SOLUTION MANUAL TO ACCOMPANY MECHANICS OF MATERIALS

Second Edition

Madhukar Vable

Michigan Technological University

EXPANDING EDUCATIONAL HORIZONS, LLC

Houghton, Michigan
madhuvable.org
2017

Paper Copy: ISBN: 978-0-9912446-6-9

Print edition: January 2nd, 2018

CONTENTS

PREFACE

The Mechanics of Materials book that is available for free for educational purposes from my website madhuvable.org has generated lot of request for this solution manual by instructors. This solution manual is designed for the instructors and may prove challenging to students. The intent was to help reduce the laborious algebra and to provide instructors with a way of checking solutions. It has been made available to students because it is next to impossible to maintain security of the manual even by large publishing companies. There are websites dedicated to obtaining a solution manuals for any course for a price. The students can use the manual as additional examples, a practice followed in many first year courses.

Most large textbooks publishers advertise by sending complimentary copies of textbooks and solution manuals to instructors. The cost of this advertisement is borne by the students, the only paying customer in this business model. The cost of textbooks has been rising 3-4 times faster than inflation. [http://www.theatlantic.com/business/archive/2013/01/why-are-college-textbooks-so-absurdly-expensive/266801/]. Students pay enough tuition to universities who can and should buy the book and other instructional material from their endowment revenues on behalf of the instructor. I hope faculty considering using my books in their courses will support my effort to keep the book cost low for the students by not providing complimentary copies.

Please see my website madhuvable.org for sample syllabus, slides, old exams, and computerized tests associated with the textbook that are all available for free for educational purposes.

ACKNOWLEDGMENTS

My thanks to Professors Ibrahim Miskioglu and Jaclyn Johnson for their corrections and suggestions.

CHAPTER 1

Sections 1.1

1. 1

Solution P=200 lbs d=1/2 in $\sigma = ?$

By equilibrium of forces we have: $N = 200 lbs$

$$A = \frac{\pi}{4}\left(\frac{1}{2}\right)^2 = 0.1963 in^2 \qquad \sigma = \frac{N}{A} = \frac{200}{0.1963} = 1018.6 psi \qquad (1)$$

ANS $\sigma = 1019\ psi(T)$

---

1. 2

Solution W=200 lbs (a) d=1/8 in σ =? (b) d=1/4 in σ =?

By equilibrium of forces we have: $N = W$

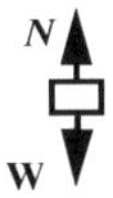

Part (a) $$A = \frac{\pi}{4}\left(\frac{1}{8}\right)^2 = 0.01227 in^2 \qquad \sigma = \frac{N}{A} = \frac{200}{0.01227} = 16297 psi \qquad (1)$$

Part (b) $$A = \frac{\pi}{4}\left(\frac{1}{4}\right)^2 = 0.04909 in^2 \qquad \sigma = \frac{N}{A} = \frac{200}{0.04909} = 4070 psi \qquad (2)$$

ANS (a) $\sigma = 16.3 ksi(T)$; (b) $\sigma = 4.1 ksi(T)$

---

1. 3

Solution d=1/5 in $\sigma_{max} \le 4 ksi(T)$ $W_{max} = ?$

From previous problem we have $N = W$

$$A = \frac{\pi}{4}\left(\frac{1}{5}\right)^2 = 31.41(10^{-3}) in^2 \qquad \sigma = \frac{N}{A} = \frac{W}{31.41(10^{-3})} \le 4(10^3) \quad or \quad W \le 125.66 lb \qquad (1)$$

ANS $W_{max} = 125.6 lb$

---

1. 4

Solution $\sigma_{max} \le 5 ksi(T)$ W = 250 lbs. d_{min} = ? to nearest 1/16th of an inch

From previous problem we have $N = W$

$$\sigma = \frac{N}{A} = \frac{W}{(\pi d^2/4)} \le 5(10^3) \quad or \quad d \ge 0.252 in \qquad (1)$$

ANS $d_{min} = (5/16)$ in

---

1. 5

Solution m = 6 kg. d = 0.75 mm $\sigma_{AB} = ?$ $\sigma_{BC} = ?$

The weight is: $W = mg = 58.86 N$. By force equilibrium in the y-direction we have the following.

$$N_{BC} = W \qquad 2N_{AB}\cos\theta = W \quad or \quad N_{AB} = \frac{W}{2\cos 23.6} = 0.5456 W \qquad A = \frac{\pi}{4}(0.00075)^2 = 0.4418(10^{-6}) m^2 \qquad (1)$$

(a)

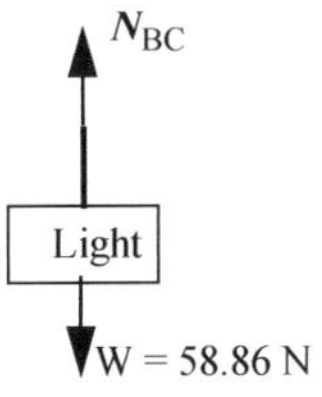

(b)

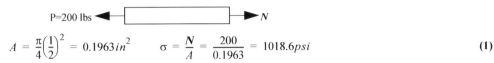

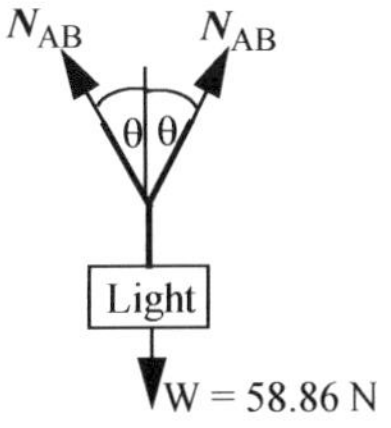

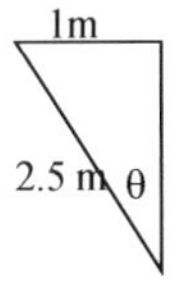

$$\theta = asin\left(\frac{1}{2.5}\right) = 23.6^o$$

$$\sigma_{BC} = \frac{N_{BC}}{A} = \frac{58.86}{0.4418(10^{-6})} = 133.2(10^6)\ N/m^2 \qquad \sigma_{AB} = \frac{N_{AB}}{A} = \frac{(0.5456)(58.86)}{0.4418(10^{-6})} = 72.7(10^6)\ N/m^2 \qquad (2)$$

ANS $\sigma_{AB} = 72.7 MPa(T)$; $\sigma_{BC} = 133.2 MPa(T)$

1.6

Solution m = 8 kg. $\sigma_{max} \le 50MPa$ d_{min} = ? nearest tenth of a millimeter

The weight is: $W = mg = 58.86N$ .By force equilibrium in the y-direction we have the following.

$$N_{BC} = W \qquad 2N_{AB}\cos\theta = W \qquad or \qquad N_{AB} = \frac{W}{2\cos 23.6} = 0.5456W. \qquad \textbf{(1)}$$

(a)

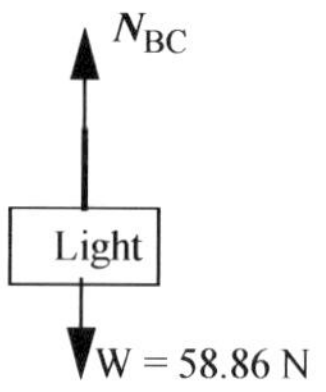

(b)

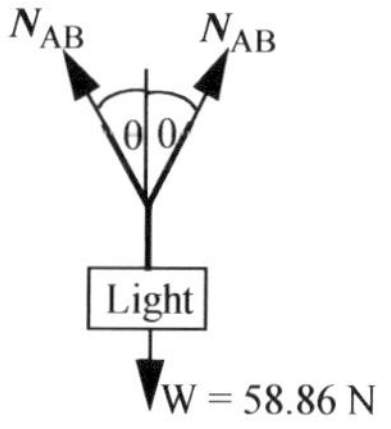

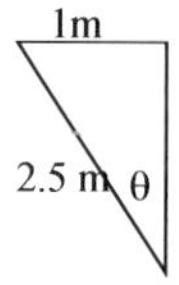

$$\theta = asin\left(\frac{1}{2.5}\right) = 23.6^o$$

$$\sigma_{max} = \frac{N_{BC}}{A} = \frac{78.48}{(\pi d^2/4)} \le 50(10^6) \qquad or \qquad d \ge 1.41(10^{-3})m \qquad \textbf{(2)}$$

ANS $d_{min} = 1.5\ mm$

1.7

Solution d = 0.5 mm $\sigma_{max} \le 80MPa$ m_{max} = ?

The weight is: $W = mg = 58.86N$. By force equilibrium in the y-direction we have the following.

$$N_{BC} = W \qquad 2N_{AB}\cos\theta = W \qquad or \qquad N_{AB} = \frac{W}{2\cos 23.6} = 0.5456W \qquad \textbf{(1)}$$

(a)

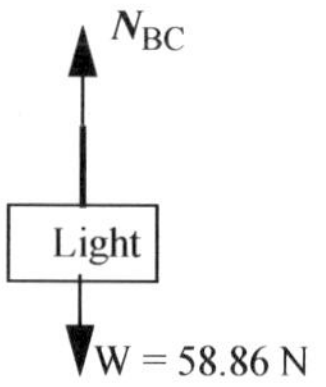

(b)

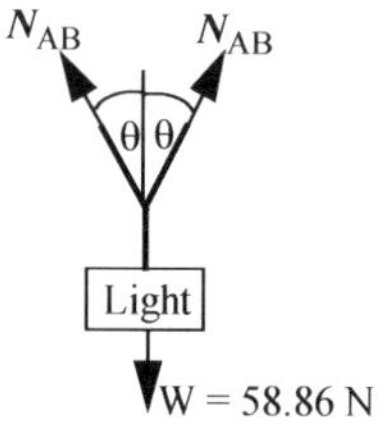

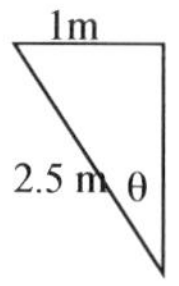

$$\theta = asin\left(\frac{1}{2.5}\right) = 23.6^o$$

$$A = \frac{\pi}{4}(0.0005)^2 = 0.1963(10^{-6})m^2 \qquad \sigma_{max} = \frac{N_{BC}}{A} = \frac{m(9.81)}{0.1963(10^{-6})} \le 80(10^6) \qquad or \qquad m \le 1.601kg \qquad \textbf{(2)}$$

ANS $m_{max} = 1.6kg$

1.8

Solution m= 3 kg d = 10 mm σ = ?

By equilibrium of forces in the y-direction we have the following equation.

$$2N\sin 54 = W \qquad or \qquad N = 0.618W = 0.618(3)(9.81) = 18.189\ N \qquad \textbf{(1)}$$

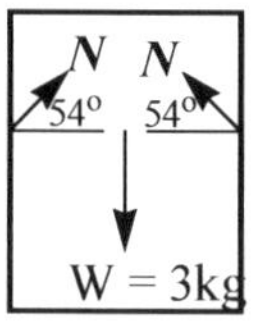

$$A = \frac{\pi}{4}(0.003)^2 = 7.0686(10^{-6})\ m^2 \qquad \sigma = \frac{N}{A} = \frac{18.189}{7.0686(10^{-6})} = 2.573(10^6)N/m^2 \qquad \textbf{(2)}$$

ANS $\sigma = 2.57MPa(T)$

1.9

Solution m= 5kg $\sigma_{max} \le 10MPa$ d_{min} = ? nearest millimeter.

From previous problem we have the following.

$$N = 0.618W \qquad or \qquad N = 0.618(5)(9.81) = 30.315\ N \qquad \textbf{(1)}$$

$$\sigma = \frac{N}{A} = \frac{30.315}{(\pi d^2/4)} \le (10)(10^6) \qquad or \qquad d \ge 1.96(10^{-3})\ m \qquad \textbf{(2)}$$

ANS $d_{min} = 2\ mm$

---

1. 10

Solution d= 0.016 inch $\sigma_{max} \le 750psi$ W_{max} =?

From previous problem we have the following.

$$N = 0.618W \quad \textbf{1}$$

$$A = \frac{\pi}{4}(0.016)^2 = 0.201(10^{-3})in^2 \qquad \sigma_{max} = \frac{N}{A} = \frac{0.618W}{0.201(10^{-3})} \le 750 \qquad or \qquad W \le 0.244lb \quad \textbf{(2)}$$

ANS $W_{max} = 0.24lbs$

---

1. 11

Solution $\sigma = f(L,\gamma,d,\theta,\alpha)$=?

By equilibrium of moment about point O we obtain the following.

$$(\gamma L)\left(\frac{L}{2}cos\theta\right) - (Nsin\alpha)L = 0 \qquad or \qquad N = \frac{\gamma Lcos\theta}{2sin\alpha} \qquad \sigma = \frac{N}{A} = \frac{(\gamma Lcos\theta/(2sin\alpha))}{(\pi d^2/4)} \quad \textbf{(1)}$$

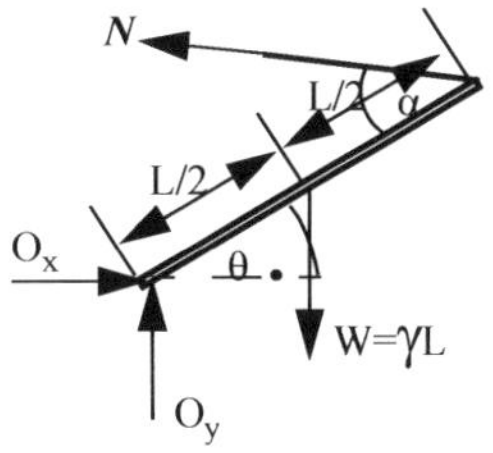

ANS $\sigma = \dfrac{2\gamma Lcos\theta}{\pi d^2 sin\alpha}$

---

1. 12

Solution d_o = 100 mm d_i = 75 mm plate: 200 mm x200 mm x 10 mm P= 800 kN σ_{col} = ? σ_b= ?

By equilibrium of forces we obtain the following equations.

(a)

(b)

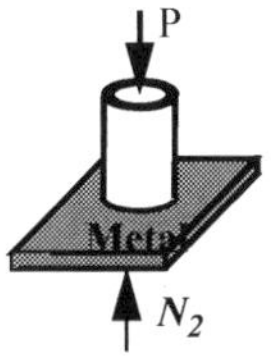

$$N_1 = P \qquad and \qquad N_2 = P \quad \textbf{(1)}$$

$$A_{col} = \frac{\pi}{4}(0.1^2 - 0.075^2) = 3.436(10^{-3}) \quad mm^2 \qquad \sigma_{col} = \frac{N_1}{A_{col}} = \frac{800(10^3)}{3.436(10^{-3})} = 232.8(10^6)N/m^2 \quad \textbf{(2)}$$

$$A_{plate} = (0.2)(0.2) = 0.04 \quad m^2 \qquad \sigma_b = \frac{N_2}{A_{plate}} = \frac{800(10^3)}{(0.04)} = 20(10^6)N/m^2 \quad \textbf{(3)}$$

ANS $\sigma_{col} = 232.8MPa(C)$; $\sigma_b = 20MPa(C)$

---

1. 13

Solution d_o = 4 in. ; d_i = 3.5 in. plate: 10 in x 10 in x 0.75 in. $\sigma_{col} \le 30ksi$; $\sigma_b \le 2ksi$ P=?

By equilibrium of forces we obtain the following equations.

$$N_1 = P \qquad and \qquad N_2 = P \quad \textbf{(1)}$$

(a)

(b)

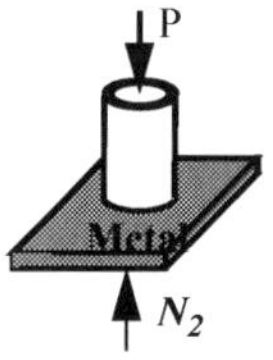

$$A_{col} = \pi(4^2 - 3.5^2)/4 = 2.945 \quad in^2 \qquad \sigma_{col} = N_1/A_{col} = (P/2.945) \le 30 \qquad or \qquad P \le 88.35kips \quad \textbf{(2)}$$

$$A_{plate} = (10)(10) = 100 \quad in^2 \qquad \sigma_b = N_2/A_{plate} = (P/100) \le 2 \qquad or \qquad P \le 200kips \quad \textbf{(3)}$$

ANS $P = 88.3kips$

---

1. 14

Solution P= 600 kN plate: 250 mm x 250 mm x 15 mm σ_{col} = ? σ_b= ?

By equilibrium of forces we obtain the following equations.

$$N_1 = P \quad and \quad N_2 = P \tag{1}$$

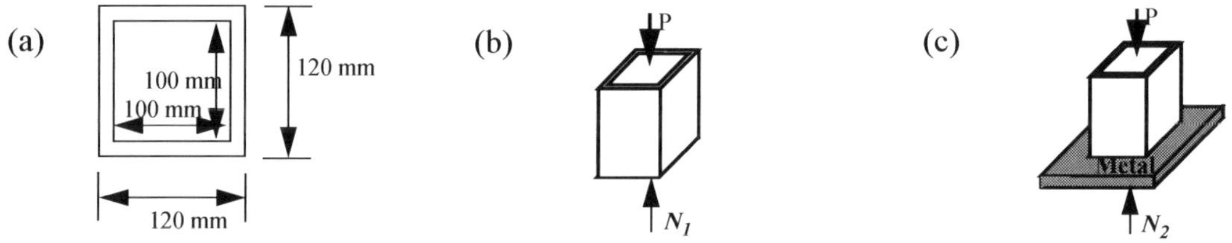

$$A_{col} = (0.12)(0.12) - (0.1)(0.1) = 4.4(10^{-3})\ m^2 \qquad \sigma_{col} = \frac{N_1}{A_{col}} = \frac{(600)(10^3)}{4.4(10^{-3})} = 136.4(10^6)\ N/m^2 \tag{2}$$

$$A_{plate} = (0.25)(0.25) = 62.5(10^{-3})\ m^2 \qquad \sigma_b = \frac{N_2}{A_{plate}} = \frac{(600)(10^3)}{62.5(10^{-3})} = 9.6(10^6)\ N/m^2 \tag{3}$$

ANS $\sigma_{col} = 136.4 MPa(C)$; $\sigma_b = 9.6 MPa(C)$

---

1. 15

Solution P = 750 kN plate: 300 x 300 x 20 mm (a) σ_{col} = ? (b) σ_p = ?

By equilibrium of forces we have:

$$N_1 = P = 750(10^3)\ N \qquad N_2 = P = 750(10^3)\ N \tag{1}$$

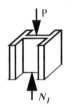

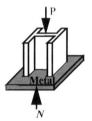

$$A_{col} = (0.160)(0.01) + (0.160)(0.01) + (0.160)(0.01) = 4.8(10^{-3})\ m^2 \qquad \sigma_{col} = \frac{N_1}{A_{col}} = \frac{750(10^3)}{4.8(10^{-3})} = 156.25(10^6) N/m^2 \tag{2}$$

$$A_{plate} = (0.3)(0.3) = 0.09\ m^2 \qquad \sigma_b = \frac{N_2}{A_{plate}} = \frac{750(10^3)}{0.09} = 8.33(10^6) N/m^2 \tag{3}$$

ANS $\sigma_{col} = 156\ MPa(C)$; $\sigma_b = 8.33\ MPa(C)$

---

1. 16

Solution m = 70 kg Scale: 150 mm x 100 mm σ_b= ?

By equilibrium of forces we obtain: ***N*** = W = (70) (9.81)=686.7N

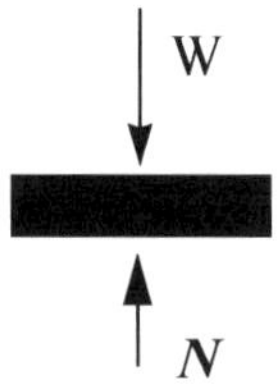

$$A = (0.15)(0.1) = 15(10^{-3})\ m^2 \qquad \sigma_b = \frac{N}{A} = \frac{686.7}{15(10^{-3})} = 45.78(10^{-3})\ N/m^2 \tag{1}$$

ANS $\sigma_b = 45.8\ kPa\ (C)$

---

1. 17

Solution L = 30 ft. d = 3 ft t = 4 in = 1/3 ft $\gamma = 80$ lbs/ft^3 σ_b = ?

The volume of the brick material and its weight can be found as shown below.

$$V = \left[\frac{\pi}{4}(d_o^2 - d_i^2)\right]L = \frac{\pi}{4}(3^2 - 2.33^2)(30) = 83.77\ ft^3 \qquad W = \gamma V = (80)(83.77) = 6702\ lbs \tag{1}$$

By force equilibrium we have: $\boldsymbol{N} = W = 6702$ lbs

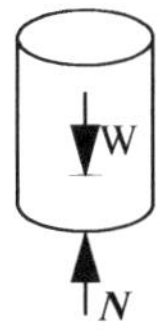

$$A = = \frac{\pi}{4}(3^2 - 2.33^2) = 2.792\ ft^2 = 402.1\ in^2 \qquad \sigma_b = \frac{N}{A} = \frac{6702}{402.1} = 16.7\ psi \qquad \textbf{(2)}$$

ANS $\sigma_b = 16.7\ psi\ (C)$

---

1. 18

Solution $\sigma_b = f(\gamma,a,h) = ?$

The volume of material and its weight can be found as shown below

$$V = [(100a + a)h/2](10h) = 505ah^2 \qquad W = \gamma V = 505\gamma ah^2 \qquad \textbf{(1)}$$

By force equilibrium we have: $\boldsymbol{N} = W = 505\gamma ah^2$

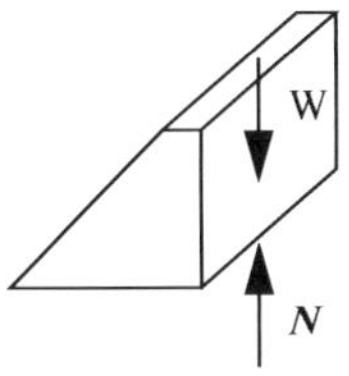

$$A = (100a)(10h) = 1000ah \qquad \sigma_b = \frac{N}{A} = \frac{505\gamma ah^2}{1000ah} = 0.505\gamma h \qquad \textbf{(2)}$$

ANS $\sigma = 0.505\gamma h\ (C)$

---

1. 19

Solution $t_0 = 4.5$ m $t_L = 2.5$ m $\gamma = 28$ kN/m^3 $\sigma_b = ?$

The outer dimension h_o and inner dimension h_i are linear functions of x and can be written as

$$h_o/2 = m_1x + C_1 \qquad and \qquad h_i/2 = m_2x + C_2 \qquad \textbf{(1)}$$

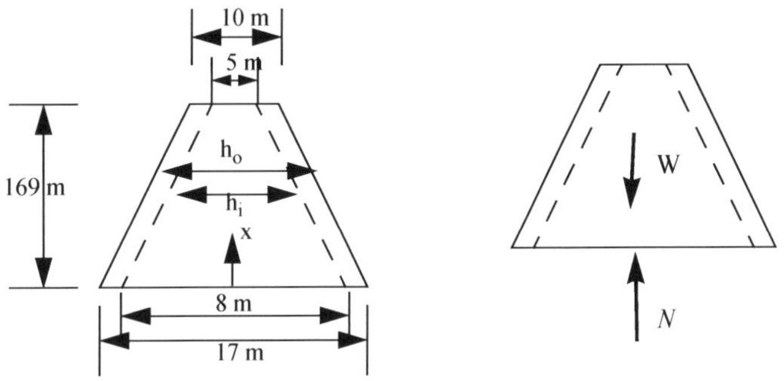

Noting that @ x = 0, $h_o = 17$ and $h_i = 8$, and @ x = 169, $h_o = 10$ and $h_i = 5$, we can solve for the constants to obtain

$$h_o = (17 - 7x/169) \qquad and \qquad h_i = (8 - 3x/169) \qquad \textbf{(2)}$$

The volume of the monument material is:

$$V = \int_0^{169} (h_o^2 - h_i^2)dx = \int_0^{169} \left[\left(17 - \frac{7x}{169}\right)^2 - \left(8 - \frac{3x}{169}\right)^2\right]dx = \frac{1}{3}\left(17 - \frac{7x}{169}\right)^3\left(-\frac{169}{7}\right)\Bigg|_0^{169} - \frac{1}{3}\left(8 - \frac{3x}{169}\right)^3\left(\frac{-169}{3}\right)\Bigg|_0^{169} = 24223.3\ m^3 \qquad \textbf{(3)}$$

The weight of the monument is: $W = \gamma V = 28\,(10^3)\,(24223.3) = 678253\,(10^3)$ N.

The force at the base of the monument is equal to the weight of the monument i.e., $\boldsymbol{N} = W = 678253\,(10^3)$N

The bearing stress is:

$$A = 17^2 - 8^2 = 225\ m^2 \qquad \sigma_b = \frac{N}{A} = \frac{678253\ (10^3)}{225} = 3014\ (10^3)\ N/m^2 \qquad \textbf{(4)}$$

ANS $\sigma_b = 3\ MPa\ (C)$

---

1. 20

Solution

By equilibrium we obtain:

$$N_1 = W_1 = \gamma(xt)(L) \qquad N_2 = W_2 = \gamma\left(\frac{1}{2}xt\right)(L) \tag{1}$$

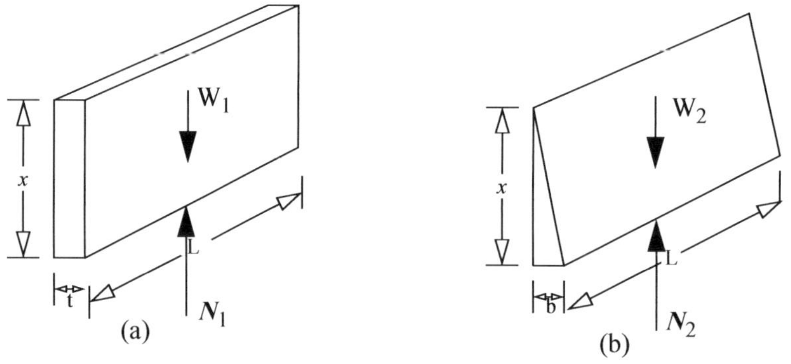

$$\sigma_1 = \frac{N_1}{A_1} = \frac{\gamma(xt)(L)}{Lt} = \gamma x \qquad \sigma_2 = \frac{N_2}{A_2} = \frac{\gamma(xb/2)(L)}{Lb} = \frac{1}{2}\gamma x \tag{2}$$

ANS $\sigma_2 = \sigma_1/2$

---

1. 21

Solution a= 757.7 ft h= 480.96 ftγ = 75 lb/ft^3 σ_b = ? σ_H = ?

The weight of the pyramid is:

$$V = \frac{1}{3}a^2h = \frac{1}{3}(757.7)^2(480.96) = 92.040(10^6)\ ft^3 \qquad W = \gamma V = (75)(92.040)(10^6) = 6.903(10^9)lb \tag{1}$$

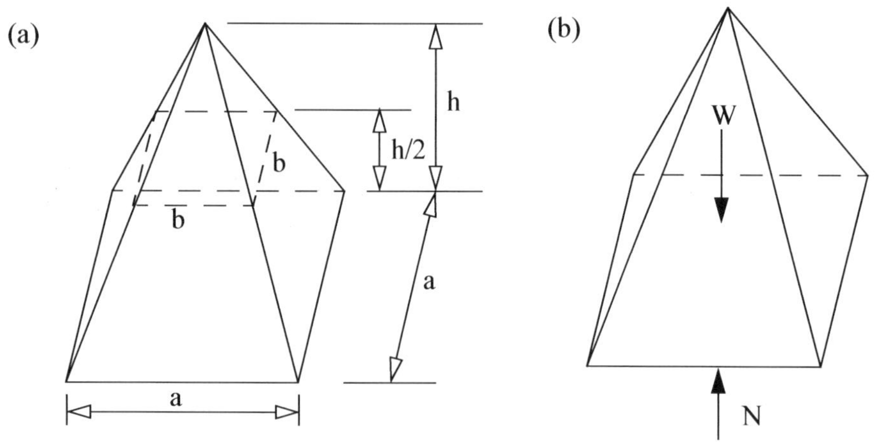

$$N = W = 6.903(10^9)lb \qquad \sigma_b = \frac{N}{a^2} = \frac{6.903(10^9)}{(757.7)^2} = 12024\ psi \tag{2}$$

The sides of square at half height will be b=a/2. The weight of top half of pyramid is:

$$V_H = \frac{1}{3}(b)^2\left(\frac{h}{2}\right) = \frac{1}{6}b^2h \qquad W_H = \gamma V_H = \frac{\gamma}{6}b^2h \tag{3}$$

$$\sigma_H = \frac{N_H}{b^2} = \frac{W_H}{b^2} = \frac{\gamma}{6}h = \frac{(75)(480.96)}{6} = 6012psi \tag{4}$$

ANS $\sigma_b = 12024\ \text{lb/ft}^2(C)\ ;\sigma_H = 6012\ \text{lb/ft}^2(C)$

---

1. 22

Solution θ_1= 54°27'44" θ_2= 43°22' g = 1200 kg/ m^3 σ = ?

Converting to degrees we obtain:

$$\theta_1 = 54 + \left(27 + \frac{44}{60}\right)/60 = 54.4622° \qquad \theta_2 = 43 + (22)/60 = 42.3667° \tag{1}$$

From geometry we obtain:

$$x_1 + x_2 = 94\ m \qquad h_1 = x_1 tan\theta_1 = 1.4x_1 \qquad h_2 = x_2 tan\theta_2 = 0.9446x_2 \qquad h_1 + h_2 = 1.4x_1 + 0.9446x_2 = 105\ m \qquad (2)$$

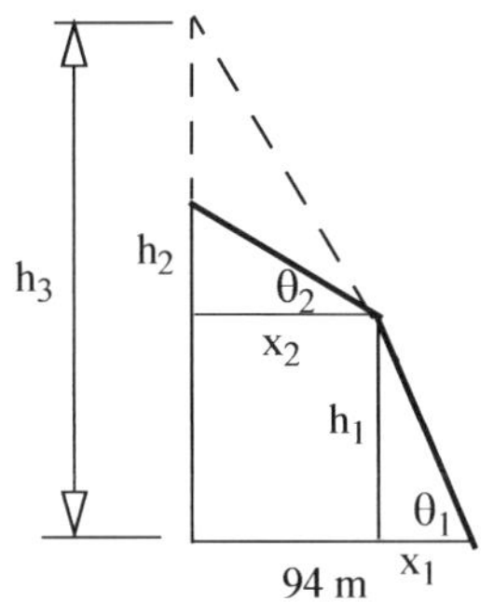

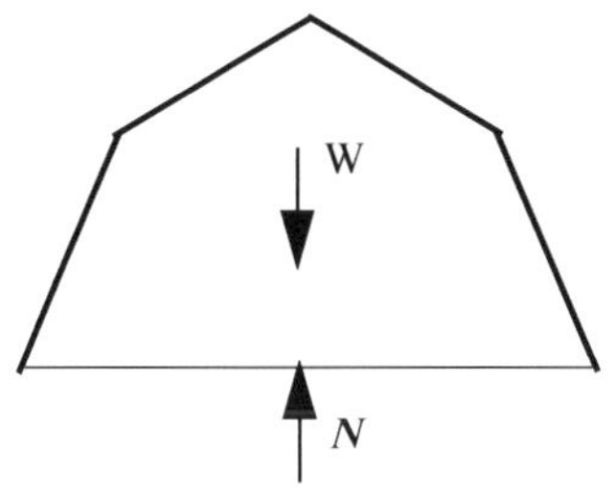

Solving we obtain

$$x_1 = 35.59\ m \qquad x_2 = 58.41\ m \qquad h_1 = 49.82\ m \qquad h_2 = 55.17\ m \qquad h_3 = 94tan\theta_1 = 131.6\ m \qquad (3)$$

The total volume of the pyramid and its weight can be found as shown below.

$$V = \frac{1}{3}(188)(188)h_3 - \frac{1}{3}(2x_2)(2x_2)(h_3 - h_1) + \frac{1}{3}(2x_2)(2x_2)(h_2) = \frac{1}{3}[188^2(h_3) - (4x_2^2)(h_3 - h_1 - h_2)] = 1.429(10^6)\ m^3 \qquad (4)$$

$$W = \gamma g V = (1200)(9.81)[1.429(10^6)] = 16.827(10^9)N \qquad (5)$$

$$\sigma = \frac{N}{A} = \frac{N}{(188)(188)} = \frac{16.827(10^9)}{(188)(188)} = 0.476(10^6)(N/m^2) \qquad (6)$$

ANS $\sigma = 0.476\ MPa(C)$

---

1. 23

Solution $d_{bolt} = 25$ mm $d_{al} = 48$ mm t= 4mm $\sigma_{bolt} = 100$ MPa(T) $\sigma_{bolt} = ?$

The cross-sectional areas of the bolt and aluminum sleeve are:

$$A_{bolt} = \frac{\pi}{4}d_{bolt}^2 = \frac{\pi}{4}(0.025)^2 = 490.87(10^{-6})m^2 \qquad A_{al} = \frac{\pi}{4}d_{bolt}^2 = \frac{\pi}{4}[(0.048)^2 - (0.04)^2] = 552.92(10^{-6})m^2 \qquad (1)$$

The internal force in the bolt can be found and by equilibrium the internal force in aluminum obtained as shown below.

$$N_{bolt} = \sigma_{bolt}A_{bolt} = 100(10^6)[490.87(10^{-6})] = 49087N \qquad N_{al} = N_{bolt} = 49087N \qquad (2)$$

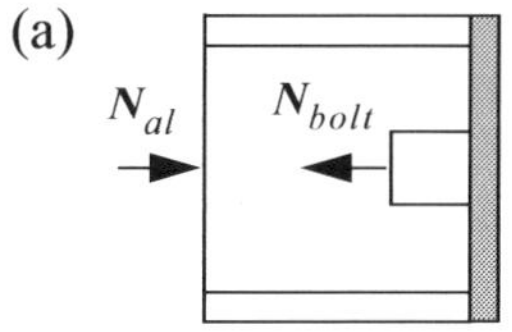

$$\sigma_{al} = \frac{N_{al}}{A_{al}} = \frac{49087}{552.92(10^{-6})} = 88.78(10^6)N/m^2 \qquad (3)$$

ANS $\sigma_{al} = 88.78\ MPa(C)$

---

1. 24

Solution P = 15 kips Wood: 6 in x 8 in x 2 in. $\tau = ?$

By equilibrium of forces we have the following equation.

$$P = \tau(2)(6) \quad or \quad \tau = P/12 = 15/12 \qquad (1)$$

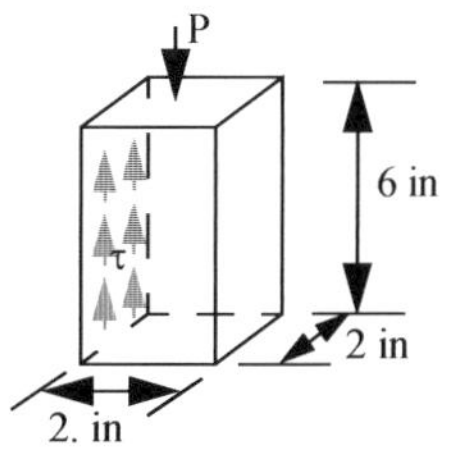

ANS $\tau = 1.25ksi$

---

1. 25

Solution Wood: 6 in x 8 in x 1.5 in. $\tau \le 1.2ksi$ P = ?

By equilibrium of forces we have the following equation.

$$P = \tau(6)(1.5) \quad or \quad \tau = P/9 \le 1.2 \quad or \quad P \le 10.8 \ kips \tag{1}$$

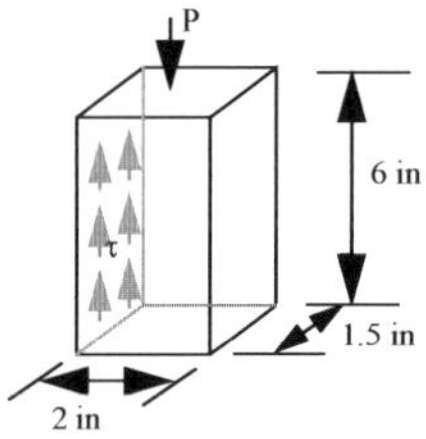

ANS $P_{max} = 10.8 \ kips$

---

1. 26

Solution P = 6 kips t = 1/8 inch d = 1 inch τ = ?

By equilibrium of forces we have the following equation.

$$P = \tau(\pi d)(t) \quad or \quad \tau = \frac{P}{\pi dt} = \frac{6}{\pi(1)(1/8)} = 15.28ksi \tag{1}$$

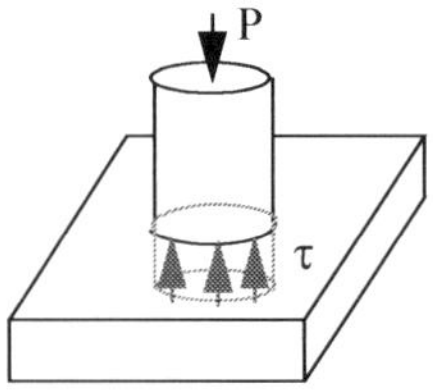

ANS $\tau = 15.3ksi$

---

1. 27

Solution t = 3 mm plate: 10 mm x 10 mm $\tau \le 200MPa$ P = ?

The free body diagram of the punch as it will go through the plate is shown below. By equilibrium of forces we have the following equation.

$$P = \tau(4a)(t) = (200)(10^6)(4)(0.01)(0.003) = 24(10^3)N \tag{1}$$

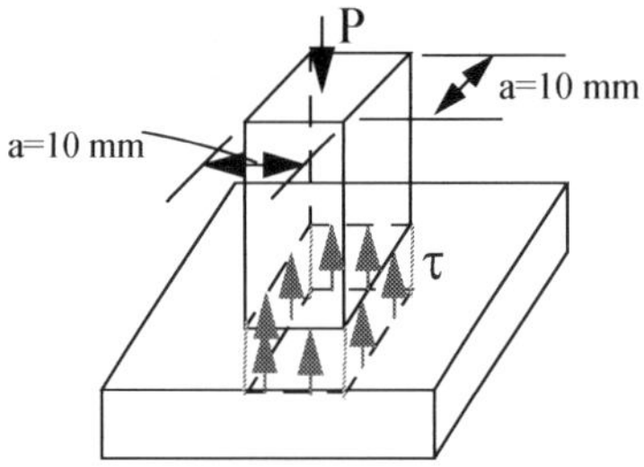

ANS $P = 24kN$

---

1. 28

Solution $P = f(\tau, t, d_i, d_o) = ?$

The surface area on which the shear stress will is the inner and outer cylindrical surface area punched through the thickness of the plate i.e., $A = \pi(d_o + d_i)t$. By equilibrium of forces we have: $P = \tau A$.

ANS $P = \tau\pi(d_o + d_i)t$

---

1. 29

Solution d = 25 mm τ_{max} = ?

From the free body diagram we see that the maximum shear force is: V_2 = 40kN.

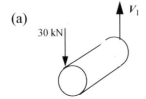

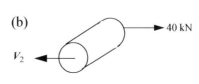

The maximum shear stress is:

$$A = \pi(0.025)^2/4 = 490.87(10^{-6})m^2 \qquad \tau_{max} = \frac{V_2}{A} = \frac{40(10^3)}{490.87(10^{-6})} = 81.48(10^6)N/m^2 \tag{1}$$

ANS $\tau_{max} = 81.5MPa$

---

1. 30

Solution W = 200lbs d_c = 1/4 in d_p = 3/8 in σ_C = ? τ_p = ?

By equilibrium of forces we have the following.

$$T = W \qquad O_x = T\cos 55 = 0.5736W = 114.7\ lb \tag{1}$$

$$O_y - T - T\sin 55 = 0 \quad or \quad O_y = T(1+\sin 55) = 1.8192W = 363.8\ lb \tag{2}$$

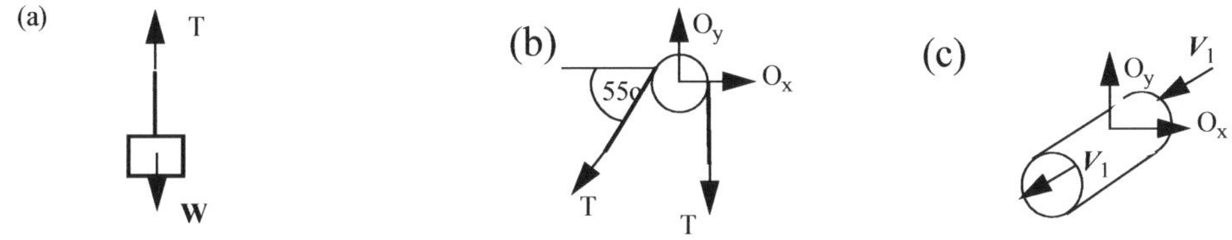

$$A_C = \pi(1/4)^2/4 = 49.08(10^{-3})in^2 \qquad \sigma_C = \frac{T}{A_C} = \frac{200}{49.08(10^{-3})} = 4074.4\ psi \tag{3}$$

From equilibrium of forces in Fig. (c) we have.

$$2V = \sqrt{O_x^2 + O_y^2} = \quad = 190.7\ lb \tag{4}$$

$$A_p = \pi(3/8)^2/4 = 110.4(10^{-3})in^2 \qquad \tau = \frac{V}{A_P} = \frac{190.7}{110.4(10^{-3})} = 1727.4psi \tag{5}$$

ANS $\sigma_C = 4074.\ psi(T)$; $\tau = 1727psi$

---

1. 31

Solution d_c = 1/5 in d_p = 3/8 in $\sigma_C \le 4ksi$ $\tau_p \le 2ksi$ W_{max} = ?

From previous problem we have

$$T = W \qquad O_x = T\cos 55 = 0.5736W \tag{1}$$

$$O_y - T - T\sin 55 = 0 \quad or \quad O_y = T(1+\sin 55) = 1.8192W \tag{2}$$

$$A_C = \pi(1/5)^2/4 = 31.42(10^{-3})in^2 \qquad \sigma_C = \frac{T}{A_C} = \frac{W}{31.42(10^{-3})} \le 4(10^3)psi \qquad W \le 125.66lbs \tag{3}$$

$$2V = \sqrt{O_x^2 + O_y^2} = W\sqrt{(0.5736)^2 + (1.8192)^2} \quad or \quad V = 0.9537W \tag{4}$$

$$A_p = \pi(3/8)^2/4 = 110.4(10^{-3})in^2 \qquad \tau_p = \frac{V}{A_P} = \frac{0.9537W}{110.4(10^{-3})} \le 2(10^3)psi \qquad W \le 231.5lb \tag{5}$$

ANS $W_{max} = 125.6lb$

---

1. 32

Solution P = 200 lbs. A_{AA} = 1.3 in² A_{AA} = 0.3 in² τ_{AA} = ? σ_{BB} = ?

By force equilibrium we obtain

$$: V_A = P/2 = 100lbs \qquad N_B = P/2 = 100lbs \tag{1}$$

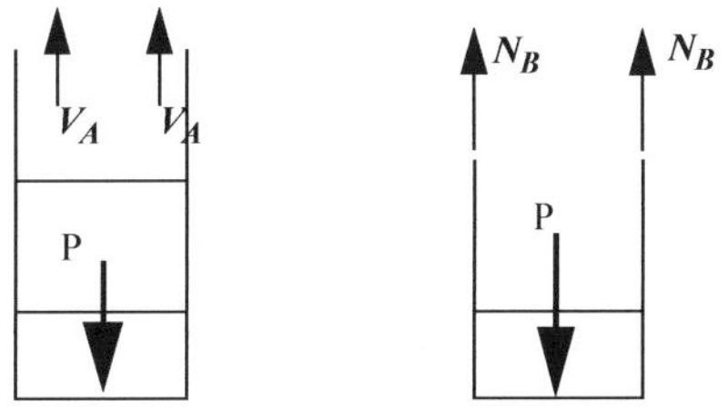

$$\tau_{AA} = \frac{V_A}{A_{AA}} = \frac{100}{1.3} = 76.92psi \qquad \sigma_{BB} = \frac{N_B}{A_{BB}} = \frac{100}{0.3} = 333.33psi \tag{2}$$

ANS $\tau_{AA} = 76.9psi$; $\sigma_{BB} = 333.3\ psi\ (T)$

---

1. 33

Solution $\sigma_{bolt} = ?$ $\tau_{bolt} = ?$ $\sigma_{bear} = ?$ $\tau_{wood} = ?$

The following free body diagrams can be drawn.

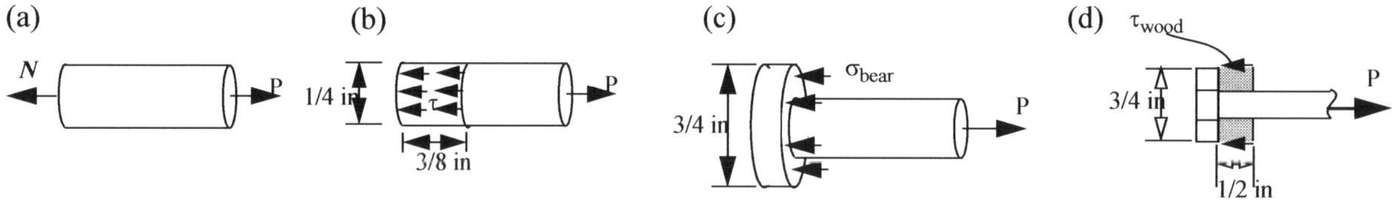

$$A_{bolt} = \pi(1/4)^2/4 = 49.09(10^{-3})in^2 \qquad N = P = 1.5\ kips \qquad \sigma_{bolt} = \frac{N}{A_{bolt}} = \frac{1.5}{49.09(10^{-3})} = 30.556ksi \tag{1}$$

$$\tau[\pi(1/4)](3/8) = P = 1.5 \quad or \quad \tau = \frac{1.5}{(3\pi/32)} = 5.093ksi \tag{2}$$

$$A_{bear} = \pi[(3/4)^2 - (1/4)^2]/4 = 0.3927in^2 \qquad \sigma_{bear}A_{bear} = P = 1.5 \quad or \quad \sigma_{bear} = \frac{1.5}{0.3927} = 3.8197ksi \tag{3}$$

$$\tau_{wood}[\pi(3/4)](1/2) = P = 1.5 \quad or \quad \tau_{wood} = \frac{1.5}{(3\pi/8)} = 1.273ksi \tag{4}$$

ANS $\sigma_{bolt} = 30.56\ ksi\ (T);\ \tau = 5.09\ ksi;\ \sigma_{bear} = 3.82\ ksi\ (C);\ \tau_{wood} = 1.27ksi$

---

1. 34

Solution P = 10 kips $\tau_r = ?$ $\sigma_{max} = ?$ $\sigma_b = ?$

By equilibrium we obtain:

$$V = P/2 = 5\ kips \qquad N = P = 10\ kips \qquad N_b = P = 10\ kips \tag{1}$$

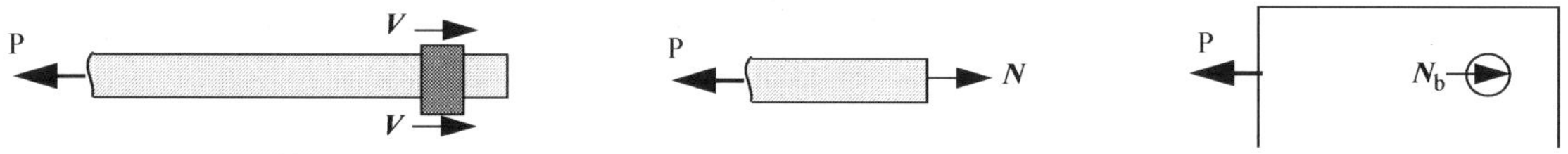

$$\tau_r = \frac{V}{A_r} = \frac{5}{(\pi/4)(0.5^2)} = \frac{5}{0.1963} = 25.46\ ksi \qquad \sigma = \frac{N}{A} = \frac{10}{(1)(0.5)} = 20\ ksi \tag{2}$$

$$\sigma_b = \frac{N_b}{A_b} = \frac{10}{(\pi/2)(t)d_r} = \frac{10}{(\pi/2)(0.5)(0.5)} = \frac{10}{0.3927} = 25.46\ ksi \tag{3}$$

ANS $\tau_r = 25.46\ ksi;\ \sigma = 20\ ksi(T);\ \sigma_b = 25.46\ ksi(C)$

---

1. 35

Solution $h = 4.375$ in $d = 1.125$ in $\tau_{BCD} = ?$ $;\sigma_{BEF} = ?$

By equilibrium we obtain:

$$V_{BCD} = 10cos30 = 8.66\ kips \qquad N_{BEF} = 10cos30 = 8.66\ kips \tag{1}$$

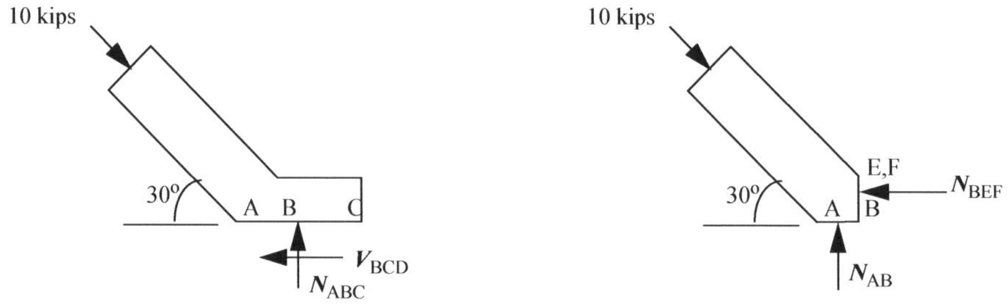

$$\tau_{BCD} = \frac{V_{BCD}}{A_{BCD}} = \frac{8.66}{(4.375)(4)} = 0.4949\ ksi \qquad \sigma_{BEF} = \frac{N_{BEF}}{A_{BEF}} = \frac{8.66}{(1.125)(4)} = 1.924\ ksi \tag{2}$$

ANS $\tau_{BCD} = 0.4949\ ksi;\sigma_{BEF} = 1.924\ ksi(C)$

---

1. 36

Solution P= 12 kips $\sigma_{bolt} = ?$ $\tau_{bolt} = ?$

By equilibrium we have:

$$N = Psin60/4 = 0.2165P = 0.2165(12) = 2.598\ kips \qquad V = Pcos60/4 = 0.125P = 0.125(12) = 1.5\ kips \tag{1}$$

$$A_{bolt} = \frac{\pi}{4}\left(\frac{1}{2}\right)^2 = 0.1963\ in^2 \qquad \sigma_{av} = \frac{N}{A_{bolt}} = \frac{2.598}{0.1963} = 13.23\ ksi \qquad \tau_{av} = \frac{V}{A_{bolt}} = \frac{1.5}{0.1963} = 7.64\ ksi \tag{2}$$

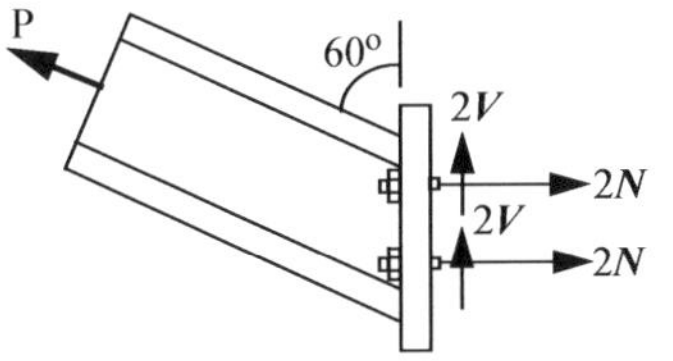

ANS $\sigma_{av} = 13.23\ ksi(T)$; $\tau_{av} = 7.64\ ksi$

---

1. 37

Solution P=50 kN σ_{allow}= 100 MPa τ_{allow}= 70 MPa d_{bolt} = ?

From previous problem we have the internal forces in terms of P and we obtain

$$N = 0.2165P = \frac{50(10^3)sin60}{4} = 10.825(10^3)\ N \qquad V = 0.125P = \frac{50(10^3)cos60}{4} = 6.25(10^3)\ N \tag{1}$$

$$\sigma_{av} = \frac{N}{A_{bolt}} = \frac{10.825(10^3)}{A_{bolt}} \le 100(10^6) \quad or \quad A_{bolt} \ge 108.25(10^{-6})\ m^2 \tag{2}$$

$$\tau_{av} = \frac{V}{A_{bolt}} = \frac{6.25(10^3)}{A_{bolt}} \le 70(10^6) \quad or \quad A_{bolt} \ge 89.29(10^{-6})\ m^2 \tag{3}$$

$$A_{bolt} = \frac{\pi}{4}d_{bolt}^2 \ge 108.25(10^{-6}) \quad or \quad d_{bolt} \ge 11.74(10^{-3})\ m \tag{4}$$

ANS $d_{bolt} = 12\ mm$

---

1. 38

Solution d= 1/2 in. σ_{allow}= 15 ksi τ_{allow}= 12 ksi P_{max} = ?

From previous problem we have the internal forces in terms of P

$$N = 0.2165P \qquad V = 0.125P \tag{1}$$

$$\sigma_{av} = \frac{N}{A_{bolt}} = \frac{0.2165P}{(\pi/4)(1/2)^2} \le 15(10^3)psi \quad or \quad P \le 13603.9\ lbs \tag{2}$$

$$\tau_{av} = \frac{V}{A_{bolt}} = \frac{0.125P}{(\pi/4)(1/2)^2} \le 12(10^3)psi \quad or \quad P \le 18849.6\ lbs \tag{3}$$

ANS $P_{max} = 13603\ lbs$

---

1. 39

Solution

By equilibrium we have:

$$V = 10/2 = 5\ kips \qquad \tau_{av} = V/A = 5/[(2)(8)] = 0.3125\ ksi \tag{1}$$

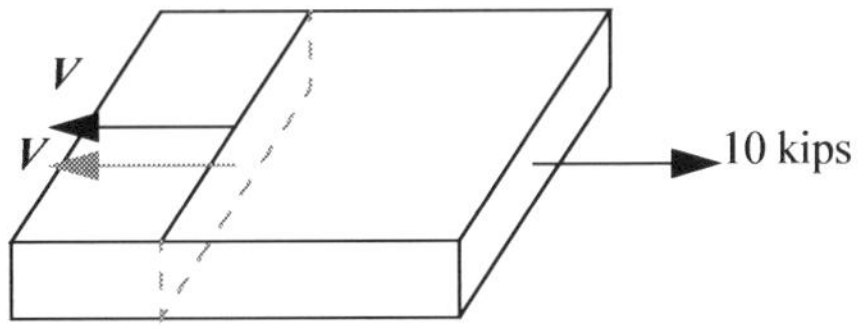

ANS $\tau_{av} = 312.5\ psi$

---

1. 40

Solution P = 20 kips L = 3 in a = 8 in h = 2 in σ_{av} = ? τ_{av} = ?

By equilibrium we have:

$$V = 20/2 = 10\ kips \qquad N = 10\ kips \tag{1}$$

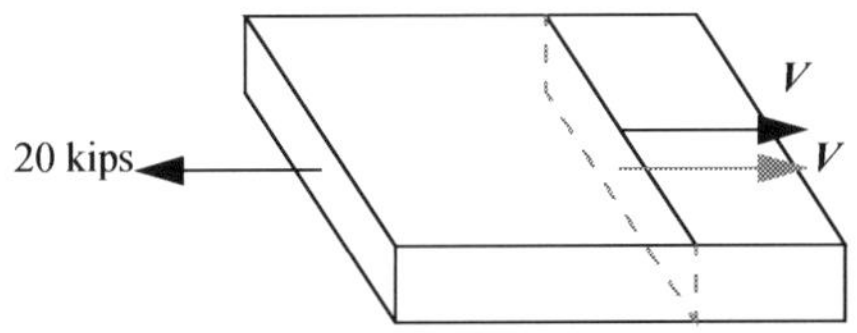

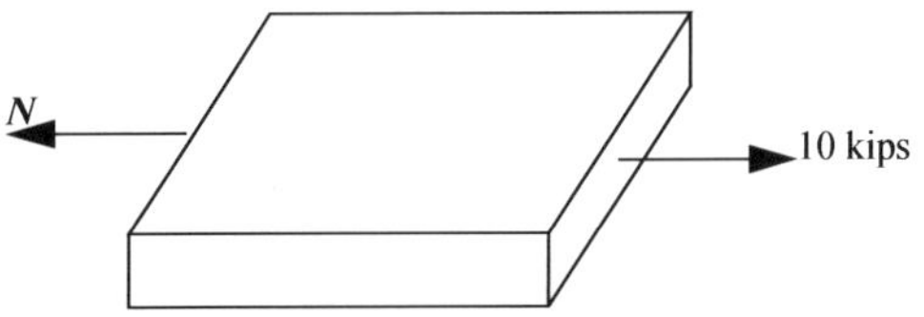

$$\tau_{av} = \frac{V}{A} = \frac{10}{(L)(a)} = \frac{10}{(3)(8)} = 0.4167\ ksi \qquad \sigma_{av} = \frac{N}{A} = \frac{10}{(a)(h/2)} = \frac{20}{(8)(1)} = 2.5\ ksi \tag{1}$$

ANS $\tau_{av} = 416.7\ psi$; $\sigma_{av} = 2500\ psi(T)$

---

1. 41

Solution σ_{allow}= 15 MPa (T) τ_{allow}= 2 MPa L = 75 mm a =200 mm h = 50 mm P_{max} = ?

By equilibrium we have:

$$V = P/2 \qquad N = P/2 \tag{1}$$

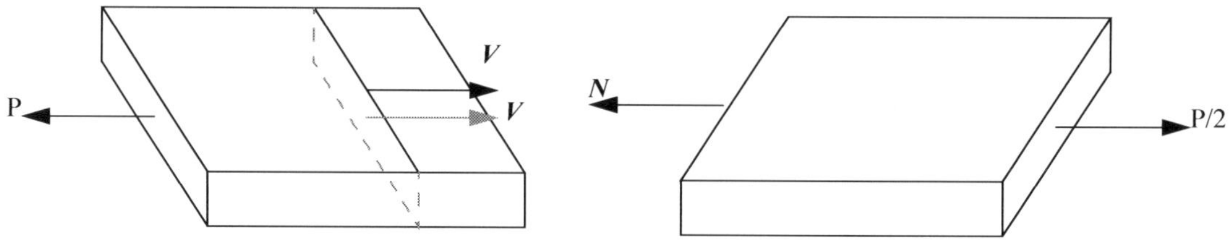

$$\tau_{av} = \frac{V}{A} = \frac{P/2}{(L)(a)} = \frac{P/2}{(0.075)(0.2)} \le 2(10^6) \quad or \quad P \le 60(10^3)\ N \tag{2}$$

$$\sigma_{av} = \frac{N}{A} = \frac{P/2}{(a)(h/2)} = \frac{P/2}{(0.2)(0.025)} \le 15(10^6) \quad or \quad P \le 150(10^3)\ N \tag{3}$$

ANS $P_{max} = 60\ kN$

---

1. 42

Solution d = 20 mm σ_{allow}= 10 MPa (T) τ_{allow}= 25 MPa σ_{br}= 18 MPa (C) P_{max} = ?

By equilibrium we have:

$$V = P/2 \qquad N = P/2 \qquad N_b = P/2 \tag{1}$$

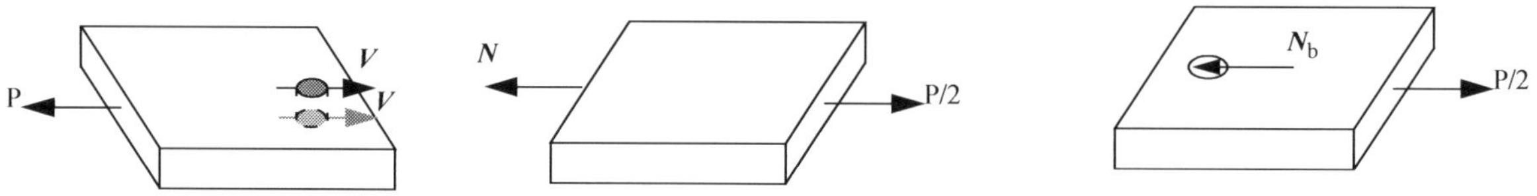

$$\tau_{av} = \frac{V}{A} = \frac{P/2}{(\pi/4)d^2} = \frac{P/2}{(314.16)(10^{-6})} \le 25(10^6) \quad or \quad P \le 15.707(10^3)\ N \tag{2}$$

$$\sigma_{av} = \frac{N}{A} = \frac{P/2}{(a)(h/2)} = \frac{P/2}{0.005} \le 10(10^6) \quad or \quad P \le 100(10^3)\ N \tag{3}$$

$$\sigma_b = \frac{N_b}{A_b} = \frac{P/2}{(\pi/2)(d)(h/2)} = \frac{P/2}{(0.7854)(10^{-3})} \le 18(10^6) \quad or \quad P \le 28.27(10^3)\ N \tag{4}$$

$P_{max} = 15.7\ kN$

---

1. 43

Solution t = 1/16 in W= 180 lb σ_A = ? σ_B = ? τ_C = ?

By equilibrium

$$N_A = 360\ lb \qquad 2N_B \sin 80 = 360\ lb \quad or \quad N_B = 182.78\ lb \qquad V_C = 180\ lb \tag{1}$$

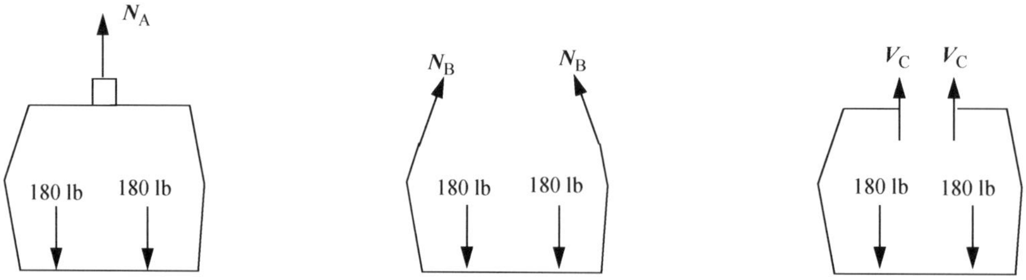

$$A_A = \frac{\pi}{4}\left[2^2 - \left(2 - \frac{2}{16}\right)^2\right] = 0.3804\ in^2 \qquad \sigma_A = \frac{N_A}{A_A} = \frac{360}{0.3804} = 946.4\ psi \tag{2}$$

$$A_B = \frac{\pi}{4}\left[1.5^2 - \left(1.5 - \frac{2}{16}\right)^2\right] = 0.2823\ in^2 \qquad \sigma_B = \frac{N_B}{A_B} = \frac{182.78}{0.2823} = 647.5\ psi \tag{3}$$

$$\tau_C = \frac{V_C}{A_C} = \frac{180}{0.4073} = 441.9\ psi \qquad A_c = (2)\left(\frac{1}{16}\right) + \frac{\pi}{4}\left[1.5^2 - \left(1.5 - \frac{2}{16}\right)^2\right] = 0.4073\ in^2 \tag{4}$$

ANS $\sigma_A = 946.4\ psi(T);\ \sigma_B = 647.5\ psi(T);\ \tau_C = 441.9\ psi$

---

1. 44

Solution P = 12 kips σ_{AA}= ? τ_{AA}= ?

By equilibrium of forces in the n and t directions we obtain:

$$N = P\sin 65 = 12\sin 65 = 10.876 kips \qquad V = P\cos 65 = 12\cos 65 = 5.0714 kips \tag{1}$$

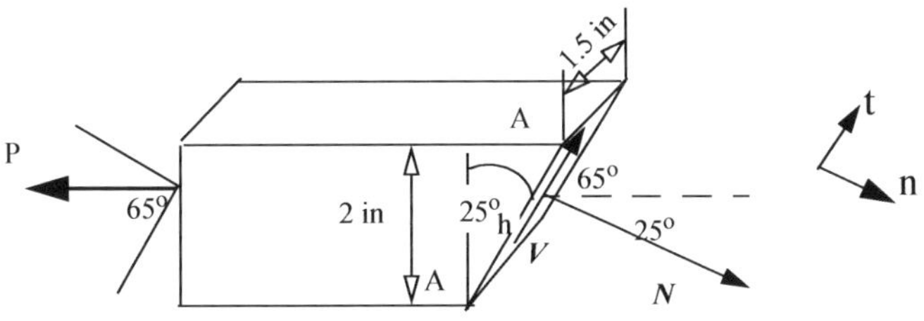

$$A_{AA} = (h)(1.5) = \left(\frac{2}{\cos 25}\right)(1.5) = 3.3107 in^2 \qquad \sigma_{AA} = \frac{N}{A_{AA}} = \frac{10.876}{3.3107} = 3.286 ksi \qquad \tau_{AA} = \frac{V}{A_{AA}} = \frac{5.0714}{3.3107} = 1.532 ksi$$

ANS $\sigma_{AA} = 3.286 ksi(T);\ \tau_{AA} = 1.53 ksi$

---

1. 45

Solution 2 in × 4 in P = ,80 kips σ_{AA}= ? τ_{AA}= ?

By equilibrium

$$N = P\cos 40 = 61.283\ kips \qquad V = P\sin 40 = 51.42\ kips \tag{1}$$

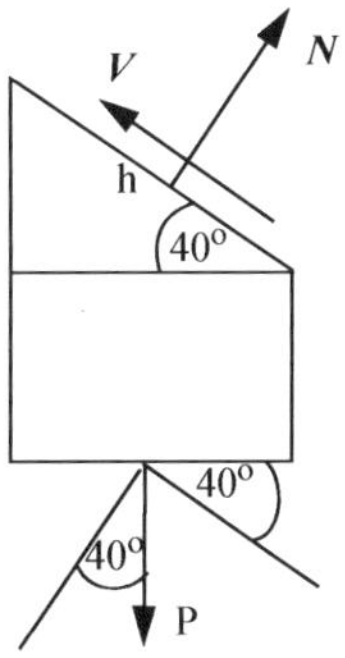

$$A_{AA} = h(2) = \left(\frac{4}{\cos 40}\right)(2) = (5.2216)(2) = 10.4433\ in^2 \tag{2}$$

$$\sigma_{AA} = \frac{N}{A_{AA}} = \frac{61.283}{10.4433} = 5.87 ksi \qquad \tau_{AA} = \frac{V}{A_{AA}} = \frac{51.42}{10.4433} = 4.924 ksi \tag{3}$$

ANS $\sigma_{AA} = 5.87 ksi(T);\ \tau_{AA} = 4.924 ksi$

---

1. 46

Solution σ_{AA}= 180 MPa (C) F_1 = ? σ_{BB}= ? τ_{BB}= ?

By equilibrium of forces in the n and t directions we obtain:

$$N_{AA} = F_1\sin 75 = 0.9659F_1 \qquad N_{BB} = F_3\sin 65 \qquad V_{BB} = F_3\cos 65 \tag{1}$$

(a)

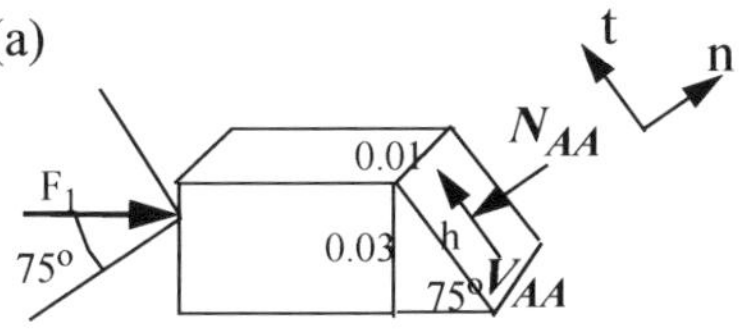

(b)

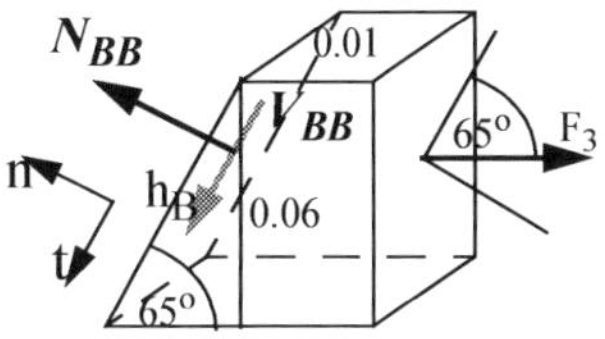

$$A_{AA} = \left(\frac{0.03}{\sin 75}\right)(0.01) = 0.3106(10^{-3})m^2 \qquad \sigma_{AA} = \frac{N_{AA}}{A_{AA}} = \frac{0.9659F_1}{0.3106(10^{-3})} = 180(10^6) \quad or \quad F_1 = 57,882N = 57.9kN \tag{2}$$

By equilibrium of forces on the bar we obtain F_3. Substituting F_3 we obtain the internal forces on BB as shown below.

$$F_1 + F_3 - 100 = 0 \quad or \quad F_3 = 42.1\ kN \tag{3}$$

$$N_{BB} = 42.1 sin 65 = 38.16 kN \qquad V_{BB} = 42.1 cos 65 = 17.79 kN \tag{4}$$

$$A_{BB} = (h_B)(1.5) = \left(\frac{0.06}{cos 65}\right)(0.01) = 0.662(10^{-3}) m^2 \tag{5}$$

$$\sigma_{BB} = \frac{N_{BB}}{A_{BB}} = \frac{(38.16)(10^3)}{0.662(10^{-3})} = 57.64(10^6) N/m^2 \qquad \tau_{BB} = \frac{V_{BB}}{A_{BB}} = \frac{(17.79)(10^3)}{0.662(10^{-3})} = 26.87(10^6) N/m^2 \tag{6}$$

ANS $F_1 = 57.9 kN$; $\sigma_{BB} = 57.6 MPa(T)$; $\tau_{BB} = 26.87 MPa$

---

1. 47

Solution P = 250 kN σ_{AA} = ? τ_{AA} = ?

By equilbrium

$$N = P sin 60 = 216.5 \ kN \qquad V = P cos 60 = 125 \ kN \tag{1}$$

The stresses are

$$A_{AA} = h(0.2) = \left(\frac{50(10^{-3})}{sin 60}\right)(0.2) = [57.735(10^{-3})](0.2) = 11.547(10^{-3}) m^2 \tag{2}$$

$$\sigma_{AA} = \frac{N}{A_{AA}} = \frac{216.5(10^3)}{11.547(10^{-3})} = 18.75(10^6) N/m^2 \qquad \tau_{AA} = \frac{V}{A_{AA}} = \frac{125}{11.547(10^{-3})} = 10.83(10^6) N/m^2 \tag{3}$$

ANS $\sigma_{AA} = 18.75 \ MPa(T)$; $\tau_{AA} = 10.83 \ MPa$

---

1. 48

Solution t= 1/4 in P = 20 kips σ_{AA} = ? τ_{AA} = ?

By equilibrium

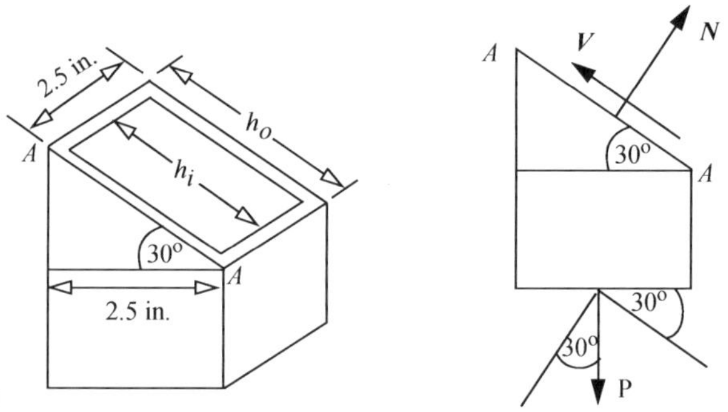

$$N = P cos 30 = 17.32 \ kips \qquad V = P sin 30 = 10 \ kips \tag{1}$$

$$A_{AA} = (2.5)h_o - \left(2.5 - \frac{2}{4}\right)h_i = (2.5)\left(\frac{2.5}{cos 30}\right) - \left(2.5 - \frac{2}{4}\right)\left(\frac{2.5 - 2/4}{cos 30}\right) = 2.598 \ in^2 \tag{2}$$

$$\sigma_{AA} = \frac{N}{A_{AA}} = \frac{17.32}{2.598} = 6.67 \ ksi \qquad \tau_{AA} = \frac{V}{A_{AA}} = \frac{10}{2.598} = 3.849 \ ksi \tag{3}$$

ANS $\sigma_{AA} = 6.67 \ ksi(T)$; $\tau_{AA} = 3.85 \ ksi$

---

1. 49

Solution $\sigma = f_1(P, a, b, \theta)$ $\tau = f_2(P, a, b, q)$ θ_σ = ? when σ is max. θ_τ = ? when τ is max.

By equilibrium of forces in the n and t direction we obtain:

$$N = P cos\theta \qquad V = -P sin\theta \tag{1}$$

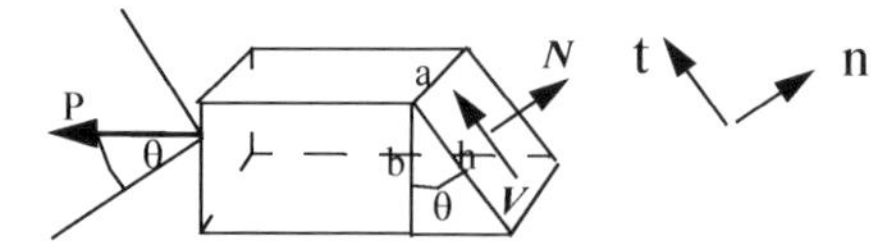

$$A = ah = a(b/cos\theta) \qquad \sigma = \frac{N}{A} = \frac{P cos\theta}{a(b/cos\theta)} = \frac{P}{ab}cos^2\theta \qquad \tau = \frac{V}{A} = \frac{-P sin\theta}{a(b/cos\theta)} = \frac{-P}{ab} sin\theta cos\theta \tag{2}$$

The minus sign in the shear stress expression indicates that is acts down the incline. The normal stress and shear stress are plotted below. From the plot it is seen that normal stress is maximum at $\theta = 0$ and shear stress magnitude is maximum at $\theta = 45^o$.

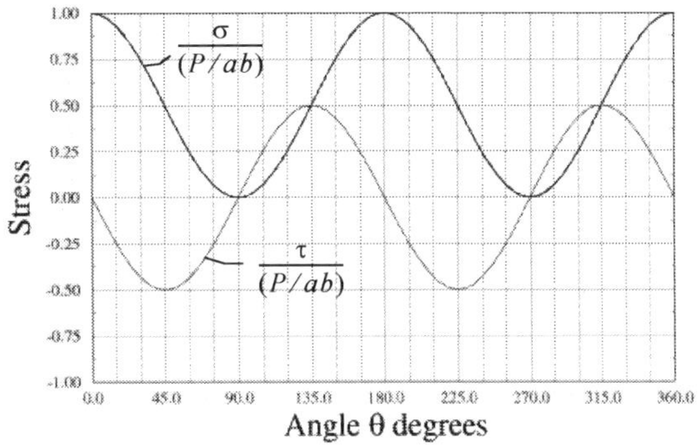

ANS $\sigma = \frac{P}{ab}cos^2\theta$; $\tau = \frac{-P}{ab}sin\theta cos\theta$; $\theta_\sigma = 0^o$ $\quad \theta_\tau = 45^o$

---

1. 50

Solution d = 1 $\tau_A = 20$ksi P = ? $\sigma_A = ?$

By equilibrium of forces in n and t direction we obtain:

$$N = Pcos45 \qquad V = Psin45 \tag{1}$$

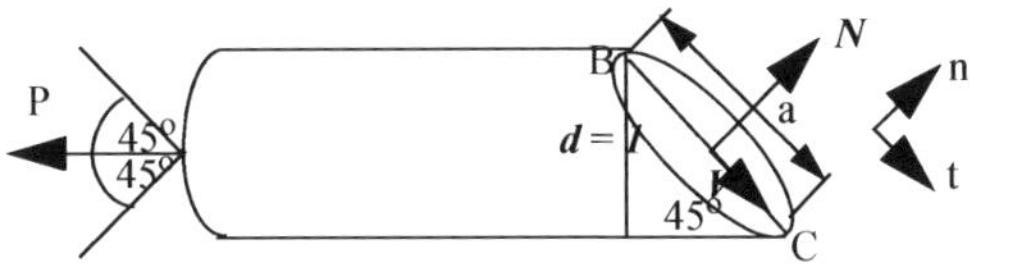

$$a = d/(sin45) = 1.414 \qquad A = \frac{\pi}{4}ad = \frac{\pi}{4}(1.414) = 1.1107in^2 \qquad V = \tau_A A = (20)(1.1107) = 22.214kips \tag{2}$$

$$P = \frac{V}{sin45} = \frac{22.214}{sin45} = 31.4\ kip \qquad N = Pcos45 = 31.4cos45 = 22.214kips \qquad \sigma = \frac{N}{A} = \frac{22.214}{1.1107} = 20 \tag{3}$$

ANS $P = 31.4\ kips$; $\sigma = 20ksi(T)$

---

1. 51

Solution A = 2in.^{2} $\sigma = ?$ $V = ?$

The angle θ can be found as $\theta = atan(6/2) = 71.56^o$

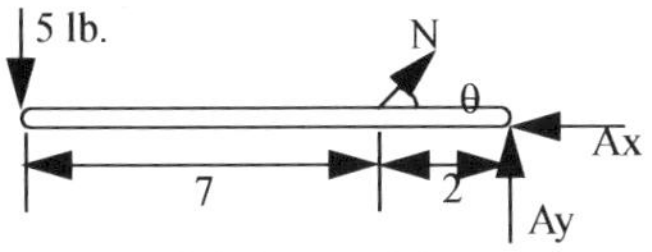

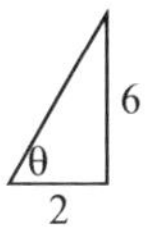

By equilibrium of moment about point A and equilibrium of forces we obtain:

$$(Nsin\theta)(2) - (5)(9) = 0 \qquad or \qquad N = \frac{45}{2sin71.56} = 23.717lbs \tag{1}$$

$$A_x = Ncos\theta = 23.717cos71.56 = 7.5lbs \qquad A_y = 5 - Nsin\theta = 5 - 23.717sin71.56 = -17.5lbs \tag{2}$$

$$\sigma = \frac{N}{A} = \frac{23.717}{2} = 11.858\ psi \qquad V = \sqrt{A_x^2 + A_y^2} = 19.04\ lbs \tag{3}$$

ANS $\sigma = 11.9psi(T)$; $V = 19lbs$

---

1. 52

Solution $A_m = 250$ mm^2 $d_p = 15$ mm $\sigma_{HA} = ?$ $\sigma_{HB} = ?$ $\sigma_{HG} = ?$ $\sigma_{HC} = ?$ $(\tau_H)_{max} = ?$

By equilibrium of moment about point E, and equilbrium of forces we obtain:

$$A_y(12) - 4(9) - 2(6) - 3(3) = 0 \qquad or \qquad A_y = 4.75\ kN \qquad E_y = 4 + 2 + 3 - A_y = 4.25\ kN \qquad E_x = 0 \tag{1}$$

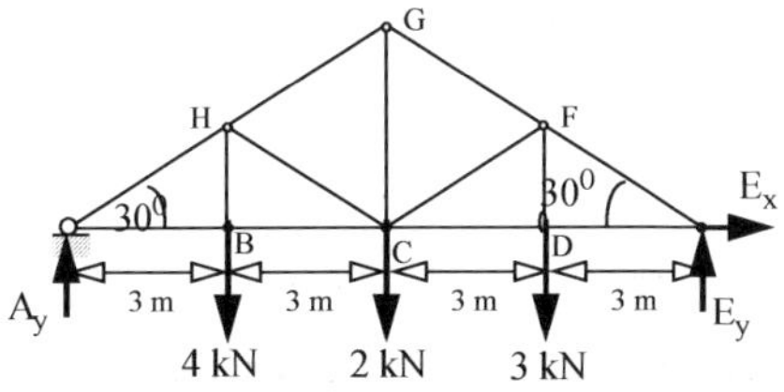

By equilibrium of moment about point C and A in Fig. (a) , we obtain:

$$(N_{HG}sin30 + A_y)(6) - 4(3) = 0 \quad or \quad N_{HG} = -5.5\ kN \qquad (N_{HC}sin30)(6) - 4(3) = 0 \quad or \quad N_{HC} = -4\ kN \quad \textbf{(2)}$$

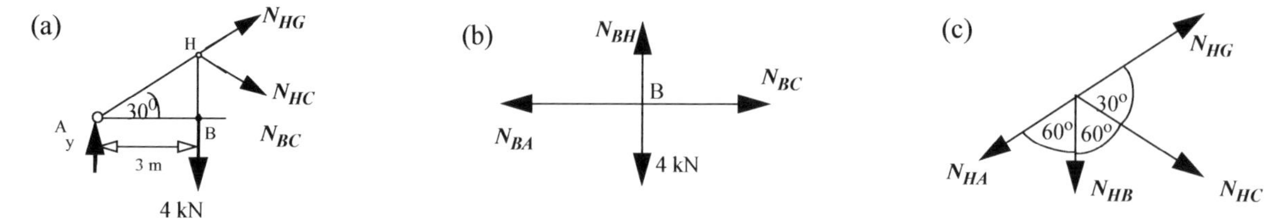

By equilibrium of forces in the y-direction in Fig. (b) and in x-direction Fig. (c) we obtain:

$$N_{HB} = 4\ kN \qquad (N_{HB} + N_{HC})cos30 - N_{HA}sin60 = 0 \quad or \quad N_{HA} = (-5.5 - 4)\frac{cos30}{sin60} = -9.5\ kN \quad \textbf{(3)}$$

$$\sigma_{HA} = \frac{N_{HA}}{A_m} = \frac{-9.5(10^3)}{250(10^{-6})} = -38(10^6)\ N/m^2 \qquad \sigma_{HB} = \frac{N_{HB}}{A_m} = \frac{4(10^3)}{250(10^{-6})} = 16(10^6)\ N/m^2 \quad \textbf{(4)}$$

$$\sigma_{HG} = \frac{N_{HG}}{A_m} = \frac{-5.5(10^3)}{250(10^{-6})} = -22(10^6)\ N/m^2 \qquad \sigma_{HC} = \frac{N_{HC}}{A_m} = \frac{(-4)(10^3)}{250(10^{-6})} = -16(10^6)\ N/m^2 \quad \textbf{(5)}$$

(b) The free body diagram of pin H is shown in Fig. (d). By making imaginary cuts at different sections of the pin the free body diagrams shown in Figs. (e), (f), and (g) are obtained. By equilibrium of forces we obtain the following

$$V_1 = 9.5\ kN \qquad V_3 = 4\ kN \quad \textbf{(6)}$$

$$V_{2x} = 9.5sin60 = 8.22\ kN \qquad V_{2y} = 9.5cos60 - 4 = 0.75\ kN \qquad V_2 = \sqrt{V_{2x}^2 + V_{2y}^2} = 8.25\ kN \quad \textbf{(7)}$$

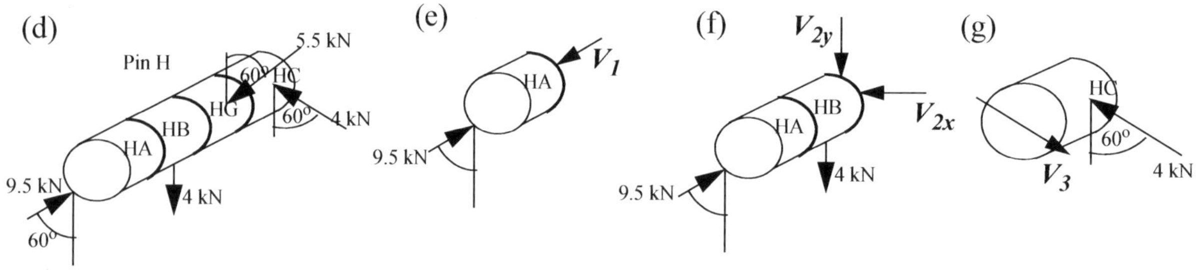

The maximum sheer stress will be between members HA and HB.

$$(\tau_H)_{max} = \frac{V_1}{A_{pin}} = \frac{9.5(10^3)}{\pi(0.015)^2/4} = 53.76(10^6)\ N/m^2 \quad \textbf{(8)}$$

ANS $\sigma_{HA} = 38\ MPa(C)$; $\sigma_{HB} = 16\ MPa(T)$; $\sigma_{HG} = 22\ MPa(C)$; $\sigma_{HC} = 16\ MPa(C)$; $(\tau_H)_{max} = 53.76\ MPa$

---

1. 53

Solution $A_m = 250\ mm^2$ $d_p = 15\ mm$ $\sigma_{FG} = ?$ $\sigma_{FC} = ?$ $\sigma_{FD} = ?$ $\sigma_{FE} = ?$ $(\tau_F)_{max} = ?$

From previous problem we have the following reaction forces.

$$A_y = 4.75\ kN \qquad E_x = 0 \qquad E_y = 4.25\ kN \quad \textbf{(1)}$$

We can make an imaginary cut through members FG, FC, and DC to obtain the free body diagram shown in Fig. (b).

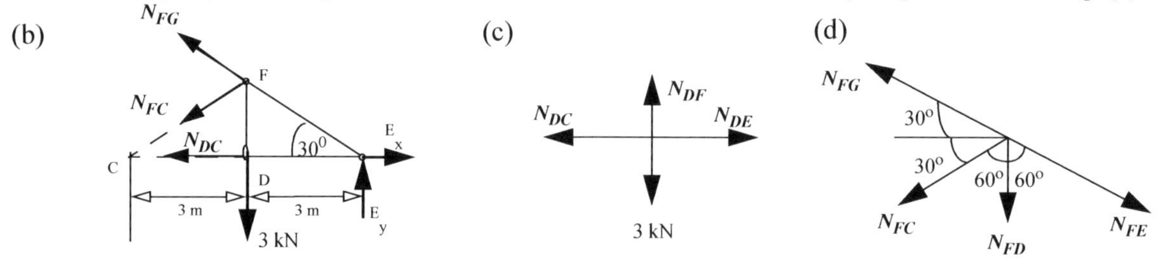

By equilibrium of moment about point C and E in Fig. (b), we obtain:

$$N_{FG}sin30(6) + 4.25(6) - 3(3) = 0 \quad or \quad N_{FG} = -5.5\ kN \qquad (N_{FC}sin30)(6) + 3(3) = 0 \quad or \quad N_{FC} = -3\ kN \quad \textbf{(2)}$$

By equilibrium of forces in the y-direction in Fig. (c) and in x-direction in Fig. (d) we obtain:

$$N_{DF} = 3\ kN \qquad N_{FE}sin60 - (N_{FG} + N_{FC})cos30 = 0 \quad or \quad N_{FE} = (-5.5 - 3)\frac{cos30}{sin60} = -8.5\ kN \quad \textbf{(3)}$$

$$\sigma_{FG} = \frac{N_{FG}}{A_m} = \frac{-5.5(10^3)}{250(10^{-6})} = -22(10^6)\ N/m^2 \qquad \sigma_{FC} = \frac{N_{FC}}{A_m} = \frac{(-3)(10^3)}{250(10^{-6})} = (-12)(10^6)\ N/m^2 \quad \textbf{(4)}$$

$$\sigma_{FD} = \frac{N_{FD}}{A_m} = \frac{3(10^3)}{250(10^{-6})} = 12(10^6)\ N/m^2 \qquad \sigma_{FE} = \frac{N_{FE}}{A_m} = \frac{(-8.5)(10^3)}{250(10^{-6})} = -34(10^6)\ N/m^2 \tag{5}$$

The free body diagram of pin F is shown in Fig. (e). By making imaginary cuts at different sections of the pin the free body diagrams shown in Figs. (f), (g), and (h) are obtained. By equilibrium of forces we obtain the following.

$$V_1 = 5.5\ kN \qquad V_3 = 8.5\ kN \tag{6}$$

$$V_{2x} = 8.5 sin 60 = 7.36\ kN \qquad V_{2y} = 8.5 cos 60 - 3 = 1.25\ kN \qquad V_2 = \sqrt{V_{2x}^2 + V_{2y}^2} = 7.47\ kN \tag{7}$$

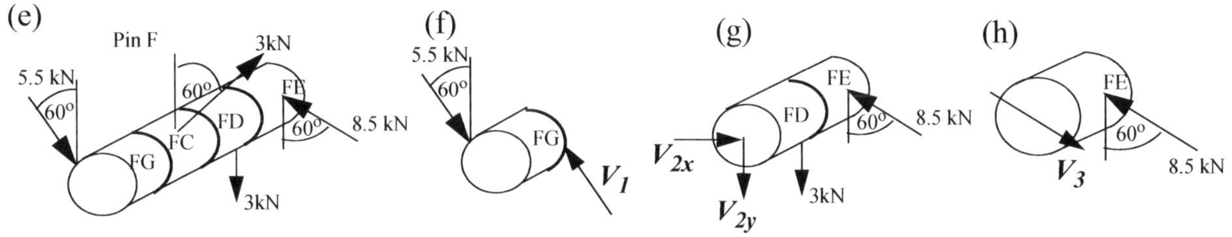

The maximum sheer stress will be between members FD and FE.

$$(\tau_F)_{max} = \frac{V_3}{A_{pin}} = \frac{8.5(10^3)}{\pi(0.015)^2/4} = 48.1(10^6)\ N/m^2 \tag{8}$$

ANS $\sigma_{FG} = 22$ MPa (C) ; $\sigma_{FC} = 12$ MPa (C) ; $\sigma_{FD} = 12$ MPa (T) ; $\sigma_{FE} = 34$ MPa (C) ; $(\tau_F)_{max} = 48.1\ MPa$

---

1. 54

Solution $A_m = 200mm^2$ $d_p = 10mm$ $\sigma_{GH} = ?$ $\sigma_{GC} = ?$ $\sigma_{GF} = ?$ $(\tau_G)_{max} = ?$

From previous problem we have the following reaction forces.

$$A_y = 4.75\ kN \qquad E_x = 0 \qquad E_y = 4.25\ kN \tag{1}$$

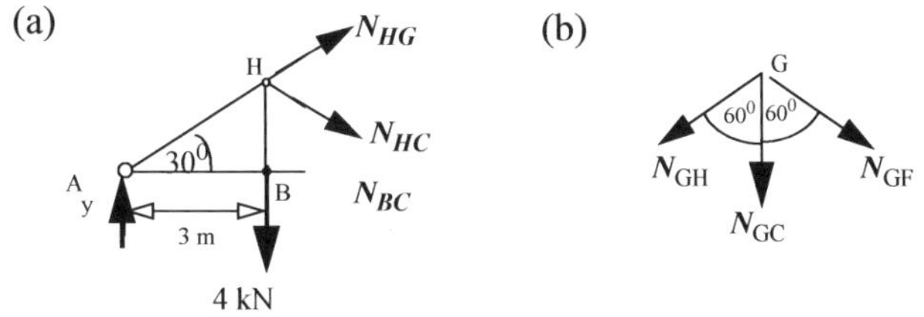

Making an imaginary cut through members HG, HC, and BC we obtain the free body diagram in Fig. (a). By equilibrium of moment about point C, and equilibrium of forces in Fig. (b) we obtain:

$$(N_{HG} sin 30 + A_y)(6) + 4 = 0 \qquad or \qquad N_{HG} = -5.5\ kN = N_{GH} \tag{2}$$

$$-N_{GH} sin 60 + N_{GF} sin 60 = 0 \quad or \quad N_{GH} = N_{GF} = -5.5\ kN \quad N_{GC} = -(N_{GH} cos 60 + N_{GF} cos 60) = 5.5\ kN \tag{3}$$

$$\sigma_{GH} = \frac{GH}{A_m} = \frac{(-5.5)(10^3)}{200(10^{-6})} = -27.5(10^6)\ N/m^2 \qquad \sigma_{GC} = \frac{N_{GC}}{A_m} = \frac{(5.5)(10^3)}{200(10^{-6})} = 27.5(10^6)\ N/m^2 \tag{4}$$

$$\sigma_{GF} = \frac{N_{GFD}}{A_m} = \frac{(-5.5)(10^3)}{200(10^{-6})} = -27.5(10^6)\ N/m^2 \tag{5}$$

The free body diagram of pin G is shown in Fig. (d). By making imaginary cuts at different sections of the pin the free body diagrams shown in Figs. (e) and (f) are obtained. By force equilibrium we obtain

$$V_1 = 5.5\ kN \qquad V_2 = 5.5\ kN \tag{6}$$

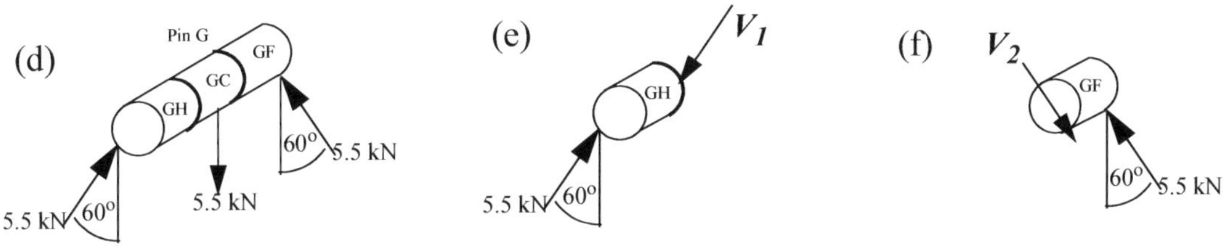

$$(\tau_G)_{max} = \frac{(5.5)(10^3)}{\pi(0.01)^2/4} = 70.03(10^6) = 70.03 MPa \tag{7}$$

ANS $\sigma_{GH} = 27.5$ MPa (C) ; $\sigma_{GC} = 27.5$ MPa (T) ; $\sigma_{GF} = 27.5$ MPa (C) ; $(\tau_G)_{max} = 70\ MPa$

---

1. 55

Solution $d_c = 1/2$ in. $A_{AB} = 2\ in^2$ $A_{BC} = 2.5\ in^2$ $\sigma_{AB} = ?$ $\tau_c = ?$

By equilibrium of moment about point C in Fig. (a) and equilibrium of forces we obtain:

$$5280(33) - N_{AB}(66 sin60) = 0 \quad or \quad N_{AB} = 3048.4 lbs \quad \textbf{(1)}$$

$$C_x = 5280 cos60 - N_{AB} = 1524.2\ lbs \quad C_y = 5280 sin60 = 2640\ lbs \quad 2V = \sqrt{C_x^2 + C_y^2} = 1524.2 \quad V = 762.1 lbs \quad \textbf{(2)}$$

(a)

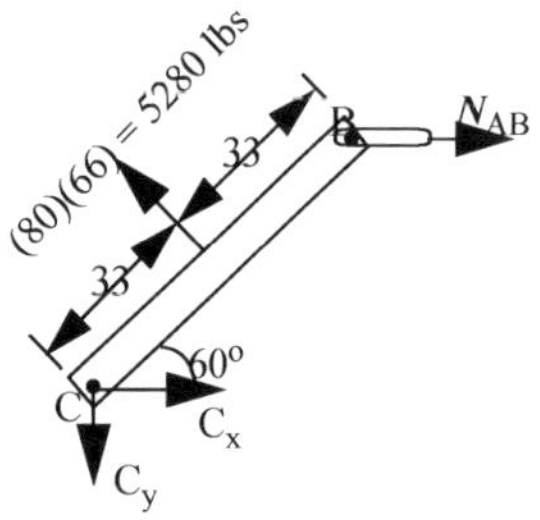

(b)

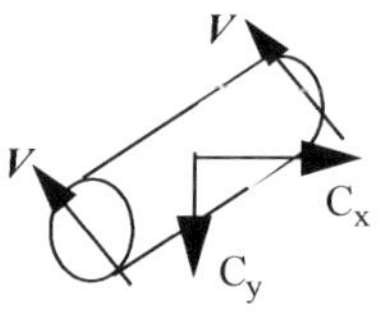

$$\sigma_{AB} = \frac{N_{AB}}{A_{AB}} = \frac{3048.4}{2} = 1524.2 psi \quad \tau_C = \frac{V}{A_C} = \frac{762.1}{(\pi(0.5)^2/4)} = 3881.3 psi \quad \textbf{(3)}$$

ANS $\sigma_{AB} = 1524\ psi(T)$; $\tau_C = 3881 psi$

---

1. 56

Solution d = 40mm $\sigma_{BD} = ?$ $\tau_{max} = ?$

By equilibrium of moment about point A in Fig. (a) and equilbrium of forces we obtain the following.

$$E_x(3) - 250(2.5) = 0 \quad or \quad E_x = 208.33\ kN \quad A_x = E_x = 208.33\ kN \quad A_y = 250\ kN \quad \textbf{(1)}$$

(a)

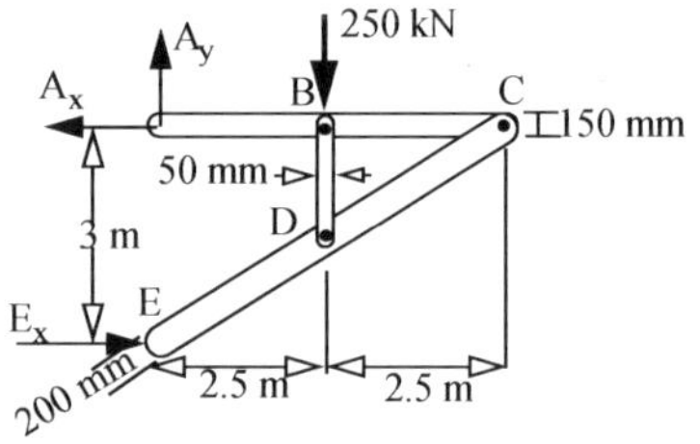

(b)

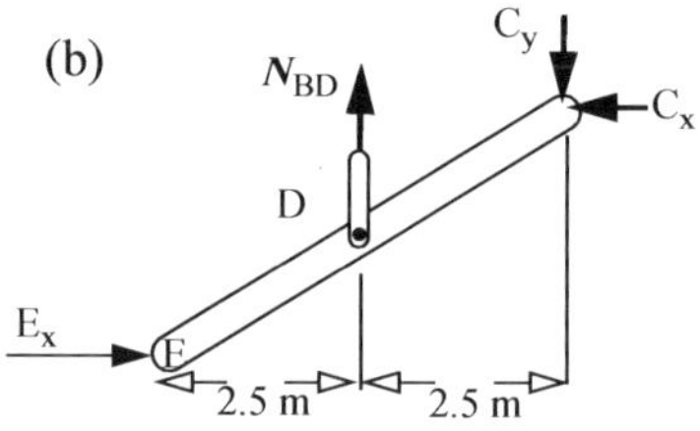

By equilibrium of moment about point C in Fig. (b) and equilibrium of forces we obtain the following.

$$N_{BD}(2.5) - 208.33(3) = 0 \quad or \quad N_{BD} = 250\ kN \quad C_x = E_x = 208.33\ kN \quad C_y = N_{BD} = 250\ kN \quad \textbf{(2)}$$

$$F_A = \sqrt{A_x^2 + A_y^2} = 325 kN \quad F_C = \sqrt{C_x^2 + C_y^2} = 325 kN$$

$$\sigma_{BD} = \frac{N_{BD}}{A_{BD}} = \frac{250(10^3)}{(0.05)(0.05)} = 100(10^6)\ N/m^2 \quad \tau_{max} = \frac{F_A}{A_A} = \frac{325(10^3)}{(\pi/4)(0.04)^2} = 258.6(10^6)\ N/m^2 \quad \textbf{(3)}$$

ANS $\sigma_{BD} = 100 MPa\ (T)$; $\tau_{max} = 259 MPa$

---

1. 57

Solution W = 36 lbs $W_A = 140$lb. $W_L = 32$ lb $\sigma = ?$

We draw the FBD and calculate various quantities as shown below.

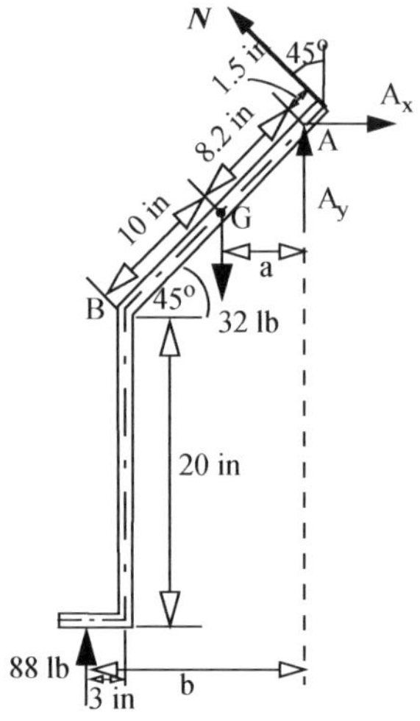

From equilibrium of moment about A

$$N(1.5) + 32a - 88b = 0 \quad \textbf{(1)}$$

From geometry

$$a = 8.2 cos45 = 5.798\ in \quad \textbf{(2)}$$

$$b = 18.2 cos45 + 3 = 15.869\ in \quad \textbf{(3)}$$

Substituting a and b we obtain

$$N = \frac{(88)(15.869) - 32(5.798)}{1.5} = 807.31\ lb \quad \textbf{(4)}$$

Thus, the stress in the muscle is:

$$\sigma = \frac{N}{A} = \frac{807.31}{1.75} = 461.3\ psi \quad \textbf{(5)}$$

ANS $\sigma = 461.3\ psi(T)$

1. 58

Solution W=15 lb $W_A = 9$ lb A=0.75 in^2

From equilibrium of moment about O

$$N\sin 15(6) - (6)W_A - (24)W = 0 \quad or \quad N = \frac{(6)(9)+(24)(15)}{6\sin 15} = 266.6\ lb \qquad \sigma = \frac{N}{A} = \frac{266.6}{0.75} = 354.7\ psi \qquad 1$$

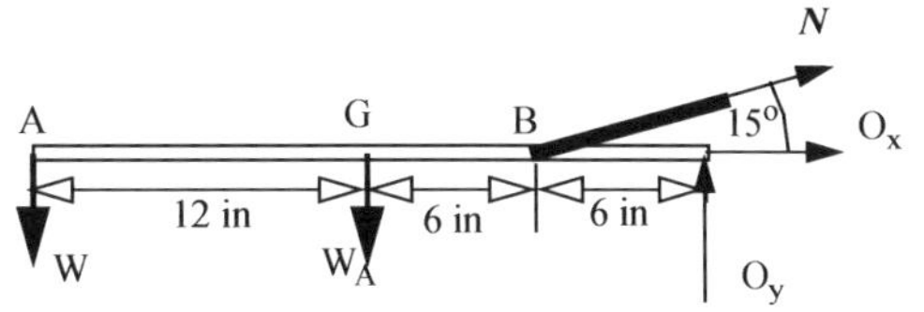

ANS $\sigma = 354.7\ psi(T)$

---

1. 59

Solution $\tau_{max} \le 50MPa$ P vs. d = ?

The shear force on the bottom screw is $V_{bottom} = 0.6P$

$$\tau_{max} = \frac{V_{bottom}}{A} = \frac{0.6P}{(\pi d^2/4)} \le 50(10^6) \quad or \quad \frac{P}{d^2} \le 65.45(10^6) \qquad (1)$$

The table below list the values of d and P that satisfies the above inequality.

d (mm)	P (Newton)
1	65
2	261
3	589
4	1047
5	1636

---

1. 60

Solution σ_{allow} - 30 ksi τ_{allow} - 15ksi σ_{bear} - 20 ksi $P_{max} = ?$

By equilibrium we obtain:

$$V = P/2 \qquad N = P \qquad N_b = P \qquad (1)$$

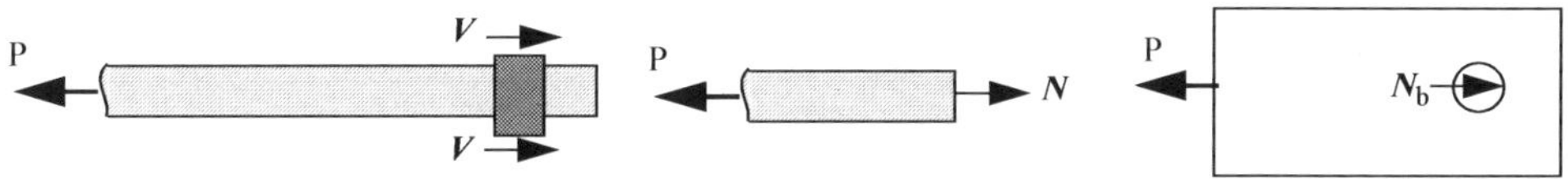

The stresses should be below the limiting values

$$\tau_r = \frac{V}{A_r} = \frac{P/2}{(\pi/4)(0.5^2)} = \frac{P/2}{0.1963} \le 20\ ksi \quad or \quad P \le 7.852\ kips \qquad (2)$$

$$\sigma = \frac{N}{A} = \frac{P}{(1)(0.5)} \le 30\ ksi \quad or \quad P \le 15\ kips \qquad (3)$$

$$\sigma_b = \frac{N_b}{A_b} = \frac{P}{(\pi/2)(t)d_r} = \frac{P}{(\pi/2)(0.5)(0.5)} = \frac{P}{0.3927} \le 15\ ksi \quad or \quad P \le 5.891\ kips \qquad (4)$$

ANS $P_{max} = 5.8\ kips$

---

1. 61

Solution $\sigma_{max} \le 10ksi$ W = 500-lb d = ? to nearest 1/16th of an inch

By force equilibrium in the y-direction on the tire and force equilibrium on the chain linl we obtain the following.

$$3T\cos 12 = 500 \quad or \quad T = 170.39\ lb \qquad N = T/2 = 85.19\ lb. \qquad (1)$$

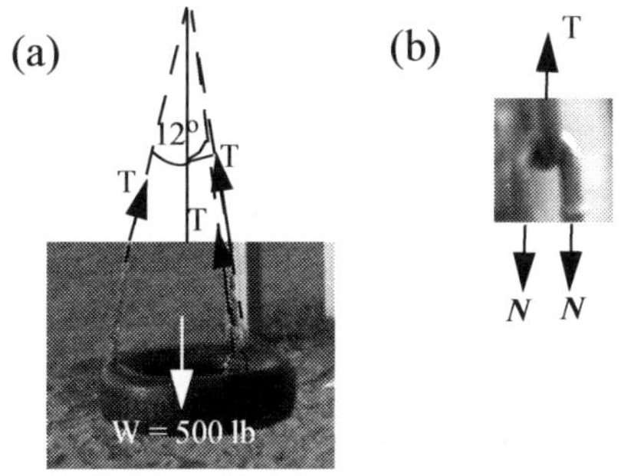

$$\sigma = \frac{N}{A} = \frac{85.195}{(\pi d^2/4)} \le 10(10^3) \qquad or \qquad d \ge 0.1041\,inch \tag{2}$$

ANS $d_{min} = (1/8)$ in

---

1. 62

Solution $\sigma_{cast} \le 150MPa$ $\tau_{bolt} \le 200MPa$ $d_{bolt} = 15mm$ $P_{max} = ?$

The smaller pipe will have higher normal stress. By force equilibrium in figure (a) and (b) we obtain:

$$N = P \qquad V = P/2 \tag{1}$$

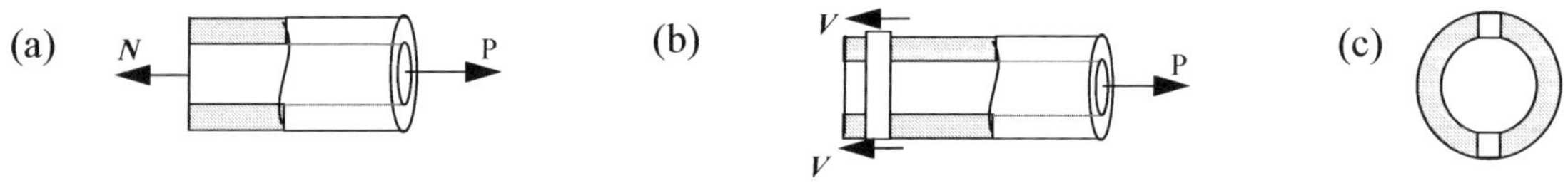

$$A_{cast} = \frac{\pi}{4}(0.05^2 - 0.03^2) - 2(0.01)(0.015) = 0.9566(10^{-3})\ m^2 \qquad A_{bolt} = \frac{\pi}{4}0.015^2 = 176.7(10^{-6})\ m^2 \tag{2}$$

$$\sigma_{cast} = \frac{N}{A_{cast}} = \frac{P}{0.9566(10^{-3})} \le 150(10^6) \qquad or \qquad P \le 143.49(10^3)\ N \tag{3}$$

$$\tau_{bolt} = \frac{V}{A_{bolt}} = \frac{(P/2)}{(176.7)(10^{-6})} \le 200(10^6) \qquad or \qquad P \le 70.68(10^3)\ N \tag{4}$$

ANS $P_{max} = 70.6kN$

---

1. 63

Solution $\sigma = 20$ ksi $d_{rivet} = 1/2$ $\tau_{rivet} \le 40ksi$ 2n= number of rivets =?

By equilibrium of forces we have:

$$nV = (20)(8)(1) \qquad or \qquad V = (160/n)\ kips \tag{1}$$

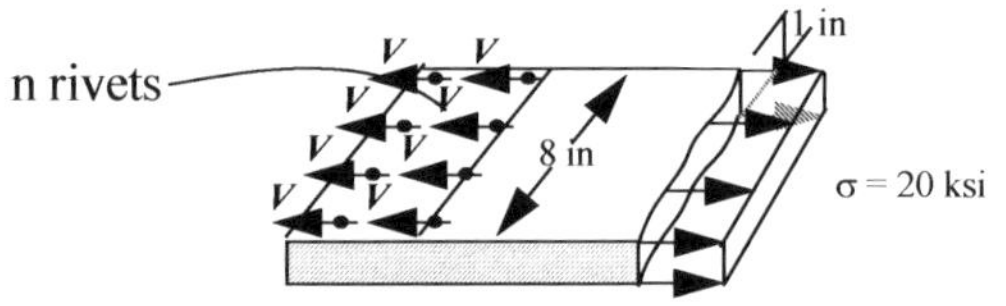

$$A_{rivet} = \frac{\pi}{4}\left(\frac{1}{2}\right)^2 = 0.1963\ in^2 \qquad \tau_{rivet} = \frac{V}{A_{rivet}} = \frac{160}{n(0.1963)} \le 40 \qquad or \qquad n \ge \frac{160}{(40)(0.1963)} \qquad or \qquad n \ge 20.37 \tag{2}$$

ANS $2n = 42\ rivets$

---

1. 64

Solution $d_p = 20$mm configuration = ?

Configuration 1

By equilibrium we obtain the shear forces and calculate the maximum shear stress.

$$V_1 = 32.68\ kN \qquad V_{2x} = 30\ kN \qquad V_{2y} = 50\ kN \qquad V_2 = \sqrt{30^2 + 50^2} = 58.31kN \qquad V_3 = 32.68\ kN \tag{1}$$

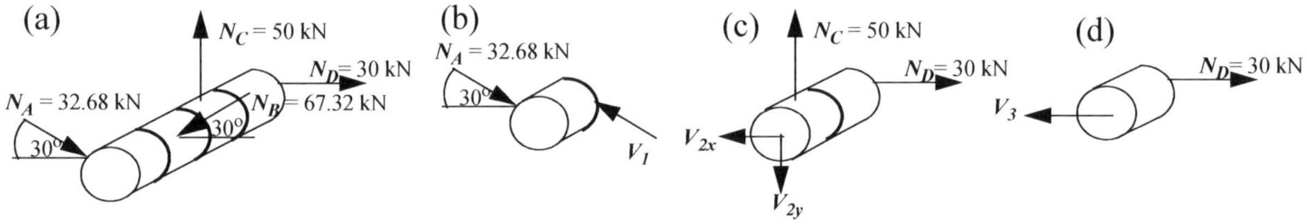

$$A = \pi(0.02^2)/4 = 0.3142(10^{-3})m^2 \qquad (\tau_1)_{max} = V_2/A = 58.31(10^3)/[0.3142(10^{-3})] = 185.6(10^6)\ N/m^2 \tag{2}$$

Configuration 2

By equilibrium we obtain the shear forces and calculate the maximum shear stress.

$$V_1 = 32.68\ kN \qquad V_{2x} = 28.3\ kN \qquad V_{2y} = 33.66\ kN \qquad V_2 = \sqrt{28.3^2 + 33.66^2} = 43.97kN \qquad V_3 = 30\ kN \tag{3}$$

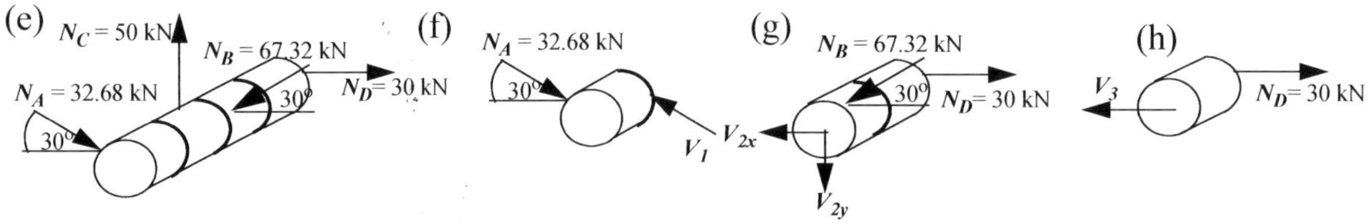

$$(\tau_2)_{max} = V_2/A = 43.97(10^3)/[0.3142(10^{-3})] = 139.98(10^6)\ N/m^2 \tag{4}$$

ANS Configuration 2 is better.

---

1. 65

Solution Sequence = ?

There are 6 possible sequences. Three of the sequences are shown below. The remaining three are the mirror image of the three shown and hence will result in the same shear stress.

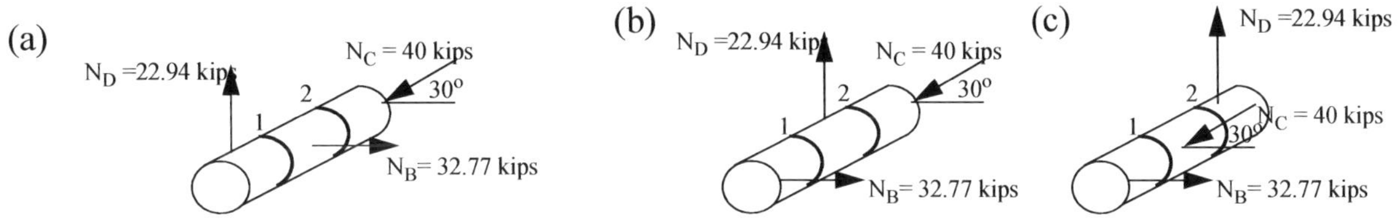

The shear force at the imaginary cuts 1 and 2 will equal to the force at the nearest end. Thus, if smaller forces are on the outside, then the shear stress will be smaller. Configuration shown in Fig.(c) will yield the smallest shear stress.

ANS Member should be assembled as in Fig.(c).

---

1. 66

Solution $N \geq 235 kips$ d_s = 1/2 in. bar = 8in x 8in σ_{conc} = 3 ksi (C) σ_s = 20 ksi (C) n_{min} = number of steel bars = ?

The cross-section of the reinforced concrete bar is shown.

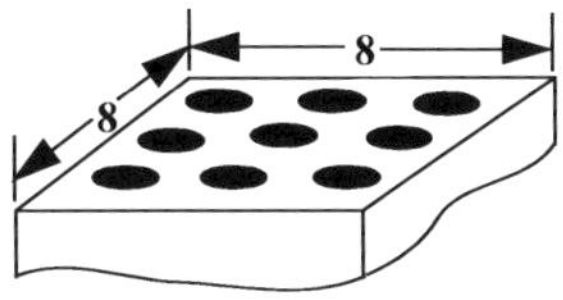

$$A_s = n(\pi(1/2)^2)/4 = 0.19635n\ in^2 \qquad A_{conc} = (8)(8) - 0.19635n = (64 - 0.19635n)\ in^2 \tag{1}$$

$$N_s = \sigma_s A_s = (20)0.19635n = 3.927n\ kips \qquad N_{conc} = \sigma_{conc} A_{conc} = (3)(64 - 0.19635n) = (192 - 0.58905n)\ kips \tag{2}$$

$$N = N_s + N_{conc} = 192 + 3.3379n \geq 235 \quad or \quad n \geq 12.88 \tag{3}$$

ANS $n_{min} = 13$

---

1. 67

Solution cross-section = 2in x 4in $\sigma \leq 800\ psi$ $\tau \leq 350\ psi$ P_{max} = ?

By equilibrium of forces in the n and t direction we have the following.

$$N = P sin 50 = 0.7660P \qquad V = P cos 50 = 0.6428P \tag{1}$$

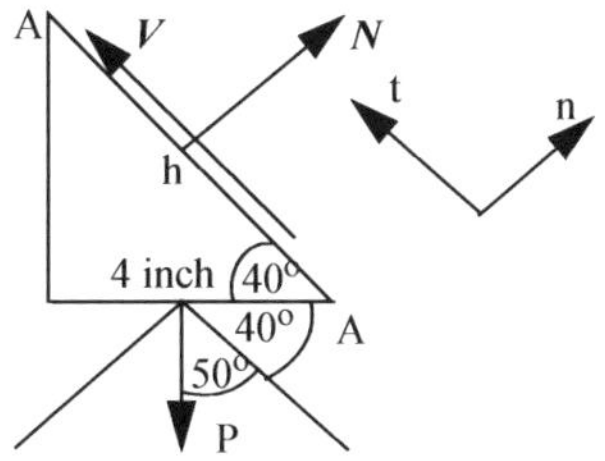

$$A = 2(h) = 2(4/(cos 40)) = 10.44\ in^2 \tag{2}$$

$$\sigma = \frac{N}{A} = \frac{0.766P}{10.44} \leq 800 \quad or \quad P \leq 10902 lbs \qquad \tau = \frac{V}{A} = \frac{0.6428P}{10.44} \leq 350 \quad or \quad P \leq 5684 lbs \tag{3}$$

ANS $P_{max} = 5684\ lb$

---

1. 68

Solution L = 5 in. $\sigma_W \leq 6000 psi$ $\tau_{glue} \leq 300 psi$ P_{max} = ?

By equilibrium of forces we have: ***N*** = P and ***V*** = P/2.

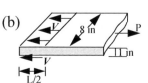

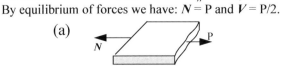

$$\sigma_W = \frac{N}{A_W} = \frac{P}{(8)(1)} \le 6000 \quad or \quad P \le 48000\ lbs \qquad \tau_{glue} = \frac{V}{A_{glue}} = \frac{P}{(2)(8)(5/2)} \le 300 \quad or \quad P \le 12000\ lbs \quad \textbf{(1)}$$

ANS $P_{max} = 12\ kips$

---

1. 69

Solution P = 25 kips $\tau_{glue} \le 300psi$ L =?

By equilibrium we have $2V = P = 25(10^3)$ or $V = 12.5(10^3)$ lbs. The shear stress can be found as shown below.

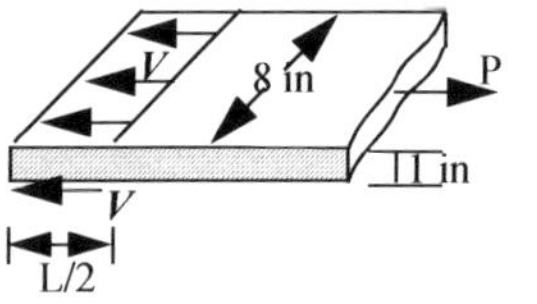

$$\tau_{glue} = \frac{V}{A_{glue}} = \frac{12.5(10^3)}{(8)(L/2)} \le 300 \quad or \quad L \ge 10.412\ in \quad \textbf{(1)}$$

ANS $L = 10.4\ in$

---

1. 70

Solution $\sigma \le 160MPa$ d_{EP} = ? d_{EG} = ? d_{EF} = ? $\tau_E \le 250MPa$ d_E = ?

By equilibrium of forces we obtain the following.

$$N_{EF} = 65\ kN \qquad N_{EG}sin\theta + N_{EF} = 0 \quad or \quad N_{EG} = -108.3\ kN \quad \textbf{(1)}$$

$$N_{ED}cos\theta + N_{EF} = 0 \quad or \quad N_{ED} = -86.67\ kN \quad \textbf{(2)}$$

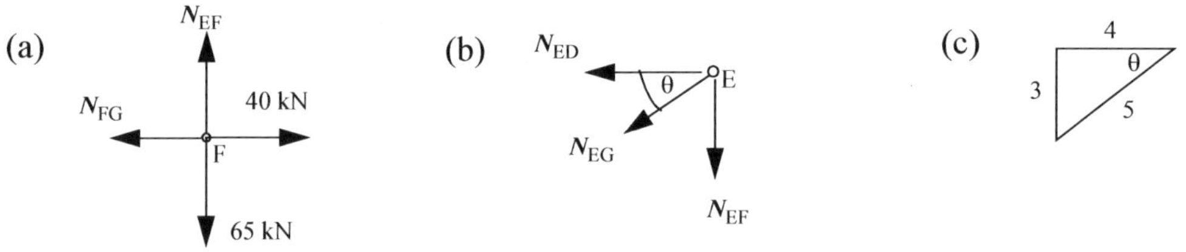

$$|\sigma_{ED}| = \left|\frac{N_{ED}}{(\pi d_{ED}^2)/4}\right| = \frac{86.67(10^3)}{(\pi d_{ED}^2)/4} \le 160(10^6) \quad or \quad d_{ED} \ge 26.26(10^{-3})m \quad \textbf{(3)}$$

$$|\sigma_{EG}| = \left|\frac{N_{EG}}{(\pi d_{EG}^2)/4}\right| = \frac{108.3(10^3)}{(\pi d_{EG}^2)/4} \le 160(10^6) \quad or \quad d_{EG} \ge 29.36(10^{-3})m \quad \textbf{(4)}$$

$$|\sigma_{EF}| = \left|\frac{N_{EF}}{(\pi d_{EF}^2)/4}\right| = \frac{65(10^3)}{(\pi d_{EF}^2)/4} \le 160(10^6) \quad or \quad d_{EF} \ge 22.7(10^{-3})m \quad \textbf{(5)}$$

There are 6 possible assembly sequences. Three of the sequences are shown below. The remaining three are mirror images and will result in the same shear stresses. The shear force at the imaginary cuts 1 and 2 will equal to the force at the nearest end. Thus, if smaller forces are on the outside, then the shear stress will be smaller. Assembly in Fig. (b) will yield the smallest shear stress.

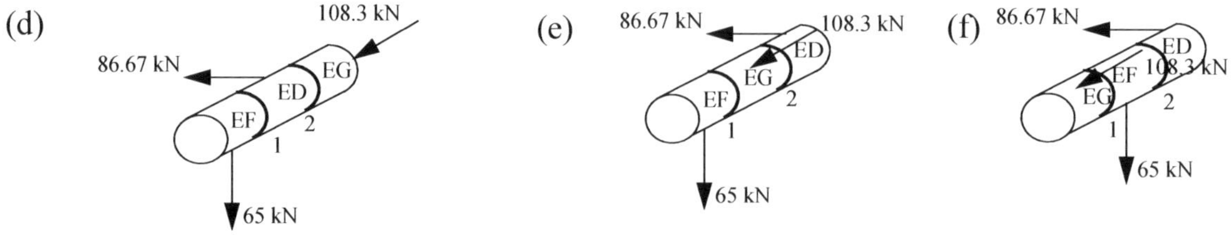

In Fig. (b) the max shear force is at cut 2 and its value is: V = 86.67 kN

$$\tau_E = \left|\frac{V}{(\pi d_E^2)/4}\right| = \frac{86.67(10^3)}{(\pi d_E^2)/4} \le 250(10^6) \quad or \quad d_E \ge 21.01(10^{-3})m \quad \textbf{(6)}$$

ANS $d_{ED} = 27\ mm$; $d_{EG} = 30\ mm$; $d_{EF} = 23\ mm$; The sequence of assembly shown in Fig. (b) should be used.; $d_E = 22\ mm$

---

1. 71

Solution $\sigma \le 160MPa$ d_{CG} = ? d_{CD} = ? d_{CB} = ? $\tau_E \le 250MPa$ d_C = ?

From previous problem we know $N_{ED} = 86.67$ kN. By equilibrium of joints D and C we obtain

$$N_{CD} = N_{ED} = 86.67\ N \qquad N_{CG}cos\theta + N_{CD} = 0 \qquad or \qquad N_{CG} = -108.3\ kN \tag{1}$$

$$N_{CG}sin\theta + N_{CD} = 0 \qquad or \qquad N_{CG} = 65.0\ kN \tag{2}$$

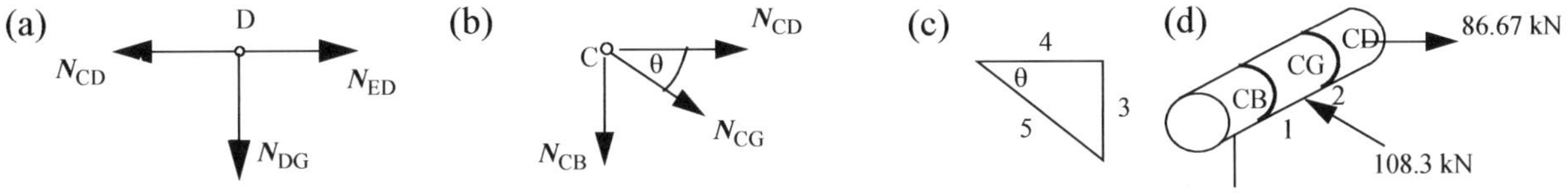

$$|\sigma_{CG}| = \left|\frac{N_{CG}}{(\pi d_{CG}^2)/4}\right| = \frac{108.3(10^3)}{(\pi d_{CG}^2)/4} \le 160(10^6) \qquad or \qquad d_{CG} \ge 29.36(10^{-3})m \tag{3}$$

$$|\sigma_{CD}| = \left|\frac{N_{CD}}{(\pi d_{CD}^2)/4}\right| = \frac{86.67(10^3)}{(\pi d_{CD}^2)/4} \le 160(10^6) \qquad or \qquad d_{CD} \ge 26.26(10^{-3})m \tag{4}$$

$$|\sigma_{CB}| = \left|\frac{N_{CB}}{(\pi d_{CB}^2)/4}\right| = \frac{65(10^3)}{(\pi d_{CB}^2)/4} \le 160(10^6) \qquad or \qquad d_{CB} \ge 22.7(10^{-3})m \tag{5}$$

The configuration that will produce the smallest maximum shear stress will be the one with members carrying the smallest forces on the outside edge of the pin. Free body diagram of pin C is shown below in Fig. (d).
In Fig. (d) the max shear force is at cut 2 and its value is: $V = 86.67$ kN

$$\tau_C = \left|\frac{V}{(\pi d_C^2)/4}\right| = \frac{86.67(10^3)}{(\pi d_C^2)/4} \le 250(10^6) \qquad or \qquad d_C \ge 21.01(10^{-3})m \tag{6}$$

ANS $d_{CG} = 30\ mm$; $d_{CD} = 27\ mm$; $d_{C;B} = 23\ mm$; The sequence of assembly shown in Fig. (d) should be used.; $d_C = 22\ mm$

---

1. 72

Solution assembly sequence = ?

There are factorial four i.e., twenty-four possible ways of assembling the four member. Twelve of the twenty-four sequence are mirror images of the other twelve. To consider the twelve assembly sequence, we consider the figure below, where a member may be in any of the four location shown.

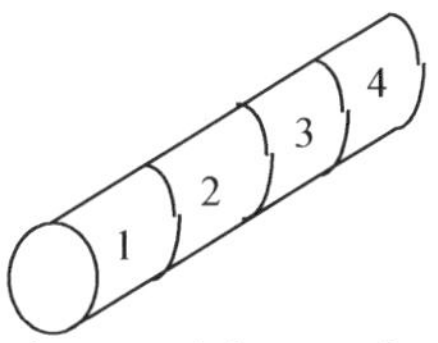

The table below describe the location of the four members and the maximum shear stress location and value of maximum shear stress for each sequence.

Sequence Number	Member Location 1	Member Location 2	Member Location 3	Member Location 4	Location of Maximum shear force between members	Maximum Shear Force (kips)
1	AB	AC	AD	AE	1 and 2	40
2	AB	AC	AE	AD	1 and 2	40
3	AC	AB	AD	AE	2 and 3	37.8
4	AC	AB	AE	AD	2 and 3	37.8
5	AB	AD	AE	AE	2 and 3	47.17
6	AB	AD	AE	AC	2 and 3	47.17
7	AD	AB	AC	AE	2 and 3	47.17
8	AD	AB	AE	AC	2 and 3	47.17
9	AB	AE	AC	AD	1 and 2	40
10	AB	AE	AD	AC	1 and 2	40
11	AE	AB	AC	AD	1 and 2	28.34
12	AE	AB	AD	AC	1 and 2	28.34

The maximum shear stress is smallest for sequence 11 and 12, thus we have the following conclusion.

The members should be assembled in the following sequence or the mirror image of the given sequence
AE, AB, AD, AC or AE, AB, AD, AC

1. 73

Solution Post 2 in x 4 in $L = 12$ in $\tau = 2$ psi $P_{min} = ?$

By equilibrium of forces the applied force should equal the average shear stress multiplied by the surface area as shown below.

$$P = (\tau)[12(2+4+2+4)] = 144\tau \quad or \quad \tau = (P/144) \le 2\ psi \quad or \quad P \le 288 \qquad \textbf{(1)}$$

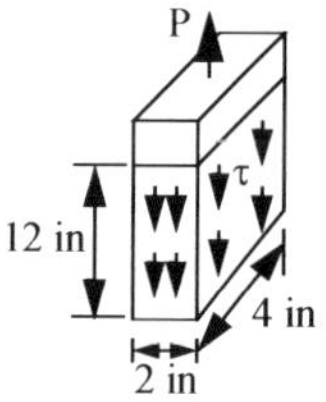

ANS $P_{min} = 288\ lbs$

---

1. 74

Solution $d = 100$ mm $= 0.1$ m $L = 400$ mm $= 0.4$ m $P = 1250$ N $\tau = ?$

By equilibrium of forces the applied force should equal the average shear stress multiplied by the surface area as shown below.

$$P = \tau(0.4)[\pi(0.1)] = 0.04\pi\tau \quad or \quad \tau = 1250/(0.04\pi) = 9947\ N/m^2 \qquad \textbf{(1)}$$

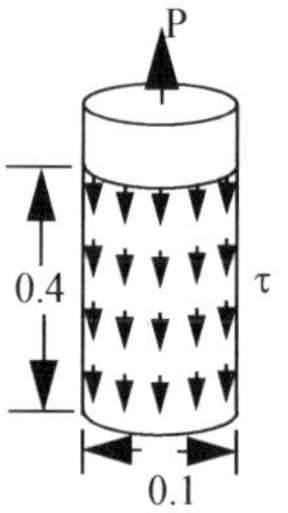

ANS $\tau = 9947\ Pa$

---

1. 75

Solution $P = f(\tau, h, a) = ?$

By equilibrium of forces the applied force should equal the average shear stress multiplied by the surface area as shown below.

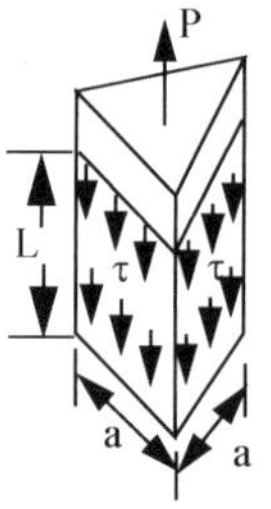

$$P = \tau(3a)(L) \qquad \textbf{(1)}$$

ANS $P = 3aL\tau$

---

1. 76

Solution $P = 10$ lbs $d = 118$ in $L = 2$ in $\tau = ?$

By moment equilibrium about point C, we obtain the following.

$$P_1(2) = P(12) \quad or \quad P_1 = 6P = 60\ lbs \qquad \textbf{(1)}$$

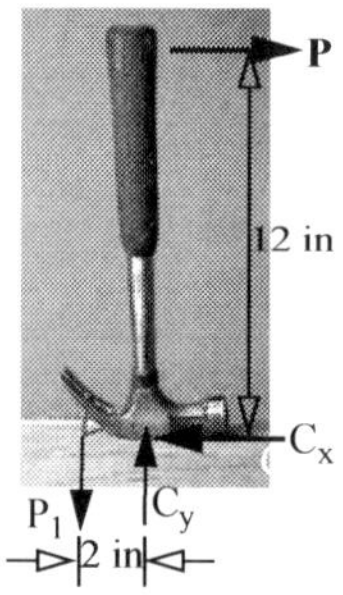

The shear force acting on the nail is $V = P_1$. The surface area of the nail in the wood and the shear stress acting on the nail can be found as shown below.

$$A = L\pi D = (2)\pi(1/8) = 0.7854\ in^2 \qquad \tau = V/A = 60/0.7854 = 76.39\ psi \qquad \textbf{(1)}$$

ANS $\tau = 76.4\ psi$

---

1. 77

Solution $(d_o)_{small} = 50$ mm $(d_i)_{small} = 30$ mm $(d_o)_{big} = 70$ mm $(d_i)_{big} = 50$ mm
P = 100 kN L = 200 mm τ = ?

By equilibrium of forces we have:

$$P = \tau[\pi(d_o)_{small}]L \quad or \quad \tau = \frac{P}{[\pi(d_o)_{small}]L} = \frac{100(10^3)}{\pi(0.05)(0.2)} = 3183(10^3)N/m^3 \quad \textbf{(1)}$$

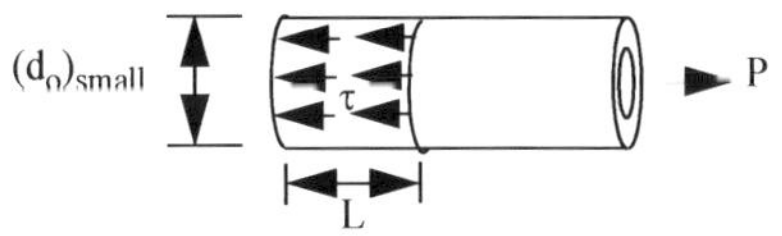

ANS $\tau = 3.18MPa$

---

1. 78

Solution $(d_o)_{small} = 50$ mm $(d_i)_{small} = 30$ mm $(d_o)_{big} = 70$ mm $(d_i)_{big} = 50$ mm
T = 2 kN-m L = 200 mm τ = ?

The free body diagram of the smaller pipe and the differential circular surface area over is shown below.

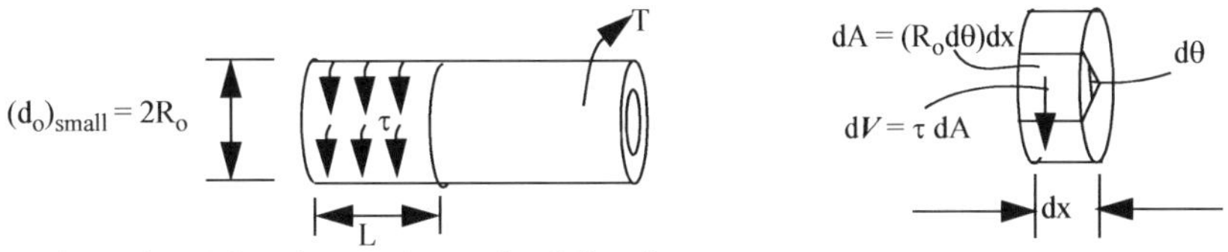

By moment equilibrium about the axis of the pipe we have the following.

$$T = \int_A R_o\ dV = \int_A R_o \tau dA = \int_0^L \int_0^{2\pi} R_o \tau(R_o d\theta)dx = \tau R_o^2 \int_0^L \left[\int_0^{2\pi} d\theta\right]dx = \tau R_o^2[2\pi](L) \text{ or}$$

$$\tau = \frac{T}{2\pi R_o^2 L} = \frac{2(10^3)}{2\pi(0.025)^2(0.2)} = 2.546(10^6)N/m^2 \quad \textbf{(1)}$$

ANS $\tau = 2.55MPa$

---

1. 79

Solution T = 2 kN-m $(d_o)_{small} = 50$ mm $d_{bolt} = 15$ mm τ = ?

By equilibrium of moment we have:

$$T = (2R_o)[V] = (2R_o)[(\tau)(\pi d_{bolt}^2/4)] \quad or \quad \tau = \frac{2T}{(\pi d_{bolt}^2)R_o} = \frac{(2)(2)(10^3)}{\pi(0.015)^2(0.025)} = 226.3(10^6)N/m^2 \quad \textbf{(1)}$$

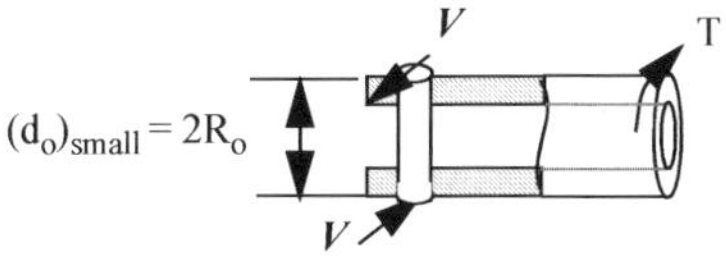

ANS $\tau = 226.3MPa$

---

1. 80

Solution t = 1/64 in a= 1/2 in b = 3 in c =1/4 in τ = 1800 psi F = ?

By equilibrium of moment about D we obtain:

$$F(b) = 2V[c + (a\cos 30)/2] = 2[(\tau)(a)(t)][c + (a\cos 30)/2] \text{ or}$$

$$F(3) = 2(1800)(1/2)(1/64)(1/4 + \cos 30/4) = 13.12 \quad or \quad F = 13.12/3 = 4.374\ lb \quad \textbf{(1)}$$

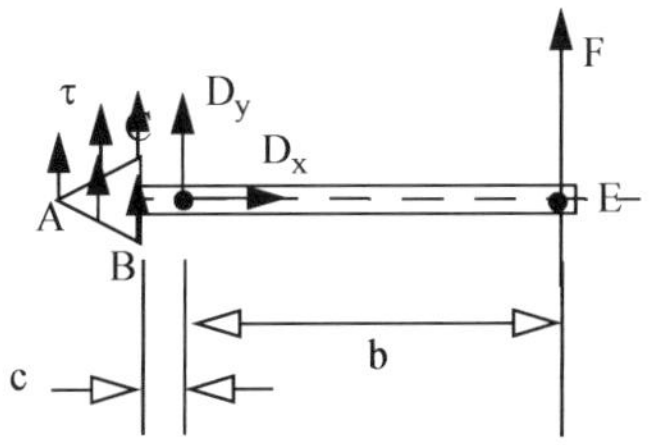

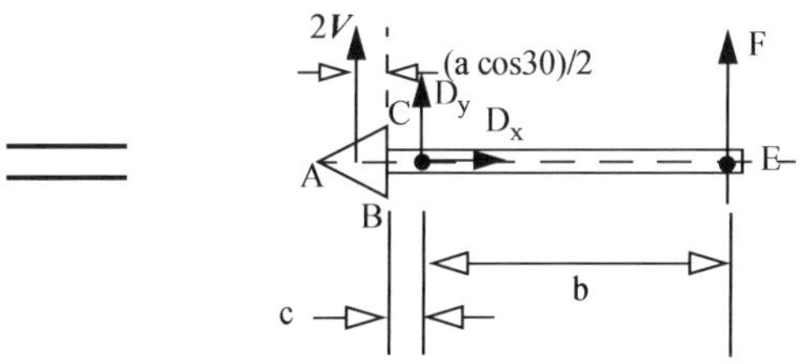

ANS $F = 4.38\ lb$

1. 81

Solution d_{bolt} = 1/2 inch T = 100 in-kips $\tau_{bolt} \le 20ksi$ n = number of bolts = ? r = ?

By equilibrium of moment about the shaft axis we obtain:

$$T = nVr \qquad or \qquad V = 100(10^3)/(nr) \tag{1}$$

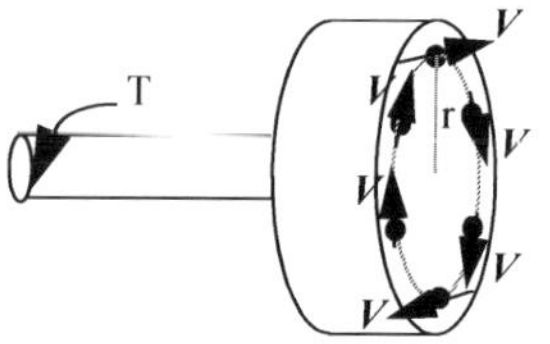

The shear stress in the bolt can be written as given below.

$$\tau_{bolt} = \frac{V}{A} = \frac{100(10^3)}{(nr)(\pi d^2/4)} \le 20(10^3) \qquad or \qquad nr \ge \frac{5}{(\pi 0.5^2/4)} \qquad or \qquad nr \ge 25.465 \tag{2}$$

The minimum value of r is r_{min} = 2+0.25=2.25 corresponding to n = 11.3 i.e., n = 12. The maximum value of r is r_{max} = 5-0.25=4.75 corresponding to n = 5.36 i.e., n = 6. The table below gives all possible answers between n = 6 to n = 12.

n (number of bolts)	6	7	8	9	10	11
r in inches	4.25	3.64	3.18	2.83	2.55	2.31

For minimizing machining and assembly cost 6 bolts should be used.

1. 82

Solution (a) T = 15 in-lb d = 1.5 inch h = 1/2 inch τ = ? (b) d = 1 inch h = 1/2 inch T = ?

The free body diagram of the bottle with lid removed is shown below.

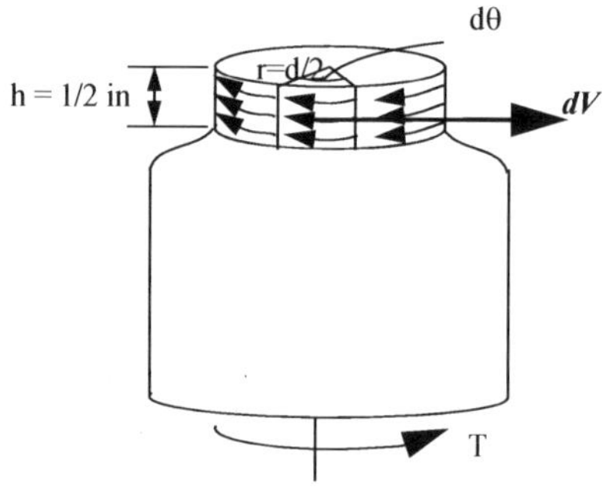

By moment equilibrium about the axis of the bottle we obtain:

$$T = \int_A (d/2)(dV) = \int_0^{2\pi}(d/2)[\tau h(d/2)(d\theta)] = (d^2/4)h\tau\int_0^{2\pi} d\theta = \tau(\pi d^2 h)/2 \tag{1}$$

(a) Substituting d= 1.5 inch, h= 0.5 inch, T= 15 in-lbs, into Eq. (1) we obtain the following.

$$15 = \tau\pi(1.5)^2(0.5)/2 \qquad or \qquad \tau = 8.49psi \tag{2}$$

(b) Substituting d= 1.0 inch, h= 0.5 inch, τ= 8.49 psi, into Eq. (1) we obtain the following

$$T = \tau\pi(1.0)^2(0.5)/2 \qquad or \qquad T = 6.67 \ in-lbs \tag{3}$$

ANS $\tau = 8.5\ psi$; $T = 6.7\ in-lb$

1. 83

Solution t = 3/8 in d = 2 1/2 in a = 4 in τ = 10 psi F = ?

We draw the FBDs as shown below

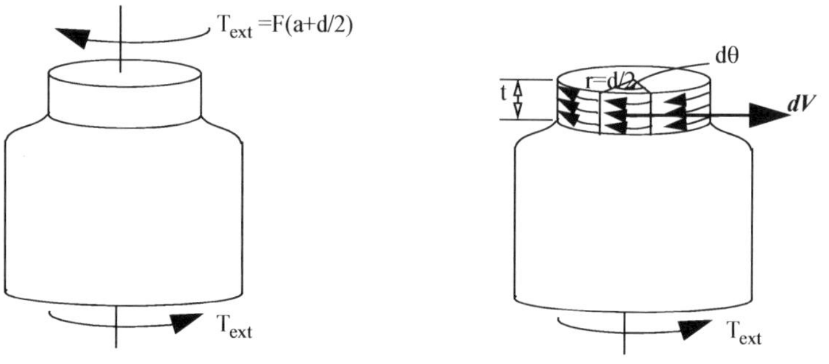

By equilibrium we obtain

$$T_{ext} = F(a+d/2) = \boldsymbol{T} = \int_A (d/2)(d\boldsymbol{V}) = \int_0^{2\pi}(d/2)[\tau t(d/2)(d\theta)] = (d^2/4)t\tau\int_0^{2\pi} d\theta = \tau(\pi d^2 t)/2 \text{ or}$$

$$F(d/2) = \tau(\pi d^2 t)/2 \quad or \quad F(4+2.5/2) = (10)(\pi(2.5)^2(0.375))/2 \quad or \quad F = 36.8155/5.25 = 7.012\ lb \quad \textbf{(1)}$$

$$F = 7.1\ lb$$

Section 1.2-1.4

1. 84

Solution

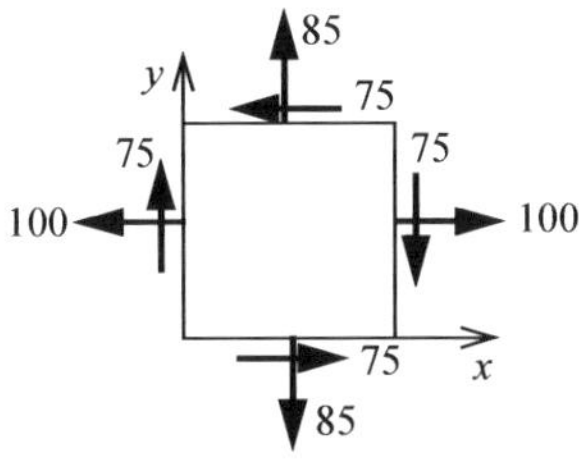

---

1. 85

Solution

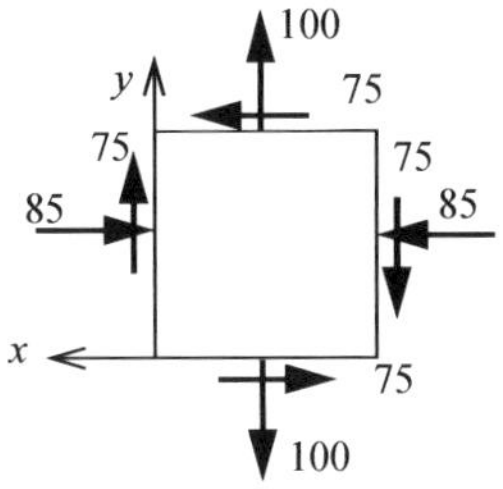

---

1. 86

Solution

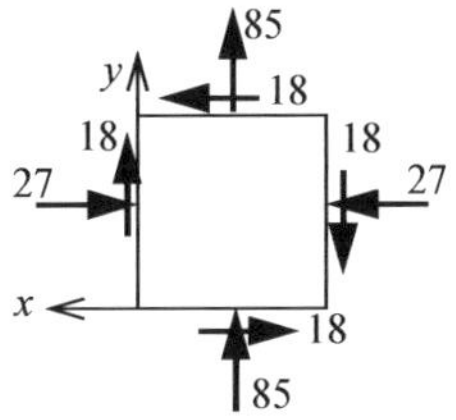

---

1. 87

Solution

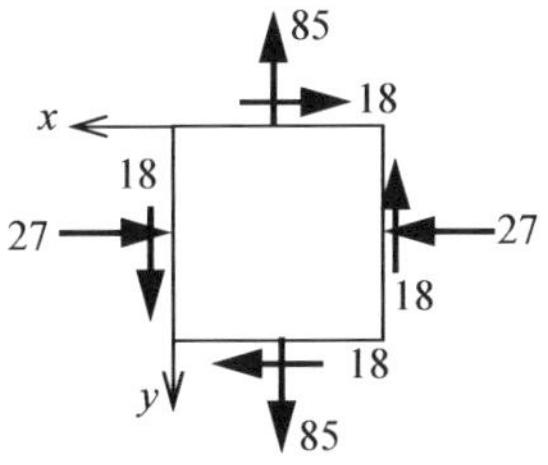

---

1. 88

Solution

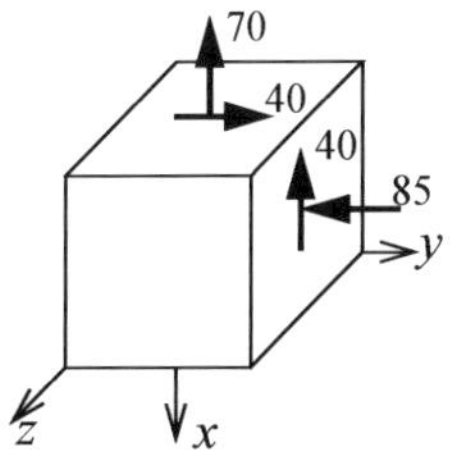

---

1. 89
Solution

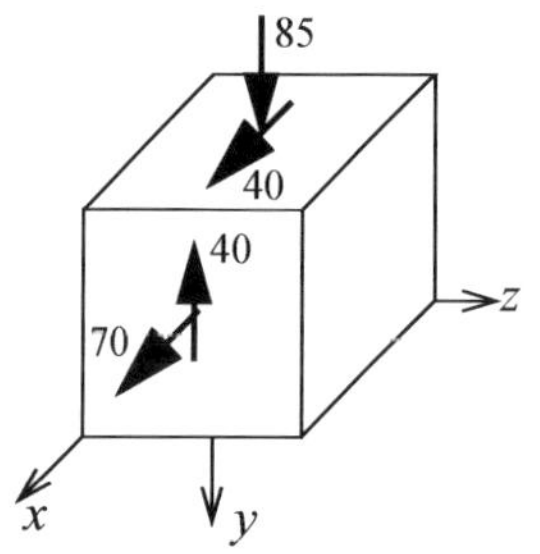

1. 90
Solution

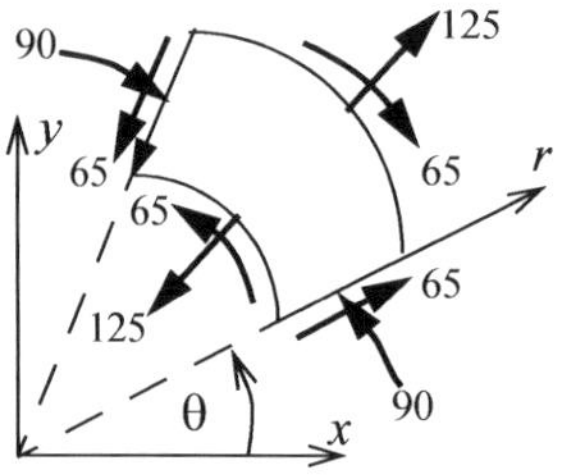

1. 91
Solution

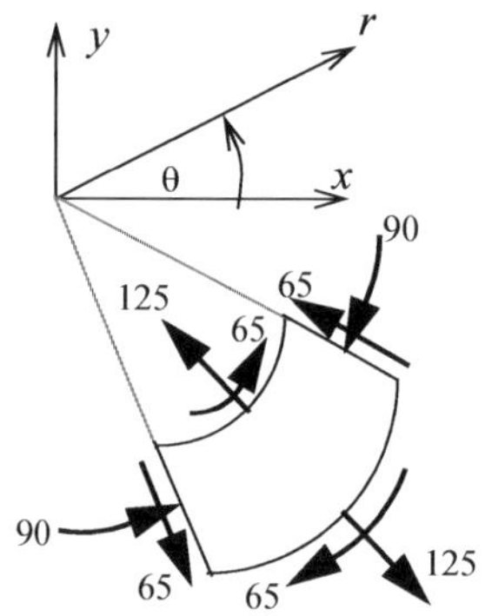

1. 92
Solution

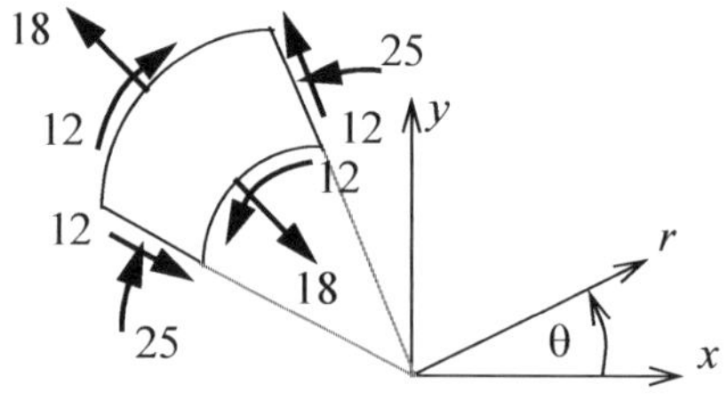

1. 93
Solution

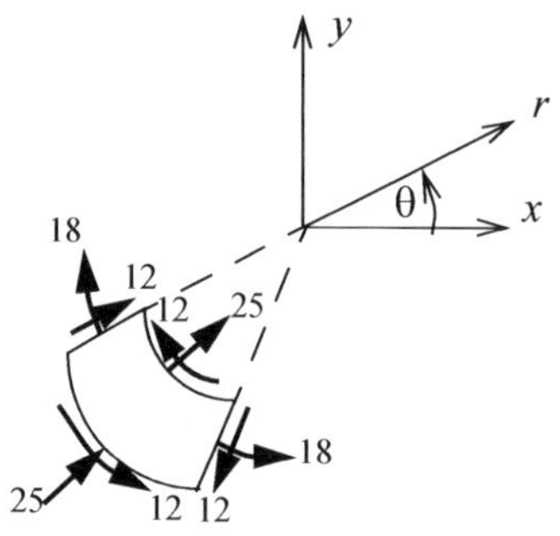

1. 94
Solution

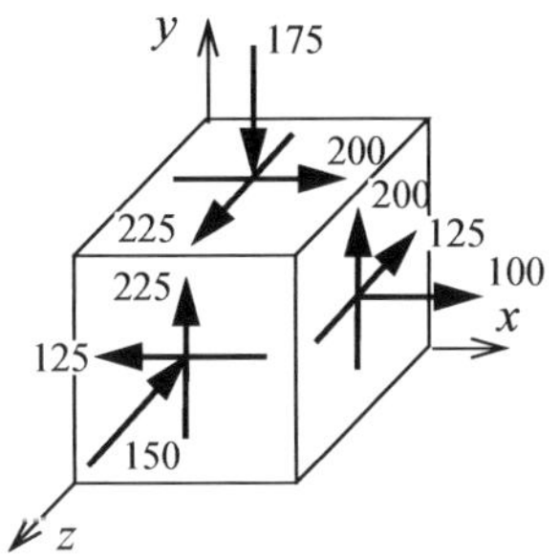

---

1. 95
Solution

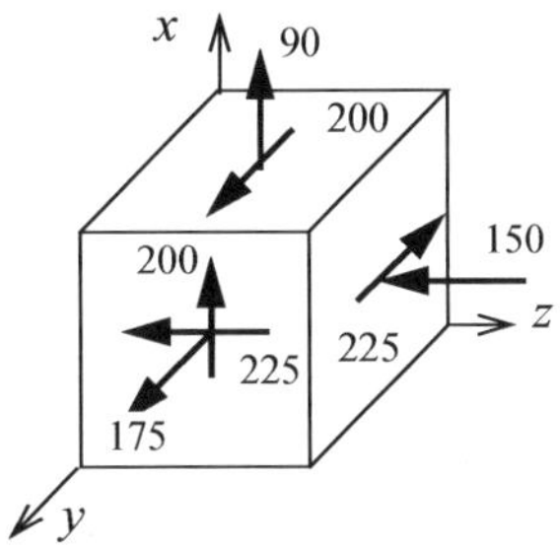

---

1. 96
Solution

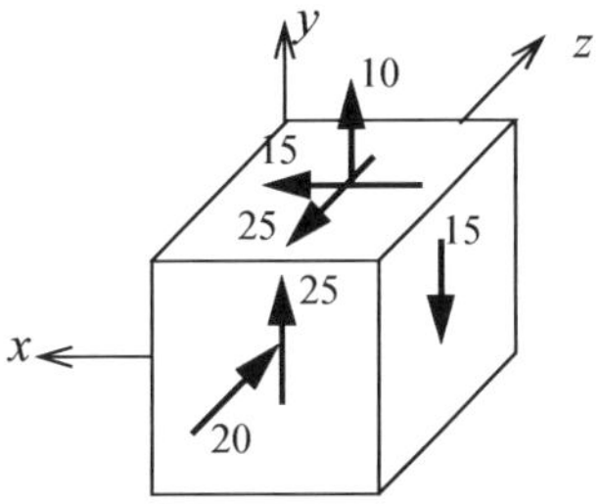

---

1. 97
Solution

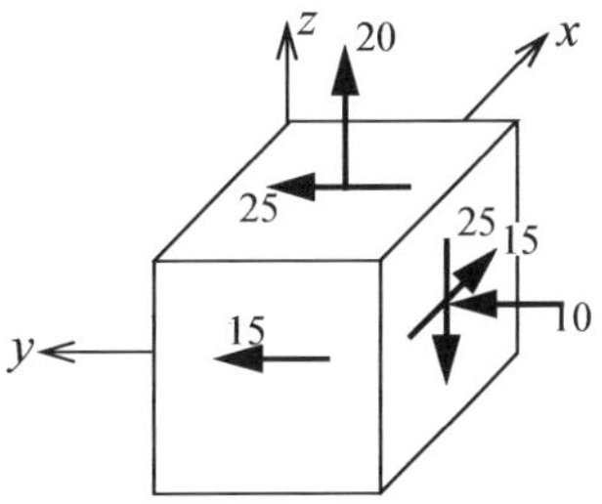

---

1. 98
Solution

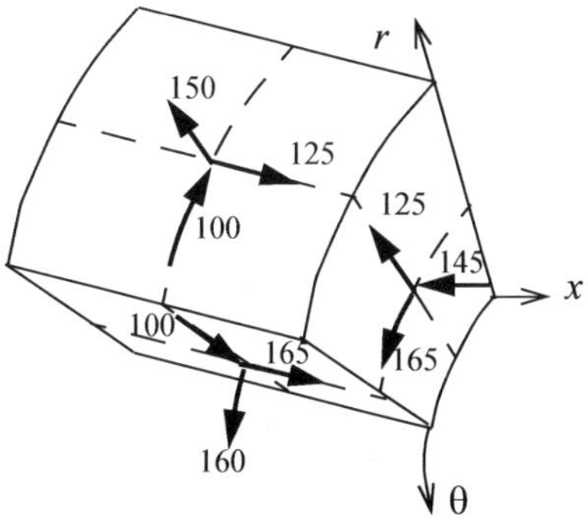

1. 99

Solution

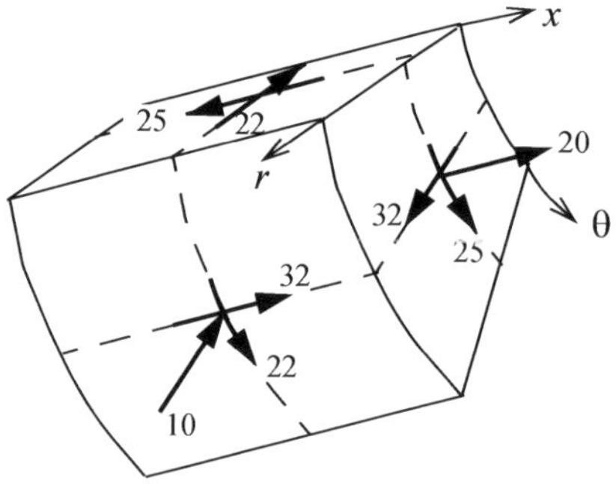

---

1. 100

Solution

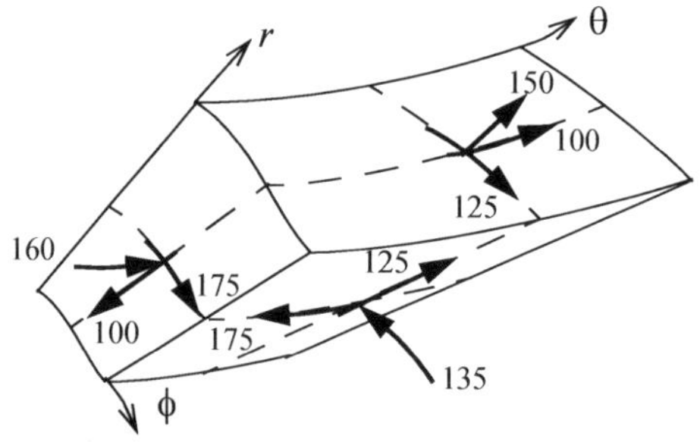

---

1. 101

Solution

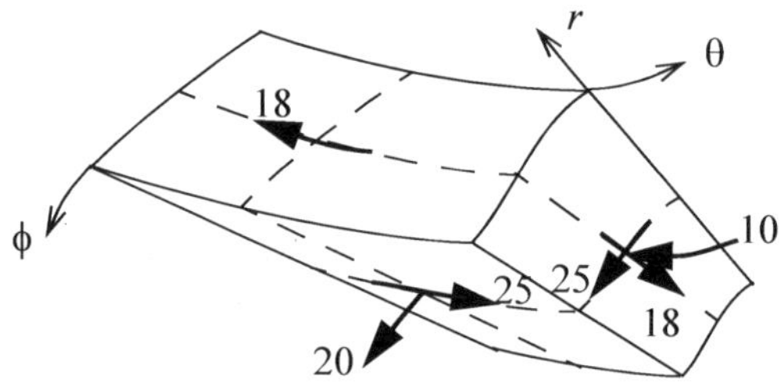

---

1. 102

Solution

Consider a differential area dA on which the normal force is $dN = \sigma_{xx}\, dA$ as shown below.

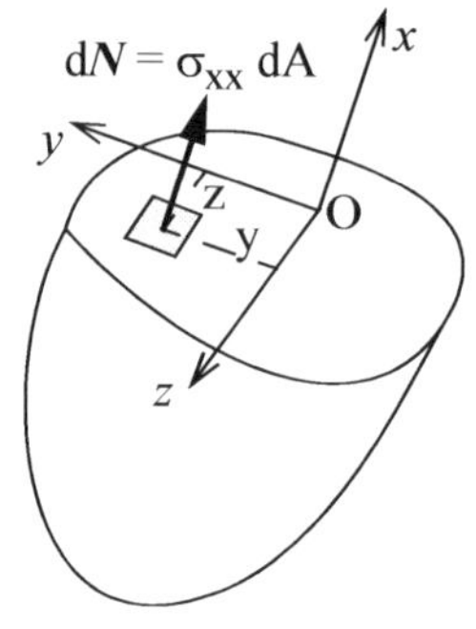

For static equivalency, the force in the x direction and the moments about the y and z axis for the above figure must be same. Thus, we obtain:

$$\boldsymbol{N} = \int_A d\boldsymbol{N} = \int_A \sigma_{xx} dA \qquad \boldsymbol{M}_y = -\int_A z\, d\boldsymbol{N} = -\int_A z\sigma_{xx} dA \qquad \boldsymbol{M}_z = -\int_A y\, d\boldsymbol{N} = -\int_A y\sigma_{xx} dA \tag{1}$$

The above equations are the desired results.

---

1. 103

Solution

Substituting $\sigma_{xx} = a + b\,y$ into the equations of the previous problem, we obtain the following:

$$N = \int_A (a+by)dA = a\int_A dA + b\int_A y\,dA \qquad and \qquad M_z = -\int_A y(a+by)dA = -a\int_A y\,dA - b\int_A y^2 dA \tag{1}$$

$\int_A y\,dA$ is zero because the origin is at the centroid. Noting that $A = \int_A dA$ and $I_{zz} = \int_A y^2 dA$ we obtain:

$$N = aA \qquad or \qquad a = N/A \qquad and \qquad M_z = -bI_{zz} \qquad or \qquad b = -(M_z/I_{zz}) \tag{2}$$

Substituting the values of a and b in $\sigma_{xx} = a + b\,y$ we obtain the desired result.

---

1. 104

Solution

Substituting $\sigma_{xx} = a + b\,y + c\,z$, into equations in the earlier problem we obtain the following.

$$N = \int_A (a+by+cz)dA = a\int_A dA + b\int_A y\,dA + c\int_A z\,dA \tag{1}$$

$$M_y = -\int_A z(a+by+cz)dA = -a\int_A z\,dA - \left(b\cdot \int_A yz\,dA\right) - c\int_A z^2 dA \tag{2}$$

$$M_z = -\int_A y(a+by+cz)dA = -a\int_A y\,dA - b\int_A y^2 dA - c\int_A yz\,dA \tag{3}$$

$\int_A y\,dA$ and $\int_A z\,dA$ are zero because the origin is at the centroid. Noting that $A = \int_A dA$, and $I_{zz} = \int_A y^2 dA$, $I_{yy} = \int_A z^2 dA$, $I_{yz} = \int_A yz\,dA$ we obtain:

$$N = aA \qquad or \qquad a = \frac{N}{A} \qquad M_y = -bI_{yz} - cI_{yy} \qquad M_z = -bI_{zz} - cI_{yz} \tag{4}$$

$$b = -\left(\frac{M_z I_{yy} - M_y I_{yz}}{I_{yy}I_{zz} - I_{yz}^2}\right) \qquad c = -\left(\frac{M_y I_{zz} - M_z I_{yz}}{I_{yy}I_{zz} - I_{yz}^2}\right) \tag{5}$$

Substituting the values of a, b, and c in $\sigma_{xx} = a + b\,y + c\,z$ we obtain the desired results.

---

1. 105

Solution

Multiplying by the stresses by areas of the plane on which they act and multiplying body force by the differential volume we obtain the free body diagram below.By force Equilibrium in the x-direction we obtain:

$$\left(\sigma_{xx} + \frac{\partial \sigma_{xx}}{\partial x}dx\right)(dydz) - (\sigma_{xx})(dydz) + \left(\tau_{yx} + \frac{\partial \tau_{yx}}{\partial y}dy\right)(dxdz) - (\tau_{yx})(dxdz) + (F_x)(dxdydz) = 0 \text{ or}$$

$$\frac{\partial \sigma_{xx}}{\partial x}(dxdydz) + \frac{\partial \tau_{yx}}{\partial y}(dxdydz) + (F_x)(dxdydz) = 0 \qquad or \qquad \frac{\partial \sigma_{xx}}{\partial x} + \frac{\partial \tau_{yx}}{\partial y} + F_x = 0. \tag{1}$$

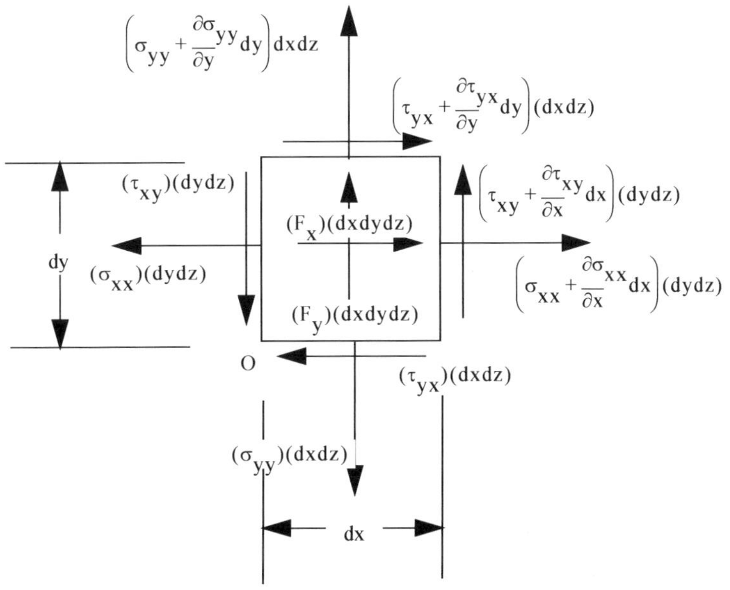

By force equilibrium in the y-direction we obtain:

$$\left(\sigma_{yy}+\frac{\partial\sigma_{yy}}{\partial y}dy\right)dxdz-(\sigma_{yy})(dxdz)+\left(\tau_{xy}+\frac{\partial\tau_{xy}}{\partial x}dx\right)(dydz)-(\tau_{xy})(dydz)+(F_y)(dxdydz) = 0 \text{ or}$$

$$\frac{\partial\sigma_{yy}}{\partial y}(dxdydz)+\frac{\partial\tau_{xy}}{\partial x}(dxdydz)+(F_y)(dxdydz) = 0 \qquad or \qquad \frac{\partial\sigma_{yy}}{\partial y}+\frac{\partial\tau_{xy}}{\partial x}+F_y = 0 \tag{2}$$

We now consider moment equilibrium about point about the center of the differential element. The body forces do not produce any moment as these pass through the center. The forces from the normal stresses do not produce any moment as these also pass through the center of the differential element. The moment from the variation of shear stresses can be written as

$$\left(\tau_{xy}+\frac{\partial\tau_{xy}}{\partial x}dx\right)(dydz)\left(\frac{dx}{2}\right)+(\tau_{xy})(dydz)\left(\frac{dx}{2}\right)-\left(\tau_{yx}+\frac{\partial\tau_{yx}}{\partial y}dy\right)(dxdz)\left(\frac{dy}{2}\right)-(\tau_{yx})(dxdz)\left(\frac{dy}{2}\right) = 0 \tag{3}$$

Neglecting the terms containing the product of four differentials as these term tend to zero faster in the limit than the terms with product of three differentials. We obtain:

$$(\tau_{xy})(dydz)(dx)-(\tau_{yx})(dxdz)(dy) = 0 \qquad or \qquad \tau_{xy} = \tau_{yx} \tag{4}$$

The above equations are the desired results.

CHAPTER 2

Sections 2.1-2.4

2. 1

Solution L_0=80cm ε=?

The final length of the cord and the average normal strain can be found as shown below.

$$L_f = AB = \sqrt{AC^2 + BC^2} = \sqrt{80^2 + 132^2} = 154.35cm \qquad \varepsilon = (L_f - L_o)/L_o = (154.35 - 80)/80 = 0.9294 \quad (1)$$

ANS $\varepsilon = 0.9294\ cm/cm$

---

2. 2

Solution d_o=250mm d_f=252mm ε=?

The average normal strain is:

$$\varepsilon = (\pi d_f - \pi d_o)/(\pi d_o) = 2/250 = 0.008 \quad (1)$$

ANS $\varepsilon = 0.008$ mm/mm

---

2. 3

Solution L_0=7in d=4.1in ε=?

The final length and the average normal strain can be found as shown below.

$$L_f = \pi d = 4.1\pi = 12.88 \qquad \varepsilon = (L_f - L_o)/L_o = (12.88 - 7)/7 = 0.8401 \quad (1)$$

ANS $\varepsilon = 0.8401\ in/in$

---

2. 4

Solution L_0=40 in ε=?

The final length of the cord and the average normal strain can be found as shown below.

$$AB = \sqrt{6^2 + 12^2} = 13.416 \qquad BC = \sqrt{5^2 + 12^2} = 13 \qquad L_f = 2(AB + BC) = 52.83 \quad (1)$$

$$\varepsilon = (L_f - L_o)/L_o = 52.83 - 40/40 = 0.3208 \quad (2)$$

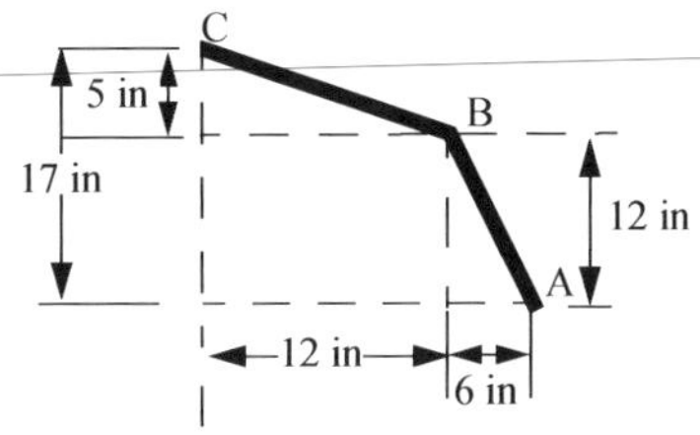

ANS $\varepsilon = 0.321\ in/in$

---

2. 5

Solution $L_o = 27$ ft

The final length and the average normal strain can be found as shown below.

$$AB = CD = \sqrt{9^2 + 1.25^2} = 9.0864 \text{ ft} \qquad L_f = AB + BC + CD = 2(9.0864) + 9 = 27.173 \text{ ft} \quad (1)$$

$$\varepsilon_{av} = (L_f - L_o)/L_o = (27.173 - 27)/27 = 6.399(10^{-3}) \quad (2)$$

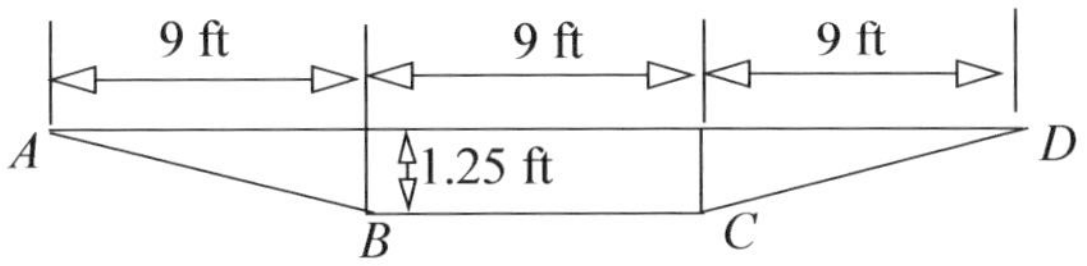

ANS $\varepsilon_{av} = 6399\ \mu\text{ft/ft}$

---

2. 6

Solution ε_{AB}=? ε_{BC}=? ε_{CD}=?

The average normal strains in the rods can be found as shown below.

$$\varepsilon_{AB} = \frac{u_B - u_A}{(x_B - x_A)} = \frac{(-1.8 - 0)}{(1.5)(10^3)} = -1.2(10^{-3}) \qquad \varepsilon_{BC} = \frac{u_C - u_B}{(x_C - x_B)} = \frac{0.7 - (-1.8)}{2.5(10^3)} = 1(10^{-3}) \quad (1)$$

$$\varepsilon_{CD} = \frac{u_D - u_C}{(x_D - x_C)} = \frac{3.7 - (0.7)}{2(10^3)} = 1.5(10^{-3}) \quad (2)$$

ANS $\varepsilon_{AB} = -1200\mu mm/mm$; $\varepsilon_{BC} = 1000\mu mm/mm$; $\varepsilon_{CD} = 1500\mu mm/mm$

---

2. 7

Solution $\varepsilon_{AB} = -800\mu$ $\varepsilon_{BC} = 600\mu$ $\varepsilon_{CD} = 1100\mu$ u_D-u_A=?

The relative displacements of the ends of each rod can be found and added as shown below.

$$u_B - u_A = \varepsilon_{AB}(x_B - x_A) = (-800)(10^{-6})(1.5)(10^3) = -1.2mm \quad (1)$$

$$u_C - u_B = \varepsilon_{BC}(x_C - x_B) = (600)(10^{-6})(2.5)(10^3) = 1.5mm \qquad u_D - u_C = \varepsilon_{CD}(x_D - x_C) = (1100)(10^{-6})(2)(10^3) = 2.2mm \quad (2)$$

$$u_D - u_A = 2.2 + 1.5 - 1.2 = 2.5mm \quad (3)$$

ANS $u_D - u_A = 2.5mm$

---

2. 8

Solution ε_A=? ε_B=?

From the exaggerated deformed geometry we obtain the following.

$$\delta_A = 0.0236 \qquad \delta_B = \delta_A - 0.02 = 0.0036(Contraction) \quad (1)$$

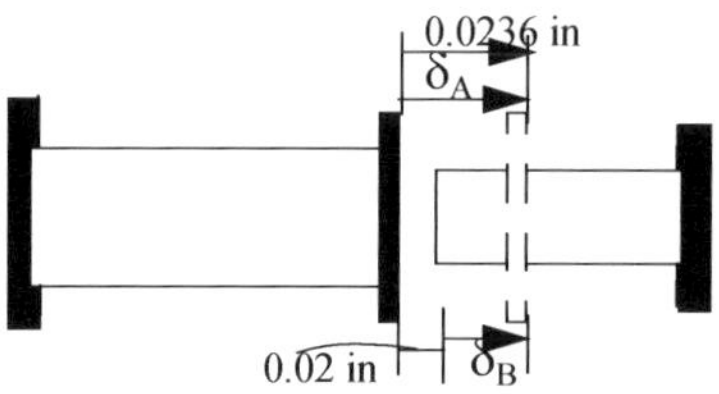

$$\varepsilon_A = \delta_A / L_A = 0.0236/60 = 393.3(10^{-6}) \qquad \varepsilon_B = \delta_B / L_B = -0.0036/24 = -150(10^{-6}) \quad (2)$$

ANS $\varepsilon_A = 393.3\mu in/in$; $\varepsilon_B = -150\mu in/in$

---

2. 9

Solution ε_A=2500μ ε_B=?

From previous problem we have $\delta_B = \delta_A - 0.02$

$$\delta_A = \varepsilon_A L_A = (2500)(10^{-6})(60) = 0.150in \qquad \delta_B = \delta_A - 0.02 = 0.13in(Contraction) \quad (1)$$

$$\varepsilon_B = \delta_B / L_B = -0.13/24 = -5.4167(10^{-3}) \quad (2)$$

ANS $\varepsilon_B = -5416.7\ \mu in/in$

---

2. 10

Solution ε_B=–4000μ ε_A=?

From previous problem we have $\delta_B = \delta_A - 0.02$.

$$\delta_B = \varepsilon_B L_B = (4000)(10^{-6})(60) = 0.24in \qquad \delta_A = \delta_B + 0.02 = 0.26in(extension) \quad (1)$$

$$\varepsilon_A = \delta_A / L_A = 0.26/60 = 4.3333(10^{-3}) \quad (2)$$

ANS $\varepsilon_A = 4333.3\ \mu in/in$

---

2. 11

Solution δ_B= 0.06 inch L_A = 24 inches ε_A = ?

Using similar triangle, the deformation equations can be written as given below.

$$\delta_A/125 = \delta_B/25 \quad or \quad \delta_A = 5\delta_B = 5(0.06) = 0.3 \; contraction \tag{1}$$

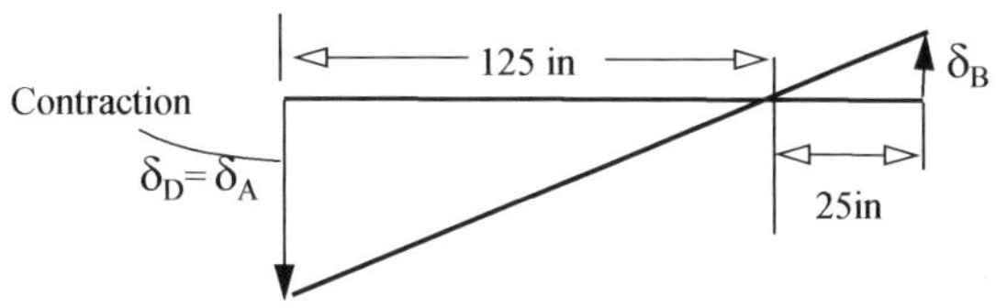

$$\varepsilon_A = \delta_A/L_A = -(0.3/24) = -0.0125 \tag{2}$$

ANS $\varepsilon_A = -0.0125 \; in/in$

---

2. 12

Solution $\varepsilon_A = -6000\mu$ $L_A = 36$ inches $\delta_B = ?$

From previous problem we have $\delta_A/125 = \delta_B/25$.

$$\delta_A = \varepsilon_A L_A = (6000)(10^{-6})(36) = 0.216 in \; contraction \qquad \delta_B = \left(\frac{25}{125}\right)\delta_A = \frac{0.216}{5} = 0.0432 inch \tag{1}$$

ANS $\delta_B = 0.0432$ in. upwards

---

2. 13

Solution $\delta_B = 0.06$ inch $L_A = 24$ inches $\varepsilon_A = ?$

The deformation equations can be written as shown below.

$$\delta_D = \delta_A + 0.04 \qquad \delta_D/125 = (\delta_A + 0.04)/125 = \delta_B/25 \tag{1}$$

$$\delta_A = 5\delta_B - 0.04 = 5(0.06) - 0.04 = 0.26 in \; contraction \qquad \varepsilon_A = \delta_A/L_A = -(0.26/24) = -0.0108 \tag{2}$$

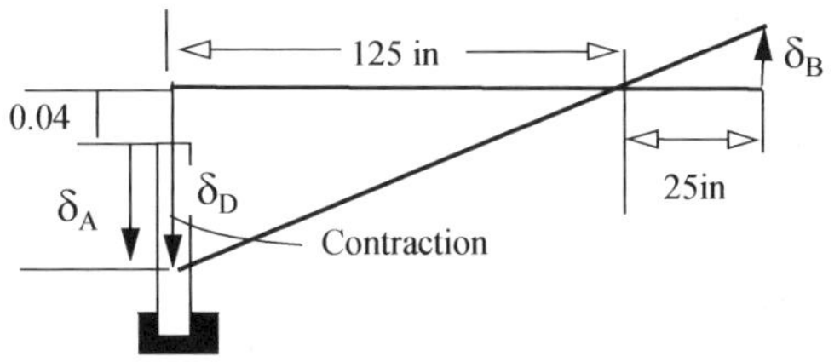

ANS $\varepsilon_A = -0.0108 \; in/in$

---

2. 14

Solution $\varepsilon_A = -6000\mu$ $L_A = 36$ inches $\delta_B = ?$

From previous problem we have $(\delta_A + 0.04)/125 = \delta_B/25$

$$\delta_A = \varepsilon_A L_A = (6000)(10^{-6})(36) = 0.216 inch \; contraction \qquad \delta_B = \frac{\delta_A + 0.04}{5} = \frac{0.216 + 0.04}{5} = 0.0512 inch \tag{1}$$

ANS $\delta_B = 0.0512$ in. upwards

---

2. 15

Solution $\delta_B = 0.06$ $L_A = 24$ inches $L_F = 24$ inches $\varepsilon_A = ?$ $\varepsilon_F = ?$

The deformation equations can be written as shown below.

$$\delta_D = \delta_A + 0.04 \qquad \delta_D/125 = \delta_E/30 = \delta_B/25 \quad or \quad (\delta_A + 0.04)/125 = \delta_F/30 = \delta_B/25 \tag{1}$$

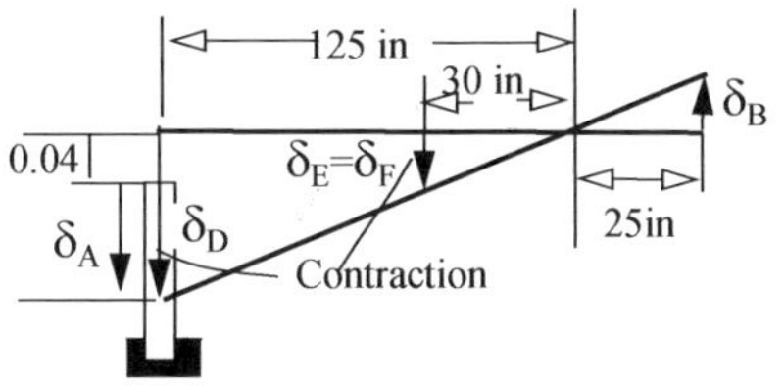

$$\delta_A = 5\delta_B - 0.04 = 5(0.06) - 0.04 = 0.26 in \; contraction \qquad \delta_F = \frac{30}{25}\delta_B = \frac{6}{5}(0.06) = 0.072 in \; contraction \tag{2}$$

$$\varepsilon_A = \delta_A/L_A = -(0.26/24) = -0.0108 \qquad \varepsilon_F = \delta_F/L_F = -(0.072/24) = -0.003 \tag{3}$$

ANS $\varepsilon_A = -0.0108 \; in/in; \varepsilon_F = -0.003 \; in/in$

---

2. 16

Solution $L_A = 36$ inches $L_F = 36$ inches $\varepsilon_A = -5000\ \mu$ $\delta_B = ?$ $\varepsilon_F = ?$

From previous problem we have $(\delta_A + 0.04)/125 = \delta_F/30 = \delta_B/25$

$$\delta_A = \varepsilon_A L_A = (5000)(10^{-6})(36) = 0.180 in\ contraction \qquad \textbf{(1)}$$

$$\delta_B = (\delta_A + 0.04)/5 = (0.18 + 0.04)/5 = 0.044 inch \qquad \delta_F = (30/25)\delta_B = (6/5)(0.044) = 0.0528 inch\ contraction \qquad \textbf{(2)}$$

$$\varepsilon_F = \frac{\delta_F}{L_F} = -\left(\frac{0.0528}{36}\right) = -1.4667(10^{-3}) \qquad \textbf{(3)}$$

ANS $\delta_B = 0.044$ in. upwards ; $\varepsilon_F = -1466.7\mu$ in/in

---

2. 17

Solution $L_A = 36$ inches $L_F = 36$ inches $\varepsilon_F = -2000\ \mu$ $\delta_B = ?$ $\varepsilon_A = ?$

From previous problem we have $(\delta_A + 0.04)/125 = \delta_F/30 = \delta_B/25$

$$\delta_F = \varepsilon_F L_F = (2000)(10^{-6})(36) = 0.072 in\ contraction \qquad \delta_D = \delta_A + 0.04 \qquad \textbf{(1)}$$

$$\delta_B = (25/30)\delta_F = 0.06 \qquad \delta_A + 0.04 = (125/30)\delta_F = (125/30)(0.072) = 0.3 \quad or \quad \delta_A = 0.26 in\ contraction \qquad \textbf{(2)}$$

$$\varepsilon_A = \delta_A/L_A = -(0.26/36) = -7.222(10^{-3}) \qquad \textbf{(3)}$$

ANS $\delta_B = 0.06$ in. upwards ; $\varepsilon_A = -7222\mu$in./in.

---

2. 18

Solution $\delta_B = 0.75$ mm $L_A = 1.2$ m $\varepsilon_A = ?$

Using similar triangle, the deformation equations can be written as given below.

$$\delta_A/2.5 = \delta_B/1.25 \quad or \quad \delta_A = 2\delta_B = 2(0.75) = 1.5 mm = 1.5(10^{-3}) m \quad contraction \qquad \textbf{(1)}$$

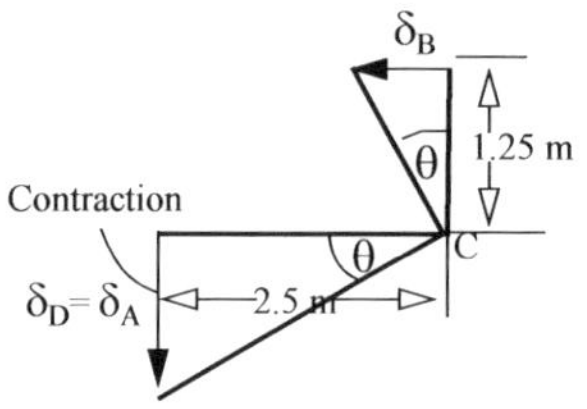

$$\varepsilon_A = \delta_A/L_A = -1.5(10^{-3})/1.2 = -1.2(10^{-3}) \qquad \textbf{(2)}$$

ANS $\varepsilon_A = -1250\mu$mm/mm

---

2. 19

Solution $\varepsilon_A = -2000\mu$ $L_A = 2$ m $\delta_B = ?$

From previous problem we have $\delta_A/2.5 = \delta_B/1.25$

$$\delta_A = \varepsilon_A L_A = (2000)(10^{-6})(2) = 4 mm\ contraction \qquad \delta_B = (1.25/2.5)\delta_A = 2 mm \qquad \textbf{(1)}$$

ANS $\delta_B = 2mm\ to\ the\ left$

---

2. 20

Solution $\delta_B = 0.75$ mm $L_A = 1.2$ m $\varepsilon_A = ?$

The deformation equations can be written as shown below.

$$\delta_D = \delta_A + 1 \qquad \delta_D/2.5 = \delta_B/1.25 \quad or \quad (\delta_A + 1)/2.5 = \delta_B/1.25 \qquad \textbf{(1)}$$

$$\delta_A = 2\delta_B - 1 = 2(0.75) - 1 = 0.5(10^{-3}) m\ contraction \qquad \varepsilon_A = \delta_A/L_A = -[0.5(10^{-3})/1.2] = -0.4167(10^{-3}) \qquad \textbf{(2)}$$

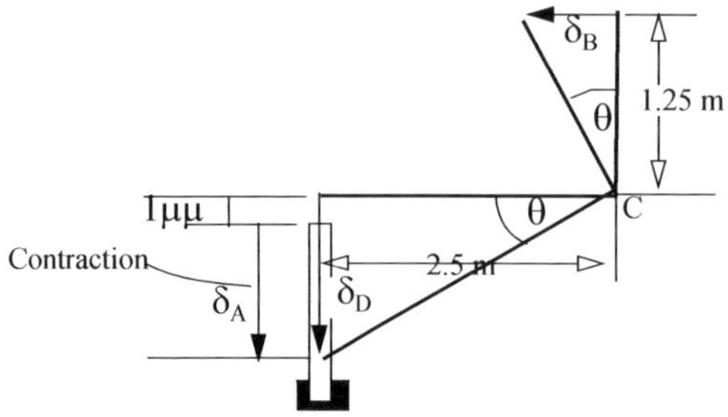

ANS $\varepsilon_A = -416.7\mu$ mm/mm

---

2. 21

Solution $\varepsilon_A = -2000\mu$ $L_A = 2$ m $\delta_B = ?$

From previous problem we have $(\delta_A + 1)/2.5 = \delta_B/1.25$

$$\delta_A = \varepsilon_A L_A = (2000)(10^{-6})(2) = 4mm\ contraction \qquad \delta_B = \left(\frac{1.25}{2.5}\right)(\delta_A + 1) = \frac{5}{2} = 2.5mm \quad \textbf{(1)}$$

ANS $\delta_B = 2.5mm\ to\ the\ left$

---

2. 22

Solution $\delta_B = 0.75$ mm $L_A = 1.2$ m $L_F = 1.2$ m $\varepsilon_A = ?$ $\varepsilon_F = ?$

The deformation equations can be written as shown below.

$$\delta_D = \delta_A + 1 \qquad \delta_D/2.5 = \delta_F/0.8 = \delta_B/1.25 \qquad or \qquad (\delta_A + 1)/2.5 = \delta_F/0.8 = \delta_B/1.25 \quad \textbf{(1)}$$

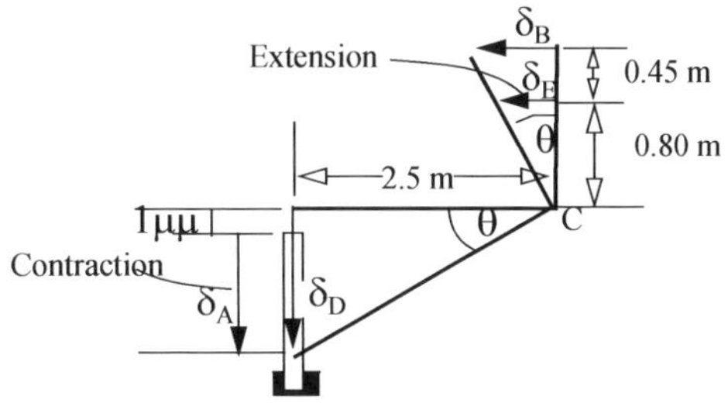

$$\delta_A = 2\delta_B - 1 = 2(0.75) - 1 = 0.5mm = 0.5(10^{-3})m\ contraction \quad \textbf{(2)}$$

$$\delta_F = 0.8\delta_B/1.25 = 0.48\ mm = 0.48(10^{-3})m\ extension \quad \textbf{(3)}$$

$$\varepsilon_A = \delta_A/L_A = -[0.5(10^{-3})]/1.2 = -0.4167(10^{-3}) \qquad \varepsilon_F = \delta_F/L_F = [0.48(10^{-3})]/1.2 = 0.4(10^{-3}) \quad \textbf{(4)}$$

ANS $\varepsilon_A = -416.7\mu\ mm/mm;\ \varepsilon_F = 400\mu\ mm/mm$

---

2. 23

Solution $\varepsilon_A = -2500\ \mu$ $L_A = 2$ m $L_F = 2$ m $\delta_B = ?$ $\varepsilon_F = ?$

From previous problem we have $(\delta_A + 1)/2.5 = \delta_F/0.8 = \delta_B/1.25$

$$\delta_A = \varepsilon_A L_A = (2500)(10^{-6})(2000) = 5mm\ contraction \qquad \delta_D = \delta_A + 1 = 6mm \quad \textbf{(1)}$$

$$\delta_B = (1.25/2.5)(6) = 3.0mm \qquad \delta_F = 0.8\delta_B/1.25 = 1.92mm = 1.92(10^{-3})m\ extension \quad \textbf{(2)}$$

$$\varepsilon_F = \delta_F/L_F = [1.92(10^{-3})]/2.0 = 0.96(10^{-3}) \quad \textbf{(3)}$$

ANS $\delta_B = 3.0$mm to the left; $\varepsilon_F = 960\mu$mm/mm

---

2. 24

Solution $\varepsilon_F = 1000\ \mu$ $L_A = 2$ m $L_F = 2$ m $\delta_B = ?$ $\varepsilon_A = ?$

From previous problem we have $(\delta_A + 1)/2.5 = \delta_F/0.8 = \delta_B/1.25$

$$\delta_F = \varepsilon_F L_F = (1000)(10^{-6})(2000) = 2mm\ extension \quad \textbf{(1)}$$

$$\delta_B = (1.25/0.8)(2) = 3.125mm \qquad \delta_A = 2.5\delta_F/0.8 - 1 = 5.25mm = 5.25(10^{-3})m\ contraction \quad \textbf{(2)}$$

$$\varepsilon_A = \delta_A/L_A = -[5.25(10^{-3})]/2.0 = -2.625(10^{-3}) \quad \textbf{(3)}$$

ANS $\delta_B = 3.125$mm to the left; ;$\varepsilon_A = -2625\mu$mm/mm

---

2. 25

Solution $L_o = 400$ mm $h = 50$ mm $\varepsilon_{AB} = 2500\ \mu$ mm/mm $\varepsilon_{CD} = 3500\ \mu$ mm/mm $R = ?$ $\psi = ?$

From geometry the arc lengths can be written and the average strain in each bar equated to the given values

$$AB = R(2\psi) \qquad CD = (R+h)(2\psi) = 2R\psi + 2h\psi \quad \textbf{(1)}$$

$$\varepsilon_{AB} = \frac{2R\psi - L_o}{L_o} = \frac{2R\psi - 400}{400} = -2500(10^{-6}) \qquad or \qquad 2R\psi - 400 = -1 \quad \textbf{(2)}$$

$$\varepsilon_{CD} = \frac{2R\psi + 2h\psi - L_o}{L_o} = \frac{2R\psi + (2)(50)\psi - 400}{400} = 3500(10^{-6}) \qquad or \qquad 2R\psi + 100\psi - 400 = 1.4 \tag{3}$$

Solving the two equations we obtain our result.

ANS $\psi = 0.024 \text{ rads} = 1.375°;\ R = 8312.5 \text{ mm}$

---

2. 26

Solution $L_o = 30$ in $\psi = 1.25$ h = 2 in $\varepsilon_{AB} = -1500\ \mu$ $\varepsilon_{CD} = ?$

From geometry the arc lengths can be written and the average strain in each bar equated to the given value.

$$\psi = \frac{1.25°}{180}\pi = 0.0218 \text{ rads} \qquad AB = R(2\psi) \qquad CD = (R+h)(2\psi) = 2R\psi + 2h\psi \tag{1}$$

$$\varepsilon_{AB} = \frac{2R\psi - L_o}{L_o} = \frac{2R\psi - 30}{30} = -1500(10^{-6}) \tag{2}$$

$$\varepsilon_{CD} = \frac{2R\psi + 2h\psi - L_o}{L_o} = \frac{2R\psi + (2)(2)\psi - 30}{30} = \frac{2R\psi - 30}{30} + \frac{4\psi}{30} = -1500(10^{-6}) + \frac{4(0.0218)}{30} = 1408.8(10^{-6}) \tag{3}$$

ANS $\varepsilon_{CD} = 1408.8\ \mu$

---

2. 27

Solution $L_o = 48$ in $\varepsilon_{AB} = -2000\ \mu$ in/in $\varepsilon_{CD} = 1500\ \mu$ in/in h = ?

From geometry the arc lengths can be written and the average strain in each bar equated to the given values.

$$AB = R(2\psi) \qquad CD = (R+4)(2\psi) = 2R\psi + 8\psi \qquad EF = (R+h)(2\psi) = 2R\psi + 2h\psi \tag{1}$$

$$\varepsilon_{AB} = (2R\psi - L_o)/L_o = (2R\psi - 48)/48 = -2000(10^{-6}) \tag{2}$$

$$\varepsilon_{CD} = \frac{2R\psi + 8\psi - L_o}{L_o} = \frac{2R\psi + 8\psi - 48}{48} = \frac{2R\psi - 48}{48} + \frac{8\psi}{48} = -2000(10^{-6}) + \frac{\psi}{6} = 1500(10^{-6}) \qquad or \qquad \psi = 0.021 \text{ rads} \tag{3}$$

$$\varepsilon_{EF} = \frac{2R\psi + 2h\psi - L_o}{L_o} = \frac{2R\psi + 2h\psi - 48}{48} = \frac{2R\psi - 48}{48} + \frac{2h\psi}{48} = -2000(10^{-6}) + \frac{h(0.021)}{24} = 0 \qquad or \qquad h = 2.2857 \text{ in.} \tag{4}$$

ANS $h = 2.286$ in.

---

2. 28

Solution $\gamma_A = ?$

From geometry, we obtain the magnitude of γ_A and noting that shear strain is positive as angle decreases, determine its sign.

$$tan|\gamma_A| = 0.840/350 \qquad or \qquad |\gamma_A| = 0.0024\ rads \tag{1}$$

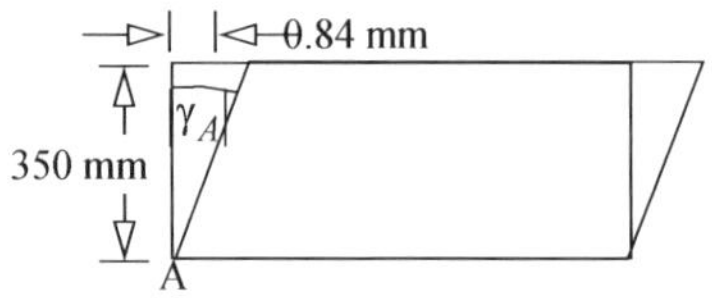

ANS $\gamma_A = 2400 \mu\text{rads}$

---

2. 29

Solution $\gamma_A = ?$

From geometry, we obtain the magnitude of γ_A and noting that shear strain is negative as angle increases, determine its sign.

$$tan|\gamma_A| = 0.0051/1.7 \qquad or \qquad |\gamma_A| = 0.003\ rads \tag{1}$$

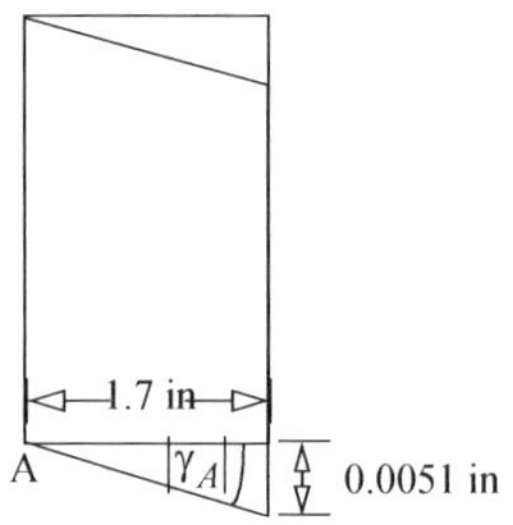

ANS $\gamma_A = -3000\mu\ rad$

---

2. 30

Solution γ_A=?

From geometry, we obtain the magnitude of γ_A and noting that shear strain is negative as angle increases, determine its sign.

$$tan|\gamma_A| = 0.007/1.4 \quad or \quad |\gamma_A| = 0.005 \ rads \tag{1}$$

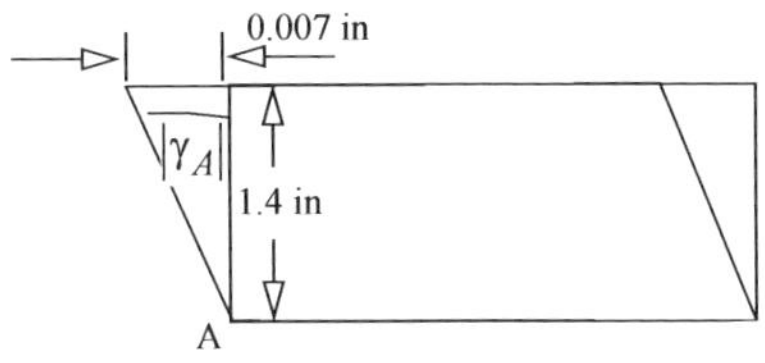

ANS $\gamma_A = -5000\mu rads$

---

2. 31

Solution γ_A=?

From geometry, we obtain the magnitude of γ_A and noting that shear strain is negative as angle increases, determine its sign.

$$tan|\gamma_A| = 0.65/250 \quad or \quad |\gamma_A| = 0.00026 \ rads \tag{1}$$

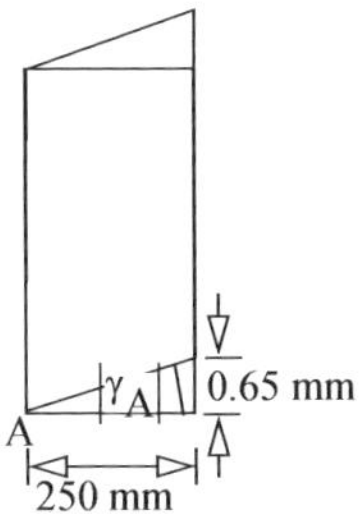

ANS $\gamma_A = 260\mu \ rad$

---

2. 32

Solution γ_A=?

From geometry, we obtain the angles ϕ_1 and ϕ_2 as shown below.

$$tan\phi_1 = 0.0042/3.0 \quad or \quad \phi_1 = 0.0014 \ rads \qquad tan\phi_2 = 0.0056/1.4 \quad or \quad \phi_2 = 0.004 \ rads \tag{1}$$

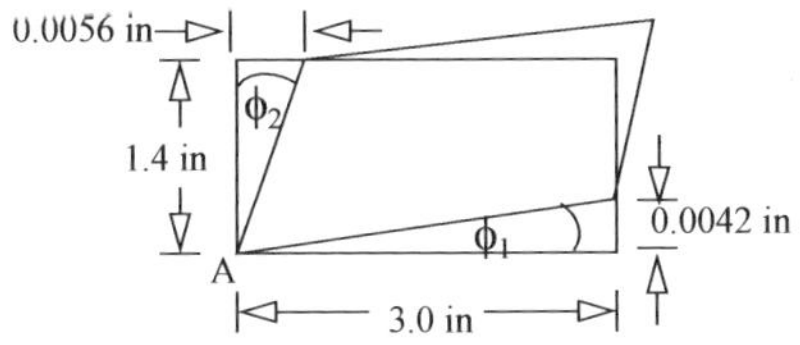

The decrease in the angle from the original right angle at A represents positive shear strain, i.e.,

$$\gamma_A = \phi_1 + \phi_2 = 0.0054 \ rads \tag{2}$$

ANS $\gamma_A = 5400\mu \ rad$

---

2. 33

Solution γ_A=?

From geometry we obtain the angles ϕ_1 and ϕ_2 as shown below.

$$tan\phi_1 = 0.6/600 \quad or \quad \phi_1 = 1.0(10^{-3}) \ rads \qquad tan\phi_2 = 0.6/350 \quad or \quad \phi_2 = 1.714(10^{-3}) \ rads \tag{1}$$

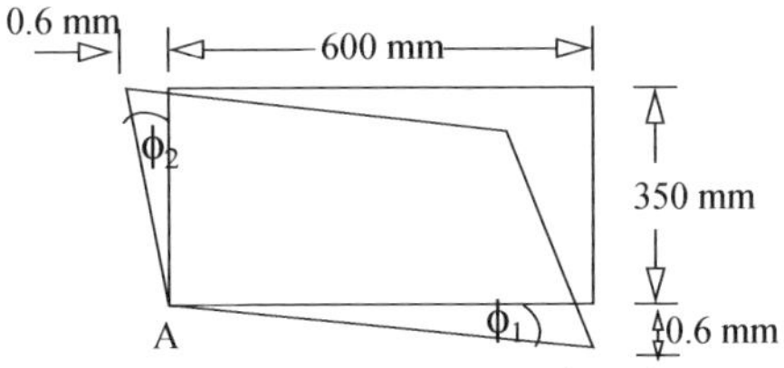

The increase in angle from the original right angle at A represents negative shear strain, i.e.,

$$\gamma_A = -(\phi_1 + \phi_2) = -2.714(10^{-3}) \ rads \tag{2}$$

ANS $\gamma_A = -2714\mu \ rad$

2. 34

Solution δ_A=0.005in γ_A=?

From geometry we can determine the following.

$$AB = 8cos25 = 7.2505 \quad BD = ABcos25 = 6.5712 \quad h = ABsin25 = 3.0642 \quad DC = BC - BD = 1.4288 \tag{1}$$

$$tan\phi_1 = \frac{BD}{h+\delta_A} = \frac{6.5712}{(3.0642+0.005)} \quad or \quad \phi_1 = 1.1338 rads \tag{2}$$

$$tan\phi_2 = \frac{DC}{h+\delta_A} = \frac{1.4288}{(3.0642+0.005)} \quad or \quad \phi_2 = 0.4357 rads \tag{3}$$

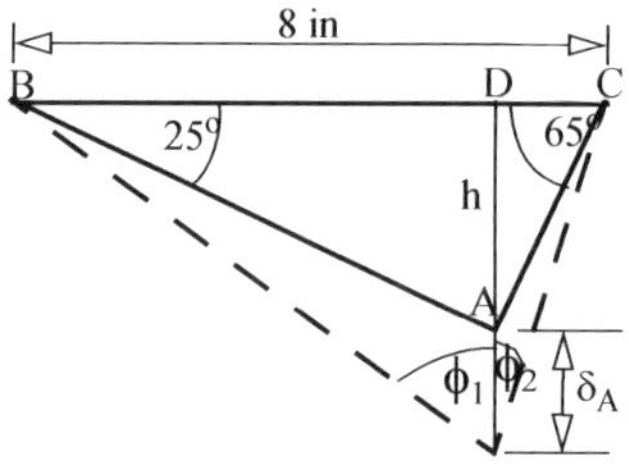

The decrease in the angle from the original right angle at A represents positive shear strain, i.e.,

$$\gamma_A = \pi/2 - (\phi_1 + \phi_2) = 1.2963(10^{-3}) rads \tag{4}$$

ANS $\gamma_A = 1296\mu rad$

---

2. 35

Solution δ_A= 0.006 γ_A= ?

From geometry we can determine the following. $AB = \sqrt{5^2 - 3^2} = 4$

$$cos\theta = 0.8 \quad sin\theta = 0.6 \quad BD = ABcos\theta = 3.2 \quad h = ABsin\theta = 2.4 \quad DC = BC - BD = 1.8 \tag{1}$$

$$tan\phi_1 = \frac{BD}{h+\delta_A} = \frac{3.2}{(2.4+0.006)} \quad or \quad \phi_1 = 0.9261 rads \tag{2}$$

$$tan\phi_2 = \frac{DC}{h+\delta_A} = \frac{1.8}{(2.4+0.006)} \quad or \quad \phi_2 = 0.6423 rads \tag{3}$$

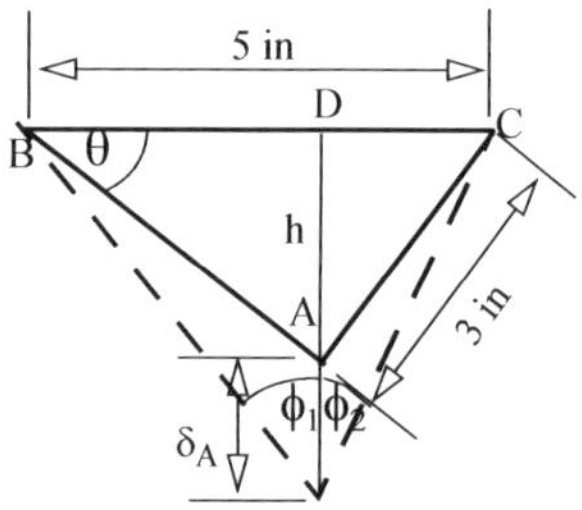

The decrease in the angle from the original right angle at A represents positive shear strain, i.e.,

$$\gamma_A = \pi/2 - (\phi_1 + \phi_2) = 2.397(10^{-3}) rads \tag{4}$$

ANS $\gamma_A = 2397\mu rads$

---

2. 36

Solution δ_A=0.75mm γ_A= ?

From geometry we can determine the following.

$$AC = \sqrt{1300^2 - 500^2} = 1200 \quad cos\theta = \frac{5}{13} \quad sin\theta = \frac{12}{13} \tag{1}$$

$$BD = ABcos\theta = 192.308 \quad h = ABsin\theta = 461.538 \quad DC = BC - BD = 1107.69 \tag{2}$$

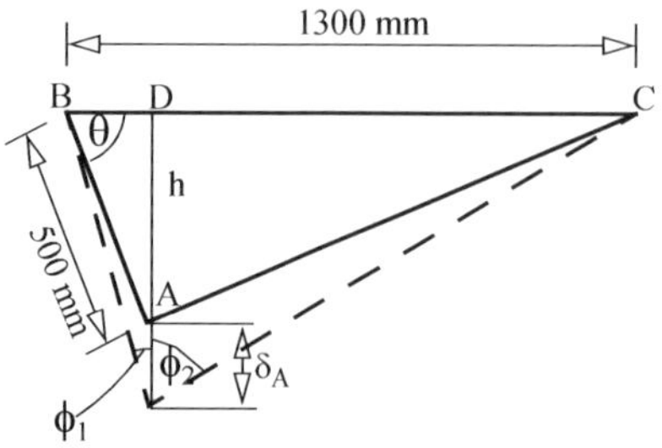

$$tan\phi_1 = \frac{BD}{h+\delta_A} = \frac{192.308}{(461.538+0.75)} \quad or \quad \phi_1 = 0.3942 rads \quad (3)$$

$$tan\phi_2 = \frac{DC}{h+\delta_A} = \frac{1107.69}{(461.538+0.75)} \quad or \quad \phi_2 = 1.1754 rads \quad (4)$$

The decrease in the angle from the original right angle at A represents positive shear strain, i.e.,

$$\gamma_A = \pi/2 - (\phi_1 + \phi_2) = 1.1529(10^{-3}) rads \quad (5)$$

ANS $\gamma_A = 1153 \mu rads$

---

2. 37

Solution δ_A=0.005 γ_A= ?

From geometry we can determine the following.

$$AB = 8cos25 = 7.2505 \quad BD = ABcos25 = 6.5712 \quad h = ABsin25 = 3.0642 \quad DC = BC - BD = 1.4288 \quad (1)$$

$$tan\phi_1 = \frac{BD - \delta_A}{h} = \frac{6.5712 - 0.005}{3.0642} \quad or \quad \phi_1 = 1.13417 rads \quad (2)$$

$$tan\phi_2 = \frac{DC + \delta_A}{h} = \frac{1.4288 + 0.005}{3.0642} \quad or \quad \phi_2 = 0.43766 rads \quad (3)$$

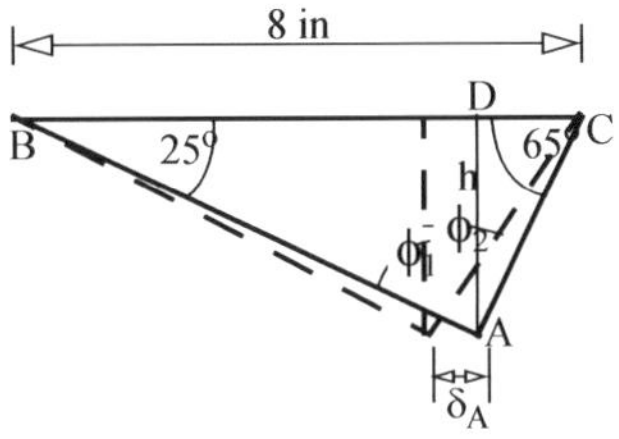

The decrease in the angle from the original right angle at A represents positive shear strain, i.e.,

$$\gamma_A = \pi/2 - (\phi_1 + \phi_2) = -1.0337(10^{-3}) rads \quad (4)$$

ANS $\gamma_A = -1034 \mu rads$

---

2. 38

Solution δ_A=0.008 γ_A= ?

From geometry we can determine the following. $AB = \sqrt{5^2 - 3^2} = 4$

$$cos\theta = 0.8 \quad sin\theta = 0.6 \quad BD = ABcos\theta = 3.2 \quad h = ABsin\theta = 2.4 \quad DC = BC - BD = 1.8 \quad (1)$$

$$tan\phi_1 = \frac{BD - \delta_A}{h} = \frac{3.2 - 0.008}{2.4} \quad or \quad \phi_1 = 0.92609 rads \quad (2)$$

$$tan\phi_2 = \frac{CD + \delta_A}{h} = \frac{1.8 + 0.008}{2.4} \quad or \quad \phi_2 = 0.64563 rads \quad (3)$$

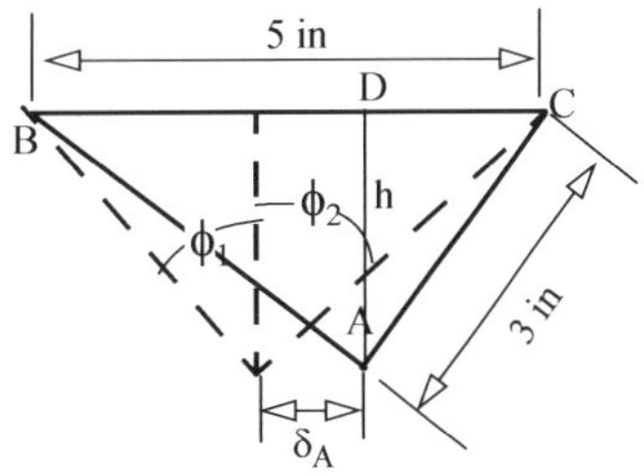

The decrease in the angle from the original right angle at A represents positive shear strain, i.e.,

$$\gamma_A = \pi/2 - (\phi_1 + \phi_2) = -0.9280(10^{-3}) rads \quad (4)$$

ANS $\gamma_A = -928 \mu rad$

---

2. 39

Solution δ_A=0.90 mm γ_A=_?

From geometry we obtain the following

$$AC = \sqrt{1300^2 - 500^2} = 1200 \quad cos\theta = 5/13 \quad sin\theta = 12/13 \quad (1)$$

$$BD = ABcos\theta = 192.308 \quad h = ABsin\theta = 461.538 \quad DC = BC - BD = 1107.69 \quad (2)$$

$$\tan\phi_1 = \frac{BD + \delta_A}{h} = \frac{192.308 + 0.90}{461.538} \quad or \quad \phi_1 = 0.39645 rads \qquad (3)$$

$$\tan\phi_2 = \frac{CD - \delta_A}{h} = \frac{1107.69 - 0.90}{461.538} \quad or \quad \phi_2 = 1.17572 rads \qquad (4)$$

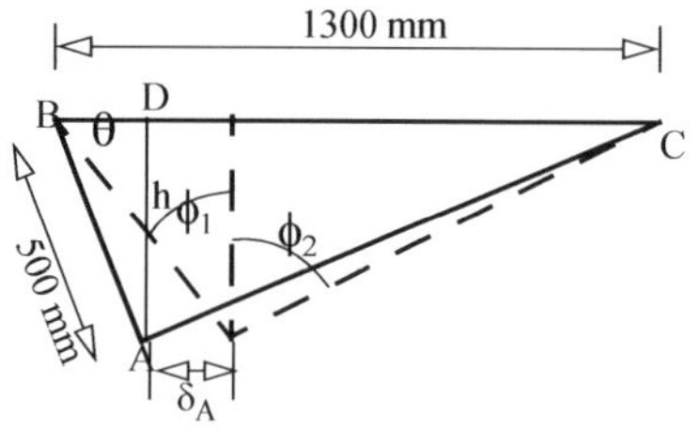

The decrease in the angle from the original right angle at A represents positive shear strain, i.e.,

$$\gamma_A = \pi/2 - (\phi_1 + \phi_2) = -1.3717(10^{-3}) rads \qquad (5)$$

ANS $\gamma_A = -1371.7 \mu rad$

---

2. 40

Solution $\varepsilon_{xx} = 500\mu \; in/in$ $\quad \varepsilon_{AB} =$

From geometry and the given strain we have:

$$\tan\theta = 5/10 \quad or \quad \theta = 26.565° \quad AB = \sqrt{5^2 + 10^2} = 11.1803 \text{ in} \qquad (1)$$

$$BB_1 = (10)\varepsilon_{xx} = 10[500(10^{-6})] = 0.005 \quad BC = BB_1 \cos\theta = 4.4721(10^{-3}) \qquad (2)$$

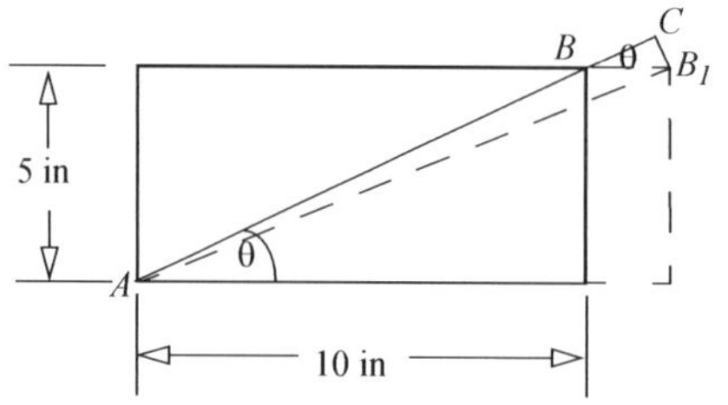

$$\varepsilon_{AB} = BC/AB = 4.4721(10^{-3})/11.1803 = 0.4(10^{-3}) \qquad (3)$$

ANS $\varepsilon_{AB} = 400 \; \mu in./in.$

---

2. 41

Solution $\varepsilon_{yy} = -1200\mu \; mm/mm$ $\quad \varepsilon_{AB} = ?$

From geometry and the given strain we have:

$$\tan\theta = 45/100 \quad or \quad \theta = 24.23° \quad AB = \sqrt{45^2 + 100^2} = 109.66 \text{ mm} \qquad (1)$$

$$BB_1 = (100)\varepsilon_{yy} = 100[120(10^{-6})] = 0.12 \quad BC = BB_1 \cos\theta = 109.43(10^{-3}) \qquad (2)$$

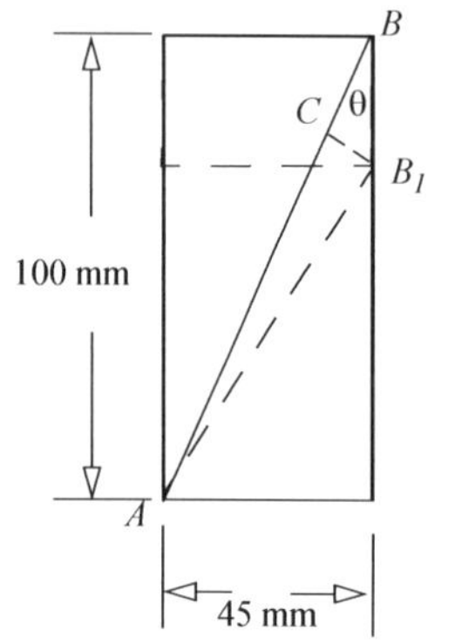

$$\varepsilon_{AB} = -BC/AB = -109.43/109.66 = -998(10^{-3}) \qquad (3)$$

ANS $\varepsilon_{AB} = -998 \; \mu mm/mm$

---

2. 42

Solution $\varepsilon_{xx} = -1000\mu\ mm/mm \qquad \varepsilon_{AB} = ?$

From geometry and given strain we have:

$$tan\theta = \frac{300}{450} \quad or \quad \theta = 33.69° \quad \alpha = 90 - \theta = 56.31° \quad AB cos\alpha = 150 \quad AB = \frac{150}{cos 56.31} = 270.42 \text{ mm} \tag{1}$$

$$AA_1 = (150)|\varepsilon_{xx}| = 150[1000(10^{-6})] = 0.15 \text{ mm} \qquad AD = AA_1 cos\alpha = 83.205(10^{-3}) \tag{2}$$

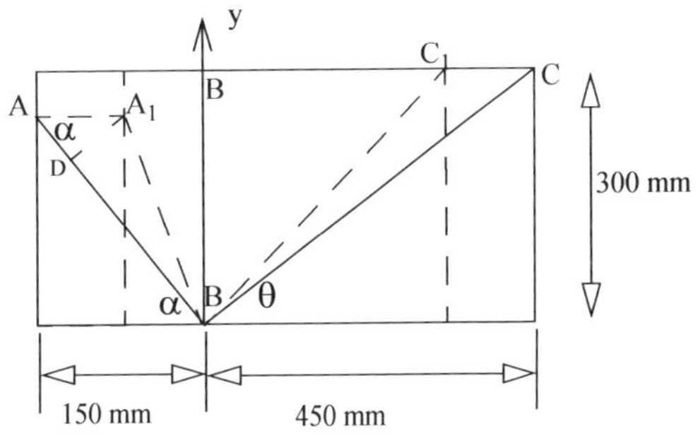

The normal strain in bar AB can be written as:

$$\varepsilon_{AB} = -AD/AB = -83.205(10^{-3})/270.42 = -307.7(10^{-3}) \tag{3}$$

ANS $\varepsilon_{AB} = -307.7\ \mu mm/mm$

---

2. 43

Solution $\varepsilon_{xx} = 700\mu\ mm/mm \qquad \varepsilon_{BC} = ?$

From geometry and the given strain we have:

$$tan\theta = \frac{300}{450} \quad or \quad \theta = 33.69° \quad BC sin\theta\theta = 300 \quad BC = \frac{300}{sin 56.31} = 540.83 \text{ mm} \tag{1}$$

$$CC_1 = (450)|\varepsilon_{xx}| = 450[700(10^{-6})] = 0.315 \text{ mm} \qquad CE = CC_1 cos\theta = 262.09(10^{-3}) \tag{2}$$

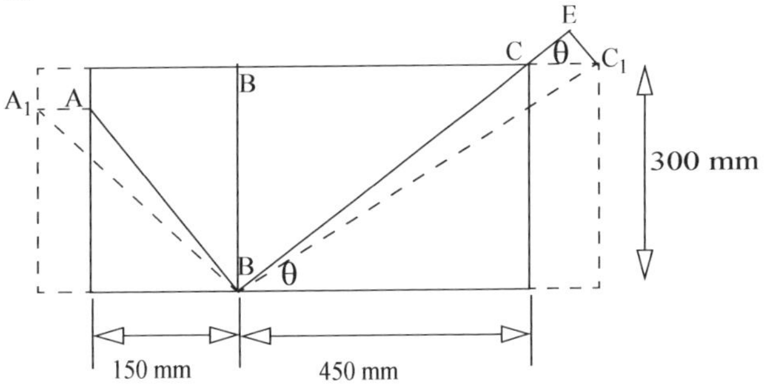

$$\varepsilon_{AB} = \frac{CE}{BC} = \frac{262.09(10^{-3})}{540.83} = 0.4846(10^{-3}) \tag{3}$$

ANS $\varepsilon_{AB} = 484.6\ \mu mm/mm$

---

2. 44

Solution $\varepsilon_{xx} = -800\mu\ mm/mm \qquad \gamma_B = ?$

From geometry and the given strain we have:

$$tan\theta = \frac{300}{450} \quad or \quad \theta = 33.69° \quad \alpha = 90 - \theta = 56.31° \tag{1}$$

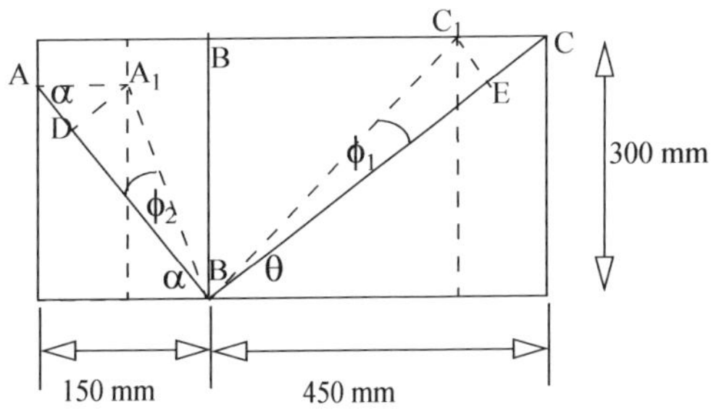

$$AB\cos\alpha = 150 \qquad AB = \frac{150}{\cos 56.31} = 270.42 \text{ mm} \qquad BC\sin\theta = 300 \qquad BC = \frac{300}{\sin 56.31} = 540.83 \text{ mm} \tag{2}$$

$$AA_1 = (150)|\varepsilon_{xx}| = 150[800(10^{-6})] = 0.12 \text{ mm} \qquad CC_1 = (450)|\varepsilon_{xx}| = 450[800(10^{-6})] = 0.360 \text{ mm} \tag{3}$$

$$AD = AA_1\cos\alpha = 66.56(10^{-3}) \qquad A_1D = AA_1\sin\alpha = 99.85(10^{-3}) \tag{4}$$

$$CE - CC_1\cos\theta = 299.54(10^{-3}) \qquad C_1E = CC_1\sin\theta = 199.69(10^{-3}) \tag{5}$$

$$\tan\phi_1 = \frac{C_1E}{BE} = \frac{C_1E}{BC-CE} = \frac{199.69(10^{-3})}{540.83 - 299.54(10^{-3})} = 0.3694(10^{-3}) \quad or \quad \phi_1 = 0.3694(10^{-3}) \text{ rads} \tag{6}$$

$$\tan\phi_2 = \frac{A_1D}{BD} = \frac{A_1D}{AB-AD} = \frac{99.85(10^{-3})}{270.42 - 66.56(10^{-3})} = 0.3692(10^{-3}) \quad or \quad \phi_2 = 0.3692(10^{-3}) \text{ rads} \tag{7}$$

$$\gamma_B = \phi_1 + \phi_2 = 0.7388(10^{-3})$$

ANS $\gamma_B = 738.8$ μrads

---

2. 45

Solution $\varepsilon_{yy} = 800\mu\ in/in \qquad \varepsilon_{AB} = ?$

From geometry and the given strain we have:

$$\tan\theta = 3/2 \quad or \quad \theta = 56.31° \qquad \alpha = 90 - \theta = 33.69° \qquad AB\sin\alpha = 1 \qquad AB = 1/(\sin 33.69) = 1.8028 \text{ in.} \tag{1}$$

$$AA_1 = (1)|\varepsilon_{yy}| = 800(10^{-6}) \text{ in.} \qquad AD = AA_1\sin\alpha = 443.76(10^{-6}) \tag{2}$$

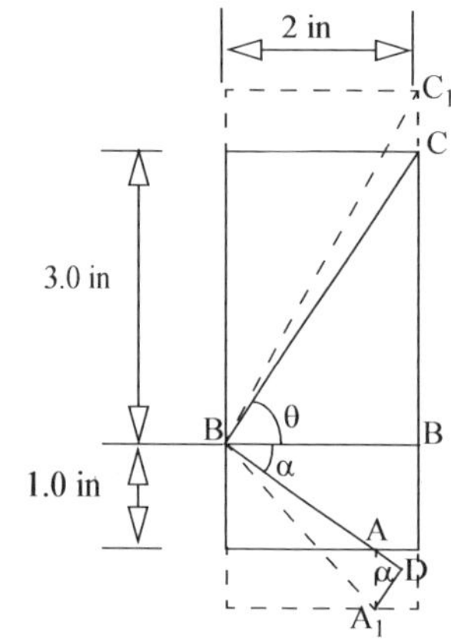

$$\varepsilon_{AB} = AD/AB = 443.76(10^{-6})/1.8028 = 246.15(10^{-6}) \tag{3}$$

ANS $\varepsilon_{AB} = 246.2$ μin./in.

---

2. 46

Solution $\varepsilon_{yy} = -500\mu\ in/in \qquad \varepsilon_{BC} = ?$

From geometry and the given strain we we have:

$$\tan\theta = 3/2 \quad or \quad \theta = 56.31° \qquad CC_1 = (3)|\varepsilon_{yy}| = 1.5(10^{-3}) \text{ in.} \qquad CE = CC_1\sin\theta = 1.2481(10^{-3}) \tag{1}$$

$$AC\sin\theta = 3 \qquad AC = 3/(\sin 56.31) = 3.6056 \text{ in.} \tag{2}$$

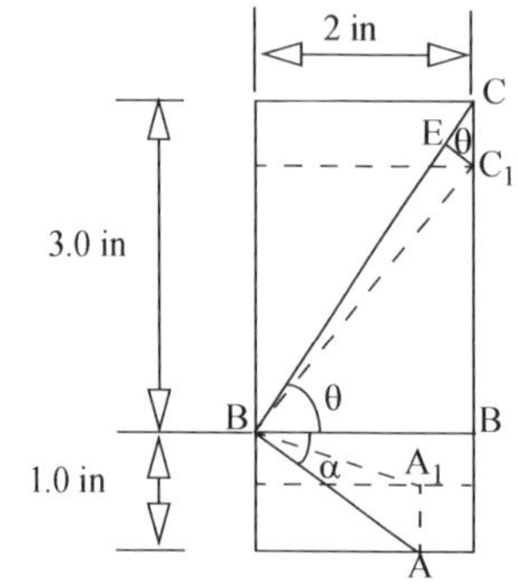

$$\varepsilon_{AC} = -CE/AC = -1.2481(10^{-3})/3.6056 = (-0.3462)(10^{-3}) \tag{3}$$

ANS $\varepsilon_{AC} = -346.2$ μin./in.

2. 47

Solution $\varepsilon_{yy} = 600\mu \; in/in$ $\gamma_B =$

From geometry and the given strain we have:

$$tan\theta = 3/2 \quad or \quad \theta = 56.31° \quad \alpha = 90 - \theta = 33.69° \quad (1)$$

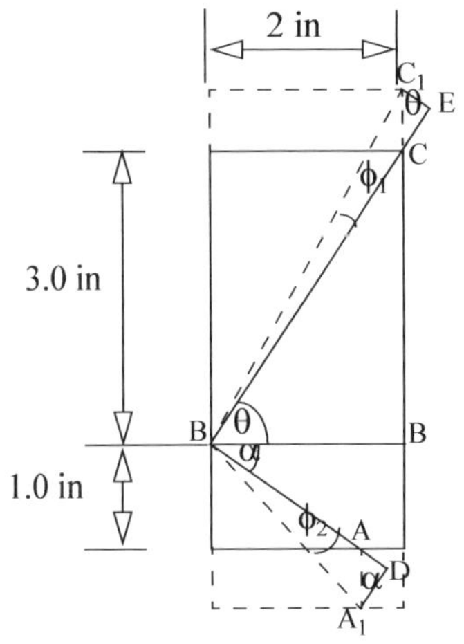

$$ABsin\alpha = 1 \quad AB = \frac{1}{sin33.69} = 1.8028 \text{ in.} \quad BCsin\theta = 3 \quad BC = \frac{3}{sin56.31} = 3.6056 \text{ in.} \quad (2)$$

$$AA_1 = (1)|\varepsilon_{yy}| = 0.6(10^{-3}) \text{ in.} \quad CC_1 = (3)|\varepsilon_{yy}| = 1.8(10^{-3}) \text{ in.} \quad (3)$$

$$AD = AA_1 sin\alpha = 0.3328(10^{-3}) \quad A_1D = AA_1 cos\alpha = 0.4992(10^{-3}) \quad (4)$$

$$CE = CC_1 sin\theta = 1.4977(10^{-3}) \quad C_1E = AA_1 cos\alpha = 0.9985(10^{-3}) \quad (5)$$

$$tan\phi_1 = \frac{C_1E}{BE} = \frac{C_1E}{BC + CE} = \frac{0.9985(10^{-3})}{3.6056 + 1.4977(10^{-3})} = 0.2768(10^{-3}) \quad or \quad \phi_1 = 0.2768(10^{-3}) \text{ rads} \quad (6)$$

$$tan\phi_2 = \frac{A_1D}{BD} = \frac{A_1D}{AB + AD} = \frac{0.4992(10^{-3})}{1.8028 + 0.3328(10^{-3})} = 0.2769(10^{-3}) \quad or \quad \phi_2 = 0.2769(10^{-3}) \text{ rads} \quad (7)$$

$$\gamma_B = -(\phi_1 + \phi_2) = -0.5537(10^{-3}) \quad (8)$$

ANS $\gamma_B = -553.7 \; \mu\text{rads}$

---

2. 48

Solution δ_A=0.4mm δ_B=0.8mm γ_A= ?

From geometry and the given values we have

$$tan\phi_1 = \frac{CD}{AD} = \frac{300 + \delta_A}{300} \quad tan\phi_2 = \frac{DE}{AD} = \frac{(CE - CD)}{AD} = \frac{(600 + \delta_B) - (300 + \delta_A)}{300} = \frac{300 + \delta_B - \delta_A}{300} \quad (1)$$

$$tan\phi_1 = 300.4/300 \quad or \quad \phi_1 = 0.78606 rads \quad tan\phi_2 = 300.4/300 \quad or \quad \phi_2 = 0.78606 rads \quad (2)$$

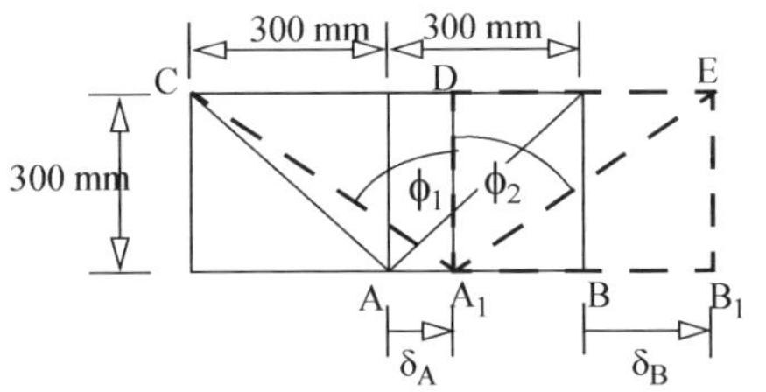

The decrease in the angle from the original right angle at A represents positive shear strain

$$\gamma_A = \pi/2 - (\phi_1 + \phi_2) = -1.332(10^{-3}) rads \quad (3)$$

ANS $\gamma_A = -1332 \mu rad$

---

2. 49

Solution δ_A=0.3 δ_B=0.9 γ_A= ?

Substituting the given values into the equation from previous problem we have

$$tan\phi_1 = (300 + \delta_A)/300 = 300.3/300 \quad or \quad \phi_1 = 0.78590 rads \quad (1)$$

$$\tan\phi_2 = (300 + \delta_B - \delta_A)/300 = 300.6/300 \qquad or \qquad \phi_2 = 0.78640 rads \qquad (2)$$

The decrease in the angle from the original right angle at A represents positive shear strain, i.e.,

$$\gamma_A = \pi/2 - (\phi_1 + \phi_2) = -1.4987(10^{-3}) rads \qquad (3)$$

ANS $\gamma_A = -1499 \mu rad$

---

2. 50

Solution $\delta_P = 0.25$ mm (a) ε_{AP}= ? using geometry (b) ε_{AP}= ? using small strain scaler approach
(c) ε_{AP}= ? using small strain vector approach.

(a) From triangle APP_1, using cosine rule we obtain the final length and then the average normal strain.

$$L_f^2 = AP^2 + PP_1^2 - 2AP(PP_1)\cos 130 = 200^2 + 0.25^2 - 2(200)(0.25)\cos 130 = 200.16079 \qquad (1)$$

$$\varepsilon_{AP} = (L_f - L_o)/L_o = 0.16079/200 = 0.80394(10^{-3}) \qquad (2)$$

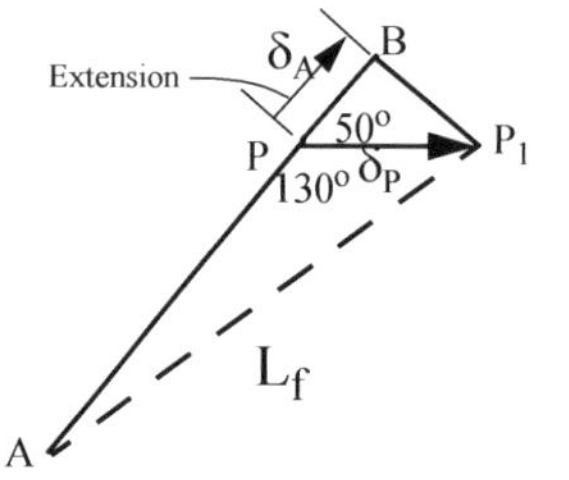

(b) From triangle PBP_1

$$\delta_{AP} = PB = PP_1 \cos 50 = 0.25\cos 50 = 0.1607 \qquad \varepsilon_{AP} = \delta_{AP}/L_o = 0.16070/200 = 0.80348(10^{-3}) \qquad (3)$$

(c) The deformation vector, the unit vector, and the deformation and strain of the bar AP can be calculated as shown below.

$$\bar{D} = 0.25\bar{i} \qquad \bar{i}_{AP} = \cos 50\bar{i} + \sin 50\bar{j} \qquad \delta_{AP} = \bar{i}_{AP} \cdot \bar{D} = 0.25\cos 50 = 0.1607 \qquad (4)$$

$$\varepsilon_{AP} = \delta_{AP}/L_o = 0.16070/200 = 0.80348(10^{-3}) \qquad (5)$$

ANS (a) $\varepsilon_{AP} = 803.9 \mu mm/mm$ (b) $\varepsilon_{AP} = 803.5 \mu mm/mm$ (c) $\varepsilon_{AP} = 803.5 \mu mm/mm$

---

2. 51

Solution $\delta_P = 0.25$ mm (a) ε_{AP}= ? using geometry
(b) ε_{AP}= ? using small strain scaler approach (c) ε_{AP}= ? using small strain vector approach.

(a) From triangle APP_1, using cosine rule we obtain the final length and then the average normal strain

$$L_f^2 = AP^2 + PP_1^2 - 2AP(PP_1)\cos 160 = 200^2 + 0.25^2 - 2(200)(0.25)\cos 160 = 200.23494 \qquad (1)$$

$$\varepsilon_{AP} = (L_f - L_o)/L_o = 0.23494/200 = 1.1747(10^{-3}) \qquad (2)$$

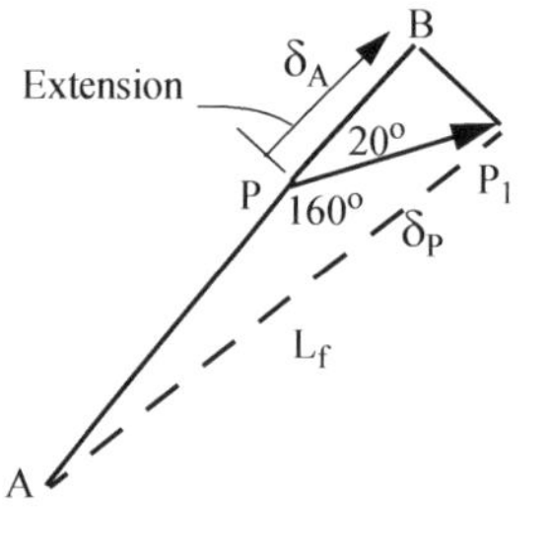

(b) From triangle PBP_1

$$\delta_{AP} = PB = PP_1 \cos 20 = 0.25\cos 20 = 0.23492 \qquad \varepsilon_{AP} = \delta_{AP}/L_o = 0.23492/200 = 01.1746(10^{-3}) \qquad (3)$$

(c) The deformation vector, the unit vector, and the deformation and strain of the bar AP can be calculated as shown below

$$\bar{D} = 0.25(\cos 30\bar{i} + \sin 30\bar{j}) \qquad \bar{i}_{AP} = \cos 50\bar{i} + \sin 50\bar{j} \qquad (4)$$

$$\delta_{AP} = \bar{i}_{AP} \cdot \bar{D} = 0.25(\cos 30\cos 50 + \sin 30\sin 50) = 0.25\cos 20 = 0.23492 \qquad (5)$$

$$\varepsilon_{AP} = \delta_{AP}/L_o = 0.23492/200 = 01.1746(10^{-3}) \qquad (6)$$

ANS (a) $\varepsilon_{AP} = 1174.7 \mu mm/mm$; (b) $\varepsilon_{AP} = 1174.6 \mu mm/mm$; (c) $\varepsilon_{AP} = 1174.6 \mu mm/mm$;

2. 52

Solution $\delta_{AP}= ?$ $\delta_{BP}= ?$

As per small strain approximation, we need component of PP_1 in the original direction of AP and BP as shown below.

$$\delta_{AP} = \delta_P = 0.25 \qquad \delta_{BP} = \delta_P cos 70 = 0.08551 \tag{1}$$

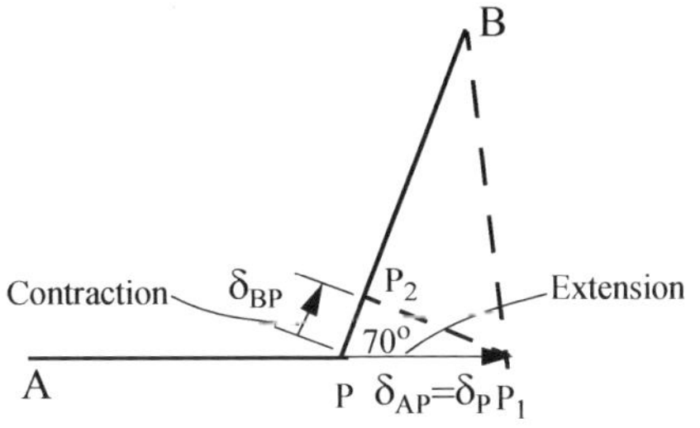

ANS $\delta_{AP} = 0.25 mm\ extension;\ \delta_{BP} = 0.0855 mm\ contraction$

---

2. 53

Solution $\delta_{AP}= ?$ $\delta_{BP}= ?$

As per small strain approximation, we need component of PP_1 in the original direction of AP and BP as shown below.

$$\delta_{AP} = \delta_P = 0.25 \qquad \delta_{BP} = \delta_P cos 60 = 0.125 \tag{1}$$

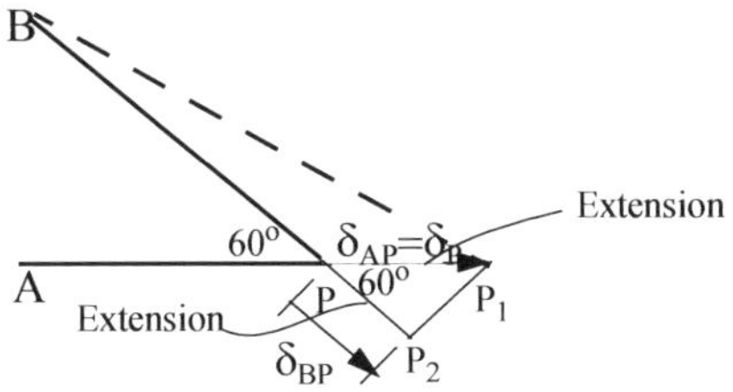

ANS δ_{AP} = 0.25 mm extension ; δ_{BP} = 0.125 mm extension

---

2. 54

Solution $\delta_{AP}= ?$ $\delta_{BP}= ?$

As per small strain approximation, we need component of PP_1 in the original direction of AP and BP as shown below.

$$\delta_{AP} = \delta_P cos 75 = 0.06470 \qquad \delta_{BP} = \delta_P cos 30 = 0.2165 \tag{1}$$

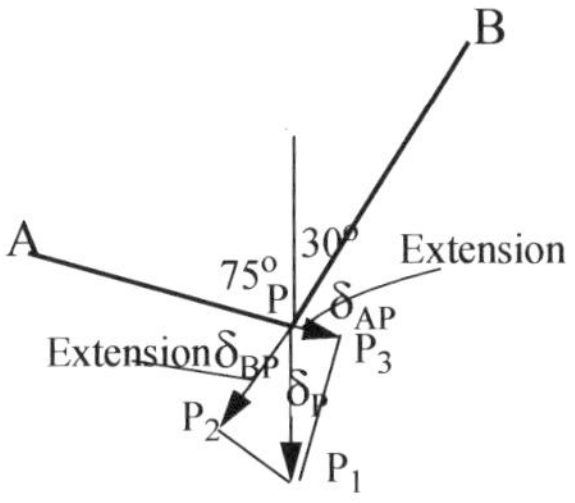

ANS $\delta_{AP} = 0.0647 mm\ extension;\ \delta_{BP} = 0.2165 mm\ extension$

---

2. 55

Solution $\delta_{AP}= ?$ $\delta_{BP}= ?$

As per small strain approximation, we need component of PP_1 in the original direction of AP and BP as shown below.

$$\delta_{AP} = \delta_P cos 40 = 0.01532 \qquad \delta_{BP} = \delta_P cos 70 = 0.00684 \tag{1}$$

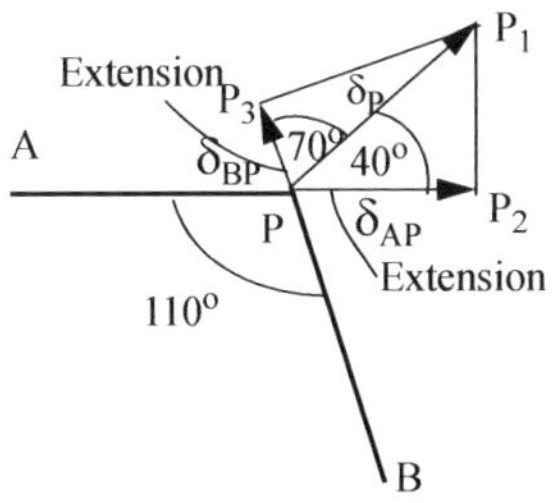

ANS δ_{AP} = 0.0153 in. extension ; δ_{BP} = 0.0068 in. extension

---

2. 56

Solution δ_{AP}= ? δ_{BP}= ?

As per small strain approximation, we need component of PP_1 in the original direction of AP and BP as shown below.

$$\delta_{AP} = \delta_P = 0.01 \qquad \delta_{BP} = \delta_P cos 25 = 0.00906 \qquad \textbf{(1)}$$

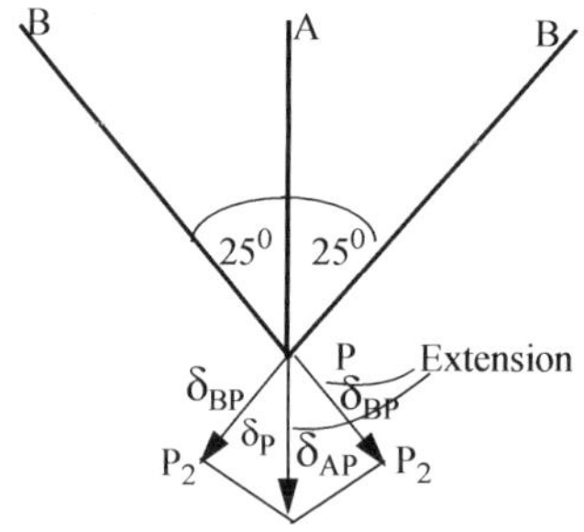

ANS $\delta_{AP} = 0.01$ in. extension ; $\delta_{BP} = 0.0091$ in. extension

---

2. 57

Solution δ_{AP}= ? δ_{BP}= ?

As per small strain approximation, we need component of PP_1 in the original direction of AP and BP as shown below.

$$\delta_{AP} = \delta_P cos 80 = 0.00347 \qquad \delta_{BP} = \delta_P cos 20 = 0.01879 \qquad \textbf{(1)}$$

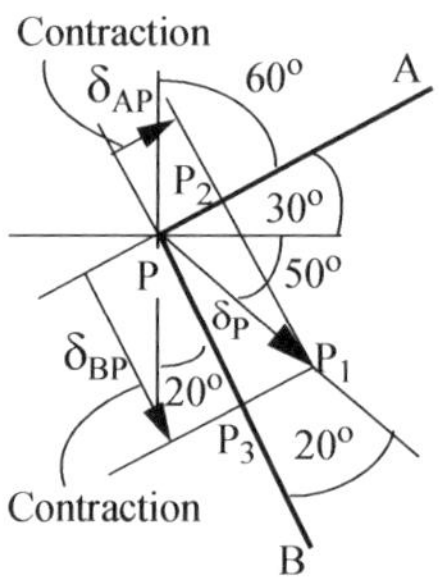

ANS $\delta_{AP} = 0.0035 inch\ contraction$; $\delta_{BP} = 0.0188 inch\ contraction$

---

2. 58

Solution ε_A=-500μ L_A=30in L_B=50in gap= 0.004 in ε_B=?

From geometry of the deformed shape and the given data we obtain:

$$\delta_D = \delta_A + gap \qquad \delta_D/36 = \delta_E/96 \qquad or \qquad (\delta_A + 0.004)/36 = \delta_E/96 \qquad \delta_B = \delta_E cos 15 \qquad \textbf{(1)}$$

$$\delta_A = (600)(10^{-6})(30) = 0.018 \qquad \delta_E = (96/36)(0.018 + 0.004) = 0.05867 in \qquad \delta_B = \delta_E cos 15 = 0.05667 \qquad \textbf{(2)}$$

$$\varepsilon_B = \delta_B/L_B = 0.05667/50 = 1.133(10^{-3}) \qquad \textbf{(3)}$$

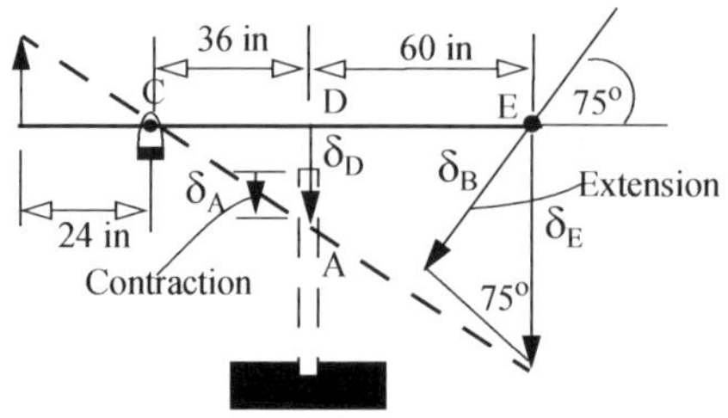

ANS $\varepsilon_B = 1133 \mu$in./in.

---

2. 59

Solution ε_B=1500μ L_A=30in L_B=50in ε_A=?

From previous problem we have the following.

$$(\delta_A + 0.004)/36 = \delta_E/96 \qquad \delta_B = \delta_E cos 15 \qquad \textbf{(1)}$$

$$\delta_B = (1500)(10^{-6})(50) = 0.075\ in \qquad \delta_E = \delta_B/(cos 15) = 0.075/(cos 15) = 0.07765\ in \qquad \textbf{(2)}$$

$\delta_A + 0.004 = (36/96)\delta_E = (36/96)(0.07765) = 0.02912 in$ or $\delta_A = \delta_D - 0.004 = 0.02512 in$ *contraction* **(3)**

$$\varepsilon_A = \delta_A/L_A = -0.02512/30 = -0.8372(10^{-3})$$ **(4)**

ANS $\varepsilon_A = -837.2 \mu in./in.$

---

2. 60

Solution ε_{AB}=? ε_{BF}=? ε_{FG}=? ε_{GB}=?

1. Deformation vector calculations

$$\bar{D}_{AB} = (u_B - u_A)\bar{i} + (v_B - v_A)\bar{j} = (12.6\bar{i} - 24.48\bar{j})mm$$ **(1)**

$$\bar{D}_{BF} = (u_F - u_B)\bar{i} + (v_F - v_B)\bar{j} = (\ 8.4 - 12.6)\bar{i} + (-28.68 + 24.48)\bar{j} = (-21\bar{i} - 4.2\bar{j})mm$$ **(2)**

$\bar{D}_{FG} = (u_G - u_F)\bar{i} + (v_G - v_F)\bar{j} = (8.4\bar{i} + 28.68\bar{j})mm$ $\bar{D}_{GB} = (u_B - u_G)\bar{i} + (v_B - v_G)\bar{j} = (12.6\bar{i} - 24.48\bar{j})mm$ **(3)**

2. Unit vector calculations

$\bar{i}_{AB} = \bar{i}$ $\bar{i}_{BF} = -\bar{j}$ $\bar{i}_{FG} = -\bar{i}$ $\bar{i}_{GB} = \cos 45\bar{i} + \sin 45\bar{j} = 0.707\bar{i} + 0.707\bar{j}$ **(4)**

3. Deformation calculation

$\delta_{AB} = \bar{D}_{AB} \cdot \bar{i}_{AB} = 12.6mm$ $\delta_{BF} = \bar{D}_{BF} \cdot \bar{i}_{BF} = 4.2mm$ $\delta_{FG} = \bar{D}_{FG} \cdot \bar{i}_{FG} = -8.4mm$ **(5)**

$$\delta_{GB} = (12.6)(0.707) - (24.48)(0.707) = -8.4004$$ **(6)**

4. Strain Calculation

$\varepsilon_{AB} = \delta_{AB}/L_{AB} = 12.6/[2(10^3)] = 6.3(10^{-3})$ $\varepsilon_{BF} = \delta_{BF}/L_{BF} = 4.2/[2(10^3)] = 2.1(10^{-3})$ **(7)**

$\varepsilon_{FG} = \delta_{FG}/L_{FG} = -8.4/(2(10^3)) = -4.2(10^{-3})$ $\varepsilon_{GB} = \delta_{GB}/L_{GB} = -8.4004/[2\sqrt{2}(10^3)] = -2.9698(10^{-3})$ **(8)**

ANS $\varepsilon_{AB} = 6300 \mu mm/mm$; $\varepsilon_{BF} = 2100 \mu mm/mm$; $\varepsilon_{FG} = -4200 \mu mm/mm$; $\varepsilon_{GB} = -2970 \mu mm/mm$

---

2. 61

Solution ε_{BC}=? ε_{CF}=? ε_{FE}=?

1. Deformation vector calculation

$$\bar{D}_{BC} = (u_C - u_B)\bar{i} + (v_C - v_B)\bar{j} = (21 - 12.6)\bar{i} + (-69.97 + 24.48)\bar{j} = (8.4\bar{i} - 45.52\bar{j})mm$$ **(1)**

$$\bar{D}_{CF} = (u_F - u_C)\bar{i} + (v_F - v_C)\bar{j} = (-8.4 - 21.0)\bar{i} + (-28.68 + 69.97)\bar{j} = (29.4\bar{i} + 41.29\bar{j})mm$$ **(2)**

$$\bar{D}_{FE} = (u_E - u_F)\bar{i} + (v_E - v_F)\bar{j} = (-12.6 + 8.4)\bar{i} + (-69.97 + 28.68)\bar{j} = (-4.2\bar{i} - 41.29\bar{j})mm$$ **(3)**

2. Unit vector calculations

$\bar{i}_{BC} = \bar{i}$ $\bar{i}_{CF} = -\cos 45\bar{i} - \sin 45\bar{j} = -0.707\bar{i} - 0.707\bar{j}$ $\bar{i}_{FE} = \bar{i}$ **(4)**

3. Deformation calculation

$\delta_{BC} = \bar{D}_{BC} \cdot \bar{i}_{BC} = 8.4mm$ $\delta_{CF} = \bar{D}_{CF} \cdot \bar{i}_{CF} = (29.4)(0.707) - (41.29)(0.707) = -8.408mm$ **(5)**

$$\delta_{FE} = \bar{D}_{FE} \cdot \bar{i}_{FE} = -4.2mm$$ **(6)**

4. Strain Calculation

$\varepsilon_{BC} = \dfrac{\delta_{BC}}{L_{BC}} = \dfrac{8.4}{2(10^3)} = 4.2(10^{-3})$ $\varepsilon_{CF} = \dfrac{\delta_{CF}}{L_{CF}} = \dfrac{-8.408}{2\sqrt{2}(10^3)} = 2.973(10^{-3})$ $\varepsilon_{FE} = \dfrac{\delta_{FE}}{L_{FE}} = \dfrac{-4.2}{2(10^3)} = -2.1(10^{-3})$ **(7)**

ANS $\varepsilon_{BC} = 4200 \mu mm/mm$; $\varepsilon_{CF} = -2973 \mu mm/mm$; $\varepsilon_{FE} = -2100 \mu mm/mm$

---

2. 62

Solution ε_{ED}=? ε_{DC}=? ε_{CE}=?

1. Deformation vector calculation

$$\bar{D}_{ED} = (u_D - u_E)\bar{i} + (v_D - v_E)\bar{j} = (-16.8 + 12.6)\bar{i} + (-119.65 + 69.97)\bar{j} = (-4.2\bar{i} - 49.98\bar{j})mm$$ **(1)**

$$\bar{D}_{DC} = (u_C - u_D)\bar{i} + (v_C - v_D)\bar{j} = (21.0 + 16.8)\bar{i} + (-69.97 + 119.65)\bar{j} = (37.8\bar{i} + 49.68\bar{j})mm$$ **(2)**

$$\bar{D}_{CE} = (u_E - u_C)\bar{i} + (v_E - v_C)\bar{j} = (-12.6 - 21)\bar{i} + (-69.97 + 69.97)\bar{j} = -33.6\bar{i}\ mm$$ **(3)**

2. Unit vector calculations

$\bar{i}_{ED} = \bar{i}$ $\bar{i}_{DC} = -\cos 45\bar{i} + \sin 45\bar{j} = -0.707\bar{i} + 0.707\bar{j}$ $\bar{i}_{CE} = \bar{j}$ **(4)**

3. Deformation calculation

$\delta_{ED} = \bar{D}_{ED} \cdot \bar{i}_{ED} = -4.2mm$ $\delta_{DC} = \bar{D}_{DC} \cdot \bar{i}_{DC} = -(37.8)(0.707) + (49.68)(0.707) = 8.4004mm$ $\delta_{CE} = \bar{D}_{CE} \cdot \bar{i}_{CE} = 0$ **(5)**

4. Strain Calculation

$$\varepsilon_{ED} = \delta_{ED}/L_{ED} = -4.2/2(10^3) = -2.1(10^{-3}) \qquad \varepsilon_{DC} = \delta_{DC}/L_{DC} = 8.4004/[2\sqrt{2}(10^3)] = 2.970(10^{-3}) \qquad \varepsilon_{CE} = 0 \quad \textbf{(6)}$$

ANS $\varepsilon_{ED} = -2100\mu mm/mm$; $\varepsilon_{DC} = 2970\mu mm/mm$; $\varepsilon_{CE} = 0$

---

2. 63

Solution ε_{AB}=? ε_{BG}=? ε_{GA}=? ε_{AH}=?

1. Deformation vector calculations

$$\bar{D}_{AB} = (u_B - u_A)\bar{i} + (v_B - v_A)\bar{j} = (7\bar{i} + 1.5\bar{j})mm \qquad \bar{D}_{BG} = (u_G - u_B)\bar{i} + (v_G - v_B)\bar{j} = (7-7)\bar{i} + (-4.125-1.5)\bar{j} = -5.625\bar{j}mm \quad \textbf{(1)}$$

$$\bar{D}_{GA} = (u_A - u_G)\bar{i} + (v_A - v_G)\bar{j} = (-7\bar{i} + 4.125\bar{j})mm \qquad \bar{D}_{AH} = (u_H - u_A)\bar{i} + (v_H - v_A)\bar{j} = (\quad \textbf{(2)}$$

2. Unit vector calculations

$$\bar{i}_{AB} = \bar{j} \qquad \bar{i}_{BG} = \bar{i} \qquad \bar{i}_{GA} = \frac{-4\bar{i} - 3\bar{j}}{\sqrt{3^2 + 4^2}} = -0.8\bar{i} - 0.6\bar{j} \qquad \bar{i}_{AH} = \bar{i} \quad \textbf{(3)}$$

3. Deformation calculation

$$\delta_{AB} = \bar{D}_{AB} \cdot \bar{i}_{AB} = 1.5mm \qquad \delta_{BG} = \bar{D}_{BG} \cdot \bar{i}_{BG} = 0 \quad \textbf{(4)}$$

$$\delta_{GA} = \bar{D}_{GA} \cdot \bar{i}_{GA} = (-7)(-0.8) + (4.125)(-0.6) = 3.125mm \qquad \delta_{AH} = 0 \quad \textbf{(5)}$$

4. Strain Calculation

$$\varepsilon_{AB} = \frac{\delta_{AB}}{L_{AB}} = \frac{1.5}{3(10^3)} = 0.5(10^{-3}) \qquad \varepsilon_{GA} = \frac{\delta_{GA}}{L_{GA}} = \frac{3.125}{5(10^3)} = 0.625(10^{-3}) \quad \textbf{(6)}$$

ANS $\varepsilon_{AB} = 500\mu mm/mm$; $\varepsilon_{BG} = 0$; $\varepsilon_{GA} = 625\mu mm/mm$; $\varepsilon_{AH} = 0$

---

2. 64

Solution ε_{BC}=? ε_{CG}=? ε_{GB}=? ε_{CD}=?

1. Deformation vector calculations

$$\bar{D}_{BC} = (u_C - u_B)\bar{i} + (v_C - v_B)\bar{j} = (17.55 - 7)\bar{i} + (3 - 1.5)\bar{j} = (10.55\bar{i} + 1.5\bar{j})mm \quad \textbf{(1)}$$

$$\bar{D}_{CG} = (u_G - u_C)\bar{i} + (v_G - v_C)\bar{j} = (7 - 17.55)\bar{i} + (-4.125 - 3)\bar{j} = (-10.55\bar{i} - 7.125\bar{j})mm \quad \textbf{(2)}$$

$$\bar{D}_{GB} = (u_B - u_G)\bar{i} + (v_B - v_G)\bar{j} = (7 - 7)\bar{i} + (1.5 + 4.125)\bar{j} = (5.625\bar{j})mm \quad \textbf{(3)}$$

$$\bar{D}_{CD} = (u_D - u_C)\bar{i} + (v_D - v_C)\bar{j} = (20.22 - 17.55)\bar{i} + (-4.125 - 3)\bar{j} = (2.67\bar{i} - 7.125\bar{j})mm \quad \textbf{(4)}$$

2. Unit vector calculations

$$\bar{i}_{BC} = \bar{j} \qquad \bar{i}_{CG} = \frac{4\bar{i} - 3\bar{j}}{5} = 0.8\bar{i} - 0.6\bar{j} \qquad \bar{i}_{GB} = -\bar{i} \qquad \bar{i}_{CD} = \bar{i} \quad \textbf{(5)}$$

3. Deformation calculation

$$\delta_{BC} = \bar{D}_{BC} \cdot \bar{i}_{BC} = 1.5mm \qquad \delta_{CG} = \bar{D}_{CG} \cdot \bar{i}_{CG} = (0.8)(-10.55) + (-0.6)(-7.125) = -4.165mm \quad \textbf{(6)}$$

$$\delta_{GB} = \bar{D}_{GB} \cdot \bar{i}_{GB} = 0 \qquad \delta_{CD} = 2.67mm \quad \textbf{(7)}$$

4. Strain Calculation

$$\varepsilon_{BC} = \frac{\delta_{BC}}{L_{BC}} = \frac{1.5}{3(10^3)} = 0.5(10^{-3}) \qquad \varepsilon_{CG} = \frac{\delta_{CG}}{L_{CG}} = \frac{-4.165}{5(10^3)} = -0.833(10^{-3}) \quad \textbf{(8)}$$

$$\varepsilon_{GB} = \frac{\delta_{GB}}{L_{GB}} = 0 \qquad \varepsilon_{CD} = \frac{\delta_{CD}}{L_{CD}} = \frac{2.67}{4(10^3)} = 0.6675(10^{-3}) \quad \textbf{(9)}$$

ANS $\varepsilon_{BC} = 500\mu\frac{mm}{mm}$; $\varepsilon_{CG} = -833\mu\frac{mm}{mm}$; $\varepsilon_{GB} = 0$; $\varepsilon_{CD} = 667.5\mu\frac{mm}{mm}$

---

2. 65

Solution ε_{GF}=? ε_{FE}=? ε_{EG}=? ε_{DE}=?

1. Deformation vector calculations

$$\bar{D}_{GF} = (u_F - u_G)\bar{i} + (v_F - v_G)\bar{j} = (9 - 7)\bar{i} + (-33.75 + 4.125)\bar{j} = (2\bar{i} - 29.625\bar{j})mm \quad \textbf{(1)}$$

$$\bar{D}_{FE} = (u_E - u_F)\bar{i} + (v_E - v_F)\bar{j} = (22.88 - 9)\bar{i} + (-32.25 + 33.75)\bar{j} = (13.88\bar{i} + 1.5\bar{j})mm \quad \textbf{(2)}$$

$$\bar{D}_{EG} = (u_G - u_E)\bar{i} + (v_G - v_E)\bar{j} = (7 - 22.88)\bar{i} + (-4.125 + 32.25)\bar{j} = (-15.88\bar{i} + 28.125\bar{j})mm \quad \textbf{(3)}$$

$$\bar{D}_{DE} = (u_E - u_D)\bar{i} + (v_E - v_D)\bar{j} = (22.88 - 20.22)\bar{i} + (-32.25 + 4.125)\bar{j} = (2.66\bar{i} - 28.125\bar{j})mm \quad \textbf{(4)}$$

2. Unit vector calculations

$$i_{GF} = i \qquad i_{EG} = \frac{-4i-3j}{5} = -0.8i-0.6j \qquad i_{FE} = j \qquad i_{DE} = i \tag{5}$$

3. Deformation calculation

$$\delta_{GF} = \bar{D}_{GF}\cdot i_{GF} = 2mm \qquad \delta_{FE} = \bar{D}_{FE}\cdot i_{FE} = 1.5mm \tag{6}$$

$$\delta_{EG} = \bar{D}_{EG}\cdot i_{EG} = (-15.88)(-0.8)+(28.125)(-0.6) = -4.171mm \qquad \delta_{DE} = 2.66mm \tag{7}$$

4. Strain Calculation

$$\varepsilon_{GF} = \delta_{GF}/L_{GF} = 2/[4(10^3)] = 0.5(10^{-3}) \qquad \varepsilon_{FE} = \delta_{FE}/L_{FE} = 1.5/[3(10^3)] = 0.5(10^{-3}) \tag{8}$$

$$\varepsilon_{EG} = \delta_{EG}/L_{EG} = -4.171/[5(10^3)] = -0.8342(10^{-3}) \qquad \varepsilon_{DE} = \delta_{DE}/L_{DE} = 2.66/[4(10^3)] = 0.665(10^{-3}) \tag{9}$$

ANS $\varepsilon_{GF} = 500\mu mm/mm$; $\varepsilon_{FE} = 500\mu mm/mm$; $\varepsilon_{EG} = -834\mu mm/mm$ $\varepsilon_{DE} = 665\mu mm/mm$

---

2. 66

Solution $\varepsilon_{AP} = ?$ $\varepsilon_{BP} = ?$ $\varepsilon_{CP} = ?$

1. Deformation vector calculations

$$\bar{D}_{AP} = 2\bar{k} \qquad \bar{D}_{BP} = 2\bar{k} \qquad \bar{D}_{CP} = 2\bar{k} \tag{1}$$

2. Unit vector calcultions:

$$\vec{r}_{AP} = -5.0\vec{i}+6\vec{k} \qquad |\vec{r}_{AP}| = \sqrt{5^2+6^2} = 7.81 \qquad \vec{i}_{AP} = \frac{\vec{r}_{AP}}{|\vec{r}_{AP}|} = -0.6402\vec{i}+0.7682\vec{k} \tag{2}$$

$$\vec{r}_{BP} = 4.0\vec{i}-6\vec{j}+6\vec{k} \qquad |\vec{r}_{BP}| = \sqrt{4^2+6^2+6^2} = 9.381 \qquad \vec{i}_{BP} = \frac{\vec{r}_{BP}}{|\vec{r}_{BP}|} = 0.4264\vec{i}-0.6396\vec{i}+0.6396\vec{k} \tag{3}$$

$$\vec{r}_{CP} = -2.0\vec{i}-3\vec{j}+6\vec{k} \qquad |\vec{r}_{BP}| = \sqrt{2^2+3^2+6^2} = 7.0 \qquad \vec{i}_{CP} = \frac{\vec{r}_{CP}}{|\vec{r}_{CP}|} = 0.2857\vec{i}+0.4286\vec{i}+0.8571\vec{k} \tag{4}$$

3. Deformation calculations

$$\delta_{AP} = \bar{D}_{AP}\bullet\vec{i}_{AP} = (2)(0.7682) = 1.5364\ inch \qquad \delta_{BP} = \bar{D}_{BP}\bullet\vec{i}_{BP} = (2)(0.6396) = 1.2792\ inch \tag{5}$$

$$\delta_{CP} = \bar{D}_{CP}\bullet\vec{i}_{CP} = (2)(0.8571) = 1.7142\ inch \tag{6}$$

4. Strain calculations: The lengths of each bar are the magnitudes of the position vectors.

$$\varepsilon_{AP} = \delta_{AP}/L_{AP} = 1.5364/[(12)(7.81)] = 16.393(10^{-3}) \qquad \varepsilon_{BP} = \delta_{BP}/L_{BP} = 1.2792/[(12)(9.381)] = 11.363(10^{-3}) \tag{7}$$

$$\varepsilon_{CP} = \delta_{CP}/L_{CP} = 1.7142/[(12)(7)] = 20.407(10^{-3}) \tag{8}$$

ANS $\varepsilon_{AP} = 16393\mu in./in.$; $\varepsilon_{BP} = 11363\mu in./in.$; $\varepsilon_{CP} = 20407\mu in./in.$

Sections 2.5-2.6

2. 67

Solution ε_{xx}=? ε_{yy}=? γ_{xy}=?

The average strains can be found as shown below.

$$\varepsilon_{xx} = \frac{\Delta u}{\Delta x} = \frac{0.0036}{3} = 0.0012 \qquad \varepsilon_{yy} = \frac{\Delta v}{\Delta y} = \frac{0.0042}{1.4} = 0.003 \qquad \gamma_{xy} = \frac{\Delta u}{\Delta y}+\frac{\Delta v}{\Delta x} = \frac{0.0056}{1.4}+\frac{0.0042}{3} = 0.0054 \tag{1}$$

ANS $\varepsilon_{xx} = 1200\mu in./in.$; $\varepsilon_{yy} = 3000\mu in./in.$; $\gamma_{xy} = 5400\mu rad$

---

2. 68

Solution ε_{xx}=? ε_{yy}=? γ_{xy}=?

The average strains can be found as shown below.

$$\varepsilon_{xx} = \frac{\Delta u}{\Delta x} = \frac{-0.32}{250} = -0.128(10^{-3}) \qquad \varepsilon_{yy} = \frac{\Delta v}{\Delta y} = \frac{-0.3}{450} = -0.6667(10^{-3}) \qquad \gamma_{xy} = \frac{\Delta u}{\Delta y}+\frac{\Delta v}{\Delta x} = \frac{0.45}{450}+\frac{0.65}{250} = 3.6(10^{-3}) \tag{1}$$

ANS $\varepsilon_{xx} = -128\mu mm/mm$; $\varepsilon_{yy} = -666.7\mu mm/mm$; $\gamma_{xy} = 3600\mu rad$

---

2. 69

Solution ε_{xx}=? ε_{yy}=? γ_{xy}=?

The average strains can be found as shown below.

$$\varepsilon_{xx} = \frac{\Delta u}{\Delta x} = \frac{-0.009}{6} = -1.5(10^{-3}) \qquad \varepsilon_{yy} = \frac{\Delta v}{\Delta y} = \frac{-0.006}{3} = -2.0(10^{-3}) \qquad \gamma_{xy} = \frac{\Delta u}{\Delta y}+\frac{\Delta v}{\Delta x} = \frac{0.033}{3}+\frac{(-0.024)}{6} = 7(10^{-3}) \tag{1}$$

ANS $\varepsilon_{xx} = -1500\mu\text{mm/mm}$; $\varepsilon_{yy} = -2000\mu\text{mm/mm}$; $\gamma_{xy} = 7000\mu\text{rad}$

---

2. 70

Solution ε_{xx}=? ε_{yy}=? γ_{xy}=?

The average strains can be found as shown below.

$$\varepsilon_{xx} = \frac{u_B - u_A}{x_B - x_A} = \frac{0.625(10^{-6})}{0.0005} = 1250(10^{-6}) \qquad \varepsilon_{yy} = \frac{v_C - v_A}{y_C - y_A} = \frac{-0.5625(10^{-6})}{0.0005} = -1125(10^{-6}) \quad \textbf{(1)}$$

$$\gamma_{xy} = \frac{u_C - u_A}{y_C - y_A} + \frac{v_B - v_A}{x_B - x_A} = \frac{-0.5(10^{-6})}{0.0005} + \frac{(-0.3125)(10^{-6})}{0.0005} = -1625(10^{-6}) \quad \textbf{(2)}$$

ANS $\varepsilon_{xx} = 1250\mu\text{mm/mm}$; $\varepsilon_{yy} = -1125\mu\text{mm/mm}$; $\gamma_{xy} = -1625\mu\text{rad}$

---

2. 71

Solution ε_{xx}=? ε_{yy}=? γ_{xy}=?

The average strains can be found as shown below.

$$\varepsilon_{xx} = \frac{u_B - u_A}{x_B - x_A} = \frac{(1.5 - 0.625)(10^{-6})}{0.0005} = 1750(10^{-6}) \qquad \varepsilon_{yy} = \frac{v_C - v_A}{y_C - y_A} = \frac{(-1.125 + 0.3125)(10^{-6})}{0.0005} = -1625(10^{-6}) \quad \textbf{(1)}$$

$$\gamma_{xy} = \frac{u_C - u_A}{y_C - y_A} + \frac{v_B - v_A}{x_B - x_A} = \frac{(0.25 - 0.625)(10^{-6})}{0.0005} + \frac{(-0.5 + 0.3125)(10^{-6})}{0.0005} \quad \textbf{(2)}$$

ANS $\varepsilon_{xx} = 1750\mu\text{mm/mm}$; $\varepsilon_{yy} = -1625\mu\text{mm/mm}$; $\gamma_{xy} = -1125\mu rad$

---

2. 72

Solution ε_{xx}=? ε_{yy}=? γ_{xy}=?

The average strains can be found as shown below.

$$\varepsilon_{xx} = \frac{u_B - u_A}{x_B - x_A} = \frac{(0.25 + 0.5)(10^{-6})}{0.0005} = 1500(10^{-6}) \qquad \varepsilon_{yy} = \frac{v_C - v_A}{y_C - y_A} = \frac{(-1.25 + 0.5625)(10^{-6})}{0.0005} = -1375(10^{-6}) \quad \textbf{(1)}$$

$$\gamma_{xy} = \frac{u_C - u_A}{y_C - y_A} + \frac{v_B - v_A}{x_B - x_A} = \frac{(-1.25 + 0.5)(10^{-6})}{0.0005} + \frac{(-1.125 + 0.5625)(10^{-6})}{0.0005} = -2625(10^{-6}) \quad \textbf{(2)}$$

ANS $\varepsilon_{xx} = 1500\mu\text{mm/mm}$; $\varepsilon_{yy} = -1375\mu\text{mm/mm}$; $\gamma_{xy} = -2625\mu\text{rad}$

---

2. 73

Solution ε_{xx}=? ε_{yy}=? γ_{xy}=?

The average strains can be found as shown below.

$$\varepsilon_{xx} = \frac{u_B - u_A}{x_B - x_A} = \frac{(1.25 - 0.25)(10^{-6})}{0.0005} = 2000(10^{-6}) \qquad \varepsilon_{yy} = \frac{v_C - v_A}{y_C - y_A} = \frac{(-2.0625 + 1.125)(10^{-6})}{0.0005} = -1875(10^{-6}) \quad \textbf{(1)}$$

$$\gamma_{xy} = \frac{u_C - u_A}{y_C - y_A} + \frac{v_B - v_A}{x_B - x_A} = \frac{(-0.375 - 0.25)(10^{-6})}{0.0005} + \frac{(-1.5625 + 1.125)(10^{-6})}{0.0005} = -2125(10^{-6}) \quad \textbf{(2)}$$

ANS $\varepsilon_{xx} = 2000\mu\text{mm/mm}$; $\varepsilon_{yy} = -1875\mu\text{mm/mm}$; $\gamma_{xy} = -2125\mu\text{rad}$

---

2. 74

Solution $u(x) = [-19.44 + 1.44x - 0.01x^2 - 933.12/(72 - x)](10^{-3})$ *inches* $\varepsilon_{xx}(24)$=?

The strain at any point x can be found as shown below.

$$\varepsilon_{xx}(x) = \frac{du}{dx} = \left[1.44 - 0.02x - \frac{933.12}{(72 - x)^2}\right](10^{-3}) \qquad \varepsilon_{xx}(24) = \frac{du}{dx} = \left[1.44 - 0.02(24) - \frac{933.12}{(72 - 24)^2}\right](10^{-3}) = 0.555(10^{-3}) \quad \textbf{(1)}$$

ANS $\varepsilon_{xx}(24) = 555\mu$

---

2. 75

Solution $u(x) = [7.5(10^{-6})x^2 - 25(10^{-6})x - 0.15\,ln(1 - 0.004x)]$ *mm* $\varepsilon_{xx}(100)$=?

The strain at any point x can be found as shown below.

$$\varepsilon_{xx}(x) = \frac{du}{dx} = \left[15(10^{-6})x - 25(10^{-6}) + \frac{(0.15)(0.004)}{(1 - 0.004x)}\right] \quad \textbf{(1)}$$

$$\varepsilon_{xx}(100) = -15(10^{-6})(100) + 25(10^{-6}) + \frac{(0.15)(0.004)}{(0.6)} = 0.0015 - 0.000025 + 0.001 \quad (2)$$

ANS $\varepsilon_{xx}(100) = 2475\mu$

---

2. 76

Solution $\varepsilon_{xx}(a) = ?$

The strain at any point x can be found as shown below.

$$\varepsilon_{xx}(x) = \frac{du}{dx} = \frac{u_1}{2a^2}[(x-2a)+(x-a)] - \frac{u_2}{a^2}[(x-2a)+x] + \frac{u_3}{2a^2}[(x-a)+x] \quad (1)$$

$$\varepsilon_{xx}(x) = \frac{u_1}{2a^2}(2x-3a) - \frac{u_2}{a^2}(2x-2a) + \frac{u_3}{2a^2}(2x-a) = \frac{u_1}{2a^2}(-a) - \frac{u_2}{a^2}(0) + \frac{u_3}{2a^2}(a) \quad (2)$$

ANS $\varepsilon_{xx}(a) = (u_3 - u_1)/(2a)$

---

2. 77

Solution $\varepsilon_{xx} = 0.2/(40-x)^2$ $\quad u(20) = ?$

Method I: Integrating we obtain: $u(x) = 0.2/(40-x) + C_1$

$$u(0) = \frac{0.2}{40} + C_1 = 0 \quad or \quad C_1 = -0.005 \quad u(x) = \frac{0.2}{(40-x)} - 0.005 \quad u(20) = \frac{0.2}{(40-20)} - 0.005 = 0.005 \quad (1)$$

Method II: Integrating from x = 0 to x = 20 we obtain the following.

$$\int_0^{u(20)} du = \int_0^{20} \frac{0.2}{(40-x)^2} dx = \left.\frac{0.2}{(40-x)}\right|_0^{20} = \frac{0.2}{40-20} = 0.005 \quad (2)$$

ANS $u(20) = 0.005in$

---

2. 78

Solution $\varepsilon_{xx} = (KL)/(4L-3x)$ $\quad u(L) - u(0) = ?$

The given strain can be integrated as shown below.

$$\int_{u(0)}^{u(L)} du = \int_0^L \frac{KL}{(4L-3x)} dx \quad or \quad u(L) - u(0) = \left.\frac{KL}{(-3)} ln(4L-3x)\right|_0^L = -\frac{KL}{2}[ln(L) - ln(4L)] = -\frac{KL}{3}(-ln4) \quad (1)$$

ANS $u(L) - u(0) = 0.4621KL$

---

2. 79

Solution $u(L) - u(0) = ?$

Integrating the strain we obtain the relative displacement as shown below.

$$\varepsilon_{xx} = \frac{du}{dx} = K\left[4L - 2x - \frac{8L^3}{(4L-2x)^2}\right] \quad or \quad \int_{u(0)}^{u(L)} du = \int_0^L K\left[4L - 2x - \frac{8L^3}{(4L-2x)^2}\right] dx \text{ or}$$

$$u(L) - u(0) = K\left.\left[4Lx - \frac{2x^2}{2} - \frac{8L^3}{(4L-2x)}\right]\right|_0^L = K\left[4L^2 - L^2 - \frac{8L^3}{4L} + \frac{8L^3}{8L}\right] = 2KL^2 \quad (1)$$

ANS $u(L) - u(0) = 2KL^2$

---

2. 80

Solution u(1250)=?

The strains can be integrated to obtain relative displacements and added. We note that $u_0 = 0$.

$$\varepsilon_{xx} = \frac{du}{dx} = \frac{1500(10^3)}{(1875-x)} \mu \quad 0 \le x \le 750 \ mm \quad \varepsilon_{xx} = \frac{du}{dx} = 1500 \ \mu \quad 750 \ mm \le x \le 1250 \ mm \quad (1)$$

$$\int_{u(0)}^{u(750)} du = \int_0^{750} \frac{1500(10^3)}{(1875-x)} dx \text{ or}$$

$$u(750) - u(0) = -1500(10^{-3}) ln(1875-x)\Big|_0^{750} = -1500(10^{-3})[ln(1875-750) - ln(1875)] = 766.24(10^{-3}) \quad (2)$$

$$\int_{u(750)}^{u(1250)} du = \int_{750}^{1250} (1500)(10^{-6}) dx \quad or \quad u(1250) - u(750) = 1500(10^{-6})x\Big|_{750}^{1250} = 750(10^{-3}) \quad (3)$$

$$u(1250) - u(0) = [u(750) - u(0)] + [u(1250) - u(750)] = 766.24(10^{-3}) + 750(10^{-3}) = 1516.24(10^{-3})m \quad (4)$$

ANS $u(1250) = 1.516mm$

---

2. 81

Solution u_N=?

The relative displacement can be found and added. We note that $u_0 = 0$.

$$\varepsilon_i = \frac{du}{dx} = a_i \qquad \int_{u_{i-1}}^{u_i} du = \int_{x_{i-1}}^{x_i} a_i dx \qquad or \qquad u_i - u_{i-1} = a_i(x_i - x_{i-1}) \qquad 1 \le i \le N \quad (1)$$

$$\sum_{i=1}^{N} (u_i - u_{i-1}) = \sum_{i=1}^{N} a_i(x_i - x_{i-1}) \qquad or \qquad \sum_{i=1}^{N} u_i - \sum_{i=1}^{N} u_{i-1} = \sum_{i=1}^{N} a_i(x_i - x_{i-1}) \text{ or}$$

$$u_N - u_0 = \sum_{i=1}^{N} a_i(x_i - x_{i-1}) \quad (2)$$

ANS $u_N = \sum_{i=1}^{N} a_i(x_i - x_{i-1})$

---

2. 82

Solution ε_T =?

Solution proceeds as follows.

$$d\varepsilon_T = (du)/(L_o + u) \qquad \varepsilon_T = \int_o^{\varepsilon_T} d\varepsilon_T = \int_0^u \frac{du}{(L_o + u)} \qquad ln(L_o + u)\Big|_0^u = ln(L_o + u) - ln(L_o) = ln\left(1 + \frac{u}{L_o}\right) = ln(1 + \varepsilon) \quad (1)$$

ANS $\varepsilon_T = ln(1 + \varepsilon)$

---

2. 83

Solution {ε_{xx}=? ε_{yy}=? γ_{xy}=?} at x = 5mm and y = 7 mm

The strains at a point can be found as shown below.

$$\varepsilon_{xx} = \frac{\partial u}{\partial x} = [0.5(2x) + 0.5y](10^{-3}) = (x + 0.5y)(10^{-3}) \qquad \varepsilon_{yy} = \frac{\partial v}{\partial y} = [0.25(-2y) - x](10^{-3}) = (-0.5y - x)(10^{-3}) \quad (1)$$

$$\gamma_{xy} = \frac{\partial u}{\partial y} + \frac{\partial v}{\partial x} = [0.5(-2y) + 0.5x](10^{-3}) + [0.25(2x) - y](10^{-3}) = \gamma_{xy} = (x - 2y)(10^{-3}) \quad (2)$$

Substituting $x = 5$ and $y = 7$ we obtain:

$$\varepsilon_{xx} = (5 + 3.5)(10^{-3}) = 8.5(10^{-3}) \qquad \varepsilon_{yy} = (-3.5 - 5)(10^{-3}) = -8.5(10^{-3}) \qquad \gamma_{xy} = (5 - 14)(10^{-3}) = -9(10^{-3}) \quad (3)$$

ANS $\varepsilon_{xx} = 8500\mu$; $\varepsilon_{yy} = -8500\mu$; $\gamma_{xy} = -9000\mu$

---

2. 84

Solution $L_o = OA = 9$ ε=?

The first derivative of y can be written and length of the deformed rod calculated as shown below.

$$\frac{dy}{dx} = (0.04)\left(\frac{3}{2}\right)x^{1/2} = 0.06x^{1/2} \quad (1)$$

$$L_f = \int ds = \int_0^9 \sqrt{1 + \left(\frac{dy}{dx}\right)^2} dx = \int_0^9 (\sqrt{1 + 0.0036x})dx = \frac{2(1 + 0.0036x)^{3/2}}{3(0.0036)}\Bigg|_0^9 = 185.19[(1 + 0.0036(9))^{3/2} - 1] = 9.07275 \quad (2)$$

$$\varepsilon = (L_f - L_o)/L_o = 0.07275/9 = 8.083(10^{-3}) \quad (3)$$

ANS $\varepsilon = 8083\mu$in./in.

---

2. 85

Solution $L_o = OA = 200mm$ ε=?

The first derivative of y can be written and length of the deformed rod calculated as shown below.

$$\frac{dy}{dx} = \left(\frac{3}{2}\right)(625)x^{1/2}(10^{-6}) = 0.9375(10^{-3})x^{1/2} \quad (1)$$

$$L_f = \int ds = \int_0^{200} \sqrt{1 + \left(\frac{dy}{dx}\right)^2} dx = \int_0^{200} \sqrt{1 + (0.9375)^2(10^{-6})x}dx = \frac{2}{3(0.9375)^2(10^{-6})}(1 + (0.9375)^2(10^{-6})x)^{3/2}\Bigg|_0^{200} \text{ or}$$

$$L_f = 711.11[(1 + (0.9375)^2(10^{-6})200)^{3/2} - 1] = 200.00844 \ mm \qquad \varepsilon = (L_f - L_o)/L_o = 42.19(10^{-6}) \quad (2)$$

ANS $\varepsilon = 42.2\mu mm/mm$

2. 86

Solution $L_o = OA = 9$ $\varepsilon = ?$

The first derivative of y can be written as:

$$\frac{dy}{dx} = \frac{3}{2}(0.04)x^{1/2} - 0.005 = 0.06x^{1/2} - 0.005 \quad (1)$$

The length of the deformed rod can be written as:

$$L_f = \int ds = \int_0^9 \sqrt{1 + \left(\frac{dy}{dx}\right)^2} dx = \int_0^9 \sqrt{1 + (0.06x^{1/2} - 0.005)^2} dx \quad (2)$$

Let $f(x) = \sqrt{1 + (0.06x^{1/2} - 0.005)^2}$, thus $L_f = \int_0^9 f(x)dx$

Using numerical integration as described in Figure B2 with a Δx_i=0.25, we obtain the following the table below.

Numerical Integration Results from Spread Sheet.

x_i	$f(x_i)$	Integral value
0	1.0000125	0.250040619
0.25	1.000312451	0.500167191
0.5	1.000700123	0.750392467
0.75	1.001102085	1.000719147
1	1.001511358	1.251148721
1.25	1.001925237	1.501682168
1.5	1.002342333	1.752320187
1.75	1.002761823	2.003063311
2	1.00318317	2.253911957
2.25	1.003605998	2.504866462
2.5	1.004030038	2.755927102
2.75	1.004455082	3.007094109
3	1.004880973	3.258367678
3.25	1.005307582	3.509747977
3.5	1.005734807	3.761235148
3.75	1.006162564	4.012829316
4	1.006590781	4.264530589
4.25	1.007019398	4.516339059
4.5	1.007448365	4.768254809

x_i	$f(x_i)$	Integral value
4.75	1.007877637	5.020277911
5	1.008307175	5.272408426
5.25	1.008736947	5.524646409
5.5	1.009166921	5.776991909
5.75	1.009597073	6.029444965
6	1.010027379	6.282005615
6.25	1.010457817	6.534673888
6.5	1.010888369	6.787449811
6.75	1.011319017	7.040333407
7	1.011749746	7.293324693
7.25	1.012180542	7.546423685
7.5	1.012611393	7.799630394
7.75	1.013042285	8.052944831
8	1.013473208	8.306367001
8.25	1.013904153	8.559896909
8.5	1.01433511	8.813534556
8.75	1.01476607	9.067279943
9	1.015197025	

The deformed length from the table and average normal strain can be found as shown below.

$$L_f = 9.0672799 \text{ in} \qquad \varepsilon = (L_f - L_o)/L_o = 0.0672799/9 = 7.4755(10^{-3}) \quad (3)$$

ANS $\varepsilon = 7475.5\mu\text{in./in.}$

2. 87

Solution L_0=40 ε=?

The deformed length of each 2 inch segment can be found using.

$$L_i = \sqrt{(x_i - x_{i-1})^2 + (y_i - y_{i-1})^2} \qquad 2 \le i \le 10 \quad (1)$$

The final length of AC can be found as $AC = \sum_{i=2}^{10} L_i$. Using Spread sheet the following table can be constructed.

Results of Calculation of the total length using spread sheet.

x_i	y_i	L_i
0	17	
2	16.9375	2.000976324
4	16.90625	2.000244126
6	16.59375	2.024266843
8	16.09375	2.061552813
10	15.5	2.086273966
12	14.75	2.136000936
14	13.875	2.18303115
16	12	2.741464025
18	0	12.16552506
	Total Length	29.39933524

From the above table: $AC = 29.399335$. The average normal strain is: $\varepsilon = [2(AC) - 40]/40 = 0.46997$

ANS $\varepsilon = 47\%$

CHAPTER 3

Sections 3.1-3.3

3. 1

Solution σ_{ult}=? σ_{frac}=? E = ? σ_{prop}= ? σ_{yield}= ? @ε_{offset} = 0.2%
E_s = ? @σ = 420 MPa E_T = ? @σ = 420 MPa

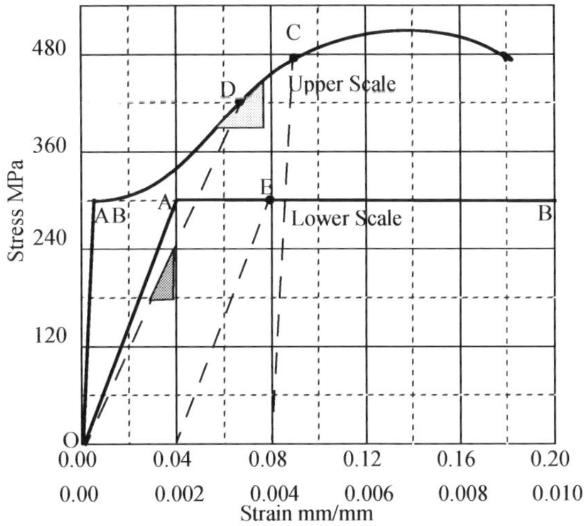

(a) By inspection σ_{ult}=480+30 = 510
(b) By inspection σ_{frac}=480

(c) From the lower plot $E = 300(10^6)/0.002 = 150(10^9)N/m^2$
(d) By inspection of the lower plot $\sigma_{prop} = 300MPa$

(e) The offset strain is: $\varepsilon_{offset} = 0.2/100 = 0.002$ Starting from a strain value of 0.002, we draw a line parallel to OA on the lower plot, intersecting the stress-strain curve at point E. The stress at point E is 300 MPa
(f) The stress value of 420 MPa corresponds to point D on the upper plot. We draw a tangent line to the stress strain curve at point D and calculate the slope as shown below:

$$E_t = [(450-400)(10^6)]/0.02 = 2.5(10^9)N/m^2$$

(g) To find the secant modulus, we draw a line from point D to point O and find the slope as shown below.

$$E_s = [(420-0)(10^6)]/[0.065-0] = 6.46(10^9)N/m^2$$

ANS $\sigma_{ult} = 510MPa$; $\sigma_{frac} = 480MPa$; $E = 150GPa$; $\sigma_{prop} = 300MPa$; $\sigma_{yield} = 300MPa$; $E_t = 2.5GPa$; $E_s = 6.5GPa$

---

3. 2

Solution d = 10 mm; L_o = 50 mm; (a) P=? @ δ = 0.2 mm (b) P=? @ δ = 4.0 mm

The area of cross-section is: $A = (\pi d^2)/4 = 78.5398mm^2$
(a) The strain in the specimen can be found as: $\varepsilon = \delta/L_0 = 0.2/50 = 0.004$
From the lower plot σ = 300 MPa

$$\sigma = P/A = (300)(10^6)(78.5398)(10^{-6}) = 23561.9\ N \qquad (1)$$

(b) The strain in the specimen can be found as: $\varepsilon = \delta/L_0 = 4.0/50 = 0.08$
From the upper plot: σ = 420 + 30 = 450MPa

$$\sigma = P/A = (450)(10^6)(78.5398)(10^{-6}) = 35342.9\ N \qquad (2)$$

ANS (a) $P = 23.56kN$ (b) $P = 35.34kN$

---

3. 3

Solution d = 10 mm L_0= 50 δ = ? @P = 33 kN

The area of cross-section from previous problem is A = 78.5398 mm^2.
The stress in the specimen is found and then from upper plot the strain is determined and then the deformation is calculated as

shown below.

$$\sigma = P/A = (33(10^3))/(78.5398(10^{-6})) = 420.1(10^6)N/m^2 = 420MPa \qquad \varepsilon = (0.064)0.005 = 0.065 \quad (1)$$

$$\delta = \varepsilon L_0 = (0.065)(50) = 3.25 \quad (2)$$

ANS $\delta = 3.25mm$

---

3. 4

Solution $\varepsilon_{total} = ?$ $\varepsilon_{elas} = ?$ $\varepsilon_{plas} = ?$ @ P = 33kN

The area of cross-section from previous problem is A = 78.5398 mm^2. The normal stress in the specimen can be found as:

$$\sigma = P/A = [33(10^6)]/[78.5398(10^{-6})] = 420.2(10^6)N/m^2 = 420.2MPa \quad (1)$$

From upper plot ε_{total}= 0.06 + 0.005 = 0.065

From previous problem we know E = 150 GPa,. We find the elastic strain and then the plastic strain as shown below.

$$\varepsilon_{elas} = \frac{\sigma}{E} = \frac{420.2(10^6)}{150(10^9)} = 2.801(10^{-3}) \qquad \varepsilon_{plas} = \varepsilon_{total} - \varepsilon_{elas} = 65(10^{-3}) - 2.801(10^{-3}) = 62.2(10^{-3}) \quad (2)$$

ANS $\varepsilon_{total} = 0.065$; $\varepsilon_{elas} = 0.0028$; $\varepsilon_{plas} = 0.062$

---

3. 5

Solution L_f = 54 mm P = ?

The plastic strain can be found as: $\varepsilon_{plas} = [L_f - L_0]/L_0 = 4/50 = 0.08$. Starting from a strain value of 0.08, we draw a line parallel to OA on the upper plot, intersecting the stress-strain curve at point C. The stress at point C is approximated as $\sigma_c = 470MPa$. The area of cross-section from previous problem is A = 78.5398 mm^2. The applied force is:

$$P = \sigma_c A = (470)(10^{-6})(78.5398)(10^{-6}) = 36913.7N \quad (1)$$

ANS $P = 36.9kN$

---

3. 6

Solution σ_{ult}=? σ_{frac}=? E = ? σ_{prop}= ? σ_{yield}= ? @ε_{offset} = 0.1%E_s = ? @σ = 72ksiE_T = ? @ σ = 72ksi

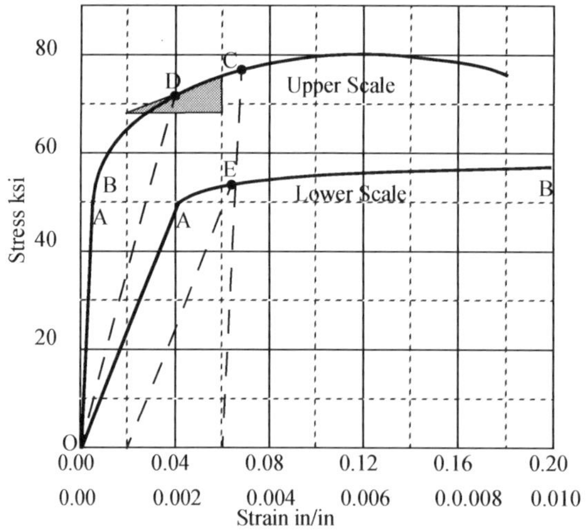

(a) By inspection we find: σ_{ult}=80 ksi

(b) By inspection we find: σ_{frac}=80 - 4=76 ksi

(c) From the lower plot $E = [50 - 0]/[0.002 - 0] = 25,000ksi$

(d) By inspection of the lower plot we obtain $\sigma_{prop} = 50ksi$

(e) The offset strain is: $\varepsilon_{offset} = 0.1/100 = 0.001$. Starting from a strain value of 0.001, we draw a line parallel to OA on the lower plot intersecting the stress-strain curve at point E. The stress at point E is 54ksi. The yield stress is: $\sigma_{yield} = 54ksi$.

(f, g) The stress value of 72 ksi corresponds to point D. On the upper plot, we draw a tangent to the stress strain curve at point D and calculate the slope for tangent modulus, while for the secant modulus, we draw a line from point D to point O, and find the slope as shown below

$E_t = [75-68]/[0.06-0.02] = 325ksi \qquad E_s = [72-0]/[0.04-0] = 1800ksi$ **(2)**

ANS $\sigma_{ult} = 80ksi$; $\sigma_{frac} = 76ksi$; $E = 25,000ksi$; $\sigma_{prop} = 50ksi$; $\sigma_{yield} = 54ksi$; $E_t = 325ksi$; $E_s = 1,800ksi$

---

3. 7

Solution $d = (5/8)inch$ $L_o = 2$ inch (a) P=? @ $\delta = 0.006$ inch (b) P =? @$\delta = 0.120$ inch

The area of cross-section can be found as: $A = (\pi d^2)/4 = 0.3068in^2$

(a) The normal strain in specimen can be found as: $\varepsilon = \delta/L_0 = 0.006/2 = 0.003$. From the lower plot, the approximate value of stress is $\sigma = 53$ ksi. The applied load can be found as: $P = \sigma A = (53)(0.3068) = 16.26kips$.

(b) The normal strain in specimen can be found as: $\varepsilon = \delta/L_0 = 0.120/2 = 0.060$. From the upper plot the approximate value of stress is $\sigma = 75$ ksi. The applied load can be found as: $P = \sigma A = (75)(0.3068) = 23.01kips$

ANS (a) $P = 16.3kips$ (b) $P = 23.0kips$

---

3. 8

Solution $d = (5/8)inch$ $L_0 = 2$ inch $\delta = ?$ @ P = 10 kips $\delta = ?$ @ P = 20 kips

The area of cross-section previous is: A=0.3068 in^2

(a) The normal stress in specimen can be found as: $\sigma = P/A = 10/0.3068 = 32.59ksi$. From the lower plot, the approximate value of strain is $\varepsilon = 0.0013$. The deformation can be found as: $\delta = \varepsilon L_0 = (0.0013)(2) = 0.0026in$.

(b) The normal stress in specimen can be found as: $\sigma = P/A = 20/0.3068 = 65.19ksi$. From the upper plot, the approximate value of strain is $\varepsilon = 0.02$. The deformation can be found as: $\delta = \varepsilon L_0 = (0.02)(2) = 0.04in$

ANS (a) $\delta = 0.0026in$ (b) $\delta = 0.04in$

---

3. 9

Solution $\varepsilon_{total} = ?$ $\varepsilon_{elas} = ?$ $\varepsilon_{plas} = ?$ @ P = 20 kips

The area of cross-section from previous is: A=0.3068 in2 . The normal stress in the specimen can be found as

$$\sigma = P/A = 20/0.3068 = 65.19ksi \qquad \textbf{(1)}$$

From the upper plot we have: $\varepsilon_{total} = 0.02$

From previous problem we know E = 25,000 ksi. We find the elastic strain and then the plastic strain as shown below.

$$\varepsilon_{elas} = \sigma/E = 65.19/25000 = 2.608(10^{-3}) \qquad \varepsilon_{plas} = \varepsilon_{total} - \varepsilon_{elas} = 20(10^{-3}) - 2.608(10^{-3}) = 17.392(10^{-3}) \qquad \textbf{(2)}$$

ANS $\varepsilon_{total} = 0.0200$; $\varepsilon_{elas} = 0.0026$; $\varepsilon_{plas} = 0.0174$

---

3. 10

Solution $L_f = 2.12$ inch P = ?

The plastic strain is: $\varepsilon_{plas} = (L_f - L_0)/L_0 = 0.12/2 = 0.06$. Starting from a strain value of 0.06, we draw a line parallel to OA on the upper plot, intersecting the stress-strain curve at point C. The stress at point C is approximated as $\sigma_c = 77ksi$. The area of cross-section from previous problem is: A=0.3068 in^2. The applied load can be found as shown below.

$$P = \sigma_c A = (77)(0.3068) = 23.62kips \qquad \textbf{(1)}$$

ANS $P = 23.6kips$

---

3. 11

Solution E = ? σ_{prop}= ? σ_{yield}= ? @ε_{offset} = 0.15% E_s = ? @σ = 72ksi E_T = ? @ σ = 72ks ε_{offset}= ?

G = 6.6 GPa ν = ? L_o = 200 mm A= 250 mm^2 P = 20 kN δ = ?

(a) Using the origin and point A for the calculation we obtain: $E = [(50-0)(10^6)]/(0.003-0) = 16.667(10^9)\ N/m^2$

(b) The end of linear region is: $\sigma_{prop} = 110$ MPa

(c) The 0.15% offset corresponds to point B. Drawing a line parallel to the linear region we intersect the curve at point C. The offset yield stress is: $\sigma_{offset} = 115$ MPa

(d) Point D corresponds to stress level of 130 MPa. Thus: $E_S = [(130-0)(10^6)]/[0.021-0] = 6.190(10^9)\ N/m^2$

(e) Using triangle DEF we obtain: $E_T = [(130-120)(10^6)]/[0.021-0.0105] = 0.952(10^9)\ N/m^2$

(f) The total strain at stress level of 130 MPa corresponds to point G. Thus. $\varepsilon_{plas} = 0.021 - [130(10^6)]/[16.667(10^9)] = 0.0132$

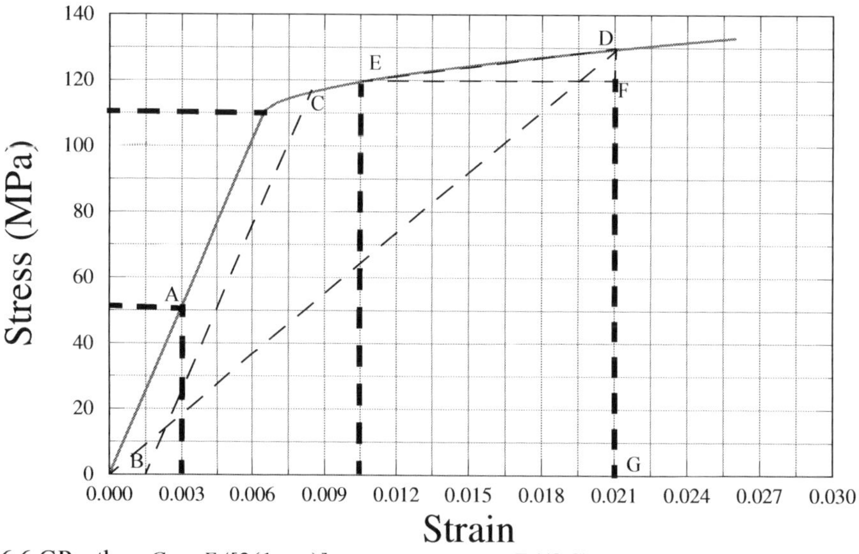

(g) We are given G = 6.6 GPa, thus $G = E/[2(1+\nu)]$ or $\nu = E/(2G) - 1 = 16.667/[2(6.6)] - 1 = 0.263$

(h) The axial stress is $\sigma = [20(10^3)]/[250(10^{-6})] = 80(10^6)\ N/m^2 = 80\ MPa$. The specimen is in linear elastic region. The strain is: $\varepsilon = \sigma/E = [80(10^6)]/[16.667(10^9)] = 0.0048$. Thus the elongation is: $\delta = \varepsilon L = (0.0048)(200) = 0.96\ mm$

ANS $E = 16.667$ GPa; $\sigma_{prop} = 110$ MPa; $\sigma_{offset} = 115$ MPa; $E_S = 6.190$ GPa; $E_T = 0.952$ GPa; $\varepsilon_{plas} = 0.0132$; $\nu = 0.263$; $\delta = 0.96$ mm

---

3. 12

Solution $L_o = 50$ mm cross-section is 12 mm x 12 mm E = ? σ_{prop} = ? σ_{yield} = ? @ε_{offset} = 0.2%
E_T = ? @ σ = 1400 MPa E_s = ? @ σ = 1400 MPa ε_{plas} = ? @ σ = 1400 MPa

The area of cross-section is A = (12)(12) = 144 mm^2=144(10^{-6}) m^2. Dividing the column of load by the area A and the column of change of length by the gage length Lo, we can obtain the columns of stress and strains on a spread sheet as shown in the table below. The stresses and strain can be plotted to obtain the stress-strain graph in figure (a). The data corresponding to first 5 points can be also plotted to get an enlarged view of region OAB as shown in figure (b).

Stress-Strain values in problem 3.17.

Strain	Stress (MPa)
0	0
0.0004	120.28
0.0014	420.7
0.0026	781.36
0.0034	1022.36
0.0106	1119.31
0.022	1168.54
0.0392	1222.43
0.0558	1269.44
0.08	1324.65

Strain	Stress (MPa)
0.0942	1342.29
0.116	1388.96
0.143	1421.18
0.1776	1458.26
0.1998	1472.64
0.2202	1473.40
0.2326	1448.89
0.2406	1423.54
0.2462	1384.31
0.2494	1334.38
0.2526	1287.92

(a)

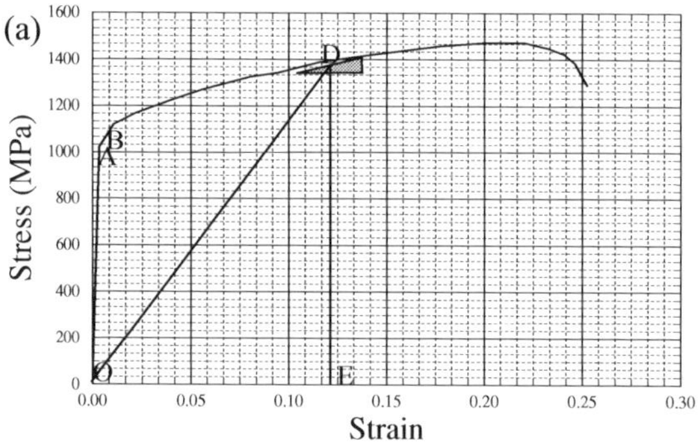

(b)

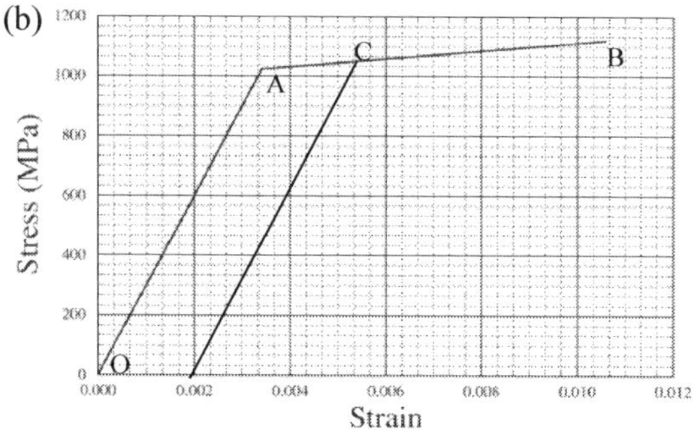

The modulus of elasticity is calculated as the average slope from the first three points as shown below:

$$E = \frac{1}{2}\left[\frac{120.28 - 0}{0.0004} + \frac{420.7 - 120.28}{0.0014 - 0.0004}\right] = 300560MPa \quad \textbf{(1)}$$

By inspection : $\sigma_{prop} = 1022 MPa$

The offset strain is: $\varepsilon_{offset} = 0.2/100 = 0.002$. On figure (b), starting at a strain value of 0.002, we draw a line parallel to OA to intersect the stress-strain curve at point C.The stress value at C is approximated as the yield stress $\sigma_{yield} = 1060 MPa$.

The stress value of 1400 MPa, corresponds to point D. in Figure (a). We draw tangent to the stress-strain curve at point D and calculate the slope as shown below:

$$E_t = [1433 - 1367]/[0.1467 - 0.1083] = 1719 MPa \tag{2}$$

To find the secant modulus we draw a line from point D to point O. and find the slope as

$$E_s = 1400/0.125 = 11200 MPa \tag{3}$$

The elastic strain at point D can be found as shown below.

$$\varepsilon_{plas} = \varepsilon_{total} - \varepsilon_{elas} = 0.125 - \sigma_D/E = [1400(10^6)]/[300(10^9)] = 0.125 - 4.667(10^{-3}) = 0.1203 \tag{4}$$

ANS \ $E = 300 Gpa$; $\sigma_{prop} = 1022 MPa$; $\sigma_{yield} = 1060 MPa$; $E_t = 1.72 GPa$; $E_s = 11.2 GPa$; $\varepsilon_{plas} = 0.1203$

---

3. 13

Solution $L_o = 2$ inch $d = 0.5$ inch $E = ?$ $\sigma_{prop} = ?$ $\sigma_{yield} = ?$ @$\varepsilon_{offset} = 0.05\%$

$E_T = ?$ @ $\sigma = 50$ ksi $E_s = ?$ @ $\sigma = 50$ ksi$\varepsilon_{plas} = ?$ @ $\sigma = 50$ ksi.

The area of cross-section is $A = \pi d^2/4 = 0.1963$ in^2.Dividing the column of load by the area A and the column of change of length by the gage length Lo, we can obtain the columns of stress and strains on a spread sheet as shown in the table below. The stress and strain can be plotted to obtain the stress-strain graph in figure (a). The data corresponding to first 5 points can be also plotted to get an enlarged view of region OAB as shown in figure (b).

The modulus of elasticity is calculated as the average slope from the first three points as shown below:

$$E = \frac{1}{2}\left[\frac{15.84 - 0}{0.00064} + \frac{36.87 - 15.84}{0.00148 - 0.00064}\right] = 24892.8 ksi \tag{1}$$

From the figure (b) and the table: $\sigma_{prop} = 36.9 ksi$

The offset strain is: $\varepsilon_{offset} = 0.05/100 = 0.0005$ on figure (b), starting at a strain value of 0.0005, we draw a line parallel to OA to intersect the stress-strain curve at point C. The stress value at C is approximated as the yield stress $\sigma_{yield} = 37.5 ksi$

Stress-Strain values in problem 3.18

Strain	Stress (ksi)
0.00000	0.00
0.00064	15.84
0.00148	36.87
0.00153	38.25
0.00438	39.22
0.00953	40.23
0.01435	41.56
0.01887	43.09
0.02359	44.92
0.02953	47.47
0.03543	50.22
0.04212	52.97
0.04893	55.11

Strain	Stress (ksi)
0.05605	56.94
0.07020	59.69
0.08061	61.06
0.09633	62.49
0.10711	63.20
0.12297	63.92
0.14174	64.68
0.15818	65.04
0.18155	65.39
0.19267	61.32
0.19802	58.26
0.20321	54.55
0.20736	50.73

(a)

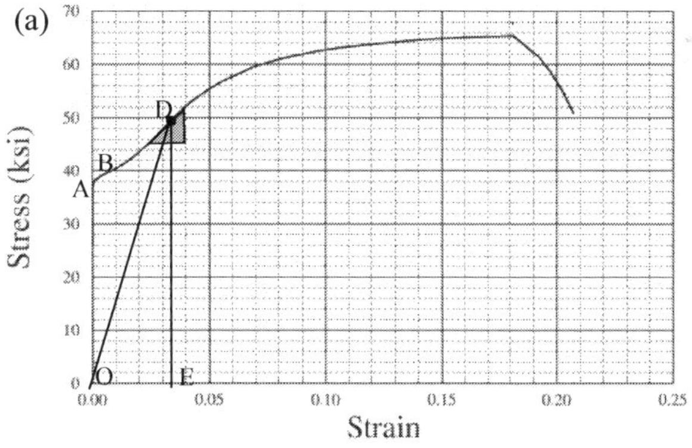

(b)

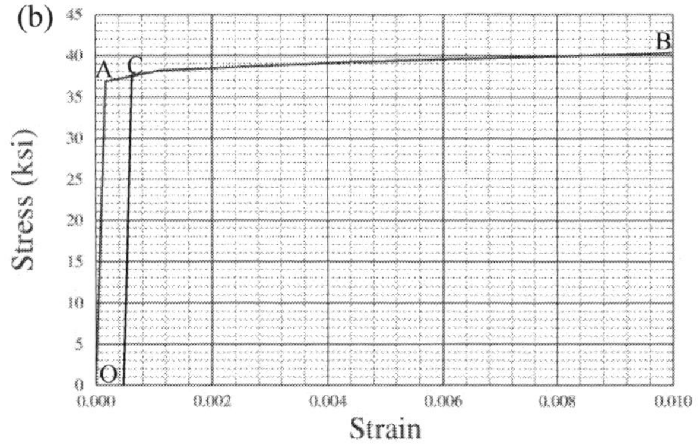

The stress value of 50ksi, corresponds to point D. in Figure (a). We draw tangent to the stress-strain curve at point D and calculate the slope as shown below:

$$E_t = [52 - 46]/[0.04 - 0.025] = 400 ksi \tag{2}$$

To find the secant modulus we draw a line from point D to point O. and find the slope as

$$E_s = 50/0.035 = 1428.57 \quad (3)$$

The plastic strain at point D can be found as shown below.

$$\varepsilon_{plas} = \varepsilon_{total} - \varepsilon_{elas} = 0.035 - \sigma_D/E = 0.035 - 0.002 = 0.033 \quad (4)$$

ANS $E = 24893ksi$; $\sigma_{prop} = 36.9ksi$; $\sigma_{prop} = 36.9ksi$; $\sigma_{yield} = 37.5ksi$; $E_t = 400ksi$; $E_s = 1428.6ksi$; $\varepsilon_{plas} = 0.033$

---

3. 14

Solution d = 1/4 in $\delta_{BC} = ?$ @P = 2 kips $(\delta_{BC})_{plas} = ?$ @ P = 0

From geometry and moment equilibrium about A we obtain the following.

$$h = 5/\tan 40 = 5.958 \text{ ft} \qquad N_{BC}h = 5P \qquad or \qquad N_{BC} = (5/5.958)P = 0.8392P \qquad A = (\pi/4)(1/4)^2 = 0.04909 \text{ in}^2 \quad (1)$$

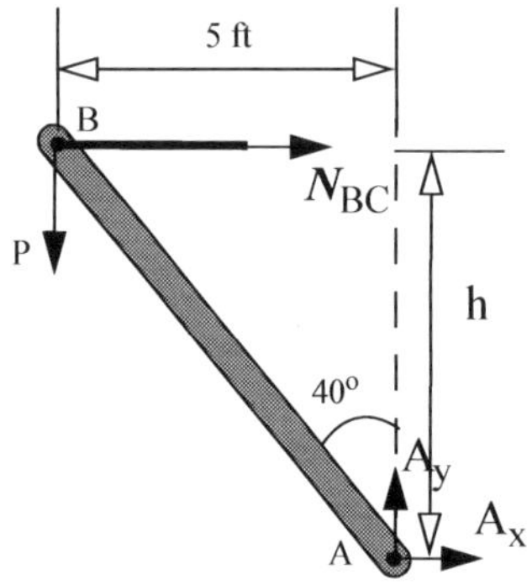

$$N_{BC} = (0.8392)(2) = 1.6782 \text{ kips} \qquad \sigma_{BC} = N_{BC}/A = 1.6782/0.04909 = 34.188 \text{ ksi} \quad (2)$$

The modulus of elasticity is $E = 50/0.002 = 25000$ ksi . The deformation can be calculated as shown below.

$$\varepsilon_{BC} = \sigma_{BC}/E = 34.188/25000 = 1.367(10^{-3}) \qquad \delta_{BC} = \varepsilon_{BC}L = (1.367)(10^{-3})(5)(12) = 0.082 \text{ in.} \quad (3)$$

The stress of 34.188 ksi is below the yield stress, hence no permanent deformation.

ANS $\delta_{BC} = 0.082$ in. ; $(\delta_{BC})_{plas} = 0$

---

3. 15

Solution d = 1/4 in $\delta_{BC} = ?$ @P = 4.25 kips $(\delta_{BC})_{plas} = ?$ @ P = 0

From previous problem we have the following.

$$h = 5.958 \text{ ft} \qquad N_{BC} = P = 0.8392P \qquad A = 0.04909 \text{ in.}^2 \quad (1)$$

$$N_{BC} = 0.8392(4.25) = 3.566 \text{ kips} \qquad \sigma_{BC} = N_{BC}/A = 3.566/0.04909 = 72.6 \text{ ksi} \quad (2)$$

From the graph the normal strain at 72.6 ksi is: $\varepsilon_{BC} = 0.04$

The deformation is: $\delta_{BC} = \varepsilon_{BC}L = (0.04)(5)(12) = 2.4$ in.

The modulus of elasticity is $E = 50/0.002 = 25000$ ksi . The plastic strain and deformation can be found as shown below

$$(\varepsilon_{BC})_{plas} = \varepsilon_{BC} - (\varepsilon_{BC})_{elas} = \varepsilon_{BC} - \sigma_{BC}/E = 0.04 - 72.6/25000 = 0.04 - 2.906(10^{-3}) = 37.094(10^{-3}) \quad (3)$$

$$(\delta_{BC})_{plas} = (\varepsilon_{BC})_{plas}L = 37.094(10^{-3})(5)(12) = 2.4 \text{ in.} = 2.22 \text{ in.} \quad (4)$$

ANS $\delta_{BC} = 2.4$ in. ; $(\delta_{BC})_{plastic} = 2.22$ in.

---

3. 16

Solution $A = 2 \text{ in}^2$ $L_0 = 5$ in $L_f = 5.005$ P = 50,000lbs E = ? ν = ?

The modulus of elasticity can be calculated as shown below.

$$\varepsilon_{long} = \frac{L_f - L_0}{L_0} = \frac{0.005}{5} = 0.001 \qquad \sigma = \frac{P}{A} = \frac{50000}{2} = 25,000psi \qquad E = \frac{\sigma}{\varepsilon_{long}} = \frac{25000}{0.001} = 25000(10^3)psi \quad (1)$$

$$\varepsilon_{trans} = -0.0004/2 = -0.0002 \qquad \nu = -\varepsilon_{trans}/\varepsilon_{long} = -(-0.0002)/0.001 = 0.2 \quad (2)$$

ANS $E = 25,000ksi$; $\nu = 0.2$

---

3. 17

Solution P = 20 kips δ=0.005 E = ?

The modulus of elasticity can be calculated as shown below.

$$\sigma = P/A = 20/4 = 5ksi \qquad \varepsilon = 0.005/10 = 0.0005 \qquad E = \sigma/\varepsilon = 5/0.0005 = 10,000ksi \qquad (1)$$

ANS $E = 10,000ksi$

---

3. 18

Solution P = 20 ksi $\delta = 0.0125$ A= 4 G = ?

The shear stress and shear strain can be found as shown below.

$$V = P \qquad \tau = V/A = 20/4 = 5ksi \qquad tan\gamma \approx \gamma = \delta/10 = 0.00125 \qquad (1)$$

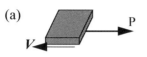

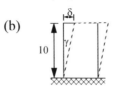

$$G = \tau/\gamma = 5/0.00125 = 4000ksi \qquad (2)$$

ANS $G = 4000ksi$

---

3. 19

Solution $\delta = 0.02$ in $W = 900$ lb $L = 12$ in, a = 3 in, and b = 2 in G = ?

The shear strain and shear stress on the rubber block can be found as shown below.

$$tan\gamma \approx \gamma = \delta/a \qquad V = W/2 \qquad (1)$$

$$\gamma = 0.02/3 = 6.667(10^{-3}) \qquad V = 900/2 = 450 \text{ lb} \qquad \tau = V/A = 450/[(12)(2)] = 18.75 \text{ psi} \qquad (2)$$

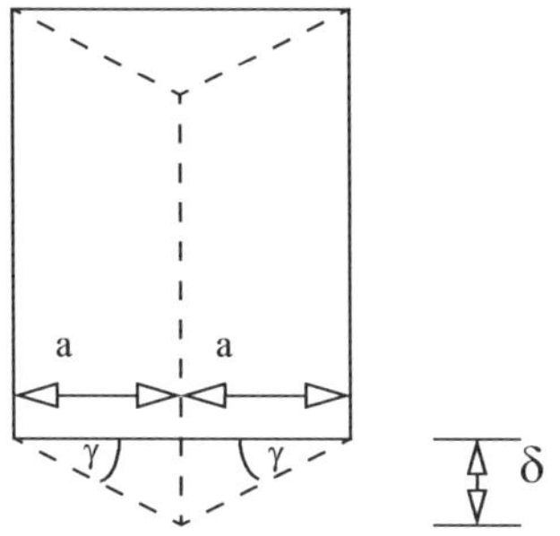

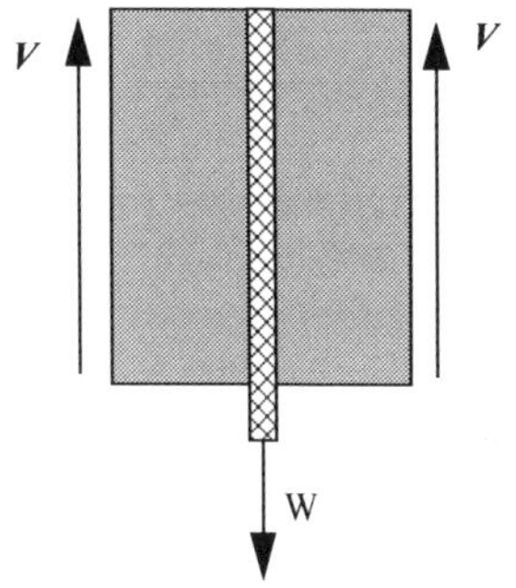

$$G = \tau/\gamma = 18.75/(6.667(10^{-3})) = 2812.5 \qquad (3)$$

ANS $G = 2812.5$ psi

---

3. 20

Solution G = 1.0 MPa W= 500 N $L = 200$ mm, $a = 45$ mm, and $b = 60$ mm $\delta = ?$

From previous problem we have the following.

$$\gamma = \delta/a \qquad V = W/2 \qquad (1)$$

$$\gamma = \delta/0.045 \qquad V = 500/2 = 250 \text{ N} \qquad \tau = V/A = 250/[(0.2)(0.06)] = 20833 \text{ N}/m^2 \qquad (2)$$

$$\gamma = \tau/G = 20833/[1(10^6)] = 20833(10^{-6}) = \delta/0.045 \quad or \quad \delta = 937.5(10^{-6}) \text{ m} \qquad (3)$$

ANS $\delta = 0.9375$ mm

---

3. 21

Solution G = 750 psi $\tau \le 15$ *psi* $\delta \le 0.03$ *in* $L = 12$ in, $a = 2$ in, and $b = 3$ in W = ?

From previous problem we have the following.

$$\gamma = \delta/a \qquad V = W/2 \qquad (1)$$

$$\gamma = \delta/2 \qquad V = W/2 \qquad \tau = V/A = W/[2(12)(3)] = W/72 \le 15 \text{ psi} \quad or \quad W \le 1080 \text{ lb} \qquad (2)$$

$$\gamma = \tau/G = (W/72)/750 = \delta/2 \quad or \quad \delta = W/27000 \le 0.03 \quad or \quad W \le 810 \text{ lb} \qquad (3)$$

ANS $W_{max} = 810$ lb

---

3. 22

Solution $\delta = f(G, L, W, a, b) = ?$

From previous problem we have the following.

$$\gamma = \delta/a \qquad V = W/2 \qquad (1)$$

$$\tau = V/A = W/[2(L)(b)] = W/(2Lb) \qquad \gamma = \tau/G = W/(2GLb) = \delta/a \quad or \quad \delta = Wa/[2GLb] \qquad (2)$$

ANS $\delta = Wa/[2GLb]$

---

3. 23

Solution L_0 = 200 mm d = 20 mm P = 77 kN δ = 4-5mm Δd = -0.162 mm
E = ? G = ?

The axial stress, axial strain and the transverse strain can be found as shown below.

$$A = (\pi d^2)/4 = \pi(0.02^2)/4 = 314.16(10^{-6})m^2 \qquad \sigma = P/A = 77(10^3)/[314.16(10^{-6})] = 245.1(10^6)N/m^2 \qquad (1)$$

$$\varepsilon_{long} = \delta/L_0 = 4.5/200 = 0.0225 \qquad \varepsilon_{tran} = \Delta d/20 = -0.162/20 = -0.0081 \qquad (2)$$

$$E = \frac{\sigma}{\varepsilon_{long}} = \frac{245.1(10^6)}{0.0225} = 10.893(10^9)\frac{N}{m^2} \qquad \nu = \frac{-\varepsilon_{tran}}{\varepsilon_{long}} = -\frac{(-0.0081)}{(0.0225)} = 0.36 \qquad G = \frac{E}{2(1+\nu)} = \frac{10.893}{2(1+0.36)} = 4.005GPa$$

ANS $E = 10.9GPa$; $G = 4.0GPa$

---

3. 24

Solution L_0 = 6 d = 1 E= 30,000ksi ν=1/3 P = 20kips Δd = ? ΔL = ?

The axial stress, axial strain and the transverse strain can be found as shown below.

$$A = \pi d^2/4 = \pi(1^2)/4 = 0.7854in^2 \qquad \sigma = P/A = 20/0.7854 = 25.465ksi \qquad (1)$$

$$\varepsilon_{long} = \sigma/E = 25.465/(30,000) = 0.8488(10^{-3}) \qquad \varepsilon_{tran} = -\nu\varepsilon_{long} = -[0.8488(10^{-3})]/3 = -0.2829(10^{-3}) \qquad (2)$$

$$\Delta L = L\varepsilon_{long} = (6)(0.8488)(10^{-3}) = 5.093(10^{-3})in \qquad \Delta d = \varepsilon_{tran}d = (1)(-0.2829)(10^{-3}) = -0.2829(10^{-3}) \qquad (3)$$

ANS $\Delta L = 0.0051in$; $\Delta d = -0.00029$

---

3. 25

Solution L_0 = 400 mm d = 20 mm E= 180 GPa ν=0.32 δ = 0.5 mm Δd = ? P = ?

The longitudinal strain, transverse strain, axial stress, applied force, and change in diameter can be found as shown below.

$$\varepsilon_{long} = \delta/L_0 = 0.5/400 = 1.25(10^{-3}) \qquad \varepsilon_{tran} = -\nu\varepsilon_{long} = -(0.32)(1.25)(10^{-3}) = -0.4(10^{-3}) \qquad (1)$$

$$\sigma = E\varepsilon_{long} = (180)(10^9)(1.25)(10^{-3}) = 225(10^6)Pa \qquad P = \sigma A = 225(10^6)\frac{\pi}{4}(0.02^2) = 70.686(10^3)N \qquad (2)$$

$$\Delta d = (d)\varepsilon_{tran} = (20)(-0.4)(10^{-3}) = -8(10^{-3}) \qquad (3)$$

ANS $P = 70.7kN$; $\Delta d = -0.008mm$

---

3. 26

Solution A = 625 mm^2 L_0 = 500 mm ν = 1/3 δ = 0.75 mm % change in V= ?

The original volume is: $V_0 = AL_0 = 312.5(10^3)$ mm^3. Let Δa represent the change in width and thickness. The volume of the deformed bar and percentage change in volume can be found as shown below.

$$\varepsilon_{long} = \delta/L_0 = 0.75/500 = 0.0015 \qquad \varepsilon_{tran} = -\nu\varepsilon_{long} = -(1/3)(0.0015) = -0.0005 \qquad \Delta a = (25)\varepsilon_{tran} = -0.0125mm \qquad (1)$$

$$V_f = FinalVolume = (25+\Delta a)(25+\Delta a)(L_0+\delta) = (25-0.0125)(25-0.0125)(500+0.75) = 312.65586\ mm^3 \qquad (2)$$

$$\frac{\Delta V}{V} \times 100 = \frac{V_f - V_0}{V_0} \times 100 = 0.04988 \qquad (3)$$

ANS 0.05%

---

3. 27

Solution d = 1 in L_0 = 15in ν = 0.32 E = 28,000ksi % change in V @P = 20 kips

The axial stress, the longitudinal strain, and the transverse strain can be found as shown below.

$$A = \pi(1^2)/4 = 0.7854\ in^2 \qquad \sigma = P/A = 20/0.7854 = 25.465ksi \qquad \varepsilon_{long} = \sigma/E = 25.465/(28,000) = 0.9095(10^{-3}) \qquad (1)$$

$$\varepsilon_{tran} = -\nu\varepsilon_{long} = -(0.32)(0.9095)(10^{-3}) = -0.2910(10^{-3}) \qquad (2)$$

$$\delta = L_0\varepsilon_{long} = (50)(0.9095)(10^{-3}) = 0.04547in \qquad \Delta d = (d)\varepsilon_{tran} = -(0.2910)(10^{-3})(1) = -0.2910(10^{-3})in \quad (3)$$

$$V_0 = AL_0 = (0.7854)(50) = 39.2699in^3 \qquad V_f = \frac{\pi}{4}(1-\Delta d)^2(L_0+\delta) = \frac{\pi}{4}(1-0.291(10^{-3}))^2(50+0.04547) = 39.2827in^3 \quad (4)$$

$$\frac{\Delta V}{V_0}\times 100 = \frac{V_f - V_0}{V_0}\times 100 = 0.0327 \quad (5)$$

ANS 0.0327%

---

3. 28

Solution E = 70GPa L_0 = 500mm ν = 0.25 % change in V @ P = 300 kN

The axial stress, the longitudinal strain , and the transverse strain can be found as shown below.

$$\sigma = P/A = 300(10^3)/[(0.05)(0.025)] = 240(10^6)N/m^2 \qquad \varepsilon_{long} = \sigma/E = 240(10^6)/[70(10^9)] = 3.4286(10^{-3}) \quad (1)$$

$$\varepsilon_{tran} = -\nu\varepsilon_{long} = -(0.25)(3.4286)(10^{-3}) = -0.85714(10^{-3}) \qquad \delta = L_0\varepsilon_{long} = (500)(3.4286)(10^{-3}) = 1.7143mm \quad (2)$$

$$\Delta a = a\varepsilon_{tran} = (25)(-0.85714)(10^{-3}) = -21.429(10^{-3})mm \qquad \Delta b = b\varepsilon_{tran} = (50)(-0.85714)(10^{-3}) = -42.857(10^{-3})mm \quad (3)$$

$$V_0 = AL_0 = (1250)(500) = 625(10^3)mm^3 \qquad V_f = (a+\Delta a)(b+\Delta b)(L_0+\delta) = 626.068(10^3)mm^3 \quad (4)$$

$$\frac{\Delta V}{V_0}\times 100 = \frac{V_f - V_0}{V_0}\times 100 = 0.1709 \quad (5)$$

ANS 0.1709%

---

3. 29

Solution %change in V = f(L,d,E,ν,P)

The axial stress, the longitudinal strain , and the transverse strain can be found as shown below.

$$A = \pi d^2/4 \qquad \sigma = P/A = 4P/(\pi d^2) \qquad \varepsilon_{long} = \sigma/E = 4P/(E\pi d^2) \qquad \varepsilon_{tran} = -\nu\varepsilon_{long} = -4\nu P/(E\pi d^2) \quad (1)$$

$$\delta = L\varepsilon_{long} = 4LP/(E\pi d^2) \qquad \Delta d = d\varepsilon_{tran} = -4\nu P/(E\pi d) \quad (2)$$

$$V_0 = \frac{\pi d^2}{4}L \qquad V_f = \frac{\pi}{4}(d+\Delta d)^2(L+\delta) = \frac{\pi}{4}d^2L\left(1+\frac{\Delta d}{d}\right)^2\left(1+\frac{\delta}{L}\right) \approx \frac{\pi}{4}d^2L\left(1+\frac{2\Delta d}{d}\right)\left(1+\frac{\delta}{L}\right) \approx \frac{\pi}{4}d^2L\left(1+\frac{2\Delta d}{d}+\frac{\delta}{L}\right) \quad (3)$$

$$\left(\frac{\Delta V}{V}\right)(100) = \left(\frac{V_f - V_0}{V_0}\right)(100) = \left(2\frac{\Delta d}{d}+\frac{\delta}{L}\right)(100) = \left(-\frac{8\nu P}{E\pi d^2}+\frac{4P}{E\pi d^2}\right)(100) \quad (4)$$

ANS $\frac{400P}{E\pi d^2}(1-2\nu)$

---

3. 30

Solution %change in V = f(L,d,E,ν,P)

The axial stress, the longitudinal strain, and the transverse strain can be found as shown below.

$$\sigma = P/A = P/(ab) \qquad \varepsilon_{long} = \sigma/E = P/(Eab) \qquad \varepsilon_{tran} = -\nu\varepsilon_{long} = -\nu P/(Eab) \quad (1)$$

$$\delta = L\varepsilon_{long} = \nu LP/(Eab) \qquad \Delta a = a\varepsilon_{tran} = -Pa\nu/(Eab) \qquad \Delta b = b\varepsilon_{tran} = -Pb\nu/(Eab)$$

The original volume is: $V_0 = abL$. The volume of deformed bar can be found as shown below.

$$V_f = (a+\Delta a)(b+\Delta b)(L+\delta) = abL\left(1+\frac{\Delta a}{a}\right)\left(1+\frac{\Delta b}{b}\right)\left(1+\frac{\delta}{L}\right) \approx (abL)\left(1+\frac{\Delta a}{a}+\frac{\Delta b}{b}\right)+\left(1+\frac{\delta}{L}\right) \approx (abL)\left(1+\frac{\Delta a}{a}+\frac{\Delta b}{b}+\frac{\delta}{L}\right) \quad (2)$$

$$\left(\frac{\Delta V}{V}\right)(100) = \left(\frac{V_f - V_0}{V_0}\right)(100) = \left(\frac{\Delta a}{a}+\frac{\Delta b}{b}+\frac{\delta}{L}\right)(100) = \left(-\frac{P\nu}{Eab}-\frac{P\nu}{Eab}+\frac{P}{Eab}\right)(100) = (100)(1-2\nu)\frac{P}{Eab}\% \quad (3)$$

ANS $(100)(1-2\nu)\frac{P}{Eab}\%$

---

3. 31

Solution P=50,000lbs A=2 in^2 L=5 in δ=0.005in U=?

The strain energy density for the bar can be found and integrated to obtain the total strain energy as shown below.

$$U_0 = \frac{1}{2}\sigma\varepsilon = \frac{1}{2}\left(\frac{P}{A}\right)\frac{\delta}{L} = \frac{1}{2}\left(\frac{50,000}{2}\right)\left(\frac{0.005}{5}\right) = 12.5in-lbs/in^3 \quad (1)$$

$$U = \int U_0 dV = 12.5\int dV = U_0 AL = (12.5)(2)(5) = 125 \quad (2)$$

ANS $U = 125in-lbs$

---

3. 32

Solution P=20 kips A=4 in^2 L=10 in δ=0.005in U=?

The strain energy density for the bar can be found and integrated to obtain the total strain energy as shown below.

$$U_0 = \frac{1}{2}\sigma\varepsilon = \frac{1}{2}\left(\frac{P}{A}\right)\frac{\delta}{L} = \frac{1}{2}\left(\frac{20(10^3)}{4}\right)\left(\frac{0.005}{10}\right) = 1.25in-lbs/in^3 \quad (1)$$

$$U = \int U_0 dV = 1.25\int dV = 1.25V = 1.25AL = (1.25)(4)(10) = 50 \quad (2)$$

ANS $U = 50in-lbs$

---

3. 33

Solution P=20 kips A=4 in^2 L=10 in δ=0.0125in U=?

The strain energy density for the bar can be found and integrated to obtain the total strain energy as shown below.

$$U_0 = \frac{1}{2}\sigma\varepsilon = \frac{1}{2}\left(\frac{P}{A}\right)\frac{\delta}{L} = \frac{1}{2}\left(\frac{20(10^3)}{4}\right)\left(\frac{0.0125}{10}\right) = 3.125in-lbs/in^3 \quad (1)$$

$$U = \int U_0 dV = 3.125\int dV = 3.125V = 3.125AL = (3.125)(4)(10) = 125 \quad (2)$$

ANS $U = 125$ in.-lbs

---

3. 34

Solution U=f(L,d,E,ν,p)=?

The strain energy density for the bar can be found and integrated to obtain the total strain energy as shown below.

$$U_0 = \frac{1}{2}\sigma\varepsilon = \frac{1}{2}\left(\frac{\sigma^2}{E}\right) = \frac{1}{2}\left(\frac{P^2}{EA^2}\right) = \frac{1}{2}\frac{p^2}{E(\pi d^2/4)^2} = \frac{8}{\pi^2 E} \quad (1)$$

$$U = \int U_0 dV = \frac{8}{\pi^2 E}\int dV = \frac{8}{\pi^2 E}V = \frac{8}{\pi^2 E}AL = \left(\frac{8P^2}{\pi^2 Ed^4}\right)\left(\frac{\pi d^2}{4}\right)L \quad (2)$$

ANS $U = (2P^2L)/(\pi Ed^2)$

---

3. 35

Solution U=f(L,a,b,E,ν,p)=?

The strain energy density for the bar can be found and integrated to obtain the total strain energy as shown below.

$$U_0 = \frac{1}{2}\sigma\varepsilon = \frac{1}{2}\left(\frac{\sigma^2}{E}\right) = \frac{1}{2}\left(\frac{P^2}{EA^2}\right) = \frac{P^2}{2Ea^2b^2} \quad (1)$$

$$U = \int U_0 dV = \frac{P^2}{2Ea^2b^2}\int dV = \frac{P^2}{2Ea^2b^2}V = \frac{P^2}{2Ea^2b^2}AL = \left(\frac{P^2}{2Ea^2b^2}\right)(ab)(L) \quad (2)$$

ANS $U = (P^2L)/(2Eab)$

---

3. 36

Solution Modulus of Resilience=? U_0= ? and $\overline{U}_0$= ? @ σ = 420 MPa Modulus of Toughness=?

From the graph we obtain the following.

$$\sigma_{prop} = 300MPa \qquad E = 300(10^6)/0.002 = 150(10^9)N/m^2 = 150GPa \quad (1)$$

The stress strain curve is represented by a series of straight line and areas calculated as shown on the right of the figure.

The Modulus of Resilience = Area of the triangle OAA_1 = 300 kN-m/m^3

Adding the area AOA_1, AA_1BB_1 and BB_1CC_1 we obtain the strain energy density at C and subtracting it from rectangle

OC_2CC_1 we obtain the complementary strain energy density at C as shown below.

$$U_O = (300 + 12160 + 9500)(10^3)\ N-m/m^3 = 21960\ (10^3)N\text{–}m/m^3 \quad (2)$$

$$\overline{U}_O = (420)(10^6)(0.065) - 21,960\ (10^3) = 5340(10^3)\ N-m/m^3 \quad (3)$$

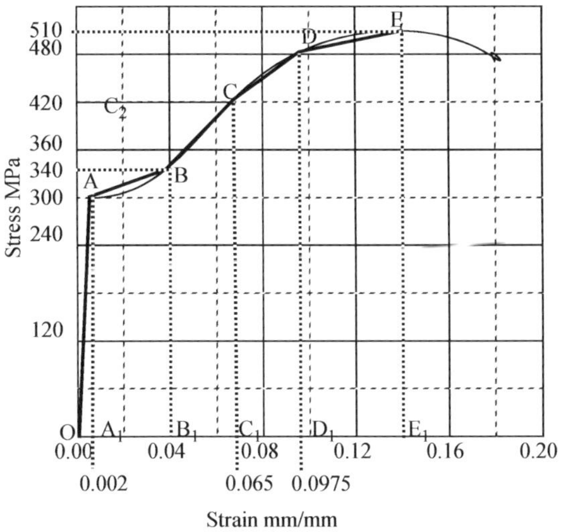

$$AOA_1 = \frac{(300)(10^6)(0.002)}{2} = 300(10^3)$$

$$AA_1BB_1 = \frac{(300+340)(10^6)(0.038)}{2} = 12160(10^3)$$

$$BB_1CC_1 = \frac{(340+420)(10^6)(0.025)}{2} = 9500(10^3)$$

$$CC_1DD_1 = \frac{(420+480)(10^6)(0.0325)}{2} = 14625(10^3)$$

$$DD_1EE_1 = \frac{(480+510)(10^6)(0.0425)}{2} = 21038(10^3)$$

The Modulus of Toughness = The sum of the areas shown on the right of the graph = $57,623(10^3)$ N-m/m^3.

ANS Modulus of Resilience = 300 kN-m/m^3 ; $U_O = 21,960\ \text{kN}-\text{m/m}^3$; $\overline{U}_O = 5,340\ \text{kN}-\text{m/m}^3$;

Modulus of Toughness = 57,623 kN-m/m^3

---

3. 37

Solution Modulus of Resilience=?U_0= ? $\overline{U}_0$= ? at 72 ksi Modulus of Toughness=?

From graph we have the following.

$$\sigma_{prop} = 50\ ksi \qquad E = 25000\ ksi \qquad \varepsilon_{prop} = \sigma_{prop}/E = 50/25000 = 0.002 \quad (1)$$

The stress strain curve is represented by a series of straight line and areas calculated as shown on the right of the figure below.

The Modulus of Resilience = Area of the triangle OAA_1 = 50 in-lbs/in^3

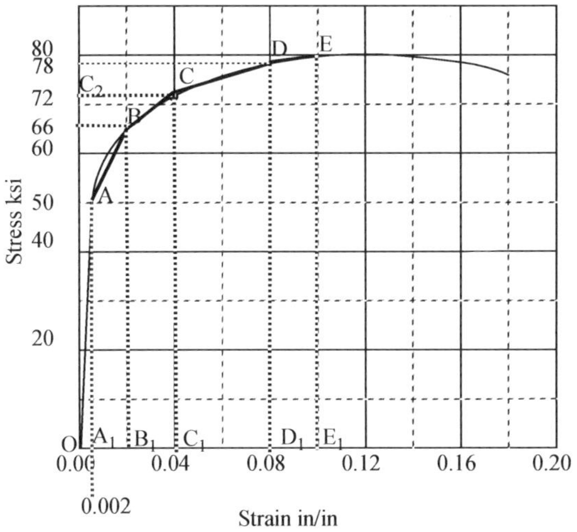

$$AOA_1 = \frac{(50)(10^3)(0.002)}{2} = 50$$

$$AA_1BB_1 = \frac{(50+66)(10^3)(0.018)}{2} = 1044$$

$$BB_1CC_1 = \frac{(66+72)(10^3)(0.02)}{2} = 1380$$

$$CC_1DD_1 = \frac{(72+78)(10^3)(0.04)}{2} = 3000$$

$$DD_1EE_1 = \frac{(78+80)(10^3)(0.02)}{2} = 1580$$

Adding the area AOA_1, AA_1BB_1 and BB_1CC_1 we obtain the strain energy density at C and subtracting it from rectangle OC_2CC_1 we obtain the complementary strain energy density at C as shown below.

$$U_O = (50 + 1044 + 1380)\ \ in-lbs/in^3 = 2474\ \text{in.-lbs/in.}^3 \qquad \overline{U}_O = (72)(10^3)(0.04) - 2474 = 406\ \ in-lbs/in^3 \quad (2)$$

The Modulus of Toughness = The sum of the areas shown on the right of the graph = 7054 in-lbs/in^3.

ANS Modulus of Resilience = 50 in-lbs/in^3 ; U_O = 2474 in.-lbs/in.^{3} ; $\overline{U}_O$ = 406 in.-lbs/in.^{3} ;

Modulus of Toughness = 7054 in-lbs/in^3

---

3. 38

Solution Modulus of Resilience=? U_o= ? and $\overline{U}_o$= ? @σ = 1400 MPa Modulus of Toughness=?

From the graph we have the following.

$$\sigma_{prop} = 1022\ MPa \qquad E = 300\ GPa \qquad \varepsilon_{prop} = \sigma_{prop}/E = 1022(10^2)/[300(10^9)] = 0.0034. \quad \textbf{(1)}$$

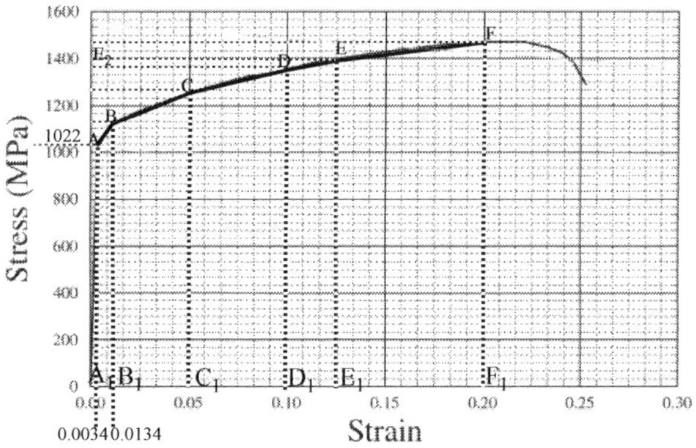

$$AOA_1 = \frac{(1022)(10^6)(0.0034)}{2} = 1734(10^3)$$

$$AA_1BB_1 = \frac{(1022+1133)(10^6)(0.01)}{2} = 10775(10^3)$$

$$BB_1CC_1 = \frac{(1133+1267)(10^6)(0.0366)}{2} = 43920(10^3)$$

$$CC_1DD_1 = \frac{(1267+1367)(10^6)(0.05)}{2} = 65850(10^3)$$

$$DD_1EE_1 = \frac{(1367+1400)(10^6)(0.025)}{2} = 34588(10^3)$$

$$EE_1FF_1 = \frac{(1400+1467)(10^6)(0.075)}{2} = 107512(10^3)$$

The stress strain curve is represented by a series of straight line and areas calculated as shown on the right of the figure below.

The Modulus of Resilience = Area of the triangle OAA_1 = 1734 kN-m/m^3

Adding the areas AOA_1 through DD_1EE_1 we obtain the strain energy density at E and subtracting it from rectangle OE_2EE_1 we obtain the complementary strain energy density at E as shown below.

$$U_O = (1734+10775+43920+65850+34588)(10^3)\ N-m/m^3 = 157\ MN-m/m^3 \quad \textbf{(2)}$$

$$\overline{U}_O = (1400)(10^6)(0.125) - 157\ (10^6) = 18(10^6)\ N-m/m^3 \quad \textbf{(3)}$$

The Modulus of Toughness = The sum of the areas shown on the right of the graph = 264(10^6) N-m/m^3.

ANS Modulus of Resilience = 1734 kN-m/m^3 ; $U_O = 157\ MN-m/m^3$; $\overline{U}_O = 18\ MN-m/m^3$;

Modulus of Toughness = 264 MN-m/m^3

---

3. 39

Solution Modulus of Resilience=? U_o= ? and $\overline{U}_o$= ? @ σ = 50 ksi Modulus of Toughness=?

From the graph we have the following.

$$\sigma_{prop} = 36.9\ ksi \qquad E = 24893\ ksi \qquad \varepsilon_{prop} = \sigma_{prop}/E = 36.9/24893 = 0.00148 \quad \textbf{(1)}$$

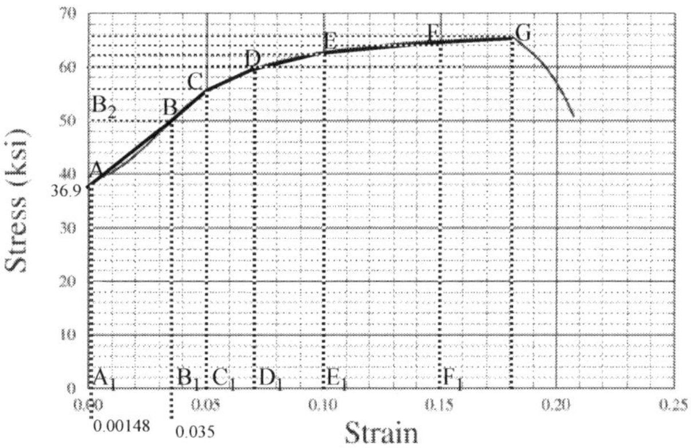

$$AOA_1 = \frac{(36.9)(10^3)(0.00148)}{2} = 27.31$$

$$AA_1BB_1 = \frac{(36.9+50)(10^3)(0.035)}{2} = 1521$$

$$BB_1CC_1 = \frac{(50+56)(10^3)(0.05)}{2} = 2650$$

$$CC_1DD_1 = \frac{(56+60)(10^3)(0.02)}{2} = 1160$$

$$DD_1EE_1 = \frac{(60+64)(10^3)(0.05)}{2} = 3100$$

$$DD_1EE_1 = \frac{(64+66)(10^3)(0.05)}{2} = 3250$$

The stress strain curve is represented by a series of straight line and areas calculated as shown on the right of the figure below.

The Modulus of Resilience = Area of the triangle OAA_1 = 727.3 in-lbs/in^3

Adding the areas AOA_1 plus area AA_1BB_1 we obtain the strain energy density at B and subtracting it from rectangle OB_2BB_1

we obtain the complementary strain energy density at B as shown below.

$$U_O = (27.31 + 1521) \quad in-lbs/in^3 = 1548.3 \text{ in.-lbs/in.}^3 \qquad \bar{U}_O = (50)(10^3)(0.035) - 1548.3 = 201.7 \quad in-lbs/in^3 \quad \textbf{(2)}$$

The Modulus of Toughness = The sum of the areas shown on the right of the graph =11708.3 in-lbs/in^3.

ANS Modulus of Resilience = 727.3 in-lbs/in^3 ; $U_O = 1548.3$ in.-lbs/in.^{3} ; $\bar{U}_O = 201.7$ in.-lbs/in.^{3} ;

Modulus of Toughness = 11,708 in-lbs/in^3

Section 3.2

3. 40

Solution $\delta_P = 0.25$ mm E = 200 GPa A = 100 mm^2 F=?

We follow the logic starting with the average strain value calculated in an earlier problem.

$$\varepsilon_{AP} = 803.5\mu \quad mm/mm \qquad \sigma_{AP} = E\varepsilon_{AP} = (200)(10^9)(803.5)(10^{-6}) = 160.7(10^6)N/m^2 \quad (T) \quad \textbf{(1)}$$

$$\boldsymbol{N}_{AP} = A\sigma_{AP} = (100)(10^{-6})(160.7)(10^6) = 16.07(10^3)N = 16.07kN \quad (T) \quad \textbf{(2)}$$

By force equilibrium in the x-direction on the free body diagram of the rigid bar we obtain the following.

$$F = \boldsymbol{N}_{AP} cos 50 = (16.07) cos 50 = 10.33 \quad \textbf{(3)}$$

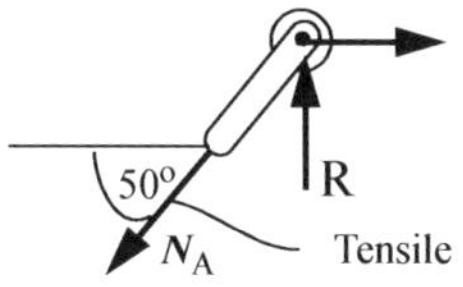

ANS $F = 10.3$ kN

---

3. 41

Solution $\delta_P = 0.25$ mm E = 200 GPa A = 100 mm^2 F=?

We follow the logic starting with the average strain value calculated in an earlier problem.

$$\varepsilon_{AP} = 1174.6\mu \quad mm/mm \qquad \sigma_{AP} = E\varepsilon_{AP} = (200)(10^9)(1174.6)(10^{-6}) = 234.9(10^6)N/m^2 \quad (T) \quad \textbf{(1)}$$

$$\boldsymbol{N}_{AP} = A\sigma_{AP} = (100)(10^{-6})(234.9)(10^6) = 23.49(10^3)N = 23.49kN \quad \textbf{(2)}$$

By force equilibrium in the x-direction on the free body diagram of the rigid bar we obtain the following.

$$F = \boldsymbol{N}_{AP} cos 20 = (23.49) cos 20 = 22.073 \quad \textbf{(3)}$$

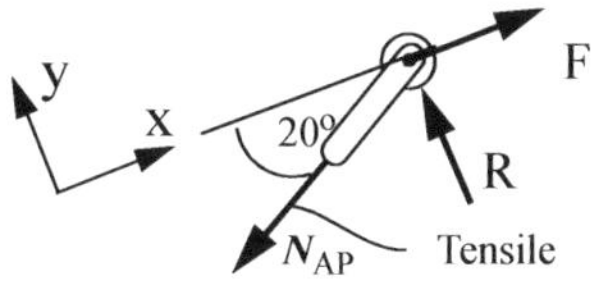

ANS $F = 22.1kN$

---

3. 42

Solution $\delta_P = 0.01$ in E = 30,000 ksi A = 0.2 in L_{AP}= 8 in. L_{BP}= 10 in. F=?

We follow the logic starting with calculation of deformation using small strain approximation as shown below.

$$\delta_{AP} = \delta_P = 0.01 in \quad extension \qquad \delta_{BP} = \delta_P cos 70 = 0.00342 in \quad contraction \quad \textbf{(1)}$$

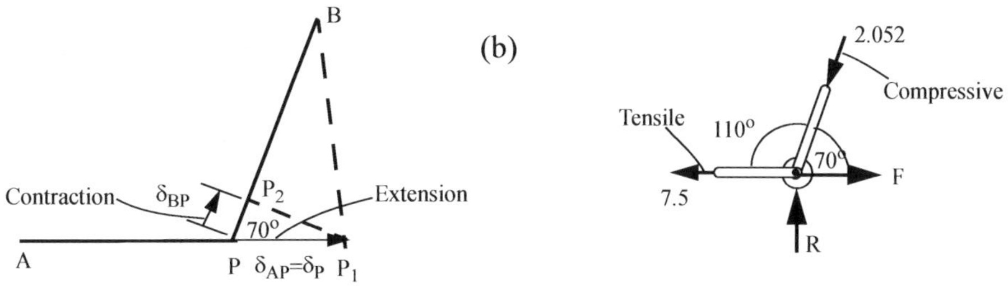

$$\varepsilon_{AP} = \delta_{AP}/L_{AP} = 0.01/8 = 1.25(10^{-3}) extension \qquad \varepsilon_{BP} = \delta_{BP}/L_{BP} = 0.00342/10 = 0.342(10^{-3}) contraction \quad \textbf{(2)}$$

$$\sigma_{AP} = E\varepsilon_{AP} = (30000)(1.25)(10^{-3}) = 37.5 ksi(T) \qquad \sigma_{BP} = E\varepsilon_{BP} = (30000)(0.342)(10^{-3}) = 10.26 ksi(C) \quad \textbf{(3)}$$

$$N_{AP} = A\sigma_{AP} = (37.5)(0.2) = 7.5kips(T) \qquad N_{BP} = A\sigma_{BP} = (10.26)(0.2) = 2.052kips(C) \quad (4)$$

By force equilibrium in the x-direction on the free body diagram we obtain the following.

$$F - 7.5 - 2.052\cos 70 = 0 \quad (5)$$

ANS $F = 8.2$ kips

---

3. 43

Solution $\delta_P = 0.25$ mm $A = 100\ mm^2$ $L_{AP} = 200$ mm $L_{BP} = 250$ mm   F=?

We follow the logic starting with the deformation values calculated in an earlier problem.

$$\delta_{AP} = 0 \qquad \delta_{BP} = 0.2165mm\ extension \qquad \varepsilon_{AP} = 0 \qquad \varepsilon_{BP} = \delta_{BP}/L_{BP} = 0.2165/250 = 0.866(10^{-3})\ extension \quad (1)$$

$$\sigma_{AP} = 0 \qquad \sigma_{BP} = E\varepsilon_{BP} = (200)(10^9)(0.866)(10^{-3}) = 173.2(10^6)N/m^2(T) \quad (2)$$

$$N_{AP} = 0 \qquad N_{BP} = A\sigma_{BP} = (173.2)(10^6)(100)(10^{-6}) = 17.32(10^3)N = 17.32kN(T) \quad (3)$$

By force equilibrium in the y-direction on the free body diagram we obtain the following.

$$F = 17.32\sin 60 = 14.999kN \quad (4)$$

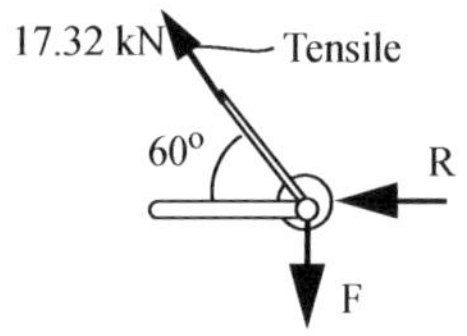

ANS $F = 15$ kN

---

3. 44

Solution $\delta_P = 0.25$ mm   E = 200 GPa $A = 100\ mm^2$ $L_{AP} = 200$ mm $L_{BP} = 250$ mm   F=?

We follow the logic starting with the deformation values calculated in earlier problem.

$$\delta_{AP} = 0.0647mm\ extension \qquad \delta_{BP} = 0.21651mm\ extension \quad (1)$$

$$\varepsilon_{AP} = \delta_{AP}/L_{AP} = 0.0647/200 = 0.3235(10^{-3})\ extension \qquad \varepsilon_{BP} = \delta_{BP}/L_{BP} = 0.21651/250 = 0.866(10^{-3})\ extension \quad (2)$$

$$\sigma_{AP} = E\varepsilon_{AP} = 200(10^9)(0.3235)(10^{-3}) = 64.7(10^6)\frac{N}{m^2}(T) \qquad \sigma_{BP} = E\varepsilon_{BP} = 200(10^9)(0.866)(10^{-3}) = 173.2(10^6)\frac{N}{m^2}(T) \quad (3)$$

$$N_{AP} = A\sigma_{AP} = 64.7(10^6)(100)(10^{-6}) = 6.47(10^3)N = 6.47kN(T)$$
$$N_{BP} = A\sigma_{BP} = 173.2(10^6)(100)(10^{-6}) = 17.32(10^3)N = 17.32kN(T) \quad (4)$$

By force equilibrium in the y-direction on the free body diagram we obtain the following.

$$F = 17.32\cos 30 + 6.47\cos 75 = 12.674 \quad (5)$$

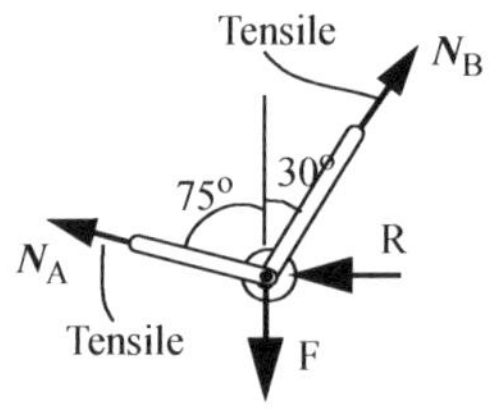

ANS $F = 16.7kN$

---

3. 45

Solution E =150 psi   A = 1/128 in $L_0 = 7$ in   θ = ?   F=?

We follow the logic starting with calculating the final length and the strain the ban.

$$L_f = 3.2 + 2.9 + 2.5 + 1 = 9.6in \qquad \varepsilon = (L_f - L_0)/L_0 = (9.6 - 7)/7 = 0.37143\ extension \quad (1)$$

$$\sigma = E\varepsilon = (150)(0.37143) = 55.71psi(T) \qquad N = A\sigma = (55.71)(1/128) = 0.43527lbs(T) \quad (2)$$

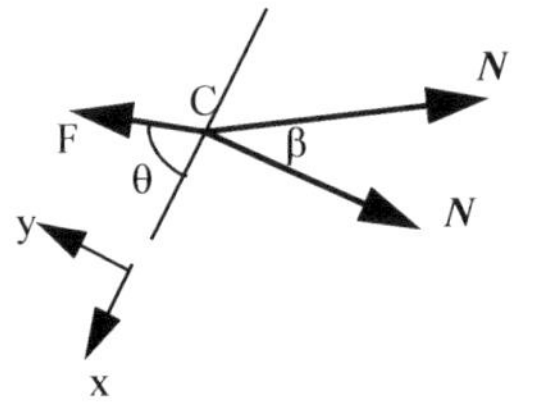

$$\cos\beta = \frac{3.2^2 + 2.9^2 - 2.5^2}{2(3.2)(2.9)} \quad \text{or} \quad \beta = 48.08^0$$

By equilibrium of forces in x and y direction on the free body diagram we obtain the following.

$$F\cos\theta - N\sin\beta = 0 \quad or \quad F\cos\theta = 0.43527(\sin 48.08) = 0.32387 \quad (3)$$

$$F\sin\theta - N - N\cos\beta = 0 \quad F\sin\theta = 0.43527(1+\cos 48.08) = 0.72608 \quad (4)$$

Solving the two equations we obtain:

$$\tan\theta = 2.24187 \quad or \quad \theta = 65.96^{\circ} \quad F = 0.32387/\cos 65.96 = 0.795 lb \quad (5)$$

ANS $\theta = 65.96^{\circ}$; $F = 0.795 lb$

---

3. 46

Solution δ_P= 2 in E =10,000ksi A = 1in^2 F–?

We follow the logic starting with the strain values calculated in earlier problem

$$\varepsilon_{AP} = 16393\mu in/in \quad \varepsilon_{BP} = 11364\mu in/in \quad \varepsilon_{CP} = 20407\mu in/in \quad (1)$$

$$\sigma_{AP} = E\varepsilon_{AP} = (10000)(16393)(10^{-6}) = 163.93 ksi(T) \quad \sigma_{BP} = E\varepsilon_{BP} = (10000)(11364)(10^{-6}) = 113.64 ksi(T)$$

$$\sigma_{CP} = E\varepsilon_{CP} = (10000)(20407)(10^{-6}) = 204.07 ksi(T) \quad (2)$$

$$N_{AP} = A\sigma_{AP} = 163.93 kips(T) \quad N_{BP} = A\sigma_{BP} = 113.64 kips(T) \quad N_{CP} = A\sigma_{CP} = 204.07 kips(T) \quad (3)$$

Noting that the unit vectors in the earlier problem were directed towards *P* and the internal forces are directed away from *P*, we can write the internal forces as vectors and then by force equilibrium in the z direction obtain the external force.

$$\overline{N_{AP}} = -N_{AP}i_{AP} = (-163.93)(-0.6402\hat{i} + 0.7682\hat{k}) \quad \overline{N_{BP}} = -N_{BP}i_{BP} = (-113.64)(-0.4264\hat{i} - 0.6396\hat{j} + 0.6396\hat{k})$$

$$\overline{N_{CP}} = -N_{CP}i_{CP} = (-204.07)(0.2857\hat{i} + 0.4286\hat{j} + 0.8571\hat{k}) \quad \overline{R} = R_x\hat{i} + R_y\hat{j} \quad (4)$$

$$\overline{N_{AP}} + \overline{N_{BP}} + \overline{N_{CP}} + F\bar{k} + \overline{R} = 0 \quad or \quad \text{z-direction: } -163.93(0.7682) - 113.64(0.6396) - 204.07(0.8571) + F = 0 \quad (5)$$

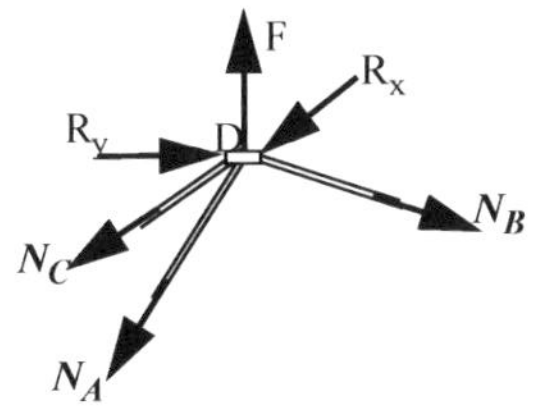

ANS $F = 373.52$ kips

---

3. 47

Solution gap= 0.004 E_A=E_B =30,000ksi L_A= 30inL_B= 50in ε_A=-500μ in/in A_A=A_B= 1in^2

We follow the logic starting with the strain value calculated in the earlier problem.

$$\varepsilon_B = 978.8\mu \quad \sigma_A = E\varepsilon_A = (30000)(-500)(10^{-6}) = -15 ksi \quad or \quad \sigma_A = 15 ksi(C) \quad (1)$$

$$\sigma_B = E\varepsilon_B = (30000)(978.8)(10^{-6}) = 29.364 ksi \quad or \quad \sigma_B = 29.364 ksi(C)$$

$$N_A = A\sigma_A = 15 kips(C) \quad N_B = A\sigma_B = 29.364 kips(T) \quad (2)$$

By moment equilibrium about point C on the free body diagram we obtain the following.

$$24F = (36)(15) + (29.364)(\sin 75)(96) \quad (3)$$

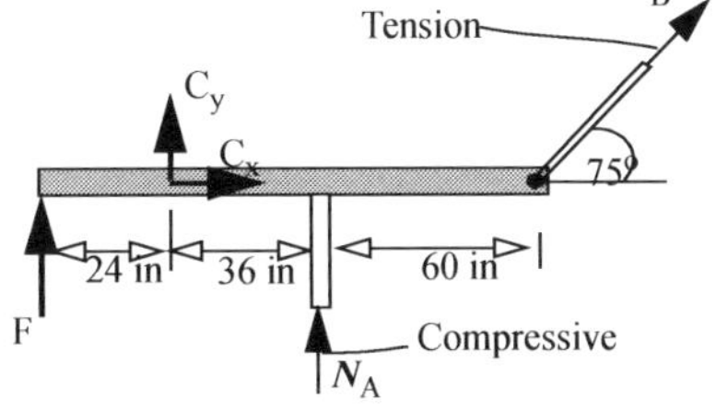

ANS $F = 136$ kips

---

3. 48

Solution d= 1/16 in E= 28,000 ksi $L_o = 27$ ft $= 324$ in. W= ?

We follow the logic starting with calculating the final length of the cable and determining the strain as shown below.

$$AB = CD = \sqrt{108^2 + 10^2} = 108.46 \text{ in} \quad L_f = AB + BC + CD = 2(108.46) + 108 = 324.9239 \text{ in} \quad \theta = atan\left(\frac{10}{108}\right) = 5.29° \quad (1)$$

$$\varepsilon_{av} = (L_f - L_o)/L_o = (324.9239 - 324)/324 = 2.8517(10^{-3}) \qquad \sigma_{av} = E\varepsilon_{av} = (28000)(2.8517)(10^{-3}) = 79.84\text{ ksi} \quad \textbf{(2)}$$

$$N = \sigma_{av}A = 79.84[(\pi(1/16)^2)/4] = 244.97(10^{-3})\text{ ksi} = 244.97\text{ psi} \quad \textbf{(3)}$$

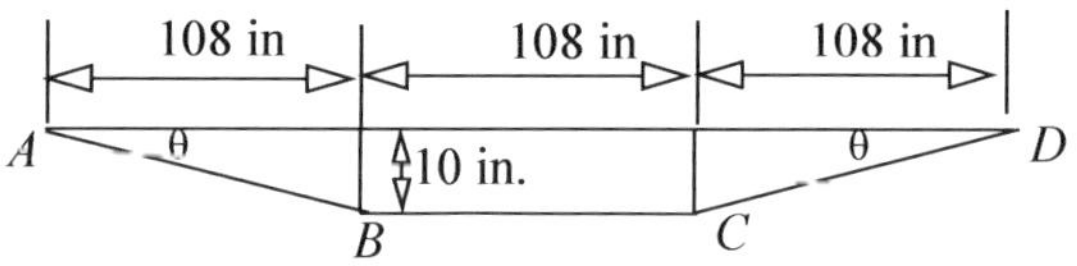

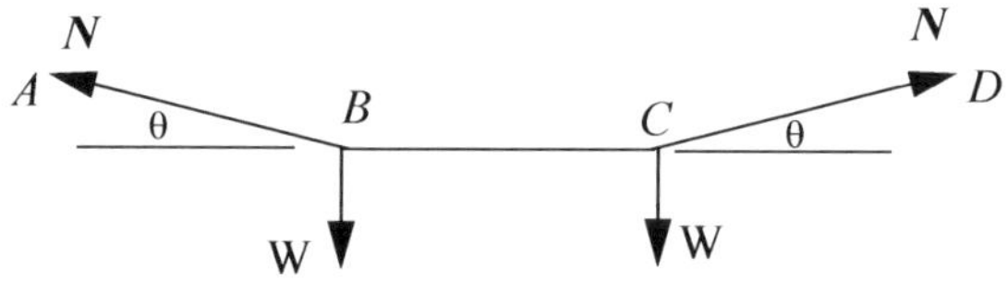

By force equilbrium in the y-direction on the free body diagram we obtain the following.

$$2N\sin\theta = 2W \qquad or \qquad W = 244.97\sin 5.29 = 22.58\text{ lb} \quad \textbf{(4)}$$

ANS $W = 22.6\text{ lb}$

---

3. 49

Solution E_s= 200 GPa E_{al} = 70 GPa) t_s=4 mm d_s = 48 mm δ_{al}= 0.75 mm σ_{al} = ? σ_s = ? δ_s = ?

The cross sectional areas, the strain, stress, and internal force in aluminum can be found as shown below.

$$A_{al} = \pi(0.048^2 - 0.04^2)/4 = 0.5529(10^{-3})\text{ m}^2 \qquad A_s = \pi(0.025^2)/4 = 0.4909(10^{-3})\text{ m}^2 \quad \textbf{(1)}$$

$$\varepsilon_{al} = \delta_{al}/L_{al} = 0.75/300 = 2.5(10^{-3}) \qquad \sigma_{Al} = E\varepsilon_{Al} = (70)(10^9)(2.5)(10^{-3}) = 175(10^6)\text{ N/m}^2(C) \quad \textbf{(2)}$$

$$N_{Al} = \sigma_{Al}A = 175(10^6)(0.5529)(10^{-3}) = 96.76(10^6)(C) \quad \textbf{(3)}$$

By equilibrium of forces we find the internal force in steel and then we find the deformation as shown below.

$$N_S = N_{Al} = 96.76(10^6)(T) \qquad \sigma_S = \frac{N_S}{A} = \frac{96.76(10^6)}{0.4909(10^{-3})} = 197.1(10^6)\text{ N/m}^2(T) \qquad \varepsilon_S = \frac{\sigma_S}{E} = \frac{197.1(10^6)}{(200)(10^9)} = 0.9856(10^{-3}) \quad \textbf{(4)}$$

$$\delta_s = L_s\varepsilon_s = (300 + 25 + 25)(0.9856)(10^{-3}) = 344.9(10^{-3})\text{ mm} \quad \textbf{(5)}$$

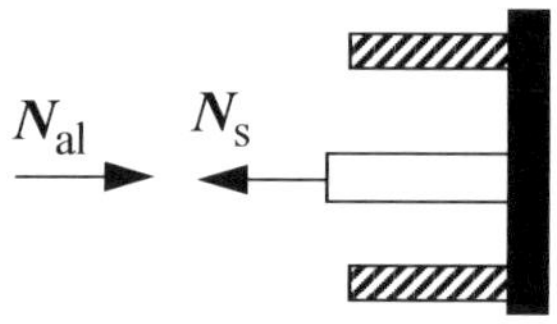

ANS $\sigma_{al} = 175\text{ MPa}(C)\,;\sigma_S = 197.1\text{ MPa}(T);\ \delta_s = 0.345\text{ mm}$

---

3. 50

Solution A=100 mm^2 E = 200GPa P_1 = ? P_2 = ?

The lengths of all members are: $L_{FC} = L_{FG} = L_{FE} = 3l\cos 30 = 3.4641m$.

For all members at joint F, we calculate the deformation vectors, the unit vectors, the deformation and then follow the logic to find the internal forces in the members as shown below.

Deformation Vector

$$\bar{D}_{FC} = (u_C - u_F)\hat{i} + (v_C - v_F)\hat{j} = (-2.0785 + 3.26)\hat{i} + (-9.7657 + 8.4118)\hat{j} = (1.1815\hat{i} - 1.3539\hat{j})\ mm \quad \textbf{(1)}$$

$$\bar{D}_{FG} = (u_G - u_F)\hat{i} + (v_G - v_F)\hat{j} = (-2.5382 + 3.26)\hat{i} + (-9.2461 + 8.4118)\hat{j} = (0.7218\hat{i} - 0.8343\hat{j})mm \quad \textbf{(2)}$$

$$\bar{D}_{FE} = (u_E - u_F)\hat{i} + (v_E - v_F)\hat{j} = (3.26\hat{i} - 8.4118\hat{j})mm \quad \textbf{(3)}$$

$$\bar{D}_{FD} = (u_D - u_F)\hat{i} + (v_D - v_F)\hat{j} = (-1.0392 + 3.26)\hat{i} + (-8.4118 + 8.4118)\hat{j} = (2.2208\hat{i})mm \quad \textbf{(4)}$$

Unit Vectors

$$\hat{i}_{FC} = -\cos 30\hat{i} - \sin 30\hat{j} = -0.866\hat{i} - 0.5\hat{j} \qquad \hat{i}_{FG} = -\cos 30\hat{i} - \sin 30\hat{j} = -0.866\hat{i} - 0.5\hat{j} \quad \textbf{(5)}$$

$$\hat{i}_{FE} = -\cos 30\hat{i} - \sin 30\hat{j} = -0.866\hat{i} - 0.5\hat{j} \qquad \hat{i}_{FD} = -\hat{j} \quad \textbf{(6)}$$

Deformation

$$\delta_{FC} = \hat{i}_{FC} \cdot \bar{D}_{FC} = (-0.866)(1.1815) + (-0.5)(-1.3539) = -0.34626m \quad \textbf{(7)}$$

$$\delta_{FG} = \hat{i}_{FG} \cdot \bar{D}_{FG} = (-0.866)(0.7218) + (0.5)(-0.8343) = -1.04225mm \quad \textbf{(8)}$$

$$\delta_{FE} = \hat{i}_{FE} \cdot \overline{D}_{FE} = (-0.866)(3.26) + (-0.5)(8.4118) = -1.38274mm \quad (9)$$

$$\delta_{FD} = \hat{i}_{FD} \cdot \overline{D}_{FD} = 0 \quad (10)$$

Strains.

$$\varepsilon_{FC} = -0.34626/[3.4641(10^3)] = -0.1(10^{-3}) \qquad \varepsilon_{FG} = -1.04225/[3.4641(10^3)] = -0.3(10^{-3}) \quad (11)$$

$$\varepsilon_{FE} = -1.38274/[3.4641(10^3)] = -0.4(10^{-3}) \qquad \varepsilon_{FD} = 0 \quad (12)$$

Stresses

$$\sigma_{FC} = E\varepsilon_{FC} = (200)(10^9)(-0.1)(10^{-3}) = -20(10^6) or \sigma_{FC} = 20(10^6)\ N/m^2(C) \quad (13)$$

$$\sigma_{FG} = E\varepsilon_{FG} = (200)(10^9)(-0.3)(10^{-3}) = -60(10^6) or \sigma_{FC} = 60(10^6)\ N/m^2(C) \quad (14)$$

$$\sigma_{FE} = E\varepsilon_{FE} = (200)(10^9)(-0.4)(10^{-3}) = -80(10^6) or \sigma_{FC} = 80(10^6)\ N/m^2(C) \qquad \sigma_{FD} = 0 \quad (15)$$

Internal Forces

$$\boldsymbol{N}_{FC} = \sigma_{FC}A = (20)(10^6)(100)(10^{-6}) = 2000N = 2kN(C) \qquad \boldsymbol{N}_{FG} = \sigma_{FG}A = (60)(10^6)(100)(10^{-6}) = 6000N = 6kN(C) \quad (16)$$

$$\boldsymbol{N}_{FE} = \sigma_{FE}A = (80)(10^6)(100)(10^{-6}) = 8000N = 8kN(C) \qquad \boldsymbol{N}_{FD} = 0 \quad (17)$$

By equilibrium of forces on the free body diagram we obtain the following.

$$-P_1 - 8cos30 + 2cos30 + 6cos30 = 0 \qquad or \qquad P_1 = 0 \quad (18)$$

$$-P_2 + 2sin30 + 8sin30 - 6sin30 = 0 \qquad or \qquad P_2 = 2\text{ kN} \quad (19)$$

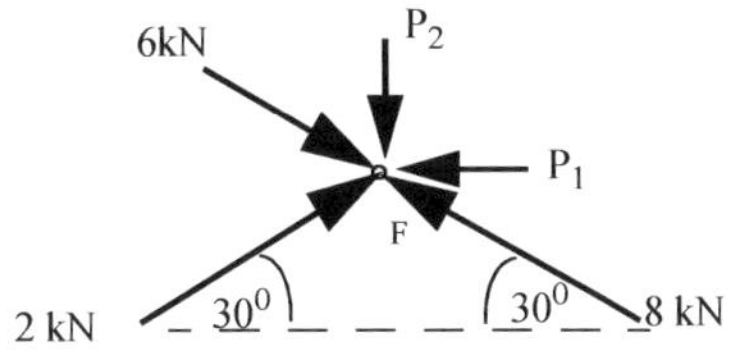

ANS $P_1 = 0;\ P_2 = 2$ kN

---

3. 51

Solution A=100 mm^2 E = 200GPa P_3 = ?

Length of all members are: $L_{GF} = L_{GN} = L_{GC} = 3/(cos30) = 3.4641m$

For all members at joint G we calculate the deformation vectors, the unit vectors, the deformation and then follow the logic to find the internal forces in the members as shown below.

Deformation vectors

$$\overline{D}_{GF} = (u_F - u_G)\hat{i} + (v_F - v_G)\hat{j} = (-3.26 + 2.5382)\hat{i} + (-8.4118 + 9.2461)\hat{j} = (-0.7218\hat{i} + 0.8343\hat{j})mm \quad (1)$$

$$\overline{D}_{GH} = (u_H - u_G)\hat{i} + (v_H - v_G)\hat{j} = (-1.55 + 2.5382)\hat{i} + (-8.8793 + 9.2461)\hat{j} = (0.98820\hat{i} + 0.3668\hat{j})mm \quad (2)$$

$$\overline{D}_{GC} = (u_C - u_G)\hat{i} + (v_C - v_G)\hat{j} = (-2.0785 + 2.5382)\hat{i} + (-9.7657 + 9.2461)\hat{j} = (0.4597\hat{i} - 0.5196\hat{j})mm \quad (3)$$

Unit vectors

$$\hat{i}_{GF} = cos30\hat{i} - sin30\hat{j} = 0.866\hat{i} - 0.5\hat{j} \qquad \hat{i}_{GH} = -cos30\hat{i} - sin30\hat{j} = -0.866\hat{i} - 0.5\hat{j} \qquad \hat{i}_{GC} = -\hat{j} \quad (4)$$

Deformations calculation

$$\delta_{GF} = \hat{i}_{GF} \cdot \overline{D}_{GF} = (0.866)(-0.7218) + (-0.5)(0.8343) = -1.0423mm \quad (5)$$

$$\delta_{GH} = \hat{i}_{GH} \cdot \overline{D}_{GH} = (-0.866)(0.9882) + (-0.5)(0.3668) = -1.03918mm \qquad \delta_{GC} = \hat{i}_{GC} \cdot \overline{D}_{GC} = 0.5196mm \quad (6)$$

Strains.

$$\varepsilon_{GF} = -\frac{1.0423}{3.4641(10^3)} = -0.3(10^{-3}) \qquad \varepsilon_{GH} = -\frac{1.03918}{3.4641(10^3)} = -0.3(10^{-3}) \qquad \varepsilon_{GC} = -\frac{0.5196}{3.4641(10^3)} = 0.15(10^{-3}) \quad (7)$$

Stresses

$$\sigma_{GF} = E\varepsilon_{GF} = (200)(10^9)(-0.3)(10^{-3}) = -60(10^6) or \sigma_{GF} = 60(10^6)\ N/m^2(C) \quad (8)$$

$$\sigma_{GH} = E\varepsilon_{GH} = (200)(10^9)(-0.3)(10^{-3}) = -60(10^6) or \sigma_{GH} = 60(10^6)\ N/m^2(C) \quad (9)$$

$$\sigma_{GC} = E\varepsilon_{GC} = (200)(10^9)(0.15)(10^{-3}) = -30(10^6) or \sigma_{GC} = 30(10^6)\ N/m^2(T) \quad (10)$$

Internal Force calculation

$$\boldsymbol{N}_{GF} = \sigma_{GF}A = (60)(10^6)(100)(10^{-6}) = 6000N = 6kN(C) \qquad \boldsymbol{N}_{GH} = \sigma_{GH}A = (60)(10^6)(100)(10^{-6}) = 6000N = 6kN(C) \quad (11)$$

$$N_{GC} = \sigma_{GC}A = (30)(10^6)(100)(10^{-6}) = 3000N = 3kN(T) \quad \textbf{(12)}$$

By equilibrium of forces in y-direction on the free body diagram we obtain the following.

$$-P_3 + 6cos60 + 6cos60 - 3 = 0 \qquad P_3 = 3 \text{ kN} \quad \textbf{(13)}$$

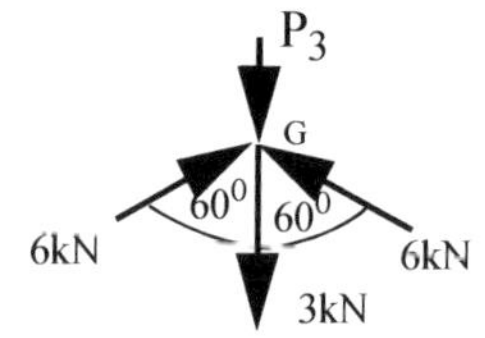

ANS $P_3 = 3$ kN

---

3. 52

Solution A=100 mm^2 E = 200GPa P_4 = ? P_5 = ?

The length of all members are: $L_{HA} = L_{HG} = L_{HC} = 3/cos30 = 3.4641m$

For all members at joint H we calculate the deformation vectors, the unit vectors, the deformation and then follow the logic to find the internal forces in the members as shown below.

Deformation vectors

$$\bar{D}_{HA} = (u_A - u_H)\hat{i} + (v_A - v_H)\hat{j} = (-4.6765 + 1.55)\hat{i} + (8.8793)\hat{j} = (-3.1265\hat{i} + 8.8793\hat{j}) \quad \textbf{(1)}$$

$$\bar{D}_{HG} = (u_G - u_H)\hat{i} + (v_G - v_H)\hat{j} = (-2.5382 + 1.55)\hat{i} + (-9.2461 + 8.8793)\hat{j} = (-0.9882\hat{i} - 0.3668\hat{j}) \quad \textbf{(2)}$$

$$\bar{D}_{HC} = (u_C - u_H)\hat{i} + (v_C - v_H)\hat{j} = (-2.0785 + 1.55)\hat{i} + (-9.7657 + 8.8793)\hat{j} = (-0.5285\hat{i} - 0.8864\hat{j}) \quad \textbf{(3)}$$

$$\bar{D}_{HB} = (u_B - u_H)\hat{i} + (v_B - v_H)\hat{j} = (-3.375 + 1.55)\hat{i} + (-8.8793 + 8.8793)\hat{j} = -1.825\hat{i} \quad \textbf{(4)}$$

Unit vectors

$$\hat{i}_{HA} = -cos30\hat{i} - sin30\hat{j} = -0.866\hat{i} - 0.5\hat{j} \qquad \hat{i}_{HG} = cos30\hat{i} + sin30\hat{j} = 0.866\hat{i} + 0.5\hat{j} \quad \textbf{(5)}$$

$$\hat{i}_{HC} = cos30\hat{i} - sin30\hat{j} = 0.866\hat{i} - 0.5\hat{j} \qquad \hat{i}_{HB} = -\hat{j} \quad \textbf{(6)}$$

Deformations

$$\delta_{HA} = \hat{i}_{HA} \cdot \bar{D}_{HA} = (-0.866)(-3.1265) + (-0.5)(8.8793) = -1.73202mm \quad \textbf{(7)}$$

$$\delta_{HG} = \hat{i}_{HG} \cdot \bar{D}_{HG} = (0.866)(-0.9882) + (0.5)(-0.3668) = -1.03918mm \quad \textbf{(8)}$$

$$\delta_{HC} = \hat{i}_{HC} \cdot \bar{D}_{HC} = (0.866)(-0.5285) + (-0.5)(-0.8864) = -0.01449mm \qquad \delta_{HB} = \hat{i}_{HB} \cdot \bar{D}_{HB} = 0 \quad \textbf{(9)}$$

Strains.

$$\varepsilon_{HA} = \delta_{HA}/L_{HA} = -1.73202/[3.4641(10^3)] = -0.5(10^{-3}) \qquad \varepsilon_{HG} = \delta_{HG}/L_{HG} = -1.03918/[3.4641(10^3)] = -0.3(10^{-3}) \quad \textbf{(10)}$$

$$\varepsilon_{HC} = \delta_{HC}/L_{HC} = -0.01449/[3.4641(10^3)] = -0.004(10^{-3}) \qquad \varepsilon_{HD} = 0 \quad \textbf{(11)}$$

Stresses

$$\sigma_{HA} = E\varepsilon_{HA} = (200)(10^9)(-0.5)(10^{-3}) = -100(10^6) = 100(10^6)\ N/m^2(C) \quad \textbf{(12)}$$

$$\sigma_{HG} = E\varepsilon_{HG} = (200)(10^9)(-0.3)(10^{-3}) = -60(10^6) = 60(10^6)\ N/m^2(C) \quad \textbf{(13)}$$

$$\sigma_{HC} = E\varepsilon_{HC} = (200)(10^9)(-0.004)(10^{-3}) = -0.8(10^6) = 0.8(10^6)\ N/m^2(C) \qquad \sigma_{HB} = 0 \quad \textbf{(14)}$$

Internal Forces

$$N_{HA} = \sigma_{HA}A = 100(10^6)(100)(10^{-6}) = 10000N = 10kN(C) \qquad N_{HG} = \sigma_{HG}A = 60(10^6)(100)(10^{-6}) = 6000N = 6kN(C) \quad \textbf{(15)}$$

$$N_{HC} = \sigma_{HC}A = (0.8)(10^6)(100)(10^{-6}) = 80N = 0.08kN(C) \qquad N_{HB} = 0 \quad \textbf{(16)}$$

By equilibrium of forces on the free body diagram we obtain the following.

$$-P_5 - 6sin60 + 10sin60 - 0.8sin60 = 0 \quad or \quad P_5 = 3.39 \text{ kN} \quad \textbf{(17)}$$

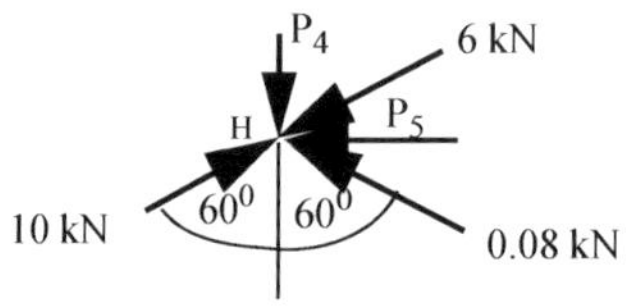

$$-P_4 + 10cos60 - 6cos60 + 0.08cos(60) = 0 \quad or \quad P_4 = 2.04 \text{ kN} \quad \textbf{(18)}$$

ANS $P_5 = 3.39$ kN ; $P_4 = 2.04$ kN

Section 3.3

3. 53

Solution K=3 $(\tau_f)_{BCD}$=1.5ksi $(\sigma_f)_{BEF}$=6ksi h=? d=? to the nearest 1/16 inch

By force equilibrium in the x-direction we obtain:

$$V_{BCD} = 10\cos 30 = 8.66 kips \qquad N_{BEF} = 10\cos 30 = 8.66 kips\ (C) \tag{1}$$

(a)

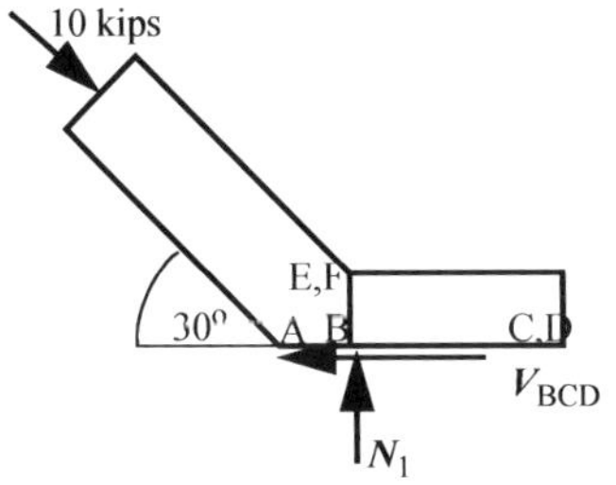

(b)

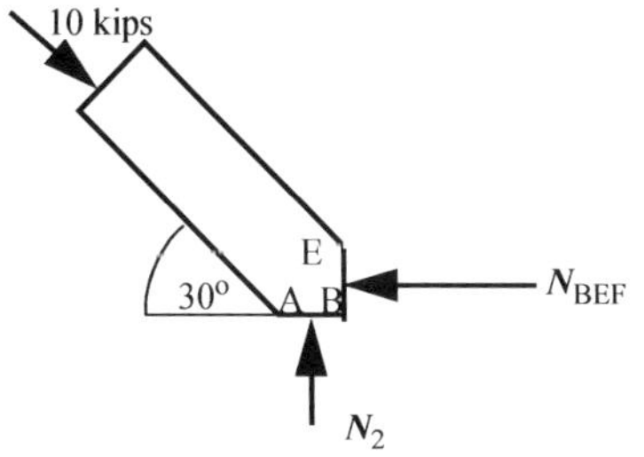

$$\tau_{BCD} = \frac{V_{BCD}}{A_{BCD}} = \frac{8.66}{4h} \qquad K = \frac{(\tau_f)_{BCD}}{\tau_{BCD}} \quad \text{or} \quad \frac{1.5}{(8.66/4h)} \quad \text{or} \quad h = 4.33 \tag{2}$$

$$\sigma_{BEF} = \frac{N_{BEF}}{A_{BEF}} = \frac{8.66}{4d}\ (C) \qquad K = \frac{(\sigma_f)_{BEF}}{\sigma_{BEF}} \quad or \quad \frac{6}{(8.66/4d)} = 3 \quad or \quad d = 1.0825\ in \tag{3}$$

ANS $h = 4\frac{3}{8}\text{in}\ ;\ d = 1\frac{1}{8} in$

---

3. 54

Solution m=125 kg K=3 σ_{fail}= 180 MPa d=? nearest millimeter

By force equilibrium we obtain the internal axial force $N = W/2$. The weight is $W = 125(9.81) = 1226.2N$

$$N = \frac{1226.2}{2} = 613.13N \qquad \sigma = \frac{N}{A} = \frac{613.13}{(\pi d^2/4)} = \frac{780.65}{d^2} \le \frac{\sigma_{fail}}{K} \quad or \quad \frac{780.65}{d^2} \le \frac{180(10^6)}{3} \quad or \quad d \ge 3.61(10^{-3})m \tag{1}$$

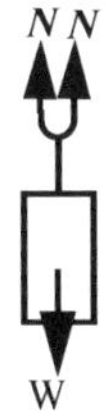

ANS $d = 4$ mm

---

3. 55

Solution W=? K=4 σ_{fail}=25ksi d= 1/8

From previous problem we have $N = W/2$.

$$\sigma = \frac{N}{A} = \frac{W}{2(\pi d^2/4)} = \frac{2W}{\pi(1/8)^2} = 40.74W \le \frac{\sigma_{fail}}{K} \quad or \quad 40.74W \le \frac{25(10^3)}{4} \quad or \quad W \le 153.4\ lb \tag{1}$$

ANS $W_{max} = 153$ lb

---

3. 56

Solution K=1.2 σ_{fail}=w.00MPa d=10mm $W_{max=?}$

By force equilibrium we obtain

$$-N_{AB}\cos 37 + N_{AC}\cos 22 = 0 \quad or \quad N_{AB} = 1.16096N_{BC} \tag{1}$$

$$N_{AB}\sin 37 + N_{BC}\sin 22 - W = 0 \quad or \quad N_{AB}\sin 37 + N_{BC}\sin 22 = W \tag{2}$$

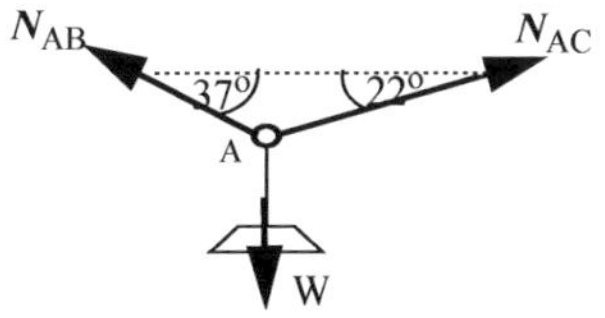

Solving we obtain the internal axial forces and also calculate cross sectional areas as shown below.

$$N_{AB} = 1.0817W \qquad N_{BC} = 0.9317W \qquad A_{AB} = A_{BC} = \frac{\pi}{4}(10)^2 = 78.54mm^2 = 78.54(10^{-6})m^2 \tag{3}$$

$$\sigma_{AB} = \frac{N_{AB}}{A_{AB}} = \frac{1.0817W}{78.54(10^{-6})} = 0.01377W(10^{-6}) \le \frac{\sigma_{fail}}{K} \quad or \quad 0.01377W(10^6) \le \frac{200(10^6)}{1.2} \quad or \quad W \le 12.103(10^3)N \tag{4}$$

$$\sigma_{BC} = \frac{N_{BC}}{A_{BC}} = \frac{0.9317W}{78.54(10^{-6})} = 0.01186W(10^6) \le \frac{\sigma_{fail}}{K} \quad or \quad 0.01186W(10^6) \le \frac{200(10^6)}{1.2} \quad or \quad W \le 14.049(10^3)N \tag{5}$$

ANS $W_{max} = 12.1$ kN

---

3. 57

Solution K=1.25 σ_{fail}=30 ksi d_{min}=? W= 2500lb

By force equilibrium we obtain

$$-N_{AB}cos37 + N_{AC}cos22 = 0 \quad or \quad N_{AB} = 1.16096N_{BC} \tag{1}$$

$$N_{AB}sin37 + N_{BC}sin22 - W = 0 \quad or \quad N_{AB}sin37 + N_{BC}sin22 = 2500\ lb \tag{2}$$

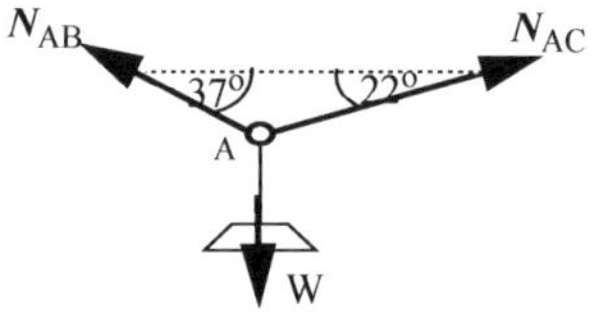

$$N_{BC} = 2329.25lb \qquad N_{AB} = 2704.17lb \qquad A_{AB} = A_{BC} = \pi d^2/4 \tag{3}$$

$$\sigma_{AB} = \frac{N_{AB}}{A_{AB}} = \frac{2704.17}{\pi d^2/4} = \frac{3443.05}{d^2} \le \frac{\sigma_{fail}}{K} \quad or \quad \frac{3443.05}{d^2} \le \frac{30(10^3)}{1.25} \quad or \quad d \ge 0.3788in \tag{4}$$

$$\sigma_{BC} = \frac{N_{BC}}{A_{BC}} = \frac{2329.25}{\pi d^2/4} = \frac{2965.7}{d^2} \le \frac{\sigma_{fail}}{K} \quad or \quad \frac{2965.7}{d^2} \le \frac{30(10^3)}{1.25} \quad or \quad d \ge 0.3515in \tag{5}$$

ANS $d_{min} = \frac{7}{16}$ in.

---

3. 58

Solution K = 1.25 τ_{fail} = 400psi σ_{fail} = 6ksi L_{min} = ? nearest1/8 inch, h_{min} = ? nearest 1/8 inch

By equilibrium we obtain the internal forces as shown below.

$$V = 10/2 = 5kips \qquad N = 10kips \tag{1}$$

(a)

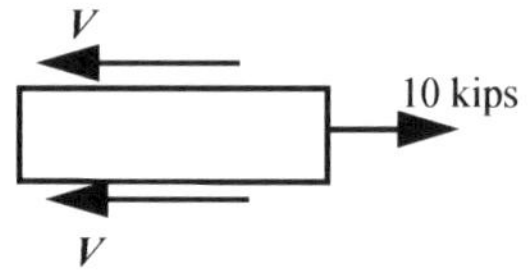

(b)

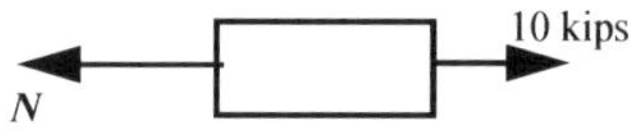

$$\sigma = \frac{N}{8h} = \frac{10}{8h} \le \frac{\sigma_{fail}}{K} \quad or \quad \frac{10}{8h} \le \frac{6}{1.25} \quad or \quad h \ge 0.2604 \tag{2}$$

$$\tau = \frac{V}{8(L/2)} = \frac{5}{4L} \le \frac{\tau_{fail}}{K} \quad or \quad \frac{5}{4L} \le \frac{0.4}{1.25} \quad or \quad L \ge 3.906 \tag{3}$$

ANS $h_{min} = (3/8)$ in. ; $L_{min} = 4$ in.

---

3. 59

Solution K=2 τ_{fail}=300MPa d_{min}=? nearest mm

By equilibrium we obtain the following.

$$V_1 = 32.68kN \qquad V_{2x} = 30kN \qquad V_{2y} = 50kN \qquad V_2 = \sqrt{30^2 + 50^2} = 58.31kN \qquad V_3 = 30kN \tag{1}$$

(a)

(b)

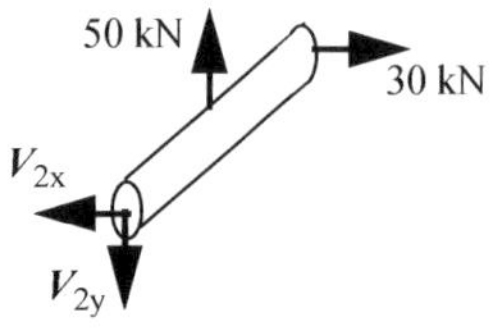

(c)

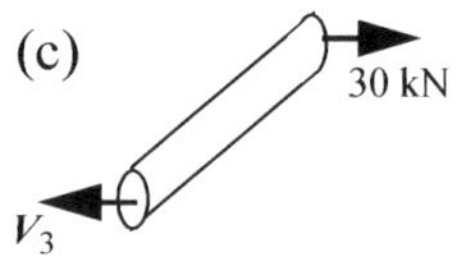

$$\tau_{max} = \frac{V_2}{A} = \frac{58.31(10^3)}{(\pi d^2/4)} \le \frac{\tau_{fail}}{K} \quad or \quad \frac{233.24(10^3)}{\pi d^2} \le \frac{300(10^6)}{2} \quad or \quad d^2 \ge 0.4949(10^{-3}) \quad d \ge 0.02225(10^{-3})m \quad \textbf{(2)}$$

ANS $d_{min} = 23mm$

---

3. 60

Solution P=1200N K=1.1 τ_{yield}=350MPa D=? for $(8\text{mm} \le d \le 16mm)$

Using the allowable maximum stress and the given stress expression we have the following.

$$\tau_{max} = \frac{\tau_{yield}}{K} = \frac{350}{1.1} = 318.18MPa = \frac{K8PC}{\pi d^2} = \frac{(KC)(8)(1200)}{\pi d^2} \text{ or}$$

$KC = 0.1041 d^2(10^6)$, where d is in meters *or* $KC = 0.1041 d^2$, where d is in millimeters **(1)**

We substitute $a = 0.1041 d^2$ and from the given Wahl factor obtain the following.

$$KC = \left(\frac{4C-1}{4C-4}\right)C + 0.615 = a \quad or \quad C^2 + (0.365 - a)C - (0.615 - a) = 0 \text{ or}$$

$$C_{1,2} = \frac{(a - 0.365) \pm \sqrt{(a-0.365)^2 + 4(0.615 - a)}}{2} \quad \textbf{(2)}$$

where, C_1 and C_2 are the two roots. Then $D_1 = C_1 d$ and $D_2 = C_2 d$. The values of D_1 & D_2 are tabulated below.

d (mm)	a	C_1	C_2	D_1 (mm)	D_2 (mm)
8	6.662	5.115	1.182	40.9	9.5
10	10.410	8.951	1.094	89.5	10.9
12	14.990	13.566	1.060	162.8	12.7
14	20.404	18.997	1.042	266.0	14.6
16	26.650	25.254	1.031	404.1	16.5

---

3. 61

Solution $(do)_1 = 2$ in $(do)_2 = 2\ 3/4$ in t=1/4 in $d_{bolt} = 1/2$ in $\sigma_{iron} = 25$ ksi
$\tau_{bolt} = 15$ ksi K= 1.2 P = ?

The area of cross section and internal forces by equilibrium can be written as shown below.

$$A_{iron} = \pi(2^2 - 1.5^2)/4 = 1.3744 \text{ in}^2 \quad A_{bolt} = (\pi/4)(1/2)^2 = 1.9635 \text{ in}^2 \quad N_{iron} = P \quad V_{bolt} = P/2 \quad \textbf{(1)}$$

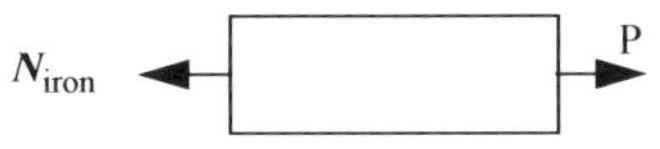

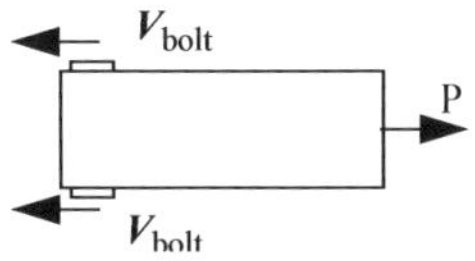

$$\sigma_{iron} = N_{iron}/A_{iron} = P/1.3744 \le \sigma_{yield}/K \quad or \quad P \le (25/1.2)(1.3744) \quad or \quad P \le 28.63 \text{ kips} \quad \textbf{(2)}$$

$$\tau_{bolt} = N_{bolt}/A_{bolt} = (P/2)/1.9635 \le \tau_{yield}/K \quad or \quad P \le (2)(15/1.2)(1.9635) \quad or \quad P \le 49.08 \text{ kips} \quad \textbf{(3)}$$

ANS $P_{max} = 28.6$ kips

---

3. 62

Solution $d_{coup} = 250$-mmn = 6 $d_{bolt} = 12.5$ mm R = 200/2=100 mm K=1.5 $\tau_{bolt} = 300$ MPa $T_{max} = ?$

The cross sectional area of the bolt and by equilibrium the shear force in the bolt can be written as shown below.

$$A_{bolt} = \pi(0.0125)^2/4 = 0.1227(10^{-3}) \text{ m}^2 \quad T = 6(VR) = 0.6V \quad or \quad V = T/0.6 \quad \textbf{(1)}$$

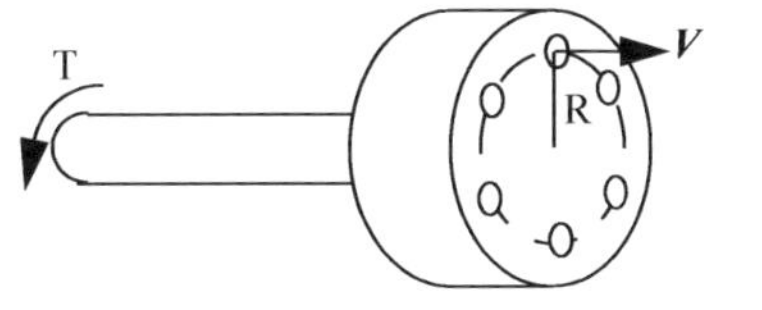

$$\tau = V/A_{bolt} = T/[(0.6)(0.1227)(10^{-3})] = 13.58T(10^3) \le (300/1.5)(10^6) \quad or \quad T \le 14.727(10^3) \text{ N-m} \quad \textbf{(2)}$$

ANS $T_{max} = 14.72$ kN-m

---

3. 63

Solution d = 15 mm E = 70 GPa p(x) = ?

We calculate strain, stress and internal force as shown below.

$$\varepsilon_{xx} = \frac{du}{dx} = 30(1-2x)(10^{-6}) \qquad \sigma_{xx} = E\varepsilon_{xx} = (70)(10^9)(30)(1-2x)(10^{-6}) = 2.1(1-2x)(10^6)\frac{N}{m^2} \quad (1)$$

$$A = (\pi/4)(0.015)^2 = 0.1767(10^{-3})\text{ m}^2 \qquad N = \sigma_{xx}A = 2.1(10^6)(0.1767)(10^{-3})(1-2x) = 371.1(1-2x)\ N \quad (2)$$

From equilibrium of free body diagram of a differential elcment we obtain the following.

$$.\ (N+dN)+p(x)dx-N = 0 \quad or \quad p(x) = -\left(\frac{dN}{dx}\right) = -371.1(-2) = 742.2\text{ N/m o} \quad (3)$$

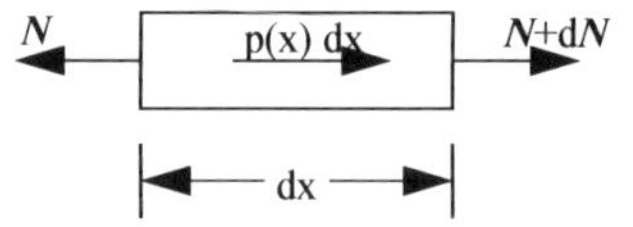

ANS $p(x) = 742.2$ N/m

---

3. 64

Solution d =15mm E = 70G Pa p(x) = ?

We calculate strain, stress and internal force as shown below.

$$\varepsilon_{xx} = \frac{du}{dx} = 50(2x-6x^2)(10^{-6}) \qquad \sigma_{xx} = E\varepsilon_{xx} = (70)(10^9)(50)(2x-6x^2)(10^{-6}) = 7.0(10^6)(x-3x^2)N/m^2 \quad (1)$$

$$A = (\pi/4)(0.015)^2 = 0.1767(10^{-3})\text{ m}^2 \qquad N = \sigma_{xx}A = 7.0(10^6)(x-3x^2)(0.1767)(10^{-3}) = 1237(x-3x^2)\ N \quad (2)$$

From equilibrium of free body diagram of a differential element we obtain the following.

$$(N+dN)+p(x)dx-N = 0 \qquad p(x) = -\left(\frac{dN}{dx}\right) = -1237(1-6x)\ N/m \quad (3)$$

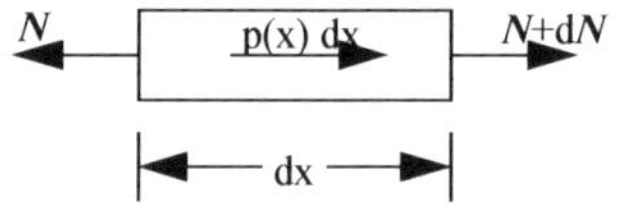

ANS $p(x) = -1.237(1-6x)$ N/m

---

3. 65

Solution E=30000ksi N=? and M_z =? @ x=20

We calculate strain and stress as shown below.

$$\varepsilon_{xx} = \frac{du}{dx} = \frac{(60+80y-2xy)}{180000} \qquad \sigma_{xx} = E\varepsilon_{xx} = 30(10^3)\frac{(60+80y-2xy)}{180000} = \frac{(30+40y-xy)}{3}ksi \quad (1)$$

$$\sigma_{xx}\big|_{x=20} = \frac{(30+40y-20y)}{3} = \frac{30+20y}{3}ksi \quad (2)$$

The stress distribution can be replaced by an equivalent force at the centroid of the distribution and then replaced by an equivalent force and moment at O as shown below.

$$N = \frac{1}{2}(20)(3)(2) = 60kips \qquad M_z = N(0.5) = 30in-kips \quad (3)$$

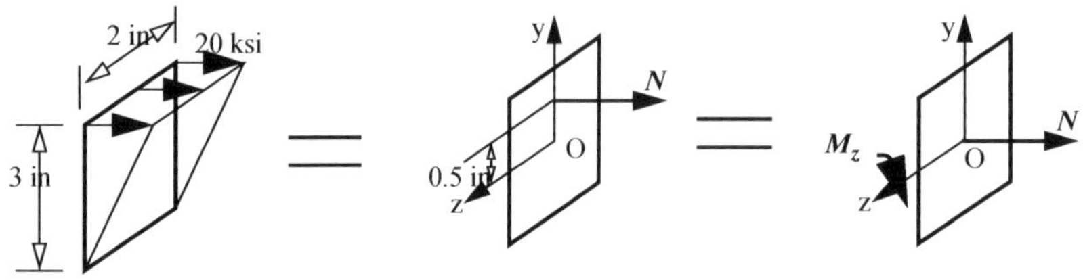

ANS $N = 60$ kips ; $M_z = 30$ in.-kips

---

3. 66

Solution a = ? b = ? c = ? U_0 = ? @ε = 0.18 E_T = ? @σ = 1400 MPa

We copy the results of stresses and strains from earlier problem in Section 3.1 and using the Least Square Method obtain the values of constants a, b, and c on a spread sheet as shown in the table below.

	Strain x_i	Stress (MPa) f_i	x_i^2	x_i^3	x_i^4	x_i*f_i	$x_i^2*f_i$
1	0.0106	1119.31	0.0001	0.0000	0.0000	11.86	0.13
2	0.0220	1168.54	0.0005	0.0000	0.0000	25.71	0.57
3	0.0392	1222.43	0.0015	0.0001	0.0000	47.92	1.88
4	0.0558	1269.44	0.0031	0.0002	0.0000	70.83	3.95
5	0.0800	1324.65	0.0064	0.0005	0.0000	105.97	8.48
6	0.0942	1342.29	0.0089	0.0008	0.0001	126.44	11.91
7	0.1160	1388.96	0.0135	0.0016	0.0002	161.12	18.69
8	0.1430	1421.18	0.0204	0.0029	0.0004	203.23	29.06
9	0.1776	1458.26	0.0315	0.0056	0.0010	258.99	46.00
10	0.1998	1472.64	0.0399	0.0080	0.0016	294.23	58.79
11	0.2202	1473.40	0.0485	0.0107	0.0024	324.44	71.44
12	0.2326	1448.89	0.0541	0.0126	0.0029	337.01	78.39
13	0.2406	1423.54	0.0579	0.0139	0.0034	342.50	82.41
14	0.2462	1384.31	0.0606	0.0149	0.0037	340.82	83.91
15	0.2494	1334.38	0.0622	0.0155	0.0039	332.79	83.00
16	0.2526	1287.92	0.0638	0.0161	0.0041	325.33	82.18
b_{ij} & r_i	2.3798	21540.14	0.4730	0.1034	0.0236	3309.21	660.77
C_{ij}	0.0005	0.0072	0.0224	0.1533	0.5288	1.90	
D	0.000770195						
a_i	1062.06	4493.16	-12993.12				

At $\sigma = 1400$ MPa we obtain the quadratic equation below and determine its root to obtain the strain at the given stress level.

$$1062.1 + 4493.3\varepsilon - 12993.1\varepsilon^2 = 1400 \qquad or \qquad \varepsilon = 0.11052 \tag{1}$$

The tangent modulus is the slope of the stress-strain curve in the non-linear region and determine its value as shown below.

$$E_T = \left.\frac{d\sigma}{d\varepsilon}\right|_{\varepsilon = 0.11052} = (b + 2c\varepsilon)|_{\varepsilon = 0.11052} = 4493.3 + 2(-12993.1)(0.11052) = 1621.3MPa \tag{2}$$

ANS $a = 1062.1MPa$; $b = 4493.3MPa$; $c = -12993.1MPa$; $E_T = 1.621\ GPa$

---

3. 67

Solution a = ? b = ? c = ? U_o = ? @ $\varepsilon = 0.15$ E_T = ? @ $\sigma = 50$ ksi

We copy the results of stresses and strains from earlier problem in Section 3.1 and using the Least Square Method obtain the values of constants a, b, and c on a spread sheet as shown in the table below..

	Strain x_i	Stress (ksi) f_i	x_i^2	x_i^3	x_i^4	x_i*f_i	$x_i^2*f_i$
1	0.00438	39.22	0.000	0.000	0.000	0.172	0.001
2	0.00953	40.23	0.000	0.000	0.000	0.383	0.004
3	0.01435	41.56	0.000	0.000	0.000	0.596	0.009
4	0.01887	43.09	0.000	0.000	0.000	0.813	0.015
5	0.02359	44.92	0.001	0.000	0.000	1.060	0.025
6	0.02953	47.47	0.001	0.000	0.000	1.402	0.041
7	0.03543	50.22	0.001	0.000	0.000	1.779	0.063
8	0.04212	52.97	0.002	0.000	0.000	2.231	0.094
9	0.04893	55.11	0.002	0.000	0.000	2.697	0.132
10	0.05605	56.94	0.003	0.000	0.000	3.191	0.179
11	0.0702	59.69	0.005	0.000	0.000	4.190	0.294
12	0.08061	61.06	0.006	0.001	0.000	4.922	0.397
13	0.09633	62.49	0.009	0.001	0.000	6.020	0.580
14	0.10711	63.2	0.011	0.001	0.000	6.769	0.725
15	0.12297	63.92	0.015	0.002	0.000	7.860	0.967
16	0.14174	64.68	0.020	0.003	0.000	9.168	1.299
17	0.15818	65.04	0.025	0.004	0.001	10.288	1.627
18	0.18155	65.39	0.033	0.006	0.001	11.872	2.155
19	0.19267	61.32	0.037	0.007	0.001	11.815	2.276
20	0.19802	58.26	0.039	0.008	0.002	11.537	2.284
21	0.20321	54.55	0.041	0.008	0.002	11.085	2.253
22	0.20736	50.73	0.043	0.009	0.002	10.519	2.181
b_{ij} & r_i	2.04273	1202.06	0.29666296	0.050327768	0.009120184	120.3690455	17.60174158
C_{ij}	0.000	0.004	0.015	0.113	0.501	2.354	
D	0.001						
a_i	35.995	455.956	-1756.978				

At $\sigma = 50$ ksi we obtain the quadratic equation below and determine its root to obtain the strain at the given stress level.

$$36.0 + 456.0\varepsilon - 1757.0\varepsilon^2 = 50 \qquad or \qquad \varepsilon = 0.03558 \tag{1}$$

The tangent modulus is the slope of the stress-strain curve in the non-linear region and can be written as $E_T = \frac{d\sigma}{d\varepsilon} = b + 2c\varepsilon$. Substituting ε_{50}=00.03558 and the values of b and c, we obtain the tangent modulus as given below.

$$E_T = \left.\frac{d\sigma}{d\varepsilon}\right|_{\varepsilon = 0.03558} = (b + 2c\varepsilon)|_{\varepsilon = 0.03558} = 455.96 + 2(-1756.96)(0.03558) = 330.94ksi \quad (2)$$

ANS $a = 36.0ksi$; $b = 456.0ksi$; $c = -1757.0ksi$ $E_T = 330.9ksi$

---

3. 68

Solution L_0= 2 in d_0= 1/2 in E = 510 psi T_i = ? P = ?

The strain ε_i in the ith segment of column three in table below is given by: $\varepsilon_i = (\Delta L_i)/L_0$, where ΔL_i is the deformed length in column two and L_0= 2 in. We find the stress in each segment $\sigma_i = E\varepsilon_i$, then the tension by multiplying stress by the area $A_i = (\pi 0.5^2)/4 = 0.19635\ in^2$. The force P exerted on the car carrier by the cord is given by the tension in segment 1.

Segment Number	Deformed Length. ΔL_i (inches)	Strain ε_i	Stress σ_i (psi)	Internal Tension T_i (lbs)
1	3.4	0.70	357	70.10
2	3.4	0.70	357	70.10
3	3.4	0.70	357	70.10
4	3.4	0.70	357	70.10
5	3.4	0.70	357	70.10
6	3.4	0.70	357	70.10
7	3.1	0.55	280.5	55.08
8	2.7	0.35	178.5	35.05
9	2.3	0.15	76.5	15.02
10	2.2	0.10	51	10.01

ANS $P = 70.1\ lb$

---

3. 69

Solution L_0= 2 in d_0= 1/2 in T_i = ? P = ?

The strain calculation is shown in the table below.

The stress σ_i in the for segments 8,9, and 10 is found using the equation $\sigma_i = 1020\varepsilon_i - 1020\varepsilon_i^2$, as the strains in these segments is less than 0.5. For segments 1 through 7, the stress is 255 psi as the strain is greater than 0.5.Multiplying the column of strain by E = 510 psi, we obtain the column of stress. The area of each segment A_i is $A_i = (\pi 0.5^2)/4 = 0.19635\ in^2$. The internal tension T_i in the ith segment is given by: $T_i = \sigma_i A_i$. Multiplying the column of stress by the area A_i we obtain the column of internal tension. The force P exerted on the car carrier by the cord is given by the tension in segment 1.

Segment Number	Deformed Length. ΔL_i (inches)	Strain ε_i	Stress σ_i (psi)	Internal Tension T_i (lbs)
1	3.4	0.70	255.00	50.07
2	3.4	0.70	255.00	50.07
3	3.4	0.70	255.00	50.07
4	3.4	0.70	255.00	50.07
5	3.4	0.70	255.00	50.07
6	3.4	0.70	255.00	50.07
7	3.1	0.55	255.00	50.07
8	2.7	0.35	232.05	45.56
9	2.3	0.15	130.05	25.54
10	2.2	0.10	91.80	18.02

ANS $P = 50.1\ lbs$

---

3. 70

Solution L_0= 2 in d_0= 1/2 in ν=1/2 T_i = ? P = ?

The longitudinal strain ε_i and the corresponding stress σ_i are calculated as in the example and are shown in the table below. The transverse strain is $-\nu\varepsilon_i$. The change in diameter is thus $-\nu\varepsilon_i d_0$. The deformed diameter d_i of the ith segment is

$d_i = d_0 - \nu\varepsilon_i d_0$. Substituting the values of ν=1/2 and d_o= 1/2 in. the entries in column five can be found using $d_i = (1-0.5\varepsilon_i)/2$. The area of each segment A_i is $A_i = \pi d_i^2/4$. The internal tension T_i in the ith segment is given by: $T_i = \sigma_i A_i$. Multiplying the column of stress by the area A_i we obtain the column of internal tension.The force P exerted on the car carrier by the cord is given by the tension in segment 1

Segment Number	Deformed Length. ΔL_i (inches)	Strain ε_i	Stress σ_i (psi)	Deformed Diameter d_i (inches)	Internal Tension T_i (lbs)
1	3.4	0.70	255.00	0.33	21.15
2	3.4	0.70	255.00	0.33	21.15
3	3.4	0.70	255.00	0.33	21 15
4	3.4	0.70	255.00	0.33	21.15
5	3.4	0.70	255.00	0.33	21.15
6	3.4	0.70	255.00	0.33	21.15
7	3.1	0.55	255.00	0.36	26.32
8	2.7	0.35	232.05	0.41	31.01
9	2.3	0.15	130.05	0.46	21.85
10	2.2	0.10	91.80	0.48	16.27

ANS $P = 21.2$ lbs

Sections 3.4-3.6

3. 71

Solution

The Generalized Hooke's Law for isotropic material in cylindrical coordinates (r, θ, z) can be written as shown below.

ANS
$$\varepsilon_{rr} = [\sigma_{rr} - \nu(\sigma_{\theta\theta} + \sigma_{zz})]/E \qquad \gamma_{r\theta} = \tau_{r\theta}/G.$$
$$\varepsilon_{\theta\theta} = [\sigma_{\theta\theta} - \nu(\sigma_{zz} + \sigma_{rr})]/E \qquad \gamma_{\theta z} = \tau_{\theta z}/G$$
$$\varepsilon_{zz} = [\sigma_{zz} - \nu(\sigma_{rr} + \sigma_{\theta\theta})]/E \qquad \gamma_{zr} = \tau_{zr}/G$$

---

3. 72

Solution

The Generalized Hooke's Law for isotropic material in cylindrical coordinates (r, θ, ϕ) can be written as shown below.

ANS
$$\varepsilon_{rr} = [\sigma_{rr} - \nu(\sigma_{\theta\theta} + \sigma_{\phi\phi})]/E \qquad \gamma_{r\theta} = \tau_{r\theta}/G$$
$$\varepsilon_{\theta\theta} = [\sigma_{\theta\theta} - \nu(\sigma_{\phi\phi} + \sigma_{rr})]/E \qquad \gamma_{\theta\phi} = \tau_{\theta\phi}/G$$
$$\varepsilon_{\phi\phi} = [\sigma_{\phi\phi} - \nu(\sigma_{rr} + \sigma_{\theta\theta})]/E \qquad \gamma_{\phi r} = \tau_{\phi r}/G$$

---

3. 73

Solution ε_{xx}=? ε_{yy}=? γ_{xy}=? ε_{zz}=? σ_{zz}=? $E = 200\ GPa$ $\nu = 0.32$ (a) plane stress (b) plane strain

$$G = E/[2(1+\nu)] = 200/[2(1.32)] = 75.76 GPa \qquad (1)$$

(a) Plane Stress: $\sigma_{zz} = 0$

$$\varepsilon_{xx} = \frac{\sigma_{xx} - \nu(\sigma_{yy} + \sigma_{zz})}{E} = \frac{[100 - 0.32(150)](10^6)}{200(10^9)} = \frac{52(10^6)}{200(10^9)} = 0.26(10^{-3}) \qquad (2)$$

$$\varepsilon_{yy} = \frac{\sigma_{yy} - \nu(\sigma_{xx})}{E} = \frac{[150 - 0.32(100)](10^6)}{200(10^9)} = \frac{118(10^6)}{200(10^9)} = 0.59(10^{-3}) \qquad (3)$$

$$\gamma_{xy} = \frac{\tau_{xy}}{G} = \frac{-125(10^6)}{75.76(10^9)} = -1.65(10^{-3}) \qquad \varepsilon_{zz} = \frac{\sigma_{zz} - \nu(\sigma_{xx} + \sigma_{yy})}{E} = -\frac{0.32(250)(10^6)}{200(10^9)} = \frac{80(10^6)}{200(10^9)} = -0.4(10^{-3}) \qquad (4)$$

ANS $\sigma_{zz} = 0$; $\varepsilon_{xx} = 260\mu$; $\varepsilon_{yy} = 590\mu$; $\gamma_{xy} = -1650\mu$ $\varepsilon_{zz} = -400\mu$

(b) Plane Strain: $\varepsilon_{zz} = 0$

$$\varepsilon_{zz} = [\sigma_{zz} - \nu(\sigma_{xx} + \sigma_{yy})]/E = 0 \quad or \quad \sigma_{zz} = \nu(\sigma_{xx} + \sigma_{yy}) = 0.32(250)(10^6) = 80(10^6)\ N/m^2 \qquad (5)$$

$$\varepsilon_{xx} = \frac{\sigma_{xx} - \nu(\sigma_{yy} + \sigma_{zz})}{E} = \frac{[100 - 0.32(230)](10^6)}{200(10^9)} = 0.132(10^{-3}) \quad (6)$$

$$\varepsilon_{yy} = \frac{\sigma_{yy} - \nu(\sigma_{xx} + \sigma_{zz})}{E} = \frac{[150 - 0.32(180)](10^6)}{200(10^9)} = 0.462(10^{-3}) \quad (7)$$

ANS $\varepsilon_{zz} = 0$; $\sigma_{zz} = 80$ MPa (T) ; $\varepsilon_{xx} = 132\mu$; $\varepsilon_{yy} = 462\mu$; $\gamma_{xy} = -1650\mu$

---

3. 74

Solution ε_{xx}=? ε_{yy}=? γ_{xy}=? ε_{zz}=? σ_{zz}=? $E = 70\ GPa$ $G = 28\ GPa$ (a) plane stress (b) plane strain

$$G = E/[2(1+\nu)] \quad or \quad \nu = E/(2G) - 1 = 70(10^9)/[(2)28(10^9)] - 1 = 0.25 \quad (1)$$

(a) Plane Stress: $\sigma_{zz} = 0$

$$\varepsilon_{xx} = \frac{\sigma_{xx} - \nu(\sigma_{yy} + \sigma_{zz})}{E} = \frac{[-225 - 0.25(125)](10^6)}{70(10^9)} = -3.661(10^{-3}) \quad (2)$$

$$\varepsilon_{yy} = \frac{\sigma_{yy} - \nu(\sigma_{xx})}{E} = \frac{[125 - 0.25(-225)](10^6)}{70(10^9)} = 2.589(10^{-3}) \quad (3)$$

$$\gamma_{xy} = \frac{\tau_{xy}}{G} = \frac{150(10^6)}{28(10^9)} = 5.357(10^{-3}) \qquad \varepsilon_{zz} = \frac{\sigma_{zz} - \nu(\sigma_{xx} + \sigma_{yy})}{E} = -\frac{0.25(-225 + 125)(10^6)}{70(10^9)} = 0.357(10^{-3}) \quad (4)$$

ANS $\sigma_{zz} = 0$; $\varepsilon_{xx} = -3661\mu$; $\varepsilon_{yy} = 2589\mu$; $\gamma_{xy} = 5357\mu rad$; $\varepsilon_{zz} = 357\mu$

(b) Plane Strain: $\varepsilon_{zz} = 0$

$$\varepsilon_{zz} = [\sigma_{zz} - \nu(\sigma_{xx} + \sigma_{yy})]/E = 0 \quad or \quad \sigma_{zz} = \nu(\sigma_{xx} + \sigma_{yy}) = 0.25(-225 + 125)(10^6) = -25(10^6)\ N/m^2 \quad (5)$$

$$\varepsilon_{xx} = \frac{\sigma_{xx} - \nu(\sigma_{yy} + \sigma_{zz})}{E} = \frac{[-225 - 0.25(125 - 25)](10^6)}{70(10^9)} = -3.571(10^{-3}) \quad (6)$$

$$\varepsilon_{yy} = \frac{\sigma_{yy} - \nu(\sigma_{xx} + \sigma_{zz})}{E} = \frac{[125 - 0.25(-225 - 25)](10^6)}{70(10^9)} = 2.679(10^{-3}) \quad (7)$$

ANS $\varepsilon_{zz} = 0$; $\sigma_{zz} = 25MPa(C)$; $\varepsilon_{xx} = -3571\mu$; $\varepsilon_{yy} = 2679\mu$; $\gamma_{xy} = 5357\mu rad$

---

3. 75

Solution ε_{xx}=? ε_{yy}=? γ_{xy}=? ε_{zz}=? σ_{zz}=? $E = 30,000\ ksi$ $\nu = 0.3$ (a) plane stress (b) plane strain

$$G = E/[2(1+\nu)] = (30(10^6))/[2(1+0.3)] = 11.54(10^6) psi \quad (1)$$

(a) Plane Stress: $\sigma_{zz} = 0$

$$\varepsilon_{xx} = \frac{\sigma_{xx} - \nu(\sigma_{yy} + \sigma_{zz})}{E} = \frac{[-22 - 0.3(-25)](10^3)}{30(10^6)} = -0.483(10^{-3}) \quad (2)$$

$$\varepsilon_{yy} = \frac{\sigma_{yy} - \nu(\sigma_{xx})}{E} = \frac{[-25 - 0.3(-22)](10^3)}{30(10^6)} = -0.613(10^{-3}) \quad (3)$$

$$\gamma_{xy} = \frac{\tau_{xy}}{G} = \frac{-15(10^3)}{11.54(10^6)} = -1.3(10^{-3}) \qquad \varepsilon_{zz} = \frac{\sigma_{zz} - \nu(\sigma_{xx} + \sigma_{yy})}{E} = -\frac{0.3(-22 - 25)(10^3)}{30(10^6)} = 0.47(10^{-3}) \quad (4)$$

ANS $\sigma_{zz} = 0$; $\varepsilon_{xx} = -483\mu$; $\varepsilon_{yy} = -613\mu$; $\gamma_{xy} = -1300\mu$; $\varepsilon_{zz} = 470\mu$

(b) Plane Strain: $\varepsilon_{zz} = 0$

$$\varepsilon_{zz} = [\sigma_{zz} - \nu(\sigma_{xx} + \sigma_{yy})]/E = 0 \quad or \quad \sigma_{zz} = \nu(\sigma_{xx} + \sigma_{yy}) = 0.3(-22 - 25)(10^3) = -14.1(10^3)\ psi \quad (5)$$

$$\varepsilon_{xx} = \frac{\sigma_{xx} - \nu(\sigma_{yy} + \sigma_{zz})}{E} = \frac{[-22 - 0.3(-25 - 14.1)](10^3)}{30(10^6)} = -0.342(10^{-3}) \quad (6)$$

$$\varepsilon_{yy} = \frac{\sigma_{yy} - \nu(\sigma_{xx} + \sigma_{zz})}{E} = \frac{[-25 - 0.3(-22 - 14.1)](10^3)}{30(10^6)} = -0.472(10^{-3}) \quad (7)$$

ANS $\varepsilon_{zz} = 0$; $\sigma_{zz} = 14.1$ ksi (C) ; $\varepsilon_{xx} = -342\mu$; $\varepsilon_{yy} = -472\mu$; $\gamma_{xy} = -1300\mu$

---

3. 76

Solution ε_{xx}=? ε_{yy}=? γ_{xy}=? ε_{zz}=? σ_{zz}=? $E = 10,000\ ksi$ $G = 3900\ ksi$ (a) plane stress (b) plane strain

$$G = E/[2(1+\nu)] \quad or \quad \nu = E/[2G]-1 = 10000(10^3)/[(2)(3900)(10^3)]-1 = 0.282 \tag{1}$$

(a) Plane Stress: $\sigma_{zz} = 0$

$$\varepsilon_{xx} = \frac{\sigma_{xx}-\nu(\sigma_{yy}+\sigma_{zz})}{E} = \frac{[15-0.282(-12)](10^3)}{10(10^6)} = 1.838(10^{-3}) \tag{2}$$

$$\varepsilon_{yy} = \frac{\sigma_{yy}-\nu(\sigma_{xx})}{E} = \frac{[-12-0.282(-15)](10^3)}{10(10^6)} = -1.623(10^{-3}) \tag{3}$$

$$\gamma_{xy} = \frac{\tau_{xy}}{G} = \frac{-10(10^3)}{3.9(10^6)} = -2.564(10^{-3}) \qquad \varepsilon_{zz} = \frac{\sigma_{zz}-\nu(\sigma_{xx}+\sigma_{yy})}{E} = -\frac{0.282(-15-12)(10^3)}{30(10^6)} = -0.085(10^{-3}) \tag{4}$$

ANS $\sigma_{zz} = 0$; $\varepsilon_{xx} = 1838\mu$; $\varepsilon_{yy} = -1623\mu$; $\gamma_{xy} = -2564\mu$; $\varepsilon_{zz} = -85\mu$

(b) Plane Strain: $\varepsilon_{zz} = 0$

$$\varepsilon_{zz} = [\sigma_{zz}-\nu(\sigma_{xx}+\sigma_{yy})]/E = 0 \quad or \quad \sigma_{zz} = \nu(\sigma_{xx}+\sigma_{yy}) = 0.282(15-12)(10^3) = 846\ psi \tag{5}$$

$$\varepsilon_{xx} = \frac{\sigma_{xx}-\nu(\sigma_{yy}+\sigma_{zz})}{E} = \frac{[15-0.282(-12+0.846)](10^3)}{10(10^6)} = 1.815(10^{-3}) \tag{6}$$

$$\varepsilon_{yy} = \frac{\sigma_{yy}-\nu(\sigma_{xx}+\sigma_{zz})}{E} = \frac{[-12-0.282(15+0.846)](10^3)}{10(10^6)} = -1.647(10^{-3}) \tag{7}$$

ANS $\varepsilon_{zz} = 0$; $\sigma_{zz} = 846$ psi (T); $\varepsilon_{xx} = 1815\mu$; $\varepsilon_{yy} = -1647\mu$; $\gamma_{xy} = -2564\mu$

---

3. 77

Solution ε_{xx}=? ε_{yy}=? γ_{xy}=? ε_{zz}=? σ_{zz}=? $G = 15\ GPa$ $\nu = 0.2$ (a) plane stress (b) plane strain

$$G = E/[2(1+\nu)] \quad or \quad E = 2G(1+\nu) = 15(10^9)(2)(1+0.2) = 36(10^9)\ N/m^2 \tag{1}$$

(a) Plane Stress: $\sigma_{zz} = 0$

$$\varepsilon_{xx} = \frac{\sigma_{xx}-\nu(\sigma_{yy}+\sigma_{zz})}{E} = \frac{[-300-0.2(300)](10^6)}{36(10^9)} = -10(10^{-3}) \tag{2}$$

$$\varepsilon_{yy} = \frac{\sigma_{yy}-\nu(\sigma_{xx}+\sigma_{zz})}{E} = \frac{[300-0.2(-300)](10^6)}{36(10^9)} = 10(10^{-3}) \tag{3}$$

$$\gamma_{xy} = \frac{\tau_{xy}}{G} = \frac{150(10^6)}{15(10^9)} = 10(10^{-3}) \qquad \varepsilon_{zz} = \frac{\sigma_{zz}-\nu(\sigma_{xx}+\sigma_{yy})}{E} = -\frac{0.2(300-300)(10^6)}{36(10^9)} = 0 \tag{4}$$

ANS $\sigma_{zz} = 0$; $\varepsilon_{xx} = -10000\mu$; $\varepsilon_{yy} = 10000\mu$; $\gamma_{xy} = 10000\mu$; $\varepsilon_{zz} = 0$

(b) Plane Strain: $\varepsilon_{zz} = 0$

As $\varepsilon_{zz} = 0$ and $\sigma_{zz} = 0$ in part (a) the results for part (b) are same.

---

3. 78

Solution ε_{xx}=? ε_{yy}=? γ_{xy}=? ε_{zz}=? σ_{zz}=? $E = 2000\ psi$ $G = 800\ psi$ (a) plane stress (b) plane strain

$$G = E/[2(1+\nu)] \quad or \quad \nu = E/[2G]-1 = 2000/[(2)(800)]-1 = 0.25 \tag{1}$$

(a) Plane Stress: $\sigma_{zz} = 0$

$$\varepsilon_{xx} = \frac{\sigma_{xx}-\nu(\sigma_{yy}+\sigma_{zz})}{E} = \frac{(-100)-0.25(150)}{2000} = -0.06875 \qquad \varepsilon_{yy} = \frac{\sigma_{yy}-\nu(\sigma_{xx}+\sigma_{zz})}{E} = \frac{150-0.25(-100)}{2000} = 0.0875 \tag{2}$$

$$\gamma_{xy} = \frac{\tau_{xy}}{G} = \frac{100}{800} = 0.125 \qquad \varepsilon_{zz} = \frac{\sigma_{zz}-\nu(\sigma_{xx}+\sigma_{yy})}{E} = -\frac{0.25(-100+150)}{2000} = -0.00625 \tag{3}$$

ANS $\sigma_{zz} = 0$; $\varepsilon_{xx} = -0.06875$; $\varepsilon_{yy} = 0.0875$; $\gamma_{xy} = 0.125$; $\varepsilon_{zz} = -0.00625$

(b) Plane Strain: $\varepsilon_{zz} = 0$

$$\varepsilon_{zz} = [\sigma_{zz}-\nu(\sigma_{xx}+\sigma_{yy})]/E = 0 \quad or \quad \sigma_{zz} = \nu(\sigma_{xx}+\sigma_{yy}) = 0.25(-100+150) = 12.5\ psi \tag{4}$$

$$\varepsilon_{xx} = \frac{\sigma_{xx} - \nu(\sigma_{yy} + \sigma_{zz})}{E} = \frac{-100 - 0.25(150 + 12.5)}{2000} = -0.0703 \quad (5)$$

$$\varepsilon_{yy} = \frac{\sigma_{yy} - \nu(\sigma_{xx} + \sigma_{zz})}{E} = \frac{150 - 0.25(-100 + 12.5)}{2000} = 0.0430 \quad (6)$$

ANS $\varepsilon_{zz} = 0$; $\sigma_{zz} = 12.50 psi(T)$; $\varepsilon_{xx} = -0.0703$; $\varepsilon_{yy} = 0.08594$; $\gamma_{xy} = 0.125$

---

3. 79

Solution $\sigma_{xx}=?$ $\sigma_{yy}=?$ $\tau_{xy}=?$ $\sigma_{zz}=?$ $\varepsilon_{zz}=?$ $E = 200\ GPa$ $\nu = 0.32$ plane stress

$$\sigma_{xx} - \nu(\sigma_{yy} + \sigma_{zz}) = E\varepsilon_{xx} = 200(10^9)500(10^{-6}) = 100000(10^3)\ N/m^2 \quad or \quad \sigma_{xx} - \nu(\sigma_{yy} + \sigma_{zz}) = 100 MPa \quad (7)$$

$$\sigma_{yy} - \nu(\sigma_{xx} + \sigma_{zz}) = E\varepsilon_{yy} = 200(10^9)400(10^{-6}) = 80000(10^3)\ N/m^2 \quad or \quad \sigma_{yy} - \nu(\sigma_{xx} + \sigma_{zz}) = 80 MPa \quad (8)$$

$$\sigma_{xx} = 139.929 MPa \qquad \sigma_{yy} = 124.78 MPa \quad (9)$$

$$\varepsilon_{zz} = \frac{\sigma_{zz} - \nu(\sigma_{xx} + \sigma_{yy})}{E} = -\frac{0.32}{200(10^9)}(139.9 + 124.8)(10^6) = -0.423(10^{-3}) \quad (10)$$

$$G = E/(2(1+\nu)) = 15(10^9)/[2(1.32)] = 75.76(10^9) \qquad \tau_{xy} = G\gamma_{xy} = 75.76(10^9)(-300)(10^{-6}) = (-22727)(10^3) N/m^2 \quad (11)$$

ANS $\sigma_{xx} = 139.93 MPa(T)$; $\sigma_{yy} = 124.78 MPa(T)$; $\varepsilon_{zz} = -423\mu$; $\tau_{xy} = -22.7 MPa$

---

3. 80

Solution $\sigma_{xx}=?$ $\sigma_{yy}=?$ $\tau_{xy}=?$ $\sigma_{zz}=?$ $\varepsilon_{zz}=?$ $E = 70\ GPa$ $G = 28\ GPa$ plane stress

$$G = E/(2(1+\nu)) \quad or \quad \nu = E/(2G) - 1 = 70(10^9)/[(2)(28)(10^9)] - 1 = 0.25 \quad (1)$$

$$\sigma_{xx} - \nu\sigma_{yy} = E\varepsilon_{xx} = 70(10^9)(2000)(10^{-6}) = 140(10^6)\ N/m^2 \quad or \quad \sigma_{xx} - \nu\sigma_{yy} = 140 MPa \quad (2)$$

$$\sigma_{yy} - \nu\sigma_{xx} = E\varepsilon_{yy} = 70(10^9)(-1000)(10^{-6}) = -70(10^6)\ N/m^2 \quad or \quad \sigma_{yy} - \nu\sigma_{xx} = -70 MPa \quad (3)$$

$$\sigma_{xx} = 130.67\ MPa \qquad \sigma_{yy} = -37.33\ MPa \quad (4)$$

$$\varepsilon_{zz} = \frac{\sigma_{zz} - \nu(\sigma_{xx} + \sigma_{yy})}{E} = -\frac{0.25}{70(10^9)}(130.67 - 37.33)(10^6) = -0.333(10^{-3}) \quad (5)$$

$$\tau_{xy} = G\gamma_{xy} = 28(10^9)(1500)(10^{-6}) = 42(10^6) \quad (6)$$

ANS $\sigma_{zz} = 0$; $\sigma_{xx} = 130.7 MPa(T)$; $\sigma_{yy} = 37.3 MPa(C)$; $\varepsilon_{zz} = -333.3\mu$; $\tau_{xy} = 42 MPa$

---

3. 81

Solution $\sigma_{xx}=?$ $\sigma_{yy}=?$ $\tau_{xy}=?$ $\sigma_{zz}=?$ $\varepsilon_{zz}=?$ $E = 30,000\ ksi$ $\nu = 0.3$ plane stress

Plane Stress : $\sigma_{zz} = 0$

$$\sigma_{xx} - \nu(\sigma_{yy} + \sigma_{zz}) = E\varepsilon_{xx} = 30(10^3)(-800)(10^{-6}) = -24 ksi \quad or \quad \sigma_{xx} - \nu\sigma_{yy} = -24 ksi \quad (1)$$

$$\sigma_{yy} - \nu(\sigma_{xx} + \sigma_{zz}) = E\varepsilon_{yy} = 30(10^3)(-1000)(10^{-6}) = -30 ksi \quad or \quad \sigma_{yy} - \nu\sigma_{xx} = -30 ksi \quad (2)$$

$$\sigma_{xx} = -36.26 ksi \qquad \sigma_{yy} = -40.88 ksi \quad (3)$$

$$\varepsilon_{zz} = \frac{\sigma_{zz} - \nu(\sigma_{xx} + \sigma_{yy})}{E} = -\frac{0.3}{30(10^3)}(-36.36 - 40.88) = 0.771(10^{-3}) \quad (4)$$

$$G = E/[2(1+\nu)] = 30000/[2(1.3)] = 11538 ksi \qquad \tau_{xy} = G\gamma_{xy} = 11538(-500)(10^{-6}) = -5.769 \quad (5)$$

ANS $\sigma_{zz} = 0$; $\sigma_{xx} = 36.26 ksi(C)$; $\sigma_{yy} = 40.9 ksi(C)$; $\varepsilon_{zz} = 771\mu$; $\tau_{xy} = -5.77 ksi$

---

3. 82

Solution $\sigma_{xx}=?$ $\sigma_{yy}=?$ $\tau_{xy}=?$ $\sigma_{zz}=?$ $\varepsilon_{zz}=?$ $E = 10,000\ ksi$ $G = 3900\ ksi$ plane stress

$$G = E/[2(1+\nu)] \quad or \quad \nu = E/[2G] - 1 = 10000/[(2)(3900)] - 1 = 0.282 \quad (1)$$

$$\sigma_{xx} - \nu(\sigma_{yy} + \sigma_{zz}) = E\varepsilon_{xx} = 10(10^3)(1500)(10^{-6}) = 15 ksi \quad or \quad \sigma_{xx} - \nu\sigma_{yy} = 15 ksi \quad (2)$$

$$\sigma_{yy} - \nu(\sigma_{xx} + \sigma_{zz}) = E\varepsilon_{yy} = 10(10^3)(-1200)(10^{-6}) = -12 ksi \quad or \quad \sigma_{yy} - \nu\sigma_{xx} = -12 ksi \quad (3)$$

$$\sigma_{xx} = 12.619 ksi \qquad \sigma_{yy} = -8.44 ksi \quad (4)$$

$$\varepsilon_{zz} = \frac{\sigma_{zz} - \nu(\sigma_{xx} + \sigma_{yy})}{E} = -\frac{0.282}{10(10^3)}(12.619 - 8.44) = -0.1178 \qquad \tau_{xy} = G\gamma_{xy} = 3900(-1000)(10^{-6}) = -3.9ksi \qquad (5)$$

ANS $\sigma_{zz} = 0$; $\sigma_{xx} = 12.62ksi(T)$; $\sigma_{yy} = 8.44ksi(C)$; $\varepsilon_{zz} = -118\mu$; $\tau_{xy} = -3.9ksi$

---

3. 83

Solution σ_{xx}=? σ_{yy}=? τ_{xy}=? σ_{zz}=? ε_{zz}=? $G = 15\ GPa$ $\nu = 0.2$ plane stress

$$G = E/[2(1+\nu)] \quad or \quad E = 2G(1+\nu) = (2)(15)(1+0.2) = 36GPa \qquad (1)$$

$$\sigma_{xx} - \nu(\sigma_{yy} + \sigma_{zz}) = E\varepsilon_{xx} = 36(10^9)(-2000)(10^{-6}) = -72MPa \quad or \quad \sigma_{xx} - \nu(\sigma_{yy} + \sigma_{zz}) = -72MPa \qquad (2)$$

$$\sigma_{yy} - \nu(\sigma_{xx} + \sigma_{zz}) = E\varepsilon_{yy} = 36(10^9)(2000)(10^{-6}) = 72MPa \qquad \sigma_{yy} - \nu\sigma_{xx} = 72MPa \qquad (3)$$

$$\sigma_{xx} = -59.98MPa \qquad \sigma_{yy} = 60.1MPa \qquad (4)$$

$$\varepsilon_{zz} = \frac{\sigma_{zz} - \nu(\sigma_{xx} + \sigma_{yy})}{E} = -\left(\frac{0.2}{36(10^9)}\right)(-60 + 60) = 0 \qquad \tau_{xy} = G\gamma_{xy} = 15(10^9)(1200)(10^{-6}) = 18(10^6) \qquad (5)$$

ANS $\sigma_{zz} = 0$; $\sigma_{xx} = 60MPa(C)$; $\sigma_{yy} = 60MPa(T)$; $\varepsilon_{zz} = 0$; $\tau_{xy} = 18MPa$

---

3. 84

Solution σ_{xx}=? σ_{yy}=? τ_{xy}=? σ_{zz}=? ε_{zz}=? $E = 2000\ psi$ $G = 800\ psi$ plane stress

$$G = E/[2(1+\nu)] \quad or \quad \nu = E/[2G] - 1 = 2000/[(2)(800)] - 1 = 0.25 \qquad (1)$$

Assuming Plane Stress: $\sigma_{zz} = 0$

$$\sigma_{xx} - \nu(\sigma_{yy} + \sigma_{zz}) = E\varepsilon_{xx} = 2000(50)(10^{-6}) = 0.1psi \quad or \quad \sigma_{xx} - \nu\sigma_{yy} = 0.1psi \qquad (2)$$

$$\sigma_{yy} - \nu(\sigma_{xx} + \sigma_{zz}) = E\varepsilon_{yy} = 2000(75)(10^{-6}) = 0.15psi \quad or \quad \sigma_{yy} - \nu(\sigma_{xx} + \sigma_{zz}) = 0.15psi \qquad (3)$$

$$\sigma_{xx} = 0.1467psi \qquad \sigma_{yy} = 0.1867psi \qquad (4)$$

$$\varepsilon_{zz} = \frac{\sigma_{zz} - \nu(\sigma_{xx} + \sigma_{yy})}{E} = -\frac{0.25}{2000}(0.1467 + 0.1867) = -41.7(10^{-6}) \qquad \tau_{xy} = G\gamma_{xy} = 800(-25)(10^{-6}) = -0.02psi \qquad (5)$$

ANS $\sigma_{zz} = 0$; $\sigma_{xx} = 0.147psi(T)$; $\sigma_{yy} = 0.187psi(T)$; $\varepsilon_{zz} = -41.7\mu$; $\tau_{xy} = -0.02psi$

---

3. 85

Solution 40 mm x 25 mm E=14GPa ν=0.3 σ_{xx}=3.2MPa(C) ΔL=?

We assume $\sigma_{yy} = 0$ and $\sigma_{zz} = 0$

$$\varepsilon_{zz} = \frac{\sigma_{zz} - \nu(\sigma_{xx} + \sigma_{yy})}{E} = \frac{-\nu\sigma_{xx}}{E} = -\left[\frac{(0.3)(-3.2)(10^6)}{14(10^9)}\right] = 0.06857(10^{-3}) \qquad (1)$$

$$\Delta L = \varepsilon_{zz}L = 0.06857(10^{-3})(125) = 8.571(10^3) \qquad (2)$$

ANS $\Delta L = 0.0086mm$

---

3. 86

Solution E=30000ksi ν=0.25 σ_{xx}=? σ_{yy}=?

Plane stress: $\sigma_{zz} = 0$

$$\varepsilon_{xx} = -0.005 = -0.0005 \qquad \sigma_{xx} - \nu\sigma_{yy} = E\varepsilon_{xx} = 30000(-0.0005) = -15ksi \qquad (1)$$

$$\varepsilon_{yy} = 0 \qquad \sigma_{yy} - \nu\sigma_{xx} = 0 \qquad \text{Solving} \qquad \sigma_{xx} = -16ksi \qquad \sigma_{yy} = -4ksi \qquad (2)$$

ANS $\sigma_{xx} = 16ksi(C)$; $\sigma_{yy} = 4ksi(C)$

---

3. 87

Solution E=30000ksi ν=0.25 σ_{xx}=-10 ksi σ_{yy}=? δ_x=?

Plane stress: $\sigma_{zz} = 0$

$$\varepsilon_{yy} = 0 \quad or \quad \sigma_{yy} - \nu\sigma_{xx} = 0 \quad or \quad \sigma_{yy} = \nu\sigma_{xx} = 0.25(-10) = -2.5ksi \qquad (1)$$

$$\varepsilon_{xx} = \frac{\sigma_{xx} - \nu\sigma_{yy}}{E} = \frac{(-10) - (0.25)(-2.5)}{30,000} = -0.3125(10^{-3}) \qquad \delta_x = \varepsilon_{xx}(10) = -3.125(10^{-3}) \quad (2)$$

ANS $\sigma_{yy} = 2.5 ksi(C)$; $\delta_x = -0.0031 in$

---

3. 88

Solution E_R=300 psi $\nu_R = 0.5$ d_R =4 in d_S =4.1 P= ?

Just closes implies: $\sigma_{yy} = \sigma_{zz} = 0$. The normal strain in radial direction (y & z) is:

$$\varepsilon_{yy} = \varepsilon_{zz} = (d_S - d_R)/d_R = (4.1 - 4.0)/4.0 = 0.025 \quad (1)$$

$$\varepsilon_{yy} = \frac{-\nu\sigma_{xx}}{E} \quad or \quad \sigma_{xx} = -\frac{E\varepsilon_{yy}}{\nu} = -\frac{(300)(0.025)}{0.5} = -15 \text{ psi} = \frac{-N}{A} = -\frac{P}{(\pi/4)(4^2)} \qquad P = 188.49 \text{ lb} \quad (2)$$

ANS $P = 188.5$ lb

---

3. 89

Solution E_R= 2.1GPa $\nu_R = 0.5$ d_S = 204 mm P = 10 kN σ_{yy}= ? σ_{zz}= ?

The compressive internal force is equal to applied force. We obtain:

$$\sigma_{xx} = -N/A = -10(10^3)/[(\pi/4)(0.02^2)] = -31.83(10^6) \text{ N/m}^2 \quad (1)$$

$$\varepsilon_{yy} = \varepsilon_{zz} = (d_S - d_R)/d_R = (204 - 200)/200 = 0.02 \qquad \sigma_{zz} = \sigma_{yy} \quad (2)$$

$$\varepsilon_{yy} = \frac{\sigma_{yy} - \nu(\sigma_{xx} + \sigma_{zz})}{E} = \frac{\sigma_{yy}(1-\nu) - \nu\sigma_{xx}}{E} = \frac{0.5\sigma_{yy} - 0.5[-31.83(10^6)]}{2.1(10^9)} = 0.02 \quad or \quad \sigma_{yy} = 52.17(10^6) \text{ N}/m^2 \quad (3)$$

ANS $\sigma_{yy} = \sigma_{zz} = 52.17$ MPa(T)

---

3. 90

Solution E=10000ksi ν=0.25 a = major axis=? b = minor axis=?

We are given: $\sigma_{xx} = 20ksi$ and $\sigma_{yy} = -10ksi$ and Plane stress $\sigma_{zz} = 0$

$$\varepsilon_{xx} = \frac{\sigma_{xx} - \nu(\sigma_{yy} + \sigma_{zz})}{E} = \frac{20 - 0.25(-10)}{10000} = 2.25(10^{-3}) \qquad \varepsilon_{yy} = \frac{\sigma_{yy} - \nu(\sigma_{xx} + \sigma_{zz})}{E} = \frac{-10 - 0.25(20)}{10000} = -1.5(10^{-3}) \quad (1)$$

The circle extends in x direction and contracts in y direction

$$\Delta a = \varepsilon_{xx}(2) = 4.5(10^{-3}) \qquad a = 2 + \Delta a = 2.0045in \qquad \Delta b = \varepsilon_{yy}(2) = -3(10^{-3}) \qquad b = 2 + \Delta b = 1.9970in \quad (2)$$

ANS $a = 2.0045in$; $b = 1.9770in$

---

3. 91

Solution E=10000ksi ν=0.25 a = major axis =? b = minor axis =?

We are given: $\sigma_{xx} = -20ksi$ and $\sigma_{yy} = -10ksi$ and plane stress $\sigma_{zz} = 0$

$$\varepsilon_{xx} = \frac{\sigma_{xx} - \nu(\sigma_{yy} + \sigma_{zz})}{E} = \frac{-20 - 0.25(-10)}{10000} = -1.75(10^{-3}) \qquad \varepsilon_{yy} = \frac{\sigma_{yy} - \nu(\sigma_{xx} + \sigma_{zz})}{E} = \frac{-10 - 0.25(-20)}{10000} = -0.5(10^{-3}) \quad (1)$$

The circle contracts more in x direction than in y direction

$$\Delta a = \varepsilon_{xx}(2) = -3.5(10^{-3}) \qquad a = 2 + \Delta a = 1.9965in \qquad \Delta b = \varepsilon_{yy}(2) = -1(10^{-3}) \qquad b = 2 + \Delta b = 1.999in \quad (2)$$

ANS $a = 1.9965in$; $b = 1.999in$

---

3. 92

Solution E=70GPa ν=0.25 a= major axis =? b = minor axis =?

We are given: $\sigma_{xx} = 154MPa$ and $\sigma_{yy} = 280MPa$ and plane stress $\sigma_{zz} = 0$

$$\varepsilon_{xx} = \frac{\sigma_{xx} - \nu(\sigma_{yy} + \sigma_{zz})}{E} = \frac{[154-0.25(280)](10^6)}{70(10^9)} = 1.2(10^{-3}) \quad (1)$$

$$\varepsilon_{yy} = \frac{\sigma_{yy} - \nu(\sigma_{xx} + \sigma_{zz})}{E} = \frac{[280-0.25(154)](10^6)}{70(10^9)} = 3.45(10^{-3}) \quad (2)$$

The circle elongates more in x direction than in y direction

$\Delta a = \varepsilon_{xx}(50) = 60(10^{-3})$ $\quad a = 50 + \Delta a = 50.06mm$ $\quad \Delta b = \varepsilon_{yy}(50) = 172.5(10^{-3})$ $\quad b = 50 + \Delta b = 50.1725mm$ **(3)**

ANS $a = 50.06mm$; $b = 50.1725mm$

---

3. 93

Solution E= 10,000 ksi, ν = 0.25 σ_{xx}=? σ_{yy}=? σ_{xy}=? plane stress.

The average strains can be found as shown below.

$$\varepsilon_{xx} = \frac{\Delta u}{\Delta x} = \frac{0.0036}{3} = 0.0012 \qquad \varepsilon_{yy} = \frac{\Delta v}{\Delta y} = \frac{0.0035}{1.4} = 0.0025 \qquad \gamma_{xy} = \frac{\Delta u}{\Delta y} + \frac{\Delta v}{\Delta x} = \frac{0.0042}{1.4} + \frac{0.0048}{3} = 0.0046 \quad (1)$$

$$G = E/[2(1+\nu)] = 10000/[2(1.25)] = 4000ksi \qquad \tau_{xy} = G\gamma_{xy} = (4000)(0.0046) = 18.4ksi \quad (2)$$

$$\sigma_{xx} - \nu\sigma_{yy} = E\varepsilon_{xx} = (10000)(0.0012) = 12ksi \quad or \quad \sigma_{xx} - \nu\sigma_{yy} = 12ksi \quad (3)$$

$$\sigma_{yy} - \nu\sigma_{xx} = E\varepsilon_{yy} = (10000)(0.0025) = 25ksi \quad or \quad \sigma_{yy} - \nu\sigma_{xx} = 50ksi \quad (4)$$

$$\sigma_{xx} = 19.47ksi \qquad \sigma_{yy} = 29.87ksi \quad (5)$$

ANS $\sigma_{xx} = 19.47ksi(T)$; $\sigma_{yy} = 29.87ksi(T)$; $\tau_{xy} = 18.4ksi$

---

3. 94

Solution E = 210 GPa ν = 0.28 σ_{xx}=? σ_{yy}=? σ_{xy}=? plane stress

The average strains can be found as shown below.

$$\varepsilon_{xx} = \frac{\Delta u}{\Delta x} = \frac{-0.075}{250} = -0.0003 \qquad \varepsilon_{yy} = \frac{\Delta v}{\Delta y} = \frac{-0.6}{450} = -0.000133 \qquad \gamma_{xy} = \frac{\Delta u}{\Delta y} + \frac{\Delta v}{\Delta x} = \frac{0.09}{450} + \frac{0.1}{250} = 0.0006 \quad (1)$$

$$G = E/[2(1+\nu)] = 210/[2(1.28)] = 82.03GPa \qquad \tau_{xy} = G\gamma_{xy} = (82.03)(10^9)(0.0006) = 49.2MPa \quad (2)$$

$$\sigma_{xx} - \nu\sigma_{yy} = E\varepsilon_{xx} = (210)(10^9)(-0.0003) = -63MPa \qquad \sigma_{xx} - \nu\sigma_{yy} = -63MPa \quad (3)$$

$$\sigma_{yy} - \nu\sigma_{xx} = E\varepsilon_{yy} = (210)(10^9)(-0.000133) = -28MPa \qquad \sigma_{yy} - \nu\sigma_{xx} = -28MPa \quad (4)$$

$$\sigma_{xx} = -76.87MPa \qquad \sigma_{yy} = -49.52MPa \quad (5)$$

ANS $\sigma_{xx} = 76.87MPa(C)$; $\sigma_{yy} = 49.52MPa(C)$; $\tau_{xy} = 49.2MPa$

---

3. 95

Solution d= 5-ft E = 30,000 ksi ν = 0.2 t = 3/4 in p = 600 psi Δd = ?

The radial stress is approximated as zero. and the hoop stress and strain found as shown below.

$$\sigma_{\theta\theta} = \sigma_{\phi\phi} = \frac{pR}{t} = \frac{(600)(30)}{(3/4)} = 24000 \text{ psi} = 24 \text{ ksi} \qquad \varepsilon_{\theta\theta} = \frac{\sigma_{\theta\theta} - \nu(\sigma_{\phi\phi} + \sigma_{rr})}{E} = \frac{24 - 0.28(24)}{30000} = 0.576(10^{-3}) \quad (1)$$

$$\varepsilon_{\theta\theta} = \frac{\pi\Delta d}{\pi d} = \frac{\Delta d}{60} = 0.576(10^{-3}) \quad or \quad \Delta d = 0.0346 \text{ in} \quad (2)$$

ANS $\Delta d = 0.0346$ in

---

3. 96

Solution E = 210 GPa ν = 0.28 d= 1 m t = 10 mm p = 250 kPa Δd = ?

The hoop and axial stress can be found as shown below.

$$\sigma_{\theta\theta} = \frac{pR}{t} = \frac{(250)(10^3)(0.5)}{(0.01)} = 12.5(10^6)\text{ N}/m^2 \qquad \sigma_{xx} = \frac{pR}{2t} = \frac{(250)(10^3)(0.5)}{2(0.01)} = 6.25(10^6)\text{ N}/m^2 \quad (1)$$

The radial stress is approximated as zero. We determine the hoop strain and change in diameter as shown below.

$$\varepsilon_{\theta\theta} = \frac{\sigma_{\theta\theta} - \nu(\sigma_{xx} + \sigma_{rr})}{E} = \frac{[12.5 - 0.28(6.25)](10^6)}{210(10^9)} = 51.19(10^{-6}) = \frac{\pi\Delta d}{\pi d} = \frac{\Delta d}{1} \quad or \quad \Delta d = 51.19(10^{-6})\text{ m} \quad (2)$$

ANS $\Delta d = 0.052$ mm

---

3. 97

Solution

Adding the three equations for normal strains in the Generalized Hooke's Law.

$$\varepsilon_{xx} + \varepsilon_{yy} + \varepsilon_{zz} = \frac{1-2\nu}{E}(\sigma_{xx} + \sigma_{yy} + \sigma_{zz}) \quad or \quad (\sigma_{xx} + \sigma_{yy} + \sigma_{zz}) = \left(\frac{E}{1-2\nu}\right)(\varepsilon_{xx} + \varepsilon_{yy} + \varepsilon_{zz}) \quad (1)$$

Substituting in Generalized Hooke's law we obtain the following.

$$\varepsilon_{xx} = \frac{\sigma_{xx}}{E} - \frac{\nu}{E}(\sigma_{yy} + \sigma_{zz}) = \sigma_{xx}\left(\frac{1+\nu}{E}\right) - \frac{\nu}{E}(\sigma_{xx} + \sigma_{yy} + \sigma_{zz}) = \sigma_{xx}\left(\frac{1+\nu}{E}\right) - \left(\frac{\nu}{E}\right)\left(\frac{E}{1-2\nu}\right)(\varepsilon_{xx} + \varepsilon_{yy} + \varepsilon_{zz}) \text{ or}$$

$$\sigma_{xx}\left(\frac{1+\nu}{E}\right) = \varepsilon_{xx}\left(1 + \frac{\nu}{1-2\nu}\right) + \frac{\nu}{1-2\nu}\varepsilon_{yy} + \frac{\nu}{1-2\nu}\varepsilon_{zz} \quad or \quad \sigma_{xx} = \frac{E}{(1+\nu)(1-2\nu)}[\varepsilon_{xx}(1-\nu) + \nu\varepsilon_{yy} + \nu\varepsilon_{zz}] \quad \text{(2)}$$

$$\varepsilon_{yy} = \frac{\sigma_{yy}}{E} - \frac{\nu}{E}(\sigma_{xx} + \sigma_{zz}) = \sigma_{yy}\left(\frac{1+\nu}{E}\right) - \frac{\nu}{E}(\sigma_{xx} + \sigma_{yy} + \sigma_{zz}) = \sigma_{yy}\left(\frac{1+\nu}{E}\right) - \left(\frac{\nu}{E}\right)\left(\frac{E}{1-2\nu}\right)(\varepsilon_{xx} + \varepsilon_{yy} + \varepsilon_{zz}) \text{ or}$$

$$\sigma_{yy}\left(\frac{1+\nu}{E}\right) = \frac{\nu}{1-2\nu}\varepsilon_{xx} + \varepsilon_{yy}\left(1 + \frac{\nu}{1-2\nu}\right) + \frac{\nu}{1-2\nu}\varepsilon_{zz} \quad \sigma_{yy} = \frac{E}{(1+\nu)(1-2\nu)}[\nu\varepsilon_{xx} + (1-\nu)\varepsilon_{yy} + \nu\varepsilon_{zz}] \quad \text{(3)}$$

$$\varepsilon_{zz} = \frac{\sigma_{zz}}{E} - \frac{\nu}{E}(\sigma_{xx} + \sigma_{yy}) = \sigma_{zz}\left(\frac{1+\nu}{E}\right) - \frac{\nu}{E}(\sigma_{xx} + \sigma_{yy} + \sigma_{zz}) = \sigma_{zz}\left(\frac{1+\nu}{E}\right) - \left(\frac{\nu}{E}\right)\left(\frac{E}{1-2\nu}\right)(\varepsilon_{xx} + \varepsilon_{yy} + \varepsilon_{zz}) \quad \text{(4)}$$

Above are the desired results.

---

3. 98

Solution

Plane Stress: $\sigma_{zz} = 0$

From generalized Hooke's Law.

$$\varepsilon_{xx} = \frac{\sigma_{xx}}{E} - \frac{\nu}{E}(\sigma_{yy} + \sigma_{zz}) \quad or \quad \sigma_{xx} - \nu\sigma_{yy} = E\varepsilon_{xx} \qquad \varepsilon_{yy} = \frac{\sigma_{yy}}{E} - \frac{\nu}{E}(\sigma_{xx} + \sigma_{zz}) \quad or \quad \sigma_{yy} - \nu\sigma_{xx} = E\varepsilon_{yy} \quad \text{(1)}$$

Multiplying by ν and adding to the other we obtain:

$$\sigma_{xx}(1-\nu^2) = E(\varepsilon_{xx} + \nu\varepsilon_{yy}) \quad or \quad \sigma_{xx} = \frac{E}{(1-\nu^2)}(\varepsilon_{xx} + \nu\varepsilon_{yy}) \quad \text{(2)}$$

$$\sigma_{yy}(1-\nu^2) = E(\varepsilon_{yy} + \nu\varepsilon_{xx}) \quad \sigma_{yy} = \frac{E}{(1-\nu^2)}(\varepsilon_{yy} + \nu\varepsilon_{xx}) \quad \text{(3)}$$

Above are the desired results.

---

3. 99

Solution

Plane Stress : $\sigma_{zz} = 0$

Adding the stress results in previous problem we obtain:

$$\sigma_{xx}(1-\nu) + \sigma_{yy}(1-\nu) = E(\varepsilon_{xx} + \varepsilon_{yy}) \quad or \quad \sigma_{xx} + \sigma_{yy} = \frac{E}{(1-\nu)}(\varepsilon_{xx} + \varepsilon_{yy}) \quad \text{(1)}$$

$$\varepsilon_{zz} = \frac{\sigma_{zz}}{E} - \frac{\nu}{E}(\sigma_{xx} + \sigma_{yy}) = 0 - \left(\frac{\nu}{E}\right)\left(\frac{E}{(1-\nu)}\right)(\varepsilon_{xx} + \varepsilon_{yy}) \quad or \quad \varepsilon_{zz} = -\left(\frac{\nu}{(1-\nu)}\right)(\varepsilon_{xx} + \varepsilon_{yy}) \quad \text{(2)}$$

Above are the desired results.

---

3. 100

Solution σ_{xx}=? σ_{yy}=? and ε_{zz}=?

Substituting $\varepsilon_{xx} = 500(10^{-6})$, $\varepsilon_{yy} = 400(10^{-6})$, $E = 200(10^9)\ N/m^2$ and $\nu = 0.32$ into the given equations we obtain:

$$\sigma_{xx} = [500 + (0.32)(400)]\frac{(10^{-6})200(10^9)}{1-0.32^2} = 139.9(10^6)\frac{N}{m^2} \qquad \sigma_{yy} = [400 + (0.32)(500)]\frac{(10^{-6})200(10^9)}{1-0.32^2} = 124.77(10^6)\frac{N}{m^2} \quad \text{(1)}$$

$$\varepsilon_{zz} = -\frac{0.25}{1-0.25}(500 + 400)(10^{-6}) = -423.5(10^{-6}) \quad \text{(2)}$$

ANS $\sigma_{xx} = 139.9MPa(T)$; $\sigma_{yy} = 124.8MPa(T)$; $\varepsilon_{zz} = -423.5\mu$

---

3. 101

Solution σ_{xx}=? σ_{yy}=? and ε_{zz}=?

Substituting $\varepsilon_{xx} = -3000(10^{-6})$, $\varepsilon_{yy} = 1500(10^{-6})$, $E = 70(10^9)\ N/m^2$ and $\nu = 0.25$ into the given equations we obtain:

$$\sigma_{xx} = [-3000 + (0.25)(1500)]\frac{(10^{-6})70(10^9)}{1-0.25^2} = -196(10^6)\frac{N}{m^2} \qquad \sigma_{yy} = [1500 + (0.25)(-3000)]\frac{(10^{-6})70(10^9)}{1-0.25^2} = 56(10^6)\frac{N}{m^2} \quad \text{(1)}$$

$$\varepsilon_{zz} = -\frac{0.25}{1-0.25}(-3000+1500)(10^{-6}) = 500(10^{-6}) \quad \textbf{(2)}$$

ANS $\sigma_{xx} = 196MPa(C)$; $\sigma_{yy} = 56MPa(T)$; $\varepsilon_{zz} = 500\mu$

---

3. 102

Solution σ_{xx}=? σ_{yy}=? and ε_{zz}=?

Substituting $\varepsilon_{xx} = -800(10^{-6})$, $\varepsilon_{yy} = -1000(10^{-6})$, $E = 30000ksi$ and $\nu = 0.3$ into the given equations we obtain:

$$\sigma_{xx} = [-800+(0.3)(-1000)]\frac{(10^{-6})30000}{1-0.3^2} = -36.26ksi \qquad \sigma_{yy} = [-1000+(0.3)(-800)]\frac{(10^{-6})30000}{1-0.3^2} - -40.88ksi \quad \textbf{(1)}$$

$$\varepsilon_{zz} = -\frac{0.3}{1-0.3}(-800-1000)(10^{-6}) = 771.4(10^{-6}) \quad \textbf{(2)}$$

ANS $\sigma_{xx} = 36.3ksi(C)$; $\sigma_{yy} = 40.9ksi(C)$; $\varepsilon_{zz} = 771.4\mu$

---

3. 103

Solution σ_{xx}=? σ_{yy}=? and ε_{zz}=?

Substituting $\varepsilon_{xx} = 1500(10^{-6})$, $\varepsilon_{yy} = -1200(10^{-6})$, $E = 10000\ ksi$ and $\nu = 0.282$ into the given equations we obtain:

$$\sigma_{xx} = [1500+(0.282)(-1200)]\frac{(10^{-6})10000}{1-0.282^2} = 12.62ksi \qquad \sigma_{yy} = [-1200+(0.282)(1500)]\frac{(10^{-6})10000}{1-0.282^2} = -8.44ksi \quad \textbf{(1)}$$

$$\varepsilon_{zz} = -\frac{0.282}{1-0.282}(1500+(-1200))(10^{-6}) = -117.8(10^{-6}) \quad \textbf{(2)}$$

ANS $\sigma_{xx} = 12.6ksi(T)$; $\sigma_{yy} = 8.44ksi(C)$; $\varepsilon_{zz} = -118\mu$

---

3. 104

Solution σ_{xx}=? σ_{yy}=? and ε_{zz}=?

Substituting $\varepsilon_{xx} = -2000(10^{-6})$, $\varepsilon_{yy} = 2000(10^{-6})$, $E = 36(10^9)N/m^2$ and $\nu = 0.2$ into the given equations we obtain:

$$\sigma_{xx} = [-2000+(0.2)(2000)]\frac{(10^{-6})(36)(10^9)}{1-0.2^2} = -60(10^6)\frac{N}{m^2} \qquad \sigma_{yy} = [2000+(0.2)(-2000)]\frac{(10^{-6})(36)(10^9)}{1-0.2^2} = 60(10^6)\frac{N}{m^2} \quad \textbf{(1)}$$

$$\varepsilon_{zz} = -\frac{0.2}{1-0.2}(-2000+2000)(10^{-6}) = 0 \quad \textbf{(2)}$$

ANS $\sigma_{xx} = 60MPa(C)$; $\sigma_{yy} = 60MPa(T)$; $\varepsilon_{zz} = 0$

---

3. 105

Solution σ_{xx}=? σ_{yy}=? and ε_{zz}=?

Substituting $\varepsilon_{xx} = 50(10^{-6})$, $\varepsilon_{yy} = 75(10^{-6})$, $E = 2000psi$ and $\nu = 0.25$ into the given equations we obtain:

$$\sigma_{xx} = [50+(0.25)(75)]\frac{(10^{-6})2000}{1-0.25^2} = 0.1467psi \qquad \sigma_{yy} = [75+(0.25)(50)]\frac{(10^{-6})2000}{1-0.25^2} = 0.187psi \quad \textbf{(1)}$$

$$\varepsilon_{zz} = -\frac{0.25}{1-0.25}(50+75)(10^{-6}) = -41.67(10^{-6}) \quad \textbf{(2)}$$

ANS $\sigma_{xx} = 0.147psi(T)$; $\sigma_{yy} = 0.187psi(T)$; $\varepsilon_{zz} = -41.67\mu$

---

3. 106

Solution

Plane strain: $\varepsilon_{zz} = 0$. From Generalized Hooke's Law:

$$\varepsilon_{zz} = [\sigma_{zz} - \nu(\sigma_{xx}+\sigma_{yy})]/E = 0 \qquad \sigma_{zz} = \nu(\sigma_{xx}+\sigma_{yy}) \quad \textbf{(1)}$$

$$\varepsilon_{xx} = \frac{\sigma_{xx}}{E} - \frac{\nu}{E}(\sigma_{yy}+\sigma_{zz}) = \frac{\sigma_{xx}}{E} - \frac{\nu}{E}(\sigma_{yy}+\nu\sigma_{xx}+\nu\sigma_{yy}) = \frac{(1-\nu^2)}{E}\sigma_{xx} - \frac{\nu(1+\nu)}{E}\sigma_{yy} = [(1-\nu)\sigma_{xx} - \nu\sigma_{yy}]\frac{(1+\nu)}{E}$$

$$\varepsilon_{yy} = \frac{\sigma_{yy}}{E} - \frac{\nu}{E}(\sigma_{xx} + \sigma_{zz}) = \frac{\sigma_{yy}}{E} - \frac{\nu}{E}(\sigma_{xx} + \nu\sigma_{xx} + \nu\sigma_{yy}) = \frac{(1-\nu^2)}{E}\sigma_{yy} - \frac{\nu(1+\nu)}{E}\sigma_{xx} = [(1-\nu)\sigma_{yy} - \nu\sigma_{xx}]\frac{(1+\nu)}{E} \text{ or}$$

Above are the desired results.

---

3. 107

Plane strain: $\varepsilon_{zz} = 0$. Multiplying one strain equation by $(1-\nu)$ and the other by ν and adding we obtain the following.

$$(1-\nu)\varepsilon_{xx} + \nu\varepsilon_{yy} = ((1-\nu)^2 - \nu^2)\sigma_{xx}\frac{(1+\nu)}{E} = (1 - 2\nu + \nu^2 - \nu^2)\sigma_{xx}\frac{(1+\nu)}{E} \text{ or}$$

$$\sigma_{xx} = [(1-\nu)\varepsilon_{xx} + \nu\varepsilon_{yy}]\frac{E}{(1-2\nu)(1+\nu)} \quad \textbf{(1)}$$

$$\nu\varepsilon_{xx} + (1-\nu)\varepsilon_{yy} = ((1-\nu)^2 - \nu^2)\sigma_{yy}\frac{(1+\nu)}{E} = (1 - 2\nu + \nu^2 - \nu^2)\sigma_{yy}\frac{(1+\nu)}{E} \text{ or}$$

$$\sigma_{yy} = [\nu\varepsilon_{xx} + (1-\nu)\varepsilon_{yy}]\frac{E}{(1-2\nu)(1+\nu)} \quad \textbf{(2)}$$

Above are the desired results.

---

3. 108

Solution

The original volume is: $V = (\Delta x)(\Delta y)(\Delta z)$. The change of volume is:

$$\Delta V = (1+\varepsilon_{xx})\Delta x(1+\varepsilon_{yy})\Delta y(1+\varepsilon_{zz})\Delta(z) - (\Delta x\Delta y\Delta z) = [1 + \varepsilon_{xx} + \varepsilon_{yy} + \varepsilon_{zz} + \varepsilon_{xx}\varepsilon_{yy} + \varepsilon_{yy}\varepsilon_{zz} + \varepsilon_{zz}\varepsilon_{xx} + \varepsilon_{xx}\varepsilon_{yy}\varepsilon_{zz} - 1](\Delta x\Delta y\Delta z) \quad \textbf{(1)}$$

For small strain, the quadratic and the cubic terms can be neglected i.e., $\Delta V = [\varepsilon_{xx} + \varepsilon_{yy} + \varepsilon_{zz}](\Delta x\Delta y\Delta z)$

$$\varepsilon_v = \Delta V/V = \varepsilon_{xx} + \varepsilon_{yy} + \varepsilon_{zz} \quad \textbf{(2)}$$

Above equation is the desired result.

---

3. 109

Solution

Adding the three equations for normal strain in generalized Hooke's law we obtain

$$\varepsilon_{xx} + \varepsilon_{yy} + \varepsilon_{zz} = (1-2\nu)(\sigma_{xx} + \sigma_{yy} + \sigma_{zz})/E \quad \textbf{(1)}$$

Substituting $\sigma_{xx} + \sigma_{yy} + \sigma_{zz} = -3p$ and $\varepsilon_{xx} + \varepsilon_{yy} + \varepsilon_{zz} = \varepsilon_v$ we obtain: $\varepsilon_v = -3(1-2\nu)(p/E)$

$$p = \frac{-E}{3(1-2\nu)}\varepsilon_v = -K\varepsilon_v \qquad where \qquad K = \frac{E}{3(1-2\nu)} \quad \textbf{(2)}$$

Above equation is the desired result.

---

3. 110

Solution $\varepsilon_{xx} = ?$ $\varepsilon_{yy} = ?$ $\gamma_{xy} = ?$

From the given stress-strain equations for orthotropic material we obtain.

$$\frac{\nu_{xy}}{E_x} = \frac{0.3}{7500} = \frac{\nu_{yx}}{E_y} \quad \textbf{(1)}$$

$$\varepsilon_{xx} = \frac{-5}{7500} - \frac{0.3}{7500}(8) = -0.9987(10^{-3}) \qquad \varepsilon_{yy} = \frac{(8)}{2500} - \frac{0.3}{7500}(-5) = 3.4(10^{-3}) \qquad \gamma_{xy} = \frac{6}{1250} = 4.8(10^{-3}) \quad \textbf{(2)}$$

ANS $\varepsilon_{xx} = -998.7\mu$; $\varepsilon_{yy} = 3400\mu$; $\gamma_{xy} = 4800\mu$

---

3. 111

Solution $\varepsilon_{xx} = ?$ $\varepsilon_{yy} = ?$ $\gamma_{xy} = ?$

From the given stress-strain equations for orthotropic material we obtain.

$$\frac{\nu_{xy}}{E_x} = \frac{0.32}{25000} = \frac{\nu_{yx}}{E_y} \quad \textbf{(1)}$$

$$\varepsilon_{xx} = \frac{(-25)}{25000} - \frac{0.32}{25000}(-5) = -0.936(10^{-3}) \qquad \varepsilon_{yy} = \frac{(-5)}{2000} - \frac{0.3}{25000}(-25) = -2.18(10^{-3}) \qquad \gamma_{xy} = \frac{-8}{1500} = -5.333(10^{-3}) \quad \textbf{(2)}$$

ANS $\varepsilon_{xx} = -936\mu$; $\varepsilon_{yy} = -2180\mu$; $\gamma_{xy} = -5333\mu$

---

3. 112

Solution $\varepsilon_{xx} = ?$ $\varepsilon_{yy} = ?$ $\gamma_{xy} = ?$

From the given stress-strain equations for orthotropic material we obtain.

$$\frac{\nu_{xy}}{E_x} = \frac{0.25}{53(10^9)} = \frac{\nu_{yx}}{E_y} \qquad \gamma_{xy} = \frac{-54(10^6)}{9(10^9)} = -6(10^{-3}) \tag{1}$$

$$\varepsilon_{xx} = \frac{(-200)(10^6)}{53(10^9)} - \frac{0.25}{53(10^9)}(-80)(10^6) = -3.396(10^{-3}) \qquad \varepsilon_{yy} = \frac{(-80)(10^6)}{18(10^9)} - \frac{0.25}{53(10^9)}(-200)(10^6) = -3.501(10^{-3}) \tag{2}$$

ANS $\varepsilon_{xx} = -3396\mu$; $\varepsilon_{yy} = -3501\mu$; $\gamma_{xy} = -6000\mu$

---

3. 113

Solution $\varepsilon_{xx} = ?$ $\varepsilon_{yy} = ?$ $\gamma_{xy} = ?$

From the given stress-strain equations for orthotropic material we obtain.

$$\frac{\nu_{xy}}{E_x} = \frac{0.28}{180(10^9)} = \frac{\nu_{yx}}{E_y} \qquad \gamma_{xy} = \frac{60(10^6)}{11(10^9)} = 5.454(10^{-3}) \tag{1}$$

$$\varepsilon_{xx} = \frac{300(10^6)}{180(10^9)} - \frac{0.28}{180(10^9)}(50)(10^6) = 1.589(10^{-3}) \qquad \varepsilon_{yy} = \frac{(50)(10^6)}{15(10^9)} - \frac{0.28}{180(10^9)}(300)(10^6) = 2.867(10^{-3}) \tag{2}$$

ANS $\varepsilon_{xx} = 1589\mu$; $\varepsilon_{yy} = 2867\mu$; $\gamma_{xy} = 5454\mu$

---

3. 114

Solution $\sigma_{xx} = ?$ $\sigma_{yy} = ?$ $\tau_{xy} = ?$

From the given stress-strain equations for orthotropic material we obtain.

$$\frac{\nu_{xy}}{E_x} = \frac{\nu_{yx}}{E_y} = \frac{0.3}{7500} = 40(10^{-6}) \qquad \frac{1}{E_x} = \frac{1}{7500} = 133.33(10^{-6}) \qquad \frac{1}{E_y} = \frac{1}{2500} = 400(10^{-6}) \tag{1}$$

Substituting the given strain values and the constants we obtain the following

$$\varepsilon_{xx} = 133.33(10^{-6})\sigma_{xx} - 40(10^{-6})\sigma_{yy} = -1000(10^{-6}) \quad or \quad 133.33\sigma_{xx} - 40\sigma_{yy} = -1000 \tag{2}$$

$$\varepsilon_{yy} = 400(10^{-6})\sigma_{yy} - 40(10^{-6})\sigma_{xx} = 500(10^{-6}) \quad or \quad 400\sigma_{yy} - 40\sigma_{xx} = 500 \tag{3}$$

$$\sigma_{xx} = -7.34ksi \qquad \sigma_{yy} = 0.515ksi \qquad \tau_{xy} = (1250)(-250)(10^{-6}) = -0.3125ksi \tag{4}$$

ANS $\sigma_{xx} = 7.34ksi(C)$; $\sigma_{yy} = 0.515ksi(T)$; $\tau_{xy} = -0.3125ksi$

---

3. 115

Solution $\sigma_{xx} = ?$ $\sigma_{yy} = ?$ $\tau_{xy} = ?$

From the given stress-strain equations for orthotropic material we obtain.

$$\frac{\nu_{xy}}{E_x} = \frac{\nu_{yx}}{E_y} = \frac{0.32}{25000} = 12.8(10^{-6}) \qquad \frac{1}{E_x} = \frac{1}{25000} = 40(10^{-6}) \qquad \frac{1}{E_y} = \frac{1}{2000} = 500(10^{-6}) \tag{1}$$

Substituting the given strain values and the constants we obtain the following

$$\varepsilon_{xx} = 40(10^{-6})\sigma_{xx} - 12.8(10^{-6})\sigma_{yy} = -750(10^{-6}) \quad or \quad 40\sigma_{xx} - 12.8\sigma_{yy} = -750$$

$$\varepsilon_{yy} = 500(10^{-6})\sigma_{yy} - 12.8(10^{-6})\sigma_{xx} = -250(10^{-6}) \quad or \quad 500\sigma_{yy} - 12.8\sigma_{xx} = -250$$

$$\sigma_{xx} = -19.07ksi \qquad \sigma_{yy} = -0.988ksi \qquad \tau_{xy} = (1500)(400)(10^{-6}) = 0.6ksi \tag{2}$$

ANS $\sigma_{xx} = 19.07ksi(C)$; $\sigma_{yy} = 0.99ksi(C)$; $\tau_{xy} = 0.6ksi$

---

3. 116

Solution $\sigma_{xx} = ?$ $\sigma_{yy} = ?$ $\tau_{xy} = ?$

From the given stress-strain equations for orthotropic material we obtain.

$$\frac{\nu_{xy}}{E_x} = \frac{\nu_{yx}}{E_y} = \frac{0.25}{53(10^9)} = 4.717(10^{-12}) \qquad \frac{1}{E_x} = \frac{1}{53(10^9)} = 18.868(10^{-12}) \qquad \frac{1}{E_y} = \frac{1}{18(10^9)} = 55.556(10^{-12}) \tag{1}$$

Substituting the given strain values and the constants, we obtain the following.

$$\tau_{xy} = 9(10^9)(600)(10^{-6}) = 5400(10^3)N/m^2 \tag{2}$$

$$\varepsilon_{xx} = 18.868(10^{-12})\sigma_{xx} - 4.717(10^{-12})\sigma_{yy} = 1500(10^{-6}) \quad or \quad 18.868\sigma_{xx} - 4.717\sigma_{yy} = 1500MPa \tag{3}$$

$$\varepsilon_{yy} = 55.556(10^{-12})\sigma_{yy} - 4.717(10^{-12})\sigma_{xx} = 800(10^{-6}) \quad or \quad 55.556\sigma_{yy} - 4.717\sigma_{xx} = 800MPa \tag{4}$$

ANS $\sigma_{xx} = 84.90MPa(T)$; $\sigma_{yy} = 21.60MPa(T)$; $\tau_{xy} = 5.4MPa$

---

3. 117

Solution σ_{xx}= ? σ_{yy} = ? τ_{xy} = ?

From the given stress-strain equations for orthotropic material we obtain.

$$\frac{\nu_{xy}}{E_x} = \frac{\nu_{yx}}{E_y} = \frac{0.28}{180(10^9)} = 1.555(10^{-12}) \qquad \frac{1}{E_x} = \frac{1}{180(10^9)} = 5.555(10^{-12}) \qquad \frac{1}{E_y} = \frac{1}{15(10^9)} = 66.667(10^{-12}) \tag{1}$$

Substituting the given strain values and the constants, we obtain the following

$$\varepsilon_{xx} = 5.555(10^{-12})\sigma_{xx} - 1.555(10^{-12})\sigma_{yy} = 1500(10^{-6}) \quad or \quad 5.555\sigma_{xx} - 1.555\sigma_{yy} = 1500MPa \tag{2}$$

$$\varepsilon_{yy} = 66.667(10^{-12})\sigma_{yy} - 1.555(10^{-12})\sigma_{xx} = -750(10^{-6}) \quad or \quad 66.667\sigma_{yy} - 1.555\sigma_{xx} = -750MPa \tag{3}$$

$$\sigma_{xx} = 268.60MPa \qquad \sigma_{yy} = -4.98MPa \qquad \tau_{xy} = 11(10^9)(-450)(10^{-6}) = -4950(10^3)N/m^2 \tag{4}$$

ANS $\sigma_{xx} = 268.6MPa(T)$; $\sigma_{yy} = 4.98MPa(C)$; $\tau_{xy} = -4.95MPa$

---

3. 118

Solution

Writing the given stress-strain equations for orthotropic material in matrix form we obtain.

$$\begin{bmatrix} \frac{1}{E_x} & -\frac{\nu_{yx}}{E_y} \\ -\frac{\nu_{xy}}{E_x} & \frac{1}{E_y} \end{bmatrix} \begin{bmatrix} \sigma_{xx} \\ \sigma_{yy} \end{bmatrix} = \begin{bmatrix} \varepsilon_{xx} \\ \varepsilon_{yy} \end{bmatrix} \tag{1}$$

Using Kramer's rule, we can write

$$D = \frac{1}{E_x E_y} - \frac{\nu_{xy}\nu_{yx}}{E_x E_y} = \frac{(1 - \nu_{xy}\nu_{yx})}{E_x E_y} \tag{2}$$

$$\sigma_{xx} = \frac{\begin{vmatrix} \varepsilon_{xx} & -\frac{\nu_{yx}}{E_y} \\ \varepsilon_{yy} & \frac{1}{E_y} \end{vmatrix}}{D} = \left(\frac{\varepsilon_{xx}}{E_y} + \frac{\nu_{yx}}{E_y}\varepsilon_{yy}\right)\frac{E_x E_y}{1 - \nu_{xy}\nu_{yx}} \qquad \sigma_{yy} = \frac{\begin{vmatrix} \frac{1}{E_x} & \varepsilon_{xx} \\ -\frac{\nu_{xy}}{E_x} & \varepsilon_{yy} \end{vmatrix}}{D} = \left(\frac{\varepsilon_{yy}}{E_x} + \frac{\nu_{xy}}{E_x}\varepsilon_{xx}\right)\frac{E_x E_y}{1 - \nu_{xy}\nu_{yx}} \tag{3}$$

ANS $\sigma_{xx} = \frac{E_x}{1 - \nu_{xy}\nu_{yx}}(\varepsilon_{xx} + \nu_{yx}\varepsilon_{yy})$; $\sigma_{yy} = \frac{E_y}{1 - \nu_{xy}\nu_{yx}}(\varepsilon_{yy} + \nu_{xy}\varepsilon_{xx})$

Sections 3.7-3.8

3. 119

Solution σ_{yield}=30ksi K=?

The nominal axial stress is: $\sigma_{nom} = 10/[(5)(0.5)] = 4ksi$

From graph, for the nominal stress and $d/h = 1/5 = 0.2$ the approximate value of stress concentration factor is

$$K_{gross} = (3 + 3.25)/2 = 3.125 \tag{1}$$

The factor of safety can be found as shown below

$$\sigma_{max} = K_{gross}\sigma_{nom} = (3.125)(4) = 12.5 \qquad K = \sigma_{yield}/\sigma_{max} = 30/12.5 = 2.4 \tag{2}$$

ANS $K = 2.4$

---

3. 120

Solution

The equation for K_{conc} can be re-written as:

$$K_{conc} = 1.970 - 0.384\left(\frac{2(r/d)}{(H/d)}\right) - 1.018\left(\frac{2(r/d)}{(H/d)}\right)^2 + 0.430\left(\frac{2(r/d)}{(H/d)}\right)^3 \tag{1}$$

Using r/d as a parameter and the above equation, the values of K_{conc} can be found for various values of H/d on a spread sheet.

Those values for which the inequality $\left(\frac{H}{d}\right) > \left(1 + 2\frac{r}{d}\right)$ does not hold can be eliminated as shown in the table below.

H/d	r/d=0.2	r/d=0.4	r/d=0.6	r/d=0.8	r/d=1.0
1.4	1.787				
1.6	1.817				
1.8	1.839	1.636			
2.0	1.856	1.681			
2.2	1.869	1.716	1.527		
2.4	1.880	1.745	1.577		
2.6	1.888	1.768	1.618	1.448	
2.8	1.896	1.787	1.652	1.498	
3.0	1.902	1.803	1.681	1.541	1.389
3.2	1.907	1.817	1.706	1.577	1.437
3.4	1.911	1.829	1.727	1.609	1.479
3.6	1.915	1.839	1.745	1.636	1.516
3.8	1.919	1.848	1.761	1.660	1.549
4.0	1.922	1.856	1.775	1.681	1.577
4.2	1.925	1.863	1.787	1.700	1.603
4.4	1.927	1.869	1.798	1.716	1.626
4.6	1.929	1.875	1.808	1.731	1.646

H/d	r/d=0.2	r/d=0.4	r/d=0.6	r/d=0.8	r/d=1.0
4.8	1.931	1.880	1.817	1.745	1.664
5.0	1.933	1.884	1.825	1.757	1.681
5.2	1.935	1.888	1.832	1.768	1.696
5.4	1.936	1.892	1.839	1.778	1.710
5.6	1.938	1.896	1.845	1.787	1.723
5.8	1.939	1.899	1.851	1.796	1.734
6.0	1.940	1.902	1.856	1.803	1.745
6.2	1.941	1.904	1.861	1.810	1.755
6.4	1.942	1.907	1.865	1.817	1.764
6.6	1.943	1.909	1.869	1.823	1.772
6.8	1.944	1.911	1.873	1.829	1.780
7.0	1.945	1.913	1.876	1.834	1.787
7.2	1.946	1.915	1.880	1.839	1.794
7.4	1.946	1.917	1.883	1.844	1.800
7.6	1.947	1.919	1.886	1.848	1.806
7.8	1.948	1.920	1.888	1.852	1.812
8.0	1.948	1.922	1.891	1.856	1.817

The values in the table can be plotted as shown in the graph below.

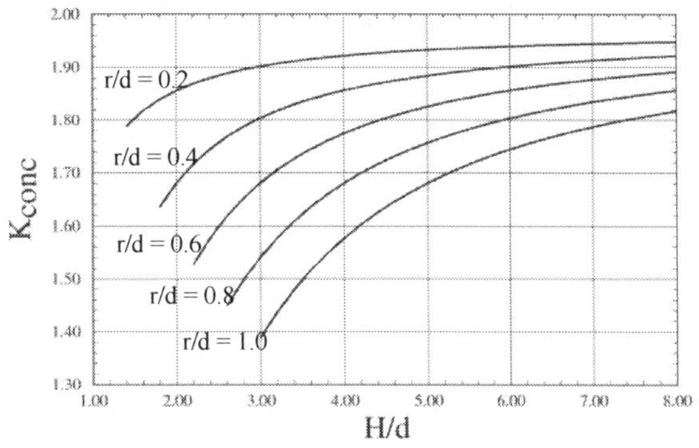

---

3. 121

Solution $\sigma_{max} = ?$

We can write the following.

$$\sigma_{nom} = P/(dt) = 9/[(3)(0.125)] = 24ksi \qquad r/H = 0.625/8 = 0.07813 \tag{1}$$

Substituting (r/H) in the expression for K_{conc} given in earlier problem we obtain

$$K_{conc} = 1.97 - 0.384[(2)(0.07813)] - 1.018[(2)(0.07813)]^2 + 0.430[(2)(0.07813)]^3 = 1.8868 \tag{2}$$

$$\sigma_{max} = K_{conc}\sigma_{nom} = (1.8868)(24) = 45.283 \tag{3}$$

ANS $\sigma_{max} = 45.3ksi$

---

3. 122

Solution P=56kN σ_{yield}=160MPa H=300mm d=100mm t=10mm K=1.6 r=?

We can write the following:

$$\sigma_{max} = \frac{\sigma_{yield}}{12} = \frac{160}{1.6} = 100MPa \qquad \sigma_{nom} = \frac{P}{dt} = \frac{56(10^3)}{(0.1)(0.01)} = 56(10^6)\frac{N}{m^2} = 56MPa \qquad K_{cone} = \frac{\sigma_{max}}{\sigma_{nom}} = \frac{100}{56} = 1.785 \tag{1}$$

From the graph in the earlier problem for $H/d = 3$ the approximate value of r/d=0.4. Thus $r = 0.4d = (0.4)(100) = 40mm$

ANS $r = 40mm$

---

3. 123

Solution

The equation for K_{conc} can be re-written as:

$$K_{conc} = 3.857 - 5.066\left(\frac{4(r/d)}{(H/d)}\right) + 2.469\left(\frac{4(r/d)}{(H/d)}\right)^2 - 0.258\left(\frac{4(r/d)}{(H/d)}\right)^3 \tag{1}$$

Using H/d as a parameter and the above equation, the values of K_{conc} can be found for various values of r/d on a spread sheet as shown in the table below.

r/d	H/d=1.25	H/d=1.50	H/d=1.75	H/d=2.00
0.05	3.109	3.225	3.310	3.375
0.06	2.974	3.109	3.208	3.284
0.07	2.843	2.996	3.109	3.195
0.08	2.718	2.886	3.012	3.109
0.09	2.597	2.780	2.917	3.024
0.10	2.480	2.677	2.825	2.940
0.11	2.368	2.577	2.735	2.859
0.12	2.261	2.480	2.648	2.780
0.13	2.158	2.387	2.563	2.702
0.14	2.060	2.296	2.480	2.626
0.15	1.966	2.209	2.400	2.552
0.16	1.876	2.125	2.322	2.480
0.17	1.790	2.044	2.246	2.410
0.18	1.709	1.966	2.173	2.341
0.19	1.632	1.890	2.101	2.274
0.20	1.558	1.818	2.032	2.209
0.21	1.489	1.749	1.966	2.146
0.22	1.424	1.683	1.901	2.084
0.23	1.363	1.619	1.839	2.024
0.24	1.306	1.558	1.778	1.966
0.25	1.252	1.501	1.720	1.909
0.26	1.203	1.445	1.664	1.854
0.27	1.157	1.393	1.610	1.801
0.28	1.114	1.343	1.558	1.749
0.29	1.076	1.297	1.509	1.699
0.30	1.041	1.252	1.461	1.651

The values in the table can be plotted as shown in the graph below.

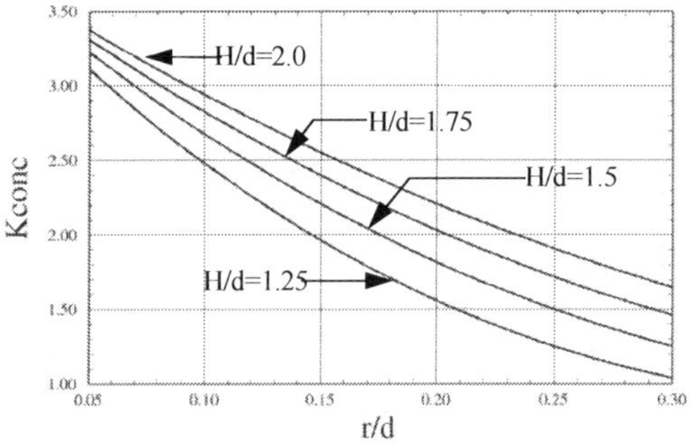

---

3. 124

Solution σ_{max}=?

We can write the following

$$\sigma_{nom} = \frac{P}{Ht} = \frac{150(10^3)}{(0.3)(0.005)} = 100(10^6)\frac{N}{m^2} = 100MPa \qquad \frac{r}{H} = \frac{15}{300} = 0.05 \quad or \quad 4\frac{r}{H} = 0.2 \quad \textbf{(1)}$$

Substituting(4r/H) in the expression for K_{conc} given in the earlier problem we obtain

$$K_{conc} = 3.857 - 5.066(0.2) + 2.469(0.2)^2 - 0.258(0.2)^3 = 2.9405 \qquad \sigma_{max} = K_{conc}\sigma_{nom} = (2.9405)(100) = 294.05 \quad \textbf{(2)}$$

ANS $\sigma_{max} = 294MPa$

---

3. 125

Solution P=18kips σ_{yield}=30ksi H=9in. t=0.25in. d=6in., K=1.4 r=?

We can write the following:

$$\sigma_{max} = \frac{\sigma_{yield}}{K} = \frac{30}{1.4} = 21.43ksi \qquad \sigma_{nom} = \frac{P}{Ht} = \frac{18}{(9)(0.25)} = 8ksi \qquad K_{conc} = \frac{\sigma_{max}}{\sigma_{nom}} = \frac{21.43}{8} = 2.68 \qquad \frac{H}{d} = \frac{9}{6} = 1.5 \quad \textbf{(1)}$$

From the graph in earlier problem, for H/d=1.5 and K_{conc}=2.68, the approximate value of r/d=0.1 or $r = 0.1d = 0.6in$.

ANS $r = 0.6inch$

3.1 An iron rim ($\alpha = 6.5\ \mu/^oF$) of diameter 35.98 inches is to be placed on a wooden cask of diameter 36 inches. Determine the minimum temperature increase needed to slip the rim onto the cask.

3. 126

Solution $\alpha=6.5\ \mu/^oF$ d_i=35.98 in d_f=36in ΔT=?

All stresses are zero. Thus, the total radial normal strain = thermal strain

$$\varepsilon = [(d_f/2)-(d_i/2)]/[d_i/2] = \alpha\Delta T \quad or \quad (36-35.98)/35.98 = (6.5)(10^{-6})\Delta T \quad or \quad \Delta T = 85.518°F \quad \textbf{(1)}$$

ANS $\Delta T = 85.52°F$

3. 127

Solution E_s=200GPa α_s=12.0 μ/^{0}C E_{al}=72GPa α_{al}=23.0 μ/^{0}C ΔT=60 θ=?

The thermal strain and the corresponding deformation in each rod can be found as shown below

$$\varepsilon_{al} = \alpha_{al}\Delta T = (23)(60)(10^{-6}) = 1.38(10^{-3}) \qquad \delta_{al} = \varepsilon_{al}L_{al} = (1.38)(10^{-3})(450) = 0.621mm \quad (1)$$

$$\varepsilon_s = \alpha_s\Delta T = (12)(60)(10^{-6}) = 0.72(10^{-3}) \qquad \delta_s = \varepsilon_s L_s = (0.72)(10^{-3})(450) = 0.324mm \quad (2)$$

The deformed geometry of the assembly can be drawn and angle calculated as shown below.

$$tan\theta = \frac{\delta_{al} - \delta_s}{50} = \frac{0.621 - 0.324}{50} = 0.00594 \quad (3)$$

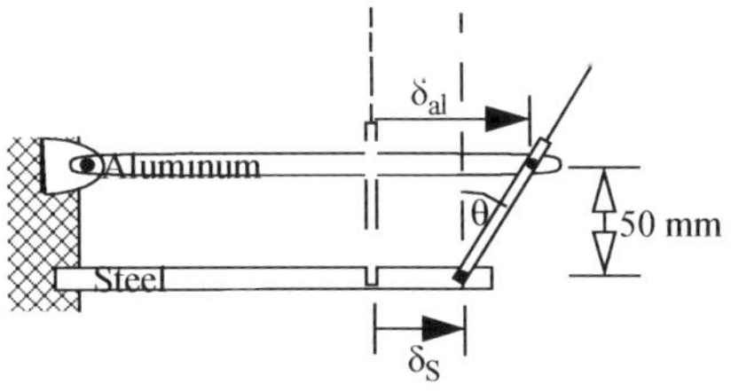

ANS $\theta = 0.34°$

---

3. 128

Solution ΔT=-25^{0}C α=11.7 μ/^{0}C {ε_{xx}=?, ε_{yy}=?, γ_{xy}=?, ε_{zz}=?, σ_{zz}=?} for plane stress and plane strain

Shear strain is unaffected from temperature change, plane stress, or plane strain. We use the value calculated in earlier problem.

(a) Plane stress: $\sigma_{zz} = 0$

Substituting $\sigma_{xx} = 100(10^6)\ N/m^2$; $\sigma_{yy} = 150(10^6)\ N/m^2$; $E = 200(10^9)\ N/m^2$; $\nu = 0.32$ we obtain the following:

$$\varepsilon_{xx} = \frac{[100 - 0.32(150)](10^6)}{200(10^9)} + (11.7)(-25)(10^{-6}) = -32.5(10^{-6}) \quad (1)$$

$$\varepsilon_{yy} = \frac{[150 - 0.32(100)](10^6)}{200(10^9)} + 11.7(-25)(10^{-6}) = 297.5(10^{-6}) \quad (2)$$

$$\varepsilon_{zz} = \frac{-0.32(100 + 150)(10^6)}{200(10^9)} + 11.7(-25)(10^{-6}) = -692.5(10^{-6}) \quad (3)$$

ANS $\varepsilon_{xx} = -32.5\mu$; $\varepsilon_{yy} = 297.5\mu$; $\gamma_{xy} = -1650\mu$; $\sigma_{zz} = 0$; $\varepsilon_{zz} = -692.5\mu$

(b) Plane strain

$$\varepsilon_{zz} = 0 = \frac{\sigma_{zz} - 0.32(100 + 150)(10^6)}{200(10^9)} + 11.7(-25)(10^{-6}) \quad or \quad \sigma_{zz} = 138.5(10^6)\frac{N}{m^2} \quad (4)$$

$$\varepsilon_{xx} = \frac{[100 - 0.32(150 + 138.5)](10^6)}{200(10^9)} + (11.7)(-25)(10^{-6}) = -254.1(10^{-6}) \quad (5)$$

$$\varepsilon_{yy} = \frac{[150 - 0.32(100 + 138.5)](10^6)}{200(10^9)} + (11.5)(-25)(10^{-6}) = 75.9(10^{-6}) \quad (6)$$

ANS $\varepsilon_{xx} = -254.1\mu$; $\varepsilon_{yy} = 75.9\mu$; $\gamma_{xy} = -1650\mu$; $\sigma_{zz} = 138.5MPa(T)$; $\varepsilon_{zz} = 0$

---

3. 129

Solution ΔT=50^0 C. α=23.6μ {ε_{xx}=?, ε_{yy}=?, γ_{xy}=?, ε_{zz}=?, σ_{zz}=?} for plane stress and plane strain

Shear strain is unaffected from temperature change, plane stress, or plane strain. We use the value obtained in earlier problem.

(a) Plane stress: $\sigma_{zz} = 0$

Substituting $\sigma_{xx} = -225(10^6)\ N/m^2$, $\sigma_{yy} = 125(10^6)\ N/m^2$, $E = 70(10^9)\ N/m^2$ and $\nu = 0.25$ we obtain the following:

$$\varepsilon_{xx} = \frac{[-225 - 0.25(125)](10^6)}{70(10^9)} + (23.6)(50)(10^{-6}) = -2481(10^{-6}) \quad (1)$$

$$\varepsilon_{yy} = \frac{[125 - 0.25(225)](10^6)}{70(10^9)} + 23.6(50)(10^{-6}) = 3769(10^{-6}) \quad (2)$$

$$\varepsilon_{zz} = \frac{-0.25(-225 + 125)(10^6)}{70(10^9)} + 23.6(50)(10^{-6}) = 1537(10^{-6}) \quad (3)$$

ANS $\varepsilon_{xx} = -2481\mu$; $\varepsilon_{yy} = 3769\mu$; $\gamma_{xy} = 5357\mu$; $\sigma_{zz} = 0$; $\varepsilon_{zz} = 1537\mu$

(b) Plane strain:

$$\varepsilon_{zz} = 0 = \frac{\sigma_{zz} - 0.25(-225 + 125)(10^6)}{70(10^9)} + 23.6(-50)(10^{-6}) \qquad \sigma_{zz} = -107.6(10^6)\frac{N}{m^2} \quad \textbf{(4)}$$

$$\varepsilon_{xx} = \frac{[-225 - 0.25(125 + 107.6)](10^6)}{70(10^9)} + (23.6)(50)(10^{-6}) = -2096.4(10^{-6}) \quad \textbf{(5)}$$

$$\varepsilon_{yy} = \frac{[125 - 0.25(-225 - 107.6)](10^6)}{70(10^9)} + (23.6)(50)(10^{-6}) = 4153.6(10^{-6}) \quad \textbf{(6)}$$

ANS $\varepsilon_{xx} = -2096\mu$; $\varepsilon_{yy} = 4154\mu$; $\gamma_{xy} = 5357\mu$; $\sigma_{zz} = 107.6MPa(C)$; $\varepsilon_{zz} = 0$

---

3. 130

Solution $\Delta T = 40^oF$ $\alpha = 6.5\mu/^oF$ $\sigma_{xx}=?$, $\sigma_{yy}=?$, $\tau_{xy}=?$, $\varepsilon_{zz}=?$ for plane stress

Shear stress is unaffected from temperature change. We use the value calculated in earlier problem.

Plane stress: $\sigma_{zz} = 0$.

$$\sigma_{xx} - \nu\sigma_{yy} = E(\varepsilon_{xx} - \alpha\Delta T) = (30,000)(-800 - 6.5(40))(10^{-6}) \qquad \sigma_{xx} - \nu\sigma_{yy} = -31.8ksi \quad \textbf{(1)}$$

$$\sigma_{yy} - \nu\sigma_{xx} = E(\varepsilon_{yy} - \alpha\Delta T) = (30,000)(-1000 - 6.5(40))(10^{-6}) \qquad \sigma_{yy} - \nu\sigma_{xx} = -37.8ksi \quad \textbf{(2)}$$

Solving with ν=0.3, we obtain:

$$\sigma_{xx} = -47.4ksi \qquad \sigma_{yy} = -52.02ksi \qquad \varepsilon_{zz} = \frac{0 - 0.3(-47.4 - 52.02)}{30,000} + (40)(6.5)(10^{-6}) = 1254.2 \quad \textbf{(3)}$$

ANS $\sigma_{xx} = 47.4ksi(C)$; $\sigma_{yy} = 52.02ksi(C)$; $\tau_{xy} = -5.77ksi$; $\varepsilon_{zz} = 1254\mu$

---

3. 131

Solution $\Delta T = -100$ $\alpha = 12.8\mu/^oF$ $\sigma_{xx}=?$, $\sigma_{yy}=?$, $\tau_{xy}=?$, $\varepsilon_{zz}=?$ for plane stress

Shear stress is unaffected from temperature change. We use the value calculated in earlier problem.

Plane stress: $\sigma_{zz} = 0$

$$\sigma_{xx} - \nu\sigma_{yy} = E(\varepsilon_{xx} - \alpha\Delta T) = (10,000)(1500 - 12.8(-100))(10^{-6}) \qquad \sigma_{xx} - \nu\sigma_{yy} = 27.8ksi \quad \textbf{(1)}$$

$$\sigma_{yy} - \nu\sigma_{xx} = E(\varepsilon_{yy} - \alpha\Delta T) = 10,000(-1200 - 12.8(-100))(10^{-6}) \qquad \sigma_{yy} - \nu\sigma_{xx} = 0.8ksi \quad \textbf{(2)}$$

Solving with ν=0.282 we obtain:

$$\sigma_{xx} = 30.45ksi \qquad \sigma_{yy} = 9.39\ ksi \qquad \varepsilon_{zz} = \frac{0 - 0.282(30.45 + 9.39)}{(10,000)} + 12.8(-100)(10^{-6}) = -2403(10^{-6}) \quad \textbf{(3)}$$

ANS $\sigma_{xx} = 30.45\ ksi(T)$; $\sigma_{yy} = 9.39\ ksi(T)$; $\tau_{xy} = -3.9ksi$; $\varepsilon_{zz} = -2403\mu$

---

3. 132

Solution $\Delta T = -75^oF$ $\alpha = 26\ \mu/^oC$ $\sigma_{xx}=?$ $\sigma_{yy}=?$ $\tau_{xy}=?$ $\varepsilon_{zz}=?$ for plane stress

Shear stress is unaffected from temperature change. We use the value calculated in earlier problem.

Plane stress: $\sigma_{zz} = 0$

$$\sigma_{xx} - \nu\sigma_{yy} = E(\varepsilon_{xx} - \alpha\Delta T) = (36)(10^9)(-2000 - 26(-75))(10^{-6}) \quad or \quad \sigma_{xx} - \nu\sigma_{yy} = -1.8MPa \quad \textbf{(1)}$$

$$\sigma_{yy} - \nu\sigma_{xx} = E(\varepsilon_{yy} - \alpha\Delta T) = 36(10^9)(-2000 - 26(-75))(10^{-6}) \quad or \quad \sigma_{yy} - \nu\sigma_{xx} = 142.2MPa \quad \textbf{(2)}$$

Solving with ν=0.2 we obtain

$$\sigma_{xx} = 27.75MPa \qquad \sigma_{yy} = 147.75MPa \qquad \varepsilon_{zz} = \frac{0 - 0.2(27.75 + 147.75)(10^6)}{(36)(10^9)} + 26(-75)(10^{-6}) = -2925(10^{-6}) \quad \textbf{(3)}$$

ANS $\sigma_{xx} = 27.8MPa(T)$; $\sigma_{yy} = 147.8MPa(T)$; $\tau_{xy} = 18MPa$; $\varepsilon_{zz} = -2925\mu$

---

3. 133

Solution E=30,000ksi ν=0.25 $\alpha = 6.5\ \mu/^oF$ $\varepsilon_{yy}=0$, $\delta_x=0.005$ $\Delta T = 100^oF$ $\sigma_{xx}=?$ $\sigma_{yy}=?$

Plane stress: $\sigma_{zz} = 0$. The normal strain in the x-direction is:

$$\varepsilon_{xx} = \delta_x/10 = 0.005/10 = 0.005 = 500(10^{-6}) \quad \textbf{(1)}$$

$$\sigma_{xx} - \nu\sigma_{yy} = E(\varepsilon_{xx} - \alpha\Delta T) = (30,000)[500 - (6.5)(100)](10^{-6}) \quad or \quad \sigma_{xx} - 0.25\sigma_{yy} = -4.5ksi \tag{2}$$

$$\sigma_{yy} - \nu\sigma_{xx} = E(\varepsilon_{yy} - \alpha\Delta T) = 30,000[0 - (6.5)(100)](10^{-6}) \quad or \quad \sigma_{yy} - 0.25\sigma_{xx} = -19.5ksi \tag{3}$$

Solving we obtain: $\sigma_{xx} = -10ksi$ and $\sigma_{yy} = -22ksi$

ANS $\sigma_{xx} = 10ksi(C)$; $\sigma_{yy} = 22ksi(C)$

---

3. 134

Solution

Adding the three normal strain equations, we obtain:

$$\varepsilon_{xx} + \varepsilon_{yy} + \varepsilon_{zz} = \frac{1-2\nu}{E}(\sigma_{xx} + \sigma_{yy} + \sigma_{zz}) + 3\alpha\Delta T \quad or \quad \sigma_{xx} + \sigma_{yy} + \sigma_{zz} = \frac{E}{1-2\nu}(\varepsilon_{xx} + \varepsilon_{yy} + \varepsilon_{zz}) - \frac{3(E\alpha\Delta T)}{1-2\nu} \tag{1}$$

We rewrite the normal strain equations and substitute the above expression to obtain the following.

$$\varepsilon_{xx} = (1+\nu)\frac{\sigma_{xx}}{E} - \frac{\nu}{E}(\sigma_{xx} + \sigma_{yy} + \sigma_{zz}) + \alpha\Delta T = \frac{1+\nu}{E}\sigma_{xx} - \frac{\nu}{1-2\nu}(\varepsilon_{xx} + \varepsilon_{yy} + \varepsilon_{zz}) + \frac{3\nu}{1-2\nu}\alpha\Delta T + \alpha\Delta T \text{ or}$$

$$\sigma_{xx} = \frac{E}{(1+\nu)}\left[\left\{\varepsilon_{xx}\left(1 + \frac{\nu}{1-2\nu}\right) + \frac{\nu\varepsilon_{yy}}{1-2\nu} + \frac{\nu\varepsilon_{zz}}{1-2\nu}\right\} - \alpha\Delta T\left(\frac{3\nu + 1 - 2\nu}{1-2\nu}\right)\right] = \frac{E}{(1+\nu)(1-2\nu)}[(1-\nu)\varepsilon_{xx} + \nu\varepsilon_{yy} + \nu\varepsilon_{zz}] - \frac{E\alpha\Delta T}{1-2\nu} \tag{2}$$

$$\varepsilon_{yy} = \frac{1+\nu}{E}\sigma_{yy} - \frac{\nu}{E}(\sigma_{xx} + \sigma_{yy} + \sigma_{zz}) + \alpha\Delta T = \frac{1+\nu}{E}\sigma_{yy} - \frac{\nu}{1-2\nu}(\varepsilon_{xx} + \varepsilon_{yy} + \varepsilon_{zz}) + \frac{3\nu}{1-2\nu}\alpha\Delta T + \alpha\Delta T \text{ or}$$

$$\sigma_{yy} = \frac{E}{(1+\nu)}\left[\left\{\frac{\nu\varepsilon_{xx}}{1-2\nu} + \varepsilon_{yy}\left(1 + \frac{\nu}{1-2\nu}\right) + \frac{\nu\varepsilon_{zz}}{1-2\nu}\right\} - \alpha\Delta T\left(\frac{3\nu + 1 - 2\nu}{1-2\nu}\right)\right] = \frac{E}{(1+\nu)(1-2\nu)}[\nu\varepsilon_{xx} + (1-\nu)\varepsilon_{yy} + \nu\varepsilon_{zz}] - \frac{E\alpha\Delta T}{1-2\nu} \tag{3}$$

$$\varepsilon_{zz} = \frac{1+\nu}{E}\sigma_{zz} - \frac{\nu}{E}(\sigma_{xx} + \sigma_{yy} + \sigma_{zz}) + \alpha\Delta T = \frac{1+\nu}{E}\sigma_{zz} - \frac{\nu}{1-2\nu}(\varepsilon_{xx} + \varepsilon_{yy} + \varepsilon_{zz}) + \frac{3\nu}{1-2\nu}\alpha\Delta T + \alpha\Delta T \tag{4}$$

$$\sigma_{zz} = \frac{E}{(1+\nu)}\left[\left\{\frac{\nu\varepsilon_{xx}}{1-2\nu} + \frac{\nu\varepsilon_{yy}}{1-2\nu} + \varepsilon_{zz}\left(1 + \frac{\nu}{1-2\nu}\right)\right\} - \alpha\Delta T\left(\frac{3\nu + 1 - 2\nu}{1-2\nu}\right)\right] = \frac{E}{(1+\nu)(1-2\nu)}[\nu\varepsilon_{xx} + \nu\varepsilon_{yy} + (1-\nu)\varepsilon_{zz}] - \frac{E\alpha\Delta T}{1-2\nu} \tag{5}$$

Above equations are the desired results.

---

3. 135

Solution

For plane stress(σ_{zz}=0) we can write

$$\sigma_{xx} - \nu\sigma_{yy} = E(\varepsilon_{xx} - \alpha\Delta T) \qquad \sigma_{yy} - \nu\sigma_{xx} = E(\varepsilon_{yy} - \alpha\Delta T) \tag{1}$$

Multiplying one equation by ν and adding it to the other we obtain:

$$\sigma_{xx}(1-\nu^2) = E[\varepsilon_{xx} + \nu\varepsilon_{yy} - \alpha\Delta T(1+\nu)] \quad or \quad \sigma_{xx} = \frac{E}{1-\nu^2}(\varepsilon_{xx} + \nu\varepsilon_{yy}) - \frac{E\alpha\Delta T}{1-\nu} \tag{2}$$

$$\sigma_{yy}(1-\nu^2) = E[\nu\varepsilon_{xx} + \varepsilon_{yy} - \alpha\Delta T(1+\nu)] \quad or \quad \sigma_{yy} = \frac{E}{1-\nu^2}(\varepsilon_{yy} + \nu\varepsilon_{xx}) - \frac{E\alpha\Delta T}{1-\nu}$$

Above equations are the desired results.

---

3. 136

Solution

Adding the stress result of previous problem we obtain

$$\sigma_{xx} + \sigma_{yy} = \frac{E}{1-\nu^2}[(1+\nu)\varepsilon_{xx} + (1+\nu)\varepsilon_{yy}] - \frac{2(E\alpha\Delta T)}{1-\nu} = \frac{E}{1-\nu}(\varepsilon_{xx} + \varepsilon_{yy}) - \frac{2(E\alpha\Delta T)}{1-\nu} \tag{1}$$

$$\varepsilon_{zz} = -\frac{\nu}{E}(\sigma_{xx} + \sigma_{yy}) + \alpha\Delta T = -\frac{\nu}{1-\nu}(\varepsilon_{xx} + \varepsilon_{yy}) + \left(\frac{2\nu}{1-\nu} + 1\right)\alpha\Delta T = -\frac{\nu}{1-\nu}(\varepsilon_{xx} + \varepsilon_{yy}) + \frac{(1+\nu)}{1-\nu}\alpha\Delta T \tag{2}$$

Above equation is the desired result.

Section 3.10-3.11

3. 137

Solution T=? for the three cases

(a) From figure for steel at 40ksi, the number of cycles to failure is n=400,000

$$T = 400,000/200 = 2,000 \ minutes = 33.33 hours \tag{1}$$

(b) At 36ksi, the number of cycles to failure is n=2(10^6)

$$T = 2(10^6)/250 = 8000\ minutes = 133.33 hours \quad (2)$$

(c) The stress value of 32 ksi is below the endurance limit and the component won't fail.

ANS (a) $T = 33.33 hours$ (b) $T = 133.33 hours$ (c) $T = \infty$

---

3. 138

Solution σ_{max}= ? for the three cases

(a) T=17 hours = 1020 minutes and n = number of cycles=(100)(1020)=102,000

From figure for aluminum, the peak stress value at 100,000 is 182 MPa and the peak stress value at 200,000 is 168 MPa. By linear interpolation we obtain a peak stress= $182 - (2000/100,000)(14) = 181.7\ MPa$

(b) T=40 hours =2400 minutes and n=(2400)(50)=120,000

By linear interpolation, the peak stress value=$182 - (20,000/100,000)(14) = 177.2$

(c) T=80 hours =4800 min and n=(4800)(20)=96,000 cycles

The peak stress value between 96,000 cycles &100,000 cycles changes slightly and is approximated at 183MPa.

ANS (a) $\sigma_{max} = 182\ MPa$ (b) $\sigma_{max} = 177\ MPa$ (c) $\sigma_{max} = 183\ MPa$

---

3. 139

Solution σ=6ksi d=3.2 H=8 n=?

From figure in Appendix C, for $d/H = 3.2/8 = 0.4$, the stress concentration factor K_{gross}=3.75

The maximum normal stress is: $\sigma_{max} = K_{gross}\sigma = (3.75)(6) = 22.5 ksi$

From S-N curves for alumina, the number of cycles at 22.5ksi are estimated at 400,000

ANS $n = 400,000 cycles$

---

3. 140

Solution n=500,000 cycles σ= 6 ksi H=8 d_{max}= ? to the nearest 1/8 inch

From S-N curves for aluminum, the peak stress at 500,000 cycles is $\sigma_{max} = 22 ksi$

The stress concentration factor is: $K_{gross} = \sigma_{max}/\sigma = 22/6 = 3.67$

From figure in Appendix C, $d/H = 0.34$ for K_{gross}=3.5 and $d/H = 0.4$ for K_{gross}=3.75, by linear interpolation the value of

$$d/H = 0.34 + (0.4 - 0.34)(3.67 - 3.5)/[3.75 - 3.5] = 0.38 \quad or \quad d = (0.38)(8) = 3.04 \quad (1)$$

ANS $d_{max} = 3.0$ in.

---

3. 141

Solution d=2.4 H=8 n=750,000 cycles σ = ?

From S-N curves for aluminum, the peak stress at 750,000 cycles is estimated as σ_{max}=21ksi

From in Appendix C, for $d/H = 0.3$ the stress concentration factor is estimated $K_{gross} = (3.25 + 3.5)/2 = 3.375$

The nominal normal stress is: $\sigma = \sigma_{max}/K_{gross} = 21/3.375 = 6.222$

ANS σ = 6.2 ksi(T)

---

3. 142

Solution σ_{yield} = 18 ksi ε_{yield} = 0.0012 σ_{ult} = 50 ksi ε_{ult} = 0.50 Constants in the three material models=?

We have coordinates of three points on the curve: P_o (σ_0 = 0.00, ε_0 = 0.000), P_1 (σ_1 = 18.0, ε_1 = 0.0012), and P_2 (σ_2 = 50, ε_2 = 0.5). Using this data we can find the various constants in the material models.

(a) The modulus of elasticity E is slope between points P_o and P_1 and can be found as shown below.

$$E_1 = \frac{\sigma_1 - \sigma_0}{\varepsilon_1 - \varepsilon_0} = \frac{18}{0.0012} = 15,000\ ksi \quad (1)$$

After yield stress, the stress is a constant. The stress strain behavior can be written as shown below.

$$\sigma = \begin{cases} 15,000\varepsilon\ ksi & |\varepsilon| \le 0.0012 \\ 18\ ksi & 0.0012 \le |\varepsilon| \le 0.5 \end{cases} \quad (2)$$

(b) In the linear strain hardening model the slope of the straight line before yield stress is as calculated in part (a). After the yield stress, the slope of the line can be found from the coordinates of point P_1 and P_2.

$$E_2 = \frac{\sigma_2 - \sigma_1}{\varepsilon_2 - \varepsilon_1} = \frac{50 - 18}{0.5 - 0.0012} = 64.15\ ksi \quad (3)$$

The stress strain behavior can be written as shown below.

$$\sigma = \begin{cases} 15,000\varepsilon \ ksi & |\varepsilon| \le 0.0012 \\ (18 + 64.15(\varepsilon - 0.0012)) \ ksi & 0.0012 \le |\varepsilon| \le 0.5 \end{cases} \quad \textbf{(4)}$$

(c) The two constants E and n in $\sigma = E\varepsilon^n$ can be found by substituting the coordinates of the two point P_1 and P_2, to generate the following two equations:

$$18 = E(0.0012)^n \qquad 50 = E(0.5)^n \quad \textbf{(5)}$$

Dividing second equation by the first and taking the logarithm of both sides, we can solve for 'n' as shown below:

$$ln\left(\frac{0.5}{0.0012}\right)^n = ln\left(\frac{50}{18}\right) \quad \text{or} \quad n\ ln(416.67) = ln(2.778) \quad \text{or} \quad n = 0.1694 \quad \textbf{(6)}$$

$$E = \frac{50}{(0.5)^n} = \frac{50}{(0.5)^{0.1694}} = 56.2 \ ksi \quad \textbf{(7)}$$

We can now write the stress-strain equations for the power law model as shown below:

$$\sigma = \begin{cases} 56.2\varepsilon^{0.1694} \ ksi & \varepsilon \ge 0 \\ -56.2(-\varepsilon)^{0.1694} \ ksi & \varepsilon < 0 \end{cases} \quad \textbf{(8)}$$

Stresses at different strains can be found plotted as shown below.

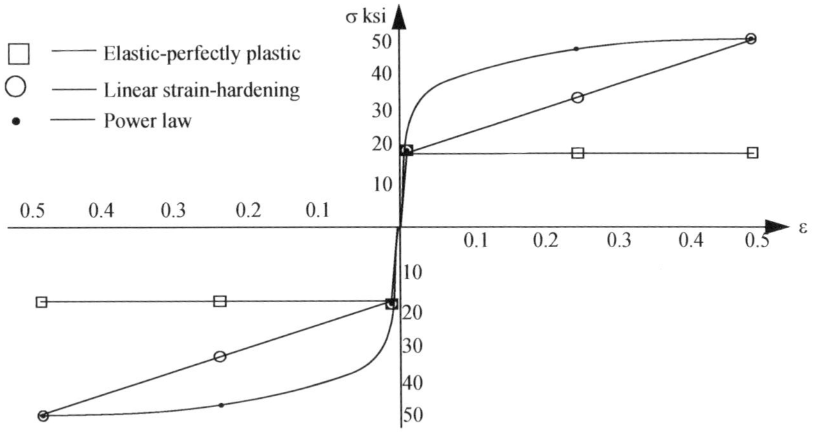

---

3. 143

Solution $\sigma_{yield} = 220$ MPa $\varepsilon_{yield} = 0.00125$ $\sigma_{ult} = 340$ MPa $\varepsilon_{ult} = 0.20$. Constants in the three material models=?

We have coordinates of three points on the curve: P_o ($\sigma_0 = 0.00$, $\varepsilon_0 = 0.000$), P_1 ($\sigma_1 = 220$, $\varepsilon_1 = 0.00125$), and P_2 ($\sigma_2 = 340$, $\varepsilon_2 = 0.2$). Using this data we can find the various constants in the material models.

(a) The modulus of elasticity E is slope between points P_o and P_1 and can be found as shown below.

$$E_1 = \frac{\sigma_1 - \sigma_0}{\varepsilon_1 - \varepsilon_0} = \frac{220}{0.00125} = 176,000 \ MPa \quad \textbf{(1)}$$

After yield stress, the stress is a constant. The stress strain behavior can be written as shown below.

$$\sigma = \begin{cases} 176,000\varepsilon \ MPa & |\varepsilon| \le 0.00125 \\ 220 \ MPa & 0.00125 \le |\varepsilon| \le 0.2 \end{cases} \quad \textbf{(2)}$$

(b) In the linear strain hardening model the slope of the straight line before yield stress is as calculated in part (a). After the yield stress, the slope of the line can be found from the coordinates of point P_1 and P_2

$$E_2 = \frac{\sigma_2 - \sigma_1}{\varepsilon_2 - \varepsilon_1} = \frac{340 - 220}{0.2 - 0.00125} = 603.8 MPa \quad \textbf{(3)}$$

The stress strain behavior can be written as shown below.

$$\sigma = \begin{cases} 176,000\varepsilon \ MPa & |\varepsilon| \le 0.00125 \\ 220 + 603.8(\varepsilon - 0.00125) \ MPa & 0.00125 \le |\varepsilon| \le 0.2 \end{cases} \quad \textbf{(4)}$$

(c) The two constants E and n in $\sigma = E\varepsilon^n$ can be found by substituting the coordinates of the two point P_1 and P_2, to generate the following two equations:

$$220 = E(0.00125)^n \qquad 340 = E(0.2)^n \quad \textbf{(5)}$$

Dividing the second equation by the first and taking the logarithm of both sides, we can solve for 'n' as shown below:

$$ln\left(\frac{0.2}{0.00125}\right)^n = ln\left(\frac{340}{220}\right) \quad or \quad n\ ln(160) = ln(1.545) \quad or \quad n = 0.0858 \quad \textbf{(6)}$$

$$E = \frac{340}{(0.2)^n} = \frac{340}{(0.2)^{0.0858}} = 390.3\ MPa \tag{7}$$

We can now write the stress-strain equations for the power law model as shown below:

$$\sigma = \begin{cases} 390.3\varepsilon^{0.0858}\ MPa & \varepsilon \geq 0 \\ -390.3(-\varepsilon)^{0.0858}\ MPa & \varepsilon < 0 \end{cases} \tag{8}$$

Stresses at different strains can be found plotted as shown below.

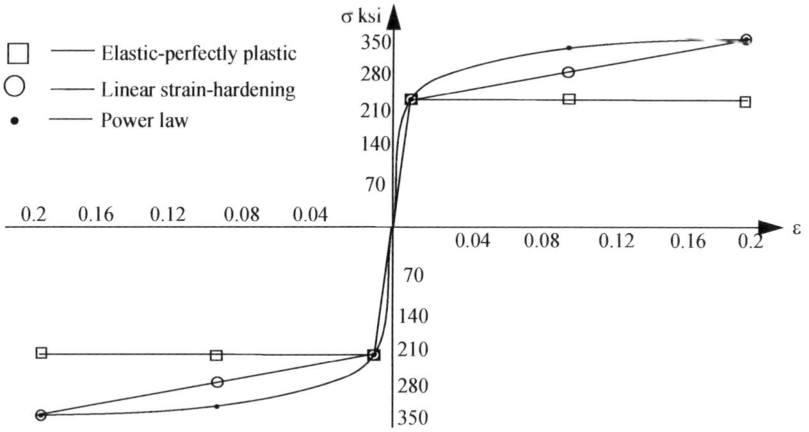

---

3. 144

Solution $\gamma_{x\theta} = 2\rho(10^{-3})$ τ_{yield}=18ksi G=12,000 ksi $\tau_{x\theta}$=f(ρ) = ? Elastic perfectly plastic

The strain at yield point is:

$$\gamma_{yield} = \frac{18}{12,000} = 0.0015 = 2\rho_{yield}(10^{-3}) \quad or \quad \rho_{yield} = 0.75 inch \tag{1}$$

$$\tau_{x\theta} = G\gamma_{x\theta} = (12000)(2\rho)(10^{-3}) = 24\rho \qquad 0 < \rho < 0.75 \tag{2}$$

The stress distribution can be written as:

$$\tau_{x\theta} = \begin{cases} 24\rho\ \text{ksi} & 0 < \rho < 0.75in \\ 18\ \text{ksi} & 0.75in < \rho < 1.5in \end{cases} \tag{3}$$

The plot of shear strain and shear stress are shown below.

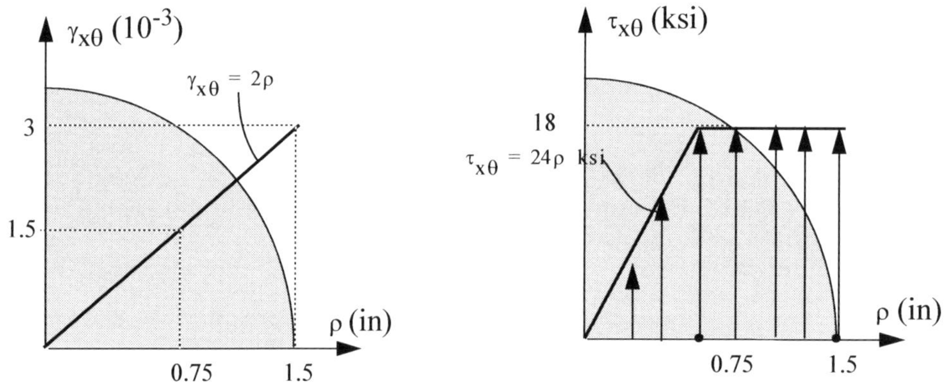

---

3. 145

Solution $\gamma\gamma_{x\theta} = 0.002\rho$) τ_{yield}=18ksi G_1=12,000ksi G_2=4,800ksi $\tau_{x\theta}$=f(ρ)=?

Before yield point we have: $\tau_1 = G_1\gamma_{x\theta} = (12,000)(2\rho)(10^{-3}) = 24\rho$

The strain at yield point is: $\gamma_{yield} = \tau_{yield}/G_1 = 18/(12,000) = 0.0015$

After yield point we can write: $\tau_2 = \tau_{yield} + G_2(\gamma_{x\theta} - \gamma_{yield}) = 18 + 4800(2\rho(10^{-3}) - 0.0015) = 10.8 + 9.6\rho$

From earlier problem we have ρ_{yield}=0.75 inch. The stress distribution can be written as:

$$\tau_{x\theta} = \begin{cases} 24\rho\ \text{ksi} & 0 < \rho < 0.75in \\ (10.8 + 9.6\rho)ksi & 0.75in < \rho < 1.5in \end{cases} \tag{1}$$

The plot of shear strain and shear stress are given below.

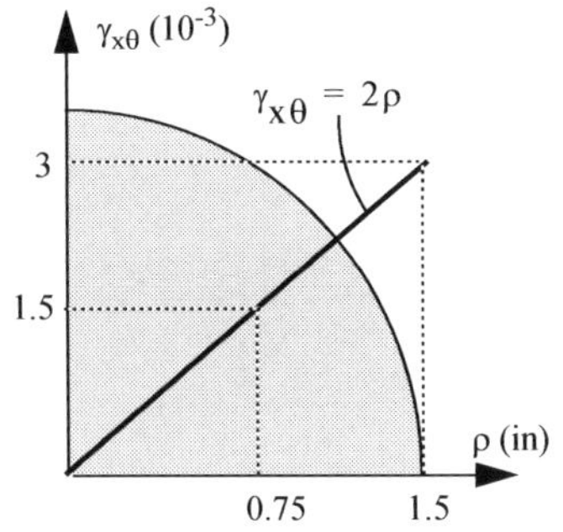

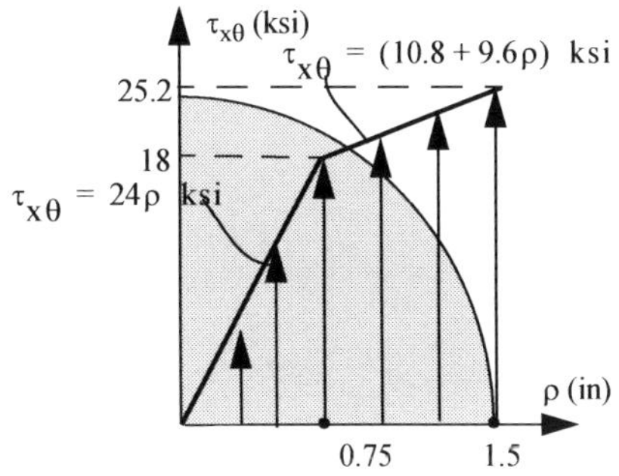

3. 146

Solution $\gamma_{x\theta} = 2\rho(10^{-3}))$ $\tau = 243\gamma^{0.4}\ ksi$ $\tau_{x\theta}= f(\rho) = ?$

The stress distribution is:

$$\tau_{x\theta} = 243\gamma_{x\theta} = 243(2\rho(10^{-3}))^{0.4} = 20.23\rho^{0.4}ksi \qquad 0 \le \rho \le 1.5 \tag{1}$$

ANS $\tau_{x\theta} = 20.23\rho^{0.4}ksi \qquad 0 \le \rho \le 1.5$

The plots of shear strain and shear stress are given below.

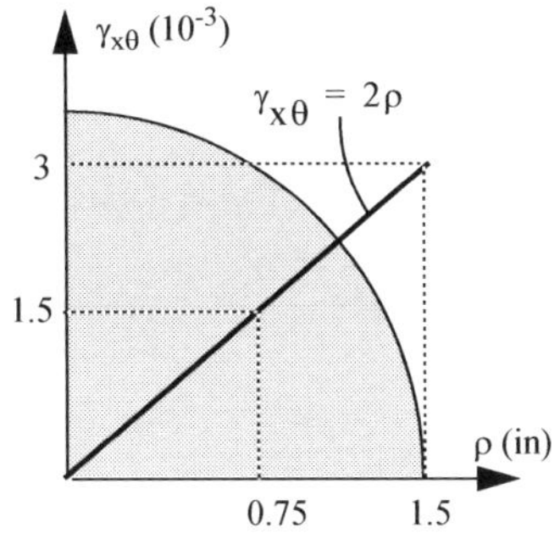

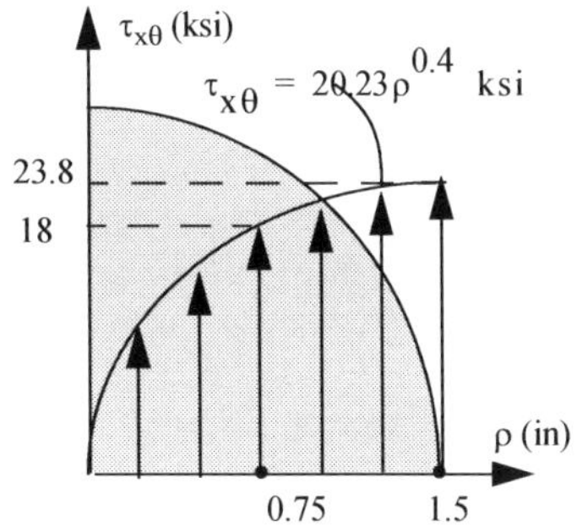

3. 147

Solution $\gamma_{x\theta} = 0.2\rho$ $\tau = (12,000\gamma - 120,000\gamma^2)\ ksi$ $\tau_{x\theta} = f(\rho) = ?$

The shear stress distribution is:

$$\tau_{x\theta} = (12,000\gamma_{x\theta} - 120,000\gamma_{x\theta}^2) = (12,000)(2\rho)(10^{-3}) - 120,000[2\rho(10^{-3})]^2 = (24\rho - 0.48\rho^2)ksi \tag{1}$$

ANS $\tau_{x\theta} = (24\rho - 0.48\rho^2)ksi \qquad 0 \le \rho \le 1.5$

The plots of shear strain and shear stress are given below.

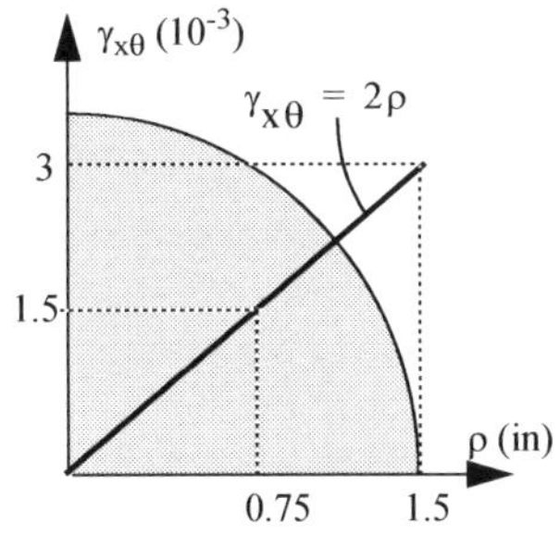

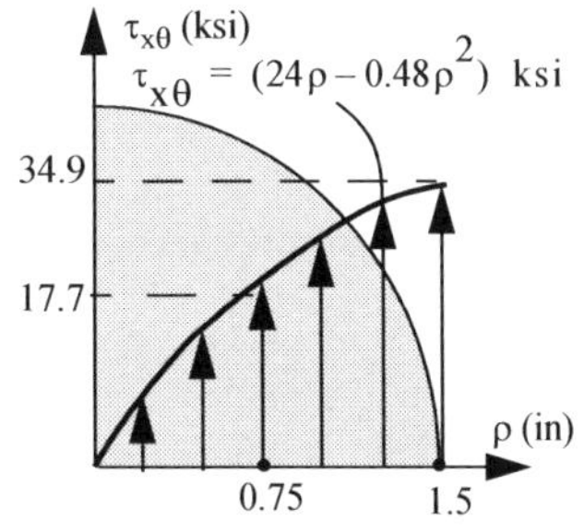

3. 148

Solution $\gamma_{x\theta} = 0.2\rho$ ρ in meters τ_{yield}=175MP G=26GPa d_i=50mm d_0=150mm $\tau_{x\theta} = f(\rho) = ?$

The strain at yield point is:

$$\gamma_{yield} = \frac{\tau_{yield}}{G} = \frac{175(10^6)}{26(10^9)} = 6.7308(10^{-3}) = 0.2\rho_{yield} \qquad or \qquad \rho_{yield} = 33.65(10^{-3})m \tag{1}$$

In the elastic region: $\tau_{x\theta} = G\gamma_{z\theta} = 26(10^9)(0.2\rho) = 5200\rho(10^6)N/m^2$ where ρ is in meters. The stress distribution can be written as:

$$\tau_{x\theta} = \begin{cases} 5200\rho MPa & 0.025m \le \rho \le 0.03365m \\ 175MPa & 0.03365m \le \rho \le 0.05m \end{cases} \tag{2}$$

The plot of shear strain and shear stress are given below

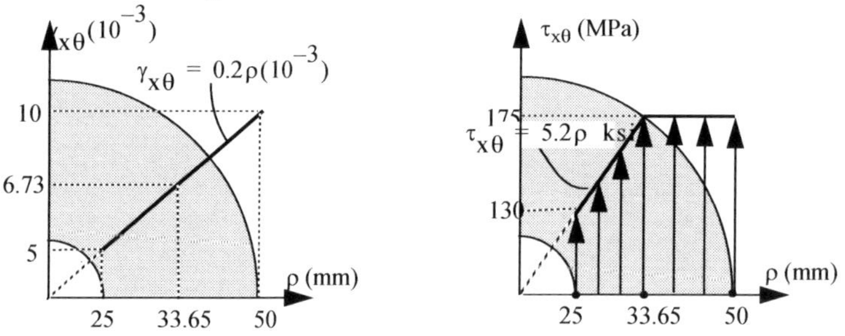

---

3. 149

Solution γ_{yield}=0.2ρ where ρ is in meters τ_{yield}=175MPa G_1=26GPa G_2=14GPa $\tau_{x\theta}$ = f(ρ) = ?

The stress before yield point is:

$$\tau_1 = G_1\gamma_{x\theta} = 26(10^9)(0.2\rho) = 5200\rho MPa \tag{1}$$

From earlier problem $\gamma_{yield} = 6.7308(10^{-3})$ and $\rho_{yield} = 33.65(10^{-3})m$

The stress after yield point is:

$$\tau_2 = \tau_{yield} + G_2(\gamma_{x\theta} - \gamma_{yield}) = 175(10^6) + 14(10^9)(0.2\rho - 6.7308(10^{-3})) = 80.77(10^6) + 2800\rho(10^6) \tag{2}$$

The stress distribution can be written as:

$$\tau_{x\theta} = \begin{cases} 5200\rho MPa & 0.025m \le \rho \le 0.03365m \\ (80.77 + 2800\rho)MPa & 0.03365m \le \rho \le 0.05m \end{cases} \tag{3}$$

The plot of shear strain and shear stress are given below.

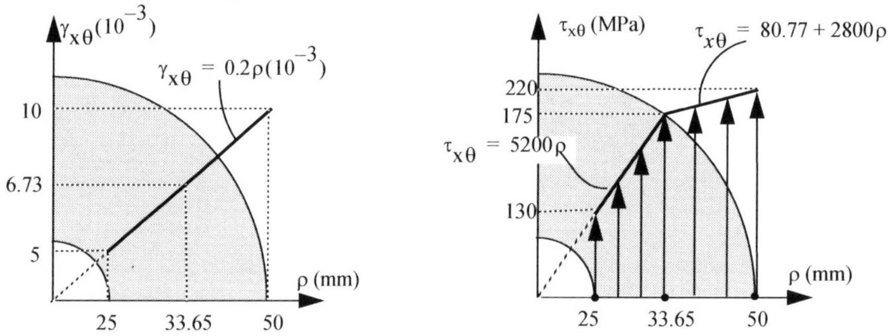

---

3. 150

Solution $\gamma_{x\theta}$=0.2ρ ρ in meters $\tau_{x\theta}$ = f(ρ) = ?

The stress distribution is:

$$\tau_{x\theta} = 3435\gamma^{0.6} = 3435(0.2\rho)^{0.6} = 1307.8\rho^{0.6} = 1307.8\rho^{0.6}MPa \tag{1}$$

ANS $\tau_{x\theta} = 1307.8\rho^{0.6}MPa$ $0.025m \le \rho \le 0.05m$

The plot of shear strain and shear stress are given below.

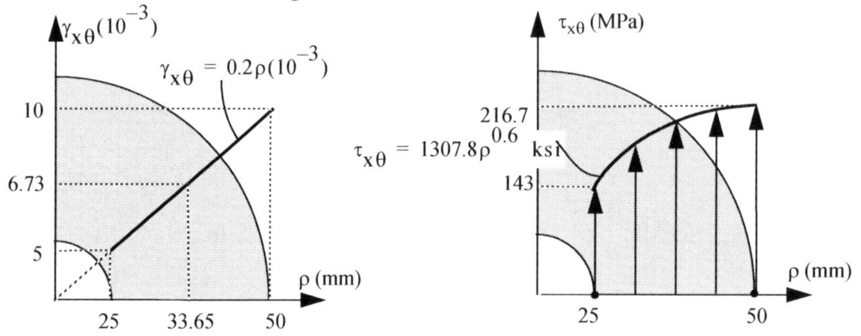

---

3. 151

Solution $\gamma_{x\theta}$=0.2ρ where ρ is in meters $\tau_{x\theta}$ = f(ρ) = ?

The shear stress distribution is:

$$\tau_{x\theta} = (26,000\gamma - 208,000\gamma^2) = 26,000(0.2\rho) - 208,000(0.2\rho)^2 = (5,200\rho - 8,320\rho^2)MPa \tag{1}$$

ANS $\tau_{x\theta} = (5,200\rho - 8,320\rho^2)MPa \qquad 0.025m \le \rho \le 0.05m$

The plots of shear strain and stress are shown below.

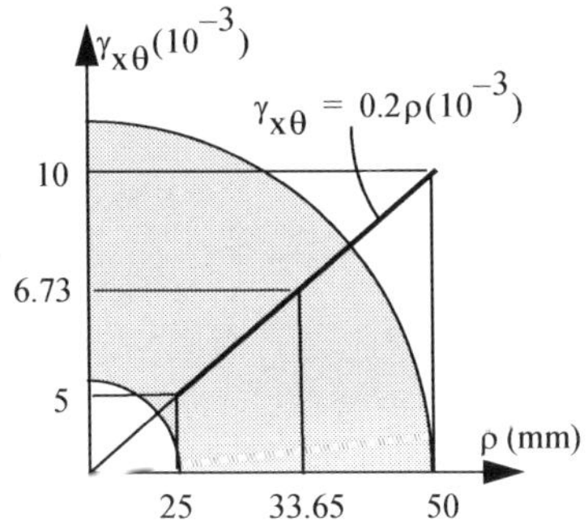

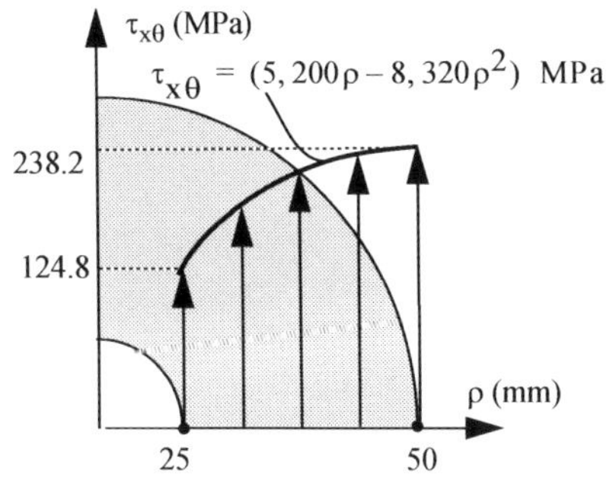

3. 152

Solution ε_{xx}=-0.01y y in meters σ_{yield}=250MPa E=200GPa σ_{xx}= f(y) Elastic-perfectly plastic

The strain at yield point is: $\varepsilon_{yield} = \sigma_{yield}/E = (250(10^6))/[200(10^9)] = 1.25(10^{-3})$

For positive values of *y* we have contraction, and for negative values of *y* we have tension. With geometry being symmetric in *y*, the elastic-plastic boundary will be on both sides

$$\varepsilon_{yield} = \pm 1.25(10^{-3}) = -0.01y_{yield} \qquad or \qquad y_{yield} = \pm 0.125m = \pm 125mm \qquad (1)$$

In the linear region

$$\sigma_{xx} = E\varepsilon_{xx} = (200)(10^9)(-0.01y) = -2000y \qquad (2)$$

$$\sigma_{xx} = \begin{cases} -250MPa & 0.125m < y < 0.150m \\ -2000yMPa & -0.125m < y < 0.125m \\ 250MPa & -0.150m < y < -0.125m \end{cases} \qquad (3)$$

The plots of normal strain and stress are shown below.

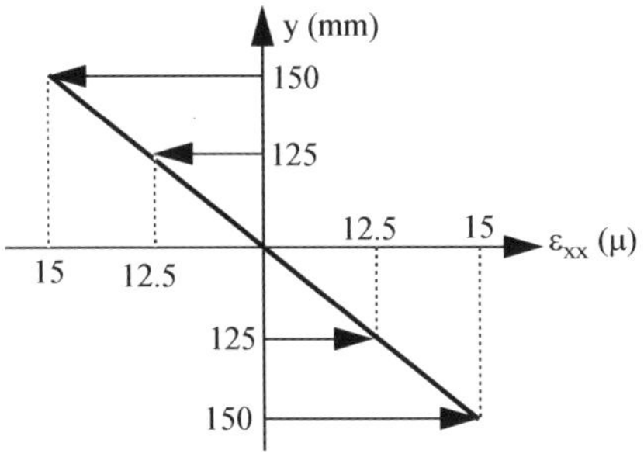

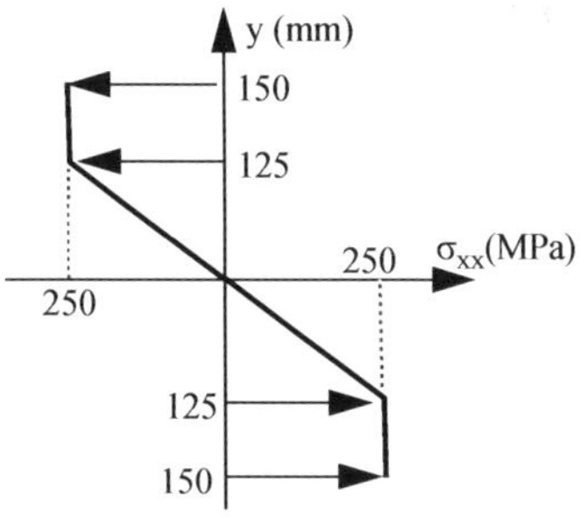

3. 153

Solution ε_{xx}=-0.01y where y is in meters σ_{yield}=250 MPa E_1=200 GPa E_2=80 GPa σ_{xx} = f(y) = ?

From earlier problem we have: $\varepsilon_{yield} = 1.25(10^{-3})$ and $y_{yield} = \pm 125mm$.

In linear region: $\sigma_1 = E_1\varepsilon_1 = (200)(10^9)(-0.01y) = -2000y \ MPa$.

For $y < -0.125m$ i.e., on the tensile side

$$\sigma_2 = \sigma_{yield} + E_2(\varepsilon_{xx} - \varepsilon_{yield}) = 250(10^6) + (80)(10^9)[-0.01y - 1.25(10^{-3})] = (150 - 800y)MPa \qquad (1)$$

For $y > 0.125m$ i.e., on the compressive side

$$\sigma_2 = -\sigma_{yield} + E_2(\varepsilon_{xx} - (-\varepsilon_{yield})) = (-250)(10^6) + (80)(10^9)[-0.01y + 1.25(10^{-3})] = (-150 - 800y)MPa \qquad (2)$$

The stress distribution can be written as:

$$\sigma_{xx} = \begin{cases} (-150 - 800y)MPa & 0.125m < y < 0.150m \\ -2000yMPa & -0.125m < y < 0.125m \\ (150 - 800y)MPa & -0.150m < y < -0.125m \end{cases} \qquad (3)$$

The plots of normal strain and stress are shown below.

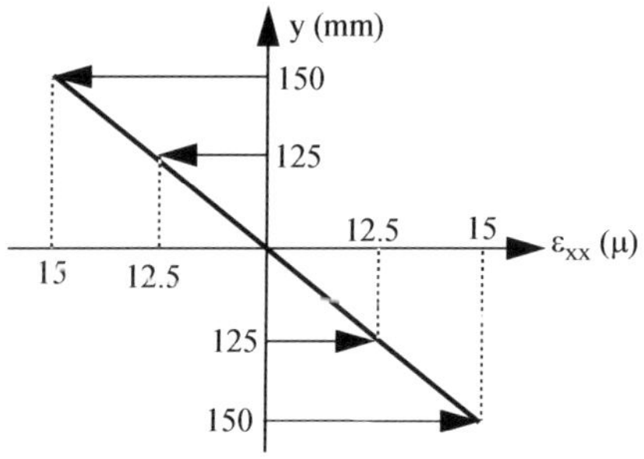

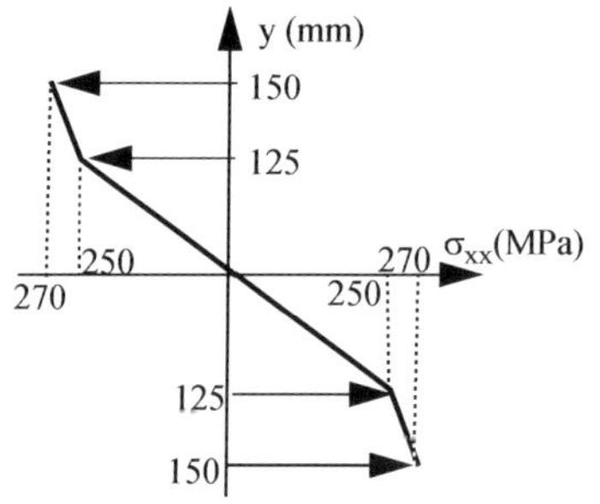

3. 154

Solution $\varepsilon_{xx} = -0.01y$ where y is in meters $\sigma = 952\varepsilon^{0.2}\ MPa$ $\sigma_{xx} = f(y) = ?$

As the material behavior in tension and compression is the same, we have:

$$\sigma_{xx} = \begin{cases} 952\varepsilon^{0.2} MPa & \varepsilon > 0 \\ -952(-\varepsilon)^{0.2} MPa & \varepsilon < 0 \end{cases} \tag{1}$$

For positive y we have contraction and for negative y we have extension, the stress distribution can be written as shown below.

$$\sigma_{xx} = \begin{cases} 952(-0.01y)^{0.2} MPa = 379(-y)^{0.2} MPa & y < 0 \\ -952(0.01y)^{0.2} MPa = -379(y)^{0.2} MPa & y > 0 \end{cases} \tag{2}$$

$$\sigma_{xx} = \begin{cases} 379(-y)^{0.2} MPa & y < 0 \\ -379(y)^{0.2} MPa & y > 0 \end{cases} \tag{3}$$

The plots of normal strain and stress are as shown below.

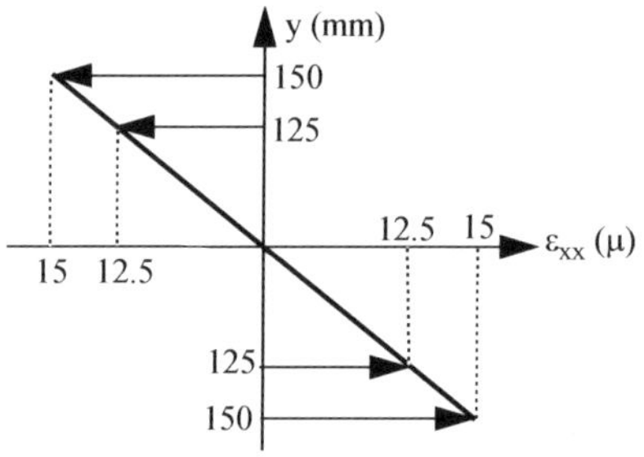

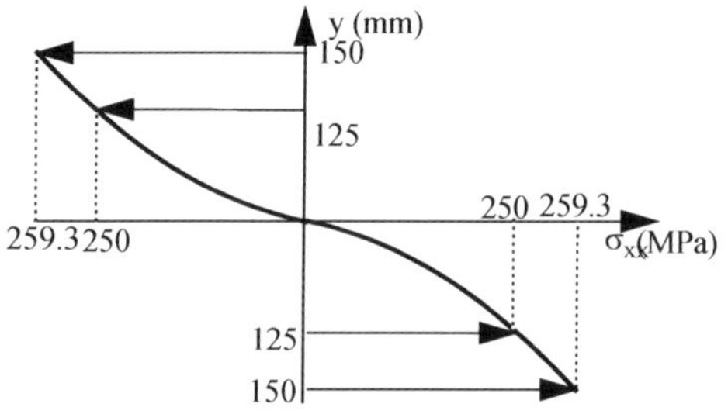

3. 155

Solution $\varepsilon_{xx} = -0.01y$ where y is in meters $\sigma = (200\varepsilon - 2,000\varepsilon^2)\ MPa$ $\sigma_{xx} = f(y) = ?$

As the material behavior in tension and compression is the same, we have:

$$\sigma_{xx} = \begin{cases} (200\varepsilon - 2,000\varepsilon^2) MPa & \varepsilon > 0 \\ (200\varepsilon + 2,000\varepsilon^2) MPa & \varepsilon < 0 \end{cases} \tag{1}$$

For positive y we have contraction and for negative y we have extension, the stress distribution can be written as shown below.

$$\sigma_{xx} = \begin{cases} [200(-0.01y) - 2,000(-0.01y)^2] MPa = (-2y - 0.2y^2) MPa & y < 0 \\ [200(-0.01y) + 2,000(-0.01y)^2] MPa = (-2y + 0.2y^2) MPa & y > 0 \end{cases} \tag{2}$$

$$\sigma_{xx} = \begin{cases} (-2y - 0.2y^2) MPa & y < 0 \\ (-2y + 0.2y^2) MPa & y > 0 \end{cases} \tag{3}$$

The plots of normal strain and stress are as shown below

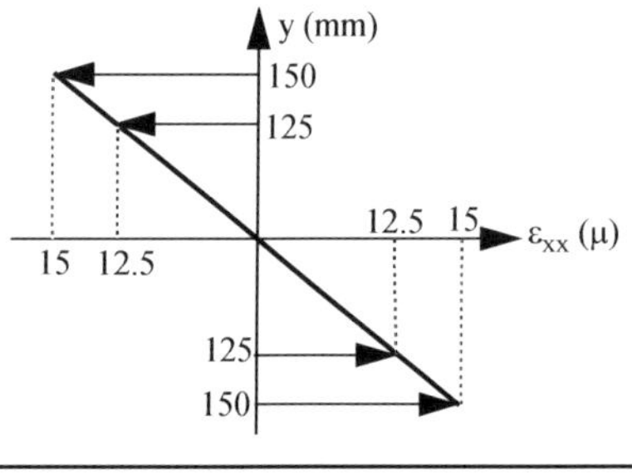

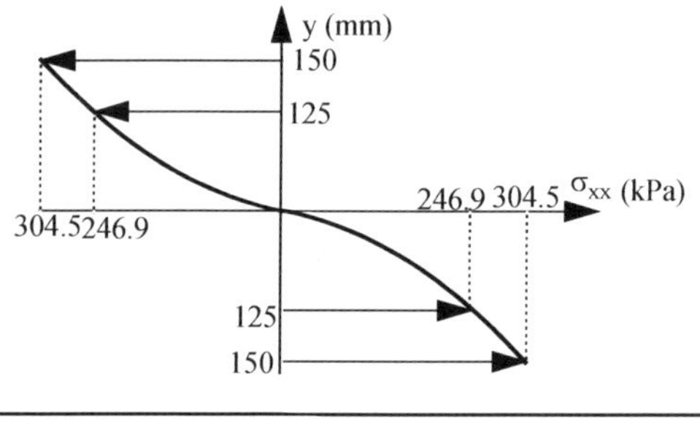

CHAPTER 4

Section 4.1

4. 1

Solution E = 30,000 ksi A = 0.5in^2 F_1 = ? F_2 = ? F_3 = ? F_4 = ?

1. Deformation calculations

$$u_D - u_C = 0.0075 - (-0.0045) = 0.0120\ in \qquad u_C - u_B = -0.0045 - 0.0080 = -0.0125\ in \tag{1}$$

$$u_B - u_A = 0.0080 - (-0.0100) = 0.0180\ in \tag{2}$$

2. Strain calculations

$$\varepsilon_{DC} = (u_D - u_C)/(x_D - x_C) = 0.012/36 = 0.3333(10^{-3}) \qquad \varepsilon_{CB} = (u_C - u_B)/(x_C - x_B) = -0.0125/50 = -0.2500(10^{-3}) \tag{3}$$

$$\varepsilon_{BA} = (u_B - u_A)/(x_B - x_A) = 0.0180/36 = 0.5000(10^{-3}) \tag{4}$$

3. Stress calculations

$$\sigma_{DC} = (30)(10^3)(0.3333(10^{-3})) = 10\ ksi = 10\ ksi(T) \qquad \sigma_{CB} = (30)(10^3)(-0.25)(10^{-3}) = -7.5\ ksi = 7.5\ ksi(C) \tag{5}$$

$$\sigma_{BA} = (30)(10^3)(0.50)(10^{-3}) = 15\ ksi = 15\ ksi(T) \tag{6}$$

4. Internal force calculations

$$\boldsymbol{N}_{DC} = (10)(0.5) = 5\ kips(T) \qquad \boldsymbol{N}_{CB} = (7.5)(0.5) = 3.75\ kips(C) \qquad \boldsymbol{N}_{BA} = (15)(0.5) = 7.5\ kips(T) \tag{7}$$

5. External force calculations

By making imaginary cuts through segments AB, BC, and CD the following free body diagrams can be obtained.

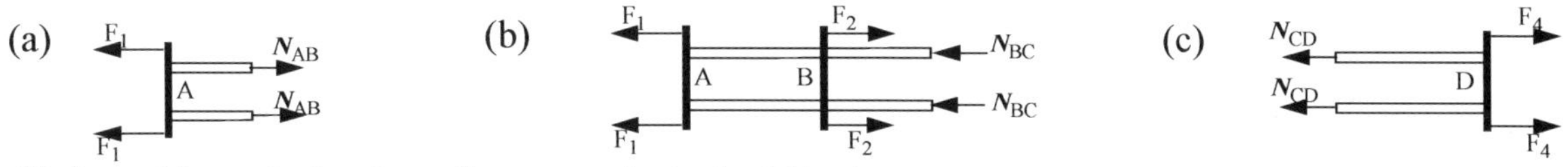

By equilibrium of forces in the above figures we obtain the following.

$$F_1 = (2\boldsymbol{N}_{BA})/2 = 7.5\text{ kips} \qquad F_2 = (2F_1 + 2\boldsymbol{N}_{BC})/2 = 7.5 + 3.75 = 5\text{ kips} \qquad F_4 = (2\boldsymbol{N}_{DC})/2 = 5\text{ kips} \tag{8}$$

$$-2F_1 + 2F_2 - 2F_3 + 2F_4 = 0 \qquad or \qquad F_3 = F_2 + F_4 - F_1 = 8.75\text{ kips} \tag{9}$$

ANS $F_1 = 7.5$ kips ; $F_2 = 11.25$ kips ; $F_3 = 8.75$ kips ; $F_4 = 5$ kips

---

4. 2

Solution F_1 = ? F_2 = ? F_3 = ?

1. Deformation calculations

$$u_D - u_C = 3.7 - 0.7 = 3.0\ mm \qquad u_C - u_B = 0.7 + 1.8 = 2.5\ mm \qquad u_B - u_A = -1.8 - 0 = -1.8\ mm \tag{1}$$

2. Strain calculations

$$\varepsilon_{DC} = (u_D - u_C)/(x_D - x_C) = 3/[2(10^3)] = 1.5(10^{-3}) \qquad \varepsilon_{CB} = (u_C - u_B)/(x_C - x_B) = 2.5/[2.5(10^3)] = 1.0(10^{-3}) \tag{2}$$

$$\varepsilon_{BD} = (u_B - u_D)/(x_B - x_D) = -1.8/[1.5(10^{-3})] = -1.2(10^{-3}) \tag{3}$$

3. Stresses

$$\sigma_{DC} = E_{DC}\varepsilon_{DC} = (200)(10^9)(1.5)(10^{-3}) = 300(10^6)\ N/m^2(T) \tag{4}$$

$$\sigma_{BC} = E_{BC}\varepsilon_{BC} = (100)(10^9)(1.0)(10^{-3}) = 100(10^6)\ N/m^2(T) \tag{5}$$

$$\sigma_{AB} = E_{AB}\varepsilon_{AB} = (70)(10^9)(-1.2)(10^{-3}) = -84(10^6) = 84(10^6)\ N/m^2(C) \tag{6}$$

4. Internal forces

$$\boldsymbol{N}_{DC} = \sigma_{DC}A_{DC} = (300)(10^6)[\pi(0.02)^2/4] = 94.25(10^3)\ N = 94.25\ kN\ (T) \tag{7}$$

$$\boldsymbol{N}_{BC} = \sigma_{BC}A_{BC} = (100)(10^6)[\pi(0.025)^2/4] = 49.09(10^3)\ N = 49.09\ kN\ (T) \tag{8}$$

$$\boldsymbol{N}_{AB} = \sigma_{AB}A_{AB} = (84)(10^6)[\pi(0.03)^2/4] = 59.38(10^3)\ N = 59.38\ kN\ (C) \tag{9}$$

5. External forces

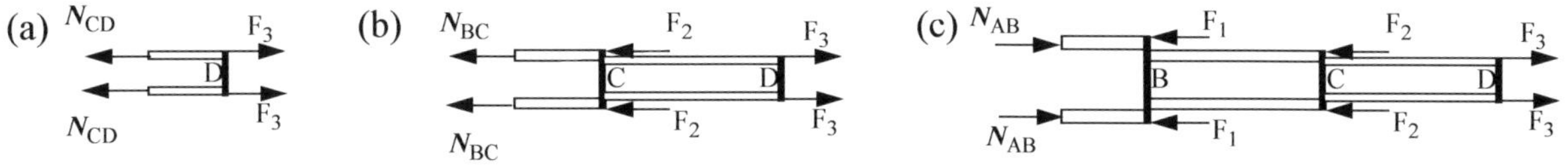

By equilibrium of forces in the above figures.

$$F_3 = 2\boldsymbol{N}_{CD}/2 = 94.3\ kN \qquad F_2 = (2F_3 - 2\boldsymbol{N}_{BC})/2 = 94.25 - 49.09 = 45.16 \tag{10}$$

$$2F_3 - 2F_2 - 2F_1 + 2N_{AB} = 0 \quad or \quad F_1 = F_3 - F_2 + N_{AB} = 94.25 - 45.16 + 59.38 = 108.47 \quad (11)$$

ANS $F_1 = 108.5\ kN;\ F_2 = 45.2\ kN;\ F_3 = 94.3\ kN$

---

4. 3

Solution E = 200 GPa d = 10 mm δ = 0.1 mm to the right F = ?

1. Deformation: From the deformed geometry we obtain the following

$$\delta_A = \delta = 0.1\ mm\ extension \qquad \delta_B = \delta = 0.1\ mm\ contraction \quad (1)$$

(a)

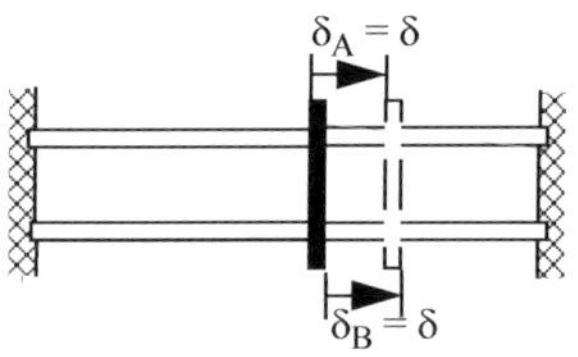

(b)

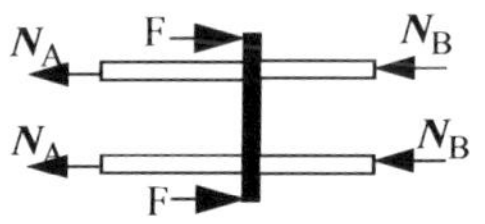

2. Strains

$$\varepsilon_A = \delta_A/L_A = 0.1/(2.5(10^3)) = 40(10^{-6})\ extension \qquad \varepsilon_B = \delta_B/L_B = 0.1/(1.5(10^3)) = 66.67(10^{-6})\ contraction \quad (2)$$

3. Stresses

$$\sigma_A = E_A\varepsilon_A = (200)(10^9)(40)(10^{-6}) = 8.0(10^6)\ N/m^2(T) \quad (3)$$

$$\sigma_B = E_B\varepsilon_B = (200)(10^9)(66.67)(10^{-6}) = 13.333(10^6)\ N/m^2(C) \quad (4)$$

4. Internal forces

$$A_A = A_B = \pi(0.01)^2/4 = 78.54(10^{-6})\ m^2 \quad (5)$$

$$N_A = \sigma_A A_A = (8.0)(10^6)(78.54)(10^{-6}) = 628.31\ N(T) \qquad N_B = \sigma_B A_B = (13.333)(10^6)(78.54)(10^{-6}) = 1047.2\ N(T) \quad (6)$$

5. External forces: By force equilibrium of free body diagram of the rigid plate we obtain the following.

$$2F - 2N_A - 2N_B = 0 \quad or \quad F = N_A + N_B = 1675.5\ N \quad (7)$$

ANS $F = 1676$ N

---

4. 4

Solution $A_A = 1\ in^2$ $E_A = 10{,}000$ksi $A_B = 0.5in^2$ $E_B = 30{,}000$ksi $\varepsilon_A = 500\mu$ F = ?

1. Deformation: From the given data and the exaggerated deformed geometry we obtain

$$\delta_A = \varepsilon_A L_A = (500)(10^{-6})(60) = 0.03\ in \qquad \delta_B = \delta_A - 0.02 = 0.01\ in\ contraction. \quad (1)$$

(a)

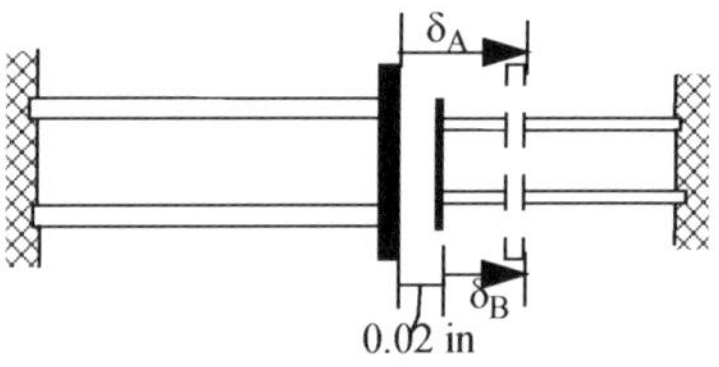

(b)

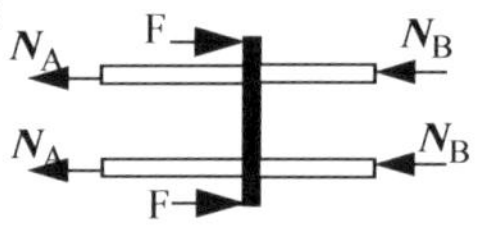

2. Strains

$$\varepsilon_A = 500(10^{-6})\ extension \qquad \varepsilon_B = \delta_B/L_B = 0.01/24 = 416.7(10^{-6})\ contraction \quad (2)$$

3. Stresses

$$\sigma_A = E_A\varepsilon_A = 10(10^3)(500)(10^{-6}) = 5\ ksi\ (T) \qquad \sigma_B = E_B\varepsilon_B = 30(10^3)(416.7)(10^{-6}) = 12.5\ ksi\ (C) \quad (3)$$

4. Internal forces

$$N_A = \sigma_A A_A = (5)(1) = 5.0\ kips\ (T) \qquad N_B = \sigma_B A_B = (12.5)(0.5) = 6.25\ kips\ (C) \quad (4)$$

5. External forces: By force equilibrium of the free body diagram of the rigid plate we obtain

$$2F - 2N_A - 2N_B = 0 \quad or \quad F = N_A + N_B = 5.0 + 6.25 = 11.25\ kips \quad (5)$$

ANS $F = 11.25\ kips$

---

4. 5

Solution $\varepsilon_{xx} = 200\mu$ $E_{al} = 100$GPa $E_w = 10$GPa $E_s = 200$GPa$\sigma_{xx} = ?$ $N = ?$ $y_N = ?$

The normal stress in steel, wood and aluminum can be found using Hooke's Law as shown below.

$$(\sigma_{xx})_s = E_s\varepsilon_{xx} = (200)(10^9)(200)(10^{-6}) = 40(10^6)N/m^2 = 40\ MPa \quad (1)$$

$$(\sigma_{xx})_w = E_s\varepsilon_{xx} = (10)(10^9)(200)(10^{-6}) = 2(10^6)N/m^2 = 2\ MPa \quad (2)$$

$$(\sigma_{xx})_{al} = E_s\varepsilon_{xx} = (100)(10^9)(200)(10^{-6}) = 20(10^6)N/m^2 = 20\ MPa \quad (3)$$

The stress distribution across the cross-section is as shown below.

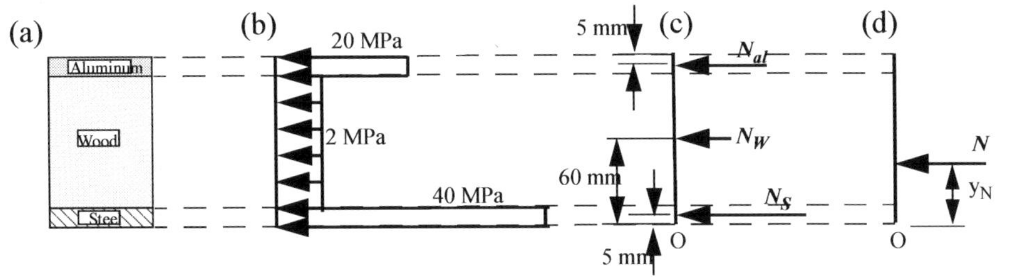

The stress distribution can be replaced by equivalent internal normal forces that act at the centroid of each distribution. The internal normal forces, can be found as shown below.

$$N_s = (\sigma_{xx})_s A_s = 40(10^6)(0.08)(0.01) = 32(10^3)\ N = 32kN \qquad (4)$$

$$N_w = (\sigma_{xx})_w A_w = 2(10^6)(0.08)(0.1) = 16(10^3)\ N = 16kN \qquad (5)$$

$$N_{A1} = (\sigma_{xx})_{A1} A_{A1} = 20(10^6)(0.08)(0.01) = 16(10^3)\ N = 16kN \qquad (6)$$

Equating the forces and the moment about point O we obtain:

$$N = N_s + N_w + N_{A1} = 64kN \qquad N(y_N) = N_s(0.005) + N_w(0.06) + (16)(0.115) \qquad (7)$$

$$(64)y_N = (32)(0.005) + (16)(0.006) + (16)(0.115) \quad or \quad y_N = \frac{2.96}{64} = 46.25(10^{-3})\ m \qquad (8)$$

ANS $N = 64kN(T);\ y_N = 46.25\ mm$

---

4. 6

Solution $\varepsilon_{xx} = -1500\mu$ $E_i = 25{,}000ksi$ $E_c = 3{,}000ksi$ $\sigma_{xx} = ?$ $N = ?$

The normal stress in iron and concrete can be found using Hooke's Law as shown below.

$$(\sigma_{xx})_i = E_i\varepsilon_{xx} = 25(10^3)(-1500)(10^{-6}) = -37.5\ ksi = 37.5\ ksi \qquad (1)$$

$$(\sigma_{xx})_c = E_c\varepsilon_{xx} = 3.0(10^3)(-1500)(10^{-6}) = -4.5\ ksi = 4.5\ ksi \qquad (2)$$

The normal stress distribution across the cross-section is as shown below.

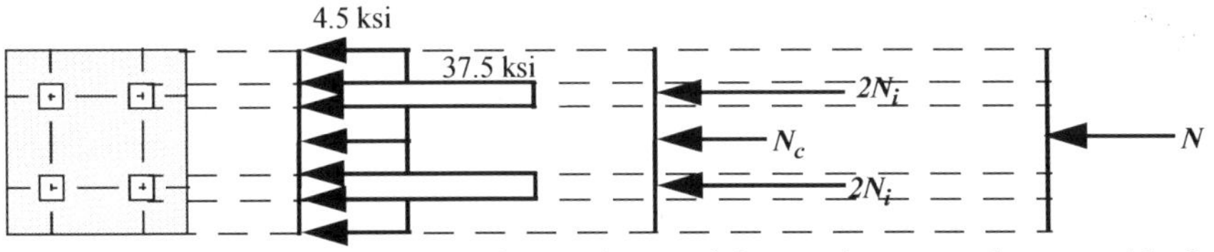

The stress distribution can be replaced by equivalent internal normal forces that act at the centroid of each distribution. The internal normal forces can be found as shown below.

$$A_i = (2)(2) = 4\ in^2 \quad and \quad A_c = (24)(24) - 4(4) = 560\ in^2 \qquad (3)$$

$$N_i = (\sigma_{xx})_i A_i = (37.5)(4) = 150\ kips\ (C) \qquad N_c = (\sigma_{xx})_c A_c = (4.5)(560) = 2520\ kips\ (C)$$

Equating forces we obtain:

$$N = 4N_i + N_c = 4(150) + 2520 = 3120\ kips \qquad (4)$$

ANS $N = 3120$ kips

Section 4.2

4. 7

Solution $m_1 = 1000kg$ $m_2 = 25kg$ $d_{AB} = 25mm$ $d_{BC} = 10mm$ $\sigma_{AB} = ?$ $\sigma_{BC} = ?$

By force equilibrium

$$N_{AB} = m_1g + m_2g = (1025)(9.81) = 10.055(10^3)\ N \qquad N_{BC} = m_1g/2 = (1000)(9.81)/2 = 4.905(10^3)\ N \qquad (1)$$

The stresses are

$$\sigma_{AB} = \frac{N_{AD}}{A_{AD}} = \frac{10.055(10^3)}{\pi(0.025)^2/4} = 20.483(10^6)\ N/m^2 \qquad \sigma_{BC} = \frac{N_{BC}}{A_{BC}} = \frac{4.905(10^3)}{\pi(0.01)^2/4} = 62045(10^6)\ N/m^2 \tag{2}$$

ANS $\sigma_{AB} = 20.5\text{ MPa(T)}$; $\sigma_{BC} = 62.5\text{ MPa(T)}$

---

4. 8

Solution d = 0.75in 2 sets of 12 cables L – soft E = 30,000ksi $\sigma_{ult} = 60\ ksi$ W = 100 ksi k_{saftey} = ?

By force equilibrium

$$24N_C = 100 \qquad or \qquad N_C = 4.167\ kips \tag{1}$$

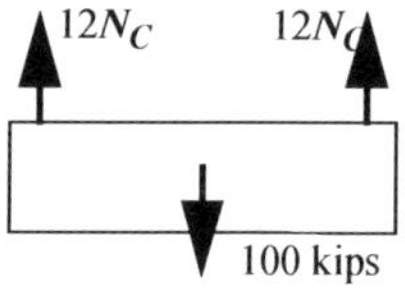

The axial stress and factor of safety can be found as shown below.

$$A_C = \pi(0.75/2)^2 = 0.4418 \qquad \sigma_C = N_C/A_C = 4.167/0.4418 = 9.43\ ksi \qquad k_{saftey} = \sigma_{ult}/\sigma_c = 60/9.43 = 6.361 \tag{2}$$

$$\varepsilon_C = \sigma_c/E = 9.43/30000 = 0.3143(10^{-3}) \qquad \delta_C = \varepsilon_C L = (0.3143)(10^{-3})(50)(12) = 0.1886\ in \tag{3}$$

ANS $k_{saftey} = 6.36$; $\delta_C = 0.19$ in.

---

4. 9

Solution N_{AB} = ? N_{BC} = ? N_{CD} = ? EA = 8,000 kips u_D-u_A = ?

We draw a template, write the template equation, and draw the axial force diagram as shown below.

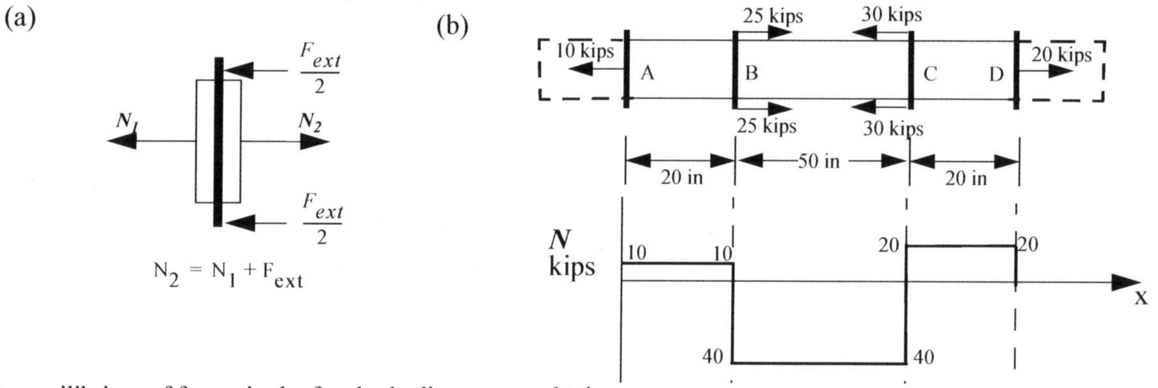

By equilibrium of forces in the free body diagram we obtain:

$$N_{AB} = 10\ kips \qquad N_{BC} = 10 - 50 = -40\ kips \qquad N_{CD} = 20\ kips \tag{1}$$

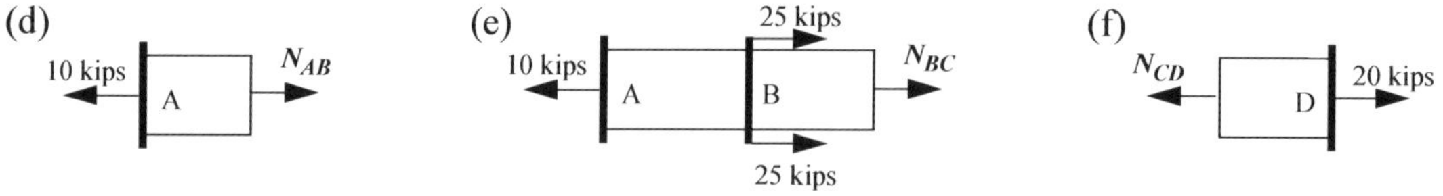

We find the relative displacements of the section ends and add to obtain relative displacement of section at D relative to A.

$$u_B - u_A = \frac{N_{AB}(x_B - x_A)}{EA} = \frac{(10)(20)}{8000} = 0.025\ in \qquad u_C - u_B = \frac{N_{BC}(x_C - x_B)}{EA} = \frac{(-40)(50)}{8000} = -0.250\ in \tag{2}$$

$$u_D - u_C = \frac{N_{CD}(x_D - x_C)}{EA} = \frac{(20)(20)}{8000} = 0.050\ in \qquad u_D - u_A = 0.025 - 0.25 + 0.05 = -0.175\ in \tag{3}$$

ANS $u_D - u_A = -0.175\ in$

---

4. 10

Solution N_{AB} = ? N_{BC} = ? N_{CD} = ? EA = 80,000 kN u_C = ?

By force equilibrium of the entire axial member, we obtain the reaction force at A as shown below.

$$R_A - 150 + 90 - 70 = 0 \qquad or \qquad R_A = 130\ kN \tag{1}$$

We draw a template, write the template equation, and draw the axial force diagram as shown below.

(a)

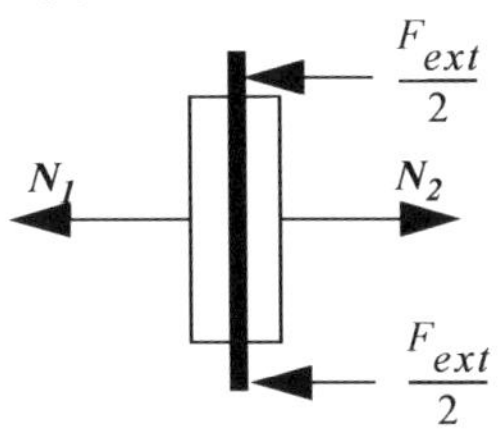

$N_2 = N_1 + F_{ext}$

(b)

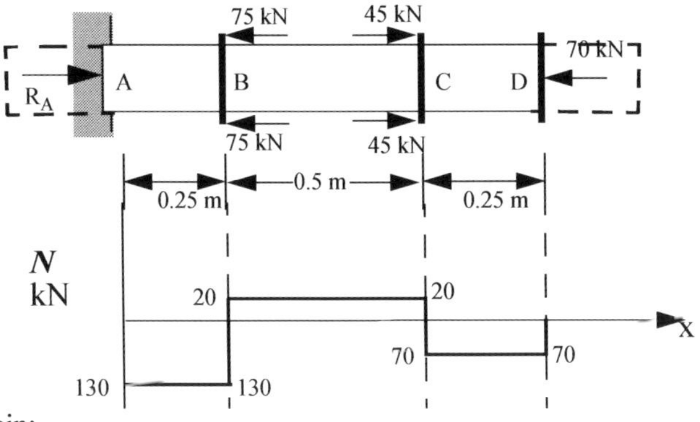

By equilibrium of forces in the free body diagram we obtain:

$$N_{AB} = -130\ kN \qquad N_{BC} = 20\ kN \qquad N_{CD} = -70\ kN \tag{2}$$

(d)

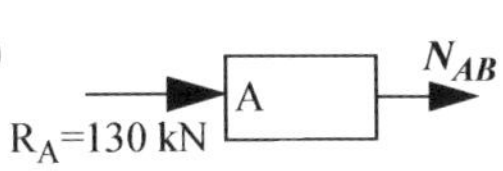

(e)

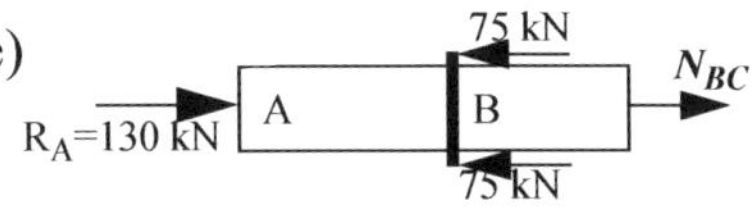

(f)

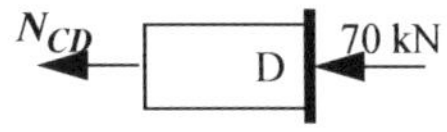

We find the relative displacements of the section ends and add to obtain relative displacement of section at C relative to A

$$u_B - u_A = \frac{N_{AB}(x_B - x_A)}{EA} = \frac{(-130)(0.25)}{80,000} = -0.4063(10^{-3})\ m \tag{3}$$

$$u_C - u_B = \frac{N_{BC}(x_C - x_B)}{EA} = \frac{(20)(0.5)}{80,000} = 0.125(10^{-3})\ m \qquad u_C - u_A = (-0.4063 + 0.125)(10^{-3}) = -0.28125(10^{-3})\ m$$

Point A is fixed to the wall, i.e., $u_A = 0$, we obtain the displacement of section at C.

ANS $u_C = -0.281$ mm

---

4. 11

Solution $N_{AB} = ?$ $N_{BC} = ?$ $N_{CD} = ?$ EA = 2,000 kips $u_B = ?$

By force equilibrium of the entire axial member, we obtain the reaction force at D as shown below.

$$1.5 + 4 - 8 + R_D = 0 \qquad or \qquad R_D = 2.5\ kips \tag{1}$$

We draw a template, write the template equation, and draw the axial force diagram as shown below.

(a)

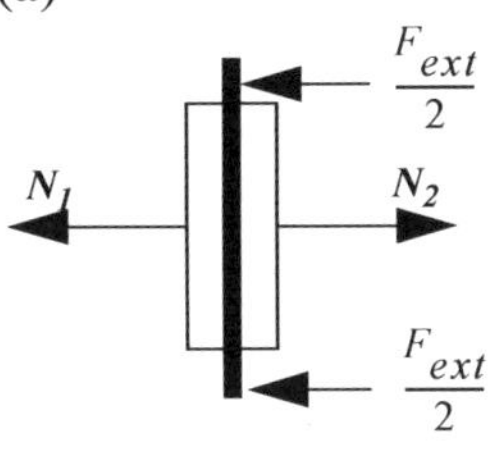

$N_2 = N_1 + F_{ext}$

(b)

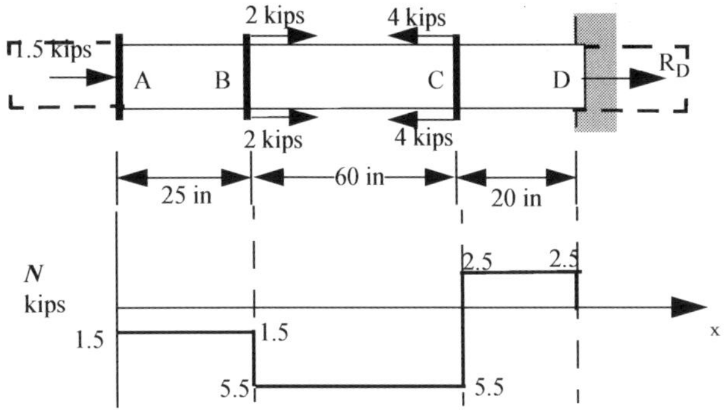

By equilibrium of forces in the free body diagram we obtain:

$$N_{AB} = -1.5\ kips \qquad N_{BC} = -5.5\ kips \qquad N_{CD} = 2.5\ kips \tag{2}$$

(d)

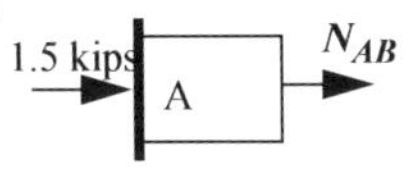

(e)

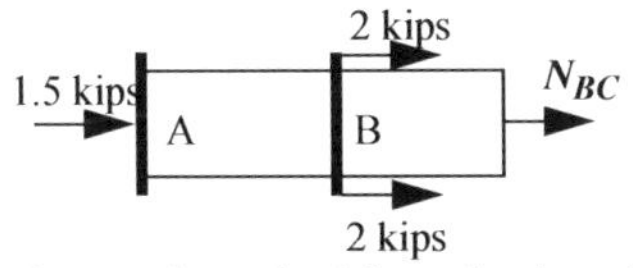

(f)

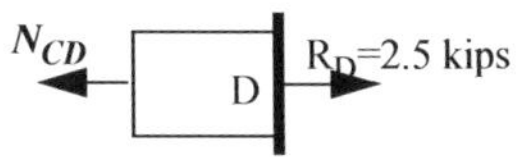

We find the relative displacements of the section ends and add to obtain relative displacement of section at B relative to D

$$u_C - u_B = \frac{N_{BC}(x_C - x_B)}{EA} = \frac{(-5.5)(60)}{2000} = -0.165\ in \tag{3}$$

$$u_D - u_C = \frac{N_{CD}(x_D - x_C)}{EA} = \frac{(2.5)(20)}{2000} = 0.025\ in \qquad u_D - u_B = -0.165 + 0.025 = -0.140\ in \tag{4}$$

Point D is fixed to the wall,i.e., $u_D = 0$, we obtain the displacement of section at B.

ANS $u_B = 0.14$ in.

---

4. 12

Solution $N_{AB} = ?$ $N_{BC} = ?$ $N_{CD} = ?$ EA = 50,000 kN $u_D - u_A = ?$

We draw a template, write the template equation, and draw the axial force diagram as shown below,

(a)   (b)

$N_2 = N_1 + F_{ext}$ Template Equation

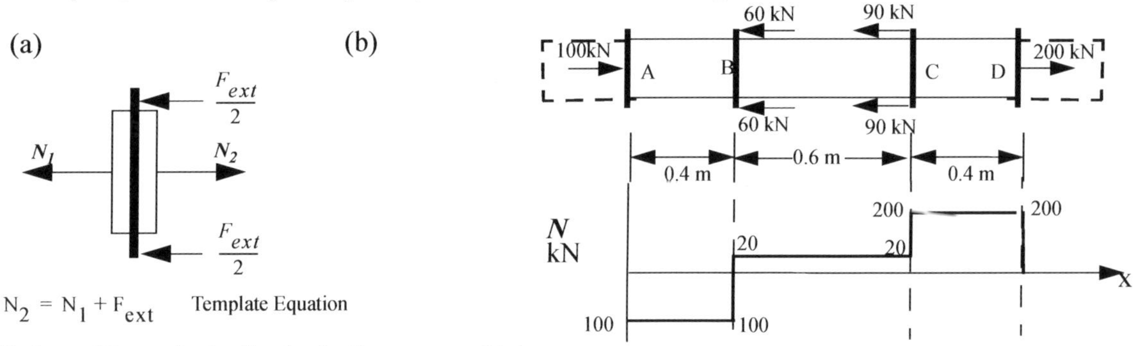

By equilibrium of forces in the free body diagram we obtain:

$$N_{AB} = -100 \ kN \qquad N_{BC} = 20 \ kN \qquad N_{CD} = 200 \ kN \qquad (1)$$

(d)   (e)   (f)We find the relative displacements of the section ends and add to obtain relative displacement of section at C relative to A.

$$u_B - u_A = \frac{N_{AB}(x_B - x_A)}{EA} = \frac{(-100)(0.4)}{50,000} = -0.80(10^{-3}) \ m \qquad u_C - u_B = \frac{N_{BC}(x_C - x_B)}{EA} = \frac{(20)(0.6)}{50,000} = 0.24(10^{-3}) \ m \qquad (2)$$

$$u_D - u_C = \frac{N_{CD}(x_D - x_C)}{EA} = \frac{(200)(0.4)}{50,000} = 1.60(10^{-3}) \ m \qquad u_C - u_A = (-0.80 + 0.24 + 1.60)(10^{-3}) = 1.04((10^{-3}) \ m) \qquad (3)$$

Point A is fixed to the wall,i.e., $u_A = 0$, we obtain the displacement of section at C

ANS $u_C = 1.04$ mm

---

4. 13

Solution   E = 1600 ksi $A = (4)(2) = 8 \ in^2$ $u_D - u_A = ?$ $\sigma_{max} = ?$

We draw a template, write the template equation, and draw the axial force diagram as shown below,

(a)   (b) Axial force diagram

$N_2 = N_1 + F_{ext}$The internal axial forces are:

$$N_{AB} = 15 \ kips \qquad N_{BC} = -30 \ kips \qquad N_{CD} = 25 \ kips \qquad (1)$$

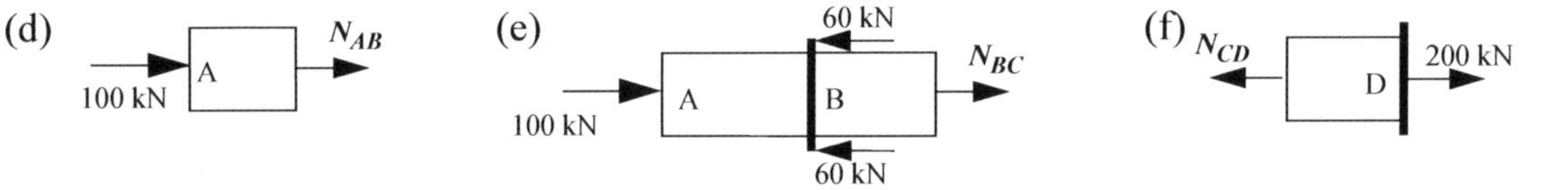

We find the relative displacements of the section ends and add to obtain relative displacement of section at D relative to A.

$$u_B - u_A = \frac{N_{AB}(x_B - x_A)}{EA} = \frac{(15)(30)}{(1600)(8)} = 0.0352 \ in \qquad u_C - u_B = \frac{N_{BC}(x_C - x_B)}{EA} = \frac{(-30)(50)}{(1600)(8)} = -0.1172 \ in \qquad (2)$$

$$u_D - u_C = \frac{N_{CD}(x_D - x_C)}{EA} = \frac{(25)(30)}{(1600)(8)} = 0.0586 \ in \qquad u_D - u_A = 0.0352 - 0.1172 + 0.0586 = -0.0234 \ in \qquad (3)$$

With all segments having the same area of cross-section, the maximum axial stress will be in the segment with the maximum internal axial force, i.e., segment BC. The axial stress in segment BC can be found as shown below.

$$\sigma_{max} = N_{BC}/A = -30/8 = -3.75 \ ksi \qquad (4)$$

ANS $u_D - u_A = -0.0234$ in. ; $\sigma_{max} = 3.75$ ksi(C)

---

4. 14

Solution $E = 30,000$ ksi $A = 0.5 \ in^2$ $F_1 = 8$ kips; $F_2 = 12$ kips; $F_3 = 9$ kips. $u_D - u_A = ?$ $\sigma_{max} = ?$

By equilibrium of the entire bar:

$$-2F_1 + 2F_2 - 2F_3 + 2F_4 = 0 \qquad or \qquad F_4 = 8 + 9 - 12 = 5 \ kips \qquad (1)$$

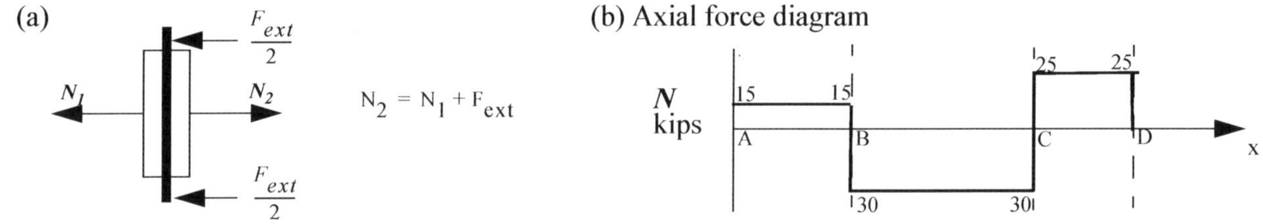

We can draw the axial force diagram as shown

(a)

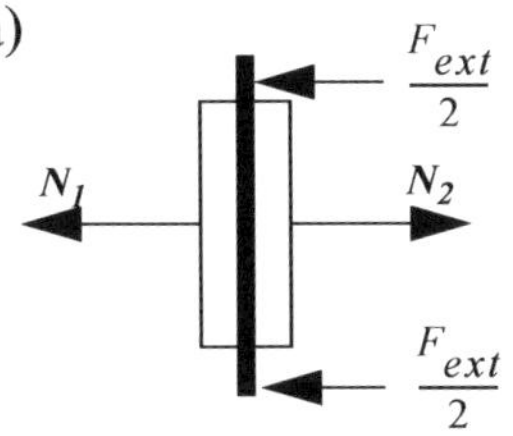

$N_2 = N_1 + F_{ext}$

(b) Axial force diagram

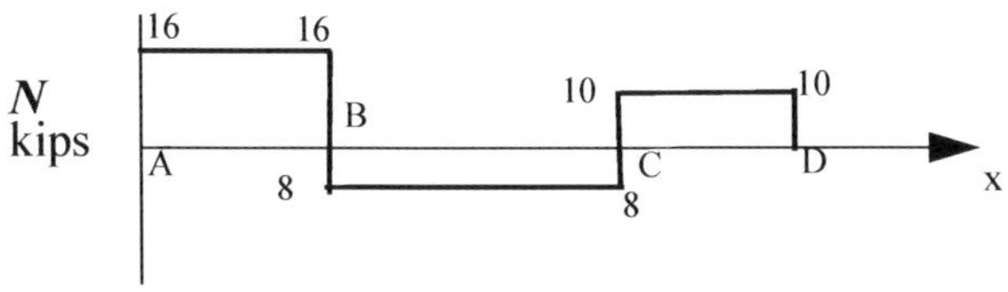

The internal axial forces are: $N_{AB} = 16\ kips$, $N_{BC} = -8\ kips$, and $N_{CD} = 10\ kips$.

We find the relative displacements of the section ends and add to obtain relative displacement of section at D relative to A.

$$u_B - u_A = \frac{N_{AB}(x_B - x_A)}{EA} = \frac{(16)(36)}{(30000)(0.5)(2)} = 19.2(10^{-3})\ in \quad (2)$$

$$u_C - u_B = \frac{N_{BC}(x_C - x_B)}{EA} = \frac{(-8)(50)}{(30000)(0.5)(2)} = -13.33(10^{-3})\ in \quad (3)$$

$$u_D - u_C = \frac{N_{CD}(x_D - x_C)}{EA} = \frac{(25)(36)}{(30000)(0.5)(2)} = 12(10^{-3})\ in \qquad u_D - u_A = (19.2 - 13.33 + 12)(10^{-3}) = 17.87(10^{-3})\ in \quad (4)$$

With all segments having the same area of cross-section, the maximum axial stress will be in the segment with the maximum internal axial force, i.e., segment AB. The axial stress in segment AB can be found as shown below.

$$\sigma_{max} = N_{AB}/A = 16/[(0.5)(2)] = 16\ ksi \quad (5)$$

ANS $u_D - u_A = 0.01787$ in. ; $\sigma_{max} = 16$ ksi(T)

---

4. 15

Solution $F_1 = 90$ kN; $F_2 = 40$ kN; $F_3 = 70$ kN. $u_D = ?$ $\sigma_{max} = ?$

We can find the reactions at the wall from equilibrium of the entire free body diagram.

$$2R_A - 2F_1 - 2F_2 + 2F_3 = 0 \quad or \quad R_A = 90 + 40 - 70 = 60 \text{ kN} \quad (1)$$

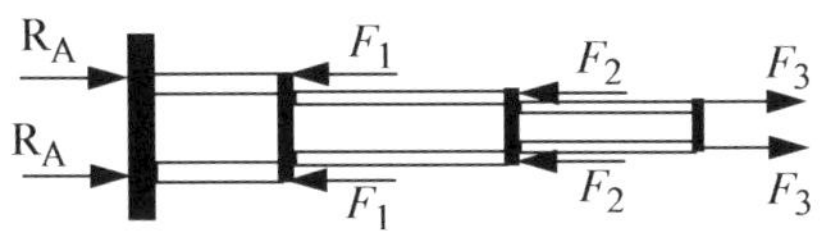

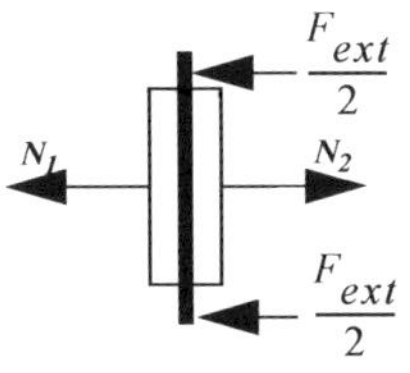

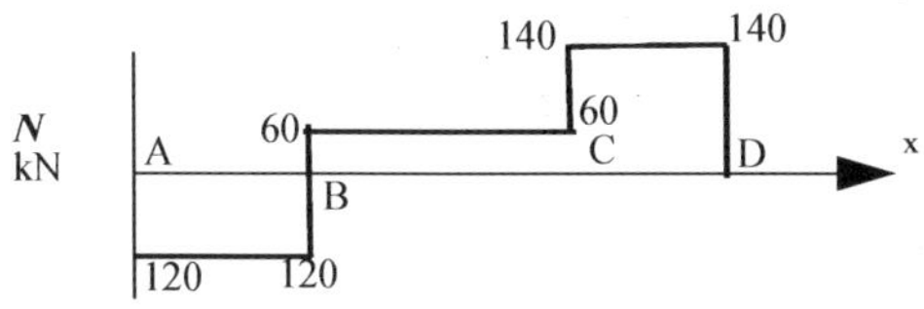

$N_2 = N_1 + F_{ext}$

The internal forces and cross sectional areas are:

$$N_{AB} = -120\ kN \qquad N_{BC} = 60\ kN \qquad N_{CD} = 140\ kN \quad (2)$$

$$A_{AB} = \frac{\pi}{4}(0.03)^2 = 0.706(10^{-3})\ \text{m}^2 \qquad A_{BC} = \frac{\pi}{4}(0.025)^2 = 0.4909(10^{-3})\ \text{m}^2 \qquad A_{CD} = \frac{\pi}{4}(0.02)^2 = 0.3142(10^{-3})\ \text{m}^2 \quad (3)$$

We find the relative displacements of the section ends and add to obtain relative displacement of section at D relative to A.

$$u_B - u_A = \frac{N_{AB}(x_B - x_A)}{E_{AB}A_{AB}} = \frac{(-120)(1.5)}{70(10^9)(0.706)(10^{-3})} = -1.819(10^{-3})\ m \quad (4)$$

$$u_C - u_B = \frac{N_{BC}(x_C - x_B)}{E_{BC}A_{BC}} = \frac{(20)(0.6)}{100(10^9)(0.4909)(10^{-3})} = 1.528(10^{-3})\ m \quad (5)$$

$$u_D - u_C = \frac{N_{CD}(x_D - x_C)}{E_{CD}A_{CD}} = \frac{(200)(0.4)}{200(10^9)(0.3142)(10^{-3})} = 2.228(10^{-3})\ m \quad (6)$$

$$u_D - u_A = (-1.819 + 1.528 + 2.228)(10^{-3}) = 1.937(10^{-3})\ m \quad (7)$$

$$\sigma_{AB} = \frac{N_{AB}}{A_{AB}} = \frac{-120(10^3)}{0.706(10^{-3})} = -169.7(10^6)\ \text{N}/m^2 \qquad \sigma_{BC} = \frac{N_{BC}}{A_{BC}} = \frac{60(10^3)}{0.4909(10^{-3})} = 61.1(10^6)\ \text{N}/m^2 \quad (8)$$

$$\sigma_{CD} = \frac{N_{CD}}{A_{CD}} = \frac{140(10^3)}{0.3142(10^{-3})} = 222.8(10^6)\ \text{N}/m^2 \quad (9)$$

ANS $u_D = 1.937$ mm ; $\sigma_{max} = 222.8$ MPa(T)

---

4. 16

Solution $E_s = 30{,}000$ ksi u_D-u_A = ? σ_{max} = ?

We can draw the axial force diagram as shown

(a)

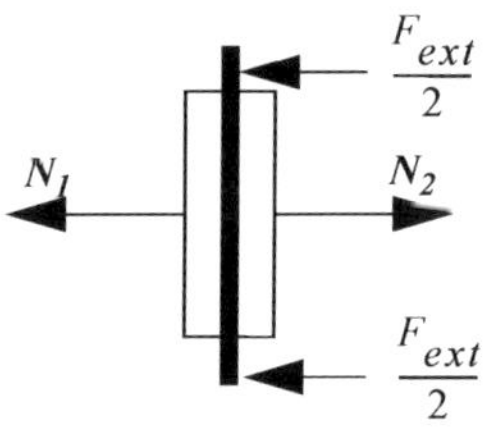

$N_2 = N_1 + F_{ext}$

(b) Axial force diagram

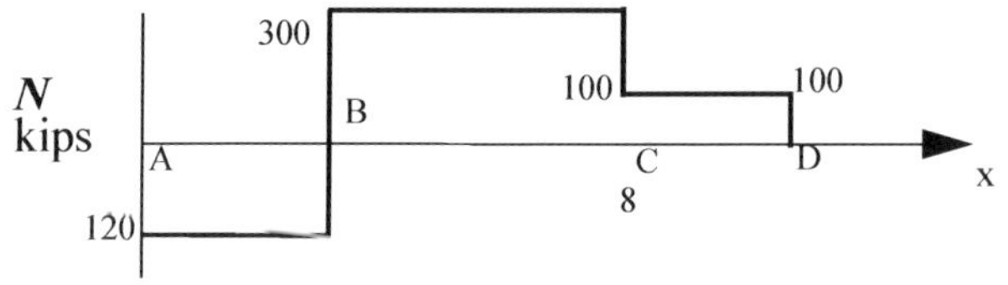

The internal axial forces and cross sectional areas are:

$$N_{AB} = -120 \;\; kips \qquad N_{BC} = 300 \;\; kips \qquad N_{CD} = 100 \;\; kips \tag{1}$$

$$A_{AB} = A_{CD} = (\pi/4)(4^2 - 2^2) = 9.428 \;\; in^2 \qquad A_{BC} = (\pi/4)(4)^2 = 12.566 \text{ in}^2 \tag{1}$$

We find the relative displacements of the section ends and add to obtain relative displacement of section at D relative to A.

$$u_B - u_A = \frac{N_{AB}(x_B - x_A)}{EA} = \frac{(-120)(24)}{(30000)(9.428)} = -10.186(10^{-3})in \qquad u_C - u_B = \frac{N_{BC}(x_C - x_B)}{EA} = \frac{(300)(36)}{(30000)(12.566)} = 28.678in \tag{2}$$

$$u_C - u_B = \frac{N_{BC}(x_C - x_B)}{EA} = \frac{(300)(36)}{(30000)(12.566)} = 28.678 \;\; in \qquad u_D - u_C = \frac{N_{CD}(x_D - x_C)}{EA} = \frac{(100)(24)}{(30000)(9.428)} = 8.488(10^{-3}) \;\; in \tag{3}$$

$$u_D - u_A = (-10.186 + 28.678 + 8.488)(10^{-3}) = 26.95(10^{-3}) \;\; in \tag{4}$$

$$\sigma_{AB} = N_{AB}/A_{AB} = -120/9.428 = -12.732 \;\; ksi \qquad \sigma_{BC} = N_{BC}/A_{BC} = 300/12.566 = 23.87 \;\; ksi$$

ANS $u_D - u_A = 0.027$ in. ; $\sigma_{max} = 23.87$ ksi(T)

---

4. 17

Solution $E_{AB} = E_{CD} = 30{,}000$ ksi $E_{BC} = 10{,}000$ ksi $\nu_{AB} = \nu_{CD} = 0.3$ $\nu_{BC} = 0.33$

$d_{AB} = d_{CD} = 2.0$ in $d_{BC} = 1.5$ in u_C-u_A = ? Δd_{max} = ?

By force equilibrium of the entire axial member, we obtain the reaction force at A as shown below.

$$R_A - 10 - 35 + 25 = 0 \qquad or \qquad R_A = 20 \;\; kips \tag{1}$$

We draw the axial force diagram and determine the internal forces as shown below.

$$N_{AB} = -20 \;\; kips \qquad N_{BC} = -10 \;\; kips \qquad N_{CD} = 25 \;\; kips \tag{2}$$

(a)

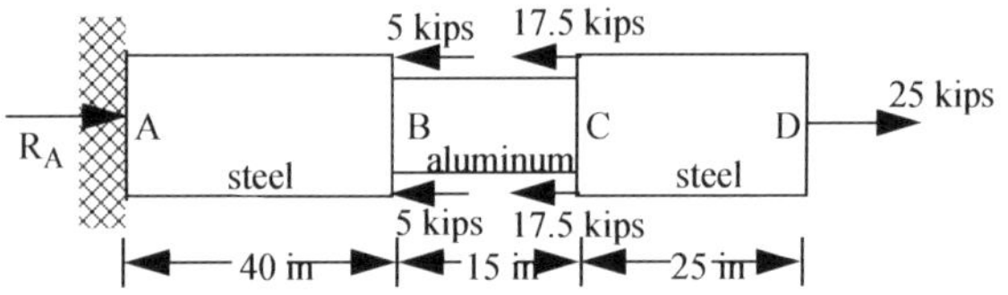

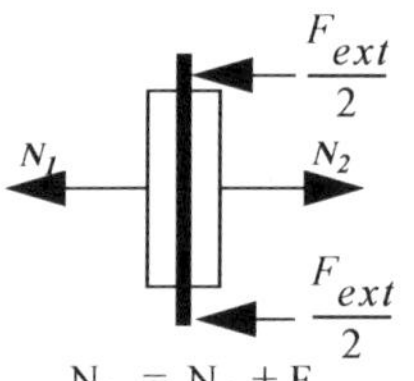

$N_2 = N_1 + F_{ext}$

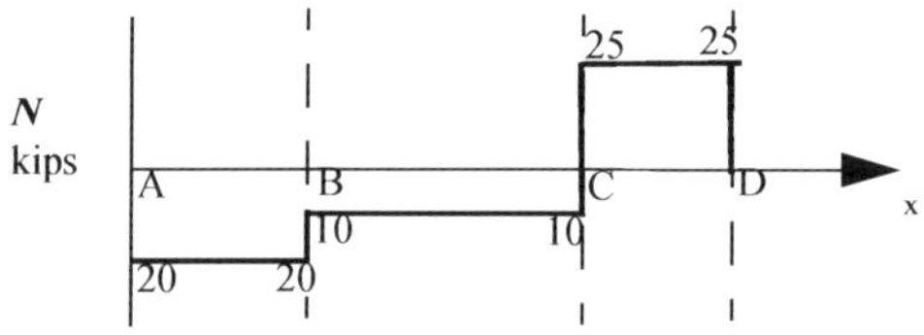

The area of cross-sections are:

$$A_{AB} = A_{CD} = \pi(2^2)/4 = 3.142 \;\; in^2 \qquad A_{BC} = \pi(1.5^2)/4 = 1.767 \;\; in^2 \tag{3}$$

We find the relative displacements of the section ends and add to obtain relative displacement of section at C relative to A.

$$u_B - u_A = \frac{N_{AB}(x_B - x_A)}{EA} = \frac{(-20)(40)}{(30,000)(3.142)} = -8.488(10^{-3}) \;\; in \tag{4}$$

$$u_C - u_B = \frac{N_{BC}(x_C - x_B)}{EA} = \frac{(-10)(15)}{(10,000)(1.767)} = -8.488(10^{-3})in \qquad u_C - u_A = (-8.488 - 8.488)(10^{-3}) = (-16.98)(10^{-3})in \tag{5}$$

$$\sigma_{BC} = N_{BC}/A_{BC} = -10/1.767 = -5.659 \;\; ksi \qquad \sigma_{CD} = N_{CD}/A_{CD} = 25/3.142 = 7.958 \;\; ksi \tag{6}$$

$$\varepsilon_{BC} = \sigma_{BC}/E_{BC} = -5.659/10,000 = (-0.5659)(10^{-3}) \qquad \varepsilon_{CD} = \sigma_{CD}/E_{CD} = 7.958/30,000 = 0.2653((10^{-3})) \tag{7}$$

$$(\varepsilon_{BC})_T = -\nu_{BC}\varepsilon_{BC} = -0.33(-0.5659)(10^{-3}) = 0.1867(10^{-3}) \qquad (\varepsilon_{CD})_T = -\nu_{CD}\varepsilon_{CD} = -0.3(0.2653)(10^{-3}) = -0.07959(10^{-3}) \tag{8}$$

$$\Delta d_{BC} = (1.5)(0.1867)(10^{-3}) = (0.2801)(10^{-3}) \;\; in \qquad \Delta d_{CD} = (2)(-0.07959)(10^{-3}) = (-0.1592)(10^{-3}) \;\; in$$

ANS $u_C - u_A = -0.017$ in. ; $\Delta d_{max} = 0.00028$ in.

---

4. 18

Solution $E = 100GP_a$ $(d_{AC})_o = 70$ mm $(d_{AC})_i = 50$ mm $(d_{DB})_o = 50$ mm
$(d_{DB})_i = 30$ mm $(d_{CD})_o = 70$ mm $(d_{CD})_i = 30$ mm $u_B\text{-}u_A = ?$

The internal axial force and material cross-section areas are:

$$N_{AC} = N_{CD} = N_{DB} = 20\ kN \qquad A_{AC} = \frac{\pi[(0.07)^2-(0.05)^2]}{4} = 1.885(10^{-3})\ m^2 \tag{1}$$

$$A_{DB} = \frac{\pi[(0.05)^2-(0.03)^2]}{4} = 1.257(10^{-3})\ m^2 \qquad A_{CD} = \frac{\pi[(0.07)^2-(0.03)^2]}{4} = 3.142(10^{-3})\ m^2 \tag{2}$$

We find the relative displacements of the section ends and add to obtain relative displacement of section at B relative to A.

$$u_C - u_A = \frac{N_{AC}(x_C-x_A)}{E_{AC}A_{AC}} = \frac{(20)(10^3)(0.5)}{(100)(10^9)(1.885)(10^{-3})} = 0.05305(10^{-3})\ m \tag{3}$$

$$u_D - u_C = \frac{N_{CD}(x_D-x_C)}{E_{CD}A_{CD}} = \frac{(20)(10^3)(0.15)}{(100)(10^9)(3.142)(10^{-3})} = 0.00955\ (10^{-3})\ m \tag{4}$$

$$u_B - u_D = \frac{N_{BD}(x_B-x_D)}{E_{BD}A_{BD}} = \frac{(20)(10^3)(0.4)}{(100)(10^9)(1.257)(10^{-3})} = 0.06364\ (10^{-3})\ m \tag{5}$$

$$u_B - u_A = (0.05305+0.00955+0.06364)(10^{-3}) = 0.12624(10^{-3})\ m \tag{6}$$

ANS $u_B - u_A = 0.126$ mm

---

4. 19

Solution $u_B\text{-}u_A = ?$

We can write the strain and integrate as shown below.

$$\frac{du}{dx} = \frac{N_{AB}}{E_{AB}A_{AB}} = \frac{P}{E[K(2L-0.25x)^2]} \quad or \quad \int_0^L du = \frac{P}{EK}\int_0^L \frac{dx}{(2L-0.25x)^2} \quad \text{or}$$

$$u = \frac{P}{EK}\left[\frac{1}{0.25(2L-0.25x)}\bigg|_0^L\right] = \frac{P}{EK}\left[\frac{1}{0.4375L} - \frac{1}{0.5L}\right] \tag{1}$$

ANS $u = 0.2857\ P/(EKL)$

---

4. 20

Solution u_B - $u_A = ?$

We can write the strain and integrate as shown below.

$$\frac{du}{dx} = \frac{N_{AB}}{E_{AB}A_{AB}} = \frac{P}{EK(4L-3x)} \quad or \quad \int_0^L du = \int_0^L \frac{P}{EK}\cdot\frac{dx}{(4L-3x)} \quad \text{or}$$

$$u = \frac{P}{3EK}[ln|4L-3x||_0^L] = -\frac{P}{3EK}[lnL - ln4L] = -\frac{P}{3EK}\left[ln\frac{1}{4}\right] \tag{1}$$

ANS $u = 0.4621P/(EK)$

---

4. 21

Solution $E_{AB} = E_{CD} = 30{,}000$ ksi $E_{BC} = 10{,}000$ ksi $d_{AB} = 2.0$ in $d_{BC} = 1.5$ in
d_{CD} varies from 1.5 in to 2 in $u_C\text{-}u_A = ?$ $\sigma_{max} = ?$

We draw the axial force diagram and determine the internal axial forces as shown below.

$$N_{AB} = 40\ kips \qquad N_{BC} = 20\ kips \qquad N_{CD} = 60\ kips \tag{1}$$

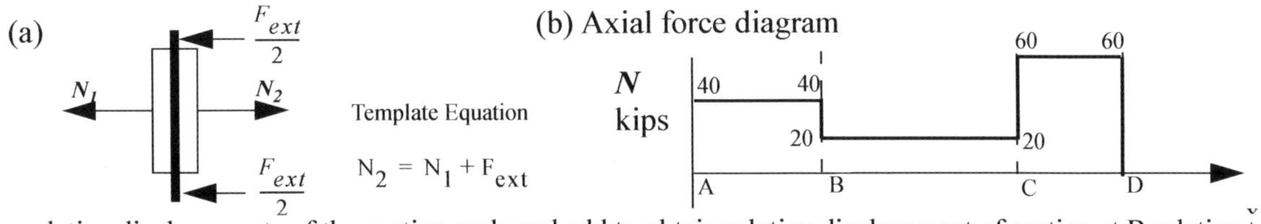

We find the relative displacements of the section ends and add to obtain relative displacement of section at B relative to A.

$$u_B - u_A = \frac{N_{AB}(x_B-x_A)}{E_{AB}A_{AB}} = \frac{(40)(40)}{(30,000)(\pi 2^2/4)} = 16.976(10^{-3})\ in \tag{2}$$

$$u_C - u_B = \frac{N_{BC}(x_C - x_B)}{E_{BC}A_{BC}} = \frac{(20)(15)}{(10,000)(\pi 1.5^2/4)} = 16.976(10^{-3})\ in \qquad u_C - u_A = (16.976 + 16.976)(10^{-3}) = 33.952(10^{-3})in \quad \textbf{(3)}$$

The maximum stress in CD will be just after C, where the diameter is 1.5 in.

$$A_{CD} = \pi 1.5^2/4 = 1.767\ \text{in}^2 \qquad \sigma_{max} = N_{CD}/A_{CD} = 60/1.767 = 33.95\ ksi \quad \textbf{(4)}$$

ANS $u_C - u_A = 0.034\ in$; $\boxed{\sigma_{max} = 33.95\ ksi\ (T)}$

---

4. 22

Solution $u_B - u_A = ?$

We can write the distributed force and integrate to obtain the internal axial force as shown below.

$$\frac{dN}{dx} = -p = -(-\gamma A) = \gamma\pi a^2 \qquad N = \gamma\pi a^2 x + c_1 \quad \textbf{(1)}$$

At $x = L$ $N = 0$ and we obtain: $c_1 = -\gamma\pi a^2 L$. Substituting N we obtain strain which we integrate to obtain displacement as shown below.

$$N = \gamma\pi a^2 x - \gamma\pi a^2 L \qquad \frac{du}{dx} = \frac{N}{EA} = \frac{\gamma\pi a^2}{E\pi a^2}(x-L) = \frac{\gamma}{E}(x-L) \qquad u_B - u_A = \int_{u_A}^{u_B} du = \frac{\gamma}{E}\int_0^L (x-L)dx = \frac{\gamma}{E}\left(\frac{x^2}{2} - Lx\right)\bigg|_0^L = \frac{-\gamma L^2}{2E} \quad \textbf{(2)}$$

ANS $u_B = -\gamma L^2/(2E)$

---

4. 23

Solution $u_B - u_A = ?$

The internal aqxial force is calculated as shown below.

$$A = \frac{1}{2}ab = \frac{\sqrt{3}}{4}a^2 \qquad p = -\gamma\left(\frac{\sqrt{3}}{4}a^2\right) \qquad \frac{dN}{dx} = -p = \frac{\sqrt{3}}{4}\gamma a^2 \qquad N = \frac{\sqrt{3}}{4}\gamma a^2 x + c_1 \quad \textbf{(1)}$$

At $x = L$, $N = 0$ and we obtain: $c_1 = (-\sqrt{3}\gamma a^2 L)/4$. Substituting ***N*** we obtain strain which we integrate to obtain displacement as shown below.

$$N = (\sqrt{3}/4)\gamma a^2(x-L) \qquad \frac{du}{dx} = \left[\frac{N}{EA}\right] = (\sqrt{3}/4)\frac{\gamma\pi a^2}{E(\sqrt{3}/4)a^2}(x-L) = \frac{\gamma}{E}(x-L) \quad \textbf{(2)}$$

$$u_B - u_A = \int_{u_A}^{u_B} du = \frac{\gamma}{E}\int_0^L (x-L)dx = \frac{\gamma}{E}\left(\frac{x^2}{2} - Lx\right)\bigg|_0^L = \frac{-\gamma L^2}{2E} \quad \textbf{(3)}$$

ANS $u_B = -\gamma L^2/(2E)$

---

4. 24

Solution $u_B - u_A = ?$

The internal aqxial force is calculated as shown below.

$$p = -\gamma A \qquad \frac{dN}{dx} = -p = \gamma A \qquad N = \gamma A x + c_1 \quad \textbf{(1)}$$

At $x = L$, $N = 0$ and we obtain $c_1 = -\gamma AL$. Substituting ***N*** we obtain strain which we integrate to obtain displacement as shown below.

$$N = \gamma A(x-L) \qquad \frac{du}{dx} = \left[\frac{N}{EA}\right] = \gamma A(x-L) \qquad \int_{u_A}^{u_B} du = \gamma A\int_0^L (x-L)dx \qquad u_B - u_A = \frac{\gamma A}{EA}\left(\frac{x^2}{2} - Lx\right)\bigg|_0^L = \frac{\gamma}{2E}\left(\frac{-L}{2}\right) \quad \textbf{(2)}$$

ANS $u_B = -\gamma L^2/(2E)$

---

4. 25

Solution $\delta = ?$

Let the total elongation $\delta = \delta_W + \delta_P$, where δ_W is the elongation due to weight and δ_P is the elongation due to the force P. From previous example we have:

$$\delta_W = \frac{7}{30}\frac{\gamma L^2}{E} \qquad R(x) = \frac{r}{L}(5L - 4x) \qquad or \qquad A(x) = \pi\frac{r^2}{L^2}(5L-4x)^2 \quad \textbf{(1)}$$

We find the internal axial force, the strain, and displacement as shown below.

$$N = P = \frac{\gamma\pi r^2 L}{5} \qquad \frac{du}{dx} = \frac{N}{EA} = \frac{\gamma\pi r^2 L}{5E\pi(r^2/L^2)(5L-4x)^2} = \frac{\gamma L^3}{5E(5L-4x)^2} \text{ or}$$

$$\delta_P = \int_{u_0}^{u_L} du = \int_0^L \frac{\gamma L^3}{5E(5L-4x)^2}dx = \frac{\gamma L^3}{5E}\left[\frac{1}{-(-4)}\cdot\frac{1}{(5L-4x)}\right]\Bigg|_0^L = \frac{\gamma L^3}{20E}\left[\frac{1}{L}-\frac{1}{5L}\right] = \frac{\gamma L^2}{25E} \quad \textbf{(2)}$$

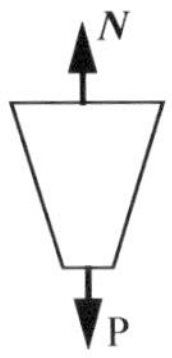

$$\delta = \delta_W + \delta_P = \frac{7}{30}\frac{\gamma L^2}{E} + \frac{\gamma L^2}{25E} = \frac{41}{150}\frac{\gamma L^2}{E} \quad \textbf{(3)}$$

ANS $\delta = 0.273\gamma L^2/E$

---

4. 26

Solution L = 20ft t = 118in d_b = 8in d_t = 2in W_{Light} = 80lbs

$\gamma = 0.1\ lb/in^3$ E = 11,000ksi G = 8000ksi σ_{max} = ? δ = ?

The outer radius R(x) of the pole at can be found as shown below.

$$R(x) = a + bx \qquad (R(x=0) = 1) \Rightarrow a = 1 \qquad (R(x=240) = 4) \Rightarrow 4 = 1 + b(240) \qquad or \qquad b = 1/80 \quad \textbf{(1)}$$

$$R(x) = \frac{(80+x)}{80}\ in \quad \textbf{(2)}$$

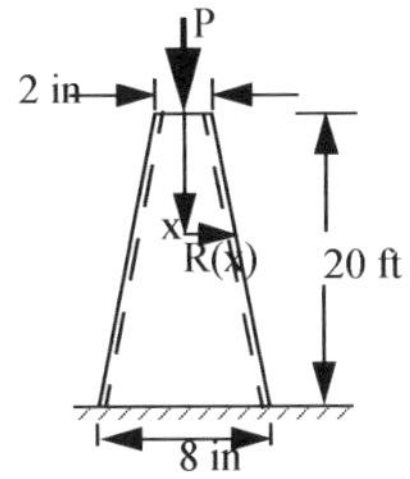

For thin-walled tubes the area of cross-section can be approximated and distributed force can be found as shown below.

$$A = 2\pi Rt = (2\pi)\frac{(80+x)}{80}\left(\frac{1}{8}\right) = 9.817(10^3)(80+x) \qquad p(x) = \gamma A = 0.9817(10^{-3})(80+x) \quad \textbf{(3)}$$

The internal axial force can be found as shown below.

$$\frac{dN}{dx} = -p = -0.9817(10^{-3})(80+x) \qquad or \qquad N = -0.4909(10^{-3})(80+x)^2 + c_1 \quad \textbf{(4)}$$

At x = 0 $\quad$ ***N*** = -80lbs, thus $\quad -80 = -0.4909(10^{-3})(80)^2 + c_1 \quad$ or $\quad c_1 = -76.858$

$$N = -0.4909(10^{-3})(80+x)^2 - 76.858 \quad \textbf{(5)}$$

The maximum axial force will be at the bottom of the pole i.e., x = 240 in

$$N_{max} = -0.4909(10^{-3})(320)^2 - 76.858 = -127.12lbs \quad \textbf{(6)}$$

The area of cross-section at the bottom and the maximum axial stress is:

$$A_b = 9.817(10^3)(320) = 3.1416\ in^2 \qquad \sigma_{max} = N_{max}/A_b = -127.12/3.1416 = -54.78\ psi \quad \textbf{(7)}$$

We can write the strain and integrate to obtain the displacement as shown below.

$$\frac{du}{dx} = \frac{N}{EA} = \frac{-0.4909(10^{-3})(80+x)^2 - 76.858}{11(10^6)L9.817(10^{-3})(80+x)} = -0.4545(10^{-9})(80+x) - \frac{0.7117(10^{-3})}{80+x} \quad \textbf{(8)}$$

$$\int_{u_0}^{u_{240}} du = -0.4545(10^{-9})\int_0^{240}(80+x)dx - 0.7117(10^{-3})\int_0^{240}\frac{dx}{80+x} \text{ or}$$

$$u_{240} - u_0 = -0.4545(10^{-9})\frac{(80+x)^2}{2}\Bigg|_0^{240} - 0.7117(10^{-3})ln(80+x)\Big|_0^{240} = -1.205(10^{-3}) \quad \textbf{(9)}$$

Now $u_{240} = 0$ as the point is fixed to the ground and we obtain the contraction $\delta = u_0 = 1.205(10^{-3})$

ANS σ_{max} = 54.8 psi(C) ; δ = 0.0012 in.

---

4. 27

Solution $\gamma = 0.28\ lb/in^3$ $E = 3,600\ ksi$ $L = 120\ in$ $R = \sqrt{240-x}$ $\delta = ?$

We can find the internal axial force and then the displacement as shown below.

$$p = -\gamma A = -\gamma\pi R^2 = -0.28\pi(240-x) \qquad \frac{dN}{dx} = -p = 0.28\pi(240-x) \tag{1}$$

$$\int_0^N dN = \int_{120}^x 0.2817(240-x)dx \quad or \quad N = -\frac{(0.28\pi)}{2}(240\ \ x)^2\Big|_{120}^x = -0.14\pi[(240-x)^2 - 120^2] \tag{2}$$

$$\frac{du}{dx} = \frac{N}{EA} = \frac{-0.14\pi[(240-x)^2-120^2]}{(3\cdot 6)(10^6)\pi(240-x)} \quad or \quad \frac{du}{dx} = -3.889(10^{-6})\left[(240-x)-\frac{120^2}{240-x}\right] \text{ or}$$

$$\int_{u_0}^{u_{120}} du = -3.889(10^{-6})\int_0^{120}\left[(240-x)-\frac{120^2}{(240-x)}\right]dx \quad or \quad u_{120}-u_0 = -3.889(10^{-6})\left[\frac{-(240-x)^2}{2}+120^2 ln(240-x)\right] \tag{3}$$

$$u_{120} = -3.889(10^{-6})\left[\frac{-120^2}{2}+\frac{240^2}{2}+120^2 ln(240-120)\right] = -45.18(10^{-3})\ inches \tag{4}$$

ANS $\delta = 0.045\ in$

---

4. 28

Solution $\gamma = 24\ kN/m^3$ $E = 25\ GPa$ $L = 10\ m$ $R = 0.5e^{-0.07x}$ $\delta = ?$

We can find the internal axial force and then the displacement as shown below.

$$p = -\gamma A = -\gamma\pi R^2 = -24(10^3)\pi(0.5e^{-0.07x})^2 = -18.85(10^3)e^{0.14x} \qquad \frac{dN}{dx} = -p = 18.85(10^3)e^{-0.14x} \tag{1}$$

$$\int_0^N dN = \int_{10}^x (18.85\pi)(10^3)e^{-0.14x}dx \quad or \quad N = -18.85\pi(10^3)\left(\frac{e^{-0.14x}}{0.14}\right)\Big|_{10}^x = -134.64(10^3)(e^{-0.14x}-e^{-1.4}) \tag{2}$$

$$\frac{du}{dx} = \frac{N}{EA} = \frac{-134.64(10^3)(e^{-0.14}-0.2466)}{25(10^9)\pi[(0.5)^2e^{-0.14x}]} = -6.857(10^{-6})(1-0.2466e^{0.14x}) \tag{3}$$

$$\int_{u_0}^{u_{10}} du = -6.857(10^{-6})\int_0^{10}(1-0.2466e^{0.14})dx \qquad u_{10}-u_0 = -6.857(10^{-6})\left(x-\frac{0.2466e^{0.14x}}{0.14}\right)\Big|_0^{10} \tag{4}$$

Point x = 0 is fixed to the ground i.e., $u_0 = 0$

$$u_{10} = -6.857(10^{-6})[10-1.761(e^{1.4}-1)] = -31.67(10^{-6})\ m \tag{5}$$

ANS $\delta = 0.317$ mm

---

4. 29

Solution F = ? $p(x) = f_{max}(x^2/L^2)$ $\delta = ?$

We can find the internal axial force and then the displacement as shown below.

$$\frac{dN}{dx} = -p(x) = -f_{max}(x^2/L^2) \quad or \quad \int_0^N dN = -f_{max}\int_L^x \frac{x^2}{L^2}dx \quad or \quad N = \frac{-f_{max}x^3}{3L^2}\Big|_L^x = \frac{f_{max}}{3L^2}(L^3-x^3) \tag{1}$$

At x = 0 N = F; Thus, $F = f_{max}L/3$

$$\frac{du}{dx} = \frac{N}{EA} = \frac{f_{max}}{3EAL^2}(L^3-x^3) \quad or \quad \int_{u_0}^{u_L} du = \frac{f_{max}}{3EAL^2}\int_0^L (L^3-x^3)dx \tag{2}$$

$$u_L - u_0 = \frac{f_{max}}{3EAL^2}\left(L^3x-\frac{x^4}{4}\right)\Big|_0^L = \frac{f_{max}}{3EAL^2}\left(L^4-\frac{L^4}{4}\right) = \frac{f_{max}L^2}{4EA} \tag{3}$$

ANS $F = f_{max}L/3$; $u_L - u_0 = f_{max}L^2/(4EA)$

---

4. 30

Solution m = 25 kg σ_{ult} = 300 MPa E = 180 GPa L_0 = 36 cm

k_{saftey} = 4 d_{min} = ? nearest millimeter d = u_L-u_0 = ?

The maximum allowable axial stress and the internal axial force can be found as shown below.

$$\sigma_{max} = \sigma_{ult}/k_{saftey} = 300/4 = 75MPa \qquad N = 25g = 245.25\ Newtons \tag{1}$$

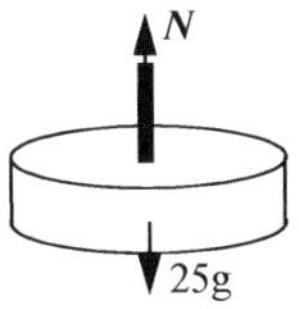

The diameter and extension can be found as shown below.

$$\sigma = \frac{N}{A} = \frac{245.25}{(\pi d^2/4)} \le \sigma_{max} \quad or \quad \frac{312.26}{d^2} \le 75(10^6) \quad or \quad d^2 \ge 4.163(10^{-6}) \quad or \quad d \ge 2.04(10^{-3})\ m$$

$$\delta = \frac{NL}{EA} = \frac{(245.25)(0.36)}{(180)(\pi 0.003^2/4)} = 0.0694(10^{-3})\ m \qquad \textbf{(2)}$$

ANS $d_{min} = 3$ mm ; $\delta = 0.069$ mm

---

4. 31

Solution E = 1800 ksi $\delta \le 0.05$ $F_{max} = ?$

The internal axial force in all segments of the joint is N = F. The area of cross-sections of the segments and relative displacements can be found as shown below.

$$A_{AB} = (4)(1) = 4\ in^2 \qquad A_{BC} = 3(4)(1) = 12\ in^2 \qquad A_{CD} = (4)(1) = 4\ in^2 \qquad \textbf{(1)}$$

We find the relative displacements of the section ends and add to obtain relative displacement of section at D relative to A.

$$u_B - u_A = \frac{N_{AB}(x_B - x_A)}{EA_{AB}} = \frac{F(36)}{(1800)(4)} = 5(10^{-3})F \qquad u_C - u_B = \frac{N_{BC}(x_C - x_B)}{EA_{BC}} = \frac{F(5)}{(1800)(12)} = 0.2318(10^{-3})F \qquad \textbf{(2)}$$

$$u_D - u_C = \frac{N_{CD}(x_D - x_C)}{EA_{CD}} = \frac{F(36)}{(1800)(4)} = 5(10^{-3})F \qquad \textbf{(3)}$$

$$u_D - u_A = F(5 + 0.2318 + 5)(10^{-3}) = 10.2318(10^{-3}) \le 0.05 \quad or \quad F \le 4.8869\ kips \qquad \textbf{(4)}$$

ANS $F_{max} = 4886\ lb$

---

4. 32

Solution L = 5 ft P=30 kips d_o= 6 in $\Delta u = 0.024$ d_i= ? nearest 1/8th inch, W=? for lightest rod

We calculate the area of cross-section for steel and aluminum to satisfy the stress and deformation requirements.

$$(\Delta u)_S = \frac{(30)(60)}{(30)(10^3)A_S} \le 0.027 \quad or \quad A_S \ge 2.22\ in^2 \qquad \sigma_S = \frac{(30)}{A_S} \le 24 \quad or \quad A_S \ge 1.25\ in^2 \qquad \textbf{(1)}$$

$$(\Delta u)_{Al} = \frac{(30)(60)}{10(10^3)A_{Al}} \le 0.027 \quad or \quad A_{Al} \ge 6.67\ in^2 \qquad \sigma_{Al} = \frac{(30)}{A_{Al}} \le 14 \quad or \quad A_{Al} \ge 2.143\ in^2 \qquad \textbf{(2)}$$

Thus if $A_S \ge 2.22\ in^2$ and $A_{Al} \ge 6.67\ in^2$ then all conditions will be met.

$$A_S = (\pi/4)(6^2 - D_S^2) \ge 2.22 \quad or \quad D_S \le 5.759\ in \qquad A_{Al} = (\pi/4)(6^2 - D_{Al}^2) \ge 6.67 \quad or \quad D_{Al} \le 5.245\ in \qquad \textbf{(3)}$$

Rounding downwards to the closest 1/8th inch, we obtain: $D_S = 5.75\ in$ $D_{Al} = 5.125\ in$

The weight of each material is:

$$W_S = (0.285)(\pi/4)(6^2 - 5.75^2)(60) = 39.45\ lbs \qquad W_{Al} = (0.1)(\pi/4)(6^2 - 5.125^2)(60) = 45.87\ lbs \qquad \textbf{(4)}$$

The rod should be of steel.

ANS $d_i = 5\frac{3}{4}\ inch$; $W_S = 39.5\ lbs$

---

4. 33

Solution P = 3,600lbs $\sigma_b \le 6\ ksi$ $\tau_p = 10\ ksi$ *double shear* $\sigma_s = 12\ ksi$ d_p = ? a_b = ? b_s = ? nearest 1/16 inch

The two-dimensional picture of the hitch and the cross-section of the tube are shown below.

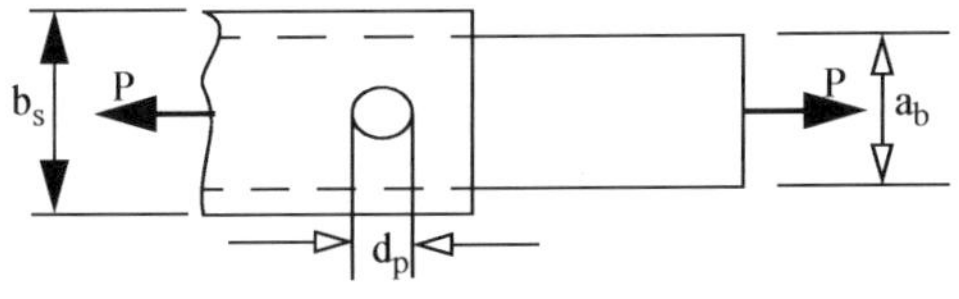

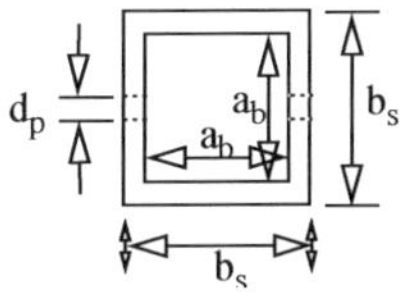

The internal axial and shear forces can be found as shown below.

$$V_{pin} = P/2 = 180\ lbs \qquad N_{bar} = P = 3600\ lbs \qquad N_s = P = 3600\ lbs \tag{1}$$

The dimensions can be found using the maximum stresses as shown below.

$$\tau_p = \frac{V_{pin}}{A_p} = \frac{1800}{(\pi dp^2/4)} \le 10,000 \quad or \quad d_p^2 \ge 0.2292 \quad or \quad d_p \ge 0.4787\ inch \quad or \quad d_p = 0.5\ in \tag{2}$$

$$\sigma_b = \frac{N_{bar}}{A_{bar}} = \frac{3600}{a_b(a_b - d_p)} \le 6000 \quad or \quad a_b^2 - 0.5a_b \ge 0.6 \quad or \quad a_b \ge 1.063 \quad or \quad a_b = 1\frac{1}{8}\ in \tag{3}$$

$$\sigma_s = \frac{N_s}{A_s} = \frac{3600}{[(b_s)(b_s - d_p) - (a_b)(a_b - d_p)]} \le 12,000 \quad or \quad b_s^2 - 0.5b_s - (1.125)(1.125 - 0.5) \ge 0.3 \text{ or}$$

$$b_s^2 - 0.5b_s - 1.003125 \ge 0 \quad or \quad b_s \ge 1.282 \quad or \quad b_s = 1\frac{5}{16}\ in \tag{4}$$

ANS $d_p = 0.5\ in;\ a_b = 1\frac{1}{8}\ in;\ b_s = 1\frac{5}{16}\ in$

---

4. 34

Solution $N(x_A) = 0$ $u_B - u_A = ?$

We first obtain internal axial force and then by integration by parts obtain our result.

$$\frac{dN}{dx} = -p(x) \quad or \quad \int_{N_A = 0}^{N} dN = -\int_{x_A}^{x} p(x)dx \quad or \quad N(x) = -\int_{x_A}^{x} p(x)dx \tag{1}$$

$$\frac{du}{dx} = \frac{N(x)}{EA} \quad or \quad \int_{u_A}^{u_B} du = \int_{x_A}^{x_A} \frac{N(x)}{EA}dx \quad or \quad u_B - u_A = \frac{1}{EA}\int_{x_A}^{x} N(x)dx \quad u_B - u_A = \frac{1}{EA}\left[xN(x)\Big|_{x_A}^{x_B} - \int_{x_A}^{x} x\frac{dN}{dx}dx\right] \text{ or}$$

$$u_B - u_A = \frac{1}{EA}\left[x_B N(x_B) - N(x_A) - \int_{x_A}^{x} x(-p(x))dx\right] = \frac{1}{EA}\left[x_B\left\{-\int_{x_A}^{x_B} p(x)dx\right\} - N(x_A) - \int_{x_A}^{x} x(-p(x))dx\right] \tag{2}$$

Noting that $N(x_A) = 0$ and x_B is a constant we obtain

$$u_B - u_A = \frac{1}{EA}\left[x_B\left\{-\int_{x_A}^{x_B} p(x)dx\right\} - 0 + \int_{x_A}^{x_B} xp(x)dx\right] = \frac{1}{EA}\left[\int_{x_A}^{x_B} (x - x_B)p(x)dx\right] \tag{3}$$

Above equation is the desired result.

---

4. 35

Solution

The axial strain, axial stress, and internal axial force can be found as shown below.

$$\varepsilon_{xx} = \frac{du}{dx}(x) \quad \sigma_{xx} = E\frac{du}{dx}(x) \quad N = \int_A \sigma_{xx} dA = \int_A E\frac{du}{dx}dA = \frac{du}{dx}\int_A E dA = \frac{du}{dx}\left[\int_{A_1} E_1 dA + \int_{A_2} E_2 dA + \cdot + \cdot + \int_{A_n} E_n dA\right] \tag{1}$$

$$N = \frac{du}{dx}[E_1A_1 + E_2A_2 + \cdot + \cdot + E_nA_n] = \frac{du}{dx}\left[\sum_{j=1}^{n} E_jA_j\right] \qquad \frac{du}{dx} = \frac{N}{\sum_{j=1}^{n} E_jA_j} \tag{2}$$

For the ith material: $(\sigma_{xx})_i = E_i\frac{du}{dx}$, where $(\sigma_{xx})_i$ is the axial stress in the ith material, we obtain:

$$(\sigma_{xx})_i = E_i\frac{N}{\sum_{j=1}^{n} E_jA_j} \tag{3}$$

Assuming all quantities on right hand side of strain equation are constant, we obtain

$$\frac{du}{dx} = \frac{u_2 - u_1}{x_2 - x_1} = \frac{N}{\sum_{j=1}^{n} E_jA_j} \quad or \quad u_2 - u_1 = \frac{N(x_2 - x_1)}{\sum_{j=1}^{n} E_jA_j} \tag{4}$$

The above two equations are the desired results. If $E_1 = E_2 = E_3 .... = E_n = E$ then

$$\sum_{j=1}^{n} E_jA_j = E\sum_{j=1}^{n} A_j = EA \tag{5}$$

Substituting we obtain:

$$(\sigma_{xx})_i = E\frac{N}{EA} = \frac{N}{A} \qquad u_2 - u_1 = \frac{N(x_2 - x_1)}{EA} \tag{6}$$

Above equations are same as those for homogeneous material.

---

4. 36

Solution $\sigma = E\varepsilon^n$ $\sigma_{xx} = ?$ $\Delta u = ?$

The axial strain, axial stress, and internal axial force can be found as shown below.

$$\varepsilon_{xx} = \frac{du}{dx}(x) \qquad \sigma_{xx} = E(\varepsilon_{xx})^n = \sigma_{xx} = E\left[\frac{du}{dx}(x)\right]^n \qquad N = \int_A \sigma_{xx} dA = \int_A E\left[\frac{du}{dx}(x)\right]^n dA = E\left[\frac{du}{dx}(x)\right]^n \int_0 dA = EA\left[\frac{du}{dx}(x)\right]^n \text{ or}$$

$$\left[\frac{du}{dx}(x)\right]^n = \frac{N}{EA} \quad or \quad \frac{du}{dx}(x) = \left[\frac{N}{EA}\right]^{1/n} = \frac{\Delta u}{L} \quad or \quad \Delta u = \left[\frac{N}{EA}\right]^{1/n} L \tag{1}$$

$$\sigma_{xx} = E\left[\frac{du}{dx}(x)\right]^n = E\frac{N}{EA} = \frac{N}{A} \tag{2}$$

ANS $\Delta u = [N/(EA)]^{1/n} L\,;\ \sigma_{xx} = N/A$

---

4. 37

Solution $\Delta u = f(\omega, \rho, L, A, E) = ?$

By equilibrium of the free body diagram of a differential element, we obtain the differential internal force which we then integrate.

$$N + dN - N = -\rho w^2 x A dx \quad or \quad dN = \rho w^2 A x dx \tag{1}$$

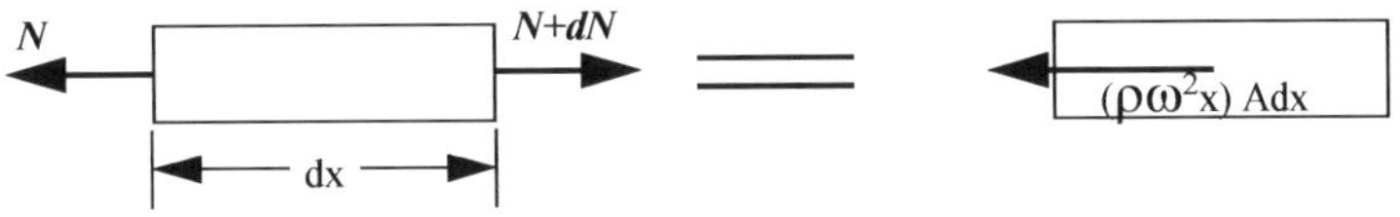

$$\int_0^N dN = \int_L^x -\rho w^2 A x dx \quad or \quad N - 0 = \rho w^2 A \frac{x^2}{2}\bigg|_L^x \quad or \quad N = \frac{\rho w^2}{2} A (L^2 - x^2) \tag{2}$$

The displacement is found as shown below.

$$\frac{du}{dx} = \frac{N}{EA} = \frac{\rho w^2}{2E}(L^2 - x^2) \quad or \quad \Delta u = u_L - u_0 = \int_{u_0}^{u_L} du = \frac{\rho w^2}{2E}\int_0^L (L^2 - x^2) dx = \frac{\rho w^2}{2E}\left(L^2 x - \frac{x^3}{3}\right)\bigg|_0^L = \frac{\rho w^2}{2E}\left(L^3 - \frac{L^3}{3}\right) \text{ or}$$

ANS $\Delta u = (\rho w^2 L^3)/(3E)$

---

4. 38

Solution

By equilibrium of forces in the given figure we obtain:

$$N + dN - N = \rho A \frac{\partial^2 u}{\partial t^2} dx \quad or \quad \frac{\partial N}{\partial x} = \rho A \frac{\partial^2 u}{\partial t^2} \tag{1}$$

$$N = EA\frac{\partial u}{\partial x} \qquad \frac{\partial N}{\partial x} = \frac{\partial}{\partial x}\left(EA\frac{\partial u}{\partial x}\right) = \rho A \frac{\partial^2 u}{\partial t^2} \quad or \quad EA\frac{\partial^2 u}{\partial x^2} = \rho A \frac{\partial^2 u}{\partial t^2} \quad or \quad \frac{\partial^2 u}{\partial t^2} = \frac{EA}{\rho A}\frac{\partial^2 u}{\partial x^2} = c^2 \frac{\partial^2 u}{\partial x^2} \qquad c = \sqrt{\frac{E}{\rho}} \tag{2}$$

Above equation is the desired result.

---

4. 39

Solution

Consider the solution: $u = f(x - ct)$. Substituting $y = x - ct$ we obtain the following.

$$\frac{\partial u}{\partial x} = \frac{\partial f}{\partial y}\frac{\partial y}{\partial x} = \frac{\partial f}{\partial y} \qquad \frac{\partial^2 u}{\partial x^2} = \frac{\partial}{\partial x}\left(\frac{\partial f}{\partial y}\right) = \frac{\partial}{\partial y}\left(\frac{\partial f}{\partial y}\right)\frac{\partial y}{\partial x} = \frac{\partial^2 f}{\partial y^2} \tag{1}$$

$$\frac{\partial u}{\partial t} = \frac{\partial f}{\partial y}\frac{\partial y}{\partial t} = -c\frac{\partial f}{\partial y} \qquad \frac{\partial^2 u}{\partial t^2} = \frac{\partial}{\partial t}\left(-c\frac{\partial f}{\partial y}\right) = \frac{\partial}{\partial y}\left(-c\frac{\partial f}{\partial y}\right)(-c) = c^2\frac{\partial^2 f}{\partial y^2} \quad or \quad \frac{\partial^2 u}{\partial t^2} = c^2\frac{\partial^2 u}{\partial x^2} \tag{2}$$

Consider now $u = g(x + ct)$. Substituting $z = x + ct$ we obtain:

$$\frac{\partial^2 u}{\partial x^2} = \frac{\partial}{\partial x}\left(\frac{\partial g}{\partial z}\right) = \frac{\partial}{\partial z}\left(\frac{\partial g}{\partial z}\right)\frac{\partial y}{\partial x} = \frac{\partial^2 g}{\partial z^2} \qquad \frac{\partial^2 u}{\partial t^2} = \frac{\partial}{\partial t}\left(c\frac{\partial g}{\partial z}\right) = \frac{\partial}{\partial z}\left(c\frac{\partial g}{\partial z}\right)(c) = c^2\frac{\partial^2 g}{\partial z^2} \quad or \quad \frac{\partial^2 u}{\partial t^2} = c^2\frac{\partial^2 u}{\partial x^2} \tag{3}$$

The above equations show that the differential equation is satisfied by $u = f(x - ct)$ and $u = g(x + ct)$.

---

4. 40

Solution

The strain, stress, and internal axial force can be written as follows.

$$\varepsilon_{xx} = \frac{du}{dx} + \frac{1}{2}\left(\frac{du}{dx}\right)^2 \qquad \sigma_{xx} = E\varepsilon_{xx} = E\left[\frac{du}{dx} + \frac{1}{2}\left(\frac{du}{dx}\right)^2\right] \qquad N = \int_A \sigma_{xx} dA = \int_A E\left[\frac{du}{dx} + \frac{1}{2}\left(\frac{du}{dx}\right)^2\right] dA = E\left[\frac{du}{dx} + \frac{1}{2}\left(\frac{du}{dx}\right)^2\right]\int_A dA \quad \textbf{(1)}$$

$$N = E\left[\frac{du}{dx} + \frac{1}{2}\left(\frac{du}{dx}\right)^2\right]A \quad or \quad \frac{du}{dx} + \frac{1}{2}\left(\frac{du}{dx}\right)^2 = \frac{N}{EA} \quad or \quad \left(\frac{du}{dx}\right)^2 + 2\left(\frac{du}{dx}\right) - 2\frac{N}{EA} = 0 \quad \textbf{(2)}$$

Solving the quadratic, we obtain

$$\frac{du}{dx} = \frac{1}{2}\left[-2 \pm \sqrt{4 + 4\left(\frac{2N}{EA}\right)}\right] \quad or \quad \frac{du}{dx} = \pm\sqrt{1 + 2\left(\frac{N}{EA}\right)} - 1 \quad \textbf{(3)}$$

The positive root is the admissible root, because with the negative root $\frac{du}{dx}$ is always negative and less that -1 irrespective of the value of N. The stress can be obtained as

$$\sigma_{xx} = E\left[\frac{du}{dx} + \frac{1}{2}\left(\frac{du}{dx}\right)^2\right] = E\left(\frac{N}{EA}\right) = \frac{N}{A} \quad \textbf{(4)}$$

ANS $\frac{du}{dx} = \sqrt{1 + \frac{2N}{EA}} - 1 \ ; \ \sigma_{xx} = \frac{N}{A}$

---

4. 41

Solution E = 100 GPa $x_A = 0$ $x_B = 1.5$ m $\Delta u = u_B - u_A = ?$ $\sigma_{max} = ?$

The internal axial force is N = 500 kN. We have the following.

$$\frac{du}{dx} = \frac{N}{EA} = \frac{(500)(10^3)}{(100)(10^9)\pi[(R(x))/1000]^2} = \frac{5(10^{-6})}{\pi R^2(x)} \quad or \quad \Delta u = \int_{u_A}^{u_B} du = \int_{x_A=0}^{x_B=1.5}\left[\frac{5}{\pi R^2(x)}\right]dx \quad \textbf{(1)}$$

Representing $f(x) = 1.5915/[R^2(x)]$ and using numerical integration as described in Appendix B, we obtain the value of the integral on a spread sheet as shown in the table below. From the results in the table we see the elongation of the bar is: $\Delta u = 0.60\ mm$. The maximum stress will occur at the cross-section where area is the smallest, i.e., just before point B and can be found as shown below.

$$\sigma_{max} = \frac{N}{A_B} = \frac{500(10^3)}{\pi(0.0506^2)} = 62.2(10^6)\ N/m^2 \quad \textbf{(2)}$$

x (m)	R(x) (mm)	f(x)	Δu (10^{-3})(m)
0.00	100.60	0.1573	0.0171
0.10	92.70	0.1852	0.0380
0.20	82.60	0.2333	0.0623
0.30	79.60	0.2512	0.0886
0.40	75.90	0.2763	0.1193
0.50	68.80	0.3362	0.1533
0.60	68.00	0.3442	0.1888
0.70	65.90	0.3665	0.2292

x (m)	R(x) (mm)	f(x)	Δu (10^{-3})(m)
0.80	60.10	0.4406	0.2731
0.90	60.30	0.4377	0.3178
1.00	59.10	0.4557	0.3678
1.10	54.00	0.5458	0.4216
1.20	54.80	0.5300	0.4753
1.30	54.10	0.5438	0.5351
1.40	49.40	0.6522	0.5988
1.50	50.60	0.6216	

ANS $\Delta u = 0.60\ mm\ ;\ \sigma_{max} = 62.2\ MPa(T)$

---

4. 42

Solution E = 100 GPa a = ? b = ? $x_A = 0$ $x_B = 1.5$ m $\Delta u = u_B - u_A = ?$

We develop the equations for the least square method for a linear representation as described in Appendix B. This difference between the radius value and the value obtained by substituting x = x_i in the equation $R(x) = a + bx$ is the error e_i that can be written as given below.

$$e_i = R_i - R(x_i) = R_i - (a + bx_i) \quad \textbf{(1)}$$

We define the error E as $E = \sum_{i=1}^{N} e_i^2$. The error E is minimized with respect to coefficients a, and b to generate a set of linear algebraic equations as shown below.

$$\frac{\partial E}{\partial a} = 0 \quad or \quad \sum 2e_i\frac{\partial e_i}{\partial a} = 0 \quad or \quad \sum 2[R_i - (a+bx_i)][-1] = 0 \qquad (2)$$

$$\frac{\partial E}{\partial b} = 0 \quad or \quad \sum 2e_i\frac{\partial e_i}{\partial b} = 0 \quad or \quad \sum 2[R_i - (a+bx_i)][-x_i] = 0 \qquad (3)$$

The above equations on the right can be rearranged and written in matrix form as shown below:

$$\begin{bmatrix} N & \sum x_i \\ \sum x_i & \sum x_i^2 \end{bmatrix}\begin{Bmatrix} a \\ b \end{Bmatrix} = \begin{Bmatrix} \sum R_i \\ \sum x_i R_i \end{Bmatrix} \quad or \quad \begin{bmatrix} b_{11} & b_{12} \\ b_{21} & b_{22} \end{bmatrix}\begin{Bmatrix} a \\ b \end{Bmatrix} = \begin{Bmatrix} r_1 \\ r_2 \end{Bmatrix} \qquad (4)$$

The coefficients of the b-matrix and the r-vector can be determined by comparison to the matrix form of the equations on the left. The coefficients a and b can be determined by Cramer's rule as shown below.

$$D = b_{11}b_{22} - b_{12}b_{21} \qquad 5$$

$$a = \left|\begin{bmatrix} r_1 & b_{12} \\ r_2 & b_{22} \end{bmatrix}\right| / D = \frac{r_1 b_{22} - r_2 b_{12}}{D} \qquad b = \left|\begin{bmatrix} b_{11} & r_1 \\ b_{21} & r_2 \end{bmatrix}\right| / D = \frac{r_2 b_{11} - r_1 b_{21}}{D} \qquad 6$$

The given data and above equations can be put in a spread sheet and the coefficients a and b can be found as shown in the table below as:

	x_i (m)	R_i (mm)	x_i^2	x_i*R_i
1	0.00	100.60	0.0000	0.000
2	0.10	92.70	0.0100	9.270
3	0.20	82.60	0.0400	16.520
4	0.30	79.60	0.0900	23.880
5	0.40	75.90	0.1600	30.360
6	0.50	68.80	0.2500	34.400
7	0.60	68.00	0.3600	40.800
8	0.70	65.90	0.4900	46.130
9	0.80	60.10	0.6400	48.080

	x_i (m)	R_i (mm)	x_i^2	x_i*R_i
10	0.90	60.30	0.8100	54.270
11	1.00	59.10	1.0000	59.100
12	1.10	54.00	1.2100	59.400
13	1.20	54.80	1.4400	65.760
14	1.30	54.10	1.6900	70.330
15	1.40	49.40	1.9600	69.160
16	1.50	50.60	2.2500	75.900
b_{ij} & r_i	12.0000	1076.50	12.4000	703.360
D	54.4			
a and b	90.226	-30.593		

Thus $R(x) = 90.226\text{-}30.593x$, where x is in meters and R is in millimeters. We obtain the elongation as shown below.

$$\Delta u = \int_0^{1.5}\left[\frac{5}{\pi(90.226-30.593x)^2}\right]dx = \left(\frac{5}{\pi}\right)\left[\frac{1}{(30.593)(90.226-30.593x)}\right]\Bigg|_0^{1.5} = 0.5968(10^{-3})\ m$$

ANS $a = 90.226\ mm;\ b = -30.593;\ \Delta u = 0.60\ \text{mm}$

---

4. 43

Solution $E = 30{,}000$ ksi $d_o = 1$ inch $d_i = 0.875$ inch $u_A = ?$ $\sigma_{max} = ?$

The equilibrium equation can be integrated from point A, where $N = 0$ to any location x, to obtain the internal axial force as a function of x as shown below.

$$\frac{dN}{dx} + p(x) = 0 \quad or \quad N(x) = \int_{N_A=0}^{N(x)} dN = -\int_{x_A=0}^{x} p(x)dx \qquad (1)$$

The area of cross-section and strain can be written as shown beloe

$$A = \pi(d_o^2 - d_i^2)/4 = 0.1841\,in^2 \qquad \frac{du}{dx} = \frac{N(x)}{EA} = \frac{N(x)}{(30)(10^6)(0.1841)} = 181.08(10^{-9})N(x) \qquad (2)$$

Integrating from point A i.e., $x_A = 0$ to point B, i.e., $x_B = 36$. We note that point B is fixed to the wall, hence $u_B = 0$. we obtain the following integral:

$$\int_{u_A}^{u_B=0} du = \int_{x_A=0}^{x_B=36}[181.08(10^{-9})N(x)]dx \quad or \quad u_A = -\int_0^{36}[181.08(10^{-9})N(x)]dx \qquad (3)$$

The axial stress σ_{xx} can be found as shown below:

$$\sigma_{xx} = \frac{N}{A} = \frac{N}{0.1841} \qquad 4$$

The internal axial force equation can be numerically integrated on a spread sheet to obtain the value of internal force at any x_i. The displacement equation then can be numerically integrated to obtain the elongation. The stress equation can be used to find the axial stress at various x_i and the maximum value chosen by inspection. These calculations can be done on a spread sheet as shown in the table below.

x_i (inches)	$p(x_i)$ (lbs./in)	$N(x_i)$ (lbs.)	$[u_A - u(x_{i+1})](10^{-3})$ (inches)	$\sigma_{xx}(x_i)$ (ksi)
0	260	0.00	0.149	0.000
3	105.5	-548.25	0.503	-2.978
6	32.0	-754.50	0.942	-4.098
9	39.5	-861.75	1.368	-4.681
12	-142.0	-708.00	1.596	-3.846
15	-242.5	-131.25	1.462	-0.713
18	-262.0	625.50	0.824	3.398
21	-470.5	1724.25	-0.548	9.366
24	-598.0	3327.00	-2.862	18.072
27	-644.5	5190.75	-6.303	28.195
30	-880.0	7477.50	-11.145	40.617
33	-1034.5	10349.25	-17.640	56.215
36	-1108.0	13563.00		73.672

ANS $u_A = 0.0176$ in. to the left ; $\sigma_{max} = 73.7$ ksi(T)

---

4. 44

Solution a = ? b = ? c = ? u_A = ?

The equilibrium equation can be integrated from point A, where $N = 0$ to any location x, to obtain the internal axial force as a function of x as shown below.

$$\frac{d\boldsymbol{N}}{dx} + p(x) = 0 \qquad \boldsymbol{N}(x) = \int_{N_A = 0}^{N(x)} d\boldsymbol{N} = -\int_{x_A = 0}^{x} p(x)dx = -\int_{x_A = 0}^{x} [a + bx + cx^2]dx = -\left[ax + \frac{bx^2}{2} + \frac{cx^3}{3}\right] \tag{1}$$

$$A = \pi(d_o^2 - d_i^2)/4 = 0.1841 in^2 \qquad \frac{du}{dx} = \frac{\boldsymbol{N}(x)}{EA} = \frac{\boldsymbol{N}(x)}{(30)(10^6)(0.1841)} = (-181.08)(10^{-9})\left[ax + \frac{bx^2}{2} + \frac{cx^3}{3}\right] \tag{2}$$

Integrating from point A i.e., $x_A = 0$ to point B, i.e., $x_B = 36$ and noting that $u_B = 0$. we obtain the following integral.

$$\int_{u_A}^{u_B = 0} du = \int_{x_A = 0}^{x_B = 36} (-181.08)(10^{-9})\left[ax + \frac{bx^2}{2} + \frac{cx^3}{3}\right]dx \text{ or}$$

$$u_A = (181.08)(10^{-9})\left(\frac{ax^2}{2} + \frac{bx^3}{6} + \frac{cx^4}{12}\right)\Bigg|_0^{36} = (181.08)(10^{-9})\left[\frac{a(36)^2}{2} + \frac{b(36)^3}{6} + \frac{c(36)^4}{12}\right] \tag{3}$$

Using the Least Square Method described in Appendix B, we obtain the value of the values of constants a, b, and c on a spread sheet as shown in the table below. We obtain the displacement of point A by substituting the values of the constants

$$u_A = (181.08)(10^{-9})\left[\frac{(224)(36)^2}{2} + \frac{(-23.60)(36)^3}{6} + \frac{(-0.40)(36)^4}{12}\right] = -0.017 \ in \tag{4}$$

	x_i	$p(x_i)$	x_i^2	x_i^3	x_i^4	$x*p_i$	$x_i^2*p_i$
1	0.0	260.0	0.0	0.000E+00	0.000E+00	0.000E+00	0.000E+00
2	3.0	105.5	9.0	2.700E+01	8.100E+01	3.165E+02	9.495E+02
3	6.0	32.0	36.0	2.160E+02	1.296E+03	1.920E+02	1.152E+03
4	9.0	39.5	81.0	7.290E+02	6.561E+03	3.555E+02	3.200E+03
5	12.0	-142.0	144.0	1.728E+03	2.074E+04	-1.704E+03	-2.045E+04
6	15.0	-242.5	225.0	3.375E+03	5.063E+04	-3.638E+03	-5.456E+04
7	18.0	-262.0	324.0	5.832E+03	1.050E+05	-4.716E+03	-8.489E+04
8	21.0	-470.5	441.0	9.261E+03	1.945E+05	-9.881E+03	-2.075E+05
9	24.0	-598.0	576.0	1.382E+04	3.318E+05	-1.435E+04	-3.444E+05
10	27.0	-644.5	729.0	1.968E+04	5.314E+05	-1.740E+04	-4.698E+05
11	30.0	-880.0	900.0	2.700E+04	8.100E+05	-2.640E+04	-7.920E+05
12	33.0	-1034.5	1089.0	3.594E+04	1.186E+06	-3.414E+04	-1.127E+06
13	36.0	-1108.0	1296.0	4.666E+04	1.680E+06	-3.989E+04	-1.436E+06
b_{ij} & r_i	234.0	-4945.0	5850.0	1.643E+05	4.918E+06	-1.513E+05	-4.531E+06
C_{ij}	1.783E+09	1.897E+08	4.216E+06	2.971E+07	7.666E+05	2.129E+04	
D	3.453E+09						
a_i	224.40	-23.60	-0.40				

ANS $a = 224.40$; $b = -23.60$; $c = -0.40$; $u_A = 0.017 in$ *to the left*

Section 4.3

4. 45

Solution E = 30,000 ksi A = 1.25 in^2 L = 24 inch δ_B = 0.002 in

From moment equilibrium about point C in the free body diagram and using the deformed shape we obtain the following.

$$125N_A = 25F \quad or \quad F = 5N_A \qquad \delta_A/125 = \delta_B/25 \quad or \quad \delta_A = 5\delta_B = 5(0.002) = 0.01in \qquad \textbf{(1)}$$

(a)

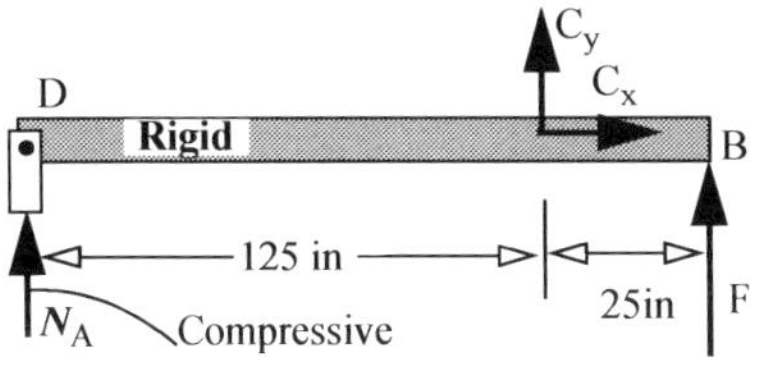

(b)

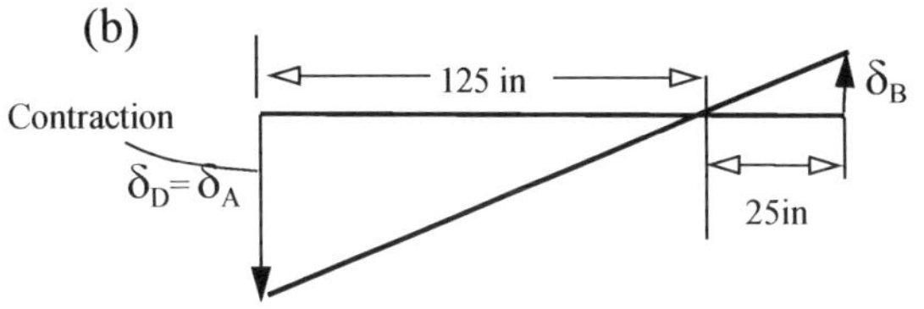

$$\delta_A = \frac{N_A L}{EA} = \frac{N_A(24)}{(30,000)(1.25)} = 0.01 \quad or \quad N_A = 15.625\ kips \qquad F = 5N_A = 5(15.625) = 78.125kips \qquad \textbf{(2)}$$

ANS $F = 78.1$ kips

---

4. 46

Solution E = 30,000 ksi A = 1.25 in^2 L = 24 inch δ_B = 0.002 in

From moment equilibrium about point C in the free body diagram and using the deformed shape we obtain the following.

$$)125N_A = 25F \quad or \quad F = 5N_A \qquad \textbf{(1)}$$

$$\delta_D = \delta_A + 0.04 \qquad \delta_D/125 = \delta_B/25 \quad or \quad \delta_A = 5\delta_B - 0.004 = 5(0.002) - 0.004 = 0.006in \qquad \textbf{(2)}$$

(a)

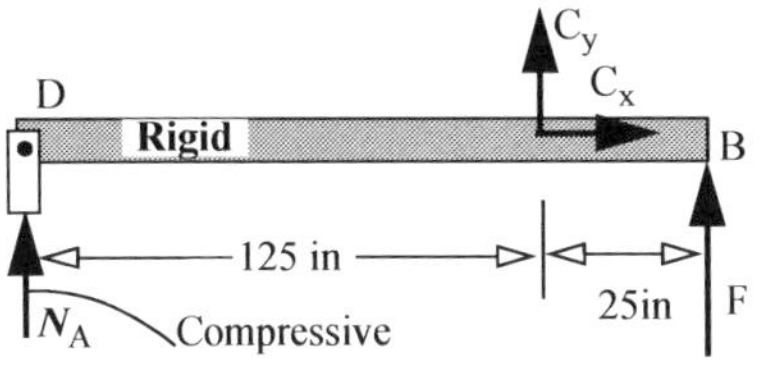

(b)

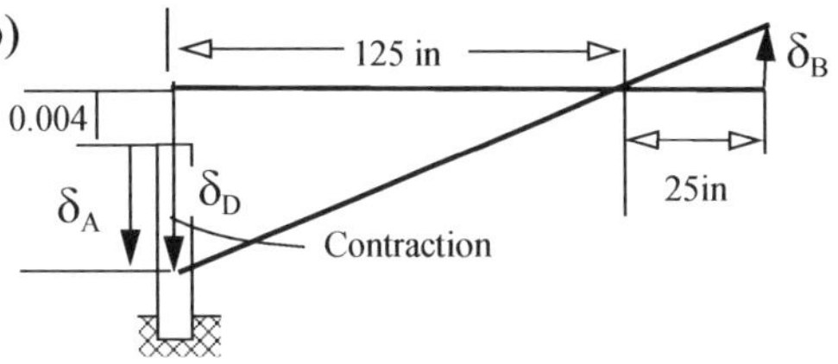

$$\delta_A = \frac{N_A L}{EA} = \frac{N_A(24)}{(30,000)(1.25)} = 0.006 \quad or \quad N_A = 9.375\ kips \qquad F = 5N_A = 5(9.375) = 46.875kips \qquad \textbf{(3)}$$

ANS $F = 46.9\ kips$

---

4. 47

Solution E = 100 GPa A = 15 mm^2 L = 1.2 m δ_B = 0.75 mm

From moment equilibrium about point C in the free body diagram and using the deformed shape we obtain the following.

$$2.5N_A = 1.25F \quad or \quad F = 2N_A \qquad \delta_A/2.5 = \delta_B/1.25 \quad or \quad \delta_A = 2\delta_B = 2(0.75) = 1.5mm = 1.5(10^{-3})m \qquad \textbf{(1)}$$

(a)

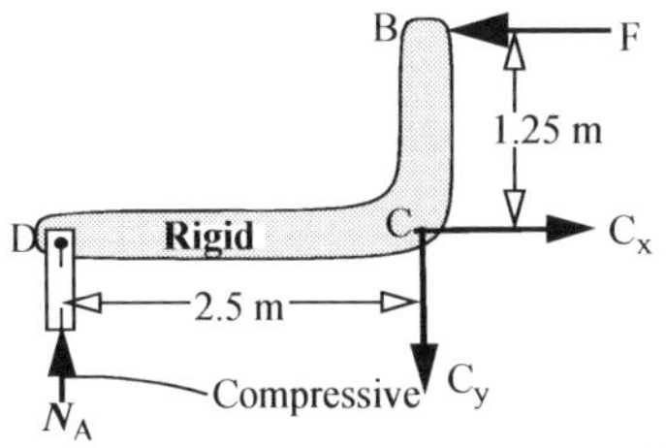

(b)

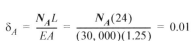

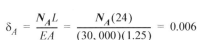

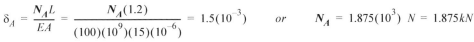

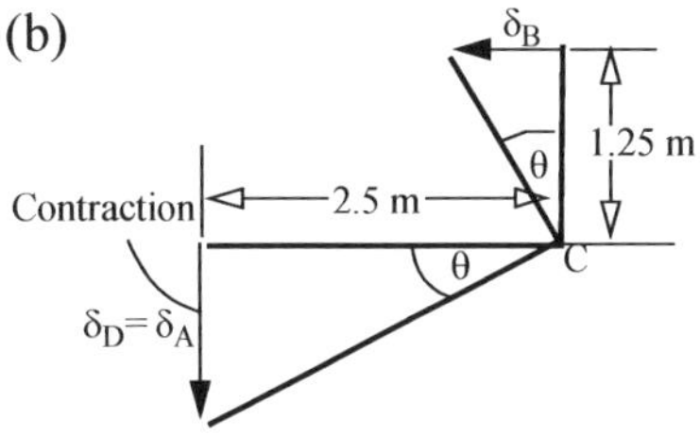

$$\delta_A = \frac{N_A L}{EA} = \frac{N_A(1.2)}{(100)(10^9)(15)(10^{-6})} = 1.5(10^{-3}) \quad or \quad N_A = 1.875(10^3)\ N = 1.875kN \qquad \textbf{(2)}$$

$$F = 2N_A = 2(1.875) = 3.75kN \qquad \textbf{(3)}$$

ANS $F = 3.75$ kN

---

4. 48

Solution E = 100 GPa A = 15 mm^2 L = 1.2 m δ_B = 0.75 mm

From moment equilibrium about point C in the free body diagram and using the deformed shape we obtain the following.

$$2.5N_A = 1.25F \quad or \quad F = 2N_A \qquad \textbf{(1)}$$

$$\delta_D = \delta_A + 1 \qquad \delta_D/2.5 = \delta_B/1.25 \qquad or \qquad \delta_A = 2\delta_B - 1 = 2(0.75) - 1 = 0.5mm = 0.5(10^{-3})m \qquad (2)$$

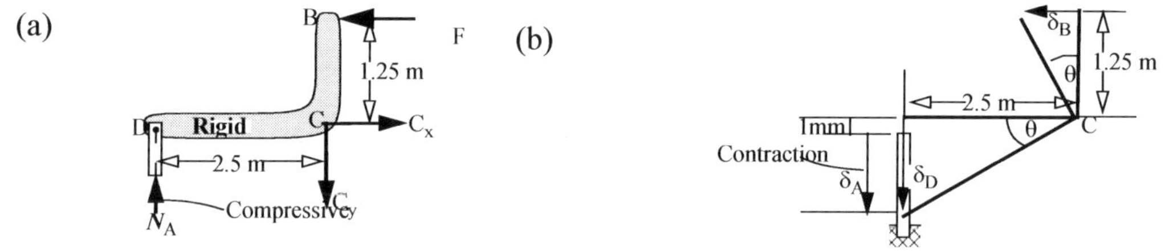

$$\delta_A = \frac{N_A L}{EA} = \frac{N_A(1.2)}{(100)(10^9)(15)(10^{-6})} = 0.5(10^{-3}) \qquad or \qquad N_A = 0.625(10^3)\ N = 0.625kN \qquad (3)$$

$$1.25F = 2.5N_A \qquad or \qquad F = 2N_A = 2(0.625) = 1.25kN \qquad (4)$$

ANS $F = 1.25$ kN

---

4. 49

Solution E = 200 GPa A = 100 mm^2 F=20 kN δ_P = ?

By force equilibrium in the x-direction on the free body diagram and the exaggerated deformed shape we obtain the following.

$$N_A cos 50 = F \qquad or \qquad N_A = 31.11kN \qquad \delta_A = \delta_P cos 50 \qquad or \qquad \delta_P = 1.556\delta_A \qquad (1)$$

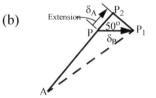

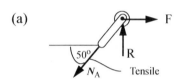

$$\delta_A = \frac{N_A L}{EA} = \frac{(31.11)(10^3)(0.2)}{(200)(10^9)(100)(10^{-6})} = 0.311(10^{-3})m = 0.311mm \qquad \delta_P = 1.556\delta_A = (1.556)(0.311) = 0.484mm \qquad (2)$$

ANS $\delta_P = 0.48$ mm

---

4. 50

Solution E = 200 GPa A = 100 mm^2 F=20 kN δ_P = ?

By force equilibrium in the x-direction on the free body diagram and the exaggerated deformed shape we obtain the following.

$$N_A cos 20 = F \qquad or \qquad N_A = 21.28kN \qquad \delta_A = \delta_P cos 20 \qquad or \qquad \delta_P = 1.064\delta_A \qquad (1)$$

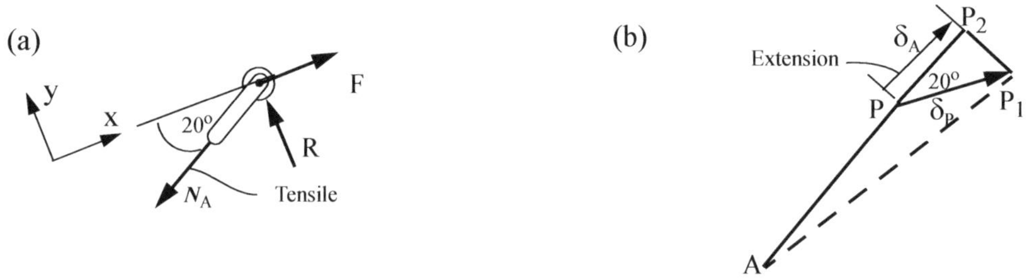

$$\delta_A = \frac{N_A L}{EA} = \frac{(21.28)(10^3)(0.2)}{(200)(10^9)(100)(10^{-6})} = 0.2128(10^{-3})m = 0.2128mm \qquad \delta_P = 1.064\delta_A = (1.064)(0.2128) = 0.2264mm \qquad (2)$$

ANS $\delta_P = 0.23mm$

---

4. 51

Solution E = 30,000 ksi A = 1.25 in^2 L = 24 inch σ_A = ? δ_D = ?

By moment equilibrium about point C on the free body diagram and the exaggerated deformed shape we obtain the following.

$$125N_A = 25F \qquad or \qquad N_A = F/5 = 10kips \qquad \delta_D = \delta_A + 0.004 \qquad \frac{\delta_D}{125} = \frac{\delta_B}{25} \qquad or \qquad \delta_B = \frac{(\delta_A + 0.004)}{5} \qquad (1)$$

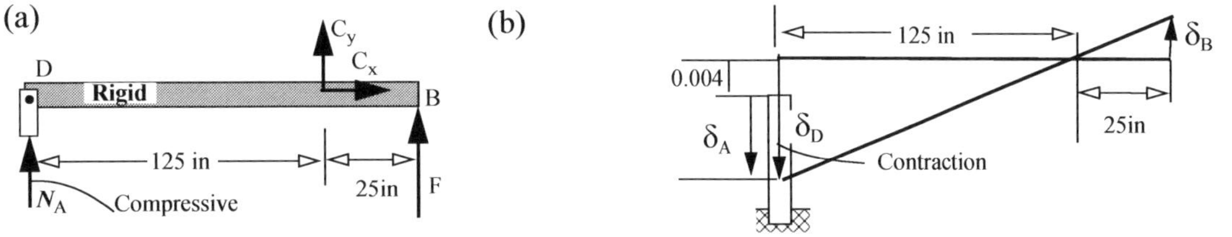

$$\sigma_A = \frac{N_A}{A} = \frac{10}{1.25} = 8.0\ ksi \qquad \delta_A = \frac{N_A L}{EA} = \frac{(10)(24)}{(30,000)(1.25)} = 0.0064\ in \qquad \delta_B = \frac{(0.0064 + 0.004)}{5} = 0.00208 in \quad \textbf{(2)}$$

ANS $\delta_B = 0.0021\ in;\ \sigma_A = 8.0\ ksi(C)$

---

4. 52

Solution E = 30,000 ksi A = 1.25 in^2 L = 24 inch σ_A = ? δ_D = ?

By moment equilibrium about point C on the free body diagram and the exaggerated deformed shape we obtain the following.

$$125 N_A = 2500 \quad or \quad N_A = 20 kips \quad \textbf{(1)}$$

$$\delta_D = \delta_A + 0.004 \qquad \delta_D/125 = \delta_B/25 \quad or \quad \delta_B = (\delta_A + 0.004)/5 \quad \textbf{(2)}$$

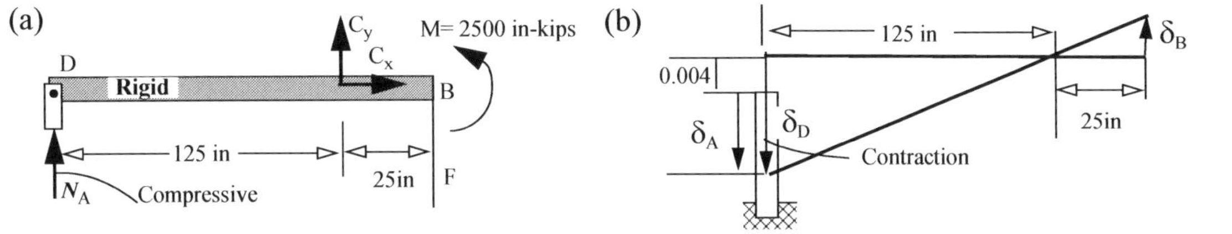

$$\delta_A = \frac{N_A L}{EA} = \frac{(20)(24)}{(30,000)(1.25)} = 0.0128\ in \qquad \delta_B = \frac{(0.0128 + 0.004)}{5} = 0.00336 in \qquad \sigma_A = \frac{N_A}{A} = \frac{20}{1.25} = 16.0\ ksi \quad \textbf{(3)}$$

ANS $\delta_B = 0.0034\ in;\ \sigma_A = 16.0\ \text{ksi}(C)$

---

4. 53

Solution E_S= 30,000 ksi, ν_S = 0.28 E_{Cu} = 15,000 ksi, ν_{Cu} = 0.35 ds= 1/2 in

d_{Cu}= 3/4 in t_{Cu} = 3/4 in P = 25 kips u_A = ? Δd_S = ?

By equilibrium of the free body diagrams and the given data we obtain the following.

$$N_S = P = 2.5\ kips \qquad N_{Cu} = -N_S = -2.5\ kips \qquad A_S = \frac{\pi}{4}\left(\frac{1}{2}\right)^2 = 0.19635\ in^2 \qquad A_{cu} = \frac{\pi}{4}(1)^2 - \frac{\pi}{4}\left(\frac{3}{4}\right)^2 = 0.3461\ \text{in}^2 \quad \textbf{(1)}$$

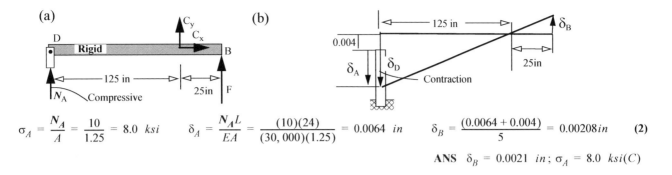

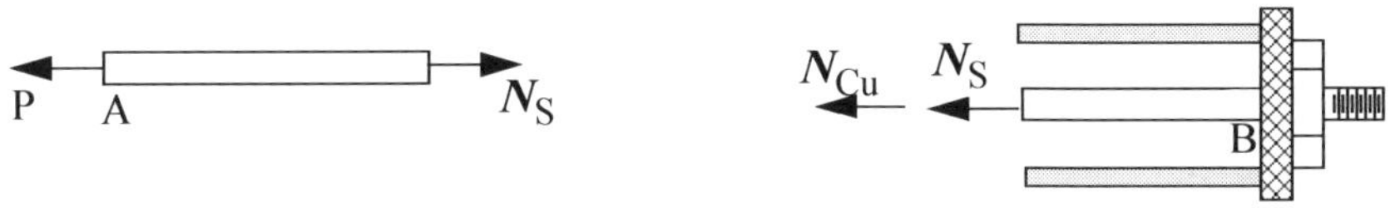

The stress calculation and displacement calculation is as follows. We note point C is fixed, i.e. $u_C = 0$

$$u_B - u_C = \frac{N_{Cu}(x_B - x_A)}{E_{Cu}A_{Cu}} = \frac{(-2.5)(16)}{(15000)(0.3461)} = -7.761(10^{-3})\ in \quad \textbf{(2)}$$

$$u_B - u_A = \frac{N_S(x_A - x_B)}{E_S A_S} = \frac{(2.5)(-24)}{(30000)(0.19635)} = -10.186\ in \qquad u_A - u_C = (-7.761 - 10.186)(10^{-3}) = (-17.947)(10^{-3})\ in \quad \textbf{(3)}$$

$$\sigma_S = \frac{N_S}{A_S} = \frac{2.5}{0.19635} = 12.732\ ksi \qquad \varepsilon_S = \frac{\sigma_S}{A_S} = \frac{12.732}{30000} = 0.4244(10^{-3}) \quad \textbf{(4)}$$

$$(\varepsilon_S)_{tran} = -\nu_S\varepsilon_S = -0.28(0.4244)(10^{-3}) = -0.1188(10^{-3}) \qquad \Delta d_S = d(\varepsilon_S)_{tran} = 0.059(10^{-3})\ \text{in} \quad \textbf{(5)}$$

ANS $u_A = 0.0018$ in. to left; $\Delta d_S = 0.059(10^{-3})$ in.

---

4. 54

Solution E = 10,000 ksi d = 1/2 in P= 5 kips θ = ?

$$A = [\pi(1/2)^2]/4 = 0.19635\ in^2 \quad \textbf{(1)}$$

By equilibrium of moment about C and force balance in y-direaction on the free body diagram we obtain

$$(8)N_1 = (5)(P) \quad or \quad N_1 = \frac{5(5)}{8} = 3.125 \text{ kips} \qquad N_1 + N_2 = P \quad or \quad N_2 = 5 - 3.125 = 1.875 \text{ kips} \tag{2}$$

(a)

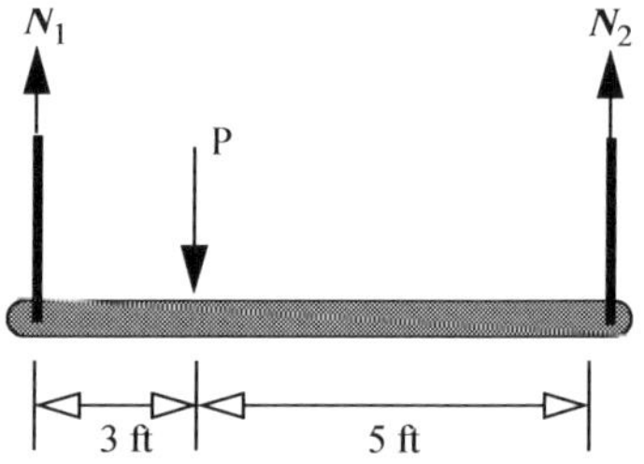

(b)

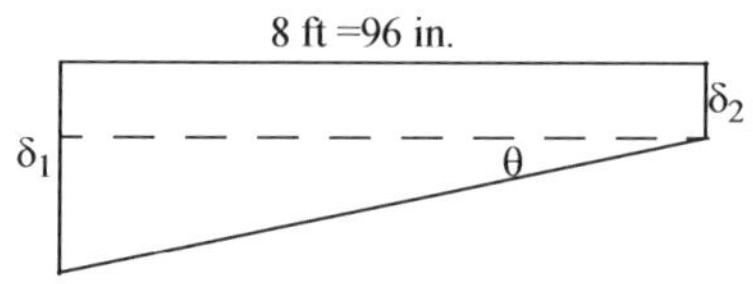

The deformations and angle of rotation can be found as shown below.are

$$\delta_1 = \frac{N_1 L}{EA} = \frac{(3.125)(5)(12)}{(10,000)(0.19635)} = 95.49(10^{-3})\ in \qquad \delta_2 = \frac{N_2 L}{EA} = \frac{(1.875)(5)(12)}{(10,000)(0.19635)} = 57.29(10^{-3})\ in \tag{3}$$

$$tan\theta = \frac{\delta_1 - \delta_2}{96} = \frac{95.49(10^{-3}) - 57.29(10^{-3})}{96} \quad or \quad \theta = atan[0.3978(10^{-3})] = 0.02279^o \tag{4}$$

ANS $\theta = 0.02279^o$ CCW

---

4. 55

Solution E = 210 GPa) d = 10 mm F = 5 kN θ = ?

$$A = \pi(0.01)^2/4 = 78.539(10^{-6})\ m^2 \tag{1}$$

By moment equilibrium about point D and force equilibrium in y direction on the free body diagram we obtain

$$(3.2)N_C = (1.2)F \quad or \quad N_C = 1.875\ kN \qquad N_D = F - N_C = 3.125\ kN \tag{2}$$

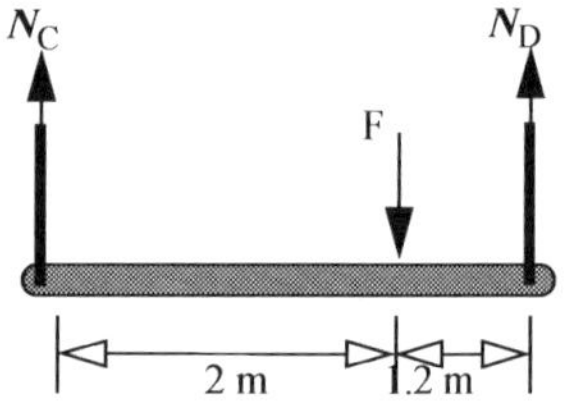

(a)

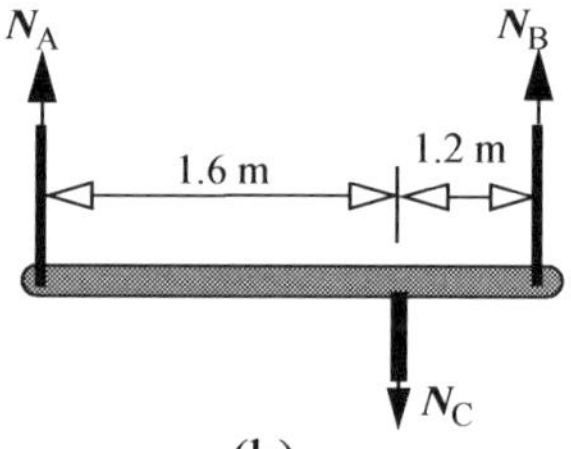

(b)

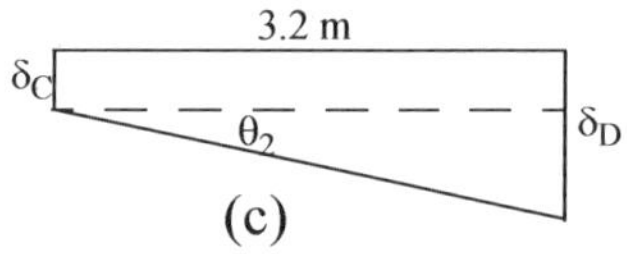

(c)

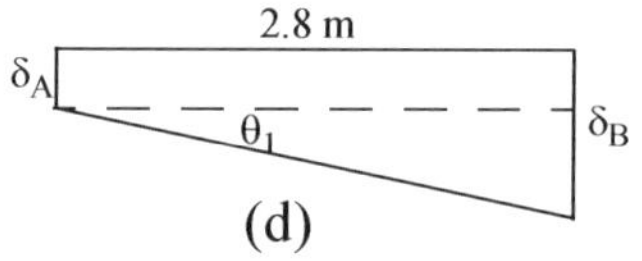

(d)

By moment equilibrium about point B in Fig. (b) and force equilibrium in y direction we obtain

$$(2.8)N_A = (1.2)N_C \quad or \quad N_A = 0.8036kN \qquad N_B = N_C - N_A = 1.0714\ kN \tag{3}$$

The deformations of the bars can be written as

$$\delta_A = \frac{N_A L_A}{EA} = \frac{(0.8036)(10^3)(2)}{(210)(10^9)(78.539)(10^{-6})} = 0.09745(10^{-3})m \qquad \delta_B = \frac{N_B L_B}{EA} = \frac{(1.0714)(10^3)(2)}{(210)(10^9)(78.539)(10^{-6})} = 0.1299(10^{-3})m \tag{4}$$

$$\delta_C = \frac{N_C L_C}{EA} = \frac{(1.875)(10^3)(1.2)}{(210)(10^9)(78.539)(10^{-6})} = 0.1364(10^{-3})m \qquad \delta_D = \frac{N_D L_D}{EA} = \frac{(3.125)(10^3)(4)}{(210)(10^9)(78.539)(10^{-6})} = 0.7579(10^{-3})m \tag{5}$$

From geometries of we obtain

$$tan\theta_2 = (\delta_D - \delta_C)/3.2 = [0.7579(10^{-3}) - 0.1364(10^{-3})]/3.2 = 0.194(10^{-3}) \quad or \quad \theta_2 = atan[0.194(10^{-3})] = 0.011^o \tag{6}$$

$$tan\theta_1 = (\delta_B - \delta_A)/2.8 = [0.1299(10^{-3}) - 0.09745(10^{-3})]/2.8 = 0.0116(10^{-3}) \quad or \quad \theta_1 = atan[0.0116(10^{-3})] = 0.0007^o \tag{7}$$

ANS $\theta_1 = 0.0007^o\ CW;\ \theta_2 = 0.011^o\ CW$

---

4. 56

Solution E = 210 GPa, σ_{yield} = 210 MP *d* = 20 mm *k* = 1.5 *F* = ?

The allowable stress and cross sectional area are

$$\sigma_{allow} = \sigma_{yield}/k = 210/1.5 = 140\ MPa \qquad A = \pi(0.02)^2/4 = 0.3142(10^{-3})\ m^2 \tag{1}$$

By moment equilibrium about point D and B and force equilibrium in y direction we obtain

$$(3.2)N_C = (1.2)F \quad or \quad N_C = 0.375F \quad N_B = N_C - N_A = 0.214F \quad (2)$$

$$(2.8)N_A = (1.2)N_C \quad or \quad N_A = 0.161F \quad N_D = F - N_C = 0.625F \quad (3)$$

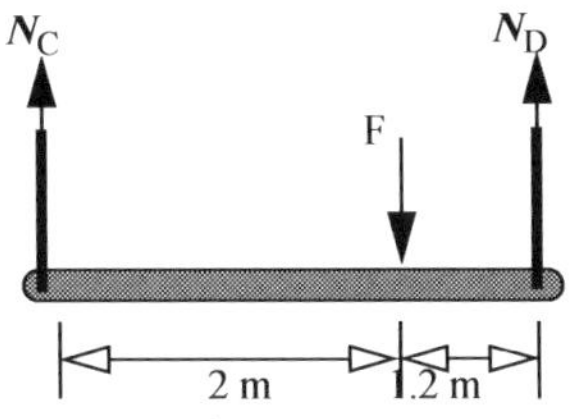

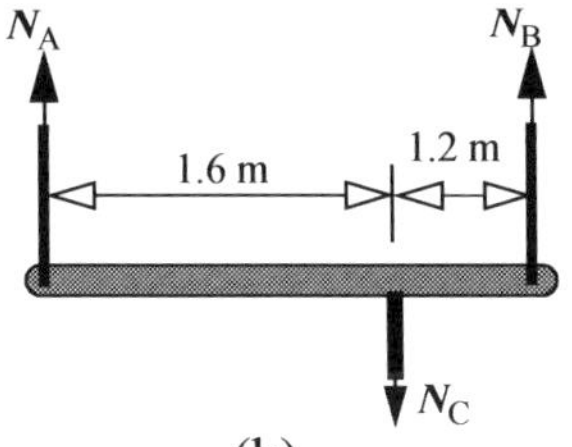

$$\sigma_D = N_D/A = (0.625F)/(0.3142(10^{-3})) \le 140(10^6) \quad or \quad F \le 70.38(10^3)\ N \quad 4$$

ANS $F = 70.3\ kN$

---

4. 57

Solution E = 10,000 ksi d = 1/2 in. P= 5 kips δ_{CE} = ? δ_{BD} =?

From geometry:

$$tan40 = \frac{CE}{(10)(12)} \quad or \quad CE = 100.69 \text{ in.} \quad BD = \frac{CE}{2} = 50.34 \text{ in.} \quad (1)$$

$$cos40 = \frac{(10)(12)}{AC} \quad or \quad AC = 156.65 \text{ in.} \quad BC = AB = \frac{AC}{2} = 78.32 \text{ in.} \quad (2)$$

$$\frac{\delta_{CE}}{AC} = \frac{\delta_{BD}}{AB} \quad or \quad \delta_{CE} = 2\delta_{BD} \quad (3)$$

By moment equilibrium about A:

$$120N_{CE} + 60N_{BD} = 100.69P = 503.45 \text{ kips} \quad (4)$$

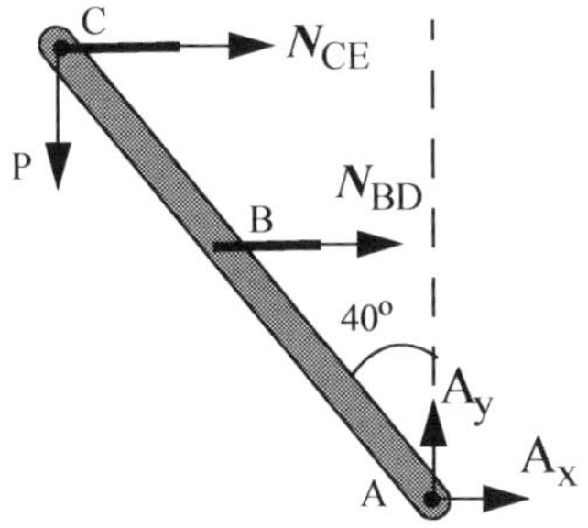

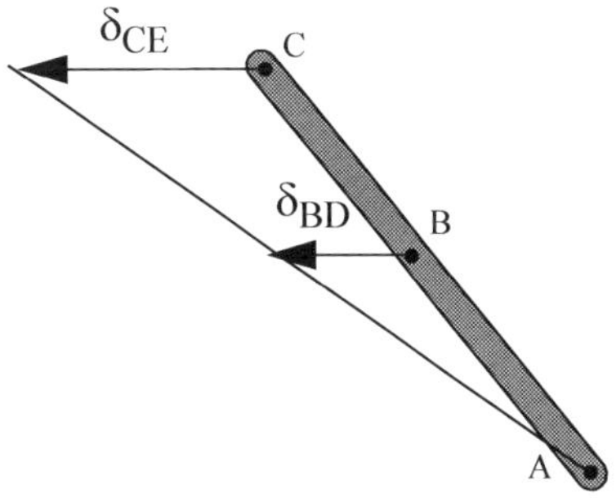

The cross sectional area is: $A = [\pi(1/2)^2]/4 = 0.1964 \text{ in.}^2$

The deformation and force calculation is as follows.

$$\delta_{CE} = \frac{N_{CE}L_{CE}}{EA} = \frac{N_{CE}(100.69)}{(10000)(0.1964)} \quad or \quad \delta_{CE} = 51.28(10^{-3})N_{CE} \quad (5)$$

$$\delta_{BD} = \frac{N_{BD}L_{BD}}{EA} = \frac{N_{BD}(50.34)}{(10000)(0.1964)} \quad or \quad \delta_{BD} = 25.64(10^{-3})N_{BD} \quad (6)$$

$$\delta_{CE} = 2\delta_{BD} \quad or \quad 51.28(10^{-3})N_{CE} = 2[25.64(10^{-3})N_{BD}] \quad or \quad N_{CE} = N_{BD} \quad (7)$$

$$20N_{CE} + 60N_{CE} = 503.45 \text{ kips} \quad N_{CE} = 2.797 \text{ kips} \quad N_{BD} = 2.797 \text{ kips} \quad (8)$$

$$\delta_{CE} = \delta_{CE} = 51.28(10^{-3})N_{CE} = 51.28(10^{-3})(2.797) = 0.1434 \text{ in.} \quad \delta_{BD} = \delta_{CE}/2 = 0.0717 \text{ in.} \quad (9)$$

ANS $\delta_{CE} = 0.1434$ in. ; $\delta_{BD} = 0.0717$ in.

---

4. 58

Solution E = 10,000 ksi σ_{yield} = 40 ksi P = 10 kips d_{min} = ? nearest 1/16 inch.

From previous problem we have the following.

$$CE = 100.69 \text{ in.} \quad BD = 50.34 \text{ in.} \quad AC = 156.65 \text{ in.} \quad BC = AB = 78.32 \text{ in.} \quad (1)$$

$$\delta_{CE} = 2\delta_{BD} \qquad 120N_{CE} + 60N_{BD} = 100.69P = 1006.9 \text{ kips} \qquad N_{CE} = N_{BD} = 5.594 \text{ kips} \quad (2)$$

$$\sigma_{CE} = \frac{N_{CE}}{A} = \frac{5.594}{(\pi d^2)/4} \le 40 \text{ ksi} \quad or \quad d^2 \ge \left(\frac{4}{\pi}\right)\left(\frac{5.594}{40}\right) \quad or \quad d \ge 0.422 \text{ in.} \quad (3)$$

ANS $d_{min} = \frac{7}{16}$ in.

---

4. 59

Solution E = 10,000 ksi d = 1/2 in. $\sigma_{yield} = 40$ ksi P = ?

From previous problem we have the following.

$$CE = 100.69 \text{ in.} \quad BD = 50.34 \text{ in.} \quad AC = 156.65 \text{ in.} \quad BC = AB = 78.32 \text{ in.} \quad A = 0.1964 \text{ in.}^2 \quad (1)$$

$$\delta_{CE} = 2\delta_{BD} \qquad 120N_{CE} + 60N_{BD} = 100.69P \qquad N_{CE} = N_{BD} = 0.5594P \quad (2)$$

The normal stress in each bar is the same as internal force and area is the same, hence

$$\sigma_{CE} = \frac{N_{CE}}{A} = \frac{0.5594P}{[0.1964]} \le 40 \text{ ksi} \quad or \quad P \le 14.0438 \text{ kips} \quad or \quad P \le 14043.8 \text{ lb} \quad (3)$$

ANS $P_{max} = 14043$ lb

---

4. 60

Solution F=20 kN E = 200 GPa A = 100 mm^2 L_{AP}= 200 mm L_{BP}= 250 mm δ_P = ? σ_A = ?

By force equilibrium of free body diagram of the rigid bar and geometry of the deformed bars we obtain the following.

$$N_A + N_B cos 70 = F \quad or \quad N_A + 0.342N_B = 20(10^3) \qquad \delta_A = \delta_P \qquad \delta_B = \delta_P cos 70 = 0.342\delta_P \quad (1)$$

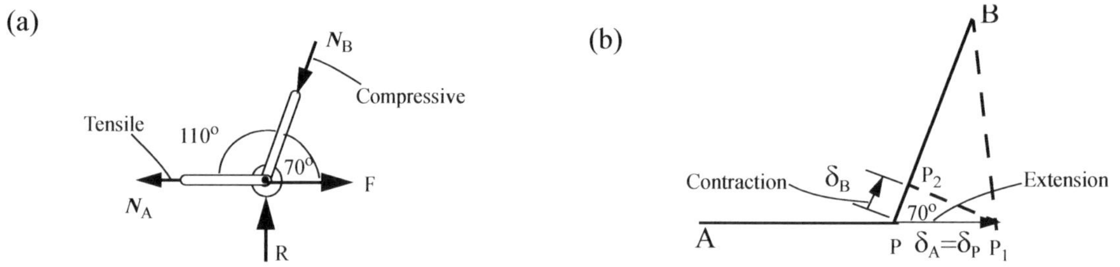

$$\delta_A = \frac{N_A L_A}{EA} = \frac{N_A(0.2)}{(200)(10^9)(100)(10^{-6})} = \delta_P \quad or \quad N_A = 100\delta_P(10^6) \quad (2)$$

$$\delta_B = \frac{N_B L_B}{EA} = \frac{N_B(0.25)}{(200)(10^9)(100)(10^{-6})} = 0.342\delta_P \quad or \quad N_B = 27.36\delta_P(10^6) \quad (3)$$

$$N_A + 0.342N_B = [(100\delta_P) + 0.342(27.36\delta_P)](10^6) = 20(10^3) \quad or \quad \delta_P = 0.1829(10^{-3})m \quad (4)$$

$$N_A = 100\delta_P(10^6) = (100)(0.1829)(10^{-3})(10^6) = 18.29(10^3)\ N \qquad \sigma_A = \frac{N_A}{A} = \frac{18.29(10^3)}{(100)(10^{-6})} = 182.9(10^6)N/m^2 \quad (5)$$

ANS $\delta_P = 0.18$ mm ; $\sigma_A = 183$ MPa(T))

---

4. 61

Solution F=20 kN E = 200 GPa A = 100 mm^2 L_{AP}= 200 mm L_{BP}= 250 mm δ_P = ? σ_A = ?

By force equilibrium of free body diagram of the rigid bar and geometry of the deformed bars we obtain the following.

$$N_A sin 30 + N_B sin 60 = F \quad or \quad 0.5N_A + 0.866N_B = 20(10^3) \qquad \delta_A = \delta_P cos 60 = 0.5\delta_P \qquad \delta_B = \delta_P sin 60 = 0.866\delta_P \quad (1)$$

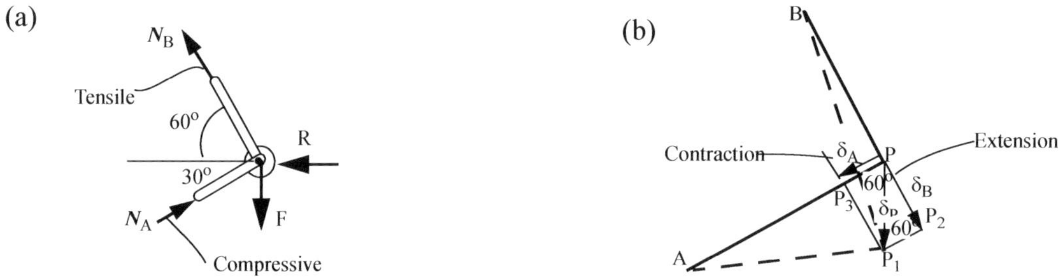

$$\delta_A = \frac{N_A L_A}{EA} = \frac{N_A(0.2)}{(200)(10^9)(100)(10^{-6})} = 0.5\delta_P \qquad or \qquad N_A = 50\delta_P(10^6) \quad \textbf{(2)}$$

$$\delta_B = \frac{N_B L_B}{EA} = \frac{N_B(0.25)}{(200)(10^9)(100)(10^{-6})} = 0.866\delta_P \qquad or \qquad N_B = 69.28\delta_P(10^6) \quad \textbf{(3)}$$

$$0.5N_A + 0.866N_B = [0.5(50\delta_P) + 0.866(69.2\delta_P)](10^6) = 20(10^3) \qquad or \qquad \delta_P = 0.2353(10^{-3})m \quad \textbf{(4)}$$

$$N_A = 50\delta_P(10^6) = (50)(0.2353)(10^{-3})(10^6) = 11.76(10^3)\ N \qquad \sigma_A = \frac{N_A}{A} = \frac{11.76(10^3)}{(100)(10^{-6})} = 117.6(10^6)N/m^2 \quad \textbf{(5)}$$

ANS $\delta_P = 0.24\ mm;\ \sigma_A = 118MPa(C)$

---

4. 62

Solution F=20 kN E = 200 GPa A = 100 mm^2 L_{AP}= 200 mm L_{BP}= 250 mm δ_P = ? σ_A = ?

By force equilibrium of free body diagram of the rigid bar and geometry of the deformed bars we obtain the following.

$$N_A cos 75 + N_B cos 30 = F \qquad or \qquad 0.2588N_A + 0.8660N_B = 20(10^3) \quad \textbf{(1)}$$

$$\delta_A = \delta_P cos 75 = 0.2588\delta_P \qquad \delta_B = \delta_P cos 30 = 0.8660\delta_P \quad \textbf{(2)}$$

(a)

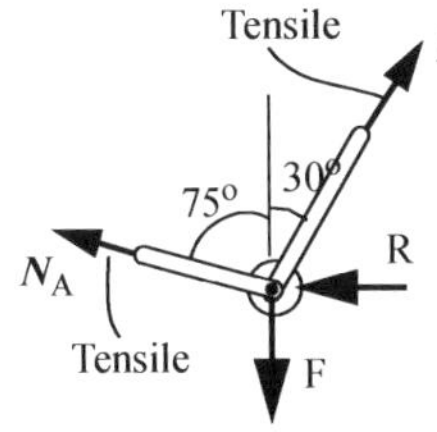

(b)

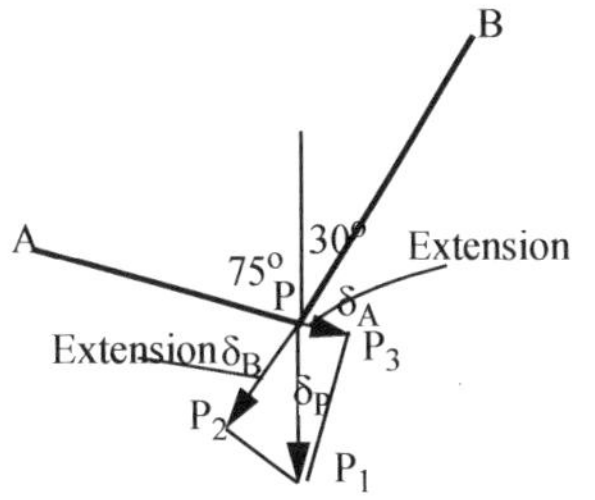

$$\delta_A = \frac{N_A L_A}{EA} = \frac{N_A(0.2)}{(200)(10^9)(100)(10^{-6})} = 0.2588\delta_P \qquad or \qquad N_A = 25.88\delta_P(10^6) \quad \textbf{(3)}$$

$$\delta_B = \frac{N_B L_B}{EA} = \frac{N_B(0.25)}{(200)(10^9)(100)(10^{-6})} = 0.866\delta_P \qquad or \qquad N_B = 69.28\delta_P(10^6) \quad \textbf{(4)}$$

$$0.2588N_A + 0.8660N_B = [0.2588(25.88\delta_P) + 0.866(69.2\delta_P)](10^6) = 20(10^3) \qquad or \qquad \delta_P = 0.3002(10^{-3})m \quad \textbf{(5)}$$

$$N_A = 25.88\delta_P(10^6) = (25.88)(0.3002)(10^{-3})(10^6) = 7.769(10^3)\ N \qquad \sigma_A = \frac{N_A}{A} = \frac{7.769(10^3)}{(100)(10^{-6})} = 77.69(10^6)N/m^2 \quad \textbf{(6)}$$

ANS δ_P = 0.30 mm ; σ_A = 77.7 MPa(T))

---

4. 63

Solution E = 70 GPa σ_{yield} = 280 MPa ν = 0.28 d = 0.5 mm m = ? %change in diameter of BC = ?

From the given data we have

$$A = \frac{\pi}{4}(0.005)^2 = 0.19635(10^{-6})\ m^2 \qquad \theta_{AB} = atan\left(\frac{200}{225}\right) = 41.63^o \qquad \theta_{CD} = atan\left(\frac{200}{225}\right) = 29.54^o \quad \textbf{(1)}$$

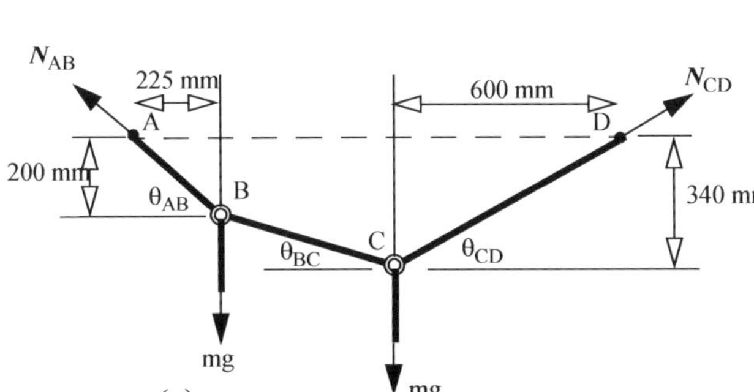

(a)

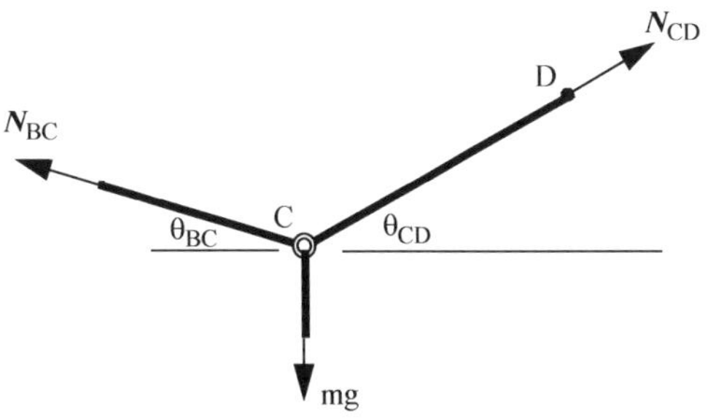

(b)

From force equilibrium of free body diagram on left we obtain two equations that can be solved.

$$N_{AB} cos\theta_{AB} = N_{CD} cos\theta_{CD} \qquad or \qquad N_{AB} = N_{CD}[cos 29.54/cos 41.63] = 1.16397N_{CD} \quad \textbf{(2)}$$

$$N_{AB}(sin\theta_{AB}) + N_{CD}(sin\theta_{CD}) = 2mg \qquad or \qquad 0.6643N_{AB} + 0.44930N_{CD} = 2mg \quad \textbf{(3)}$$

$$N_{AB} = 1.83844mg \qquad N_{CD} = 1.5795mg \tag{4}$$

From force equilibrium of free body diagram on right we obtain two equations that can be solved.

$$N_{BC}cos\theta_{BC} = N_{CD}cos\theta_{CD} = 1.5795mgcos29.54 = 1.3741mg \tag{5}$$

$$N_{BC}sin\theta_{BC} + N_{CD}sin\theta_{CD} = mg \quad or \quad N_{BC}sin\theta_{BC} = mg - 1.5795mgsin29.54 = 0.22126mg \tag{6}$$

$$tan\theta_{BC} = \frac{0.22126mg}{1.3741mg} = 0.161 \quad or \quad \theta_{BC} = 9.147^o \quad N_{BC} = \frac{0.22126mg}{sin9.147} = 1.3918mg \tag{7}$$

The maximum axial stress will be in CD and change in diameter in BC can be found as shown below.

$$\sigma_{CD} = N_{CD}/A = (1.5795mg)/[(0.19635)(10^{-6})] \le 280(10^6) \quad or \quad mg \le 34.887 \quad or \quad m \le 3.5481kg \tag{8}$$

$$N_{BC} = 1.3918(3.548)(9.81) = 48.443\ N \qquad \sigma_{BC} = N_{BC}/A = 48.443/[(0.19635)(10^{-6})] = 246.7(10^6)\ N/m^2 \tag{9}$$

$$\varepsilon_{BC} = \frac{\sigma_{BC}}{E} = \frac{246.7(10^6)}{70(10^9)} = 3.525(10^{-3}) \qquad (\varepsilon_{BC})_{tran} = -\nu\varepsilon_{BC} = -0.28(3.525)(10^{-3}) = -0.98687(10^{-3}) \tag{10}$$

$$\%\text{change in diameter of BC} = (\Delta d/d)x100 = (\varepsilon_{BC})_{tran}(100) = -0.0987 \tag{11}$$

ANS $m = 3548g$; %change in diameter of BC $= -0.0987\%$

---

4. 64

Solution E = 70 GPa σ_{yield} = 280 MPa ν = 0.28) d = ? nearest 1/10 of a millimeter

For a mass of 5 kg the internal axial force in CD can be found from the results of previous problem as shown below.

$$N_{CD} = 1.5795mg = 1.5795(5)(9.81) = 77.47\ N \tag{1}$$

$$\sigma_{CD} = N_{CD}/A = 77.47/[\pi d^2/4] \le 280(10^6) \quad or \quad d^2 \ge (4)(77.47)/[\pi(280)(10^6)] \quad or \quad d \ge 0.5935(10^{-3})m \tag{2}$$

ANS $d = 0.6\ mm$

---

4. 65

Solution E_{al} = 10,000ksi ν_{al} = 0.25 E_s = 30,000ksi ν_s = 0.28 t = 1/8 in $(d_{al})_0$ = 4in $(d_s)_0$ = 3in

P = 20 kips δp = ? Δd_s = ? Δd_{al} = ?

From the given data we can find the following.

$$(d_s)_i = (d_s)_0 - 2t = 2.75in \qquad (d_{al})_i = (d_{al})_0 - 2t = 3.75in \tag{1}$$

$$A_s = (\pi/4)(3^2 - 2.75^2) = 1.129in^2 \qquad A_{al} = (\pi/4)(4^2 - 3.75^2) = 1.522in^2 \tag{2}$$

From force equilibrium of the free body diagram and deformed shape we have the following.

$$N_{al} + N_s = 2P = 40kips \qquad \delta_s = \delta_p \qquad \delta_{al} = \delta_p \tag{3}$$

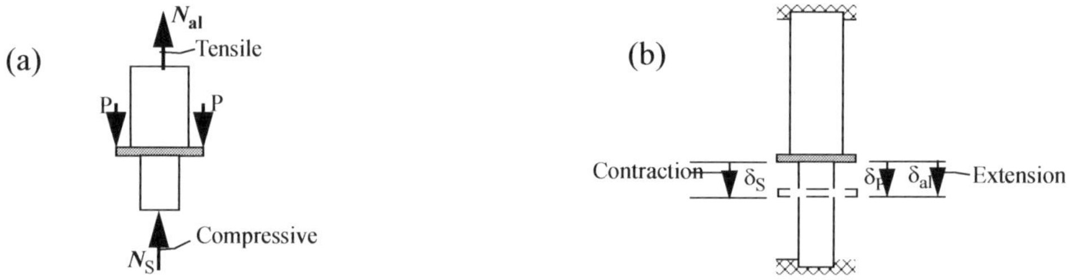

The solution proceed as follows.

$$\delta_{al} = \frac{N_{al}L_{al}}{E_{al}A_{al}} = \frac{N_{al}(40)}{10(10^3)(1.522)} = 2.628(10^{-3})N_{al} = \delta_p \quad or \quad N_{al} = 0.3804\delta_p(10^3) \tag{4}$$

$$\delta_s = \frac{N_sL_s}{E_sA_s} = \frac{N_s(30)}{30(10^3)(1.129)} = 0.8857(10^{-3})N_s = \delta_p \quad or \quad N_s = 1.129\delta_p(10^3) \tag{5}$$

$$N_{al} + N_s = (0.3804 + 1.129)\delta_p(10^3) = 40 \quad or \quad \delta_p = 26.5(10^{-3})in = \delta_{al} = \delta_s \tag{6}$$

$$\varepsilon_s = \delta_s/L_s = 26.5(10^{-3})/30 = 0.8833(10^{-3}) \qquad \varepsilon_{al} = \delta_{al}/L_{al} = 26.5(10^{-3})/40 = 0.6625(10^{-3}) \tag{7}$$

$$\Delta d_s = (d_s)_0[-\nu_s\varepsilon_s] = 3[-0.28(-0.8833)(10^{-3})] = 0.7420(10^{-3}) \tag{8}$$

$$\Delta d_{al} = (d_{al})_0[-\nu_{al}\varepsilon_{al}] = 4[-0.25(0.6625)(10^{-3})] = -0.6625(10^{-3}) \tag{9}$$

ANS $\delta_p = 0.0265$ in. ; $\Delta d_s = 0.00074$ in. ; $\Delta d_{al} = -0.00066$ in.

---

4. 66

Solution $E_{al} = 10{,}000 ksi$ $\nu_{al} = 0.25$ $E_s = 30{,}000 ksi$ $\nu_s = 0.28$ $t = 1/8$ in $(d_{al})_0 = 4 in$ $(d_s)_0 = 3 in$

$\sigma_{al} \le 10 ksi$ $\sigma_s \le 25 ksi$ $P_{max} = ?$

From previous problem we have the following.

$$N_{al} + N_s = 2P \qquad N_{al} = 0.3804\delta_p(10^3) \qquad N_s = 1.129\delta_p(10^3) \tag{1}$$

The solution proceeds as follows.

$$N_{al} + N_s = [0.3804 + 1.129]\delta_p(10^3) = 2P \quad or \quad \delta_p = 1.325P(10^{-3}) \qquad N_{al} = 0.504P \qquad N_s = 1.496P \tag{2}$$

$$\sigma_{al} = \frac{N_{al}}{A_{al}} = \frac{0.504P}{1.522} \le 10 \quad or \quad P \le 30.19 kips \qquad \sigma_s = \frac{N_s}{A_s} = \frac{1.496P}{1.129} \le 25 \quad or \quad P \le 18.87 kips \tag{3}$$

ANS $P_{max} = 18.8$ kips

---

4. 67

Solution $E_A = E_B = 30{,}000 ksi$ $L_A = 30$ in $L_B = 50$ in $A_A = A_B = 1 in^2$ $P = 100$ kips $\sigma_A = ?$ $\sigma_B = ?$

Assume gap closes at equilibrium. By moment equilibrium about point C on the free body diagram of the rigid body and from geometry of the exaggerated deformed shape we obtain the following.

$$F(24) - N_A(36) - N_B \sin 75(96) = 0 \quad or \quad 36N_A + 92.73N_B = 24P = 2400 \tag{1}$$

$$\delta_A = \delta_D - 0.004 \qquad \delta_B = \delta_E \sin 75 \qquad (\delta_D/36) = (\delta_E/96) \tag{2}$$

(a)

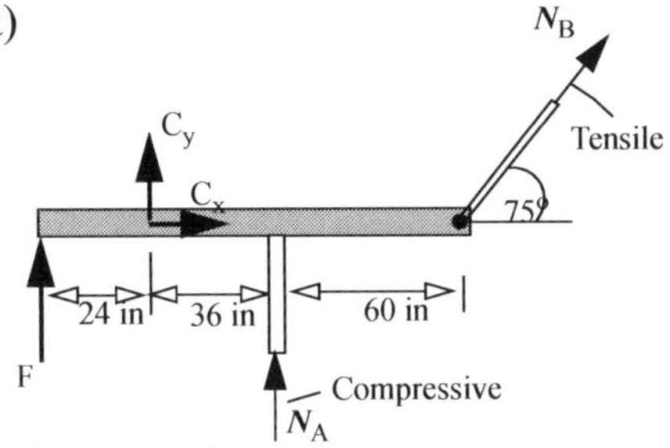

(b)

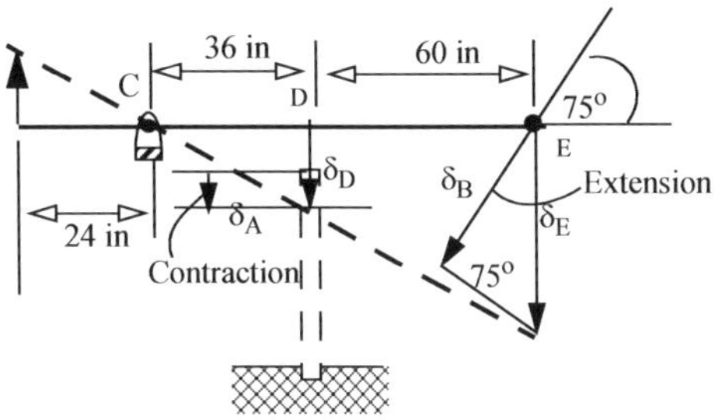

The solution proceed as follows.

$$\delta_A = \frac{N_A L_A}{E_A A_A} = \frac{N_A(30)}{(30)(10^3)(1)} = N_A(10^{-3}) = \delta_D - 0.004 \quad or \quad N_A = 10^3\delta_D - 4 \tag{3}$$

$$\delta_B = \frac{N_B L_B}{E_B A_B} = \frac{N_B(50)}{(30)(10^3)(1)} = 1.667N_B(10^{-3}) = \left(\frac{96}{36}\delta_D\right)\sin 75 \quad or \quad N_B = 0.647\delta_D(10^3) \tag{4}$$

$$36N_A + 92.73N_B = 36[10^3(\delta_D) - 4] + 92.73[0.647\delta_D(10^3)] = 2400 \quad or \quad \delta_D = 2544(10^{-3})/96 = 0.0265 in \tag{5}$$

As δ_D is greater than the gap, the assumption of gap closing is correct.

$$N_A = (10^3)(0.0265) - 4 = 22.5 kips \qquad \sigma_A = N_A/A_A = 22.5/1 = 22.5 ksi \tag{6}$$

$$N_B = (0.647)(0.0265)(10^3) = 17.15 kips \qquad \sigma_B = N_B/A_B = 17.15/1 = 17.15 ksi \tag{7}$$

ANS $\sigma_A = 22.5 ksi(C)$; $\sigma_B = 17.2 ksi(T)$

---

4. 68

Solution $E_A = E_B = 30{,}000 ksi$ $L_A = 30$ in $L_B = 50$ in$A_A = A_B = 1 in^2$ $\sigma_A \le 20 ksi$ $\sigma_B \le 20 ksi$ $P_{max} = ?$

From previous problem we have the following.

$$36N_A + 92.73N_B = 24P \qquad N_A = (10^3)\delta_D - 4 \qquad N_B = 0.647\delta_D(10^3) \tag{1}$$

The solution proceeds as follows.

$$36N_A + 92.73N_B = 36[(10^3)\delta_D - 4] + 92.73[0.647\delta_D(10^3)] = 24P = 96(10^3)\delta_D = 24P + 144 \text{ or}$$

$$\delta_D = (0.25P + 1.5)(10^{-3}) \qquad N_A = 0.25P - 2.5 \qquad N_B = 0.1618P + 0.9705 \tag{2}$$

$$\sigma_A = \frac{N_A}{A_A} = (0.25P - 2.5) \le 20 \quad or \quad P \le 90 kips \qquad \sigma_B = \frac{N_B}{A_B} = (0.1618P + 0.9705) \le 20 \quad or \quad P \le 117.6 kips \tag{3}$$

ANS $P_{max} = 90$ kips

---

4. 69

Solution $E_s = 200GPa$ $\nu = 0.28$ $L_A = 1$ m $L_B = 1.5$m $F = 75$ kN
$d_A = 50$ mm $d_B = 30$mm $\delta_A = ?$ $\delta_B = ?$ $\Delta d_A = ?$ $\Delta d_B = ?$

Assume gap closes at equilibrium. By moment equilibrium about point C on the free body diagram of the rigid body and from geometry of the exaggerated deformed shape we obtain the following.

$$F(0.4) - N_A(0.9) - (N_B sin40)(0.9) = 0 \quad or \quad 0.9N_A + 0.5785N_B = 0.4F = 30(10^3) \tag{1}$$

$$\delta_A = \delta_D - 0.0002 \qquad \delta_B = \delta_D cos50 = 0.6428\delta_D \tag{2}$$

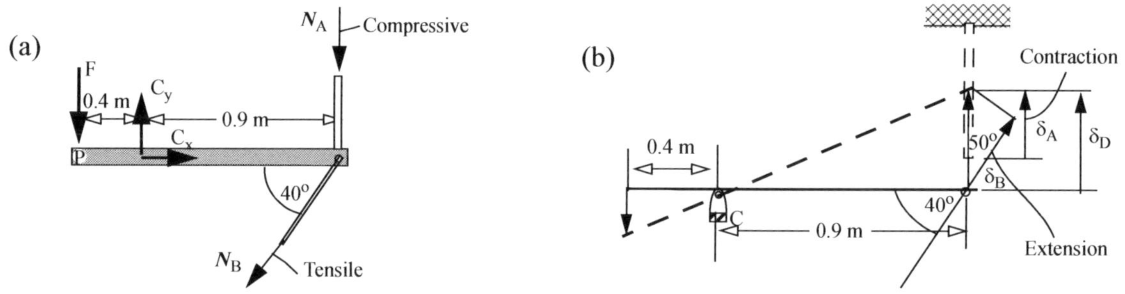

The solution proceeds as follows.

$$A_A = \frac{\pi}{4}(0.05)^2 = 1.963(10^{-3})m^2 \qquad A_B = \frac{\pi}{4}(0.03)^2 = 0.7068(10^{-3})m^2 \tag{3}$$

$$\delta_A = \frac{N_A L_A}{E_A A_A} = \frac{N_A(1)}{(200)(10^9)(1.963)(10^{-3})} = 2.547N_A(10^{-9}) = \delta_D - 0.0002 \quad or \quad N_A = (392.6\delta_D - 0.0785)(10^6) \tag{4}$$

$$\delta_B = \frac{N_B L_B}{E_B A_B} = \frac{N_B(1.5)}{(200)(10^9)(0.7068)(10^{-3})} = 10.611N_B(10^{-9}) = 0.6428\delta_D \quad or \quad N_B = 60.576\delta_D(10^6) \tag{5}$$

$$0.9N_A + 0.5785N_B = 0.9(392.6\delta_D - 0.0785)(10^6) + 0.5785(60.576)\delta_D(10^6) = 30(10^3) \quad or \quad \delta_D = \frac{0.10065}{388.4} = 0.2591(10^{-3})m$$

As δ_D is greater than the gap, the assumption of gap closing is correct.

$$\delta_A = 0.2591(10^{-3}) - 0.2(10^{-3}) = 0.0591(10^{-3})m \qquad \delta_B = 0.2591(0.6428)(10^{-3}) = 0.1666(10^{-3})m \tag{6}$$

The axial strains in A and B can be found as

$$\varepsilon_A = \delta_A/L_A = 0.0591(10^{-3}) \qquad \Delta d_A = d_A[-\nu\varepsilon_A] = 50[-0.28(-0.0591)(10^{-3})] = 0.828(10^{-3})mm$$

$$\varepsilon_B = \delta_B/L_B = 0.166(10^{-3})/1.5 = 0.1106(10^{-3}) \qquad \Delta d_B = d_B[-\nu\varepsilon_B] = 30[-0.28(0.1106)(10^{-3})] = -0.929(10^{-3})mm \tag{7}$$

ANS $\delta_A = 0.059$ mm ; $\delta_B = 0.166$ mm ; $\Delta d_A = 0.00083$ mm ; $\boxed{\Delta d_B = -0.00093 \text{ mm}}$

---

4. 70

Solution $E_s = 200$ GPa $L_A = 1$ m $L_B = 1.5$ m $d_A = 50$ mm $d_B = 30$ mm
$\nu = 0.28$ $\sigma_A \le 110MPa$ $\sigma_B \le 125MPa$ $F_{max} = ?$

From previous problem we have the following.

$$0.9N_A + 0.5785N_B = 0.4F \qquad N_A = (392.6\delta_D - 0.0785)(10^6) \qquad N_B = 63.581\delta_D(10^6) \tag{1}$$

The solution proceeds as follows.

$$0.9N_A + 0.5785N_B = [0.9(392.6\delta_D - 0.0785) + (0.5785)(63.581)\delta_D](10^6) = (390.1\delta_D - 0.07065)(10^6) = 0.4F \text{ or}$$

$$\delta_D = 1.025F(10^{-9}) + 0.1811(10^{-3}) \tag{2}$$

$$N_A = 402.4F(10^{-3}) - 7.400(10^3) \qquad N_B = 65.17F(10^{-3}) - 11.514(10^3) \tag{3}$$

$$\sigma_A = \frac{N_A}{A_A} = \frac{402.4F(10^{-3}) - 7.4(10^3)}{1.963(10^{-3})} \le 110(10^6) \quad or \quad F \le 555.0(10^3)Newtons \tag{4}$$

$$\sigma_B = \frac{N_B}{A_B} = \frac{65.17F(10^{-3}) + 11.514(10^3)}{0.7068(10^{-3})} \le 125(10^6) \quad or \quad F \le 1179(10^3)Newtons \tag{5}$$

ANS $F_{max} = 555kN$

---

4. 71

Solution E_{al} = 10,000ksi E_s = 30,000ksi E_{bl} = 15,000ksi t = 0.5 in P = 15 kips σ_s = ? δ_p = ?

$$A_{br} = (2)(0.5) = 1in^2 \qquad A_s = (2)(0.5) = 1in^2 \qquad A_{al} = (6)(0.5) = 3in^2 \qquad (1)$$

Assume that the gap closes at equilibrium. By force equilibrium and geometry we obtain the following.

$$N_{al} + N_{br} + N_s = 2P = 30kips \qquad \delta_{al} = \delta_p \qquad \delta_s = \delta_p - 0.02 \qquad \delta_{br} = \delta_p - 0.02 \qquad (2)$$

(a)

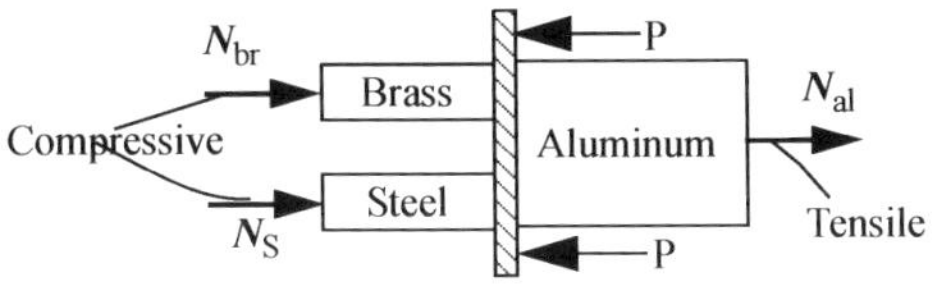

(b)

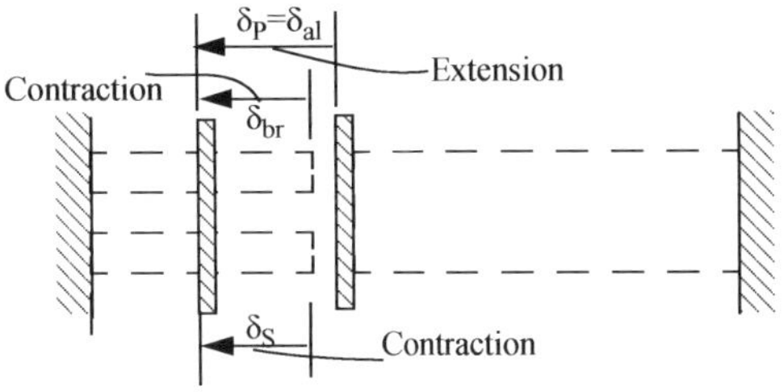

The solution proceeds as follows.

$$\delta_{al} = \frac{N_{al}L_{al}}{E_{al}A_{al}} = \frac{N_{al}(60)}{(10)(10^3)(3)} = 2N_{al}(10^{-3}) = \delta_p \quad or \quad N_{al} = 0.5\delta_p(10^3) \qquad (3)$$

$$\delta_s = \frac{N_s L_s}{E_s A_s} = \frac{N_s(30)}{(30)(10^3)(1)} = N_s(10^{-3}) = \delta_p - 0.02 \quad or \quad N_s = (10^3)\delta_p - 20 \qquad (4)$$

$$\delta_{br} = \frac{N_{br}L_{br}}{E_{br}A_{br}} = \frac{N_{br}(30)}{(15)(10^3)(1)} = 2N_{br}(10^{-3}) = \delta_p - 0.02 \quad or \quad N_{br} = 0.5(10^3)\delta_p - 10 \qquad (5)$$

$$0.5\delta_p(10^3) + (10^3)\delta_p - 20 + 0.5(10^3)\delta_p - 10 = 30 \quad or \quad 2.0\delta_p(10^3) = 60 \quad or \quad \delta_p = 30(10^{-3})inch \qquad (6)$$

As $\delta_p > 0.02$ the assumption gap closes is correct.

$$N_s = (10^3)(30)(10^{-3}) - 20 = 10kips \qquad \sigma_s = N_s/A_s = 10/1 = 10ksi \qquad (7)$$

ANS $\delta_p = 0.03$ in. ; $\sigma_s = 10$ ksi(C)

---

4. 72

Solution E_{al} = 10,000ksi E_s = 30,000ksi E_{bl} = 15,000ksi t = 0.5 in $\sigma_{br} \le 8ksi$ $\sigma_s \le 15ksi$ $\sigma_{al} \le 10ksi$

From previous problem we have

$$N_{al} + N_s + N_{br} = 2P \qquad N_{al} = 0.5(10^3)\delta_p \qquad N_s = (10^3)\delta_p - 20 \qquad N_{br} = 0.5(10^3)\delta_p - 10 \qquad (1)$$

The solution proceeds as follows.

$$0.5(10^3)\delta_p + (10^3)\delta_p - 20 + 0.5(10^3)\delta_p - 10 = 2P \quad or \quad 2.0\delta_p(10^3) = 2P + 30 \quad or \quad \delta_p = P(10^{-3}) + 0.015 \qquad (2)$$

$$N_{al} = 0.5P + 7.5 \qquad N_s = P - 5 \qquad N_{br} = 0.5P - 2.5 \qquad (3)$$

$$\sigma_{al} = N_{al}/A_{al} = (0.5P + 7.5)/3 \le 10ksi \quad or \quad P \le 45kips \qquad \sigma_s = N_s/A_s = (P-5)/1 \le 15ksi \quad or \quad P \le 20kips \qquad (4)$$

$$\sigma_{br} = N_{br}/A_{br} = (0.5P - 2.5)/1 \le 8ksi \quad or \quad P \le 21kips \qquad (5)$$

ANS P_{max} = 20 kips

---

4. 73

Solution $A_A = A_B = 400\ mm^2$ $E_A = E_B = 200$ GPa F = 10 kN δ_A = ? σ_B =

Assume that the gap closes at equilibrium. By moment equilibrium about point C and geometry we obtain the following.

$$N_A(5) + N_B(3) - F(7) = 0 \quad or \quad 5N_A + 3N_B = 7F = 70(10^3) \qquad \frac{\delta_D}{5} = \frac{\delta_B}{3} or, \delta_B = \frac{3}{5}\delta_D \qquad \delta_A = \delta_D - 0.0005 \qquad (1)$$

The solution proceeds as follows.

$$\delta_A = \frac{N_A L_A}{E_A A_A} = \frac{N_A(1.5)}{(200)(10^9)(400)(10^{-6})} = 18.75N_A(10^{-9}) = \delta_D - 0.0005 \quad or \quad N_A = (53.33\delta_D - 0.02667)(10^6) \qquad (2)$$

$$\delta_B = \frac{N_B L_B}{E_B A_B} = \frac{N_B(2)}{(200)(10^9)(400)(10^{-6})} = 25N_B(10^{-9}) = \frac{3}{5}\delta_D \quad or \quad N_B = 24\delta_D(10^6) \qquad (3)$$

$$5N_A + 3N_B = (5)(53.33\delta_D - 0.02667)(10^6) + (3)(24)(\delta_D)(10^6) = 70(10^3) \quad or \quad \delta_D = \frac{203.33}{338.65(10^6)} = 0.6004(10^{-3})m \qquad (4)$$

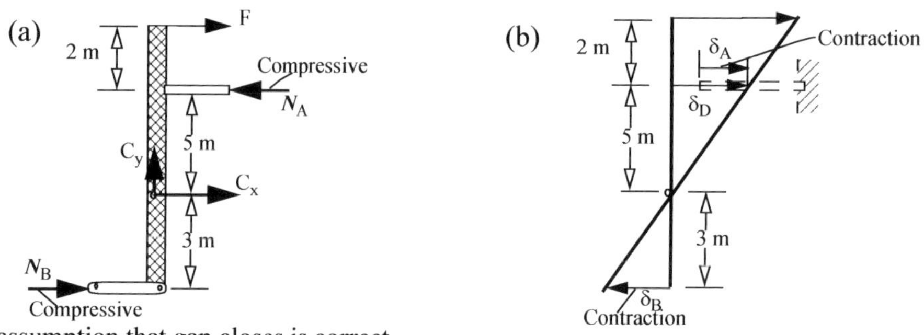

As $\delta_D > 0.0005$ the assumption that gap closes is correct.

$$\delta_A = (0.6004 - 0.5)(10^{-3}) = 0.1004(10^{-3})m \quad \mathbf{(5)}$$

$$N_B = 24\delta_D(10^6) = (24)(0.6004)(10^{-3})(10^6) = 14.41(10^3)N \qquad \sigma_B = \frac{N_B}{A_B} = \frac{14.41(10^3)}{400(10^{-6})} = 36.03(10^6)N/m^2 \quad \mathbf{(6)}$$

ANS $\delta_A = 0.1\text{ mm}$; $\sigma_B = 36\text{ MPa(C)}$

---

4. 74

Solution $A_A = A_B = 400\text{ mm}^2$ $E_A = E_B = 200\text{ GPa}$ $P_{max} = ?$ $\sigma_B \le 120MPa(C)$ $\delta_A \le 0.25mm$

From previous problem we have

$$5N_A + 3N_B = 7F \qquad N_A = (53.33\delta_D - 0.02667)(10^6) \qquad N_B = 24\delta_D(10^6) \quad \mathbf{(1)}$$

The solution proceeds as follows.

$$5N_A + 3N_B = 5(53.33\delta_D - 0.02667)(10^6) + (3)(24)\delta_D(10^6) = 7F \quad or \quad 338.65\delta_D = 7F(10^{-6}) + 0.1333 \text{ or}$$

$$\delta_D = 20.67F(10^{-9}) + 0.3936(10^{-3}) \quad \mathbf{(2)}$$

$$\delta_A = \delta_D - 0.5(10^{-3}) = 20.67F(10^{-9}) - 0.1064(10^{-3}) \le 0.25(10^{-3}) \quad or \quad F \le 17.24(10^3)N \quad \mathbf{(3)}$$

$$N_B = 0.496F + 9.446(10^3) \qquad \sigma_B = \frac{N_B}{A_B} = \frac{0.496F + 9.446(10^3)}{400(10^{-6})} \le 120(10^6) \quad or \quad F \le 77.73 \quad \mathbf{(4)}$$

ANS $F_{max} = 17.2kN$

---

4. 75

Solution $E = 30{,}000$ $\nu = 0.25$ $t = 0.5$ $\delta c = ?$ Δd_{CD}

Assume gap is closed at equilibrium. By force equilibrium and geometry we obtain the following.

$$N_{AB} = R_A \qquad N_{BC} = R_A - 25 \qquad N_{CD} = 60 - R_A \qquad \delta_{AB} + \delta_{BC} - \delta_{CD} = 0.01 \quad \mathbf{(1)}$$

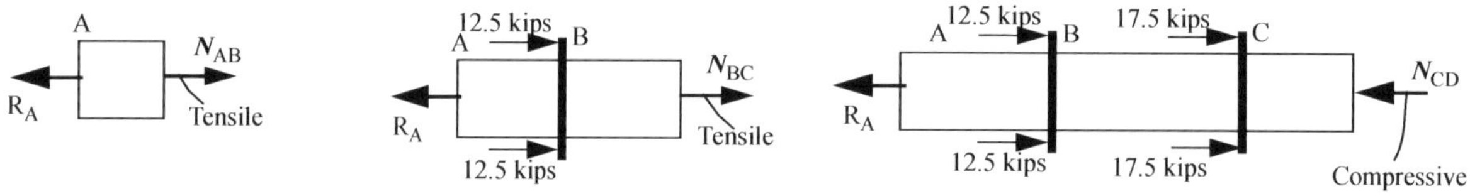

The solution proceeds as follows.

$$\delta_{AB} = \frac{N_{AB}L_{AB}}{EA} = \frac{(R_A)(18)}{(30)(10^3)(0.5)(3)} = 0.40R_A(10^{-3}) \qquad \delta_{BC} = \frac{N_{BC}L_{BC}}{EA} = \frac{(R_A - 25)(24)}{(30)(10^3)(0.5)(3)} = 0.533(R_A - 25)(10^{-3}) \quad \mathbf{(2)}$$

$$\delta_{CD} = \frac{N_{CD}L_{CD}}{EA} = \frac{(60 - R_A)(36)}{(30)(10^3)(0.5)(3)} = 0.80(60 - R_A)(10^{-3}) \quad \mathbf{(3)}$$

$$[0.4R_A + 0.533(R_A - 25) - 0.8(60 - R_A)](10^{-3}) = 0.01 \quad or \quad 1.733R_A = 71.325 \quad or \quad R_A = 41.157kips \quad \mathbf{(4)}$$

$$\delta_C = \delta_{AB} + \delta_{BC} = [(0.4)(0.533)(41.157 - 25)](10^{-3}) = 25.07(10^{-3})inch \quad \mathbf{(5)}$$

$$\varepsilon_{CD} = \frac{\delta_{CD}}{L_{CD}} = \frac{0.8(60 - 41.157)(10^{-3})}{36} = 0.4187(10^{-3}) \qquad \Delta d_{CD} = -\nu\varepsilon_{CD}d_{CD} = -\nu(-0.4187)(10^{-3})(3) = 0.314(10^{-3})in \quad \mathbf{(6)}$$

ANS $\delta_C = 0.025\text{ in.}$; $\Delta d_{CD} = 0.00031\text{ in.}$

---

4. 76

Solution $E_A = 1.5GPa$ $E_B = 2.0GPa$ $w = 20\ MPa$ $\sigma_A = ?$ $\sigma_B = ?$

Assume that the gap closes at equilibrium. By force equilibrium and geometry we obtain the following.

$$2N_A + N_B = 0.002w = 40(10^3)N/m^2 \qquad \delta_A = \delta_B + 0.0005 \qquad \textbf{(1)}$$

The solution proceeds as follows.

$$\delta_A = \frac{N_A L_A}{E_A A_A} = \frac{N_A(0.08)}{(1.5)(10^9)(0.01)(0.02)} = 0.2667(10^{-6})N_A \qquad \delta_B = \frac{N_B L_B}{E_B A_B} = \frac{N_B(0.08)}{(2.0)(10^9)(0.01)(0.02)} = 0.2000(10^{-6})N_B \qquad \textbf{(2)}$$

$$N_A = 3.75(10^6)\delta_A \qquad N_B = 5.00(10^6)(\delta_A - (0.0005)) \qquad \textbf{(3)}$$

$$2N_A + N_B = [2(3.75\delta_A) + 5(\delta_A - 0.0005)](10^6) = 40(10^3) \quad or \quad 12.5\delta_A = 0.0425 \quad or \quad \delta_A = 0.0034 \qquad \textbf{(4)}$$

$$N_A = (3.75)(0.0034)(10^6) = 12.75(10^3)N/m^2 \qquad N_B = (5.0)(0.0034 - 0.0005)(10^6) = 14.5(10^3)N/m^2 \qquad \textbf{(5)}$$

$$\sigma_A = \frac{N_A}{A_A} = \frac{(12.75)(10^3)}{(0.01)(0.02)} = 63.75(10^6)N/m^2 \qquad \sigma_B = \frac{N_B}{A_B} = \frac{(14.5)(10^3)}{(0.01)(0.02)} = 72.5(10^6)N/m^2 \qquad \textbf{(6)}$$

ANS $\sigma_A = 63.75\ MPa(C)$; $\sigma_B = 72.5\ MPa(C)$

---

4. 77

Solution $E_A = 1.5GPa$ $E_B = 2.0GPa$ $w_{max} = ?$ $\sigma_A \le 50MPa$ $\sigma_B = 30MPa$

From previous problem we have the following.

$$2N_A + N_B = 0.002w \qquad N_A = 3.75(10^6)\delta_A \qquad N_B = 5.0(10^6)(\delta_A - 0.0005) \qquad \textbf{(1)}$$

$$2N_A + N_B = [2(3.75)\delta_A + 5(\delta_A - 0.0005)](10^6) = 0.002w \quad or \quad \delta_A = \frac{2w(10^{-9}) + 2.5(10^{-3})}{12.5} = 0.16w(10^{-9}) + 0.2(10^{-3}) \qquad \textbf{(2)}$$

$$N_A = 3.75(10^6)\delta_A = 0.6w(10^{-3}) + 0.75(10^3) \qquad N_B = 5.0(10^6)(\delta_A - 0.0005) = 0.8w(10^{-3}) - 1.5(10^3) \qquad \textbf{(3)}$$

$$\sigma_A = \frac{N_A}{A_A} = \frac{0.6w(10^{-3}) + 0.75(10^3)}{(0.01)(0.02)} \le 50(10^6)N/m^2 \quad or \quad w \le 17.92(10^6)N/m^2 \qquad \textbf{(4)}$$

$$\sigma_B = \frac{N_B}{A_B} = \frac{0.8w(10^{-3}) - 1.5(10^3)}{(0.01)(0.02)} \le 30(10^6)N/m^2 \quad or \quad w \le 9.375(10^6)N/m^2 \qquad \textbf{(5)}$$

ANS $w_{max} = 9.4MPa$

---

4. 78

Solution $E_{al} = 70\ GPa$ $\nu_{al} = 0.25$ $E_s = 210\ GPa$ $\nu_s = 0.28$ $P = 200\ kN$ $\sigma_s = ?$ $\sigma_{al} = ?$ $\Delta d_{al} = ?$

$$A_{al} = \pi(0.04)^2/4 = 1.2566(10^{-3})m^2 \qquad A_s = (\pi[(0.07)^2 - (0.05)^2])/4 = 1.8849(10^{-3})m^2 \qquad \textbf{(1)}$$

From force equilibrium and deformed shape we obtain the following.

$$N_{al} + N_s = P = 200(10^3) \qquad \delta_{al} = \delta_s + 0.00015 \qquad \textbf{(2)}$$

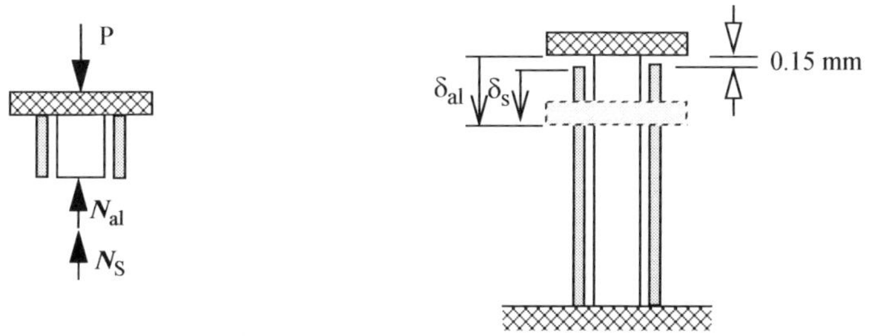

The solution proceeds as follows.

$$\delta_{al} = \frac{N_{al}L_{al}}{E_{al}A_{al}} = \frac{N_{al}(0.25)}{(70)(10^9)(1.2566)(10^{-3})} = 2.842(10^{-9})N_{al} \qquad \delta_s = \frac{N_sL_s}{E_sA_s} = \frac{N_s(0.25)}{(210)(10^9)(1.8849)(10^{-3})} = 0.6316(10^{-9})N_s \quad (3)$$

$$2.842(10^{-9})N_{al} = 0.6316(10^{-9})N_s + 0.00015 \qquad or \qquad 2.842N_{al} - 0.6316N_s = 150(10^3) \quad (4)$$

$$N_s = 120.45(10^3)N \qquad N_{al} = 79.55(10^3)N \quad (5)$$

$$\sigma_s = \frac{N_s}{A_s} = \frac{120.45(10^3)}{(1.8849)(10^{-3})} = 63.9(10^6)\ N/m^2 \qquad \sigma_{al} = \frac{N_{al}}{A_{al}} = \frac{79.55(10^3)}{(1.2566)(10^{-3})} = 63.3(10^6)(\ N/m^2) \quad (6)$$

$$\varepsilon_{al} = \frac{\sigma_{al}}{E_{al}} = \frac{(-63.3)(10^6)}{(70)(10^9)} = -0.904(10^{-3}) \qquad \Delta d_{al} = -\nu_{al}\varepsilon_{al}d_{al} = -0.25[-0.904(10^{-3})](40) = 90.43(10^{-3})mm \quad (7)$$

ANS $\sigma_s = 63.9\ MPa(C)\ ;\ \sigma_{al} = 63.3\ MPa(C)\ ;\ \Delta d_{al} = 0.0904mm$

---

4. 79

Solution $E_{al} = 70$ GPa $(\sigma_{yield})_{al} = 280$ MPa $E_s = 210$ GPa $(\sigma_{yield})_s = 210$ MPa $P = ?$

From previous problem we have the following.

$$A_{al} = \frac{\pi}{4}(0.04)^2 = 1.2566(10^{-3})m^2 \qquad A_s = \frac{\pi}{4}(0.07)^2 - \frac{\pi}{4}(0.05)^2 = 1.8849(10^{-3})m^2 \quad (1)$$

$$N_{al} + N_s = P \qquad 2.842N_{al} - 0.6316N_s = 150(10^3) \quad (2)$$

The solution proceeds as follows.

$$N_s = 0.8182P - 43.183(10^3) \qquad N_{al} = 0.1818P - 43.183(10^3) \quad (3)$$

$$\sigma_s = \frac{N_s}{A_s} = \frac{0.8182P - 43.183(10^3)}{(1.8849)(10^{-3})} \le 210(10^6) \quad or \quad 0.8182P \le (395.83 + 43.183)(10^3) \quad or \quad P \le 536.6(10^3)\ N \quad (4)$$

$$\sigma_{al} = \frac{N_{al}}{A_{al}} = \frac{0.1818P - 43.183(10^3)}{(1.2566)(10^{-3})} \le 280(10^6) \quad or \quad 0.1818P \le (351.8 + 43.183)(10^3) \quad or \quad P \le 1697.8(10^3)\ N \quad (5)$$

ANS $P_{max} = 536.6\ kN$

---

4. 80

Solution $E = 10{,}000$ ksi $d = 1/4$ in. $P = 10$kips $\theta = ?$

The cross sectional area is: $A = \pi(1/4)^2/4 = 0.04909\ in^2$

By equilibrium of moment about point D and geometry we obtain

$$13N_A + 5N_C = 10P = 100\ kips \qquad tan\theta = (\delta_A/13) = (\delta_C/5) \quad (1)$$

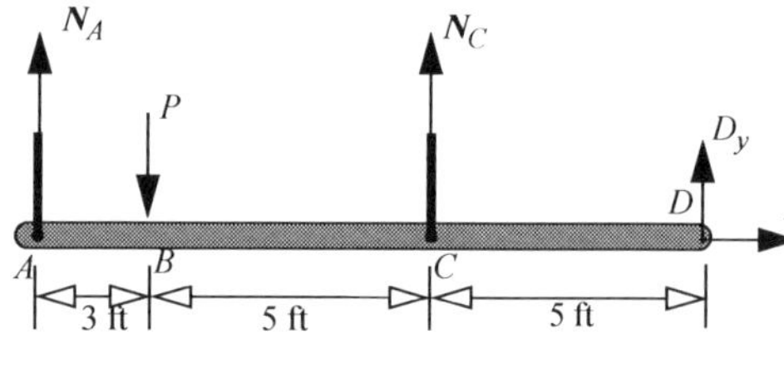

(a)

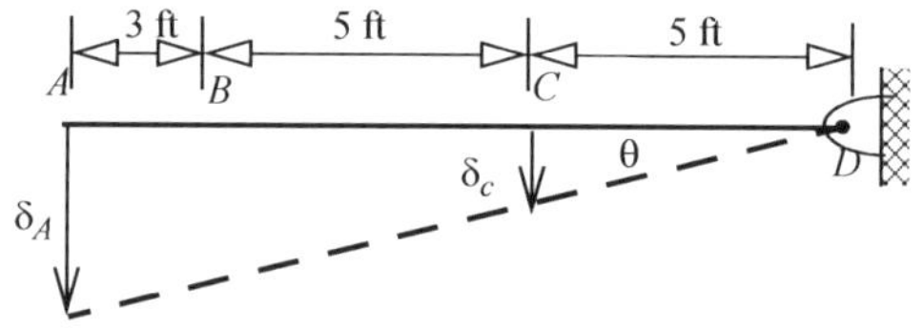

(b)

The solution proceeds as follows.

$$\delta_A = \frac{N_AL_A}{EA} = \frac{N_A(60)}{(10000)(0.04909)} = 122.23(10^{-3})N_A \qquad or \qquad N_A = 8.181\delta_A \quad (2)$$

$$\delta_C = \frac{N_CL_C}{EA} = \frac{N_C(60)}{(10000)(0.04909)} = 122.23(10^{-3})N_C = \frac{5}{13}\delta_A \qquad or \qquad N_C = 3.1466\delta_A \quad (3)$$

$$13N_A + 5N_C = (13)(8.181\delta_A) + 5(3.1466\delta_A) = 100 \qquad or \qquad \delta_A = 0.8191\ in \quad (4)$$

$$tan\theta = \frac{\delta_A}{13(12)} = \frac{0.8191}{156} = 0.00525 \quad (5)$$

ANS $\theta = 0.3^o$

4. 81

Solution E =30,000 ksi) d = 2 in $A_{bearing}$=4 in.^{2} w = 725lb/ft. σ_{AB} = ? σ_{CD} =? $\sigma_A = \sigma_D$ = ?

The cross sectional area of the bars is: $A = \pi(2)^2/4 = 3.1416\ in^2$

By equilibrium of moment about point E and geometry we obtain

$$10N_A + 30N_C = (15)(40w) = 600 = 435000\ lb = 435\ kip \qquad \frac{\delta_{AB}}{10} = \frac{\delta_{CD}}{30} \quad (1)$$

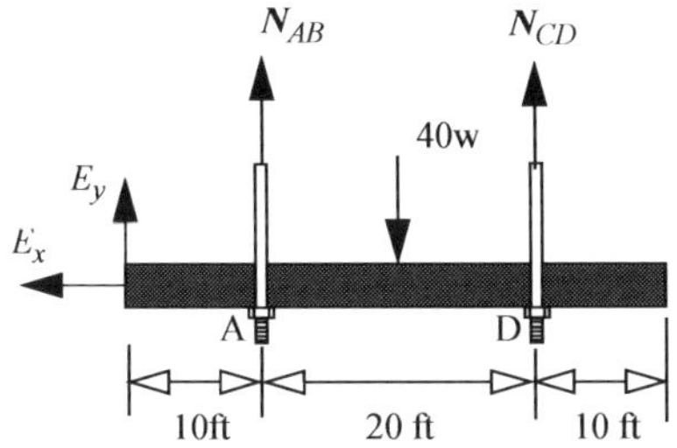

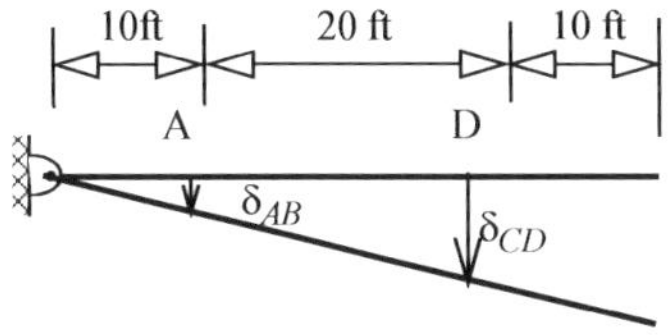

The solution proceeds as follows.

$$\delta_{AB} = \frac{N_{AB}L_A}{EA} = \frac{N_{AB}(36)(12)}{(30000)(3.1416)} = 4.5836(10^{-3})N_{AB} \quad or \quad N_{AB} = 218.17\delta_{AB} \quad (2)$$

$$\delta_{CD} = \frac{N_{CD}L_{CD}}{EA} = \frac{N_{CD}(36)(12)}{(30000)(3.1416)} = 4.4836(10^{-3})N_{CD} = \frac{30}{10}\delta_{AB} \quad or \quad N_{CD} = 654.45\delta_{AB} \quad (3)$$

$$10N_A + 30N_C = (10)(218.17\delta_{AB}) + 30(654.45)\delta_{AB} = 435 \quad or \quad \delta_{AB} = 19.939(10^{-3})\ in \quad (4)$$

$$N_{AB} = 218.17(19.939)(10^{-3}) = 4.35\ kips \qquad N_{CD} = 654.45(19.939)(10^{-3}) = 13.05\ kips \quad (5)$$

$$\sigma_{AB} = \frac{N_{AB}}{A} = \frac{4.35}{3.1416} = 1.38\ ksi \qquad \sigma_{CD} = \frac{N_{CD}}{A} = \frac{13.05}{3.1416} = 4.15\ ksi \quad (6)$$

$$\sigma_A = \frac{N_{AB}}{A_{bearing}} = \frac{4.35}{4} = 1.0875\ ksi \qquad \sigma_D = \frac{N_{CD}}{A_{bearing}} = \frac{13.05}{4} = 3.26\ ksi \quad (7)$$

ANS $\sigma_{AB} = 1.38\ ksi(T)$; $\sigma_{CD} = 4.15\ ksi(T)$; $\sigma_A = 1.0875\ ksi(C)$; $\sigma_D = 3.26\ ksi(C)$

---

4. 82

Solution E_{al} = 70GPa E_s = 200GPa δ_{AB} = ? $(\sigma_S)_{max}$ = ?

The radius of the bar BC varies linearly: $R(x) = a + bx$. At x=0.5m; R=0.025m and x=2.5m; R=0.05. Using these values we obtain two equations that can be solved as shown below and cross section area calculated.

$$a + b(0.5) = 0.025 \qquad a + b(2.5) = 0.050 \qquad a = 18.75(10^{-3}) \qquad b = 12.5(10^{-3}) \qquad A_{BC} = \pi(18.75 + 12.5x)^2(10^{-6}) \quad (1)$$

By force equilibrium and geometry we obtain the following.

$$N_{AB} = R_A \qquad N_{BC} = 100(10^3) - R_A \qquad \delta_{AB} = \delta_{BC} \quad (2)$$

(a)

(b)

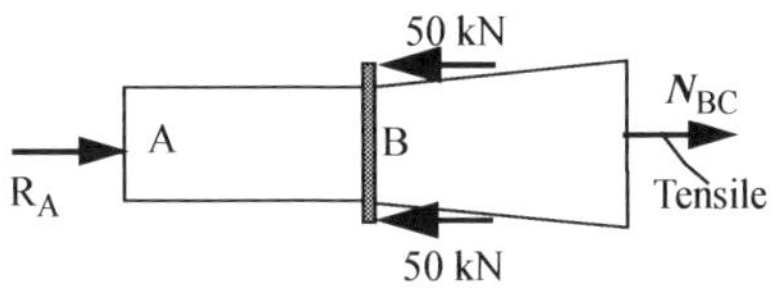

The solution proceeds as follows.

$$\delta_{AB} = \frac{N_{AB}L_{AB}}{E_{AB}A_{AB}} = \frac{R_A(0.5)}{(70)(10^9)\pi(0.025)^2} = 3.638R_A(10^{-9}) \qquad \frac{du}{dx} = \frac{N_{BC}}{E_{BC}A_{BC}} = \frac{[100(10^3) - R_A]}{(200)(10^9)\pi(18.75 + 12.5x)^2(10^{-6})} \quad (3)$$

$$\delta_{BC} = \int_{u_B}^{u_C} du = \frac{[100(10^3) - R_A]}{(200\pi)(10^3)}\int_{0.5}^{2.5}\frac{dx}{(18.75 + 12.5x)^2} = \frac{[100(10^3) - R_A]}{(200\pi)(10^3)}\frac{1}{(-12.5)}\frac{1}{(18.75 + 12.5x)}\Bigg|_{0.5}^{2.5} \quad or$$

$$\delta_{BC} = 2.546[(100)(10^3) - R_A](10^{-9}) \quad (4)$$

$$3.638R_A(10^{-9}) = 2.546[100(10^3) - R_A](10^{-9}) \quad or \quad R_A = 41.17(10^3)N \quad (5)$$

$$\delta_{AB} = 3.638R_A(10^{-9}) = (3.638)(41.17)(10^3)(10^{-9}) = 0.1498(10^{-3})m \quad (6)$$

The maximum axial stress in the segment BC will be just after B, as the area of the cross-section is smallest there.

$$N_{BC} = 58.83(10^3)N \qquad (\sigma_S)_{max} = \frac{58.83(10^3)}{\pi(0.025)^2} = 29.96(10^6)N/m^2 \qquad (7)$$

ANS $\delta_{AB} = 0.15$ mm to the left ; $(\sigma_S)_{max} = 30$ MPa(T)

---

4. 83

Solution $E = 30000$ $\quad A_A = 1\,in^2$ $\quad A_B = 2\,in^2$ $\quad \delta_C \le 0.01 in$ $\quad \sigma_A \le 25ksi$ $\quad \sigma_B \le 25ksi$ $\quad P_{max} = ?$

Assume gap closes. By moment equilibrium about point O and geometry we obtain the following.

$$N_A(30) + N_B(72) - P(24) = 0 \quad or \quad 30N_A + 72N_B = 24P \qquad \frac{\delta_C}{24} = \frac{\delta_A}{30} = \frac{\delta_D}{72} \qquad \delta_D = \delta_B + 0.005 \qquad (1)$$

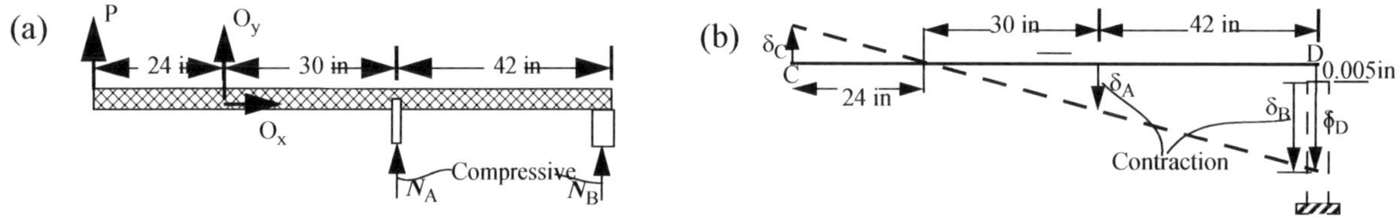

The solution proceeds as follows.

$$\delta_A = \frac{N_A L_A}{E_A A_A} = \frac{N_A(36)}{(30000)(1)} = \frac{5}{4}\delta_C \quad or \quad N_A = 1.0417\delta_C(10^3) \qquad (2)$$

$$\delta_B = \frac{N_B L_B}{E_B A_B} = \frac{N_B(48)}{(30000)(2)} = 3\delta_C - 0.005 \quad or \quad N_B = 3.75\delta_C(10^3) - 6.25 \qquad (3)$$

$$30N_A + 72N_B = 30(1.0417\delta_C)(10^3) + 72(3.75\delta_C(10^3) - 6.25) = 24P \quad or \quad 301.25(10^3)\delta_C = 24P + 450 \qquad (4)$$

$$\delta_C = (0.0791P + 1.4938)(10^{-3}) \le 0.01 \qquad P \le 106.73kips \qquad (5)$$

$$N_A = (0.083P + 1.556)kips \qquad N_B = (0.2988P - 0.6483)kips \qquad (6)$$

$$\sigma_A = \frac{N_A}{A_A} = \frac{0.083P + 1.556}{1} \le 25 \quad or \quad P \le 282.5 \qquad \sigma_B = \frac{N_B}{A_B} = \frac{0.2988P - 0.6483}{2} \le 25 \quad or \quad P \le 169.5 \qquad (7)$$

ANS $P_{max} = 106.7kips$

---

4. 84

Solution $\quad \sigma_{AC} \le 15ksi \quad \sigma_{BC} = 15ksi \quad v_C \le 0.1in \quad A_{AC} = ? \quad A_{BC} = ?$

The length $L_{BC} = 12/\sin 52 = 15.23$ ft.

By force equilibrium we obtain

$$N_{BC}\cos 52 = 75\cos 55 \quad or \quad N_{BC} = 69.87kips \qquad N_{AC} = N_{BC}\sin 52 + 75\sin 55 \quad or \quad N_{AC} = 116.49kips \qquad (1)$$

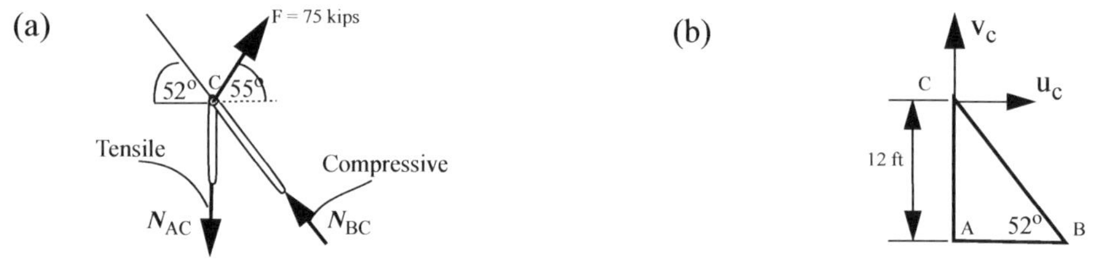

The solution proceeds as follows.

$$\sigma_{BC} = \frac{N_{BC}}{A_{BC}} = \frac{69.87}{A_{BC}} \le 15 \quad or \quad A_{BC} \ge 4.658in^2 \qquad \sigma_{AC} = \frac{N_{AC}}{A_{AC}} = \frac{116.49}{A_{AC}} \le 15 \quad or \quad A_{AC} \ge 7.766in^2 \qquad (2)$$

$$\delta_{AC} = \frac{N_{AC}L_{AC}}{EA_{AC}} = \frac{(69.87)(12)(12)}{(30000)A_{AC}} = \frac{0.3354}{A_{AC}}in^2 \qquad \delta_{BC} = \frac{N_{BC}L_{BC}}{EA_{BC}} = \frac{(116.49)(15.23)(12)}{(30000)A_{BC}} = \frac{0.7097}{A_{BC}}in^2 \qquad (3)$$

Let the pin move by u_C and v_C in the x and y-direction, respectively. Assuming small strain.

$$\delta_{AC} = v_C \qquad v_C = \frac{0.3354}{A_{AC}} \le 0.1 \qquad A_{AC} \ge 3.35in^2 \qquad (4)$$

ANS $A_{AC} = 7.77\ in^2$; $A_{BC} = 4.66\ in^2$

4. 85

Solution $\sigma_{BC} \le 30ksi$ $\tau_C \le 12ksi$ $d_c = ?$ $A_{BC} = ?$

By moment equilibrium about point A we obtain the following.

$$18\sin 20(44) - (N_{BC}\cos 36.5)(12) + (N_{BC}\sin 36.5)(2) = 0 \qquad or \qquad N_{BC} = \frac{270.9}{8.457} = 32.03kips \qquad \textbf{(1)}$$

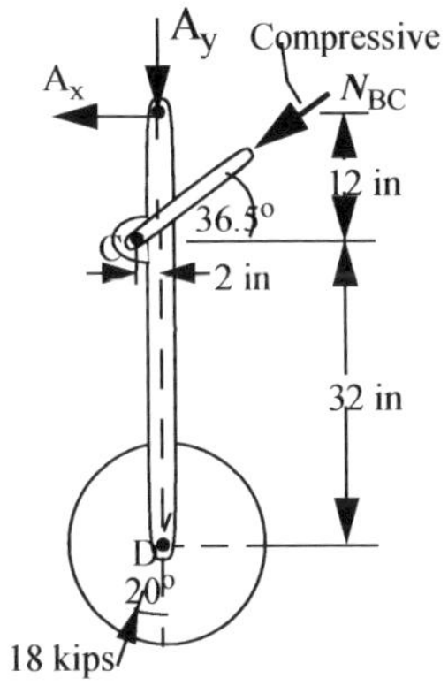

The axial stress in BC and shear stress in pin C that is in double shear can be found as shown below.

$$\sigma_{BC} = N_{BC}/A_{BC} = 32.03/A_{BC} \le 30 \qquad or \qquad A_{BC} \ge 1.068in^2 \qquad \textbf{(2)}$$

$$V_C = N_{BC}/2 = 16.015kips \qquad \tau_C = V_C/A_C = 16.015/[\pi d^2/4] \le 12 \qquad or \qquad d^2 \ge 1.699in \qquad or \qquad d \ge 1.303in \qquad \textbf{(3)}$$

ANS $A_{BC} = 1.1in^2$; $d = 1.31in$

---

4. 86

Solution $\sigma_{max} \le 20ksi$ $r = ?$

The axial stress in BC and maximum stress can can be found as shown below.

$$\sigma_{BC} = 10/[\pi(1^2)/4] = 12.73ksi \qquad \sigma_{max} = k_{conc}\sigma_{BC} = 12.73k_{conc} \le 20 \qquad or \qquad k_{conc} \le 1.57 \qquad \textbf{(1)}$$

From Appendix C we obtain the approximate value of r/d corresponding to D/d = 2 and $k_{conc} = 1.5$

$$r/d = 0.26 \qquad or \qquad r = (0.26)(1) = 0.26inch \qquad \textbf{(2)}$$

ANS $r = 0.26$ in.

---

4. 87

Solution $\sigma_{max} \le 120\ MPa$ r = 6 mm E = 70GPa $F_{max} = ?$

The area of cross-sections of various segments can be found as shown below.

$$A_{AB} = (\pi/4)(0.06)^2 = 2.83(10^{-3})m^2 \qquad A_{BC} = A_{CD} = (\pi/4)(0.048)^2 = 1.81(10^{-3})m^2 \qquad \textbf{(1)}$$

For the ratio's $r/d = 6/48 = 0.125$ and $D/d = 60/48 = 1.25$ we obtain the stress concentration factor as $K_{conc} = 1.7$

By force equilibrium and geometry we obtain

$$N_{AB} = R_A \qquad N_{BC} = R_A \qquad N_{CD} = 2F - R_A \qquad \delta_{AB} + \delta_{BC} - \delta_{CD} = 0 \qquad \textbf{(2)}$$

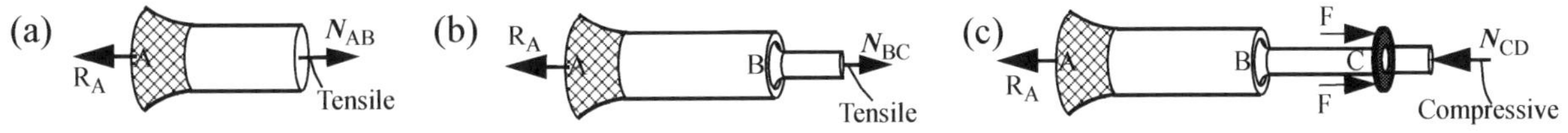

The solution proceeds as follows.

$$\delta_{AB} = \frac{N_{AB}L_{AB}}{E_{AB}A_{AB}} = \frac{R_A(0.9)}{70(10^9)(2.83)(10^{-3})} = 4.543R_A(10^{-6}) \qquad \delta_{BC} = \frac{N_{BC}L_{BC}}{E_{BC}A_{BC}} = \frac{R_A(0.75)}{70(10^9)(1.81)(10^{-3})} = 5.919R_A(10^{-6}) \qquad \textbf{(3)}$$

$$\delta_{CD} = \frac{N_{CD}L_{CD}}{E_{CD}A_{CD}} = \frac{(2F - R_A)(1)}{70(10^9)(1.81)(10^{-3})} = 7.893(2F - R_A)(10^{-6}) \qquad \textbf{(4)}$$

$$\delta_{AB} + \delta_{BC} - \delta_{CD} = [4.543R_A + 5.919R_A - 7.893(2F - R_A)](10^{-6}) = 0 \qquad or \qquad R_A = \frac{15.786F}{18.355} = 0.86F \qquad \textbf{(5)}$$

The nominal axial stress in each member can be found as

$$\sigma_{AB} = \frac{N_{AB}}{A_{AB}} = \frac{0.86F}{2.83(10^{-3})} = 303.9F \qquad \sigma_{BC} = \frac{N_{BC}}{A_{BC}} = \frac{0.86F}{1.81(10^{-3})} = 475.1F \qquad \sigma_{CD} = \frac{N_{CD}}{A_{CD}} = \frac{(2 - 0.86)F}{2.83(10^{-3})} = 629.8F \qquad \textbf{(6)}$$

The maximum stress in BC is used to find F as shown below

$$(\sigma_{BC})_{max} = k_{conc}\sigma_{BC} = (1.7)(475.1F) = 807.7F \qquad 807.7F \le 120(10^6) \qquad or \qquad F \le 148.6(10^3)N \qquad \textbf{(7)}$$

ANS $F_{max} = 148.6$ kN

---

4. 88

Solution r = 0.2 n = 0.5(10^6) steel F_{max} = ?

From Figure 3.36 the peak stress for half a million cycles is $\sigma_{max} = 40ksi$. For r/d = 0.2 and d/D = 2, the stress concentration factor is K_{conc} = 1.65. The nominal stress in BC is as shown below.

$$\sigma_{BC} = \sigma_{max}/K_{conc} = 40/1.65 = F/[\pi(1)^2/4] \qquad or \qquad F = 19.04kips \qquad \textbf{(1)}$$

ANS F_{max} = 19 kips

---

4. 89

Solution n = 10^6 Aluminum F_{max} = ?

From S-N curves the peak stress for a million cycles in Aluminum is approximately $\sigma_{max} = 147MPa$

From earlier problem, we have

$$(\sigma_{BC})_{max} = 807.7F \le 147(10^6) \qquad or \qquad F \le 181.998(10^3)N \qquad \textbf{(1)}$$

ANS $F_{max} = 181.9kN$

Section 4.4

4. 90

Solution E = 10,000ksi d = 1inch σ_A =? σ_B =?

By moment equilibrium about point C and geometry , we obtain

$$0.4N_A = 0.6N_B \qquad \frac{\delta_D}{0.4} = \frac{\delta_B}{0.6} \qquad or \qquad \delta_B = 1.5\delta_D \qquad \delta_A = 0.05 - \delta_D \qquad \textbf{(1)}$$

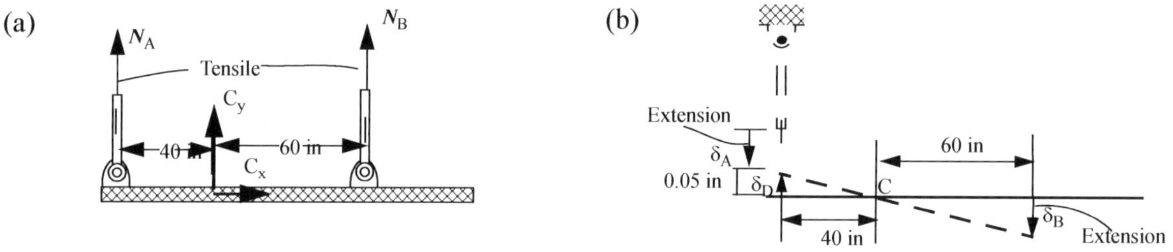

Solution proceeds as shown below.

$$\delta_A = \frac{N_A L_A}{E_A A_A} = \frac{N_A(60)}{(10,000)(\pi(1)^2/4)} = 7.639N_A(10^{-3}) = 0.05 - \delta_D \qquad or \qquad N_A = 6.545 - 130.91\delta_D \qquad \textbf{(2)}$$

$$\delta_B = \frac{N_B L_B}{E_B A_B} = \frac{N_B(80)}{(10,000)(\pi(1)^2/4)} = 10.186N_B(10^{-3}) = 1.5\delta_D \qquad or \qquad N_B = 147.26\delta_D \qquad \textbf{(3)}$$

$$0.4(6.545 - 130.91\delta_D) = 0.6(147.26\delta_D) \qquad or \qquad \delta_D = \frac{2.618}{140.72} = 18.6(10^{-3}) \qquad N_A = 4.11kips \qquad N_B = 2.74kips \qquad \textbf{(4)}$$

$$\sigma_A = N_A/A_A = 4.11/[\pi(1)^2/4] = 5.233ksi \qquad \sigma_B = N_B/A_B = 2.74/[\pi(1)^2/4] = 3.488ksi \qquad \textbf{(5)}$$

ANS $\sigma_A = 5.2ksi(T)$; $\sigma_B = 3.5ksi(T)$

---

4. 91

Solution E = 70GPa d = 25mm assembly = ? σ_A =? σ_B =?

The area cross-section is $A = \pi(0.025)^2/4 = 0.4909(10^{-3})$

Assembly 1 Analysis: By moment equilibrium about point C and geometry we obtain the following.

$$N_A(3) = N_B(1.5) \qquad or \qquad 2N_A - N_B = 0 \qquad \delta_D/3 = \delta_B/1.5 \qquad or \qquad \delta_B = \delta_D/2 \qquad \delta_A = 0.002 - \delta_D \qquad \textbf{(1)}$$

Solution proceeds as follows.

$$\delta_A = \frac{N_A L_A}{E_A A_A} = \frac{N_A(2)}{(70)(10^9)(0.4909)(10^{-3})} = 58.2N_A(10^{-9}) = 0.002 - \delta_D \qquad or \qquad N_A = 34.36(10^3) - 17.18\delta_D(10^6) \qquad \textbf{(2)}$$

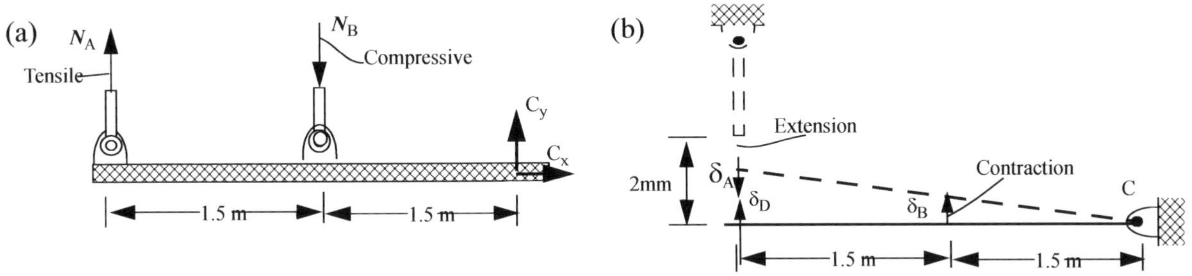

$$\delta_B = \frac{N_B L_B}{E_B A_B} = 58.2 N_B(10^{-9}) = \frac{\delta_D}{2} \quad or \quad N_B = 8.591\delta_D(10^6) \tag{3}$$

$$2N_A - N_B = 2[34.36 - 17.18\delta_D(10^6)] - 8.591\delta_D(10^6) = 0 \quad or \quad \delta_D = \frac{68.72}{42.95}(10^{-3}) = 1.60(10^{-3}) \tag{4}$$

$$N_A = 34.36(10^3) - 17.18\delta_D(10^6) = 6.872(10^3) \quad N_B = 8.591\delta_D(10^6) = 13.75(10^3) \tag{5}$$

$$\sigma_A = \frac{N_A}{A_A} = \frac{6.872(10^3)}{(0.4909)(10^{-3})} = 14.0(10^6) \quad \sigma_B = \frac{N_B}{A_B} = \frac{13.75(10^3)}{(0.4909)(10^{-3})} = 28.0(10^6) \tag{6}$$

ANS $\sigma_A = 14.0 \text{ MPa}(T); \sigma_B = 28.0 \text{ MPa}(C)$

Assembly 2 Analysis: By moment equilibrium about point C and from geometery we obtain the following.

$$N_B(3) = 1.5N_A \quad or \quad N_A - 2N_B = 0 \quad \frac{\delta_B}{3} = \frac{\delta_D}{1.5} \quad or \quad \delta_B = 2\delta_D \quad \delta_A = 0.002 - \delta_D \tag{7}$$

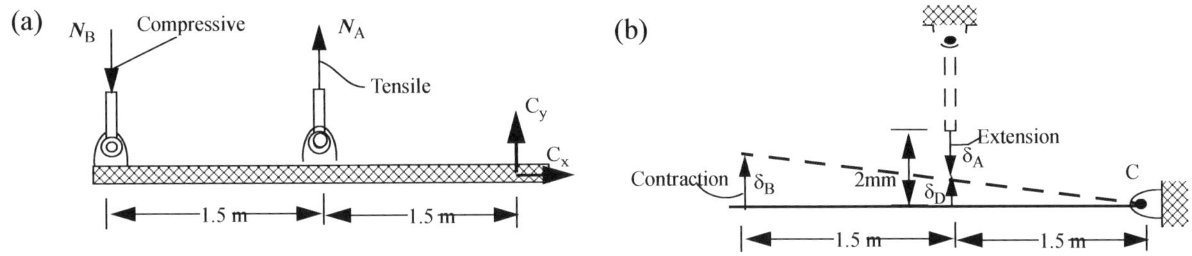

Solution proceeds as follows.

$$\delta_B = (N_B L_B)/(E_B A_B) = 58.2N_B(10^{-9}) = 2\delta_D \quad or \quad N_B = 34.36\delta_D(10^6) \tag{8}$$

$$\delta_A = 58.2N_A(10^{-9}) = 0.002 - \delta_D \quad or \quad N_A = 34.36(10^3) - 17.18\delta_D(10^6) \tag{9}$$

$$N_A - 2N_B = [34.36(10^3) - 17.18\delta_D(10^6)] - 2(34.36)\delta_D(10^6) = 0 \quad or \quad \delta_D = 34.36(10^3)/[85.908(10^6)] = 0.4(10^{-3}) \tag{10}$$

$$N_A = 34.36(10^3) - 17.18\delta_D(10^6) = 27.49(10^3) \quad N_B = 34.36\delta_D(10^6) = 13.74(10^3) \tag{11}$$

$$\sigma_A = N_A/A_A = (27.49(10^3))/[0.4909(10^{-3})] = 56(10^6) \quad \sigma_B = N_B/A_B = 13.74((10^3)/[0.4909(10^{-3})]) = 27.989 \tag{12}$$

ANS $\sigma_A = 56 \text{ MPa}(T); \sigma_B = 28 \text{ MPa}(C)$

Assembly 1 is preferred as the initial stress in member A is significantly smaller.

---

4. 92

Solution p =0.3 mm n=1/4 E_{st} = 200 GPa E_{al} = 70 GPa A_{st} = 500 mm^2 A_{al} = 1100 mm^2

By force equilibrium and geometry we have the following.

$$N_{al} = N_s \quad \delta_{al} + \delta_s = np = 0.75(10^{-3})m \tag{1}$$

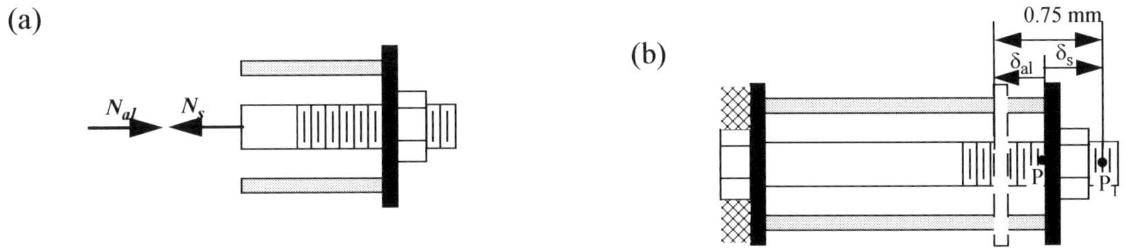

Solution proceeds as follows.

$$\delta_{al} = \frac{N_{al}L_{al}}{E_{al}A_{al}} = \frac{N_{al}(0.3)}{(70)(10^9)(1100)(10^{-6})} = 3.896N_{al}(10^{-9}) \quad \delta_s = \frac{N_s L_s}{E_s A_s} = \frac{N_s(0.3 + 0.025 + 0.025)}{(200)(10^9)(500)(10^{-6})} = 3.5N_s(10^{-9}) \tag{2}$$

$$\delta_{al} + \delta_s = (3.896 + 3.5)N_{al}(10^{-9}) = 0.75(10^{-3}) \qquad or \qquad N_{al} = 101.4(10^3)N \qquad N_s = 101.4(10^3)N \tag{3}$$

$$\sigma_{al} = N_{al}/A_{al} = 101.4(10^3)/[1100(10^{-6})] = 92.2(10^6) \qquad \sigma_s = N_s/A_s = 101.4(10^3)/[500(10^{-6})] = 202.8(10^6) \tag{4}$$

ANS $\sigma_{al} = 92.2 \text{ MPa(C)}; \sigma_s = 202.8 \text{ MPa}(T)$

---

4. 93

Solution p = 0.125 inch n = number of quarter turns = ? σ_A = ? σ_B = ?

By moment equilibrium about point C and geometry

$$N_A(5) = 15N_B \qquad or \qquad N_A - 3N_B = 0 \tag{1}$$

$$\frac{\delta_A}{5} = \frac{\delta_D}{15} \qquad or \qquad \delta_A = \frac{\delta_D}{3} \qquad \delta_B + \delta_D = \frac{nP}{4} \qquad or \qquad \delta_B = 0.03125n - \delta_D \tag{1}$$

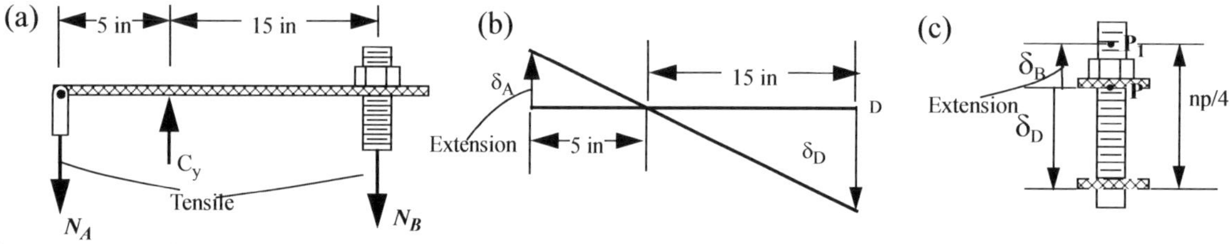

Solution proceed as follows.

$$\delta_A = \frac{N_A L_A}{E_A A_A} = \frac{N_A(50)}{(10)(10^3)(0.5)} = 0.01N_A = \frac{\delta_D}{3} \qquad or \qquad N_A = 33.33\delta_D \tag{2}$$

$$\delta_B = \frac{N_B L_B}{E_B A_B} = \frac{N_B(50)}{(30)(10^3)(0.75)} = 2.222(10^{-3})N_B = 0.03125n - \delta_D \qquad or \qquad N_B = 14.0625n - 450\delta_D \tag{3}$$

$$N_A - 3N_B = 33.33\delta_D - 3(14.0625n - 450\delta_D) = 0 \qquad or \qquad \delta_D = 42.1875n/1383.33 = 0.0305n \tag{4}$$

$$N_A = 33.33\delta_D = 1.0166n \ kips \qquad N_B = 14.0625n - 450\delta_D = 0.3388n \tag{5}$$

$$\sigma_A = N_A/A_A = 1.0166n/0.5 = 2.033n \qquad \sigma_B = N_B/A_B = 0.3388n/0.75 = 0.4518n \tag{6}$$

The stress in A exceeds material yield stress when n = 12. The initial stress for 11 quarter turns are given in the table below.

n	σ_A	σ_B
1	2.03	0.45
2	4.07	0.90
3	6.10	1.36
4	8.13	1.81
5	10.17	2.26
6	12.20	2.71
7	14.23	3.16
8	16.26	3.61
9	18.30	4.07
10	20.33	4.52
11	22.36	4.97

---

4. 94

Solution $\Delta T = T_L(x^2/L^2)$ σ_{xx}=? $u(L/2)$=?

As the right side is free to expand and there are no external forces. $\sigma_{xx} = 0$

The axial strain can be integrated to give displacement as shown below.

$$\varepsilon_{xx} = \frac{du}{dx} = \alpha\Delta T = \alpha T_L \frac{x^2}{L^2} \qquad or \qquad \int_0^{u(L/2)} du = \int_0^{L/2} \alpha T_L \frac{x^2}{L^2} dx \qquad or \qquad u(L/2) = \frac{\alpha T_L}{3L^2} x^3 \Big|_0^{L/2} = \frac{\alpha T_L L^3}{24L^2} \tag{1}$$

ANS $\sigma_{xx} = 0; u(L/2) = (\alpha T_L L)/24$

---

4. 95

Solution $\Delta T = T_L(x^2/L^2)$ σ_{xx}=? $u(L/2)$=?

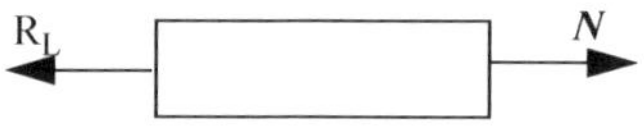

We can write the following

$$N = R_L \qquad \sigma_{xx} = N/A = R_L/A \tag{1}$$

The axial strain can be integrated as shown below.

$$\varepsilon_{xx} = \frac{du}{dx} = \frac{\sigma_{xx}}{E} + \alpha\Delta T \qquad or \qquad \frac{du}{dx} = \frac{R_L}{EA} + \alpha T_L(x^2/L^2) \qquad u(x) = \frac{R_L}{EA}x + \frac{\alpha T_L x^3}{3L^2} + C_1 \tag{2}$$

Noting u(0) = 0 and u(L) = 0, we obtain the following.

$$u(0) = C_1 = 0 \qquad u(L) = \frac{R_L L}{EA} + \frac{\alpha T_L L^3}{3L^2} = 0 \qquad or \qquad R_L = -\left(\frac{\alpha T_L EA}{3}\right) \tag{3}$$

$$\sigma_{xx} = -\left(\frac{\alpha T_L EA}{3}\right)\frac{1}{A} = -\left(\frac{E\alpha T_L}{3}\right) \qquad u(L/2) = -\frac{\alpha T_L L}{6} + \frac{\alpha T_L L^3}{24L^2} = -\left(\frac{\alpha T_L L}{8}\right) \tag{4}$$

ANS $\sigma_{xx} = \frac{E\alpha T_L}{3}(C)$; $u(L/2) = -\frac{\alpha T_L L}{8}$

---

4. 96

Solution $A = K(L - 0.5\ x)^2$ $\qquad \Delta T = T_L(x^2/L^2)$ $\qquad \sigma_{xx}(L/2) = ?$

We can write the following.

$$N = R \qquad \sigma_{xx} = \frac{N}{A} = \frac{R}{K(L-0.5x)^2} \tag{1}$$

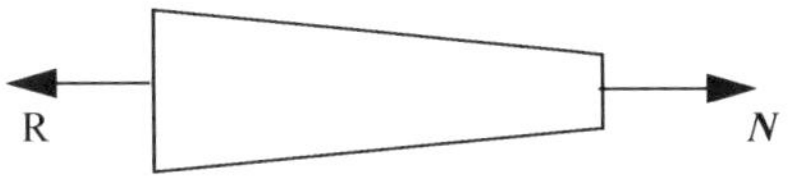

The strain can be integrated as shown below.

$$\varepsilon_{xx} = \frac{du}{dx} = \frac{\sigma_{xx}}{E} + \alpha\Delta T = \frac{R}{EK(L-0.5x)^2} + \alpha T_L \frac{x^2}{L^2} \tag{2}$$

$$\int_{u_0=0}^{u_L=0} du = \int_0^L \left(\frac{R}{EK(L-0.5x)^2} + \frac{\alpha T_L x^2}{L^2}\right)dx = \left(\frac{R}{(0.5EK)(L-0.5x)} + \frac{\alpha T_L x^3}{3L^2}\right)\Bigg|_0^L = 0 \text{ or}$$

$$\frac{R}{(0.5EK)}\left[\frac{1}{0.5L} - \frac{1}{L}\right] + \frac{\alpha T_L L}{3} = 0 \qquad \frac{2R}{EKL} = \frac{\alpha T_L L}{3} \qquad or \qquad R = -\frac{EK\alpha T_L L^2}{6} \tag{3}$$

$$\sigma_{xx} = \frac{R}{K[L-0.5(L/2)]^2} = \frac{R}{K(0.75L)^2} = -\frac{E\alpha T_L}{6(0.75)^2} \tag{4}$$

ANS $\sigma_{xx} = \frac{8}{27}E\alpha T_L(C)$

---

4. 97

Solution $\Delta T = -100°F$ $\qquad \delta_P =$

By force equilibrium and geometry we have the following

$$N_{a1} + N_{s1} + N_{s2} = 13000 lbs = 13 kips \qquad \delta_{a1} = \delta_P \qquad \delta_{s1} = \delta_P \qquad \delta_{s2} = \delta_P \tag{1}$$

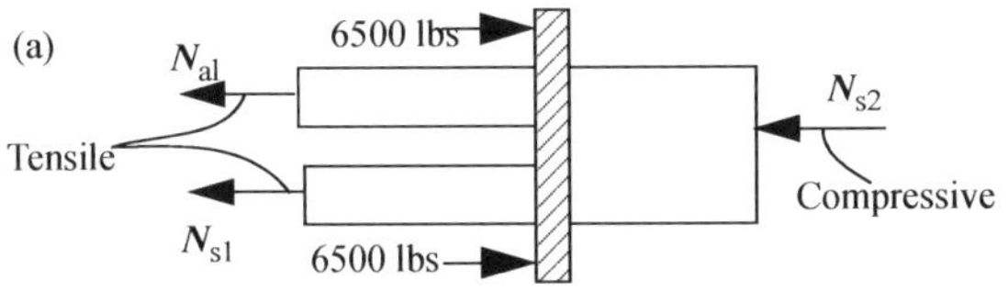

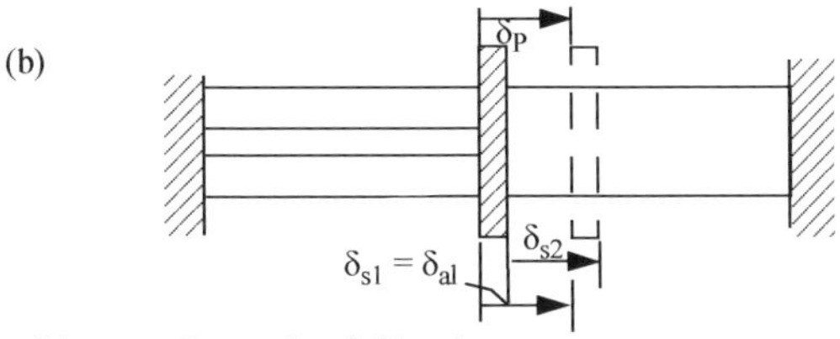

Noting that the deformation shown in Fig (b) are considered positive we have the following

$$\delta_{a1} = \frac{N_{a1}(100)}{10(10^3)(4)} - (100)(12.5)(10^{-6})(100) = \delta_P \qquad or \qquad N_{a1} = 400\delta_P + 50 \tag{2}$$

$$\delta_{s1} = \frac{N_{s1}(100)}{30(10^3)(4)} - (100)(6.6)(10^{-6})(100) = \delta_P \qquad or \qquad N_{s1} = 1200\delta_P + 79.2 \tag{3}$$

$$\delta_{s2} = \frac{N_{s2}(100)}{30(10^3)(12)} + (100)(6.6)(10^{-6})(100) = \delta_P \qquad or \qquad N_{s2} = 3600\delta_P - 237.6 \tag{4}$$

$$N_{a1} + N_{s1} + N_{s2} = (400 + 1200 + 3600)\delta_P + 50 + 79.2 - 237.6 = 13 \quad or \quad \delta_P = 121.4/5200 = 23.35(10^{-3}) \quad (5)$$

ANS $\delta_P = 0.0233$ in. to the right

---

4. 98

Solution $\Delta T = 40°C$ $\alpha_{st} = 12\mu/°C$ $\alpha_{al} = 22.5\mu/°C$

From earlier problem we have

$$N_{al} = N_s \qquad \delta_{al} + \delta_s = 0.75(10^{-3})m \quad (1)$$

$$\delta_{al} = 3.896N_{al}(10^{-9}) - (40)(22.5)(10^{-6})(0.3) \quad or \quad \delta_{al} = 3.896N_{al}(10^{-9}) - 270(10^{-6}) \quad (2)$$

$$\delta_s = 3.5N_s(10^{-9}) - (40)(12)(10^{-6})(0.3 + 0.025 + 0.025) \quad or \quad \delta_s = 3.5N_s(10^{-9}) - 168(10^{-6}) \quad (3)$$

$$\delta_{al} + \delta_s = (3.896 + 3.5)N_{al}(10^{-9}) + (-270 + 168)(10^{-6}) = 0.75(10^{-3}) \quad or \quad N_{al} = \frac{0.852(10^{-3})}{7.396(10^{-9})} = 115.2(10^3)N = N_s \quad (4)$$

$$\sigma_{al} = N_{al}/A_{al} = 115.2(10^3)/[1100(10^{-6})] = 104.7(10^6) \qquad \sigma_s = N_s/A_s = 115.2(10^3)/[500(10^{-6})] = 230.4(10^6) \quad (5)$$

ANS $\sigma_{al} = 104.7MPa(C)$; $\sigma_s = 230.4$ MPa(T)

---

4. 99

Solution p = 0.125 inch n = 4 quarter turns $\Delta T_A = -80°F$ $\alpha_{st} = 22.5\ \mu/°F$ $\sigma_A = ?$ $\sigma_B = ?$

From earlier problem we have the following.

$$N_A(5) = 15N_B \quad or \quad N_A - 3N_B = 0 \quad (1)$$

$$\delta_A/5 = \delta_D/15 \quad or \quad \delta_A = \delta_D/3 \qquad \delta_B + \delta_D = nP/4 \quad or \quad \delta_B = 0.03125n - \delta_D \quad (2)$$

$$\delta_A = \frac{N_A(50)}{(10)(10^3)(0.5)} - (80)(22.5)(10^{-6})(50) = 0.01N_A - 90(10^{-3}) = \frac{\delta_D}{3} \quad or \quad N_A = 33.33\delta_D + 9 \quad (3)$$

$$\delta_B = \frac{N_B L_B}{E_B A_B} = \frac{N_B(50)}{(30)(10^3)(0.75)} = 2.222(10^{-3})N_B = 0.03125n - \delta_D \quad or \quad N_B = 14.0625n - 450\delta_D \text{ or}$$

$$N_B = 14.0625(4) - 450\delta_D = 56.25 - 450\delta_D \quad (4)$$

$$N_A - 3N_B = 33.33\delta_D + 9 - 3(56.25 - 450\delta_D) = 0 \quad or \quad \delta_D = 159.75/1383.33 = 0.1155 \quad (5)$$

$$N_A = 33.33\delta_D + 9 = 12.849 \ kips \qquad \sigma_A = N_A/A_A = 12.849/0.5 = 25.70ksi \quad (6)$$

ANS $\sigma_A = 25.70ksi(T)$

Section 4.6-4.7

4. 100

Solution n = 50 $d_r = 10$ mm d = 1000 mm t = 10 mm p = 200kPa $\sigma_{\theta\theta} = ?$ $\tau_r = ?$

The hoop stress can be found as shown below:

$$\sigma_{\theta\theta} = \frac{pr}{t} = \frac{(200)(10^3)(0.500)}{(0.01)} = 10(10^6)\ N/m^2 \quad (1)$$

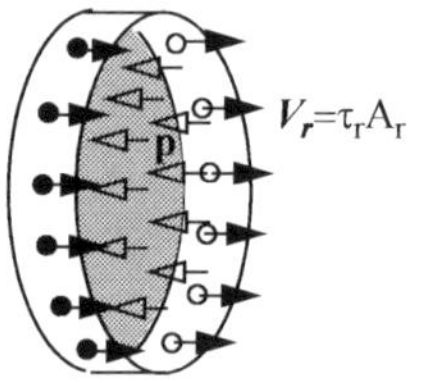

The shear force on each rivet is $V_r = \tau_r A_r$, where Ar is the area of cross-section of each rivet. By force equilibrium of the free body diagram of the cap we obtain

$$p(\pi r^2) = (50)(\tau_r A_r) \quad or \quad \tau_r = \frac{(200)(10^3)\pi(0.5)^2}{(50)\pi(0.01/2)^2} = 40(10^6)\ N/m^2 \quad (2)$$

ANS $\sigma_{\theta\theta} = 10MPa(T)$; $\tau_r = 40MPa$

---

4. 101

Solution L = 15 ft. d = 40 inch t = 0.5 inch W = 8 inch p = 500 psi $\tau_a = ?$

The hoop stress can be found as

$$\sigma_{\theta\theta} = pr/t = (75)(20)/0.5 = 3000psi \tag{1}$$

.By force equilibrium we obtain the following

$$\tau_a(W/2)(L) + \sigma_{\theta\theta}(L)(t) - (p)(L)(d) = 0 \quad or \quad \tau_a = \frac{(p)(d) - \sigma_{\theta\theta}(t)}{(W/2)} = \frac{75(40) - (3000)(0.5)}{(8/2)} = 375psi \tag{2}$$

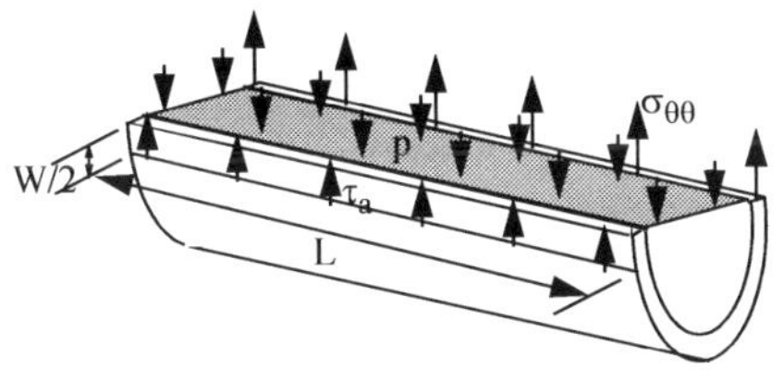

ANS $\tau_a = 375$ psi

---

4. 102

Solution d = 5 ft. t = 0.75 inch $\sigma_{max} \le 10\ ksi$ $p_{max} = ?$

The maximum pressure can be found as shown below.

$$\sigma = pr/t = p(5)(12)/0.75 \le 10 \quad or \quad p \le 0.125ksi \tag{1}$$

ANS $p_{max} = 125psi$

---

4. 103

Solution r = 500 mm t = 40 mm $d_{plug} = 50$ mm $\tau_a \le 1.2MPa$ $p_{max} = ?$ $\sigma = ?$

By force equilibrium of the free body diagram of the plug we obtain the following

$$p(\pi d_{plug}^2/4) = (\tau_a)(\pi d_{plug})(t) \quad or \quad (\tau_a) = \frac{pd_{plug}}{4t} = \frac{p(0.05)}{(4)(0.04)} \le 1.2(10^6) \quad or \quad p \le 3.84(10^6) \tag{1}$$

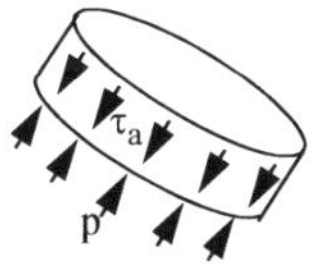

The hoop stress can be found as shown below.

$$\sigma = pr/t = 3.84(10^6)(0.5)/0.04 = 48(10^6)\ N/m^2 \tag{2}$$

ANS $p_{max} = 3.8$ MPa ; $\sigma = 48$ MPa(T)

---

4. 104

Solution d = 20 inch p = 15 psi $\sigma_{max} \le 3\ ksi$ $t_{min} = ?$ W = 0.5 lb $d_{noz} = ?$

The hoop stress in the cylinder is greater than the axial stress and is used to determine the minimum thickness as shown below.

$$\sigma_{\theta\theta} = pr/t = \frac{(15)(10)}{t} \le 3(10^3) \quad or \quad t \ge 0.05\ in \tag{1}$$

By force equilibrium of the weight on the nozzle we obtain:

$$p\pi(d_{noz}^2/4) = 0.5 \quad or \quad d_{noz} = \sqrt{\frac{2}{\pi p}} = \sqrt{\frac{2}{\pi(15)}} = d_{noz} = 0.206\ in \tag{2}$$

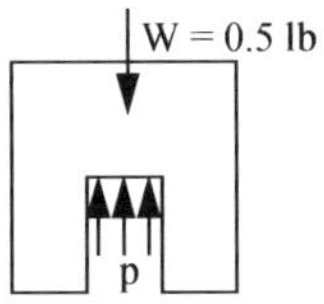

ANS $t_{min} = 0.05in$; $d_{noz} = 0.206\ in$

---

4. 105

Solution t = 8 mm $\sigma_{max} \le 100MPa$ p vs. d

Assuming d is measured in millimeters and p in MPa, we can write the hoop stress expression as shown below:

$$\sigma_{\theta\theta} = \frac{pr}{t} = \frac{(p)(10^6)(d/2)(10^{-3})}{0.008} \le 100(10^6) \qquad or \qquad pd \le 1600 \qquad (1)$$

The above equation is used to determine the values of p and d in the table below. All pressure values are rounded downwards,

d (mm	p (MPa)
400	4.00
500	3.20
600	2.66
700	2.28
800	2.00
900	1.77

---

4. 106

Solution $L = 15$ ft. $d = 40$ inch $t = 0.5$ inch $W_{plate} = 8$ inch $t_{plate} = 0.5$ inch

$n_{rivets}=90$ $\sigma_{max} \le 20\ ksi$ $\tau_{riv} \le 36\ ksi$ $p_{max} = ?$ $(d_{riv})_{min} = ?$

The hoop stress in the cylinder is greater than the axial stress and is used to determine the maximum pressure as shown below.

$$\sigma_{\theta\theta} = \frac{pr}{t} = \frac{(p)(20)}{0.5} \le 20(10^3) \qquad or \qquad p \le 500psi \qquad (1)$$

By force equilibrium we obtain:

$$n_{riv}\tau_r(\pi d_{riv}^2)/4 + \sigma_{\theta\theta}(L)(t) - (p)(L)(d) = 0 \qquad or \qquad \tau_{riv} = \frac{(p)(L)(d) - (\sigma_{\theta\theta})(L)(t)}{n_{riv}\pi(d_{riv}^2/4)} \qquad or$$

$$\tau_{riv} = \frac{(500)(15)(12)(40) - (20000)(15)(12)(0.5)}{(90)\pi(d_{riv}^2/4)} \le (36)(10^3) \qquad or \qquad d_{riv} \ge \sqrt{\frac{1800(10^3)}{810\pi(10^3)}} \qquad or \qquad d_{riv} \ge 0.841 inch \qquad (2)$$

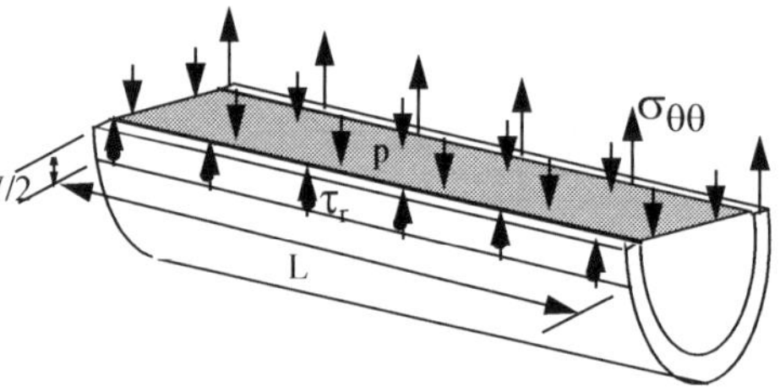

ANS $p_{max} = 500psi$; $(d_{riv})_{min} = 0.85in$

---

4. 107

Solution $L = 5$ m $d = 1$ m $t = 0.01$m $W = 0.2$ m $n_{riv}=100$

$k_{safety}= 2$ $\sigma_{yield} = 200MPa$ $\tau_{riv} = 300MPa$ $p = ?$ $d_r = ?$

The hoop stress can be found as

$$\sigma_{\theta\theta} = \frac{pr}{t} = \frac{p(0.5)}{(0.01)} = 50p \le \frac{200}{2} \qquad or \qquad p \le 2MPa \qquad (1)$$

By force equilibrium we obtain the following.

$$n_{riv}\tau_r(\pi d_{riv}^2)/4 + \sigma_{\theta\theta}(L)(t) - (p)(L)(d) = 0 \qquad \tau_{riv} = \frac{(p)(L)(d) - (\sigma_{\theta\theta})(L)(t)}{n_{riv}\pi(d_{riv}^2/4)} \qquad or$$

$$\tau_{riv} = \frac{(4)(10^6)(5)(1) - (200)(10^6)(5)(0.01)}{(100)\pi(d_{riv}^2/4)} \le \frac{300}{2}(10^6) \qquad or \qquad d_{riv} \ge \sqrt{\frac{10(10^6)}{3750\pi(10^6)}} \qquad or \qquad d_{riv} \ge 0.02913m \qquad (2)$$

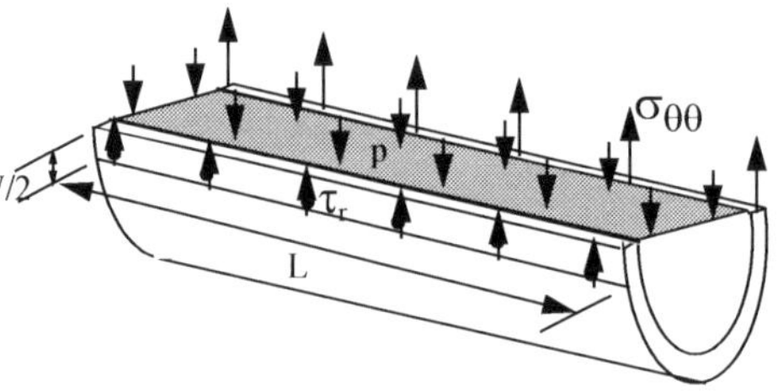

ANS $p_{max} = 2$ MPa ; $(d_{riv})_{min} = 30$ mm

CHAPTER 5

Section 5.1

5. 1

Solution $\gamma_A = 3000\ \mu$ r= 2 inch. $\gamma_D = ?$

From triangles ABB_1 and DCC_1 we have the following:

$$tan\gamma_A \approx \gamma_A = BB_1/AB = BB_1/48 = 3000(10^{-6}) \quad or \quad BB_1 = 0.144\ in = CC_1 \qquad (1)$$

$$tan\gamma_D \approx \gamma_D = CC_1/CD = BB_1/CD \quad or \quad \gamma_D = 0.144/60 = 2.400(10^{-3}) \qquad (2)$$

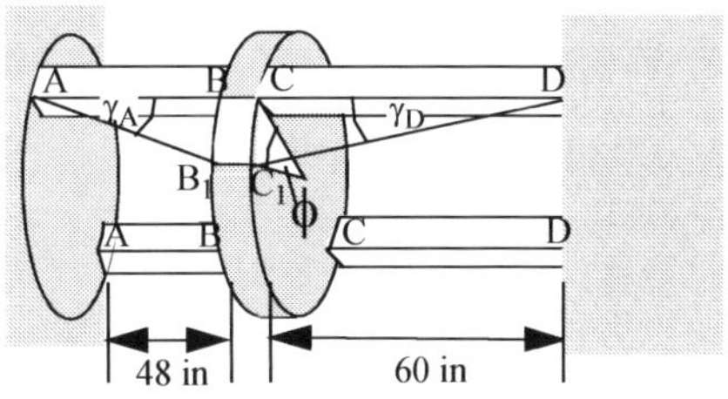

ANS $\gamma_D = 2400\ \mu rad$

5. 2

Solution G = 12,000 ksi A = 0.25in^2 T = ?

From previous problem we have

$$\gamma_A = 3.000(10^{-3}) \qquad \gamma_D = 2.400(10^{-3}) \qquad (1)$$

The solution proceeds as follows.

$$\tau_{AB} = G\gamma_A = (12,000)(3.000)(10^{-3}) = 36\ ksi \qquad \tau_{CD} = G\gamma_D = (12,000)(2.400)(10^{-3}) = 28.8\ ksi \qquad (2)$$

$$V_{AB} = \tau_{AB}A = (36)(0.25) = 9.0\ kips \qquad V_{CD} = \tau_{CD}A = (28.8)(0.25) = 7.2\ kips \qquad (3)$$

By equilibrium of moment about the axis of the rigid disc, we obtain:

$$T = (2)(V_{AB})(2) + (2)(V_{CD})(2) = (4)(V_{AB}) + (4)(V_{CD}) \quad or \quad T = (4)(9) + (4)(7.2) = 64.8\ in-kips \qquad (4)$$

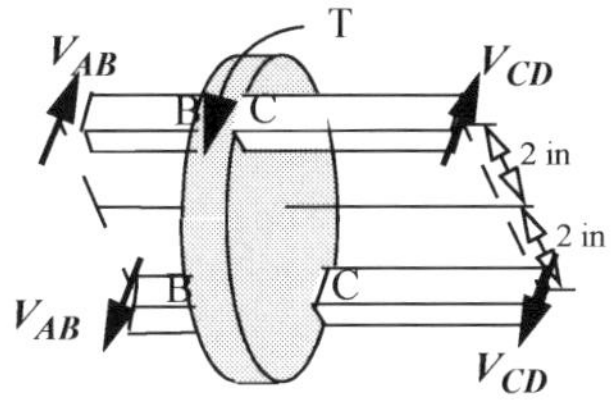

ANS $T = 64.8\ in-kips$

5. 3

Solution G_{AB}= 4,000 ksi G_{CD}= 6,500 ksi A = 0.25in^2 T = ?

From previous problem we have

$$\gamma_A = 3.000(10^{-3}) \qquad \gamma_D = 2.400(10^{-3}) \qquad (1)$$

The solution proceeds as follows.

$$\tau_{AB} = G_{AB}\gamma_A = (4,000)(3.000)(10^{-3}) = 12.0\ ksi \qquad \tau_{CD} = G_{CD}\gamma_D = (6,500)(2.400)(10^{-3}) = 15.6\ ksi \qquad (2)$$

$$V_{AB} = \tau_{AB}A = (12.0)(0.25) = 3.0\ kips \qquad V_{CD} = \tau_{CD}A = (15.6)(0.25) = 3.9\ kips \qquad (3)$$

From previous problem, by equilibrium we have:

$$T = (4)(V_{AB}) + (4)(V_{CD}) = (4)(3) + (4)(3.9) = 27.6 \text{ in.-kips} \qquad (4)$$

ANS $T = 27.6$ in.-kips

5. 4

Solution $\phi_1 = 1.5°$ $\phi_2 = 3.0°$ $\phi_3 = 2.5°$ G = 40 ksi A = 0.04 in^2 $T_1 = ?$ $T_2 = ?$ $T_3 = ?$

Converting the given angles from degrees to radians and using geometry we obtain the following.

$$\phi_1 = \left(\frac{1.5°}{180}\right)\pi = 0.0262rads \qquad \phi_2 = \left(\frac{3.0°}{180}\right)\pi = 0.0524rads \qquad \phi_2 = \left(\frac{2.5°}{180}\right)\pi = 0.0436rads \tag{1}$$

$$tan|\gamma_{AB}| = \frac{BB_1}{AB} \quad or \quad |\gamma_{AB}| = \frac{1.25\phi_1}{25} = \frac{1.25(0.0262)}{25} = 1.31(10^{-3}) \tag{2}$$

$$tan|\gamma_{CD}| = \frac{D_2D_1}{C_1D} \quad or \quad |\gamma_{CD}| = \frac{D_2D + DD_1}{CD} = \frac{2\phi_1 + 2\phi_2}{CD} = \frac{2(0.0262 + 0.0524)}{40} = 3.93(10^{-3}) \tag{3}$$

$$tan|\gamma_{EF}| = \frac{F_2F_1}{E_1F_2} \quad or \quad |\gamma_{EF}| = \frac{F_2F + FF_1}{EF} = \frac{1.5\phi_2 + 1.5\phi_3}{EF} = \frac{2(0.0524 + 0.0436)}{30} = 4.8(10^{-3}) \tag{4}$$

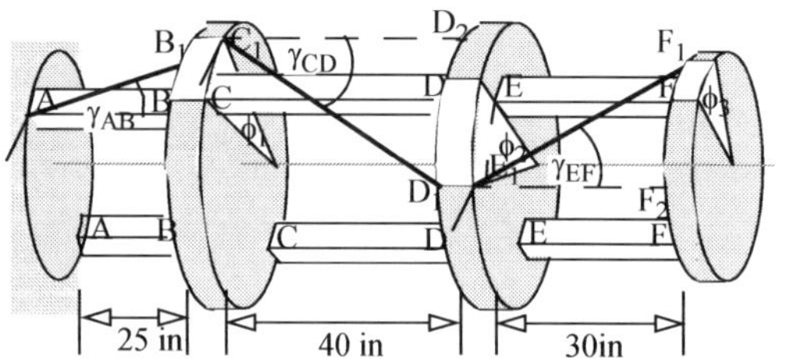

Using the sign convention that positive shear strain result in decrease in angle from right angle, we write the shear strains as:

$$\gamma_{AB} = -1.31(10^{-3}) \qquad \gamma_{CD} = 3.93(10^{-3}) \qquad \gamma_{EF} = -4.8(10^{-3}) \tag{5}$$

The shear stress and internal force in each bar can be found as shown below.

$$\tau_{AB} = G\gamma_{AB} = (40)(-1.31)(10^{-3}) = -52.4(10^{-3})ksi = -52.4\ psi \qquad V_{AB} = \tau_{AB}A = (-52.4)(0.04) = -2.096\ lbs \tag{6}$$

$$\tau_{CD} = G\gamma_{CD} = (40)(3.93)(10^{-3}) = 157.2(10^{-3})ksi = 157.2\ psi \qquad V_{CD} = \tau_{CD}A = (157.2)(0.04) = 6.288\ lbs \tag{7}$$

$$\tau_{EF} = G\gamma_{EF} = (40)(-4.80)(10^{-3}) = -192.0(10^{-3})ksi = -192.0\ psi \qquad V_{EF} = \tau_{EF}A = (-192.0)(0.04) = -7.680\ lbs \tag{8}$$

From moment equilibrium we obtain the following.

$$T_3 = (2)(V_{AB})(1.5) = (3)(7.628) = 23.04 \tag{9}$$

$$T_2 - T_3 - (2)(V_{CD})(2.0) = 0 \quad or \quad T_2 = 23.04 + (4)(6.288) = 48.192 \tag{10}$$

$$T_1 - T_2 + T_3 + (2)(V_{AB})(1.25) = 0 \quad or \quad T_1 = 48.2 - 23.04 + (2.5)(2.096) = 30.4 \tag{11}$$

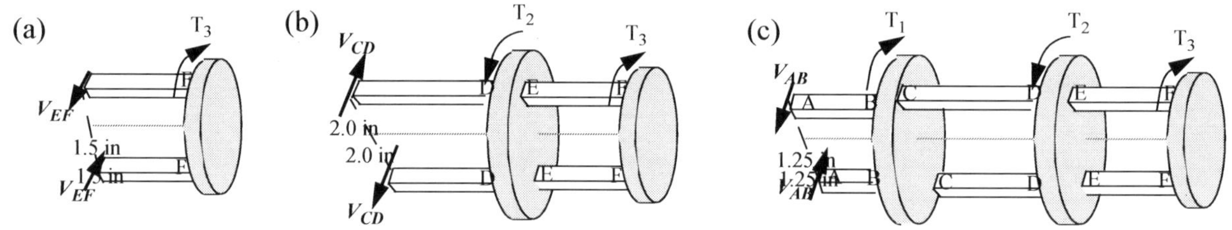

ANS $T_1 = 30.4$ in.-lbs ; $T_2 = 48.2$ in.-lbs ; $T_3 = 23.0$ in.-lbs

---

5. 5

Solution $\gamma = f(r, \Delta x, \Delta\phi) = ?$

From triangle AB_2B_1 and small strain approximation, we have the following:

$$tan\gamma \approx \gamma = \frac{B_2B_1}{A_1B_2} = \frac{BB_1 - BB_2}{AB} = \frac{BB_1 - AA_1}{AB} = \frac{r\phi_2 - r\phi_1}{\Delta x} = r\frac{\Delta\phi}{\Delta x} \tag{1}$$

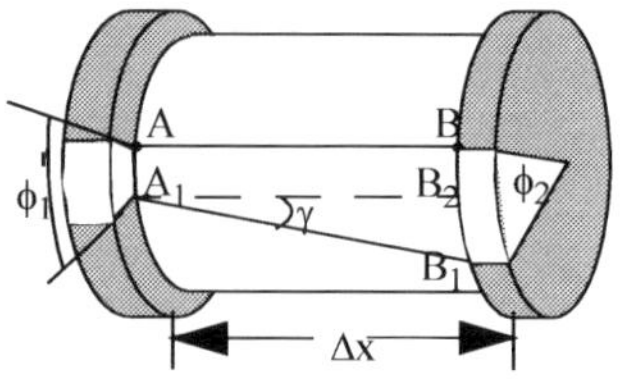

ANS $\gamma = r\Delta\phi/\Delta x$

---

5. 6

Solution $d_o = 4$ inch $\quad d_i = 1.5$ inch $\quad \gamma_A = 4000\ \mu$ rads $\gamma_C = ?$

From triangles ABB_1 and CDD_1 we obtain the following:

$$\tan\gamma_A \approx \gamma_A = \frac{BB_1}{AB} = \frac{(d_o/2)\phi}{AB} \quad or \quad 4000(10^{-6}) = \frac{(4/2)\phi}{36} \quad or \quad \phi = 0.072\ rads \tag{1}$$

$$\tan\gamma_C \approx \gamma_C = \frac{DD_1}{CD} = \frac{(d_i/2)\phi}{AB} = \frac{(1.5/2)(0.072)}{36} = 1.500(10^{-3}) \tag{2}$$

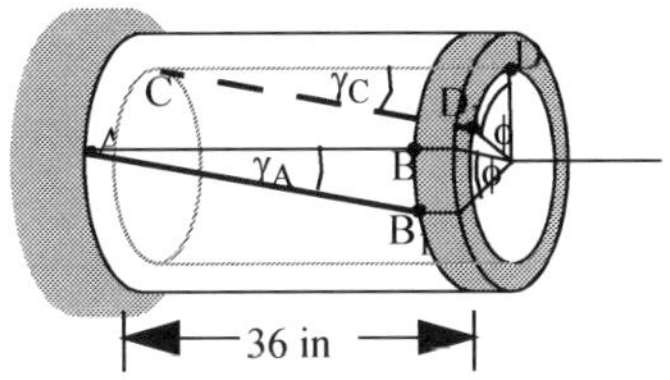

ANS $\gamma_C = 1500\ \mu$ rads

---

5. 7

Solution $\gamma_{AB} = 3000\ \mu rads$ $\gamma_{CD} = 2500\ \mu rads$ $\gamma_{EF} = 6000\ \mu rads$ $r_{AB} = 150$ mm

$r_{CD} = 70$ mm $r_{EF} = 60$ mm $\phi_1 = ?$ $\phi_2 = ?$ $\phi_3 = ?$

The shear strains can be found using the geometry and small strain approximation as shown below.

$$\tan\gamma_{AB} \approx \gamma_{AB} = \frac{BB_1}{AB} = \frac{r_{AB}\phi_1}{AB} = \frac{(0.150)\phi_1}{2} = 3000(10^{-6}) \tag{1}$$

$$\tan\gamma_{CD} \approx \gamma_{CD} = \frac{D_2D_1}{C_1D_2} = \frac{D_2D + DD_1}{CD} = \frac{r_{CD}\phi_1 + r_{CD}\phi_2}{CD} = \frac{0.07(0.04 + \phi_2)}{1.8} = 2500(10^{-6}) \tag{2}$$

$$\tan\gamma_{CD} \approx \gamma_{CD} = \frac{F_2F_1}{E_1F_2} = \frac{F_2F + FF_1}{EF} = \frac{r_{EF}\phi_2 + r_{EF}\phi_3}{EF} = \frac{0.06(0.0243 + \phi_3)}{1.2} = 6000(10^{-6}) \tag{3}$$

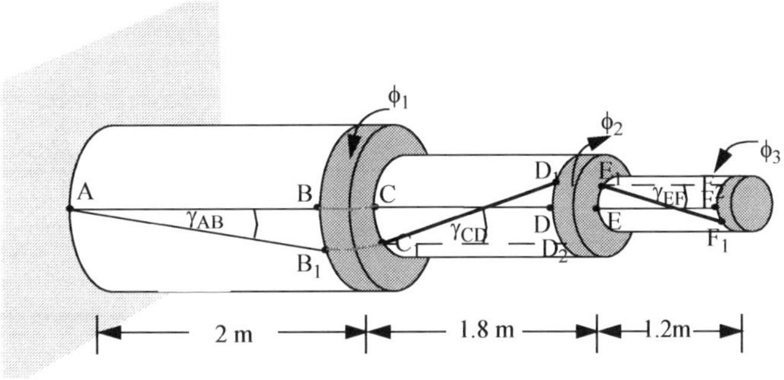

ANS $\phi_1 = 0.0400$ rad ; $\phi_2 = 0.0243$ rad; ; $\phi_3 = 0.0957$ rad

---

5. 8

Solution G = 26 GPa $\gamma_{x\theta} = -0.06\rho$ $d_i = 30$ mm $d_o = 50$ mm $T = ?$

We can write the stress and calculate the internal torque as shown below.

$$\tau_{x\theta} = G\gamma_{x\theta} = (26)(10^9)(-0.06\rho) = -1.56\rho(10^9)\ N/m^2 \tag{1}$$

$$T = \int_{0.015}^{0.025} \rho\tau_{x\theta}(2\pi\rho d\rho) = \int_{0.015}^{0.025} \rho(-1.56\rho)(10^9)(2\pi\rho)d\rho = \left[(-3.12\pi)\frac{\rho^4}{4}\Big|_{0.015}^{0.025}\right](10^9) = \left[\left(\frac{-3.12\pi}{4}\right)(0.025^4 - 0.015^4)\right](10^9) \tag{2}$$

ANS $T = -833.1$ N-m

---

5. 9

Solution G = 26 GPa $\gamma_{x\theta} = 0.05\rho$ $d_i = 40$ mm $d_o = 120$ mm $T = ?$

We can write the stress and calculate the internal torque as shown below.

$$\tau_{x\theta} = G\gamma_{x\theta} = (26)(10^9)(0.05\rho) = 1.3\rho(10^9)\ N/m^2 \tag{1}$$

$$T = \int_{0.02}^{0.06} \rho\tau_{x\theta}(2\pi\rho)d\rho = \int_{0.02}^{0.06} \rho(1.3\rho)(10^9)(2\pi\rho)d\rho = \left[(2.6\pi)\frac{\rho^4}{4}\Big|_{0.02}^{0.06}\right](10^9) = \left[\left(\frac{2.6\pi}{4}\right)(0.06^4 - 0.02^4)\right](10^9) = 26.1(10^3)\text{ N-m} \tag{2}$$

ANS $T = 26.1$ kN-m

---

5. 10

Solution $G_B = 6500$ ksi $G_S = 13{,}000$ ksi) $\gamma_{x\theta} = 0.001\rho$ $d_B = 4$ in $d_S = 2$ in. $T = ?$

We can write the stress and calculate the internal torque as shown below.

$$(\tau_{x\theta})_{steel} = G_S\gamma_{x\theta} = (13,000)(0.001\rho) = 13\rho\ ksi \qquad (\tau_{x\theta})_{brass} = G_B\gamma_{x\theta} = (6,500)(0.001\rho) = 6.5\rho\ ksi \quad \textbf{(1)}$$

$$\tau_{x\theta} = \begin{cases} 13\rho\ ksi & 0.0 \le \rho < 1.0 \\ 6.5\rho\ ksi & 1.0 < \rho \le 2.0 \end{cases} \quad \textbf{(2)}$$

$$T = \int_0^2 \rho\tau_{x\theta}(2\pi\rho)d\rho = \int_0^1 \rho(13\rho)(2\pi\rho)d\rho + \int_1^2 \rho(6.5\rho)(2\pi\rho)d\rho = (26\pi)\frac{\rho^4}{4}\bigg|_0^1 + (13\pi)\frac{\rho^4}{4}\bigg|_1^2 = \left(\frac{26\pi}{4}\right) + \left(\frac{13\pi}{4}\right)(2^4 - 1^4) = 173.57 \quad \textbf{(3)}$$

ANS $T = 173.6$ in.-kips

---

5. 11

Solution $G_B = 6500$ ksi $G_S = 13{,}000$ ksi) $\gamma_{x\theta} = -0.0005\rho$ $d_B = 6$ in $d_S = 4$ in.

We can write the stress and calculate the internal torque as shown below.

$$(\tau_{x\theta})_{brass} = G_B\gamma_{x\theta} = (6,500)(-0.0005\rho) = -3.25\rho\ ksi \qquad (\tau_{x\theta})_{steel} = G_S\gamma_{x\theta} = (13,000)(-0.0005\rho) = -6.50\rho\ ksi \quad \textbf{(1)}$$

$$\tau_{x\theta} = \begin{cases} -3.25\rho\ ksi & 0.0 \le \rho < 2.0 \\ -6.50\rho\ ksi & 2.0 < \rho \le 3.0 \end{cases} \quad \textbf{(2)}$$

$$T = \int_0^3 \rho\tau_{x\theta}(2\pi\rho)d\rho = \int_0^2 \rho(-6.50\rho)(2\pi\rho)d\rho + \int_2^3 \rho(-3.25\rho)(2\pi\rho)d\rho = (-13\pi)\frac{\rho^4}{4}\bigg|_0^2 + (-6.5\pi)\frac{\rho^4}{4}\bigg|_2^3 \text{ or}$$

$$T = \left(\frac{-13\pi}{4}\right)(2^4 - 0) + \left(\frac{-6.5\pi}{4}\right)(3^4 - 2^4) = -495.19\ in-kips \quad \textbf{(3)}$$

ANS $T = -495.2$ in.-kips

---

5. 12

Solution $G_B = 6500$ ksi $G_S = 13{,}000$ ksi) $\gamma_{x\theta} = 0.002\rho$ $d_B = 3$ in $d_S = 1$ in. $T = ?$

We can write the stress and calculate the internal torque as shown below.

$$(\tau_{x\theta})_{steel} = G_S\gamma_{x\theta} = (13,000)(0.002\rho) = 26\rho\ ksi \qquad (\tau_{x\theta})_{brass} = G_B\gamma_{x\theta} = (6,500)(0.002\rho) = 13\rho\ ksi \quad \textbf{(1)}$$

$$\tau_{x\theta} = \begin{cases} 26\rho\ ksi & 0.0 \le \rho < 0.5 \\ 13\rho\ ksi & 0.5 < \rho \le 1.5 \end{cases} \quad \textbf{(2)}$$

$$T = \int_0^{1.5} \rho\tau_{x\theta}(2\pi\rho)d\rho = \int_0^{0.5} \rho(26\rho)(2\pi\rho)d\rho + \int_{0.5}^{1.5} \rho(13\rho)(2\pi\rho)d\rho = (52\pi)\frac{\rho^4}{4}\bigg|_0^{0.5} + (26\pi)\frac{\rho^4}{4}\bigg|_{0.5}^{1.5} \text{ or}$$

$$T = (52\pi/4)(0.5^4 - 0) + (26\pi/4)(1.5^4 - 0.5^4) = 104.65\ in-kips \quad \textbf{(3)}$$

ANS $T = 104.65$ in.-kips

---

5. 13

Solution $G_{Ti} = 36$ GPa $G_{Al} = 26$ GPa $\gamma_{x\theta} = 0.04\rho$ $d_i = 50$ mm $d_{Al} = 90$ mm $d_{Ti} = 100$ mm $T = ?$

We can write the stress and calculate the internal torque as shown below.

$$(\tau_{x\theta})_{Al} = G_{Al}\gamma_{x\theta} = (26)(10^9)(0.04\rho) = 1.04\rho(10^9)\ N/m^2 \qquad (\tau_{x\theta})_{Ti} = G_{Ti}\gamma_{x\theta} = (36)(10^9)(0.04\rho) = 1.44\rho(10^9)\ N/m^2 \quad \textbf{(1)}$$

$$\tau_{x\theta} = \begin{cases} 1.04\rho(10^9)(\ N/m^2) & 0.025 \le \rho < 0.045 \\ 1.44\rho(10^9)\ N/m^2 & 0.045 < \rho \le 0.050 \end{cases} \quad \textbf{(2)}$$

$$T = \int_{0.025}^{0.050} \rho\tau_{x\theta}(2\pi\rho)d\rho = \int_{0.025}^{0.045} \rho(1.04\rho)(10^9)(2\pi\rho)d\rho + \int_{0.045}^{0.05} \rho(1.44\rho)(10^9)(2\pi\rho)d\rho = \left[(2.08\pi)\frac{\rho^4}{4}\bigg|_{0.025}^{0.045} + (2.88\pi)\frac{\rho^4}{4}\bigg|_{0.045}^{0.050}\right](10^9) \text{ or}$$

$$T = [(2.08\pi/4)(0.045^4 - 0.025^4) + (2.88\pi/4)(0.05^4 - 0.045^4)](10^9) = 10.92(10^3)\ N-m \quad \textbf{(3)}$$

ANS $T = 10.9$ kN-m

---

5. 14

Solution $T = ?$

We can write the stress and calculate the internal torque as shown below.

$$\tau_{x\Theta} = \begin{cases} \dfrac{24}{0.3}\rho = 80\rho \ ksi & 0 \le \rho \le 0.3 \\ 24 & 0.3 \le \rho \le 0.6 \end{cases} \quad (1)$$

$$T = \int_0^{0.6} \rho\tau_{x\theta}(2\pi\rho)d\rho = \int_0^{0.3} \rho(80\rho)(2\pi\rho)d\rho + \int_{0.3}^{0.6} \rho(24)(2\pi\rho)d\rho = \left[(160\pi)\frac{\rho^4}{4}\bigg|_0^{0.3} + (48\pi)\frac{\rho^3}{3}\bigg|_{0.3}^{0.6}\right] = 1.0179 + 9.50 = 10.52 \ in-kips$$

ANS $T = 10.52 \ in-kips$

---

5. 15

Solution $\gamma_{x\theta}=0.002\rho$ $\tau_{yield}=18ksi$ $G=12{,}000$ ksi $T = ?$ Elastic perfectly plastic

The strain at yield point and stress distribution can be written and internal torque calculate as shown below.

$$\gamma_{yield} = 18/12,000 = 0.0015 = 0.002\rho_{yield} \quad or \quad \rho_{yield} = 0.75 inch \quad (1)$$

For $0 < \rho < 0.75$, $\tau_{x\theta} = G\gamma_{x\theta} = (12000)(0.002\rho) = 24\rho$. The stress distribution can be written as:

$$\tau_{x\theta} = \begin{cases} 24\rho \ \text{ksi} & 0 < \rho < 0.75in \\ 18 \ \text{ksi} & 0.75in < \rho < 1.5in \end{cases} \quad (2)$$

$$T = \int_A \rho\tau_{x\theta}dA = \int_0^{0.75} (\rho)(24\rho)(2\pi\rho)d\rho + \int_{0.75}^{1.5} (\rho)(18)(2\pi\rho)d\rho = (48\pi)\frac{\rho^4}{4}\bigg|_0^{0.75} + (36\pi)\frac{\rho^3}{3}\bigg|_{0.75}^{1.5} = 11.93 + 111.33 = 123.26 \quad (3)$$

ANS $T = 123.3$ in.-kips

---

5. 16

Solution $\gamma_{x\theta}=0.002\rho$ $\tau_{yield}=18ksi$ $G_1=12{,}000ksi$ $G_2=4{,}800ksi$ $\tau_{x\theta}= f(\rho) = ?$ $T = ?$

The strain at yield point and stress distribution can be written and internal torque calculate as shown below.

$$\gamma_{yield} = 18/12,000 = 0.0015 = 0.002\rho_{yield} \quad or \quad \rho_{yield} = 0.75 inch \quad (1)$$

$$\tau_1 = G_1\gamma_{x\theta} = 12000(0.002\rho) = 24\rho \qquad \tau_2 = \tau_{yield} + G_2(\gamma_{x\theta} - \gamma_{yield}) = 18 + 4800(2\rho(10^{-3}) - 0.0015) = 10.8 + 9.6\rho \quad (2)$$

$$\tau_{x\theta} = \begin{cases} 24\rho \ \text{ksi} & 0 < \rho < 0.75in \\ (10.8 + 9.6\rho)ksi & 0.75in < \rho < 1.5in \end{cases} \quad (3)$$

$$T = \int_A \rho\tau_{x\theta}dA = \int_0^{0.75} \rho(24\rho)(2\pi\rho)d\rho + \int_{0.75}^{1.5} \rho(10.8 + 9.6\rho)(2\pi\rho)d\rho = 48\pi\frac{\rho^4}{4}\bigg|_0^{0.75} + 2\pi\left(10.8\frac{\rho^3}{3} + 9.6\frac{\rho^4}{4}\right)\bigg|_{0.75}^{1.5} = 11.93 + 138.37 \quad (4)$$

ANS $T = 150.3$ in.-kips

---

5. 17

Solution $\gamma_{x\theta}= 0.002\rho$ $\tau = 243\gamma^{0.4}$ ksi $T = ?$

Stress distribution can be written and internal torque calculate as shown below.

$$\tau_{x\theta} = 243\gamma_{x\theta} = 243(0.002\rho)^{0.4} = 20.23\rho^{0.4}ksi \qquad 0 \le \rho \le 1.5 \quad (1)$$

$$T = \int_A \rho\tau_{x\theta}dA = \int_0^{1.5} (\rho)(20.23\rho^{0.4})(2\pi\rho)d\rho = (40.46\pi)\frac{\rho^{3.4}}{3.4}\bigg|_0^{1.5} = 148.39 \ in-kips \quad (2)$$

ANS $T = 148.4$ in.-kips

---

5. 18

Solution $\gamma_{x\theta}=0.002\rho$ $\tau = (12,000\gamma - 120,000\gamma^2) \ ksi$ $T = ?$

Stress distribution can be written and internal torque calculate as shown below.

$$\tau_{x\theta} = (12,000\gamma_{x\theta} - 120,000\gamma_{x\theta}^2) = (12,000)(0.002\rho) - 120,000(0.002\rho)^2 = (24\rho - 0.48\rho^2)ksi \qquad 0 \le \rho \le 1.5 \quad (1)$$

$$T = \int_A \rho\tau_{x\theta}dA = \int_0^{1.5}(\rho)(24\rho - 0.48\rho^2)(2\pi\rho)d\rho = (2\pi)\left(24\frac{\rho^4}{4} - 0.48\frac{\rho^5}{5}\right)\Big|_0^{1.5} = 186.27\ in-kips \quad (2)$$

ANS $T = 186.3$ in.-kips

Section 5.2

5. 19

Solution $\tau_A = 120$ MPa $\gamma_{max} = ?$

Using similar triangles on the torsional shear stress distribution across the cross-section we obtain the following.

$$.(\tau_{max}/0.1) = (\tau_A/0.06) \quad or \quad \tau_{max} = 200(0.1/0.06) = 200\ MPa \quad (1)$$

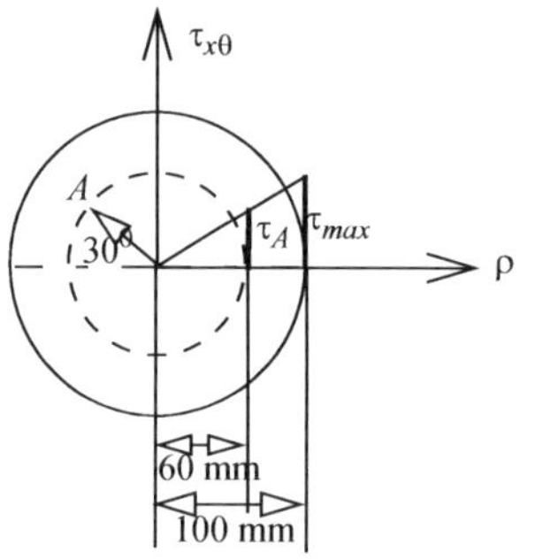

ANS $\tau_{max} = 200\ MPa$

---

5. 20

Solution $\gamma_A = 900\ \mu$ $\gamma_B = ?$

Using similar triangles on the torsional shear strain distribution we obtain the following.

$$(\gamma_B/2.5) = (\gamma_A/1.5) \quad or \quad \gamma_B = 900(2.5/1.5) = 1500\ \mu \quad (1)$$

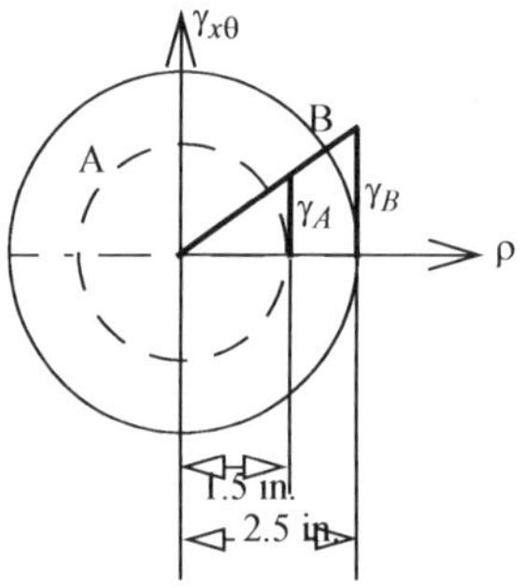

ANS $\gamma_B = 1500\ \mu$

---

5. 21

Solution $\gamma_{al} = 1500\ \mu$ $G_{al} = 28$ GPa $G_S = 82$ GPa $\gamma_s = ?$

Using similar triangles on the torsional shear strain distribution we obtain the following.

$$(\gamma_s/0.1) = (\gamma_{al}/0.06) \quad or \quad \gamma_s = 1500(0.1/0.06) = 2500\ \mu \quad (1)$$

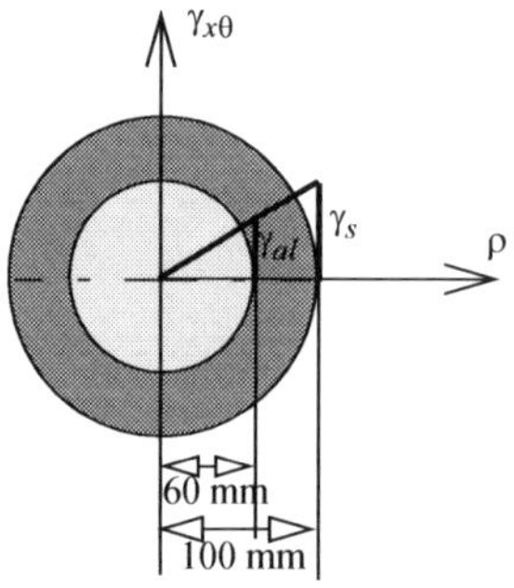

ANS $\gamma_s = 2500\ \mu$

5. 22

Solution $\tau_{al} = 1500\ MPa$ Gal= 28 GPa GS=82 GPa $\tau_s = ?$

From previous problem we have the following.

$$(\gamma_s/0.1) = (\gamma_{al}/0.06) \quad \textbf{(1)}$$

$$\gamma_{al} = \tau_{al}/G_{al} = [21(10^6)]/[28(10^9)] = 0.75(10^{-3}) \qquad \gamma_s = 0.75(10^{-3})(0.1/0.06) = 1.25(10^{-3}) \quad \textbf{(2)}$$

$$\tau_s = G_s\gamma_s = 82(10^9)(1.25)(10^{-3}) = 102.5(10^6)\ N/m^2 \quad \textbf{(3)}$$

ANS $\tau_s = 102.5\ MPa$

---

5. 23

Solution

By Inspection: The reaction torque at the wall will be counter-clockwise with respect to the x-axis. Thus, the segment of the shaft near the wall would rotate counter-clockwise, that is point A would move downwards and point B upwards. To oppose this imaginary motion, the shear stress at point A will be upward and at point B it will be downwards as shown in Figs (a) and (b). The shear stress on the rest of the surfaces can be drawn using the observation that a symmetric pair of shear stress points towards the corner or away from the corner.

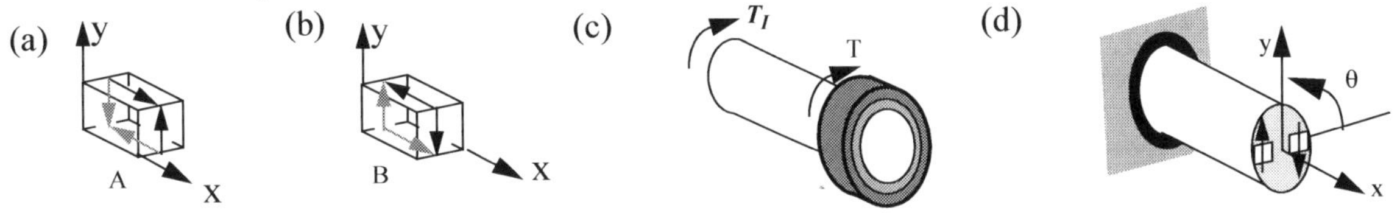

By Subscripts: We can make an imaginary cut through the cross-section containing points A and B and draw the free body diagram as shown in Fig. (c). the internal torque is drawn as per our sign convention. By moment equilibrium we obtain $T_I = -T$. Thus from the torsional stress formula we obtain $\tau_{x\theta} < 0$. Noting that the outward normal of the cross-section is in the positive x-direction, a negative $\tau_{x\theta}$ requires the stress to be in the negative θ-direction as shown in Fig. (d). We then write the stresses in the x-y coordinate system.

ANS $(\tau_{xy})_A > 0$; $(\tau_{xy})_B < 0$

---

5. 24

Solution

By Inspection: The reaction torque at the wall will be counter-clockwise with respect to the x-axis. Thus, the segment of the shaft near the wall would move counter-clockwise, that is point A would move upwards and point B would downwards. To oppose this imaginary motion, the shear stress at point A will be downwards and at point B it will be upwards as shown in Figs (a) and (b). The shear stress on the rest of the surfaces can be drawn using the observation that a symmetric pair of shear stress points towards the corner or away from the corner.

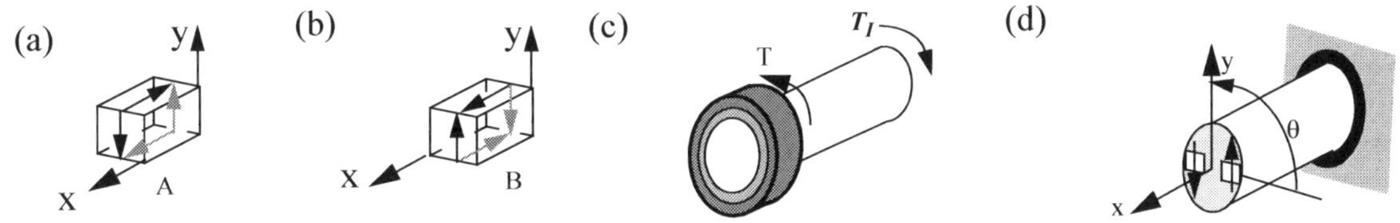

By Subscripts: We can make an imaginary cut through the cross-section containing points A and B and draw the free body diagram as shown in Fig. (c). the internal torque is drawn as per our sign convention. By moment equilibrium we obtain $T_I = T$. Thus from the torsional stress formula we obtain $\tau_{x\theta} > 0$. Noting that the outward normal of the cross-section is in the positive x-direction, a positive $\tau_{x\theta}$ requires the stress to be in the negative θ-direction as shown in Fig. (d). We write the stresses in the x-y coordinate system.

ANS $(\tau_{xy})_A < 0$; $(\tau_{xy})_B > 0$

---

5. 25

Solution

By Inspection: Due to the applied torque, the segment of the shaft containing the points A and B would rotate clockwise, that is point A would move downwards and point B upwards. To oppose this imaginary motion, the shear stress at point A will be upward and at point B it will be downwards as shown in Figs (a) and (b). The shear stress on the rest of the surfaces can be drawn using the observation that a symmetric pair of shear stress points towards the corner or away from the corner.

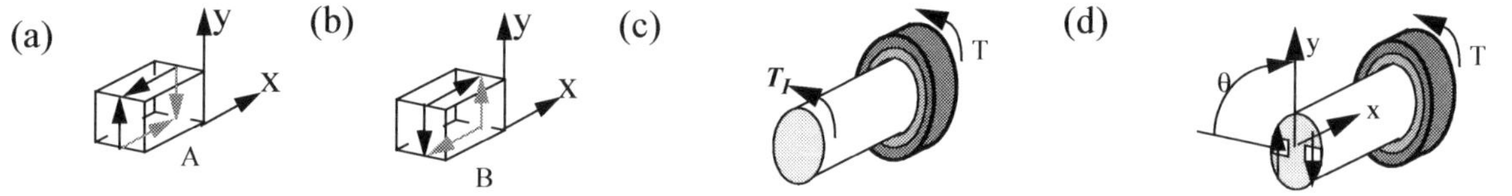

By Subscripts: We can make an imaginary cut through the cross-section containing points A and B and draw the free body diagram as shown in Fig. (c). the internal torque is drawn as per our sign convention. By moment equilibrium we obtain $T_I = -T$. Thus from the torsional stress formula we obtain $\tau_{x\theta} < 0$. Noting that the outward normal of the cross-section is in the negative x-direction, a negative $\tau_{x\theta}$ requires the stress to be in the positive θ-direction as shown in Fig. (d). We write the stresses in the x-y coordinate system.

ANS $(\tau_{xy})_A < 0$; $(\tau_{xy})_B > 0$

---

5. 26

Solution

By Inspection: Due to the applied torque, the segment of the shaft containing the points A and B would rotate in the direction of the torque, that is point A would move upwards and point B would move downwards. To oppose this imaginary motion, the shear stress at point A will be downward and at point B it will be upwards as shown in Figs (a) and (b). The shear stress on the rest of the surfaces can be drawn using the observation that a symmetric pair of shear stress points towards the corner or away from the corner.

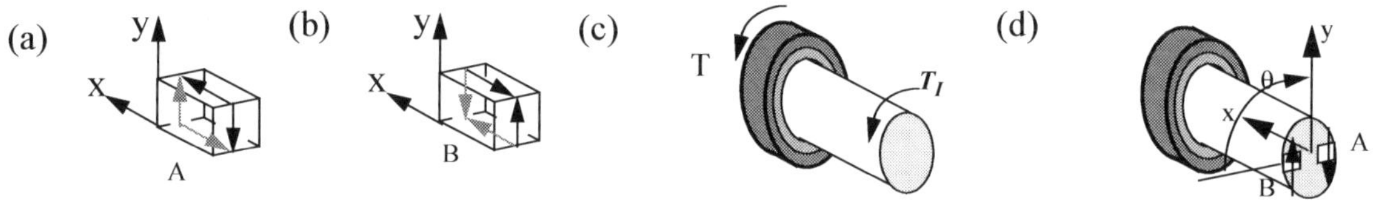

By Subscripts: We can make an imaginary cut through the cross-section containing points A and B and draw the free body diagram as shown in Fig. (c). The internal torque is drawn as per our sign convention. By moment equilibrium we obtain $T_I = -T$. Thus from the torsional stress formula we obtain $\tau_{x\theta} < 0$. Noting that the outward normal of the cross-section is in the negative x-direction, a negative $\tau_{x\theta}$ requires the stress to be in the positive θ-direction as shown in Fig. (d). In the x-y coordinate system we obtain:

ANS $(\tau_{xy})_A > 0$; $(\tau_{xy})_B < 0$

---

5. 27

Solution

We calculate the radius R_H and R_S in terms of the cross-sectional area A as shown below

$$A_H = \pi[(\alpha R_H)^2 - R_H^2] = A \quad or \quad R_H^2 = \frac{A}{\pi(\alpha^2 - 1)} \qquad A_S = \pi R_S^2 = A \quad or \quad R_S^2 = \frac{A}{\pi} \tag{1}$$

The polar area moment of inertia for a hollow shaft with inside radius R_i and outside radius R_o is $J = \pi(R_o^4 - R_i^4)/2$. For the hollow shaft $R_o = \alpha R_H$ and $R_i = R_H$, while for solid shaft $R_o = R_S$ and $R_i = 0$. Substituting these values we obtain the two polar area moments shown below:

$$J_H = \frac{\pi}{2}[(\alpha R_H)^4 - (R_H)^4] = \frac{\pi}{2}(\alpha^4 - 1)(R_H)^4 = \frac{\pi}{2}(\alpha^4 - 1)\left(\frac{A}{\pi(\alpha^2 - 1)}\right)^2 = \left(\frac{\alpha^2 + 1}{\alpha^2 - 1}\right)\left(\frac{A^2}{2\pi}\right) \qquad J_S = \frac{\pi}{2}R_S^4 = \frac{\pi}{2}\left(\frac{A}{\pi}\right)^2 = \frac{A^2}{2\pi} \tag{2}$$

$$\frac{J_H}{J_S} = \frac{[(\alpha^2 + 1)/(\alpha^2 - 1)](A^2/(2\pi))}{A^2/(2\pi)} = (\alpha^2 + 1)/(\alpha^2 - 1) \tag{3}$$

ANS $J_H/J_S = (\alpha^2 + 1)/(\alpha^2 - 1)$

---

5. 28

Solution

The polar moment of inertia of a hollow shaft in terms of outer radius R_o and inner radius R_i can be written as shown below:

$$J = \frac{\pi}{2}(R_o^4 - R_i^4) = \frac{\pi}{2}(R_o^2 + R_i^2)(R_o^2 - R_i^2) = \frac{\pi}{2}(R_o^2 + R_i^2)(R_o + R_i)(R_o - R_i) \tag{1}$$

Substituting $R_o = R + t/2$ and $R_i = R - t/2$ in the equation above we obtain:

$$J = \frac{\pi}{2}\left(R^2 + Rt + \frac{t^2}{4} + R^2 - Rt + \frac{t^2}{4}\right)(R + t/2 + R - t/2)(R + t/2 - R + t/2) = \frac{\pi}{2}\left(2R^2 + \frac{t^2}{2}\right)(2R)(t) \quad \textbf{(2)}$$

The t^2 term is two orders of magnitude smaller than the R^2 term and can be neglected to obtain the following expression.

$$J \approx (\pi/2)(2R^2)(2R)(t) = 2\pi R^3 t \quad \textbf{(3)}$$

ANS $J = 2\pi R^3 t$

---

5. 29

Solution GJ = 90,000 kips-in^2 $\phi_D - \phi_A = ?$

The torque diagram can be drawn using the template and template equation shown below..

(a) Template

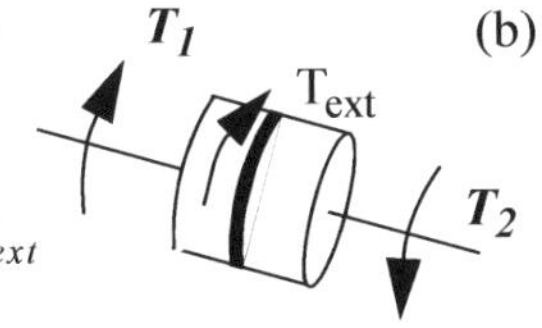

(b) Torque Diagram

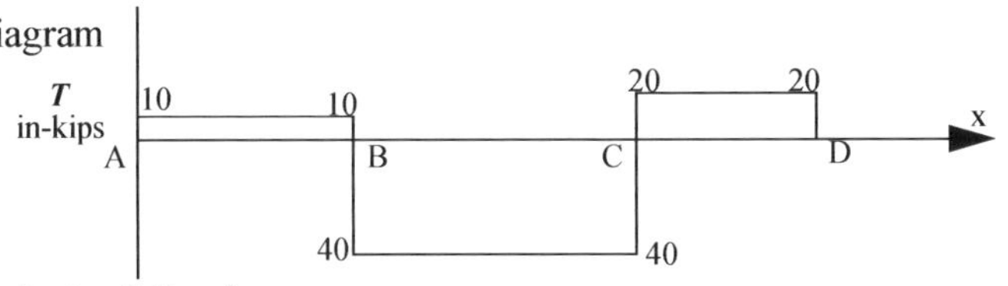

By equilibrium of moment about the shaft axis in we obtain the following.

$$T_{AB} = 10in-kips \qquad T_{BC} - 10 + 50 = 0 \quad or \quad T_{BC} = -40in-kips \qquad T_{CD} = 20in-kips \quad \textbf{(1)}$$

(c)

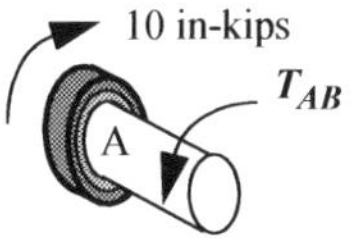

(d)

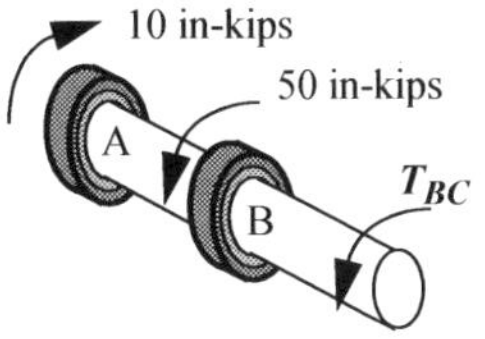

(e)

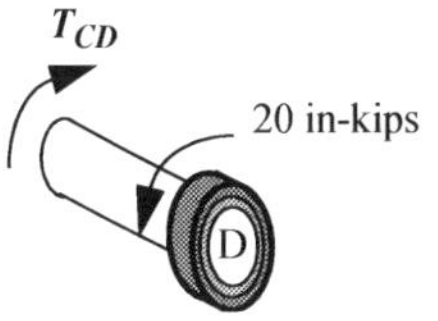

The relative rotation of the ends of each segment can be found and added as shown below:

$$\phi_B - \phi_A = \frac{T_{AB}(x_B - x_A)}{GJ} = \frac{(10)(20)}{90,000} = 2.222(10^{-3})\ rads \qquad \phi_C - \phi_B = \frac{T_{BC}(x_C - x_B)}{GJ} = \frac{(-40)(36)}{90,000} = -16.0(10^{-3})\ rads \quad \textbf{(2)}$$

$$\phi_D - \phi_C = \frac{T_{CD}(x_C - x_B)}{GJ} = \frac{(20)(30)}{90,000} = 6.667(10^{-3})\ rads \qquad \phi_D - \phi_A = (2.222 - 16.0 + 6.667)(10^{-3}) = -7.111(10^{-3})\ rads \quad \textbf{(3)}$$

ANS $\phi_D - \phi_A = 0.00711$ rads CW

---

5. 30

Solution GJ = 1,270 kN-m^2 $\phi_D - \phi_A = ?$

The torque diagram can be drawn using the template and template equation shown.

(a) Template

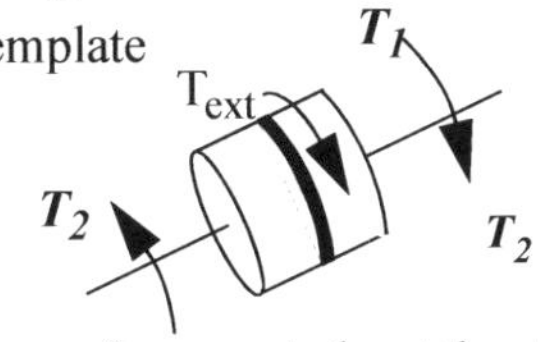

(b) Torque Diagram

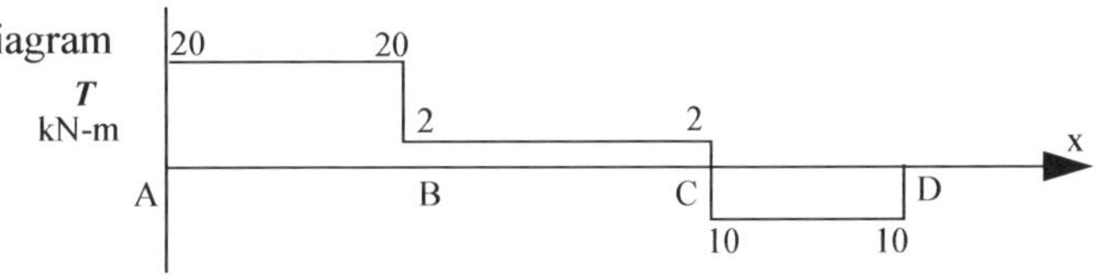

By equilibrium of moment about the shaft axis we obtain the following.

$$T_{AB} = 20\ kN-m \qquad T_{BC} - 20 + 18 = 0 \quad or \quad T_{BC} = 2\ kN-m \qquad T_{CD} = -10\ kN-m \quad \textbf{(1)}$$

(c)

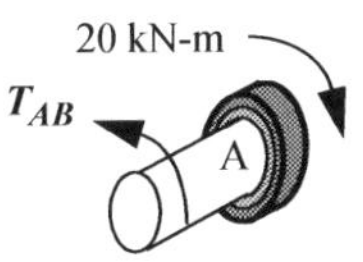

(d)

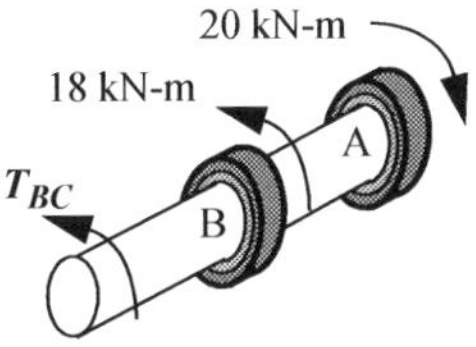

(e)

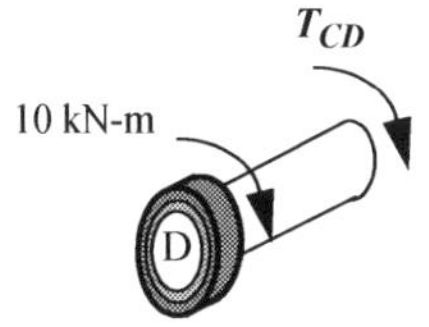

The relative rotation of the ends of each segment can be found and added as shown below:

$$\phi_B - \phi_A = \frac{T_{AB}(x_B - x_A)}{GJ} = \frac{(20)(0.4)}{1,270} = 6.299(10^{-3})\ rads \qquad \phi_C - \phi_B = \frac{T_{BC}(x_C - x_B)}{GJ} = \frac{(2)(1)}{1,270} = 1.575(10^{-3})\ rads \quad \textbf{(2)}$$

$$\phi_D - \phi_C = \frac{T_{CD}(x_C - x_B)}{GJ} = \frac{(-10)(0.5)}{1,270} = -3.937(10^{-3})rads \qquad \phi_D - \phi_A = (6.299 + 1.575 - 3.937)(10^{-3}) = 3.937(10^{-3})rads \quad \textbf{(3)}$$

ANS $\phi_D - \phi_A = 0.00394$ rads CCW

---

5. 31

Solution G = 80 GPa d = 150 mm ϕ_D = ? $(\tau_{x\theta})_E$ = ? γ_{max} = ?

The wall reaction torque T_A can be found from the free body diagram of the entire shaft as shown below.

$$T_A - 150 + 90 - 70 = 0 \quad or \quad T_A = 130\ kN-m \tag{1}$$

The torque diagram can be drawn using the template and template equation as shown below.

(a) Template

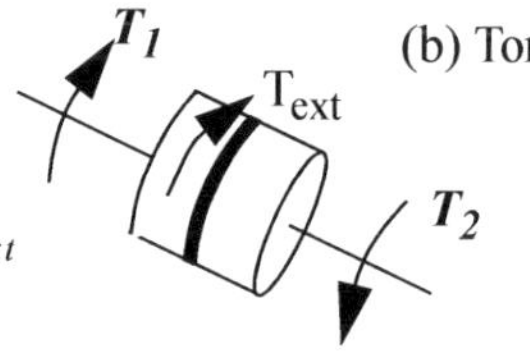

$T_2 = T_1 + T_{ext}$

(b) Torque Diagram

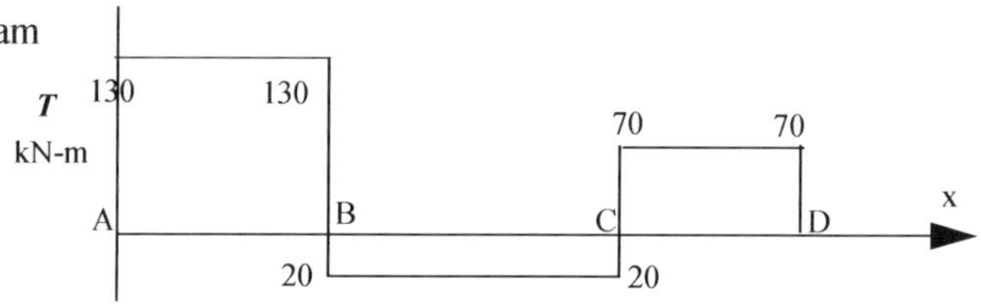

$$T_{AB} = 130\ kN-m \qquad T_{BC} = -20\ kN-m \qquad T_{CD} = 70\ kN-m \tag{2}$$

$$J = \pi(0.15^4)/32 = 49.701(10^{-6})\ m^4 \qquad GJ = (80)(10^9)(49.701)(10^{-6}) = 3.976(10^6)N-m^2 = 3.976(10^3)kN-m^2 \tag{3}$$

The relative rotation of the ends of each segment can be found and added as shown below. Note $\phi_A = 0$.

$$\phi_B - \phi_A = \frac{T_{AB}(x_B - x_A)}{GJ} = \frac{(130)(0.25)}{3.976(10^3)} = 8.174(10^{-3})\ rads \qquad \phi_C - \phi_B = \frac{T_{BC}(x_C - x_B)}{GJ} = \frac{(-20)(0.5)}{3.976(10^3)} = -2.515(10^{-3})\ rads \tag{4}$$

$$\phi_D - \phi_C = \frac{T_{CD}(x_C - x_B)}{GJ} = \frac{(70)(0.3)}{3.976(10^3)} = 5.282(10^{-3})\ rads \qquad \phi_D - \phi_A = (8.174 - 2.515 + 5.282)(10^{-3}) = 10.94(10^{-3})\ rads \tag{5}$$

$$(\tau_{x\theta})_E = \frac{T_{BC}\rho_E}{J} = \frac{(-20)(10^3)(0.075)}{49.701(10^{-6})} = -30.18(10^6)\ N/m^2 \tag{6}$$

The outward normal is in the positive x-direction, thus the shear stress must be in the negative θ-direction at point E as shown below.

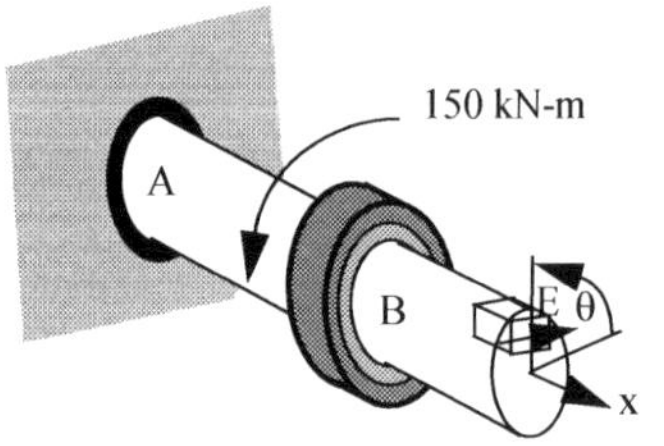

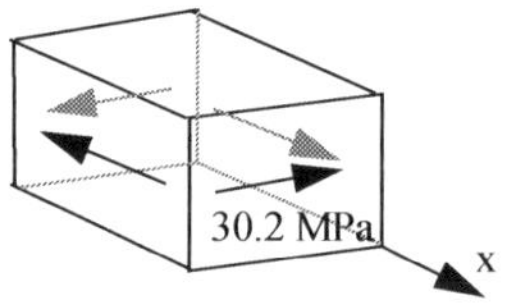

The maximum shear stress will be in segment AB as the internal torque is maximum.

$$\tau_{max} = \frac{T_{AB}\rho_{max}}{J} = \frac{130(10^3)(0.075)}{49.701(10^{-6})} = 196.2(10^6)\ N/m^2 \qquad \gamma_{max} = \frac{\tau_{max}}{G} = \frac{196.2(10^6)}{80(10^9)} = 2.452(10^{-3}) \tag{7}$$

ANS $\phi_D = 0.0109$ rads CCW ; $(\tau_{x\theta})_E = -30.2$ MPa ; $\gamma_{max} = 2452\mu$

---

5. 32

Solution G = 4000 ksi d = 4 in ϕ_D = ? $(\tau_{x\theta})_E$ = ? γ_{max} = ?

The wall reaction torque T_A can be found from the free body diagram of the entire shaft as shown below.

$$T_A - 80 + 40 + 15 = 0 \quad or \quad T_A = 25\ in-kips \tag{1}$$

The torque diagram can be drawn using the template and template equation shown.

(a) Template

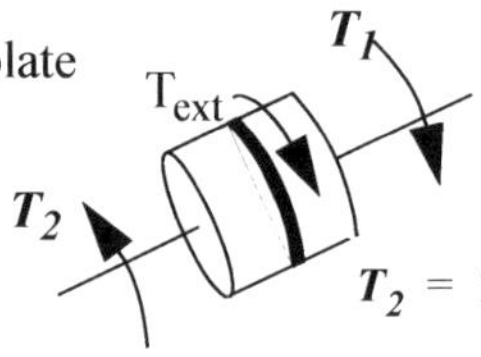

$T_2 = T_1 + T_{ext}$

(b) Torque Diagram

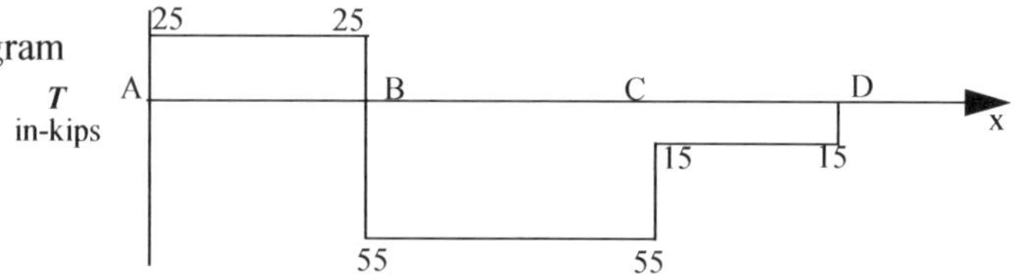

$$T_{AB} = 25\ in-kips \qquad T_{BC} = -55\ in-kips \qquad T_{CD} = -15\ in-kips \tag{2}$$

$$J = \pi(4^4)/32 = 25.133\ in^4 \qquad GJ = (4000)(25.133) = 100.53(10^3)kips-in^2 \tag{3}$$

The relative rotation of the ends of each segment can be found and added as shown below: Note $\phi_A = 0$

$$\phi_B - \phi_A = \frac{T_{AB}(x_B - x_A)}{GJ} = \frac{(25)(20)}{(100.53)(10^3)} = 4.97(10^{-3})rads \qquad \phi_C - \phi_B = \frac{T_{BC}(x_C - x_B)}{GJ} = \frac{(-55)(32)}{(100.53)(10^3)} = -17.51(10^{-3})rads \tag{4}$$

$$\phi_D - \phi_C = \frac{T_{CD}(x_C - x_B)}{GJ} = \frac{(-15)(25)}{(100.53)(10^3)} = -3.730(10^{-3})rads \qquad \phi_D - \phi_A = (4.97 - 17.51 - 3.73)(10^{-3}) = -16.26(10^{-3})rads \tag{5}$$

$$(\tau_{x\theta})_E = \frac{T_{BC}\rho_E}{J} = \frac{(-55)(2)}{25.133} = -4.377\ ksi \qquad \gamma_{max} = \frac{\tau_{max}}{G} = \frac{-4.377}{4,000} = -1.094(10^{-3}) \qquad \textbf{(6)}$$

The outward normal is in the positive x-direction, the shear stress must be in the negative θ-direction at point E as shown below.

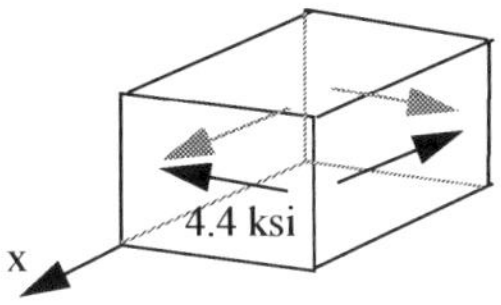

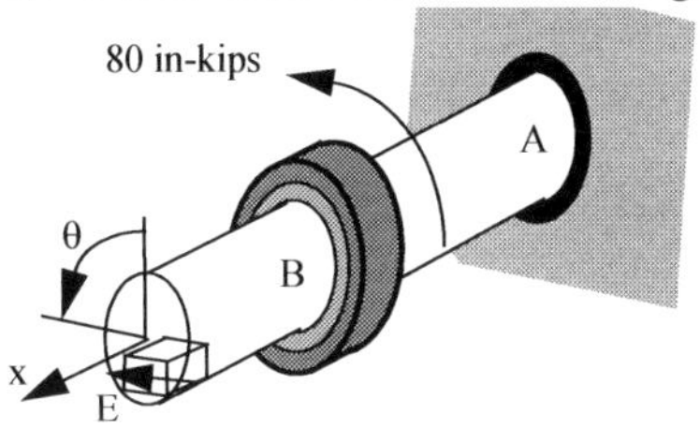

ANS $\phi_D = 0.0163$ rads CW; ; $(\tau_{x\theta})_E = -4.4$ ksi ; $\gamma_{max} = -1094\mu$

---

5. 33

Solution G =12,000 ksi $d_s = 2$ in $d_{al} = 1.5$ in $\phi_D - \phi_A = ?$ $\tau_{max} = ?$

$$J_{AB} = J_{CD} = \frac{\pi}{32}(2)^4 = 1.5708\text{ in}^4 \qquad J_{BC} = \frac{\pi}{32}(1.5)^4 = 0.497\text{ in}^4 \qquad \textbf{(1)}$$

We find the wall reaction torque and draw the torque diagram and determine the internal torques as shown below.

$$T_A = 12 + 25 - 15 = 22\text{ in-kips} \qquad T_{AB} = 22\text{ in-kips} \qquad T_{BC} = 10\text{ in-kips} \qquad T_{CD} = -15\text{ in-kips} \qquad \textbf{(2)}$$

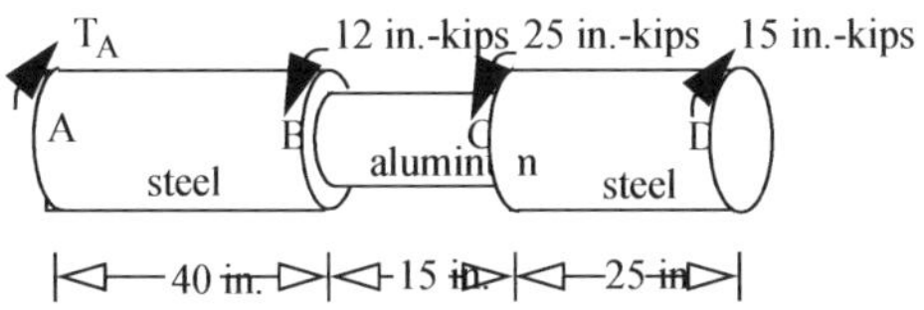

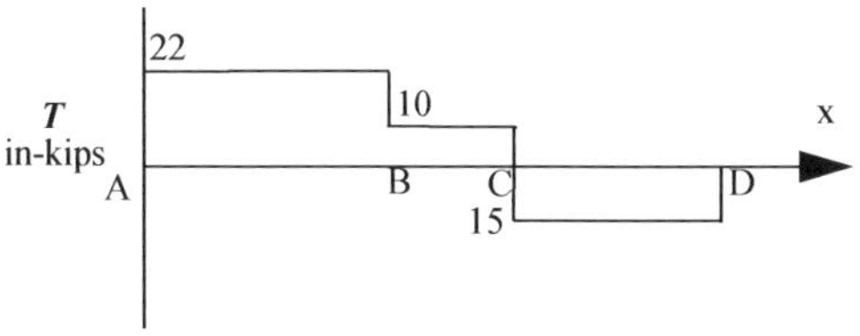

The relative rotations of each section and maximum torsional shear stress in each section can be found as shown below.

$$\phi_B - \phi_A = \frac{T_{AB}(x_B - x_A)}{G_{AB}J_{AB}} = \frac{(22)(40)}{(12000)(1.5708)} = 46.68(10^{-3})rads \qquad \tau_{AB} = \frac{T_{AB}(\rho_{AB})_{max}}{J_{AB}} = \frac{22(1)}{1.5708} = 14.01\text{ ksi} \qquad \textbf{(3)}$$

$$\phi_C - \phi_B = \frac{T_{BC}(x_C - x_B)}{G_{BC}J_{BC}} = \frac{(10)(15)}{(4000)(0.497)} = 75.45(10^{-3})rads \qquad \tau_{BC} = \frac{T_{BC}(\rho_{BC})_{max}}{J_{BC}} = \frac{(10)(0.75)}{1.5708} = 15.09\text{ ksi} \qquad \textbf{(4)}$$

$$\phi_D - \phi_C = \frac{T_{CD}(x_C - x_B)}{G_{CD}J_{CD}} = \frac{(-15)(25)}{(12000)(1.5708)} = -19.89(10^{-3})\ rads \qquad \tau_{CD} = \frac{T_{AB}(\rho_{AB})_{max}}{J_{AB}} = \frac{(-15)(1)}{1.5708} = -9.55\text{ ksi} \qquad \textbf{(5)}$$

$$\phi_D - \phi_A = (46.68 + 75.45 - 19.89)(10^{-3}) = 102.2(10^{-3})\ rads \qquad \textbf{(6)}$$

ANS $\phi_D - \phi_A = 0.102$ rads CCW ; $\tau_{max} = 15.09$ ksi

---

5. 34

Solution G = 12,000 ksi $\phi_D - \phi_A = ?$ $\tau_{max} = ?$ $(\tau_{x\theta})_E = ?$

$$J_{AB} = J_{AB} = \frac{\pi}{32}(4^4 - 2^4) = 23.562\ in^4 \qquad J_{BC} = \frac{\pi}{32}(4^4) = 25.133\ in^4 \qquad \textbf{(1)}$$

The torque diagram can be drawn using the template and intrnal torques determined as shown below.

$$T_{AB} = 120\ in-kips \qquad T_{BC} = -300\ in-kips \qquad T_{CD} = -100\ in-kips. \qquad \textbf{(2)}$$

(a) Template

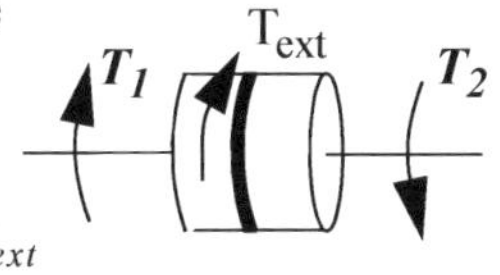

$T_2 = T_1 + T_{ext}$

(b) Torque Diagram

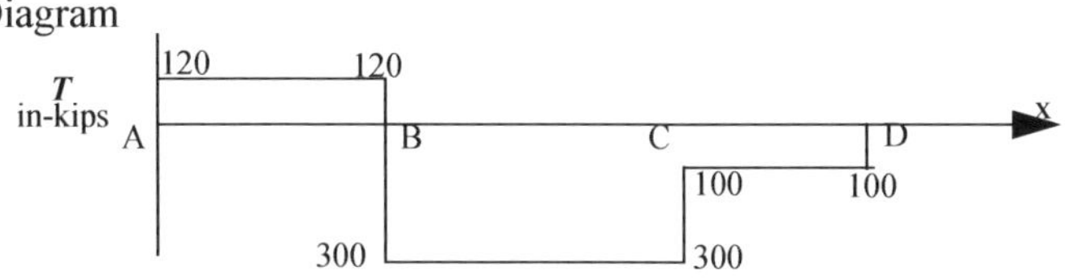

The relative rotation of the ends of each segment can be found and added as shown below:

$$\phi_B - \phi_A = \frac{T_{AB}(x_B - x_A)}{G_{AB}J_{AB}} = \frac{(120)(24)}{(12,000)(23.562)} = 10.186(10^{-3})\ rads \qquad \textbf{(3)}$$

$$\phi_C - \phi_B = \frac{T_{BC}(x_C - x_B)}{G_{BC}J_{BC}} = \frac{(-300)(36)}{(12,000)(25.133)} = -35.810(10^{-3})\ rads \qquad \textbf{(4)}$$

$$\phi_D - \phi_C = \frac{T_{CD}(x_D - x_C)}{G_{CD}J_{CD}} = \frac{(-100)(24)}{(12,000)(23.562)} = -8.488(10^{-3})\ rads \quad (5)$$

$$\phi_D - \phi_A = (10.186 - 35.810 - 8.488)(10^{-3}) = -34.11(10^{-3})\ rads \quad (6)$$

The maximum torsional shear stress will be in BC and at point E can be found and shown on a stress cube as shown below.

$$\tau_{max} = \left|\frac{T_{BC}\rho_{max}}{J_{BC}}\right| = \frac{(300)(2)}{25.133} = 23.87\ ksi \qquad (\tau_{x\theta})_E = \frac{T_{CD}\rho_E}{J_{CD}} = \frac{(-100)(1)}{23.562} = -4.244\ ksi \quad (7)$$

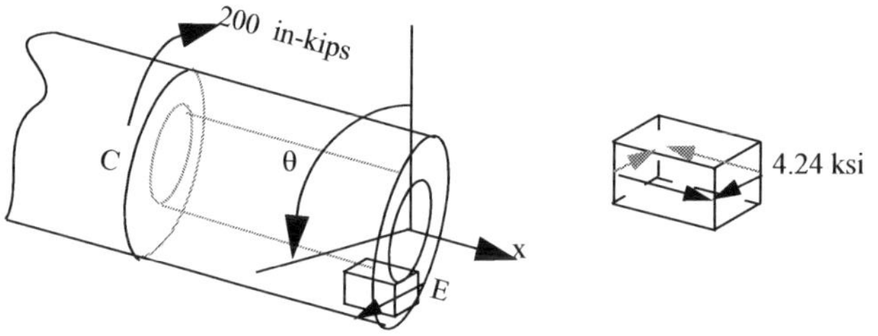

ANS $\phi_D - \phi_A = 0.0341$ rads CW ; $\tau_{max} = 23.87$ ksi ; $(\tau_{x\theta})_E = -4.24$ ksi

---

5. 35

Solution G=80GPa $\phi_A = ?$ $(\tau_{x\theta})_E = ?$

$$J_{AB} = J_{BC} = \frac{\pi}{32}(0.3^4) = 0.7952(10^{-3})\ mm^4 \qquad J_{CD} = \frac{\pi}{32}(0.5^4) = 6.1359(10^{-3})\ mm^4 \quad (1)$$

The wall reaction torque T_D can be found from the free body diagram of the entire shaft and torque diagram drawn to determine the internal torques.

$$T_D - 80 + 120 - 160 = 0 \quad or \quad T_D = 120kN-m \quad T_{AB} = 80kN-m \quad T_{BC} = -40kN-m \quad T_{CD} = 120kN-m. \quad (2)$$

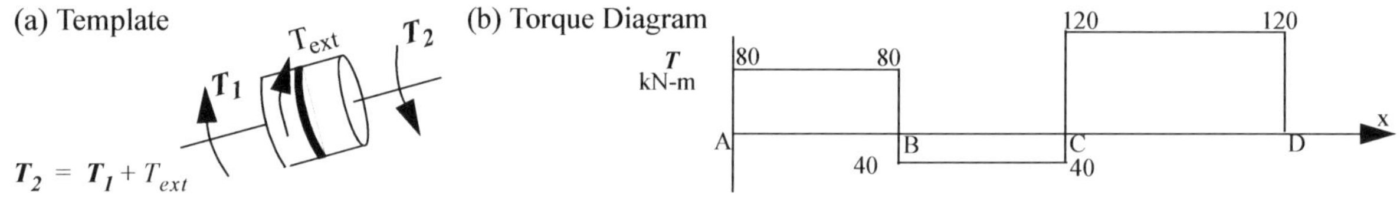

The relative rotation of the ends of each segment can be found and added as shown below.

$$\phi_B - \phi_A = \frac{T_{AB}(x_B - x_A)}{G_{AB}J_{AB}} = \frac{(80)(10^3)(2)}{(80)(10^9)(0.7952)(10^{-3})} = 2.515(10^{-3})\ rads \quad (3)$$

$$\phi_C - \phi_B = \frac{T_{BC}(x_C - x_B)}{G_{BC}J_{BC}} = \frac{(-40)(10^3)(2.5)}{(80)(10^9)(0.7952)(10^{-3})} = -1.572(10^{-3})\ rads \quad (4)$$

$$\phi_D - \phi_C = \frac{T_{CD}(x_D - x_C)}{G_{CD}J_{CD}} = \frac{(120)(10^3)(3.0)}{(80)(10^9)(6.1359)(10^{-3})} = 0.7334(10^{-3})\ rads \quad (5)$$

$$\phi_D - \phi_A = (2.515 - 1.572 + 0.7334)(10^{-3}) = 1.676(10^{-3})\ rads \quad (6)$$

Note ϕ_D is zero. The torsional shear stress at point E can be found and shown on a stress cube as shown below.

$$(\tau_{x\theta})_E = \frac{T_{AB}\rho_E}{J_{AB}} = \frac{(80)(10^3)(0.150)}{(0.7952)(10^{-3})} = 15.09(10^6)\ N/m^2. \quad (7)$$

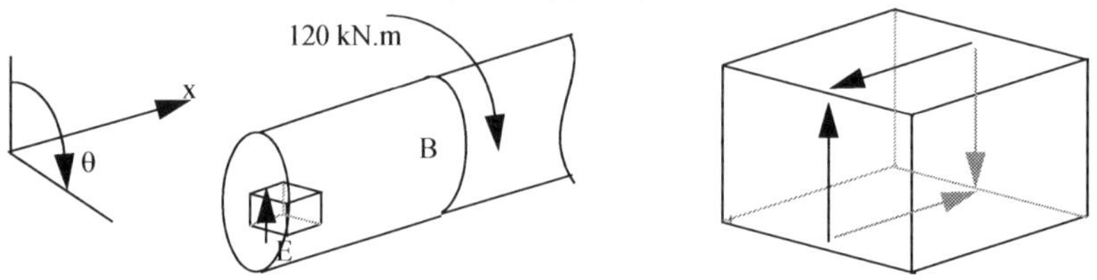

ANS $\phi_A = 1676\ \mu$ rads CW ; $(\tau_{x\theta})_E = 15.1$ MPa

---

5. 36

Solution R_A = 200 mm R_B = 50 mm G = 40 GPa $\phi_C = ?$ $\gamma_{max} = ?$

We note that R in segment AB is a linear function of x i.e., $R(x) = a + bx$. The constants can be found as shown below.

$$R(x=0) = R_A = a = 0.2 \qquad R(x=7.5) = R_B = 0.2 + b(7.5) = 0.05 \qquad or \qquad b = -0.02 \qquad R(x) = 0.2 - 0.02x \qquad \textbf{(1)}$$

From moment equilibrium we obtain:

$$\boldsymbol{T_{AB}} = -7.5kN-m \qquad \boldsymbol{T_{BC}} = 2.5kN-m \qquad J_{AB} = \frac{\pi}{2}(0.2-0.02x)^4\ m^4 \qquad J_{BC} = \frac{\pi}{2}(0.05)^4 = 9.817(10^{-6})\ m^4 \qquad \textbf{(2)}$$

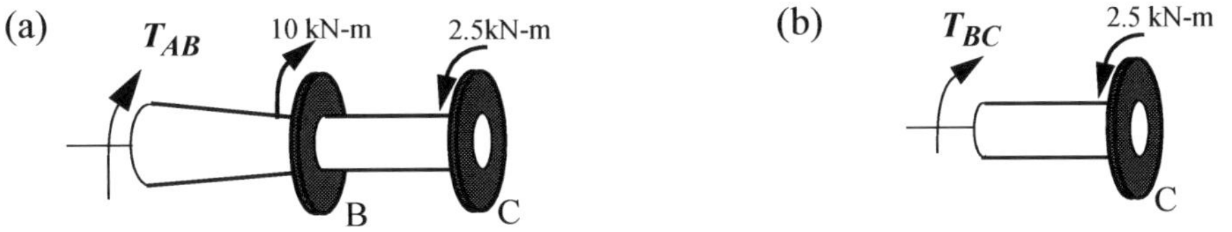

The relative rotation of the ends of segment AB can be found as shown below

$$\phi_C - \phi_B = \frac{\boldsymbol{T_{BC}}(x_C - x_B)}{G_{BC}J_{BC}} = \frac{(2.5)(10^3)(2)}{(40)(10^9)(9.817)(10^{-6})} = 12.732(10^{-3})rads \qquad \textbf{(3)}$$

$$\left(\frac{d\phi}{dx}\right)_{AB} = \frac{\boldsymbol{T_{AB}}}{G_{AB}J_{AB}} = \frac{-(7.5)(10^3)}{(40)(10^9)[\pi(0.2-0.02x)^4/2]} \qquad or \qquad \int_{\phi_A}^{\phi_B} d\phi = -\frac{15(10^{-6})}{40\pi}\int_0^{7.5}\frac{dx}{(0.2-0.02x)^4} \quad \text{or}$$

$$\phi_B - \phi_A = -\frac{0.1194(10^{-6})}{(-3)(-0.02)}\left[\frac{1}{(0.2-0.02x)^3}\right]\Bigg|_0^{7.5} = -(1.989)(10^{-6})\left[\frac{1}{(0.05)^3} - \frac{1}{(0.20)^3}\right] = -15.668(10^{-3})\ rads \qquad \textbf{(4)}$$

$$\phi_C - \phi_A = (-15.668 + 12.732)(10^{-3}) = -2.934(10^{-3})\ rads \qquad or \qquad \phi_C = -2.934(10^{-3})\ rads \qquad \textbf{(5)}$$

The maximum shear stress will be on a cross-section just left of where $J_{AB} = 9.817(10^{-6})\ m^4$ and $\rho_{max} = 0.05m$. The magnitude of maximum torsional shear stress in the shaft can be found as shown below.

$$\tau_{max} = \left|\frac{\boldsymbol{T_{AB}}\rho_{max}}{J_{AB}}\right| = \left|\frac{(7.5)(10^3)(0.05)}{9.817(10^{-6})}\right| = 38.199(10^6)\ N/m^2 \qquad \gamma_{max} = \frac{\tau_{max}}{G} = \frac{38.199(10^6)}{(40)(10^9)} = 0.955(10^{-3}) \qquad \textbf{(6)}$$

ANS $\phi_C = 0.0029$ rads CW ; $\gamma_{max} = 955\ \mu$

---

5. 37

Solution $R = Ke^{-ax}$ $\qquad \phi_B = f(T_{ext}, L, G, r) = ?$ $\qquad \tau_{max} = g(T_{ext}, L, G, r) = ?$

The internal torque and the polar moment of the cross-section can be calculated as shown below.

$$\boldsymbol{T_{AB}} = T_{ext} \qquad J_{AB} = \pi K^4 e^{-4ax}/2 \qquad \textbf{(1)}$$

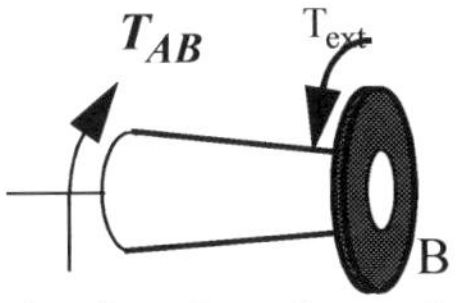

The relative rotation of the ends of segment AB can be found as shown below

$$\left(\frac{d\phi}{dx}\right)_{AB} = \frac{\boldsymbol{T_{AB}}}{G_{AB}J_{AB}} = \frac{2T_{ext}e^{4ax}}{\pi GK^4} \qquad or \qquad \phi_B - \phi_A = \int_{\phi_A}^{\phi_B} d\phi = \frac{2T_{ext}}{\pi GK^4}\int_0^L e^{4ax}dx = \left(\frac{2T_{ext}}{\pi GK^4}\right)\left[\frac{e^{4ax}}{4a}\right]\Bigg|_0^L = \left(\frac{T_{ext}}{2\pi aGK^4}\right)[e^{4aL} - 1] \qquad \textbf{(2)}$$

ANS $\phi_B = \left(\frac{T_{ext}}{2\pi aGK^4}\right)[e^{4aL} - 1]$ CCW

---

5. 38

Solution $R = r\sqrt{(2 - 0.25x/L)}$ $\qquad \phi_B = f(T_{ext}, L, G, r) = ?$ $\qquad \tau_{max} = g(T_{ext}, L, G, r) = ?$

From previous problem and the given data we obtain the following.

$$\boldsymbol{T_{AB}} = T_{ext} \qquad J_{AB} = (\pi r^4)(2 - 0.25\ x/L)^2/2 = (\pi r^4)(2L - 0.25\ x)^2/(2L^2). \qquad \textbf{(1)}$$

The relative rotation of the ends of segment AB can be found as shown below

$$\left(\frac{d\phi}{dx}\right)_{AB} = \frac{\boldsymbol{T_{AB}}}{G_{AB}J_{AB}} = \frac{T_{ext}}{G[(\pi r^4)(2L-0.25x)^2/(2L^2)]} = \frac{2T_{ext}L^2}{\pi Gr^4(2L-0.25x)^2} \qquad or \qquad \int_{\phi_A}^{\phi_B} d\phi = \left(\frac{2T_{ext}L^2}{\pi Gr^4}\right)\int_0^L \frac{dx}{(2L-0.25x)^2} \quad \text{or}$$

$$\phi_B - \phi_A = \left(\frac{2T_{ext}L^2}{\pi G r^4}\right)\frac{1}{(-1)(-0.25)}\left[\frac{1}{(2L-0.25x)}\right]\Bigg|_0^L = \left(\frac{8T_{ext}L^2}{\pi G r^4}\right)\left[\frac{1}{(1.75L)} - \frac{1}{(2L)}\right] = 0.1819\left(\frac{T_{ext}L}{Gr^4}\right) \quad \textbf{(2)}$$

The maximum shear stress will be on a cross-section just left of B where $R = (r)\sqrt{(2-0.25)} = 1.3229r$.

$$\tau_{max} = \left|\frac{T_{AB}\rho_{max}}{J_{AB}}\right| = \frac{T_{ext}(1.3229r)}{\pi(1.3229r)^4/2} = \frac{0.275T_{ext}}{r^3} \quad \textbf{(3)}$$

ANS $\phi_B = 0.1819\left(\frac{T_{ext}L}{Gr^4}\right)$ CCW; $\tau_{max} = \frac{0.275T_{ext}}{r^3}$

---

5. 39

Solution $\phi_B - \phi_A = f(q, L, G, r) = ?$ $t(x) = q(x^2/L^2)$ $\tau_{max} = g(q, L, G, r) = ?$

From equilibrium equation we obtain

$$\frac{d\boldsymbol{T}}{dx} + q\left(\frac{x^2}{L^2}\right) = 0 \quad or \quad \boldsymbol{T} = -q\left(\frac{x^3}{3L^2}\right) + c \qquad \boldsymbol{T}(x = L) = -q\left(\frac{L^3}{2L^2}\right) + c = 0 \quad or \quad c = \frac{qL}{3} \qquad \boldsymbol{T} = \frac{q}{3L^2}(L^3 - x^3) \quad \textbf{(1)}$$

The maximum internal torque will exist at x = 0 where $\boldsymbol{T}_{max} = qL/3$. The maximum shear stress is

$$\tau_{max} = \left|\frac{T_{AB}\rho_{max}}{J}\right| = \frac{(qL/3)(d/2)}{[\pi d^4/32]} = \frac{16qL}{3\pi d^3} \quad \textbf{(2)}$$

$$\frac{d\phi}{dx} = \frac{\frac{q}{3L}(L^3 - x^3)}{G(\pi d^4/32)} \quad or \quad \phi_B - \phi_A = \int_{\phi_A}^{\phi_B} d\phi = \left(\frac{32q}{3\pi G L^2 d^4}\right)\int_{x_A = 0}^{x_B = L}(L^3 - x^3)dx = \left(\frac{32q}{3\pi G L^2 d^4}\right)\left(L^3x - \frac{x^4}{4}\right)\Bigg|_0^L = \left(\frac{8qL^2}{\pi G d^4}\right) \quad \textbf{(3)}$$

ANS $\tau_{max} = \frac{16qL}{3\pi d^3}$; $\phi_B - \phi_A = \left(\frac{8qL^2}{\pi G d^4}\right)$ CCW

---

5. 40

Solution $\phi_A = f(q, L, G, J) = ?$

From moment equilibrium we obtain the following.

$$\boldsymbol{T}_{AB} - 2qL + qx = 0 \quad or \quad \boldsymbol{T}_{AB} = 2qL - qx \qquad \boldsymbol{T}_{BC} - 2qL + qx + qL = 0 \quad or \quad \boldsymbol{T}_{BC} = qL - qx \quad \textbf{(1)}$$

(a)

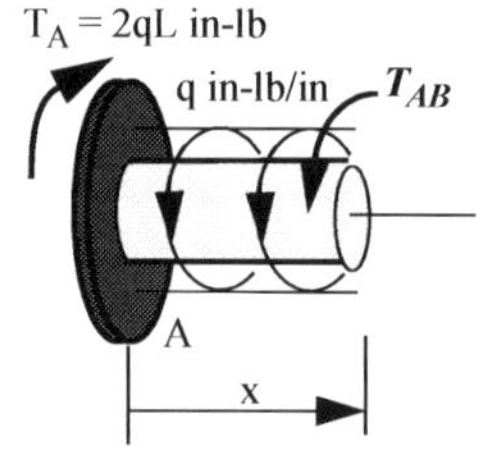

(b)

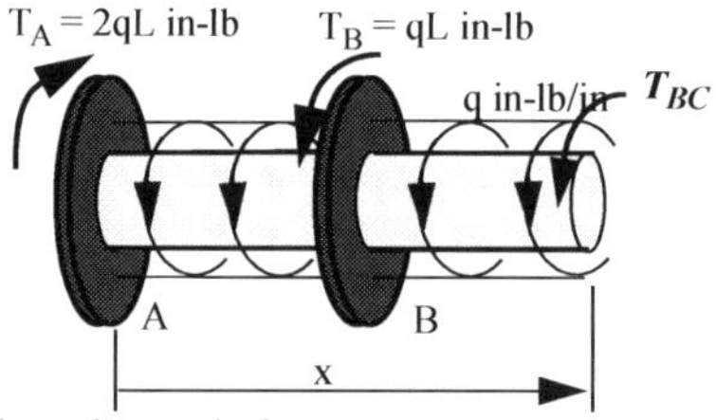

The relative rotation for the ends of segments AB and BC can be found as shown below

$$\left(\frac{d\phi}{dx}\right)_{AB} = \frac{T_{AB}}{G_{AB}J_{AB}} = \frac{2qL - qx}{GJ} \quad or \quad \phi_B - \phi_A = \int_{\phi_A}^{\phi_B} d\phi = \int_0^{0.5L}\left(\frac{2qL - qx}{GJ}\right)dx = \frac{1}{GJ}\left(2qLx - \frac{qx^2}{2}\right)\Bigg|_0^{0.5L} = \frac{7}{8}\left(\frac{qL^2}{GJ}\right) \quad \textbf{(2)}$$

$$\left(\frac{d\phi}{dx}\right)_{BC} = \frac{T_{BC}}{G_{BC}J_{BC}} = \frac{qL - qx}{GJ} \quad or \quad \phi_C - \phi_B = \int_{\phi_B}^{\phi_C} d\phi = \int_{0.5L}^{L}\left(\frac{qL - qx}{GJ}\right)dx = \frac{1}{GJ}\left(qLx - \frac{qx^2}{2}\right)\Bigg|_{0.5L}^{L} = \frac{1}{8}\left(\frac{qL^2}{GJ}\right) \quad \textbf{(3)}$$

$$\phi_C - \phi_A = \frac{7}{8}\left(\frac{qL^2}{GJ}\right) + \frac{1}{8}\left(\frac{qL^2}{GJ}\right) = \left(\frac{qL^2}{GJ}\right) \quad \textbf{(4)}$$

ANS $\phi_A = \left(\frac{qL^2}{GJ}\right)$ CW

---

5. 41

Solution: G = 12,000 ksit = 1/8 inch d = 6 inch L = 36 inch $\tau_{max} \le 10ksi$ $\Delta\phi \le 0.015rads$ T_{max} =?

The outer and inner diameters of the tube are: d_o = 6 + (1/8) = 6.125 in. and d_i = 6 - (1/8) = 5.875 in.

The maximum stress and the relative rotation of the two ends are as shown below.

$$J = \frac{\pi}{32}[(6.125)^4 - (5.875)^4] = 21.215 \qquad \tau_{max} = \frac{T(d_o)/2}{J} = \frac{T(6.125/2)}{21.215} \le 10 \qquad or \qquad T \le 69.27 in-kips \qquad (1)$$

$$\Delta\phi = \frac{TL}{GJ} = \frac{T(36)}{(12000)(21.211)} \le 0.015 \qquad or \qquad T \le 106.1 in-kips \qquad (2)$$

ANS $T = 69.2$ in-kips

---

5. 42

Solution: d = 2 inch. G = 4,000 ksi L = 4 feet $\tau_{max} \le 18ksi$ $\Delta\phi \le 0.2rads$ T_{max} =?

The maximum shear stress and the relative rotation of the two ends are as shown below.

$$J = \frac{\pi}{32}(2)^4 = 1.5708\ in^4 \qquad \tau_{max} = \frac{T(d/2)}{J} = \frac{T(1)}{1.5708} \le 18 \qquad or \qquad T \le 28.27 in-kips \qquad (1)$$

$$\Delta\phi = \frac{TL}{GJ} = \frac{T(4)(12)}{(4000)(1.5708)} \le 0.2 \qquad or \qquad T \le 26.18 in-kips \qquad (2)$$

ANS $T = 26.1$ in.-kips

---

5. 43

Solution: G = 80 GPa r_o = 3.0 mm T = 2700 m L = 1 m. $\tau_{max} \le 120MPa$ $\Delta\phi \le 0.1rads$ $(r_i)_{max}$ =?nearest mm.

The maximum shear stress and the relative rotation of the two ends are as shown below.

$$\tau_{max} = \frac{Tr_o}{J} = \frac{(2700)(30)(10^{-3})}{J} \le 120(10^6) \qquad or \qquad J \ge 675(10^{-9})m^4 \qquad (1)$$

$$\Delta\phi = \frac{TL}{GJ} = \frac{2700(1)}{(80)(10^9)(J)} \le 0.1 \qquad or \qquad J \ge 337.5(10^{-9})m^4 \qquad (2)$$

$$(\pi/2)(r_o^4 - r_i^4) \ge 675(10^{-9}) \qquad or \qquad r_i^4 \le (0.03)^4 - (1350/\pi)(10^{-9}) \qquad or \qquad r_i \le 0.0248m \qquad (3)$$

ANS $(r_i)_{max} = 24mm$

---

5. 44

Solution L = 5 ft. T= 200 in-kips d_o = 6 inch $\Delta\phi \le 0.05\ rads$ $(d_i)_{max}$ = ? nearest 1/8 inch W = ?

The relative rotation of the ends and maximum shear stress in steel and aluminum can be found as shown below.

$$(\Delta\phi)_S = \frac{(200)(5)(12)}{12(10^3)J_S} \le 0.05 \qquad or \qquad J_S \ge 20\ in^4 \qquad (\tau_{max})_S = \frac{(200)(3)}{J_S} \le 18 \qquad or \qquad J_S \ge 33.33\ in^4 \qquad (1)$$

$$(\Delta\phi)_{Al} = \frac{(200)(5)(12)}{(4000)J_{Al}} \le 0.05 \qquad or \qquad J_{Al} \ge 60\ in^4 \qquad (\tau_{max})_{Al} = \frac{(200)(3)}{J_{Al}} \le 10 \qquad or \qquad J_{Al} \ge 60\ in^4 \qquad (2)$$

The internal diameters d_S and d_{Al} can be found and rounded downwards as shown below:

$$J_S = (\pi/32)(6^4 - d_S^4) \ge 33.33\ in^4 \qquad or \qquad d_S \le 5.561\ in \qquad J_{Al} = (\pi/32)(6^4 - d_{Al}^4) \ge 60\ in^4 \qquad or \qquad d_{Al} \le 5.116\ in \qquad (3)$$

$$d_S = 5.5\ in \qquad or \qquad d_{Al} = 5.0\ in \qquad (4)$$

The weight of each material by taking the product of the material density and the volume of a hollow cylinder as shown below:

$$W_S = (0.285)(\pi/4)(6^2 - 5.5^2)(60) = 77.224\ lbs \qquad W_{Al} = (0.1)(\pi/4)(6^2 - 5.0^2)(60) = 51.8\ lbs \qquad (5)$$

The aluminum shaft is lighter.

ANS $(d_i)_{max} = 5.0$ in. ; $W = 51.8$ lbs

---

5. 45

Solution: P = 100 hp w = 3600 rpm $\tau_{max} \le 10ksi$ d_{min} =? nearest 1/8 inch.

The rotational speed in rads/sec and the torque on the shaft can be found as shown below.

$$\omega = (3600)(2\pi)/60 = 376.99(rad)/(sec) \qquad T = P/\omega = (100)(6600)/376.99 = 1751 in-lbs \qquad (1)$$

$$\tau_{max} = \frac{T(d/2)}{(\pi d^4)/32} = \frac{16}{\pi}\frac{1751}{d^3} \le 10(10^3) \qquad or \qquad d \ge 0.9625 \qquad (2)$$

ANS $d_{min} = 1.0$ in.

---

5. 46

Solution: $d_{bolt} = 1/4$ inch $\tau_{bolt} \le 12ksi$ R = 5/8 inch n_{min} =?

From earlier problem we have T = 1751 in - kips. The area of cross section of a bolt is: $A = \pi d_{bolt}^2/4 = 0.4909in^4$

By moment equilibrium about the shaft axis we obtain

$$T = nVR = (nR)(\tau_{bolt})(A) \quad or \quad \tau_{bolt} = \frac{T}{nRA} = \frac{1751}{n(5/8)(0.04909)} \le 12(10^3) \quad or \quad n \ge 4.76 \tag{1}$$

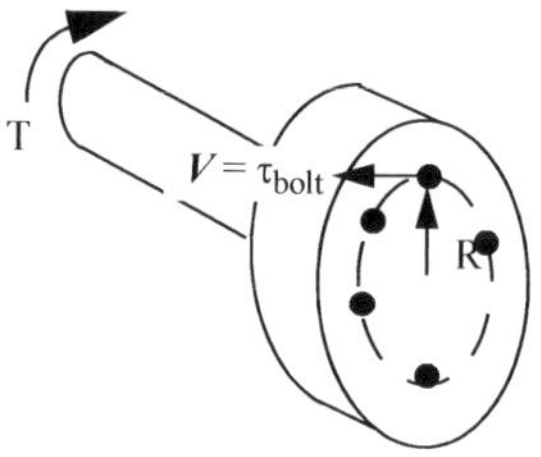

ANS $n_{min} = 5$

---

5. 47

Solution: P = 20 kW f = 20 Hz. P_A = 8kW P_B = 7KN P_C = 5 kW

τ_{yield} = 145 MPa k = 1.5 d_{min} =? to the nearest millimeter τ_{AB} =?

The torque delivered by the motor and the torques transferred by the gears can be found as shown below.

$$T = \frac{P}{2\pi f} = \frac{(20)(10^3)}{(2\pi)(20)} = 159.2N-m \qquad T_A = \frac{P_A}{2\pi f} = \frac{(8)(10^3)}{(2\pi)(20)} = 63.7N-m \tag{1}$$

$$T_B = \frac{P_B}{2\pi f} = \frac{(7)(10^3)}{(2\pi)(20)} = 55.7N-m \qquad T_C = \frac{P_C}{2\pi f} = \frac{(5)(10^3)}{(2\pi)(20)} = 39.8N-m \tag{2}$$

The shaft with the torques and the associated torque diagram shown below and internal torques calculated.

$$\boldsymbol{T}_{MA} = 159.2N-m \qquad \boldsymbol{T}_{AB} = 95.5N-m \qquad \boldsymbol{T}_{BC} = 39.8N-m$$

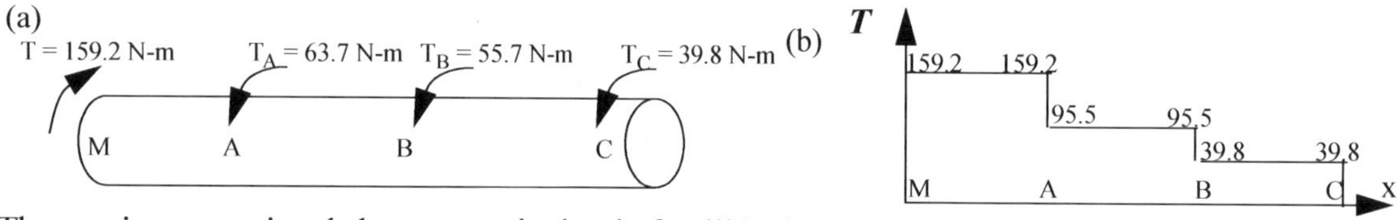

The maximum torsional shear stress in the shaft will be in segment MA

$$\tau_{max} = \frac{\boldsymbol{T}_{MA}(d/2)}{(\pi d^4)/32} = \frac{16(159.2)}{\pi d^3} \le \frac{145(10^6)}{1.5} \quad or \quad d \ge 0.0203m \tag{3}$$

$$\tau_{AB} = \frac{\boldsymbol{T}_{AB}(d/2)}{(\pi d^4)/32} = \frac{16\boldsymbol{T}_{AB}}{\pi d^3} = \frac{16(95.5)}{\pi(0.021)^3} = 52.5(10^6) \tag{4}$$

ANS $d_{min} = 21mm$; $\tau_{AB} = 52.5MPa$

---

5. 48

Solution $\boldsymbol{T}(x_A) = 0$ $\phi_B - \phi_A = ?$

The solution proceeds as follows.

$$\frac{d\boldsymbol{T}}{dx} = -t(x) \qquad \int_{T_A=0}^{T} d\boldsymbol{T} = -\int_{x_A}^{x} t(x)dx \quad or \quad \boldsymbol{T}(x) = -\int_{x_A}^{x} t(x)dx \tag{1}$$

$$\phi_B - \phi_A = \int_{\phi_A}^{\phi_B} d\phi = \int_{x_A}^{x_A} \frac{\boldsymbol{T}(x)}{GJ}dx = \frac{1}{GJ}\int_{x_A}^{x} \boldsymbol{T}(x)dx = \frac{1}{GJ}\left[x\boldsymbol{T}(x)\Big|_{x_A}^{x_B} - \int_{x_A}^{x} x\frac{d\boldsymbol{T}}{dx}dx\right] = \frac{1}{GJ}\left[x_B\boldsymbol{T}(x_B) - \boldsymbol{T}(x_A) - \int_{x_A}^{x} x(-t(x))dx\right] \tag{2}$$

Noting that $\boldsymbol{T}(x_A) = 0$, we obtain

$$\phi_B - \phi_A = \frac{1}{GJ}\left[x_B\left\{-\int_{x_A}^{x_B} t(x)dx\right\} - 0 + \int_{x_A}^{x_B} xt(x)dx\right] = \frac{1}{GJ}\left[\int_{x_A}^{x_B} (x - x_B)t(x)dx\right] \tag{3}$$

The above equation is the desired result.

---

5. 49

Solution

Solution proceeds as follows.

$$\gamma_{x\theta} = \rho\frac{d\phi}{dx}(x) \qquad \tau_{x\theta} = G\rho\frac{d\phi}{dx}(x) \qquad T = \int_A \rho\tau_{x\theta}dA = \int_A \rho\left(G\rho\frac{d\phi}{dx}(x)\right)(2\pi\rho)dA = \frac{d\phi}{dx}\int_A G\rho^2\, dA \quad (1)$$

$$T = \frac{d\phi}{dx}\left[\int_{A_1} G_1\rho^2 dA + \int_{A_2} G_2\rho^2 dA + \cdot + \cdot + \int_{A_n} G_n\rho^2 dA\right] = \frac{d\phi}{dx}[G_1J_1 + G_2J_2 + \cdot + \cdot + G_NJ_N] \quad (2)$$

$$T = \frac{d\phi}{dx}\left[\sum_{j=1}^{n} G_jJ_j\right] \qquad \frac{d\phi}{dx} = \frac{T}{\sum_{j=1}^{n} G_jJ_j} = \frac{\phi_2 - \phi_1}{x_2 - x_1} \qquad or \qquad \phi_2 - \phi_1 = \frac{T(x_2 - x_1)}{\sum_{j=1}^{n} G_jJ_j} \quad (3)$$

We can write for the ith material

$$(\tau_{x\theta})_i = G_i\rho\frac{d\phi}{dx} = G_i\rho\frac{T}{\sum_{j=1}^{n} G_jJ_j} \quad (4)$$

The above equations are the desired results.

If $G_1=G_2=G_3....=G_n=G$, then $\sum_{j=1}^{n} G_jJ_j = G\sum_{j=1}^{n} J_j = GJ$. Substituting we obtain:

$$(\tau_{x\theta})_i = G\rho\frac{T}{GJ} = \frac{T\rho}{J} \qquad \phi_2 - \phi_1 = \frac{T(x_2 - x_1)}{GJ} \quad (5)$$

Above equations are same as those for homogeneous materials.

---

5. 50

Solution

We follow the logic and determine the internal torque as shown below.

$$\gamma_{x\theta} = \rho\frac{d\phi}{dx} \qquad \tau_{x\theta} = G\gamma_{x\theta}^{0.5} = G\rho^{0.5}\left(\frac{d\phi}{dx}\right)^{0.5} \quad (1)$$

$$T = \int_A \rho\tau_{x\theta}dA = \int_0^R \rho\left[G\rho^{0.5}\left(\frac{d\phi}{dx}\right)^{0.5}\right](2\pi)d\rho = 2\pi G\left(\frac{d\phi}{dx}\right)^{0.5}\int_0^R \rho^{2.5}d\rho = 2\pi G\left(\frac{d\phi}{dx}\right)^{0.5}\frac{\rho^{3.5}}{3.5}\bigg|_0^R = 1.795G\left(\frac{d\phi}{dx}\right)^{0.5}R^{3.5} \text{ or}$$

$$\left(\frac{d\phi}{dx}\right)^{0.5} = \frac{T}{1.795GR^{3.5}} = 0.577\frac{T}{GR^{3.5}} \qquad \tau_{x\theta} = G\rho^{0.5}\left(\frac{d\phi}{dx}\right)^{0.5} = 0.577\frac{T\rho^{0.5}}{R^{3.5}} \qquad \tau_{max} = 0.577\frac{TR^{0.5}}{R^{3.5}} = 0.577\frac{T}{R^3} \quad (2)$$

$$\left(\frac{d\phi}{dx}\right) = \left[0.577\frac{T}{GR^{3.5}}\right]^2 \qquad or \qquad \frac{\phi_B - \phi_A}{L} = 0.3103\frac{T^2}{G^2R^7} \quad (3)$$

ANS $\tau_{max} = 0.577(T/R^3)$; $\phi_B = 0.3103T^2L/(G^2R^7)$

---

5. 51

Solution $\tau = G\gamma^2$ $\qquad \tau_{max}$ =? $\qquad \phi_2 - \phi_1$ =?

We follow the logic and determine the internal torque as shown below.

$$\gamma_{x\theta} = \rho\frac{d\phi}{dx} \qquad \tau_{x\theta} = G\gamma_{x\theta}^2 = G\rho^2\left(\frac{d\phi}{dx}\right)^2 \qquad T = \int_A \rho\tau_{x\theta}dA = \int_R^{2R} \rho\left[G\rho^2\left(\frac{d\phi}{dx}\right)^2\right](2\pi)d\rho \quad (1)$$

$$T = 2\pi G\left(\frac{d\phi}{dx}\right)^2\int_R^{2R} \rho^4 d\rho = 2\pi G\left(\frac{d\phi}{dx}\right)^2\frac{\rho^5}{5}\bigg|_R^{2R} = \frac{62}{5}\pi G\left(\frac{d\phi}{dx}\right)^2 R^5 \qquad or \qquad \left(\frac{d\phi}{dx}\right)^2 = \frac{5T}{62\pi GR^5} = 0.02567\frac{T}{GR^5} \quad (2)$$

$$\tau_{x\theta} = G\rho^2\left(\frac{d\phi}{dx}\right)^2 = 0.02567\frac{T\rho^2}{R^5} \qquad \tau_{max} = 0.02567\frac{TR^2}{R^5} = 0.02567\frac{T}{R^3} \quad (3)$$

$$\left(\frac{d\phi}{dx}\right) = \sqrt{0.02567\frac{T}{GR^5}} \qquad or \qquad \frac{\phi_2 - \phi_1}{x_2 - x_1} = 0.1602\sqrt{\frac{T}{GR^5}} \quad (4)$$

ANS $\tau_{max} = 0.02567(T/R^3)$; $\phi_2 - \phi_1 = 0.1602(x_2 - x_1)\sqrt{T/(GR^5)}$

---

5. 52

Solution $\tau = G\gamma^n$ $\tau_{max} = ?$ $\Delta\phi = ?$

The solution follows as shown below.

$$\gamma_{x\theta} = \rho\frac{d\phi}{dx}(x) \qquad \tau_{x\theta} = G\rho^n\left[\frac{d\phi}{dx}(x)\right]^n \qquad \boldsymbol{T} = \int_A \rho\tau_{x\theta}dA = \int_0^R \rho\left\{G\rho^n\left[\frac{d\phi}{dx}(x)\right]^n\right\}(2\pi\rho)d\rho \tag{1}$$

$$\boldsymbol{T} = 2\pi G\left[\frac{d\phi}{dx}(x)\right]^n \int_0^R \rho^{n+2}d\rho = 2\pi G\left[\frac{d\phi}{dx}(x)\right]^n\left(\frac{R^{n+3}}{n+3}\right) \quad or \quad \left[\frac{d\phi}{dx}(x)\right]^n = \frac{\boldsymbol{T}(n+3)}{2\pi GR^{n+3}} \tag{2}$$

$$\tau_{x\theta} = G\rho^n\left[\frac{d\phi}{dx}(x)\right]^n = G\rho^n\frac{\boldsymbol{T}(n+3)}{2\pi GR^{n+3}} \qquad \tau_{max} = GR^n\frac{\boldsymbol{T}(n+3)}{2\pi GR^{n+3}} \qquad \frac{d\phi}{dx} = \left[\frac{\boldsymbol{T}(n+3)}{2\pi GR^{n+3}}\right]^{1/n} = \frac{\Delta\phi}{L} \tag{3}$$

$$\textbf{ANS} \quad \tau_{max} = \frac{\boldsymbol{T}(n+3)}{2\pi R^3};\ \Delta\phi = \left[\frac{(n+3)\boldsymbol{T}}{2\pi GR^{(3+n)}}\right]^{1/n}L$$

For n=1 we obtain $\tau_{max} = 2\boldsymbol{T}/\pi R^3$ and $\Delta\phi = 2TL/(\pi GR^4)$, which are the same results as our classical theories.

---

5. 53

Solution $u = \psi(y, z)\frac{d\phi}{dx}$ $v = -xz\frac{d\phi}{dx}$ $w = xy\frac{d\phi}{dx}$ $\boldsymbol{T} = \int_A (y\tau_{xz} - z\tau_{xy})dA$ $\tau_{xy} = ?$ $\tau_{xz} = ?$

The solution follows as shown below.

$$\gamma_{xy} = \frac{\partial u}{\partial y} + \frac{\partial v}{\partial x} = \frac{\partial}{\partial y}\left[\psi(y,z)\frac{d\phi}{dx}\right] + \frac{\partial}{\partial x}\left[-xz\frac{d\phi}{dx}\right] = \frac{d\phi}{dx}\left[\frac{\partial\psi}{\partial y} - z\right] \qquad \gamma_{xz} = \frac{\partial w}{\partial x} + \frac{\partial u}{\partial z} = \frac{\partial}{\partial x}\left[xy\frac{d\phi}{dx}\right] + \frac{\partial}{\partial z}\left[\psi(y,z)\frac{d\phi}{dx}\right] = \frac{d\phi}{dx}\left[y + \frac{\partial\psi}{\partial z}\right] \tag{1}$$

$$\tau_{xy} = G\gamma_{xy} = G\left(\frac{\partial\psi}{\partial y} - z\right)\frac{d\phi}{dx} \qquad \tau_{xz} = G\gamma_{xz} = \tau_{xz} = G\left(\frac{\partial\psi}{\partial z} + y\right)\frac{d\phi}{dx} \tag{2}$$

The above equation is the desired result.

---

5. 54

Solution

Substituting $\psi(x, y) = 0$ into results of previous problem we obtain.

$$\tau_{xy} = -Gz\frac{d\phi}{dx} \qquad \tau_{xz} = Gy\frac{d\phi}{dx} \tag{1}$$

Substituting

$$\boldsymbol{T} = \int_A\left[y\left(Gy\frac{d\phi}{dx}\right) - z\left(-Gz\frac{d\phi}{dx}\right)\right]dA = G\frac{d\phi}{dx}\int_A[y^2 + z^2]dA = G\frac{d\phi}{dx}\int_A[\rho^2]dA = GJ\frac{d\phi}{dx} \tag{2}$$

The above equation is the desired result.

---

5. 55

Solution

By moment equilibrium in the given figure , we obtain the following.

$$\boldsymbol{T} + d\boldsymbol{T} - \boldsymbol{T} = \rho J\frac{\partial^2\phi}{\partial t^2}dx \quad or \quad \frac{\partial\boldsymbol{T}}{\partial x} = \rho J\frac{\partial^2\phi}{\partial t^2} \tag{1}$$

Substituting $\boldsymbol{T} = GJ\frac{\partial\phi}{\partial x}$ and assuming GJ as constant, we obtain

$$\frac{\partial}{\partial x}\left[GJ\frac{\partial\phi}{\partial x}\right] = \rho J\frac{\partial^2\phi}{\partial t^2} \quad or \quad GJ\left[\frac{\partial^2\phi}{\partial x^2}\right] = \rho J\frac{\partial^2\phi}{\partial t^2} \quad or \quad \frac{\partial^2\phi}{\partial t^2} = \left(\frac{G}{\rho}\right)\frac{\partial^2\phi}{\partial x^2} = c^2\frac{\partial^2\phi}{\partial x^2} \qquad c = \sqrt{\frac{G}{\rho}} \tag{2}$$

Above equation is the desired result.

---

5. 56

Solution

The solution follows as shown below.

$$\frac{\partial\phi}{\partial x} = [C\cos\omega t + D\sin\omega t]\frac{\partial}{\partial x}\left[A\cos\frac{\omega x}{c} + B\sin\frac{\omega x}{c}\right] = [C\cos\omega t + D\sin\omega t]\left[-\left(\frac{\omega}{c}\right)A\sin\frac{\omega x}{c} + \left(\frac{\omega}{c}\right)B\cos\frac{\omega x}{c}\right] \tag{1}$$

$$\frac{\partial^2\phi}{\partial x^2} = [C\cos\omega t + D\sin\omega t]\frac{\partial}{\partial x}\left[-\left(\frac{\omega}{c}\right)A\sin\frac{\omega x}{c} + \left(\frac{\omega}{c}\right)B\cos\frac{\omega x}{c}\right] = [C\cos\omega t + D\sin\omega t]\left[-\left(\frac{\omega}{c}\right)^2 A\cos\frac{\omega x}{c} - \left(\frac{\omega}{c}\right)^2 B\sin\frac{\omega x}{c}\right] \text{ or}$$

$$\frac{\partial^2\phi}{\partial x^2} = -\left(\frac{\omega}{c}\right)^2[C\cos\omega t + D\sin\omega t]\left(A\cos\frac{\omega x}{c} + B\sin\frac{\omega x}{c}\right) = -\left(\frac{\omega}{c}\right)^2\phi \quad or \quad c^2\frac{\partial^2\phi}{\partial x^2} = -\omega^2\phi \tag{2}$$

$$\frac{\partial\phi}{\partial t} = \left[A\cos\frac{\omega x}{c} + B\sin\frac{\omega x}{c}\right]\frac{\partial}{\partial t}[C\cos\omega t + D\sin\omega t] = \left[A\cos\frac{\omega x}{c} + B\sin\frac{\omega x}{c}\right][-\omega C\sin\omega t + \omega D\cos\omega t] \tag{3}$$

$$\frac{\partial^2\phi}{\partial t^2} = \left[A\cos\frac{\omega x}{c} + B\sin\frac{\omega x}{c}\right]\frac{\partial}{\partial t}[-\omega C\sin\omega t + \omega D\cos\omega t] = \left[A\cos\frac{\omega x}{c} + B\sin\frac{\omega x}{c}\right][-\omega^2 C\cos\omega t - \omega^2 B\sin\omega t] = -\omega^2\phi = c^2\frac{\partial^2\phi}{\partial x^2} \tag{4}$$

The above equation is the desired result.

---

5. 57

Solution L = 5 ft T = 200 in-kips R_i = 1 inch $\tau_{max} \le 10 \ ksi$ R_o = ? nearest 1/8 inch

The maximum torsional shear stress is as shown below.

$$\tau_{max} = \frac{T\rho}{J} = \frac{(200)(R_o)}{\pi[R_o^4 - 1^4]/2} \le 10 \quad or \quad (R_o^4 - 12.732R_o - 1) \ge 0 \tag{1}$$

The minimum value of R_o corresponds to the root of the left hand side i.e., $f(R_o) = R_o^4 - 12.732R_o - 1 = 0$. We find the root on the spread sheet as described in Appendix B. The calculations are shown in the table below.

R_o	$f(R_o)$
1.1	-13.542
1.2	-14.205
1.3	-14.696
1.4	-14.984
1.5	-15.036
1.6	-14.818
1.7	-14.293
1.8	-13.421
1.9	-12.159
2	-10.465
2.1	-8.290
2.2	-5.586
2.3	-2.300
2.4	1.620
2.5	6.232

R_o	$f(R_o)$
2.3	-2.300
2.31	-1.938
2.32	-1.569
2.33	-1.194
2.34	-0.812
2.35	-0.423
2.36	-0.028
2.37	0.374
2.38	0.782
2.39	1.198
2.4	1.620

From the above table, we see that the root of $f(R_o)$ is between 2.36 and 2.37. The value of R_o to the nearest 1/8 is 2.375.

ANS $R_o = 2\frac{3}{8} \ in$

---

5. 58

Solution L = 4 ft. T= 100 in-kips $\Delta\phi \le 0.06 \ rads$ R_i = 1 inch R_o = ? to the nearest 1/8 inch W = ?

The relative rotation of the ends and maximum shear stress in steel can be found as shown below..

$$(\Delta\phi)_S = \frac{(100)(4)(12)}{12(10^3)(\pi/2)[R_o^4 - 1^4]} \le 0.06 \quad or \quad R_o \ge 1.513 \ in \tag{1}$$

$$(\tau_{max})_S = \frac{(100)(R_o)}{(\pi/2)[R_o^4 - 1^4]} \le 18 \quad or \quad R_o^4 - 3.5368R_o - 1 \ge 0 \tag{2}$$

Substituting the value of $R_o = 1.513 \ in$ into the left hand side of second equation we obtain a negative value, which implies R_o must be greater than 1.513. The value of R_o corresponds to the root of the left hand side of second equation. We find the root on the spread sheet as described in Appendix B. The calculations are shown in the table below.

R_0	$f(R_0)$
1.5	-1.243
1.6	-0.105
1.7	1.340
1.8	3.131
1.9	5.312
2	7.926

R_0	$f(R_0)$
1.6	-0.105
1.61	0.025
1.62	0.158
1.63	0.294
1.64	0.434
1.65	0.576

From the above table, we see that the root of $f(R_0)$ is between 1.6 and 1.61. Rounding to the nearest 1/8 inch, we obtain for steel. $R_O = 1.625 inch$

Using similar calculations for Aluminum shaft we obtain the following.

$$(\Delta\phi)_{Al} = \frac{(100)(4)(12)}{(4000)(\pi/2)[r_o^4 - 1^4]} \le 0.06 \qquad or \qquad r_o \ge 1.925\ in \tag{3}$$

$$(\tau_{max})_{Al} = \frac{(100)(r_o)}{(\pi/2)[r_o^4 - 1^4]} \le 10 \qquad or \qquad r_o^4 - 6.366 r_o - 1 \ge 0 \tag{4}$$

Substituting the value of $r_o = 1.925\ in$ into the left hand side of the second equation above we obtain a positive value, which implies that $r_0 = 1.925$inch satisfies both Eqs. (4) and (5). Rounding to the nearest 1/8 inch, we obtain for aluminum $r_o = 2.0 inch$. We can find the weight of each material by taking the product of the material density and the volume of a hollow cylinder as shown below:

$$W_S = (0.285)(\pi)(1.625^2 - 1^2)(48) = 70.51\ lbs \qquad W_{Al} = (0.1)(\pi)(2^2 - 1^2)(48) = 45.24\ lbs \tag{5}$$

We see that the aluminum shaft is lighter.

ANS $r_o = 2.0$ in. ; $W = 45.24$lbs

---

5. 59

Solution G = 28 GPa $x_A = 0$ $x_B = 1.5$ m $\Delta\phi = \phi_B - \phi_A = ?$ $\tau_{max} = ?$

The internal torque is $\boldsymbol{T} = 35$ kN-m.

$$\frac{d\phi}{dx} = \frac{T}{GJ} = \frac{(35)(10^3)}{(28)(10^9)(\pi/2)(R/1000)^4} = \frac{2.5(10^6)}{\pi R^4(x)} \qquad or \qquad \Delta\phi = \int_{\phi_A}^{\phi_B} d\phi = \int_{x_A = 0}^{x_B = 1.5}\left[\frac{2.5(10^6)}{\pi R^4(x)}\right]dx \tag{1}$$

Representing $f(x) = 2.5(10^6)/[\pi R^4(x)]$ and using numerical integration as described in Appendix B, we obtain the value of the integral on a spread sheet as shown below.

x	R(x)	f(x)	$\Delta\phi$
0.00	100.60	0.0078	0.0009
0.10	92.70	0.0108	0.0023
0.20	82.60	0.0171	0.0042
0.30	79.60	0.0198	0.0064
0.40	75.90	0.0240	0.0093
0.50	68.80	0.0355	0.0130
0.60	68.00	0.0372	0.0169
0.70	65.90	0.0422	0.0221

x	R(x)	f(x)	$\Delta\phi$
0.80	60.10	0.0610	0.0282
0.90	60.30	0.0602	0.0344
1.00	59.10	0.0652	0.0424
1.10	54.00	0.0936	0.0515
1.20	54.80	0.0882	0.0605
1.30	54.10	0.0929	0.0718
1.40	49.40	0.1336	0.0846
1.50	50.60	0.1214	

The maximum shear stress will occur at the cross-section just before point B and can be found as shown below.

$$\tau_{max} = \frac{T\rho}{J_B} = \frac{35(10^3)(0.0506)}{(\pi/2)(0.0506^4)} = 171.99(10^6)\ N/m^2 \tag{2}$$

ANS $\Delta\phi = 0.085$ rad; ; $\tau_{max} = 25.8$ ksi

---

5. 60

Solution G = 28 GPa a = ? b = ? $x_A = 0$ $x_B = 1.5$ m $\Delta\phi = \phi_B - \phi_A = ?$

The error e_i that can be written as given below.

$$e_i = R_i - R(x_i) = R_i - (a + bx_i) \tag{1}$$

We define the error as $E = \sum_{i=1}^{N} e_i^2$. The error E is minimized with respect to coefficients a, and b to generate a set of linear algebraic equations as shown below.

$\frac{\partial E}{\partial a} = \sum 2e_i\frac{\partial e_i}{\partial a} = 0 \quad or \quad \sum 2[R_i-(a+bx_i)][-1] \qquad \frac{\partial E}{\partial b} = \sum 2e_i\frac{\partial e_i}{\partial b} = 0 \quad or \quad \sum 2[R_i-(a+bx_i)][-x_i] = 0$ **(2)**

The above equations on the right can be rearranged and written in matrix form as shown below:

$$\begin{bmatrix} N & \sum x_i \\ \sum x_i & \sum x_i^2 \end{bmatrix}\begin{Bmatrix} a \\ b \end{Bmatrix} = \begin{Bmatrix} \sum R_i \\ \sum x_i R_i \end{Bmatrix} \quad or \quad \begin{bmatrix} b_{11} & b_{12} \\ b_{21} & b_{22} \end{bmatrix}\begin{Bmatrix} a \\ b \end{Bmatrix} = \begin{Bmatrix} r_1 \\ r_2 \end{Bmatrix} \tag{3}$$

The coefficients a and b can be determined by Cramer's rule as shown below.

$$D = b_{11}b_{22} - b_{12}b_{21} \tag{4}$$

$$a = \left\| \begin{bmatrix} r_1 & b_{12} \\ r_2 & b_{22} \end{bmatrix} \right\| / D = \frac{r_1 b_{22} - r_2 b_{12}}{D} \qquad b = \left\| \begin{bmatrix} b_{11} & r_1 \\ b_{21} & r_2 \end{bmatrix} \right\| / D = \frac{r_2 b_{11} - r_1 b_{21}}{D} \tag{5}$$

The given data and the above equations are put in a spread sheet and the coefficients a and b found as shown in the table below.

	x_i (m)	R_i (mm)	x_i^2	x_i*R_i
1	0.00	100.60	0.0000	0.000
2	0.10	92.70	0.0100	9.270
3	0.20	82.60	0.0400	16.520
4	0.30	79.60	0.0900	23.880
5	0.40	75.90	0.1600	30.360
6	0.50	68.80	0.2500	34.400
7	0.60	68.00	0.3600	40.800
8	0.70	65.90	0.4900	46.130
9	0.80	60.10	0.6400	48.080
10	0.90	60.30	0.8100	54.270

	x_i (m)	R_i (mm)	x_i^2	x_i*R_i
11	1.00	59.10	1.0000	59.100
12	1.10	54.00	1.2100	59.400
13	1.20	54.80	1.4400	65.760
14	1.30	54.10	1.6900	70.330
15	1.40	49.40	1.9600	69.160
16	1.50	50.60	2.2500	75.900
b_{ij} & r_i	12.0000	1076.50	12.4000	703.360
D	54.4			
a and b	90.226	-30.593		

Thus R(x) = 90.226-30.593x, where x is in meters and R is in millimeters. Substituting we obtain the rotation as shown below.

$$\Delta\phi = \int_0^{1.5}\left[\frac{2.5(10^6)}{\pi(90.226-30.593x)^4}\right]dx = \left(\frac{2.5(10^6)}{\pi}\right)\left[\frac{1}{(3)(30.593)(90.226-30.593x)^3}\right]\Bigg|_0^{1.5} = 0.0877 \tag{6}$$

ANS $a = 90.226\ mm;\ b = -30.593;\ \Delta\phi = 0.088\ rad$

---

5. 61

Solution G = 12,000 ksi d= 1 inch $u_A = ?$ $\sigma_{max} = ?$

The solution proceeds as shown below.

$$J = \pi d^4/32 = 0.0981 in^4 \qquad \frac{d\boldsymbol{T}}{dx} + t(x) = 0 \quad or \quad \boldsymbol{T}(x) = \int_{\boldsymbol{T}_A=0}^{\boldsymbol{T}(x)} d\boldsymbol{T} = -\int_{x_A=0}^{x} t(x)dx \tag{1}$$

$$\frac{d\phi}{dx} = \frac{\boldsymbol{T}}{GJ} = \frac{\boldsymbol{T}}{(12)(10^6)(0.0981)} = 848.8(10^{-9})\boldsymbol{T} \quad or \quad -\phi_A = \int_{\phi_A}^{\phi_B=0} d\phi = 848.8(10^{-9})\int_{x_A=0}^{x_B=36}\boldsymbol{T}dx \tag{2}$$

$$\tau_{x\theta} = \boldsymbol{T}\rho/J = \boldsymbol{T}(0.5)/0.0981 = 5.0929\boldsymbol{T} \tag{3}$$

The first equation is numerically integrated to obtain $\boldsymbol{T}(x_i)$. Then the next equation is numerically integrated to obtain the rotation. The stress equation is used to find the torsional shear stress at various x_i and the maximum value chosen by inspection.

x_i (inches)	$t(x_i)$ (in-lbs./in)	$\boldsymbol{T}(x_i)$ (in-lbs.)	$[\phi_A-\phi(x_{i+1})](10^{-3})$ (rads)	$\tau_{x\theta}(x_i)$ (ksi)
0	93	0.00	0.456	0.000
3	146.0	-358.50	2.057	-1.826
6	214.1	-898.65	5.251	-4.577
9	260.0	-1609.80	10.487	-8.199
12	335.0	-2502.30	18.310	-12.744
15	424.7	-3641.85	29.334	-18.548
18	492.0	-5016.90	44.174	-25.551
21	588.8	-6638.10	63.539	-33.808
24	700.1	-8571.45	88.211	-43.654
27	789.6	-10806.00	118.970	-55.034
30	907.4	-13351.50	156.689	-67.999
33	1040.3	-16273.05	202.314	-82.878
36	1151.4	-19560.60		-99.621

ANS $\phi_A = 0.202$ rads ; $\tau_{max} = 99.6$ ksi

5. 62

Solution a = ? b = ? c = $\phi_A = ?$

The solution proceeds as follows.

$$\frac{\mathrm{d}\boldsymbol{T}}{\mathrm{d}x} + t(x) = 0 \qquad or \qquad \boldsymbol{T}(x) = \int_{\boldsymbol{T}_A = 0}^{\boldsymbol{T}(x)} d\boldsymbol{T} = -\int_{x_A = 0}^{x} t(x)dx = -\int_{x_A = 0}^{x} [a + bx + cx^2]dx = -\left[ax + \frac{bx^2}{2} + \frac{cx^3}{3}\right] \quad (1)$$

$$J = \pi d^4/32 = 0.0981 in^4 \qquad \frac{\mathrm{d}\phi}{\mathrm{d}x} = \frac{\boldsymbol{T}(x)}{GJ} = \frac{\boldsymbol{T}(x)}{(12)(10^6)(0.0981)} = -848.82(10^{-9})\left[ax + \frac{bx^2}{2} + \frac{cx^3}{3}\right] \quad (2)$$

$$\int_{\phi_A}^{\phi_B = 0} d\phi = \int_{x_A = 0}^{x_B = 36} (-848.82)(10^{-9})\left[ax + \frac{bx^2}{2} + \frac{cx^3}{3}\right]dx \text{ or}$$

$$\phi_A = (848.82)(10^{-9})\left(\frac{ax^2}{2} + \frac{bx^3}{6} + \frac{cx^4}{12}\right)\Bigg|_0^{36} = (848.82)(10^{-9})\left[\frac{a(36)^2}{2} + \frac{b(36)^3}{6} + \frac{c(36)^4}{12}\right] \quad (3)$$

Using the Least Square Method described in Appendix B, we obtain the value of the values of constants a, b, and c on a spread sheet as shown in the table below.

	x_i	$t(x_i)$	x_i^2	x_i^3	x_i^4	$x*t_i$	$x_i^2*t_i$
1	0.0	93	0.0	0.000E+00	0.000E+00	0.000E+00	0.000E+00
2	3.0	146.0	9.0	2.700E+01	8.100E+01	4.380E+02	1.314E+03
3	6.0	214.1	36.0	2.160E+02	1.296E+03	1.285E+03	7.708E+03
4	9.0	260.0	81.0	7.290E+02	6.561E+03	2.340E+03	2.106E+04
5	12.0	335.0	144.0	1.728E+03	2.074E+04	4.020E+03	4.824E+04
6	15.0	424.7	225.0	3.375E+03	5.063E+04	6.371E+03	9.556E+04
7	18.0	492.0	324.0	5.832E+03	1.050E+05	8.856E+03	1.594E+05
8	21.0	588.8	441.0	9.261E+03	1.945E+05	1.236E+04	2.597E+05
9	24.0	700.1	576.0	1.382E+04	3.318E+05	1.680E+04	4.033E+05
10	27.0	789.6	729.0	1.968E+04	5.314E+05	2.132E+04	5.756E+05
11	30.0	907.4	900.0	2.700E+04	8.100E+05	2.722E+04	8.167E+05
12	33.0	1040.3	1089.0	3.594E+04	1.186E+06	3.433E+04	1.133E+06
13	36.0	1151.4	1296.0	4.666E+04	1.680E+06	4.145E+04	1.492E+06
b_{ij} &r_i	234.0	7142.4	5850.0	1.643E+05	4.918E+06	1.768E+05	5.014E+06
C_{ij}	1.783E+09	1.897E+08	4.216E+06	2.971E+07	7.666E+05	2.129E+04	
D	3.453E+09						
a_i	96.36	15.45	0.39				

Substituting the values of the constants we obtain:

$$\phi_A = (848.82)(10^{-9})\left[\frac{(96.36)(36)^2}{2} + \frac{(15.45)(36)^3}{6} + \frac{(0.39)(36)^4}{12}\right] = 0.2013 \quad (4)$$

ANS $a = 96.36$; $b = 15.45$; $c = 0.39$; $\phi_A = 0.2013$ rads

---

Section 5.3

5. 63

Solution $G_{AB} = 12000\ ksi$ $G_{BC} = 5600 ksi$ $T = 50 in\text{–}kips$ $\tau_{max} = ?$ $\phi_B = ?$

By moment equilibrium about the shaft axis we obtain the internal torques.

$$J = (\pi 2^4)/32 = 1.5708 in^4 \qquad \boldsymbol{T}_{AB} = T_A \qquad \boldsymbol{T}_{BC} = T_A - T \quad (1)$$

(a) 

(b) 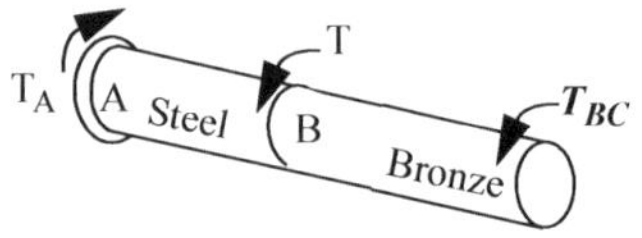

The relative rotations of the ends of the each segment can be written and added and equated to zero as $\phi_C = 0$ and $\phi_A = 0$

$$\phi_B - \phi_A = \frac{\boldsymbol{T}_{AB}(x_B - x_A)}{G_{AB}J} = \frac{T_A(2)(12)}{(12000)(1.5708)} = 1.273(10^{-3})T_A \quad (2)$$

$$\phi_C - \phi_B = \frac{\boldsymbol{T}_{BC}(x_C - x_B)}{G_{BC}J} = \frac{(T_A - 50)(4)(12)}{(5600)(1.5708)} = 5.456(10^{-3})(T_A - T) \quad (3)$$

$$\phi_C - \phi_A = [1.273T_A + 5.456(T_A - T)](10^{-3}) = 0 \qquad or \qquad T_A = \frac{5.456}{(1.273 + 5.456)}T = (0.8108)(50) = 40.541 in - kips \quad (4)$$

$$T_{AB} = 40.541 \qquad T_{BC} = -9.459 in-kips \qquad \phi_B = 1.273(10^{-3})(40.541) = 51.608(10^{-3}) rads \tag{5}$$

$$\tau_{max} = \left|\frac{T_{AB}\rho_{max}}{J}\right| = \frac{(40.541)(1)}{1.5708} \tag{6}$$

ANS $\phi_B = 0.0516$ rads CCW ; $\tau_{max} = 25.8$ ksi

---

5. 64

Solution $\phi_B = 0.02\ rads \qquad \gamma_{max} = ? \qquad T = ?$

The rate of twist in each segment can be found as shown below:

$$\left(\frac{d\phi}{dx}\right)_{AB} = \frac{\phi_B - \phi_A}{x_B - x_A} = \frac{0.02 - 0}{24} = 0.833(10^{-3})\ \frac{rads}{in} \qquad \left(\frac{d\phi}{dx}\right)_{BC} = \frac{\phi_C - \phi_B}{x_C - x_B} = \frac{0 - 0.02}{48} = -0.4167(10^{-3})\ \frac{rads}{in} \tag{1}$$

The maximum torsional shear strain will be in AB and can be found as shown

$$\gamma_{max} = \rho_{max}\left(\frac{d\phi}{dx}\right)_{AB} = (1)(0.833)(10^{-3}) \tag{2}$$

Substituting $\phi_B = 0.02$, $\phi_A = 0$, and $\phi_C = 0$ in equations in previous problem we obtain the following.

$$\phi_B - \phi_A = 1.273(10^{-3})T_A \qquad or \qquad T_A = \frac{0.02}{1.273(10^{-3})} = 15.711\ in-kips \tag{3}$$

$$\phi_C - \phi_B = 5.456(10^{-3})(T_A - T) \qquad or \qquad (15.711 - T) = \frac{-0.02}{5.456(10^{-3})} \qquad or \qquad T = 19.377\ in-kips \tag{4}$$

ANS $\gamma_{max} = 833$ μrads ; $T = 19.4$ in-kips

---

5. 65

Solution $G = 10000\ ksi \qquad T = 300\ in\text{–}kips \qquad \phi_C - \phi_A = \ ? \qquad \gamma_E = \ ?$

By moment equilibrium about the shaft axis we obtain the internal torques.

$$J_{AB} = J_{CD} = \frac{\pi}{32}(4^4 - 2^4) = 23.562\ in^4 \qquad J_{BC} = \frac{\pi}{32}4^4 = 25.133\ in^4 \qquad T_{AB} = T_A \qquad T_{BC} = T_A - T \qquad T_{CD} = T_A - T \tag{1}$$

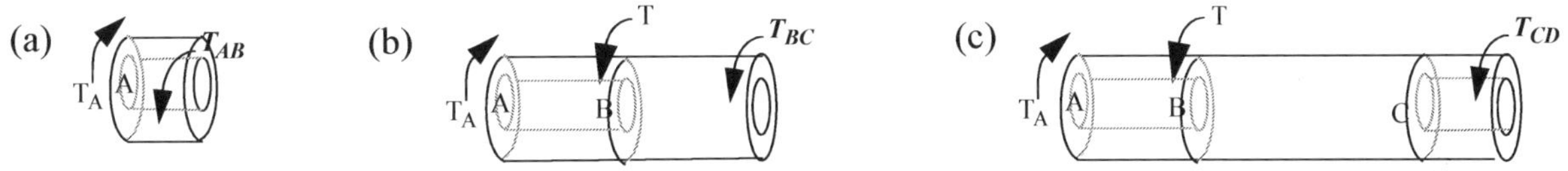

The relative rotation of the ends of each segment can be written, added and equated to zero as $\phi_D = 0$ and $\phi_A = 0$.

$$\phi_B - \phi_A = \frac{T_{AB}(x_B - x_A)}{GJ_{AB}} = \frac{(T_A)(24)}{(10000)(23.562)} = 0.1019(10^{-3})T_A \tag{2}$$

$$\phi_C - \phi_B = \frac{T_{BC}(x_C - x_B)}{GJ_{BC}} = \frac{(T_A - T)(36)}{(10000)(25.133)} = 0.1432(10^{-3})(T_A - T) \tag{3}$$

$$\phi_D - \phi_C = \frac{T_{CD}(x_D - x_C)}{GJ_{CD}} = \frac{(T_A - T)(24)}{(10000)(23.562)} = 0.1019(10^{-3})(T_A - T) \tag{4}$$

$$\phi_D - \phi_A = [0.1019T_A + 0.1432(T_A - T) + 0.1019(T_A - T)](10^{-3}) = 0 \textbf{ or} \tag{5}$$

$$T_A = \frac{(0.1527 + 0.1019)T}{0.1019 + 0.1432 + 0.1019} = 0.7064T = 211.93 \tag{6}$$

$$\phi_C - \phi_A = [0.1019(211.93) + 0.1432(211.93 - 300)](10^{-3}) = 8.984(10^{-3}) \qquad T_{CD} = -88.07\ in-kips \tag{7}$$

$$\tau_E = \frac{T_{CD}\rho_E}{J_{CD}} = \frac{(-88.07)(1)}{23.562} = -3.738\ ksi \qquad \gamma_E = \frac{\tau_E}{G} = \frac{-3.738}{10000} = -0.3738(10^{-3}) \tag{8}$$

ANS $\phi_C - \phi_A = 0.00898$ rads CCW ; $\gamma_E = -374$ μrads

---

5. 66

Solution $G = 10000\ ksi \qquad \gamma_E = -250\ \mu rads \qquad \phi_C = ? \qquad T = ?$

The rotation of the section at C can be found as shown below

$$\gamma_E = \rho_E\left(\frac{d\phi}{dx}\right)_{CD} = (1)\frac{(\phi_D - \phi_C)}{x_D - x_C} = (1)\frac{(0 - \phi_C)}{24} = (-250)(10^{-6}) \qquad or \qquad \phi_C = 6(10^{-3}) rads \tag{1}$$

Substituting $\phi_D = 0$ and $\phi_C = 0.006$ into equation of previous problem, we obtain

$$T_A = 0.7064\ T \qquad T_A - T = \frac{\phi_D - \phi_C}{0.1019(10^{-3})} = -\left(\frac{0.006}{0.1019(10^{-3})}\right) = -58.88 \qquad or \qquad T = \frac{-58.88}{(0.7064-1)} = 200.5\ \text{in.-kips} \quad \textbf{(2)}$$

ANS $\phi_C = 0.006$ rads CCW; ; $T = 200.5$ in.-kips

---

5. 67

Solution $G_{AB} = G_{BC} = 80\ GPa$ $G_{CD} = 40\ GPa$ $T = 10\ kN-m$ $\phi_B = ?$

By moment equilibrium about the shaft axis we obtain the internal torques.

$$J = \pi(0.1)^4/32 = 9.817(10^{-6})\ m^4 \qquad \boldsymbol{T}_{AB} = -T_A \qquad \boldsymbol{T}_{BC} = T - T_A \qquad \boldsymbol{T}_{CD} = T - T_A \quad \textbf{(1)}$$

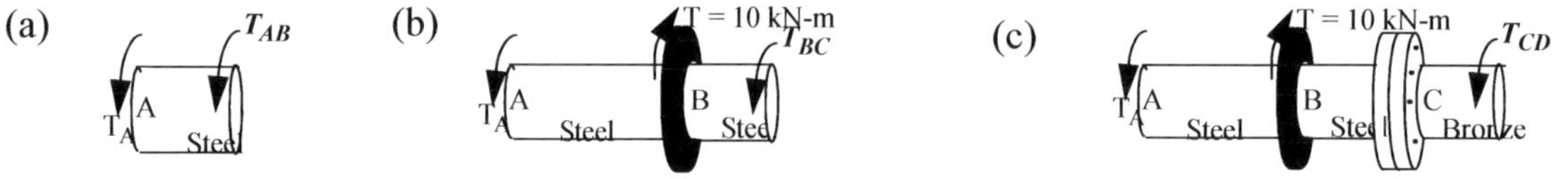

The relative rotation of the ends of the each segment can be written, added and equated to zero as $\phi_D = 0$ and $\phi_A = 0$.

$$\phi_B - \phi_A = \frac{\boldsymbol{T}_{AB}(x_B - x_A)}{G_{AB}J} = \frac{\boldsymbol{T}_A(5)}{(80)(10^9)(9.817)(10^{-6})} = 6.366(10^{-6})\boldsymbol{T}_A \quad \textbf{(2)}$$

$$\phi_C - \phi_B = \frac{\boldsymbol{T}_{BC}(x_C - x_B)}{G_{BC}J} = \frac{(T-\boldsymbol{T}_A)(3)}{(80)(10^9)(9.817)(10^{-6})} = 3.8199(10^{-6})(T - \boldsymbol{T}_A) \quad \textbf{(3)}$$

$$\phi_D - \phi_C = \frac{\boldsymbol{T}_{CD}(x_D - x_C)}{G_{CD}J} = \frac{(T-\boldsymbol{T}_A)(4)}{(40)(10^9)(9.817)(10^{-6})} = 10.1864(10^{-6})(T - \boldsymbol{T}_A) \quad \textbf{(4)}$$

$$\phi_D - \phi_A = [-6.3661T_A + 3.8199(T - T_A) + 10.1864(T - T_A)](10^{-6}) = 0 \text{ or}$$

$$T_A = \frac{(3.8199 + 10.1864)T}{(6.3661 + 3.8199 + 10.1864)} = 0.6875T = 6.875\ kN-m \quad \textbf{(5)}$$

$$\phi_B = 6.366(10^{-6})\boldsymbol{T}_A = (-6.366)(10^{-6})(6.875)(10^3) = -43.77(10^{-3}) \quad \textbf{(6)}$$

ANS $\phi_B = 0.0438$ rads CW

---

5. 68

Solution $G_{AB} = G_{BC} = 80\ GPa$ $G_{CD} = 40\ GPa$ $T = 10\ kN-m$ $|\phi_{C_2} - \phi_{C_1}| = 0.5^o$ [$\phi_B = ?$

The coupling plate C_1 will rotate clockwise relative to the coupling plate C_2 ,hence the relative rotation of plate C_2 with respect to plate C_1 will be positive.We can write the following

$$\phi_{C_2} - \phi_{C_1} = (0.5\pi)/180 = 8.7267(10^{-3}) \quad \textbf{(1)}$$

From equations in previous problem we have the following.

$$\phi_B - \phi_A = 6.366(10^{-6})\boldsymbol{T}_A \qquad \phi_{C1} - \phi_B = 3.8199(10^{-6})(T - \boldsymbol{T}_A) \qquad \phi_D - \phi_{C2} = 10.1864(10^{-6})(T - \boldsymbol{T}_A) \quad \textbf{(2)}$$

Adding the above four equations and noting that $\phi_D = 0$ and $\phi_A = 0$, we obtain

$$\phi_D - \phi_A = [-6.3661\boldsymbol{T}_A + 3.8199(T - \boldsymbol{T}_A) + 10.1864(T - \boldsymbol{T}_A)](10^{-6}) + 8.7267(10^3) = 0 \text{ or}$$

$$\boldsymbol{T}_A = \frac{(3.8199 + 10.1864)T + 8.7267(10^3)}{(6.3661 + 3.8199 + 10.1864)} = \frac{(3.8199 + 10.1864)10(10^3) + 8.7267(10^3)}{(6.3661 + 3.8199 + 10.1864)} = 7.303(10^3)\ N-m \quad \textbf{(3)}$$

$$\phi_B = 6.366(10^{-6})\boldsymbol{T}_A = (-6.366)(10^{-6})(7.303)(10^3) = -46.497(10^{-3})\ rads \quad \textbf{(4)}$$

ANS $\phi_B = -0.0465$ rads CW

---

5. 69

Solution $G_{AB} = 40GPa$ $G_{BC} = 80\ GPa$ $T = 10\ kN-m$ $\tau_{max} = ?$ $\phi_1 = ?$

By moment equilibrium about the shaft axis we obtain the internal torques.

$$J_{AB} = \frac{\pi}{32}(0.12^4) = 20.358(10^{-6})\ m^4 \qquad J_{BC} = \frac{\pi}{32}(0.08)^4 = 4.021(10^{-6})\ m^4 \qquad T_{AB} = T_A \qquad T_{BC} = T_A - T \qquad (1)$$

The relative rotation of the ends of the each segment can be written, added and equated to zero as $\phi_C = 0$ and $\phi_A = 0$.

$$\phi_B - \phi_A = \frac{T_{AB}(x_B - x_A)}{G_{AB}J_{AB}} = \frac{T_A(2)}{(40)(10^9)(20.358)(10^{-6})} = 2.456(10^{-6})T_A \qquad (2)$$

$$\phi_C - \phi_B = \frac{T_{BC}(x_C - x_B)}{G_{BC}J_{BC}} = \frac{(T - T_A)(1)}{(80)(10^9)(4.021)(10^{-6})} = 3.109(10^{-6})(T_A - T) \qquad (3)$$

$$\phi_C - \phi_A = [2.456T_A + 3.109(T_A - T)](10^{-6}) = 0 \qquad or \qquad T_A = \frac{3.109T}{2.456 + 3.109} = 0.5586T = 5.586(10^3)N-m \text{ or}$$

$$T_{AB} = 5.586(10^3)N-m \qquad (\tau_{AB})_{max} = \frac{T_{AB}(\rho_{AB})_{max}}{J_{AB}} = \frac{(5.586)(10^3)(0.06)}{20.358(10^{-6})} = 16.46(10^6)\frac{N}{m^2} \qquad (4)$$

$$T_{BC} = -4.413(10^3)N-m \qquad (\tau_{BC})_{max} = \frac{T_{BC}(\rho_{BC})_{max}}{J_{BC}} = \frac{(-4.413)(10^3)(0.04)}{4.021(10^{-6})} = -43.9(10^6)\frac{N}{m^2} \qquad (5)$$

$$\phi_1 - \phi_A = \frac{T_{AB}(x_1 - x_A)}{G_{AB}J_{AB}} = \frac{(5.586)(10^3)(1)}{(40)(10^9)(20.358)(10^{-6})} = 6.86(10^{-3})\ rads \qquad (6)$$

ANS $\tau_{max} = 43.9$ MPa ; $\phi_1 = 0.0069$ rads CCW

---

5. 70

Solution $G_{AB} = 80GPa \quad G_{BC} = 40\ GPa \quad \phi_B = 0.05\ rads \qquad \tau_{max} = ? \qquad T = ?$

The rate of twist in the segment AB and BC can be found as

$$\left(\frac{d\phi}{dx}\right)_{AB} = \frac{\phi_B - \phi_A}{x_B - x_A} = \frac{0.05}{2} = 0.025\ \frac{rads}{m} \qquad \left(\frac{d\phi}{dx}\right)_{BC} = \frac{\phi_C - \phi_B}{x_C - x_B} = \frac{0 - 0.05}{1} = -0.05\ \frac{rads}{m} \qquad (1)$$

$$(\gamma_{AB})_{max} = (\rho_{AB})_{max}\left(\frac{d\phi}{dx}\right)_{AB} = (0.06)(0.025) = 1.5(10^{-3}) \qquad (\gamma_{BC})_{max} = (\rho_{BC})_{max}\left(\frac{d\phi}{dx}\right)_{BC} = (0.04)(-0.05) = -2.0(10^{-3}) \qquad (2)$$

Substituting $\phi_B = 0.05$ and $\phi_A = 0$ in equations of previous problem we obtain the following.

$$T_A = \frac{\phi_B - \phi_A}{2.456(10^{-6})} = \frac{0.05}{2.456(10^{-6})} = 20.36(10^3)N-m \qquad T = \frac{T_A}{0.5586} = \frac{20.36(10^3)}{0.5586} = 36.44(10^3)N-m \qquad (3)$$

ANS $\gamma_{max} = -2000$ μ rads ; $T = 36.44$ kN-m

---

5. 71

Solution $\tau_{max} = f(T, L, G, d) = ?$ $\qquad \phi_B = g(T, L, G, d) = ?$

By moment equilibrium about the shaft axis we have the following equations

$$T_{AB} = T_A \qquad T_{BC} = T_A - T \qquad (1)$$

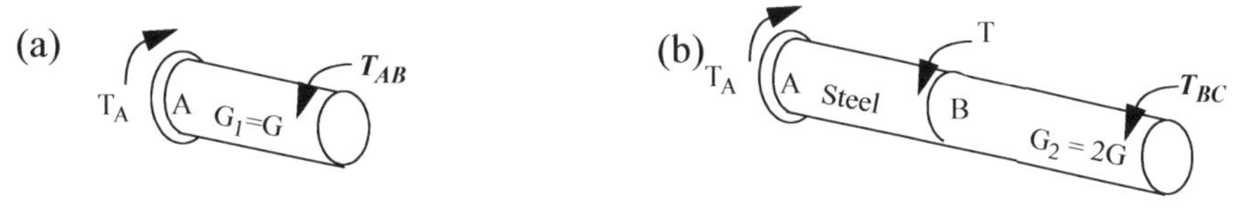

The relative rotations of each segment can be written, added, and equated to zero as $\phi_C = 0$ and $\phi_A = 0$.

$$\phi_B - \phi_A = \frac{T_{AB}(x_B - x_A)}{G_{AB}J} = \frac{T_A(L)}{GJ} \qquad \phi_C - \phi_B = \frac{T_{BC}(x_C - x_B)}{G_{BC}J} = \frac{(T_A - T)(2.5L)}{2GJ} = \frac{1.25}{GJ}(T_A - T)L \qquad (2)$$

$$\phi_C - \phi_A = \frac{L}{GJ}[T_A + 1.25(T_A - T)] = 0 \qquad or \qquad T_A = \frac{1.25}{1.25 + 1}T = \frac{5}{9}T \qquad (3)$$

$$T_{AB} = \frac{5}{9}T \qquad T_{BC} = -\frac{4}{9}T \qquad \phi_B = \frac{5TL}{9GJ}\ CCW \qquad \tau_{max} = \frac{T_{AB}(\rho_{AB})_{max}}{J} = \frac{(5T/9)(d/2)}{(\pi d^3/32)} = \frac{80}{9\pi}\frac{T}{d^3} \qquad (4)$$

ANS $\phi_B = 5.659\frac{TL}{Gd^4}CCW$; $\tau_{max} = 2.83\ \frac{T}{d^3}$

---

5. 72

Solution $\tau_{max} = f(q, L, G, r) = ?$ $\phi_B = g(q, L, G, r) = ?$

By moment equilibrium about the shaft axis we have the following equations.

$$T_{AB} + qx - T_A = 0 \quad or \quad T_{AB} = T_A - qx \qquad T_{BC} + qx - T_A - 3qL = 0 \quad or \quad T_{BC} = T_A - qx + 3qL \tag{5}$$

(a)

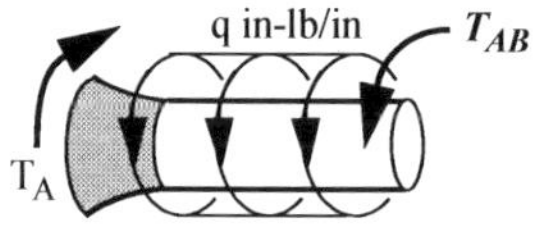

(b)

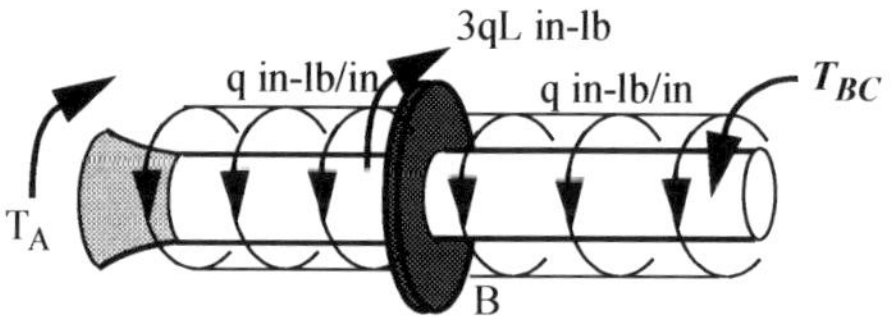

The solution proceeds as follows.

$$\left(\frac{d\phi}{dx}\right)_{AB} = \frac{T_{AB}}{GJ} = \frac{(T_A - qx)}{GJ} \qquad \phi_B - \phi_A = \int_{\phi_A}^{\phi_B} d\phi = \int_{x_B=0}^{x_B=L} \frac{(T_A - qx)}{GJ}dx = \frac{1}{GJ}\left(T_A x - \frac{q}{2}x^2\right)\Big|_0^L = \frac{1}{GJ}\left(T_A L - \frac{q}{2}L^2\right) \tag{6}$$

$$\left(\frac{d\phi}{dx}\right)_{BC} = \frac{T_{BC}}{GJ} = \frac{T_A - qx + 3qL}{GJ} \qquad \phi_c - \phi_B = \int_{\phi_B}^{\phi_C} d\phi = \int_{x_B=L}^{x_C=3L} \frac{T_A - qx + 3qL}{GJ}dx = \frac{1}{GJ}\left(T_A x - \frac{q}{2}x^2 + 3qLx\right)\Big|_L^{3L} = \frac{2T_A L + 2qL^2}{GJ} \tag{7}$$

$$\phi_C - \phi_A = \frac{L}{GJ}\left(T_A L - \frac{q}{2}L^2 + 2T_A L + 2qL^2\right) = 0 \quad or \quad T_A = -q\frac{L}{2} \qquad \phi_B = \frac{1}{GJ}\left(-\frac{q}{2}L^2 - \frac{q}{2}L^2\right) = \frac{-qL^2}{G(\pi r^4/2)} \tag{8}$$

$$T_{AB} = -\frac{q}{2}\left(x + \frac{L}{2}\right) \qquad T_{BC} = q\left(\frac{5}{2}L - x\right) \qquad (T_{AB})_{max} = -\left(\frac{3}{2}qL\right) \qquad (T_{BC})_{max} = \frac{3}{2}qL \tag{9}$$

$$\tau_{max} = \frac{(T_{BC})_{max}(\rho_{BC})_{max}}{J_{BC}} = \frac{((3/2)qL)r}{(\pi r^4/2)} \tag{10}$$

ANS $\phi_B = \dfrac{2qL^2}{\pi G r^4}$ CW ; $\tau_{max} = \dfrac{3qL}{\pi r^3}$

---

5. 73

Solution $G_{AB} = 40 GPa$ $G_{BC} = 80\ GPa$ $(\tau_{AB})_{max} \le 60\ MPa$ $(\tau_{BC})_{max} \le 160\ MPa$ $T_{ext} = ?$

By moment equilibrium about the shaft axis we obtain the internal torques.

$$J_{AB} = \pi(0.12^4)/32 = 20.358(10^{-6})\ m^4 \qquad J_{BC} = \pi(0.08)^4/32 = 4.021(10^{-6})\ m^4 \qquad T_{AB} = T_A \qquad T_{BC} = T_A - T \tag{1}$$

(a)

(b)

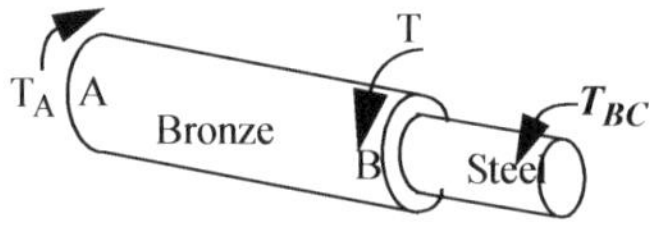

The relative rotation of the ends of the each segment can be written, added, and equated to zero as $\phi_C = 0$ and $\phi_A = 0$

$$\phi_B - \phi_A = \frac{T_{AB}(x_B - x_A)}{G_{AB}J_{AB}} = \frac{T_A(2)}{(40)(10^9)(20.358)(10^{-6})} = 2.456(10^{-6})T_A \tag{2}$$

$$\phi_C - \phi_B = \frac{T_{BC}(x_C - x_B)}{G_{BC}J_{BC}} = \frac{(T - T_A)(1)}{(80)(10^9)(4.021)(10^{-6})} = 3.109(10^{-6})(T_A - T) \tag{3}$$

$$\phi_C - \phi_A = [2.456T_A + 3.109(T_A - T)](10^{-6}) = 0 \qquad T_A = \frac{3.109T}{2.456 + 3.109} = 0.5586T$$

$$T_{AB} = 0.5586T \qquad (\tau_{AB})_{max} = \left|\frac{T_{AB}(\rho_{AB})_{max}}{J_{AB}}\right| = \frac{(0.5586T)(0.06)}{20.358(10^{-6})} \le 60(10^6) \quad or \quad T \le 36444.6 \text{ N-m} \tag{4}$$

$$T_{BC} = -0.4414T \qquad (\tau_{BC})_{max} = \left|\frac{T_{BC}(\rho_{BC})_{max}}{J_{BC}}\right| = \frac{(0.4414T)\ (0.04)}{4.021(10^{-6})} \le 160(10^6) \quad or \quad T \le 36438.6 \text{ N-m} \tag{5}$$

ANS $T_{max} = 36438$ N-m

---

5. 74

Solution $G_{AB} = 80$ GPa $G_{BC} = 40$ GPa $(\tau_{AB})_{max} \le 160\ MPa$ $(\tau_{BC})_{max} \le 60\ MPa$

$\phi_B \le 0.05\ rads$ d = 100 mm T_{max} =? nearest kN τ_{max} = ? ϕ_B =?

By moment equilibrium about the shaft axis we have the following equations.

$$J = \pi(0.1)^4/32 = 9.817(10^{-6})\ m^4 \qquad \boldsymbol{T}_{AB} = T_A \qquad \boldsymbol{T}_{BC} = T_A - T \tag{1}$$

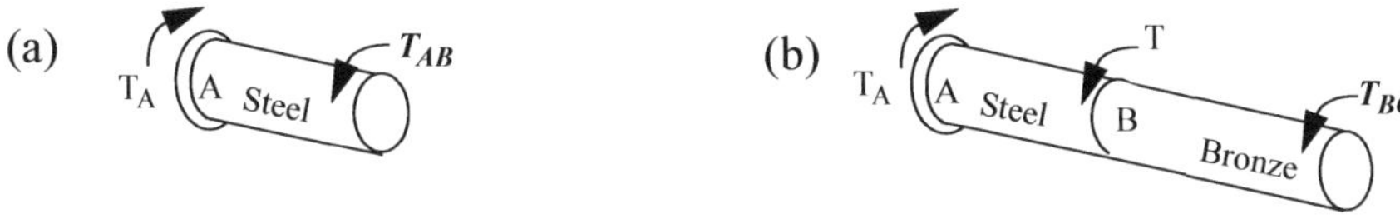

The relative rotation of the ends of each segment can be written, added, and equated to zero as ϕ_C=0 and ϕ_A=0

$$\phi_B - \phi_A = \frac{\boldsymbol{T}_{AB}(x_B - x_A)}{G_{AB}J} = \frac{1.5T_A}{80(10^9)J} \qquad \phi_C - \phi_B = \frac{\boldsymbol{T}_{BC}(x_C - x_B)}{G_{BC}J} = \frac{3(T_A - T)}{40(10^9)J} \tag{2}$$

$$\phi_C - \phi_A = \frac{1}{40(10^9)J}\left[\frac{1.5}{2}T_A + 3(\boldsymbol{T}_A - T)\right] = 0 \qquad or \qquad T_A = \frac{3}{0.75+3}T = 0.8T \tag{3}$$

$$\phi_B = \frac{1.5T_A}{80(10^9)J} = \frac{(1.5)(0.8T)}{(80)(10^9)(9.817)(10^{-6})} = 1.528T(10^{-6}) \le 0.05 \qquad or \qquad T \le 32.72(10^3)\ N-m \tag{4}$$

$$\boldsymbol{T}_{AB} = 0.8T \qquad (\tau_{AB})_{max} = \frac{(0.8T)(0.05)}{9.817(10^{-6})} = 4.074(10^3)T \le 160(10^6) \qquad or \qquad T \le 39.27(10^3)\ N-m \tag{5}$$

$$\boldsymbol{T}_{BC} = -0.2T \qquad (\tau_{BC})_{max} = \frac{(0.2T)(0.05)}{9.817(10^{-6})} = 1.018(10^3)T \le 60(10^6) \qquad or \qquad T \le 58.90(10^3)\ N-m \tag{6}$$

$$T_{max} = 32(10^3)\text{ N-m} \qquad \phi_B = 1.528T(10^{-6}) = 48.896(10^{-3})\ rads \tag{7}$$

$$(\tau_{AB})_{max} = 4.074(10^3)T = 130.37(10^6)\ N/m^2 \qquad (\tau_{BC})_{max} = 1.018(10^3)T = 32.6(10^6)\ N/m^2 \tag{8}$$

ANS T_{max} = 32 kN-m ; ϕ_B = 0.048 rads CCW ; τ_{max} = 130.4 MPa

---

5. 75

Solution G_{AB} = 80 GPa G_{BC} = 40 GPa $(\tau_{AB})_{max} \le 160\ MPa$ $(\tau_{BC})_{max} \le 60\ MPa$ $\phi_B \le 0.05\ rads$

T = 20 kN-m d_{min} =? nearest mm τ_{max} = ? ϕ_B =?

From previous problem we have

$$T_A = 0.8T = 16\ kN-m \qquad \phi_B = \frac{(1.5)(16)(10^3)}{(80)(10^9)\pi d^4/32} = \frac{3.056}{d^4}(10^{-6}) \le 0.05 \qquad or \qquad d \ge 88.417(10^{-3})\ m \tag{1}$$

$$\boldsymbol{T}_{AB} = T_A = 16\ kN-m \qquad \boldsymbol{T}_{BC} = T_A - T = 16 - 20 = -4kN-m \tag{2}$$

$$(\tau_{AB})_{max} = \left|\frac{(16)(10^3)(\ )}{(\pi d^4/32)}\right| = \frac{81.487(10^3)}{d^3} \le 160(10^6) \qquad or \qquad d \ge 79.86(10^{-3})\ m \tag{3}$$

$$(\tau_{BC})_{max} = \left|\frac{(-4)(10^3)(d/2)}{(\pi d^4/32)}\right| = \frac{20.37(10^3)}{d^3} \le 60(10^6) \qquad or \qquad d \ge 69.76(10^{-3})\ m \tag{4}$$

$$d_{min} = 89\text{ mm} \qquad \phi_B = \frac{3.056}{d^4}(10^{-6}) = \frac{3.056(10^{-6})}{(0.089)^4} = 48.71(10^{-3}) \tag{5}$$

$$(\tau_{AB})_{max} = \frac{81.487(10^3)}{d^3} = \frac{81.487(10^3)}{(0.089)^3} = 115.6(10^6)\ N/m^2 \qquad (\tau_{BC})_{max} = \frac{20.37(10^3)}{d^3} = \frac{20.37(10^3)}{(0.089)^3} = 39.8(10^6)\ N/m^2 \tag{6}$$

ANS d_{min} = 89 mm ; ϕ_B = 0.0487 rads CCW ; τ_{max} = 116 MPa

---

5. 76

Solution G = 80 GPa $|\tau_{max}| \le 60\ MPa$ $|\phi_{max}| \le 0.03\ rads$ T_1 = 10 kN-m

T_2 = 25 kN-m d_{min} =? nearest mm τ_{max} = ? ϕ_{max} =?

By moment equilibrium about the shaft axis we obtain the internal torque.

$$\boldsymbol{T}_{AB} = T_A \qquad \boldsymbol{T}_{BC} = T_A + T_1 \qquad \boldsymbol{T}_{CD} = T_A + T_1 - T_2 \tag{1}$$

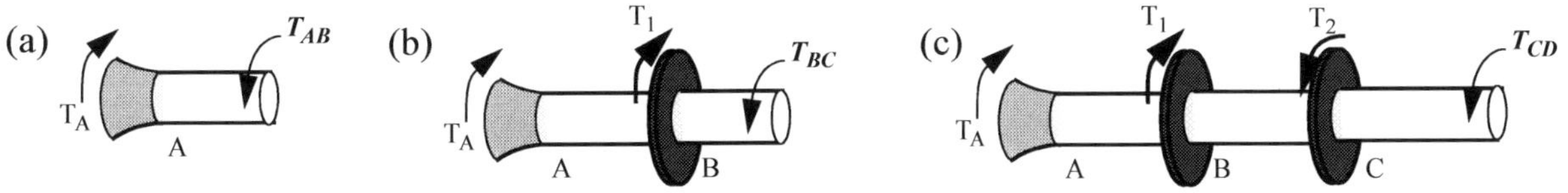

The relative rotation of the ends of each segment can be written, added and equated to zero as $\phi_A=0$ and $\phi_D=0$.

$$\phi_B - \phi_A = \frac{T_{AB}(x_B - x_A)}{GJ} = \frac{(1)T_A}{GJ} \qquad \phi_C - \phi_B = \frac{T_{BC}(x_C - x_B)}{GJ} = \frac{(T_A + T_1)(1.5)}{GJ} \quad (2)$$

$$\phi_D - \phi_C = \frac{T_{CD}(x_D - x_C)}{GJ} = \frac{(T_A + T_1 - T_2)(2.5)}{GJ} \quad (3)$$

$$\phi_D - \phi_A = \frac{1}{GJ}[T_A + 1.5(T_A + T_1) + 2.5(T_A + T_1 - T_2)] = 0 \quad or \quad T_A = \frac{2.5T_2 - 4T_1}{5} = 0.5T_2 - 0.8T_1 \quad (4)$$

$$T_{AB} = 0.5T_2 - 0.8T_1 = 4.5\ kN-m \qquad T_{BC} = 0.5T_2 + 0.2T_1 = 14.5\ kN-m \qquad T_{CD} = -0.5T_2 + 0.2T_1 = -10.5kN-m \quad (5)$$

$$-\phi_C = \frac{(-10.5)(10^3)(2.5)}{(80)(10^9)\pi(d^4/32)} = \frac{-3.342}{d^4}(10^{-6}) \quad or \quad \phi_{max} = \frac{3.342}{d^4}(10^{-6}) \le 0.03 \quad or \quad d \ge 0.1027\ m$$

$$\tau_{max} = \frac{(14.5)(10^3)(d/2)}{\pi(d^4/32)} = \frac{73.847(10^3)}{d^3} \le 60(10^6) \quad or \quad d \ge 0.1072\ m \quad (6)$$

$$d_{\min} = 108\text{ mm} \qquad \phi_{max} = \frac{3.342}{(0.108)^4}(10^{-6}) = 0.0246\ rads \qquad \tau_{max} = \frac{73.847(10^3)}{(0.108)^3} = 58.62(10^6)\ N/m^2 \quad (7)$$

ANS $d_{\min} = 108$ mm ; $\phi_B = 0.025$ rads CCW ; $\tau_{\max} = 58.62$ MPa

---

5. 77

Solution G = 80 GPa $|\tau_{max}| \le 60\ MPa$ $|\phi_{max}| \le 0.03\ rads$ d = 80 mm T_1 =? T_2 =? nearest kN-m

From equations in the previous problems we have

$$T_{AB} = 0.5T_2 - 0.8T_1 \qquad T_{BC} = 0.5T_2 + 0.2T_1 \qquad T_{CD} = -0.5T_2 + 0.2T_1 \qquad J = \pi(0.08)^4/32 = 4.021(10^{-6})\ m^4 \quad (1)$$

$$\tau_{max} = \frac{T_{BC}(\rho_{BC})_{max}}{J} = \frac{(0.5T_2 + 0.2T_1)(0.04)}{4.021(10^{-6})} = 60(10^6) \quad or \quad 0.5T_2 + 0.2T_1 = 6.032(10^3)\ N-m \quad (2)$$

$$\phi_B - \phi_A = \frac{T_{AB}(1)}{(80)(10^9)(4.021)(10^{-6})} \quad or \quad \phi_B = (0.5T_2 - 0.8T_1)(3.109)(10^{-6}) \quad (3)$$

$$\phi_D - \phi_C = \frac{T_{CD}(2.5)}{(80)(10^9)(4.021)(10^{-6})} \quad or \quad -\phi_C = (-0.5T_2 + 0.2T_1)7.772(10^{-6}) \quad (4)$$

Assuming ϕ_B is greater in magnitude than ϕ_C, and ϕ_B is clockwise, we obtain

$$(0.5T_2 - 0.8T_1)(3.109)(10^{-6}) = 0.03 \quad or \quad 0.5T_2 - 0.8T_1 = -9.649(10^3) \quad (5)$$

$$T_1 = 15.68(10^3)\ N-m \quad and \quad T_2 = 5.792(10^3)\ N-m \quad (6)$$

Substituting we obtain $\phi_C = 0.0019$, which is less than 0.03, Hence our assumption that ϕ_B is greater in magnitude is correct, rounding downwards to the nearest kN, we obtain the maximum values of the torques.

ANS $T_1 = 15$ kN-m ; $T_2 = 5.0$ kN-m

---

5. 78

Solution $t = 5\ mm$ $d_{al} = 125\ mm$ $d_{cu} = 50\ mm$ $L = 1.5\ m$ $G_{al} = 28\ GPa$ $G_{cu} = 40\ GPa$ $\phi_1 = 0.03\ rads$ $(\tau_{al})_{max}$ =? $(\tau_{cu})_{max}$ =? $T = Fd$ =?

T_{al} and T_{cu} represents the internal torque on each material cross section. We can write the following.

$$T_{al} + T_{cu} = T \qquad J_{al} = \frac{\pi}{32}[(0.130)^4 - (0.120^4)] = 7.6820(10^{-6})\ m^4 \qquad J_{cu} = \frac{\pi}{32}[(0.055)^4 - (0.045^4)] = 0.4958(10^{-6})\ m^4 \quad (1)$$

The relative rotation of the ends of the each segment can be written and $\phi_A = 0$, $\phi_B = 0.03$ be substituted as shown below.

$$(\phi_B)_{al} - (\phi_A)_{al} = \frac{T_{al}(x_B - x_A)}{G_{al}J_{al}} = \frac{T_{al}(1.5)}{(28)(10^9)(7.682)(10^{-6})} = 6.973(10^{-6})T_{al} = 0.03 \quad or \quad T_{al} = 4.302(10^3)\ N-m \quad (2)$$

$$(\phi_B)_{cu} - (\phi_A)_{cu} = \frac{T_{cu}(x_C - x_B)}{G_{cu}J_{cu}} = \frac{T_{cu}(1.5)}{(40)(10^9)(0.4958)(10^{-6})} = 75.63(10^{-6})T_{cu} = 0.03 \quad or \quad T_{cu} = 0.397(10^3)\ N-m \quad (3)$$

$$T = T_{al} + T_{cu} = (4.302 + 0.397)(10^3) = 4.699(10^3)\ N-m \quad (4)$$

$$(\tau_{al})_{max} = \frac{T_{al}(\rho_{al})_{max}}{J_{al}} = \frac{(4.302)(10^3)(0.065)}{7.682(10^{-6})} = 36.4(10^6)\frac{N}{m^2} \quad (5)$$

$$(\tau_{cu})_{max} = \frac{T_{cu}(\rho_{cu})_{max}}{J_{cu}} = \frac{(0.397)(10^3)(0.0275)}{0.4958(10^{-6})} = 22.0(10^6)\ \frac{N}{m^2} \quad (6)$$

ANS $T = 4.699$ kN-m ; $(\tau_{cu})_{max} = 22.0$ MPa ; $(\tau_{al})_{max} = 36.4$ MPa

---

5. 79

Solution $t = 5\ mm$ $d_{al} = 125\ mm$ $d_{cu} = 50\ mm$ $L = 1.5\ m$ $\phi_1 = 0.03\ rads$

$G_{al} = 28\ GPa$ $G_{cu} = 40\ GPa$ $(\tau_{al})_{max} = ?$ $(\tau_{cu})_{max} = ?$ $T = Fd = ?$

The solution proceeds as follows.

$$J_{al} = \frac{\pi}{32}[(0.130)^4 - (0.120^4)] = 7.6820(10^{-6})\ m^4 \qquad J_{cu} = \frac{\pi}{32}[(0.055)^4 - (0.045^4)] = 0.4958(10^{-6})\ m^4 \quad (1)$$

$$\Sigma G_j J_j = (28)(10^9)7.6820(10^{-6}) + (40)(10^9)(0.4958)(10^{-6}) = 234.93(10^3)\ N-m^2 \quad (2)$$

$$\phi_B - \phi_A = \frac{T_{AB}(x_B - x_A)}{\Sigma G_j J_j} = \frac{T(1.5)}{234.93(10^3)} \le 0.03 \quad or \quad T \le 4.669(10^3)\ N-m \quad (3)$$

$$\tau_{x\theta} = G_i\frac{T_{AB}\rho}{\Sigma G_j J_j} = \frac{4.669(10^3)}{234.93(10^3)}G_i\rho = 0.02G_i\rho \quad (4)$$

The maximum torsional shear stress in copper will be at $\rho = (d_{cu}/2 + t/2) = 0.0275\ m$ and in aluminum will be at $\rho = (d_{al}/2 + t/2) = 0.065\ m$. We obtain

$$(\tau_{cu})_{max} = (0.02)(40)(10^9)(0.0275) = 22(10^6)\ N/m^2 \qquad (\tau_{al})_{max} = (0.02)(28)(10^9)(0.065) = 36.4(10^6)\ N/m^2 \quad (5)$$

ANS $T = 4.7$ kN-m ; $(\tau_{cu})_{max} = 22$ MPa ; $(\tau_{al})_{max} = 36.4$ MPa

---

5. 80

Solution $t = 5\ mm$ $d_{al} = 125\ mm$ $d_{cu} = 50\ mm$ $L = 1.5\ m$ $G_{al} = 28\ GPa$

$G_{cu} = 40\ GPa$ $T = Fd = 10kN-m$ $(\tau_{al})_{max} = ?$ $(\tau_{cu})_{max} = ?$ $\phi_B = ?$

$$J_{al} = (\pi/32)[(0.130)^4 - (0.120^4)] = 7.6820(10^{-6})\ m^4 \qquad J_{cu} = (\pi/32)[(0.055)^4 - (0.045^4)] = 0.4958(10^{-6})\ m^4 \quad (1)$$

The relative rotation of the ends of the each segment can be written and equated to each other as shown below.

$$(\phi_B)_{al} - (\phi_A)_{al} = \frac{T_{al}(x_B - x_A)}{G_{al}J_{al}} = \frac{T_{al}(1.5)}{(28)(10^9)(7.682)(10^{-6})} = 6.973(10^{-6})T_{al} \quad (2)$$

$$(\phi_B)_{cu} - (\phi_A)_{cu} = \frac{T_{cu}(x_C - x_B)}{G_{cu}J_{cu}} = \frac{T_{cu}(1.5)}{(40)(10^9)(0.4958)(10^{-6})} = 75.63(10^{-6})T_{cu} \quad (3)$$

$$6.973(10^{-6})T_{al} = 75.63(10^{-6})T_{cu} \quad or \quad T_{al} = 10.846T_{cu} \text{ or}$$

$$T = T_{al} + T_{cu} = 10(10^3)\ N-m \quad or \quad T_{cu} = \frac{10(10^3)}{(10.846+1)} = 0.8442(10^3)\ N-m \qquad T_{al} = 9.1558(10^3)\ N-m \quad (4)$$

$$\phi_B = (6.973)(10^{-6})(9.1558)(10^{-6}) = 63.85(10^{-3}) \quad (5)$$

$$(\tau_{al})_{max} = \frac{(9.1558)(10^3)(0.065)}{7.682(10^{-6})} = 77.47(10^6)\ N/m^2 \qquad (\tau_{cu})_{max} = \frac{(0.8442)(10^3)(0.065)}{0.4958(10^{-6})} = 46.8(10^6)\ N/m^2 \quad (6)$$

ANS $\phi_B = 0.064$ rads CCW ; $(\tau_{cu})_{max} = 46.8$ MPa ; $(\tau_{al})_{max} = 77.5$ MPa

---

5. 81

Solution $t = 5\ mm$ $d_{al} = 125\ mm$ $d_{cu} = 50\ mm$ $L = 1.5\ m$ $G_{al} = 28\ GPa$

$G_{cu} = 40\ GPa$ $T = Fd = 10kN-m$ $(\tau_{al})_{max} = ?$ $(\tau_{cu})_{max} = ?$ $\phi_B = ?$

$$J_{al} = (\pi/32)[(0.130)^4 - (0.120^4)] = 7.6820(10^{-6})\ m^4 \qquad J_{cu} = (\pi/32)[(0.055)^4 - (0.045^4)] = 0.4958(10^{-6})\ m^4 \quad (1)$$

$$\Sigma G_j J_j = (28)(10^9)7.6820(10^{-6}) + (40)(10^9)(0.4958)(10^{-6}) = 234.93(10^3)\ N-m^2 \quad (2)$$

$$\phi_B - \phi_A = \phi_B = \frac{T_{AB}(x_B - x_A)}{\Sigma G_j J_j} = \frac{(10)(10^3)(1.5)}{234.93(10^3)} = 63.85(10^{-3})$$

$$\tau_{x\theta} = G_i \frac{T_{AB}\rho}{\Sigma G_j J_j} = \frac{(10)(10^3)}{234.93(10^3)} G_i \rho = 0.0426 G_i \rho \quad \textbf{(3)}$$

The maximum torsional shear stress in copper will be at $\rho = (d_{cu}/2 + t) = 0.0275 \ m$ and at $\rho = (d_{al}/2 + t) = 0.0675 \ m$ in aluminum.

$$(\tau_{cu})_{max} = (0.0426)(40)(10^9)(0.0275) = 46.86(10^6) \ \frac{N}{m^2} \qquad (\tau_{al})_{max} = (0.0426)(28)(10^9)(0.065) = 77.53(10^6) \ \frac{N}{m^2} \quad \textbf{(4)}$$

ANS $\phi_B = 0.0639$ rads CCW ; $(\tau_{cu})_{max} = 46.9$ MPa ; $(\tau_{al})_{max} = 77.5$ MPa

---

5. 82

Solution

The solution proceeds as follows.

$$J_S = (\pi/32)(0.08)^4 = 4.02(10^{-6}) \ m^4 \qquad J_{Br} = (\pi/32)(0.12^4 - 0.08^4) = 16.33(10^{-6}) \ m^4 \quad \textbf{(1)}$$

$$\sum_{j=1}^{2} G_j J_j = G_S J_S + G_{Br} J_{Br} = 321.6(10^3) + 653.2(10^3) = 974.8(10^3) \ N \cdot m^2 \quad \textbf{(2)}$$

$$\boldsymbol{T_{AB}} = 75 \ kN - m \qquad \phi_B - \phi_A = \boldsymbol{T_{AB}}(x_B - x_A)/[\textstyle\sum_{j=1}^{2} G_j J_j] = \frac{(75)(10^3)(2)}{974.8(10^3)} = 0.1538 \ rad \quad \textbf{(3)}$$

$$\tau_i = (G_i \rho \boldsymbol{T_{AB}})/[\textstyle\sum_{j=1}^{2} G_j J_j] = \frac{G_i \rho (75)(10^3)}{974.8(10^3)} = (76.9)(10^{-3}) G_i \rho \quad \textbf{(4)}$$

$$(\tau_{Br})_{max} = (76.9)(10^{-3})(40)(10^9)(0.06) = 184.6(10^6) \qquad (\tau_s)_{max} = (76.9)(10^{-3})(80)(10^9)(0.04) = 246.2(10^6) \quad \textbf{(5)}$$

ANS $\phi_B - \phi_A = 0.1538$ rads ; $\tau_{max} = 246.2$ MPa

---

5. 83

Solution $G_{al} = 4000$ ksi $G_{br} = 6000$ ksi $G_{st} = 12{,}000$ ksi

The solution proceeds as follows.

$$J_s = \frac{\pi}{32}(2^4 - 1.5^4) = 1.0738 \ in^4 \qquad J_{br} = \frac{\pi}{32}(1.5^4 - 1^4) = 0.3988 \ in^4 \qquad J_{al} = \frac{\pi}{32}(1^4) = 0.0982 \ in^4 \quad \textbf{(1)}$$

$$\phi_B - \phi_A = (\phi_B)_{al} - (\phi_A)_{al} = \frac{\boldsymbol{T}_{al}(x_B - x_A)}{G_{al} J_{al}} = \frac{\boldsymbol{T}_{al}(25)}{(4000)(0.0982)} = 0.06365 \boldsymbol{T}_{al} \quad \textbf{(2)}$$

$$\phi_B - \phi_A = (\phi_B)_{br} - (\phi_A)_{br} = \frac{\boldsymbol{T}_{br}(x_B - x_A)}{G_{br} J_{br}} = \frac{\boldsymbol{T}_{br}(25)}{(6000)(0.3988)} = 0.01045 \boldsymbol{T}_{br} \quad \textbf{(3)}$$

$$\phi_B - \phi_A = (\phi_B)_s - (\phi_A)_s = \frac{\boldsymbol{T}_s(x_B - x_A)}{G_s J_s} = \frac{T_s(25)}{(12000)(1.0738)} = 0.00194 \boldsymbol{T}_s \quad \textbf{(4)}$$

$$\boldsymbol{T}_{br} = \frac{0.06365}{0.01045} \boldsymbol{T}_{al} = 6.0916 \boldsymbol{T}_{al} \qquad \boldsymbol{T}_s = \frac{0.06365}{0.00194} \boldsymbol{T}_{al} = 32.8045 \boldsymbol{T}_{al} \quad \textbf{(5)}$$

$$\boldsymbol{T}_{al} + \boldsymbol{T}_{br} + \boldsymbol{T}_s = 30 \quad or \quad \boldsymbol{T}_{al} = \frac{30}{1 + 6.0916 + 32.8045} = 0.7519 \ in-kips \quad \textbf{(6)}$$

$$\boldsymbol{T}_{br} = 6.0916(0.7519) = 4.5806 \ in-kips \qquad \boldsymbol{T}_s = 32.8045(0.7519) = 24.6674 \ in-kips \quad \textbf{(7)}$$

$$\phi_B - \phi_A = (\phi_B)_{al} - (\phi_A)_{al} = 0.06365(0.7519) = 0.04786 \quad \textbf{(8)}$$

$$(\tau_{al})_{max} = \frac{(0.7519)(0.5)}{0.0982} = 3.83 \ ksi \qquad (\tau_{br})_{max} = \frac{(4.5806)(0.75)}{0.3988} = 8.61 \ ksi \qquad (\tau_s)_{max} = \frac{(24.6674)(1)}{1.0738} = 22.97 \ ksi \quad \textbf{(9)}$$

ANS $\phi_B = 0.0479$ rads CCW **;** $(\tau_{al})_{max} = 3.83$ ksi **;** $(\tau_{br})_{max} = 8.61$ ksi ; $(\tau_s)_{max} = 22.97$ ksi

---

5. 84

Solution $G_{al} = 4000$ ksi $G_{br} = 6000$ ksi $G_s = 12000$ ksi $\phi_B - \phi_A = ?$ $\gamma_{x\theta}$ vs. $\rho = ?$ $\tau_{x\theta}$ vs. $\rho = ?$

The solution proceeds as follows.

$$J_s = (\pi/32)(2^4 - 1.5^4) = 1.0738 \ in^4 \qquad J_{br} = (\pi/32)(1.5^4 - 1^4) = 0.3988 \ in^4 \qquad J_{al} = (\pi/32)(1^4) = 0.0982 \ in^4 \quad \textbf{(1)}$$

$$\Sigma G_j J_j = (4000)(0.0982) + (6000)(0.3988) + (12000)(1.0738) = 15.671(10^3)\ kips - in^2 \quad (2)$$

$$\phi_B - \phi_A = \phi_B = \frac{T_{AB}(x_B - x_A)}{\Sigma G_j J_j} = \frac{(30)(25)}{15.671(10^3)} = 47.861(10^{-3}) \qquad \tau_{x\theta} = G_i \frac{T_{AB}\rho}{\Sigma G_j J_j} = 1.914(10^{-3}) G_i \rho \quad (3)$$

$$(\tau_{Al})_{max} = 1.914(10^{-3})(4000)(0.5) = 3.828\ ksi \qquad (\tau_{br})_{max} = 1.914(10^{-3})(6000)(0.75) = 8.613\ ksi \quad (4)$$

$$(\tau_s)_{max} = 1.914(10^{-3})(12000)(1) = 22.968\ ksi \quad (5)$$

ANS $\phi_B = 0.0479$ rads CCW **;** $(\tau_{al})_{max} = 3.83$ ksi **;** $(\tau_{br})_{max} = 8.61$ ksi ; $(\tau_s)_{max} = 22.97$ ksi

---

5. 85

Solution $G_{ir} = 70\ GPa$ $G_{cu} = 40\ GPa$ $(d_{ir})_o = 70\ mm$ $(d_{ir})_i = 50\ mm$ $(d_{cu})_o = 50\ mm$

$(d_{cu})_i = 30\ mm$ $T = 1500\ N-m$ $(\tau_{ir})_{max} = ?$ $(\tau_{cu})_{max} = ?$ $\phi_D - \phi_A = ?$

The solution proceeds as shown below.

$$J_{ir} = (\pi/32)[(0.07)^4 - (0.05)^4] = 1.7436(10^{-6})m^4 \qquad J_{cu} = (\pi/32)[(0.05)^4 - (0.03)^4] = 0.5341(10^{-6})m^4 \quad (1)$$

$$\Sigma G_j J_j = (70)(10^9)(1.7436)(10^{-6}) + (40)(10^9)(0.5341)(10^{-6}) = 143.41(10^3)\ N-m^2 \quad (2)$$

$$T_{AB} = -T = -1500\ N-m \qquad \phi_B - \phi_A = \frac{T_{AB}(x_B - x_A)}{G_{ir}J_{ir}} = \frac{(-1500)(0.5)}{(70)(10^9)(1.7436)(10^{-6})} = -6.145(10^{-3}) \quad (3)$$

$$T_{BC} = -T = -1500\ N-m \qquad \phi_C - \phi_B = \frac{T_{BC}(x_C - x_B)}{\Sigma G_j J_j} = \frac{(-1500)(0.15)}{(143.4)(10^{-3})} = -1.569(10^{-3}) \quad (4)$$

$$T_{CD} = -T = -1500\ N-m \qquad \phi_D - \phi_C = \frac{T_{CD}(x_D - x_C)}{G_{cu}J_{cu}} = \frac{(-1500)(0.4)}{(40)(10^9)(0.5341)(10^{-6})} = -28.08(10^{-3}) \quad (5)$$

$$\phi_D - \phi_A = -(0.145 + 1.569 + 28.08)(10^{-3}) = 35.798(10^{-3}) \quad (6)$$

$$(\tau_{AB})_{max} = \left|\frac{T_{AB}(\rho_{AB})_{max}}{J_{ir}}\right| = \left|\frac{(-1500)(0.035)}{1.7436(10^{-6})}\right| = 30.1(10^6)\ \frac{N}{m^2} \quad (7)$$

$$(\tau_{CD})_{max} = \left|\frac{T_{CD}(\rho_{CD})_{max}}{J_{cu}}\right| = \left|\frac{(-1500)(0.025)}{0.5341(10^{-6})}\right| = 70.2(10^6)\ \frac{N}{m^2} \quad (8)$$

$$\tau_{x\theta} = \left|G_i \frac{T_{BC}\rho}{\Sigma G_j J_j}\right| = \left|\frac{(-1500)}{143.41(10^3)} G_i \rho\right| = 10.46(10^{-3}) G_i \rho \quad (9)$$

At ρ=0.035, G_i=G_{ir}, we obtain $\tau_{x\theta} = 25.62(10^6)$ and at ρ=0.025, G_i=G_{cu}, we obtain $\tau_{x\theta} = 10.46(10^6)$ which are less than those obtained earlier in segments *AB* and *CD*.

ANS $\phi_D - \phi_A = 0.0358$ rads CCW ; $(\tau_{cu})_{max} = 70.2$ MPa ; $(\tau_{ir})_{max} = 30.1$ MPa

---

5. 86

Solution $G = 80$ GPa $d_{AB} = d_{CD} = 40$ mm $R_B = 250$ mm $R_C = 200$ mm

$T = 1.5$ kN-m $(\tau_{AB})_{max} = ?$ $\phi_D = ?$

The contact force between the gears can be calculated as shown below.

$$FR_C = T_{CD} = T \quad or \quad F = \frac{T}{R_C} = \frac{1.5(10^3)}{0.2} = 7.5(10^3)N-m \qquad T_{AB} = FR_B = 7.5(10^3)(0.25) = 1.875(10^3)N-m. \quad (1)$$

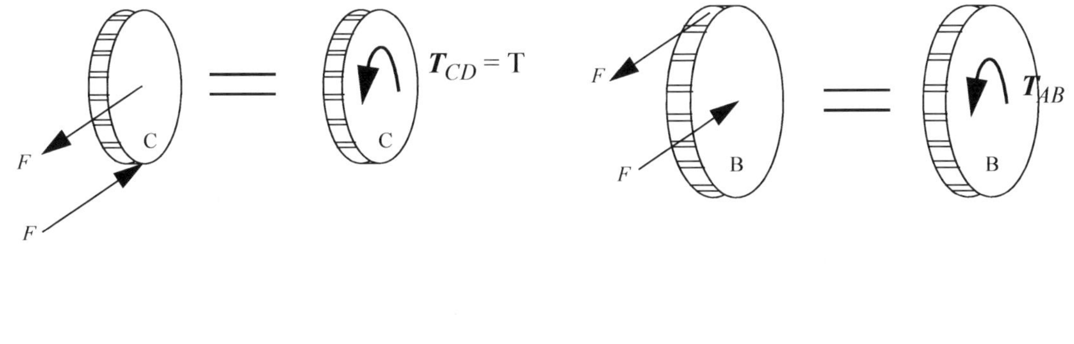

$J = \pi(0.04)^4/32 = 0.2513(10^{-6})\ m^4 \qquad (\tau_{AB})_{max} = \frac{T_{AB}(\rho_{AB})_{max}}{J} = \frac{1.875(10^3)(0.02)}{0.2513(10^{-6})} = 149.2(10^6)\ \frac{N}{m^2}$ **(2)**

$\phi_B - \phi_A = \frac{T_{AB}(x_B - x_A)}{GJ} = \frac{1.875(10^3)(1.2)}{(80)(10^9)(0.2513)(10^{-6})} = 0.1119\ rads$ **(3)**

$R_C\phi_C = R_B\phi_B \quad or \quad \phi_C = \frac{R_B\phi_B}{R_C} = \frac{0.1119(0.25)}{0.2} = 0.1399\ rads$ **(4)**

$\phi_D - \phi_C = \frac{T_{CD}(x_D - x_C)}{GJ} = \frac{1.5(10^3)(1.5)}{(80)(10^9)(0.2513)(10^{-6})} = 0.1119 \quad or \quad \phi_D = 0.1119 + \phi_C = 0.1119 + 0.1399 = 0.252 rads$ **(5)**

ANS $(\tau_{AB})_{max} = 149.2\ MPa;\ \phi_D = 0.252\ rads\ CCW$

---

5. 87

Solution $G = 80$ GPa $\quad d_{AB} = d_{CD} = 40$ mm $\quad R_B = 250$ mm $\quad R_C = 200$ mm $\quad \tau_{max} = 120$ MPa $\quad T_{max} = ?$

From previous problem we have the contact force between the gears as shown below.

$FR_C = T_{CD} = T \quad or \quad F = T/R_C = 5T \quad T_{AB} = FR_B = (5T)(0.25) = 1.25T$ **(1)**

$J = \pi(0.04)^4/32 = 0.2513 \qquad (\tau_{AB})_{max} = \frac{T_{AB}(\rho_{AB})_{max}}{J} = \frac{1.25T(0.02)}{0.2513(10^{-6})} \le 120(10^6) \quad or \quad T \le 1.20(10^3)N-m$ **(2)**

ANS $T_{max} = 1.2\ kN-m$

---

5. 88

Solution $G = 12000$ ksi $\quad d_{AB} = d_{CDE} = 1.5$ in $\quad R_B = 9$ in $\quad R_D = 5$ in

$T = 800$ ft-lb $\quad (\tau_{AB})_{max} = ? \quad \phi_E = ?$

The contact force between the gears can be calculated as shown below.

$FR_D + T_{CD} = T = 800(12) = 9600\ in-lb = 9.6\ in-kips \qquad T_{AB} = FR_B$ **(1)**

$T_{CD} + T_{AB}(R_D/R_B) = 9.6\ in-kips \quad or \quad T_{CD} + T_{AB}(5/9) = 9.6\ in-kips$ **(2)**

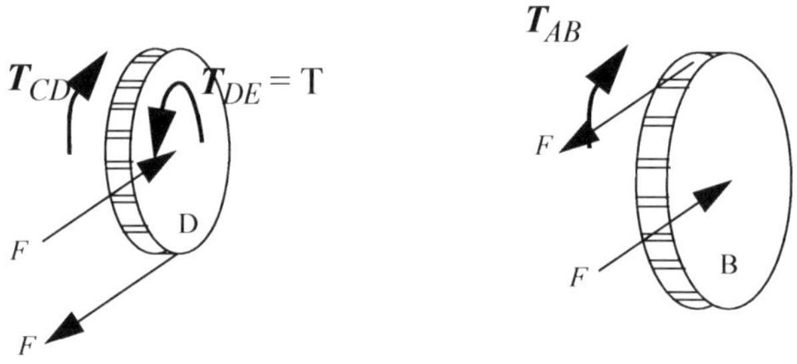

$J = \pi(1.5)^4/32 = 0.497\ in^4 \qquad \phi_D - \phi_C = \phi_D = \frac{T_{CD}(x_D - x_C)}{GJ} = \frac{T_{CD}(4)(12)}{(12000)(0.497)} = 8.048(10^{-3})T_{CD}$ **(3)**

$\phi_B - \phi_A = \phi_B = \frac{T_{AB}(x_B - x_A)}{GJ} = \frac{T_{AB}(4)(12)}{(12000)(0.497)} = 8.048(10^{-3})T_{AB}$ **(4)**

$R_D\phi_D = R_B\phi_B \quad or \quad (5)(8.048)(10^{-3})T_{CD} = (9)(8.048)(10^{-3})T_{AB} \quad or \quad 5T_{CD} = 9T_{AB}$ **(5)**

$T_{AB} = 4.075\ in-kips \qquad T_{CD} = 7.336\ in-kips$ **(6)**

$(\tau_{AB})_{max} = \frac{T_{AB}(\rho_{AB})_{max}}{J} = \frac{4.075(0.75)}{0.497} = 6.15\ ksi \qquad \phi_D = 8.048(10^{-3})(7.336) = 0.05904\ rads$ **(7)**

$\phi_E - \phi_D = \frac{T_{DE}(x_E - x_D)}{GJ} = \frac{(9.6)(5)(12)}{(12000)(0.497)} = 0.09658\ rads \quad or \quad \phi_E = 0.05904 + 0.09658 = 0.1556\ rads$

ANS $(\tau_{AB})_{max} = 6.15\ ksi;\ \phi_E = 0.1556\ rads\ CCW$

---

5. 89

Solution $G = 80$ GPa $\quad d_{AB} = d_{CD} = 60$ mm $\quad R_B = 175$ mm $\quad R_D = 125$ mm

$T = 2$ kN-m $\quad (\tau_{AB})_{max} = ? \quad \phi_D = ?$

The contact force between the gears can be calculated as shown below.

$$FR_D + \boldsymbol{T}_{CD} = T \qquad \boldsymbol{T}_{AB} = FR_B \qquad \boldsymbol{T}_{CD} + \frac{\boldsymbol{T}_{AB}R_D}{R_B} = 2000\ N-m \quad or \quad \boldsymbol{T}_{CD} + \frac{0.125\boldsymbol{T}_{AB}}{0.175} = T = 2000\ N-m \qquad (1)$$

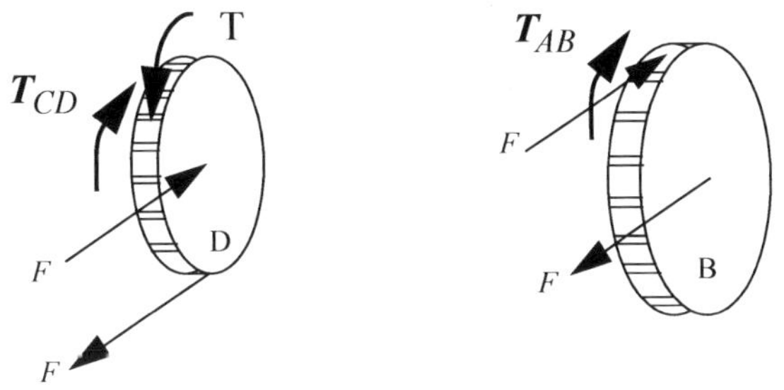

$$J = \pi(0.06)^4/32 = 1.272(10^{-6})\ m^4 \qquad \phi_D - \phi_C = \phi_D = \frac{\boldsymbol{T}_{CD}(x_D - x_C)}{GJ} = \frac{\boldsymbol{T}_{CD}(1.5)}{(80)(10^9)(1.272)(10^{-6})} = 14.737(10^{-6})\boldsymbol{T}_{CD} \qquad (2)$$

$$\phi_B - \phi_A = \phi_B = \frac{\boldsymbol{T}_{AB}(x_B - x_A)}{GJ} = \frac{\boldsymbol{T}_{AB}(4)(1.5)}{(80)(10^9)(1.272)(10^{-6})} = 14.737(10^{-6})\boldsymbol{T}_{AB} \qquad (3)$$

$$R_D\phi_D = R_B\phi_B \quad or \quad (0.125)14.737(10^{-6})\boldsymbol{T}_{CD} = (0.175)14.737(10^{-6})\boldsymbol{T}_{AB} \quad or \quad \boldsymbol{T}_{CD} = 1.4\boldsymbol{T}_{AB} \text{ or}$$

$$\boldsymbol{T}_{AB} = 945.9\ N-m \qquad \boldsymbol{T}_{CD} = 1324.3\dot{N}-m \qquad (4)$$

$$(\tau_{CD})_{max} = \frac{\boldsymbol{T}_{CD}(\rho_{CD})_{max}}{J} = \frac{1324.3(0.03)}{(1.272)(10^{-6})} = 31.23(10^6)N/m^2 \qquad \phi_D = 14.737(10^{-6})(1324.3) = 0.0195\ rads \qquad (5)$$

ANS $(\tau_{CD})_{max} = 31.2\ MPa;\ \phi_D = 0.0195\ rads\ CCW$

---

5. 90

Solution $G = 80$ GPa $\quad d_{AB} = d_{CD} = 60$ mm $\quad R_B = 175$ mm $\quad R_D = 125$ mm $\quad \tau_{max} = 120$ MPa $\quad T_{max} = ?$

From previous problem we have

$$J = \pi(0.06)^4/32 = 1.272(10^{-6})\ m^4 \qquad \boldsymbol{T}_{CD} + (0.125/0.175)\boldsymbol{T}_{AB} = T \qquad \boldsymbol{T}_{CD} = 1.4\boldsymbol{T}_{AB} \qquad (1)$$

Solving for internal torques we obtain

$$\boldsymbol{T}_{AB} = 0.47297T \qquad \boldsymbol{T}_{CD} = 0.66216T \qquad (2)$$

$$(\tau_{CD})_{max} = \frac{\boldsymbol{T}_{CD}(\rho_{CD})_{max}}{J} = \frac{0.66216T(0.03)}{(1.272)(10^{-6})} \le 120(10^6) \quad or \quad T \le 7.684(10^3)N-m \qquad (3)$$

ANS $T_{max} = 7684\ N-m$

---

5. 91

Solution

From two problem before we have

$$\boldsymbol{T}_{AB} = 945.9\ N-m \qquad \boldsymbol{T}_{CD} = 1324.3N-m \qquad (1)$$

$$(\tau_{CD})_{max} = \frac{\boldsymbol{T}_{CD}(\rho_{CD})_{max}}{J} = \frac{1324.3(d/2)}{(\pi d^4/32)} \le 120(10^6) \quad or \quad \frac{1}{d^3} \le 17.792(10^{-3}) \quad or \quad d \ge 0.0383\ m \qquad (2)$$

ANS $d_{min} = 38\ mm$

---

5. 92

Solution $\tau_{max} \le 17\ ksi$ $\qquad$ r =?

The maximum shear stress in section BC can be found as shown below.

$$(\tau_{BC})_{max} = \frac{T\rho_{max}}{J_{BC}} = \frac{(2.5)(0.5)}{\pi(1)^4/32} = 12.73\ ksi \qquad \tau_{max} = K_{conc}(\tau_{BC})_{max} = 12.73K_{conc} \le 17 \qquad K_{conc} \le 1.335 \qquad (1)$$

From Section C.4.3, we obtain the approximate value of r/d corresponding to D/d = 2 and $K_{conc} = 1.335$

$$r/d = 0.135 \quad or \quad r = (0.135)(1) \qquad (2)$$

ANS $r = 0.135$ in.

---

5. 93

Solution r = 6 mm $\tau_{max} \le 80\ MPa$ G = 28 MPa T_{max} =?

The polar moment of cross-sections in various segments can be found as

$$J_{AB} = \frac{\pi}{32}(0.06)^4 = 1.272(10^{-6})\ m^4 \qquad J_{BC} = J_{CD} = \frac{\pi}{32}(0.048)^4 = 0.521(10^{-6})\ m^4 \tag{1}$$

By moment equilibrium about the shaft axis we have the following equations.

$$T_{AB} = T_A \qquad T_{BC} = T_A \qquad T_{CD} = T_A - T \tag{2}$$

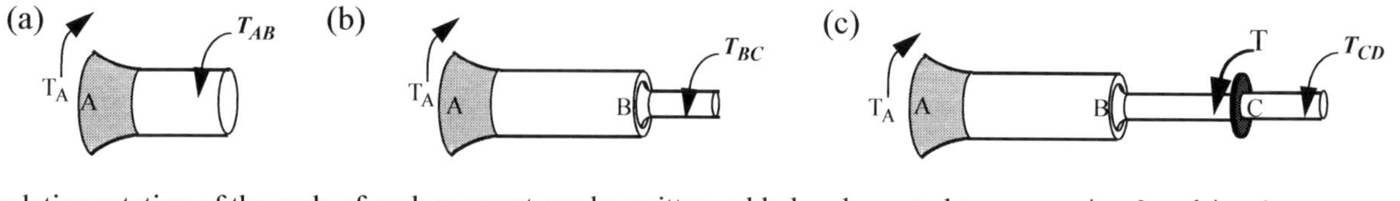

The relative rotation of the ends of each segment can be written, added and equated to zero as ϕ_A=0 and ϕ_D=0.

$$\phi_B - \phi_A = \frac{T_{AB}(x_B - x_A)}{G_{AB}J_{AB}} = \frac{T_A(0.9)}{(28)(10^9)(1.272(10^{-6}))} = 25.26(10^{-6})T_A \tag{3}$$

$$\phi_C - \phi_B = \frac{T_{BC}(x_C - x_B)}{G_{BC}J_{BC}} = \frac{(T_A)(0.75)}{(28)(10^9)(0.521)(10^{-6})} = 51.40(10^{-6})T_A \tag{4}$$

$$\phi_D - \phi_C = \frac{T_{CD}(x_D - x_C)}{G_{CD}J_{CD}} = \frac{(T_A - T)(1)}{(28)(10^9)(0.521)(10^{-6})} = 68.53(10^{-6})(T_A - T) \tag{5}$$

$$\phi_D - \phi_A = [25.26T_A + 51.40T_A + 68.53(T_A - T)] = 0 \quad or \quad T_A = \frac{68.53T}{25.26 + 51.40 + 68.53} = 0.472T \tag{6}$$

$$T_{AB} = 0.472T \qquad T_{BC} = 0.472T \qquad T_{CD} = -0.528T \tag{7}$$

The nominal torsional shear stresses in segment BC can be found as shown below

$$(\tau_{BC})_{max} = \frac{T_{BC}(\rho_{BC})_{max}}{J_{BC}} = \frac{(0.472T)(0.04)}{0.521(10^{-6})} = 21.74T(10^3) \tag{8}$$

Now $r/d = 6/48 = 0.125$ and $D/d = 60/48 = 1.25$. From Section C.4.3, we obtain the stress concentration factor as K_{conc}= 1.29.

$$(\tau_{max})_{BC} = 1.29(\tau_{BC})_{max} = (1.29)(21.74)T(10^3) \le 80(10^6) \quad or \quad T \le 2.853(10^3)\ N-m \tag{9}$$

$$(\tau_{CD})_{max} = \left|\frac{T_{CD}(\rho_{CD})_{max}}{J_{CD}}\right| = \frac{(0.528T)(0.024)}{0.521(10^{-6})} = 24.32T(10^3) \le 80(10^6) \quad or \quad T \le 3.29(10^3)\ N-m \tag{10}$$

ANS T_{max} = 2.853 kN-m

Section 5.4

5. 94

Solution T=100 in-kips τ_{max} = ?

The solution proceeds as follows.

$$A_E = 2(\pi/2)2^2 + (4)(4) = 28.566\ in^2 \qquad \tau_{max} = \frac{T}{2tA_E} = \frac{100}{2(1/4)(28.566)} = 7.001\ ksi \tag{1}$$

ANS $|\tau_{max}|$ = 7.0 ksi

---

5. 95

Solution T = 900 N-m τ_{max} = ?

The solution proceeds as follows.

$$A_E = (\pi/2)(0.05)^2 + (0.1)(0.1) = 0.0139\ m^2 \qquad \tau_{max} = \frac{T}{2tA_E} = \frac{900}{2(0.003)(0.0139)} = 10.77(10^6)N/m^2$$

ANS $|\tau_{max}|$ = 10.8 MPa

---

5. 96

Solution T = 15 kN-m τ_{max} = ?

The solution proceeds as follows.

$$A_E = \frac{1}{4}\pi(0.1)^2 = 7.854(10^{-3})\ m^2 \qquad \tau_{max} = \frac{\boldsymbol{T}}{2tA_E} = \frac{15(10^3)}{2(0.006)(7.854)(10^{-3})} = 159.1(10^6)N/m^2 \qquad \textbf{(1)}$$

ANS $|\tau_{max}| = 159\text{ MPa}$

---

5. 97

Solution τ_{max} =f(t, a, T)= ?

The solution proceeds as follows.

$$A_E = \frac{1}{2}(a)\left(\frac{a}{2}tan60\right) = \frac{\sqrt{3}}{4}a^2 \qquad \tau_{max} = \frac{\boldsymbol{T}}{2tA_E} = \frac{T}{2(t)(\sqrt{3}a^2)/4} \qquad \textbf{(1)}$$

ANS $\tau_{max} = 2T/(\sqrt{3}a^2t)$

---

5. 98

Solution τ_{max} =f(t, a, T)= ?

The solution proceeds as follows.

$$A_E = (a)(a) = a^2 \qquad \tau_{max} = \frac{\boldsymbol{T}}{2tA_E} = \frac{T}{2(t)(a^2)} \qquad \textbf{(1)}$$

ANS $\tau_{max} = T/(2a^2t)$

---

5. 99

Solution τ_{max} =f(t, a, T)= ?

The solution proceeds as follows.

$$A_E = \pi a^2 \qquad \tau_{max} = \frac{\boldsymbol{T}}{2tA_E} = \frac{T}{2(t)(\pi a^2)} \qquad \textbf{(2)}$$

ANS $\tau_{max} = T/(2\pi a^2t)$

---

5. 100

Solution τ_{max} =f(t, a, b, T)= ?

The solution proceeds as follows.

$$A_E = \pi ab \qquad \tau_{max} = \frac{\boldsymbol{T}}{2tA_E} = \frac{T}{2(t)(\pi ab)} \qquad \textbf{(1)}$$

ANS $\tau_{max} = T/(2\pi abt)$

---

5. 101

Solution τ_{max} = ?

The torque diagram is drawn and the maximum internal torque in segment AB calculated as shown below.

$$\boldsymbol{T}_{max} = -2250\ N-m \qquad A_E = (6)\left[\frac{1}{2}(0.1)\left(\frac{0.1}{2}sin60\right)\right] = 12.99(10^{-3})\ m^2. \qquad \textbf{(1)}$$

(a) Template

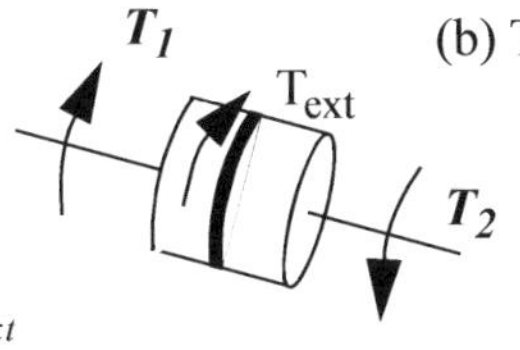

$\boldsymbol{T_2} = \boldsymbol{T_1} + T_{ext}$

(b) Torque Diagram

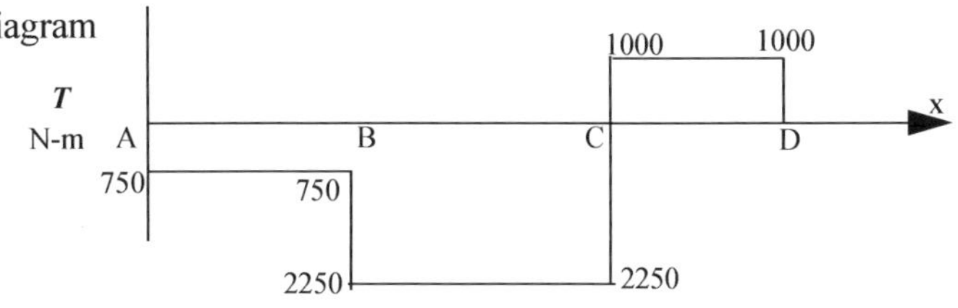

$$\tau_{max} = \frac{\boldsymbol{T}_{max}}{2tA_E} = \frac{(-2250)}{2(0.004)(12.99)(10^{-3})} = -21.65(10^6)N/m^2 \qquad \textbf{(2)}$$

ANS $|\tau_{max}| = 21.65\text{ MPa}$

---

5. 102

Solution $\tau_{max} = ?$

The torque diagram is drawn and maximum internal torque is in segment calculated as shown below.

$$T_{max} = 5 \ in-kips \qquad A_E = (4)(6) = 24 \ in^2 \tag{1}$$

(a) Template

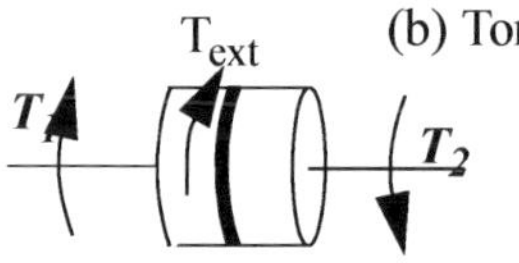

$T_2 = T_1 + T_{ext}$

(b) Torque Diagram

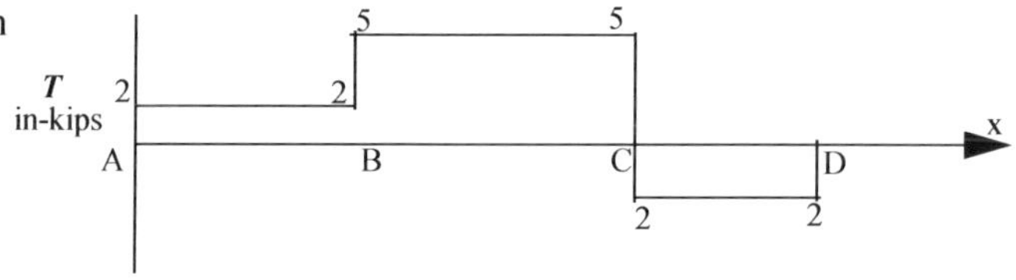

$$\tau_{max} = \frac{T_{max}}{2tA_E} = \frac{(5)}{2(1/8)(24)} = 0.833 \ ksi \tag{2}$$

ANS $|\tau_{max}| = 833$ psi

---

5. 103

Solution Shape= Triangle or Square or Circle =? and percentage torque =?

Let a_T, a_S and a_C represent the dimension of the triangle, square and circle. We find the dimension of each shape in terms of the material area and use the results of earlier problems as shown below.

Triangle:

$$A = (3a_T)t \quad or \quad a_T = \frac{A}{3t} \quad \tau_{max} = \frac{2}{\sqrt{3}}\frac{T_T}{a_T^2 t} = \frac{2}{\sqrt{3}}\frac{T_T}{(A/(3t))^2 t} = 10.392\frac{T_T}{A^2 t} \quad or \quad T_T = \frac{1}{10.392}A^2 t\tau_{max} \tag{1}$$

Square:

$$A = (4a_S)t \quad or \quad a_S = \frac{A}{4t} \quad \tau_{max} = \frac{T_S}{2a_S^2 t} = \frac{T_S}{2(A/(4t))^2 t} = 8.0\frac{T_S}{A^2 t} \quad or \quad T_S = \frac{1}{8.0}A^2 t\tau_{max} \tag{2}$$

Circle:

$$A = (2\pi a_C)t \quad or \quad a_C = \frac{A}{2\pi t} \quad \tau_{max} = \frac{T_C}{2\pi a_C^2 t} = \frac{T_C}{2\pi(A/(2\pi t))^2 t} = 6.283\frac{T_C}{A^2 t} \quad or \quad T_C = \frac{1}{6.283}A^2 t\tau_{max} \tag{3}$$

The circular shape will support the maximum internal torque.

$$\text{\% torque in triangle relative to circle} = \left(\frac{T_T}{T_C}\right)(100) = \frac{(A^2 t\tau_{max})/8}{(A^2 t\tau_{max})/6.283}(100) = 60.46 \tag{4}$$

$$\text{\% torque in square relative to circle} = \left(\frac{T_S}{T_C}\right)(100) = \frac{(A^2 t\tau_{max})/10.392}{(A^2 t\tau_{max})/6.283}(100) = 78.54 \tag{5}$$

ANS Circular shape sould be used. ; %t orque in triangle relative to circle is 60.46% ; %t orque in square relative to circle is 78.54%

CHAPTER 6

Section 6.1

6. 1

Solution $\varepsilon_1 = 2000\ \mu\ in/in$ $\varepsilon_2 = -1500\ \mu\ in/in$ $\psi = ?$

From the deformed geometry we obtain two equations that can be solved as shown below.

$$\varepsilon_1 = \frac{AB_1 - AB}{AB} = \frac{R\psi - 48}{48} = 2000(10^{-6}) \qquad \varepsilon_2 = \frac{CD_1 - CD}{CD} = \frac{(R-4)\psi - 48}{48} = -1500(10^{-6}) \quad \textbf{(1)}$$

$$\text{Subtracting} \qquad (4\psi)/48 = 3500(10^{-6}) \qquad or \qquad \psi = 0.042\ rads = 2.41° \quad \textbf{(2)}$$

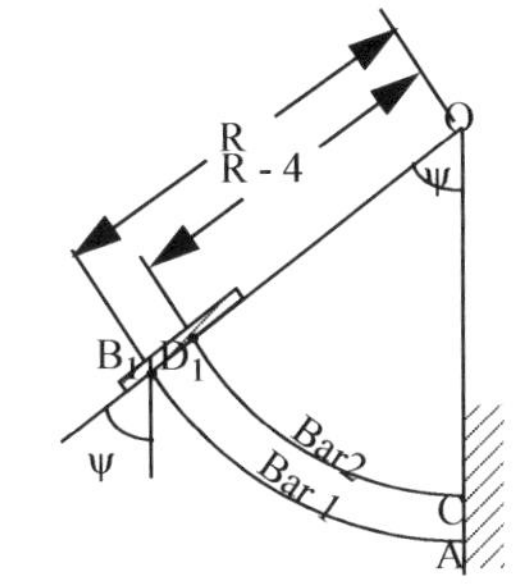

ANS $\psi = 2.41°$

6. 2

Solution $\varepsilon_1 = 2000\ \mu\ in/in$ $\varepsilon_2 = -1500\ \mu\ in/in$ $h = ?$ where $\varepsilon_3 = 0$

From previous problem and the deformed geometry we obtain the following.

$$\frac{R\psi - 48}{48} = 2000(10^{-6}) \qquad \psi = 0.042\ rads \qquad \varepsilon_3 = \frac{EF_1 - EF}{EF} = \frac{(R-h)\psi - 48}{48} = 2000(10^{-6}) - h\psi/48 = 0 \text{ or}$$

$$h\psi/48 = 2000(10^{-6}) \qquad or \qquad h = 0.096/0.042 = 2.2857\ in \quad \textbf{(1)}$$

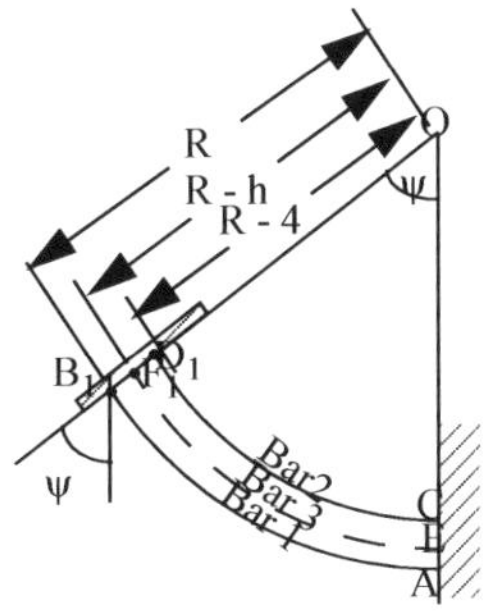

ANS $h = 2.29$ in.

6. 3

Solution $\varepsilon_2 = 0$ $\varepsilon_5 = 0$ $\varepsilon_1 = ?$ $\varepsilon_3 = ?$ $\varepsilon_4 = ?$ $\varepsilon_6 = ?$

The rotation of the rigid plates can be written in radians as shown below.

$$\psi_1 = (1.25/180)\pi = 0.02182\ rads \qquad \psi_2 = (2.5/180)\pi = 0.04363\ rads \quad \textbf{(1)}$$

From the deformed geometry, we obtain the following.

$$\varepsilon_2 = \frac{CD_1 - CD}{CD} = \frac{R_2\psi_1 - 3}{3} = 0 \qquad or \qquad R_2\psi_1 = 3 \quad \textbf{(2)}$$

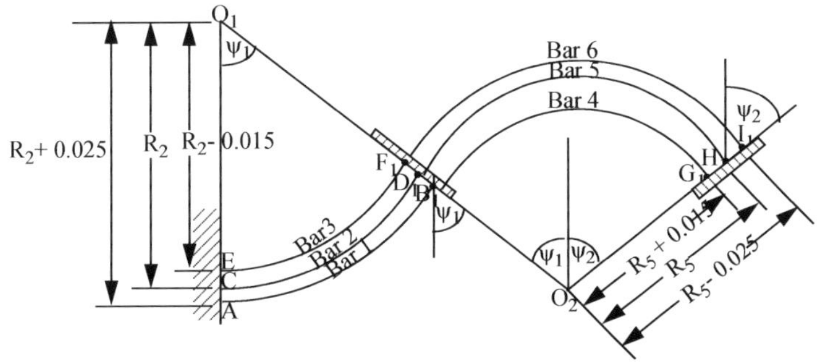

$$\varepsilon_1 = \frac{AB_1 - AB}{AB} = \frac{(R_2 + 0.025)\psi_1 - 3}{3} = \frac{(0.025)\psi_1}{3} = \frac{(0.025)(0.02182)}{3} = 181.8(10^{-6}) \quad \textbf{(3)}$$

$$\varepsilon_3 = \frac{EF_1 - EF}{EF} = \frac{(R_2 - 0.015)\psi_1 - 3}{3} = \frac{-(0.015)\psi_1}{3} = -\frac{(0.015)(0.02182)}{3} = -109.1(10^{-6}) \quad \textbf{(4)}$$

$$\varepsilon_5 = \frac{D_1H_1 - DH}{DH} = \frac{R_5(\psi_1 + \psi_2) - 2.5}{2.5} = 0 \quad or \quad R_5(\psi_1 + \psi_2) = 2.5 \quad \textbf{(5)}$$

$$\varepsilon_4 = \frac{B_1G_1 - BG}{BG} = \frac{(R_5 - 0.025)(\psi_1 + \psi_2) - 2.5}{2.5} = \frac{-(0.025)(\psi_1 + \psi_2)}{2.5} = -\frac{(0.025)(0.02182 + 0.04363)}{2.5} = -654.49(10^{-6}) \quad \textbf{(6)}$$

$$\varepsilon_6 = \frac{F_1I_1 - FI}{FI} = \frac{(R_5 + 0.015)(\psi_1 + \psi_2) - 2.5}{2.5} = \frac{(0.015)(\psi_1 + \psi_2)}{2.5} = \frac{(0.015)(0.02182 + 0.04363)}{2.5} = 392.7(10^{-6}) \quad \textbf{(7)}$$

ANS $\varepsilon_1 = 182\mu\ m/m$; $\varepsilon_3 = -109.1\mu\ m/m$; $\varepsilon_4 = -654\ \mu\ m/m$; $\varepsilon_6 = 393\ \mu\ m/m$

---

6. 4

Solution $\varepsilon_1 = 800\ \mu$ $\quad\varepsilon_3 = 500\ \mu$ $\quad\varepsilon_2 = ?$ $\quad\varepsilon_4 = ?$

The rotation of the rigid plates can be written in radians as shown below.

$$\psi_1 = (2.0/180)\pi = 0.03491\ rads \qquad \psi_2 = (3.5/180)\pi = 0.06109\ rads \quad \textbf{(1)}$$

From the deformed geometry we obtain the following.

$$\varepsilon_1 = \frac{AB_1 - AB}{AB} = \frac{R_1\psi_1 - 3}{3} = 800(10^{-6}) \quad \textbf{(2)}$$

$$\varepsilon_2 = \frac{CD_1 - CD}{CD} = \frac{(R_1 - 0.025)\psi_1 - 3}{3} = \frac{R_1\psi_1 - 3}{3} - \frac{0.025\psi_1}{3} = 800(10^{-6}) - \frac{0.025(0.03491)}{3} = 509.11(10^{-6}) \quad \textbf{(3)}$$

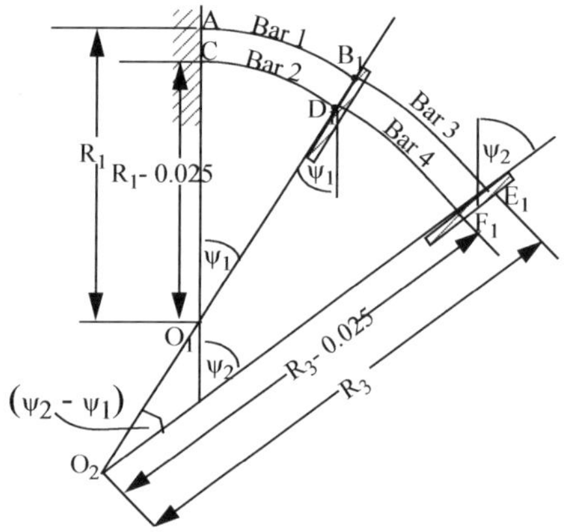

$$\varepsilon_3 = \frac{B_1E_1 - BE}{BE} = \frac{R_3(\psi_2 - \psi_1) - 2.5}{2.5} = 500(10^{-6}) \quad \textbf{(4)}$$

$$\varepsilon_4 = \frac{D_1F_1 - DF}{DF} = \frac{(R_3 - 0.025)(\psi_2 - \psi_1) - 2.5}{2.5} = \frac{R_3(\psi_2 - \psi_1) - 2.5}{2.5} - \frac{0.025(\psi_2 - \psi_1)}{2.5} \quad \text{or}$$

$$\varepsilon_4 = 500(10^{-6}) - \frac{0.025(0.06109 - 0.03491)}{2.5} = 238.2(10^{-6}) \quad \textbf{(5)}$$

ANS $\varepsilon_2 = 509\ \mu m/m$; $\varepsilon_4 = 238\ \mu m/m$

---

6. 5

Solution $\varepsilon_1 = 2000\ \mu\ in/in$ $\quad\psi = 2^o$ $\quad E = 30{,}000$ ksi $\quad A = 1/2\ in^2$ $\quad M_z = ?$ $\quad P = ?$

1. Strain The rotation of the rigid plates can be written in radians as shown below.

$$\psi = 2.0\pi/180 = 0.03491\ rads \quad \textbf{(1)}$$

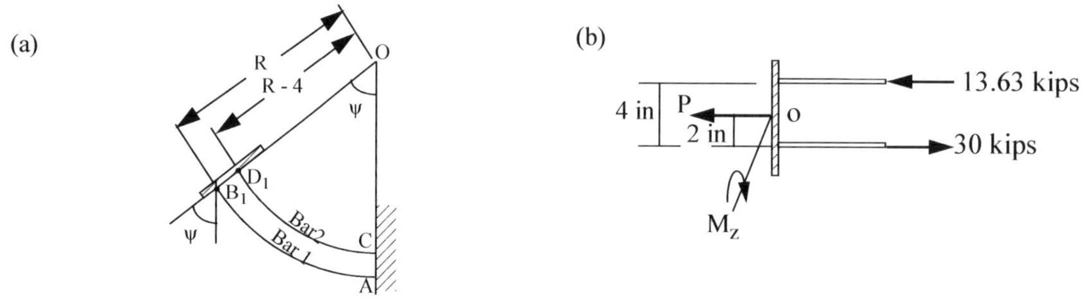

From the deformed geometry we obtain the following.

$$\varepsilon_1 = \frac{AB_1 - AB}{AB} = \frac{R\psi - 48}{48} = 2000(10^{-6}) \quad \textbf{(2)}$$

$$\varepsilon_2 = \frac{CD_1 - CD}{CD} = \frac{(R-4)\psi - 48}{48} = \frac{R\psi - 48}{48} - \frac{4\psi}{48} = 2000(10^{-6}) - \frac{4(0.03491)}{48} = -908.88(10^{-6}) \quad \textbf{(3)}$$

2. Stresses

$$\sigma_1 = E\varepsilon_1 = (30,000)(2000)(10^{-6}) = 60\ ksi \quad or \quad \sigma_1 = 60\ ksi\ (T) \quad \textbf{(4)}$$

$$\sigma_2 = E\varepsilon_2 = (30,000)(-908.88)(10^{-6}) = -27.27\ ksi \quad or \quad \sigma_2 = 27.27\ ksi\ (C) \quad \textbf{(5)}$$

3. Internal forces

$$N_1 = \sigma_1 A = (60)(1/2) = 30\ kips(T) \qquad N_2 = \sigma_2 A = (27.27)(1/2) = 13.63\ kips(C) \quad \textbf{(6)}$$

4. External forces and moments: By equilibrium of forces and moment about point O we obtain the following.

$$P + 13.63 - 30 = 0 \quad or \quad P = 16.37\ kips \qquad M_z - (13.63)(2) - (30)(2) = 0 \quad or \quad M_z = 87.26\ in-kips \quad \textbf{(7)}$$

ANS $P = 16.37\ kips;\ M_z = 87.26\ in-kips$

---

6. 6

Solution $\varepsilon_2 = 0$ $\psi = 1.25^o$ E = 200 GPa A = 100 mm^2 M_z = ? P= ?

1. Strains

$$\psi = 1.25(\pi/180) = 0.02182\ rads \quad \textbf{(1)}$$

We can draw an exaggerated approximate deformed geometry as shown in figure (a).

(a)

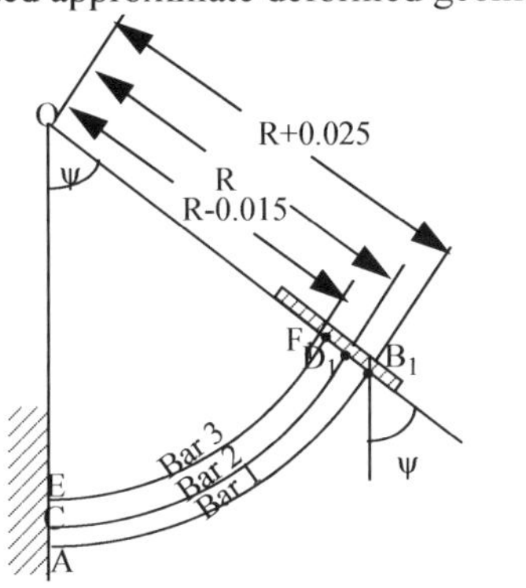

(b)

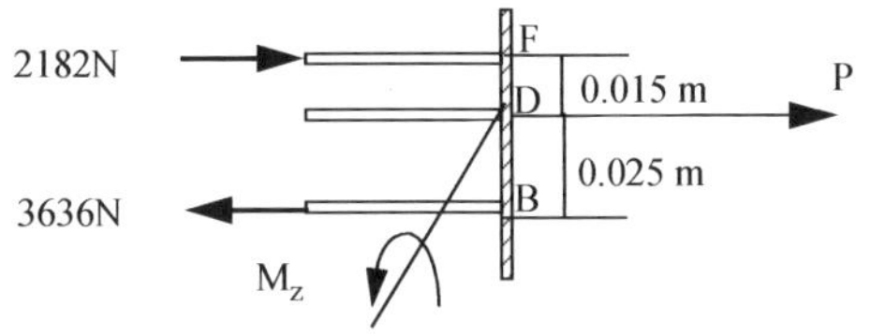

From the deformed geometry, we obtain the following.

$$\varepsilon_2 = \frac{CD_1 - CD}{CD} = \frac{R\psi - 3}{3} = 0 \quad or \quad R\psi = 3 \quad \textbf{(2)}$$

$$\varepsilon_1 = \frac{AB_1 - AB}{AB} = \frac{(R+0.025)\psi - 3}{3} = \frac{R\psi + (0.025)\psi - 3}{3} = \frac{(0.025)\psi}{3} = \frac{(0.025)(0.02182)}{3} = 181.8(10^{-6}) \quad \textbf{(3)}$$

$$\varepsilon_3 = \frac{EF_1 - EF}{EF} = \frac{(R-0.015)\psi - 3}{3} = \frac{R\psi - (0.015)\psi - 3}{3} = \frac{-(0.015)\psi}{3} = -\frac{(0.015)(0.02182)}{3} = -109.1(10^{-6}) \quad \textbf{(4)}$$

2. Stress calculations:

$$\sigma_1 = E\varepsilon_1 = (200)(10^9)(181.8)(10^{-6}) = 36.36(10^6)\ N/m^2 \quad or \quad \sigma_1 = 36.36(10^6)\ N/m^2(T) \quad \textbf{(5)}$$

$$\sigma_2 = E\varepsilon_2 = (200)(10^9)(-109.1)(10^{-6}) = -21.82(10^6) N/m^2 \quad or \quad \sigma_2 = 21.82(10^6)\ N/m^2(C) \quad \textbf{(6)}$$

3. Internal forces

$$N_1 = \sigma_1 A = (36.36)(10^6)(100)(10^{-6}) = 3636N\ (T) \qquad N_2 = \sigma_2 A = (21.82)(10^6)(100)(10^{-6}) = 2182N\ (C) \quad \textbf{(7)}$$

4. External forces and moments :By equilibrium of forces and moment about point D we obtain the following.

$$P + 2182 - 3636 = 0 \quad or \quad P = 1454\ N \qquad M_z - (2182)(0.015) - (3636)(0.025) = 0 \quad or \quad M_z = 123.6\ N-m \quad \textbf{(8)}$$

ANS $P = 1454\ N;\ M_z = 123.6\ N-m$

---

6. 7

Solution $\varepsilon_1 = 800\ \mu$ $\varepsilon_3 = 500\ \mu$ $\psi_1 = 2^o$ $\psi_1 = 3.5^o$ $E_1 = E_3$ =200 GPa

$E_2 = E_4$ = 70 GPa A = 125 mm^2 M_1 = ? P_1= ? M_2 = ? P_2= ?

1. Strain calculations: From earlier problem we have:

$$\varepsilon_4 = 238.2(10^{-6}) \qquad \varepsilon_2 = 509.11(10^{-6}) \quad \textbf{(1)}$$

2. Stresses

$$\sigma_1 = E_1\varepsilon_1 = (200)(10^9)(800)(10^{-6}) = 160(10^6)\frac{N}{m^2}(T) \qquad \sigma_2 = E_2\varepsilon_2 = (70)(10^9)(509.11)(10^{-6}) = 35.64(10^6)\frac{N}{m^2}(T) \quad \textbf{(2)}$$

$$\sigma_3 = E_3\varepsilon_3 = (200)(10^9)(500)(10^{-6}) = 100(10^6)\frac{N}{m^2}(T) \qquad \sigma_4 = E_4\varepsilon_4 = (70)(10^9)(238.2)(10^{-6}) = 16.67(10^6)\frac{N}{m^2}(T) \quad \textbf{(3)}$$

3. Internal force calculations: The internal normal force in each bar can be found as shown below.

$$N_1 = 160(10^6)(125)(10^{-6}) = 20(10^3)N = 20kN(T) \qquad N_2 = 35.64(10^6)(125)(10^{-6}) = 4.46(10^3)N = 4.46kN(T) \quad \textbf{(4)}$$

$$N_3 = 100(10^6)(125)(10^{-6}) = 12.5(10^3)N = 12.5kN(T) \qquad N_4 = 16.67(10^6)(125)(10^{-6}) = 2.08(10^3)N = 2.08kN(T) \quad \textbf{(5)}$$

4. External forces and moments: By equilibrium of forces and moments we obtain the following.

(a)

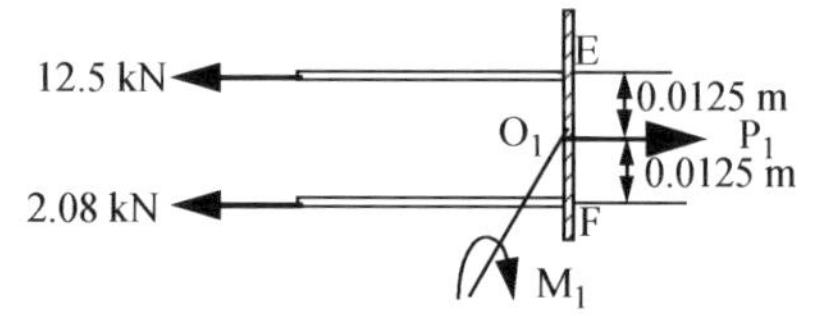

(b)

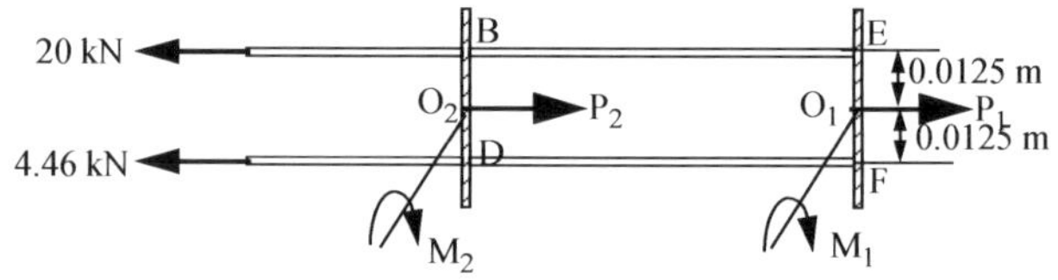

$$P_1 - 12.5 - 2.08 = 0 \quad or \quad P_1 = 14.58kN \qquad M_1 = (12.5)(0.0125) - (2.08)(0.0125) = 0.1303 \ \ kN-m \quad \textbf{(6)}$$

$$P_2 + P_1 - 20 - 4.46 = 0 \quad or \quad P_2 = 9.88kN \qquad M_2 = (20)(0.0125) - (4.46)(0.0125) - M_1 = 0.06395 \ \ kN-m \quad \textbf{(7)}$$

ANS $P_1 = 14.58kN$; $M_1 = 130.3N-m$; $P_2 = 9.88kN$; $M_2 = 64.0N-m$

---

6. 8

Solution $E_{wood} = 10$ GPa $\varepsilon_{xx} = -0.012y$ $t_W = 20$ mm h = 250 mm $t_F = 20$ mm $d_f = 125$ mm. $\boldsymbol{M_z} = ?$

The differential area in the flange is $dA = 0.25dy$ and in the web is $dA = 0.02dy$. From Hooke's law we can write:

$$\sigma_{xx} = E_{wood}\varepsilon_{xx} = (10)(10^9)(-0.012y) = -120y(10^6) \ \ N/m^2 \quad \textbf{(1)}$$

$$\sigma_{xx} = \begin{cases} -120y(10^6) \ \ N/m^2 & dA = 0.25dy \quad 0.250 < y \le 0.270 \\ -120y(10^6) \ \ N/m^2 & dA = 0.02dy \quad -0.250 < y < 0.250 \\ -120y(10^6) \ \ N/m^2 & dA = 0.25dy \quad -0.270 \le y < -0.250 \end{cases} \quad \textbf{(2)}$$

By symmetry the moment equation is twice the sum of integrals over the upper half (y>0) of the cross-section.

$$\boldsymbol{M_z} = -\int_A y\sigma_{xx}dA = -2\left[\int_0^{0.250} y(-120y)(0.02)dy + \int_{0.250}^{0.270} y(-120y)(0.25)dy\right](10^6) \ \ \text{or}$$

$$\boldsymbol{M_z} = -2\left[-2.4\frac{y^3}{3}\Bigg|_0^{0.250} - 30\frac{y^3}{3}\Bigg|_{0.250}^{0.270}\right](10^6) = -2[-12.5 - 40.58](10^3) = 106.2(10^3) \ \ N-m \quad \textbf{(3)}$$

ANS $\boldsymbol{M_z} = 106.2$ kN-m

---

6. 9

Solution $E_{wood} = 10$ GPa $\varepsilon_{xx} = -0.15y$ $t_W = 10$ mm h = 50 mm $t_F = 10$ mm $d_f = 25$ mm. $\boldsymbol{M_z} = ?$

The differential area in the flange is $dA = 0.05dy$ and in the web is $dA = 0.01dy$. From Hooke's law we can write:

$$\sigma_{xx} = E_{wood}\varepsilon_{xx} = (10)(10^9)(-0.15y) = -1.5y(10^9) \ \ N/m^2 \quad \textbf{(1)}$$

$$\sigma_{xx} = \begin{cases} -1.5y(10^9) \ \ N/m^2 & dA = 0.05dy \quad 0.05 < y \le 0.06 \\ -1.5y(10^9) \ \ N/m^2 & dA = 0.01dy \quad -0.05 < y < 0.05 \\ -1.5y(10^9) \ \ N/m^2 & dA = 0.05dy \quad -0.06 \le y < -0.05 \end{cases} \quad \textbf{(2)}$$

By symmetry the moment equation is twice the sum of integrals in the upper half (y>0) of the cross-section.

$$\boldsymbol{M_z} = -\int_A y\sigma_{xx}dA = -2\left[\int_0^{0.05} y(-1.5y)(0.01)dy + \int_{0.05}^{0.06} y(-1.5y)(0.05)dy\right](10^9) \ \ \text{or}$$

$$\boldsymbol{M_z} = -2\left[-0.015\frac{y^3}{3}\Bigg|_0^{0.05} - 0.075\frac{y^3}{3}\Bigg|_{0.05}^{0.06}\right](10^9) = -2[-625 - 2275] \ \ N-m = 5800 \ \ N-m \quad \textbf{(3)}$$

ANS $\boldsymbol{M_z} = 5.8$ kN-m

6. 10

Solution $E_{wood} = 10$ GPa$\varepsilon_{xx} = 0.02y$ $t_W = 15$ mm h =200 mm $t_F = 20$ mm d= 150 mm. $M_z = ?$

The differential area in the flange is $dA = 0.3dy$ and in the web is $dA = 0.015dy$. From Hooke's law we can write:

$$\sigma_{xx} = E_{wood}\varepsilon_{xx} = (10)(10^9)(0.02y) = 0.2y(10^9)\ N/m^2 \quad (1)$$

$$\sigma_{xx} = \begin{cases} 0.2y(10^9)\ N/m^2 & dA = 0.3dy & 0.20 < y \le 0.22 \\ 0.2y(10^9)\ N/m^2 & dA = 0.015dy & -0.20 < y < 0.20 \\ 0.2y(10^9)\ N/m^2 & dA = 0.3dy & -0.22 \le y < -0.20 \end{cases} \quad (2)$$

By symmetry the moment equation is twice the sum of integrals over the upper half (y>0) of the cross-section.

$$M_z = -\int_A y\sigma_{xx}dA = -2\left[\int_0^{0.20} y(0.2y)(0.015)dy + \int_{0.20}^{0.22} y(0.2y)(0.3)dy\right](10^9) \text{ or}$$

$$M_z = -2\left[0.003\frac{y^3}{3}\bigg|_0^{0.20} + 0.06\frac{y^3}{3}\bigg|_{0.20}^{0.22}\right](10^9) = -2[8 + 52.96](10^3) = -121.92\ N-m \quad (3)$$

ANS $M_z = -121.9$ kN-m

---

6. 11

Solution $E_{steel} = 30{,}000$ ksi $E_{wood} = 2{,}000$ ksi $\varepsilon_{xx} = -100y\ \mu$. d =2 in $h_W = 4$ in $h_S = (1/8)$ in. $M_z = ?$

The differential area is $dA = 2dy$. From Hooke's law we can write:

$$(\sigma_{xx})_{steel} = E_{steel}\varepsilon_{xx} = (30000)(-100y)(10^{-6}) = -3.0y\ ksi \qquad (\sigma_{xx})_{wood} = E_{wood}\varepsilon_{xx} = (2000)(-100y)(10^{-6}) = -0.2y\ ksi \quad (1)$$

$$\sigma_{xx} = \begin{cases} -3.0y\ ksi & 2.0 < y \le 2.125 \\ -0.2y\ ksi & -2.0 < y < 2.0 \\ -3.0y\ ksi & -2.125 \le y < -2.0 \end{cases} \quad (2)$$

$$M_z = -\int_A y\sigma_{xx}dA = -\left[\int_{-2.125}^{-2.0} y(-3.0y)(2)dy + \int_{-2.0}^{2.0} y(-0.2y)(2)dy + \int_{2.0}^{2.125} y(-3.0y)(2)dy\right] \text{ or}$$

$$M_z = -\left[-6\frac{y^3}{3}\bigg|_{-2.125}^{-2.0} - 0.4\frac{y^3}{3}\bigg|_{-2.0}^{2.0} - 6\frac{y^3}{3}\bigg|_{2.0}^{2.125}\right] = -[-3.191 - 2.133 - 3.191] = 8.52 \text{ in-kips} \quad (3)$$

ANS $M_z = 8.52$ in-kips

---

6. 12

Solution $E_{steel} = 30{,}000$ ksi $E_{wood} = 2{,}000$ ksi $\varepsilon_{xx} = -50\ y\ \mu$ d =1 in $h_W = 6$ in $h_S = (1/4)$ in. $M_z = ?$

The differential area is $dA = (1)dy$. From Hooke's law we can write:

$$(\sigma_{xx})_{steel} = E_{steel}\varepsilon_{xx} = (30000)(-50y)(10^{-6}) = -1.5y\ ksi \qquad (\sigma_{xx})_{wood} = E_{wood}\varepsilon_{xx} = (2000)(-50y)(10^{-6}) = -0.1y\ ksi \quad (1)$$

$$\sigma_{xx} = \begin{cases} -1.5y\ ksi & 3.0 < y \le 3.25 \\ -0.1y\ ksi & -3.0 < y < 3.0 \\ -1.5y\ ksi & -3.25 \le y < -3.0 \end{cases} \quad (2)$$

$$M_z = -\int_A y\sigma_{xx}dA = -\left[\int_{-3.25}^{-3.0} y(-1.5y)(1)dy + \int_{-3.0}^{3.0} y(-0.1y)(1)dy + \int_{3.0}^{3.25} y(-1.5y)(1)dy\right] \text{ or}$$

$$M_z = -\left[-1.5\frac{y^3}{3}\bigg|_{-3.25}^{-3.0} - 0.1\frac{y^3}{3}\bigg|_{-3.0}^{3.0} - 1.5\frac{y^3}{3}\bigg|_{3.0}^{3.25}\right] = -[-3.664 - 1.800 - 3.664] = 9.13 \text{ in-kips} \quad (3)$$

ANS $M_z = 9.13$ in-kips

---

6. 13

Solution $E_{steel} = 30{,}000$ ksi $E_{wood} = 2{,}000$ ksi $\varepsilon_{xx} = 200\ y\ \mu$. d =1 in $h_W = 2$ in $h_S = (1/16)$ in. $M_z = ?$

The differential area is $dA = (1)dy$. From Hooke's law we can write:

$$(\sigma_{xx})_{steel} = E_{steel}\varepsilon_{xx} = (30000)(200y)(10^{-6}) = 6y\ ksi \qquad (\sigma_{xx})_{wood} = E_{wood}\varepsilon_{xx} = (2000)(200y)(10^{-6}) = 0.4y\ ksi \quad (1)$$

$$\sigma_{xx} = \begin{cases} 6y \ \ ksi & 1.0 < y \le 1.0625 \\ 0.4y \ \ ksi & -1.0 < y < 1.0 \\ 6y \ \ ksi & -1.0625 \le y < -1.0 \end{cases} \quad (2)$$

$$M_z = -\int_A y\sigma_{xx}dA = -\left[\int_{-1.0625}^{-1.0} y(6y)(1)dy + \int_{-1.0}^{1.0} y(0.4y)(1)dy + \int_{1.0}^{1.0625} y(6y)(1)dy\right] \text{ or}$$

$$M_z = -\left[6\frac{y^3}{3}\bigg|_{-1.0625}^{-1.0} + 0.4\frac{y^3}{3}\bigg|_{-1.0}^{1.0} + 6\frac{y^3}{3}\bigg|_{1.0}^{1.0625}\right] = -[0.3989 + 0.2667 + 0.3989] = -1.06 \text{ in.-kips} \quad (3)$$

ANS $M_z = -1.06$ in.-kips

---

6. 14

Solution $E_{steel} = 200$ GPa $E_{wood} = 10$ GPa$\varepsilon_{xx} = 0.02y$ $t_W = 15$ mm $h_W = 200$ mm $t_f = 20$ mm $d_f = 150$ mm. $M_z = ?$

The differential area in steel is $dA = 0.3dy$ and in wood is $dA = 0.015dy$ .From Hooke's law we can write:

$$(\sigma_{xx})_{steel} = E_{steel}\varepsilon_{xx} = (200)(10^9)(0.02y) = 4y(10^9)\frac{N}{m^2} \qquad (\sigma_{xx})_{wood} = E_{wood}\varepsilon_{xx} = (10)(10^9)(0.02y) = 0.2y(10^9)\frac{N}{m^2} \quad (1)$$

$$\sigma_{xx} = \begin{cases} 4y(10^9) \ N/m^2 & dA = 0.3dy & 0.20 < y \le 0.22 \\ 0.2y(10^9) \ N/m^2 & dA = 0.015dy & -0.20 < y < 0.20 \\ 4y(10^9) \ N/m^2 & dA = 0.3dy & -0.22 \le y < -0.20 \end{cases} \quad (2)$$

By symmetry the integral is written as twice the sum of integrals over steel and wood in the upper half (y>0) of the cross-section.

$$M_z = -\int_A y\sigma_{xx}dA = -2\left[\int_0^{0.20} y(0.2y)(0.015)dy + \int_{0.20}^{0.22} y(4y)(0.3)dy\right](10^9) \text{ or}$$

$$M_z = -2\left[0.003\frac{y^3}{3}\bigg|_0^{0.20} + 1.2\frac{y^3}{3}\bigg|_{0.20}^{0.22}\right](10^9) = -2[8 + 1059.2](10^3) \quad (3)$$

ANS $M_z = -2134 kN - m$

---

6. 15

Solution $M_z = ?$

The stress distribution in the linear region is: $\sigma = -(30/1.5)y = -20y \ \ ksi$ as shown below.

(a)

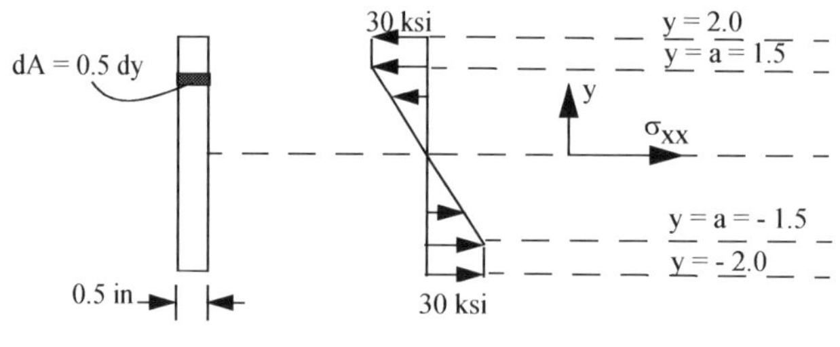

$$\begin{aligned} \sigma_{xx} &= 30 \ \ ksi & dA &= 0.5dy & -2 \le y < -1.5 \\ \sigma_{xx} &= -20y \ \ ksi & dA &= 0.5dy & -1.5 < y < 1.5 \\ \sigma_{xx} &= -30 \ \ ksi & dA &= 0.5dy & 1.5 < y < 2 \end{aligned} \quad (1)$$

By symmetry the integral is twice the integral in the positive y half as shown below.

$$M_z = -\int_A y\sigma_{xx}dA = -2\left[\int_0^{1.5} y[\sigma_{xx}](0.5dy) + \int_{1.5}^{2} y[\sigma_{xx}](0.5dy)\right] = -2\left[\int_0^{1.5} y[-20y](0.5dy) + \int_{1.5}^{2} y[-30](0.5dy)\right] \text{ or}$$

$$M_z = -2\left[-10\frac{y^3}{3}\bigg|_0^{1.5} - 15\frac{y^2}{2}\bigg|_{1.5}^{2}\right] = 2[11.25 + 13.125] = 48.75 \text{ in.-kips} \quad (2)$$

ANS $M_z = 48.75$ in.-kips

6. 16

Solution ε_{xx}=-0.01y y in meters σ_{yield}=250MPa E=200GPa M_z = ?

With geometry being symmetric in y, the elastic-plastic boundary will be on both sides. The strain at yield point is:

$$\varepsilon_{yield} = \frac{\sigma_{yield}}{E} = \frac{250(10^6)}{200(10^6)} = 1.25(10^{-3}) \qquad \varepsilon_{yield} = \pm 1.25(10^{-3}) = -0.01 y_{yield} \quad or \quad y_{yield} = \pm 0.125m = \pm 125mm \quad (1)$$

In the linear region $\sigma_{xx} = E\varepsilon_{xx} = (200)(10^9)(-0.01y) = -2000y$ y in meters.

$$\sigma_{xx} = \begin{cases} -250\ MPa & 0.125m < y < 0.150m \\ -2000y\ MPa & -0.125m < y < 0.125m \\ 250\ MPa & -0.150m < y < -0.125m \end{cases} \quad (2)$$

$$M_z = -\int_A y\sigma_{xx} dA = -[\int_{-0.15}^{-0.125} y(250)(0.2dy) + \int_{-0.125}^{0.125} y(-2000y)(0.2dy) + \int_{0.125}^{0.15} y(-250)(0.2dy)](10^6) \text{ or}$$

$$M_z = -\left[50\frac{y^2}{2}\Bigg|_{-0.15}^{-0.125} - 400\frac{y^3}{3}\Bigg|_{-0.125}^{0.125} - 50\frac{y^2}{2}\Bigg|_{0.125}^{0.15}\right](10^6) = 864.5(10^3)\ N-m \quad (3)$$

ANS M_z = 864.5 kN-m

---

6. 17

Solution ε_{xx}=-0.01y y in meters σ_{yield}=250MPa E_1 = 200 GPa E_2= 80 GPa M_z = ?

With geometry being symmetric in y, the elastic-plastic boundary will be on both sides. The strain at yield point is:

$$\varepsilon_{yield} = \frac{\sigma_{yield}}{E} = \frac{250(10^6)}{200(10^6)} = 1.25(10^{-3}) \qquad \varepsilon_{yield} = \pm 1.25(10^{-3}) = -0.01 y_{yield} \quad or \quad y_{yield} = \pm 0.125m = \pm 125mm \quad (1)$$

In linear region: . $\sigma_1 = E_1\varepsilon_1 = (200)(10^9)(-0.01y) = -2000y\ MPa$

For $y < -0.125m$ i.e., on the tensile side

$$\sigma_2 = \sigma_{yield} + E_2(\varepsilon_{xx} - \varepsilon_{yield}) = 250(10^6) + (80)(10^9)[-0.01y - 1.25(10^{-3})] = (150 - 800y)MPa \quad (2)$$

For $y > 0.125m$ i.e., on the compressive side

$$\sigma_2 = -\sigma_{yield} + E_2(\varepsilon_{xx} - (-\varepsilon_{yield})) = (-250)(10^6) + (80)(10^9)[-0.01y + 1.25(10^{-3})] = (-150 - 800y)MPa$$

$$\sigma_{xx} = \begin{cases} (-150 - 800y)MPa & 0.125m < y < 0.150m \\ -2000yMPa & -0.125m < y < 0.125m \\ (150 - 800y)MPa & -0.150m < y < -0.125m \end{cases} \quad (3)$$

$$M_z = -[\int_{-0.15}^{-0.125} y(150 - 800y)(0.2dy) + \int_{-0.125}^{0.125} y(-2000y)(0.2dy) + \int_{0.125}^{0.15} y(-150 - 800y)(0.2dy)](10^6) \text{ or}$$

$$M_z = -\left[\left(30\frac{y^2}{2} - 40\frac{y^3}{3}\right)\Bigg|_{-0.15}^{-0.125} - 400\frac{y^3}{3}\Bigg|_{-0.125}^{0.125} + \left(-30\frac{y^2}{2} - 40\frac{y^3}{3}\right)\Bigg|_{0.125}^{0.15}\right](10^6) = 765.0(10^3)\ N-m \quad (4)$$

ANS M_z = 765 kN-m

---

6. 18

Solution ε_{xx}=-0.01y y in meters $\sigma = 952\varepsilon^{0.2}\ MPa$ M_z = ?

As the material behavior in tension and compression is the same, we have:

$$\sigma_{xx} = \begin{cases} 952\varepsilon^{0.2} MPa & \varepsilon > 0 \\ -952(-\varepsilon)^{0.2} MPa & \varepsilon < 0 \end{cases} \quad or \quad \sigma_{xx} = \begin{cases} 379(-y)^{0.2} MPa & y < 0 \\ -379(y)^{0.2} MPa & y > 0 \end{cases} \quad (1)$$

By symmetry the integral in moment equation is twice the integral in the positive y half as shown below.

$$M_z = -\int_A y\sigma_{xx} dA = -[\int_{-0.15}^{0} y[\sigma_{xx}](0.2dy) + \int_0^{0.15} y[\sigma_{xx}](0.2dy)] = -2\int_0^{0.15} y[\sigma_{xx}](0.2dy) \text{ or}$$

$$M_z = -2\left[\int_0^{0.15} y[-379(y)^{0.2}](0.2dy)\right](10^6) = 151.6\left[\int_0^{0.15} y^{1.2}dy\right](10^6) = 151.6\left(\frac{y^{2.2}}{2.2}\right)\Bigg|_0^{0.15}(10^6) = 1061(10^3) \quad (2)$$

ANS M_z = 1061 kN-m

Section 6.2

6. 19

Solution

The lengths a_S and a_H in terms of area can be found as shown below:

$$A_S = a_S^2 = A \quad or \quad a_S = \sqrt{A} \qquad A_H = (\alpha a_H)^2 - a_H^2 = (\alpha^2 - 1)a_H^2 = A \quad or \quad a_H = \sqrt{\frac{A}{(\alpha^2 - 1)}} \qquad (1)$$

$$I_S = \frac{1}{12}a_S a_S^3 = \frac{1}{12}A^2 \qquad I_H = \frac{1}{12}(\alpha a_H)(\alpha a_H)^3 - \frac{1}{12}a_H a_H^3 = \frac{(\alpha^4 - 1)}{12}a_H^4 = \frac{(\alpha^4 - 1)}{12}\left[\frac{A}{(\alpha^2 - 1)}\right]^2 = \frac{1}{12}\left(\frac{\alpha^2 + 1}{\alpha^2 - 1}\right)A^2 \qquad (2)$$

ANS $I_H/I_S = (\alpha^2 + 1)/(\alpha^2 - 1)$

---

6. 20

Solution

The total area moment of inertia for separate beams is four times the area moment of inertia for the individual beam.

$$I_S = 4\left[\frac{1}{12}ab^3\right] = \frac{ab^3}{3} \qquad I_G = \frac{1}{12}a(4b)^3 = \frac{16ab^3}{3} \qquad I_G = 16I_S \qquad (1)$$

The maximum bending normal stress for separate beams will be $y_{max} = b/2$ while glued beam will be at $y_{max} = 2b$.

$$\sigma_S = \frac{M_Z(b/2)}{I_S} = \frac{M_Z(b/2)}{(ab^3/3)} = \frac{3}{2}\left(\frac{M_Z}{ab^2}\right) \qquad \sigma_G = \frac{M_Z(2b)}{I_G} = \frac{M_Z(2b)}{(16ab^3/3)} = \frac{3}{8}\left(\frac{M_Z}{ab^2}\right) \qquad \sigma_G = \sigma_S/4 \qquad (2)$$

ANS $I_G = 16I_S;\ \sigma_G = \sigma_S/4$

---

6. 21

Solution $\quad I_T = ?$

The height of the triangle is $h = (a)sin60 = (\sqrt{3}/2)a$. Using Table A2 and neglecting t^2 terms as t<<a, we obtain the following.

$$I_T = \frac{1}{36}a\left[\left(\frac{\sqrt{3}}{2}\right)a\right]^3 - \frac{1}{36}(a-t)\left[\left(\frac{\sqrt{3}}{2}\right)(a-t)\right]^3 = \left(\frac{\sqrt{3}}{24}\right)[a^4 - (a-t)^4] = \left(\frac{\sqrt{3}}{24}\right)[a^4 - (a^2 - 2at + t^2)^2] \qquad (1)$$

$$(I_T = (\sqrt{3}/24)[a^4 - (a^2 - 2at)^2] \approx (\sqrt{3}/24)[a^4 - (a^4 - 4a^3t + t^2)]) \approx (\sqrt{3}/6)a^3t = a^3t/(2\sqrt{3}) \qquad (2)$$

ANS $I_T \approx a^3t/(2\sqrt{3})$

---

6. 22

Solution $\quad I_S = ?$

As t<<a, the t^2 term can be neglected and the remainder expanded as shown below.

$$I_S = \frac{1}{12}aa^3 - \frac{1}{12}(a-t)(a-t)^3 = \frac{1}{12}[a^4 - (a-t)^4] = \frac{1}{12}[a^4 - (a^2 - 2at + t^2)^2] \approx \frac{1}{12}[a^4 - (a^2 - 2at)^2] \text{ or}$$

$$I_S \approx \frac{1}{12}[a^4 - (a^4 - 4a^3t + 4a^2t^2)] \approx \frac{1}{12}[a^4 - a^4 + 4a^3t] \approx \frac{1}{3}a^3t \qquad (1)$$

ANS $I_S \approx a^3t/3$

---

6. 23

Solution $\quad I_C = ?$

As t<<a, the t^2 term can be neglected and the remainder expanded as shown below.

$$I_C = \frac{\pi}{4}(a)^4 - \frac{\pi}{4}(a-t)^4 = \frac{\pi}{4}[a^4 - (a^2 - 2at + t^2)^2] \approx \frac{\pi}{4}[a^4 - (a^2 - 2at)^2] = \frac{\pi}{4}[a^4 - (a^4 - 4a^3t + t^2)] \approx \frac{\pi}{4}[a^4 - a^4 + 4a^3t] \approx a^3t \qquad (1)$$

ANS $I_C \approx \pi a^3t/3$

---

6. 24

Solution $\quad \sigma_T : \sigma_S : \sigma_C = ?$

The dimensions of the figures a_T, a_S, and a_C can be found in terms of the material area as shown below.

$$(3a_T)t = A \quad or \quad a_T = \frac{A}{3t} \qquad (4a_S)t = A \quad or \quad a_S = \frac{A}{4t} \qquad (2\pi a_C)t = A \quad or \quad a_C = \frac{A}{2\pi t} \qquad (1)$$

The maximum values of y for the three cases are: $y_T = 2a_T/3$, $y_S = a_S/2$, and $y_C = a_C$.

$$\sigma_T = \frac{M_z y_T}{I_T} = \frac{M_z(2a_T/3)}{(a_T^3 t/2\sqrt{3})} = \frac{4\sqrt{3}}{3}\frac{M_z}{(a_T^2 t)} = \frac{4\sqrt{3}}{3}\frac{M_z}{[(A/3t)^2 t]} = 12\sqrt{3}\frac{M_z t}{A^2} = 20.78\left(\frac{M_z t}{A^2}\right) \quad (2)$$

$$\sigma_S = \frac{M_z y_S}{I_S} = \frac{M_z(a_S/2)}{(a_S^3 t/3)} = \frac{3}{2}\frac{M_z}{(a_S^2 t)} = \frac{3}{2}\frac{M_z}{[(A/4t)^2 t]} = 24\frac{M_z t}{A^2} \quad (3)$$

$$\sigma_C = \frac{M_z y_C}{I_C} = \frac{M_z(a_C)}{(\pi a_C^2 t/3)} = \frac{3}{\pi}\frac{M_z}{(a_C^2 t)} = \frac{3}{\pi}\frac{M_z}{[(A/2\pi t)^2 t]} = 12\pi\frac{M_z t}{A^2} = 37.7\left(\frac{M_z t}{A^2}\right) \quad (4)$$

ANS $\sigma_T : \sigma_S : \sigma_C = 20.78 : 24 : 37.7$

---

6. 25

Solution $\sigma_A = ?$ $\sigma_T = ?$ $\sigma_C = ?$

The bending normal stress at point A and the location of the centroid from the bottom of the cross-section is as shown below.

$$\sigma_A = E\varepsilon_A = (8000)(200)(10^{-6}) = 1.6\ ksi(T) \qquad \eta_c = \frac{\sum_i \eta_i A_i}{\sum_i A_i} = \frac{(4)(1)(2)+(4)(1)(4.5)}{(4)(1)+(4)(1)} = 3.25\ in \quad (1)$$

The bending normal stress is zero at the centroid C and is a linear function of y as shown below..

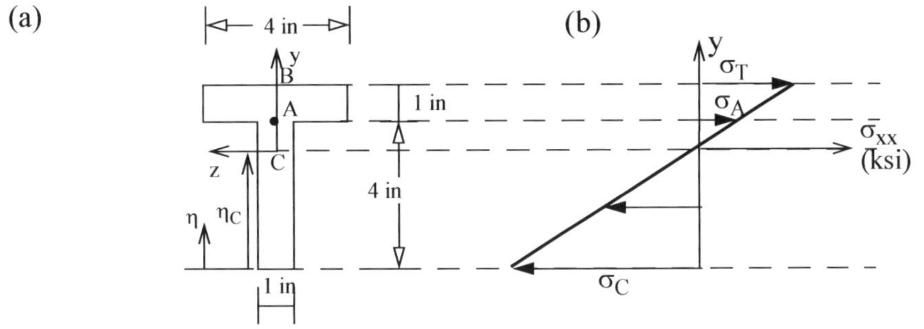

$$\frac{\sigma_T}{(5-3.25)} = \frac{\sigma_C}{3.25} = \frac{\sigma_A}{(4-3.25)} \quad or \quad \sigma_T = \frac{(1.75)(1.6)}{0.75} = 3.73 \qquad \sigma_C = \frac{(3.25)(1.6)}{0.75} = 6.93 \quad (2)$$

ANS $\sigma_T = 3.73 ksi(T); \sigma_C = 6.93 ksi(C)$

---

6. 26

Solution $\sigma_{max} = 40ksi(C$ $\varepsilon_A = ?$ $\sigma_T =$

The location of the centroid from the bottom of the cross-section can be found as shown below.

$$\eta_c = \frac{\sum_i \eta_i A_i}{\sum_i A_i} = \frac{2[(2.0)(0.5)(1.0)]+(4)(0.5)(2.25)}{2[(2.0)(0.5)]+(4)(0.5)} = 1.625\ in \quad (1)$$

The bending normal stress is zero at the centroid C and is a linear function of y as shown below..

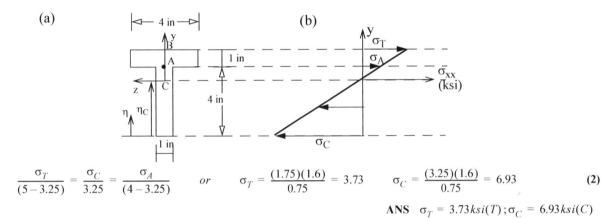

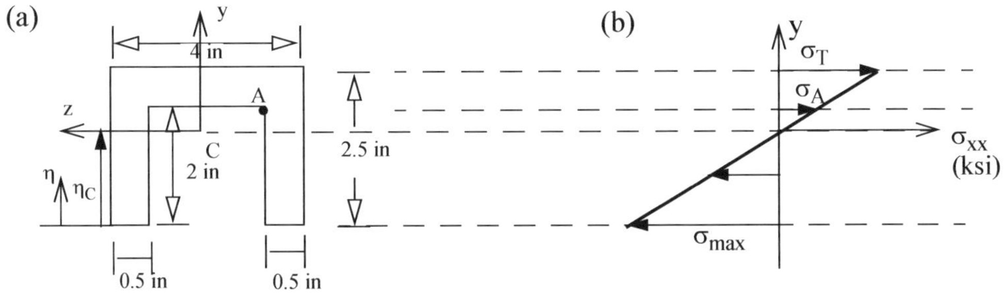

$$\frac{\sigma_T}{(2.5-1.625)} = \frac{\sigma_{max}}{1.625} = \frac{\sigma_A}{(2-1.625)} \qquad \sigma_T = \frac{0.875(40)}{1.625} = 21.54ksi \qquad \sigma_A = \frac{0.375(40)}{1.625} = 9.23ksi \quad (2)$$

$$\varepsilon_A = \sigma_A/E = 9.23/30000 = 307.69(10^{-6}) \quad (3)$$

ANS $\varepsilon_A = 307.69\ \mu;\ \sigma_T = 21.54$ ksi(T)

---

6. 27

Solution $(\varepsilon_{xx})_A = -200\mu$ $E_1 = 200$ GPa $E_2 = 70$ GPa $(\sigma_1)_{max} = ?$ $(\sigma_2)_{max} = ?$

At $y_A = 0.025$ m. the strain is -200μ and is zero at C. Noting the bending normal strain varies linearly with y, we obtain:

$$\varepsilon_{xx} = (-200/0.025)y = -8000y \ \mu \qquad (\varepsilon_{xx})_B = -8000(\pm 0.035) = \mp 280 \ \mu \tag{1}$$

The strain at top or bottom surface is:
The maximum stress in each material is:

$$(\sigma_1)_{max} = E_1(\varepsilon_{xx})_B = (200)(10^9)(\mp 280)(10^{-6}) = (\mp 56)(10^6) N/m^2 \tag{2}$$

$$(\sigma_2)_{max} = E_2(\varepsilon_{xx})_A = (70)(10^9)(\mp 200)(10^{-6}) = (\mp 14)y(10^6) N/m^2 \tag{3}$$

ANS $(\sigma_1)_{max} = 56$ MPa(T) or (C) ; $(\sigma_2)_{max} = 14$ MPa(T) or (C)

---

6. 28

Solution $(\varepsilon_{xx})_A = 300\mu$ $E_1 = 30{,}000$ ksi $E_2 = 20{,}000$ ksi $(\sigma_1)_{max} = ?$ $(\sigma_2)_{max} = ?$

At $y_A = 0.25$ in. the strain is 30μ and is zero at C. Noting the bending normal strain varies linearly with y, we obtain:

$$\varepsilon_{xx} = (300/0.25)y = 1200y \ \mu \qquad (\varepsilon_{xx})_B = 1200(-1.75) \ \mu = 2100 \ \mu \qquad (\varepsilon_{xx})_D = 1200(0.75) \ \mu = 900 \ \mu \tag{1}$$

The maximum stress in each material is:

$$(\sigma_1)_{max} = E_1(\varepsilon_{xx})_D = (30000)(10^3)(900)(10^{-6}) = 27000 \ psi = 27 \ ksi \tag{2}$$

$$(\sigma_2)_{max} = E_2(\varepsilon_{xx})_B = (20000)(10^3)(2100)(10^{-6}) = 42000 \ psi = 42 \ ksi \tag{3}$$

ANS $(\sigma_1)_{max} = 27$ ksi(T) ; $(\sigma_2)_{max} = 42$ ksi(C)

---

6. 29

Solution $M_z = 20 \ in-kips$ $\sigma_A = ?$ $\sigma_B = ?$ $\sigma_D = ?$

The second area moment of inertia about the z-axis can be found and substituted in the stress equation as shown below.

$$I_{zz} = \frac{1}{12}(2)(1)^3 + (2)(1)(3)^2 + \frac{1}{12}(1)(4)^3 + (1)(4)(0.5)^2 + \frac{1}{12}(4)(1)^3 + (4)(1)(2)^2 = 40.833 in^4 \tag{1}$$

$$\sigma_{xx} = -\left(\frac{M_z y}{I_{zz}}\right) = -\left(\frac{20y}{40.833}\right) = -0.4898y \tag{2}$$

$$\sigma_A = -0.4898(2.5) = -1.224 ksi \qquad \sigma_B = -0.4898(1.5) = -0.735 ksi \qquad \sigma_D = -0.4898(-3.5) = 1.714 ksi \tag{3}$$

ANS $\sigma_A = 1224$ psi(C) ; $\sigma_B = 735$ psi (C); ; $\sigma_D = 1714$ ksi (T)

---

6. 30

Solution $M_z = 10 \ kN-m$ $\sigma_A = ?$ $\sigma_B = ?$ $\sigma_D = ?$

From symmetry, the centroid is at 35 mm from the bottom. The second area moment of inertia about the z-axis can be found and substituted in the stress equation as shown below.

$$I_{zz} = \frac{1}{12}(0.01)(0.05)^3 + 2\left[\frac{1}{12}(0.05)(0.01)^3 + (0.05)(0.01)(0.005 - 0.035)^2\right] = 1.0125(10^{-6})m^4 \tag{1}$$

$$\sigma_{xx} = -\frac{M_z y}{I_{zz}} = -\frac{(10)(10^3)y}{1.0125(10^{-6})} = -9.8765(10^9)y \tag{2}$$

$$\sigma_A = -9.8765(10^9)(0.035) = -345.7(10^6) \ N/m^2 \qquad \sigma_B = -9.8765(10^9)(0.025) = -246.9(10^6) \ N/m^2 \tag{3}$$

$$\sigma_D = -9.8765(10^9)(-0.035) = 345.7(10^6) \ N/m^2 \tag{4}$$

ANS $\sigma_A = 346$ MPa(C) ;$\sigma_B = 247$ MPa(C) ; $\sigma_D = 346$ MPa(T)

---

6. 31

Solution $M_z = -12 \ kN-m$ $\sigma_A = ?$ $\sigma_B = ?$ $\sigma_D = ?$

The second area moment of inertia about the z-axis can be found and substituted into stress equation as shown below.

$$I_{zz} = 2\left[\frac{1}{12}(0.01)(0.10)^3 + (0.01)(0.1)(0.0206)^2\right] + \frac{1}{12}(0.12)(0.01)^3 + (0.12)(0.01)(0.0344)^2 = 3.9463(10^{-6})m^4 \tag{1}$$

$$\sigma_{xx} = -\frac{M_z y}{I_{zz}} = -\frac{(-12)(10^3)y}{3.9463(10^{-6})} = 3.0408(10^9)y \tag{2}$$

$$\sigma_A = 3.0408(10^9)(0.0394) = 119.8(10^6) \ N/m^2 \qquad \sigma_B = 3.0408(10^9)(0.0294) = 89.44(10^6) \ N/m^2 \tag{3}$$

$$\sigma_D = 3.0408(10^9)(-0.0706) = -214.6(10^6)\ N/m^2 \quad (4)$$

ANS $\sigma_A = 120$ MPa(T) ; $\sigma_B = 89.4$ MPa(T) ; $\sigma_D = 214.6$ MPa(C)

---

6. 32

Solution $V_y = ?$ $M_z = ?$ for the three cases

The internal shear force and internal bending moment are drawn as per the sign convention. The distributed loads are replace by an equivalent load at the centroid of the distribution as shown below. By equilibrium of forces in the y-direction and moment about point A we obtain:

ANS Case I: $V_y = 0.25$ kN ; $M_Z = 0.0625$ kN-m ; Case II; $V_y = 0.25$ kN ; $M_Z = 0.0625$ kN-m ; Case III: $V_y = -0.25$ kN ; $M_Z = 0.0625$ kN-m

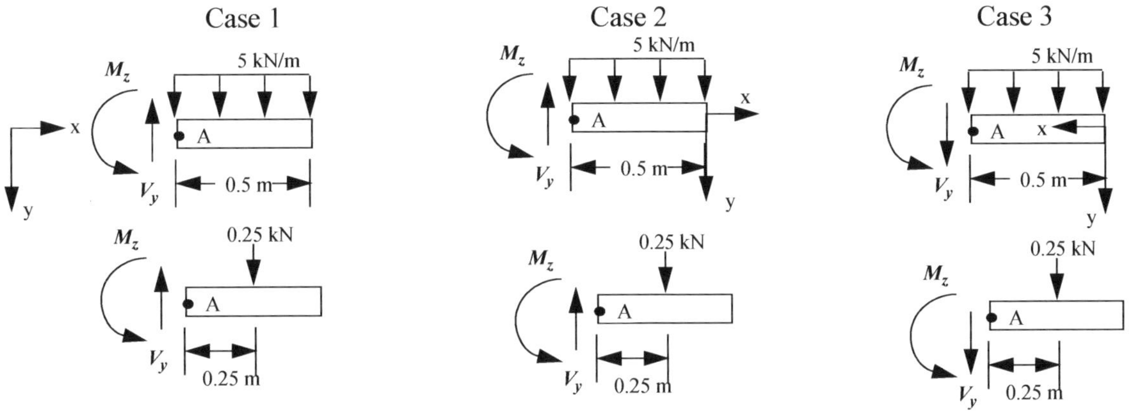

---

6. 33

Solution $V_y = ?$ $M_z = ?$ for the three cases

By moment equilibrium at the right end of the entire beam we obtain: $R_L(1) - 20 = 0$ *or* $R_L = 20\ kN$

By equilibrium of forces in the y-direction and moment about point A we obtain the shear force and moment for each case..

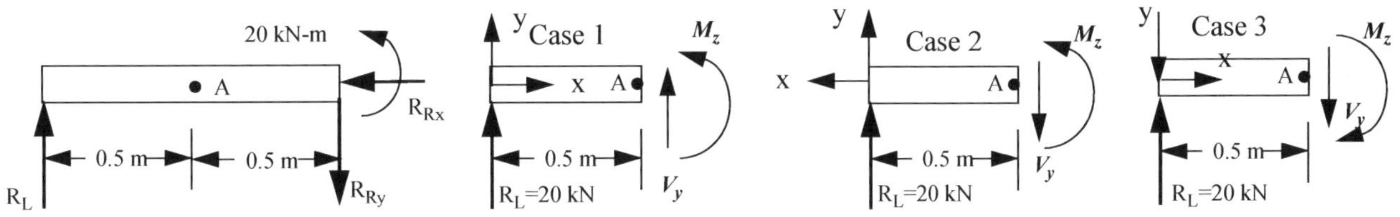

ANS Case I $V_y = -20\ kN$; $M_Z = 10\ kN-m$; Case II $V_y = 20\ kN$; $M_Z = 10\ kN-m$; Case III: $V_y = 20\ kN$; $M_Z = -10\ kN-m$

---

6. 34

Solution $V_y = ?$ $M_z = ?$ for the three cases

By moment equilibrium at the right support for the entire beam we obtain: $R_L(1) - (10)(0.5) = 0$ *or* $R_L = 5\ kN$

By equilibrium of forces in the y-direction and moment about point A we obtain the shear force and moment for each case..

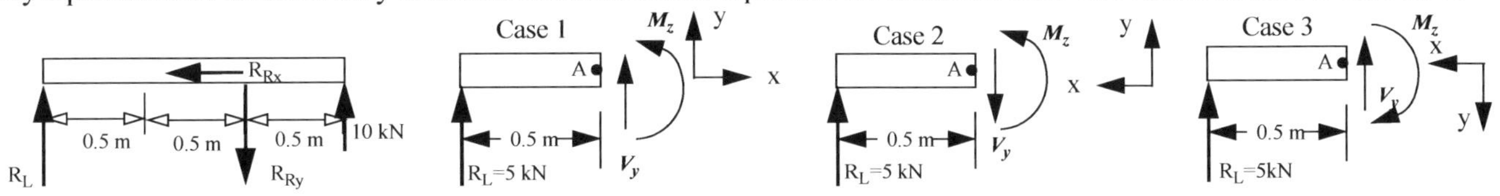

ANS Case I: $V_y = -5$ kN ; $M_Z = 2.5$ kN-m ; Case II: $V_y = 5$ kN ; $M_Z = 2.5$ kN-m ; Case III: $V_y = -5$ kN ; $M_Z = -2.5$ kN-m

---

6. 35

Solution $\sigma_A =$ (T) or (C) ? $\sigma_B =$ (T) or (C) ?

An approximate deformed shape of the beam is shown below.By inspection, the bending normal stresses are determined.

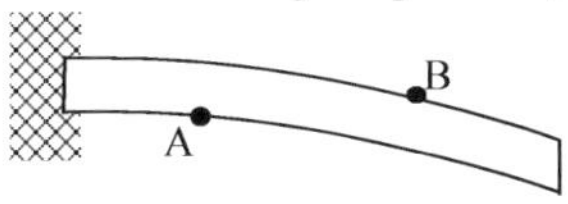

ANS σ_A *is* (C) ; σ_B *is* (T)

---

6. 36

Solution σ_A= (T) or (C) ? σ_B= (T) or (C) ?

An approximate deformed shape of the beam is shown below.By inspection, the bending normal stresses are determined.

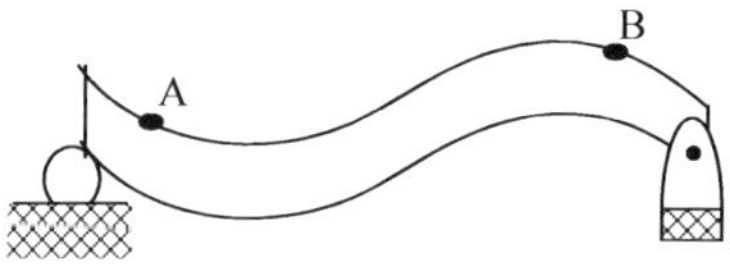

ANS σ_A is (C); ; σ_B is (T);

---

6. 37

Solution σ_A= (T) or (C) ? σ_B= (T) or (C) ?

There are two possibilities, depending upon the length of the overhang. Approximate deformed shape of the beam for both possibilities are shown below. By inspection, the bending normal stresses are determined.

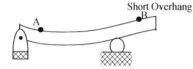

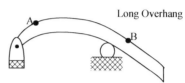

ANS *short overhang* : σ_A is (C); ; σ_B is (T);

ANS *long overhang* are: σ_A is (T); ; σ_B is (T);

---

6. 38

Solution σ_A= (T) or (C) ? σ_B= (T) or (C) ?

An approximate deformed shape of the beam is shown below.By inspection, the bending normal stresses are determined.

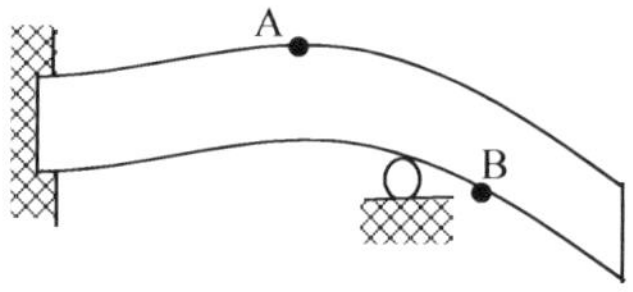

ANS σ_A is (T); ; σ_B is (C);

---

6. 39

Solution σ_A= (T) or (C) ? σ_B= (T) or (C) ?

An approximate deformed shape of the beam is shown below.By inspection, the bending normal stresses are determined.

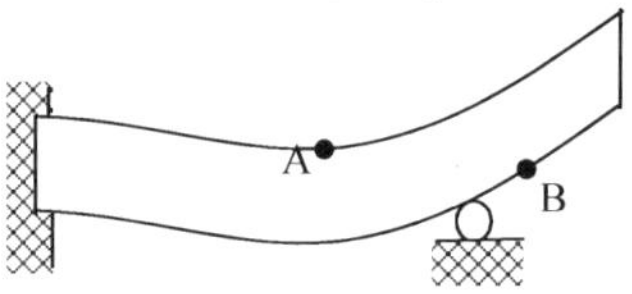

ANS σ_A is (C); ; $\sigma_B = 0$

---

6. 40

Solution σ_A= (T) or (C) ? σ_B= (T) or (C) ?

An approximate deformed shape of the beam is shown below.By inspection, the bending normal stresses are determined.

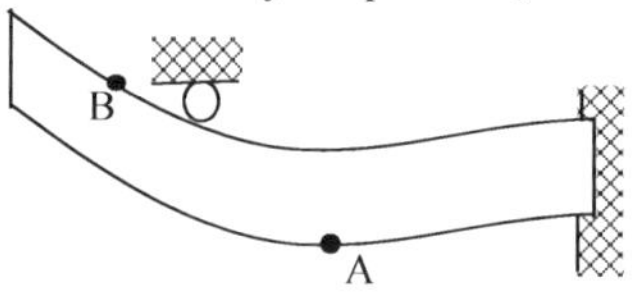

ANS σ_A is (T); ; σ_B is (C)

---

6. 41

Solution W 150 x 24 L = 4 m p = 2 kN/m σ_{40} = ? and σ_{max} = ? @ x=L/2

A simply supported beam with uniform distributed load will have a reaction force at each end whose value is half of the total force acting on the beam, i.e.,R = (pL)/2 = 4kN. By equilibrium of moment about point O in we obtain:

$$M_z - (4)(2) + (4)(1) = 0 \quad or \quad M_z = 4 \ kN-m \tag{1}$$

From Table E.2, we obtain the geometry and dimension of W150 x 24 beam cross-section shown above and $I_{zz} = 13.36(10^6)\ mm^4$. Substituting $y_A = -0.04$ m, $y_{max} = \pm 0.08$ m, M_z I_{zz} we obtain the following.

$$\sigma_{40} = -\frac{M_z y_A}{I_{zz}} = -\frac{(4)(10^3)(-0.04)}{13.36(10^{-6})} = 11.98(10^6)\ \frac{N}{m^2} \qquad \sigma_{max} = -\frac{M_z y_{max}}{I_{zz}} = -\frac{(4)(10^3)(\pm 0.08)}{13.36(10^{-6})} = \mp 23.95(10^6)\ \frac{N}{m^2} \qquad \textbf{(2)}$$

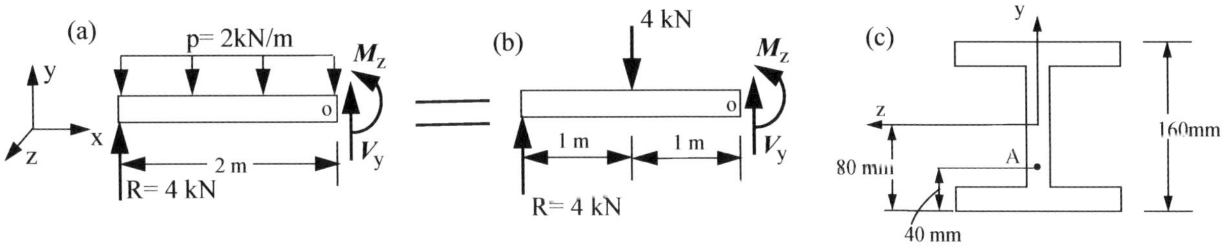

ANS $\sigma_{40} = 12.0\ MPa(T)$; $\sigma_{max} = 24.0$ MPa (T) or (C)

---

6. 42

Solution W10 x 30 L = 10 ft p = 1.5 kips /ft $\sigma_{3.0}$ = ? and σ_{max} = ? @ x=L/2

A simply supported beam with uniform distributed load will have a reaction force at each end whose value is half of the total force acting on the beam, i.e., R = (pL)/2 = 7.5 kipsBy equilibrium of moment about point O in Fig. (b) we obtain:

$$M_z - (7.5)(5) + (7.5)(2.5) = 0 \qquad or \qquad M_z = 18.75\ ft-kips \qquad \textbf{(1)}$$

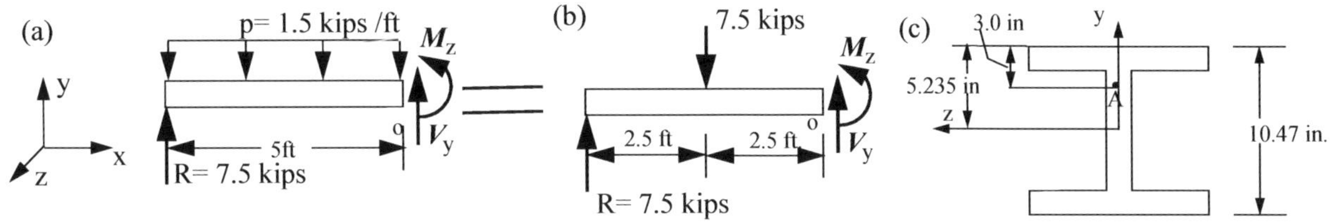

From Table E.2, we obtain the geometry and dimension of W10 x 30 beam cross-section shown above and $I_{zz} = 170\ in^4$. Substituting y_A = 2.235 inch, $y_{max} = \pm 5.235$ inch, I_{zz}, and $M_z = (18.75)(12) = 225 in-lbs$ we obtain the following.

$$\sigma_{3.0} = -\frac{M_z y_A}{I_{zz}} = -\frac{(225)(2.235)}{170} = -2.958\ ksi \qquad \sigma_{max} = -\frac{M_z y_{max}}{I_{zz}} = -\frac{(225)(\pm 5.235)}{170} = \mp 6.928\ ksi \qquad \textbf{(2)}$$

ANS $\sigma_{3.0} = 2.96\ ksi(C)$; $\sigma_{max} = 6.93\ ksi\ (C)\ or\ (T)$

---

6. 43

Solution S12 x 35 L = 20 ft F = 3 kips $\sigma_{2.0}$ = ? and σ_{max} = ? @ x=0

By equilibrium of moment about point O in Fig. (b) we obtain:

$$M_z + (3)(20) = 0 \qquad or \qquad M_z = -60\ ft-kips = -720 in-lbs \qquad \textbf{(1)}$$

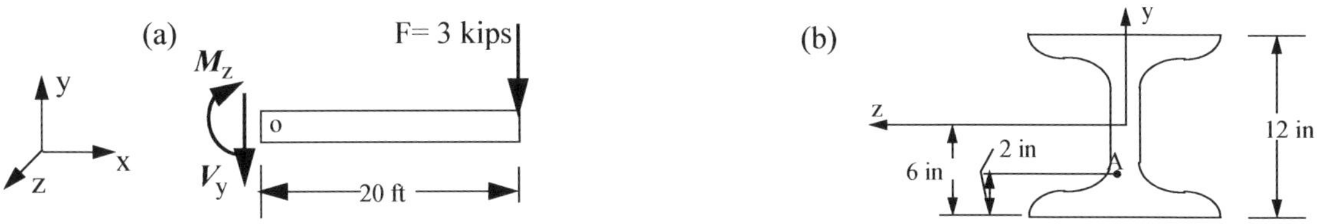

From Table E.2, we obtain the geometry and dimension of S12 x 35 beam cross-section shown above and $I_{zz} = 229\ in^4$. Substituting y_A= -4 inch, $y_{max} = \pm 6$ inch, I_{zz}, and M_z, we obtain the following.

$$\sigma_{2.0} = -\frac{M_z y_A}{I_{zz}} = -\frac{(-720)(-4)}{229} = -12.58\ ksi \qquad \sigma_{max} = -\frac{M_z y_{max}}{I_{zz}} = -\frac{(-720)(\pm 6)}{229} = \pm 18.86\ ksi \qquad \textbf{(2)}$$

ANS $\sigma_{2.0}$ = 12.58 ksi (C) ; σ_{max} = 18.86 ksi (T) or (C)

---

6. 44

Solution S250 x 52 L = 5 m F = 15 kN σ_{30}= ? and σ_{max} = ? @ x=0

By equilibrium of moment about point O in Fig. (b) we obtain:

$$M_z + (5)(15) = 0 \qquad or \qquad M_z = -75\ kN-m = -75(10^3)N-m \qquad \textbf{(1)}$$

From Appendix Table E.2, we obtain the geometry and dimension of S250 x 52 beam cross-section shown above and

$I_{zz} = 61.20(10^6)\ mm^4 = 61.20(10^{-6})\ m^4$. Substituting y_A= = 0.097 m, $y_{max} = \pm 0.127$ m I_{zz} ,and M_z we obtain the following.

$$\sigma_{30} = -\frac{M_z y_A}{I_{zz}} = -\frac{(-75)(10^3)(0.097)}{61.20(10^{-6})} = 118.87(10^6)\frac{N}{m^2} \qquad \sigma_{max} = -\frac{M_z y_{max}}{I_{zz}} = \frac{(-75)(10^3)(\pm 0.127)}{61.20(10^{-6})} = \pm 155.63\ (10^6)\frac{N}{m^2} \quad \textbf{(2)}$$

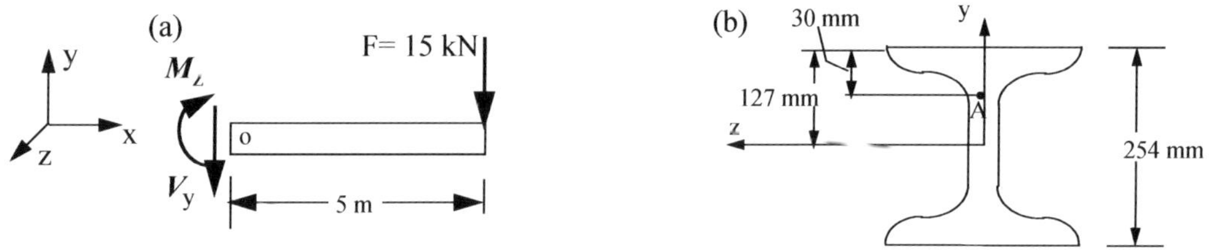

ANS $T\sigma_{30} = 118.9$ MPa(T) ; $\sigma_{max} = 155.6$ MPa (T) or (C)

---

6. 45

Solution: $\sigma_A = ?$ $\sigma_{max} = ?$

By equilibrium of moment about point C and moment about point A we obtain:

$$R_B(20) - 5000(5) = 0 \quad or \quad R_B = 1250\ lbs \qquad M_z = 10R_B = 12500 ft - lbs = 150\ in - kips \quad \textbf{(1)}$$

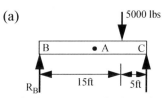

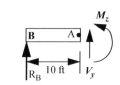

Substituting y_A = 1 inch, $y_{max} = \pm 3$ inch, and M_z we obtain the following.

$$I_{zz} = \frac{1}{12}(2)(6^3) = 36\ in^4 \qquad \sigma_A = \frac{-M_z y_A}{I_{zz}} = -\frac{(150)(1)}{36} = -4.17\ ksi \qquad \sigma_{max} = -\frac{(150)(\pm 3)}{36} = \mp 12.5\ ksi \quad \textbf{(2)}$$

ANS $\sigma_A = 4.17$ ksi (C); ; $\sigma_{max} = 12.5$ ksi (C) or (T)

---

6. 46

Solution: σ_A =?, σ_{max} =?

By equilibrium of moment about point C and moment about point A we obtain:

$$R_B(1) - 20 = 0 \quad or \quad R_B = 20\ kN \qquad M_z = (0.5)R_B = 10\ kN - m = 10,000\ N - m \quad \textbf{(1)}$$

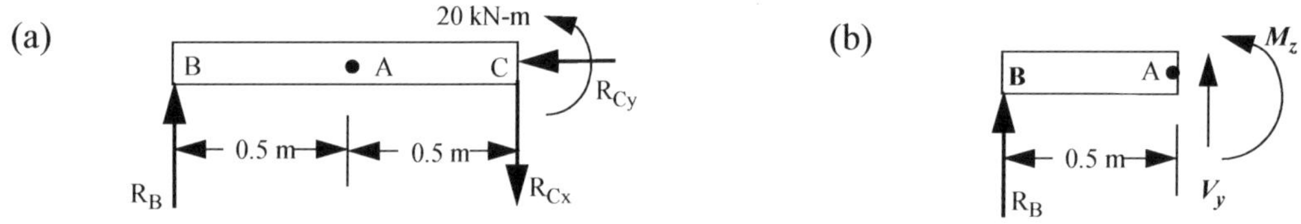

Substituting y_A = 0.03 m, $y_{max} = \pm 0.05$, M_z we obtain.

$$I_{zz} = \frac{1}{12}(0.06)(0.1^3) - \frac{1}{12}(0.05)(0.09^3) = 1.9625(10^{-6})\ m^4 \quad \textbf{(2)}$$

$$\sigma_A = \frac{-M_z y_A}{I_{zz}} = -\frac{(10000)(-0.03)}{1.9625(10^{-6})} = 152.87(10^6)\frac{N}{m^2} \qquad \sigma_{max} = -\frac{(10000)(\pm 0.05)}{1.9625(10^{-6})} = \mp 254.77(10^6)\frac{N}{m^2} \quad \textbf{(3)}$$

Top and bottom surfaces are equally distant from the centroid. Substituting Eqs. (1), (2) and in Eq. 6-12, we obtain

ANS $\sigma_A = 152.9$ MPa (T) ; $\sigma_{max} = 254.8$ MPa (C) ot (T)

---

6. 47

Solution: σ_A =? σ_{max} =?

By equilibrium of moment about point A, we obtain.

$$M_z = -(15)(1.5) = -22.5KN - m = -22500\ N - m \qquad I_{zz} = \pi[(0.1)^4 - (0.08^4)]/64 = 2.8916(10^{-6})\ m^4 \quad \textbf{(1)}$$

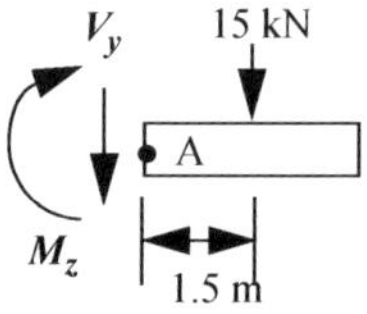

Substituting $y_A = -0.04$ m, $y_{max} = \pm 0.05$, I_{zz}, and M_z we obtain

$$\sigma_A = \frac{-M_z y_A}{I_{zz}} = -\frac{(-22500)(-0.04)}{2.8916(10^{-6})} = -310.5(10^6)\frac{N}{m^2} \qquad \sigma_{max} = -\frac{(-22500)(\pm 0.05)}{2.8916(10^{-6})} = \pm 388.18(10^6)\frac{N}{m^2} \quad \textbf{(2)}$$

ANS $\sigma_A = 310.5$ MPa (C) ; $\sigma_{max} = -388.2$ MPa (T) or (C)

---

6. 48

Solution: $I_{zz} = 1.01(10^6)mm^4$ $\sigma_A = ?$ $\sigma_{max} = ?$

By symmetry the reaction force of each support is half the total force. By equilibrium of moment about point O.

$$R_A = R_C = \frac{1}{2}\left[\frac{1}{2}(5)(1.0)\right] = 1.25kN \qquad M_z - (1.25)(0.5) + (1.25)\left(\frac{0.5}{3}\right) = 0 \quad or \quad M_z = 0.4167kN-m \quad \textbf{(1)}$$

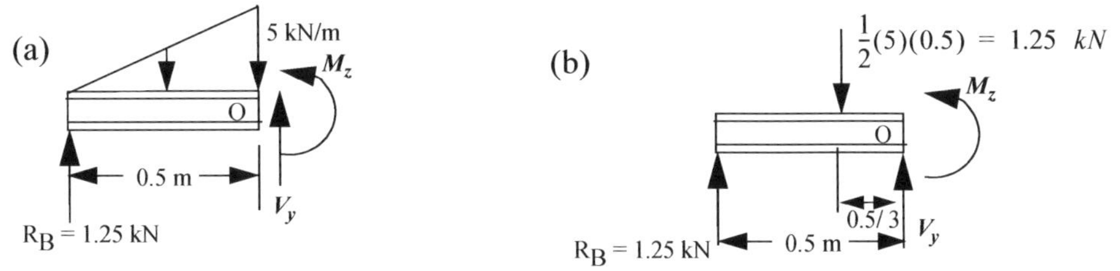

Substituting $y_A = -0.025$ m, $y_{max} = \pm 0.035$, I_{zz}, and M_z we obtain the following.

$$\sigma_A = -\frac{M_z y_A}{I_{zz}} = -\frac{(0.4167)(10^3)(-0.025)}{1.01(10^{-6})} = 10.31(10^6)\frac{N}{m^2} \qquad \sigma_A = -\frac{M_z y_{max}}{I_{ZZ}} = -\frac{(0.4167)(10^3)(\pm 0.035)}{1.01(10^{-6})} = \mp 14.44(10^6)\frac{N}{m^2} \quad \textbf{(2)}$$

ANS $\sigma_A = 10.3$ MPa (T) ; $\sigma_{max} = 14.4$ MPa (C) or (T)

---

6. 49

Solution: $I_{zz} = 18.2\ in^4$, $\sigma_a = ?$, $\sigma_{max} = ?$

By force and moment equilibrium in of the entire beam we obtain the wall reactions. By equilibrium of moment about point O we obtain the internal moment.

$$R_B = 10,800\ lbs \qquad M_B = (48)(10800) = 518.4(10^3)in-lbs \qquad M_z = (2700)(12) + M_B - R_B(36) = 162000\ in-lbs \quad \textbf{(1)}$$

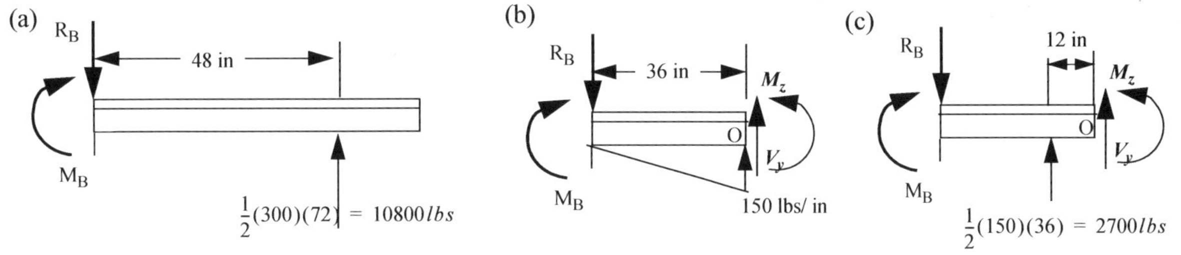

Substituting $y_A = 0.75$ inch, $y_{max} = -3.25$ inch, I_{zz}, and M_z we obtain the following.

$$\sigma_A = -\frac{M_t y_A}{I_{zz}} = -\frac{(162000)(0.75)}{18.2} = -6675.8\ psi \qquad \sigma_{max} = -\frac{M_{max} y_{max}}{I_{zz}} = -\frac{(162000)(-3.25)}{18.2} = 28929\ psi \quad \textbf{(2)}$$

ANS $\sigma_A = 6.68$ ksi (C); ; $\sigma_{max} = 28.9$ ksi (T)

---

6. 50

Solution: $I_{zz} = 2.27in^4$ $\sigma_a = ?$, $\sigma_{max} = ?$

By equilibrium of force we obtain we obtain the reaction force and by equilibrium of moment about point O we obtain the internal moment as shown below.

$$R_B = 2\ kips \qquad M_z + 10 - (24)(R_B) = 0 \quad or \quad M_z = 38\ in-lbs \quad \textbf{(1)}$$

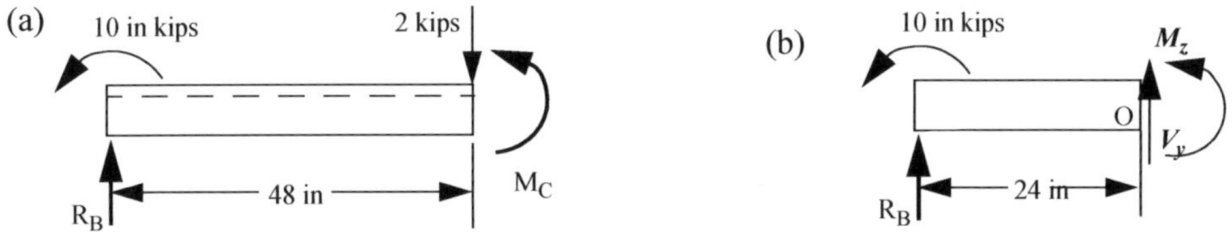

Substituting $y_A = 0.375$ inch, $y_{max} = -1.625$ inch, I_{zz}, and M_z we obtain the following.

$$\sigma_A = -\frac{M_z y_A}{I_{zz}} = -\frac{(38)(0.375)}{2.27} = -6.28\ ksi \qquad \sigma_{max} = -\frac{M_z y_{max}}{I_{zz}} = -\frac{(38)(-1.625)}{2.27} = 27.2\ ksi \qquad \textbf{(2)}$$

ANS $\sigma_A = 6.3$ ksi (C) ; $\sigma_{max} = 27.2$ ksi (T)

---

6. 51

Solution: $E = 10\ GPa$ $\qquad w = 5\ kN/m$ $\qquad \varepsilon_a = ?$

By equilibrium of moment about point O.

$$M_z + (0.25)(0.5w) = 0 \quad or \quad M_z = -0.125w \qquad I_{zz} = (0.025)(0.1^3)/12 = 2.083(10^{-6})\ m^4 \qquad \textbf{(1)}$$

(a)

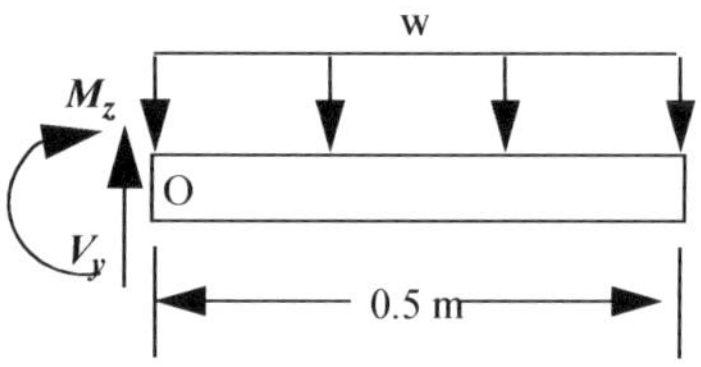

(b)

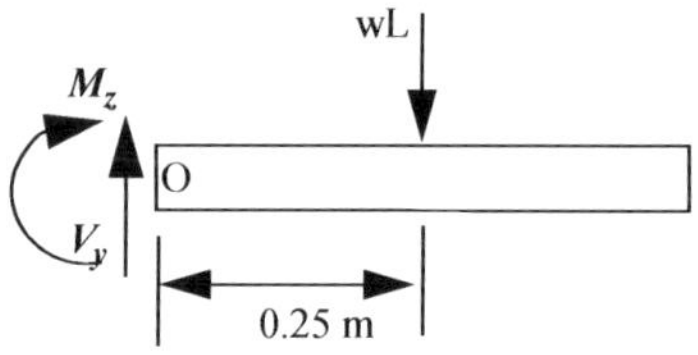

Substituting $w = 5\ kN/m$ we obtain $M_z = -0.625\ kN-m = -625\ N-m$. The bending normal stress at and strain at A are

$$\sigma_A = -\frac{M_z y_A}{I_{zz}} = -\frac{(-625)(-0.05)}{2.083(10^{-6})} = -15(10^6)\ N/m^2 \qquad \varepsilon_A = \frac{\sigma_A}{E} = -\frac{(15)(10^6)}{(10)(10^9)} = -1.5(10^{-3}) \qquad \textbf{(2)}$$

ANS $\varepsilon_A = -1500\ \mu$

---

6. 52

Solution: $E = 10\ GPa$ $\qquad \varepsilon_A = -600\ \mu$ $\qquad w = ?$

From previous problem we have $I_{zz} = 2.083(10^{-6})\ m^4$ and $M_z = -0.125w\ kN-m = -125w\ N-m$. The bending normal stress and strain at A are

$$\sigma_A = -\frac{M_z y_A}{I_{zz}} = -\frac{(-125w)(-0.05)}{2.083(10^{-6})} = -3.0w(10^6)\ \frac{N}{m^2} \qquad \varepsilon_A = \frac{\sigma_A}{E} = -\frac{-3.0w(10^6)}{(10)(10^9)} = -600(10^{-6}) \quad or \quad w = 2\text{ kN/m} \qquad \textbf{(1)}$$

ANS $w = 2$ kN/m

---

6. 53

Solution: $E = 8000\ ksi$ $\qquad I_{zz} = 95.47\ in^4$ $\qquad P = 6\ kips$ $\qquad \varepsilon_a = ?$

By equilibrium about point B, we obtain the reaction force. By equilibrium of moment about point O, we obtain,

$$R_B(6) - P(4) = 0 \quad or \quad R_B = 2P/3 \qquad M_z = -(2)R_B = -4P/3\ ft-kips = -8\ ft-kips = -96\ in-kips \qquad \textbf{(1)}$$

(a)

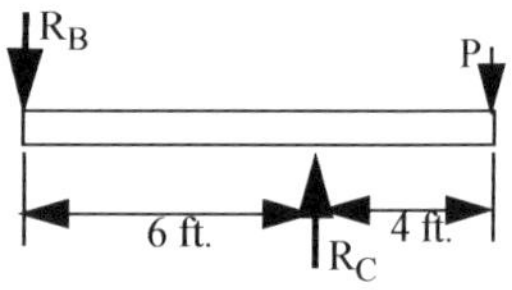

(b)

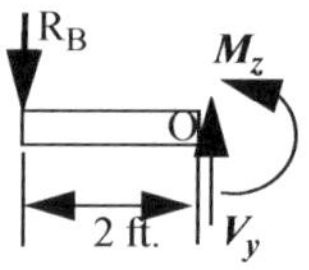

The bending normal stress and strain at A are

$$\sigma_A = -\left[\frac{M_z y_A}{I_{zz}}\right] = -\left[\frac{(-96)(-2.6)}{95.47}\right] = -2.614\ ksi \qquad \varepsilon_A = \frac{\sigma_A}{E} = \frac{-2.614}{(8000)} = -0.3268(10^{-3}) \qquad \textbf{(2)}$$

ANS $\varepsilon_A = -327\ \mu$

---

6. 54

Solution: $E = 8000\ ksi$ $\qquad I_{zz} = 95.47\ in^4$ $\qquad \varepsilon_A = -250\ \mu$ $\qquad P = ?$

From previous problem we have

$$M_z = -4P/3\ ft-kips = -16P\ in-kips \qquad \textbf{(1)}$$

The bending normal stress and strain at A are

$$\sigma_A = -\frac{M_z y_A}{I_{zz}} = -\frac{(-16P)(-2.6)}{95.47} = -0.4357P \qquad \varepsilon_A = \frac{\sigma_A}{E} = \frac{-0.4357P}{8000} = -250(10^{-6}) \quad or \quad P = 4.59\text{ kips} \qquad \textbf{(2)}$$

ANS $P = 4.59$ kips

---

6. 55

Solution

In the condition of zero axial force we substitute the new coordinate shown in the figure to obtain the following.

$$\int_A yE dA = 0 \qquad y = \eta - \eta_c \qquad \int_A \eta E\ dA - \int_A \eta_c E\ dA = 0 \qquad or \qquad \eta_c = \left(\int_A \eta E\ dA\right) / \left(\int_A E\ dA\right) \tag{1}$$

Writing the integration over the area as the sum of integration over each material we obtain the following.

$$\eta_c = \frac{\sum_{i=1}^{n} \int_{A_i} \eta E_i dA}{\sum_{i=1}^{n} \int_{A_i} E_i dA} = \frac{\sum_{i=1}^{n} E_i \int_{A_i} \eta dA}{\sum_{i=1}^{n} E_i \int_{A_i} dA} = \frac{\sum_{i=1}^{n} E_i \eta_i A_i}{\sum_{i=1}^{n} E_i A_i} \tag{2}$$

The above equation is the desired result.

---

6. 56

Solution

The solution proceeds as follows.

$$\varepsilon_{xx} = -y \frac{d^2 v}{dx^2}(x) \qquad \sigma_{xx} = -Ey \frac{d^2 v}{dx^2} \qquad \boldsymbol{M}_z = \int_A Ey^2 \frac{d^2 v}{dx^2} dA = \frac{d^2 v}{dx^2} \int_A Ey^2 dA \text{ or}$$

$$\boldsymbol{M}_z = \frac{d^2 v}{dx^2}\left[\int_{A_1} E_1 y^2 dA + \int_{A_2} E_2 y^2 dA + \cdot + \cdot + \int_{A_n} E_n y^2 dA\right] = \frac{d^2 v}{dx^2}\left[E_1 \int_{A_1} y^2 dA + E_2 \int_{A_2} y^2 dA + \cdot + \cdot + E_n \int_{A_n} y^2 dA\right] \text{ or}$$

$$\boldsymbol{M}_z = \frac{d^2 v}{dx^2}[E_1 (I_{zz})_1 + E_2 (I_{zz})_2 + \cdot + \cdot + E_N (I_{zz})_N] = \frac{d^2 v}{dx^2}\left[\sum_{j=1}^{n} E_j (I_{zz})_j\right] \tag{1}$$

$$\frac{d^2 v}{dx^2} = \frac{\boldsymbol{M}_z}{\sum_{j=1}^{n} E_j (I_{zz})_j} \qquad \sigma_{xx} = Ey \frac{d^2 v}{dx^2} = -Ey \frac{\boldsymbol{M}_z}{\sum_{j=1}^{n} E_j (I_{zz})_j} \qquad or \qquad (\sigma_{xx})_i = -E_i y \frac{\boldsymbol{M}_z}{\sum_{j=1}^{n} E_j (I_{zz})_j} \tag{2}$$

The above equation is the desired result.

---

6. 57

Solution $\sigma = K\,\varepsilon^{0.5}$ $\sigma_{xx} = f(y, h, b, \boldsymbol{M}_z) = ?$

The solution proceeds as follows.

$$\varepsilon_{xx} = -y \frac{d^2 v}{dx^2}(x) \qquad \sigma_{xx} = \begin{cases} K\varepsilon_{xx}^{0.5} & \varepsilon_{xx} \ge 0 \\ -K(-\varepsilon_{xx})^{0.5} & \varepsilon_{xx} \le 0 \end{cases} \qquad or \qquad \sigma_{xx} = \begin{cases} K\left(-y\frac{d^2 v}{dx^2}\right)^{0.5} & y < 0 \\ -K\left(y\frac{d^2 v}{dx^2}\right)^{0.5} & y > 0 \end{cases} \tag{1}$$

By symmetry the integral can be written as twice the integral in the positive y half as shown below.

$$\boldsymbol{M}_z = -\int_A y\sigma_{xx} dA = -2\left\{\int_0^{h/2} y\left[-K\left(y\frac{d^2 v}{dx^2}\right)^{0.5}\right](bdy)\right\} = 2bK\left(\frac{d^2 v}{dx^2}\right)^{0.5} \int_0^{h/2} y^{1.5} dy = 2bK\left(\frac{d^2 v}{dx^2}\right)^{0.5} \frac{y^{2.5}}{2.5}\Bigg|_0^{h/2} = 2bK\left(\frac{d^2 v}{dx^2}\right)^{0.5} \frac{(h/2)^{2.5}}{2.5} \text{ or}$$

$$\boldsymbol{M}_z = \frac{2}{2.5(2^{2.5})} bK\left(\frac{d^2 v}{dx^2}\right)^{0.5} h^{2.5} = \frac{1}{5\sqrt{2}} bK\left(\frac{d^2 v}{dx^2}\right)^{0.5} h^{2.5} \qquad or \qquad \left(\frac{d^2 v}{dx^2}\right)^{0.5} = \frac{5\sqrt{2}\boldsymbol{M}_z}{bh^{2.5}K} \tag{2}$$

$$\sigma_{xx} = \begin{cases} K\left(-y\frac{d^2 v}{dx^2}\right)^{0.5} = \left(\frac{5\sqrt{2}\boldsymbol{M}_z}{bh^{2.5}K}\right)(-y)^{0.5} & y < 0 \\ -K\left(y\frac{d^2 v}{dx^2}\right)^{0.5} = -\left(\frac{5\sqrt{2}\boldsymbol{M}_z}{bh^{2.5}K}\right)(y)^{0.5} & y > 0 \end{cases} \qquad or \qquad \sigma_{xx} = \begin{cases} \left(\frac{5\sqrt{2}}{bh^2}\right)\left(\frac{-y}{h}\right)^{0.5} \boldsymbol{M}_z & y < 0 \\ -\left(\frac{5\sqrt{2}}{bh^2}\right)\left(\frac{y}{h}\right)^{0.5} \boldsymbol{M}_z & y > 0 \end{cases} \tag{3}$$

The above equation is the desired result.

---

6. 58

Solution $\sigma = K\varepsilon^{0.4}$ $\sigma_A = f(K, L, a, M_{ext}) = ?$

The solution proceeds as follows.

$$\varepsilon_{xx} = -y\frac{d^2v}{dx^2}(x) \qquad \sigma_{xx} = \begin{cases} K\varepsilon_{xx}^{0.4} & \varepsilon_{xx} \geq 0 \\ -K(-\varepsilon_{xx})^{0.4} & \varepsilon_{xx} \leq 0 \end{cases} \qquad or \qquad \sigma_{xx} = \begin{cases} K\left(-y\frac{d^2v}{dx^2}\right)^{0.4} & y < 0 \\ -K\left(y\frac{d^2v}{dx^2}\right)^{0.4} & y > 0 \end{cases} \qquad \textbf{(1)}$$

By symmetry the integral can be written as twice the integral in the positive y half as shown below.

$$\boldsymbol{M}_z = -\int_A y\sigma_{xx}dA = -2\left[\int_0^{a/2} y\sigma_{xx}(ady) + \int_{a/2}^{a} y\sigma_{xx}(2ady)\right] = -2\left[\int_0^{a/2} y\left[-K\left(y\frac{d^2v}{dx^2}\right)^{0.4}\right](ady) + \int_{a/2}^{a} y\left[-K\left(y\frac{d^2v}{dx^2}\right)^{0.4}\right](2ady)\right] \text{ or}$$

$$\boldsymbol{M}_z = 2aK\left(\frac{d^2v}{dx^2}\right)^{0.4}\left[\int_0^{a/2} y^{1.4}dy + 2\int_{a/2}^{2} y^{1.4}dy\right] = 2aK\left(\frac{d^2v}{dx^2}\right)^{0.4}\left[\frac{y^{2.4}}{2.4}\bigg|_0^{a/2} + 2\frac{y^{2.4}}{2.4}\bigg|_{a/2}^{a}\right] \text{ or}$$

$$\boldsymbol{M}_z = 2aK\left(\frac{d^2v}{dx^2}\right)^{0.4}\left[\frac{(a/2)^{2.4}}{2.4} + 2\left(\frac{(a)^{2.4}}{2.4} - \frac{(a/2)^{2.4}}{2.4}\right)\right] = 1.509a^{3.4}K\left(\frac{d^2v}{dx^2}\right)^{0.4} \qquad or \qquad \left(\frac{d^2v}{dx^2}\right)^{0.4} = \frac{\boldsymbol{M}_z}{1.509a^{3.4}K} \qquad \textbf{(2)}$$

$$\sigma_{xx} = \begin{cases} K\left(-y\frac{d^2v}{dx^2}\right)^{0.4} = \left(\frac{\boldsymbol{M}_z}{1.509a^{3.4}}\right)(-y)^{0.4} & y < 0 \\ -K\left(y\frac{d^2v}{dx^2}\right)^{0.4} = -\left(\frac{\boldsymbol{M}_z}{1.509a^{3.4}}\right)(y)^{0.4} & y > 0 \end{cases} \qquad \textbf{(3)}$$

Substituting $\boldsymbol{M}_z = M_{ext}$ and $y_A = a$ we obtain the stress at point A as shown below.

$$\sigma_A = -\left(\frac{M_{ext}}{1.509a^{3.4}}(a)^{0.4}\right) = -\left(\frac{M_{ext}}{1.509a^3}\right) \qquad \textbf{(4)}$$

ANS $\sigma_A = \left(\frac{M_{ext}}{1.509a^3}\right)$ *compression*

Section 6.3-6.5

6. 59

Solution $V_y(x) = ?$ $\qquad$ $\boldsymbol{M}_z(x) = ?$

By equilibrium of moment about point B we obtain.

$$A_y(3) - 15(1.5) = 0 \qquad or \qquad A_y = 7.5 \ kN \qquad \textbf{(1)}$$

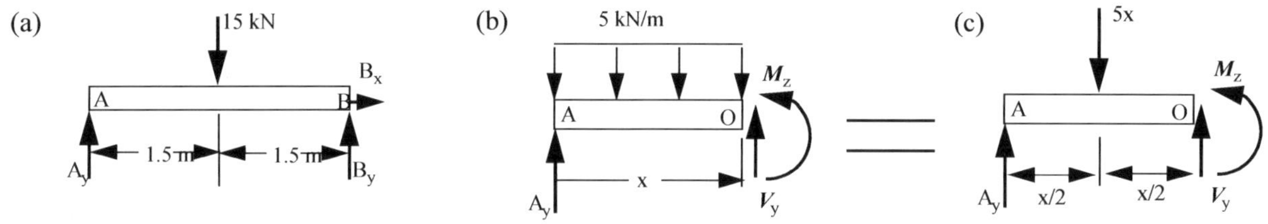

By equilibrium of forces in the y-direction and moment about point O we obtain.

$$V_y + A_y - 5x = 0 \quad or \quad V_y = (5x - 7.5) \text{ kN} \qquad \boldsymbol{M}_z - A_y(x) + (5x)\left(\frac{x}{2}\right) = 0 \quad or \quad \boldsymbol{M}_z = (7.5x - 2.5x^2) \text{ kN-m} \qquad \textbf{(2)}$$

Taking the derivative of shear force and moment we confirm the equilibrium equations are satisfied as shown below.

$$\frac{d\boldsymbol{V}_y}{dx} = 5 = -p_y \qquad \frac{d\boldsymbol{M}_z}{dx} = 7.5 - (2)(2.5)x = (7.5 - 5x) = -\boldsymbol{V}_y \qquad \textbf{(3)}$$

ANS $\boldsymbol{V}_y = (5x - 7.5)$ kN ; $\boldsymbol{M}_z = (7.5x - 2.5x^2)$ kN-m

---

6. 60

Solution $V_y(x) = ?$ $\qquad$ $\boldsymbol{M}_z(x) = ?$

By equilibrium of forces in the y-direction and equilibrium of moment about point O we obtain:

$$-V_y + 3(72-x) = 0 \quad or \quad V_y = 3(72-x)\text{ kips} \qquad M_z - 3(72-x)\left(\frac{72-x}{2}\right) = 0 \quad or \quad M_z = 1.5(72-x)^2\text{ in-kips} \quad \textbf{(1)}$$

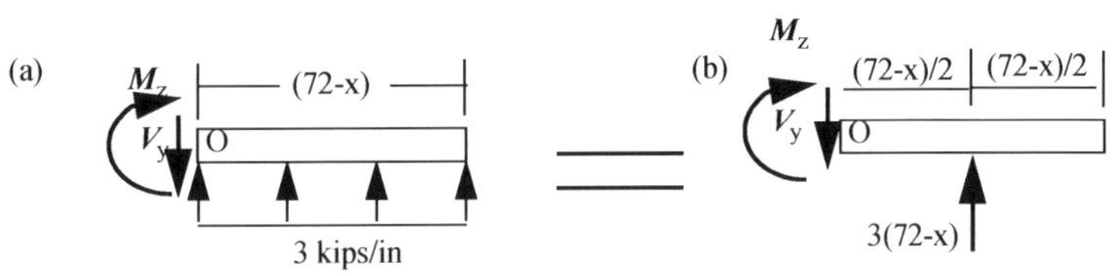

Taking the derivative of shear force and moment we confirm the equilibrium equations are satisfied as shown below.

$$\frac{dV_y}{dx} = 3(-1) = -3 = -p_y \qquad \frac{dM_z}{dx} = (1.5)(2)(-1)(72-x) = -3(72-x) = -V_y \quad \textbf{(2)}$$

ANS $V_y = 3(72-x)$ kips ; $M_z = 1.5(72-x)^2$ in-kips

---

6. 61

Solution $V_y(x) = ?$ $M_z(x) = ?$

By equilibrium of moment about point A we obtain the reaction force and by similar triangle we obtain the intensity of distributed load at the imaginary cut.

$$B_y(3) - (7.5)(1) = 0 \quad or \quad B_y = 2.5\ kN \qquad p/(3-x) = 5/3 \quad or \quad p = (5/3)(3-x) \quad \textbf{(1)}$$

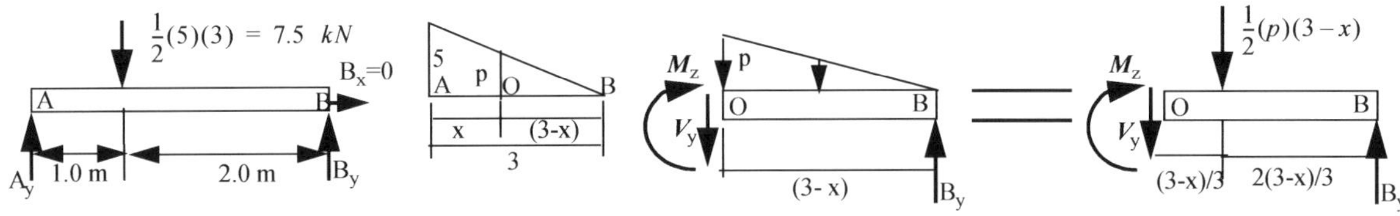

By equilibrium of forces in the y-direction and equilibrium of moment about point O we obtain the following.

$$-V_y + B_y - \frac{1}{2}(p)(3-x) = 0 \quad or \quad V_y = \left[2.5 - \frac{5}{6}(3-x)^2\right]\text{ kN} \quad \textbf{(2)}$$

$$M_z - B_y(3-x) + \left[\frac{1}{2}(p)(3-x)\right]\left(\frac{3-x}{3}\right) = 0 \quad or \quad M_z = B_y(3-x) - \frac{1}{6}p(3-x)^2 = \left[2.5(3-x) - \frac{5}{18}(3-x)^3\right]\text{ kN-m} \quad \textbf{(3)}$$

Taking the derivative of shear force and moment we confirm the equilibrium equations are satisfied as shown below.

$$\frac{dV_y}{dx} = -\left(\frac{5}{6}\right)(2)(-1)(3-x) = \frac{5}{3}(3-x) = -p_y \qquad \frac{dM_z}{dx} = 2.5(-1) - \frac{5}{18}(3)(-1)(3-x)^2 = -\left[2.5 - \frac{5}{6}(3-x)^2\right] = -V_y \quad \textbf{(4)}$$

ANS $V_y = \left[2.5 - \frac{5}{6}(3-x)^2\right]$ kN ; $M_z = \left[2.5(3-x) - \frac{5}{18}(3-x)^3\right]$ kN-m

---

6. 62

Solution $V_y(x) = ?$ $M_z(x) = ?$

By equilibrium of forces moment about point A we obtain the wall reactions and by similar triangles we obtain the intensity of distributed load at the imaginary cut.

$$A_y = 108\ kips \qquad M_A - (108)(48) = 0 \quad or \quad M_A = 5184\ in-kips \qquad p/x = 3/72 \quad or \quad p = x/24 \quad \textbf{(1)}$$

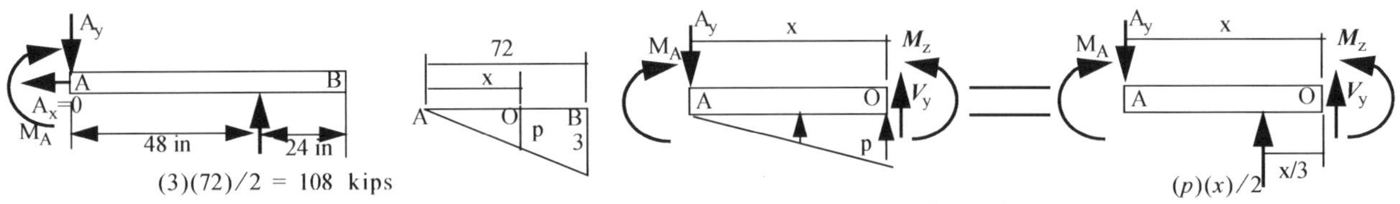

By equilibrium of forces in the y-direction and moment about point O we obtainthe following.

$$V_y - A_y + \frac{1}{2}px = 0 \quad or \quad V_y = \left[108 - \frac{1}{48}x^2\right]\text{ kips}; \qquad M_z = M_A + \frac{1}{2}px\left(\frac{x}{3}\right) - A_y(x) = \left[5184 - 108x + \frac{1}{144}x^3\right]\text{ in-kips} \quad \textbf{(1)}$$

Taking the derivative of shear force and moment we confirm the equilibrium equations are satisfied as shown below.

$$\frac{dV_y}{dx} = -\left(\frac{1}{48}\right)(2)(x) = -\frac{1}{24}(x) = -p_y \qquad \frac{dM_z}{dx} = -108 + \frac{1}{144}(3)x^2 = -\left[108 - \frac{1}{48}x^2\right] = -V_y \quad \textbf{(2)}$$

ANS $V_y = [108 - \frac{1}{48}x^2]$ kips; ; $M_z = \left[5184 - 108x + \frac{1}{144}x^3\right]$ in-kips

---

6. 63

Solution $V_y(x) = ?$ $M_z(x) = ?$

From the symmetry in the problem we conclude that the reaction forces at A and C in the y-direction are equal. By similar triangles we obtain the intensity of distributed load at the imaginary cut.

$$A_y = C_y = 1.25\ kN \qquad \frac{p}{x} = \frac{5}{0.5} \qquad or \qquad p = 10x \tag{1}$$

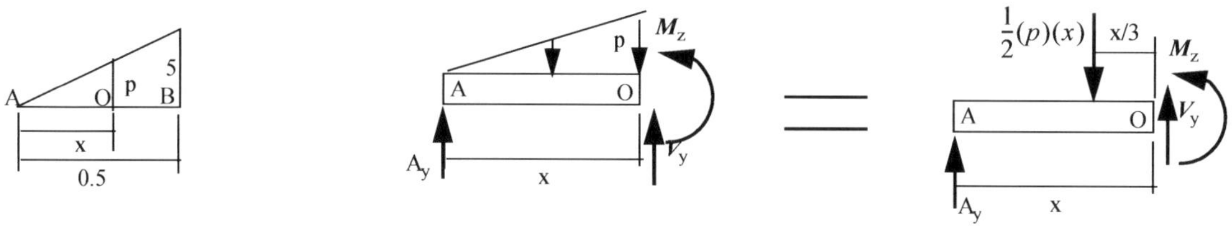

By equilibrium of forces in the y-direction and moment about point O we obtain the following.

$$V_y = \frac{1}{2}(p)(x) - A_y = [5x^2 - 1.25]\text{ kN} \qquad M_z = A_y(x) - \frac{1}{2}(p)(x)\left(\frac{x}{3}\right) = \left[1.25x - \frac{5}{3}x^3\right]\text{ kN-m} \qquad 0 \le x < 0.5 \tag{2}$$

Taking the derivative of shear force and moment we confirm the equilibrium equations are satisfied as shown below.

$$\frac{dV_y}{dx} = 10x = -p_y \qquad \frac{dM_z}{dx} = 1.25 - \left(\frac{5}{3}\right)(3)x^2 = -V_y \tag{3}$$

By symmetry the shear force and moment in right side can be obtained by replacing x by (1-x).

ANS $V_y = [5x^2 - 1.25]\text{ kN}$ $\quad 0 \le x < 0.5$; $M_z = \left[1.25x - \frac{5}{3}x^3\right]\text{ kN-m}$ $\quad 0 \le x < 0.5$;

$V_y = [5(1-x)^2 - 1.25]\text{ kN}$ $\quad 0.5 < x \le 1$; $M_z = \left[1.25(1-x) - \frac{5}{3}(1-x)^3\right]\text{ kN-m}$ $\quad 0.5 < x \le 1$

---

6. 64

Solution $V_y(x) = ?$ $M_z(x) = ?$

By equilibrium of forces in the y-direction and moment about point O_1 we obtain the following.

$$A_y = wL\ kips \qquad V_y = -A_y = -wL\text{ kips} \qquad M_z = A_y(x) - wL^2 = (wLx - wL^2)\text{ in-kips} \quad 0 \le x < L; \tag{1}$$

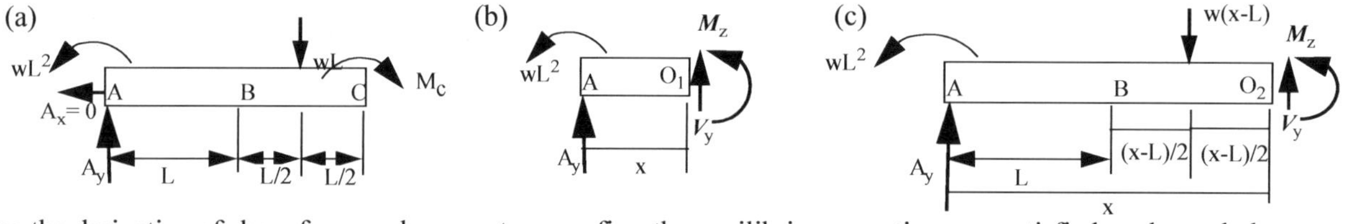

Taking the derivative of shear force and moment we confirm the equilibrium equations are satisfied as shown below.

$$\frac{dV_y}{dx} = 0 \qquad \frac{dM_z}{dx} = wL = -V_y \tag{2}$$

By equilibrium of forces in the y-direction and moment about point O_2 we obtain the following.

$$V_y = w(x-L) - A_y = [w(x-L) - wL]\text{ kips} \qquad L < x \le 2L \tag{3}$$

$$M_z + (w)(x-L)\left(\frac{x-L}{2}\right) - A_y(x) + wL^2 = 0 \qquad or \qquad M_z = [wLx - \tfrac{w}{2}(x-L)^2 - wL^2]\text{ in-kips} \quad L < x \le 2L \tag{4}$$

Taking the derivative of shear force and moment we confirm the equilibrium equations are satisfied as shown below.

$$\frac{dV_y}{dx} = w = -p_y \qquad \frac{dM_z}{dx} = wL - \frac{w}{2}(2)(x-L) = -V_y \qquad L < x \le 2L \tag{5}$$

Taking the first derivative of Eq. (7) we obtain:

ANS $V_y = -wL\text{ kips}$ $\quad 0 \le x < L$; $M_z = (wLx - wL^2)\text{ in-kips}$ $\quad 0 \le x < L$;

$V_y = [w(x-L) - wL]\text{ kips}$ $\quad L < x \le 2L$; $M_z = [wLx - \tfrac{w}{2}(x-L)^2 - wL^2]\text{ in-kips}$ $\quad L < x \le 2L$

---

6. 65

Solution $V_y(x) = ?$ $M_z(x) = ?$

By equilibrium of moment about point C in Fig.(a) we obtain.

$$A_y(2L) - wL(L/2) = 0 \qquad or \qquad A_y = wL/4\ kips \tag{1}$$

By equilibrium of forces in the y-direction and moment about point O_1 we obtain the following.

$$V_y = -A_y = -(wL/4)\text{ kips} \qquad M_z = A_y(x) = (wLx/4)\text{ in.-kips} \qquad 0 \le x < L \tag{2}$$

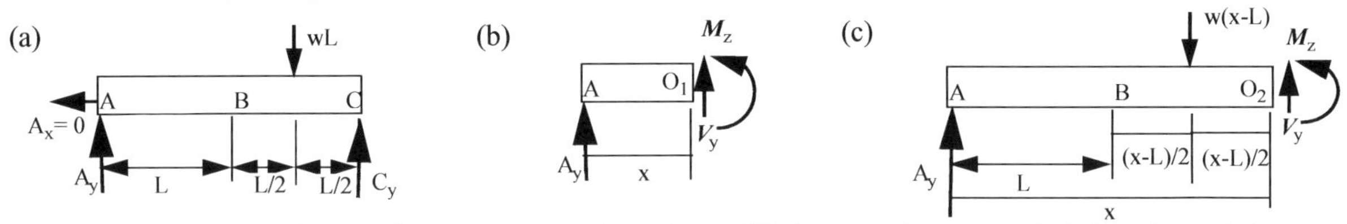

Taking the derivative of shear force and moment we confirm the equilibrium equations are satisfied as shown below.

$$\frac{dV_y}{dx} = 0 \qquad \frac{dM_z}{dx} = \frac{wL}{4} = -V_y \tag{3}$$

By equilibrium of forces in the y-direction and moment about point O_2 we obtain the following.

$$V_y = w(x-L) - A_y = \left[w(x-L) - \frac{wL}{4}\right]\text{ kips} \qquad M_z = A_y(x) - w(x-L)\left(\frac{x-L}{2}\right) = \left[\frac{wL}{4}x - \frac{w}{2}(x-L)^2\right]\text{ in.-kips} \qquad L < x \le 2L \tag{4}$$

Taking the derivative of shear force and moment we confirm the equilibrium equations are satisfied as shown below.

$$\frac{dV_y}{dx} = w = -p_y \qquad \frac{dM_z}{dx} = \frac{wL}{4} - \frac{w}{2}(2)(x-L) = -V_y \tag{5}$$

ANS $V_y = -\left(\frac{wL}{4}\right)$ kips; $M_z = \left(\frac{wL}{4}x\right)$ in.-kips $\qquad 0 \le x < L$

$V_y = \left[w(x-L) - \frac{wL}{4}\right]$ kips; $M_z = \left[\frac{wL}{4}x - \frac{w}{2}(x-L)^2\right]$ in.-kips $\qquad L < x \le 2L$;

---

6. 66

Solution $\qquad V_y(x) = ?$ $\qquad M_z(x) = ?$

By equilibrium of forces and moment about point A we obtain the following.

$$A_y = wL\ kips \qquad M_A = wL\left(\frac{3L}{2}\right) = \frac{3}{2}wL^2 \tag{1}$$

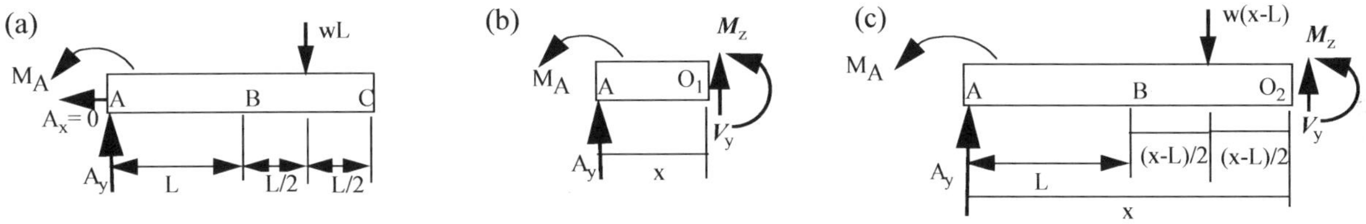

By equilibrium of forces in the y-direction and moment about point O_1 we obtain the following.

$$V_y = -A_y = wL\text{ kips} \qquad M_z = A_y(x) - M_A = \left(wLx - \frac{3}{2}wL^2\right)\text{ in.- kips} \qquad 0 \le x < L \tag{2}$$

Taking the derivative of shear force and moment we confirm the equilibrium equations are satisfied as shown below.

$$\frac{dV_y}{dx} = 0 \qquad \frac{dM_z}{dx} = wL = -V_y \tag{3}$$

By equilibrium of forces in the y-direction and moment about point O_2 we obtain the following.

$$V_y = w(x-L) - A_y = [w(x-L) - wL]\text{ kips} \qquad L < x \le 2L \tag{4}$$

$$M_z + (w)(x-L)\left(\frac{x-L}{2}\right) - A_y(x) + M_A = 0 \quad or \quad M_z = \left[wLx - \frac{w}{2}(x-L)^2 - \frac{3}{2}wL^2\right]\text{ in.- kips} \qquad L < x \le 2L \tag{5}$$

Taking the derivative of shear force and moment we confirm the equilibrium equations are satisfied as shown below.

$$\frac{dV_y}{dx} = w = -p_y \qquad \frac{dM_z}{dx} = wL - \frac{w}{2}(2)(x-L) = -V_y \tag{6}$$

ANS $V_y = -wL$ kips; $M_z = \left(wLx - \frac{3}{2}wL^2\right)$ in.- kips $\qquad 0 \le x < L$

$V_y = [w(x-L) - wL]$ kips; $M_z = \left[wLx - \frac{w}{2}(x-L)^2 - \frac{3}{2}wL^2\right]$ in.- kips $\qquad L < x \le 2L$

---

6. 67

Solution $\qquad V_y(x) = ?$ $\qquad M_z(x) = ?$

By equilibrium of moment about point E and equilibrium of forces we obtain the following.

$$B_y(9) - (36)(10.5) + 10 + (48)(5) - (12)(3) - 16 = 0 \quad or \quad B_y = 20\ kN \tag{1}$$

$$B_y - E_y - 36 + 48 - 12 = 0 \quad or \quad E_y = 20\ kN \tag{2}$$

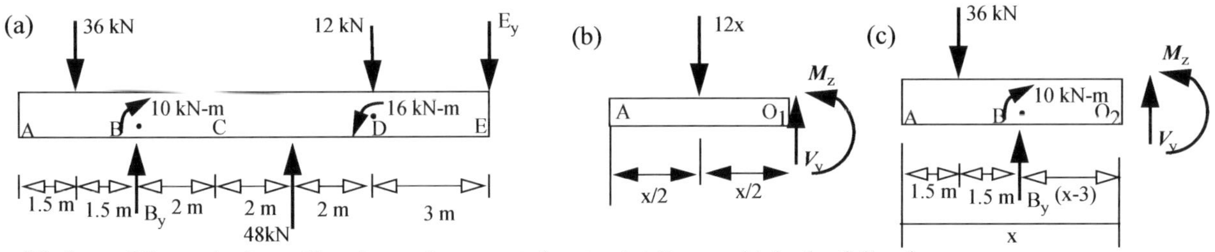

By equilibrium of forces in the y-direction and moment about point O_1 we obtain the following

$$V_y = 12x \text{ kN} \qquad M_z + (12x)\left(\frac{x}{2}\right) = 0 \quad or \quad M_z = -6x^2 \text{ kN-m} \qquad 0 \le x < 3m \tag{3}$$

Taking the derivative of shear force and moment we confirm the equilibrium equations are satisfied as shown below.

$$\frac{dV_y}{dx} = 12 = -p_y \qquad \frac{dM_z}{dx} = -12x = -V_y \tag{4}$$

By equilibrium of forces in the y-direction and equilibrium of moment about point O_2 we obtain the following.

$$V_y = 36 - B_y = 16 \text{ kN} \qquad M_z + (36)(x - 1.5) - B_y(x-3) - 10 = 0 \quad or \quad M_z = (4 - 16x) \text{ kN-m} \qquad 3m < x < 5m \tag{5}$$

Taking the derivative of shear force and moment we confirm the equilibrium equations are satisfied as shown below.

$$\frac{dV_y}{dx} = 0 \qquad \frac{dM_z}{dx} = -16 = -V_y \tag{6}$$

Substituting x = 3 in expressions for $0 \le x < 3m$ and in expressions for $3m < x < 5m$ we obtain the value of V_y and M_z just before and just after point B as shown below.

$$V_y(3^-) = 36\ kN \qquad M_z(3^-) = 54\ kN-m \qquad V_y(3^+) = 16\ kN \qquad M_z(3^+) = -44\ kN-m \tag{7}$$

The differences in values for shear force and bending moment just before and after B are equal to the value of concentrated force By and the applied moment of 10 kN-m at B.

ANS $V_y = 12x$ kN; $M_z = -6x^2$ kN-m $\quad 0 \le x < 3m$; ; $V_y = 16$ kN; $M_z = (4 - 16x)$ kN-m $\quad 3m < x < 5m$;

---

6. 68

Solution $\quad V_y(x) = ? \qquad M_z(x) = ?$

From previous problem we have the following values for reactions.

$$B_y = 20\ kN \qquad E_y = 20\ kN \tag{1}$$

We can make an imaginary cut in CD. That is the location of the imaginary cut is defined by: $5m < x < 9m$. We take the right part of the cut and after replacing the distributed force, draw the free body diagram. The internal shear force V_y and internal bending moment M_z are drawn as per our sign convention and shown in Fig. (a).

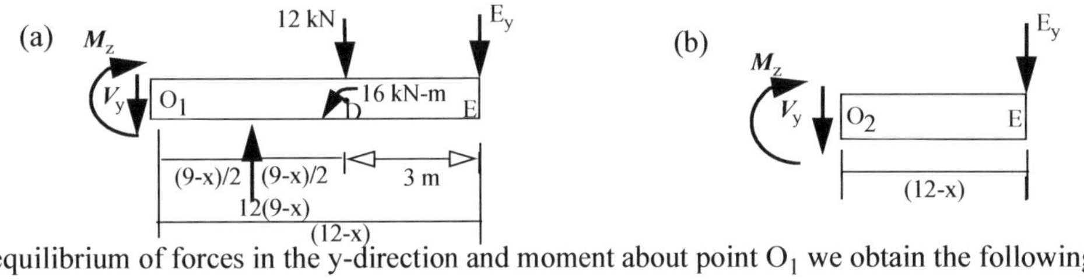

By equilibrium of forces in the y-direction and moment about point O_1 we obtain the following.

$$-V_y - E_y - 12 + 12(9-x) = 0 \quad or \quad V_y = (76 - 12x) \text{ kN} \quad 5 \text{ m} < x < 9 \text{ m}; \tag{2}$$

$$M_z - 12(9-x)\frac{(9-x)}{2} + 12(9-x) + E_y(12-x) - 16 = 0 \quad or \quad M_z = (6x^2 - 76x + 154) \text{ kN-m} \qquad 5m < x < 9m \tag{3}$$

Taking the derivative of shear force and moment we confirm the equilibrium equations are satisfied as shown below.

$$\frac{dV_y}{dx} = -12 = -p_y \qquad \frac{dM_z}{dx} = (12x - 76) = -V_y \tag{4}$$

By equilibrium of forces in the y-direction and moment about point O_2 we obtain the following.

$$V_y = -E_y = -20 \text{ kN} \qquad M_z = -E_y(12-x) = M_z = (20x - 240) \text{ kN-m} \qquad 9m < x < 12m \tag{5}$$

Taking the derivative of shear force and moment we confirm the equilibrium equations are satisfied as shown below.

$$\frac{dV_y}{dx} = 0 \qquad \frac{dM_z}{dx} = 20 = -V_y \qquad (6)$$

Substituting x = 9 in expressions for $5m < x < 9m$ and in expressions for $9m < x < 12m$ we obtain the value of V_y and M_z just before and after point D as shown below.

$$V_y(9^-) = -32\ kN \qquad M_z(9^-) = -44\ kN-m \qquad V_y(9^+) = -20\ kN \qquad M_z(9^+) = -60\ kN-m \qquad (7)$$

The differences in values for shear force and bending moment just before and after D are equal to the value of concentrated force of 12 kN and the applied moment of 16 kN-m at D.

ANS $V_y = (76 - 12x)$ kN; $M_z = (6x^2 - 76x + 154)$ kN-m $\quad 5\ m < x < 9$ m; ;

$V_y = -20$ kN; $M_z = (20x - 240)$ kN-m $\quad 9m < x < 12m$

---

6. 69

Solution $\quad V_y(x) = ?\qquad M_z(x) = ?$

By equilibrium of forces in the y-direction and moment about point O_1 we obtain the following.

$$V_y = -6x\text{ kN} \qquad M_z - (6x)\frac{x}{2} = 0 \qquad or \qquad M_z = 3x^2\text{ kN-m} \qquad 0 \le x < 3m \qquad (1)$$

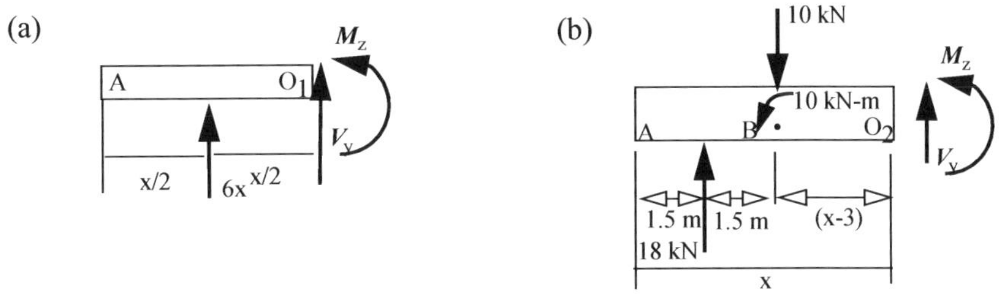

Taking the derivative of shear force and moment we confirm the equilibrium equations are satisfied as shown below.

$$\frac{dV_y}{dx} = -6 = -p_y \qquad \frac{dM_z}{dx} = 6x = -V_y \qquad (2)$$

By equilibrium of forces in the y-direction and moment about point O_2 we obtain the following.

$$V_y = 10 - 18 = -8\text{ kN} \qquad M_z - (18)(x - 1.5) + 10(x - 3) + 10 = 0 \qquad or \qquad M_z = (8x - 7)\text{ kN-m} \qquad 3m < x < 5m$$

Taking the derivative of shear force and moment we confirm the equilibrium equations are satisfied as shown below.

$$\frac{dV_y}{dx} = 0 \qquad \frac{dM_z}{dx} = 8 = -V_y \qquad (3)$$

Substituting x = 3 in expressions for $0 \le x < 3m$ and in expressions for $3m < x < 5m$ we obtain the value of V_y and M_z just before and after point D as shown below.

$$V_y(3^-) = -18\ kN \qquad M_z(3^-) = 27\ kN-m \qquad V_y(3^+) = -8\ kN \qquad M_z(3^+) = 17\ kN-m \qquad (4)$$

The differences in values for shear force and bending moment just before and after B are equal to the value of concentrated force of 10 kN and the applied moment of 10 kN-m at B.

ANS $V_y = -6x$ kN; $M_z = 3x^2$ kN-m $\quad 0 \le x < 3m$; $V_y = -8$ kN; $M_z = (8x - 7)$ kN-m $\quad 3m < x < 5m$

---

6. 70

Solution $\quad V_y(x) = ?\qquad M_z(x) = ?$

By equilibrium of forces and moment about point E we obtain the following.

$$E_y = 10 + 24 - 6 - 18 = 10\ kN \qquad M_E + (18)(10.5) - (10)(9) - 10 - (24)(5) + (6)(3) = 0 \qquad or \qquad M_E = 13\ kN-m \qquad (1)$$

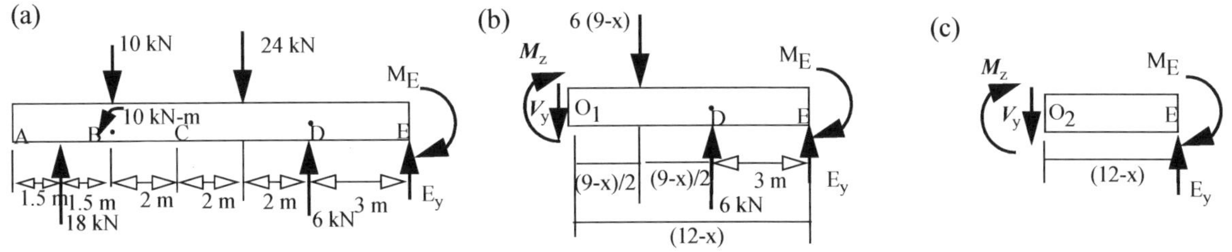

By equilibrium of forces in the y-direction and moment about point O_1 we obtain the following.

$$-V_y + E_y - 6(9 - x) + 6 = 0 \qquad or \qquad V_y = (6x - 38)\text{ kN} \qquad 5m < x < 9m \qquad (2)$$

$$M_z + 6(9-x)\frac{(9-x)}{2} - 6(9-x) - E_y(12-x) + M_E = 0 \quad or \quad M_z = (-3x^2 + 38x - 82)\ \text{kN-m} \quad 5m < x < 9m \quad (3)$$

Taking the derivative of shear force and moment we confirm the equilibrium equations are satisfied as shown below.

$$\frac{dV_y}{dx} = 6 = -p_y \qquad \frac{dM_z}{dx} = (-6x + 38) = -V_y \quad (4)$$

By equilibrium of forces in the y-direction and moment about point O_2 we obtain the following.

$$V_y = E_y = 10\ \text{kN} \qquad M_z + M_E - E_y(12-x) = 0 \quad or \quad M_z = (107 - 10x)\ \text{kN-m} \quad 9m < x < 12m \quad (5)$$

Taking the derivative of shear force and moment we confirm the equilibrium equations are satisfied as shown below.

$$\frac{dV_y}{dx} = 0 \qquad \frac{dM_z}{dx} = 10 = -V_y \quad (6)$$

Substituting x = 3 in expressions for $0 \le x < 3m$ and in expressions for $3m < x < 5m$ we obtain the value of V_y and M_z just before and after point D as shown below.

$$V_y(9^-) = 16\ kN \qquad M_z(9^-) = 17\ kN-m \qquad V_y(9^+) = 10\ kN \qquad M_z(9^+) = 17\ kN-m \quad (7)$$

The difference in values for shear force just before and after D are equal to the value of concentrated force of 6 kN. As no moment is applied at D, the internal moment are continuous.

ANS $V_y = (6x - 38)$ kN; ($M_z = (-3x^2 + 38x - 82)$ kN-m) $5m < x < 9m$;

$V_y = 10$ kN; ($M_z = (107 - 10x)$ kN-m) $9m < x < 12m$

---

6. 71

Solution $V_y(x) = ?$ $M_z(x) = ?$ $\sigma_{max} = ?$

By moment equilibrium about point C and equilibrium of forces we obtain the following.

$$[(p)(1.5)/2](1.4) + [(p)(0.9)/2]0.6 - (792)(0.9) = 0 \quad or \quad p = 540\ N/m \quad (1)$$

$$P - 792 + (p)(1.5)/2 + (p)(0.9)/2 = 0 \quad or \quad P = 144\ N \quad (2)$$

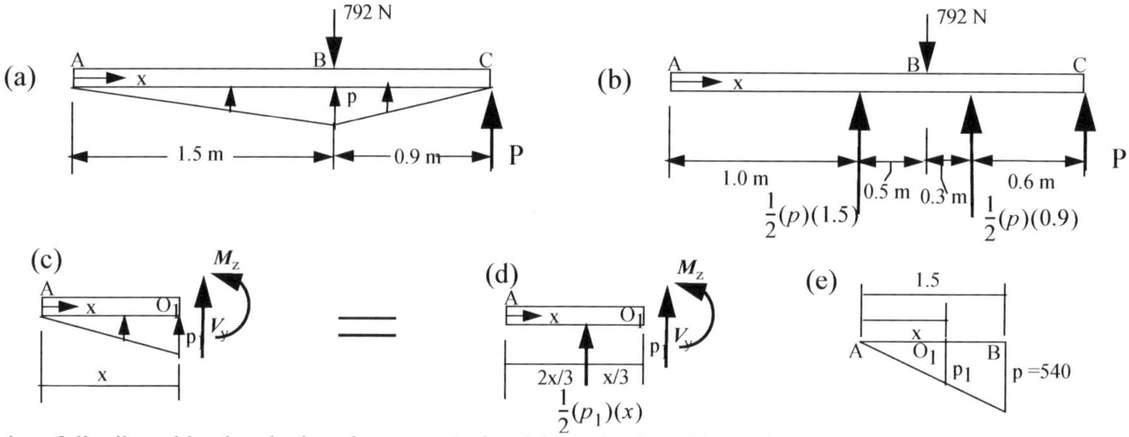

The intensity of distributed load at the imaginary cut (point O_1) can be found from similar triangle as shown below.

$$p_1/x = 540/1.5 \quad or \quad p_1 = 360x\ N/m \quad (3)$$

By equilibrium of forces in the y-direction and moment about point O_1 we obtain the following.

$$V_y + p_1x/2 = 0 \quad or \quad V_y = -180x^2\ \text{N} \qquad M_z - [p_1x/2](x/3) = 0 \quad or \quad M_z = 60x^3\ \text{N-m} \quad 0 \le x < 1.5 \quad (4)$$

The intensity of distributed load at the imaginary cut (point O_2) can be found from similar triangle shown in Fig. (h) as:

$$p_2/(2.4-x) = 540/0.9 \quad or \quad p_2 = 600(2.4-x) \quad (5)$$

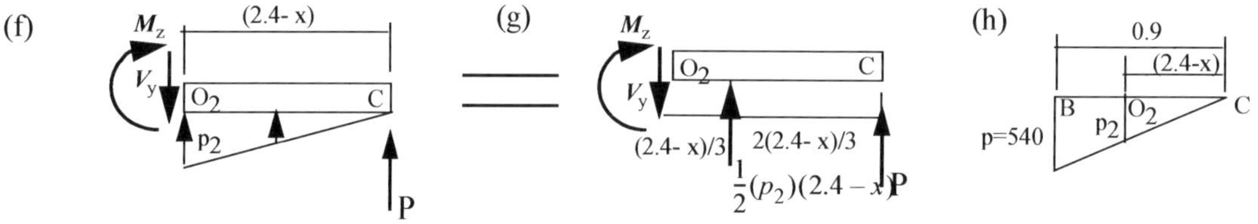

By equilibrium of forces in the y-direction and moment about point O_2 we obtain the following.

$$-V_y + \frac{1}{2}(p_2)(2.4-x) + P = 0 \quad or \quad V_y = [300(2.4-x)^2 - 144]\ \text{N} \quad 1.5 < x < 2.4 \quad (6)$$

$$M_z - \left[\frac{1}{2}(p_2)(2.4-x)\right]\left(\frac{2.4-x}{3}\right) - 144(2.4-x) = 0 \quad or \quad M_z = [100(2.4-x)^3 + 144(2.4-x)]\ \text{N-m} \quad 1.5 < x < 2.4 \quad (7)$$

The thickness t(x) of the ski, the area moment of inertia and section modulus can be written as shown below.

$$t(x) = \begin{cases} 0.007 + 0.012x & 0 \le x < 1.5 \\ 0.055 - 0.020x & 1.5 < x < 2.4 \end{cases} \qquad I_{zz} = \frac{1}{12}b[t(x)]^3 \qquad S(x) = \frac{I_{zz}}{y_{max}} = \frac{I_{zz}}{t(x)/2} = \frac{1}{6}b[t(x)]^2 = \frac{0.05}{6}[t(x)]^2 \qquad (8)$$

The maximum bending normal stress at any section is: $\sigma = M_z/S(x)$ and can be written as follows.

$$\sigma = \frac{60x^3}{0.05[0.007 + 0.012x]^2/6} = \frac{7200x^3}{[0.007 + 0.012x]^2}\ N/m^2 \qquad 0 \le x < 1.5 \qquad (9)$$

$$\sigma = \frac{100(2.4-x)^3 + 144(2.4-x)}{0.05[0.055 - 0.020x]^2/6} = \frac{12000(2.4-x)^3 + 17280(2.4-x)}{[0.055 - 0.020x]^2}\ N/m^2 \qquad 1.5 < x < 2.4 \qquad (10)$$

The maximum bending normal stress in the above equations is determined on a spread sheet as shown in the table below.

x (m)	σ (N/m²)	x (m)	σ (N/m²)	x (m)	σ (N/m²)	x (m)	σ (N/m²)
0	0.000E+00	0.6	7.713E+06	1.2	2.717E+07	1.8	8.975E+06
0.1	1.071E+05	0.7	1.041E+07	1.3	3.097E+07	1.9	6.488E+06
0.2	6.519E+05	0.8	1.338E+07	1.4	3.488E+07	2	4.267E+06
0.3	1.730E+06	0.9	1.657E+07	1.5	3.888E+07	2.1	2.396E+06
0.4	3.309E+06	1	1.994E+07	1.6	1.452E+07	2.2	9.917E+05
0.5	5.325E+06	1.1	2.349E+07	1.7	1.167E+07	2.3	1.852E+05
						2.4	0.000

ANS $V_y = -180x^2$ N; $M_z = 60x^3$ N-m $\quad 0 \le x < 1.5$;

$V_y = [300(2.4-x)^2 - 144]$ N; $M_z = [100(2.4-x)^3 + 144(2.4-x)]$ N-m $\quad 1.5 < x < 2.4 \quad \sigma_{max} = 38.9$ MPa (T) or (C)

---

6. 72

Solution Plot of V_y and M_z $\quad (V_y)_{max} = ?$ $\quad (M_z)_{max} = ?$

By symmetry the reactions at the support are equal. The shear force and bending moment diagrams are drawn as shown below.

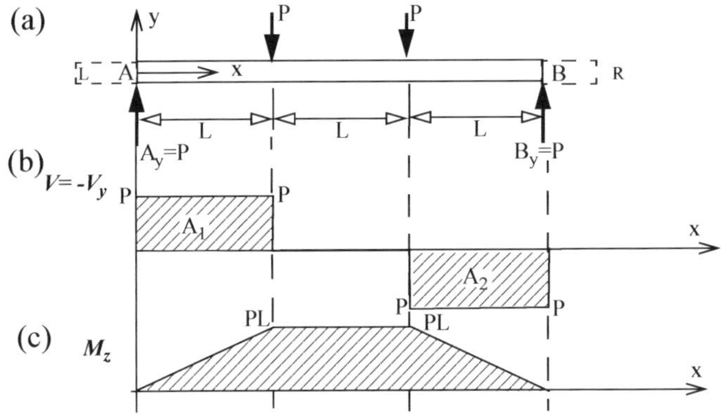

$$A_1 = A_2 = PL$$

ANS $(V_y)_{max} = \pm P$; $(M_z)_{max} = PL$

---

6. 73

Solution Plot of V_y and M_z $\quad (V_y)_{max} = ?$ $\quad (M_z)_{max} = ?$

By equilibrium of moment about point B and force equilibrium in the y direction we obtain the following.

$$3A_yL = 2M \quad or \quad A_y = 2M/3L \qquad B_y = A_y = 2M/3L \qquad (1)$$

The shear force and bending moment diagrams are drawn as shown below

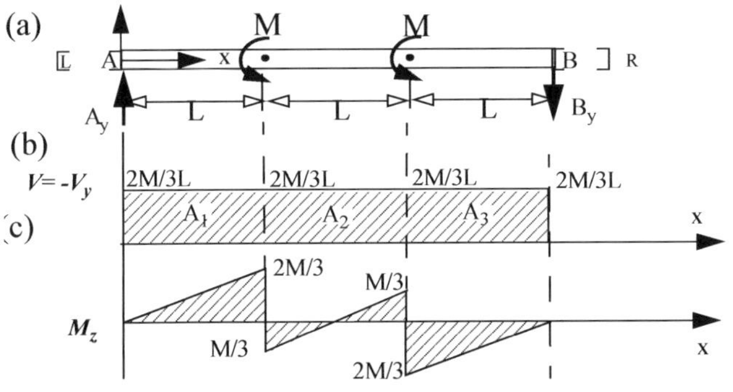

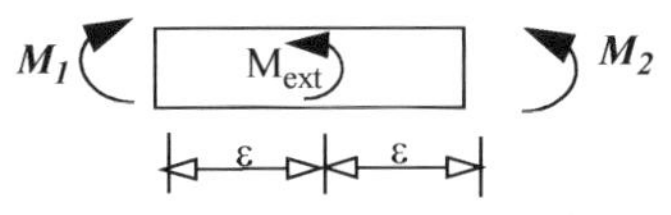

$M_2 = M_1 - M_{ext}$ Template Equations

$$A_1 = A_2 = A_3 = 2M/3$$

ANS $(V_y)_{max} = -2M/3L$; $(M_z)_{max} = 2M/3$

6. 74

Solution Plot of V_y and M_z $(V_y)_{max} = ?$ $(M_z)_{max} = ?$

By symmetry the reactions at the support are equal. The shear force and bending moment diagrams are drawn as shown below.

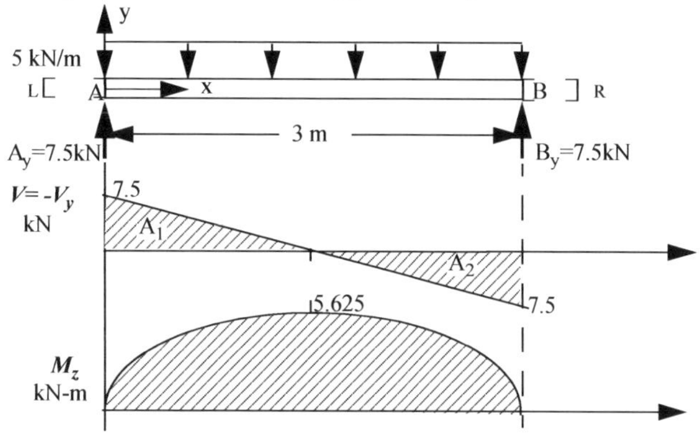

$$A_1 = A_2 = \frac{1}{2}(7.5)(1.5) = 5.625$$

ANS $(V_y)_{max} = \pm 7.5 \text{ kN}$; $(M_z)_{max} = 5.625 \text{ kN-m}$

---

6. 75

Solution Plot of V_y and M_z $(V_y)_{max} = ?$ $(M_z)_{max} = ?$

By equilibrium of forces in the y-direction and moment about point A we obtain the following.

$$A_y = 216 \ kips \qquad M_A = (216)(36) = 7776 \ in\text{–}kips \tag{1}$$

The shear force and bending moment diagrams are drawn as shown below.

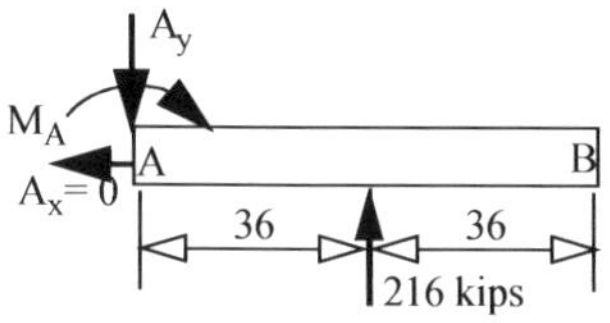

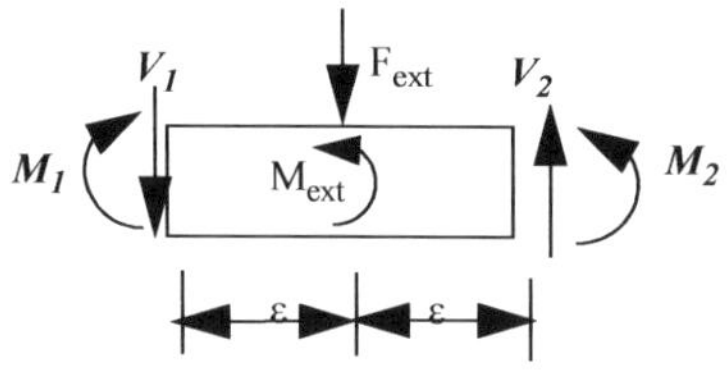

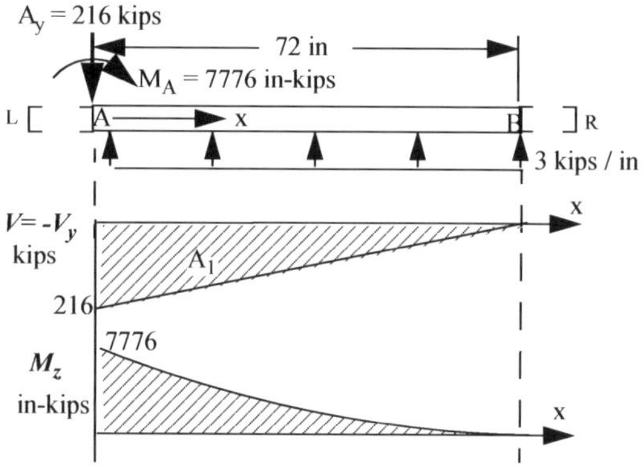

$$A_1 = \frac{1}{2}(216)(772) = 776$$

Template Equations

$$V_2 = V_1 + F_{ext}$$

$$M_2 = M_1 - M_{ext}$$

ANS $(V_y)_{max} = 216$ kips ; $(M_z)_{max} = 7776$ in-kips

---

6. 76

Solution m= 80 kg w_{boat} =130 N/m Plot of V_y and M_z

The water pressure can be found from equilibrium and shear force and bending moment diagrams drawn as shown below.

$$p(4.5) = 784.8 + 130(4.5) \qquad or \qquad p = 304.4 \tag{1}$$

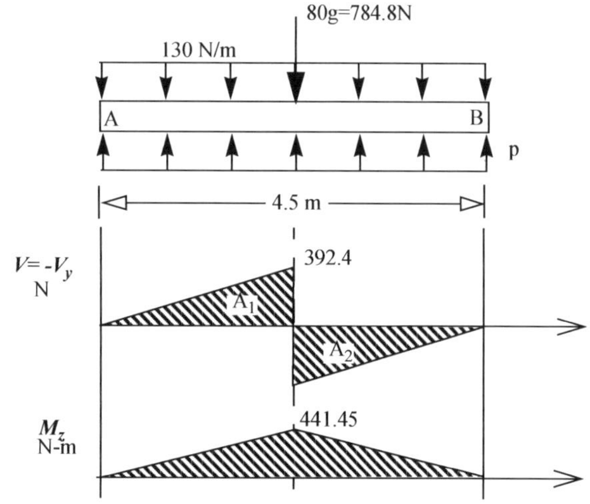

$$A_1 = A_2 = \frac{1}{2}(392.4)\left(\frac{4.5}{2}\right) = 441.45$$

ANS $(V_y)_{max} = \pm 392.4 \ N$; $(M_z)_{max} = 441.45 \ N-m$

6.77

Solution Plot of V_y and M_z $(V_y)_{max} = ?$ $(M_z)_{max} = ?$

By equilibrium of moment about point B and equilibrium of forces in the y-direction we obtain the following.

$$A_y(3L) - (wL)3(L/2) = 0 \quad or \quad A_y = wL/2 \qquad A_y + B_y - wL = 0 \quad or \quad B_y = wL/2 \tag{1}$$

The shear force and bending moment diagrams are drawn as shown below.

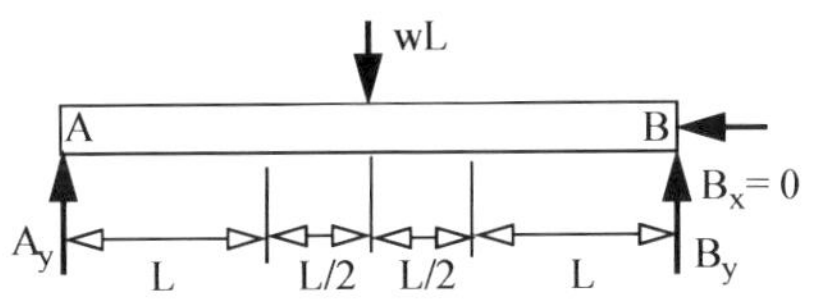

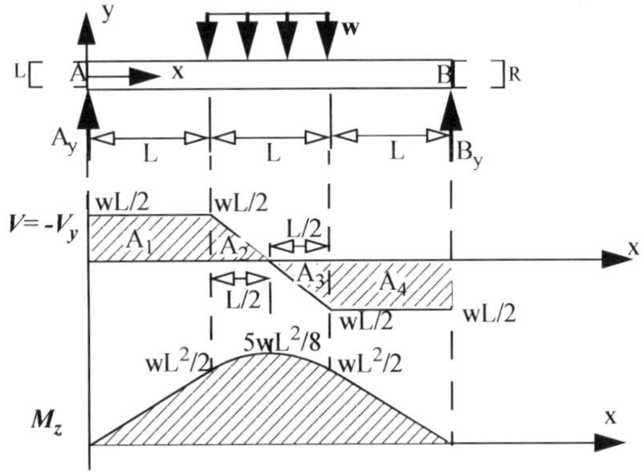

$$A_1 = A_4 = \frac{wL^2}{2}$$

$$A_2 = A_3 = \frac{1}{2}\left(\frac{wL}{2}\right)\left(\frac{L}{2}\right) = \frac{wL^2}{8}$$

ANS $(V_y)_{max} = \pm wL/2$; $(M_z)_{max} = 5wL^2/8$

6.78

Solution Plot of V_y and M_z $(V_y)_{max} = ?$ $(M_z)_{max} = ?$

By equilibrium of moment about point B and equilibrium of forces in the y-direction we obtain the following.

$$A_y(3L) - (wL)(5L/2) - (wL)(L/2) = 0 \quad or \quad A_y = wL \qquad A_y + B_y - 2wL = 0 \quad or \quad B_y = wL \tag{1}$$

The shear force and bending moment diagrams are drawn as shown below.

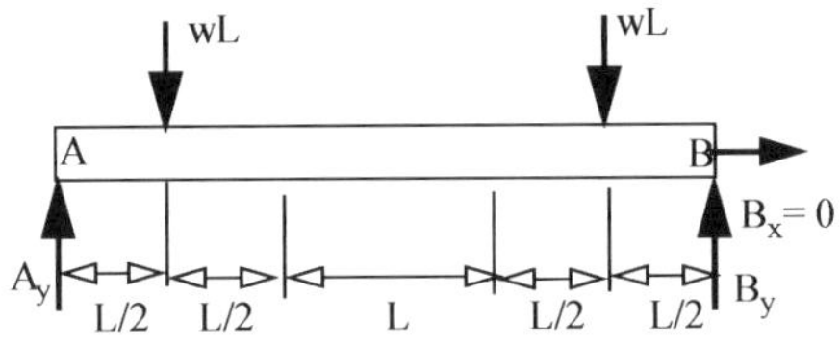

$$A_1 = A_2 = \frac{1}{2}wL^2$$

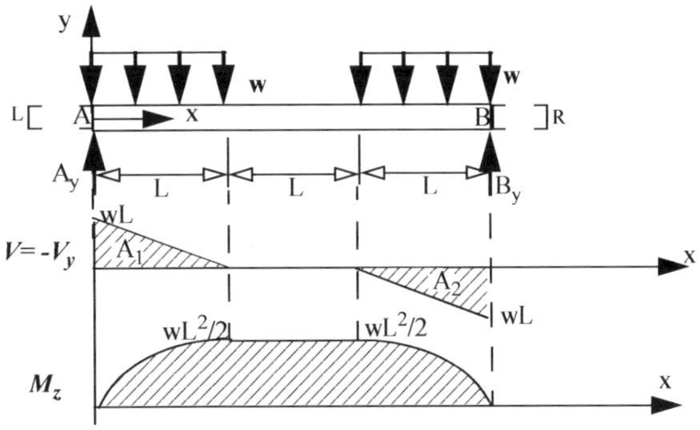

ANS $(V_y)_{max} = \pm wL$; $(M_z)_{max} = wL^2/2$

6.79

Solution Plot of V_y and M_z $(V_y)_{max} = ?$ $(M_z)_{max} = ?$

The reactions at the support were calculated in previous problem and are shown on the beam. The shear force and bending moment diagrams are drawn as shown below.

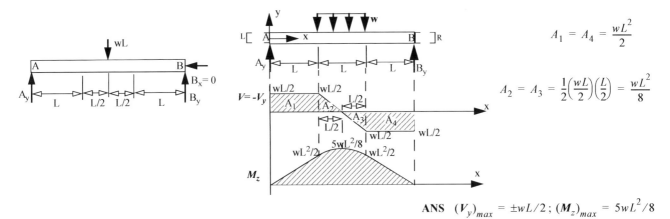

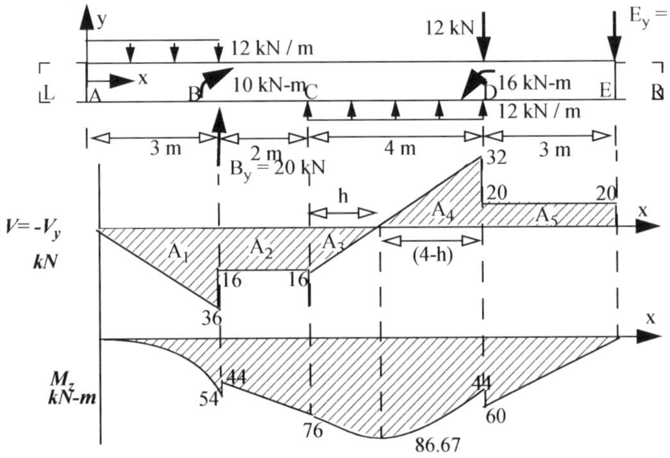

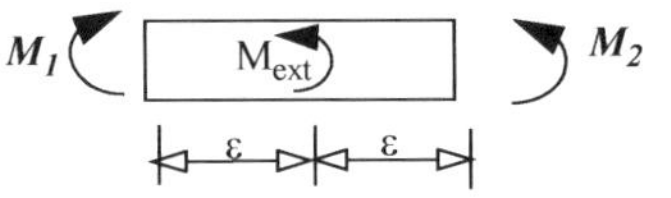

$$M_2 = M_1 - M_{ext}$$ Template Equations

$$\frac{h}{16} = \frac{4-h}{32} \quad or \quad h = 1.333m$$

$$A_1 = \frac{1}{2}(36)(3) = 54$$

$$A_2 = (16)(2) = 32$$

$$A_3 = \frac{1}{2}(16)(h) = 10.67$$

$$A_4 = \frac{1}{2}(32)(4-h) = 42.67$$

$$A_5 = (20)(3) = 60$$

ANS $(V_y)_{max} = 36$ kN; ; $(M_z)_{max} = -86.67$ kN-m

6. 80

Solution Plot of V_y and M_z $(V_y)_{max} = ?$ $(M_z)_{max} = ?$

The reactions at the support were calculated in earlier problem and are shown on the beam. The shear force and bending moment diagrams are drawn as shown below.

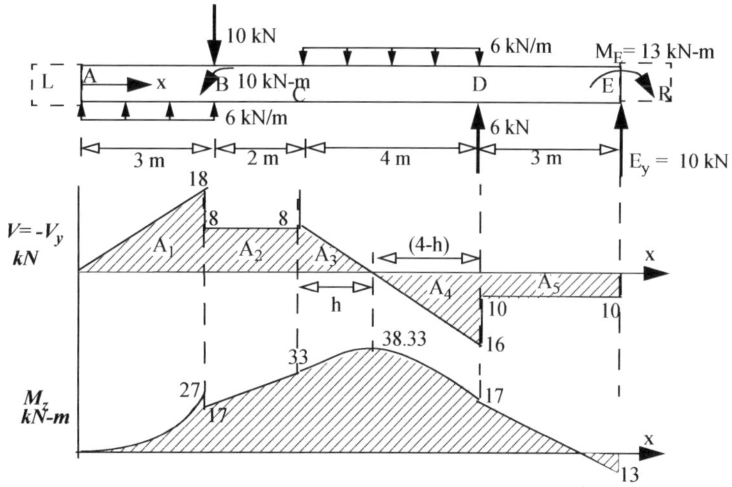

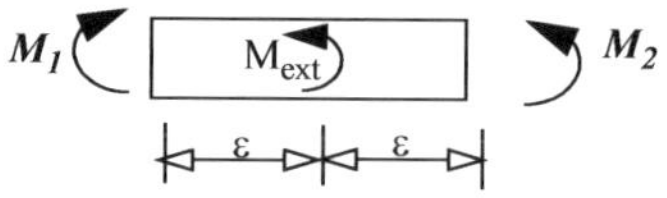

$M_2 = M_1 - M_{ext}$ Template Equations

$$\frac{h}{8} = \frac{4-h}{16} \quad or \quad h = 1.333m$$

$$A_1 = \frac{1}{2}(18)(3) = 27$$

$$A_2 = (8)(2) = 16$$

$$A_3 = \frac{1}{2}(8)(h) = 5.333$$

$$A_4 = \frac{1}{2}(16)(4-h) = 21.333$$

$$A_5 = (10)(3) = 30$$

ANS $(V_y)_{max} = -18$ kN ; $(M_z)_{max} = 38.33$ kN-m

---

6. 81

Solution Plot of V_y and M_z $(V_y)_{max} = ?$ $(M_z)_{max} = ?$

The shear force and bending moment diagrams are drawn as shown below.

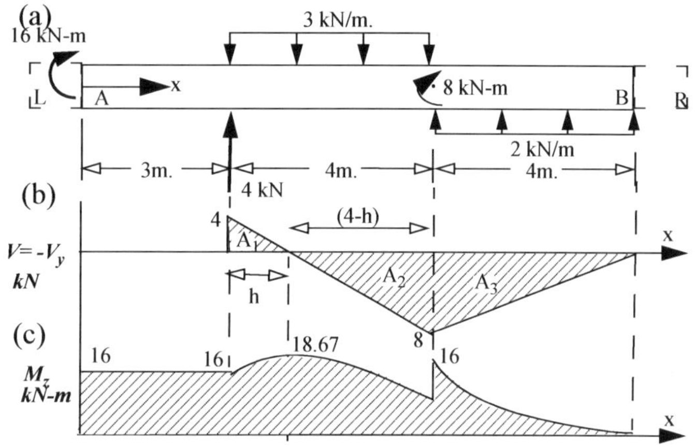

(d)

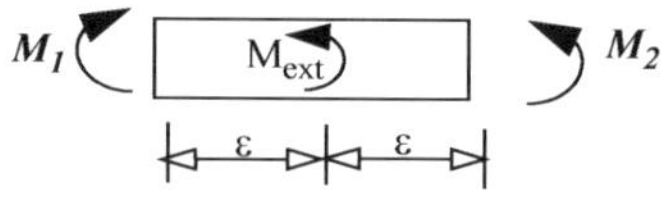

$M_2 = M_1 - M_{ext}$ Template Equations

$$\frac{h}{4} = \frac{4-h}{8} \quad or \quad h = 1.333m$$

$$A_1 = \frac{1}{2}(4)(h) = 2.667$$

$$A_2 = \frac{1}{2}(8)(4-h) = 10.667$$

$$A_3 = \frac{1}{2}(8)(4) = 16$$

ANS $(V_y)_{max} = 8$ kN ; $(M_z)_{max} = 18.67$ kN-m

---

6. 82

Solution Plot of V_y and M_z $(V_y)_{max} = ?$ $(M_z)_{max} = ?$

The shear force and bending moment diagrams are drawn as shown below.

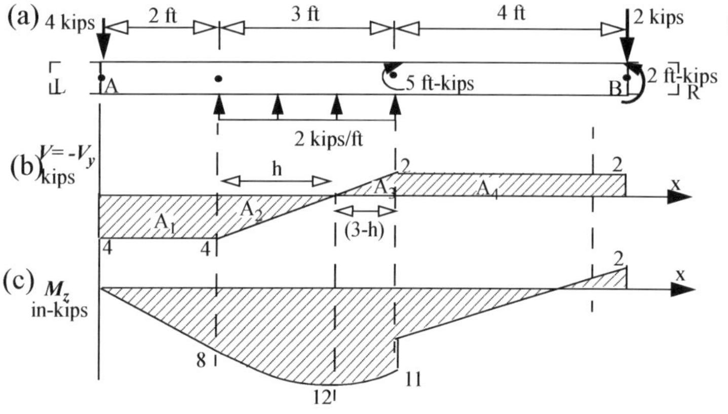

(d)

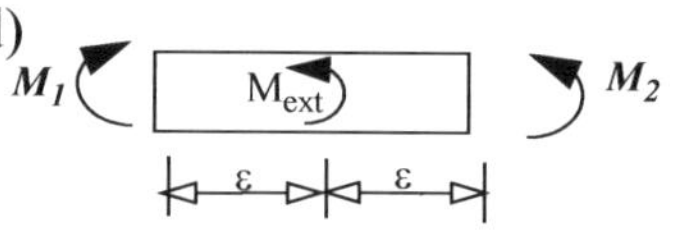

$M_2 = M_1 - M_{ext}$ Template Equations

$$\frac{h}{4} = \frac{3-h}{2} \quad or \quad h = 2ft$$

$$A_1 = (4)(2) = 8$$

$$A_2 = \frac{1}{2}(4)(h) = 4$$

$$A_3 = \frac{1}{2}(2)(3-h) = 1$$

$$A_4 = (2)(4) = 8$$

ANS $(V_y)_{max} = 4$ kips ; $(M_z)_{max} = -12$ in.-kips

---

6. 83

Solution Plot of V_y and M_z $(V_y)_{max} = ?$ $(M_z)_{max} = ?$

The shear force and bending moment diagrams are drawn as shown below.

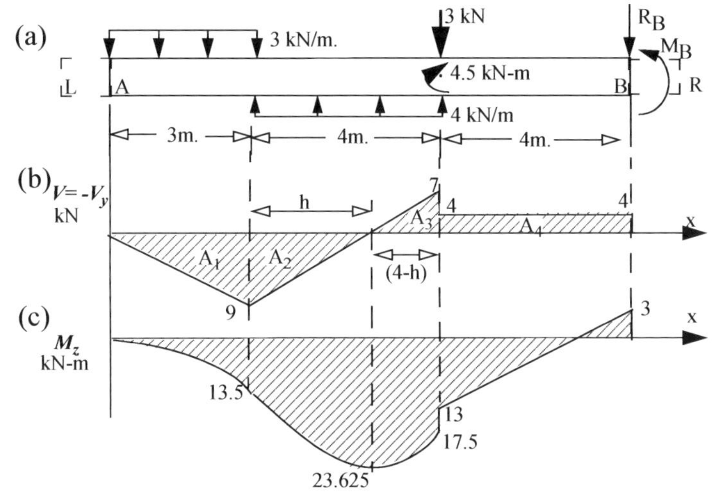

(d)

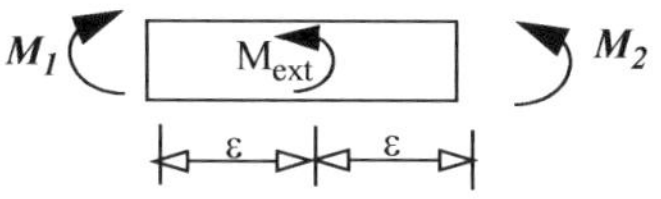

$M_2 = M_1 - M_{ext}$ Template Equations

$$\frac{h}{9} = \frac{4-h}{7} \quad or \quad h = 2.25m$$

$$A_1 = \frac{1}{2}(9)(3) = 13.5$$

$$A_2 = \frac{1}{2}(9)(h) = 10.125$$

$$A_3 = \frac{1}{2}(7)(4-h) = 6.125$$

$$A_4 = (4)(4) = 16$$

To check the results we draw the free body diagram of the entire beam with distributed forces replaced by equivalent forces. By equilibrium of forces in the y direction and moment equilibrium about point B we obtain the values of the wall reaction as shown below.

$$-R_B - 9 + 16 - 3 = 0 \quad or \quad R_B = 4kN \qquad M_B + (9)(9.5) - (16)(6) + (3)(4) - 4.5 = 0 \quad or \quad M_B = 3kN-m. \quad (1)$$

(e)

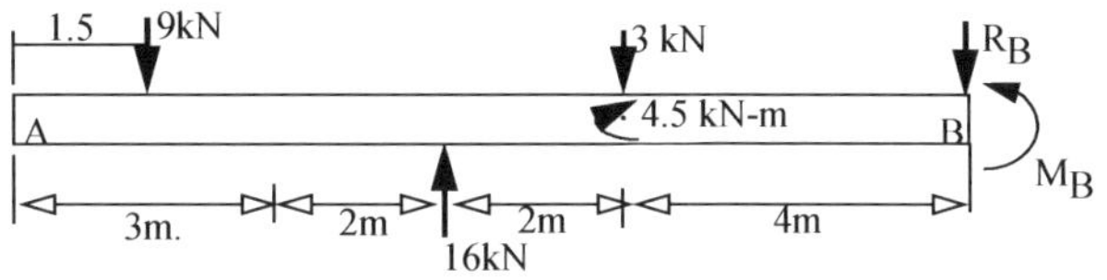

ANS $(V_y)_{max} = 9$ kN ; $(M_z)_{max} = -23.625$ kN-m

---

6. 84

Solution Plot of V_y and M_z $(V_y)_{max} = ?$ $(M_z)_{max} = ?$

The shear force and bending moment diagrams are drawn as shown below.

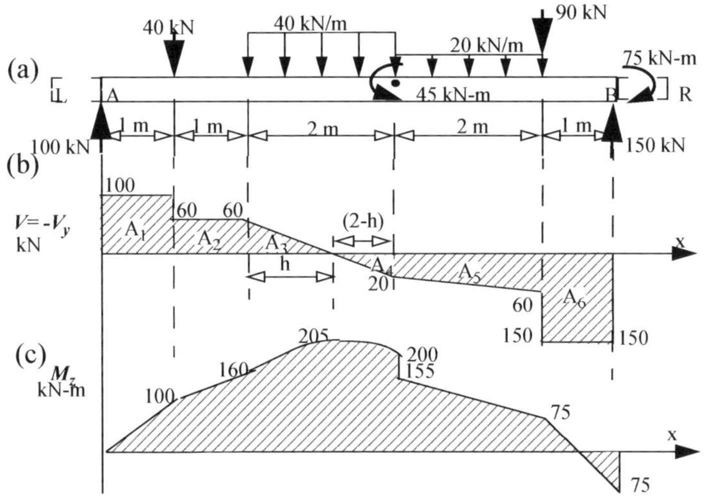

(d)

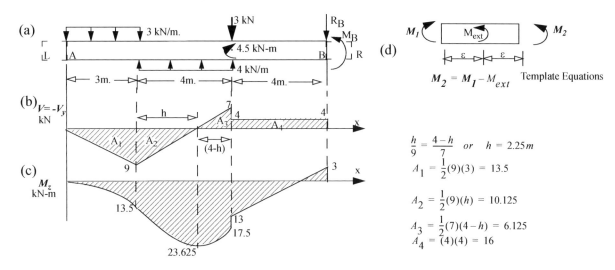

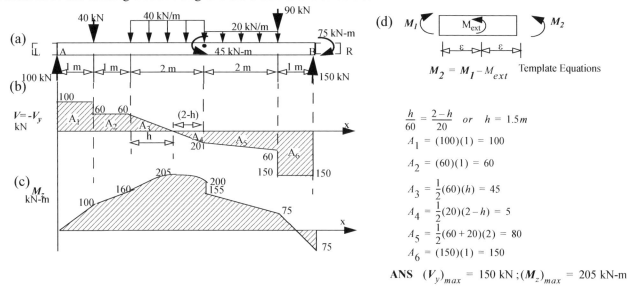

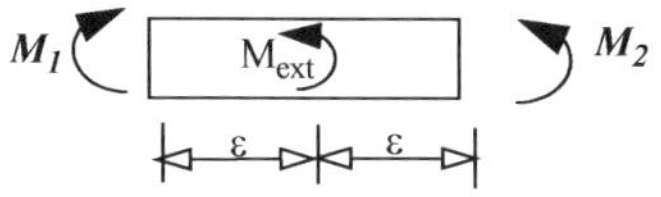

$M_2 = M_1 - M_{ext}$ Template Equations

$$\frac{h}{60} = \frac{2-h}{20} \quad or \quad h = 1.5m$$

$$A_1 = (100)(1) = 100$$

$$A_2 = (60)(1) = 60$$

$$A_3 = \frac{1}{2}(60)(h) = 45$$

$$A_4 = \frac{1}{2}(20)(2-h) = 5$$

$$A_5 = \frac{1}{2}(60+20)(2) = 80$$

$$A_6 = (150)(1) = 150$$

ANS $(V_y)_{max} = 150$ kN ; $(M_z)_{max} = 205$ kN-m

---

6. 85

Solution

By moment equilibrium about E and equilibrium of forces in the y-direction we obtain the following.

$$4(5) - 6(4.5) - 4(2) + R_B(4) = 0 \quad or \quad R_B = 3.75kips \qquad 4 + R_B - 6 - 4 + R_E = 0 \quad or \quad R_E = 2.25kips \quad (1)$$

The two forces of 4 kips acting on the rigid bracket are replaced by an equivalent moment and force acting at A and D. The shear force and bending moment diagrams are drawn as shown below.

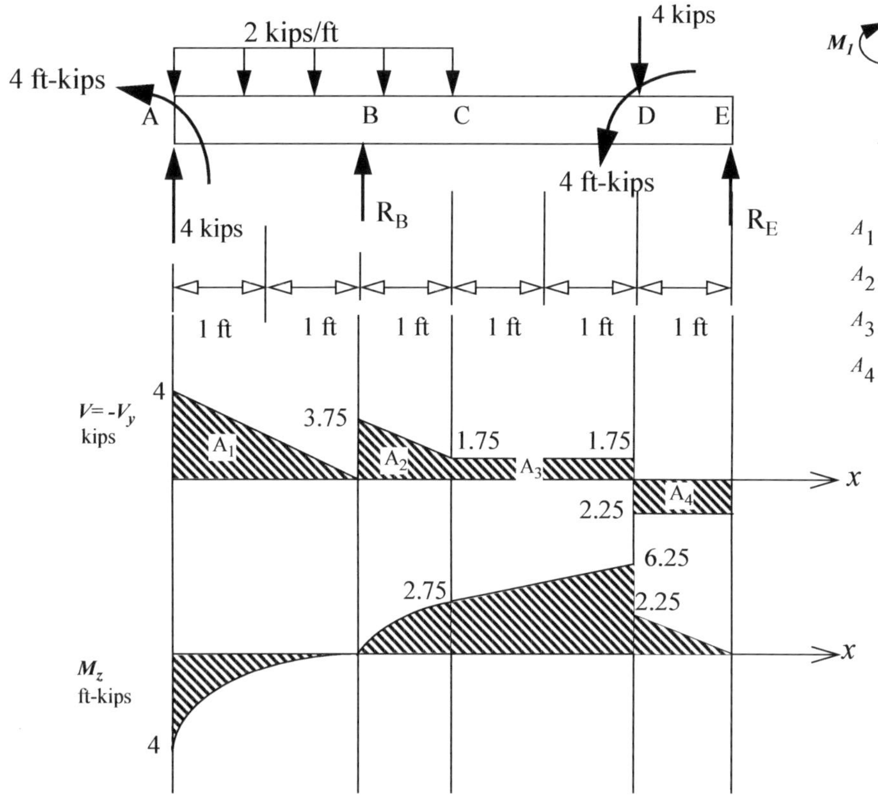

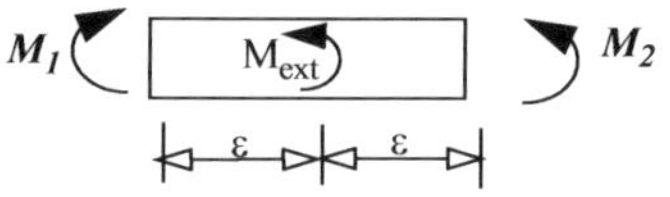

$M_2 = M_1 - M_{ext}$ Template Equations

$A_1 = (4)(2)/2 = 4$

$A_2 = (3.75 + 1.75)(2)/1 = 2.75$

$A_3 = (1.75)(2) = 3.5$

$A_4 = (2.25)(1) = 2.25$

ANS $(V_y)_{max} = -4 \text{ kips}; (M_z)_{max} = 6.25 \text{ ft-kips}$

---

6. 86

Solution Plot of V_y and M_z $(V_y)_{max} = ?$ $(M_z)_{max} = ?$

The shear force and bending moment diagrams are drawn as shown below.

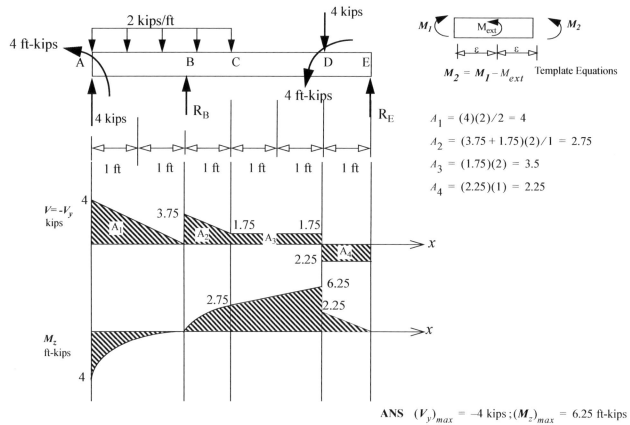

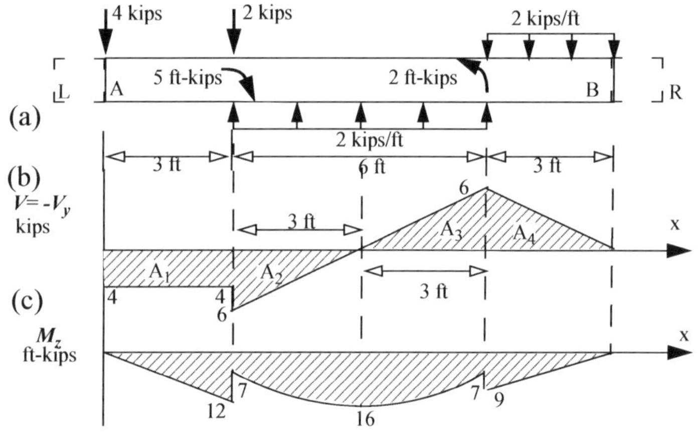

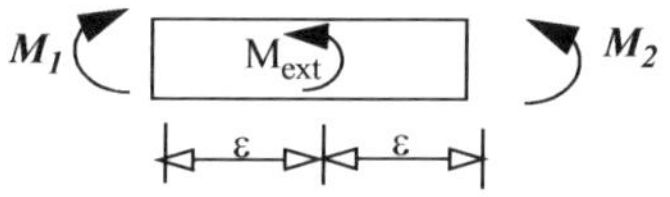

$M_2 = M_1 - M_{ext}$ Template Equations

$A_1 = (4)(3) = 12$

$A_2 = \frac{1}{2}(6)(3) = 9$

$A_3 = \frac{1}{2}(6)(3) = 9$

$A_4 = \frac{1}{2}(6)(3) = 9$

ANS $(V_y)_{max} = \pm 6 \text{ kips}; ; (M_z)_{max} = -16 \text{ in-kips}$

---

6. 87

Solution W= 200 lb cross section = 16 in.x 1 in. σ_{max}=?

The maximum moment will be at B.The area moment of inertia is:

$$M_{max} = (200)(64) = 12800 \ in-lb \qquad I_{zz} = (16)(1)^3/12 = 1.333 in^4 \qquad |\sigma_{max}| = \frac{M_{max} y_{max}}{I_{zz}} = \frac{(12800)(0.5)}{1.333} = 4800 psi \quad \textbf{(1)}$$

ANS $\sigma_{max} = 4800$ psi (T) or (C)

---

6. 88

Solution W= 200 lb cross section = 16 in.x 1 in. p = 0.5 lb/in σ_{max}=?

By equilibrium of moment about point A and equilibrium of forces in y direction, we obtain the reaction forces. The shear force and bending moment diagrams are drawn as shown below.

$$R_B(56)-200(120)-60(60) = 0 \quad or \quad R_B = 492.86\ lb \tag{1}$$

$$R_B-200-60-R_A = 0 \quad or \quad R_A = 232.86\ lb \tag{2}$$

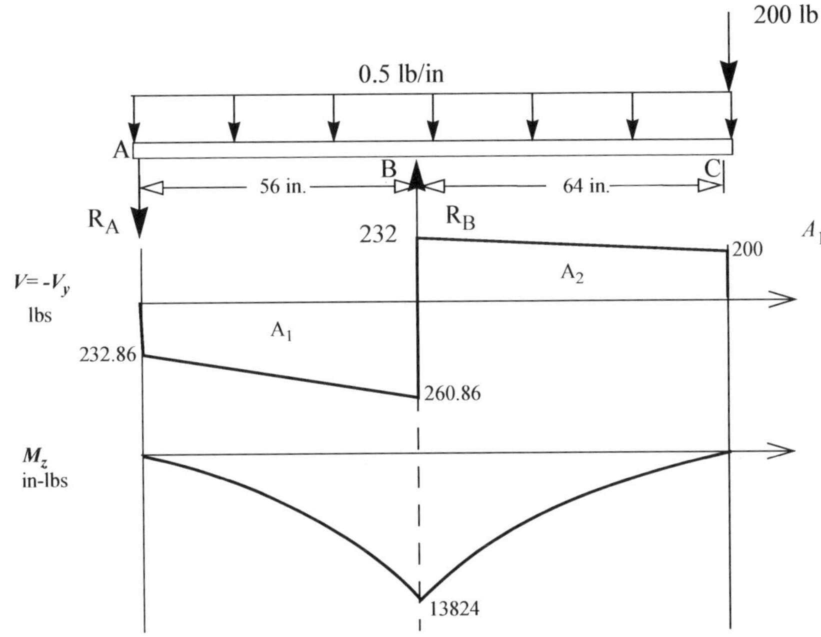

$$A_1 = \frac{1}{2}(232.86+260.86)(56) = 13824$$

$$A_2 = \frac{1}{2}(232+200)(64) = 13824$$

$$M_{max} = -13824\ in-lb \qquad I_{zz} = (16)(1)^3/12 = 1.333\ in^4 \qquad \sigma_{max} = -\left(\frac{M_{max}y_{max}}{I_{zz}}\right) = \frac{(13824)(\pm 0.5)}{1.333} = \pm 5184\ psi \tag{3}$$

ANS $\sigma_{max} = 5184$ psi (T) or (C)

---

6. 89

Solution cross section 18 in.x 1 in $|\sigma_{max}| = 10\ ksi$ $F_{max} = ?$

The maximum internal bending moment will be at B.

$$M_{max} = (F)(64) = 64F \qquad I_{zz} = (18)(1)^3/12 = 1.5\ in^4 \tag{1}$$

$$|\sigma_{max}| = \frac{M_{max}y_{max}}{I_{zz}} = \frac{(64F)(0.5)}{1.5} \le 10000 \quad or \quad F \le 468.7\ lbs \tag{2}$$

ANS $F_{max} = 468\ lbs$

---

6. 90

Solution $W_F = 225$ lb $W_s = 80$ lb 12ft x 10 in.x 1.5 in. $\sigma_{max} = ?$

By equilibrium of moment about point O we obtain

$$(W_F - m_F a)(6) - (W_s + m_s a)(6) = 0 \quad or \quad a = g\frac{(W_F - W_s)}{(W_F + W_s)} = g\left(\frac{225-80}{225+80}\right) = 0.4754g \tag{1}$$

$$m_F a = \frac{W_F}{g}0.4754g = 0.4754W_F \quad and \quad m_s a = \frac{W_s}{g}0.4754g = 0.4754W_s \tag{2}$$

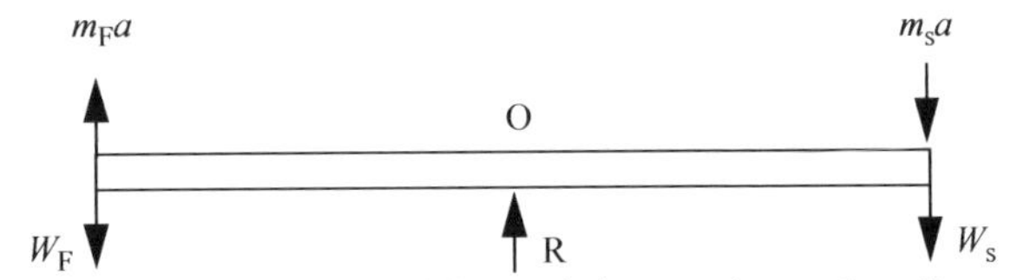

The maximum internal bending moment will be in the middle, and the maximum bending stress can be found as shown below.

$$|M_{max}| = (m_F a - W_F)(6) = 354.09\ ft-lbs = 4249.08\ in-lb \tag{3}$$

$$I_{zz} = \frac{1}{12}(10)(1.5)^3 = 2.8125\ in^4 \qquad |\sigma_{max}| = \frac{M_{max}y_{max}}{I_{zz}} = \frac{(4249.08)(0.75)}{2.8125} = 1133\ psi$$

ANS $\sigma_{max} = 1133\ psi\ (T)\ or\ (C)$

---

6. 91

Solution m_m= 70 kg m_s= 80 kg m_d= 40 kg3.5 m x 250 mm x 40 mm σ_{max} =?

By equilibrium of moment about point O we obtain

$$(m_m g - m_F a)(1.75) - (m_d g + m_d a)(1.75) = 0 \quad or \quad a = g\frac{(m_m - m_d)}{(m_m + m_d)} = g\left(\frac{70-40}{70+40}\right) = 0.2727g \qquad (1)$$

$$m_m g - m_m a = 0.7273 m_m g = 499.4 \ N \qquad m_d g + m_d a = 1.2727 m_d g = 499.4 \ N \qquad (2)$$

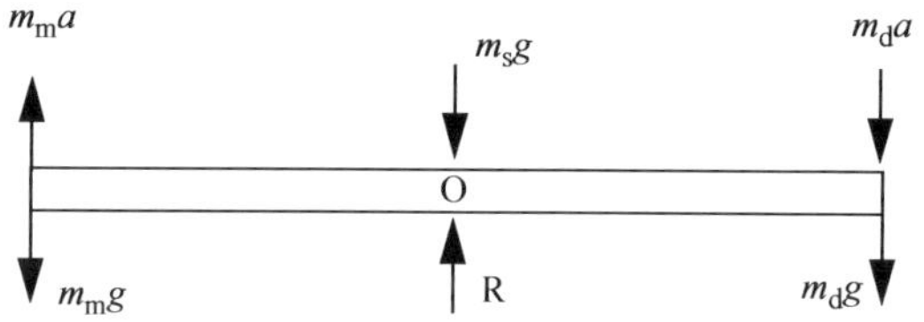

The maximum internal bending moment will be in the middle, and the maximum bending stress can be found as shown below.

$$|M_{max}| = (m_m g - m_m a)(1.75) = (499.4)(1.75) = 873.98 \ N-m \qquad (3)$$

$$I_{zz} = \frac{1}{12}(0.25)(0.04)^3 = 1.333(10^{-6}) \ m^4 \qquad |\sigma_{max}| = \frac{M_{max} y_{max}}{I_{zz}} = \frac{(873.98)(0.02)}{1.333(10^{-6})} = 13.1(10^6) \ N/m^2$$

ANS $\sigma_{max} = 13.1 \ MPa \ (T) \ or \ (C)$

---

6. 92

Solution $\sigma_C \le 10 \ ksi(C)$ $\sigma_T \le 6 \ ksi(T)$ w_{max} = ?

By equilibrium of moment about point B and equilibrium of forces in the y direction we obtain the reaction forces. The shear force and bending moment diagrams are drawn as shown below.

$$A_y(100) - (50w)(75) = 0 \quad or \quad A_y = 37.5w \ kips \qquad A_y + B_y - (50w) = 0 \quad or \quad B_y = 12.5w \ kips \qquad (1)$$

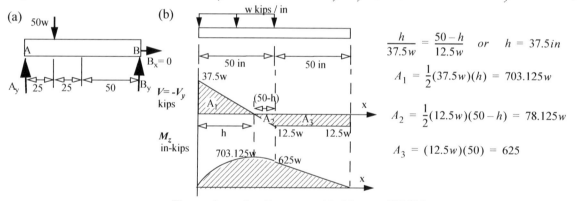

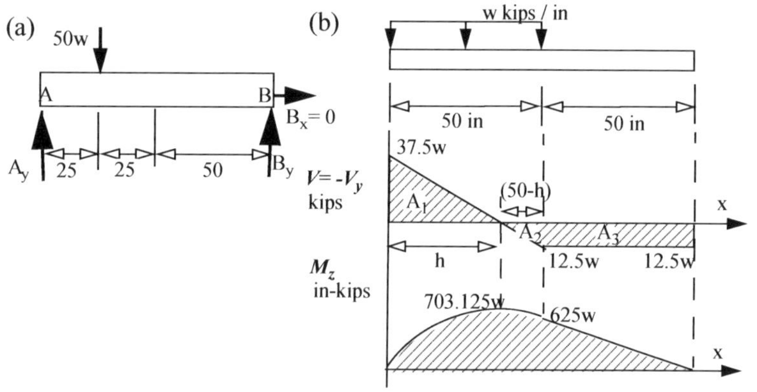

$$\frac{h}{37.5w} = \frac{50-h}{12.5w} \quad or \quad h = 37.5in$$

$$A_1 = \frac{1}{2}(37.5w)(h) = 703.125w$$

$$A_2 = \frac{1}{2}(12.5w)(50-h) = 78.125w$$

$$A_3 = (12.5w)(50) = 625$$

The maximum bending moment is M_{max} = 703.125w .

$$\sigma_D = \frac{-M_{max} y_D}{I_{zz}} = -\frac{(703.125w)(4.4)}{47.73} = -64.82w \qquad \sigma_D = 64.82w \ ksi(C) \le 10 \ ksi(C) \qquad w \le 0.1543 \ kips/(in) \qquad (2)$$

$$\sigma_E = \frac{-M_{max} y_E}{I_{zz}} = -\frac{(703.125w)(-2.6)}{47.73} = -38.3w \qquad \sigma_E = 38.3w \ ksi(T) \le 6ksi(T) \qquad w \le 0.1566 kips/in \qquad (3)$$

ANS $w_{max} = 154.3 \ \text{lb/in}$

---

6. 93

Solution: $\sigma_G \le 800 \ psi(T)$ $\sigma_w \le 1200 \ psi$ w_{max} =?

By equilibrium of moment about point B and equilibrium of forces in the y direction we obtain the reaction forces.The shear force and bending moment diagrams are drawn as shown below.

$$A_y(70) - (100w)(50) = 0 \quad or \quad A_y = 71.43w \ kips \qquad A_y + B_y - (100w) = 0 \quad or \quad B_y = 28.57w \ kips \qquad (1)$$

The maximum bending moment o and area moment of Inertia are

$$M_{max} = -450w \qquad I_{zz} = (2)(5)^3/12 = 20.83 \ in^4 \qquad (2)$$

The maximum bending normal stresses in wood and glue are

$$\sigma_W = \frac{-M_{max} y_{max}}{I_{zz}} = -\frac{(-450w)(\pm 2.5)}{20.83} = \pm 54w \ psi \qquad |\sigma_w| = 54w \le 1200 \qquad w \le 22.2 \ lbs/in \qquad (3)$$

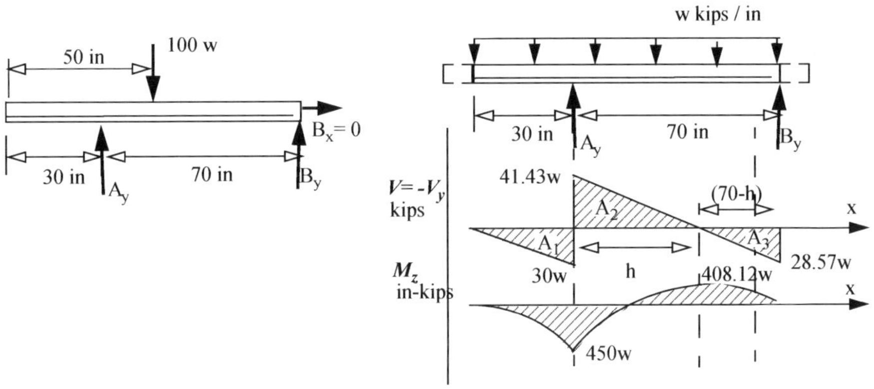

$$\frac{h}{41.43w} = \frac{70-h}{28.57w} \quad or \quad h = 41.43in$$

$$A_1 = \frac{1}{2}(30w)(30) = 450w$$

$$A_2 = \frac{1}{2}(41.43w)(h) = 858.22w$$

$$A_3 = \frac{1}{2}(28.57w)(70-h) = 408.12w$$

$$\sigma_G = \frac{-M_{max}y_G}{I_{zz}} = \frac{(-450w)(-1.5)}{20.83} = -43.2w \qquad (4)$$

The normal stress in glue above is compressive and our limit is only on tensile normal stress. We consider the maximum positive moment i.e. M = 408.12 w. The maximum tensile stress in glue is

$$\sigma_G = \frac{-My_G}{I_{zz}} = \frac{(-408.12w)(-1.5)}{20.83} = 29.38w \qquad \sigma_G = 29.38w\ psi(T) \le 800\ psi(T) \qquad w \le 27.2\ lbs/in \qquad (5)$$

ANS $w_{max} = 22.2$ lbs/in.

---

6. 94

Solution: $w = 25\ lbs/in$ $L = 72$ $|\sigma_{max}| \le 12\ ksi$ Lightest W or S shape =?

The reaction forces were found out in earlier problem. The shear and moment diagrams can be drawn as shown below.

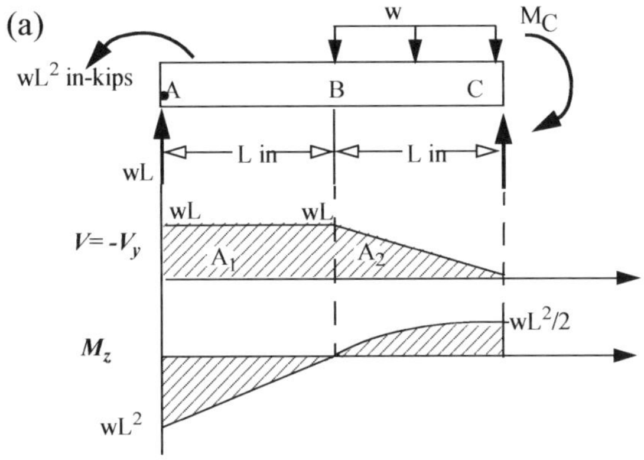

$$A_1 = wL(L) = wL^2$$

$$A_2 = \frac{1}{2}(wL)(L) = \frac{1}{2}(wL^2)$$

The magnitude of maximum bending normal stress is:

$$M_{max} = wL^2 = -(25)(72^2) = 129.6(10^3)\ in-lbs \qquad |\sigma_{max}| = \left|\frac{M_{max}}{S}\right| = \frac{129.6(10^3)}{S} \le 12(10^3) \qquad or \qquad S \ge 10.8\ in^3 \qquad (1)$$

Section modulus with shapes just greater than 10.8 in^3 are W8 x 15 and S7 x20.

ANS W8x15

---

6. 95

Solution: $w = 0.4\ (kips)/in$, $L = 48$, $|\sigma_{max}| \le 16\ ksi$ Lightest W or S shape =?

The reaction forces were found out in the earlier problem. The shear and moment diagrams can be drawn as shown below.

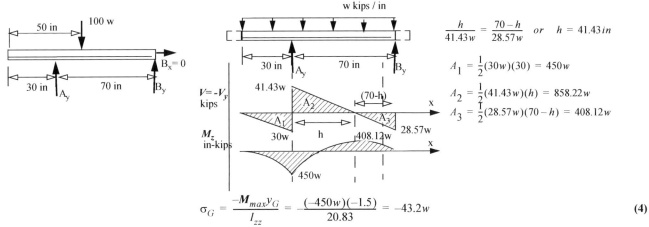

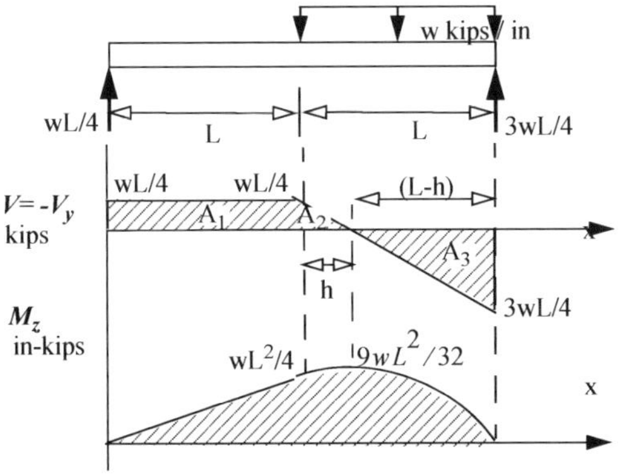

$$\frac{h}{(wL/4)} = \frac{L-h}{3wL/4} \quad or \quad h = L/4$$

$$A_1 = \left(\frac{wL}{4}\right)(L) = \frac{wL^2}{4}$$

$$A_2 = \frac{1}{2}\left(\frac{wL}{4}\right)\left(\frac{L}{4}\right) = \frac{wL^2}{32}$$

$$A_3 = \frac{1}{2}\left(\frac{3wL}{4}\right)\left(3\frac{L}{4}\right) = \frac{9wL^2}{32}$$

$$\boldsymbol{M}_{max} = \frac{9wL^2}{32} = \frac{9(0.4)(48^2)}{32} = 259.2 \ in-kips \qquad |\sigma_{max}| = \left|\frac{\boldsymbol{M}_{max}}{S}\right| = \frac{259.2}{S} \le 16 \qquad or \qquad S \ge 16.2 \ in^3 \qquad \textbf{(1)}$$

Section modulus with shapes just greater than16.2 in^3 are W 10 x 22and S 8x23.

ANS W10 x22

---

6. 96

Solution $w = 0.15 \ (kips)/in$ $L = 48$ $|\sigma_{max}| \le 21 \ ksi$ Lightest W or S shape =?

The maximum moment will be at the built in end. The maximum bending normal stress calculated as shown below.

$$\boldsymbol{M}_{max} = \frac{3wL^2}{2} = \frac{3(0.15)(48^2)}{2} = 518.4 \ in-kips \qquad |\sigma_{max}| = \left|\frac{\boldsymbol{M}_{max}}{S}\right| = \frac{518.4}{S} \le 21 \qquad or \qquad S \ge 24.68 \ in^3 \qquad \textbf{(1)}$$

Section modulus with shapes just greater than 24.68 in^3 are W10 x 30 and S10 x25.4.

ANS $S10x25.4$

---

6. 97

Solution: $|\sigma_{max}| \le 180 \ MPa$ Lightest W or S shape =?

From the solution in earlier problem we have the maximum internal moment.

$$\boldsymbol{M}_{max} = -86.68 \ kN-m \qquad |\sigma_{max}| = \left|\frac{\boldsymbol{M}_{max}}{S}\right| = \frac{86.67(10^3)}{S} \le 180(10^6) \qquad S \ge 0.4815(10^{-3}) \ m^3 \qquad S \ge 481.5(10^3) \ mm^3 \quad \textbf{(1)}$$

Section modulus with shapes just greater than 481.5(10^3) mm^3 are W250 x 44.8 and S250 x 52.

ANS W250x44.8

---

6. 98

Solution: $|\sigma_{max}| \le 225 \ MPa$ Lightest W or S shape =?

From the solution in earlier problem we have the maximum internal moment.

$$M_{max} = 38.33 \ kN-m \qquad |\sigma_{max}| = \left|\frac{M_{max}}{S}\right| = \frac{38.33(10^3)}{S} \le 225(10^6) \qquad S \ge 0.1704(10^{-3}) \ m^3 \qquad S \ge 170.4(10^3) \ mm^3 \quad \textbf{(1)}$$

Section modulus with shapes just greater than 170.4(10^3) mm^3 are W 150 x 29.8 and S 180 x 22.8.

ANS $S180x22.8$

---

6. 99

Solution: $p = 33 \ lbs/(ft^2)$, $|\sigma_{max}| \le 24 \ ksi$, $a_0 = 12 \ in$, a_i =?nearest 118 inch

The maximum bending moment will be at the base of the sign post. By equilibrium of moment about point O, we obtain

$$\boldsymbol{M}_z + (42.5p)(9.5) + 7p(3.5) = 0 \qquad or \qquad \boldsymbol{M}_z = (428.25)p = 14132.25 \ ft-lbs = 169.6 \ in-kips \qquad \textbf{(1)}$$

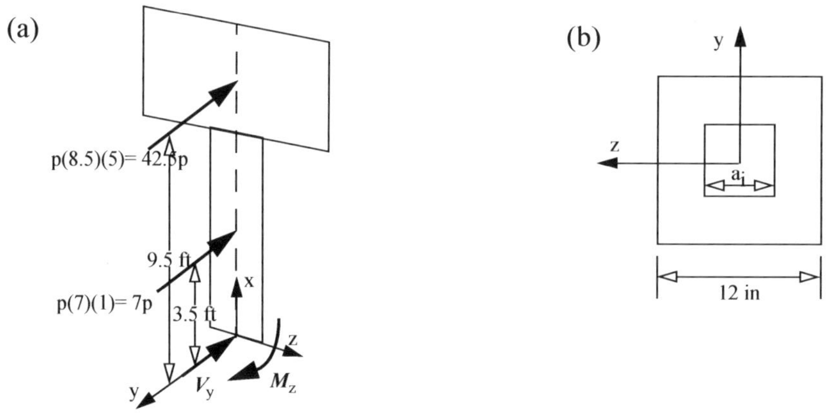

$$|\sigma_{max}| = \left|\frac{M_{max}y_{max}}{I_{zz}}\right| = \frac{169.6(6)}{I_{zz}} \le 24 \qquad or \qquad I_{zz} \ge 42.397 \qquad \textbf{(2)}$$

$$I_{zz} = [(12)(12)^3 - (a_i)(a_i)^3]/12 \ge 42.397 \qquad or \qquad (a_i)^4 \le (12)^4 - (12)(42.397) \qquad or \qquad (a_i) \le 11.925 \qquad \textbf{(3)}$$

ANS $a_i = 11\frac{7}{8}$ in.

---

6. 100

Solution: $\sigma_{max} = 200\ MPa$ $P = 200\ N$ r =?

By moment equilibrium about point B, we obtain

$$M_z = -P(1) = -200\ N-m \qquad I_{zz} = \pi(0.025)^4/64 = 19.17(10^{-9})\ m^4 \quad \textbf{(1)}$$

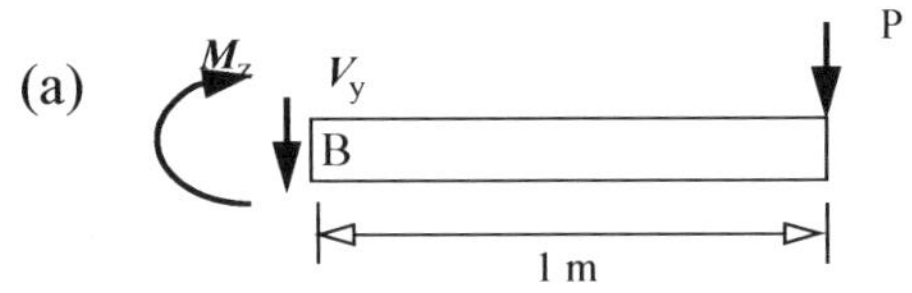

The maximum bending normal stress is

$$|(\sigma_{BC})_{max}| = \left|\frac{-(-200)(12.5)(10^{-3})}{19.17(10^{-9})}\right| = 130.4(10^6)\ N/m^2 = 130.4\ MPa \qquad K_{conc} = \frac{|\sigma_{max}|}{|(\sigma_{BC})_{max}|} \le 1.53 \quad \textbf{(2)}$$

From Appendix C we obtain the approximate value of r/d corresponding to D/d = 2 and K_{cone} = 1.53

$$r/d = 0.15 \quad or \quad r = (0.15)(25) = 3.75\text{ mm} \quad \textbf{(3)}$$

ANS $r = 3.75$ mm

---

6. 101

Solution $\sigma_{max} = 48\ ksi$ (a) r = 0.3 w_{max} =? (b) r = 0.5 w_{max} =?

By equilibrium of moment about point B we obtain

$$M_z + (24w)(12) = 0 \quad or \quad M_z = -288w\ in-lb \quad \textbf{(1)}$$

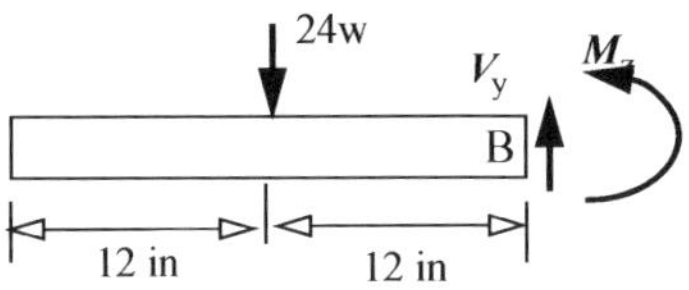

The magnitude of maximum bending normal stress is

$$I_{zz} = \frac{\pi}{64}(2)^4 = 0.7854\ in^4 \qquad |(\sigma_{BC})_{max}| = \left|\frac{-(-288w)(1)}{0.7854}\right| = 366.7w \quad \textbf{(2)}$$

(a) From Appendix C for $r/d = 0.3/2 = 0.15$ and $D/d = 2$ we obtain an approximate value of K_{conc} = 1.55. Thus we have

$$|\sigma_{max}| = K_{conc}|(\sigma_{BC})_{max}| = (1.55)(366.7)w \quad or \quad \sigma_{max} = 568.37w \le 48(10^3)\ psi \quad or \quad w \le 84.45\ lbs/in \quad \textbf{(3)}$$

(b) From Appendix C for $r/d = 0.25$ and $D/d = 2$ we obtain an approximate value of K_{conc} = 1.35. Thus we have

$$|\sigma_{max}| = K_{conc}|(\sigma_{BC})_{max}| = (1.35)(366.7)w \quad or \quad \sigma_{max} = 495w \le 48(10^3)\ psi \quad or \quad w \le 96.96\ lbs/in \quad \textbf{(4)}$$

ANS (a) w_{max} = 84.4 lbs/in. (b) w_{max} = 96.96 lbs/in.

---

6. 102

Solution: $r = 5\ mm$ $n = (0.5)(10)^6$ $Aluminium\ \sigma_{max} = 200\ MPa$

From the S-N curves the peak stress for half a million cycles in aluminum is approximately $\sigma_{max} = 154\ MPa$**.**

For $r/d = 5/25 = 0.2$ and $D/d = 2$, the stress concentration factor is approximately $K_{conc} = (1.4 + 1.45)/2 = 1.425$

From problem earlier problem we have $M_z = -P$ and $I_{zz} = 19.17(10^{-9})\ m^4$. The nominal stress in BC can be written as

$$|(\sigma_{BC})_{max}| = \frac{\sigma_{max}}{K_{cone}} = \frac{154}{1.425} = 108.1MPa \qquad |(\sigma_{BC})_{max}| = \left|\frac{-(-P)(12.5)(10^{-3})}{19.17(10^{-9})}\right| = 0.652P(10^6) \le 108.1(10^6) \qquad P \le 165.74 \quad \textbf{(1)}$$

ANS $P = 165.7$ N

---

6. 103

Solution $r = 0.36in$ $w = 80\ lbs/in$ $Steel$ n =?

From earlier problem we have $I_{zz} = 0.7854in^4$ and $M_z = -288w = -(288)(80) = -23040\ in-lb$

The nominal stress in BC is:

$$|(\sigma_{BC})_{max}| = \left|\frac{-M_z y_{max}}{I_{zz}}\right| = \left|\frac{-(-23040)(1)}{0.7854}\right| = 29.34(10^3)psi = 29.34\ ksi \quad \textbf{(1)}$$

From Appendix C for $r/d = 0.36/2 = 0.18$ and $D/d = 2$, the stress concentration factor can be found and maximum stress determine. as shown below.

$K_{conc} = (1.4+1.5)/2 = 1.5$ $\quad |\sigma_{max}| = K_{conc}|(\sigma_{BC})_{max}| = (1.475)(29.34) = 43.3\ ksi$.

From S-N curves for a peak stress of 43.3 ksi in steel the approximate number of cycles is $n \approx 175,000$

ANS $n = 175,000$

---

6. 104

Solution L = 3m p = 10 kN/mE$_{al}$ = 70 GPaE$_s$ = 200 GPa $(\sigma_{al})_{max} = ?$ $\quad (\sigma_w)_{max} = ?$ $\quad (\sigma_s)_{max} = ?$

By symmetry the reaction on each support of the simply supported beam is half the total load on the beam, i.e. $R_A = (3)(10)/(2) = 15kN$. The maximum moment will be at the center of the beam and can be found by equilibrium of moment about point O as shown below.

$$\boldsymbol{M}_{max} + 15(0.75) - 15(1.5) = 0 \quad or \quad \boldsymbol{M}_{max} = 11.25kN-m \tag{1}$$

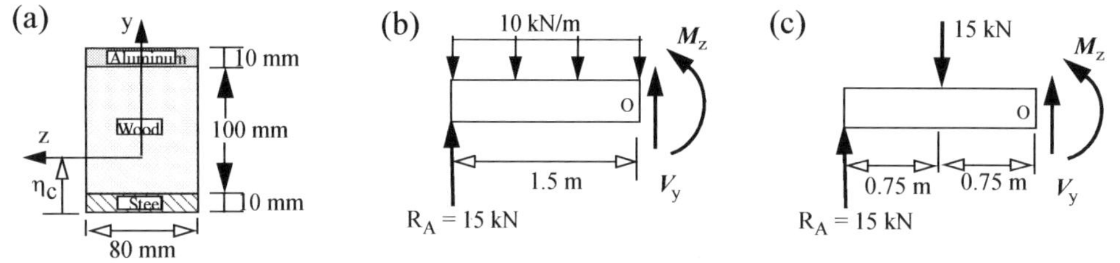

The location of the neutral axis can be found as shown below.

$$\eta_c = \frac{\sum E_i A_i \eta_{ci}}{\sum E_i A_i} = \frac{(200)(10)(80)(5) + (10)(100)(80)(60) + (70)(10)(80)(115)}{(200)(10)(80) + (10)(100)(80) + (70)(10)(80)} = 40.7mm \tag{2}$$

The area moment of inertias for each material and the bending rigidity of the cross-section can be found as shown below.

$$(I_{zz})_{al} = \frac{1}{12}(0.08)(0.01)^3 + (0.08)(0.01)(0.115-0.0407)^2 = 4.426(10^{-6})m^4 \tag{3}$$

$$(I_{zz})_w = \frac{1}{12}(0.08)(0.1)^3 + (0.08)(0.1)(0.06-0.0407)^2 = 9.654(10^{-6})m^4 \tag{4}$$

$$(I_{zz})_s = \frac{1}{12}(0.08)(0.01)^3 + (0.08)(0.01)(0.0407-0.005)^2 = 1.025(10^{-6})m^4 \tag{5}$$

$$\sum E_j(I_{zz})_j = [(70)(4.426) + (10)(9.654) + (200)(1.025))](10^3) = 611.3(10^3)N-m^2 \tag{6}$$

The maximum bending normal stress in aluminum will be at $y = (0.12-0.0407) = 0.0793m$ in wood will be at $y = (0.11-0.0407) = 0.0693m$, and in steel will be at $y = -0.0407m$.

$$(\sigma_{al})_{max} = \frac{-E_{al}(\boldsymbol{M}_{max})y}{\sum E_j(I_{zz})_j} = -\frac{(70)(10^9)(11.25)(10^3)(0.0793)}{611.3(10^3)} = -102.1(10^6)\frac{N}{m^2} \tag{7}$$

$$(\sigma_w)_{max} = \frac{-E_w(\boldsymbol{M}_{max})y}{\sum E_j(I_{zz})_j} = -\frac{(10)(10^9)(11.25)(10^3)(0.0693)}{611.3(10^3)} = -12.75(10^6)\frac{N}{m^2} \tag{8}$$

$$(\sigma_s)_{max} = \frac{-E_s(\boldsymbol{M}_{max})y}{\sum E_j(I_{zz})_j} = -\frac{(200)(10^9)(11.25)(10^3)(-0.0407)}{611.3(10^3)} = 149.7(10^6)\frac{N}{m^2} \tag{9}$$

ANS $(\sigma_{al})_{max}$ =102.1 MPa (C) ; $(\sigma_w)_{max}$ = 12.75 MPa (C); $(\sigma_s)_{max}$ = 149.7 MPa (T)

---

6. 105

Solution: E$_s$ = 200 GPa E$_{br}$ = 100 GPa $(\sigma_s)_{max} = ?$ $\quad (\sigma_{br})_{max} = ?$ $\quad |\tau|_{max} = ?$

The area moment of inertia and bending rigidity of the cross section can be found as shown below.

$$(I_{zz})_s = \frac{\pi}{64}(0.24^2 - 0.20^2) = 84.32(10^{-6})m^4 \qquad (I_{zz})_{br} = \frac{\pi}{64}(0.20^2 - 0.16^2) = 46.37(10^{-6})m^4 \tag{1}$$

$$\sum E_j(I_{zz})_j = [(200)(84.32) + (100)(46.37)](10^3) = 21.5(10^6)N-m \tag{2}$$

By equilibrium of forces in the y- direction and equilibrium of moment about point A we obtain the wall reactions. The shear force and bending moment diagrams can be drawn as shown below.

$$R_A = 12-8 = 4kN \qquad \boldsymbol{M}_A + (12)(5) + (8) - (8)(9) = 0 \quad or \quad \boldsymbol{M}_A = 4kN-m \tag{3}$$

The maximum bending normal stress in steel will be at y = $\pm$ 0.12 m, and in aluminum will be at y = $\pm$ 0.10 m.

$$M_{max} = 18.667 \; kN-m \qquad \left|(\sigma_s)_{max}\right| = \left|\frac{-E_s(M_{max})y}{\sum E_j(I_{zz})_j}\right| = \frac{(200)(10^9)(18.666)(10^3)(0.12)}{21.5(10^6)} = 21.5(10^6)(N/m^2) \qquad \textbf{(4)}$$

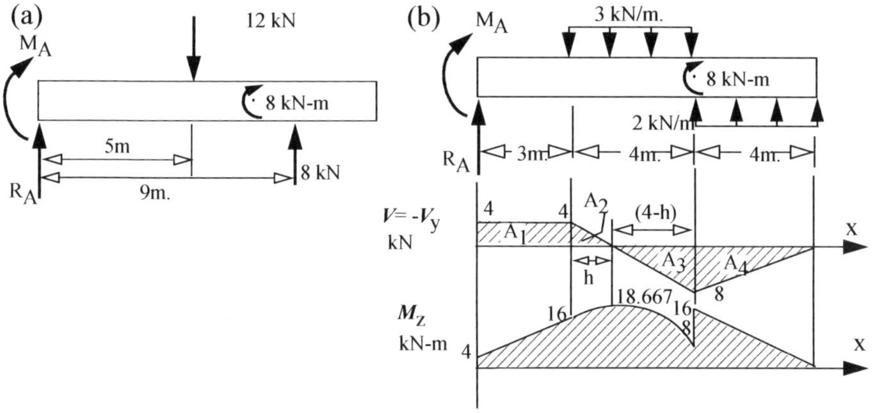

$$\frac{h}{4} = \frac{(4-h)}{8}$$

$or \qquad h = 1.333$

$A_1 = (4)(3) = 12$

$A_2 = \frac{1}{2}(4)(h) = 2.667$

$A_3 = \frac{1}{2}(4)(4-h) = 10.667$

$A_4 = \frac{1}{2}(8)(4) = 16$

$$\left|(\sigma_{al})_{max}\right| = \left|\frac{-E_{al}(M_{max})y}{\sum E_j(I_{zz})_j}\right| = \frac{(100)(10^9)(18.666)(10^3)(0.10)}{21.5(10^6)} = 8.681(10^6)(N/m^2) \qquad \textbf{(5)}$$

ANS $\left|(\sigma_s)_{max}\right| = 21.5$ MPa (T) or (C) ; $\left|(\sigma_{al})_{max}\right| = 8.7$ MPa (T) or (C)

Section 6.6

6. 106

Solution Shear flow sketch=? τ_{xy} or τ_{xz} at points A, B, C, and D =?

The shear flow is as shown below..

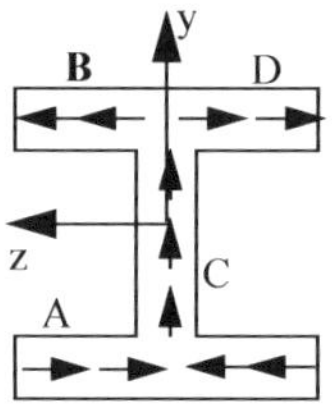

ANS Point A: negative τ_{xz} ; Point B: positive τ_{xz} ; Point C: positive τ_{xy} ; Point D: negative τ_{xz}

---

6. 107

Solution Shear flow sketch=? τ_{xy} or τ_{xz} at points A, B, C, and D =?

The shear flow is as shown below..

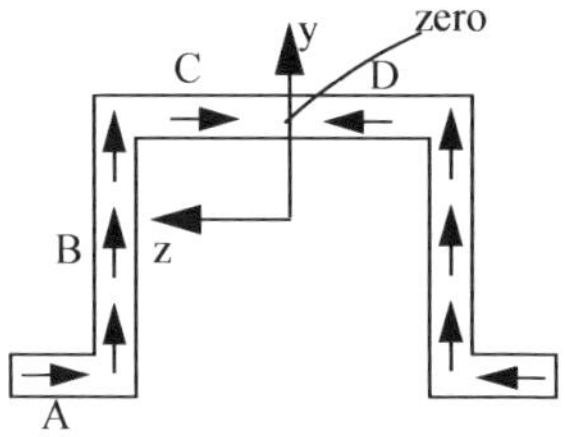

ANS Point A: negative τ_{xz} ; Point B: positive τ_{xy} ; Point C: negative τ_{xz} ; Point D: positive τ_{xz}

---

6. 108

Solution Shear flow sketch=? τ_{xy} or τ_{xz} at points A, B, C, and D =?

The shear flow is as shown below.

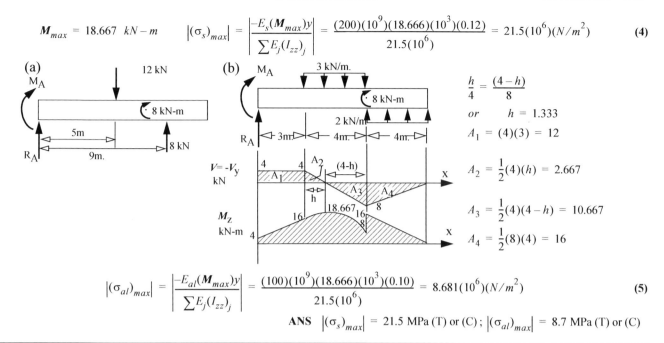

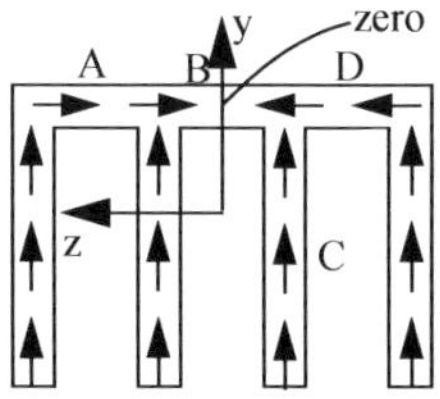

ANS Point A: negative τ_{xz} ; Point B: negative τ_{xz} ; Point C: positive τ_{xy} ; Point D: positive τ_{xz}

---

6. 109

Solution Shear flow sketch=? τ_{xy} or τ_{xz} at points A, B, C, and D =?

The shear flow is as shown.

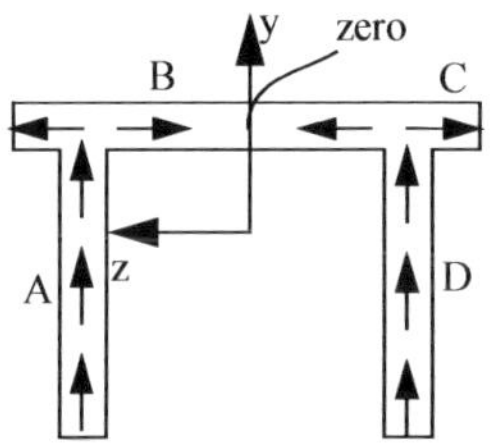

ANS Point A: positive τ_{xy} ; Point B: negative τ_{xz} ; Point C: negative τ_{xz} ; Point D: positive τ_{xy}

---

6. 110

Solution Shear flow sketch=? τ_{xy} or τ_{xz} at points A, B, C, and D =?

The shear flow is as shown.

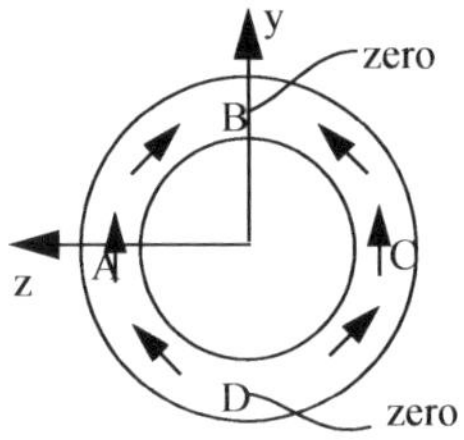

ANS Point A: positive τ_{xy} ; Point B: zero shear stress ; Point C: positive τ_{xy} ; Point D: zero shear stress

---

6. 111

Solution Shear flow sketch=? τ_{xy} or τ_{xz} at points A, B, C, and D =?

The shear flow is as shown

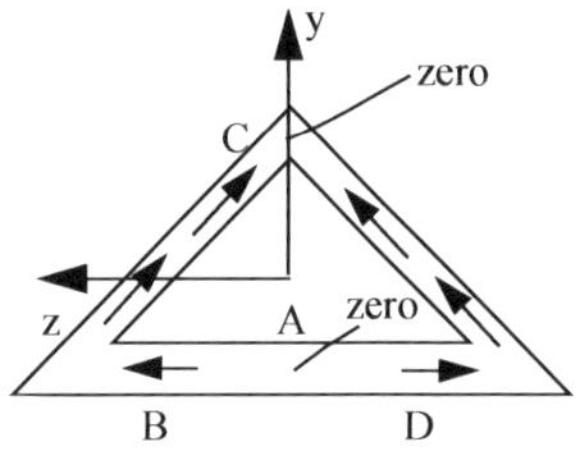

ANS Point A: zero shear stress ; Point B: positive τ_{xz} ; Point C: zero shear stress ; Point D: negative τ_{xz}

---

6. 112

Solution Shear flow sketch=? τ_{xy} or τ_{xz} at points A, B, C, and D =?

The shear flow is as shown.

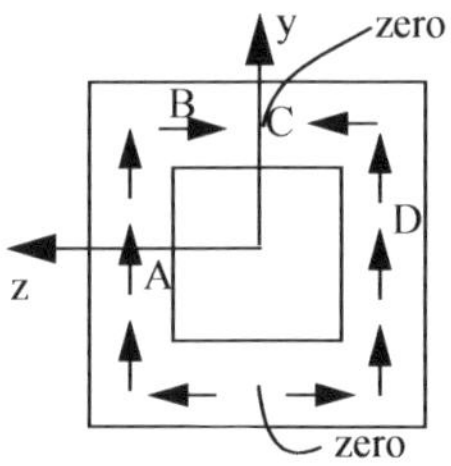

ANS Point A: positive τ_{xy} ; Point B: negative τ_{xz} ; Point C: zero shear stress ; Point D: positive τ_{xy}

---

6. 113

Solution $V_y = 5$ kips τ_{xy} or τ_{xz} at points A, B, C, and D =?

$$I_{zz} = \frac{1}{12}(2)(1)^3 + (2)(1)(3)^2 + \frac{1}{12}(1)(4)^3 + (1)(4)(0.5)^2 + \frac{1}{12}(4)(1)^3 + (4)(1)(2)^2 = 40.833 \text{ in}^4 \quad \textbf{(1)}$$

The first moment of area As for various points can be found and shear stress calculated as shown below.

$$(Q_z)_A = (1)(1)(2) = 2 \text{ in}^3 \qquad (\tau_{xs})_A = -\frac{(5)(2)}{(40.833)(1)} = -0.2449 \text{ ksi} \quad \textbf{(2)}$$

$$(Q_z)_B = (4)(1)(2) = 8 \text{ in}^3 \qquad (\tau_{xs})_B = -\frac{(5)(8)}{(40.833)(1)} = -0.980 \text{ ksi} \tag{3}$$

$$(Q_z)_C = (Q_z)_B + (1.5)(1)(0.75) = 9.125 \text{ in}^3 \qquad (\tau_{xs})_C = -\frac{(5)(9.125)}{(40.833)(1)} = -1.117 \text{ ksi} \tag{4}$$

$$(Q_z)_D = (Q_z)_C + (1.5)(1)(-0.75) = 8 \text{ in}^3 \qquad (\tau_{xs})_A = -\frac{(5)(8)}{(40.833)(1)} = -0.980 \text{ ksi}. \tag{5}$$

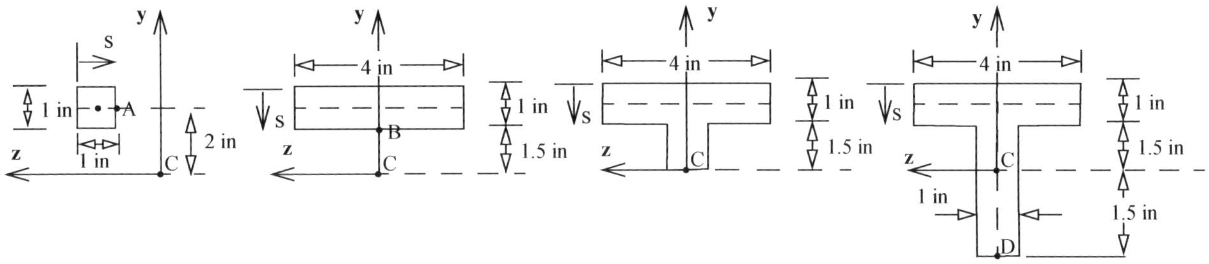

ANS $(\tau_{xz})_A = 0.2449$ ksi ; $(\tau_{xy})_B = 0.980$ ksi ; $(\tau_{xy})_C = 1.117$ ksi ; $(\tau_{xy})_D = 0.980$ ksi

---

6. 114

Solution $V_y = -10$ kN $\quad \tau_{xy}$ or τ_{xz} at points A, B, C, and D =?

$$I_{zz} = \frac{1}{12}(50)(70^3) - \frac{1}{12}(40)(50^3) = 1.0125(10^6)\ mm^4 = 1.0125(10^{-6})\ m^4 \tag{1}$$

The first moment of area As for various points can be found and shear stress calculated as shown below.

$$(Q_z)_A = (0.01)(0.01)(0.03) = 3(10^{-6}) \text{ m}^3 \qquad (\tau_{xs})_A = \frac{(10)(10^3)(3)(10^{-6})}{(1.0125)(10^{-6})(0.01)} = 2.963(10^6)N/m^2 \tag{2}$$

$$(Q_z)_B = (0.01)(0.05)(0.03) = 15(10^{-6}) \text{ m}^3 \qquad (\tau_{xs})_B = \frac{(10)(10^3)(15)(10^{-6})}{(1.0125)(10^{-6})(0.01)} = 14.81(10^6)N/m^2 \tag{3}$$

$$(Q_z)_C = (Q_z)_B + (0.01)(0.025)(0.0125) = 18.125(10^{-6}) \text{ m}^3 \qquad (\tau_{xs})_C = \frac{(10)(10^3)(18.125)(10^{-6})}{(1.0125)(10^{-6})(0.01)} = 17.9(10^6)N/m^2 \tag{4}$$

$$(Q_z)_D = (0.01)(0.01)(-0.03) = (-3)(10^{-6}) \text{ m}^3 \qquad (\tau_{xs})_A = -\frac{(-10)(10^3)(-3)(10^{-6})}{(1.0125)(10^{-6})(0.01)} = -2.963(10^6)N/m^2 \tag{5}$$

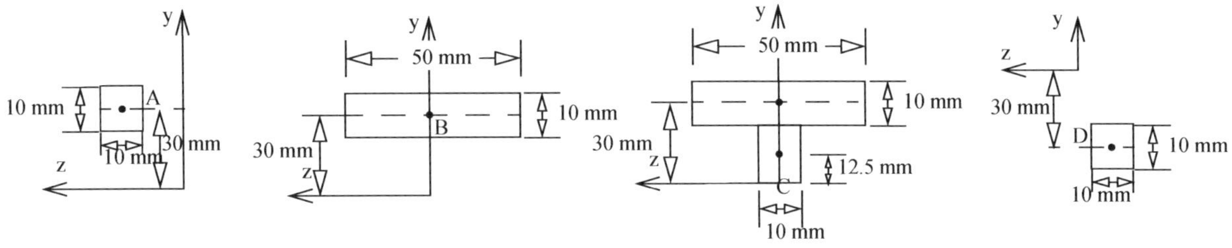

$(\tau_{xz})_A = -2.96\ MPa$; $(\tau_{xy})_B = -14.8\ MPa$; $(\tau_{xy})_C = -17.9\ MPa$ $\quad (\tau_{xz})_D = 2.96\ MPa$

---

6. 115

Solution $M_z = 50$ in.-kips $V_y = 10$ kips, $\quad \sigma_{xx}$ and τ_{xy} or τ_{xz} at points A, B, C, and D =?

$$I_{zz} = \left[\frac{(8)(0.5)^3}{12} + (8)(0.5)(3.21 - 0.25)^2\right] + \left[\frac{(0.5)(7)^3}{12} + (0.5)(7)(3.21 - 4)^2\right] + \left[\frac{(4)(0.5)^3}{12} + (4)(0.5)(7.75 - 3.21)^2\right] = 92.871 in^4 \tag{1}$$

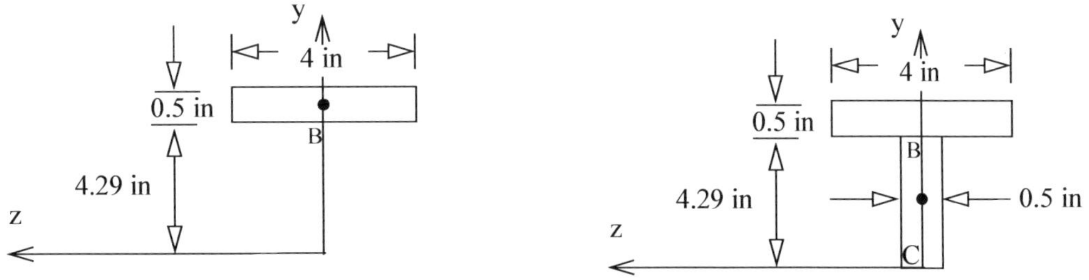

The first moment of areas can be found and stresses calculated as shown belowas shown below.

$$(Q_z)_B = (0.5)(4)(4.29 + 0.25) = 9.08 \ in^3 \qquad (Q_z)_C = (Q_z)_B + (0.5)(4.29)\left(\frac{4.29}{2}\right) = 13.681 \text{ in}^3 \tag{2}$$

$$(\sigma_{xx})_A = -\left(\frac{M_z y_A}{I_{zz}}\right) = -\frac{(50)(-3.21)}{92.871} = 1.73 \ ksi \qquad (\sigma_{xx})_B = -\left(\frac{M_z y_B}{I_{zz}}\right) = -\frac{(50)(4.29)}{92.871} = -2.309 \ ksi \tag{3}$$

$$(\tau_{xs})_B = -\frac{(10)(9.08)}{(92.871)(0.5)} = -1.955 \ ksi \qquad (\tau_{xs})_C = -\frac{(10)(13.681)}{(92.871)(0.5)} = -2.94 \ ksi \tag{4}$$

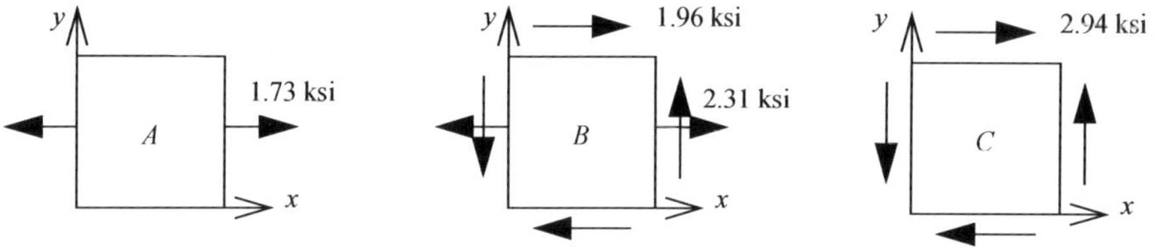

ANS $(\sigma_{xx})_A = 1.73 \ ksi(T)$; $(\sigma_{xx})_B = 2.31 \ ksi(C)$; $(\tau_{xy})_B = 1.96 \ ksi$; $(\tau_{xy})_C = 2.94 \ ksi$

---

6. 116

Solution| $|\sigma_{max}| = ?$ $|\tau_{max}| = ?$

The shear and moment diagrams are drawn as shown below

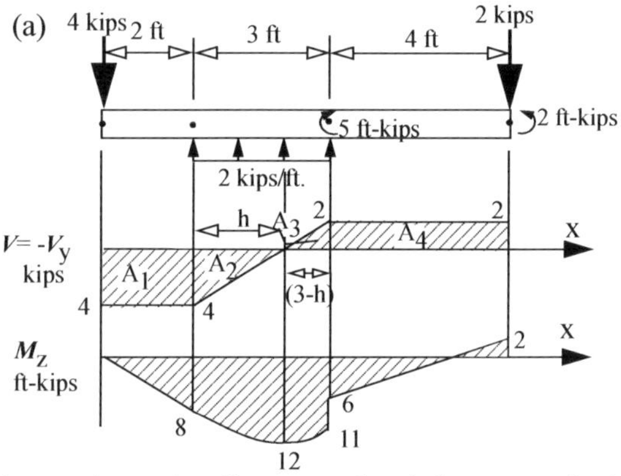

$$\frac{h}{4} = \frac{(3-h)}{2} \qquad or \qquad h = 2$$

$$A_1 = (4)(2) = 8$$

$$A_2 = \frac{1}{2}(4)(h) = 4$$

$$A_3 = \frac{1}{2}(2)(3-h) = 1$$

$$A_4 = (4)(2) = 8$$

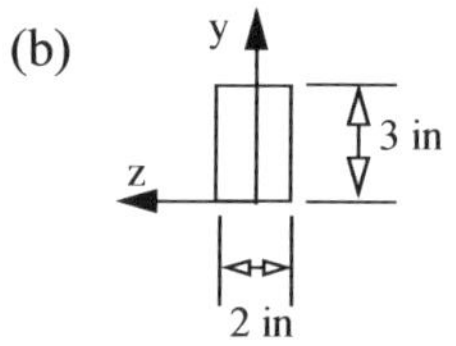

The maximum bending normal and shear stress in the beam can be calculated as shown below.

$$M_{max} = -12 \ ft-kips = -144 \ in-kips \qquad I_{zz} = \frac{1}{12}(2)(6^3) = 36 \ in^4 \qquad \sigma_{max} = -\frac{M_{max} y_{max}}{I_{zz}} = -\frac{(-144)(\pm 3)}{36} = \pm 12 ksi \tag{1}$$

$$(V_y)_{max} = 4 \ kips \qquad Q_{max} = (3)(2)(1.5) = 9 \ in^3 \qquad |\tau_{max}| = \left|-\frac{(V_y)_{max} Q_{max}}{I_{zz}(t)}\right| = \frac{(4)(9)}{(36)(2)} = 0.5 \ ksi \tag{2}$$

ANS $|\sigma_{max}| = 12$ ksi ; $|\tau_{max}| = 500$ psi

---

6. 117

Solution $I_{zz} = 3.6(10^6) mm^4$, $|\tau_{max}| = ?$ $|\sigma_{max}| = ?$ $(\sigma_{xx})_A = ?$ $(\tau_{xy})_A = ?$

The shear and moment diagrams are drawn as shown below.

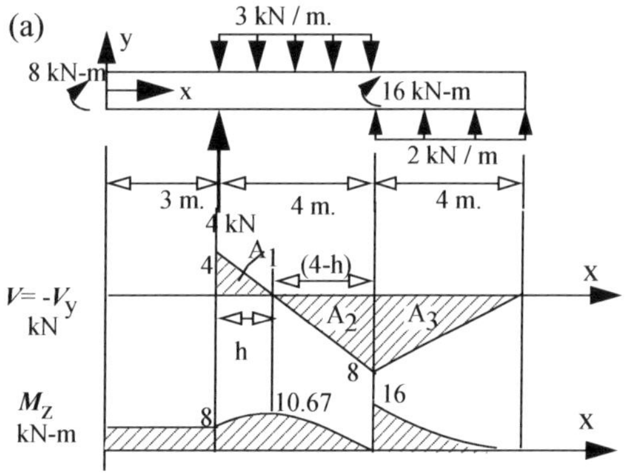

$$\frac{h}{4} = \frac{(4-h)}{8} \qquad or \ h = 1.333$$

$$A_1 = \frac{1}{2}(4)(h) = 2.67$$

$$A_2 = \frac{1}{2}(8)(4-h) = 10.67$$

$$A_3 = \frac{1}{2}(8)(4) = 16$$

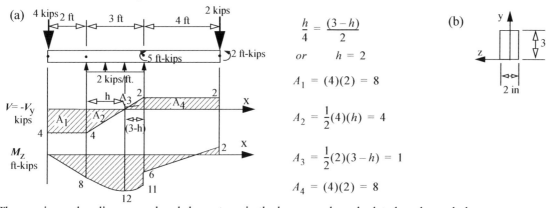

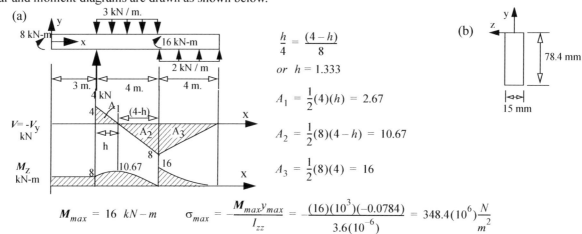

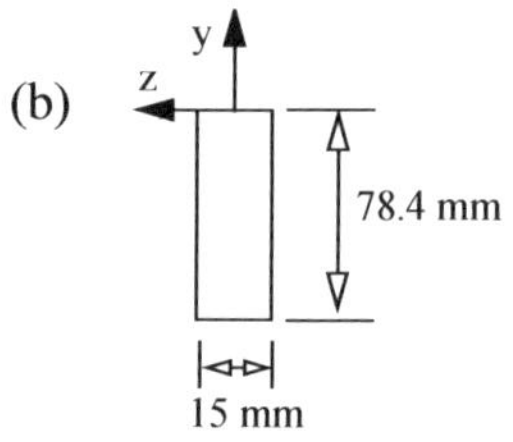

$$M_{max} = 16 \ kN-m \qquad \sigma_{max} = -\frac{M_{max} y_{max}}{I_{zz}} = -\frac{(16)(10^3)(-0.0784)}{3.6(10^{-6})} = 348.4(10^6)\frac{N}{m^2} \tag{1}$$

$$Q_{max} = (0.0784)(0.015)\left(-\frac{0.0784}{2}\right) = -46.10(10^{-6})m^3 \qquad (V_y)_{max} = 8kN \tag{2}$$

$$|\tau_{max}| = \left|-\frac{Q_{max}V_{ymax}}{I_{zz}\ t}\right| = -\frac{8(10^3)(46.10)(10^{-6})}{3.6(10^{-6})(0.015)} = 6.84(10^6)\frac{N}{m^2} \tag{3}$$

Note that s and y are in opposite direction in the calculation of shear stress at point A.

$$M_A = 8kN-m \qquad (\sigma_{xx})_A = -\frac{M_A y_A}{I_{zz}} = -\frac{(8)(10^3)(0.0216)}{3.6(10^{-6})} = -48(10^6)\frac{N}{m^2} \tag{4}$$

$$Q_A = (0.160)(0.010)(0.0266) = 42.56(10^{-6})m^3 \qquad (V_y)_A = -4kN \tag{5}$$

$$(\tau_{xs})_A = -\frac{Q_A(V_y)_A}{I_{zz}\ t} = -\frac{(-4)(10^3)42.56(10^{-6})}{3.6(10^{-6})(0.015)} = 3.15(10^6)\frac{N}{m^2} \quad or \quad (\tau_{xy})_A = -(\tau_{xs})_A = -3.15MPa. \tag{6}$$

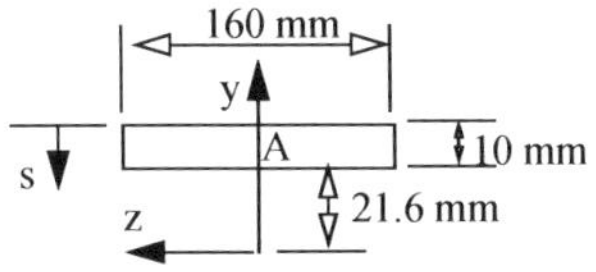

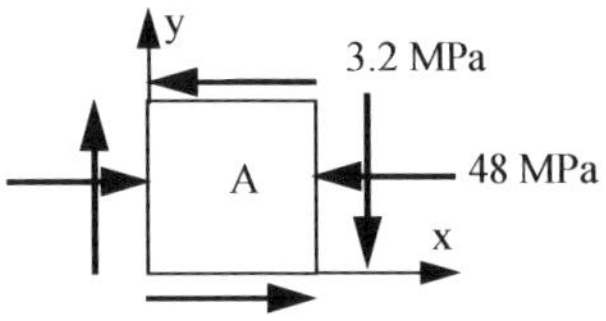

ANS $|\sigma_{max}| = 348.4MPa$; $|\tau_{max}| = 6.84MPa$; $(\sigma_{xx})_A = 48MPa(C)$; $(\tau_{xy})_A = -3.2MPa$

---

6. 118

Solution $I_{zz} = 453(10^6)mm^4$, $|\tau_{max}| = ?$ $|\sigma_{max}| = ?$ $(\sigma_{xx})_A = ?$ $(\tau_{xy})_A = ?$

The shear and moment diagrams are drawn as shown below.

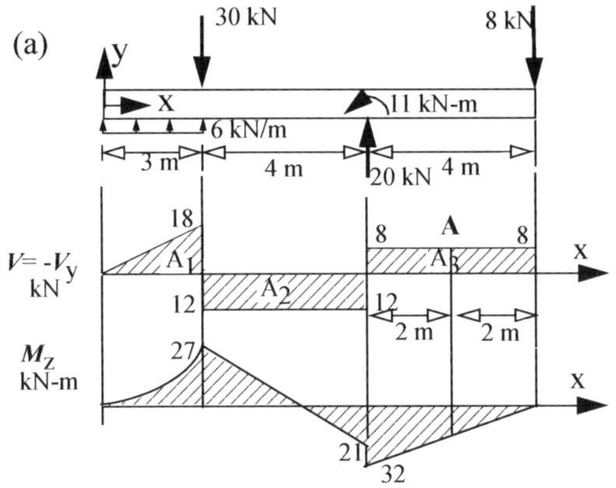

$$A_1 = \frac{1}{2}(18)(3) = 27$$

$$A_2 = (12)(4) = 48$$

$$A_3 = (8)(4) = 32$$

(b)

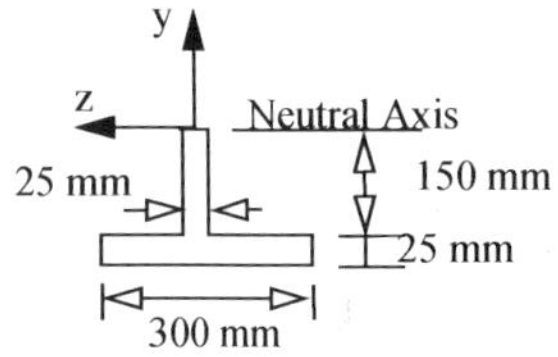

$$M_{max} = -32\ kN-m \qquad \sigma_{max} = -\frac{M_{max}y_{max}}{I_{zz}} = -\frac{(-32)(10^3)(\pm 0.175)}{453(10^{-6})} = 12.36(10^6)\frac{N}{m^2} \tag{1}$$

$$V_y = -18kN \qquad Q_{max} = (0.3)(0.025)(-0.1625) + (0.15)(0.025)(-0.075) = -1.5(10^{-3})m^3 \tag{2}$$

$$|\tau_{max}| = \left|-\frac{Q_{max}V_{ymax}}{I_{zz}\ t}\right| = \frac{18(10^3)(1.5)(10^{-3})}{(453)(10^{-6})(0.025)} = 2.38(10^6)\frac{N}{m^2} \tag{3}$$

Note that s and y are in opposite direction in the calculation of shear stress at point A.

$$M_A = -16kN-m \qquad (\sigma_{xx})_A = -\frac{M_A y_A}{I_{zz}} = -\frac{(-16)(10^3)(-0.055)}{(453)(10^{-6})} = -1.943(10^6)\frac{N}{m^2} \tag{4}$$

$$(V_y)_A = -8kN \qquad Q_A = (0.3)(0.025)(-0.1625) + (0.095)(0.025)(-0.1025) = -1.46(10^{-3})m^3 \tag{5}$$

$$(\tau_{xs})_A = -\frac{Q_A(V_y)_A}{I_{zz}\ t} = -\frac{(-8)(10^3)(-1.46)(10^{-3})}{(453)(10^{-6})(0.025)} = -1.03(10^6)\frac{N}{m^2} \qquad (\tau_{xy})_A = (\tau_{xs})_A = -1.03(10^6)\frac{N}{m^2} \tag{6}$$

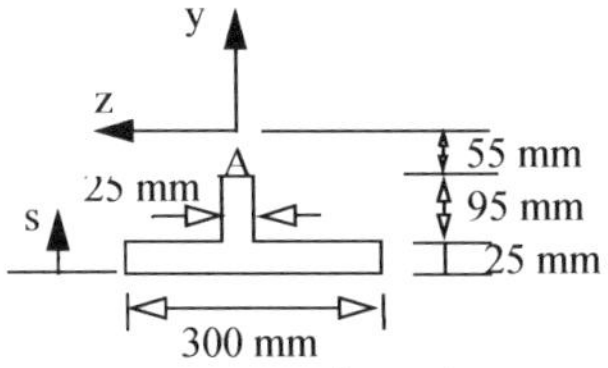

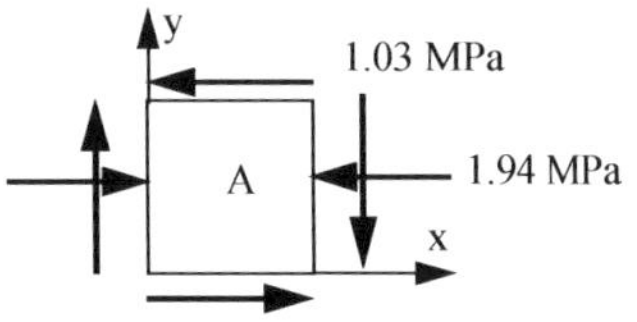

ANS $|\sigma_{max}| = 12.4$ MPa; $|\tau_{max}| = 2.38$ MPa; $(\sigma_{xx})_A = 1.94$ MPa (C); $(\tau_{xy})_A = -1.03$ MPa

6. 119

Solution $\sigma_{max} = ?$ $\tau_{max} = ?$

$$I_{zz} = (12)(36^3)/12 - (9)(34^3)/12 = 17.178(10^3)in^4 \qquad (1)$$

The shear and moment diagrams are drawn as shown below..

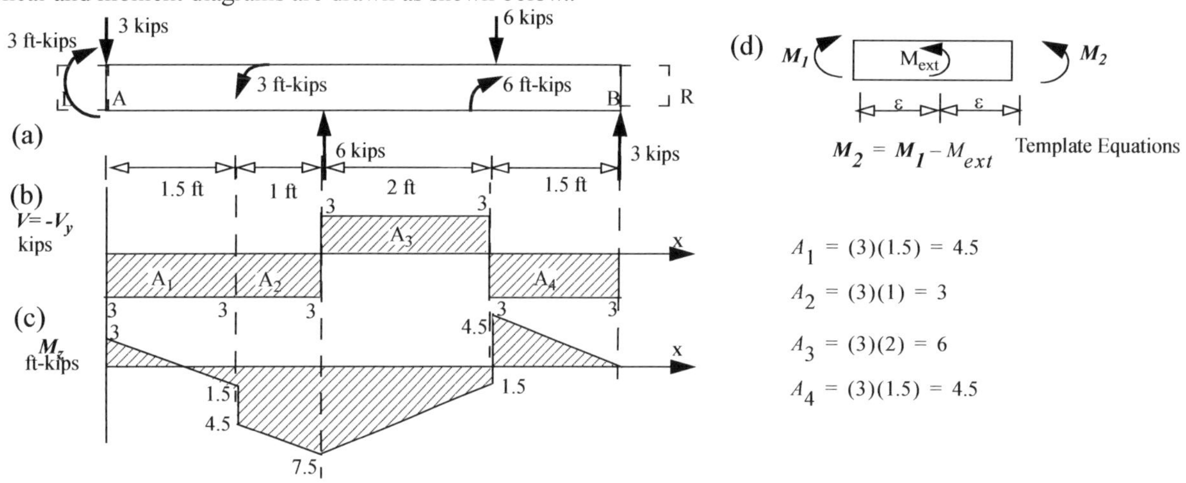

$A_1 = (3)(1.5) = 4.5$

$A_2 = (3)(1) = 3$

$A_3 = (3)(2) = 6$

$A_4 = (3)(1.5) = 4.5$

$$(M_z)_{max} = -7.5\ ft-kips = -90(10^3)\ in-lbs \qquad \sigma_{max} = \left|-\frac{M_{max}y_{max}}{I_{zz}}\right| = \left|-\frac{(-90)(10^3)(\pm 18)}{17.178(10^3)}\right| = 94.3\ psi \qquad (2)$$

$$(V_y)_{max} = \pm 3\ kips \qquad Q_z = (12)(1)(17.5) + 3(17)(1)(8.5) = 643.5\ in^3 \qquad (3)$$

$$|\tau_{max}| = \left|-\frac{Q_{max}V_{ymax}}{I_{zz}\ t}\right| = \frac{(3)(10^3)(643.5)}{17.178(10^3)(3)} = 37.4\ psi \qquad (4)$$

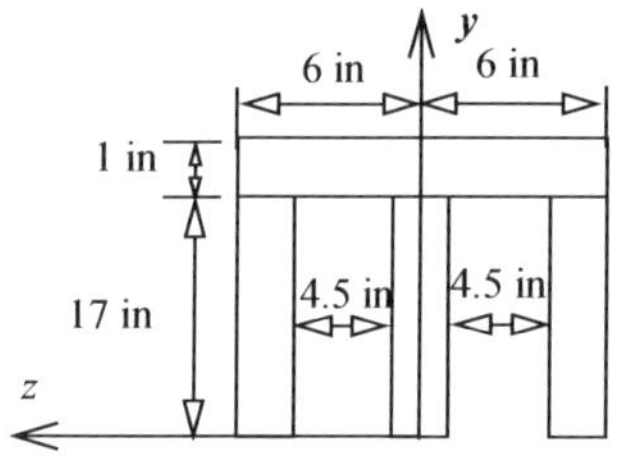

ANS $|\sigma_{max}| = 94.3$ psi ; $|\tau_{max}| = 37.4$ psi

---

6. 120

Solution $\Delta s = 10in$, $V_{AB} = ?$ $V_{BC} = ?$

By equilibrium of moment about point C we obtain the reaction forces and then draw the shear force diagram as shown below.

$$R_A(10) - 800(4) = 0 \quad or \quad R_A = 320lb \quad R_B = 480lb \quad I_{zz} = (2)(5^3)/12 = 20.83in^4 \qquad (1)$$

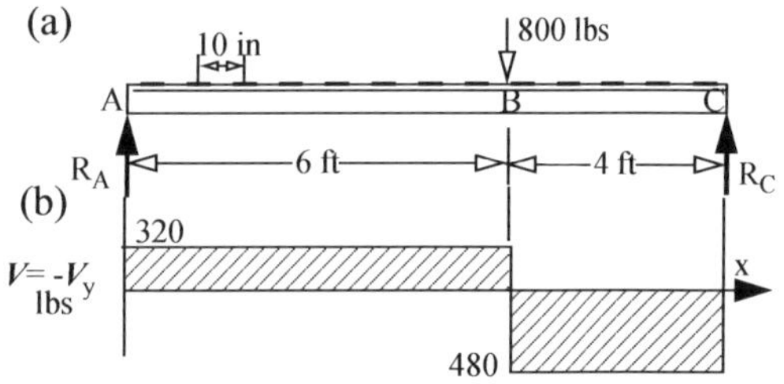

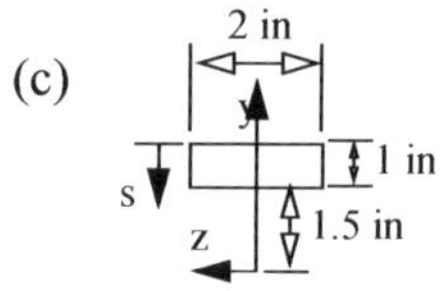

$Q_z = (2)(1)(2) = 4in^3$ is the value at the line junction of the wood pieces. The shear flow in each segment is

$$(V_y)_{AB} = -320lb \qquad q_{AB} = \left|\frac{(V_y)_{AB}Q_z}{I_{zz}}\right| = \frac{(320)(4)}{20.83} = 61.44\frac{lbs}{in} \qquad (2)$$

$$(V_y)_{BC} = -480lb \qquad q_{BC} = \left|\frac{(V_y)_{BC}Q_z}{I_{zz}}\right| = \frac{(480)(4)}{20.83} = 92.16\frac{lbs}{in} \qquad (3)$$

$$V_{AB} = q_{AB}\Delta s = (61.44)(10) \qquad V_{BC} = q_{BC}\Delta s = (92.16)(10) \qquad (4)$$

ANS $V_{AB} = 614.4$ lbs; ; $V_{BC} = 921.6$ lbs

6. 121

Solution $\Delta s = 75mm$ $V_{nail} = ?$

The location of the centroid is shown and area moment of inertia can be found out as shown below.

$$\eta_c = \frac{(2)(200)(25)(100) + (200)(25)(212.5)}{(2)(200)(25) + (200)(25)} = 137.5mm \quad (1)$$

$$I_{zz} = 2[(25)(200^3)/12 + (25)(200)(137.5 - 100)^2] + [(200)(25^3)/12 + (200)(25)(212.5 - 137.5)^2] = 75.78(10^6)mm^4 \quad (2)$$

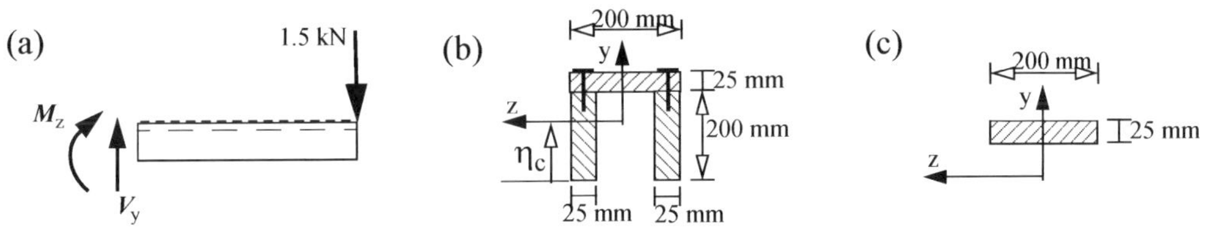

$$V_y = -1.5kN \qquad Q_t = (200)(25)(212.5 - 137.5) = 375(10)^3 mm^3 \quad (3)$$

$$q = \left|\frac{V_y Q_t}{I_{zz}}\right| = \frac{1.5(10^3)(375)(10^{-6})}{(75.78)(10^{-6})} = 7.42(10^3)N/m \qquad V_{nail} = (q/2)\Delta s = (7.42/2)(10^3)(0.075) = 278.25N \quad (4)$$

ANS $V_{nail} = 278$ N

---

6. 122

Solution $\Delta s = 75mm$, $V_{nail} = ?$

The location of the centroid is shown and area moment of inertia can be found out as shown below.

$$\eta_c = \frac{(2)(200)(75)(100) + (200)(25)(187.5)}{(2)(200)(25) + (200)(25)} = 129.17mm \quad (1)$$

$$I_{zz} = 2\left[\frac{1}{12}(25)(200^3) + (25)(200)(129.17 - 100)^2\right] + \left[\frac{1}{12}(200)(25^3) + (200)(25)(187.5 - 129.17)^2\right] = 45.66(10^6)mm^4 \quad (2)$$

$$V_y = -1.5kN \qquad Q_z = (200)(25)(187.5 - 129.17) = 291.65(10)^3 mm^3 \quad (3)$$

$$q = \left|\frac{V_y Q_t}{I_{zz}}\right| = \frac{(1.5)(10^3)(291.65)(10^{-6})}{115.66(10^{-6})} = 9.58(10^3)N/m \qquad V_{nail} = (q/2)\Delta s = (9.58/2)(10^3)(0.075) = 359.2N \quad (4)$$

ANS $V_{nail} = 359$ N

As the force on the nail in the previous problem is smaller, the joining method of the previous problem is better.

---

6. 123

Solution m = 100 kg $|\sigma_{max}| \leq 10$MPa $|\tau_{max}| \leq 2$MPa

12cm $\leq$ d $\leq$ 20cm in steps of 2 cm. Lightest beam t =? nearest cm

By equilibrium of moment about point B, we obtain the reaction force and draw the shear-moment diagram as shown below.

$$W = (100)(9.81) = 981 \ N \qquad R_A(1) - W(0.7) - W(0.3) = 0 \qquad R_A = W = 981 \ N \quad (1)$$

$$I_{zz} = dt^3/12 \qquad Q_{max} = (d)(t/2)(t/4) = dt^2/8 \quad (2)$$

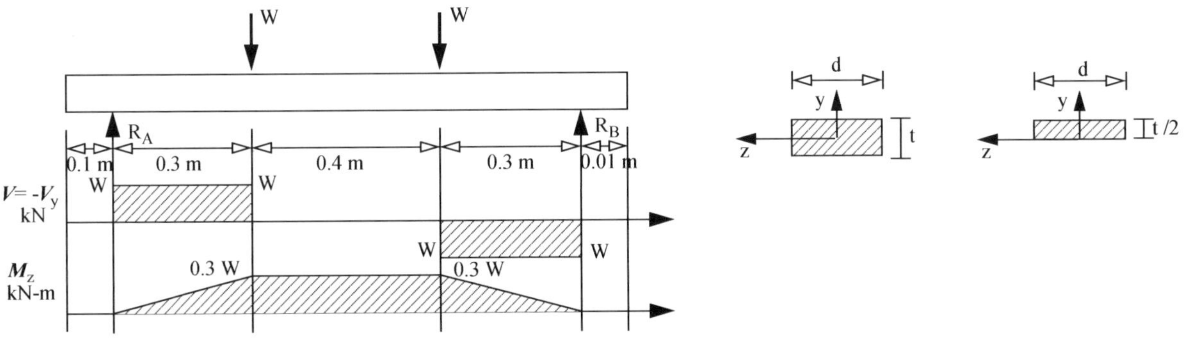

$$M_{max} = 0.3W = 294.3N-m \qquad |\sigma_{max}| = \left|\frac{M_{max}y_{max}}{I_{zz}}\right| = \frac{(294.3)(t/2)}{dt^3/12} = \frac{1765.8}{dt^2} \le 10(10^6) \qquad or \qquad t \ge \sqrt{\frac{1765.8(10^{-6})}{d}} \tag{3}$$

$$(V_y)_{max} = \pm W = \pm 981N \qquad |\tau_{max}| = \left|\frac{Q_{max}(V_y)_{max}}{I_{zz}\, t_{ma}}\right| = -\frac{(981)(dt^2/8)}{(dt^3/12)(d)} = \frac{1471.5}{dt} \le 2000 \qquad or \qquad t \ge \left[\frac{735.75(10^{-6})}{d}\right] \tag{4}$$

Table below shows the calculation for the value of t for each corresponding value of d. The minimum value of t is rounded upward to the nearest cm.

d (m)	t from Eq. 1 (m)	t from Eq.2 m	t_{min} cm
0.12	0.0384	0.0061	4
0.14	0.0355	0.0053	4
0.16	0.0332	0.0046	4
0.18	0.0313	0.0041	4
0.20	0.0297	0.0037	3

6. 124

Solution $|\sigma_W| \le 3ksi$ $|\tau_W| \le 1ksi$ $|\sigma_G| \le 600psi(T)$ $|\tau_G| \le 250psi$ $M_{ext} = ?$

By equilibrium of moment about point B we obtain the reaction force and draw the shear-moment diagram as shown below.

$$R_A(100) - M_{ext} = 0 \qquad R_A = M_{ext}/100 \qquad I_{zz} = (2)(5^3)/12 = 20.83in^4 \tag{1}$$

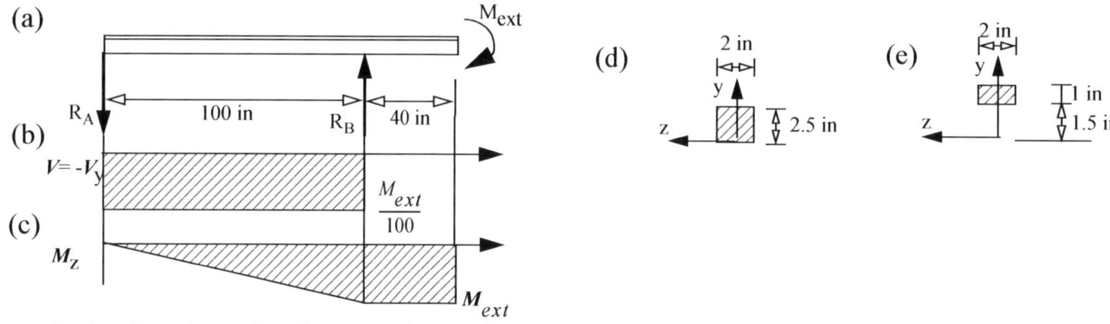

The magnitude of maximum bending stress in wood and glue is

$$M_{max} = -M_{ext} \qquad \sigma_W = \left|\frac{M_{max}y_{max}}{I_{zz}}\right| = \frac{(M_{ext})(2.5)}{20.83} \le 3000psi \qquad or \qquad M_{ext} \le 25000in-lbs \tag{2}$$

$$\sigma_G = \frac{-M_{max}y_G}{I_{zz}} = \frac{-(-M_{ext})(1.5)}{20.83} = \frac{M_{ext}(1.5)}{20.83} \le 600 \qquad or \qquad M_{ext} \le 8333.33in-lbs \tag{3}$$

The Q_z maximum and at glue can be found as shown below.

$$Q_{max} = (2.5)(2)(1.25) = 6.25in^3 \qquad Q_G = (2)(1)(2) = 4in^3 \tag{4}$$

The magnitude of maximum shear stress in wood and glue can be found as shown below.

$$(V_y)_{max} = M_{ext}/100 \qquad |\tau_W| = \left|\frac{Q_{max}(V_y)_{max}}{I_{zz}\; t}\right| = -\frac{(M_{ext}/100)(6.25)}{(20.83)(2)} \le 1000 \qquad or \qquad M_{ext} \le 666.6(10^3)in-lbs \tag{5}$$

$$|\tau_G| = \left|\frac{Q_G(V_y)_{max}}{I_{zz}\; t}\right| = -\frac{(M_{ext}/100)(4)}{(20.83)(2)} \le 250 \qquad or \qquad M_{ext} \le 260.3(10^3)in-lbs \tag{6}$$

ANS $M_{ext} = 8333.33$ in.-lbs

6. 125

Solution $|\sigma_W| \le 7MPa$ $|\tau_W| \le 1.5MPa$ $|V_{nail}| \le 300$ $P_{max} = ?$ $\Delta s = ?$

The centroid and area moment of inertia.

$$\eta_c = \frac{(120)(20)(60) + (80)(20)(130)}{(120)(20) + (80)(20)} = 88mm \tag{1}$$

$$I_{zz} = [(20)(120^3)/12 + (20)(120)(88-60)^2] + [(80)(20^3)/12 + (80)(20)(130-88)^2] = 7.63(10^6)mm^4 \tag{2}$$

The maximum bending moment will be at the wall. The maximum bending normal stress in wood is

$$|M_{max}| = 3P \qquad |\sigma_{max}| = \left|\frac{-M_{max}y_{max}}{I_{zz}}\right| = -\frac{(3P)(-0.088)}{7.63(10^{-6})} = 34.6P(10^3) \le 7(10^6) \qquad or \qquad P \le 202.3N \tag{3}$$

The shear force is uniform across the beam.

$$|V_y| = P \qquad Q_{max} = (0.088)(0.02)\left(-\frac{0.088}{2}\right) = -77.44(10^{-6})m^3 \qquad Q_{nail} = (0.08)(0.02)(-0.042) = 67.2(10^{-6})m^3 \quad \textbf{(4)}$$

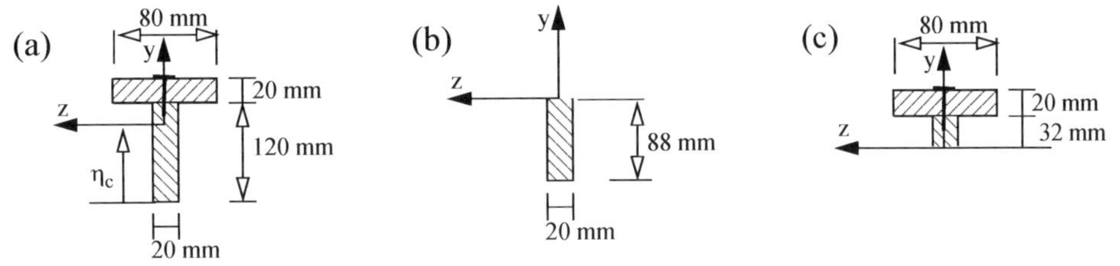

$$\tau_w = \left|\frac{-V_y Q_{max}}{I_{zz} t}\right| = \frac{(P)(77.44)(10^{-6})}{(7.63)(10^{-6})(0.02)} = 507.5P \le 1.5(10^6) \quad or \quad P \le 2956N \quad \textbf{(5)}$$

$$q = \left|\frac{Q_{nail} V_y}{I_{zz}}\right| = \frac{(P)(67.2)(10^{-6})}{(7.63)(10^{-6})} = 8.807P \le 1.5(10^6) \qquad V_{nail} = q\Delta s = 8.807P\Delta s \le 300 \qquad P\Delta s \le 34.06 \quad \textbf{(6)}$$

ANS $P_{max} = 202$ N ; $\Delta s = 16$ cm

---

6. 126

Solution $|\sigma_W| \le 730$psi $|\tau_W| \le 150$ psi $|V_{nail}| \le 100$lbs P_{max} =? configuration =?

The configuration that gives the largest values of P_{max} and Δs is the better joining method.

The area moment of inertias and section modulus for the joining methods are as follow.

$$(I_{zz})_1 = \frac{1}{12}(6)(8^3) - \frac{1}{12}(4)(6^3) = 184in^4 \qquad (I_{zz})_2 = \frac{1}{12}(8)(6^3) - \frac{1}{12}(6)(4^3) = 112in^4 \quad \textbf{(1)}$$

$$S_1 = \frac{(I_{zz})_1}{(y_{max})_1} = \frac{184}{4} = 46in^3 \qquad S_2 = \frac{(I_{zz})_2}{(y_{max})_2} = \frac{112}{4} = 37.33in^3 \quad \textbf{(2)}$$

The stresses in configuration 1 will be smaller because the section modulus and area moment of inertia are larger.The maximum bending moment will be at the wall and the shear force is uniform across the beam.

$$|M_{max}| = 240P \ in-lb \qquad |\sigma_w| = \left|\frac{M_{max}}{S_1}\right| = \frac{240P}{46} = 5.218P \le 750 \quad or \quad P \le 143.7lbs \quad \textbf{(3)}$$

$$|V_y| = P \qquad Q_{max} = (2)(3)(1)(1.5) + (6)(1)(3.5) = 30in^3 \qquad Q_{Nail} = (6)(1)(3.5) = 21in^3 \quad \textbf{(4)}$$

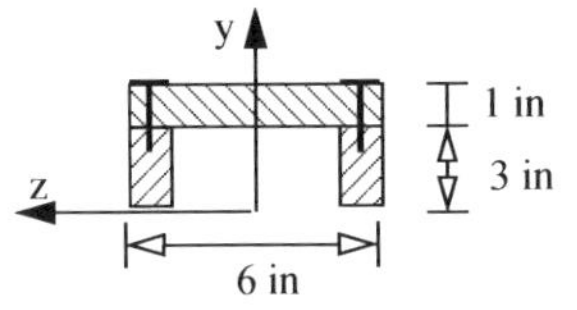

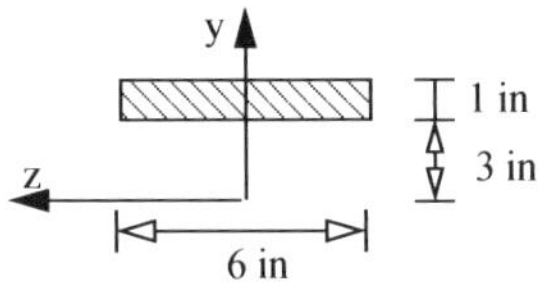

$$|\tau_W| = \left|\frac{Q_{max} V_y}{(I_{zz})_1 t}\right| = \frac{P(30)}{(184)(2)} = 81.52P(10^{-3}) \le 150 \quad or \quad P \le 1840lbs \quad \textbf{(5)}$$

$$q = \left|\frac{Q_{Nail} V_y}{(I_{zz})_1}\right| = \frac{P(21)}{184} = 0.114P \qquad |V_{Nail}| = \frac{q\Delta S}{2} = 0.0570P\Delta S \le 100 \qquad \Delta S \le \frac{1752}{P} \qquad \Delta S \le \frac{1752}{143} \qquad \Delta S \le 12.25inch \quad \textbf{(6)}$$

ANS Use joining method 1 ; ΔS = 12 in. ; P_{max} = 143 lbs

---

6. 127

Solution: Show $W_2 = \alpha W$

By symmetry each support reaction is half the load on the beam. The shear and moment diagrams can be drawn as shown below.

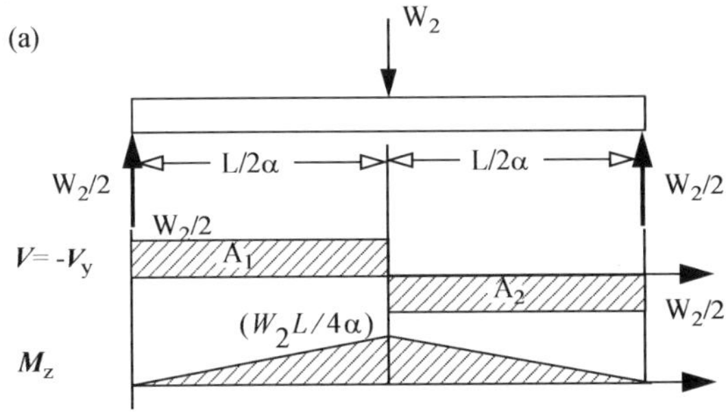

$$A_1 = A_2 = \left(\frac{W_2}{2}\right)\left(\frac{L}{2\alpha}\right) = \frac{W_2 L}{4\alpha}$$

The maximum bending normal stress is

$$M_{max} = W_2L/4\alpha \qquad \sigma_{max} = \frac{M_{max}}{S} = \frac{W_2L}{4\alpha S} \tag{1}$$

When $\alpha = 1$, $W_2 = W$ and we obtain

$$\sigma_{max} = \frac{WL}{4S} = \frac{W_2L}{4\alpha S} \quad or \quad W_2 = \alpha W \tag{2}$$

ANS $W_2 = \alpha W$

---

6. 128

Solution: Show $\sigma_{correct} = 3\sigma_{Galileo}$

By moment equilibrium about point B, we obtain

$$(\sigma_{Galileo})(bh)\left(\frac{h}{2}\right) = (P)(L) \quad or \quad \sigma_{Galileo} = \frac{2PL}{bh^2} \tag{1}$$

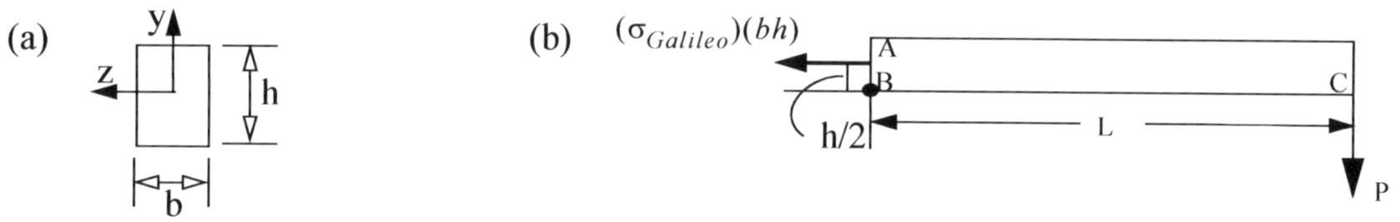

The correct stress at B is

$$I_{zz} = \frac{1}{12}(b)(h^3) \qquad |M_z| = PL \qquad \sigma_{correct} = \left|\frac{y_B M_z}{I_{zz}}\right| = \frac{(PL)(h/2)}{(b)(h^3)/12} = \frac{6PL}{bh^2} \qquad \frac{\sigma_{Galileo}}{\sigma_{correct}} = \frac{1}{3} \tag{2}$$

ANS $\sigma_{correct} = 3\sigma_{Galileo}$

---

6. 129

Solution: $M_z = f(\gamma, A, L) = ?$

The distributed force due to gravity acts in the negative y- direction and its magnitude is γA force / length as shown below. By equilibrium of moment about point O we obtain

$$M_z + (\gamma AL)\left(\frac{L}{2}\right) = 0 \tag{1}$$

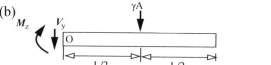

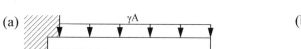

ANS $M_z = -\left(\frac{\gamma AL^2}{2}\right)$

The moment varies with the square of the length as concluded by Galileo before Newton's law were formulated.

---

6. 130

Solution: Show ***M*** is maximum when a = b

By equilibrium of moment about point C we obtain the reaction force and then we draw the shear-moment diagram.

$$R_A(a+b) - P(b) = 0 \quad or \quad R_A = \frac{b}{(a+b)}P \tag{1}$$

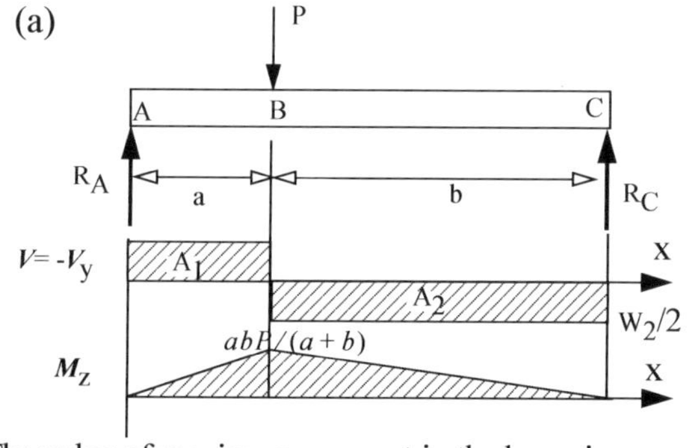

$$A_1 = \left(\frac{bP}{a+b}\right)(a) = \frac{abP}{a+b}$$

$$A_2 = \left(\frac{bP}{a+b}\right)(b) = \frac{abP}{a+b}$$

The value of maximum moment in the beam is

$$M = \frac{Pab}{(a+b)} \qquad L = a+b \qquad M = \frac{Pa(L-a)}{L} = \frac{P(aL-a^2)}{L} \tag{1}$$

Taking the first derivative with respect to a and setting it to zero we obtain

$$\frac{dM}{da} = \frac{P(L-2a)}{L} = 0 \qquad a = (L/2) \tag{2}$$

Thus the bending moment is maximum when a = b = L/2 as concluded by Galileo before calculus.

---

6. 131

Solution: Show $\sigma_{correct} = 2\sigma_{Mariotte}$

In earlier problem the correct value of stress at the bottom of the beam was found as

$$\sigma_{correct} = \frac{6PL}{bh^2} \tag{1}$$

Figures below shows Mariotte's assumed stress distribution and the equivalent force diagram. By equilibrium of moment about point B we obtain

$$\frac{1}{2}(\sigma_{Mariotte})(bh)\left(\frac{2h}{3}\right) = PL \qquad or \qquad \sigma_{Mariotte} = \frac{3PL}{bh^2} \qquad \frac{\sigma_{correct}}{\sigma_{Mariotte}} = 2 \tag{2}$$

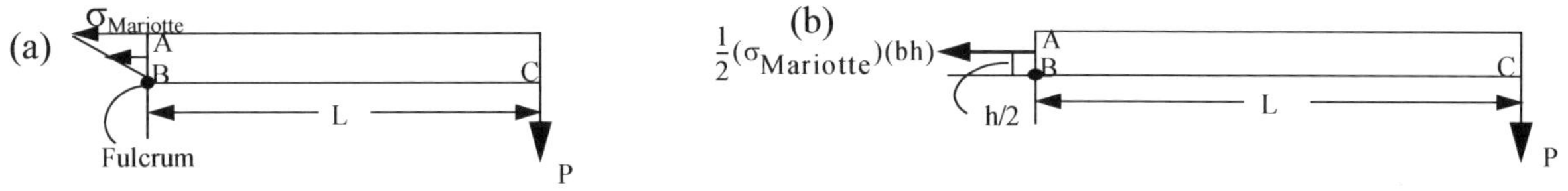

ANS $\sigma_{correct} = 2\sigma_{Mariotte}$

---

6. 132

Solution

The solution proceeds as follows.

$$\frac{dV_y}{dx} = -p(x) \qquad \int_{V_A}^{V_y} dV_y = -\int_{x_A}^{x} p(x)dx \qquad or \qquad V_y = V_A - \int_{x_A}^{x} p(x)dx \tag{1}$$

$$\frac{dM_z}{dx} = -V_y \qquad or \qquad \int_{M_A}^{M_B} dM_z = -\int_{x_A}^{x_B} V_y dx \qquad or \qquad M_B = M_A - \int_{x_A}^{x_B} V_y dx \text{ or}$$

$$M_B = M_A - (x-x_B)V_y\Big|_{x_A}^{x_B} + \int_{x_A}^{x_B}(x-x_B)\frac{dV_y}{dx}dx = M_A - [(x_B-x_B)V_B - (x_A-x_B)V_A] + \int_{x_A}^{x_B}(x-x_B)[-p(x)]dx \text{ or}$$

$$M_B = M_A + (x_A-x_B)V_A + \int_{x_A}^{x_B}(x_B-x)[p(x)]dx \tag{2}$$

The above equation is the desired result.

---

6. 133

Solution

The solution proceeds as follows.

$$\varepsilon_{xx} = \frac{\partial u}{\partial x} = \frac{du_o}{dx} - y\frac{d^2v}{dx^2} \qquad \sigma_{xx} = E\varepsilon_{xx} = E\left(\frac{du_o}{dx} - y\frac{d^2v}{dx^2}\right) \tag{1}$$

$$N = \int_A \sigma_{xx}dA = \int_A E\left(\frac{du_o}{dx} - y\frac{d^2v}{dx^2}\right)dA = E\left(\frac{du_o}{dx}\right)\int_A dA - E\left(\frac{d^2v}{dx^2}\right)\int_A ydA = EA\left(\frac{du_o}{dx}\right) - Ey_cA\left(\frac{d^2v}{dx^2}\right) \tag{2}$$

$$M_z = -\int_A y\sigma_{xx}dA = -\int_A Ey\left(\frac{du_o}{dx} - y\frac{d^2v}{dx^2}\right)dA = -\left[E\left(\frac{du_o}{dx}\right)\int_A ydA - E\left(\frac{d^2v}{dx^2}\right)\int_A y^2dA\right] = -Ey_cA\left(\frac{du_o}{dx}\right) + EI_{zz}\left(\frac{d^2v}{dx^2}\right) \tag{3}$$

The expressions of N and M_z are the desired results.

---

6. 134

Solution

The solution proceeds as follows.

$$\varepsilon_{xx} = -y\frac{d^2v}{dx^2} \qquad \varepsilon_{xx} = \frac{\sigma_{xx}}{E} + \alpha\Delta T(x,y) = -y\frac{d^2v}{dx^2} \qquad or \qquad \sigma_{xx} = -\left[Ey\frac{d^2v}{dx^2} + E\alpha\Delta T(x,y)\right] \tag{1}$$

$$M_z = -\int_A y\sigma_{xx}dA = \int_A y\left[Ey\frac{d^2v}{dx^2} + E\alpha\Delta T(x,y)\right]dA = E\frac{d^2v}{dx^2}\int_A y^2 dA + E\alpha\int_A y\Delta T(x,y)dA \text{ or}$$

$$M_z = EI_{zz}\frac{d^2v}{dx^2} + M_T \qquad or \qquad \frac{d^2v}{dx^2} = \frac{(M_z - M_T)}{EI_{zz}} \qquad M_T = E\alpha\int_A y\Delta T(x,y)dA \tag{2}$$

$$\sigma_{xx} = -\left[Ey\frac{d^2v}{dx^2} + E\alpha\Delta T(x,y)\right] = -\left[Ey\frac{(M_z - M_T)}{EI_{zz}} + E\alpha\Delta T(x,y)\right] = -\left[\frac{M_z y}{I_{zz}} - \frac{M_T y}{I_{zz}} + E\alpha\Delta T(x,y)\right] \tag{3}$$

Above equation is the desired result.

---

6. 135

Solution

The solution proceeds as follows.

$$\varepsilon_{xx} = \frac{\partial u}{\partial x} = -y\frac{d^2v}{dx^2} - z\frac{d^2w}{dx^2} \qquad \sigma_{xx} = E\varepsilon_{xx} = E\left(-y\frac{d^2v}{dx^2} - z\frac{d^2w}{dx^2}\right) \tag{1}$$

$$M_z = -\int_A y\sigma_{xx}dA = \int_A y\left[E\left(y\frac{d^2v}{dx^2} + z\frac{d^2w}{dx^2}\right)\right]dA = E\frac{d^2v}{dx^2}\int_A y^2 dA + E\frac{d^2w}{dx^2}\int_A yzdA \tag{2}$$

$$M_y = -\int_A z\sigma_{xx}dA = \int_A z\left[E\left(y\frac{d^2v}{dx^2} + z\frac{d^2w}{dx^2}\right)\right]dA = E\frac{d^2v}{dx^2}\int_A yzdA + E\frac{d^2w}{dx^2}\int_A z^2 dA \tag{3}$$

Substituting $I_{zz} = \int_A y^2 dA$, $I_{yy} = \int_A z^2 dA$, and $I_{yz} = \int_A yzdA$ we obtain:

$$M_z = EI_{zz}\frac{d^2v}{dx^2} + EI_{yz}\frac{d^2w}{dx^2} \qquad and \qquad M_y = EI_{yz}\frac{d^2v}{dx^2} + EI_{yy}\frac{d^2w}{dx^2} \tag{4}$$

Solving for $\frac{d^2v}{dx^2}$ and $\frac{d^2w}{dx^2}$ using Cramer's rule we obtain the following.

$$\frac{d^2v}{dx^2} = \frac{1}{E}\left(\frac{I_{yy}M_z - I_{yz}M_y}{I_{yy}I_{zz} - I_{yz}^2}\right) \qquad and \qquad \frac{d^2w}{dx^2} = \frac{1}{E}\left(\frac{I_{zz}M_y - I_{yz}M_z}{I_{yy}I_{zz} - I_{yz}^2}\right) \tag{5}$$

$$\sigma_{xx} = E\left(-y\frac{d^2v}{dx^2} - z\frac{d^2w}{dx^2}\right) = E\left[-\frac{y}{E}\left(\frac{I_{yy}M_z - I_{yz}M_y}{I_{yy}I_{zz} - I_{yz}^2}\right) - \frac{z}{E}\left(\frac{I_{zz}M_y - I_{yz}M_z}{I_{yy}I_{zz} - I_{yz}^2}\right)\right] = -\left(\frac{I_{yy}M_z - I_{yz}M_y}{I_{yy}I_{zz} - I_{yz}^2}\right)y - \left(\frac{I_{zz}M_y - I_{yz}M_z}{I_{yy}I_{zz} - I_{yz}^2}\right)z \tag{6}$$

The above equations are the desired results.

---

6. 136

Solution

Substituting $\sigma_{xx} = -\left(\frac{M_z y}{I_{zz}}\right)$ into $\frac{\partial\sigma_{xx}}{\partial x} + \frac{\partial\tau_{yx}}{\partial y} = 0$ we obtain the following:

$$-\frac{\partial}{\partial x}\left[\frac{M_z y}{I_{zz}}\right] + \frac{\partial\tau_{yx}}{\partial y} = 0 \qquad or \qquad -\left(\frac{y}{I_{zz}}\right)\left[\frac{\partial M_z}{\partial x}\right] + \frac{\partial\tau_{yx}}{\partial y} = 0 \tag{1}$$

Substituting $\frac{dM_z}{dx} = -V_y$ in the above equation we obtain:

$$\frac{\partial\tau_{yx}}{\partial y} = \left(\frac{y}{I_{zz}}\right)\left[\frac{\partial M_z}{\partial x}\right] = -\left(\frac{y}{I_{zz}}\right)V_y \tag{2}$$

We note that V_y and I_{zz} are not functions of y and the shear stress at the bottom is zero.

$$\int_{\tau_{yx}=0}^{\tau_{yx}} d\tau_{yx} = -\left(\frac{V_y}{I_{zz}}\right)\int_{-b/2}^{y} ydy \qquad or \qquad \tau_{yx} = -\left(\frac{V_y}{I_{zz}}\right)\frac{y^2}{2}\bigg|_{-b/2}^{y} = -\left(\frac{V_y}{2I_{zz}}\right)\left(y^2 - \frac{b^2}{4}\right) = \left(\frac{V_y}{2(tb^3/12)}\right)\left(y^2 - \frac{b^2}{4}\right) = \frac{6V_y(b^2/4 - y^2)}{b^3 t} \tag{3}$$

ANS $\tau_{yx} = \frac{6V_y(b^2/4 - y^2)}{b^3 t}$

---

6. 137

Solution $R_i = 1$ inch $L = 5$ft. $P = 1200$ lbs $\sigma_{max} \le 10ksi$ $R_o = ?$ nearest 1/16th in.

The maximum moment will be at the built in wall. Its magnitude is:

$$M_{max} = (1200)(5) = 6000\ ft-lbs = 72,000\ in-lbs \quad (1)$$

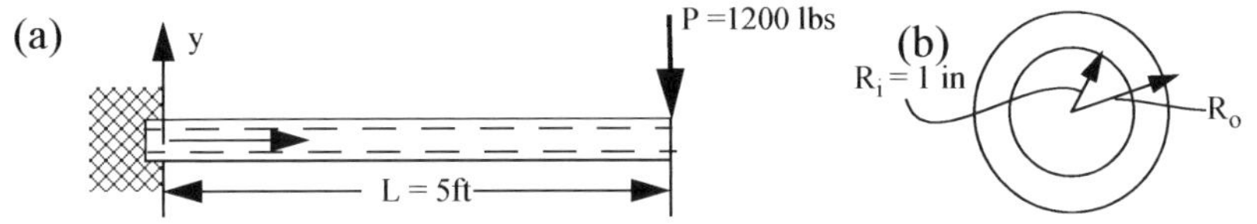

$$I_{zz} = \frac{\pi}{4}(R_o^4 - 1) \qquad \sigma_{max} = \left|\frac{M_{max}y_{max}}{I_{zz}}\right| = \frac{(72000)(R_o)}{(\pi/4)(R_o^4 - 1)} \le 10,000psi \qquad or \qquad R_o^4 - 1 - 9.167R_o \ge 0 \quad (2)$$

The value of R_o corresponds to the root of the left hand side of the above equation, i.e., $f(R_o) = R_o^4 - 9.167R_o - 1 = 0$. We find the root on the spreadsheet as described in Appendix B. The calculations are shown in the table below.

R_o	$f(R_o)$	R_o	$f(R_o)$
1.100	-9.620	2.300	5.899
1.200	-9.927	2.100	-0.803
1.300	-10.061	2.105	-0.663
1.400	-9.993	2.110	-0.522
1.500	-9.688	2.115	-0.379
1.600	-9.114	2.120	-0.235
1.700	-8.232	2.125	-0.090
1.800	-7.004	2.130	0.057
1.900	-5.386	2.135	0.205
2.000	-3.335	2.140	0.355
2.100	-0.803	2.145	0.506
2.200	2.257	2.150	0.658

The value of R_o is between 2.125 and 2.130. Rounding upwards to the nearest 1/16 inch, we obtain the minimum value of R_o.

ANS $R_O = 2\frac{3}{16}$ in.

---

6. 138

Solution $|\sigma_{max}| = ?$ $|\tau_{max}| = ?$

The solution proceeds as follows.

$$\frac{dV_y}{dx} = -p(x) \qquad \int_0^{V_y} dV_y = -\int_L^x p(x)dx \qquad or \qquad V_y(x) = -\int_L^x p(x)dx\ lbs \quad (1)$$

$$\frac{dM_z}{dx} = -V_y \qquad \int_0^{M_z} dM_z = -\int_L^x V_y dx \qquad or \qquad M_z(x) = -\int_L^x V_y(x)dx\ ft-lbs \quad (2)$$

$$I_{zz} = \frac{(2)(8)^3}{12} = 85.33in^4 \qquad \sigma_{xx}(x_i) = \left|\frac{M_z(x_i)y_{max}}{I_{zz}}\right| = \left|\frac{M_z(x_i)(12)(4)}{85.33}\right| = 0.562522|M_z(x_i)| \quad (3)$$

$$Q_{max} = (4)(2)(2) = 16in^3 \qquad \tau(x_i) = \left|\frac{V_y(x_i)Q_{max}}{I_{zz}t}\right| = \left|\frac{V_y(x_i)(16)}{(85.33)(2)}\right| = 0.093754|V_y(x_i)| \quad (4)$$

We first obtain $V_y(x_i)$ and then $M_z(x_i)$ by numerical integration on a spread sheet. We then find $\sigma_{xx}(x_i)$ and $\tau(x_i)$ on the spread sheet as shown in the Table below.

x_i	$p(x_i)$	$-V_y(x_i)$ (lbs)	$M_z(x_i)$ ft-lbs	$\sigma_{xx}(x_i)$ psi	$\tau(x_i)$ psi
10	0	0.00	0.00	0.00	0.00
9	128	-180.50	90.25	50.77	16.92
8	233	-455.00	408.00	229.51	42.66
7	316	-801.50	1036.25	582.91	75.14
6	377	-1198.00	2036.00	1145.29	112.32
5	416	-1622.00	3446.00	1938.45	152.07
4	432	-2051.00	5282.50	2971.52	192.29
3	426	-2463.00	7539.50	4241.13	230.92
2	398	-2836.00	10189.00	5731.54	265.89
1	348	-3147.50	13180.75	7414.46	295.09
0	275	-3147.50	16328.25	9185.00	295.09

ANS $|\sigma_{max}| = 9185$ psi; ; $|\tau_{max}| = 295$ psi

6. 139

Solution a = ? b = ? c = $|\sigma_{max}|$ = ? $|\tau_{max}|$ = ?

The solution proceeds as follows.

$$\frac{dV_y}{dx} = -p(x) \qquad \int_0^{V_y} dV_y = -\int_L^x p(x)dx \qquad or \qquad V_y(x) = -\int_L^x [a + bx + cx^2]dx = -\left[ax + \frac{bx^2}{2} + \frac{cx^3}{3}\right]_L^x \quad \text{or}$$

$$V_y(x) = -\left[a(x-L) + \frac{b}{2}(x^2 - L^2) + \frac{c}{3}(x^3 - L^3)\right] \tag{1}$$

$$\frac{dM_z}{dx} = -V_y \qquad \int_0^{M_z} dM_z = -\int_L^x V_y dx \qquad or \qquad M_z(x) = \int_L^x \left[a(x-L) + \frac{b}{2}(x^2 - L^2) + \frac{c}{3}(x^3 - L^3)\right]dx \ ft-lbs \quad \text{or}$$

$$M_z(x) = a\left(\frac{x^2}{2} - Lx\right) + \frac{b}{2}\left(\frac{x^3}{3} - L^2x\right) + \frac{c}{3}\left(\frac{x^4}{4} - L^3x\right)\Bigg|_L^x = a\left(\frac{x^2}{2} - Lx + \frac{L^2}{2}\right) + \frac{b}{2}\left(\frac{x^3}{3} - L^2x + \frac{2L^3}{3}\right) + \frac{c}{3}\left(\frac{x^4}{4} - L^3x + \frac{3L^4}{4}\right) \tag{2}$$

The maximum value of shear force and bending moment is at the wall i.e., at x = 0.

$$V_y(x=0) = \left(aL + \frac{bL^2}{2} + \frac{cL^3}{3}\right) \ lbs \qquad M_z(x=0) = \left(\frac{aL^2}{2} + \frac{bL^3}{3} + \frac{cL^4}{4}\right) \ ft-lbs \tag{3}$$

Using the Least Square Method described in Appendix B, we obtain the value of the values of constants a, b, and c on a spread sheet as shown in the table below.

	x_i	$p(x_i)$	x_i^2	x_i^3	x_i^4	$x*p_i$	$x_i^2*p_i$
1	10	0	100.0	1.000E+03	1.000E+04	0.000E+00	0.000E+00
2	9	128	81.0	7.290E+02	6.561E+03	1.152E+03	1.037E+04
3	8	233	64.0	5.120E+02	4.096E+03	1.864E+03	1.491E+04
4	7	316	49.0	3.430E+02	2.401E+03	2.212E+03	1.548E+04
5	6	377	36.0	2.160E+02	1.296E+03	2.262E+03	1.357E+04
6	5	416	25.0	1.250E+02	6.250E+02	2.080E+03	1.040E+04
7	4	432	16.0	6.400E+01	2.560E+02	1.728E+03	6.912E+03
8	3	426	9.0	2.700E+01	8.100E+01	1.278E+03	3.834E+03
9	2	398	4.0	8.000E+00	1.600E+01	7.960E+02	1.592E+03
10	1	348	1.0	1.000E+00	1.000E+00	3.480E+02	3.480E+02
11	0	275	0.0	0.000E+00	0.000E+00	0.000E+00	0.000E+00
Sums (b_{ij} &r_i)	55.0	3349.0	385.0	3025.0	25333.0	13720.0	77422.0
C_{ij}	6.026E+05	2.287E+05	1.815E+04	1.304E+05	1.210E+04	1.210E+03	
D	1.038E+06						
a_i	275.119	83.724	-11.122				

The values of the constants are: $a = 275.119$, $b = 83.724$, and $c = -11.122$

Substituting the above values of a, b, and c and L = 10 we obtain:

$$V_y(x=0) = aL + \frac{bL^2}{2} + \frac{cL^3}{3} = 3230 \ lbs \qquad M_z(x=0) = \frac{aL^2}{2} + \frac{bL^3}{3} + \frac{cL^4}{4} = 13858.95 \ ft-lbs \tag{4}$$

Substituting the shear force and moment into the equations of previous problem we obtain:

$$|\sigma_{max}| = \sigma_{xx}(x=0) = 0.562522|M_z(x=0)| = (0.562522)(13858.95) = 7795.96 \ psi \tag{5}$$

$$|\tau_{max}| = \tau(x=0) = 0.093754|V_y(x=0)| = (0.093754)(3230) = 302.83 \ psi \tag{6}$$

ANS $a = 275.119$; $b = 83.724$; $c = -11.122$; $|\sigma_{max}| = 7796$ psi; $|\tau_{max}| = 303$ psi

CHAPTER 7

Section 7.1

7. 1

Solution $v(x) = ?$ $v(L) = ?$

By equilibrium of moment about point O, we obtain:

$$M_z + P(L-x) = 0 \qquad or \qquad M_z = -P(L-x) \tag{1}$$

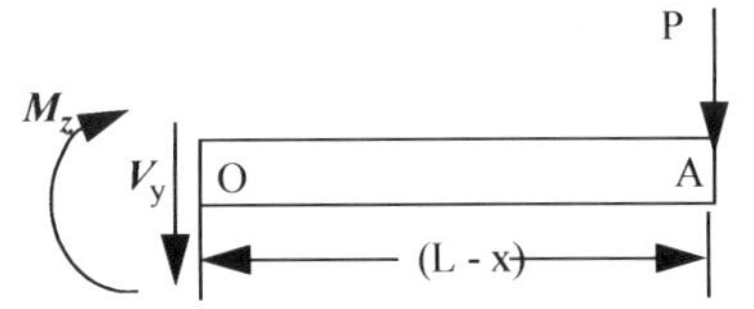

The boundary value problem statement is

$$EI\frac{d^2v}{dx^2} = -PL + Px \qquad v(0) = 0 \qquad \frac{dv}{dx}(0) = 0 \tag{2}$$

Integrating and using the boundary conditions we obtain the deflection as shown below.

$$EI\frac{dv}{dx} = -PLx + \frac{Px^2}{2} + C_1 \qquad EIv = \frac{-PLx^2}{2} + \frac{Px^3}{6} + C_1x + C_2 \tag{3}$$

$$EI\frac{dv}{dx}(0) = C_1 = 0 \qquad EIv(0) = C_2 = 0 \tag{4}$$

$$v(x) = \frac{P}{EI}\left(\frac{-3Lx^2 + x^3}{6}\right) \qquad v(L) = \frac{PL^2}{6EI}(-2L) \tag{5}$$

ANS $v(x) = \frac{Px^2}{6EI}(x-3L)$; $v(L) = -\frac{PL^3}{3EI}$

---

7. 2

Solution $v(x) = ?$ $v(L/2) = ?$

The reaction at each support is half the total load by symmetry. By equilibrium of moment about point O, we obtain

$$M_z + wx\left(\frac{x}{2}\right) - \frac{wL}{2}(x) = 0 \qquad or \qquad M_z = \frac{wLx}{2} - \frac{wx^2}{2} \tag{1}$$

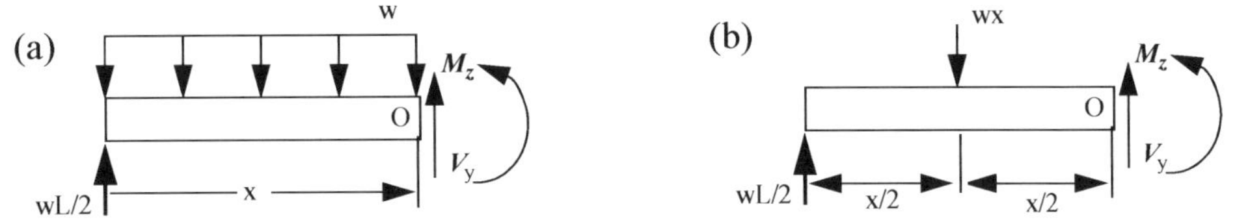

The boundary value statement is

$$EI\frac{d^2v}{dx^2} = \frac{wLx}{2} - \frac{wx^2}{2} \qquad v(0) = 0 \qquad v(L) = 0 \tag{2}$$

Integrating and using the boundary conditions we obtain the deflection as shown below.

$$EI\frac{dv}{dx} = \frac{wLx^2}{4} - \frac{wx^3}{6} + C_1 \qquad EIv = \frac{wLx^3}{12} - \frac{wx^4}{24} + C_1x + C_2 \tag{3}$$

$$EIv(0) = C_2 = 0 \qquad EIv(L) = \frac{wL^4}{12} - \frac{wL^4}{24} + C_1L = 0 \qquad or \qquad \frac{wL^4}{24}(2-1) + C_1L = 0 \qquad or \qquad C_1 = -\frac{wL^3}{24} \tag{4}$$

$$v(x) = \frac{w}{EI}\left(\frac{2Lx^3 - x^4 - L^3x}{24}\right) \qquad v\left(\frac{L}{2}\right) = \frac{-wL}{48EI}\left(\frac{L^3}{8} - \frac{L^3}{2} + L^3\right) \tag{5}$$

ANS $v(x) = \frac{-wx}{24EI}(x^3 - 2Lx^2 + L^3)$; $v\left(\frac{L}{2}\right) = -\left(\frac{5wL^4}{384EI}\right)$

---

7. 3

Solution $v(x) = ?$ $v(L/2) = ?$

By equilibrium of moment about point C, we obtain the reaction force. By equilibrium of moment about point O we obtain the internal moment as shown below.

$$R_B(L) - PL = 0 \quad or \quad R_B = P \qquad \boldsymbol{M}_z - R_B x = 0 \quad or \quad \boldsymbol{M}_z = Px \tag{1}$$

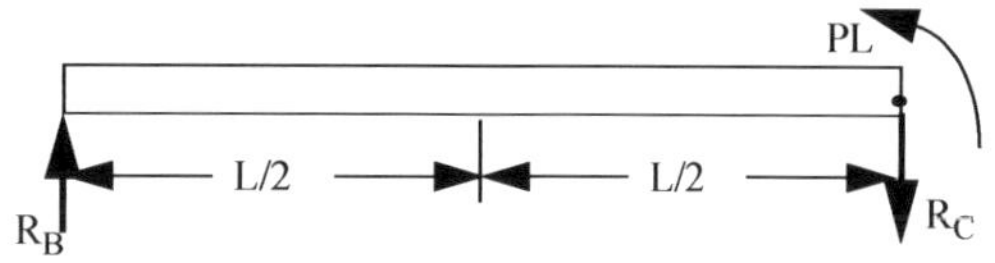

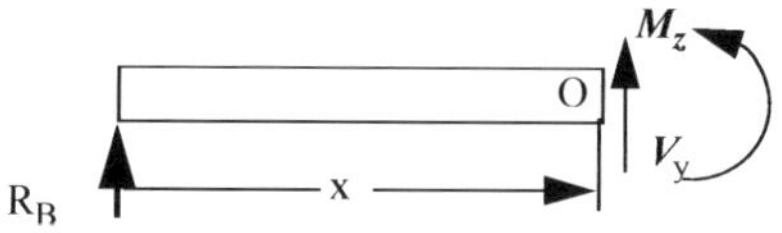

The boundary value problem statement is,

$$EI\frac{d^2v}{dx^2} = Px \qquad v(0) = 0 \qquad v(L) = 0 \tag{2}$$

Integrating and using the boundary conditions we obtain the deflection as shown below.

$$EI\frac{dv}{dx} = \frac{Px^2}{2} + C_1 \qquad EIv = \frac{Px^3}{6} + C_1x + C_2 \tag{3}$$

$$EIv(0) = C_2 = 0 \qquad EIv(L) = \frac{PL^3}{6} + C_1L = 0 \quad or \quad C_1 = -\frac{PL^2}{6} \tag{4}$$

$$v(x) = \frac{P}{6EI}x^3 - \frac{P}{6EI}L^3x \qquad v\left(\frac{L}{2}\right) = \frac{P}{6EI}\left(\frac{L}{2}\right)\left(\frac{L^2}{4} - L^2\right) \tag{5}$$

ANS $v(x) = \frac{Px}{6EI}(x^2 - L^2)$; $v\left(\frac{L}{2}\right) = -\left(\frac{PL^3}{16EI}\right)$

---

7. 4

Solution v(x) =? v(L) =?

By equilibrium of moment about point O, we obtain,

$$\boldsymbol{M}_z + w(L-x)\left(\frac{L-x}{2}\right) = 0 \quad or \quad \boldsymbol{M}_z = -\frac{w}{2}(L-x)^2 = -\frac{w}{2}(L^2 - 2Lx + x^2) \tag{1}$$

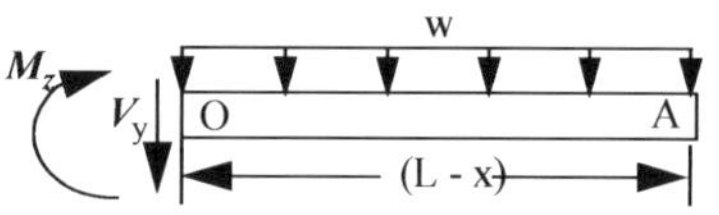

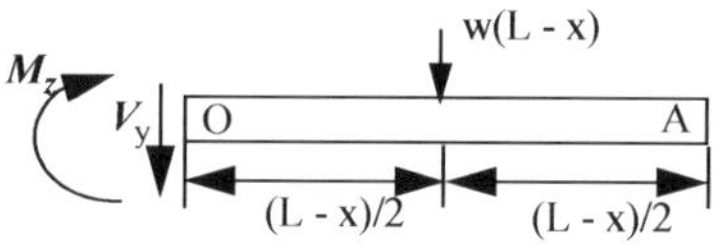

The boundary value statement is

$$EI\frac{d^2v}{dx^2} = -\frac{wL^2}{2} + wLx - \frac{wx^2}{2} \qquad v(0) = 0 \qquad \frac{dv}{dx}(0) = 0 \tag{2}$$

Integrating and using the boundary conditions we obtain the deflection as shown below.

$$EI\frac{dv}{dx} = -\frac{wL^2}{2}x + wL\frac{x^2}{2} - \frac{wx^3}{6} + C_1 \qquad EIv = -\frac{wL^2x^2}{4} + \frac{wLx^3}{6} - \frac{wx^4}{24} + C_1x + C_2 \tag{3}$$

$$EI\frac{dv}{dx}(0) = C_1 = 0 \qquad EIv(0) = C_2 = 0 \tag{4}$$

$$v(x) = \frac{w}{EI}\left(\frac{-6L^2x^2 + 4Lx^3 - x^4}{24}\right) \qquad v(L) = \frac{-w(L^2)}{24EI}(L^2 - 4L^2 + 6L^2) \tag{5}$$

ANS $v(x) = \frac{-wx^2}{24EI}(x^2 - 4Lx + 6L^2)$; $v(L) = -\left(\frac{wL^4}{8EI}\right)$

---

7. 5

Solution v(x) =? v(0) =?

By equilibrium of moment about point B and O, we obtain the reaction moment and the internal moment as shown below.

$$M_A - wL(L/2) = 0 \quad or \quad M_A = wL^2/2 \qquad \boldsymbol{M}_z + wx(x/2) - M_A = 0 \quad or \quad \boldsymbol{M}_z = wL^2/2 - wx^2/2 \tag{1}$$

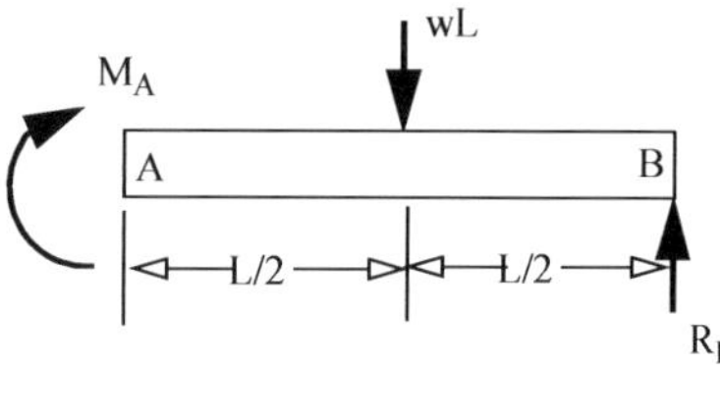

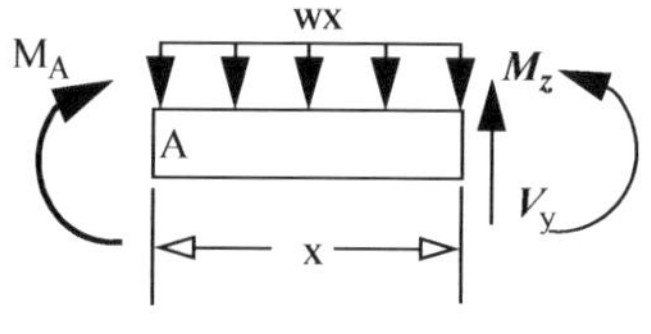

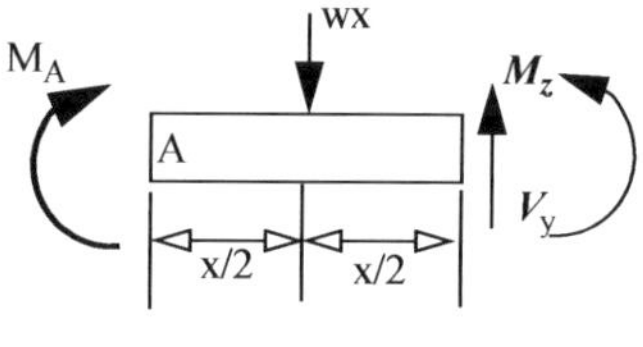

The boundary value statement is

$$EI\frac{d^2v}{dx^2} = \frac{wL^2}{2} - \frac{wx^2}{2} \qquad \frac{dv}{dx}(0) = 0 \qquad v(L) = 0 \tag{2}$$

Integrating and using the boundary conditions we obtain the deflection as shown below.

$$EI\frac{dv}{dx} = \frac{wL^2x}{2} - \frac{wx^3}{6} + C_1 \qquad EIv = \frac{wL^2x^2}{4} - \frac{wx^4}{24} + C_1x + C_2 \tag{3}$$

$$EI\frac{dv}{dx}(0) = C_1 = 0 \qquad EIv(L) = \frac{wL^4}{12} - \frac{wL^4}{24} + C_2 = 0 \qquad or \qquad C_2 = -\frac{5wL^4}{24} \tag{4}$$

$$v(x) = \frac{w}{EI}\left(\frac{6L^2x^2 - x^4 - 5L^4}{24}\right) \qquad v(0) = -\left(\frac{5wL^4}{24EI}\right) \tag{5}$$

ANS $v(x) = \frac{w}{24EI}(6L^2x^2 - x^4 - 5L^4)$; $v(0) = -\left(\frac{5wL^4}{24EI}\right)$

---

7. 6

Solution v(x) =? v(L) =?

By equilibrium of forces in the y-direction, we obtain the reaction force. By equilibrium of moment about point O, we obtain the internal moment as shown below.

$$R_B = P \qquad \boldsymbol{M_z} + PL - R_Bx = 0 \qquad or \qquad \boldsymbol{M_z} = Px - PL \tag{1}$$

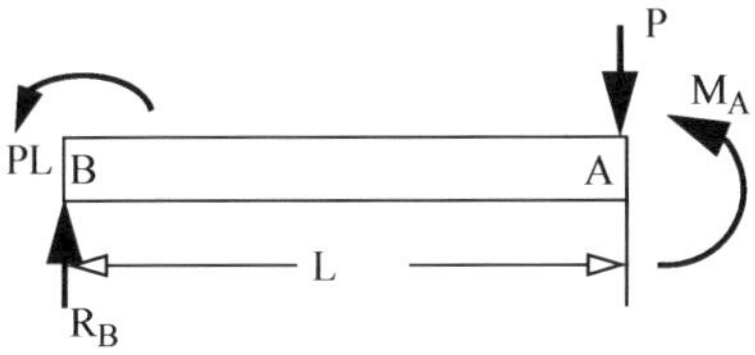

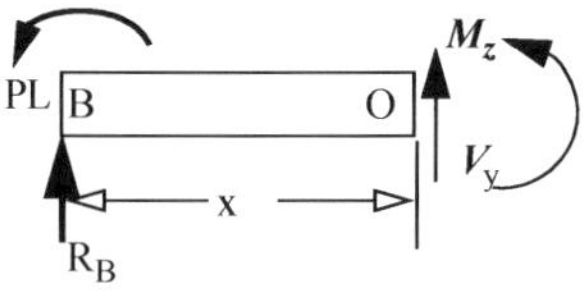

The boundary value problem statement is,

$$EI\frac{d^2v}{dx^2} = Px - PL \qquad v(0) = 0 \qquad \frac{dv}{dx}(L) = 0 \tag{2}$$

Integrating and using the boundary conditions we obtain the deflection as shown below.

$$EI\frac{dv}{dx} = \frac{Px^2}{2} - PLx + C_1 \qquad EIv = \frac{Px^3}{6} - \frac{PLx^2}{2} + C_1x + C_2 \tag{3}$$

$$EIv(0) = C_2 = 0 \qquad EI\frac{dv}{dx}(L) = \frac{PL^2}{2} - PL^2 + C_1 = 0 \qquad or \qquad C_1 = \frac{PL^2}{2} \tag{4}$$

$$v(x) = \frac{P}{6EI}(x^3 - 3Lx^2 + 3L^2x) \qquad v(L) = \frac{PL^3}{6EI} \tag{5}$$

ANS $v(x) = \frac{Px}{6EI}(x^2 - 3Lx + 3L^2)$; $v(L) = \frac{PL^3}{6EI}$

---

7. 7

Solution v(x) =?

By equilibrium of moment about point O, we obtain

$$\boldsymbol{M_z} + m(L - x) = 0 \qquad or \qquad \boldsymbol{M_z} = -m(L - x) \tag{1}$$

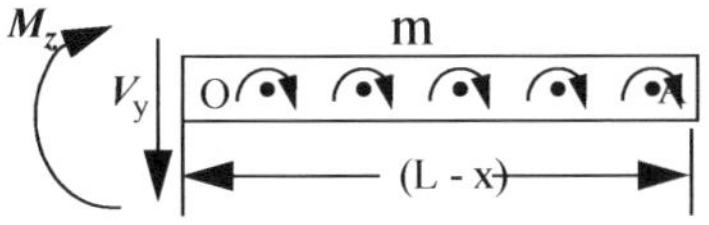

The boundary value problem statement is

$$EI\frac{d^2v}{dx^2} = -mL + mx \qquad \frac{dv}{dx}(0) = 0 \qquad v(0) = 0 \tag{2}$$

Integrating and using the boundary condition we obtain the deflection as shown below.

$$EI\frac{dv}{dx} = -mLx + \frac{mx^2}{2} + C_1 \qquad EIv = \frac{-mLx^2}{2} + \frac{mx^3}{6} + C_1x + C_2 \tag{3}$$

$$EI\frac{dv}{dx}(0) = C_1 = 0 \qquad EIv(0) = C_2 = 0 \tag{4}$$

$$v(x) = \frac{m}{EI}\left(\frac{-3Lx^2 + x^3}{6}\right) \qquad v(L) = -\left(\frac{mL^3}{3EI}\right) \tag{5}$$

ANS $v(x) = \frac{mx^2}{6EI}(x - 3L)$; $v(L) = -\left(\frac{mL^3}{3EI}\right)$

---

7. 8

Solution v_A =?

The elastic curve in the region BC is same as in earlier problem and is given below.

$$v(x) = \frac{-wx^2}{24EI}(x^2 - 4Lx + 6L^2) \tag{1}$$

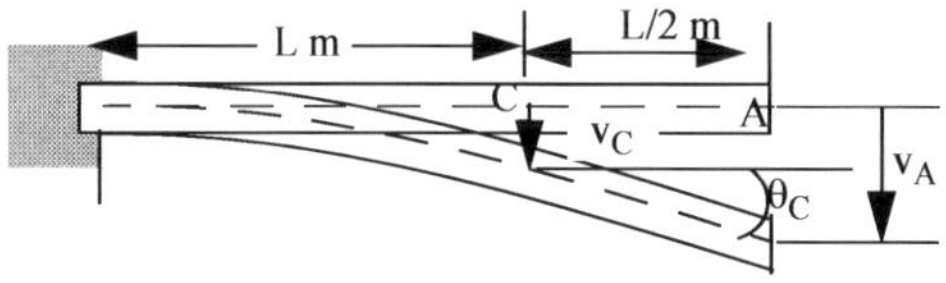

The elastic curve in the region CA is a straight line as the internal moment is zero. We obtain the following.

$$v_A = v_C + \frac{L}{2}\tan\theta_C \qquad \frac{dv}{dx} = \frac{-w}{24EI}(4x^3 - 12Lx^2 + 12L^2x) \tag{2}$$

$$v_C = v(L) = \frac{-wL^4}{8EI} \qquad \tan\theta_C = \frac{dv}{dx}(L) = \frac{-wL^3}{24EI}(4 - 12 + 12) = \frac{-wL^3}{6EI} \qquad v_A = \frac{-wL^4}{8EI} + \frac{L}{2}\left(\frac{-wL^3}{6EI}\right) \tag{3}$$

ANS $v_A = -5wL^4/(24EI)$

---

7. 9

Solution v_A =?

From earlier problem the elastic curve in BC is

$$v(x) = \frac{P}{6EI}(x^3 - L^2x) \tag{1}$$

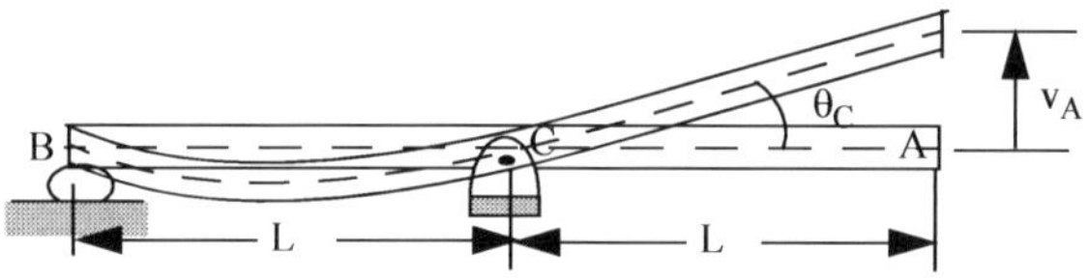

The elastic curve in the region CA is a straight line as the internal moment is zero. We obtain the following.

$$v_A = L\tan\theta_C \qquad \frac{dv}{dx} = \frac{P}{6EI}(3x^2 - L^2) \qquad \tan\theta_C = \frac{dv}{dx}(L) = \frac{P}{6EI}(3L^2 - L^2) = \frac{PL^2}{3EI} \qquad v_A = L\left(\frac{PL^2}{3EI}\right) \tag{2}$$

ANS $v_A = PL^3/(3EI)'$

---

7. 10

Solution v_A =?

From earlier problem the elastic curve in CB is

$$v(x) = \frac{Px^2}{6EI}(x - 3L) \tag{1}$$

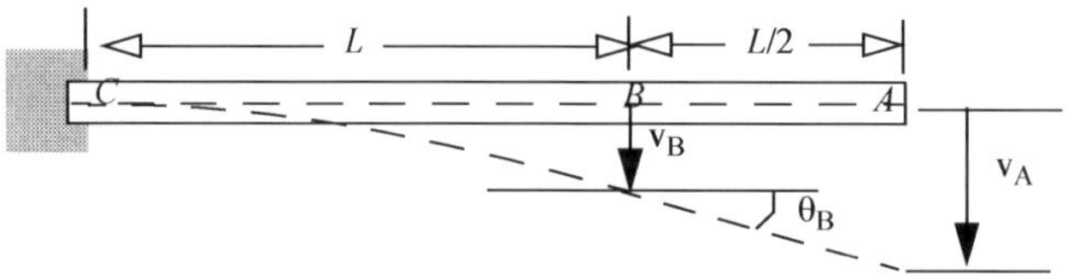

The elastic curve in the region BA is a straight line as the internal moment is zero. We obtain the following.

$$v_B = v(L) = -\frac{PL^3}{3EI} \qquad \theta_B = \frac{dv}{dx}(L) = \frac{P}{6EI}(3x^2 - 6xL)\bigg|_{x=L} = -\frac{PL^2}{2EI} \qquad v_A = v_B + \theta_B\frac{L}{2} = -\frac{PL^3}{3EI} - \frac{PL^3}{4EI} = -\frac{7PL^3}{12EI} \tag{2}$$

ANS $v_A = -7PL^3/(12EI)$

---

7. 11

Solution v_A =?

From earlier problem the elastic curve in BC is

$$v(x) = \frac{-wx}{24EI}(x^3 - 2Lx^2 + L^3) \quad \textbf{(1)}$$

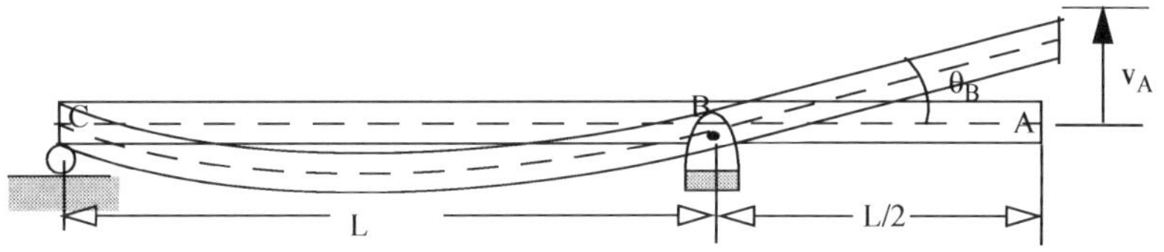

The elastic curve in the region BA is a straight line as the internal moment is zero. We obtain the following.

$$\theta_B = \frac{dv}{dx}(L) = \left.\frac{-w}{24EI}(4x^3 - 6Lx^2 + L^3)\right|_{x=L} = \frac{wL^3}{24EI} \qquad v_A = \theta_B\frac{L}{2} = \frac{wL^4}{48EI} \quad \textbf{(2)}$$

ANS $v_A = \dfrac{wL^4}{48EI}$

---

7. 12

Solution v_A =?

From earlier problem the elastic curve in BC is

$$v(x) = \frac{w}{24EI}(6L^2x^2 - x^4 - 5L^4) \quad \textbf{(1)}$$

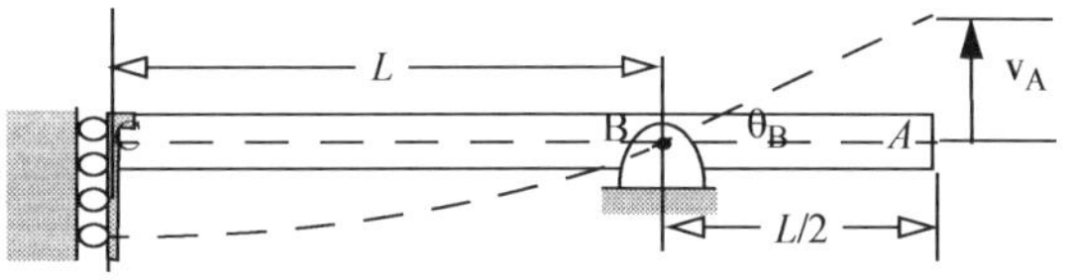

The elastic curve in the region BA is a straight line as the internal moment is zero. We obtain the following.

$$\theta_B = \frac{dv}{dx}(L) = \left.\frac{w}{24EI}(12L^2x - 4x^3)\right|_{x=L} = \frac{wL^3}{3EI} \qquad v_A = \theta_B\frac{L}{2} = \frac{wL^4}{6EI} \quad \textbf{(2)}$$

ANS $v_A = wL^4/(6EI)$

---

7. 13

Solution

ANS The applicable conditions are: (f), (h), (j), and (l).

---

7. 14

Solution

Three sets of conditions are possible answers.

ANS The applicable conditions are: {i} (b), (c),(f), and (k). **or** {ii} (b),(f), (i) and (k). **or** {iii} (c), (f),(i), and (k).

---

7. 15

Solution

ANS The applicable conditions are: (a), (g),(i), and (k).

---

7. 16

Solution

Three sets of conditions are possible answers.

ANS The applicable conditions are: {i} (a), (d),(e), and (l). **or** {ii} (a), (d),(j), and (l). **or** {iii} (a),(e), (j)and (l).

---

7. 17

Solution $v(x) = ?$ $v(L) = ?$

By equilibrium of moment about point C, we obtain the reaction force. By equilibrium of moment about points O_1 and O_2 we obtain the internal moments as shown below.

$$R_A(3L) - wL(2L) = 0 \quad or \quad R_A = 2wL/3 \tag{1}$$

$$\boldsymbol{M_1} - R_A x = 0 \quad or \quad \boldsymbol{M_1} = \frac{2}{3}wLx \qquad \boldsymbol{M_2} - R_A x + wL(x-L) = 0 \quad or \quad \boldsymbol{M_2} = \frac{2}{3}wLx - wL(x-L) \tag{2}$$

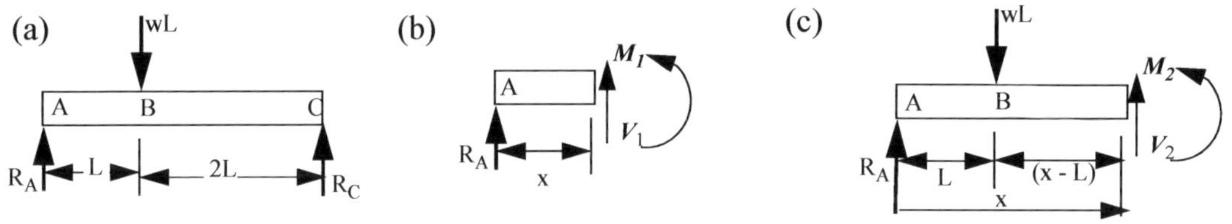

The boundary value problem statement is

$$EI\frac{d^2v_1}{dx^2} = \frac{2}{3}wLx \quad 0 \le x \le L \qquad EI\frac{d^2v_2}{dx^2} = \frac{2}{3}wLx - wL(x-L) \quad L \le x \le 3L \tag{3}$$

$$v_1(0) = 0 \qquad v_2(3L) = 0 \qquad v_1(L) = v_2(L) \qquad \frac{dv_1}{dx}(L) = \frac{dv_2}{dx}(L) \tag{4}$$

Integrating and using the boundary and continuity conditions we obtain the deflection as shown below.

$$EI\frac{dv_1}{dx} = \frac{1}{3}wLx^2 + C_1 \qquad EI\frac{dv_2}{dx} = \frac{1}{3}wLx^2 - \frac{wL}{2}(x-L)^2 + C_2 \tag{5}$$

$$EIv_1 = \frac{1}{9}wLx^3 + C_1x + C_3 \qquad EIv_2 = \frac{1}{9}wLx^3 - \frac{wL}{6}(x-L)^3 + C_1x + C_4 \tag{6}$$

$$EI\frac{dv_1}{dx} = \frac{1}{3}wL^3 + C_1 = EI\frac{dv_2}{dx} = \frac{1}{3}wL^3 + C_2 \quad or \quad C_1 = C_2 \tag{7}$$

$$v_1(L) = \frac{1}{9}wL^3 + C_1L + C_3 = v_2(L) = \frac{1}{9}wL^3 + C_1L + C_4 \quad or \quad C_3 = C_4 \tag{8}$$

$$EIv_1(0) = C_3 = 0 = C_4 \tag{9}$$

$$EIv_2(3L) = \frac{1}{9}wL(27L^3) - \frac{wL}{6}(8L^3) + C_1(3L) = C_1(3L) + \frac{wL^4(54-24)}{18} = 0 \quad or \quad C_1 = -\frac{5}{9}wL^3 = C_2 \tag{10}$$

$$v_1 = \frac{w}{EI}\frac{(Lx^3 - 5L^3x)}{9} = \frac{wLx}{9EI}(x^2 - 5L^2) \qquad v_2 = \frac{w}{9EI}(Lx^3 - 5L^3x) - \frac{w}{6EI}(x-L)^3 \tag{11}$$

$$v(L) = v\ (L) = \frac{wL^2}{9EI}(-4L^2) \tag{12}$$

$$\textbf{ANS} \quad v(x) = \begin{cases} \dfrac{wLx}{9EI}(x^2 - 5L^2) & 0 \le x \le L \\ \dfrac{wLx}{9EI}(x^2 - 5L^2) - \dfrac{wL}{6EI}(x-L)^3 & L \le x \le 3L \end{cases} \; ; \; v(L) = -\left(\frac{wL^4}{9EI}\right)$$

---

7. 18

Solution $v(x) = ?$ $v(L) = ?$

By equilibrium of forces in the y-direction, we obtain the reaction force. By equilibrium of moment about points O_1 and O_2 we obtain the internal moments as shown below.

$$R_A - wL = 0 \quad or \quad R_A = wL \tag{1}$$

$$\boldsymbol{M_1} + wL^2 - R_A x = 0 \quad or \quad \boldsymbol{M_1} = wLx - wL^2 \tag{2}$$

$$\boldsymbol{M_2} + wL^2 - R_A x + w(x-L)(x-L)/2 = 0 \quad or \quad \boldsymbol{M_2} = wLx - wL^2 - w(x-L)^2/2 \tag{3}$$

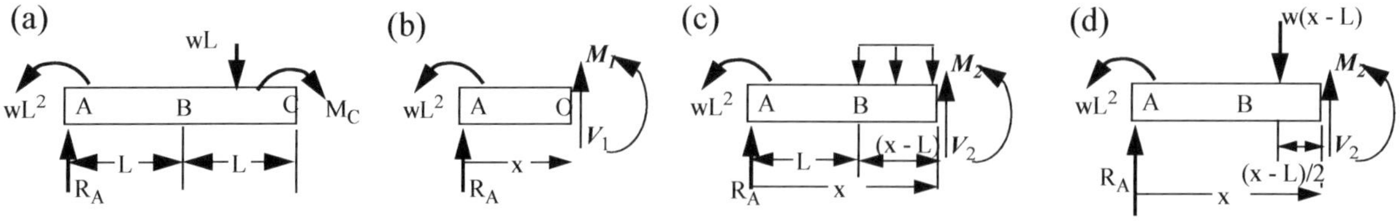

The boundary value problem statement is,

$$EI\frac{d^2v_1}{dx^2} = wLx - wL^2 \qquad 0 \le x \le L \qquad EI\frac{d^2v_2}{dx^2} = wLx - wL^2 - w(x-L)^2/2 \qquad L \le x \le 2L \tag{4}$$

$$v_1(0) = 0 \qquad \frac{dv_2}{dx}(2L) = 0 \qquad v_1(L) = v_2(L) \qquad \frac{dv_1}{dx}(L) = \frac{dv_2}{dx}(L) \tag{5}$$

Integrating and using the boundary and continuity conditions we obtain the deflection as shown below.

$$EI\frac{dv_1}{dx} = \frac{1}{2}wLx^2 - wL^2x + C_1 \qquad EI\frac{dv_2}{dx} = \frac{1}{2}wLx^2 - wL^2x - \frac{w}{6}(x-L)^3 + C_2 \tag{6}$$

$$EIv_1 = \frac{wLx^3}{6} - \frac{wL^2x^2}{2} + C_1x + C_3 \qquad EIv_2 = \frac{wLx^3}{6} - \frac{wL^2x^2}{2} - \frac{w(x-L)^4}{24} + C_1x + C_4 \tag{7}$$

$$EI\frac{dv_1}{dx}(L) = \frac{1}{2}wL^3 - wL^3 + C_1 = EI\frac{dv_2}{dx}(L) = \frac{1}{2}wL^3 - wL^3 + C_2 \qquad or \qquad C_1 = C_2 \tag{8}$$

$$EIv_1(L) = \frac{1}{6}wL^4 - \frac{w}{2}L^4 + C_1L + C_3 = EIv_2(L) = \frac{1}{6}wL^4 - \frac{w}{2}L^4 + C_1x + C_4 \qquad or \qquad C_3 = C_4 \tag{9}$$

$$EIv_1(0) = C_3 = 0 = C_4 \tag{10}$$

$$EI\frac{dv_2}{dx}(2L) = \frac{1}{2}wL(4L^2) - wL^2(2L) - \frac{1}{6}wL^3 + C_2 = 0 \qquad or \qquad C_2 + \frac{wL^3(12-12-1)}{6} = 0 \qquad or \qquad C_2 = \frac{1}{6}wL^3 = C_1 \tag{11}$$

$$v_1 = \frac{w}{EI}\frac{(Lx^3 - 3L^2x^2 + L^3x)}{6} = \frac{wLx}{6EI}(x^2 - 3Lx + L^2) \qquad v_2 = \frac{w}{EI}\frac{(Lx^3 - 3L^2x^2 + L^3x)}{6} - \frac{w}{24}(x-L)^4 \tag{12}$$

$$v(L) = v_1(L) = \frac{wL^2}{6EI}(L^2 - 3L^2 + L^2) \tag{13}$$

ANS $$v(x) = \begin{cases} \frac{wLx}{6EI}(x^2 - 3Lx + L^2) & 0 \le x \le L \\ \frac{w}{EI}\frac{(Lx^3 - 3L^2x^2 + L^3x)}{6} - \frac{w}{24}(x-L)^4 & L \le x \le 2L \end{cases} \quad ; \; v(L) = -\left(\frac{4wL^4}{6EI}\right)$$

---

7. 19

Solution v(x) =? v(L) =?

By equilibrium of moment about point C in Fig.(a) we obtain the reaction force. By equilibrium of moment about points O_1 and O_2 we obtain the internal moments as shown below.

$$A_y(2L) - wL\left(\frac{L}{2}\right) = 0 \qquad or \qquad A_y = \frac{wL}{4}\ kips \tag{1}$$

$$\boldsymbol{M}_1 - A_y(x) = 0 \qquad or \qquad \boldsymbol{M}_1 = \frac{wL}{4}x\ in-kips \tag{2}$$

$$\boldsymbol{M}_2 + (w)(x-L)\left(\frac{x-L}{2}\right) - A_y(x) = 0 \qquad or \qquad \boldsymbol{M}_2 = \left[\frac{wL}{4}x - \frac{w}{2}(x-L)^2\right]\ in-kips \tag{3}$$

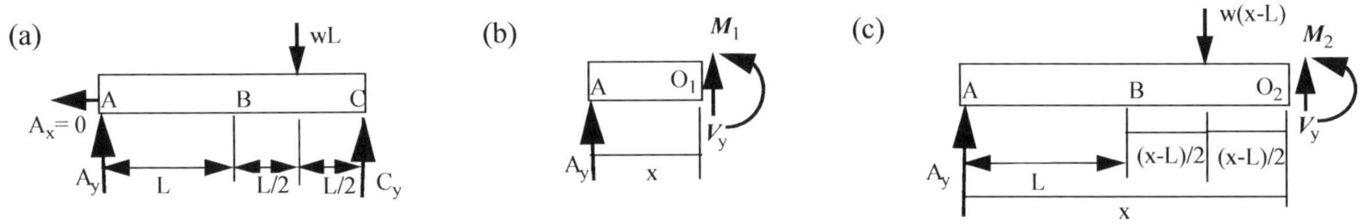

The boundary value problem statement is,

$$EI\frac{d^2v_1}{dx^2} = \frac{wLx}{4} \qquad 0 \le x \le L \qquad EI\frac{d^2v_2}{dx^2} = \frac{wLx}{4} - \frac{w}{2}(x-L)^2 \qquad L \le x \le 2L \tag{4}$$

$$v_1(0) = 0 \qquad v_2(2L) = 0 \qquad v_1(L) = v_2(L) \qquad \frac{dv_1}{dx}(L) = \frac{dv_2}{dx}(L) \tag{5}$$

Integrating and using the boundary and continuity conditions we obtain the deflection as shown below.

$$EI\frac{dv_1}{dx} = \frac{1}{8}wLx^2 + C_1 \qquad EI\frac{dv_2}{dx} = \frac{1}{8}wLx^2 - \frac{w}{6}(x-L)^3 + C_2 \tag{6}$$

$$EIv_1 = \frac{wLx^3}{24} + C_1x + C_3 \qquad EIv_2 = \frac{wLx^3}{24} - \frac{w(x-L)^4}{24} + C_1x + C_4 \tag{7}$$

$$EI\frac{dv_1}{dx}(L) = \frac{1}{8}wL^3 + C_1 = EI\frac{dv_2}{dx}(L) = \frac{1}{8}wL^3 + C_2 \qquad or \qquad C_1 = C_2 \tag{8}$$

$$EIv_1(L) = \frac{wL^4}{24} + C_1L + C_3 = EIv_2(L) = \frac{wL^4}{24} + C_1x + C_4 \qquad or \qquad C_3 = C_4 \tag{9}$$

$$EIv_1(0) = C_3 = 0 = C_4 \tag{10}$$

$$EIv_2(2L) = \frac{w}{24}L(8L^3) - \frac{wL^4}{24} + C_1(2L) = 0 \qquad or \qquad C_1(2L) + \frac{wL^4(8-1)}{24} = 0 \qquad or \qquad C_1 = -\frac{7wL^3}{48} = C_2 \text{ or} \tag{11}$$

$$v_1 = \frac{w}{EI}\frac{(2Lx^3 - 7L^3x)}{48} = \frac{wLx}{48EI}(2x^2 - 7L^2) \qquad v_2 = \frac{w}{EI}\frac{(2Lx^3 - 7L^3x)}{48} - \frac{w}{24EI}(x-L)^4 \tag{12}$$

$$v_1(L) = \frac{wL^2}{48EI}(2L^2 - 7L^2) = \frac{-5wL^4}{48EI} \tag{13}$$

$$\textbf{ANS} \quad v(x) = \begin{cases} \dfrac{wLx}{48EI}(2x^2 - 7L^2) & 0 \le x \le L \\ \dfrac{w}{EI}\dfrac{(2Lx^3 - 7L^3x)}{48} - \dfrac{w}{24EI}(x-L)^4 & L \le x \le 2L \end{cases} \quad ; v(L) = \frac{-5wL^4}{48EI}$$

---

7. 20

Solution v(x) =? v(L) =?

By equilibrium of forces and moment about point A we obtain the reaction force. By equilibrium of moment about points O_1 and O_2 we obtain the internal moments as shown below.

$$A_y = wL \ kips \qquad M_A = wL\left(\frac{3L}{2}\right) = \frac{3}{2}wL^2 \tag{1}$$

$$\boldsymbol{M}_1 = A_y(x) - M_A = \left(wLx - \frac{3}{2}wL^2\right) \text{ in.- kips} \tag{2}$$

$$\boldsymbol{M}_2 + (w)(x-L)\left(\frac{x-L}{2}\right) - A_y(x) + M_A = 0 \qquad or \qquad \boldsymbol{M}_2 = \left[wLx - \frac{w}{2}(x-L)^2 - \frac{3}{2}wL^2\right] \text{ in.- kips} \tag{3}$$

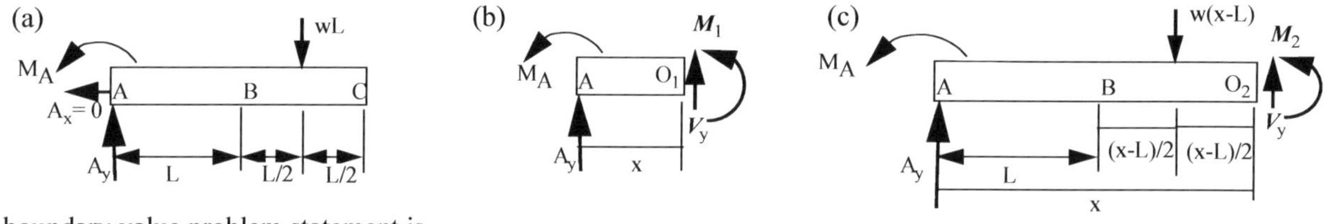

The boundary value problem statement is

$$EI\frac{d^2v_1}{dx^2} = wLx - \frac{3}{2}wL^2 \qquad 0 \le x \le L \qquad EI\frac{d^2v_2}{dx^2} = wLx - \frac{3}{2}wL^2 - \frac{w}{2}(x-L)^2 \qquad L \le x \le 2L \tag{4}$$

$$v_1(0) = 0 \qquad \frac{dv_1}{dx}(0) = 0 \qquad v_1(L) = v_2(L) \qquad \frac{dv_1}{dx}(L) = \frac{dv_2}{dx}(L) \tag{5}$$

Integrating and using the boundary and continuity conditions we obtain the deflection as shown below.

$$EI\frac{dv_1}{dx} = \frac{1}{2}wLx^2 - \frac{3}{2}wL^2x + C_1 \qquad EI\frac{dv_2}{dx} = \frac{1}{2}wLx^2 - \frac{3}{2}wL^2x - \frac{w}{6}(x-L)^3 + C_2 \tag{6}$$

$$EIv_1 = \frac{wLx^3}{6} - \frac{3wL^2x^2}{4} + C_1x + C_3 \qquad EIv_2 = \frac{wLx^3}{6} - \frac{3wL^2x^2}{4} - \frac{w(x-L)^4}{24} + C_1x + C_4 \tag{7}$$

$$EI\frac{dv_1}{dx}(L) = \frac{1}{2}wL^3 - \frac{3}{2}wL^3 + C_1 = EI\frac{dv_2}{dx}(L) = \frac{1}{2}wL^3 - \frac{3}{2}wL^3 + C_2 \qquad or \qquad C_1 = C_2 \tag{8}$$

$$EIv_1(L) = \frac{1}{6}wL^4 - \frac{3}{4}wL^4 + C_1L + C_3 = EIv_2(L) = \frac{1}{6}wL^4 - \frac{3}{4}wL^4 + C_1x + C_4 \qquad or \qquad C_3 = C_4 \tag{9}$$

$$EI\frac{dv_1}{dx}(0) = C_1 = 0 = C_2 \qquad EIv_1(0) = C_3 = 0 = C_4 \tag{10}$$

$$v_1 = \frac{w}{EI}\frac{(2Lx^3 - 9L^2x^2)}{12} = \frac{wLx^2}{12EI}(2x - 9L) \qquad v_2 = \frac{w}{EI}\frac{(2Lx^3 - 9L^2x^2)}{12} - \frac{w}{24EI}(x - L)^4 \tag{11}$$

$$v_1(L) = \frac{wL^3}{12EI}(-7L) = \frac{-7wL^4}{12EI} \tag{12}$$

$$\textbf{ANS} \quad v(x) = \begin{cases} \frac{wLx^2}{12EI}(2x - 9L) & 0 \le x \le L \\ \frac{w}{EI}\frac{(2Lx^3 - 9L^2x^2)}{12} - \frac{w}{24EI}(x - L)^4 & L \le x \le 2L \end{cases} \quad ; \; v(L) = \frac{-7wL^4}{12EI}$$

---

7. 21

Solution v(x) =? v(L) =?

By equilibrium of moment about point C we obtain the reaction force. By equilibrium of moment about points O_1 and O_2 we obtain the internal moments as shown below.

$$A_y(2L) - mL = 0 \quad or \quad A_y = \frac{m}{2} \; kips \tag{1}$$

$$\boldsymbol{M}_1 - A_y(x) = 0 \quad or \quad \boldsymbol{M}_1 = \frac{m}{2}x \; in-kips \tag{2}$$

$$\boldsymbol{M}_2 + (m)(x - L) - A_y(x) = 0 \quad or \quad \boldsymbol{M}_2 = \left[\frac{m}{2}x - m(x - L)\right] \; in-kips \tag{3}$$

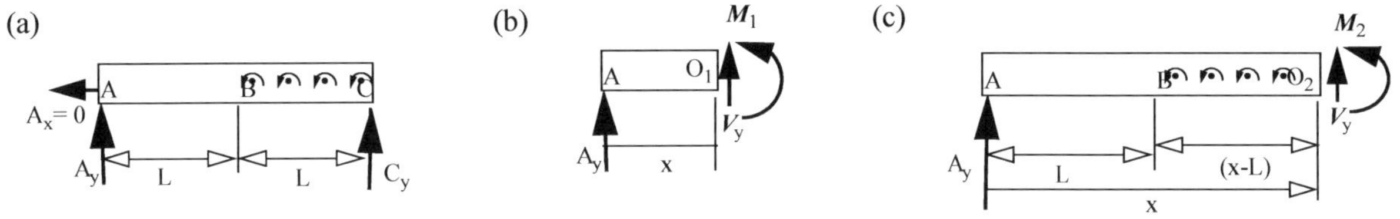

The boundary value problem statement is,

$$EI\frac{d^2v_1}{dx^2} = \frac{m}{2}x \quad 0 \le x \le L \qquad EI\frac{d^2v_2}{dx^2} = \frac{m}{2}x - m(x - L) \quad L \le x \le 2L \tag{4}$$

$$v_1(0) = 0 \qquad v_2(2L) = 0 \qquad v_1(L) = v_2(L) \qquad \frac{dv_1}{dx}(L) = \frac{dv_2}{dx}(L) \tag{5}$$

Integrating and using the boundary and continuity conditions we obtain the deflection as shown below.

$$EI\frac{dv_1}{dx} = \frac{1}{4}mx^2 + C_1 \qquad EI\frac{dv_2}{dx} = \frac{1}{4}mx^2 - \frac{m}{2}(x - L)^2 + C_2 \tag{6}$$

$$EIv_1 = \frac{mx^3}{12} + C_1x + C_3 \qquad EIv_2 = \frac{mx^3}{12} - \frac{m(x - L)^3}{6} + C_1x + C_4 \tag{7}$$

$$EI\frac{dv_1}{dx}(L) = \frac{1}{4}mL^2 + C_1 = EI\frac{dv_2}{dx}(L) = \frac{1}{4}mL^2 + C_2 \quad or \quad C_1 = C_2 \tag{8}$$

$$EIv_1(L) = \frac{mL^3}{12} + C_1L + C_3 = EIv_2(L) = \frac{mL^3}{12} + C_1x + C_4 \quad or \quad C_3 = C_4 \tag{9}$$

$$EIv_1(0) = C_3 = 0 = C_4 \tag{10}$$

$$EIv_2(2L) = \frac{m}{12}(8L^3) - \frac{mL^3}{6} + C_1(2L) = 0 \quad ore \quad C_1(2L) + \frac{mL^3(8 - 2)}{24} = 0 \quad or \quad C_1 = -\frac{mL^2}{12} = C_2 \tag{11}$$

$$v_1 = \frac{m}{EI}\frac{(x^3 - L^2x)}{12} = \frac{mx}{12EI}(x^2 - L^2) \qquad v_2 = \frac{mx}{12EI}(x^2 - L^2) - \frac{m}{6EI}(x - L)^3 \tag{12}$$

$$v_1(L) = \frac{mL}{12EI}(L^2 - L^2) = 0 \tag{13}$$

$$\textbf{ANS} \quad v(x) = \begin{cases} \frac{mx}{12EI}(x^2 - L^2) & 0 \le x \le L \\ \frac{mx}{12EI}(x^2 - L^2) - \frac{m}{6EI}(x - L)^3 & L \le x \le 2L \end{cases} \quad ; \; v(L) = 0$$

---

7. 22

Solution $v_{max} = ?$

$$I_{zz} = (16)(1)^3/12 = 1.333\ in^4 \qquad EI = (15)(10^6)(1.333) = 20(10^6)\ lb-in^2 \tag{1}$$

By equilibrium of forces we obtain the reaction force. By equilibrium of moment about points O_1 and O_2 we obtain the internal moments as shown below.

$$R_B = \frac{(64)(200)}{56} = 492.86\ lb \qquad R_A = R_B - 200 = 232.86\ lb \tag{2}$$

$$\boldsymbol{M}_1 = -232.86x\ in-lb \qquad \boldsymbol{M}_2 = -232.86x + 492.86(x-56)\ in-lb \tag{3}$$

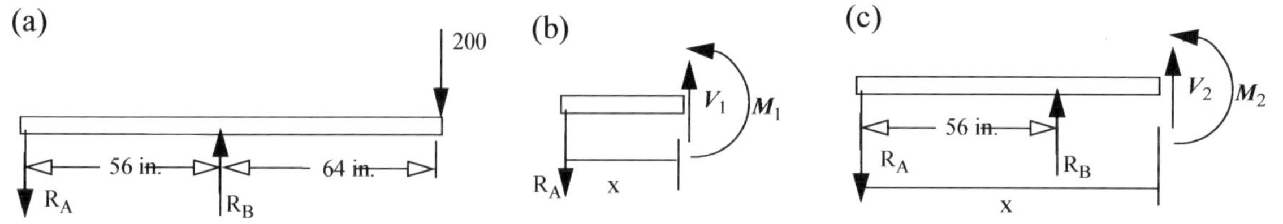

The boundary value problem can be writen as:

$$EI\frac{d^2v_1}{dx^2} = -232.86x \qquad 0 \le x \le 56 \qquad EI\frac{d^2v_2}{dx^2} = -232.86x + 492.86(x-56) \qquad 56 \le x \le 120 \tag{4}$$

$$v_1(0) = 0 \qquad v_1(56) = 0 \qquad v_2(56) = 0 \qquad \frac{dv_1}{dx}(56) = \frac{dv_2}{dx}(56) \tag{5}$$

Integrating and using the boundary and continuity conditions we obtain the deflection as shown below.

$$EI\frac{dv_1}{dx} = -232.86\frac{x^2}{2} + C_1 \qquad EI\frac{dv_2}{dx} = -232.86\frac{x^2}{2} + 492.86\frac{(x-56)^2}{2} + C_2 \tag{6}$$

$$EIv_1 = -232.86\frac{x^3}{6} + C_1x + C_3 \qquad EIv_2 = -232.86\frac{x^3}{6} + 492.86\frac{(x-56)^3}{6} + C_2x + C_4 \tag{7}$$

$$EIv_1(0) = C_3 = 0 \qquad EIv_1(56) = -232.86\frac{(56)^3}{6} + C_1(56) = 0 \qquad or \qquad C_1 = 121.71(10^3) \tag{8}$$

$$EI\frac{dv_1}{dx}(56) = -232.86\frac{(56)^2}{2} + C_1 = EI\frac{dv_2}{dx}(56) = -232.86\frac{(56)^2}{2} + C_2 \qquad or \qquad C_2 = C_1 = 121.71(10^3) \tag{9}$$

$$EIv_2(56) = -232.86\frac{(56)^3}{6} + C_2(56) + C_4 = 0 \qquad or \qquad C_4 = 0 \tag{10}$$

$$EIv_2(120) = -232.86\frac{(120)^3}{6} + 492.86\frac{(64)^3}{6} + (121.71)(10^3)(120) = -30.92(10^6) \text{ or}$$

$$v_2(120) = -\frac{30.92(10^6)}{20(10^6)} = -1.546\ in \tag{11}$$

ANS $v_{max} = 1.55$ in.

---

7. 23

Solution Boundary Value Problem =?

By equilibrium of moment about points O_1, O_2, and O_3we obtain the following.

$$\boldsymbol{M_1} - wx\left(\frac{x}{2}\right) = 0 \qquad or \qquad \boldsymbol{M_1} = \frac{wx^2}{2} \qquad \boldsymbol{M_2} - wL\left(x - \frac{L}{2}\right) = 0 \qquad or \qquad \boldsymbol{M_2} = wL\left(x - \frac{L}{2}\right) \tag{1}$$

$$\boldsymbol{M_3} - wL\left(x - \frac{L}{2}\right) + wL(x-2L) + wL^2 = 0 \qquad or \qquad \boldsymbol{M_3} - wLx + \frac{wL^2}{2} + wLx - 2wL^2 + wL^2 = 0 \qquad or \qquad \boldsymbol{M_3} = \frac{wL^2}{2} \tag{2}$$

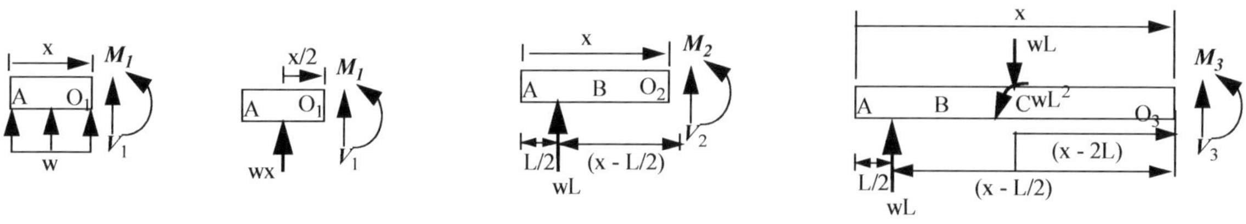

The boundary value problem statement is as shown below.

$$EI\frac{d^2v_1}{dx^2} = \frac{wx^2}{2} \quad 0 \le x < L \qquad EI\frac{d^2v_2}{dx^2} = wL\left(x-\frac{L}{2}\right) \quad L < x < 2L \qquad EI\frac{d^2v_3}{dx^2} = \frac{wL^2}{2} \quad 2L < x \le 4L \tag{3}$$

$$v_3(4L) = 0 \qquad \frac{dv_3}{dx}(4L) = 0 \qquad v_1(L) = v_2(L) \qquad \frac{dv_1}{dx}(L) = \frac{dv_2}{dx}(L) \qquad v_2(2L) = v_3(2L) \qquad \frac{dv_1}{dx}(2L) = \frac{dv_2}{dx}(2L) \tag{4}$$

---

7. 24

Solution Boundary Value Problem =?

By equilibrium of moment about point D and quilibrium of forces in y-direction we obtain the reaction forces. By equilibrium of moment about points O_1, O_2, and O_3we obtain the internal moments.

$$R_B(3L) - wL(3L+L/2) + wL^2 - 2wL^2 + wL(L) = 0 \quad or \quad R_B(3L) - 7wL^2/2 = 0 \qquad R_B = 7wL/6 \tag{1}$$

$$R_B - R_D + wL - wL = 0 \quad or \quad R_B = R_D = 7wL/6 \tag{2}$$

$$\boldsymbol{M_1} + wx(x/2) = 0 \quad or \quad \boldsymbol{M_1} = -w(x^2/2) \tag{3}$$

$$\boldsymbol{M_2} + wL\left(x-\frac{L}{2}\right) - wL^2 - R_B(x-L) = 0 \quad or \quad \boldsymbol{M_2} + wLx - \frac{wL^2}{2} - wL^2 - \frac{7w}{6}Lx + \frac{7}{6}wL^2 = 0 \quad or \quad \boldsymbol{M_2} = \frac{wLx}{6} + \frac{wL^2}{3} \tag{4}$$

$$\boldsymbol{M_3} + R_D(4L-x) = 0 \quad or \quad \boldsymbol{M_3} = \frac{7wLx}{6} - \frac{14wL^2}{3} \tag{5}$$

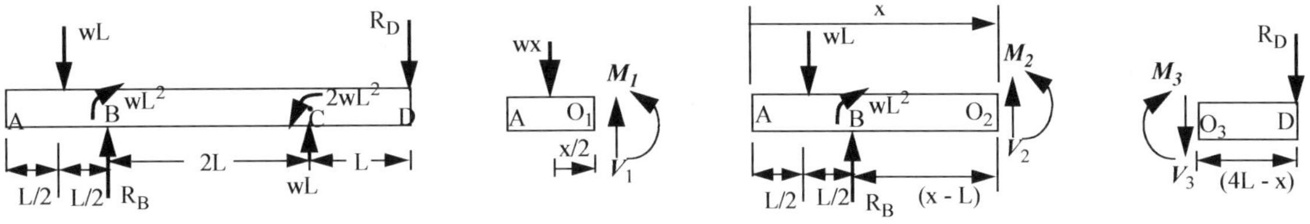

The boundary value problem statement is

$$EI\frac{d^2v_1}{dx^2} = \frac{-wx^2}{2} \quad 0 \le x < L \qquad EI\frac{d^2v_2}{dx^2} = \frac{wLx}{6} + \frac{wL^2}{3} \quad L < x < 3L \qquad EI\frac{d^2v_3}{dx^2} = \frac{7wLx}{6} - \frac{14wL^2}{3} \quad 3L < x \le 4L \tag{6}$$

$$v_1(L) = 0 \qquad v_2(L) = 0 \qquad v_3(4L) = 0 \qquad \frac{dv_1}{dx}(L) = \frac{dv_2}{dx}(L) \qquad v_2(3L) = v_3(3L) \qquad \frac{dv_2}{dx}(3L) = \frac{dv_3}{dx}(3L) \tag{7}$$

---

7. 25

Solution h(x) = ? v_{max} =?

By equilibrium of moment about point O_1, we obtain,

$$\boldsymbol{M_z} + Px = 0 \quad or \quad \boldsymbol{M_z} = -Px \tag{1}$$

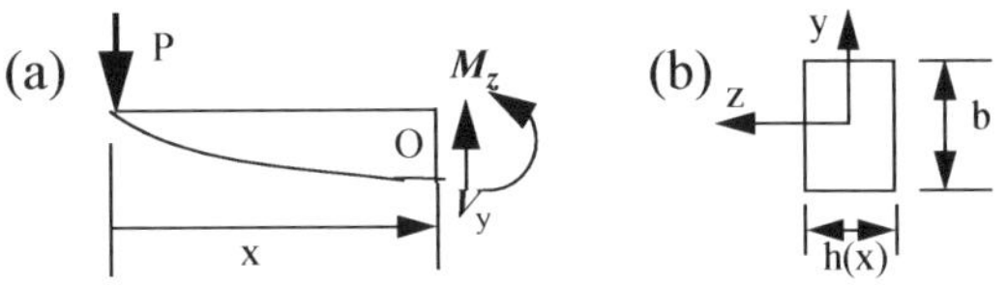

The maximum bending normal stress will be at $y = \mp h(x)/2$. and its magnitude is

$$\boldsymbol{I_{zz}} = \frac{1}{12}bh^3(x) \qquad \sigma_{max} = \left|-\frac{\boldsymbol{M_z}y}{\boldsymbol{I_{zz}}}\right| = \frac{(Px)(h(x)/2)}{bh^3(x)/12} = \sigma \quad or \quad \frac{6Px}{bh^2(x)} = \sigma \quad or \quad h(x) = \sqrt{\frac{6Px}{b\sigma}} \tag{2}$$

The moment curvature equation can be written as

$$\boldsymbol{I_{zz}} = \frac{1}{12}b\left(\frac{6Px}{b\sigma}\right)^{3/2} \qquad \boldsymbol{M_z} = EI_{zz}\frac{d^2v_2}{dx^2} = -Px \quad or \quad \frac{d^2v_2}{dx^2} = -\frac{Px}{E\sqrt{3P^3/2b\sigma^3}x^{3/2}} = -\sqrt{\frac{2b\sigma^3}{3PE^2}}\frac{1}{\sqrt{x}} \tag{3}$$

The boundary value problem statement is,

$$\frac{d^2v}{dx^2} = -\left(\frac{k}{\sqrt{x}}\right) \quad where\ k = \sqrt{\frac{2b\sigma^3}{3PE^2}} \qquad v(L) = 0 \qquad \frac{dv}{dx}(L) = 0 \tag{4}$$

Integrating and using the boundary conditions we obtain the displacement as shown below.

$$\frac{dv}{dx} = -2kx^{1/2} + C_1 \qquad v = -\frac{4}{3}kx^{3/2} + C_1x + C_2 \tag{5}$$

$$\frac{dv}{dx}(L) = -2kL^{1/2} + C_1 = 0 \qquad or \qquad C_1 = 2kL^{1/2} \tag{6}$$

$$v(L) = -k\,\frac{4}{3}L^{3/2} + 2k\;L^{1/2}L + C_2 = kL^{3/2}\left(\frac{-4+6}{3}\right) + C_2 = 0 \qquad or \qquad C_2 = -\left(\frac{2}{3}kL^{3/2}\right) \tag{7}$$

$$v_{max} = v(0) = C_2 = -\frac{2}{3}\sqrt{\frac{2b\sigma^3}{3PE^2}}L^{3/2} \tag{8}$$

ANS $h(x) = \sqrt{\dfrac{6Px}{b\sigma}}$; $v_{max} = -\sqrt{\dfrac{8b\sigma^3L^3}{27PE^2}}$

---

7. 26

Solution R(x) =? v_{max} =?

By equilibrium of moment about point O, we obtain,

$$\boldsymbol{M}_z = -Px \tag{1}$$

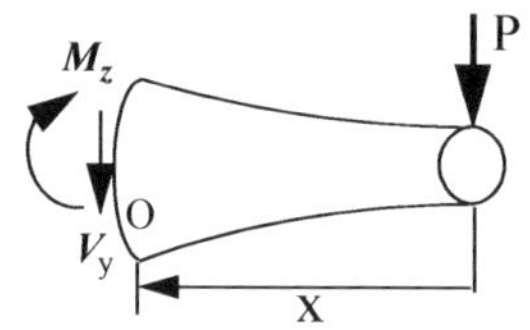

The maximum bending normal stress will be at $y = \mp R(x)$ .The magnitude of maximum bending stress is

$$\boldsymbol{I}_{zz} = \frac{\pi}{4}R^4(x) \qquad \sigma_{max} = \left|-\frac{\boldsymbol{M}_z y}{\boldsymbol{I}_{zz}}\right| = \frac{(Px)R}{\pi R^4/4} = \sigma \qquad or \qquad R = \left(\frac{4Px}{\pi\sigma}\right)^{1/3} \tag{2}$$

The moment curvature relationship can be written as

$$\boldsymbol{I}_{zz} = \frac{\pi}{4}\left(\frac{4Px}{\pi\sigma}\right)^{4/3} = \left(\frac{4P^4}{\pi\sigma^4}\right)^{1/3}x^{4/3} \qquad \boldsymbol{M}_z = EI_{zz}\frac{d^2v}{dx^2} = -Px \qquad or \qquad \frac{d^2v}{dx^2} = -\frac{Px}{E[4P^4/(3\sigma^4)]^{1/3}x^{4/3}} = -\left(\frac{3\sigma^4}{4PE^3}\right)^{1/3}\frac{1}{x^{1/3}} \tag{3}$$

The boundary value problem statement is,

$$\frac{d^2v}{dx^2} = -\frac{k}{x^{1/3}} \qquad where\ k = \left(\frac{3\sigma^4}{4PE^3}\right)^{1/3} \qquad v(L) = 0 \qquad \frac{dv}{dx}(L) = 0 \tag{4}$$

Integrating and using the boundary conditions we obtain the displacement as shown below.

$$\frac{dv}{dx} = -k\left(\frac{3}{2}x^{2/3}\right) + C_1 \qquad v = -k\left(\frac{9}{10}x^{5/3}\right) + C_1x + C_2 \tag{5}$$

$$\frac{dv}{dx}(L) = -k\left(\frac{3}{2}L^{2/3}\right) + C_1 = 0 \qquad or \qquad C_1 = \frac{3}{2}kL^{2/3} \tag{6}$$

$$v(L) = -\frac{9}{10}kL^{5/3} + \frac{3}{2}kL^{2/3}L + C_2 \qquad or \qquad kL^{5/3}\left(\frac{-9+15}{10}\right) + C_2 = 0 \qquad or \qquad C_2 = -\frac{3}{5}kL^{5/3} \tag{7}$$

$$v_{max} = v(0) = C_2 = -\frac{3}{5}\left(\frac{3\sigma^4}{4PE^3}\right)^{1/3}L^{5/3} \tag{8}$$

ANS $R = \left(\dfrac{4Px}{\pi\sigma}\right)^{1/3}$; $v_{max} = -\left(\dfrac{81\sigma^4L^5}{500PE^3}\right)^{1/3}$

---

7. 27

Solution σ_{max}=? v_{max} =?

By similar triangle and by equilibrium of moment about point O, we obtain the following.

$$h = (h_o/L)x \qquad \boldsymbol{M}_z = -wx^2/2 \tag{1}$$

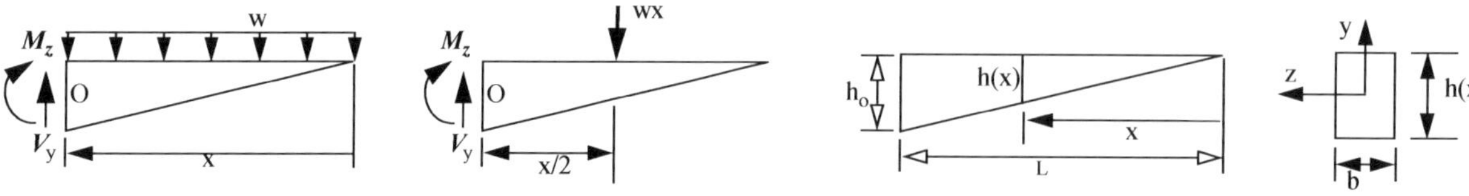

The maximum bending normal stress will be at $y = \mp h(x)/2$. The magnitude of maximum bending stress is

$$I_{zz} = \frac{1}{12}bh^3 = \frac{1}{12}b\frac{h_o^3}{L^3}x^3 \qquad \sigma_{max} = \left|-\frac{M_z y}{I_{zz}}\right| = \frac{(-wx^2/2)(h/2)}{(bh^3)/12} = \frac{3wx^2}{bh^2} = \frac{3wx^2}{b(h_o^2/L^2)x^2} \quad or \quad \sigma_{max} = \frac{3wL^2}{bh_o^2} \qquad (2)$$

The moment curvature relationship can be written as

$$M_z = EI_{zz}\frac{d^2v}{dx^2} = -\frac{wx^2}{2} \quad or \quad E\left(\frac{3bh_o^3}{L^3}\right)x^3\frac{d^2v}{dx^2} = -\frac{wx^2}{2} \quad or \quad \frac{d^2v}{dx^2} = -\left(\frac{wL^3}{6Ebh_o^3}\right)\frac{1}{x} \qquad (3)$$

The boundary value problem statement is:

$$\frac{d^2v}{dx^2} = -\frac{k}{x} \quad where\ k = \frac{wL^3}{6Ebh_o^3} \quad v(L) = 0 \quad \frac{dv}{dx}(L) = 0 \qquad (4)$$

Integrating and using the boundary conditions we obtain the displacement as shown below.

$$\frac{dv}{dx} = -k\ln(x) + C_1 \qquad v = -k(x\ln(x) - x) + C_1x + C_2 \qquad (5)$$

$$\frac{dv}{dx}(L) = -k\ln(L) + C_1 = 0 \quad or \quad C_1 = k\ln(L) \qquad (6)$$

$$v(L) = -k[L\ln(L) - L] + [k\ln(L)]L + C_2 \quad or \quad C_2 = -kL \qquad (7)$$

$$v_{max} = v(0) = C_2 = -kL \qquad (8)$$

ANS $\sigma_{max} = \dfrac{3wL^2}{bh_o^2}$; $v_{max} = -\left(\dfrac{wL^4}{6Ebh_o^3}\right)$

---

7. 28

Solution σ_{max}=? v_{max} =?

The diameter is a linear function of x and by equilibrium of moment about point O, we obtain the following.

$$d(x) = a + bx \quad d(0) = a = d_o \quad d(L) = d_o + bL \quad or \quad b = d_o/L \quad d = d_o + (d_o/L)x = d_o(x+L)/L \qquad (1)$$

$$M_z = -Px \qquad (2)$$

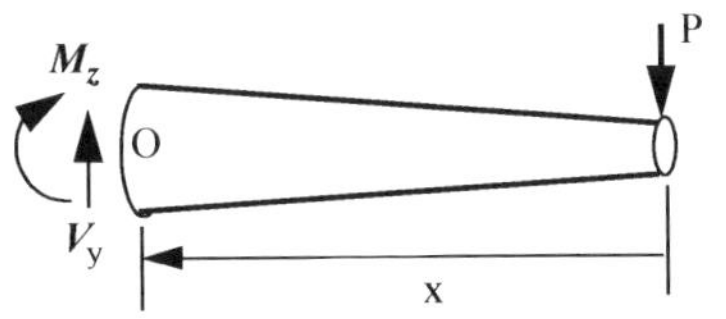

The maximum bending normal stress will be at $y = \pm d/2$. The magnitude of maximum bending stress is

$$I_{zz} = \frac{\pi}{64}d^4 = \frac{\pi}{64}(d_o^4/L^4)(x+L)^4 \qquad \sigma = \left|-\frac{M_z y}{I_{zz}}\right| = \frac{(Px)\frac{d}{2}}{\pi d^4/64} = \frac{32}{\pi}P\frac{x}{d^3} \quad or \quad \sigma = \frac{32PL^3}{\pi d_o^3}\frac{x}{(x+L)^3} \qquad (3)$$

Taking the first derivative of the above equation with respect to x and setting it to zero at x_1, we determine the maximum bending normal stress as shown below.

$$\left.\frac{d\sigma}{dx}\right|_{x_1} = \left(\frac{32PL^3}{\pi d_o^3}\right)\left[\frac{1}{(x_1+L)^3} - \frac{3x_1}{(x_1+L)^4}\right] = 0 \quad or \quad (x_1 + L) - 3x_1 = 0 \quad or \quad x_1 = \frac{L}{2} \qquad (4)$$

$$\sigma_{max} = \frac{32PL^3}{\pi d_o^3}\left[\frac{L/2}{(3L/2)^3}\right] = \frac{128PL}{27\pi d_o^3} \qquad (5)$$

$$M_z = EI_{zz}\frac{d^2v}{dx^2} = -Px \quad or \quad \frac{d^2v}{dx^2} = -\frac{Px}{EI_{zz}} = -\frac{64PL^4}{E\pi d_o^4}\left[\frac{x}{(x+L)^4}\right] \qquad (6)$$

We can write $x = x + L - L$ in the numerator and state the boundary value problem as shown below,

$$\frac{d^2v}{dx^2} = -k\left[\frac{1}{(x+L)^3} - \frac{L}{(x+L)^4}\right] \quad where\ k = \frac{64PL^4}{E\pi d_o^4} \quad v(L) = 0 \quad \frac{dv}{dx}(L) = 0 \qquad (7)$$

Integrating and using the boundary conditions we obtain the displacement as shown below.

$$\frac{dv}{dx} = -k\left[-\frac{1}{2}\frac{1}{(x+L)^2} + \frac{1}{3}\frac{L}{(x+L)^3}\right] + C_1 \qquad v = -k\left[\frac{1}{2}\frac{1}{(x+L)} - \frac{1}{6}\frac{L}{(x+L)^2}\right] + C_1x + C_2 \qquad (8)$$

$$\frac{dv}{dx}(L) = -k\left[-\frac{1}{2}\left(\frac{1}{4L^2}\right)+\frac{1}{3}\left(\frac{L}{8L^3}\right)\right]+C_1 = 0 \quad or \quad -k\left[\frac{-3+1}{24L^2}\right]+C_1 = 0 \quad or \quad C_1 = -\left(\frac{k}{12L^2}\right) \tag{9}$$

$$v(L) = -k\left[\frac{1}{2}\left(\frac{1}{2L}\right)-\frac{L}{6}\left(\frac{1}{4L^2}\right)\right]-\frac{kL}{12L^2}+C_2 \quad or \quad -\frac{k}{24L}[6-1+2]+C_2 = 0 \quad or \quad C_2 = \frac{7k}{24L} \tag{10}$$

$$v_{max} = v(0) = -k\left[\frac{1}{2L}-\frac{1}{6L}\right]+\frac{7k}{24L} = \frac{k}{24L}[-12+4+7] = -\frac{k}{24L} \tag{11}$$

ANS $\sigma_{max} = \dfrac{128PL}{27\pi d_o^3}$; $v_{max} = -\dfrac{8PL^3}{3E\pi d_o^4}$

---

7. 29

Solution E = 2000 psi σ_{max}=? v_{max}=?

The length of the beam is 108 inch. By equilibrium of moment about point O, we obtain,

$$\boldsymbol{M}_z = -800(108-x) \tag{1}$$

The area moment of inertia and bending rigidity in the segment AB and BC are

$$I_{AB} = (2)(10)^3/12 = 166.66\ in^4 \quad 0 \le x \le 36\ in \qquad I_{BC} = (2)(8)^3/12 = 85.33\ in^4 \quad 36 \le x \le 108\ in \tag{2}$$

$$EI_{AB} = 2(10^6)(166.66) = 333.33(10^6)\ lbs-in^2 \qquad EI_{BC} = 2(10^6)(85.33) = 170.67(10^6)\ lbs-in^2 \tag{3}$$

The maximum bending normal stress in BC will be on a section just right of B i.e. at x = 36 inch and at $y = \pm 4\ in$. The maximum bending normal stress in AB will be at the wall i.e. at x = 0 inch and at $y = \pm 5\ in$

$$\sigma_{BC} = \left|-\frac{\boldsymbol{M}_z y}{\boldsymbol{I}_{zz}}\right| = \frac{(800)(72)(4)}{85.33} = 2700\ psi \qquad \sigma_{AB} = \left|-\frac{\boldsymbol{M}_z y}{\boldsymbol{I}_{zz}}\right| = \frac{(800)(108)(5)}{166.66} = 2592\ psi \tag{4}$$

The boundary value problem is

$$\frac{d^2 v_1}{dx^2} = \frac{\boldsymbol{M}_z}{EI_{AB}} = -2.4(10^{-6})(108-x) \quad 0 \le x \le 36\ in \qquad \frac{d^2 v_2}{dx^2} = \frac{\boldsymbol{M}_z}{EI_{BC}} = -4.6875(10^{-6})(108-x) \quad 36 \le x \le 108\ in \tag{5}$$

$$v_1(0) = 0 \qquad \frac{dv_1}{dx}(0) = 0 \qquad v_1(36) = v_2(36) \qquad \frac{dv_1}{dx}(36) = \frac{dv_2}{dx}(36) \tag{6}$$

Integrating and using boundary and continuity conditions we determine the deflection as shown below.

$$\frac{dv_1}{dx} = -2.4(10^{-6})\left(108x-\frac{x^2}{2}\right)+C_1 \qquad \frac{dv_2}{dx} = -4.6875(10^{-6})\left(108x-\frac{x^2}{2}\right)+C_2 \tag{7}$$

$$v_1 = -2.4(10^{-6})\left(54x^2-\frac{x^3}{6}\right)+C_1x+C_3 \qquad v_2 = -4.6875(10^{-6})\left(54x^2-\frac{x^3}{6}\right)+C_2x+C_4 \tag{8}$$

$$\frac{dv_1}{dx}(0) = C_1 = 0 \qquad v_1(0) = C_3 = 0 \tag{9}$$

$$\frac{dv_1}{dx}(36) = -2.4(10^{-6})\left(108(36)-\frac{(36)^2}{2}\right) = \frac{dv_2}{dx}(36) = -4.6875(10^{-6})\left(108(36)-\frac{(36)^2}{2}\right)+C_2 \quad or \quad C_2 = 7.406(10^{-3}) \tag{10}$$

Substituting x = 36 in Eq.(16) and (17) and equating to satisfy Eq.(10) we obtain,

$$v_1(36) = -2.4(10^{-6})\left(54(36)^2-\frac{(36)^3}{6}\right) = v_2(36) = -4.6875(10^{-6})\left(54x^2-\frac{x^3}{6}\right)+7.406(10^{-3})x+C_4 \text{ or}$$

$$C_4 = -(2.4-4.6875)(10^{-6})(54432)-266.62(10^{-3}) = -142.11(10^{-3}) \tag{11}$$

$$v_{max} = v(108) = -4.6875(10^{-6})(54(108)^2-108^3/6)+7.406(10^{-3})(108)+-142.11(10^{-3}) = -1389.3(10^{-3})\ in$$

ANS $\sigma_{max} = 2.7$ ksi ; $v_{max} = -1.4$ in.

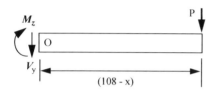

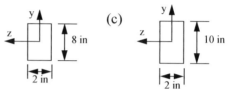

7.30

Solution σ_{max}= ? v_{max}=v(L/2)=?

By symmetry each supports has half the load. By equilibrium of moment about point O, we obtain,

$$R_A = R_B = 800\ lb \qquad \boldsymbol{M_z} = 800x \tag{1}$$

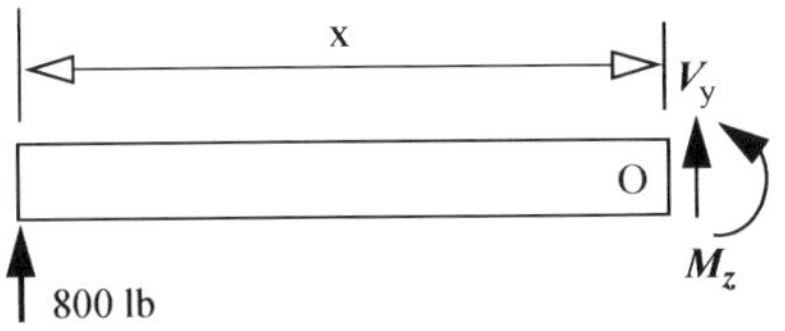

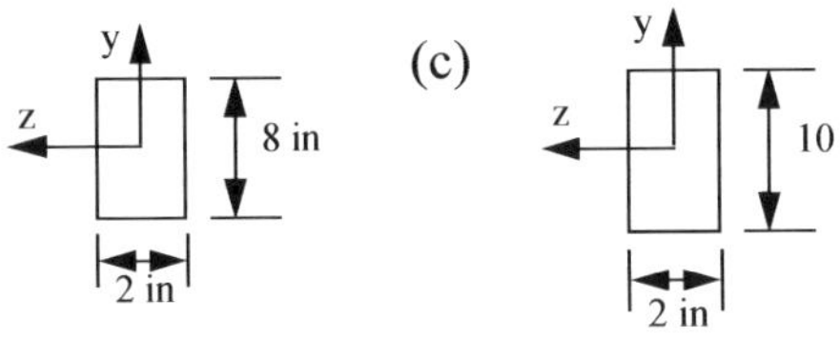

The area moment of inertia and bending rigidity in the segment AB and BC are

$$I_{AB} = \frac{1}{12}(2)(8)^3 = 85.33\ in^4 \quad 0 \le x < 72\ in \qquad I_{BC} = \frac{1}{12}(2)(10)^3 = 166.66\ in^4 \quad 72 < x \le 108\ in \tag{2}$$

$$EI_{AB} = 2(10^6)(85.33) = 170.67(10^6)\ lbs-in^2 \qquad EI_{BC} = 2(10^6)(166.66) = 333.33(10^6)\ lbs-in^2 \tag{3}$$

The maximum bending normal stress in AB will be on a section just left of B, i.e., at x = 72 inch and at $y = \pm 4\ in$. The maximum bending normal stress in BC will be at the wall, i.e., at x = 108 inch and at $y = \pm 5\ in$

$$\sigma_{AB} = \left|-\frac{\boldsymbol{M_z}y}{\boldsymbol{I_{zz}}}\right| = \frac{(800)(72)(4)}{85.33} = 2700\ psi \qquad \sigma_{BC} = \left|-\frac{\boldsymbol{M_z}y}{\boldsymbol{I_{zz}}}\right| = \frac{(800)(108)(5)}{166.66} = 2592\ psi \tag{4}$$

We solve for elastic curve for half the beam as it is symmetric. The boundary value problem is

$$\frac{d^2v_1}{dx^2} = \frac{\boldsymbol{M_z}}{EI_{AB}} = 4.6875(10^{-6})x \quad 0 \le x \le 72\ in \qquad \frac{d^2v_2}{dx^2} = \frac{\boldsymbol{M_z}}{EI_{BC}} = 2.4(10^{-6})x \quad 72 \le x \le 108\ in \tag{5}$$

$$v_1(0) = 0 \qquad v_1(72) = v_2(72) \qquad \frac{dv_1}{dx}(72) = \frac{dv_2}{dx}(72) \qquad \frac{dv_2}{dx}(108) = 0 \tag{6}$$

Integrating and using boundary and continuity conditions we determine the deflection as shown below.

$$\frac{dv_1}{dx} = 4.6875(10^{-6})\frac{x^2}{2} + C_1 \qquad \frac{dv_2}{dx} = 2.4(10^{-6})\frac{x^2}{2} + C_2 \tag{7}$$

$$v_1 = 4.6875(10^{-6})\frac{x^3}{6} + C_1x + C_3 \qquad v_2 = 2.4(10^{-6})\frac{x^3}{6} + C_2x + C_4 \tag{8}$$

$$v_1(0) = C_3 = 0 \qquad \frac{dv_2}{dx}(108) = 2.4(10^{-6})\frac{(108)^2}{2} + C_2 = 0 \quad or \quad C_2 = -0.014 \tag{9}$$

$$\frac{dv_1}{dx}(72) = 4.6875(10^{-6})\frac{(72)^2}{2} + C_1 = \frac{dv_2}{dx}(72) = 2.4(10^{-6})\frac{(72)^2}{2} + C_2 \quad or \quad C_1 = -0.01993 \tag{10}$$

$$v_1(72) = 4.6875(10^{-6})\frac{(72)^3}{6} + C_1(72) + C_3 = v_2(72) = 2.4(10^{-6})\frac{(72)^3}{6} + C_2(72) + C_4 \quad or \quad C_4 = -0.2846 \tag{11}$$

$$v_{max} = v_2(108) = 2.4(10^{-6})\frac{(108)^3}{6} + C_2(108) + C_4 = -1.293\ in$$

ANS $\sigma_{max} = 2.59$ ksi ; $v_{max} = -1.29$ in.

7.31

Solution: E_s = 200 GPa E_{al} = 70 GPa v_{max} = ?

The area moment of inertias and o bending rigidity of the cross-section are as follows.

$$(I_{zz})_s = 2\left[\frac{1}{12}(0.07)(0.01)^3 + (0.07)(0.01)(0.04)^2\right] = 2.252(10^{-6})m^4 \qquad (I_{zz})_{al} = \frac{1}{12}(0.01)(0.07)^3 = 0.2858(10^{-6})m^4 \tag{1}$$

$$\sum E_j(I_{zz})_j = [(200)(2.252) + (70)(0.2858)](10^3) = 470.5(10^3)N-m^2 \tag{2}$$

By equilibrium of moment about point O, we obtain

$$\boldsymbol{M_z} + P(L-x) = 0 \quad or \quad \boldsymbol{M_z} = -P(L-x) \tag{3}$$

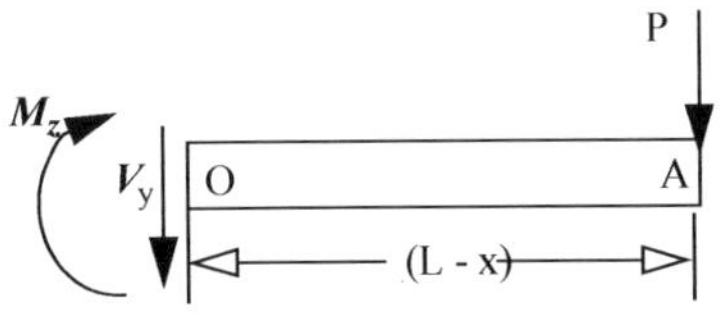

The boundary value problem statement is

$$[\sum E_j(I_{zz})_j]\frac{d^2v}{dx^2} = -PL + Px \qquad v(0) = 0 \qquad \frac{dv}{dx}(0) = 0 \tag{4}$$

Integrating and using boundary conditions we obtain the deflection as shown below.

$$[\sum E_j(I_{zz})_j]\frac{dv}{dx} = -PLx + \frac{Px^2}{2} + C_1 \qquad [\sum E_j(I_{zz})_j]v = \frac{-PLx^2}{2} + \frac{Px^3}{6} + C_1x + C_2 \tag{5}$$

$$[\sum E_j(I_{zz})_j]\frac{dv}{dx}(0) = C_1 = 0 \qquad \sum E_j(I_{zz})_j v(0) = C_2 = 0 \tag{6}$$

$$v(x) = P\frac{-3Lx^2 + x^3}{6[\sum E_j(I_{zz})_j]} \qquad v_{max} = v(L) = \frac{PL^2}{6[\sum E_j(I_{zz})_j]}(-2L) = -\frac{PL^3}{3[\sum E_j(I_{zz})_j]} = -\frac{(5)(10^3)(2)^3}{3(470.5)(10^3)} = 0.02834m \tag{7}$$

ANS $v_{max} = -28.3$ mm

Section 7.2

7. 32

Solution $v(x) = (20x^3 - 40x^2)(10)^{-6}$ *in* EI = $135(10^6)$ lb $R_A = ?$ $M_A = ?$

By equilibrium of forces and the moment we obtain,

$$R_A = -\boldsymbol{V}_y(0) \qquad M_A = \boldsymbol{M}_z(0) \tag{1}$$

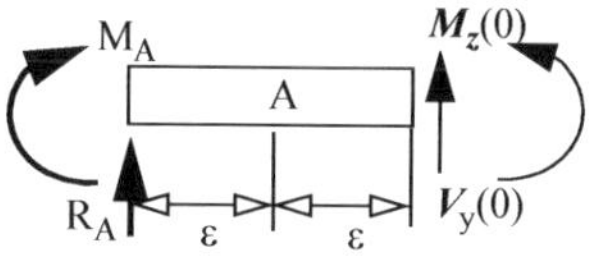

The moment and shear force in AB can be found as shown

$$\boldsymbol{M}_z(x) = EI\frac{d^2v}{dx^2} = EI(120x - 80)(10^{-6}) \qquad \boldsymbol{V}_y(x) = -\frac{d\boldsymbol{M}_z}{dx} = -120EI(10^{-6}) \tag{2}$$

$$R_A = -\boldsymbol{V}_y(0) = -120EI(10^{-6}) = -(120)(135)(10^6)(10^{-6}) = -16200\ lb \tag{3}$$

$$M_A = \boldsymbol{M}_z(0) = -80EI(10^{-6}) = -(80)(135)(10^6)(10^{-6}) = -10800 in\ lb \tag{4}$$

ANS R_A = 16.2 kips up; ; M_A = 10.8 in.-kips CCW

---

7. 33

Solution $v_1(x) = -3(x^4 - 20x^3)(10^{-6})$ *in* $v_2(x) = -8(x^2 - 100x + 1600)(10^{-3})$ *in* EI = $135(10^6)$ lb-10^2 $M_B = ?$ $R_B = ?$

By equilibrium of forces and the moment, we obtain,

$$R_B = \boldsymbol{V}_1(20) - \boldsymbol{V}_2(20) \qquad or \qquad M_B = \boldsymbol{M}_1(20) - \boldsymbol{M}_2(20) \tag{1}$$

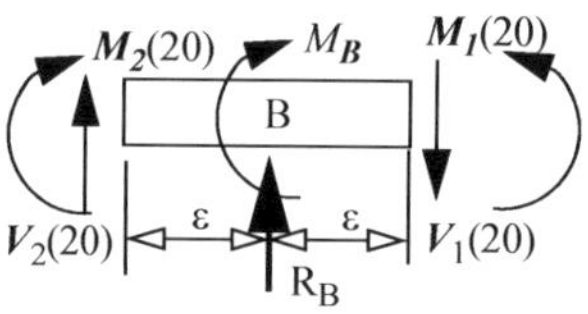

The moment and shear force in AB and BC can be found as shown below.

$$\boldsymbol{M}_1(x) = EI\frac{d^2v_1}{dx^2} = EI[-3(12x^2 - 120x)(10^{-6})] \qquad \boldsymbol{V}_1(x) = -\frac{d}{dx}(\boldsymbol{M}_1) = EI[-3(24x - 120)(10^{-6})] \tag{2}$$

$$\boldsymbol{M}_2(x) = EI\frac{d^2v_2}{dx^2} = EI[-8(2)](10^{-3}) = -16EI(10^{-3}) \qquad \boldsymbol{V}_2(x) = -\frac{d\boldsymbol{M}_2}{dx} = 0 \tag{3}$$

$$R_B = \boldsymbol{V}_1(20) - \boldsymbol{V}_2(20) = EI[-3(24(20) - 120)] = -(1080(120))(135)(10^6)(10^{-6}) = 145.8(10^3)\ lb \tag{4}$$

$$M_B = \boldsymbol{M}_1(20) - \boldsymbol{M}_2(20) = EI[-3(12(20)^2 - 120(20))(10^{-6})] - [-16EI(10^{-3})] = (-7.2 + 16)EI(10^{-3}) = 1180(10^3)\ in\ lb \tag{5}$$

ANS R_B = 145.8 kips up ; M_B = 1180 in.-kips

---

7. 34

Solution v(x) = ? R_A = ?

By equilibrium of moment about points O_1 and O_2 we obtain

$$M_1 + P(2L-x) - R_A(L-x) = 0 \quad or \quad M_1 = P(x-2L) - R_A(x-L) \tag{1}$$

$$M_2 + P(2L-x) = 0 \quad or \quad M_2 = P(x-2L) \tag{2}$$

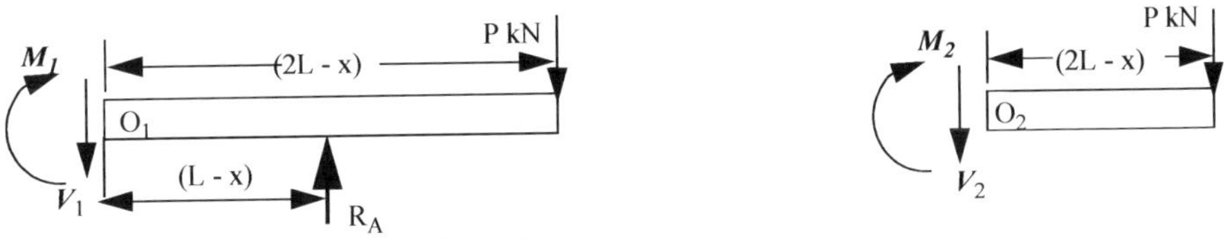

The boundary value problem statement can be written as

$$EI\frac{d^2 v_1}{dx^2} = P(x-2L) - R_A(x-L) \qquad EI\frac{d^2 v_2}{dx^2} = P(x-2L) \tag{3}$$

$$v_1(0) = 0 \qquad \frac{dv_1}{dx}(0) = 0 \qquad v_1(L) = 0 \qquad v_2(L) = 0 \qquad \frac{dv_1}{dx}(L) = \frac{dv_2}{dx}(L) \tag{4}$$

Integrating and using all the conditions we obtain the displacement as shown below.

$$EI\frac{dv_1}{dx} = \frac{P}{2}(x-2L)^2 - \frac{R_A}{2}(x-L)^2 + C_1 \qquad EI\frac{dv_2}{dx} = \frac{P}{2}(x-2L)^2 + C_2 \tag{5}$$

$$EIv_1 = \frac{P}{6}(x-2L)^3 - \frac{R_A}{6}(x-L)^3 + C_1x + C_3 \qquad EIv_2 = \frac{P}{6}(x-2L)^3 + C_1x + C_4 \tag{6}$$

$$EI\frac{dv_1}{dx}(0) = 2PL^2 - \frac{R_A}{2}L^2 + C_1 = 0 \quad or \quad C_1 = \frac{(R_A - 4P)}{2}L^2 \tag{7}$$

$$\frac{dv_1}{dx}(L) = \frac{PL^2}{2} + C_1 = \frac{dv_2}{dx}(L) = \frac{PL^2}{2} + C_2 \quad or \quad C_1 = C_2 \tag{8}$$

$$EIv_1(0) = -\frac{8P}{6}L^3 - \frac{R_A}{6}L^3 + C_3 \quad or \quad C_3 = \frac{(8P - R_A)}{6}L^3 \tag{9}$$

$$EIv_1(L) = \frac{-P}{6}L^3 + C_1L + C_3 = -\frac{P}{6}L^3 + \frac{(R_A - 4P)}{2}L^3 + \frac{(8P - R_A)}{6}L^3 = 0 \quad or \quad R_A = \frac{5P}{2} \tag{10}$$

$$C_1 = C_2 = \frac{(R_A - 4P)}{2}L^2 = -\frac{3}{4}PL^2 \qquad C_3 = \frac{(8P - R_A)}{6}L^3 = \frac{11}{12}PL^3 \tag{11}$$

$$EIv_2(L) = \frac{-P}{6}L^3 + C_1L + C_4 = 0 \quad or \quad \frac{-P}{6}L^3 + \left(\frac{-3P}{4}L^2\right)L + C_4 = 0 \quad or \quad C_4 = \frac{11}{12}PL^3 \tag{12}$$

$$v_1 = \frac{P}{12EI}[2(x-2L)^3 - 5(x-L)^3 - 9L^2x + 11L^3] \qquad v_2 = \frac{P}{12EI}[2(x-2L)^3 - 9L^2x + 11L^3] \tag{13}$$

$$\textbf{ANS} \quad R_A = \frac{5P}{2}; \; v(x) = \begin{cases} \frac{P}{12EI}[2(x-2L)^3 - 5(x-L)^3 - 9L^2x + 11L^3] & 0 \le x \le L \\ \frac{P}{12EI}[2(x-2L)^3 - 9L^2x + 11L^3] & L \le x \le 2L \end{cases}$$

---

7. 35

Solution v(x) = ? R_A = ? M_A = ?

The boundary value problem is

$$\frac{d^2}{dx^2}\left(EI\frac{d^2 v}{dx^2}\right) = -w \qquad v(0) = 0 \qquad \frac{dv}{dx}(0) = 0 \qquad v(L) = 0 \qquad \frac{dv}{dx}(L) = 0 \tag{1}$$

Integrating and using the boundary conditions we obtain the displacement

$$\frac{d}{dx}\left(EI\frac{d^2 v}{dx^2}\right) = -wx + C_1 \qquad EI\frac{d^2 v}{dx^2} = -\frac{wx^2}{2} + C_1x + C_2 \tag{2}$$

$$EI\frac{dv}{dx} = -\frac{wx^3}{6} + \frac{C_1x^2}{2} + C_2x + C_3 \qquad EIv = -\frac{wx^4}{24} + \frac{C_1x^3}{6} + \frac{C_2x^2}{2} + C_3x + C_4 \tag{3}$$

$$EI\frac{dv}{dx}(0) = C_3 = 0 \qquad EIv(0) = C_4 = 0 \tag{4}$$

$$EI\frac{dv}{dx}(L) = -\frac{wL^3}{6} + \frac{C_1L^2}{2} + C_2L \quad or \quad \frac{C_1}{2}L + C_2 = \frac{wL^2}{6} \tag{5}$$

$$EIv(L) = -\frac{wL^4}{24} + \frac{C_1L^3}{6} + \frac{C_2L^2}{2} = 0 \quad or \quad \frac{C_1}{6}L + \frac{C_2}{2} = \frac{wL^2}{24} \tag{6}$$

$$C_1 = \frac{wL}{2} \qquad C_2 = -\frac{wL^2}{12} \qquad v = \frac{w}{EI}\frac{(-x^4 + 2x^3L - L^2x^2)}{24} = -\frac{wx^2}{24EI}(x^2 - 2xL - L^2) = -\frac{wx^2}{24EI}(x-L)^2 \tag{7}$$

The moment and shear force can be found as shown below.

$$\boldsymbol{M}_z = EI\frac{d^2v}{dx^2} = \frac{w}{24}(-12x^2 + 12xL - 2L^2) \qquad \boldsymbol{M}_z(L) = -\frac{wL^2}{12} \qquad \boldsymbol{V}_y = -\frac{d\boldsymbol{M}_z}{dx} = -\frac{w}{24}(-24x + 12L) \qquad \boldsymbol{V}_y(L) = \frac{wL}{2} \tag{8}$$

By equilibrium of forces and the moment we obtain,

$$R_A = \boldsymbol{V}_y(L) = wL/2 \qquad M_A = -\boldsymbol{M}_z(L) = wL^2/12 \tag{9}$$

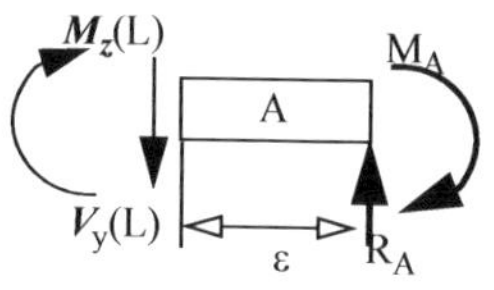

ANS $v(x) = -\frac{wx^2}{24EI}(x-L)^2$; $R_A = \frac{wL}{2}$ up ; $M_A = \frac{wL^2}{12}$ CW

---

7. 36

Solution $\frac{dv}{dx}(L) = ??$ $\quad R_A = ?$ $\quad M_A = ?$

The boundary value problem can be written as:

$$\frac{d^2}{dx^2}\left(EI\frac{d^2v}{dx^2}\right) = -w\left(1 - \frac{x^2}{L^2}\right) \qquad v(0) = 0 \qquad \frac{dv}{dx}(0) = 0 \qquad v(L) = 0 \qquad EI\frac{d^2v}{dx^2}(L) = 0 \tag{1}$$

Integrating and using the boundary conditions we obtain the displacement.

$$\frac{d}{dx}\left(EI\frac{d^2v}{dx^2}\right) = -w\left(x - \frac{x^3}{3L^2}\right) + C_1 \qquad EI\frac{d^2v}{dx^2} = -w\left(\frac{x^2}{2} - \frac{x^4}{12L^2}\right) + C_1x + C_2 \tag{2}$$

$$EI\frac{dv}{dx} = -w\left(\frac{x^3}{6} - \frac{x^5}{60L^2}\right) + \frac{C_1x^2}{2} + C_2x + C_3 \qquad EIv = -w\left(\frac{x^4}{24} - \frac{x^6}{360L^2}\right) + \frac{C_1x^3}{6} + \frac{C_2x^2}{2} + C_3x + C_4 \tag{3}$$

$$EI\frac{dv}{dx}(0) = C_3 = 0 \qquad EIv(0) = C_4 = 0 \tag{4}$$

$$EI\frac{d^2v}{dx^2}(L) = -w\left(\frac{L^2}{2} - \frac{L^2}{12}\right) + C_1L + C_2 \quad or \quad C_1L + C_2 = \frac{5wL^2}{12} \tag{5}$$

$$EIv(L) = -w\left(\frac{L^4}{24} - \frac{L^4}{360}\right) + \frac{C_1L^3}{6} + \frac{C_2L^2}{2} = 0 \quad or \quad \frac{C_1}{6}L + \frac{C_2}{2} = \frac{7wL^2}{180} \tag{6}$$

$$C_1 = \frac{61wL}{120} \quad and \quad C_2 = -\frac{11wL^2}{120} \qquad EI\frac{dv}{dx}(L) = -w\left(\frac{L^3}{6} - \frac{L^3}{60}\right) + \frac{61wL^3}{240} - \frac{11wL^3}{120} = \frac{wL^3}{80} \tag{7}$$

By equilibrium of forces and the moment, we obtain,

$$R_A = -\boldsymbol{V}_y(0) = \left.\frac{d}{dx}\left(EI\frac{d^2v}{dx^2}(L)\right)\right|_{x=0} = C_1 = \frac{61wL}{120} \qquad M_A = -\boldsymbol{M}_z(0) = \left.-EI\frac{d^2v}{dx^2}(L)\right|_{x=0} = -C_2 = \frac{11wL^2}{120} \tag{8}$$

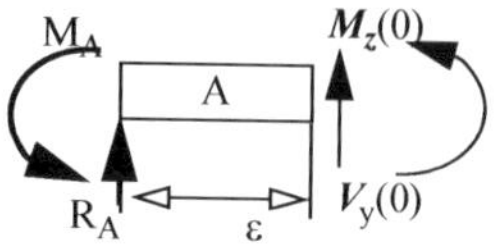

ANS $\frac{dv}{dx}(L) = \frac{wL^3}{80EI}$; $R_A = \frac{61wL}{120}$ up; ; $M_A = \frac{11wL^2}{120}$ CW

---

7. 37

Solution v(L) = ? M_B = ?

The boundary value problem is

$$\frac{d^2}{dx^2}\left(EI\frac{d^2v}{dx^2}\right) = -\frac{wx}{L} \qquad v(0) = 0 \qquad \frac{dv}{dx}(0) = 0 \qquad \frac{dv}{dx}(L) = 0 \qquad V_y(L) = -\frac{d}{dx}\left(EI\frac{d^2}{dx^2}v(L)\right)\Bigg|_{x=L} = 0 \tag{1}$$

Integrating and using the boundary conditions we obtain the displacement.

$$\frac{d}{dx}\left(EI\frac{d^2v}{dx^2}\right) = -\frac{wx^2}{2L} + C_1 \qquad EI\frac{d^2v}{dx^2} = -\frac{wx^3}{6L} + C_1x + C_2 \tag{2}$$

$$EI\frac{dv}{dx} = -\frac{wx^4}{24L} + \frac{C_1x^2}{2} + C_2x + C_3 \qquad EIv = -\frac{wx^5}{120L} + \frac{C_1x^3}{6} + \frac{C_2x^2}{2} + C_3x + C_4 \tag{3}$$

$$V_y(L) = -\left[-\frac{wL}{2} + C_1\right] = 0 \qquad or \qquad C_1 = \frac{wL}{2} \tag{4}$$

$$EI\frac{dv}{dx}(0) = C_3 = 0 \qquad EIv(0) = C_4 = 0 \tag{5}$$

$$EI\frac{dv}{dx}(L) = -\frac{wL^3}{24} + \left(\frac{wL}{2}\right)\frac{L^2}{2} + C_2L = wL^3\frac{(-1+6)}{24} + C_2L = 0 \qquad or \qquad C_2 = -\frac{5}{24}wL^2 \tag{6}$$

$$v = \frac{w}{240EIL}[-2x^5 + 20L^2x^3 - 25L^3x^2] \qquad v(L) = \frac{w(L^5)}{240EIL}[-2+20-25] = \frac{-7wL^4}{240EI} \tag{7}$$

By equilibrium of moment we obtain:

$$M_B = -\boldsymbol{M}_z(L) = -EI\frac{d^2v}{dx^2}(L) = \frac{-w}{240EIL}[-40L^3 + 120L^3 - 50L^3] = \frac{-30wL^3}{240EIL} \tag{8}$$

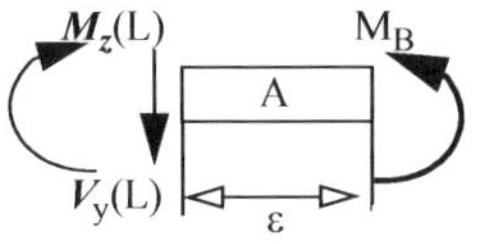

ANS $v = \dfrac{-7wL^4}{240EI}$; $M_B = \dfrac{wL^2}{8EI}$ CW

---

7. 38

Solution v(L) = ? $\frac{dv}{dx}(L) = ?$

By equilibrium of forces and the moment, we obtain

$$V_y(L) = wL \qquad \boldsymbol{M}_z(L) = -wL^2 \tag{1}$$

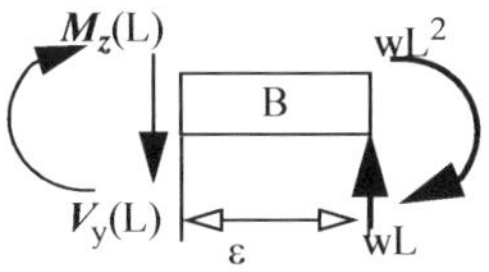

The boundary value problem is,

$$\frac{d^2}{dx^2}\left(EI\frac{d^2v}{dx^2}\right) = -\frac{wx^2}{L^2} \qquad v(0) = 0 \qquad \frac{dv}{dx}(0) = 0 \qquad \boldsymbol{M}_z(L) = EI\frac{d^2v}{dx^2}(L) = -wL^2 \qquad V_y(L) = -\frac{d}{dx}\left(EI\frac{d^2v}{dx^2}\right)\Bigg|_{x=L} = wL \tag{2}$$

Integrating and using the boundary conditions we obtain the displacement.

$$\frac{d}{dx}\left(EI\frac{d^2v}{dx^2}\right) = -\frac{wx^3}{3L^2} + C_1 \qquad EI\frac{d^2v}{dx^2} = -\frac{wx^4}{12L^2} + C_1x + C_2 \tag{3}$$

$$EI\frac{dv}{dx} = -\frac{wx^5}{60L^2} + C_1\frac{x^2}{2} + C_2x + C_3 \qquad EIv = -\frac{wx^6}{360L^2} + C_1\frac{x^3}{6} + C_2\frac{x^2}{2} + C_3x + C_4 \tag{4}$$

$$V_y(L) = -\left[-\frac{wL}{3} + C_1\right] = wL \qquad or \qquad C_1 = \frac{-2wL}{3} \tag{5}$$

$$\boldsymbol{M}_z(L) = -\frac{wL^2}{12} + C_1L + C_2 = -\frac{wL^2}{12} - \frac{2wL^2}{3} + C_2 = -wL^2 \qquad or \qquad C_2 = \frac{-wL^2}{4} \tag{6}$$

$$EI\frac{dv}{dx}(0) = C_3 = 0 \qquad EIv(0) = C_4 = 0 \tag{7}$$

$$EI\frac{dv}{dx}(L) = -\frac{wL^3}{60} + \frac{(-2wL)L^2}{6} + \frac{(-wL^2)L}{4} = \frac{wL^3}{60}(-1-20-15) = -\frac{3wL^3}{5} \tag{8}$$

$$EIv(L) = -\frac{wL^4}{360} + \left(\frac{-2wL}{3}\right)\frac{L^3}{6} + \left(\frac{-wL^2}{4}\right)\frac{L^2}{2} = \frac{wL^4}{360}(-1-40-45) = -\frac{43wL^4}{180} \tag{9}$$

ANS $\frac{dv}{dx}(L) = -\frac{3wL^3}{5EI}$; $v(L) = -\frac{43wL^4}{180EI}$

---

7. 39

Solution $v(L) = ?$ $\frac{dv}{dx}(L) = ?$

The boundary value problem can be written as:

$$\frac{d^2}{dx^2}\left(EI\frac{d^2v}{dx^2}\right) = -w\cos\frac{\pi x}{2L} \qquad v(0) = 0 \qquad \frac{dv}{dx}(0) = 0 \qquad \left.\frac{d}{dx}\left(EI\frac{d^2v}{dx^2}\right)\right|_{x=L} = 0 \qquad EI\frac{d^2v}{dx^2}(L) = 0 \tag{1}$$

Integrating and using the boundary conditions we obtain the displacement.

$$\frac{d}{dx}\left(EI\frac{d^2v}{dx^2}\right) = -w\frac{2L}{\pi}\sin\frac{\pi x}{2L} + C_1 \qquad EI\frac{d^2v}{dx^2} = w\left(\frac{2L}{\pi}\right)^2\cos\frac{\pi x}{2L} + C_1x + C_2 \tag{2}$$

$$EI\frac{dv}{dx} = w\left(\frac{2L}{\pi}\right)^3\sin\frac{\pi x}{2L} + \frac{C_1x^2}{2} + C_2x + C_3 \qquad EIv = -w\left(\frac{2L}{\pi}\right)^4\cos\frac{\pi x}{2L} + \frac{C_1x^3}{6} + \frac{C_2x^2}{2} + C_3x + C_4 \tag{3}$$

$$EI\frac{dv}{dx}(0) = C_3 = 0 \qquad EIv(0) = -w\left(\frac{2L}{\pi}\right)^4 + C_4 = 0 \quad or \quad C_4 = w\left(\frac{2L}{\pi}\right)^4 = 16w\left(\frac{L}{\pi}\right)^4 \tag{4}$$

$$\left.\frac{d}{dx}\left(EI\frac{d^2v}{dx^2}\right)\right|_{x=L} = -w\frac{2L}{\pi}\sin\frac{\pi}{2} + C_1 = 0 \quad or \quad C_1 = \frac{2wL}{\pi} \tag{5}$$

$$EI\frac{d^2v}{dx^2}(L) = w\left(\frac{2L}{\pi}\right)^2\cos\frac{\pi}{2} + C_1L + C_2 = 0 \quad or \quad C_2 = -C_1L = -\left(\frac{2wL^2}{\pi}\right) \tag{6}$$

$$EIv(L) = -w\left(\frac{2L}{\pi}\right)^4\cos\frac{\pi}{2} + \frac{C_1L^3}{6} + \frac{C_2L^2}{2} + C_3L + C_4 = \frac{2wL^4}{3\pi} - \frac{2wL^4}{\pi} + 16w\left(\frac{L}{\pi}\right)^4 = -1.109wL^4 \tag{7}$$

$$EI\frac{dv}{dx}(L) = w\left(\frac{2L}{\pi}\right)^3\sin\frac{\pi}{2} + \frac{C_1L^2}{2} + C_2L + C_3 = 8w\left(\frac{L}{\pi}\right)^3 + \frac{wL^3}{\pi} - \frac{2wL^3}{\pi} = -0.0603wL^3 \tag{8}$$

ANS $v(L) = -1.109\frac{wL^4}{EI}$; $\frac{dv}{dx}(L) = -0.0603\frac{wL^3}{EI}$

---

7. 40

Solution $\frac{dv}{dx}(L) = ?$ $R_A = ?$ $M_A = ?$

The boundary value problem can be written as:

$$\frac{d^2}{dx^2}\left(EI\frac{d^2v}{dx^2}\right) = -w\cos\frac{\pi x}{2L} \qquad v(0) = 0 \qquad \frac{dv}{dx}(0) = 0 \qquad v(L) = 0 \qquad EI\frac{d^2v}{dx^2}(L) = 0 \tag{1}$$

Integrating and using the boundary conditions we obtain the displacement.

$$\frac{d}{dx}\left(EI\frac{d^2v}{dx^2}\right) = -w\frac{2L}{\pi}\sin\frac{\pi x}{2L} + C_1 \qquad EI\frac{d^2v}{dx^2} = w\left(\frac{2L}{\pi}\right)^2\cos\frac{\pi x}{2L} + C_1x + C_2 \tag{2}$$

$$EI\frac{dv}{dx} = w\left(\frac{2L}{\pi}\right)^3\sin\frac{\pi x}{2L} + \frac{C_1x^2}{2} + C_2x + C_3 \qquad EIv = -w\left(\frac{2L}{\pi}\right)^4\cos\frac{\pi x}{2L} + \frac{C_1x^3}{6} + \frac{C_2x^2}{2} + C_3x + C_4 \tag{3}$$

$$EI\frac{dv}{dx}(0) = C_3 = 0 \qquad EIv(0) = -w\left(\frac{2L}{\pi}\right)^4 + C_4 = 0 \quad or \quad C_4 = w\left(\frac{2L}{\pi}\right)^4 = \frac{16wL^4}{\pi^4} \tag{4}$$

$$v(L) = \frac{C_1 L^3}{6} + \frac{C_2 L^2}{2} + C_4 = 0 \qquad or \qquad C_1 L + 3C_2 = -\left(\frac{96 w L^2}{\pi^4}\right) \tag{5}$$

$$EI\frac{d^2v}{dx^2}(L) = w\left(\frac{2L}{\pi}\right)^2 cos\frac{\pi}{2} + C_1 L + C_2 = 0 \qquad or \qquad C_2 = -C_1 L \tag{6}$$

$$C_1 = \frac{48wL}{\pi^4} \qquad C_2 = -\frac{48wL^2}{\pi^4} \tag{7}$$

$$EI\frac{dv}{dx}(L) = w\left(\frac{2L}{\pi}\right)^3 sin\frac{\pi}{2} + \frac{C_1 L^2}{2} + C_2 L + C_3 = \frac{8wL^3}{\pi^3} + \frac{24wL^3}{\pi^4} - \frac{48wL^3}{\pi^4} = -0.0116wL^3 \tag{8}$$

By equilibrium of forces and the moment, we obtain,

$$R_A = -V_y(0) = \left.\frac{d}{dx}\left(EI\frac{d^2v}{dx^2}(L)\right)\right|_{x=0} = C_1 = \frac{48wL}{\pi^4} = 0.4928wL \tag{9}$$

$$M_A = -M_z(0) = \left.-EI\frac{d^2v}{dx^2}(L)\right|_{x=0} = -\left[w\left(\frac{2L}{\pi}\right)^2 + C_2\right] = -\left[\frac{4wL^2}{\pi^2} - \frac{48wL^2}{\pi^4}\right] = -0.0875wL^2 \tag{10}$$

(a) 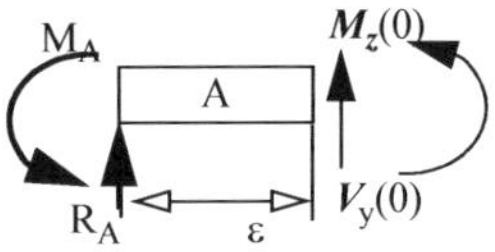

ANS $\frac{dv}{dx}(L) = -0.0116\frac{wL^3}{EI}$; $R_A = 0.4928wL$ up ; $M_A = 0.0875wL^2$ CW

---

7. 41

Solution $v(L/2)$**=?**

The boundary value problem can be written as:

$$\frac{d^2}{dx^2}\left(EI\frac{d^2v}{dx^2}\right) = -w\,sin\frac{\pi x}{L} \qquad v(0) = 0 \qquad EI\frac{d^2v}{dx^2}(0) = 0 \qquad v(L) = 0 \qquad EI\frac{d^2v}{dx^2}(L) = 0 \tag{1}$$

Integrating and using the boundary conditions we obtain the displacement.

$$\frac{d}{dx}\left(EI\frac{d^2v}{dx^2}\right) = w\frac{L}{\pi}cos\frac{\pi x}{L} + C_1 \qquad EI\frac{d^2v}{dx^2} = w\left(\frac{L}{\pi}\right)^2 sin\frac{\pi x}{L} + C_1 x + C_2 \tag{2}$$

$$EI\frac{dv}{dx} = -w\left(\frac{L}{\pi}\right)^3 cos\frac{\pi x}{L} + \frac{C_1 x^2}{2} + C_2 x + C_3 \qquad EIv = -w\left(\frac{L}{\pi}\right)^4 sin\frac{\pi x}{L} + \frac{C_1 x^3}{6} + \frac{C_2 x^2}{2} + C_3 x + C_4 \tag{3}$$

$$EI\frac{dv}{dx}(0) = -w\left(\frac{L}{\pi}\right)^3 + C_3 = 0 \qquad or \qquad C_3 = \frac{wL^3}{\pi^3} \qquad EIv(0) = C_4 = 0 \qquad or \qquad C_4 = 0 \tag{4}$$

$$v(L) = \frac{C_1 L^3}{6} + \frac{C_2 L^2}{2} + C_3 L + C_4 = 0 \qquad or \qquad C_1 L + 3C_2 = -\left(\frac{6wL^2}{\pi^3}\right) \tag{5}$$

$$EI\frac{dv}{dx}(L) = -w\left(\frac{L}{\pi}\right)^3 cos\,\pi + \frac{C_1 L^2}{2} + C_2 L + C_3 = 0 \qquad or \qquad C_1 L + 2C_2 = \frac{4wL^2}{\pi^3} \tag{6}$$

$$C_1 = \frac{24wL}{\pi^3} \qquad C_2 = -\frac{10wL^2}{\pi^3} \tag{7}$$

$$EIv(L/2) = -w\left(\frac{L}{\pi}\right)^4 sin\frac{\pi}{2} + \frac{C_1(L/2)^3}{6} + \frac{C_2(L/2)^2}{2} + C_3(L/2) + C_4 = -\left(\frac{wL^4}{\pi^4}\right) + \frac{wL^4}{2\pi^3} - \frac{5wL^4}{4\pi^3} + \frac{wL^4}{2\pi^3} \tag{8}$$

ANS $v(L/2) = -0.01833\left(\frac{wL^4}{EI}\right)$

---

7. 42

Solution $v(L)$**=?**

The boundary value problem can be written as:

$$\frac{d^2}{dx^2}\left(EI\frac{d^2v}{dx^2}\right) = -w\sin\frac{\pi x}{L} \qquad v(0) = 0 \qquad \frac{dv}{dx}(0) = 0 \qquad \left.\frac{d}{dx}\left(EI\frac{d^2v}{dx^2}\right)\right|_{x=L} = 0 \qquad EI\frac{d^2v}{dx^2}(L) = 0 \tag{1}$$

Integrating and using the boundary conditions we obtain the displacement.

$$\frac{d}{dx}\left(EI\frac{d^2v}{dx^2}\right) = w\frac{L}{\pi}\cos\frac{\pi x}{L} + C_1 \qquad EI\frac{d^2v}{dx^2} = w\left(\frac{L}{\pi}\right)^2\sin\frac{\pi x}{L} + C_1x + C_2 \tag{2}$$

$$EI\frac{dv}{dx} = -w\left(\frac{L}{\pi}\right)^3\cos\frac{\pi x}{L} + \frac{C_1x^2}{2} + C_2x + C_3 \qquad EIv = -w\left(\frac{L}{\pi}\right)^4\sin\frac{\pi x}{L} + \frac{C_1x^3}{6} + \frac{C_2x^2}{2} + C_3x + C_4 \tag{3}$$

$$EI\frac{dv}{dx}(0) = -w\left(\frac{L}{\pi}\right)^3 + C_3 = 0 \quad or \quad C_3 = \frac{wL^3}{\pi^3} \qquad EIv(0) = C_4 = 0 \quad or \quad C_4 = 0 \tag{4}$$

$$\frac{d}{dx}\left(EI\frac{d^2v}{dx^2}\right) = w\frac{L}{\pi}\cos\pi + C_1 = 0 \quad or \quad C_1 = \frac{wL}{\pi} \tag{5}$$

$$EI\frac{d^2v}{dx^2} = w\left(\frac{L}{\pi}\right)^2\sin\pi + C_1L + C_2 = 0 \quad or \quad C_2 = -C_1L = -\left(\frac{wL^2}{\pi}\right) \tag{6}$$

$$EIv(L) = -w\left(\frac{L}{\pi}\right)^4\sin\pi + \frac{C_1L^3}{6} + \frac{C_2L^2}{2} + C_3L + C_4 = \left(\frac{wL^4}{6\pi}\right) - \frac{wL^4}{2\pi} + \frac{wL^4}{\pi^3} \tag{7}$$

ANS $v(L) = -0.0738\left(\frac{wL^4}{EI}\right)$

---

7. 43

Solution $\delta = ?$

Figure below shows the free body diagram of an infinitesimal element of the free end. By equilibrium we obtain:

$$\boldsymbol{M}_z(L) = 0 \qquad \boldsymbol{V}_y(L) = -kv(L) \tag{1}$$

(a)

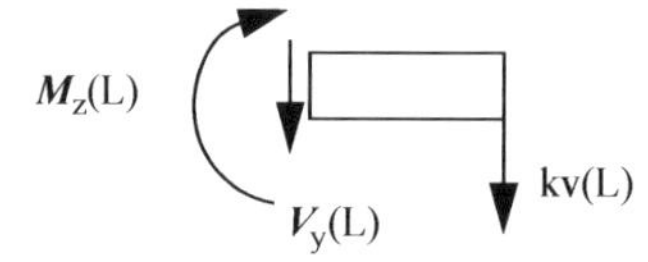

The boundary value problem can be written as:

$$\frac{d^2}{dx^2}\left(EI\frac{d^2v}{dx^2}\right) = -w \qquad v(0) = 0 \qquad \frac{dv}{dx}(0) = 0 \qquad -\left.\frac{d}{dx}\left(EI\frac{d^2v}{dx^2}\right)\right|_{x=L} = -\alpha\frac{EI}{L^3}v(L) \qquad EI\frac{d^2v}{dx^2}(L) = 0 \tag{2}$$

Integrating and using the boundary conditions we obtain the displacement.

$$\frac{d}{dx}\left(EI\frac{d^2v}{dx^2}\right) = -wx + C_1 \qquad EI\frac{d^2v}{dx^2} = -\frac{wx^2}{2} + C_1x + C_2 \tag{3}$$

$$EI\frac{dv}{dx} = -\frac{wx^3}{6} + \frac{C_1x^2}{2} + C_2x + C_3 \qquad EIv = -\frac{wx^4}{24} + \frac{C_1x^3}{6} + \frac{C_2x^2}{2} + C_3x + C_4 \tag{4}$$

$$EI\frac{dv}{dx}(0) = C_3 = 0 \qquad EIv(0) = C_4 = 0 \tag{5}$$

$$\left.\frac{d}{dx}\left(EI\frac{d^2v}{dx^2}\right)\right|_{x=L} = -wL + C_1 = \alpha\frac{EI}{L^3}v(L) \quad or \quad C_1 = wL + \alpha\frac{EI}{L^3}v(L) \tag{6}$$

$$EI\frac{d^2v}{dx^2}(L) = -\frac{wL^2}{2} + C_1L + C_2 = 0 \quad or \quad C_2 = -\frac{wL^2}{2} - \alpha\frac{EI}{L^2}v(L) \tag{7}$$

$$EIv(L) = -\frac{wL^4}{24} + \frac{C_1L^3}{6} + \frac{C_2L^2}{2} + C_3L + C_4 = -\frac{wL^4}{24} + \frac{(wL + \alpha[EI/L^3]v(L))L^3}{6} + \frac{(-wL^2/2 - \alpha[EI/L^2]v(L))L^2}{2} \quad \text{or}$$

$$EIv(L)\left[1 - \frac{\alpha}{6} + \frac{\alpha}{2}\right] = (wL^4)\left[-\frac{1}{24} + \frac{1}{6} - \frac{1}{4}\right] \quad or \quad EIv(L)\left[1 + \frac{\alpha}{3}\right] = (wL^4)\left[\frac{-1 + 4 - 6}{24}\right] = -\frac{wL^4}{8} \tag{8}$$

$$\delta = v(L) = -\frac{3}{8(3+\alpha)}\left(\frac{wL^4}{EI}\right) \qquad (9)$$

ANS $\delta = \frac{3}{8(3+\alpha)}\left(\frac{wL^4}{EI}\right)$

---

7. 44

Solution Boundary Value Problem

By equilibrium of forces and the moment , we obtain,

$$\boldsymbol{V}_2(L) - \boldsymbol{V}_1(L) + wL = 0 \quad or \quad \boldsymbol{V}_2(L) - \boldsymbol{V}_1(L) = -wL \qquad \boldsymbol{M}_2(L) - \boldsymbol{M}_1(L) - wL^2 = 0 \quad or \quad \boldsymbol{M}_2(L) - \boldsymbol{M}_1(L) = wL^2 \qquad (1)$$

$$\boldsymbol{V}_3(3L) = -Kv_3(3L) \qquad \boldsymbol{M}_3(3L) = 0 \qquad (2)$$

(a) 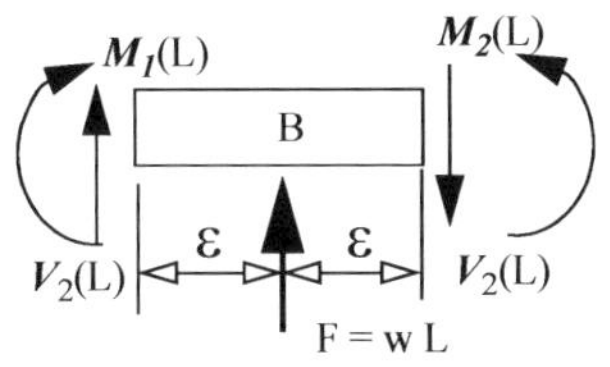

(b) 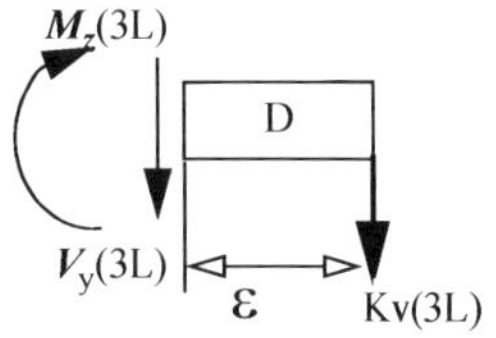

The boundary value problem is

$$\frac{d^2}{dx^2}\left(EI\frac{d^2v_1}{dx^2}\right) = -w\sin\frac{\pi x}{2L} \quad 0 \le x < L \qquad \frac{d^2}{dx^2}\left(EI\frac{d^2v_2}{dx^2}\right) = 0 \quad L < x < 2L \qquad \frac{d^2}{dx^2}\left(EI\frac{d^2v_3}{dx^2}\right) = -w \quad 2L < x \le 3L \qquad (3)$$

$$v_1(0) = 0 \qquad \frac{dv_1}{dx}(0) = 0 \qquad v_1(L) = v_2(L) \qquad \frac{dv_1}{dx}(L) = \frac{dv_2}{dx}(L) \qquad EI\frac{d^2v_2}{dx^2}(L) - EI\frac{d^2v_1}{dx^2}(L) = wL^2 \qquad (4)$$

$$-\frac{d}{dx}\left(EI\frac{d^2}{dx^2}v_2(L)\right)\bigg|_{x=L} + \frac{d}{dx}\left(EI\frac{d^2}{dx^2}v_1(L)\right)\bigg|_{x=L} = -wL \qquad v_2(2L) = 0 \qquad v_3(2L) = 0 \qquad \frac{dv_2}{dx}(2L) = \frac{dv_3}{dx}(2L) \qquad (5)$$

$$EI\frac{d^2v_2}{dx^2}(2L) = EI\frac{d^2v_3}{dx^2}(2L) \qquad EI\frac{d^2v_3}{dx^2}(3L) = 0 \qquad -\frac{d}{dx}\left(EI\frac{d^2v_3}{dx^2}\right)\bigg|_{x=3L} = -Kv_3(3L) \qquad (6)$$

---

7. 45

Solution

By equilibrium of moment about points $O_1$and O_2 we obtain,

$$\boldsymbol{M}_1 + P(a-x) - R_A(L-x) = 0 \quad or \quad \boldsymbol{M}_1 = P(x-a) - R_A(x-L) \qquad \boldsymbol{M}_2 - R_A(L-x) = 0 \quad or \quad \boldsymbol{M}_2 = -R_A(x-L) \qquad (1)$$

(a) 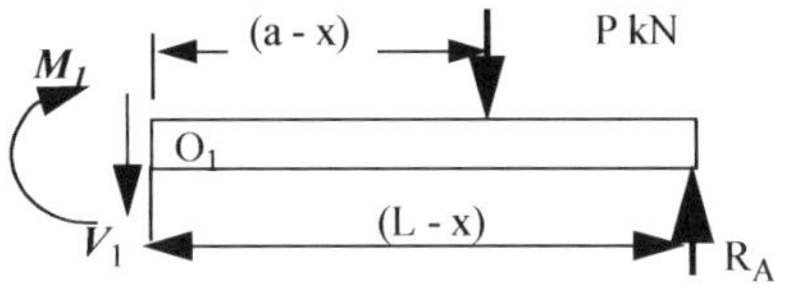

(b) 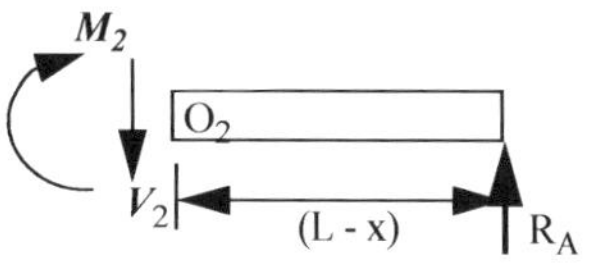

The boundary value problem statement can be written as,

$$EI\frac{d^2v_1}{dx^2} = P(x-a) - R_A(x-L) \qquad EI\frac{d^2v_2}{dx^2} = -R_A(x-L) \qquad (2)$$

$$v_1(0) = 0 \qquad \frac{dv_1}{dx}(0) = 0 \qquad v_2(L) = 0 \qquad v_1(a) = v_2(a) \qquad \frac{dv_1}{dx}(a) = \frac{dv_2}{dx}(a) \qquad (3)$$

Integrating and using the boundary conditions we obtain the displacement.

$$EI\frac{dv_1}{dx} = \frac{P}{2}(x-a)^2 - \frac{R_A}{2}(x-L)^2 + C_1 \qquad EI\frac{dv_2}{dx} = \frac{R_A}{2}(x-L)^2 + C_2 \qquad (4)$$

$$EIv_1 = \frac{P}{6}(x-a)^3 - \frac{R_A}{6}(x-L)^3 + C_1x + C_3 \qquad EIv_2 = -\frac{R_A}{6}(x-L)^3 + C_1x + C_4 \qquad (5)$$

$$EI\frac{dv_1}{dx}(a) = -\frac{R_A}{2}(a-L)^2 + C_1 = EI\frac{dv_2}{dx}(a) = -\frac{R_A}{2}(a-L)^2 + C_2 \quad or \quad C_1 = C_2 \qquad (6)$$

$$EIv_1(a) = -\frac{R_A}{2}(a-L)^3 + C_1a + C_3 = EIv_2(a) = -\frac{R_A}{2}(a-L)^3 + C_1a + C_4 \quad or \quad C_3 = C_4 \qquad (7)$$

$$EI\frac{dv_1}{dx}(0) = \frac{P}{2}a^2 - \frac{R_A}{2}L^2 + C_1 = 0 \qquad or \qquad C_1 = \frac{R_A L^2 - Pa^2}{2} \tag{8}$$

$$EIv_1(0) = -\frac{P}{6}a^3 + \frac{R_A}{6}L^3 + C_3 = 0 \qquad or \qquad C_3 = \frac{Pa^3 - R_A L^3}{6} \tag{9}$$

$$v_2(L) = C_1 L + C_3 = 0 \qquad or \qquad \frac{R_A L^2 - Pa^2}{2}L + \frac{Pa^3 - R_A L^3}{6} = 0 \qquad or \qquad 3R_A L^3 + R_A L^3 = 3Pa^3 L - Pa^3 \tag{10}$$

ANS $R_A = \frac{Pa^2}{2L^3}(3L - a)$

---

7. 46

Solution

Let R_c represents the correct radius of curvature. Noting that curvature is positive concave upwards, we write the following.

$$R_c = -\frac{d^2 v}{dx^2} \qquad I_{zz} = \frac{1}{12}bh^3 \tag{1}$$

By equilibrium of moment about point O in Fig.(a), we obtain,

$$\boldsymbol{M}_z + Px = 0 \qquad or \qquad \boldsymbol{M}_z = -Px \tag{2}$$

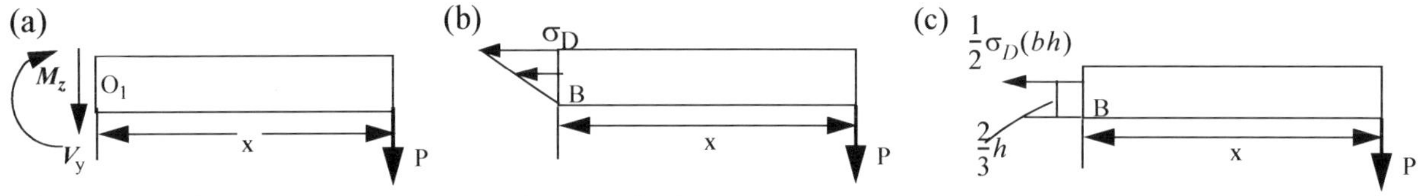

Substituting moment in moment curvature equation, we obtain

$$EI_{zz}\frac{d^2 v}{dx^2} = \boldsymbol{M}_z \qquad or \qquad \frac{Ebh^3}{12}\frac{1}{R_c} = Px \qquad R_c = \frac{Ebh^3}{12Px} \tag{3}$$

To derive Bernoulli's equation we consider AB = R dθ and CD = (R+h)dθ. The strain in AB is zero, hence AB represents the original length that was equal to the original length of CD. Thus the strain in CD is,

$$\varepsilon_D = \frac{CD - AB}{AB} = \frac{(R+h)d\theta - Rd\theta}{Rd\theta} = \frac{h}{R} \qquad \sigma_D = E\varepsilon_D = \frac{Eh}{R} \tag{4}$$

As plane section remain plane, the bending normal strain and stress vary linearly as shown in figures above. By moment equilibrium about point B we have

$$\left(\frac{1}{2}\sigma_D bh\right)\left(2\frac{h}{3}\right) = Px \qquad or \qquad \frac{1}{3}bh^2\left(\frac{Eh}{R}\right) = Px \qquad or \qquad \left(\frac{Ebh^3}{3R}\right) = Px \qquad or \qquad R = \frac{Ebh^3}{3Px} \tag{5}$$

The above equations are the desired equation. We note that $R_c/R = 1/4$.

---

7. 47

Solution

By equilibrium of f moment about point O, we obtain,

$$(\boldsymbol{M}_z)_i = Rx - P_1(x - x_1) - P_2(x - x_2) \ldots\ldots - P_i(x - x_i) \qquad or \qquad (\boldsymbol{M}_z)_i = Rx - \sum_{j=1}^{i} P_i \cdot (x - x_i) \tag{1}$$

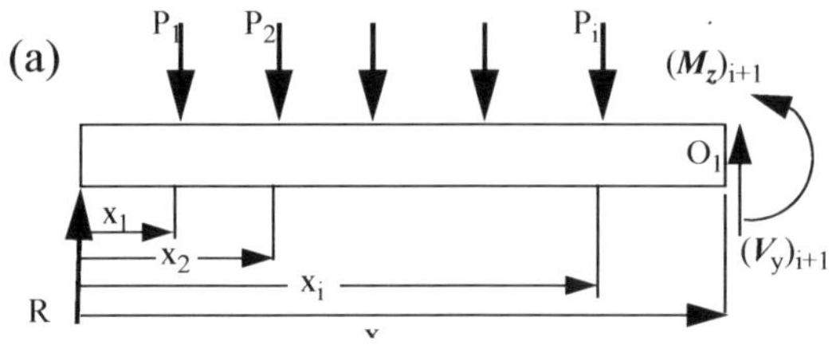

The boundary value problem is,

$$EI\frac{d^2 v_i}{dx^2} = Rx - \sum_{j=1}^{i} P_j(x - x_i) \qquad v_{i-1}(x_i) = v_i(x_i) \qquad \frac{dv_{i-1}}{dx}(x_i) = \frac{dv_i}{dx}(x_i) \qquad i \geq 1 \tag{2}$$

Integrating and using the boundary conditions we obtain the displacement.

$$EI\frac{dv_i}{dx}(x) = \frac{Rx^2}{2} - \sum_{j=1}^{i} P_j\frac{(x - x_i)^2}{2} + C_i \qquad EIv_i(x) = \frac{Rx_i^3}{6} - \sum_{j=1}^{i} P_j\frac{(x - x_i)^3}{6} + C_i x + D_i \qquad i \geq 1 \tag{3}$$

$$EI\frac{dv_{i-1}}{dx}(x_i) = \frac{Rx_i^2}{2} - \sum_{j=1}^{i-1} P_j\frac{(x_i-x_j)^2}{2} + C_{i-1} = EI\frac{dv_i}{dx}(x_i) = \frac{Rx_i^2}{2} - \sum_{j=1}^{i} P_j\frac{(x-x_i)^2}{2} + C_i \quad or \quad C_{i-1} = C_i \quad i \geq 1 \quad (4)$$

$$EIv_{i-1}(x_i) = \frac{Rx_i^3}{6} - \sum_{j=1}^{i-1} P_j\frac{(x_i-x_j)^3}{6} + C_{i-1}x_i + D_{i-1} \text{ or}$$

$$EIv_i(x_i) = \frac{Rx_i^3}{6} - \sum_{j=1}^{i} P_j\frac{(x_i-x_j)^3}{6} + C_ix_i + D_i = \frac{Rx_i^3}{6} - \sum_{j=1}^{i-1} P_j\frac{(x_i-x_j)^3}{6} + C_ix_i + D_i \quad (5)$$

$$\frac{dv_{i-1}}{dx}(x_i) = \frac{dv_i}{dx}(x_i) \quad or \quad C_{i-1}x_i + D_{i-1} = C_ix_i + D_i \quad or \quad D_{i-1} = D_i \quad i \geq 1 \quad (6)$$

The above equations are the desired results.

---

7. 48

Solution

By the force equilibrium in the y-direction and moment equilibrium about point O we obtain the following.

$$d\boldsymbol{V}_y + pdx - (kdx)v = 0 \quad or \quad \frac{d\boldsymbol{V}_y}{dx} - kv = -p \quad (1)$$

$$\boldsymbol{M}_z + d\boldsymbol{M}_z + (\boldsymbol{V}_y + d\boldsymbol{V}_y)\frac{dx}{2} + \boldsymbol{V}_y\frac{dx}{2} - \boldsymbol{M}_z = 0 \quad or \quad d\boldsymbol{M}_z + \boldsymbol{V}_ydx = 0 \quad or \quad \boldsymbol{V}_y = -\frac{d\boldsymbol{M}_z}{dx} \quad (2)$$

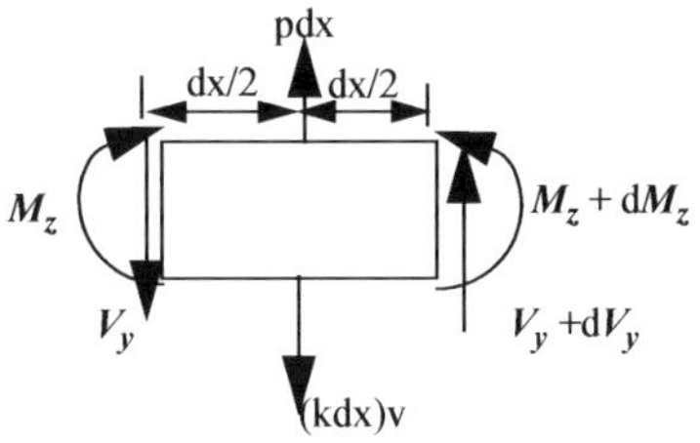

$$\boldsymbol{M}_z = EI_{zz}\frac{d^2v}{dx^2} \quad \boldsymbol{V}_y = -\frac{d}{dx}\left[EI_{zz}\frac{d^2v}{dx^2}\right] \quad \frac{d}{dx}\left\{-\frac{d}{dx}\left[EI_{zz}\frac{d^2v}{dx^2}\right]\right\} - kv = -p \quad or \quad -\frac{d^2}{dx^2}\left(EI_{zz}\frac{d^2v}{dx^2}\right) - kv = -p \quad (3)$$

$$\textbf{ANS} \quad \frac{d^2}{dx^2}\left(EI_{zz}\frac{d^2v}{dx^2}\right) + kv = p$$

---

7. 49

Solution $u = -y\psi(x) \quad v = v(x)$

The solution proceeds as follows.

$$\varepsilon_{xx} = \frac{\partial u}{\partial x} = -y\frac{d}{dx}\psi(x) \quad \gamma_{xy} = \frac{\partial u}{\partial y} + \frac{\partial v}{\partial x} = \frac{dv}{dx} - \psi(x) \quad \sigma_{xx} = E\varepsilon_{xx} = -Ey\frac{d\psi}{dx} \quad \tau_{xy} = G\gamma_{xy} = G\left(\frac{dv}{dx} - \psi(x)\right) \quad (1)$$

$$\boldsymbol{V}_y = \int_A \tau_{xy}dA = \int_A G\left(\frac{dv}{dx} - \psi(x)\right)dA = GA\left[\frac{dv}{dx} - \psi(x)\right] \quad \boldsymbol{M}_z = -\int_A y\sigma_{xx}dA = \int_A y\left[(Ey)\frac{d\psi}{dx}\right]dA = E\frac{d\psi}{dx}\int_A y^2dA = EI_{zz}\frac{d\psi}{dx} \quad (2)$$

$$\frac{d\boldsymbol{V}_y}{dx} = \frac{d}{dx}\left[GA\left(\frac{dv}{dx} - \psi(x)\right)\right] = -p \quad or \quad \frac{d\boldsymbol{M}_z}{dx} = \frac{d}{dx}\left[EI_{zz}\frac{d\psi}{dx}\right] = -\boldsymbol{V}_y = -GA\left[\frac{dv}{dx} - \psi(x)\right] \quad (3)$$

$$\textbf{ANS} \quad \frac{d}{dx}\left[GA\left(\frac{dv}{dx} - \psi(x)\right)\right] = -p\,; \quad \frac{d}{dx}\left[EI_{zz}\frac{d\psi}{dx}\right] = -GA\left[\frac{dv}{dx} - \psi(x)\right]$$

---

7. 50

Solution

By the force equilibrium and moment equilibrium about the center in the given figure we obtain the following.

$$\boldsymbol{V}_y + \Delta\boldsymbol{V}_y - \boldsymbol{V}_y = \rho A\frac{\partial^2}{\partial t^2}(v)\Delta x \quad or \quad \lim_{\Delta x \to 0}\left(\frac{\Delta\boldsymbol{V}_y}{\Delta x}\right) = \rho A\frac{\partial^2}{\partial t^2}(v) \quad or \quad \frac{\partial\boldsymbol{V}_y}{\partial x} = \rho A\frac{\partial^2}{\partial t^2}(v) \quad (1)$$

$$\boldsymbol{M}_z + \Delta\boldsymbol{M}_z + (\boldsymbol{V}_y + \Delta\boldsymbol{V}_y)\frac{\Delta x}{2} + \boldsymbol{V}_y\frac{\Delta x}{2} - \boldsymbol{M}_z = 0 \quad \lim_{\Delta x \to 0}\left(\frac{\Delta\boldsymbol{M}_z}{\Delta x}\right) = -\boldsymbol{V}_y \quad or \quad \frac{\partial\boldsymbol{M}_z}{\partial x} = -\boldsymbol{V}_y \quad (2)$$

$$M_z = EI_{zz}\frac{\partial^2 v}{\partial x^2} \qquad V_y = -\frac{\partial M_z}{\partial x} = -\frac{\partial}{\partial x}\left[EI_{zz}\frac{\partial^2 v}{\partial x^2}\right] \qquad \frac{\partial V_y}{\partial x} = -\frac{\partial^2}{\partial x^2}\left(EI_{zz}\frac{\partial^2 v}{\partial x^2}\right) = \rho A\frac{\partial^2}{\partial t^2}(v) \text{ or}$$

$$\frac{\partial^2}{\partial x^2}\left(EI_{zz}\frac{\partial^2 v}{\partial x^2}\right) + \rho A\frac{\partial^2}{\partial t^2}(v) = 0 \quad or \quad \frac{\partial^2}{\partial x^2}\left(\frac{\partial^2 v}{\partial x^2}\right) + \frac{\rho A}{EI_{zz}}\frac{\partial^2}{\partial t^2}(v) = 0 \quad or \quad \frac{\partial^2 v}{\partial t^2} + c^2\frac{\partial^4 v}{\partial x^4} = 0 \quad c = \sqrt{\frac{EI_{zz}}{\rho A}} \quad (3)$$

Above arc the desired equations.

---

7. 51

Solution

The x- derivatives of functions G are

$$\frac{\partial G}{\partial x} = -A\omega\cos\omega x + B\omega\sin\omega x + C\omega\cosh\omega x + D\omega\sinh\omega x \qquad \frac{\partial^2 G}{\partial x^2} = -A\omega^2\cos\omega x - B\omega^2\sin\omega x + C\omega^2\cosh\omega x + D\omega^2\sinh\omega x \quad (1)$$

$$\frac{\partial^3 G}{\partial x^3} = A\omega^3\cos\omega x - B\omega^3\sin\omega x + \omega\cosh\omega x + D\omega^3\sinh\omega x \quad (2)$$

$$\frac{\partial^4 G}{\partial x^4} = A\omega^4\cos\omega x + B\omega^4\sin\omega x + C\omega^4\cosh\omega x + D\omega^4\sinh\omega x = \omega^4 G \quad (3)$$

Taking the time derivatives of the function H, we obtain

$$\frac{\partial H}{\partial t} = -E(c\omega^2)\cos(c\omega^2)t + D(c\omega^2)\sin(c\omega^2)t \quad (4)$$

$$\frac{\partial^2 H}{\partial t^2} = -E(c\omega^2)^2\cos(c\omega^2)t - D(c\omega^2)^2\sin(c\omega^2)t = -c^2\omega^4[E\cos(c\omega^2) + D\sin(c\omega^2)t] = -c^2\omega^4 H(t) \quad (5)$$

$$v(x,t) = G(x)H(t) \qquad \frac{\partial^2 v}{\partial t^2} + c^2\frac{\partial^4 v}{\partial x^4} = G(x)\frac{\partial^2 H}{\partial t^2} + c^2 H(t)\frac{\partial^4 G}{\partial x^4} = G(x)(-c^2\omega^4)H(t) + c^2\omega^4 H(t)G(x) = 0 \quad (6)$$

Above equation is the desired result.

---

7. 52

Solution

The boundary value problem for the cantilever beam shown in the given figure is shown below.

$$\frac{d^2}{dx^2}\left(EI\frac{d^2 v}{dx^2}\right) = p \qquad v(0) = 0 \qquad \frac{dv}{dx}(0) = 0 \quad (1)$$

$$V_y(L) = -\frac{d}{dx}\left(EI\frac{d^2 v}{dx^2}\right) = -EI\frac{d^3 v}{dx^3}\bigg|_{x=L} = 0 \quad or \quad \frac{d^3 v}{dx^3}(L) = 0 \qquad M_z(L) = \left(EI\frac{d^2 v}{dx^2}\right)\bigg|_{x=L} = 0 \quad or \quad \frac{d^2 v}{dx^2}(L) = 0 \quad (2)$$

The given solution is:

$$v(x) = \frac{1}{6EI}[R_A x^3 + 3M_A x^2 + \textstyle\int_0^x (x-x_1)^3 p(x_1)dx_1] \quad (3)$$

We record the following formula for the derivative of the integral below and take the derivatives of the above equations

$$\frac{d}{dx}[\textstyle\int_0^x (x-x_1)^n p(x_1)dx_1] = (x-x_1)^n p(x_1)\big|_{x_1=x} + n\int_0^x (x-x_1)^{n-1}p(x_1)dx_1 = n\int_0^x (x-x_1)^{n-1}p(x_1)dx_1 \quad (4)$$

Taking the derivatives of displacement we obtain the following.

$$\frac{dv}{dx} = \frac{1}{6EI}[3R_A x^2 + 6M_A x + 3\textstyle\int_0^x (x-x_1)^2 p(x_1)dx_1] = \frac{1}{2EI}[R_A x^2 + 2M_A x + \int_0^x (x-x_1)^2 p(x_1)dx_1] \quad (5)$$

$$\frac{d^2 v}{dx^2} = \frac{1}{EI}[R_A x + 2M_A + \textstyle\int_0^x (x-x_1)p(x_1)dx_1] \qquad \frac{d^3 v}{dx^3} = \frac{1}{EI}[R_A + \int_0^x p(x_1)dx_1] \qquad \frac{d^4 v}{dx^4} = \frac{1}{EI}p(x) \quad (6)$$

For constant EI, the above equation shows the differential equation is satisfied. The boundary conditions on displacement and slopes are satisfied as shown below.

$$v(0) = \frac{1}{6EI}\left[\int_0^0 (0-x_1)^3 p(x_1)dx_1\right] = 0 \qquad \frac{dv}{dx}(0) = \frac{1}{2EI}\left[\int_0^0 (0-x_1)^2 p(x_1)dx_1\right] = 0 \quad (7)$$

The value of R_A and M_A are given and used to show the boundary conditions on moment and shear force are satisfied as shown below.

$$R_A = -\int_0^L p(x_1)dx_1 \qquad \frac{d^3v}{dx^3}(L) = \frac{1}{EI}[R_A + \int_0^L p(x_1)dx_1] = 0 \tag{8}$$

$$M_A = \int_0^L x_1 p(x_1)dx_1 \qquad R_A L + M_A = \int_0^L x_1 p(x_1)dx_1 - \int_0^L p(x_1)dx_1 = \int_0^L (L-x_1)p(x_1)dx_1 \tag{9}$$

$$\frac{d^2v}{dx^2}(L) = \frac{1}{EI}[R_A L + M_A + \int_0^L (L-x_1)p(x_1)dx_1] = 0 \tag{10}$$

Thus all equations of the boundary value problem are satisfied by the given displacement solution.

---

7. 53

Solution: E = 2000 ksi $\quad v_B = ? \quad \frac{dv_B}{dx} = ?$

Using the results of previous problem the solution proceeds as follows.

$$v(x) = \frac{1}{6EI}[R_A x^3 + 3M_A x^2 + \int_0^x (x-x_1)^3 p(x_1)dx_1] \qquad R_A = -\int_0^L p(x_1)dx_1 \qquad M_A = \int_0^L x_1 p(x_1)dx_1 \tag{1}$$

$$\frac{dv}{dx} = \frac{1}{2EI}[R_A x^2 + 2M_A x + \int_0^x (x-x_1)^2 p(x_1)dx_1] \tag{2}$$

$$I_{zz} = \left(\frac{1}{12}\right)(2)(8^3) = 85.333 in^4 \qquad EI_{zz} = (2)(10^6)(85.333) = 170.667(10^6) lb-in^2 \tag{3}$$

$$v(120) = \frac{1}{1024(10^6)}[R_A(120)^3 + 3M_A(120)^2 + \int_0^{120}(120-x_1)^3 p(x_1)dx_1] \qquad R_A = -\int_0^{120} p(x_1)dx_1 \qquad M_A = \int_0^{120} x_1 p(x_1)dx_1 \tag{4}$$

$$\frac{dv}{dx} = \frac{1}{(341.33)(10^6)}[R_A(120)^2 + 2M_A(120) + \int_0^{120}(120-x_1)^2 p(x_1)dx_1] \tag{5}$$

We define $J_1 = \int_0^{120}(120-x_1)^3 p(x_1)dx_1$ and $J_2 = \int_0^{120}(120-x_1)^2 p(x_1)dx_1$. The integrals J_1, J_2 and the integrals for R_A and M_A are evaluated using numerical integration on a spreadsheet as described in Appendix B and shown on table below.

x (m)	p(x) (lb/in)	R_A (lb)	M_A(in-lb)	J_1	J_2
0	22.917	-3.115E+02	2.088E+03	4.568E+08	4.010E+06
12	29.000	-6.845E+02	8.952E+03	8.520E+08	7.873E+06
24	33.167	-1.097E+03	2.140E+04	1.154E+09	1.121E+07
36	35.500	-1.526E+03	3.943E+04	1.361E+09	1.383E+07
48	36.000	-1.950E+03	6.228E+04	1.487E+09	1.570E+07
60	34.667	-2.346E+03	8.833E+04	1.553E+09	1.688E+07
72	31.417	-2.693E+03	1.152E+05	1.581E+09	1.752E+07
84	26.333	-2.967E+03	1.396E+05	1.590E+09	1.780E+07
96	19.417	-3.148E+03	1.577E+05	1.591E+09	1.787E+07
108	10.667	-3.212E+03	1.646E+05	1.592E+09	1.788E+07
120	0.000				

From the table above we have the following:

$$R_A = -3.212(10^3)\ lb \qquad M_A = 164.6(10^3)\ in-lb \qquad J_1 = 1.592(10^9) \qquad J_2 = 17.88(10^6) \tag{6}$$

Substituting we obtain:

$$v(120) = \frac{1}{1024(10^6)}[(-3.212)(10^3)(120)^3 + 3(164.6)(10^3)(120)^2 + 1.592(10^9)] = 3.078 in \tag{7}$$

$$\frac{dv}{dx}(120) = \frac{1}{(341.33)(10^6)}[(-3.212)(10^3)(120)^2 + 2(164.6)(10^3)(120) + 17.88(10^6)] = 0.0326 \tag{8}$$

ANS $v(120) = 3.1$ in. ; $\frac{dv}{dx}(120) = 0.033$

---

7. 54

Solution: E = 2000 ksi, $\quad p(x) = a + bx + cx^2 \quad$ a =?, b =?, c=?, v_B =? $\frac{dv_B}{dx}$ =?

Using the least square method described in Appendix B, we obtain the value of constants a, b, and c on a spread sheet as shown in the table below.

	x_i	$p(x_i)$	x_i^2	x_i^3	x_i^4	$x*p_i$	$x_i^2*p_i$
1	0	22.917	0.0	0.000E+00	0.000E+00	0.000E+00	0.000E+00
2	12	29.000	144.0	1.728E+03	2.074E+04	3.480E+02	4.176E+03
3	24	33.167	576.0	1.382E+04	3.318E+05	7.960E+02	1.910E+04
4	36	35.500	1296.0	4.666E+04	1.680E+06	1.278E+03	4.601E+04
5	48	36.000	2304.0	1.106E+05	5.308E+06	1.728E+03	8.294E+04
6	60	34.667	3600.0	2.160E+05	1.296E+07	2.080E+03	1.248E+05
7	72	31.417	5184.0	3.732E+05	2.687E+07	2.262E+03	1.629E+05
8	84	26.333	7056.0	5.927E+05	4.979E+07	2.212E+03	1.858E+05
9	96	19.417	9216.0	8.847E+05	8.493E+07	1.864E+03	1.789E+05
10	108	10.667	11664.0	1.260E+06	1.360E+08	1.152E+03	1.244E+05
11	120	0.000	14400.0	1.728E+06	2.074E+08	0.000E+00	0.000E+00
b_{ij} & r_i	660.0	279.1	55440.0	5227200.0	525305088.0	13720.0	929064.0
C_{ij}	1.799E+12	5.691E+10	3.764E+08	2.705E+09	2.091E+07	1.742E+05	
D	3.100E+12						
a_i	22.927	0.581	-0.006				

The value of the constants are, $a = 22.927$ $b = 0.581$, and $c = -0.006$.

The boundary value problem for the cantilever beam shown in the given figure is given below.

$$EI = 170.667(10^6) lb-in^2 \qquad EI\frac{d^4v}{dx^4} = a + bx + cx^2 \qquad v(0) = 0 \qquad \frac{dv}{dx}(0) = 0 \tag{1}$$

$$V_y(120) = -\frac{d}{dx}\left(EI\frac{d^2v}{dx^2}\right)\bigg|_{x=120} = -EI\frac{d^3v}{dx^3}\bigg|_{x=120} = 0 \quad or \quad \frac{d^3v}{dx^3}(120) = 0 \tag{2}$$

$$M_z(120) = \left(EI\frac{d^2v}{dx^2}\right)\bigg|_{x=120} = 0 \quad or \quad \frac{d^2v}{dx^2}(120) = 0 \tag{3}$$

Integrating and using the boundary conditions we obtain the deflection and slope as shown below.

$$EI\frac{d^3v}{dx^3} = ax + \frac{bx^2}{2} + \frac{cx^3}{3} + c_1 \qquad EI\frac{d^2v}{dx^2} = \frac{ax^2}{2} + \frac{bx^3}{6} + \frac{cx^4}{12} + c_1x + c_2 \tag{4}$$

$$EI\frac{dv}{dx} = \frac{ax^3}{6} + \frac{bx^4}{24} + \frac{cx^5}{60} + \frac{c_1x^2}{2} + c_2x + c_3 \qquad EIv = \frac{ax^4}{24} + \frac{bx^5}{120} + \frac{cx^6}{360} + \frac{c_1x^3}{6} + \frac{c_2x^2}{2} + c_3x + c_4 \tag{5}$$

$$v(0) = c_4 = 0 \qquad \frac{dv}{dx}(0) = c_3 = 0 \tag{6}$$

$$EI\frac{d^3v}{dx^3}(120) = a(120) + \frac{b(120)^2}{2} + \frac{c(120)^3}{3} + c_1 = 0 \quad or \quad c_1 = -3478.4 \tag{7}$$

$$EI\frac{d^2v}{dx^2}(120) = \frac{a(120)^2}{2} + \frac{b(120)^3}{6} + \frac{c(120)^4}{12} + c_1(120) + c_2 = 0 \quad or \quad c_2 = 188.69(10^3) \tag{8}$$

$$EIv(120) = \frac{a(120)^4}{24} + \frac{b(120)^5}{120} + \frac{c(120)^6}{360} + \frac{c_1(120)^3}{6} + \frac{c_2(120)^2}{2} = 3.665 in \tag{9}$$

$$EI\frac{dv}{dx}(120) = \frac{a(120)^3}{6} + \frac{b(120)^4}{24} + \frac{c(120)^5}{60} + \frac{c_1(120)^2}{2} + c_2(120) = 0.03945 \tag{10}$$

ANS $a = 22.927$; $b = 0.581$; $c = -0.006$; $v(120) = 3.67$ in. ; $\frac{dv}{dx}(120) = 0.039$

---

7. 55

Solution E = 28 GPa , v_B = ? $\frac{dv_B}{dx}$ =?

The solution proceeds as follows.

$$\frac{d^2v}{dx^2} = \frac{M_z}{EI} \qquad \frac{dv}{dx}(x) = \int_0^x \frac{M_z}{EI}dx \qquad v = \int_{v(0)=0}^{v} dv = \int_0^x \frac{dv}{dx_1}dx_1 = x\frac{dv}{dx_1}\bigg|_0^x - \int_0^x x_1\frac{d^2v}{dx^2}dx_1 \text{ or} \qquad (1)$$

$$v(x) = x\frac{dv}{dx}(x) - \int_0^x x_1\frac{d^2v}{dx^2}dx_1 = x\int_0^x \frac{M_z}{EI}dx - \int_0^x x_1\frac{M_z}{EI}dx_1 = \int_0^x (x-x_1)\frac{M_z}{EI}dx_1 \qquad (2)$$

By equilibrium of moment about point O we obtain:

$$M_z = -P(1.5-x) \qquad I = (\pi R^4/4)mm^4 = (\pi R^4/4)(10^{-12})m^4 \qquad (3)$$

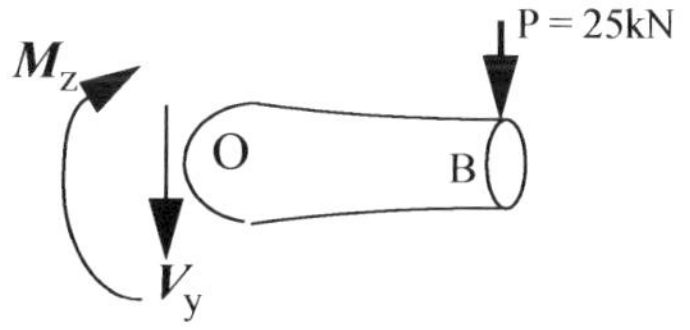

Substituting x = 1.5m, E = 28(10^9) N/m^2, and P = 25 (10^3) N we obtain the following.

$$v_B = \int_0^{1.5} \frac{(-25)(10^3)(1.5-x)^2}{(28)(10^9)(\pi R^4/4)(10^{-12})}dx = -1.1368(10^6)\int_0^{1.5}\frac{(1.5-x)^2}{R^4}dx = \int_0^{1.5} F_1 dx \qquad F_1 = -1.1368(10^6)\frac{(1.5-x)^2}{R^4} \qquad (4)$$

$$\frac{dv_B}{dx} = \int_0^{1.5} \frac{(-25)(10^3)(1.5-x)}{(28)(10^9)(\pi R^4/4)(10^{-12})}dx = -1.1368(10^6)\int_0^{1.5}\frac{(1.5-x)}{R^4}dx = \int_0^{1.5} F_2 dx \qquad F_2 = -1.1368(10^6)\frac{(1.5-x)}{R^4} \qquad (5)$$

The integrals above are evaluated using numerical integration on a spreadsheet as described in Appendix B and the results are given in the table below.

x	R(x)	$F_1(x)$	v(x) (m)	$F_2(x)$	$\frac{dv}{dx}(x)$
0.00	100.60	-2.497E-02	-2.757E-03	-1.665E-02	-1.910E-03
0.10	92.70	-3.017E-02	-6.330E-03	-2.155E-02	-4.575E-03
0.20	82.60	-4.127E-02	-1.043E-02	-3.175E-02	-7.861E-03
0.30	79.60	-4.078E-02	-1.454E-02	-3.398E-02	-1.144E-02
0.40	75.90	-4.145E-02	-1.915E-02	-3.768E-02	-1.587E-02
0.50	68.80	-5.074E-02	-2.384E-02	-5.074E-02	-2.079E-02
0.60	68.00	-4.307E-02	-2.792E-02	-4.785E-02	-2.560E-02
0.70	65.90	-3.858E-02	-3.199E-02	-4.822E-02	-3.106E-02
0.80	60.10	-4.270E-02	-3.567E-02	-6.099E-02	-3.669E-02
0.90	60.30	-3.095E-02	-3.838E-02	-5.159E-02	-4.160E-02
1.00	59.10	-2.330E-02	-4.062E-02	-4.659E-02	-4.660E-02
1.10	54.00	-2.139E-02	-4.225E-02	-5.348E-02	-5.117E-02
1.20	54.80	-1.134E-02	-4.309E-02	-3.782E-02	-5.438E-02
1.30	54.10	-5.308E-03	-4.345E-02	-2.654E-02	-5.666E-02
1.40	49.40	-1.909E-03	-4.354E-02	-1.909E-02	-5.762E-02
1.50	50.60	0.000E+00		0.000E+00	

From the table above we obtain the value of deflection and slope at point B as

$$v_B = -43.54(10^{-3})m = -43.5 \text{ mm} \qquad \frac{dv_B}{dx} = -57.62(10^{-3}) = -0.058 \qquad (6)$$

ANS $v_B = -43.5$ mm ; $\frac{dv_B}{dx} = -0.058$

---

7. 56

Solution: E = 28 GPa , a = ? , b = ? , v_B = ? $\frac{dv_B}{dx}$ =?

In an earlier problem in Chapter 4 we obtained: $a = 90.226mm$ **,** $b = -30.593$ **.** Thus R(x) can be written as:

$$R(x) = 90.226 - 30.593x \qquad (1)$$

From results of previous problem we have

$$v_B = -1.1368(10^6)\int_0^{1.5}\frac{(1.5-x)^2}{R^4}dx = \int_0^{1.5}F_1(x)dx \qquad F_1 = -1.1368(10^6)\frac{(1.5-x)^2}{R^4} \quad (2)$$

$$\frac{dv_B}{dx} = -1.1368(10^6)\int_0^{1.5}\frac{(1.5-x)}{R^4}dx = \int_0^{1.5}F_2(x)dx \qquad F_2 = -1.1368(10^6)\frac{(1.5-x)}{R^4} \quad (3)$$

The above integrals are evaluated using numerical integration on a spreadsheet as described in Appendix B and the results are given in the table below.

x	R(x)	$F_1(x)$	v(x)	$F_2(x)$	$\frac{dv}{dx}(x)$
0.00	90.23	-3.860E-02	-3.860E-03	-2.573E-02	-2.665E-03
0.10	87.17	-3.860E-02	-7.709E-03	-2.757E-02	-5.520E-03
0.20	84.11	-3.839E-02	-1.153E-02	-2.953E-02	-8.577E-03
0.30	81.05	-3.794E-02	-1.528E-02	-3.162E-02	-1.185E-02
0.40	77.99	-3.718E-02	-1.894E-02	-3.380E-02	-1.534E-02
0.50	74.93	-3.606E-02	-2.247E-02	-3.606E-02	-1.906E-02
0.60	71.87	-3.451E-02	-2.582E-02	-3.835E-02	-2.301E-02
0.70	68.81	-3.245E-02	-2.893E-02	-4.056E-02	-2.716E-02
0.80	65.75	-2.980E-02	-3.175E-02	-4.258E-02	-3.150E-02
0.90	62.69	-2.649E-02	-3.420E-02	-4.415E-02	-3.596E-02
1.00	59.63	-2.247E-02	-3.621E-02	-4.495E-02	-4.042E-02
1.10	56.57	-1.776E-02	-3.772E-02	-4.439E-02	-4.472E-02
1.20	53.51	-1.248E-02	-3.869E-02	-4.158E-02	-4.856E-02
1.30	50.46	-7.017E-03	-3.916E-02	-3.508E-02	-5.144E-02
1.40	47.40	-2.253E-03	-3.927E-02	-2.253E-02	-5.256E-02
1.50	44.34	0.000E+00		0.000E+00	

From the table above we obtain the value of deflection and slope at point B as

$$v_B = -39.27(10^{-3})m = -39.3 \text{ mm} \qquad \frac{dv_B}{dx} = -52.56(10^{-3}) = -0.053 \quad (4)$$

ANS $v_B = -39.3$ mm ; $\frac{dv_B}{dx} = -0.053$

Section 7.3

7. 57

Solution: $v_A = v(2L) = ?$

The loading on the beam can be considered as the sum of two loading shown below:

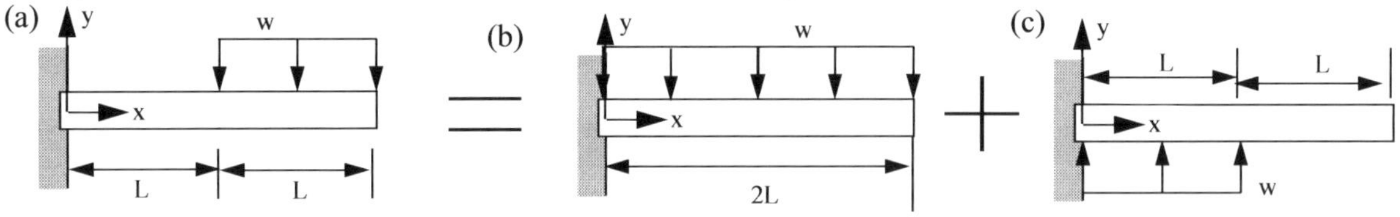

Comparing Figs.(b) and (c) with case 3 of Table C.3 and superposing the solution we obtain the following.

$$a = 2L \qquad b = 0 \qquad p_o = -w \qquad v_{A1} = \frac{-w(8L^3)(6L)}{24EI} = \frac{-2wL^4}{EI} \quad (1)$$

$$a = L \qquad b = L \qquad p_o = w \qquad v_{A2} = \frac{w(7L^3)(6L)}{24EI} = \frac{7wL^4}{24EI} \quad (2)$$

$$v_A = v_{A1} + v_{A2} = \frac{wL^4}{24EI}(-48+7) = \frac{-41wL^4}{24EI} \quad (3)$$

ANS $v_A = \frac{-41wL^4}{24EI}$

7. 58

Solution: $R_A = ?$

The loading on the beam can be considered as the sum of two loading shown below;

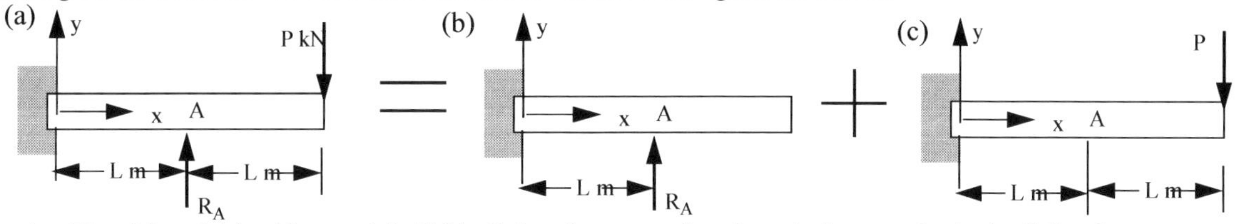

Comparing Figs.(b) and (c) with case 1 in Table C.3 and superposing the solution we obtain the following.

$$a = L \qquad b = 0 \qquad P = R_A \qquad v_{A1} = \frac{R_A L^3}{3EI} \tag{1}$$

$$a = 2L \qquad b = 0 \qquad P = -P \qquad x = L \qquad v_{A2} = \frac{-PL^2}{6EI}(6L - L) = \frac{-5PL^3}{6EI} \tag{2}$$

$$v = v_{A1} + v_{A2} = \frac{R_A L^3}{3EI} - \frac{5PL^3}{6EI} = 0 \qquad or \qquad R_A = \frac{5}{2}P \tag{3}$$

ANS $R_A = 5P/2$

---

7. 59

Solution $v_A = ?$

The beam and loading given is the sum of three loadings as shown below.

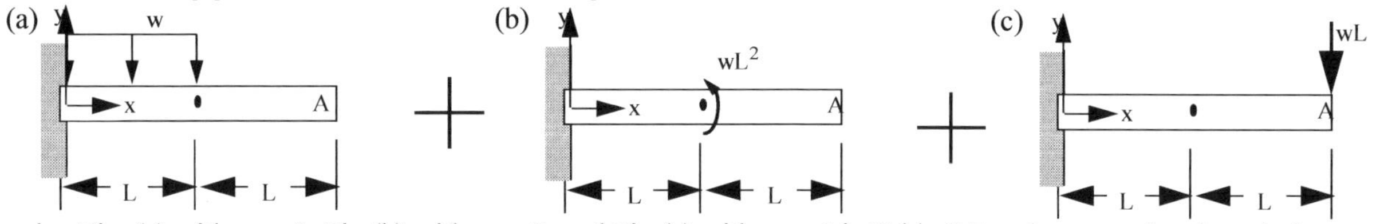

Comparing Fig. (a) with case 3, Fig.(b) with case 2, and Fig.(c) with case 1 in Table C.3 and superposing the solution we obtain the following.

$$a = L \qquad b = L \qquad p_o = -w \qquad v_{A1} = \frac{-7wL^4}{24EI} \tag{1}$$

$$a = L \qquad b = L \qquad M = wL^2 \qquad v_{A2} = \frac{wL^2}{2EI}(3L) = \frac{3wL^4}{2EI} \tag{2}$$

$$a = 2L \qquad b = 0 \qquad P = -wL \qquad v_{A3} = \frac{(-wL)}{6EI}(4L^2)(4L) = -\frac{8wL^3}{3EI} \tag{3}$$

$$v = v_{A1} + v_{A2} + v_{A3} = \frac{wL^3}{EI}\left(\frac{-7 + 36 - 64}{6EI}\right) = \frac{-35wL^3}{24EI} \tag{4}$$

ANS $v_A = \frac{-35wL^3}{24EI}$

---

7. 60

Solution $R_A = ?$ $\qquad \frac{dv_A}{dx} = ?$

The beam and loading in the given figure can be considered as the sum of the two loadings shown below.

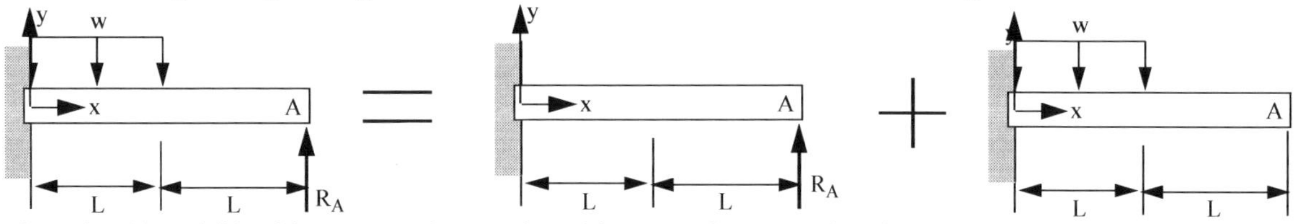

Comparing Figs.(a) and (b) with case 3 and case 1 in Table C.3 and superposing the solution we obtain the following.

$$a = L \qquad b = L \qquad p_o = -w \qquad v_{A1} = \frac{(-wL^2)}{24EI}(7L) = -\frac{7wL^4}{24EI} \qquad \frac{dv_{A1}}{dx} = \left(-\frac{wL^3}{6EI}\right) \tag{1}$$

$$a = 2L \qquad b = 0 \qquad P = R_A \qquad v_{A2} = (R_A)\frac{(4L^2)}{6EI}(4L) = \frac{8R_AL^3}{3EI} \qquad \frac{dv_{A2}}{dx} = \frac{R_A(4L^2)}{2EI} = \frac{2R_AL^2}{EI} \tag{2}$$

$$v_A = v_{A1} + v_{A2} = -\frac{7wL^4}{24EI} + \frac{8R_AL^3}{3EI} = 0 \qquad or \qquad R_A = \frac{7}{64}wL \tag{3}$$

$$\frac{dv_A}{dx} = \frac{dv_{A1}}{dx} + \frac{dv_{A2}}{dx} = -\frac{wL^3}{6EI} + \frac{2R_AL^2}{EI} = -\frac{wL^3}{6EI} + \frac{7wL^3}{32EI} = \frac{5wL^3}{96EI} \tag{4}$$

ANS $R_A = 7wL/64$; $\frac{dv_A}{dx} = 5wL^3/(96EI)$

---

7. 61

Solution: v_A=?

The beams and loading in the given figure is the sum of beams and loading shown below:

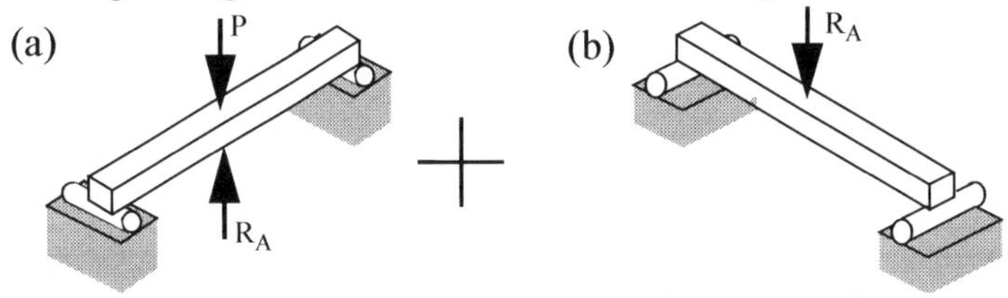

Comparing Figs.(a) and (b) with case 4 in Table C.3 1 and superposing the solutions to obtain the following.

$$L = L \qquad P = R_A - P \qquad v_{A1} = \frac{(R_A - P)}{48EI}L^3 \tag{1}$$

$$L = L \qquad P = -R_A \qquad v_{A2} = -\frac{R_AL^3}{48EI} \tag{2}$$

$$v_{A1} = v_{A2} \qquad or \qquad \frac{(R_A - P)}{48EI}L^3 = -\frac{R_AL^3}{48EI} \qquad or \qquad R_A = \frac{P}{2} \qquad v_A = v_{A2} = -\frac{R_AL^3}{48EI} = -\frac{PL^3}{96EI} \tag{3}$$

ANS $R_A = P/2$; $v_A = -PL^3/(96EI)$

---

7. 62

Solution: v_A=?

The beams in the given figure can be considered as the sum of three beams shown below:

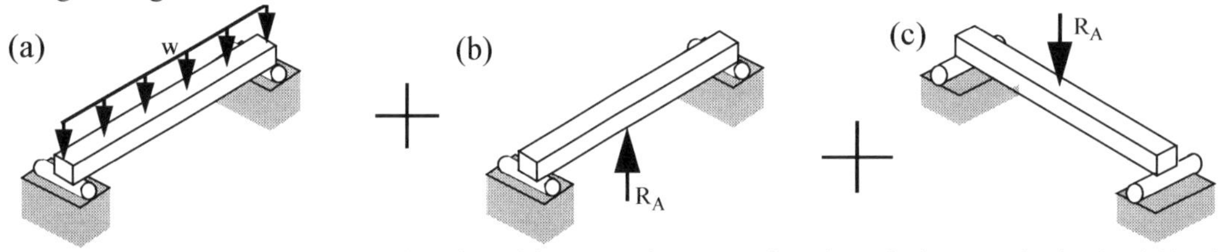

Comparing Fig.(a) with case 5 and Figs.(b) and (c) in Table C.3 and superposing the solutions to obtain the following.

$$L = L \qquad p_o = -w \qquad v_{A1} = \frac{-5wL^4}{384EI} \tag{1}$$

$$L = L \qquad P = R_A \qquad v_{A2} = \frac{R_AL^3}{48EI} \tag{2}$$

$$L = L \qquad P = -R_A \qquad v_{A3} = -\frac{R_AL^3}{48EI} \tag{3}$$

The deflection of the top beam is the sum of deflection of beams in Figs.(a) and (b) which must be equal to deflection of the bottom beam shown in Fig. (c). We obtain

$$v_{A1} + v_{A2} = \frac{-5wL^4}{384EI} + \frac{R_AL^3}{48EI} = v_{A3} = -\frac{R_AL^3}{48EI} \qquad or \qquad \frac{-5wL}{8} + 2R_A = 0 \qquad or \qquad R_A = \frac{5wL}{16} \tag{4}$$

$$v_A = v_{A3} = -\frac{R_AL^3}{48EI} = \frac{-5wL^4}{758EI} \tag{5}$$

ANS $v_A = -5wL^4/(758EI)$

7. 63

Solution: v_A=?, R_C=?, M_C=?

The beam and loading given can be considered as the sum of the two beams shown below

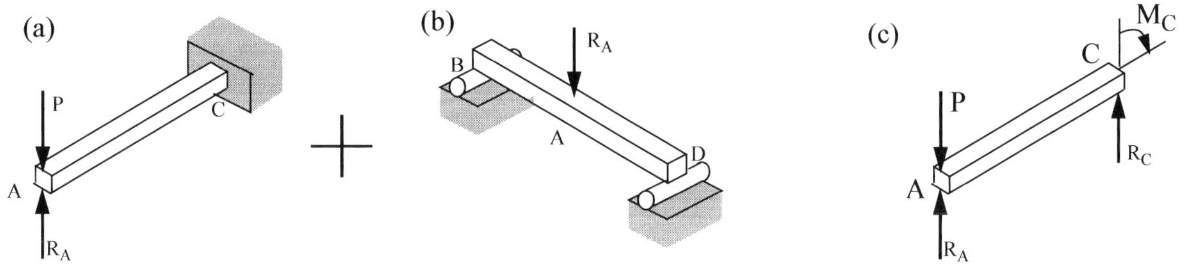

Comparing Figures (a) and (b) with case 1 in Table C.3 and superposing the solutions to obtain the following.

$$a = L \qquad b = 0 \qquad P = R_A - P \qquad v_{A1} = \frac{(R_A - P)}{3EI}L^3 \tag{1}$$

$$L = L \qquad P = -R_A \qquad v_{A2} = -\frac{R_A L^3}{48EI} \tag{2}$$

$$v_{A1} = \frac{(R_A - P)}{3EI}L^3 = v_{A2} = -\frac{R_A L^3}{48EI} \quad or \quad R_A = \frac{16}{17}P \qquad v_A = v_{A2} = -\frac{R_A L^3}{48EI} = -\frac{PL^3}{51EI} \tag{3}$$

By equilibrium of forces and equilibrium of moment about point C in Fig. (c) we obtain: R_C+R_A-P=0 or

$$R_C = P - R_A = \frac{P}{17} \qquad M_c = (P - R_A)L = \frac{PL}{17} \tag{4}$$

ANS $R_A = \frac{16}{17}P$; $v_A = -\frac{PL^3}{51EI}$; $M_c = \frac{PL}{17}$; $R_C = \frac{P}{17}$

---

7. 64

Solution: v_A=?, R_C = ?, M_C = ?

The beams and loading given can be considered as the sum of beams shown in figures (a), (b)

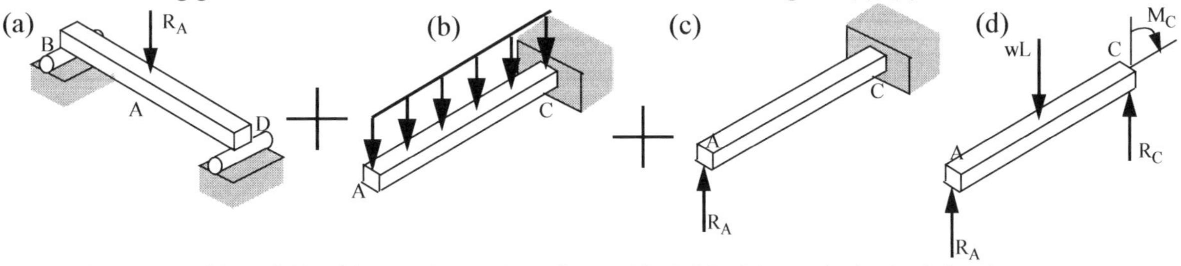

Comparing Figures (a), (b), and (c) with case 4, case 3, and case 1 in Table C.3 we obtain the following.

$$L = L \qquad P = -R_A \qquad v_{A1} = -\frac{R_A L^3}{48EI} \tag{1}$$

$$a = L \qquad b = 0 \qquad p_o = -w \qquad v_{A2} = \frac{-wL^4}{8EI} \tag{2}$$

$$a = L \qquad b = 0 \qquad P = R_A \qquad v_{A3} = \frac{R_A L^3}{3EI} \tag{3}$$

The deflection of the top beam is the sum of deflection of beams in figures (b) and (c) which must be equal to deflection of the bottom beam given in figure (a). We obtain

$$v_{A2} + v_{A3} = \frac{-wL^4}{8EI} + \frac{R_A L^3}{3EI} = v_{A1} = -\frac{R_A L^3}{48EI} \quad or \quad -6wL + 16R_A = -R_A \quad or \quad R_A = \frac{6wL}{17} \tag{4}$$

$$v_A = v_{A1} = -\frac{R_A L^3}{48EI} = \frac{-wL^4}{136EI} \tag{5}$$

By equilibrium of forces in the y-direction and equilibrium of moment about point C in figure (d) , we obtain:

$$R_C = wL - R_A = 11wL/17 \qquad M_C = wL(L/2) - R_A L = 5wL^2/34 \tag{6}$$

ANS $v_A = -wL^4/(136EI)$; $R_C = 11wL/17$; $M_C = 5wL^2/34$

---

7. 65

Solution: v_A=?, R_C = ?, M_C = ?

The beams and loading given can be considered as the sum of beams shown in figures below.)

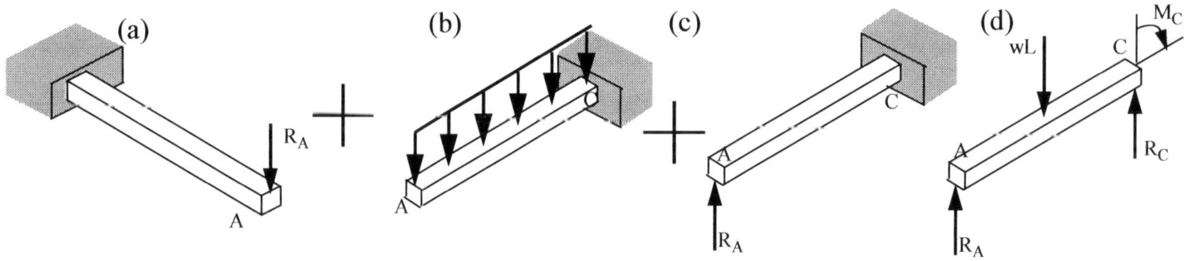

Comparing figures (a), (b), and (c) with case 1, case 3, and case 1 in Table C.3 we obtain the following.

$$a = L \qquad b = 0 \qquad P = -R_A \qquad v_{A1} = -\frac{R_A L^3}{3EI} \tag{1}$$

$$a = L \qquad b = 0 \qquad p_o = -w \qquad v_{A2} = \frac{-wL^4}{8EI} \tag{2}$$

$$a = L \qquad b = 0 \qquad P = R_A \qquad v_{A3} = \frac{R_A L^3}{3EI} \tag{3}$$

The deflection of the top beam is the sum of deflection in figures (b) and (c) which must be equal to deflection of the bottom beam in figure (a) as shown below.

$$v_{A2} + v_{A3} = \frac{-wL^4}{8EI} + \frac{R_A L^3}{3EI} = v_{A1} = -\frac{R_A L^3}{3EI} \quad or \quad R_A = \frac{3wL}{16} \qquad v_A = v_{A1} = -\frac{R_A L^3}{3EI} = \frac{-wL^4}{16EI} \tag{4}$$

By equilibrium of forces in the y-direction and equilibrium of moment about point C in figure (d), we obtain

$$R_C = wL - R_A = \frac{13wL}{16} \qquad M_C = wL\left(\frac{L}{2}\right) - R_A L = \frac{5wL^2}{8} \tag{5}$$

ANS $v_A = \frac{-wL^4}{16EI}$; $R_C = \frac{13wL}{16}$; $M_C = \frac{5wL^2}{8}$

Section 7.4

7. 66

Solution m = 60 kg E= 12.6 GPa (a) simply supported (b) built in ends. $v(3) = ?$

The area moment of inertia can be calculated using formula in Table C2.

$$I_{zz} = \frac{(150)^3[(80)^2 + 4(80)(120) + (120)^2]}{36(80+120)} = 27.75(10^6)mm^4 = 27.75(10^{-6})m^4 \tag{1}$$

$$EI = 12.6(10^9)(27.75)(10^{-6}) = 349.65(10^3) \qquad W = (60)(9.81) = 588.6N \tag{2}$$

The differential equation can be written and integrated as shown below.

$$\frac{d^2}{dx^2}\left(EI\frac{d^2v}{dx^2}\right) = -W\langle x-3\rangle^{-1} \tag{3}$$

$$\frac{d}{dx}\left(EI\frac{d^2v}{dx^2}\right) = -W\langle x-3\rangle^0 + C_1 \qquad EI\frac{d^2v}{dx^2} = -W\langle x-3\rangle^1 + C_1 x + C_2 \tag{4}$$

$$EI\frac{dv}{dx} = \frac{-W}{2}\langle x-3\rangle^2 + \frac{C_1 x^2}{2} + C_2 x + C_3 \qquad EIv = \frac{-W}{6}\langle x-3\rangle^3 + \frac{C_1 x^3}{6} + \frac{C_2 x^2}{2} + C_3 x + C_4 \tag{5}$$

(a) Simply supported: The boundary conditions are

$$v(0) = 0 \qquad EI\frac{d^2v}{dx^2}(0) = 0 \qquad v(6) = 0 \qquad EI\frac{d^2v}{dx^2}(6) = 0 \tag{6}$$

The constants can be solved and displacement obtained as shown below.

$$EIv(0) = C_4 = 0 \qquad EI\frac{d^2v}{dx^2}(0) = C_2 = 0 \tag{7}$$

$$EIv(6) = \frac{-W}{6}\langle 3\rangle^3 + \frac{C_1 6^3}{6} + C_3(6) = 0 \quad or \quad C_3 = -6C_1 + 0.75W \tag{8}$$

$$EI\frac{d^2v}{dx^2}(6) = -W\langle 3\rangle^1 + C_1(6) = 0 \quad or \quad C_1 = \frac{W}{2} \tag{9}$$

$$EIv(3) = \frac{C_1(3)^3}{6} + C_3(3) = \frac{9}{4}W + 3(-3W + 0.75W) = -4.5W \quad or \quad v(3) = -4.5\frac{(588.6)}{349.65(10^3)} = -7.575(10^{-3})m \tag{10}$$

(b) Built in End: The boundary conditions are

$$v(0) = 0 \quad \frac{dv}{dx}(0) = 0 \quad v(6) = 0 \quad \frac{dv}{dx}(6) = 0 \tag{11}$$

(12)

The constants can be solved and displacement obtained as shown below.

$$EIv(0) = C_4 = 0 \quad \frac{dv}{dx}(0) = C_3 = 0 \tag{13}$$

$$EIv(6) = \frac{-W}{6}\langle 3\rangle^3 + \frac{C_1 6^3}{6} + \frac{C_2(6)^2}{2} = 0 \quad or \quad C_2 = -2C_1 + 0.25W \tag{14}$$

$$EI\frac{d^2v}{dx^2}(6) = -W\langle 3\rangle^1 + C_1(6) + C_2 = 4C_1 - 2.75W = 0 \quad or \quad C_1 = 0.6875W \tag{15}$$

$$EIv(3) = \frac{C_1(3)^3}{6} + \frac{C_2(3)^2}{2} = 4.5C_1 + 4.5(-2C_1 + 0.25W) = -4.5(0.6875W) + 1.125W = -1.968W \text{ or}$$

$$v(3) = -1.968\frac{W}{EI} = -1.968\frac{(588.6)}{349.65(10^3)} = -3.314(10^{-3})m \tag{16}$$

ANS (a) $v(3) = -7.575$ mm (b) $v(3) = -3.314$ mm

---

7. 67

Solution: v(x) = ? v(L) = ?

The reaction force R_A=2wL/3 was found in earlier problem. Using templates the moment equation can be written as

$$M_z = \frac{2}{3}wLx - wL\langle x - L\rangle^1 \tag{1}$$

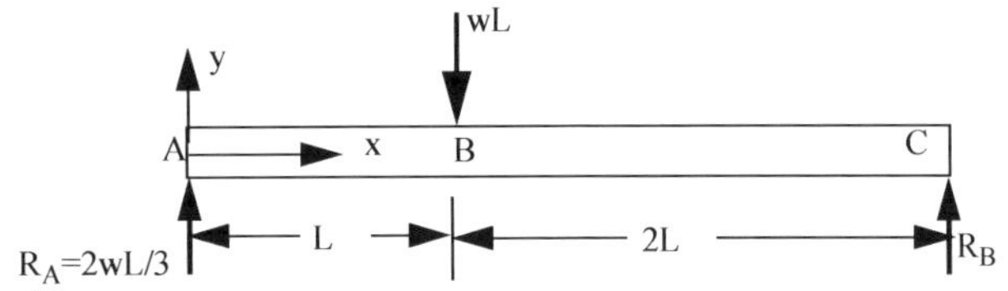

The boundary value problem can be written as

$$EI\frac{d^2v}{dx} = \frac{2}{3}wLx - wL\langle x - L\rangle^1 \quad v(0) = 0 \quad v(3L) = 0 \tag{2}$$

Integrating and using the boundary conditions we obtain the displacements as shown below.

$$EI\frac{dv}{dx} = \frac{1}{3}wLx^2 - \frac{wL}{2}\langle x - L\rangle^2 + C_1 \qquad EIv = \frac{1}{9}wLx^3 - \frac{wL}{6}\langle x - L\rangle^3 + C_1x + C_2 \tag{3}$$

$$EIv(0) = C_2 = 0 \qquad EIv(3L) = \frac{1}{9}wL(3L)^3 - \frac{wL}{6}(8L^3) + C_1(3L) = 0 \quad or \quad C_1 = -\frac{5}{9}wL^3 \tag{4}$$

$$v(x) = \frac{w}{18EI}[2Lx^3 - 3L\langle x - L\rangle^3 - 10L^3x] \qquad v(L) = \frac{w}{18EI}[2L^4 - 10L^4] = \frac{-4wL^4}{9EI} \tag{5}$$

ANS $v(x) = \frac{w}{18EI}[2Lx^3 - 3L\langle x - L\rangle^3 - 10L^3x]$; $v(L) = \frac{-4wL^4}{9EI}$

---

7. 68

Solution: v(x) = ?, v(L) = ?

The reaction force R_A= wL was found in earlier problem.Using templates the moment equation can be written as

$$M_z = wLx - wL^2 - \frac{w}{2}\langle x-L\rangle^2 \qquad \textbf{(1)}$$

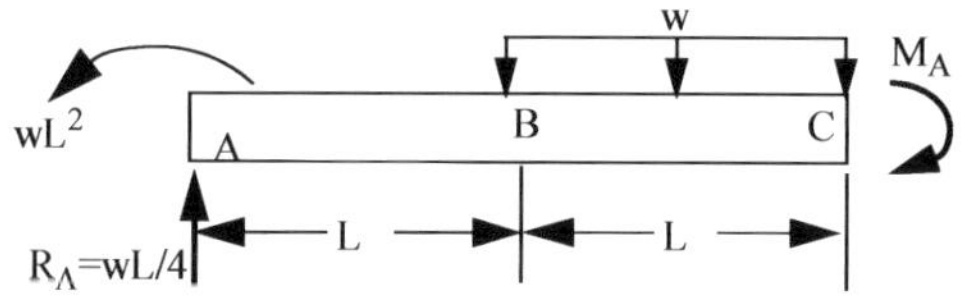

The boundary value problem can be written as

$$EI\frac{d^2v}{dx^2} = wLx - wL^2 - \frac{w}{2}\langle x-L\rangle^2 \qquad v(0) = 0 \qquad \frac{dv}{dx}(2L) = 0 \qquad \textbf{(2)}$$

Integrating and using the boundary conditions we obtain the displacements as shown below.

$$EI\frac{dv}{dx} = \frac{1}{2}wLx^2 - wL^2x - \frac{w}{6}\langle x-L\rangle^3 + C_1 \qquad EIv = \frac{1}{6}wLx^3 - \frac{wL^2x^2}{2} - \frac{w}{24}\langle x-L\rangle^4 + C_1x + C_2 \qquad \textbf{(3)}$$

$$EIv(0) = C_2 = 0 \qquad EI\frac{dv}{dx}(2L) = \frac{w}{2}L(4L^2) - wL^2(2L) - \frac{w}{6}L^3 + C_1 = 0 \qquad or \qquad C_1 = \frac{wL^3}{6} \qquad \textbf{(4)}$$

$$v(x) = \frac{w}{24EI}[4Lx^3 - 12L^2x^2 + 4L^3x - \langle x-L\rangle^4] \qquad v(L) = \frac{w}{24EI}[4L^4 - 12L^4 + 4L^4] = \frac{-wL^4}{6EI} \qquad \textbf{(5)}$$

ANS $v(x) = \frac{w}{24EI}[4Lx^3 - 12L^2x^2 + 4L^3x - \langle x-L\rangle^4]; v(L) = \frac{-wL^4}{6EI}$

---

7. 69

Solution v(x) = ? v(L) = ?

The reaction force R_A= wL/4 was found in earlier problem. Using templates the moment equation can be written as

$$M_z = \frac{wLx}{4} - \frac{w}{2}\langle x-L\rangle^2 \qquad \textbf{(1)}$$

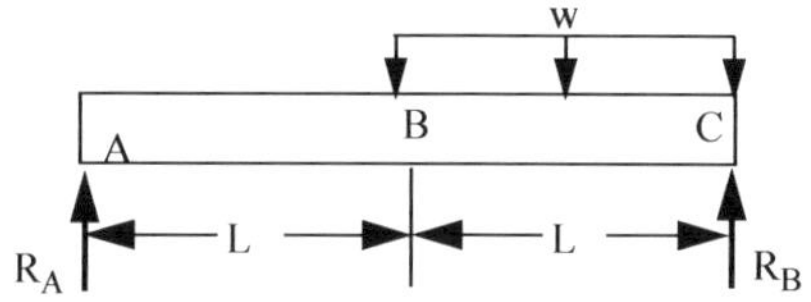

The boundary value problem can be written as

$$EI\frac{d^2v}{dx^2} = \frac{1}{4}wLx - \frac{w}{2}\langle x-L\rangle^2 \qquad v(0) = 0 \qquad v(2L) = 0 \qquad \textbf{(2)}$$

Integrating and using the boundary conditions we obtain the displacements as shown below.

$$EI\frac{dv}{dx} = \frac{1}{8}wLx^2 - \frac{w}{6}\langle x-L\rangle^3 + C_1 \qquad EIv = \frac{1}{24}wLx^3 - \frac{w}{24}\langle x-L\rangle^4 + C_1x + C_2 \qquad \textbf{(3)}$$

$$EIv(0) = C_2 = 0 \qquad EIv(2L) = \frac{1}{24}wL(8L^3) - \frac{wL^4}{24} + C_1(2L) = 0 \qquad or \qquad C_1 = -\frac{7}{48}wL^3 \qquad \textbf{(4)}$$

$$v(x) = \frac{w}{48EI}[2Lx^3 - 7L^3x - 2\langle x-L\rangle^4] \qquad v(L) = \frac{w}{48EI}[2L^4 - 7L^4] = \frac{-5wL^4}{48EI} \qquad \textbf{(5)}$$

ANS $v(x) = \frac{w}{48EI}[2Lx^3 - 7L^3x - 2\langle x-L\rangle^4]; \; v(L) = \frac{-5wL^4}{48EI}$

---

7. 70

Solution: v(x) = ?, v(L) = ?

The reactions A_y = wL and M_A = $3wL^2/2$ were found in earlier problem. Using templates the moment equation can be written in terms of discontinuity function as

$$M_z = wLx - 3wL^2/2 - (w\langle x-L\rangle^2)/2 \qquad \textbf{(1)}$$

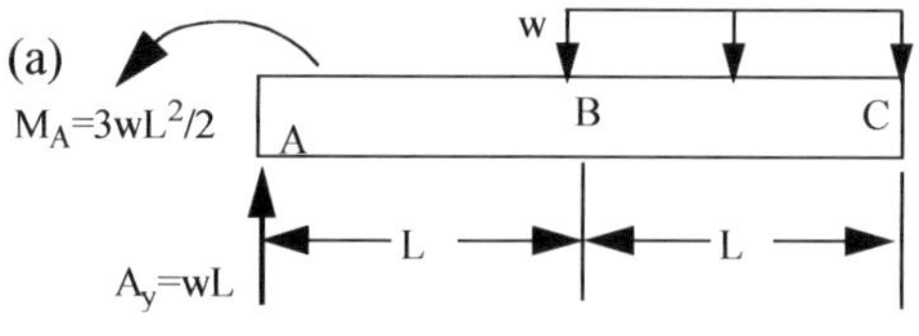

The boundary value problem can be written as

$$EI\frac{d^2v}{dx^2} = wLx - \frac{3}{2}wL^2 - \frac{w}{2}\langle x-L\rangle^2 \qquad v(0) = 0 \qquad \frac{dv}{dx}(0) = 0 \tag{2}$$

Integrating and using the boundary conditions we obtain the displacements as shown below.

$$EI\frac{dv}{dx} = \frac{1}{2}wLx^2 - \frac{3}{2}wL^2x - \frac{w}{6}\langle x-L\rangle^3 + C_1 \qquad EIv = \frac{1}{6}wLx^3 - \frac{3}{4}wL^2x^2 - \frac{w}{24}\langle x-L\rangle^4 + C_1x + C_2 \tag{3}$$

$$EI\frac{dv}{dx}(0) = C_1 = 0 \qquad EIv(0) = C_2 = 0 \tag{4}$$

$$v = \frac{w}{24EI}[4Lx^3 - 18L^2x^2 - \langle x-L\rangle^4] \qquad v(L) = \frac{w}{24EI}[4L^4 - 18L^4] = \frac{-7wL^4}{12EI} \tag{5}$$

ANS $v = \frac{w}{24EI}[4Lx^3 - 18L^2x^2 - \langle x-L\rangle^4]$; $v(L) = \frac{-7wL^4}{12EI}$

---

7. 71

Solution: v(x) = ?, v(2L) = ?

The moment equation can be written in as

$$M_z = \frac{wx^2}{2} - \frac{w}{2}\langle x-L\rangle^2 - wL\langle x-2L\rangle^1 - wL^2\langle x-2L\rangle^0 \tag{1}$$

The boundary value problem can be written as

$$EI\frac{d^2v}{dx^2} = \frac{wx^2}{2} - \frac{w}{2}\langle x-L\rangle^2 - wL\langle x-2L\rangle^1 - wL^2\langle x-2L\rangle^0 \qquad v(3L) = 0 \qquad \frac{dv}{dx}(3L) = 0 \tag{2}$$

Integrating and using the boundary conditions we obtain the displacements as shown below.

$$EI\frac{dv}{dx} = \frac{wx^3}{6} - \frac{w}{6}\langle x-L\rangle^3 - \frac{wL}{2}\langle x-2L\rangle^2 - wL^2\langle x-2L\rangle^1 + C_1 \tag{3}$$

$$EIv = \frac{wx^4}{24} - \frac{w}{24}\langle x-L\rangle^4 - \frac{wL}{6}\langle x-2L\rangle^3 - \frac{wL^2}{2}\langle x-2L\rangle^2 + C_1x + C_2 \tag{4}$$

$$EI\frac{dv}{dx}(3L) = \frac{w(27L^3)}{6} - \frac{w}{6}(8L^3) - \frac{wL^3}{2} - wL^3 + C_1 = 0 \quad or \quad C_1 = \frac{-5wL^3}{3} \tag{5}$$

$$EIv(3L) = \frac{w}{24}(81L^4) - \frac{w}{24}(16L^4) - \frac{wL^4}{6} - \frac{wL^4}{2} - \frac{5}{3}wL^3(3L) + C_2 = 0 \quad or \quad C_2 = \frac{71wL^4}{24} \tag{6}$$

$$v(x) = \frac{w}{24EI}[x^4 - \langle x-L\rangle^4 - 4L\langle x-2L\rangle^3 - 12L^2\langle x-2L\rangle^2 - 40L^3x + 71L^4] \qquad v(L) = \frac{w}{24EI}[16L^4 - L^4 - 80L^4 + 71L^4] = \frac{wL^4}{4EI} \tag{7}$$

ANS $v(x) = \frac{w}{24EI}[x^4 - \langle x-L\rangle^4 - 4L\langle x-2L\rangle^3 - 12L^2\langle x-2L\rangle^2 - 40L^3x + 71L^4]$; $v(2L) = \frac{wL^4}{4EI}$

---

7. 72

Solution: v(x) = ? v(2L) = ?

By equilibrium of forces in the y-direction and equilibrium of moment about point B we obtain

$$R_A = P - R_B \qquad M_B = 2PL - R_AL = PL + R_BL \tag{1}$$

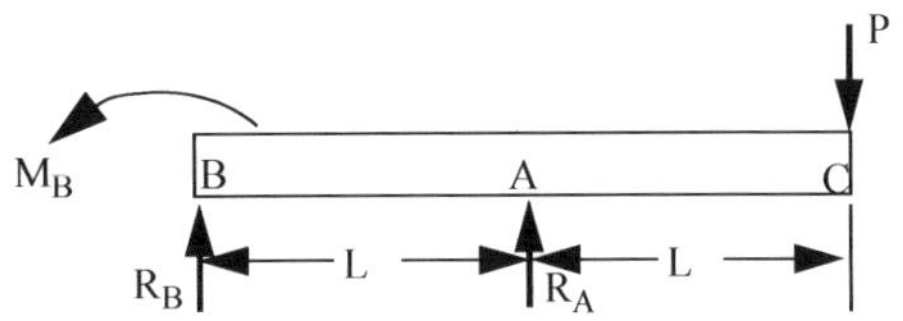

Using templates, the moment equation can be written as

$$M_z = -M_B + R_Bx + R_A\langle x-L\rangle^1 \tag{2}$$

The boundary value problem can be written as

$$EI\frac{d^2v}{dx^2} = -M_B + R_Bx + R_A\langle x-L\rangle^1 \qquad \frac{dv}{dx}(0) = 0 \qquad v(0) = 0 \qquad v(L) = 0 \tag{3}$$

Integrating and using the boundary conditions we obtain the displacements as shown below.

$$EI\frac{dv}{dx} = -M_B x + \frac{R_B x^2}{2} + \frac{R_A}{2}\langle x-L\rangle^2 + C_1 \qquad EIv = -\frac{M_B}{2}x^2 + \frac{R_B x^3}{6} + \frac{R_A}{6}\langle x-L\rangle^3 + C_1 x + C_2 \tag{4}$$

$$EI\frac{dv}{dx}(0) = C_1 = 0 \qquad EIv(0) = C_2 = 0 \qquad EIv(L) = -\frac{M_B}{2}L^2 + \frac{R_B L^3}{6} = 0 \quad or \quad R_B L = 3M_B \tag{5}$$

$$M_B - PL + R_B L - PL + 3M_B \quad or \quad M_B = -PL/2 \qquad R_B = -3P/2 \qquad R_A = 5P/2 \tag{6}$$

$$v(x) = \frac{P}{EI}\left[\frac{x^2 L}{4} - \frac{x^3}{4} + \frac{5}{12}P\langle x-L\rangle^3\right] = \frac{P}{12EI}[3x^2L - 3x^3 + 5\langle x-L\rangle^3] \qquad v(2L) = \frac{P}{12EI}[12L^3 - 24L^3 + 5L^3] = -\left(\frac{7PL^3}{12EI}\right) \tag{7}$$

ANS $v(x) = \frac{P}{12EI}[3x^2L - 3x^3 + 5\langle x-L\rangle^3]$; $v(2L) = -\left(\frac{7PL^3}{12EI}\right)$

---

7. 73

Solution $\delta = ?$

Figure below shows the free body diagram of an infinitesimal element of the free end. By equilibrium we obtain:

$$\boldsymbol{M}_z(L) = 0 \qquad \boldsymbol{V}_y(L) = -kv(L) \tag{1}$$

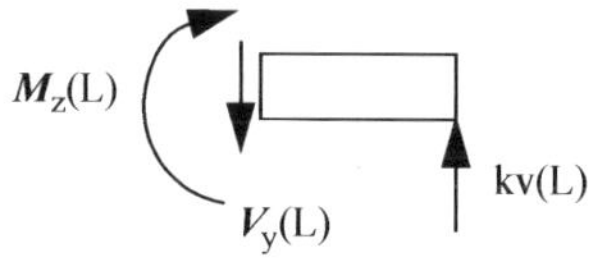

The boundary value problem can be written as:

$$\frac{d^2}{dx^2}\left(EI\frac{d^2v}{dx^2}\right) = R_B\left\langle x-\frac{L}{2}\right\rangle^{-1} - w\left\langle x-\frac{L}{2}\right\rangle^0 \tag{2}$$

$$v(0) = 0 \qquad EI\frac{d^2v}{dx^2}(0) = 0 \qquad v\left(\frac{L}{2}\right) = 0 \qquad -\frac{d}{dx}\left(EI\frac{d^2v}{dx^2}\right)\bigg|_{x=L} = -\alpha\frac{EI}{L^3}v(L) \qquad EI\frac{d^2v}{dx^2}(L) = 0 \tag{3}$$

Integrating and using the boundary conditions we obtain the displacements as shown below.

$$\frac{d}{dx}\left(EI\frac{d^2v}{dx^2}\right) = R_B\left\langle x-\frac{L}{2}\right\rangle^0 - w\left\langle x-\frac{L}{2}\right\rangle^1 + C_1 \qquad EI\frac{d^2v}{dx^2} = R_B\left\langle x-\frac{L}{2}\right\rangle^1 - \frac{w}{2}\left\langle x-\frac{L}{2}\right\rangle^2 + C_1x + C_2 \tag{4}$$

$$EI\frac{dv}{dx} = \frac{R_B}{2}\left\langle x-\frac{L}{2}\right\rangle^2 - \frac{w}{6}\left\langle x-\frac{L}{2}\right\rangle^3 + \frac{C_1x^2}{2} + C_2x + C_3 \qquad EIv = \frac{R_B}{6}\left\langle x-\frac{L}{2}\right\rangle^3 - \frac{w}{24}\left\langle x-\frac{L}{2}\right\rangle^4 + \frac{C_1x^3}{6} + \frac{C_2x^2}{2} + C_3x + C_4 \tag{5}$$

$$EIv(0) = C_4 = 0 \qquad EI\frac{d^2v}{dx^2}(0) = C_2 = 0 \qquad EIv\left(\frac{L}{2}\right) = \frac{C_1(L/2)^3}{6} + C_3(L/2) = 0 \quad or \quad C_3 = -\frac{C_1L^2}{24} \tag{6}$$

$$EI\frac{d^2v}{dx^2}(L) = R_B\left\langle\frac{L}{2}\right\rangle^1 - \frac{w}{2}\left\langle\frac{L}{2}\right\rangle^2 + C_1L + C_2 = 0 \quad or \quad R_B = -2C_1 + \frac{wL}{4} \tag{7}$$

$$\frac{d}{dx}\left(EI\frac{d^2v}{dx^2}\right)\bigg|_{x=L} = R_B\left\langle\frac{L}{2}\right\rangle^0 - w\left\langle\frac{L}{2}\right\rangle^1 + C_1 = \left(\alpha\frac{EI}{L^3}\right)v(L) \quad or \quad C_1 = -\frac{wL}{4} - \left(\alpha\frac{EI}{L^3}\right)v(L) \tag{8}$$

$$EIv(L) = \frac{R_BL^3}{48} - \frac{wL^4}{384} + \frac{C_1L^3}{6} + C_3L = \frac{L^3}{48}\left[-2C_1 + \frac{wL}{4}\right] - \frac{wL^4}{384} + \frac{C_1L^3}{6} - \frac{C_1L^3}{24} = wL^4\left[\frac{1}{192} - \frac{1}{384}\right] + L^3\left[-\frac{1}{24} + \frac{1}{6} - \frac{1}{24}\right]C_1 \text{ or}$$

$$EIv(L) = \frac{wL^4}{384} + \frac{C_1L^3}{12} = \frac{wL^4}{384} - \frac{wL^4}{48} - \left(\alpha\frac{EI}{12L^3}\right)v(L) \qquad EIv(L)\left[1 + \frac{\alpha}{12}\right] = -\frac{7wL^4}{384} \quad or \quad v(L) = -\frac{7wL^4}{32(12+\alpha)} = \delta \tag{9}$$

ANS $\delta = \frac{7wL^4}{32(12+\alpha)}$

Section 7.5

7.74

Solution $v_A = v(L/2) = ?$

The shear-moment diagrams, the area under the moment curve and the location of the centroid are shown below.The deflection at C can be written as:

$$v(x_C) = v(x_B) + v'(x_B)(x_C - x_B) + \frac{1}{EI}[A_1(x_C - \bar{x}_1) + A_2(x_C - \bar{x}_2)] \quad (1)$$

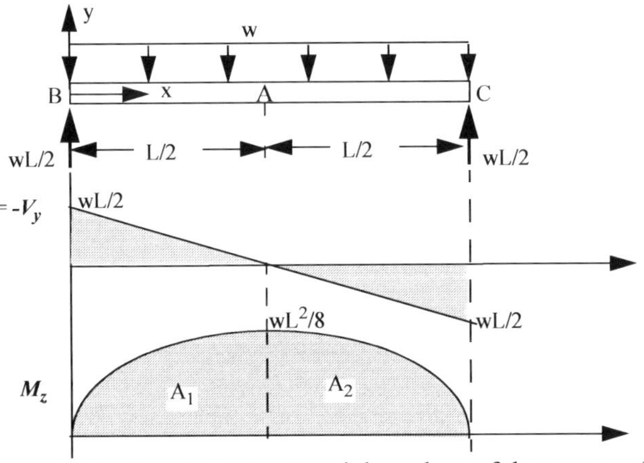

$$A_1 = A_2 = \frac{2}{3}\left(\frac{L}{2}\right)\left(\frac{wL^2}{8}\right) = \frac{wL^3}{24}$$

$$\bar{x}_1 = \frac{5}{8}\left(\frac{L}{2}\right) = \frac{5L}{16}$$

$$\bar{x}_2 = \frac{L}{2} + \frac{3}{8}\left(\frac{L}{2}\right) = \frac{11L}{16}$$

Substituting $v(x_C) = 0$ and $v(x_B) = 0$ and the values of the areas and centroids, the slope at B can be found as shown below.

$$v'(x_B)(L) + \frac{1}{EI}\left[\frac{wL^3}{24}\left(L - \frac{5L}{16}\right) + \frac{wL^3}{24}\left(L - \frac{11L}{16}\right)\right] = 0 \quad or \quad v'(x_B) = -\left(\frac{wL^3}{24EI}\right) \quad (2)$$

The deflection at A can be found as shown below.

$$v(x_A) = v(x_B) + v'(x_B)(x_A - x_B) + \frac{1}{EI}[A_1(x_A - \bar{x}_1)] = -\left(\frac{wL^3}{24EI}\right)\left(\frac{L}{2}\right) + \frac{1}{EI}\left[\frac{wL^3}{24}\left(\frac{L}{2} - \frac{5L}{16}\right)\right] = -\left(\frac{5wL^4}{384EI}\right) \quad (3)$$

ANS $v(x_A) = -\left(\frac{5wL^4}{384EI}\right)$

7.75

Solution $v(L/2) = ?$

By equilibrium of moment about point C and equilibrium of forces in the y-direction we obtain $R_B = P$ and $R_C = P$. The shear-moment diagrams, the area under the moment curve and the location of the centroid are shown below..

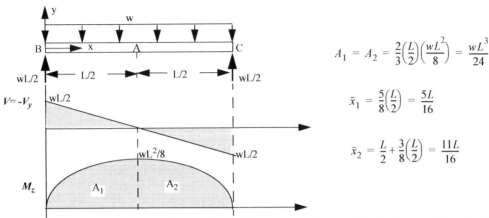

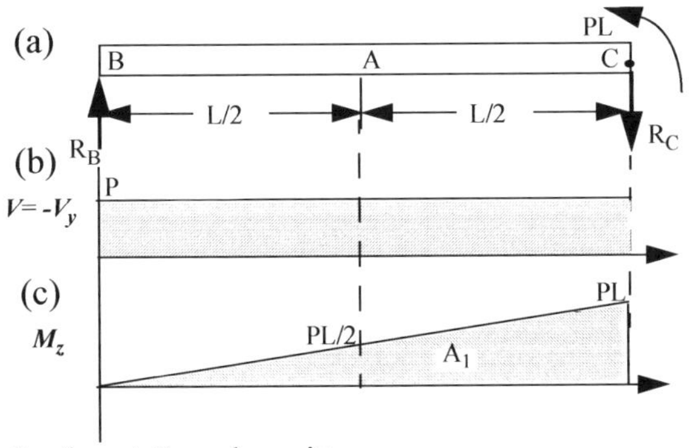

$$A_1 = \frac{1}{2}(PL)(L) = \frac{PL^2}{2}$$

$$\bar{x}_1 = \frac{2L}{3}$$

The deflection at C can be written as:

$$v(x_C) = v(x_B) + v'(x_B)(x_C - x_B) + \frac{1}{EI}[A_1(x_C - \bar{x}_1)] \quad (1)$$

Substituting $v(x_C) = 0$ and $v(x_B) = 0$ and the values of the area and centroids, the slope at B can be found as shown below.

$$v'(x_B)(L) + \frac{1}{EI}\left[\frac{PL^2}{2}\left(L - \frac{2L}{3}\right)\right] = 0 \quad or \quad v'(x_B) = -\left(\frac{PL^2}{6EI}\right) \quad (2)$$

The area under the moment curve up to the middle of the beam is $A_2 = PL^2/8$ and its centroid is at $\bar{x}_2 = L/3$.

$$v(x_A) = v(x_B) + v'(x_B)(x_A - x_B) + \frac{1}{EI}[A_2(x_A - \bar{x}_2)] = -\left(\frac{PL^2}{6EI}\right)\left(\frac{L}{2}\right) + \frac{1}{EI}\left[\frac{PL^2}{8}\left(\frac{L}{2} - \frac{L}{3}\right)\right] = -\left(\frac{PL^3}{16EI}\right) \quad (3)$$

ANS $v(x_A) = -\left(\frac{PL^3}{16EI}\right)$

---

7. 76

Solution $v_A = ?$, $v'(x_A) = ?$

The shear-moment diagrams, the area under the moment curve and the location of the centroid are shown below..

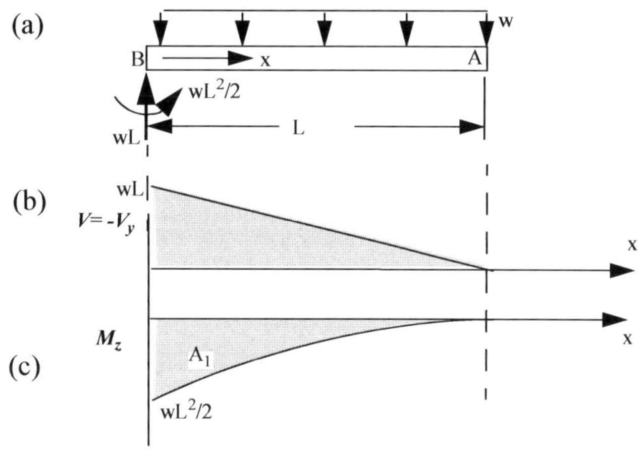

$$A_1 = -\frac{1}{3}\left(\frac{wL^2}{2}\right)(L) = -\left(\frac{wL^3}{6}\right)$$

$$\bar{x}_1 = \frac{L}{4}$$

Noting that the slope at B is zero, the slope at A can be found as shown below.

$$v'(x_A) = v'(x_B) + \frac{A_1}{EI} \quad (1)$$

Noting that the deflection and slope at B is zero, the deflection at A can be found as shown below.

$$v(x_A) = v(x_B) + v'(x_B)(x_A - x_B) + \frac{1}{EI}[A_1(x_A - \bar{x}_1)] = \frac{1}{EI}\left[\frac{wL^3}{6}\left(L - \frac{L}{4}\right)\right] \quad (2)$$

ANS $v(x_A) = -\left(\frac{wL^4}{8EI}\right)$; $v'(x_A) = -\left(\frac{wL^3}{6EI}\right)$

---

7. 77

Solution $v'(x_B) = v'(0) = ?$ $v(L) = v_A = ?$

By equilibrium of forces in the y-direction and moment about point A we obtain $R_B = P$ and $M_A = 0$. The shear-moment diagrams, the area under the moment curve and the location of the centroid are shown below.

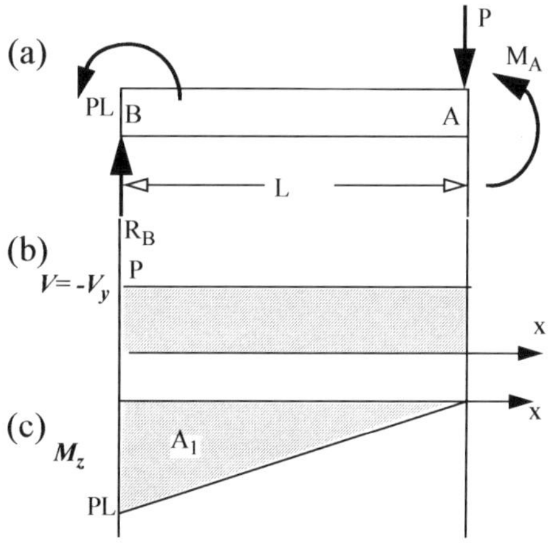

$$A_1 = -\left(\frac{1}{2}\right)(PL)(L) = -\left(\frac{PL^2}{2}\right)$$

$$\bar{x}_1 = \frac{L}{3}$$

Noting that the slope at A is zero the slope at B can be found as shown below.

$$v'(x_A) = v'(x_B) + \frac{A_1}{EI} \quad or \quad v'(x_B) = -\left(\frac{A_1}{EI}\right) = \frac{PL^2}{2EI} \quad (1)$$

Noting that the deflection B is zero, the deflection at A can be can be found as shown below.

$$v(x_A) = v(x_B) + v'(x_B)(x_A - x_B) + \frac{1}{EI}[A_1(x_A - \bar{x}_1)] = 0 + \left(\frac{PL^2}{2EI}\right)(L) + \frac{1}{EI}\left(-\frac{PL^2}{2}\right)\left(L - \frac{L}{3}\right) = \frac{PL^3}{6EI} \quad (2)$$

ANS $v'(x_B) = \frac{PL^2}{2EI}$; $v(x_A) = \frac{PL^3}{6EI}$

---

7. 78

Solution $v'(x_A) = v'(0) = ?$ $v(L) = v_B = ?$

By equilibrium of moment about point C and equilibrium of forces in the y-direction we obtain $R_B = 2wL/3$ and $R_C = wL/3$. The shear-moment diagrams, the area under the moment curve and the location of the centroid are shown below.

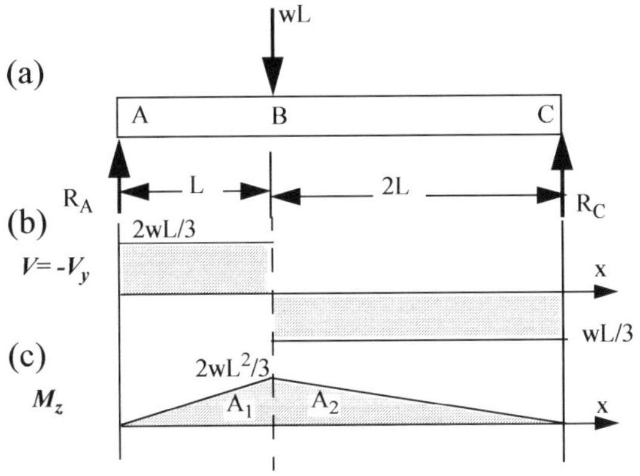

$$A_1 = \left(\frac{1}{2}\right)\left(\frac{2wL^2}{3}\right)(L) = \left(\frac{wL^3}{3}\right)$$

$$A_2 = \left(\frac{1}{2}\right)\left(\frac{2wL^2}{3}\right)(2L) = \left(\frac{2wL^3}{3}\right)$$

$$\bar{x}_1 = \frac{2L}{3}$$

$$\bar{x}_2 = L + \frac{2L}{3} = \frac{5L}{3}$$

The deflection at C can be written as:

$$v(x_C) = v(x_A) + v'(x_A)(x_C - x_A) + \frac{1}{EI}[A_1(x_C - \bar{x}_1) + A_2(x_C - \bar{x}_2)] \quad (1)$$

Substituting $v(x_C) = 0$ and $v(x_A) = 0$ and the values of the area and centroids, the slope at A can be found as shown below.

$$v'(x_A)(3L) + \frac{1}{EI}\left[\left(\frac{wL^3}{3}\right)\left(3L - \frac{2L}{3}\right) + \frac{2wL^3}{3}\left(3L - \frac{5L}{3}\right)\right] = 0 \quad or \quad v'(x_A) = -\left(\frac{5wL^3}{9EI}\right) \quad (2)$$

The deflection at B can be written as:

$$v(x_B) = v(x_A) + v'(x_A)(x_B - x_A) + \frac{1}{EI}[A_1(x_B - \bar{x}_1)] = -\left(\frac{5wL^3}{9EI}\right)(L) + \frac{1}{EI}\left[\frac{wL^3}{3}\left(L - \frac{2L}{3}\right)\right] = -\left(\frac{4wL^4}{9EI}\right) \quad (3)$$

ANS $v'(x_A) = -\left(\frac{5wL^3}{9EI}\right)$; $v(x_B) = -\left(\frac{4wL^4}{9EI}\right)$

---

7. 79

Solution $v'(x_A) = v'(0) = ?$ $v(L) = v(x_B) = ?$

By equilibrium of forces in the y-direction and equilibrium of moment about point C we obtain $R_A = wL$and $M_C = wL^2/2$. The shear-moment diagrams, the area under the moment curve and the location of the centroid are shown below.

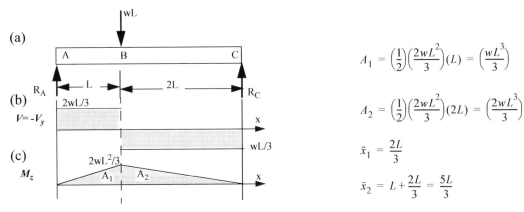

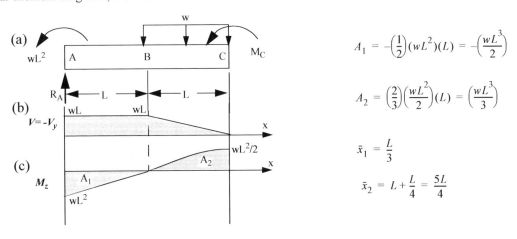

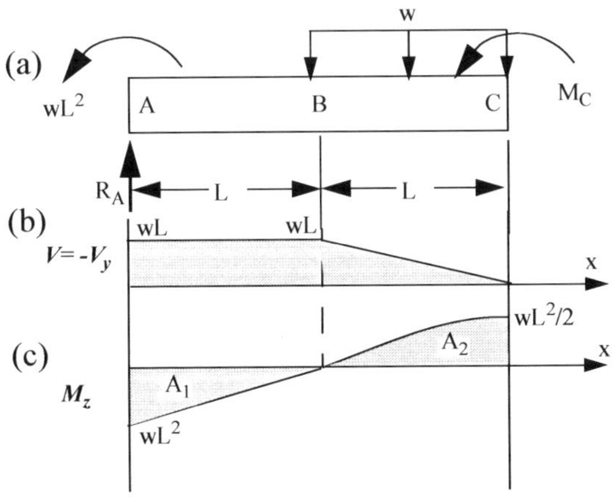

$$A_1 = -\left(\frac{1}{2}\right)(wL^2)(L) = -\left(\frac{wL^3}{2}\right)$$

$$A_2 = \left(\frac{2}{3}\right)\left(\frac{wL^2}{2}\right)(L) = \left(\frac{wL^3}{3}\right)$$

$$\bar{x}_1 = \frac{L}{3}$$

$$\bar{x}_2 = L + \frac{L}{4} = \frac{5L}{4}$$

Noting that the slope at C is zero, the slope at A can be found as shown below.

$$v'(x_C) = v'(x_A) + \frac{1}{EI}(A_1 + A_2) \qquad or \qquad v'(x_A) = -\frac{1}{EI}\left(-\frac{wL^3}{2} + \frac{wL^3}{3}\right) = \left(\frac{wL^3}{6EI}\right) \qquad \textbf{(1)}$$

The deflection at B can be written as:

$$v(x_B) = v(x_A) + v'(x_A)(x_B - x_A) + \frac{1}{EI}[A_1(x_B - \bar{x}_1)] = \left(\frac{wL^3}{6EI}\right)(L) + \frac{1}{EI}\left[-\left(\frac{wL^3}{2}\right)\left(L - \frac{L}{3}\right)\right] \qquad \textbf{(2)}$$

ANS $v'(x_A) = \left(\frac{wL^3}{6EI}\right)$; $v(x_B) = -\left(\frac{wL^4}{6EI}\right)$

---

7. 80

Solution $v'(x_C) = v'(2L) = ?$ $\qquad v(2L) = v(x_C) = ?$

By equilibrium of forces in the y-direction and equilibrium of moment about point A we obtain $R_A = wL$ and $M_A = 3wL^2/4$. The shear-moment diagrams, the area under the moment curve and the location of the centroid are shown below.

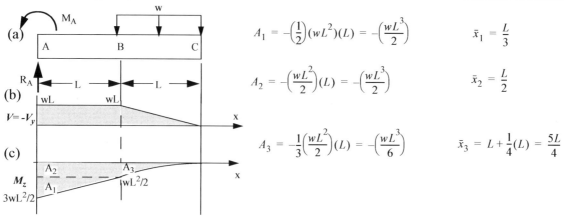

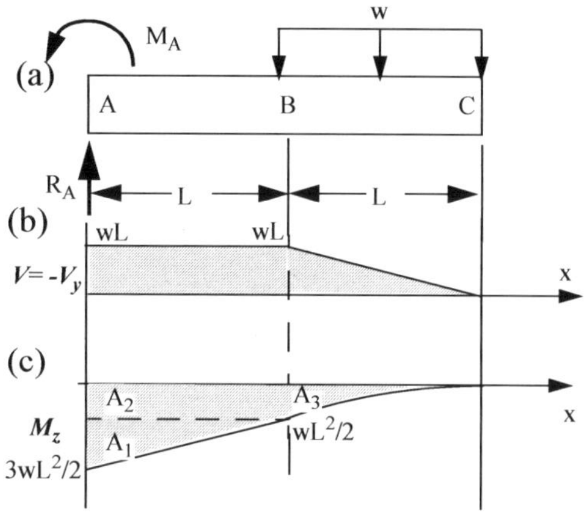

$$A_1 = -\left(\frac{1}{2}\right)(wL^2)(L) = -\left(\frac{wL^3}{2}\right) \qquad \bar{x}_1 = \frac{L}{3}$$

$$A_2 = -\left(\frac{wL^2}{2}\right)(L) = -\left(\frac{wL^3}{2}\right) \qquad \bar{x}_2 = \frac{L}{2}$$

$$A_3 = -\frac{1}{3}\left(\frac{wL^2}{2}\right)(L) = -\left(\frac{wL^3}{6}\right) \qquad \bar{x}_3 = L + \frac{1}{4}(L) = \frac{5L}{4}$$

The slope at point C can be written as:

$$v'(x_C) = v'(x_A) + \frac{1}{EI}[A_1 + A_2 + A_3] \qquad \textbf{(1)}$$

Substituting $v'(x_A) = 0$ and the values of the areas, the slope at A can be found as shown below.

$$v'(x_C) = \frac{1}{EI}\left[-\frac{wL^3}{2} - \frac{wL^3}{2} - \frac{wL^3}{6}\right] = -\left(\frac{7wL^3}{6EI}\right) \qquad \textbf{(2)}$$

The deflection at C can be written as:

$$v(x_C) = v(x_A) + v'(x_A)(x_C - x_A) + \frac{1}{EI}[A_1(x_C - \bar{x}_1) + A_2(x_C - \bar{x}_2) + A_3(x_C - \bar{x}_3)] \qquad \textbf{(3)}$$

Substituting $v(x_A) = 0$, $v'(x_A) = 0$ and the values of the area and centroids, the deflection at C can be found as shown below.

$$v(x_C) = \frac{1}{EI}\left[\left(-\frac{wL^3}{2}\right)\left(2L - \frac{L}{3}\right) + \left(-\frac{wL^3}{2}\right)\left(2L - \frac{L}{2}\right) + \left(-\frac{wL^3}{6}\right)\left(2L - \frac{5L}{4}\right)\right] = \frac{1}{EI}\left[\left(-\frac{5wL^4}{6}\right) + \left(-\frac{3wL^4}{4}\right) + \left(-\frac{wL^4}{8}\right)\right]$$

ANS $v'(x_C) = -\left(\frac{7wL^3}{6EI}\right)$; $v(x_C) = -\left(\frac{41wL^4}{24EI}\right)$

---

7. 81

Solution: $E_w = 2000$ ksi $\quad E_s = 3000$ ksi $\quad v(L) = ?$

The location of neutral axis can be found and the bending rigidity of the cross section can be found as shown below.

$$\eta_c = \frac{\sum E_i A_i \eta_{ci}}{\sum E_i A_i} = \frac{(2000)(4)(2)(2) + (30000)(0.25)(2)(4.125)}{(2000)(4)(2) + (30000)(0.25)(2)} = 3.028in \qquad \textbf{(1)}$$

$$(I_{zz})_s = \frac{1}{12}(2)(0.25)^3 + (2)(0.25)(4.125 - 3.028)^2 = 0.6043in^4 \qquad (I_{zz})_w = \frac{1}{12}(2)(4)^3 + (2)(4)(3.028 - 2)^2 = 19.12in^4 \qquad \textbf{(2)}$$

$$\sum E_j(I_{zz})_j = [(30000)(0.6043) + (2000)(19.12)] = 56369 kip-in^2 = 56369(10^3) lb-in^2 \quad (3)$$

The reaction force R_A= wL/4 can be obtained from moment equilibrium. The moment equation can be written in terms of discontinuity function as shown below.

$$M_z = \frac{wLx}{4} - \frac{w}{2}\langle x-L\rangle^2 \quad (4)$$

(a)

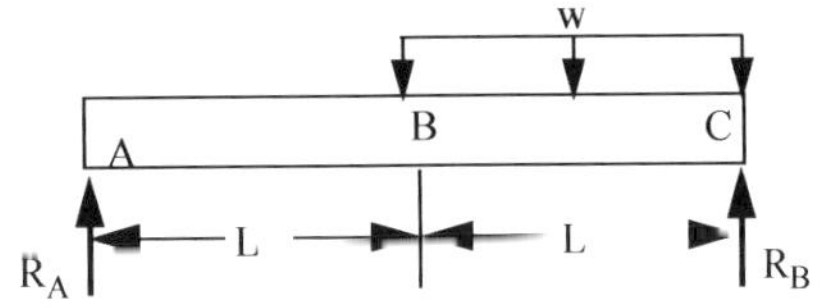

The boundary value problem can be written as

$$\left(\sum E_j(I_{zz})_j\right)\frac{d^2v}{dx^2} = \frac{1}{4}wLx - \frac{w}{2}\langle x-L\rangle^2 \qquad v(0) = 0 \qquad v(2L) = 0 \quad (5)$$

Integrating and using the boundary conditions we obtain the deflection of the beam as shown below.

$$\left(\sum E_j(I_{zz})_j\right)\frac{dv}{dx} = \frac{1}{8}wLx^2 - \frac{w}{6}\langle x-L\rangle^3 + C_1 \qquad \left(\sum E_j(I_{zz})_j\right)v = \frac{1}{24}wLx^3 - \frac{w}{24}\langle x-L\rangle^4 + C_1x + C_2 \quad (6)$$

$$\left(\sum E_j(I_{zz})_j\right)v(0) = C_2 = 0 \qquad \left(\sum E_j(I_{zz})_j\right)v(2L) = \frac{1}{24}wL(8L^3) - \frac{wL^4}{24} + C_1(2L) = 0 \qquad or \qquad C_1 = -\frac{7}{48}wL^3 \quad (7)$$

$$v(x) = \frac{w[2Lx^3 - 7L^3x - 2\langle x-L\rangle^4]}{48\left(\sum E_j(I_{zz})_j\right)} \qquad v(L) = \frac{w[2L^4 - 7L^4]}{48\sum E_j(I_{zz})_j} = \frac{-5wL^4}{48\sum E_j(I_{zz})_j} = \frac{-5(20)(72)^4}{48(56369)(10^3)} = -0.993 in \quad (8)$$

ANS $v(L) = -0.993$ in.

CHAPTER 8

Section 8.1

8. 1

Solution

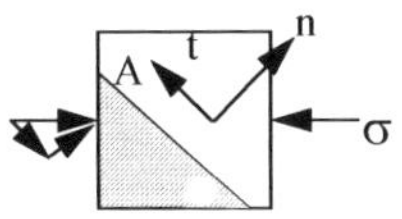

ANS $\sigma_{nn} \Rightarrow (C)$; $\tau_{nt} \Rightarrow positive$

---

8. 2

Solution

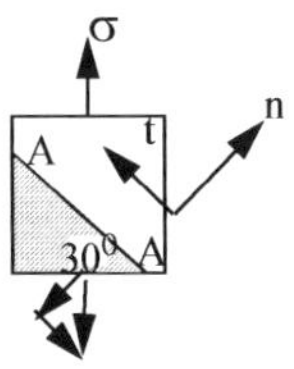

ANS $\sigma_{nn} \Rightarrow (T)$; $\tau_{nt} \Rightarrow positive$

---

8. 3

Solution

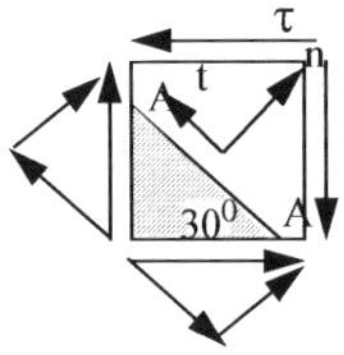

ANS $\sigma_{nn} \Rightarrow (C)$; $\tau_{nt} \Rightarrow Can't\ Say$

---

8. 4

Solution

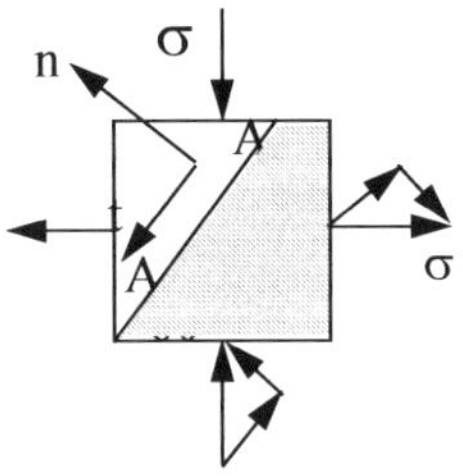

ANS $\sigma_{nn} \Rightarrow Can't\ Say$; $\tau_{nt} \Rightarrow positive$

---

8. 5

Solution

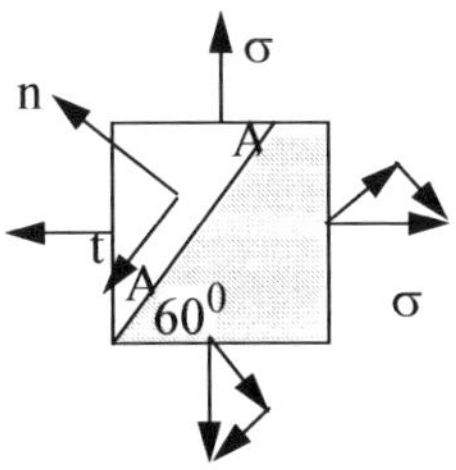

ANS $\sigma_{nn} \Rightarrow (T)$; $\tau_{nt} \Rightarrow Can't\ Say$

8. 6

Solution

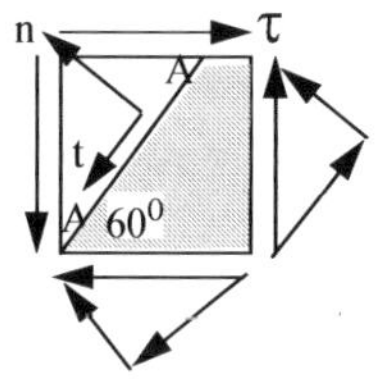

ANS $\sigma_{nn} \Rightarrow (C)$; $\tau_{nt} \Rightarrow Can't\ Say$

---

8. 7

Solution

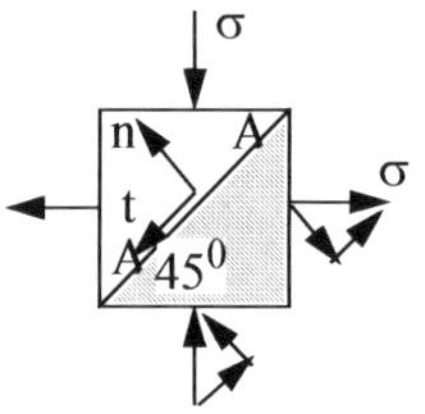

ANS $\sigma_{nn} \Rightarrow zero\ \ or\ \ Can't\ Say$; $\tau_{nt} \Rightarrow positive$

---

8. 8

Solution

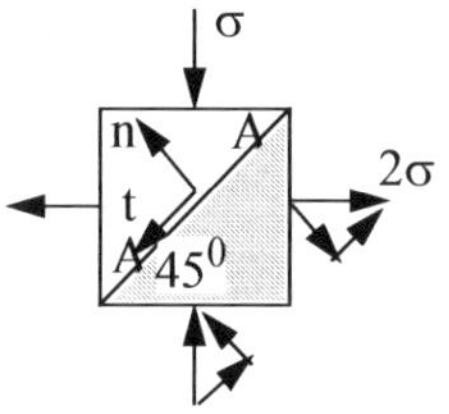

ANS $\sigma_{nn} \Rightarrow (T)\ \ or\ \ Can't\ Say$; $\tau_{nt} \Rightarrow positive$

---

8. 9

Solution

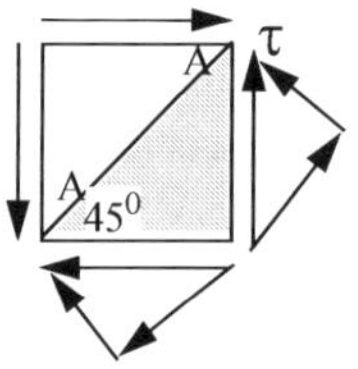

ANS $\sigma_{nn} \Rightarrow (C)$; $\tau_{nt} \Rightarrow zero\ \ or\ \ Can't\ Say$

---

8. 10

Solution $\sigma = 10$ ksi $\sigma_{nn} = ?$ $\tau_{nt} = ?$

The stress cube is as given. The stress wedge, the local coordinate system, and the force wedge are as shown below.

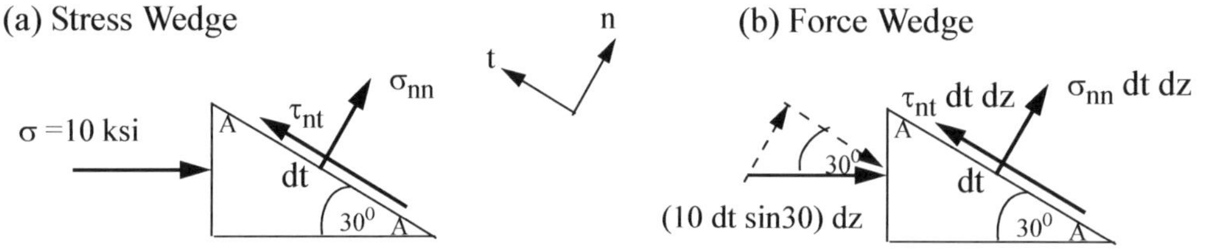

By equilibrium of forces in the *n* and *t* directions we obtain the following.

$$(10\ dt\ \sin 30)dz\ \sin 30 + \sigma_{nn}\ dt\ dz = 0 \qquad or \qquad \sigma_{nn} = -10\sin^2 30 = -2.5 \text{ ksi} \tag{1}$$

$$-(10\ dt\ \sin 30)dz\ \cos 30 + \tau_{nt}\ dt\ dz = 0 \qquad or \qquad \tau_{nt} = 10\sin 30\cos 30 = 4.33 \text{ ksi} \tag{2}$$

ANS $\sigma_{nn} = 2.5$ ksi(C); $\tau_{nt} = 4.33$ ksi

---

8. 11

Solution $\sigma = 10$ ksi $\sigma_{nn} = ?$ $\tau_{nt} = ?$

The stress cube is as given. The stress wedge, the local coordinate system, and the force wedge are as shown below.

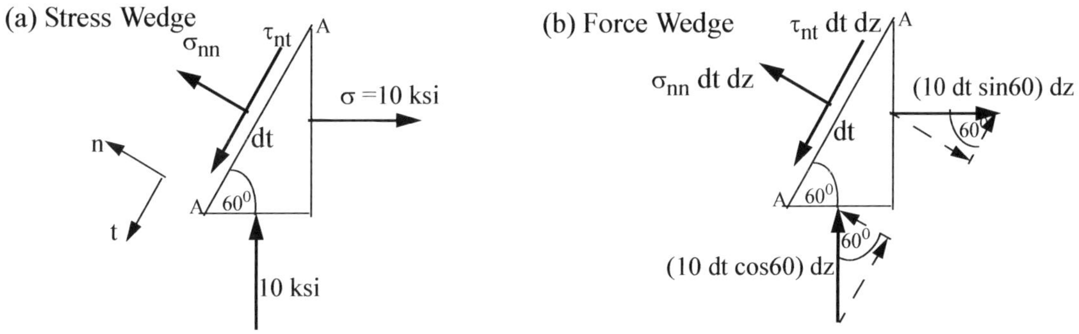

By equilibrium of forces in the *n* and *t* directions we obtain the following.

$$-(10\ dt\ \sin 60)dz\sin 60 + (10\ dt\ \cos 60)dz\cos 60 + \sigma_{nn}\ dt\ dz = 0 \quad or \quad \sigma_{nn} = 10\sin^2 60 - 10\cos^2 60 = 5.0 \text{ ksi} \quad \textbf{(1)}$$

$$-(10\ dt\ \sin 60)dz\cos 60 - (10dt\cos 60)dz\sin 60 + \tau_{nt}\ dt\ dz = 0 \quad or \quad \tau_{nt} = 10\sin 60\cos 60 + 10\cos 60\sin 60 = 8.66 \text{ ksi} \quad \textbf{(2)}$$

ANS $\sigma_{nn} = 5.0$ ksi (T) ; $\tau_{nt} = 8.66$ ksi

---

8. 12

Solution $\sigma = 10$ ksi $\sigma_{nn} = ?$ $\tau_{nt} = ?$

The stress cube is as given. The stress wedge, the local coordinate system, and the force wedge are as shown below.

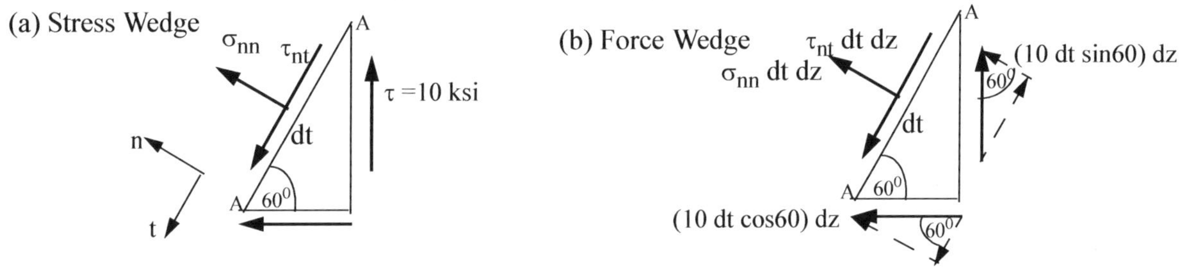

By equilibrium of forces in the *n* and *t* directions we obtain the following.

$$(10dt\sin 60)dz\cos 60 + (10dt\cos 60)dz\sin 60 + \sigma_{nn}dtdz = 0 \quad or \quad \sigma_{nn} = -10\sin 60\cos 60 - 10\cos 60\sin 60 = -8.66 \text{ ksi} \quad \textbf{(1)}$$

$$-(10dt\sin 60)dz\sin 60 + (10dt\cos 60)dz\cos 60 + \tau_{nt}dtdz = 0 \quad \tau_{nt} = 10\sin^2 60 - 10\cos^2 60 = 5.0 \text{ ksi} \quad \textbf{(2)}$$

ANS $\sigma_{nn} = 8.66$ ksi (C) ; $\tau_{nt} = 5.0$ ksi

---

8. 13

Solution $\sigma = 60$ MPa $\sigma_{nn} = ?$ $\tau_{nt} = ?$

The stress cube is as given. The stress wedge, the local coordinate system, and the force wedge are as shown below.

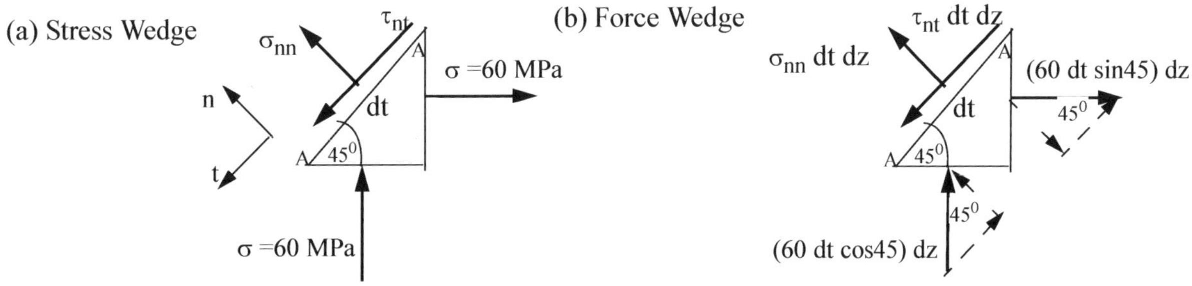

By equilibrium of forces in the *n* and *t* directions we obtain the following.

$$-(60\ dt\ \sin 45)dz\ \sin 45 + (60\ dt\ \cos 45)dz\ \cos 45 + \sigma_{nn}\ dt\ dz = 0 \quad \sigma_{nn} = 60\sin^2 45 - 60\cos^2 45 = 0 \quad \textbf{(1)}$$

$$-(60\ dt\ \sin 45)dz\ \cos 45 - (60\ dt\ \cos 45)dz\ \sin 45 + \tau_{nt}\ dt\ dz = 0 \quad \tau_{nt} = 60\sin 45\cos 45 + 60\cos 45\sin 45 = 60 \text{ MPa} \quad \textbf{(2)}$$

ANS $\sigma_{nn} = 0$; $\tau_{nt} = 60$ MPa

---

8. 14

Solution $\sigma = 60$ MPa $\sigma_{nn} = ?$ $\tau_{nt} = ?$

The stress cube is as given. The stress wedge, the local coordinate system, and the force wedge are as shown below

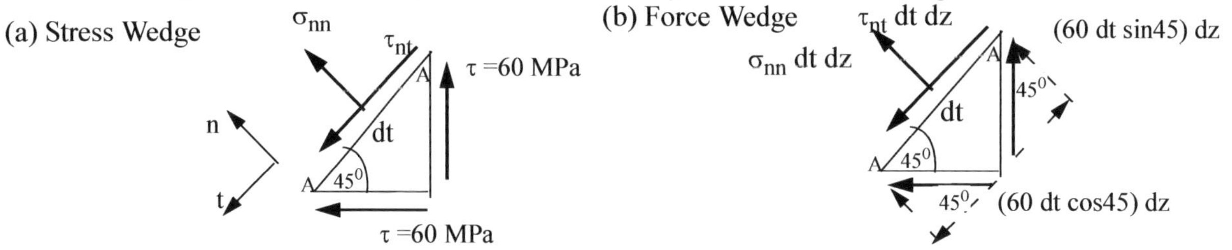

By equilibrium of forces in the *n* and *t* directions we obtain the following.

$$(60\ dt\ sin45)dz\ cos45 + (60\ dt\ cos45)dz\ sin45 + \sigma_{nn}\ dt\ dz = 0 \qquad \sigma_{nn} = -60sin45cos45 - 60cos45sin45 = 60\text{ MPa}$$

$$-(60\ dt\ sin45)dz\ sin45 + (60\ dt\ cos45)dz\ cos45 + \tau_{nt}\ dt\ dz = 0 \qquad \tau_{nt} = 60sin^2 45 - 60cos^2 45 = 0$$

ANS $\sigma_{nn} = 60$ MPa (C) ; $\tau_{nt} = 0$

---

8. 15

Solution

The upper surface will be pressed into the lower surface. The normal stress in adhesive is compression.

ANS *Compression*

---

8. 16

Solution

The upper surface will be move away from the lower surface. The normal stress in adhesive is tension.

ANS *Tension*

---

8. 17

Solution

The upper surface will be move away from the lower surface. The normal stress in adhesive is tension.

ANS *Tension*

---

8. 18

Solution

The lower surface will be pressed into the upper surface. The normal stress in adhesive is:

ANS *Compression*

---

8. 19

Solution $\sigma_{nn} = ?$ $\tau_{nt} = ?$

The stress cube is as given. The stress wedge, the local coordinate system, and the force wedge are as shown below.

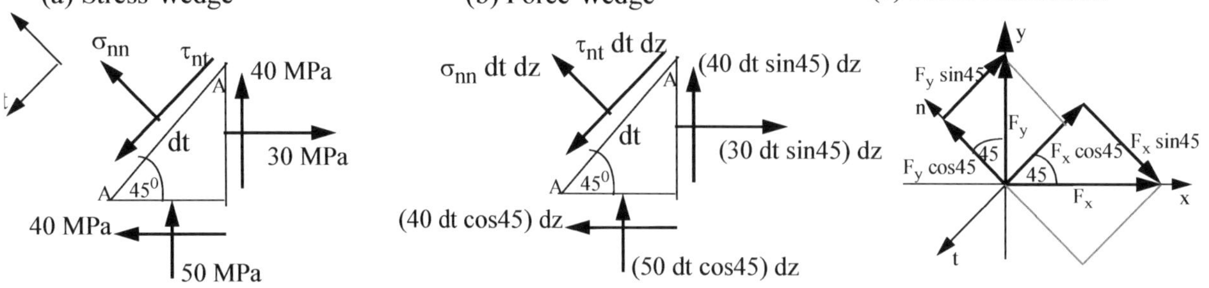

The forces in *x*, and *y* direction are given below and resolved into *n* ant *t* direction using Figure (c) for purpose of equilibrium.

$$F_x = [(30dtsin45)dz - (40dtcos45)dz] \quad and \quad F_y = [(40dtsin45)dz + (50dtcos45)dz] \tag{1}$$

$$\sigma_{nn}dtdz - F_x sin45 + F_y cos45 = \sigma_{nn}dtdz - [30dtsin45dz - 40dtcos45dz]sin45 + [40dtsin45dz + 50dtcos45dz]cos45 = 0 \text{ or}$$

$$\sigma_{nn} = 30sin^2 45 - 50cos^2 45 - 2(40)sin45cos45 = 50\text{ MPa} \tag{2}$$

$$\tau_{nt}dtdz - F_x cos45 - F_y sin45 = \tau_{nt}dtdz - [30dt sin45dz - 40dt cos45dz]cos45 - [40dt sin45dz + 50dt cos45dz]sin45 = 0 \text{ or}$$

$$\tau_{nt} = 30 sin45 cos45 - 40 cos^2 45 + 40 sin^2 45 + 50 cos45 sin45 = 40 \text{ MPa} \quad (3)$$

ANS $\sigma_{nn} = 50$ MPa (C); ; $\tau_{nt} = 40$ MPa

---

8. 20

Solution $\sigma_{nn} = ?$ $\tau_{nt} = ?$

The stress cube is as given. The stress wedge, the local coordinate system, and the force wedge are as shown below

(a) Stress Wedge

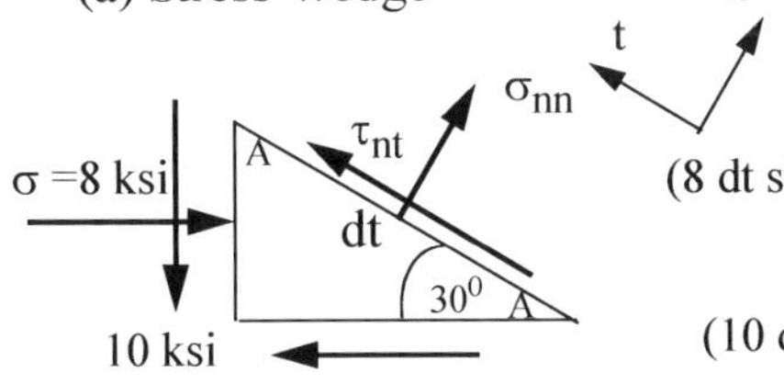

(b) Force Wedge

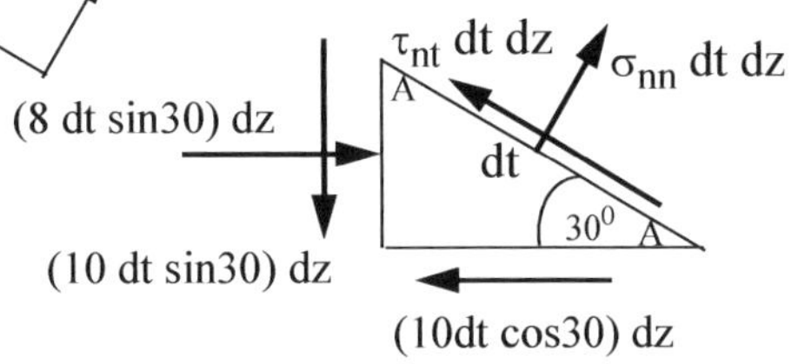

(c) Force Resolution

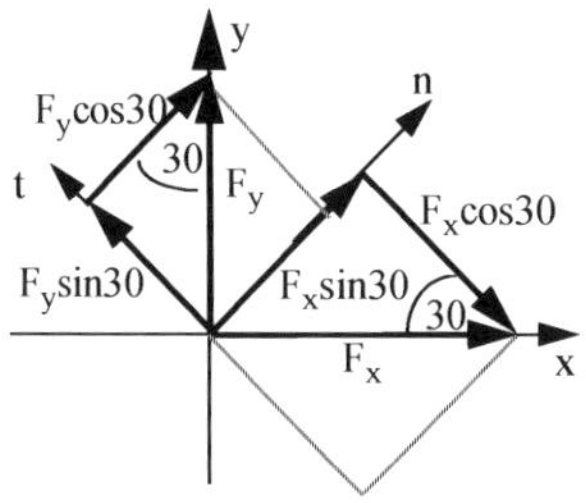

The forces in x, and y direction are given below and resolved into n ant t direction using Figure (c) for purpose of equilibrium.

$$F_x = [(8dt sin30)dz - (10dt cos30)dz] \quad and \quad F_y = -(10dt sin30)dz \quad (1)$$

$$\sigma_{nn}dtdz + F_x sin30 + F_y cos30 = \sigma_{nn}dtdz + [8dt sin30dz - 10dt cos30dz]sin30 + [-10dt sin30dz]cos30 = 0 \text{ or}$$

$$\sigma_{nn} = -8 sin^2 30 + 2(10) sin30 cos30 = 6.66 \text{ ksi} \quad (2)$$

$$\tau_{nt}dtdz - F_x cos30 + F_y sin30 = \tau_{nt}dtdz - [(8dt sin30)dz - (10dt cos30)dz]cos30 + [-(10dt sin30)dz]sin30 = 0 \text{ or}$$

$$\tau_{nt} = 8 sin30 cos30 - 10 cos^2 30 + 10 sin^2 30 = -1.53 \text{ ksi} \quad (3)$$

ANS $\sigma_{nn} = 6.66$ ksi (T) ; $\tau_{nt} = -1.53$ ksi

---

8. 21

Solution $\sigma_{nn} = ?$ $\tau_{nt} = ?$

The stress cube is as given. The stress wedge, the local coordinate system, and the force wedge are as shown below.

(a) Stress Wedge

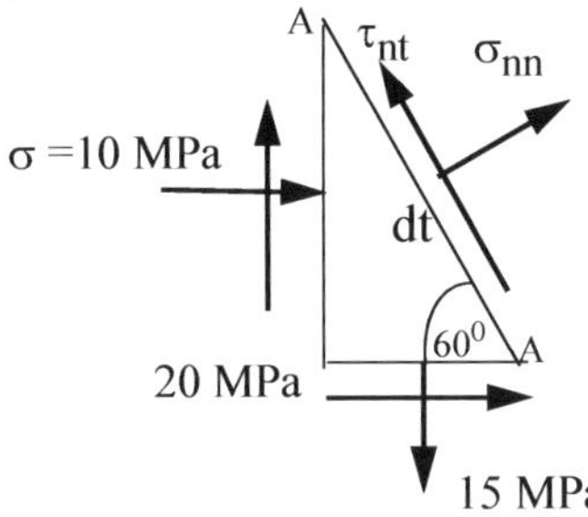

(b) Force Wedge

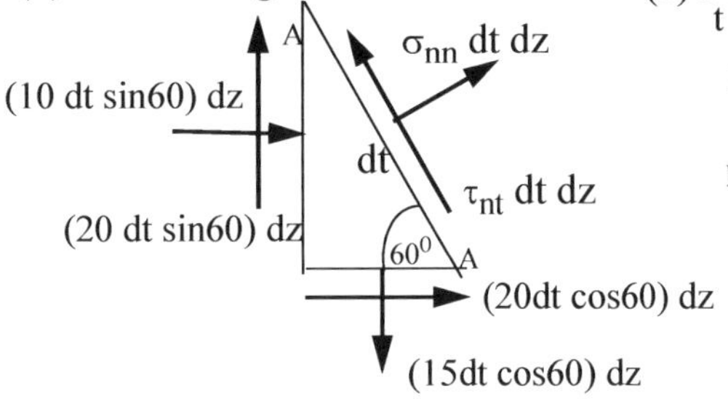

(c) Force Resolution

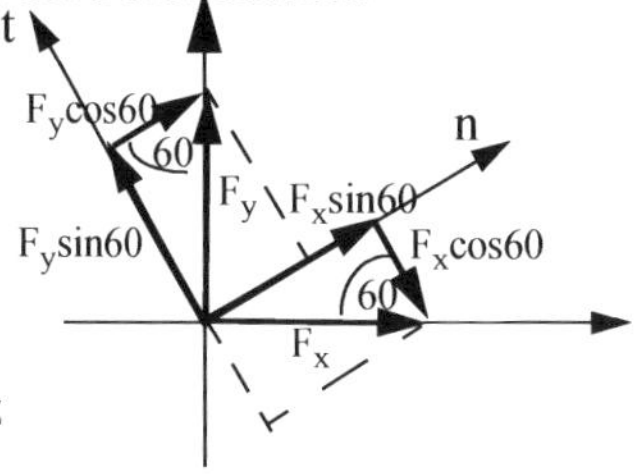

The forces in x, and y direction are given below and resolved into n ant t direction using Figure (c) for purpose of equilibrium.

$$F_x = [(10dt sin60)dz + (20dt cos60)dz] \quad and \quad F_y = [(20dt sin60)dz - (15dt cos60)dz] \quad (1)$$

$$\sigma_{nn}dtdz + F_x sin60 + F_y cos60 = \sigma_{nn}dtdz + [10dt sin60dz + 20dt cos60dz]sin60 + [20dt sin60dz - 15dt cos60dz]cos60 = 0 \text{ or}$$

$$\sigma_{nn} = -10 sin^2 60 + 15 cos^2 60 - 2(20) sin60 cos60 = 21.07 \text{ MPa} \quad (2)$$

$$\tau_{nt}dtdz - F_x cos60 + F_y sin60 = \tau_{nt}dtdz - [10dt sin60dz + 20dt cos60dz]cos60 + [20dt sin60dz - 15dt cos60dz]sin60 = 0 \text{ or}$$

$$\tau_{nt} = 10 sin60 cos60 + 20 cos^2 60 - 20 sin^2 60 + 15 sin60 cos60 = 0.825 \text{ MPa} \quad (3)$$

ANS $\sigma_{nn} = 21.07$ MPa (C) ; $\tau_{nt} = 0.825$ MPa

---

8. 22

Solution $\sigma_{nn} = ?$ $\tau_{nt} = ?$

The stress cube is as given. The stress wedge, the local coordinate system, and the force wedge are as shown below.
The forces in x, and y direction are given below and resolved into n ant t direction using Figure (c) for purpose of equilibrium.

$$F_x = [(45dt sin28)dz + (20dt cos28)dz] \quad and \quad F_y = [-(20dt sin28)dz - (15dt cos28)dz] \quad (1)$$

$$\sigma_{nn}dtdz - F_x sin28 + F_y cos28 = \sigma_{nn}dtdz - [45dtsin28dz + 20dtcos28dz]sin28 + [-20dtsin28dz - 15dtcos28dz]cos28 = 0 \text{ or}$$

$$\sigma_{nn} = 45sin^2 28 + 15cos^2 28 + 2(20)sin28cos28 = 38.19 \text{ MPa} \qquad (2)$$

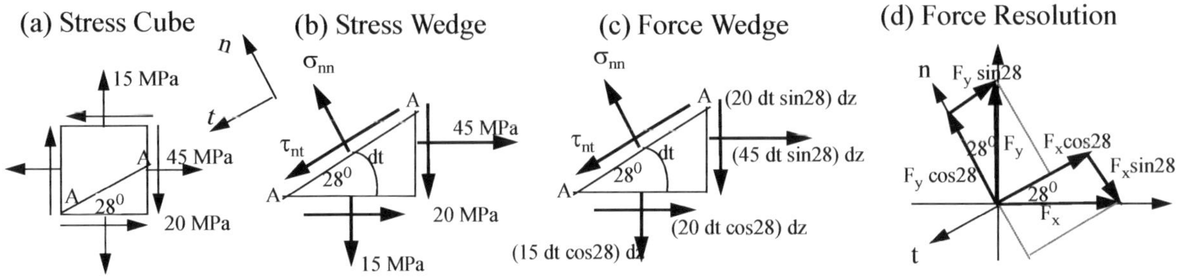

$$\tau_{nt}dtdz - F_x cos28 - F_y sin28 = \tau_{nt}dtdz - [45dtsin28dz + 20dtcos28dz]cos28 - [-20dtsin28dz - 15dtcos28dz]sin28 = 0 \text{ or}$$

$$\tau_{nt} = 45sin28cos28 + 20cos^2 28 - 20sin^2 28 - 15cos28sin28 = 23.62 \text{ MPa} \qquad (3)$$

ANS $\sigma_{nn} = 38.19$ MPa (T) ; $\tau_{nt} = 23.62$ MPa

---

8. 23

Solution $\sigma_{nn} = ?$ $\tau_{nt} = ?$

The stress cube is as given. The stress wedge, the local coordinate system, and the force wedge are as shown below.

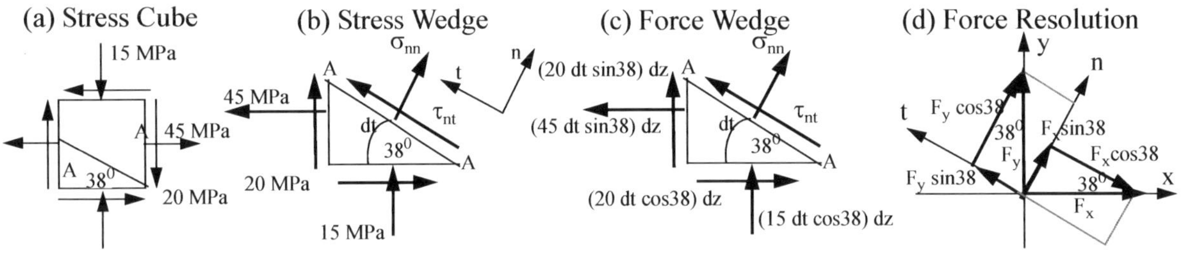

The forces in *x*, and *y* direction are given below and resolved into *n* ant *t* direction using Figure (c) for purpose of equilibrium.

$$F_x = [(-45dtsin38)dz + (20dtcos38)dz] \quad and \quad F_y = [(20dtsin38)dz + (15dtcos38)dz] \qquad (1)$$

$$\sigma_{nn}dtdz + F_x sin38 + F_y cos38 = \sigma_{nn}dtdz + [-45dtsin38dz + 20dtcos38dz]sin38 + [20dtsin38dz + 15dtcos38dz]cos38 = 0 \text{ or}$$

$$\sigma_{nn} = 45sin^2 38 - 15cos^2 38 - 2(20)sin38cos38 = 11.66 \text{ MPa} \qquad (2)$$

$$\tau_{nt}dtdz - F_x cos38 + F_y sin38 = \tau_{nt}dtdz - [-45dtsin38dz + 20dtcos38dz]cos38 + [20dtsin38dz + 15dtcos38dz]sin38 = 0 \text{ or}$$

$$\tau_{nt} = -45sin38cos38 + 20cos^2 38 - 20sin^2 38 - 15cos38sin38 = -24.27 \text{ MPa} \qquad (3)$$

ANS $\sigma_{nn} = 11.66$ MPa (C) ; $\tau_{nt} = -24.27$ MPa

---

8. 24

Solution $\sigma_{nn} = ?$ $\tau_{nt} = ?$

The stress cube is as given. The stress wedge, the local coordinate system, and the force wedge are as shown below.

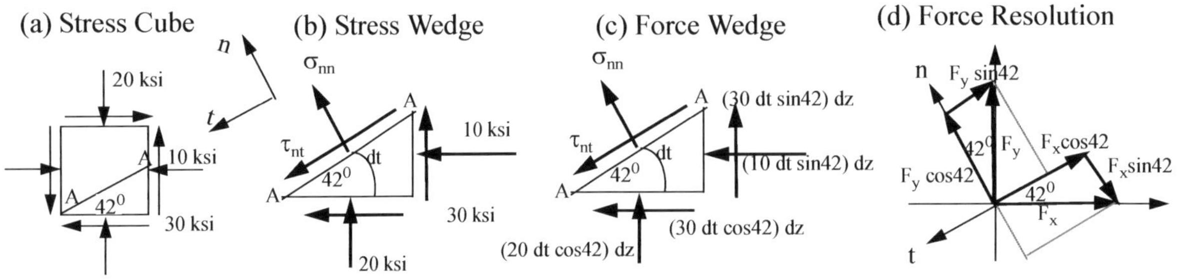

The forces in *x*, and *y* direction are given below and resolved into *n* ant *t* direction using Figure (c) for purpose of equilibrium.

$$F_x = [-(10dtsin42)dz - (30dtcos42)dz] \quad and \quad F_y = [(30dtsin42)dz + (20dtcos42)dz] \qquad (1)$$

$$\sigma_{nn}dtdz - F_x sin42 + F_y cos42 = \sigma_{nn}dtdz - [-10dtsin42dz - 30dtcos42dz]sin42 + [30dtsin42dz + 20dtcos42dz]cos42 = 0 \text{ or}$$

$$\sigma_{nn} = -10sin^2 42 - 20cos^2 42 - 2(30)sin42cos42 = 45.36 \text{ ksi} \qquad (2)$$

$$\tau_{nt}dtdz - F_x\cos 42 - F_y\sin 42 = \tau_{nt}dtdz - [-10dt\sin 42dz - 30dt\cos 42dz]\cos 42 - [30dt\sin 42dz + 20dt\cos 42dz]\sin 42 = 0 \text{ or}$$

$$\tau_{nt} = -10\sin 42\cos 42 - 30\cos^2 42 + 30\sin^2 42 + 20\cos 42\sin 42 = 1.84 \text{ ksi} \quad (3)$$

ANS $\sigma_{nn} = 45.36$ ksi (C); $\tau_{nt} = 1.84$ ksi

---

8. 25

Solution d = 25 mm σ = 330 MPa T = ?

The stress wedge, the local coordinate system, and the force wedge are as shown below.

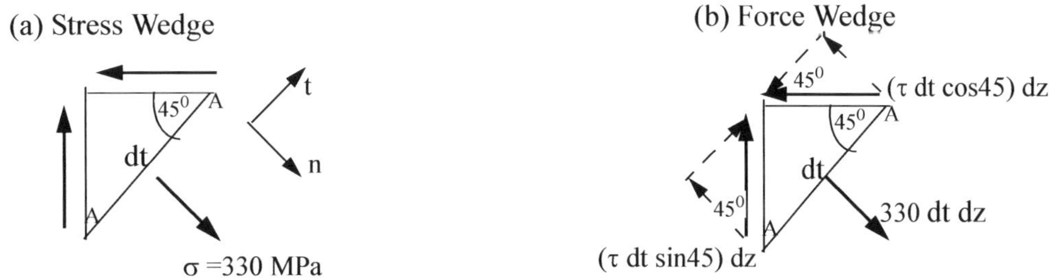

By equilibrium of forces in the *n* and *t* direction we obtain the following.

$$330dtdz - [\tau dt\cos 45dz]\sin 45 - [\tau dt\sin 45dz]\cos 45 = 0 \quad or \quad 2\tau\cos 45\sin 45 = 330 \quad or \quad \tau = 330 \ MPa \quad (1)$$

From torsional shear stress formula we obtain:

$$J = \pi(0.025^4/32) = 38.35(10^{-9})m^4 \qquad \tau = \frac{T\rho}{J} = \frac{T(0.0125)}{38.35(10^{-9})} = 330(10^6) \quad or \quad T = 1012 \text{ N-m} \quad (2)$$

ANS $T = 1012$ N-m

---

8. 26

Solution $\sigma_a \le 500psi$ $\tau_a \le 200psi$ P = ?

The stress cube is as given. The stress wedge, the local coordinate system, and the force wedge are as shown below.

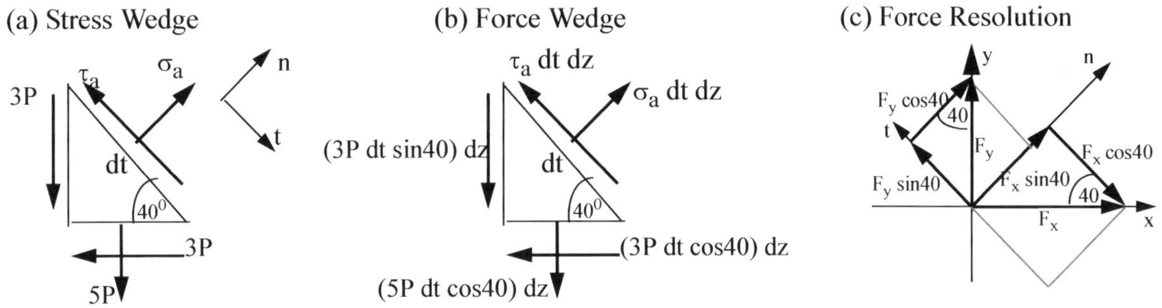

The forces in *x*, and *y* direction are given below and resolved into *n* ant *t* direction using Figure (c) for purpose of equilibrium.

$$F_x = -(3Pdt\cos 40)dz \quad and \quad F_y = [(-3Pdt\sin 40)dz - (5Pdt\cos 40)dz] \quad (1)$$

$$\sigma_a dtdz + F_x\sin 40 + F_y\cos 40 = \sigma_a dtdz + [-3Pdt\cos 40dz]\sin 40 + [-3Pdt\sin 40dz - 5Pdt\cos 40dz]\cos 40 = 0 \text{ or}$$

$$\sigma_a = 6P\sin 40\cos 40 + 5P\cos^2 40 = 5.888P \le 500 \quad or \quad P \le 84.9 \ lb \quad (2)$$

$$\tau_a dtdz - F_x\cos 40 + F_y\sin 40 = \tau_a dtdz - [-3Pdt\cos 40dz]\cos 40 + [-3Pdt\sin 40dz - 5Pdt\cos 40dz]\sin 40 = 0 \text{ or}$$

$$\tau_a = -3P\cos^2 40 + 3P\sin^2 40 + 5P\cos 40\sin 40 = 1.941P \le 200 \quad or \quad P \le 103 \ lb \quad (3)$$

ANS $P_{max} = 84.9$ lb

---

8. 27

Solution $\theta_x = 60^o$ $\theta_y = -60^o$ $\theta_z = 45^o$ $\sigma_{nn} = ?$

The unit normal vector to the inclined plane and the projected areas in the coordinate directions are as shown below.

$$\bar{i}_n = \cos\theta_x\bar{i} + \cos\theta_y\bar{j} + \cos\theta_z\bar{k} = 0.5\bar{i} + 0.5\bar{j} + 0.7071\bar{k} \quad (1)$$

$$A_x = \cos\theta_x A = 0.5A \qquad A_y = \cos\theta_y A = 0.5A \qquad A_z = \cos\theta_z A = 0.7071A \quad (2)$$

Figure (a) shows the stress wedge. Figure (b) shows the force wedge. The forces on the coordinate planes can be written as a force vector shown below.

$$\bar{F} = -8A_x\bar{i} - 12A_y\bar{j} + 8A_z\bar{k} = (-4\bar{i} - 6\bar{j} + 5.657\bar{k})A \quad (3)$$

(a) Stress Wedge

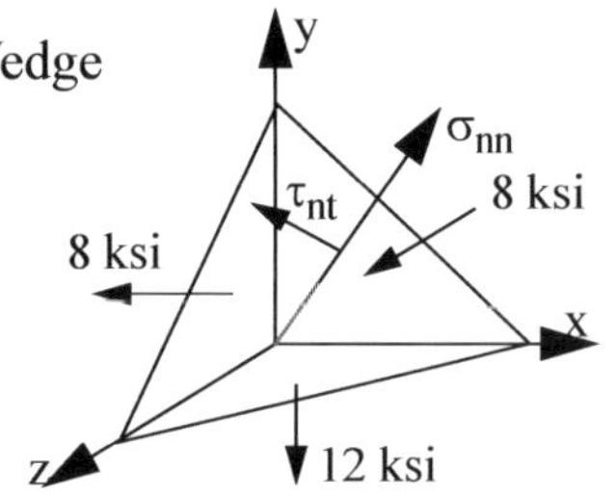

(a) Force Wedge

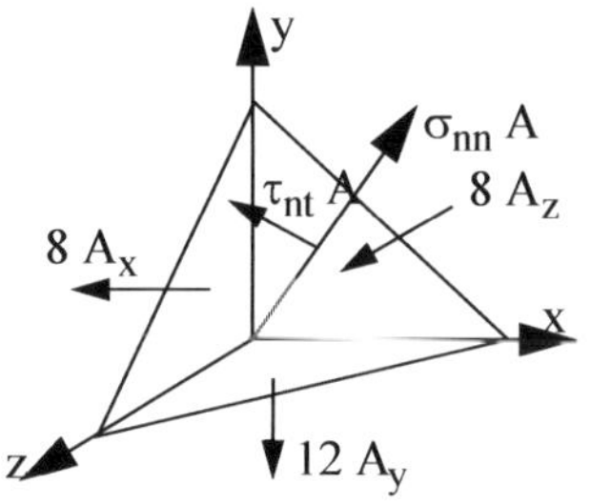

By equilibrium of forces in the n direction we obtain the following.

$$\sigma_{nn}A + \bar{F}\bullet\bar{i}_n = 0 \quad or \quad \sigma_{nn}A + (-4\bar{i} - 6\bar{j} + 5.657\bar{k})A \bullet 0.5\bar{i} + 0.5\bar{j} + 0.7071\bar{k} = 0 \text{ or}$$

$$\sigma_{nn} - (4)(0.5) - (6)(0.5) + (5.657)(0.7071) = 0 \quad or \quad \sigma_{nn} = 1.0 \text{ ksi} \quad (4)$$

ANS $\sigma_{nn} = 1.0$ ksi (T)

---

8. 28

Solution $\theta_{\mathbf{x}} = 72.54^{\circ}$ $\theta_{\mathbf{y}} = 120^{\circ}$ $\theta_{\mathbf{z}} = 35.67^{\circ}$ $\sigma_{nn} = ?$

The unit normal vector to the inclined plane is:

$$\bar{i}_n = cos\theta_x\bar{i} + cos\theta_y\bar{j} + cos\theta_z\bar{k} = 0.3\bar{i} - 0.5\bar{j} + 0.8123\bar{k} \quad (1)$$

The projected areas in the coordinate directions are:

$$A_x = cos\theta_x A = 0.3A \qquad A_y = cos\theta_y A = -0.5A \qquad A_z = cos\theta_z A = 0.8123A \quad (2)$$

Figure (a) shows the stress wedge. Figure (b) shows the force wedge.

(a) Stress Wedge

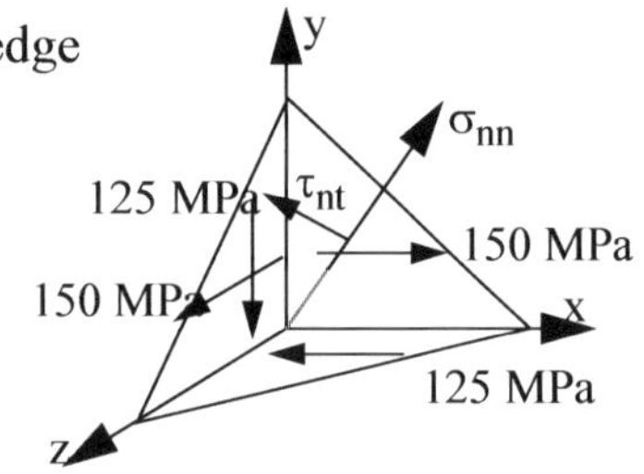

(a) Force Wedge

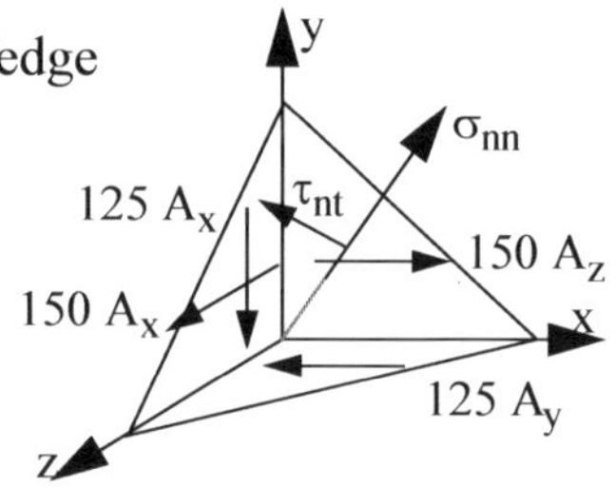

The forces on the coordinate planes can be written as a force vector shown below.

$$\bar{F} = (150A_z - 125A_y)\bar{i} - 125A_x\bar{j} + 150A_x\bar{k} = (184.345\bar{i} - 37.5\bar{j} + 45\bar{k})A \quad (3)$$

By equilibrium of forces in the n direction we obtain the following.

$$\sigma_{nn}A + \bar{F}\bullet\bar{i}_n = 0 \quad or \quad \sigma_{nn}A + (184.345\bar{i} - 37.5\bar{j} + 45\bar{k})A \bullet (0.3\bar{i} - 0.5\bar{j} + 0.8123\bar{k}) = 0 \text{ or}$$

$$\sigma_{nn} + (184.345)(0.3) - (37.5)(0.5) + (45)(0.8123) = 0 \quad (4)$$

ANS $\sigma_{nn} = 73.1$ ksi (C)

Sections 8.2-8.3

8. 29

Solution

From the given figure we have:

$$R = \sqrt{\left(\frac{\sigma_{xx} - \sigma_{yy}}{2}\right)^2 + \tau_{xy}^2} \qquad sin2\theta_{1,2} = \pm\tau_{xy}/R \qquad cos2\theta_{1,2} = \pm\left(\frac{\sigma_{xx} - \sigma_{yy}}{2}\right)/R \quad (1)$$

$$\sigma_{1,2} = \frac{\sigma_{xx} + \sigma_{yy}}{2} + \frac{\sigma_{xx} - \sigma_{yy}}{2}cos\,2\theta_{1,2} + \tau_{xy}sin\,2\theta_{1,2} = \frac{(\sigma_{xx} + \sigma_{yy})}{2} + \frac{(\sigma_{xx} - \sigma_{yy})}{2}\left[\pm\left(\frac{\sigma_{xx} - \sigma_{yy}}{2}\right)/R\right] + \tau_{xy}[\pm\tau_{xy}/R] \text{ or}$$

$$\sigma_{1,2} = \frac{(\sigma_{xx} + \sigma_{yy})}{2} \pm \frac{1}{R}\left[\left(\frac{\sigma_{xx} - \sigma_{yy}}{2}\right)^2 + \tau_{xy}^2\right] = \frac{(\sigma_{xx} + \sigma_{yy})}{2} \pm \frac{1}{R}[R^2] = \frac{(\sigma_{xx} + \sigma_{yy})}{2} + R = \frac{(\sigma_{xx} + \sigma_{yy})}{2} \pm \sqrt{\left(\frac{\sigma_{xx} - \sigma_{yy}}{2}\right)^2 + \tau_{xy}^2} \quad (2)$$

$$\tau_{1,2} = -\frac{\sigma_{xx} - \sigma_{yy}}{2}sin\,2\theta_{1,2} + \tau_{xy}cos2\theta_{1,2} = -\frac{(\sigma_{xx} - \sigma_{yy})}{2}[\pm\tau_{xy}/R] + \tau_{xy}\left[\pm\left(\frac{\sigma_{xx} - \sigma_{yy}}{2}\right)/R\right] = 0 \quad (3)$$

The above equations are the desired results.

---

8. 30

Solution

From given figure we have:

$$R = \sqrt{\{(\sigma_{xx}-\sigma_{yy})/2\}^2+\tau_{xy}^2} \qquad sin2\bar{\theta}_{1,2} = \mp\{(\sigma_{xx}-\sigma_{yy})/2\}/R \qquad cos2\bar{\theta}_{1,2} = \pm\tau_{xy}/R \tag{1}$$

$$\sigma_{av} = \frac{\sigma_{xx}+\sigma_{yy}}{2}+\frac{\sigma_{xx}-\sigma_{yy}}{2}cos\,2\theta_{1,2}+\tau_{xy}sin\,2\theta_{1,2} = \frac{(\sigma_{xx}+\sigma_{yy})}{2}+\frac{(\sigma_{xx}-\sigma_{yy})}{2}[\pm\tau_{xy}/R]+\tau_{xy}\left[\mp\left(\frac{\sigma_{xx}-\sigma_{yy}}{2}\right)/R\right] = \frac{(\sigma_{xx}+\sigma_{yy})}{2} \tag{2}$$

$$\tau_p = -\{(\sigma_{xx}-\sigma_{yy})/2\}\,sin\,2\theta_{1,2}+\tau_{xy}cos2\theta_{1,2} = -\{(\sigma_{xx}-\sigma_{yy})/2\}[\mp\{(\sigma_{xx}-\sigma_{yy})/2\}/R]+\tau_{xy}[\pm\tau_{xy}/R] \text{ or}$$

$$\tau_p = \pm[\{(\sigma_{xx}-\sigma_{yy})/2\}^2+\tau_{xy}^2]/R = \pm R = \pm[\sigma_1-\sigma_2] \tag{3}$$

The above equations are the desired results.

---

8. 31

Solution

$$\sigma_{tt} = \sigma_{xx}sin^2\theta+\sigma_{yy}cos^2\theta-2\tau_{xy}cos\theta sin\theta = \{(\sigma_{xx}+\sigma_{yy})/2\}+\{(\sigma_{xx}-\sigma_{yy})/2\}cos2\theta-\tau_{xy}sin2\theta \tag{1}$$

By differentiating with respect to θ and letting its value be zero at $\theta = \theta_t$, we obtain

$$\left.\frac{d\sigma_{tt}}{d\theta}\right|_{\theta=\theta_t} = \left.\left(-2\frac{(\sigma_{yy}-\sigma_{xx})sin2\theta}{2}-2\tau_{xy}cos2\theta\right)\right|_{\theta=\theta_t} = 0 \quad or \quad tan2\theta_t = \frac{-2\tau_{xy}}{(\sigma_{yy}-\sigma_{xx})} = \frac{2\tau_{xy}}{(\sigma_{xx}-\sigma_{yy})} = tan2\theta_p \tag{2}$$

As θ_t and θ_p are equivalent, we obtain

$$R = \sqrt{\{(\sigma_{xx}-\sigma_{yy})/2\}^2+\tau_{xy}^2} \qquad sin2\theta_{1,2} = \pm\tau_{xy}/R \qquad cos2\theta_{1,2} = \pm\{(\sigma_{xx}-\sigma_{yy})/2\}/R \tag{3}$$

$$\sigma_{1,2} = \frac{(\sigma_{xx}+\sigma_{yy})}{2}+\frac{(\sigma_{yy}-\sigma_{xx})}{2}\left[\pm\left(\frac{\sigma_{xx}-\sigma_{yy}}{2}\right)/R\right]-\tau_{xy}[\pm\tau_{xy}/R] = \frac{(\sigma_{xx}+\sigma_{yy})}{2}-\left[\pm\frac{1}{R}\left[\left(\frac{\sigma_{xx}-\sigma_{yy}}{2}\right)^2+\tau_{xy}^2\right]\right] \text{ or}$$

$$\sigma_{1,2} = \{(\sigma_{xx}+\sigma_{yy})/2\}\mp[R^2]/R = \{(\sigma_{xx}+\sigma_{yy})/2\}\mp R = \{(\sigma_{xx}+\sigma_{yy})/2\}\mp\sqrt{\{(\sigma_{xx}-\sigma_{yy})/2\}^2+\tau_{xy}^2} \tag{4}$$

The above equations are the desired results.

---

8. 32

Solution $\sigma_A = ?$ $\tau_A = ?$

The Mohr's circle with the inclined plane are drawn below. Using the local coordinate system the signs of shear stress is determined as shown below.

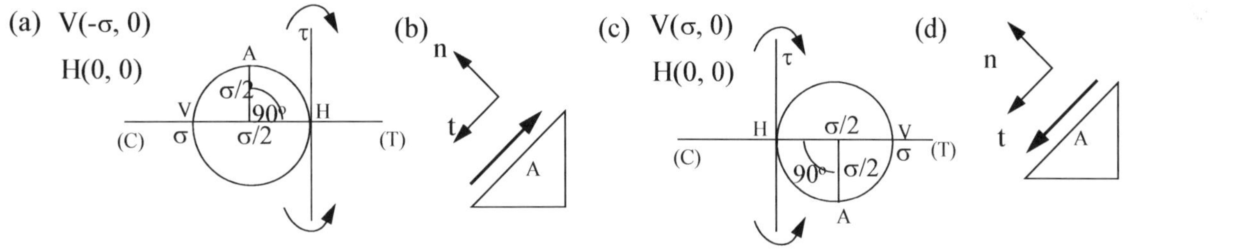

ANS (a) $\sigma_A = (\sigma/2)\,(C)$; $\tau_A = -(\sigma/2)$ (b) $\sigma_A = (\sigma/2)\,(T)$; $\tau_A = (\sigma/2)$

---

8. 33

Solution $\sigma_A = ?$ $\tau_A = ?$

(a) The Mohr's circle associated is a point. Thus all planes have the same normal stress s and zero shear stress.

(b) The Mohr's circle with the inclined plane are drawn below. Using the local coordinate system the signs of shear stress is determined as shown below.

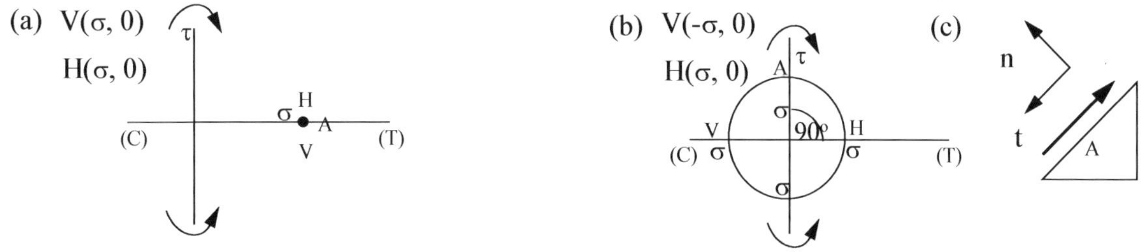

ANS (a) $\sigma_A = \sigma\,(T)$; $\tau_A = 0$ (b) $\sigma_A = 0$; $\tau_A = -\sigma$

---

8. 34

Solution $\sigma_A = ?$ $\tau_A = ?$

The Mohr's circle are drawn and stresses determined by inspection.

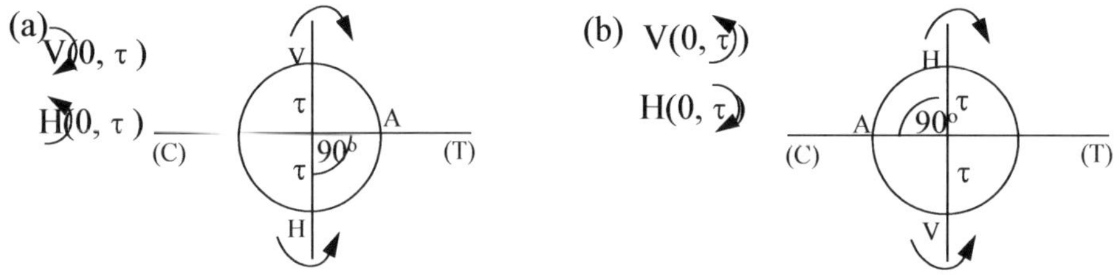

ANS (a) $\sigma_A = \tau\,(T)$; $\tau_A = 0$ (b) $\sigma_A = \tau\,(C)$; $\tau_A = 0$

---

8. 35

Solution

Figure (a) shows the stress cube with the torsional shear stress as shown in the given figure. Figure (b) shows the Mohr's circle associated with the stress cube. Figure (c) shows the principal element.

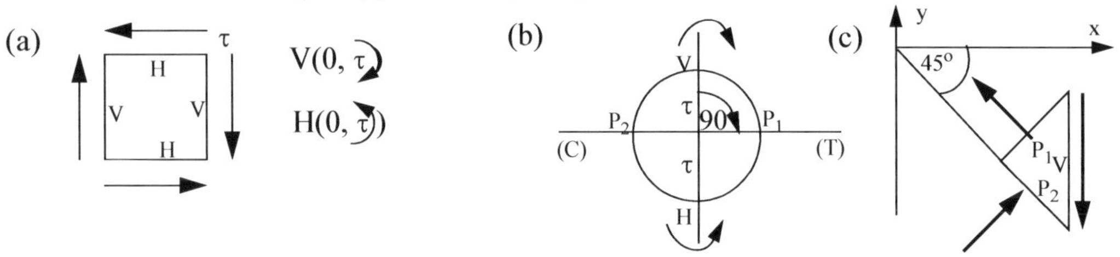

Cast iron being a brittle material will fail from maximum tensile stress that is from principle stress one. From the principal element in Figure (c) we see that the orientation of the principal plane one (P_1) is as shown in the given figure for cast iron under torsion. Aluminum being a ductile material will fail from maximum shear stress which is on the vertical (V) plane of the principal element. The orientation is the same as shown in the given figure for aluminum bar under torsion.

---

8. 36

Solution $\sigma_A = ?$ $\tau_A = ?$

From Figures (a) and (b) we obtain the following.

$$\sigma_{xx} = 30MPa \qquad \sigma_{yy} = -50MPa \qquad \tau_{xy} = 40MPa \qquad \theta = 135^o \tag{1}$$

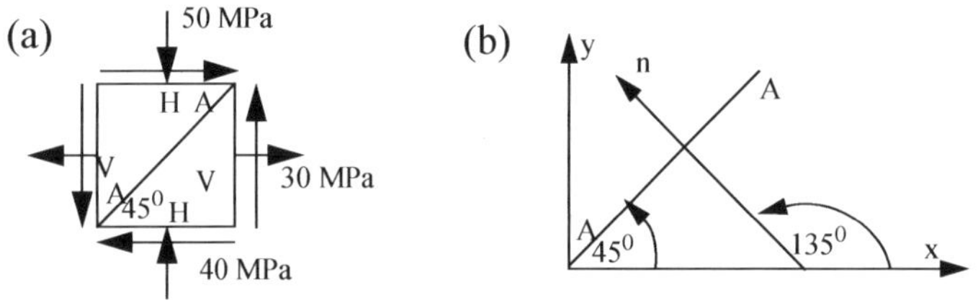

$$\sigma_A = (30)cos^2 135 + (-50)sin^2 135 + 2(40)sin 135 cos 135 = 50MPa \tag{1}$$

$$\tau_A = -(30)cos 135 sin 135 + (-50)sin 135 cos 135 + (40)(cos^2 135 - sin^2 135) = 40MPa \tag{2}$$

ANS $\sigma_A = 50MPa(C)$; $\tau_A = 40MPa$

---

8. 37

Solution $\sigma_A = ?$ $\tau_A = ?$

The Mohr's circle and the inclined plane with local coordinates are drawn and used to calculate the stresses on the inclined plane as shown below.

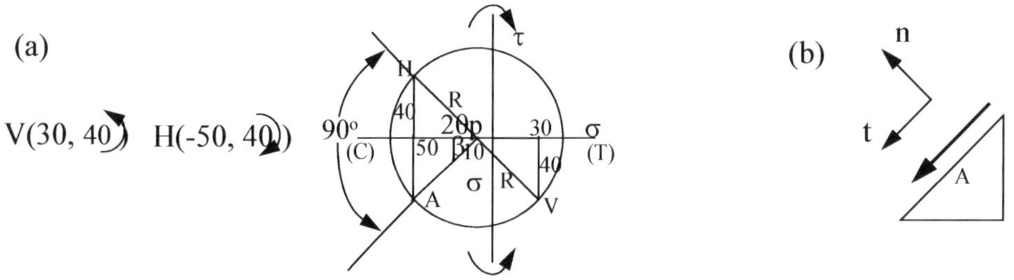

$$R = \sqrt{40^2 + 40^2} = 56.57MPa \qquad 2\theta_p = atan(40/40) = 45^o \qquad \beta = 90^o - 45^o = 45^o \quad \textbf{(1)}$$

$$\sigma_A = -10 - Rcos\beta = -50 \qquad |\tau_A| = Rsin\beta = 40 \quad \textbf{(2)}$$

ANS $\sigma_A = 50MPa(C)$; $\tau_A = 40MPa$

---

8. 38

Solution $\sigma_A = ?$ $\tau_A = ?$

From Figures (a) and (b) we obtain the following.

$$\sigma_{xx} = -8ksi \qquad \sigma_{yy} = 0 \qquad \tau_{xy} = 10ksi \qquad \theta = 60^o \quad \textbf{(1)}$$

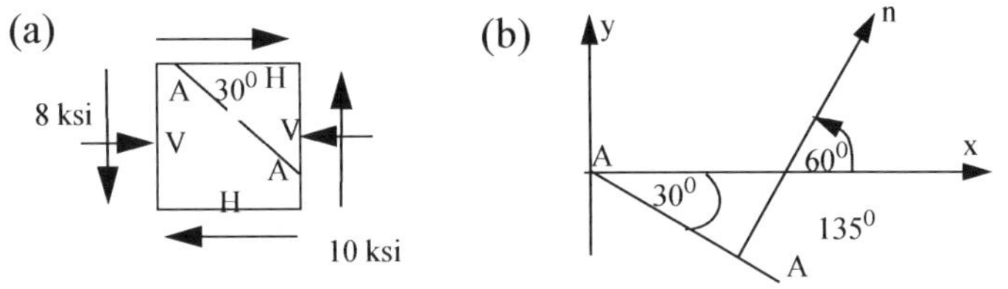

$$\sigma_A = (-8)cos^2 60 + (0)sin^2 60 + 2(10)sin60cos60 = 6.66 \text{ ksi} \quad \textbf{(2)}$$

$$\tau_A = -(30)cos60sin60 + (-50)sin60cos60 + (40)(cos^2 60 - sin^2 60) \quad \textbf{(3)}$$

ANS $\sigma_A = 6.66$ ksi (T) ; $\tau_A = -1.54$ ksi

---

8. 39

Solution $\sigma_A = ?$ $\tau_A = ?$

The Mohr's circle and the inclined plane with local coordinates are drawn and used to calculate the stresses on the inclined plane as shown below

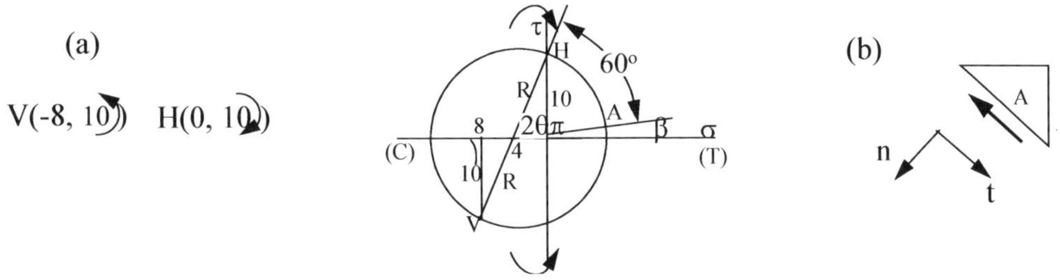

$$R = \sqrt{4^2 + 10^2} = 10.77ksi \qquad 2\theta_p = atan(10/4) = 68.2^o \qquad \beta = 68.2^o - 60^o = 8.2^o \quad \textbf{(4)}$$

$$\sigma_A = -4 + Rcos\beta = 6.66 \qquad |\tau_A| = Rsin\beta = 1.54 \quad \textbf{(5)}$$

ANS $\sigma_A = 6.66$ ksi (T) ; $\tau_A = -1.54$ ksi

---

8. 40

Solution $\sigma_A = ?$ $\tau_A = ?$

From Figures (a) and (b) we obtain the following.

$$\sigma_{xx} = -10MPa \qquad \sigma_{yy} = 15MPa \qquad \tau_{xy} = -20MPa \qquad \theta = 30^o \quad \textbf{(1)}$$

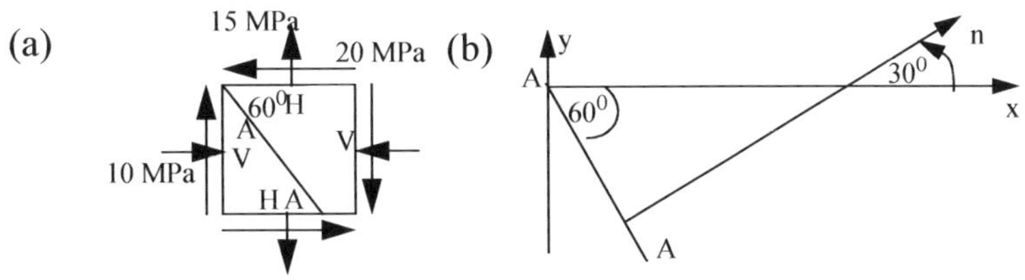

$$\sigma_A = (-10)cos^2 30 + (15)sin^2 30 + 2(-20)sin30cos30 = 21.1 \text{ MPa} \quad \textbf{(2)}$$

$$\tau_A = -(-10)cos30sin30 + (15)sin30cos30 + (-20)(cos^2 30 - sin^2 30) = 0.83 \text{ MPa} \quad \textbf{(3)}$$

ANS $\sigma_A = 21.1$ MPa (C) ; $\tau_A = 0.83$ MPa

8. 41

Solution $\sigma_A = ?$ $\tau_A = ?$

The Mohr's circle and the inclined plane are drawn and used to calculate the stresses on the inclined plane as shown below

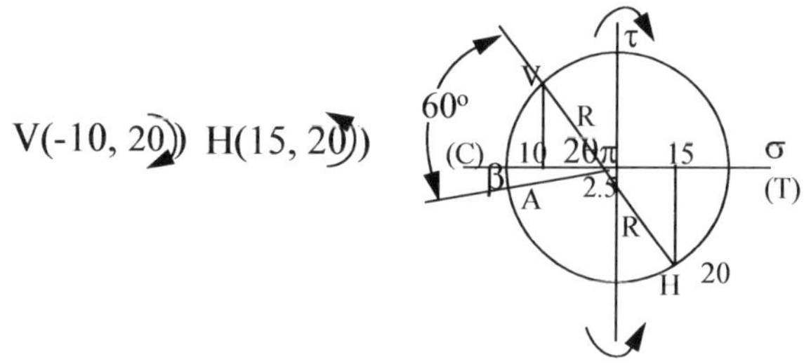

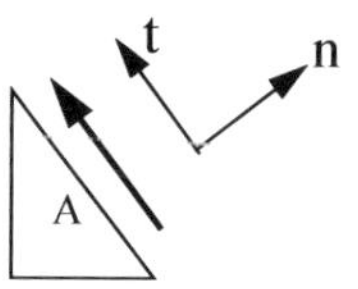

$$R = \sqrt{17.5^2 + 20^2} = 23.58MPa \qquad 2\theta_p = atan(20/17.5) = 57.995^o \qquad \beta = 60^o - 57.995^o = 2.005^o \tag{1}$$

$$\sigma_A = -2.5 - Rcos\beta = -21.1 \text{ MPa} \qquad |\tau_A| = Rsin\beta = 0.825 \text{ MPa} \tag{2}$$

ANS $\sigma_A = 21.1$ MPa (C) ; $\tau_A = 0.83$ MPa

---

8. 42

Solution Plane Stress $\sigma_A = ?$ $\tau_A = ?$ $\sigma_1 = ?$ $\sigma_2 = ?$ $\sigma_3 = ?$ $\theta_1 = ?$ $\tau_{max} = ?$

From the stress cube given and Figure (a) we obtain the following.

$$\sigma_{xx} = -30MPa \qquad \sigma_{yy} = 60MPa \qquad \tau_{xy} = 40MPa \qquad \theta = 118^o \tag{1}$$

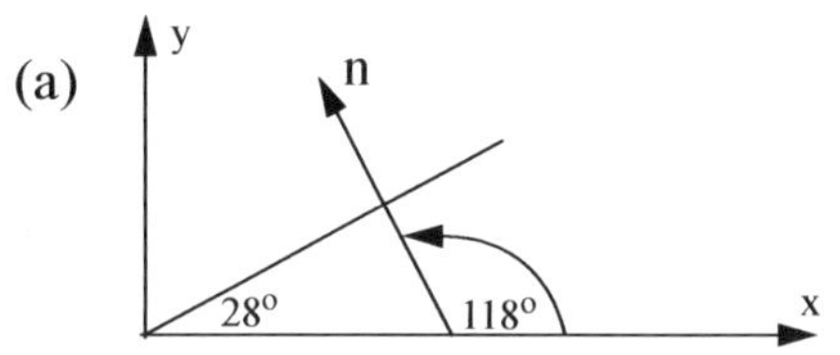

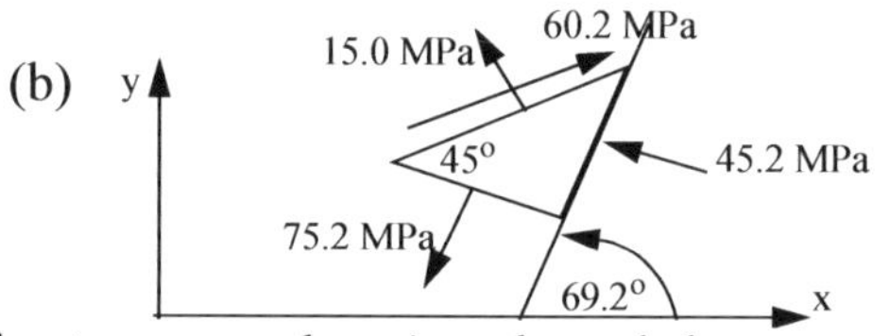

(a) Substituting in the normal and shear stress equations we obtain the stresses on plane A as shown below.

$$\sigma_A = (-30)cos^2 118 + (60)sin^2 118 + 2(40)sin 118 cos 118 = 7 \text{ MPa} \tag{2}$$

$$\tau_A = -(-30)cos 118 sin 118 + (60)sin 118 cos 118 + (40)(cos^2 118 - sin^2 118) = -59.7 \text{ MPa} \tag{3}$$

(b) The principal angle θ_p is calculated and substituted into the normal stress equation to determine one principal stress.

$$\theta_p = 0.5 \ atan\left(\frac{(2)(40)}{-30-60}\right) = -20.82^o \qquad \sigma_p = -30cos^2(-20.82) + 60sin^2(-20.82) + 2(40)sin(-20.82)cos(-20.82) = -45.2MPa \tag{4}$$

The other principal stress is: $-30 + 60 - (45.2) = 75.2MPa$. As this stress is greater than the stress found above, therefore this is principal stress one. Thus principal angle 1 is $\theta_1 = \theta_p \pm 90$ For plane stress we have the following answers:

$$\sigma_1 = 75.2 \text{ MPa (T)} \qquad \sigma_2 = 45.2 \text{ MPa (C)} \qquad \sigma_3 = 0 \qquad \theta_1 = 69.2° \ or \ -110.8^o \tag{5}$$

(c) The largest difference between principal stresses is between σ_1 and σ_2.

$$\tau_{max} = |(\sigma_1 - \sigma_2)/2| = |(75.2 + 45.2)/2| \tag{6}$$

(d) The average normal stress is: $\sigma_{avg} = (\sigma_{xx} + \sigma_{yy})/2 = 15MPa(T)$. The principal element is shown in Figure (b).

ANS $\sigma_A = 7$ MPa (T) ; $\tau_A = -59.7$ MPa ; $\sigma_1 = 75.2$ MPa (T) ; $\sigma_2 = 45.2$ MPa (C) ; $\sigma_3 = 0$; $\theta_1 = 69.2°$ or -110.8^o ;

$\tau_{max} = 60.2$ MPa

---

8. 43

Solution Plane Stress $\sigma_A = ?$ $\tau_A = ?$ $\sigma_1 = ?$ $\sigma_2 = ?$ $\sigma_3 = ?$ $\theta_1 = ?$ $\tau_{max} = ?$

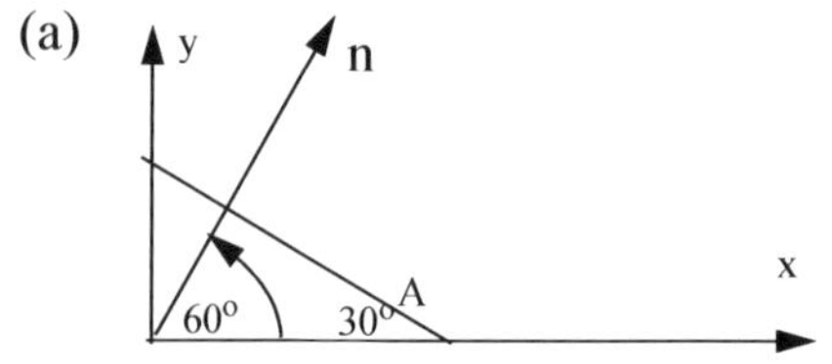

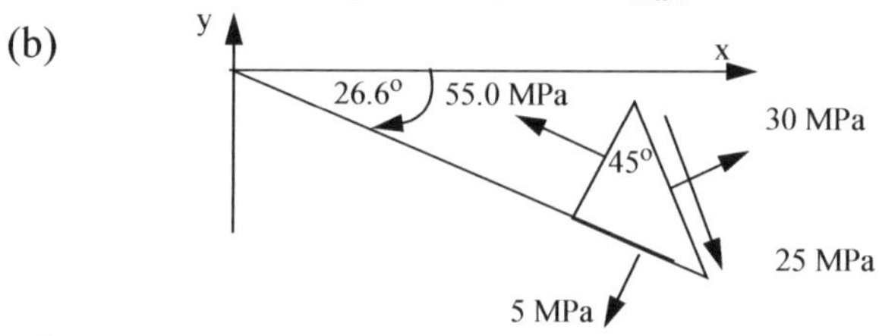

From the stress cube given and Figure (a) we obtain the following.

$$\sigma_{xx} = 45MPa \qquad \sigma_{yy} = 15 \ MPa \qquad \tau_{xy} = -20MPa \qquad \theta = 60^o \tag{1}$$

(a) Substituting in the normal and shear stress equations we obtain the stresses on plane A as shown below.

$$\sigma_A = (45)cos^2 60 + (15)sin^2 60 + 2(-20)sin 60 cos 60 = 5.2 \text{ MPa} \quad \textbf{(2)}$$

$$\tau_A = -(45)cos 60 sin 60 + (15)sin 60 cos 60 + (-20)(cos^2 60 - sin^2 60) = -3.0 \text{ MPa} \quad \textbf{(3)}$$

(b) The principal angle θ_p is calculated and substituted into the normal stress equation to determine one principal stress.

$$\theta_p = 0.5\, atan\left(\frac{2(-20)}{45-15}\right) = -26.56^o \qquad \sigma_p = 45 cos^2(-26.56^o) + 15 sin^2(-26.56^o) + 2(-20) sin(-26.56^o) cos(-26.56^o) = 55 MPa \quad \textbf{(4)}$$

The other principal stress as: $45 + 15 - (55) = 5.0 MPa$. As this stress is smaller than the stress above, therefore the stress above is principal stress one. Thus principal angle 1 is θ_p. For plane stress we have the following answers;

$$\sigma_1 = 55 \text{ MPa (T)} \qquad \sigma_2 = 5 \text{ MPa (T)} \qquad \sigma_3 = 0 \qquad \theta_1 = -26.6^o \; or \; 153.4^o \quad \textbf{(5)}$$

(c) The difference between σ_1 and σ_3 is the largest difference between principal stresses.

$$\tau_{max} = |(\sigma_1 - \sigma_3)/2| = |(55-0)/2| = 27.5 MPa \quad \textbf{(6)}$$

(d) The average normal stress is: $\sigma_{avg} = (\sigma_{xx} + \sigma_{yy})/2 = 15 MPa(T)$. The inplane maximum shear stress is: $\tau_p = |(\sigma_1 - \sigma_2)/2| = |(55-5)/2| = 25$. The principal element is shown in Figure (b).

ANS $\sigma_A = 5.2$ MPa (T) ; $\tau_A = -3.0$ MPa ; $\sigma_1 = 55$ MPa (T) ; $\sigma_2 = 5$ MPa (T) ; $\sigma_3 = 0$; $\theta_1 = -26.6^o$ or 153.4^o ; $\tau_{max} = 27.5 MPa$

---

8. 44

Solution Plane Stress $\sigma_A = ?$ $\tau_A = ?$ $\sigma_1 = ?$ $\sigma_2 = ?$ $\sigma_3 = ?$ $\theta_1 = ?$ $\tau_{max} = ?$

From the stress cube given and Figure (a) we obtain the following.

$$\sigma_{xx} = -10 ksi \qquad \sigma_{yy} = -20 ksi \qquad \tau_{xy} = 30 ksi \qquad \theta = 132^o \quad \textbf{(1)}$$

(a)

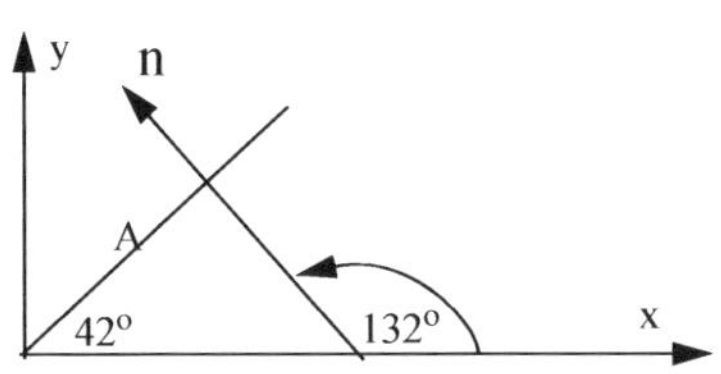

(b)

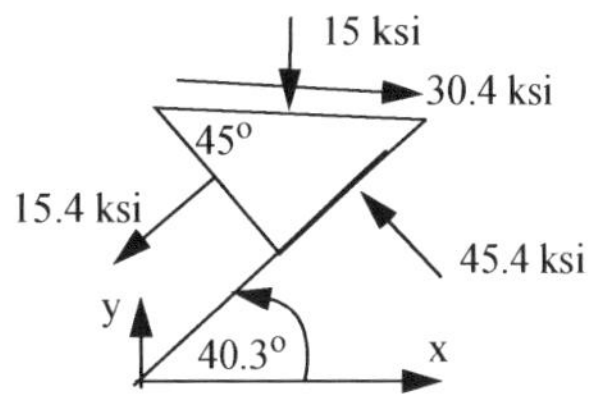

(a) Substituting in the normal and shear stress equations we obtain the stresses on plane A as shown below.

$$\sigma_A = (-10)cos^2 132 + (-20)sin^2 132 + 2(30) sin 132 cos 132 = -45.36 ksi \quad \textbf{(2)}$$

$$\tau_A = -(-10)cos 132 sin 132 + (-20) sin 132 cos 132 + (30)(cos^2 132 - sin^2 132) = 1.84 ksi \quad \textbf{(3)}$$

(b) The principal angle θ_p is calculated and substituted into the normal stress equation to determine one principal stress.

$$\theta_p = 0.5 \; atan\left(\frac{2(30)}{-10-(-20)}\right) = 40.27^o \qquad \sigma_p = -10 cos^2(40.27) + (-20) sin^2(40.27) + 2(30) sin(40.27) cos(40.27) = 15.41 ksi \quad \textbf{(4)}$$

The other principal stress is: $-10 - 20 - (15.414) = -45.414 ksi$. As this stress is smaller than the stress above, therefore the stress above is principal stress one. Thus principal angle 1 is θ_p. For plane stress we have the following answers:

$$\sigma_1 = 15.4 ksi(T) \qquad \sigma_2 = 45.4 ksi(C) \qquad \sigma_3 = 0 \qquad \theta_1 = 40.3^o \; or \; -139.7^o \quad \textbf{(5)}$$

(c) The difference between σ_1 and σ_2 is the largest difference between principal stresses.

$$\tau_{max} = |(\sigma_1 - \sigma_2)/2| = |(15.4 - (-45.4))/2| = 30.4 ksi \quad \textbf{(6)}$$

(d) The average normal stress is: $\sigma_{avg} = (\sigma_{xx} + \sigma_{yy})/2 = 15 ksi(C)$. The principal stress element is shown in Figure (b).

ANS $\sigma_A = 45.36 ksi(C)$; $\tau_A = 1.84 ksi$; $\sigma_1 = 15.4 ksi(T)$; $\sigma_2 = 45.4 ksi(C)$; $\sigma_3 = 0$; $\theta_1 = 40.3^o$ or -139.7^o ; $\tau_{max} = 30.4 ksi$

---

8. 45

Solution Plane Strain $\nu = 0.3$ $\sigma_A = ?$ $\tau_A = ?$ $\sigma_1 = ?$ $\sigma_2 = ?$ $\sigma_3 = ?$ $\theta_1 = ?$ $\tau_{max} = ?$

(a)

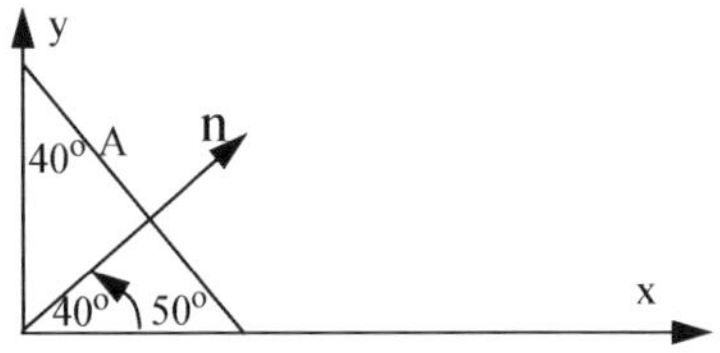

(b)

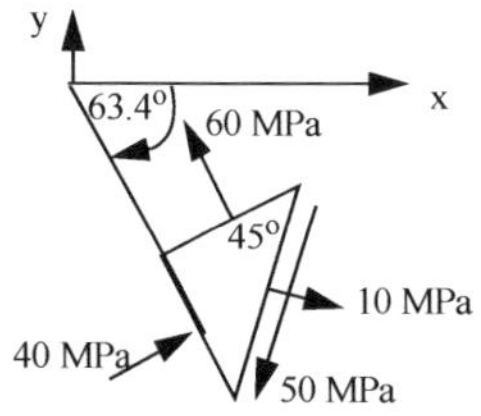

From the stress cube given and Figure (a) we obtain the following.

$$\sigma_{xx} = -20MPa \qquad \sigma_{yy} = 40MPa \qquad \tau_{xy} = -40MPa \qquad \theta = 40^o \tag{1}$$

(a) Substituting in the normal and shear stress equations we obtain the stresses on plane A as shown below.

$$\sigma_A = (-20)cos^2 40 + (40)sin^2 40 + 2(-40)sin40cos40 = 34.6 \text{ MPa} \tag{2}$$

$$\tau_A = -(-20)cos40sin40 + (40)sin40cos40 + (-40)(cos^2 40 - sin^2 40) = 22.6 \text{ MPa} \tag{3}$$

(b) The principal angle θ_p is calculated and substituted into the normal stress equation to determine one principal stress as shown below.

$$\theta_p = 0.5 \ atan\left(\frac{2(-40)}{-20-(40)}\right) = 26.56^o \qquad \sigma_p = -20cos^2(26.56) + 40sin^2(26.56) + 2(-40)sin(26.56)cos(26.56) = -40MPa \tag{4}$$

The other principal stress is: $-20 + 40 - (-40) = 60MPa$. As this stress is greater than the stress above, therefore this stress is principal stress one. Thus principal angle 1 is $\theta_1 = \theta_p \pm 90$. For plane strain $\sigma_3 = \sigma_{zz} = \nu(\sigma_{xx} + \sigma_{yy}) = 0.3(-20 + 40) = 6$.

$$\sigma_1 = 60 \text{ MPa (T)} \qquad \sigma_2 = 40 \text{ MPa (C)} \qquad \sigma_3 = 6 \text{ MPa (T)} \qquad \theta_1 = 116.6^o \ or \ -63.4^o \tag{5}$$

(c) The difference between σ_1 and σ_2 is the largest difference between principal stresses. Thus, the magnitude of the maximum shear stress is

$$\tau_{max} = |(\sigma_1 - \sigma_2)/2| = |(60-(-40))/2| = 50MPa \tag{6}$$

(d) The average normal stress is: $\sigma_{avg} = (\sigma_{xx} + \sigma_{yy})/2 = 10MPa(T)$. The principal stress element is shown in Figure (b).

ANS $\sigma_A = 34.6$ MPa (C); $\tau_A = 22.6$ MPa; $\sigma_1 = 60$ MPa (T); $\sigma_2 = 40$ MPa (C); $\sigma_3 = 6$ MPa (T); $\theta_1 = 116.6^o$ or -63.4^o;
$\tau_{max} = 50MPa$

---

8. 46

Solution Plane Strain $\nu = 0.3$ $\sigma_A = ?$ $\tau_A = ?$ $\sigma_1 = ?$ $\sigma_2 = ?$ $\sigma_3 = ?$ $\theta_1 = ?$ $\tau_{max} = ?$

From the stress cube given and Figure (a) we obtain the following.

$$\sigma_{xx} = 35MPa \qquad \sigma_{yy} = 25MPa \qquad \tau_{xy} = -20MPa \qquad \theta = 150^o \tag{1}$$

(a) Substituting in the normal and shear stress equations we obtain the stresses on plane A as shown below.

$$\sigma_A = (35)cos^2 150 + (25)sin^2 150 + 2(-20)sin150cos150 = 49.8 \text{ MPa} \tag{2}$$

$$\tau_A = -(35)cos150sin150 + (25)sin150cos150 + (-20)(cos^2 150 - sin^2 150) = -5.67 \text{ MPa}; \tag{3}$$

(a)

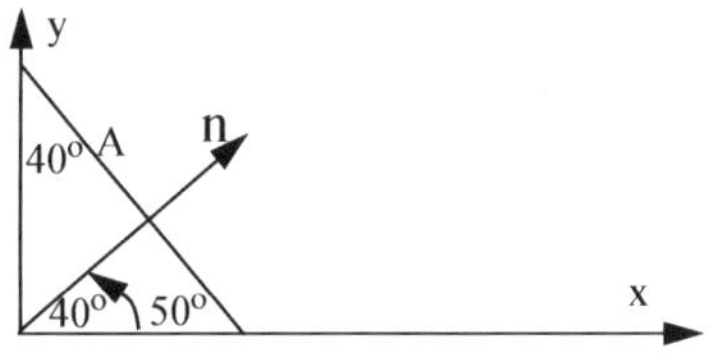

(b)

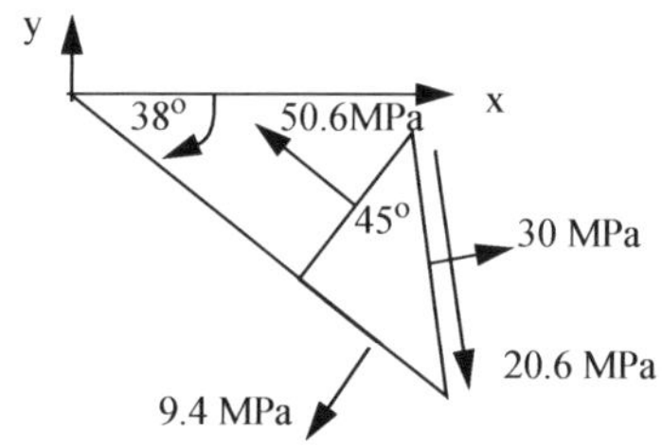

(b) The principal angle θ_p is calculated and substituted into the normal stress equation to determine one principal stress as shown below.

$$\theta_p = 0.5 \ atan\left(\frac{2(-20)}{35-25}\right) = -37.98^o \qquad \sigma_p = 35cos^2(-37.98) + 25sin^2(-37.98) + 2(-20)sin(-37.98)cos(-37.98) = 50.62MPa \tag{4}$$

The other principal stress is: $35 + 25 - 50.62 = 9.38MPa$. As this stress is smaller than the stress above, therefore the stress above is principal stress one. Thus principal angle 1 is θ_p. For plane strain $\sigma_3 = \sigma_{zz} = \nu(\sigma_{xx} + \sigma_{yy}) = 0.3(35 + 25) = 18MPa$.

$$\sigma_1 = 50.6 \text{ MPa (T)} \qquad \sigma_2 = 9.4 \text{ MPa (T)} \qquad \sigma_3 = 18 \text{ MPa (T)} \qquad \theta_1 = -38^o \ or \ 142^o \tag{5}$$

(c) The difference between σ_1 and σ_2 is the largest difference between principal stresses. Thus, the magnitude of the maximum shear stress is same as maximum inplane shear stress and is given by:

$$\tau_{max} = |(\sigma_1 - \sigma_2)/2| = |(50.6 - 9.4)/2| = 20.6 \text{ MPa} \tag{6}$$

(d) The average normal stress is: $\sigma_{avg} = (\sigma_{xx} + \sigma_{yy})/2 = 30MPa(T)$. The principal stress element is shown in Figure (b).

ANS $\sigma_A = 49.8$ MPa (T); $\tau_A = -5.67$ MPa; $\sigma_1 = 50.6$ MPa (T); $\sigma_2 = 9.4$ MPa (T); $\sigma_3 = 18$ MPa (T); $\theta_1 = -38^o$ or 142°;
$\tau_{max} = 20.6$ MPa

---

8. 47

Solution Plane Strain $\nu = 0.3$ $\sigma_A = ?$ $\tau_A = ?$ $\sigma_1 = ?$ $\sigma_2 = ?$ $\sigma_3 = ?$ $\theta_1 = ?$ $\tau_{max} = ?$

From the stress cube given and Figure (a) we obtain the following.

$$\sigma_{xx} = -25ksi \qquad \sigma_{yy} = -15ksi \qquad \tau_{xy} = -20ksi \qquad \theta = 138^o \tag{1}$$

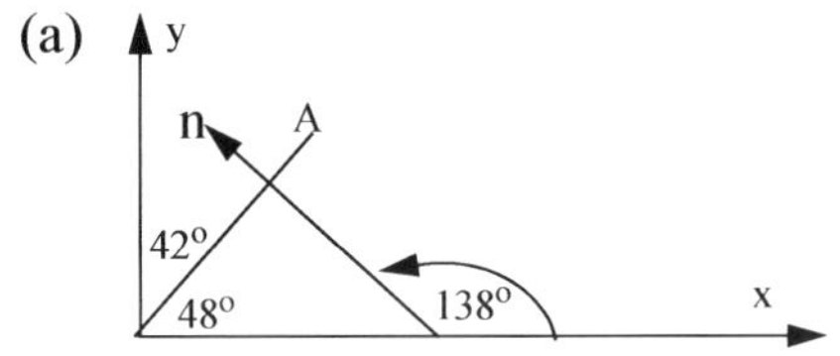

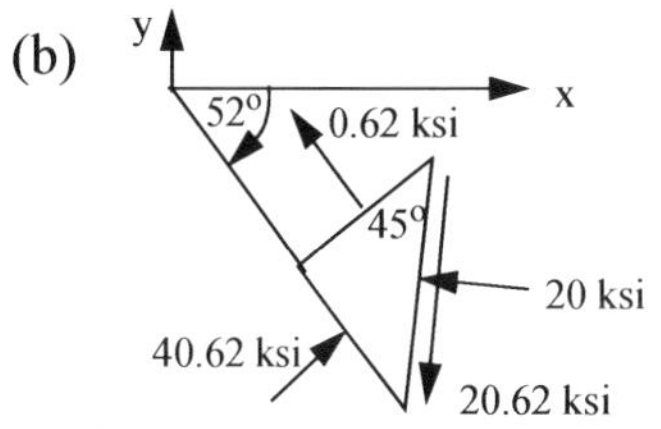

(a) Substituting in the normal and shear stress equations we obtain the stresses on plane A as shown below.

$$\sigma_A = (-25)cos^2 138 + (-15)sin^2 138 + 2(-20)sin 138 cos 138 = -40.62ksi \tag{2}$$

$$\tau_A = -(-25)cos 138 sin 138 + (-15)sin 138 cos 138 + (-20)(cos^2 138 - sin^2 138) = -7.06ksi \tag{3}$$

(b) The principal angle θ_p is calculated and substituted into the normal stress equation to determine one principal stress as shown below.

$$\theta_p = 0.5 atan\left(\frac{2(-20)}{-25-(-15)}\right) = 37.98^o \qquad \sigma_p = (-25)cos^2(37.98) + (-15)sin^2(37.98) + 2(-20)sin(37.98)cos(37.98) = -40.62 \tag{4}$$

The other principal stress is: $-25-15-(-40.616) = 0.616ksi$. As this stress is greater than the stress above, therefore this is the principal stress one. The principal angle 1 is $\theta_1 = \theta_p \pm 90$. For plane strain $\sigma_3 = \nu(\sigma_{xx} + \sigma_{yy}) = 0.3(-25-15) = -12ksi$.

$$\sigma_1 = 0.62ksi(T) \qquad \sigma_2 = 40.62ksi(C) \qquad \sigma_3 = 12ksi(C) \qquad \theta_1 = 128^o \; or \; -52^o \tag{5}$$

(c) The difference between σ_1 and σ_2 is the largest difference between principal stresses. Thus, the maximum shear stress is

$$\tau_{max} = |(\sigma_1 - \sigma_2)/2| = |(0.62-(-40.62))/2| = 20.62ksi \tag{6}$$

(d) The average normal stress is: $\sigma_{avg} = (\sigma_{xx} + \sigma_{yy})/2 = 20MPa(C)$. The principal stress element is shown in Figure (b).

ANS $\sigma_A = 0.63ksi(C)$; $\tau_A = -7.06ksi$; $\sigma_1 = 0.62ksi(T)$; $\sigma_2 = 40.62ksi(C)$; $\sigma_3 = 12ksi(C)$; $\theta_1 = 128^o \; or \; -52^o$; $\tau_{max} = 20.62ksi$

---

8. 48

Solution E=30000ksi ν=0.25 σ_{xx}=-10 ksi τ_{max}=?

Plane stress: $\sigma_{zz} = 0$

$$\varepsilon_{yy} = 0 \quad or \quad \sigma_{yy} - \nu\sigma_{xx} = 0 \quad or \quad \sigma_{yy} = \nu\sigma_{xx} = 0.25(-10) = -2.5ksi \tag{1}$$

The shear stress τ_{xy} =0, hence σ_{xx} and σ_{xx} are the principal stresses. The maximum shear stress is

$$\tau_{max} = \left|\frac{\sigma_1 - \sigma_2}{2}\right| = \left|\frac{\sigma_{xx} - \sigma_{yy}}{2}\right| = \left|\frac{(-10)-(-2.5)}{2}\right| = 3.75ksi \tag{2}$$

ANS $\tau_{max} = 3.75$ ksi

---

8. 49

Solution Plane stress $\sigma_1 = ?$ $\sigma_2 = ?$ $\sigma_3 = ?$ $\theta_1 = ?$ $\tau_{max} = ?$

(1)

Plane Stress: $\sigma_{zz} = 0$

$$\sigma_{xx} - \nu(\sigma_{yy} + \sigma_{zz}) = E\varepsilon_{xx} = 200(10^9)500(10^{-6}) = 100000(10^3)\ N/m^2 \quad or \quad \sigma_{xx} - \nu(\sigma_{yy} + \sigma_{zz}) = 100MPa \tag{2}$$

$$\sigma_{yy} - \nu(\sigma_{xx} + \sigma_{zz}) = E\varepsilon_{yy} = 200(10^9)400(10^{-6}) = 80000(10^3)\ N/m^2 \quad or \quad \sigma_{yy} - \nu(\sigma_{xx} + \sigma_{zz}) = 80MPa \tag{3}$$

$$\sigma_{xx} = 139.93MPa \qquad \sigma_{yy} = 124.78MPa \tag{4}$$

$$G = \frac{E}{2(1+\nu)} = \frac{15(10^9)}{2(1.32)} = 75.76(10^9)\ N/m^2 \qquad \tau_{xy} = G\gamma_{xy} = 75.76(10^9)(-300)(10^{-6}) = -22727(10^3)N/m^2 \tag{5}$$

The principal angle θ_p is calculated and substituted into the normal stress equation to determine one principal stress.

$$\theta_p = 0.5\ atan\left(\frac{2(-22.7)}{139.93-124.78}\right) = -35.78^o \tag{6}$$

$$\sigma_p = 139.93cos^2(-35.78) + 124.78sin^2(-35.78) + 2(-22.7)sin(-35.78)cos(-35.78) = 156.31MPa \quad (7)$$

The other principal stress as: $139.93 + 124.78 - 156.31 = 108.40MPa$. As this stress is smaller than the stress in above, therefore the above stress is principal stress one. Thus principal angle 1 is θ_p.

$$\sigma_1 = 156.3MPa(T) \quad \sigma_2 = 108.4MPa(T) \quad \sigma_3 = 0 \quad \theta_1 = -35.78^o \; or \; 144.22^o \quad (8)$$

The difference between σ_1 and σ_3 is the largest difference between principal stresses.

$$\tau_{max} = |(\sigma_1 - \sigma_3)/2| = |(156.3 - 0)/2| = 78.2 \text{ MPa} \quad (9)$$

ANS $\sigma_1 = 156.3MPa(T)$; $\sigma_2 = 108.4MPa(T)$; $\sigma_3 = 0$; $\theta_1 = -35.78^o$ or 144.22^o; $\tau_{max} = 78.2$ MPa

---

8. 50

Solution Plane stress $\sigma_1 = ?$ $\sigma_2 = ?$ $\sigma_3 = ?$ $\theta_1 = ?$ $\tau_{max} = ?$

Assuming Plane Stress

$$\sigma_{xx} - \nu(\sigma_{yy} + \sigma_{zz}) = E\varepsilon_{xx} = 70(10^9)(-3000)(10^{-6}) = -210(10^6)\frac{N}{m^2} \quad or \quad \sigma_{xx} - \nu(\sigma_{yy} + \sigma_{zz}) = -210MPa \quad (1)$$

$$\sigma_{yy} - \nu(\sigma_{xx} + \sigma_{zz}) = E\varepsilon_{yy} = 70(10^9)1500(10^{-6}) = 105(10^6)\frac{N}{m^2} \quad or \quad \sigma_{yy} - \nu(\sigma_{xx} + \sigma_{zz}) = 105MPa \quad (2)$$

$$\nu = \frac{E}{2G} - 1 = \frac{70(10^9)}{2(28)(10^9)} - 1 = 0.25 \quad \sigma_{xx} = 196MPa(C) \quad \sigma_{yy} = 56MPa(T) \quad (3)$$

$$\tau_{xy} = G\gamma_{xy} = 28(10^9)(2000)(10^{-6}) = 56(10^6) = 56MPa \quad (4)$$

The principal angle θ_p is calculated and substituted into the normal stress equation to determine one principal stress.

$$\theta_p = 0.5 \; atan\left(\frac{(2)(56)}{((-196) - 56)}\right) = -11.98^o \quad (5)$$

$$\sigma_p = (-196)cos^2(-11.98) + (56)sin^2(-11.98) + 2(56)sin(-11.98)cos(-11.98) = -207.88MPa \quad (6)$$

The other principal stress is: $(-196) + 56 - (-207.88) = 67.88MPa$. As this stress is greater than the above stress, therefore this stress is principal stress one. The principal angle 1 is $\theta_1 = \theta_p \pm 90$.

$$\sigma_1 = 67.9 \text{ MPa (T)}; \quad \sigma_2 = 207.9 \text{ MPa(C)} \quad \sigma_3 = 0 \quad \theta_1 = 78.02^o \; or \; -101.98^o \quad (7)$$

The difference between σ_1 and σ_2 is the largest difference between principal stresses.

$$\tau_{max} = |(\sigma_1 - \sigma_2)/2| = |(67.88 - (-207.88))/2| = 137.9 \text{ MPa} \quad (8)$$

ANS $\sigma_1 = 67.9$ MPa (T); $\sigma_2 = 207.9$ MPa(C); $\sigma_3 = 0$; $\theta_1 = 78.02^o$ or -101.98^o; $\tau_{max} = 137.9$ MPa

---

8. 51

Solution Plane stress $\sigma_1 = ?$ $\sigma_2 = ?$ $\sigma_3 = ?$ $\theta_1 = ?$ $\tau_{max} = ?$

Plane Stress: $\sigma_{zz} = 0$

$$\sigma_{xx} - \nu(\sigma_{yy} + \sigma_{zz}) = E\varepsilon_{xx} = 30(10^3)(-800)(10^{-6}) = -24ksi \quad or \quad \sigma_{xx} - \nu\sigma_{yy} = -24ksi \quad (1)$$

$$\sigma_{yy} - \nu(\sigma_{xx} + \sigma_{zz}) = E\varepsilon_{yy} = 30(10^3)(-1000)(10^{-6}) = -30ksi \quad or \quad \sigma_{yy} - \nu\sigma_{xx} = -30ksi \quad (2)$$

$$\sigma_{xx} = -36.26ksi \quad \sigma_{yy} = -40.88ksi \quad (3)$$

$$G = \frac{E}{2(1+\nu)} = \frac{30000}{2(1.3)} = 11538ksi \quad \tau_{xy} = G\gamma_{xy} = 11538(-500)(10^{-6}) = -5.769 \quad (4)$$

The principal angle θ_p is calculated and substituted into the normal stress equation to determine one principal stress.

$$\theta_p = 0.5 \; atan\left(\frac{(2)(-5.77)}{(-36.26) - (-40.9)}\right) = -34.10^o \quad (5)$$

$$\sigma_p = (-36.26)cos^2(-34.10) + (-40.9)sin^2(-34.10) + 2(-5.77)sin(-34.10)cos(-34.10) = -32.36ksi \quad (6)$$

The other principal stress is: $(-36.26) + (-40.9) - (-32.36) = -44.79ksi$. As this stress is smaller than the stress above, therefore the above stress is principal stress one. Thus principal angle 1 is $\theta_1 = \theta_p \pm 90$.

$$\sigma_1 = 32.4ksi(C) \quad \sigma_2 = 44.8ksi(C) \quad \sigma_3 = 0 \quad \theta_1 = -34.10^o \; or \; 145.9^o \quad (7)$$

The difference between σ_2 and σ_3 is the largest difference between principal stresses.

$$\tau_{max} = |(\sigma_2 - \sigma_3)/2| = |(-44.8 - 0)/2| = 22.4 \text{ ksi} \quad \textbf{(8)}$$

$$\textbf{ANS} \quad \sigma_1 = 32.4 ksi(C); \sigma_2 = 44.8 ksi(C); \sigma_3 = 0; \theta_1 = -34.10^o \; or \; 145.9^o; \tau_{max} = 22.4 \text{ ksi}$$

---

8. 52

Solution σ_{xx}=? σ_{yy}=? σ_{xy}=?

The solution proceeds as follows.

$$\varepsilon_{xx} = \frac{\Delta u}{\Delta x} = \frac{0.0036}{3} = 0.0012 \qquad \varepsilon_{yy} = \frac{\Delta v}{\Delta y} = \frac{0.0035}{1.4} = 0.0025 \qquad \gamma_{xy} = \frac{\Delta u}{\Delta y} + \frac{\Delta v}{\Delta x} = \frac{0.0042}{1.4} + \frac{0.0048}{3} = 0.0046 \quad \textbf{(1)}$$

$$G = \frac{E}{2(1+\nu)} = \frac{10000}{2(1.25)} = 4000 ksi \qquad \tau_{xy} = G\gamma_{xy} = (4000)(0.0046) = 18.4 ksi \quad \textbf{(2)}$$

Plane Stress: $\sigma_{zz} = 0$

$$\sigma_{xx} - \nu\sigma_{yy} = E\varepsilon_{xx} = (10000)(0.0012) = 12 ksi \qquad \sigma_{yy} - \nu\sigma_{xx} = E\varepsilon_{yy} = (10000)(0.0025) = 25 ksi \quad \textbf{(3)}$$

$$\sigma_{xx} = 19.47 ksi \qquad \sigma_{yy} = 29.87 ksi \quad \textbf{(4)}$$

The principal angle θ_p is calculated and substituted into the normal stress equation to determine one principal stress.

$$\theta_p = 0.5 \; atan\left(\frac{(2)(18.4)}{(19.47) - (29.87)}\right) = -37.11^o \quad \textbf{(5)}$$

$$\sigma_p = (19.47)cos^2(-74.22) + (29.87)sin^2(-74.22) + 2(18.4)sin(-74.22)cos(-74.22) = 5.55 ksi \quad \textbf{(6)}$$

The other principal stress is: $(19.47) + (29.87) - (5.55) = 43.79 ksi$. As this stress is greater than the stress above, therefore this stress is principal stress one. Thus principal angle 1 is $\theta_1 = \theta_p \pm 90$.

$$\sigma_1 = 43.79 \text{ ksi (T)} \qquad \sigma_2 = 5.55 \text{ ksi (T)} \qquad \sigma_3 = 0 \qquad \theta_1 = 52.89^o \; or -127.1° \quad \textbf{(7)}$$

The difference between σ_1 and σ_3 is the largest difference between principal stresses.

$$\tau_{max} = |(\sigma_1 - \sigma_3)/2| = |(43.79 - 0)/2| = 21.9 \text{ ksi} \quad \textbf{(8)}$$

$$\textbf{ANS} \quad \tau_{max} = 21.9 \text{ ksi}$$

---

8. 53

Solution plane stress $\sigma_{xx} = ?$ $\sigma_{yy} = ?$ $\tau_{xy} = ?$

Solving the two equations $\sigma_1 - \sigma_2 = 10 ksi$ and $\sigma_1 + \sigma_2 = 6 ksi$ we obtain: $\sigma_1 = 8 ksi$ and $\sigma_2 = -2 ksi$.

Figure (a) shows the Mohr's circle for the given state of stress. The center is at 3 ksi and the radius is R = 5 ksi. Figure (b) shows the orientation of principal planes P_1 and P_2 at 15° clockwise from the x axis as per the value of θ_1. To get to the vertical plane V and horizontal plane H, we must rotate counter-clockwise from the principal planes. Starting from P_1 and P_2 on the Mohr's circle we rotate 30° counter-clockwise to get planes V and H as shown in Figure (a).The normal stresses in the x and y coordinate system are:

$$\sigma_{xx} = 3 + Rcos30 = 7.33 \text{ ksi} \qquad \sigma_{yy} = 3 - Rcos30 = -1.33 \text{ ksi} \qquad |\tau_{xy}| = Rsin30 = 2.5 \text{ ksi} \quad \textbf{(1)}$$

(a)

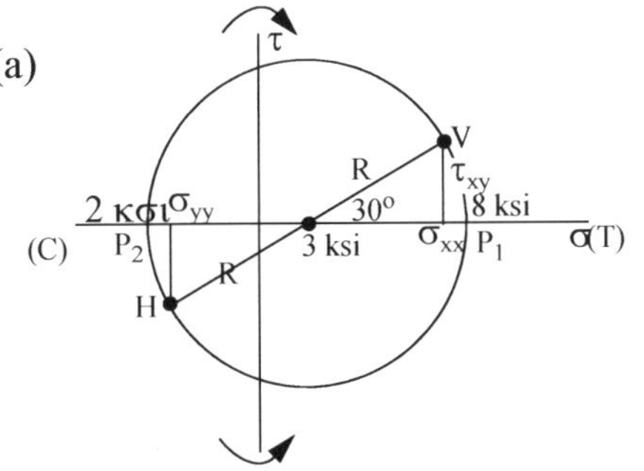

(b)

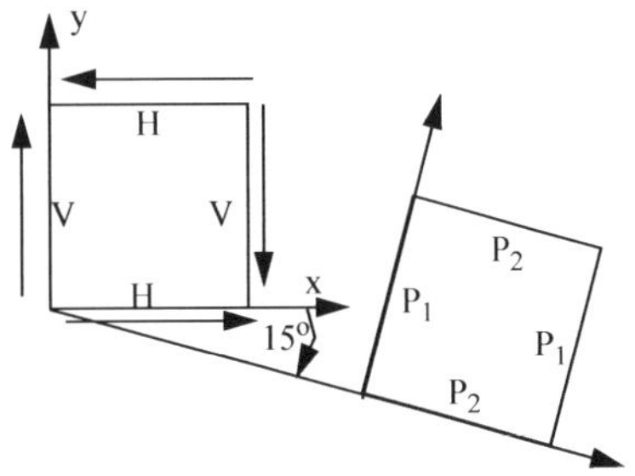

$$\textbf{ANS} \quad \sigma_{xx} = 7.33 \text{ ksi (T)}; \sigma_{yy} = 1.33 \text{ ksi (C)}; \tau_{xy} = -2.5 \text{ ksi}$$

---

8. 54

Solution plane stress $\sigma_{xx} = ?$ $\sigma_{yy} = ?$ $\tau_{xy} = ?$

Solving the two equations $\sigma_1 - \sigma_2 = 3 ksi$ and $\sigma_1 + \sigma_2 = -17 ksi$ we obtain: $\sigma_1 = -7 ksi$ and $\sigma_2 = -10 ksi$. Figure (a) shows the Mohr's circle for the given state of stress. The center is at -8.5 ksi and the radius is R = 1.5 ksi. Figure (b) shows the orientation of principal planes P_1 and P_2 at 25° counter-clockwise from the x axis as per the value of θ_1. To get to the vertical plane

V and horizontal plane H, we must rotate clockwise from the principal planes. Starting from P_1 and P_2 on the Mohr's circle we rotate 50° clockwise to get planes V and H as shown in Figure (a).

(a)

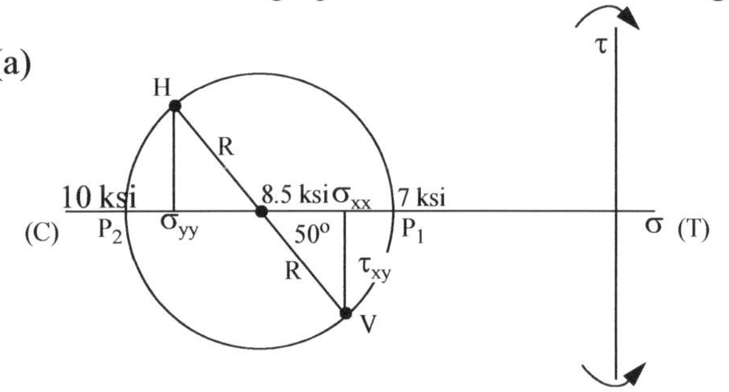

(b)

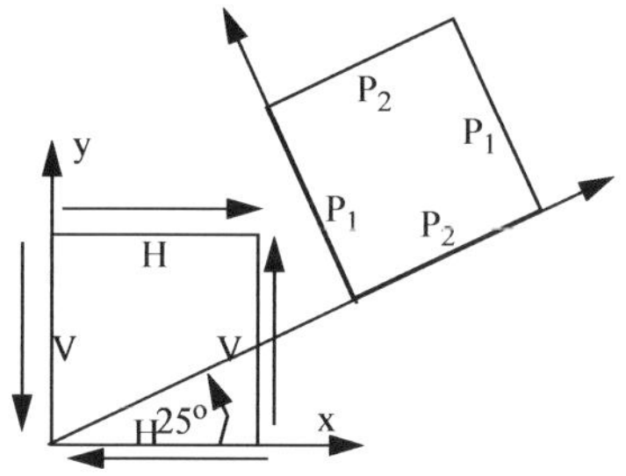

The normal stresses in the x and y coordinate system are:

$$\sigma_{xx} = -8.5 + R\cos 50 = -7.54 \text{ ksi} \qquad \sigma_{yy} = -8.5 - R\cos 50 = -9.46 \text{ ksi} \qquad |\tau_{xy}| = R\sin 50 = 1.15 \text{ ksi} \qquad (1)$$

ANS $\sigma_{xx} = 7.54 \text{ ksi(C)}$; $\sigma_{yy} = 9.46 \text{ ksi (C)}$; $\tau_{xy} = 1.15 \text{ ksi}$

---

8. 55

Solution plane stress $\sigma_{xx} = ?$ $\sigma_{yy} = ?$ $\tau_{xy} = ?$

Solving the two equations $\sigma_1 - \sigma_2 = 5ksi$ and $\sigma_1 + \sigma_2 = 5ksi$ we obtain: $\sigma_1 = 5ksi$ and $\sigma_2 = 0$. Figure (a) shows the Mohr's circle for the given state of stress. The center is at 2.5 ksi and the radius is R = 2.5 ksi. Figure (b) shows the orientation of principal planes P_1 and P_2 at 35° counter-clockwise from the x axis as per the value of θ_1. To get to the vertical plane V and horizontal plane H, we must rotate clockwise from the principal planes. Starting from P_1 and P_2 on the Mohr's circle we rotate 50° clockwise to get planes V and H as shown in Figure (a). The normal stresses in the x and y coordinate system are:

$$\sigma_{xx} = 2.5 + R\cos 70 = 3.36 \text{ ksi} \qquad \sigma_{yy} = 2.5 - R\cos 70 = 1.64 \text{ ksi} \qquad |\tau_{xy}| = R\sin 70 = 2.35 \text{ ksi} \qquad (1)$$

(a)

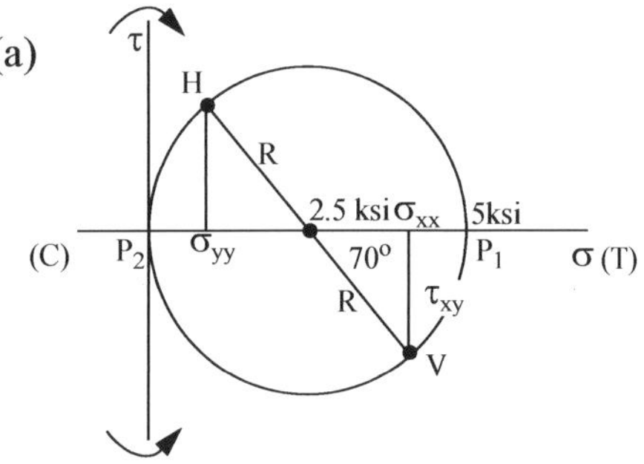

(b)

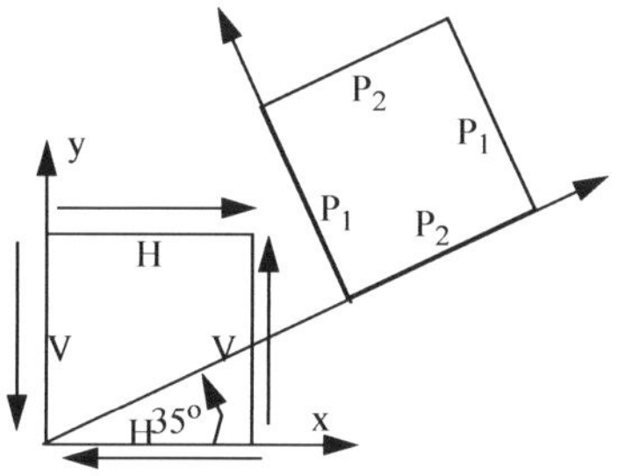

ANS $\sigma_{xx} = 3.36 \text{ ksi (T)}$; $\sigma_{yy} = 1.64 \text{ ksi (C)}$; $\tau_{xy} = 2.35 \text{ ksi}$

---

8. 56

Solution $\sigma_A = ?$ $\tau_A = ?$

From Figures (a) and (b) we obtain:

$$\sigma_{xx} = 12/[(2)(6)] = 1ksi(T) \qquad \sigma_{yy} = 0 \qquad \tau_{xy} = 0 \qquad \theta = 150^o \qquad (1)$$

(a)

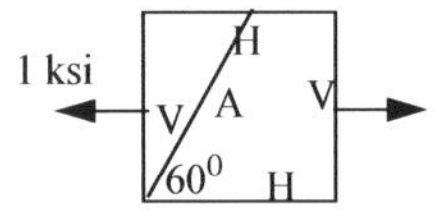

(b)

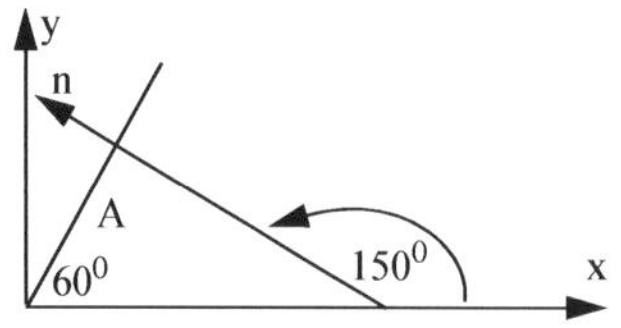

Substituting in the normal and shear stress equations we obtain the stresses on plane A as shown below.

$$\sigma_A = (1)\cos^2 150 + (0)\sin^2 150 + 2(0)\sin 150\cos 150 = 0.75ksi \qquad (2)$$

$$\tau_A = -(1)\cos 150\sin 150 + (0)\sin 150\cos 150 + (0)(\cos^2 150 - \sin^2 150) = 0.433ksi \qquad (3)$$

ANS $\sigma_A = 750 \text{ psi (T)}$; $\tau_A = 433 \text{ psi}$

---

8. 57

Solution W= 10-lbs d = 1/8 inch $\tau_{max} = ?$

By equilibrium of forces in the y-direction we have the following equation.

$$2N\sin 57 = W \quad or \quad N = 0.5962W \quad (1)$$

(a)

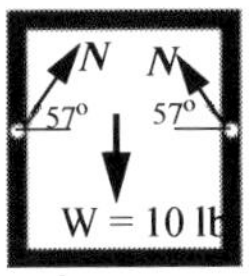

The average axial stress and maximum shear stress in the wire are as shown below.

$$A = \frac{\pi}{4}\left(\frac{1}{8}\right)^2 = 12.27(10^{-3})in^2 \quad \sigma = \frac{N}{A} = \frac{(0.5962)(10)}{12.27(10^{-3})} = 485.9psi \quad \tau_{max} = \frac{\sigma}{2} = 242.9 \quad (2)$$

ANS $\tau_{max} = 242.9$ psi

---

8. 58

Solution σ_{AA}= 150 MPa (C) F_1 = ? σ_{BB}= 1? τ_{BB}= ?

By equilibrium we obtain the following.The area of cross-section of the bar in the two sections are:

$$F_3 = 100 - 25 = 75kN \quad N_A = -F_1 = -25kN \quad N_B = F_3 = 75kN \quad (1)$$

$$A_A = (0.03)(0.01) = 0.3(10^{-3})m^2 \quad A_B = (0.06)(0.01) = 0.6(10^{-3})m^2 \quad (2)$$

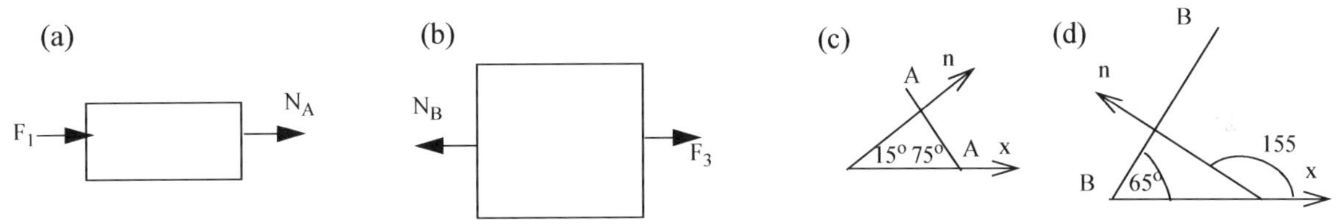

$$(\sigma_{xx})_A = \frac{N_A}{A_A} = \frac{-25(10^3)}{0.3(10^{-3})} = -83.33MPa \quad (\sigma_{xx})_B = \frac{N_B}{A_B} = \frac{75(10^3)}{0.6(10^{-3})} = 125MPa \quad (3)$$

$$\sigma_{AA} = (-83.33)cos^2 15 = -77.75 \text{ MPa} \quad \tau_{AA} = -(-83.33)cos 15 sin 15 = 20.83 \text{ MPa} \quad (4)$$

$$\sigma_{BB} = (125)cos^2 155 = 102.67 \text{ MPa} \quad \tau_{BB} = -(125)cos 155 sin 155 = -47.88 \text{ MPa} \quad (5)$$

ANS σ_{AA} = 77.75 MPa (C) ; τ_{AA} = 20.83 MPa ; σ_{BB} = 102.67 MPa (C) ; τ_{BB} = –47.88 MPa

---

8. 59

Solution E_{al} = 70 GPa E_{steel} = 210 GPa d_{al} = 20 mm d_s = 10 mm τ_{al}=? τ_s=? P=25 kN

By equilibrium of forces and exaggerated deformed shape. we obtain the following.

$$N_S + N_{al} = 2P \quad \delta_{al} = \delta_S \quad (1)$$

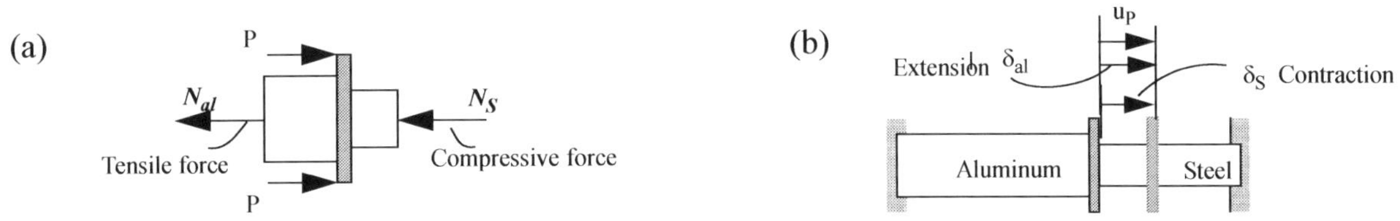

$$A_s = \pi(0.01)^2/4 = 78.54(10^{-6})m^2 \quad \delta_S = \frac{N_S L_S}{E_S A_S} = \frac{N_S(1.2)}{210(10^9)78.54(10^{-6})} = 0.07277N_S(10^{-6}) \quad (2)$$

$$A_{al} = \pi(0.02)^2/4 = 314.6(10^{-6})m^2 \quad \delta_{al} = \frac{N_{al}L_{al}}{E_{al}A_{al}} = \frac{N_{al}(1)}{70(10^9)314.6(10^{-6})} = 0.04547N_{al}(10^{-6}) \quad (3)$$

$$\delta_{al} = 0.04547N_{al}(10^{-6}) = \delta_S = 0.07277N_S(10^{-6}) \quad or \quad N_{al} = 1.6N_S \quad (4)$$

$$N_S + N_{al} = N_S(1 + 1.6) = 2P \quad or \quad N_S = 0.7691P \quad and \quad N_{al} = 1.2306P \quad (5)$$

$$\sigma_S = 0.7691P/[78.54(10^{-6})] = 9792.4P \quad \tau_s = \sigma_S/2 = 4896.2P = 4896.2(25)(10^3) = 122.4(10^6)(N/m^2) \quad (6)$$

$$\sigma_{al} = 1.2306P/[314.6(10^{-6})] = 3911.64P \quad \tau_{al} = \sigma_{al}/2 = 1955.82P = 1955.82(25)(10^3) = 48.89((10^6)(N/m^2)) \quad (7)$$

ANS τ_s = 122.4 MPa ; τ_{al} = 48.9 MPa

8. 60

Solution $\sigma_A = ?$ $\tau_A = ?$

The magnitude of torsional shear stress is at outer surface is: $|\tau_{x\theta}| = \dfrac{30(1)}{\pi 2^4/32} = 19.1\,ksi$. Figure (a) shows the direction of torsional shear stress determined by inspection. Figure (b) shows the stress cube. Figure (c) shows the orientation of the plane of the seam and the angle of the outward normal to it.

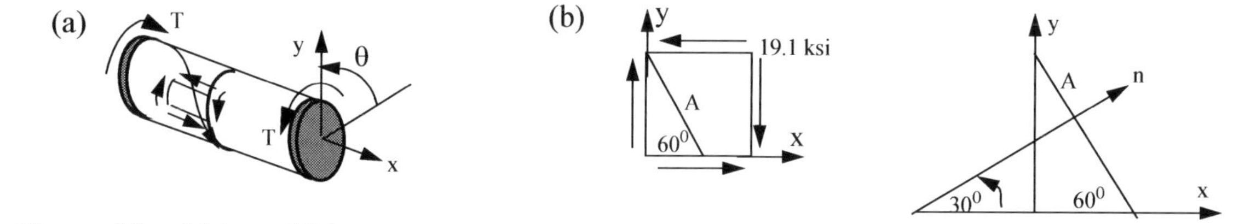

From Figures (b) and (c) we obtain:

$$\sigma_{xx} = 0 \qquad \sigma_{yy} = 0 \qquad \tau_{xy} = -19.1\,ksi \qquad \theta = 30^o \tag{1}$$

$$\sigma_A = (0)cos^2 30 + (0)sin^2 30 + 2(-19.1)sin30\,cos30 = 16.5 \text{ ksi} \tag{2}$$

$$\tau_A = -(0)cos30\,sin30 + (0)sin30\,cos30 + (-19.1)(cos^2 30 - sin^2 30) = -9.55 \text{ ksi} \tag{3}$$

ANS $\sigma_A = 16.5$ ksi (C); $\tau_A = -9.55$ ksi

---

8. 61

Solution $G_s = 12{,}000$ ksi $d_s = 2$ in $G_{al} = 4{,}000$ ksi $d_{al} = 1.5$in $\sigma_{max} = ?$ $\sigma_E = ?$ $\tau_E = ?$

We find the reaction torque at the wall and draw the torque diagram to determine the internal torques as shown below.

$$T_A = 12 + 25 - 15 = 22 \text{ in-kips} \qquad J_{AB} = J_{CD} = \frac{\pi}{32}(2)^4 = 1.5708 \text{ in}^4 \qquad J_{BC} = \frac{\pi}{32}(1.5)^4 = 0.497 \text{ in}^4 \tag{1}$$

$$\boldsymbol{T}_{AB} = 22 \text{ in-kips} \qquad \boldsymbol{T}_{BC} = 10 \text{ in-kips} \qquad \boldsymbol{T}_{CD} = -15 \text{ in-kips} \tag{2}$$

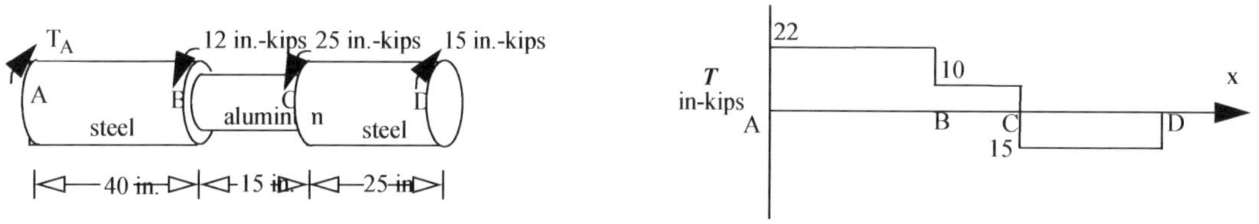

The maximum torsional shear stress in each segment can be found as:

$$\tau_{AB} = \frac{\boldsymbol{T}_{AB}(\rho_{AB})_{max}}{J_{AB}} = \frac{22(1)}{1.5708} = 14.01 \text{ ksi} \qquad \tau_{BC} = \frac{\boldsymbol{T}_{BC}(\rho_{BC})_{max}}{J_{BC}} = \frac{(10)(0.75)}{1.5708} = 15.09 \text{ ksi} \tag{3}$$

$$(\tau_{x\theta})_E = \tau_{CD} = \frac{\boldsymbol{T}_{AB}(\rho_{AB})_{max}}{J_{AB}} = \frac{(-15)(1)}{1.5708} = -9.55 \text{ ksi} = -\tau_{xy} \quad or \quad \tau_{xy} = 9.55 \text{ ksi} \tag{4}$$

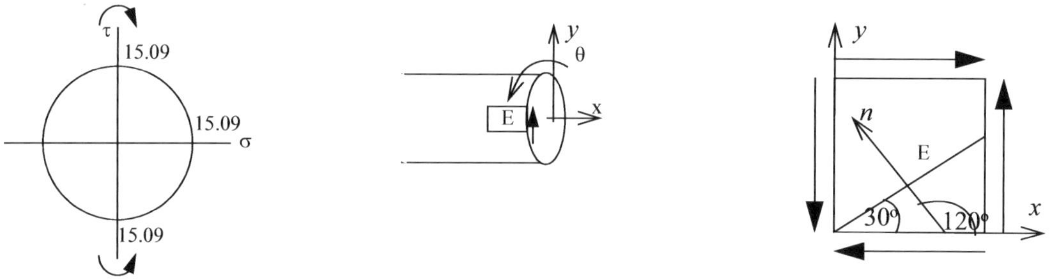

From Mohr's circle we see that the principal stress is equal in magnitude to the maximum torsional shear stress. The torsional shear stress at E is $\tau_{xy} = 9.55$ ksi. Noting that the angle of the seam from form x-axis is 120° we obtain:

$$\sigma_E = 2\tau_{xy}cos\theta\,sin\theta = 2(9.55)cos120\,sin120 = -8.27 \text{ ksi} \qquad \tau_E = \tau_{xy}(cos^2\theta - sin^2\theta) = 9.55(cos^2 120 - sin^2 120) = -4.78 \text{ ksi} \tag{5}$$

ANS $\sigma_{max} = 15.09$ ksi ; $\sigma_E = -8.27$ ksi ; $\tau_E = -4.78$ ksi

---

8. 62

Solution G = 80 GPa d = 75 mm $T = 8$ kN-m $\sigma_{max} = ?$

By moment equilibrium we obtain the internal torques and also write the polar moment of inertia.

$$T_{AB} = T_A \qquad T_{BC} = T_A - T \qquad J = \pi(0.075)^4/32 = 3.106(10^{-6})m^4 \qquad (1)$$

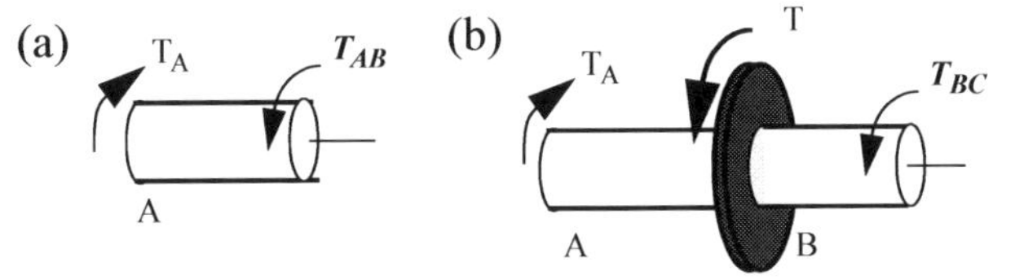

The relative rotation of the ends can be written and added to obtain the following.

$$\phi_B - \phi_A = \frac{T_{AB}(x_B - x_A)}{GJ} = \frac{(T_A)(0.75)}{GJ} \qquad \phi_C - \phi_B = \frac{T_{BC}(x_C - x_B)}{GJ} = \frac{(T_A - T)(2)}{GJ} \qquad (2)$$

$$\phi_C - \phi_A = \frac{(T_A)(0.75)}{GJ} + \frac{(T_A - T)(2)}{GJ} = 0 \quad or \quad T_A = \frac{2T}{2.75} = 0.7273T \qquad T_{AB} = 0.7273T \qquad T_{BC} = -0.2727T \qquad (3)$$

The maximum torsional shear stress will be in segment AB. The principal stresses are equal to the torsional shear stress.

$$T_{AB} = 0.7273T = 0.7273(8)(10^3) = 5.82(10^3)N-m \qquad \tau_{AB} = \frac{T_{AB}\rho_{max}}{J} = \frac{5.82(10^3)(0.0375)}{3.106(10^{-6})} = 70.25(10^6)(N/m^2) = \sigma_{max} \qquad (4)$$

ANS $\sigma_{max} = 70.25$ MPa

---

8. 63

Solution P = 1.8 kN $\sigma_1 = ?$ $\sigma_2 = ?$ $\sigma_3 = ?$ $\tau_{max} = ?$ @A, B, and C

By equilibrium we obtain the internal shear force and moment. We also write the area moment of inertia.

$$V_y = -1.8kN \qquad M_z = -(1.8)(0.4) = -0.72kN-m \qquad I_{zz} = (0.012)(0.060)^3/12 = 216(10^{-9})\ m^4 \qquad (1)$$

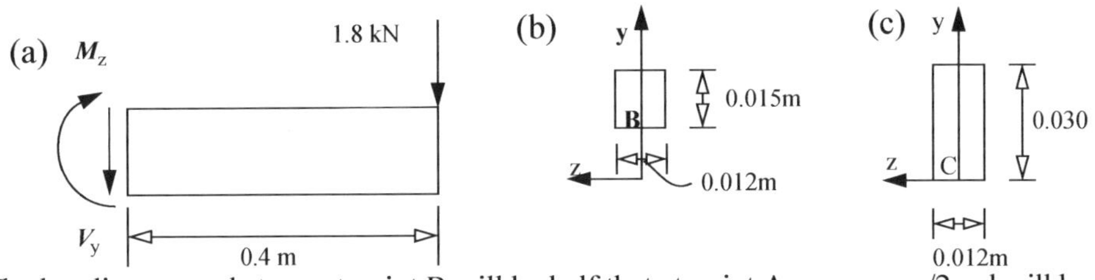

The bending normal stress at point B will be half that at point A as $y_B = y_A/2$and will be zero at point C.

$$(\sigma_{xx})_A = -\frac{M_z y_A}{I_{zz}} = -\frac{(-0.72)(10^3)(0.03)}{216(10^{-9})} = 100MPa(T) \qquad (\sigma_{xx})_B = 50MPa(T) \qquad (\sigma_{xx})_C = 0 \qquad (2)$$

$$(Q_z)_B = (0.012)(0.015)(0.015 + 0.0075) = 4.05(10^{-6})m^3 \qquad (Q_z)_C = (0.012)(0.030)(0.015) = 5.40(10^{-6})m^3 \qquad (3)$$

The bending shear stress at A is zero and at points B and C will be negative as the shear force V_y is negative.

$$(\tau_{xy})_A = 0 \qquad (\tau_{xy})_B = -\left|\frac{V_y(Q_z)_B}{I_{zz}t}\right| = -\left|\frac{1.8(10^3)(4.05)(10^{-6})}{216(10^{-9})(0.012)}\right| = -2.8125MPa \qquad (4)$$

$$(\tau_{xy})_C = -\left|\frac{V_y(Q_z)_C}{I_{zz}t}\right| = -\left|\frac{1.8(10^3)(5.40)(10^{-6})}{216(10^{-9})(0.012)}\right| = -3.75MPa \qquad (5)$$

The third principal stress at the three points is zero as these points are on free surfaces. At point A the bending normal stress is the principal stress 1 as all other stress components in the x, y, z coordinate system are zero. The maximum shear stress is half the principal stress 1 as other principal stresses are zero.

ANS $(\sigma_1)_A = 100$ MPa (T) $(\sigma_2)_A = 0$ $(\sigma_3)_A = 0$ $(\tau_{max})_A = 50$ MPa

Noting σ_{yy} is zero, we obtain the principal stress 1 and 2 at point B as:

$$(\sigma_{1,2})_B = \left(\frac{50+0}{2}\right) \pm \sqrt{\left(\frac{50-0}{2}\right)^2 + (2.8125)^2} = 25 \pm 25.16 \quad or \quad (\sigma_1)_B = 50.16 \text{ MPa}) \qquad (\sigma_2)_B = 0.16 \text{ MPa} \qquad (6)$$

ANS $(\sigma_1)_B = 50.16$ MPa (T) $(\sigma_2)_B = 0.16$ MPa (C) $(\sigma_3)_B = 0$ $(\tau_{max})_B = 25.16$ MPa

Noting that σ_{xx} and σ_{yy} are zero, we obtain the principal stress 1 and 2 at point C as:

$$(\sigma_{1,2})_C = 0 \pm \sqrt{0 + (3.75)^2} = 0 \pm 3.75 \qquad (7)$$

ANS $(\sigma_1)_C = 3.75$ MPa (T) $(\sigma_2)_C = 3.75$ MPa (C) $(\sigma_3)_C = 0$ $(\tau_{max})_C = 3.75$ MPa

---

8. 64

Solution P = 1.8 kN $\sigma_1 = ?$ $\sigma_2 = ?$ $\sigma_3 = ?$ $\tau_{max} = ?$ @A, B, and C

By equilibrium we obtain the internal shear force and moment. We also write the area moment of inertia.

$$V_y = 1.8kN \qquad M_z = (1.8)(0.4) = 0.72kN-m \qquad I_{zz} = (0.060)(0.012)^3/12 = 8.64(10^{-9})\ m^4 \quad \textbf{(1)}$$

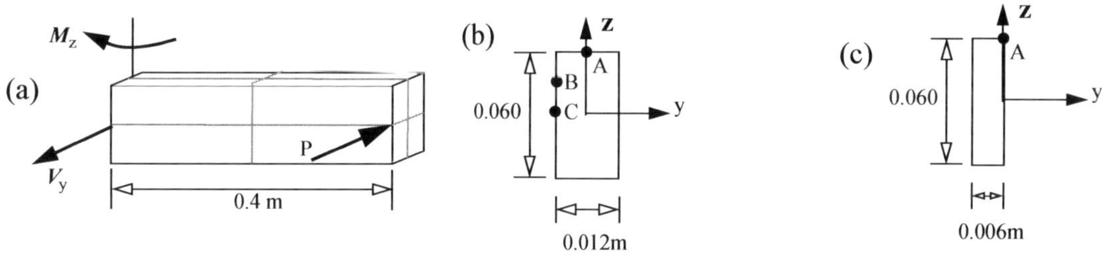

The bending normal stress at point B and C will be same as $y_B = y_C = -0.006$ m and will be zero at point A.

$$(\sigma_{xx})_A = 0 \qquad (\sigma_{xx})_B = (\sigma_{xx})_C = -\frac{M_z y_{B,C}}{I_{zz}} = -\frac{(0.72)(10^3)(-0.006)}{8.64(10^{-9})} = 500MPa(T) \quad \textbf{(2)}$$

$$(Q_z)_A = (0.060)(0.006)(0.003) = 1.08(10^{-6})m^3 \quad \textbf{(3)}$$

The bending shear stress τ_{xy} at points B and C will be zero as these points are on the free surface. The bending shear stress at A is positive as the shear force V_y is positive.The bending shear stress at A is:

$$(\tau_{xy})_A = \left|\frac{V_y(Q_z)_A}{I_{zz}t}\right| = \left|\frac{1.8(10^3)(1.08)(10^{-9})}{8.64(10^{-9})(0.06)}\right| = 3.75MPa \qquad (\tau_{xy})_B = (\tau_{xy})_C = 0 \quad \textbf{(4)}$$

The third principal stress at the three points is zero as these points are on free surfaces. Noting that σ_{xx} and σ_{yy} are zero at point A the principal stress 1 and 2 are:

$$(\sigma_{1,2})_A = 0 \pm \sqrt{0+(3.75)^2} = 0 \pm 3.75 \quad \textbf{(5)}$$

ANS $(\sigma_1)_A = 3.75$ MPa (T) $(\sigma_2)_A = 3.75$ MPa (C) $(\sigma_3)_A = 0$ $(\tau_{max})_A = 3.75$ MPa

At points B and C the bending normal stress is the principal stress 1 as all other stress components in the x, y, z coordinate system are zero. The maximum shear stress is half the principal stress 1 as other principal stresses are zero.

ANS $(\sigma_1)_B = 500$ MPa (T) $(\sigma_2)_B = 0$ $(\sigma_3)_B = 0$ $(\tau_{max})_B = 250$ MPa

ANS $(\sigma_1)_C = 500$ MPa (T) $(\sigma_2)_C = 0$ $(\sigma_3)_C = 0$ $(\tau_{max})_C = 250$ MPa

---

8. 65

Solution: $w = 25$ lb/in $\tau_{max} = ?$ $(\sigma_G)_{max} = ?$

By equilibrium of moment about point B and equilibrium of forces in the y direction we obtain the following.

$$A_y(70) - (100w)(50) = 0 \quad or \quad A_y = 71.43w\ kips \qquad A_y + B_y - (100w) = 0 \quad or \quad B_y = 28.57w\ kips \quad \textbf{(1)}$$

The shear force and bending moment diagrams are shown below

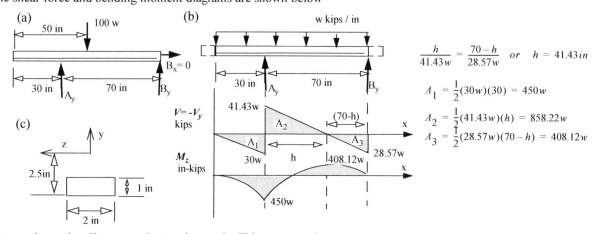

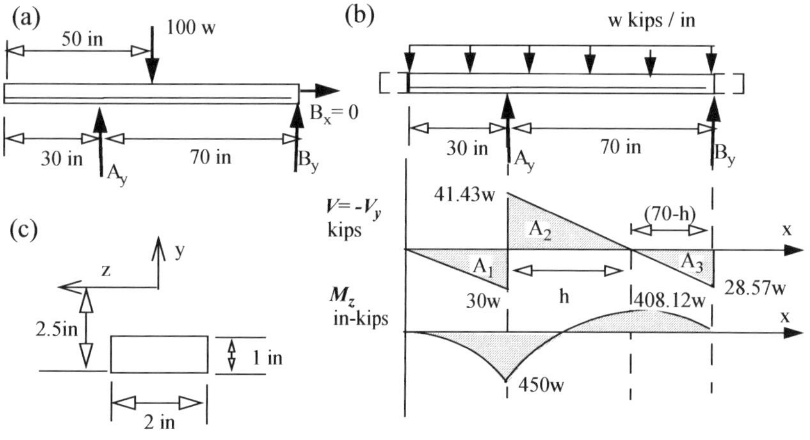

$$\frac{h}{41.43w} = \frac{70-h}{28.57w} \quad or \quad h = 41.43in$$

$$A_1 = \frac{1}{2}(30w)(30) = 450w$$

$$A_2 = \frac{1}{2}(41.43w)(h) = 858.22w$$

$$A_3 = \frac{1}{2}(28.57w)(70-h) = 408.12w$$

The maximum bending normal stress in wood will be at top or bottom:

$$M_{max} = -450w \qquad I_{zz} = \frac{1}{12}(2)(5)^3 = 20.83\ in^4 \qquad (\sigma_{xx})_W = \frac{-M_{max}y_{max}}{I_{zz}} = -\frac{(-450w)(\pm 2.5)}{20.83} = \pm 54w\ psi \quad \textbf{(2)}$$

The maximum shear stress in the beam is:

$$\tau_{max} = |(\sigma_{xx})_W|/2 = 27w = 27(25) = 675 \text{ psi} \quad (3)$$

The maximum bending normal and shear stress in glue can be calculated as shown below.

$$(\sigma_{xx})_G = \frac{-M_{max}y_G}{I_{zz}} = -\frac{(-450w)(-1.5)}{20.83} = -32.4w \quad (4)$$

$$(V_y)_A = -41.43w \qquad Q_z = (2)(1)(-2) = -4 \ in^3 \qquad (\tau_{xy})_G = -\left|\frac{(V_y)_A Q_z}{I_{zz}t}\right| = -\frac{(41.43w)(4)}{(20.83)(2)} = 3.978w \quad (5)$$

The maximum normal stress in glue will be principal stress 2.

$$(\sigma_{max})_G = \frac{(\sigma_{xx})_G}{2} - \sqrt{\left(\frac{(\sigma_{xx})_G}{2}\right)^2 + (\tau_{xy})_G^2} = -32.4w - \sqrt{\left(\frac{-32.4w}{2}\right)^2 + (3.978w)^2} = -32.88w = -32.88(25) = -822 \text{ psi} \quad (6)$$

ANS $\tau_{max} = 675$ psi ; $(\sigma_{max})_G = 822$ psi (C)

---

8. 66

Solution $M_{ext} = 9$ in-kips $\qquad \tau_{max} = ?$ $\qquad (\sigma_G)_{max} = ?$

By equilibrium of moment about point B we obtain the reaction force and draw the shear-moment diagram.

$$R_A(100) - M_{ext} = 0 \quad or \quad R_A = M_{ext}/100 \quad (1)$$

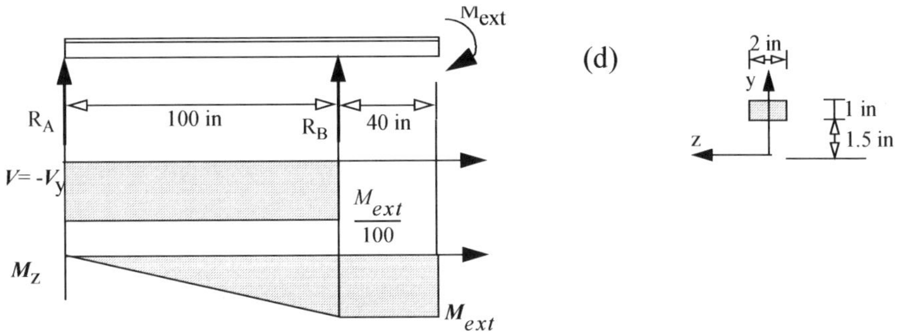

The maximum bending stress and the maximum shear stress in the beam can be calculated as shown below.

$$M_{max} = -M_{ext} \qquad I_{zz} = \frac{1}{12}(2)(5^3) = 20.83 in^4 \qquad (\sigma_{xx})_W = \left|\frac{M_{max}y_{max}}{I_{zz}}\right| = \frac{(M_{ext})(2.5)}{20.83} = 0.12M_{ext} \quad (2)$$

$$\tau_{max} = \left|\frac{(\sigma_{xx})_W}{2}\right| = 0.06M_{ext} = 0.06(9000) = 540 psi \quad (3)$$

The maximum normal stress and shear stress in glue are as shown below.

$$(\sigma_{xx})_G = \frac{-M_{max}y_G}{I_{zz}} = \frac{-(-M_{ext})(1.5)}{20.83} = \frac{M_{ext}(1.5)}{20.83} = 0.072M_{ext} \quad (4)$$

The magnitude of maximum shear stress in glue is

$$(V_y)_{max} = M_{ext}/100 \qquad Q_G = (2)(1)(2) = 4in^3 \qquad (\tau_{xy})_G = \left|\frac{Q_G(V_y)_{max}}{I_{zz}\ t}\right| = \frac{(M_{ext}/100)(4)}{(20.83)(2)} = 0.96(10^{-3})M_{ext} \quad (5)$$

The maximum normal stress in will be principal stress 1.

$$(\sigma_{max})_G = \frac{0.072M_{ext}}{2} - \sqrt{\left(\frac{0.072M_{ext}}{2}\right)^2 + (0.96(10^{-3})M_{ext})^2} = 0.07201M_{ext} = 0.07201(9000) = 648.11 \text{ psi} \quad (6)$$

ANS $\tau_{max} = 540$ psi ; $(\sigma_{max})_G = 648.11$ psi (T)

---

8. 67

Solution $\sigma_A \le 600 psi(T) \qquad \tau_A \le 400 psi \qquad F_{max} = ?$

The axial stress is given below. All other stress components in the x-y coordinate system are zero.

$$\sigma_{xx} = F/[(2)(6)] = F/12 \qquad \sigma_{yy} = 0 \qquad \tau_{xy} = 0 \qquad \theta = 150^o \quad (1)$$

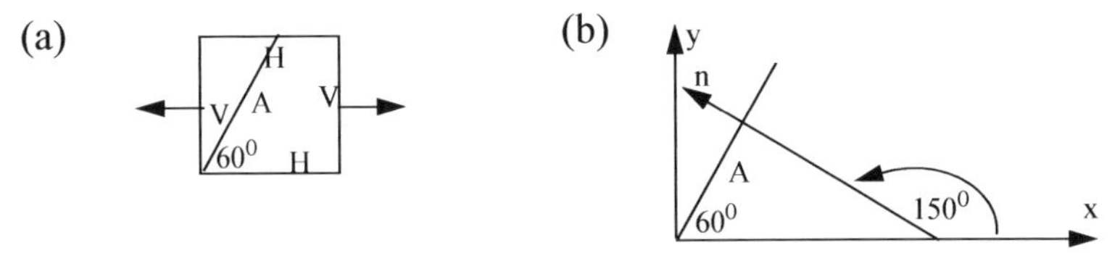

The stresses in the glue are:

$$\sigma_A = \left(\frac{F}{12}\right)\cos^2 150 = 0.0625F \le 600 \qquad or \qquad F \le 9600lb \tag{2}$$

$$|\tau_A| = \left|-\left(\frac{F}{12}\right)\cos 150 \sin 150\right| = 0.03608F \le 400 \qquad or \qquad F \le 11085 \tag{3}$$

ANS $F_{max} = 9600$ lb

---

8. 68

Solution E = 10,000 ksi $\tau_{max} \le 20$ ksi $P = 10$ kips d_{min} =?

From geometry we obtain the following.

$$CE = 120\tan 40 = 100.69 \text{ in} \qquad BD = \frac{CE}{2} = 50.34 \text{ in} \qquad AC = 120\cos 40 = 156.65 \text{ in} \qquad BC = AB = \frac{AC}{2} = 78.32 \text{ in} \tag{1}$$

By moment equilibrium about A and geometry we obtain the following.

$$120N_{CE} + 60N_{BD} = (156.65\sin 40)P = 1006.9 \text{ kips} \qquad \frac{\delta_{CE}}{AC} = \frac{\delta_{BD}}{AB} \qquad or \qquad \delta_{CE} = 2\delta_{BD} \tag{2}$$

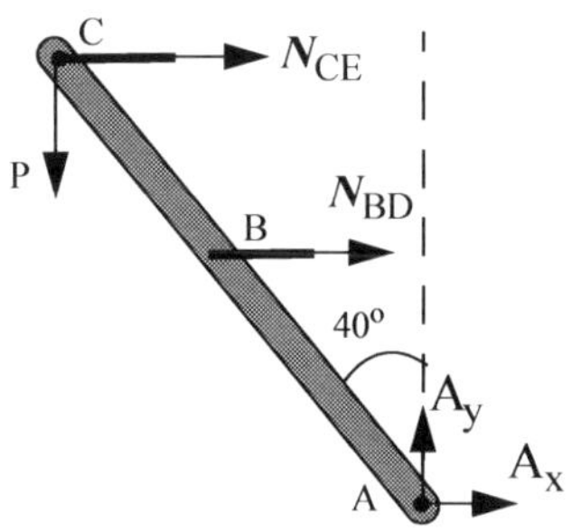

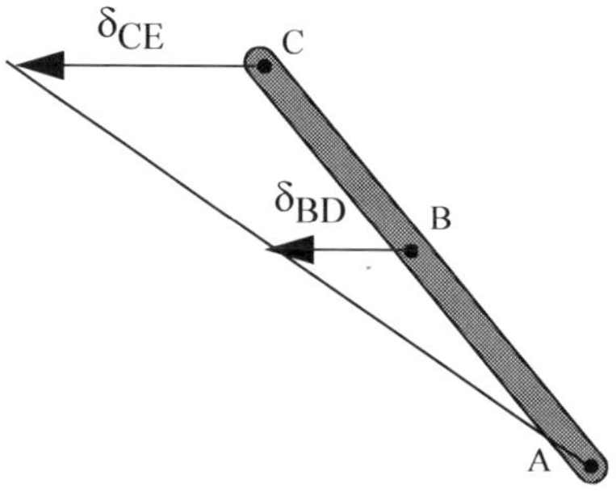

$$\delta_{CE} = \frac{N_{CE}L_{CE}}{EA} = \frac{N_{CE}(100.69)}{EA} \qquad \delta_{BD} = \frac{N_{BD}L_{BD}}{EA} = \frac{N_{BD}(50.34)}{EA} \tag{3}$$

$$\delta_{CE} = \frac{N_{CE}(100.69)}{EA} = 2\delta_{BD} = 2\left[\frac{N_{BD}(50.34)}{EA}\right] \qquad or \qquad N_{CE} = N_{BD} \tag{4}$$

$$120N_{CE} + 60N_{CE} = 1006.9 \text{ kips} \qquad N_{CE} = 5.594 \text{ kips} \tag{5}$$

The maximum shear stress is half of the axial stress in *CE* (or *BD*) on a 45° plane from the axis, hence

$$\tau_{max} = \frac{\sigma_{CE}}{2} = \frac{1}{2}\left(\frac{N_{CE}}{A}\right) = \frac{5.594}{2[\pi d^2/4]} \le 20 \text{ ksi} \qquad or \qquad d^2 \ge \left(\frac{4}{\pi}\right)\left(\frac{5.594}{40}\right) \qquad or \qquad d \ge 0.422 \text{ in.} \tag{6}$$

ANS $d_{min} = \frac{7}{16}$ in.

---

8. 69

Solution t= 1/8 in $d_{mean} = 6$ in $\sigma_2 \le 10ksi(C)$ T_{max} = ?

The magnitude of maximum torsional shear stress is:

$$J = \pi(6.125^4 - 5.875^4)/32 = 21.21in^4 \qquad \rho_{max} = \frac{d_{mean}}{2} + \frac{t}{2} = 3.0625 \qquad \tau = \frac{T\rho_{max}}{J} = \frac{T(3.0625)}{21.21} \tag{1}$$

As the normal stresses on the cross-section plane are zero, the principal stresses are equal to the shear stress. Thus:

$$\sigma_2 = \tau = \frac{T(3.0625)}{21.21} \le 10 \qquad or \qquad T \le 69.27 \ \ in-kip \tag{2}$$

ANS $T_{max} = 69.2$ in-kips

---

8. 70

Solution L = 30 in T= 25π in-kips t= 1/16 in G=12000ksi $\sigma_S \le 10ksi(C)$

$\tau_W \le 12ksi$ $\sigma_W \le 20ksi(T)$ $\Delta\phi \le 3^o$ minimum R_o = ? to nearest 1/16 in

By inspection we see that weld would be subjected to compressive stress, hence we do not have to account for the limitation on tensile stress in the weld. The maximum compressive stress in steel is equal to the torsional shear stress τ_{tor}.

$$\sigma_{xx} = 0 \qquad \sigma_{yy} = 0 \qquad \tau_{xy} = -\tau_{tor} \qquad \theta = 65^o \qquad \sigma_2 = \tau_{tor} \le 10ksi \qquad (1)$$

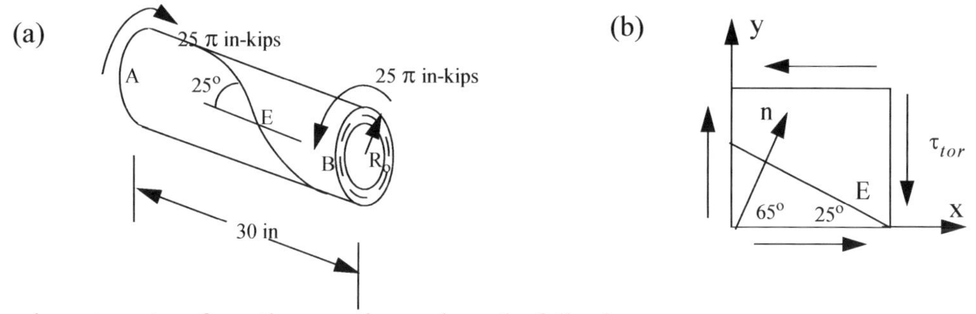

From the shear stress transformation equation we have the following.

$$\tau_W = -(0)cos65\,sin65 + (0)sin65\,cos65 + (-\tau_{tor})(cos^2 65 - sin^2 65) = 0.6428\tau_{tor} \le 12 \qquad or \qquad \tau_{tor} \le 18.7ksi \qquad (2)$$

$$\tau_{tor} = \frac{T\rho_{max}}{J} = \frac{(25\pi)R_o}{\pi[R_o^4 - (R_o - 0.0625)^4]/2} \le 10 \qquad or \qquad R_o^4 - (R_o - 0.0625)^4 - 5R_o \ge 0 \qquad (3)$$

$$\Delta\phi = \frac{TL}{GJ} = \frac{(25\pi)(30)}{(12000)\pi[R_o^4 - (R_o - 0.0625)^4]/2} \le \frac{3\pi}{180} \qquad or \qquad R_o^4 - (R_o - 0.0625)^4 - 2.3873 \ge 0 \qquad (4)$$

$$(f_1(R_o) = R_o^4 - (R_o - 0.0625)^4 - 5R_o) \ge 0 \qquad (f_2(R_o) = R_o^4 - (R_o - 0.0625)^4 - 2.3873) \ge 0 \qquad (5)$$

We consider the values of the two functions for R_o starting at 4 in. in steps of 1/16th in. The calculations are done on a spreadsheet and results reported below.

R_o	$f_1(R_o)$	$f_2(R_o)$
4.0000	-4.37	13.34
4.0625	-3.93	14.09
4.1250	-3.47	14.87
4.1875	-2.99	15.66
4.2500	-2.48	16.48
4.3125	-1.94	17.33
4.3750	-1.38	18.20
4.4375	-0.80	19.10
4.5000	-0.19	20.02
4.5625	0.45	20.97
4.6250	1.11	21.95
4.6875	1.80	22.95
4.7500	2.52	23.98

$f_2(R_o)$ is always positive. The smallest value of R_o for which $f_1(R_o)$ becomes positive is $R_o = 4.5625$ in.

ANS $R_o = 4\frac{9}{16}$ in.

---

8. 71

Solution $E_{al} = 70$ GPa $E_{steel} = 210$ GPa $d_{al} = 20$ mm $d_s = 10$ mm

$\tau_{al} \le 120MPa$ $\tau_s \le 150MPa$ $P_{max} = ?$

By equilibrium of forces and geometry we obtain the following.

$$N_S + N_{al} = 2P \qquad \delta_{al} = \delta_S \qquad (1)$$

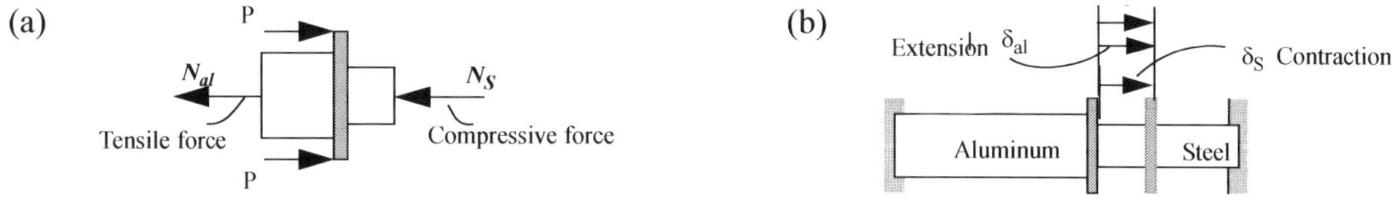

The deformations of the bars can be written as

$$A_s = \pi(0.01)^2/4 = 78.54(10^{-6})m^2 \qquad \delta_S = \frac{N_S L_S}{E_S A_S} = \frac{N_S(1.2)}{210(10^9)78.54(10^{-6})} = 0.07277N_S(10^{-6}) \qquad (2)$$

$$A_{al} = \pi(0.02)^2/4 = 314.6(10^{-6})m^2 \qquad \delta_{al} = \frac{N_{al}L_{al}}{E_{al}A_{al}} = \frac{N_{al}(1)}{70(10^9)314.6(10^{-6})} = 0.04547N_{al}(10^{-6}) \qquad (3)$$

$$\delta_{al} = 0.04547N_{al}(10^{-6}) = \delta_S = 0.07277N_S(10^{-6}) \quad or \quad N_{al} = 1.6N_S \tag{4}$$

$$N_S + N_{al} = (1+1.6)N_S = 2P \quad or \quad N_S = 0.7691P \quad and \quad N_{al} = 1.2306P \tag{5}$$

$$\sigma_{al} = \frac{1.2306P}{314.6(10^{-6})} = 3911.64P \qquad \tau_{al} = \sigma_{al}/2 = 1955.82P \le 120(10^6) \quad or \quad P \le 61.35(10^3)N$$

$$\sigma_S = \frac{0.7691P}{78.54(10^{-6})} = 9792.4P \qquad \tau_s = \sigma_S/2 = 4896.2P \le 150(10^6) \quad or \quad P \le 30.64(10^3)N \tag{6}$$

ANS $P_{max} = 30.6$ kN

---

8. 72

Solution $\sigma_A \le 15ksi(T)$ $\tau_A \le 10ksi$ $T_{ext} = ?$

From Figures (b) and (c) we obtain the following.

$$\sigma_{xx} = 0 \qquad \sigma_{yy} = 0 \qquad |\tau_{x\theta}| = \frac{T_{ext}(1)}{\pi 2^4/32} = 0.6366T_{ext} \qquad \tau_{xy} = 0.6366T_{ext} \qquad \theta = 30^o \tag{1}$$

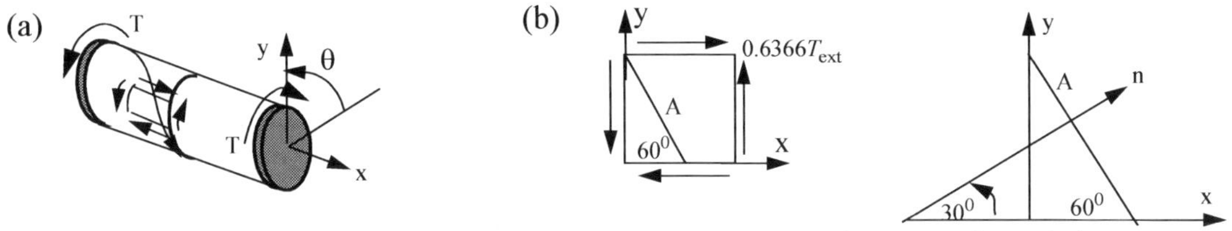

Substituting in the normal and shear stress equations we obtain the stresses on plane A as shown below.

$$\sigma_A = 2(0.6366T_{ext})sin30cos30 = 0.55133T_{ext} \le 15 \quad or \quad T_{ext} \le 27.2 \;\; in-kips \tag{2}$$

$$\tau_A = (0.6366T_{ext})(cos^2 30 - sin^2 30) = 0.3183T_{ext} \le 10 \quad or \quad T_{ext} \le 31.42 in-kips \tag{3}$$

ANS $T_{ext} = 27.2$ in.-kips

---

8. 73

Solution G = 80 GPa d = 75 mm $\sigma_{max} \le 90MPa$ $T_{max} = ?$

The internal torques and the polar moment of inertia can be written as shown below.

$$T_{AB} = T_A \qquad T_{BC} = T_A - T \qquad J = \pi(0.075)^4/32 = 3.106(10^{-6})m^4 \tag{1}$$

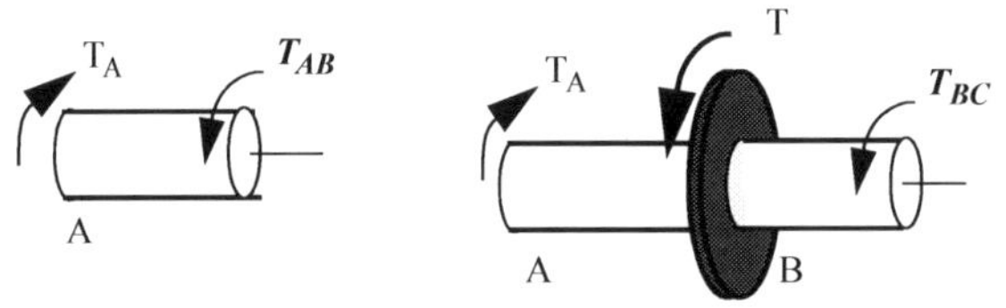

The relative rotation of the ends can be found, added and equated to zero as shown below.

$$\phi_B - \phi_A = \frac{T_{AB}(x_B - x_A)}{GJ} = \frac{(T_A)(0.75)}{GJ} \qquad \phi_C - \phi_B = \frac{T_{BC}(x_C - x_B)}{GJ} = \frac{(T_A - T)(2)}{GJ} \tag{2}$$

$$\phi_C - \phi_A = \frac{(T_A)(0.75)}{GJ} + \frac{(T_A - T)(2)}{GJ} = 0 \quad or \quad T_A = \frac{2T}{2.75} = 0.7273T \qquad T_{AB} = 0.7273T \qquad T_{BC} = -0.2727T \tag{3}$$

The maximum torsional shear stress will be in segment AB. The principal stresses are equal to the torsional shear stress.

$$\tau_{AB} = \frac{T_{AB}\rho_{max}}{J} = \frac{0.7273T(0.0375)}{3.106(10^{-6})} = 8.7807T(10^3) \tag{4}$$

$$\sigma_{AB} = \tau_{AB} = 8.7807T(10^3) \le 90(10^6) \quad or \quad T \le 10.25(10^3)N-m \tag{5}$$

ANS $T_{max} = 10.2$ kN-m

---

8. 74

Solution $\tau_G \le 300psi$ $\sigma_G \le 200psi(T)$ $\sigma_W \le 2000psi$ $w_{max} = ?$

The reactions can be found and shear moment diagram drawn as shown below.

$$R = 4w \qquad M = 24w \tag{1}$$

The maximum bending normal and shear stress will occur at the wall. By inspection the glue on the top junction will be in tensile and the shear stress τ_{xy} will be negative as shear force is negative. The normal and shear stress in the glue are as follows.

$$I_{zz} = (4)(10)^3/12 = 333.3\ in^4 \qquad M_z = -24w \qquad (\sigma_{xx})_G = -\frac{M_z y_G}{I_{zz}} = -\frac{(-24w)(12)(3)}{333.3} = 2.592w \qquad (2)$$

$$Q_z = (4)(2)(4) = 32 in^3 \qquad V_y = -4w \qquad (\tau_{xy})_G = -\left|\frac{V_y Q_z}{I_{zz} t}\right| = -\left|\frac{(4w)(32)}{(333.3)(4)}\right| = 0.096w \qquad (3)$$

$$(\sigma_{1,2})_G = \left(\frac{2.592w + 0}{2}\right) \pm \sqrt{\left(\frac{2.592w - 0}{2}\right)^2 + (0.096w)^2} = 1.296w \pm 1.300w \qquad (\sigma_1)_G = 2.596w \qquad (\sigma_2)_G = -0.004w \qquad (4)$$

$$(\sigma_1)_G = 2.596w \le 200 \qquad or \qquad w \le 77.04\ lb/ft \qquad (5)$$

$$\tau_{max} = \left|\frac{(\sigma_1)_G - (\sigma_2)_G}{2}\right| = 1.300w \le 300 \qquad or \qquad w \le 230.7\ lb/ft \qquad (6)$$

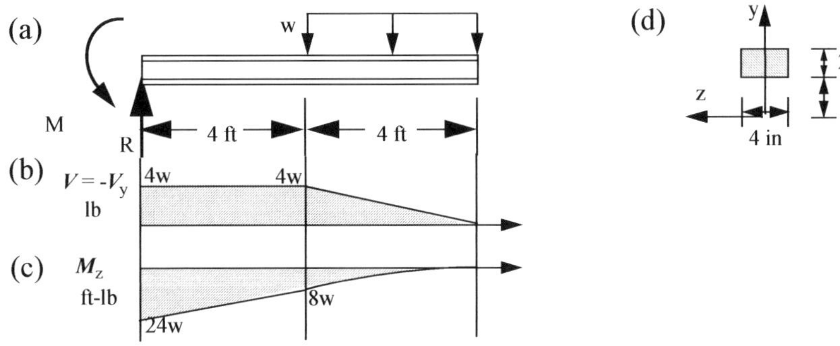

The maximum normal stress in wood is the maximum bending normal stress.

$$\sigma_W = -\frac{(-24w)(12)(5)}{333.3} = 4.32w \le 2000 \qquad or \qquad w \le 462.9\ lb/ft \qquad (7)$$

ANS $w_{max} = 77\ lb/ft$

---

8. 75

Solution τ_{max}=500 psi $(\sigma_G)_{max}$ =800 psi (T) M_{ext} =?

From earlier problem we have the following.

$$(V_y)_{max} = M_{ext}/100 \qquad M_{max} = -M_{ext} \qquad I_{zz} = (2)(5^3)/12 = 20.83 in^4 \qquad Q_G = 4 in^3 \qquad (1)$$

$$(\sigma_{xx})_W = \left|\frac{M_{max} y_{max}}{I_{zz}}\right| = \frac{(M_{ext})(2.5)}{20.83} = 0.12 M_{ext} \qquad \tau_{max} = \left|\frac{(\sigma_{xx})_W}{2}\right| = 0.06 M_{ext} \le 600 psi \qquad or \qquad M_{ext} \le 10000 in - lb \qquad (2)$$

$$(\sigma_{xx})_G = \frac{-M_{max} y_G}{I_{zz}} = \frac{M_{ext}(1.5)}{20.83} = 0.072 M_{ext} \qquad (\tau_{xy})_G = \left|\frac{Q_G (V_y)_{max}}{I_{zz}\ t}\right| = \frac{(M_{ext}/100)(4)}{(20.83)(2)} = 0.96(10^{-3}) M_{ext} \qquad (3)$$

$$(\sigma_{max})_G = \frac{0.072 M_{ext}}{2} + \sqrt{\left(\frac{0.072 M_{ext}}{2}\right)^2 + [0.96(10^{-3}) M_{ext}]^2} = 0.07201 M_{ext} \le 650 \qquad or \qquad M_{ext} \le 9026.5 in - lb$$

ANS M_{ext} = 9026 in.-lb

---

8. 76

Solution d = 2.5 m $\sigma_{max} \le 100 MPa(T)$ $\tau_{max} \le 60 MPa$ p = 1800 kPa t = ?

The axial stress and hoop stress can be found as shown below.

$$\sigma_{xx} = \frac{pr}{2t} = \frac{(1800)(10^3)(1.25)}{2t} = \frac{1.125(10^6)}{t} \qquad \sigma_{yy} = \frac{pr}{t} = \frac{(1800)(10^3)(1.25)}{t} = \frac{2.25(10^6)}{t} \qquad (1)$$

The shear stress $\tau_{xy} = 0$. The maximum normal and shear stress can be written as:

$$\sigma_{max} = 2.25(10^6)/t \le 100(10^6) \qquad or \qquad t \ge 0.0225 m \qquad (2)$$

$$\tau_{max} = |2.25(10^6)/t - 1.125(10^6)/t|/2 = 0.5625(10^6)/t \le 60(10^6) \qquad or \qquad t \ge 0.0094 m \qquad (3)$$

ANS t_{min} = 22.5 mm

---

8. 77

Solution d = 4 ft = 48 in t = 1/2 in. $\sigma_{max} \le 30ksi(T)$ $\sigma_W \le 25ksi$ $\tau_W \le 18ksi$ $p_{max} = ?$

The axial stress and hoop stress can be found as shown below.

$$\sigma_{xx} = \frac{pr}{2t} = \frac{p(24)}{2(0.5)} = 24p \qquad \sigma_{yy} = \frac{pr}{t} = \frac{p(24)}{0.5} = 48p \tag{1}$$

The shear stress $\tau_{xy} = 0$. The angle of the outward normal to the weld can be found.

$$\sigma_{max} = 48p \le 30 \quad or \quad p \le 0.625ksi \quad \theta = 40°. \tag{2}$$

(a)

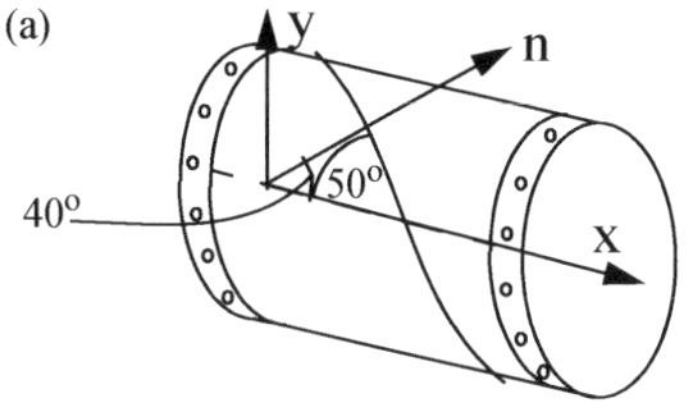

Substituting in the normal and shear stress equations we obtain the stresses on plane *W* as shown below.

$$\sigma_W = (24p)cos^2 40 + (48p)sin^2 40 = 33.9162p \le 25 \quad or \quad p \le 0.737ksi \tag{3}$$

$$\tau_W = -(24p)cos40sin40 + (48p)sin40cos40 = 11.818p \le 18 \quad or \quad p \le 1.523ksi \tag{4}$$

ANS $p_{max} = 625$ psi

---

8. 78

Solution

The solution proceeds as follows.

$$[\sigma]_{nt} = [T]^T[\sigma][T] = [T]^T \begin{bmatrix} \sigma_{xx} & \tau_{xy} \\ \tau_{yx} & \sigma_{yy} \end{bmatrix} \begin{bmatrix} cos\theta & -sin\theta \\ sin\theta & cos\theta \end{bmatrix} = [T]^T \begin{bmatrix} \sigma_{xx}cos\theta + \tau_{xy}sin\theta & -\sigma_{xx}sin\theta + \tau_{xy}cos\theta \\ \tau_{yx}cos\theta + \sigma_{yy}sin\theta & -\tau_{yx}sin\theta + \sigma_{yy}cos\theta \end{bmatrix} \tag{1}$$

$$[\sigma]_{nt} = \begin{bmatrix} cos\theta & sin\theta \\ -sin\theta & cos\theta \end{bmatrix} \begin{bmatrix} \sigma_{xx}cos\theta + \tau_{xy}sin\theta & -\sigma_{xx}sin\theta + \tau_{xy}cos\theta \\ \tau_{yx}cos\theta + \sigma_{yy}sin\theta & -\tau_{yx}sin\theta + \sigma_{yy}cos\theta \end{bmatrix} = \begin{bmatrix} \sigma_{nn} & \tau_{nt} \\ \tau_{tn} & \sigma_{tt} \end{bmatrix} \tag{2}$$

We evaluate each term and use the identity $\tau_{xy} = \tau_{yx}$ to obtain the following:

$$\sigma_{nn} = (\sigma_{xx}cos\theta + \tau_{xy}sin\theta)cos\theta + (\tau_{yx}cos\theta + \sigma_{yy}sin\theta)sin\theta = \sigma_{xx}cos^2\theta + \sigma_{yy}sin^2\theta + 2\tau_{xy}sin\theta cos\theta \tag{3}$$

$$\tau_{nt} = (-\sigma_{xx}sin\theta + \tau_{xy}cos\theta)cos\theta + (-\tau_{yx}sin\theta + \sigma_{yy}cos\theta)sin\theta = (-\sigma_{xx} + \sigma_{yy})cos\theta sin\theta + \tau_{xy}(cos^2\theta - sin^2\theta) \tag{4}$$

$$\tau_{tn} = -(\sigma_{xx}cos\theta + \tau_{xy}sin\theta)sin\theta + (\tau_{yx}cos\theta + \sigma_{yy}sin\theta)cos\theta = (-\sigma_{xx} + \sigma_{yy})cos\theta sin\theta + \tau_{xy}(cos^2\theta - sin^2\theta) \tag{5}$$

$$\sigma_{tt} = -(-\sigma_{xx}sin\theta + \tau_{xy}cos\theta)sin\theta + (-\tau_{yx}sin\theta + \sigma_{yy}cos\theta)cos\theta = \sigma_{xx}sin^2\theta + \sigma_{yy}cos^2\theta - 2\tau_{xy}cos\theta sin\theta \tag{6}$$

The above equations are the desired results.

---

8. 79

Solution

Let λ be the eigenvalue of the matrix $[\sigma]$. Thus the following determinant must be zero. We note $\tau_{xy} = \tau_{yx}$.

$$\begin{vmatrix} \sigma_{xx} - \lambda & \tau_{xy} \\ \tau_{yx} & \sigma_{yy} - \lambda \end{vmatrix} = 0 \quad or \quad (\sigma_{xx} - \lambda)(\sigma_{yy} - \lambda) - \tau_{xy}\tau_{yx} = 0 \quad or \quad \sigma_{xx}\sigma_{yy} - \lambda(\sigma_{xx} + \sigma_{yy}) + \lambda^2 - \tau_{xy}^2 = 0 \tag{1}$$

The roots of the above equation are as follows.

$$\lambda_{1,2} = \frac{(\sigma_{xx} + \sigma_{yy}) \pm \sqrt{(\sigma_{xx} + \sigma_{yy})^2 - 4(\sigma_{xx}\sigma_{yy} - \tau_{xy}^2)}}{2} = \frac{(\sigma_{xx} + \sigma_{yy})}{2} \pm \sqrt{\frac{\sigma_{xx}^2 + \sigma_{yy}^2 - 2\sigma_{xx}\sigma_{yy} + 4\tau_{xy}^2}{4}} \quad \text{or}$$

$$\lambda_{1,2} = \frac{(\sigma_{xx} + \sigma_{yy})}{2} \pm \sqrt{\frac{(\sigma_{xx} - \sigma_{yy})^2}{4} + \tau_{xy}^2} = \frac{(\sigma_{xx} + \sigma_{yy})}{2} \pm \sqrt{\left(\frac{\sigma_{xx} - \sigma_{yy}}{2}\right)^2 + \tau_{xy}^2} \tag{2}$$

λ_1 and λ_2 represents the principal stresses.

---

8. 80

Solution

From the given figure we have the following relation between the area A of the inclined plane and the area of the Cartesian plane on the wedge.

$$A_x = n_x A \qquad A_y = n_y A \qquad A_z = n_z A \tag{1}$$

(a) Force Wedge

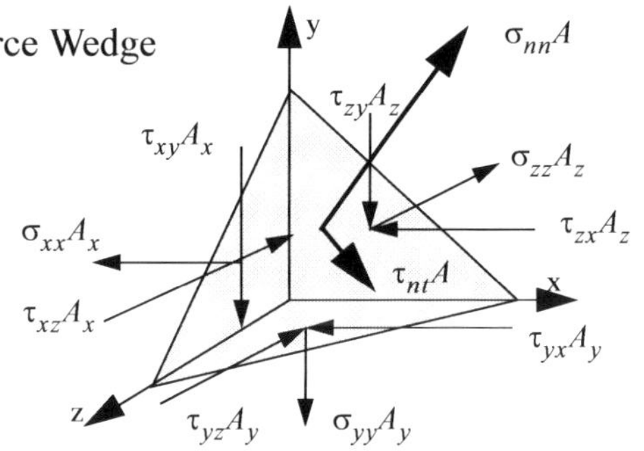

Figure (a) shows the force wedge obtained by multiplying the stress components with the area of the plane on which they act. We can find the components of the force vectors in the n-direction by taking the dot product of the force vector and the unit vector in the n-direction. By force equilibrium in the n-direction we obtain the following.

$$\sigma_{nn}A + (n_x\bar{i} + n_y\bar{j} + n_z\bar{k}) \bullet [(-\sigma_{xx}A_x - \tau_{yx}A_y - \tau_{zx}A_z)\bar{i} + (-\tau_{xy}A_x - \sigma_{yy}A_y - \tau_{zy}A_z)\bar{j} + ((-\tau_{xz}A_x - \tau_{yz}A_y - \sigma_{zz}A_z)\bar{k}] = 0) \text{ or}$$

$$\sigma_{nn}A + (-\sigma_{xx}n_xA - \tau_{yx}n_yA - \tau_{zx}n_zA)n_x + (-\tau_{xy}n_xA - \sigma_{yy}n_yA - \tau_{zy}n_zA)n_y + (-\tau_{xz}n_xA - \tau_{yz}n_yA - \sigma_{zz}n_zA)n_z = 0 \text{ or}$$

$$\sigma_{nn} = \sigma_{xx}n_x^2 + \sigma_{yy}n_y^2 + \sigma_{zz}n_z^2 + (\tau_{xy} + \tau_{yx})n_xn_y + (\tau_{yz} + \tau_{zy})n_yn_z + (\tau_{zx} + \tau_{zx})n_zn_x \text{ or}$$

$$\sigma_{nn} = \sigma_{xx}n_x^2 + \sigma_{yy}n_y^2 + \sigma_{zz}n_z^2 + 2\tau_{xy}n_xn_y + 2\tau_{yz}n_yn_z + 2\tau_{zx}n_zn_x \tag{2}$$

The above equation is the desired result.

---

8. 81

Solution

In terms of principal stresses the result of previous problem can be written as:

$$\sigma_{nn} = \sigma_1 n_x^2 + \sigma_2 n_y^2 + \sigma_3 n_z^2 \tag{1}$$

where n_x, n_y, and n_z, are now measured from the principal planes. Substituting $|n_x| = |n_y| = |n_z| = 1/\sqrt{3}$ we obtain:

$$\sigma_{nn} = \sigma_1\left(\frac{1}{\sqrt{3}}\right)^2 + \sigma_2\left(\frac{1}{\sqrt{3}}\right)^2 + \sigma_3\left(\frac{1}{\sqrt{3}}\right)^2 = \frac{(\sigma_1 + \sigma_2 + \sigma_3)}{3} \tag{2}$$

The above equation are the desired results.

---

8. 82

Solution Plane stress E= 30,000 ksi, G = 11,600 ksi $\varepsilon_{xx} = [100(2x + y) + 50]\mu$, $\varepsilon_{yy} = -100(2x + y)\mu$, $\gamma_{xy} = 200(x - 2y)\mu$

$\sigma_1 = ?$ $\sigma_2 = ?$ $\sigma_3 = ?$ $\theta_1 = ?$ $\tau_{max} = ?$ every 30° with r =1in

The Poisson's ratio is: $\nu = E/(2G) - 1 = 30000/[2(11600)] - 1 = 0.2931$.

θ	x	y	ε_{xx} (μ)	ε_{yy} (μ)	γ_{xy} (μ)	σ_{xx} (psi)	σ_{yy} (psi)	τ_{xy} (psi)	σ_1 (psi)	σ_2 (psi)	θ_p	σ_p (psi)	θ_1	τ_{max} (psi)
0	1.000	0.000	250	-200	200	6281	-4159	2320	6773	-4651	12.0	6773	12.0	5712
30	0.866	0.500	273	-223	-27	6819	-4697	-311	6828	-4706	-1.5	6828	-1.5	5767
60	0.500	0.866	237	-187	-246	5970	-3848	-2858	6742	-4620	-15.1	6742	-15.1	5681
90	0.000	1.000	150	-100	-400	3961	-1839	-4640	6533	-4411	-29.0	6533	-29.0	5472
120	-0.500	0.866	37	13	-446	1330	792	-5178	6246	-4124	-43.5	6246	-43.5	5185
150	-0.866	0.500	-73	123	-373	-1217	3339	-4329	5953	-3831	31.1	-3831	121.1	4892
180	-1.000	0.000	-150	200	-200	-2999	5121	-2320	5737	-3615	14.9	-3615	104.9	4676
210	-0.866	-0.500	-173	223	27	-3537	5659	311	5670	-3548	-1.9	-3548	88.1	4609
240	-0.500	-0.866	-137	187	246	-2688	4810	2858	5775	-3654	-18.7	-3654	71.3	4715
270	-0.000	-1.000	-50	100	400	-679	2801	4640	6016	-3895	-34.7	-3895	55.3	4956
300	0.500	-0.866	63	-13	446	1952	170	5178	6315	-4193	40.1	6315	40.1	5254
330	0.866	-0.500	173	-123	373	4499	-2377	4329	6589	-4468	25.8	6589	25.8	5528

Starting with $\theta = 0$ in steps of 30° we find the following on a spreadsheet as shown in the table above.

1. The coordinates: $x = r\cos\theta = \cos\theta \qquad y = r\sin\theta = \sin\theta$
2. The strains: $\varepsilon_{xx} = [100(2x + y) + 50]\ \mu \qquad \varepsilon_{yy} = -100(2x + y)\ \mu \qquad \gamma_{xy} = 200(x - 2y)\ \mu$
3. The stresses: $\sigma_{xx} = \dfrac{E[\varepsilon_{xx} + \nu\varepsilon_{yy}]}{(1 - \nu^2)} \qquad \sigma_{yy} = \dfrac{E[\varepsilon_{yy} + \nu\varepsilon_{xx}]}{(1 - \nu^2)} \qquad \tau_{xy} = G\gamma_{xy}$

4. The principal stresses: $\sigma_{1,2} = \frac{(\sigma_{xx} + \sigma_{yy})}{2} \pm \sqrt{\left(\frac{\sigma_{xx} - \sigma_{yy}}{2}\right)^2 + \tau_{xy}^2}$

5. The principal angle: $\theta_p = atan[2\tau_{xy}/(\sigma_{xx} - \sigma_{yy})]/2$

6. The normal stress on principal plane: $\sigma_p = \sigma_{xx}\cos^2\theta_p + \sigma_{yy}\sin^2\theta_p + 2\tau_{xy}\sin\theta_p\cos\theta_p$.

7. The principal angle 1: If σ_p is equal to σ_1 then $\theta_1 = \theta_p$ otherwise $\theta_1 = \theta_p + 90$.

8. The maximum shear stress: $\tau_{max} = (\sigma_1 - \sigma_2)/2$

---

8. 83

Solution a= 12 in $\sigma_1 = ?$ $\sigma_2 = ?$ every 15° with r =18in

On a spreadsheet for θ starting from zero in steps of 15° the following is calculated using a = 12 and r = 18.

1. The stresses:

$$\sigma_{rr} = \frac{\sigma}{2}\left(1 - \frac{a^2}{r^2}\right) - \frac{\sigma}{2}\left(1 - \frac{4a^2}{r^2} + \frac{3a^4}{r^4}\right)\cos 2\theta \qquad \sigma_{\theta\theta} = \frac{\sigma}{2}\left(1 + \frac{a^2}{r^2}\right) + \frac{\sigma}{2}\left(1 + \frac{3a^4}{r^4}\right)\cos 2\theta \qquad \tau_{r\theta} = \frac{\sigma}{2}\left(1 + \frac{2a^2}{r^2} - \frac{3a^4}{r^4}\right)\sin 2\theta \qquad (1)$$

2. The principal stresses: $\sigma_{1,2} = \frac{(\sigma_{xx} + \sigma_{yy})}{2} \pm \sqrt{\left(\frac{\sigma_{xx} - \sigma_{yy}}{2}\right)^2 + \tau_{xy}^2}$

The calculations done on a spreadsheet are reported in the table below.

θ	σ_{rr} (ksi)	$\sigma_{\theta\theta}$ (ksi)	$\tau_{r\theta}$ (ksi)	σ_1 (ksi)	σ_2 (ksi)
0	3.70	15.19	0.00	15.19	3.70
15	3.58	14.12	3.24	15.04	2.66
30	3.24	11.20	5.61	14.10	0.34
45	2.78	7.22	6.48	11.85	-1.85
60	2.31	3.24	5.61	8.41	-2.85
75	1.98	0.33	3.24	4.50	-2.19
90	1.85	-0.74	0.00	1.85	-0.74
105	1.98	0.33	-3.24	4.50	-2.19
120	2.31	3.24	-5.61	8.41	-2.85
135	2.78	7.22	-6.48	11.85	-1.85
150	3.24	11.20	-5.61	14.10	0.34
165	3.58	14.12	-3.24	15.04	2.66
180	3.70	15.19	-0.00	15.19	3.70
195	3.58	14.12	3.24	15.04	2.66
210	3.24	11.20	5.61	14.10	0.34
225	2.78	7.22	6.48	11.85	-1.85
240	2.31	3.24	5.61	8.41	-2.85
255	1.98	0.33	3.24	4.50	-2.19
270	1.85	-0.74	0.00	1.85	-0.74
285	1.98	0.33	-3.24	4.50	-2.19
300	2.31	3.24	-5.61	8.41	-2.85
315	2.78	7.22	-6.48	11.85	-1.85
330	3.24	11.20	-5.61	14.10	0.34
345	3.58	14.12	-3.24	15.04	2.66

CHAPTER 9

Section 9.1

9. 1

Solution $\varepsilon_{xx} = 500\mu$ $\varepsilon_{OP} = ?$ $\phi = ?$

From geometry we can write:

$$\Delta x = \Delta n \cos 50 \qquad PP_1 = \varepsilon_{xx}\Delta x = 500(10^{-6})\Delta n\ \cos 50 = 321.39(\Delta n)(10^{-6}) \tag{1}$$

$$PP_n = PP_1 \cos 50 = 206.59(\Delta n)(10^{-6}) \qquad P_nP_1 = PP_1 \sin 50 = 246.20(\Delta n)(10^{-6}) \qquad \varepsilon_{OP} = \frac{PP_n}{OP} = \frac{206.59(\Delta n)(10^{-6})}{\Delta n} \tag{2}$$

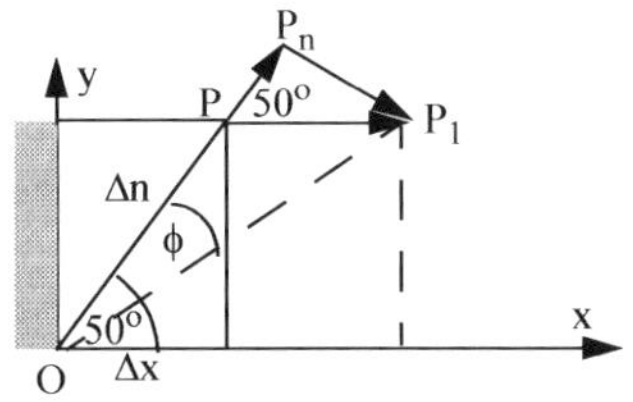

$$OP_n = OP + PP_n = (\Delta n)[1 + 206.59(10^{-6})] \approx (\Delta n) \qquad \tan\phi \approx \phi = P_nP_1/OP_n \approx 246.20(10^{-6}) \tag{3}$$

ANS $\varepsilon_{OP} = 206.59\mu$; $\phi = 246.20\mu$ rads CW

9. 2

Solution $\gamma_{yx} = 300\mu$ $\varepsilon_{OP} = ?$ $\phi = ?$

From geometry we can write:

$$\Delta x = \Delta n \cos 50 \qquad PP_1 = AP\tan\gamma_{xy} \approx \Delta x\gamma_{yx} = 300(10^{-6})\Delta n\ \cos 50 = 192.84(\Delta n)(10^{-6}) \tag{1}$$

$$PP_n = PP_1 \cos 40 = 147.72(\Delta n)(10^{-6}) \qquad P_nP_1 = PP_1 \sin 40 = 123.95(\Delta n)(10^{-6}) \qquad \varepsilon_{OP} = PP_n/OP = 147.72(10^{-6}) \tag{2}$$

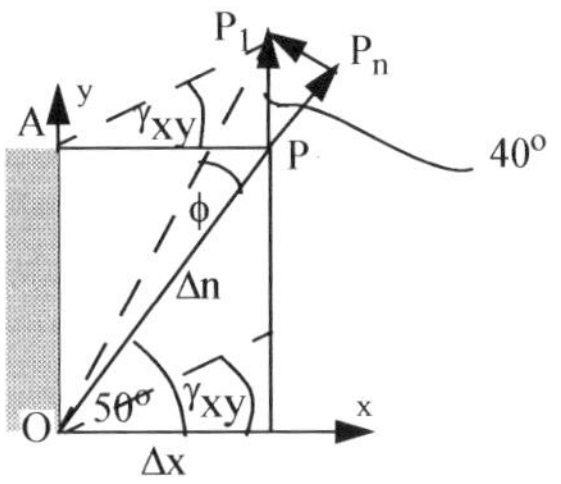

$$OP_n = OP + PP_n = (\Delta n)[1 + 147.72(10^{-6})] \approx (\Delta n) \qquad \tan\phi \approx \phi = P_nP_1/OP_n \approx 123.95(10^{-6}) \tag{3}$$

ANS $\varepsilon_{OP} = 147.72\mu$; $\phi = 123.95\mu$ rads CCW

9. 3

Solution $\varepsilon_{yy} = -400\mu$ $\varepsilon_{OP} = ?$ $\phi = ?$

From geometry we can write:

$$\Delta y = \Delta n \sin 50 \qquad PP_1 = |\varepsilon_{yy}\Delta x| = 400(10^{-6})\Delta n\ \sin 50 = 306.42(\Delta n)(10^{-6}) \tag{1}$$

$$PP_n = PP_1 \cos 40 = 234.72(\Delta n)(10^{-6}) \qquad P_nP_1 = PP_1 \sin 40 = 196.96(\Delta n)(10^{-6}) \qquad \varepsilon_{OP} = -PP_n/OP = -234.72(10^{-6}) \tag{2}$$

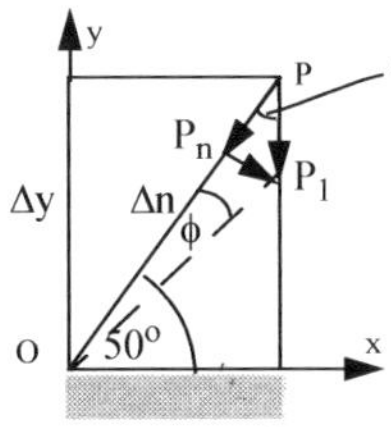

$$OP_n = OP - PP_n = (\Delta n)[1 - 234.72(10^{-6})] \approx (\Delta n) \qquad \tan\phi \approx \phi = P_nP_1/OP_n \approx 196.96(10^{-6}) \tag{3}$$

ANS $\varepsilon_{OP} = -234.72\mu$; $\phi = 196.96\ \mu$ rads CW

9. 4

Solution $\gamma_{xy} = 300\mu$ $\varepsilon_{OP} = ?$ $\phi = ?$

From geometry we can write:

$$\Delta y = \Delta n \sin 50 \qquad PP_1 = \gamma_{xy}\Delta y = 300(10^{-6})\Delta n \ \sin 50 = 229.81(\Delta n)(10^{-6}) \quad \textbf{(1)}$$

$$PP_n = PP_1 \cos 50 = 147.72(\Delta n)(10^{-6}) \qquad P_nP_1 = PP_1 \sin 50 = 176.05(\Delta n)(10^{-6}) \qquad \varepsilon_{OP} = PP_n/OP = 147.72(10^{-6}) \quad \textbf{(2)}$$

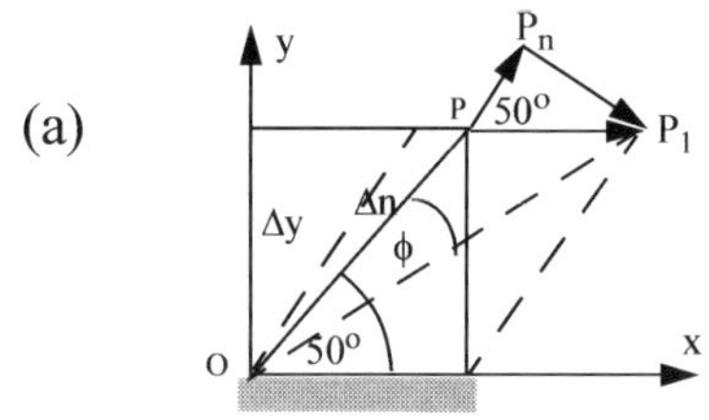

$$OP_n = OP + PP_n = (\Delta n)[1 + 147.72(10^{-6})] \approx (\Delta n) \qquad \tan\phi \approx \phi = P_nP_1/OP_n \approx 176.05(10^{-6}) \quad \textbf{(3)}$$

ANS $\varepsilon_{OP} = 147.72\mu$; $\phi = 176.05\mu$ rads CW

---

9. 5

Solution $\varepsilon_{xx} = -400\ \mu$ $\varepsilon_{nn} = ?$ $\varepsilon_{tt} = ?$ $\gamma_{nt} = ?$

From geometry in Figure (b) we obtain:

$$\Delta x = \Delta n \cos 30 \qquad PP_1 = |\varepsilon_{xx}|\Delta x = 400(10^{-6})\Delta n \ \cos 30 = 346.41(\Delta n)(10^{-6}) \quad \textbf{(1)}$$

$$PP_n = PP_1 \cos 30 = 300(\Delta n)(10^{-6}) \qquad P_nP_1 = PP_1 \sin 30 = 173.2(\Delta n)(10^{-6}) \qquad \varepsilon_{nn} = -PP_n/OP = (-300)(10^{-6}) \quad \textbf{(2)}$$

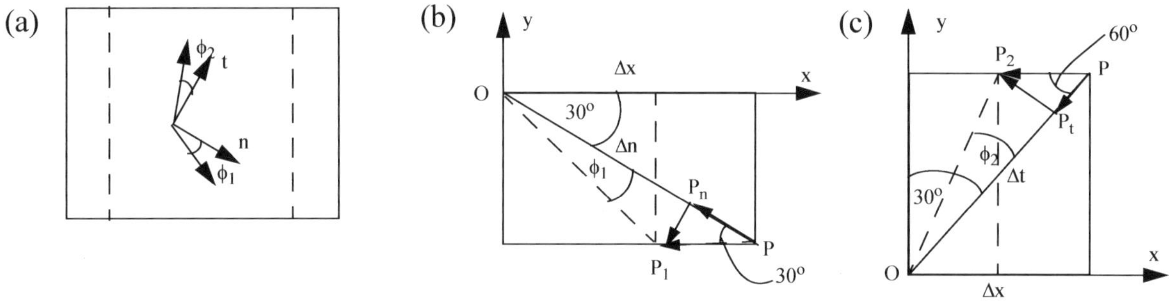

$$OP_n = OP - PP_n = (\Delta n)[1 - 300(10^{-6})] \approx (\Delta n) \qquad \tan\phi_1 \approx \phi_1 = P_nP_1/OP_n \approx 173.2(10^{-6}) \quad or \quad \phi_1 = 173.2\mu rads \quad \textbf{(3)}$$

From geometry in Figure (c) we obtain

$$\Delta x = \Delta t \sin 30 \qquad PP_2 = |\varepsilon_{xx}|\Delta x = 400(10^{-6})\Delta t \ \sin 30 = 200(\Delta t)(10^{-6}) \quad \textbf{(4)}$$

$$PP_t = PP_2 \cos 60 = 100(\Delta t)(10^{-6}) \qquad P_tP_2 = PP_2 \sin 60 = 173.2(\Delta t)(10^{-6}) \qquad \varepsilon_{tt} = -PP_t/OP = (-100)(10^{-6}) \quad \textbf{(5)}$$

$$OP_t = OP - PP_t = (\Delta t)[1 - 100(10^{-6})] \approx (\Delta t) \qquad \tan\phi_2 \approx \phi_2 = P_tP_2/OP_t \approx 173.2(10^{-6}) \quad or \quad \phi_2 = 173.2\mu rads \quad \textbf{(6)}$$

$$\gamma_{nt} = -(\phi_1 + \phi_2) = -(173.2 + 173.2) \quad \textbf{(7)}$$

ANS $\varepsilon_{nn} = -300\mu$; $\varepsilon_{tt} = -100\mu$; $\gamma_{nt} = -346.4\mu$

---

9. 6

Solution $\varepsilon_{yy} = 600\ \mu$ $\varepsilon_{nn} = ?$ $\varepsilon_{tt} = ?$ $\gamma_{nt} = ?$

From geometry in Figure (b) we obtain the following.

$$\Delta y = \Delta n \sin 30 \qquad PP_1 = |\varepsilon_{yy}|\Delta y = 600(10^{-6})\Delta n \ \sin 30 = 300(\Delta n)(10^{-6}) \quad \textbf{(1)}$$

$$PP_n = PP_1 \cos 60 = 150(\Delta n)(10^{-6}) \qquad P_nP_1 = PP_1 \sin 60 = 259.8(\Delta n)(10^{-6}) \qquad \varepsilon_{nn} = PP_n/OP = (150)(10^{-6}) \quad \textbf{(2)}$$

$$OP_n = OP + PP_n = (\Delta n)[1 + 300(10^{-6})] \approx (\Delta n) \qquad \tan\phi_1 \approx \phi_1 = P_nP_1/OP_n \approx 259.8(10^{-6}) \quad or \quad \phi_1 = 259.8\mu rads \quad \textbf{(3)}$$

From geometry in Figure (c) we obtain the following.

$$\Delta y = \Delta t \cos 30 \qquad PP_2 = |\varepsilon_{yy}|\Delta y = 600(10^{-6})\Delta t \ \cos 30 = 519.62(\Delta t)(10^{-6}) \quad \textbf{(4)}$$

$$PP_t = PP_2 \cos 30 = 450(\Delta t)(10^{-6}) \qquad P_tP_2 = PP_2 \sin 30 = 259.8(\Delta t)(10^{-6}) \qquad \varepsilon_{tt} = PP_t/OP = (450)(10^{-6}) \quad \textbf{(5)}$$

$$OP_t = OP + PP_t = (\Delta t)[1 + 450(10^{-6})] \approx (\Delta t) \qquad tan\phi_2 \approx \phi_2 = P_tP_2/OP_t \approx 259.8(10^{-6}) \qquad or \qquad \phi_2 = 259.8\mu rads \quad \textbf{(6)}$$

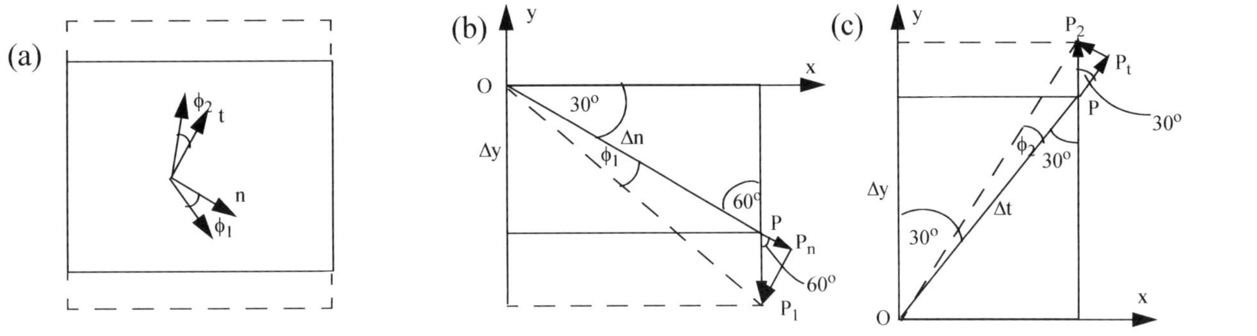

$$\gamma_{nt} = -(\phi_1 + \phi_2) = -(259.8 + 259.8) = -519.6 \quad \textbf{(7)}$$

ANS $\varepsilon_{nn} = 150\mu$; $\varepsilon_{tt} = 450\mu$; $\gamma_{nt} = -519.6\mu rad$

---

9. 7

Solution $\gamma_{xy} = -500\ \mu$ $\varepsilon_{nn} = ?$ $\varepsilon_{tt} = ?$ $\gamma_{nt} = ?$

From geometry in Figure (b) we obtain the following.

$$\Delta x = \Delta n cos 30 \qquad PP_1 = AP tan|\gamma_{xy}| \approx \Delta x|\gamma_{xy}| = 500(10^{-6})\Delta n\ cos 30 = 433.01(\Delta n)(10^{-6}) \quad \textbf{(1)}$$

$$PP_n = PP_1 cos 60 = 216.5(\Delta n)(10^{-6}) \qquad P_nP_1 = PP_1 sin 60 = 375(\Delta n)(10^{-6}) \qquad \varepsilon_{nn} = PP_n/OP = 216.5(10^{-6}) \quad \textbf{(2)}$$

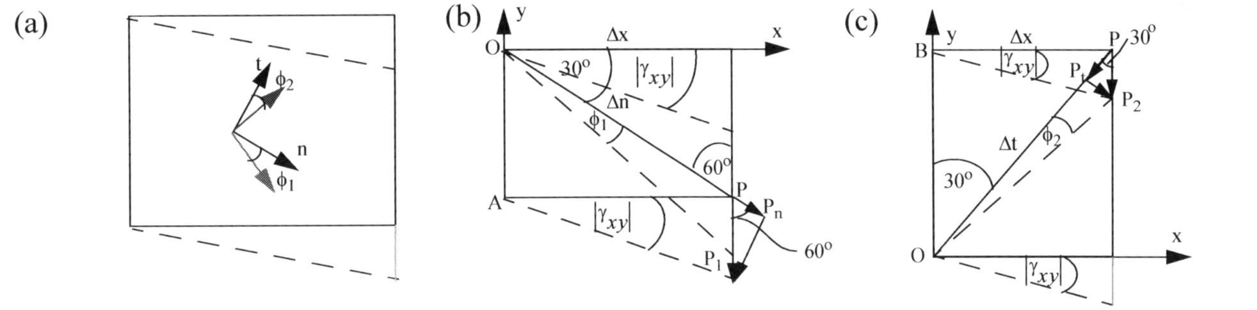

$$OP_n = OP + PP_n = (\Delta n)[1 + 216.5(10^{-6})] \approx (\Delta n) \qquad tan\phi_1 \approx \phi_1 = P_nP_1/OP_n \approx 375(10^{-6}) \qquad or \qquad \phi_1 = 375\mu rads \quad \textbf{(3)}$$

From geometry in Figure (c) we obtain the following.

$$\Delta x = \Delta t sin 30 \qquad PP_2 = BP tan|\gamma_{xy}| \approx \Delta x|\gamma_{xy}| = 500(10^{-6})\Delta t\ sin 30 = 250(\Delta t)(10^{-6}) \quad \textbf{(4)}$$

$$PP_t = PP_2 cos 30 = 216.5(\Delta t)(10^{-6}) \qquad P_tP_2 = PP_2 sin 30 = 125(\Delta t)(10^{-6}) \qquad \varepsilon_{tt} = -PP_t/OP = -216.5(10^{-6}) \quad \textbf{(5)}$$

$$OP_t = OP - PP_t = (\Delta t)[1 - 216.5(10^{-6})] \approx (\Delta t) \qquad tan\phi_2 \approx \phi_2 = P_tP_2/OP_t \approx 125(10^{-6}) \qquad or \qquad \phi_2 = 125\mu rads \quad \textbf{(6)}$$

$$\gamma_{nt} = -(\phi_1 - \phi_2) = -(375 - 125) = -250 \quad \textbf{(7)}$$

ANS $\varepsilon_{nn} = 216.5\mu$; $\varepsilon_{tt} = -216.5\mu$; $\gamma_{nt} = -250\mu$

---

9. 8

Solution $\varepsilon_{xx} = -600\ \mu$ $\varepsilon_{nn} = ?$ $\varepsilon_{tt} = ?$ $\gamma_{nt} = ?$

Figure (a) shows the rotation of the n and t lines, while Figures (b) and (c) show rectangles with diagonals in the n and t directions deforming due to ε_{xx} .

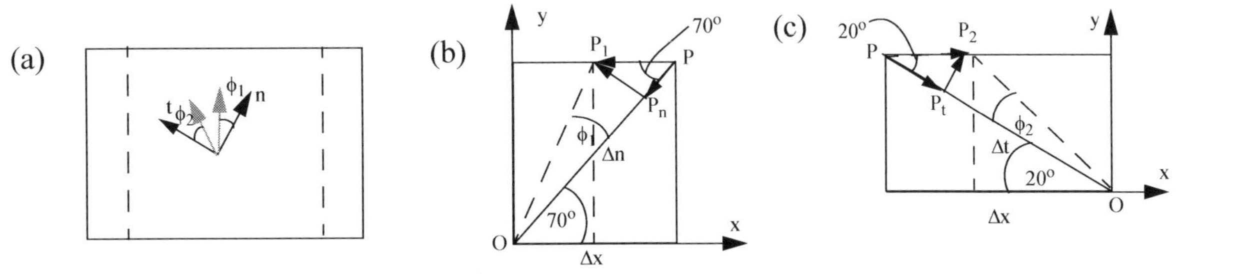

From geometry in Figure (b) we obtain the following.

$$\Delta x = \Delta n cos 70 \qquad PP_1 = |\varepsilon_{xx}|\Delta x = 600(10^{-6})\Delta n\ cos 70 = 205.21(\Delta n)(10^{-6}) \quad \textbf{(1)}$$

$$PP_n = PP_1\cos 70 = 70.19(\Delta n)(10^{-6}) \qquad P_nP_1 = PP_1\sin 70 = 192.84(\Delta n)(10^{-6}) \qquad \varepsilon_{nn} = -PP_n/OP = (-70.19)(10^{-6}) \quad (2)$$

$$OP_n = OP - PP_n = (\Delta n)[1 - 70.19(10^{-6})] \approx (\Delta n) \qquad \tan\phi_1 \approx \phi_1 = P_nP_1/OP_n \approx 192.84(10^{-6}) = 192.84\mu rads \quad (3)$$

From geometry in Figure (c) we obtain the following.

$$\Delta x = \Delta t\cos 20 \qquad PP_2 = |\varepsilon_{xx}|\Delta x = 600(10^{-6})\Delta t\ \cos 20 = 563.82(\Delta t)(10^{-6}) \quad (4)$$

$$PP_t = PP_2\cos 20 = 529.81(\Delta t)(10^{-6}) \qquad P_tP_2 = PP_2\sin 20 = 192.84(\Delta t)(10^{-6}) \qquad \varepsilon_{tt} = -PP_t/OP = -529.81(10^{-6}) \quad (5)$$

$$OP_t = OP - PP_t = (\Delta t)[1 - 529.81(10^{-6})] \approx (\Delta t) \qquad \tan\phi_2 \approx \phi_2 = P_tP_2/OP_t \approx 192.84(10^{-6}) = 192.84\mu rads \quad (6)$$

$$\gamma_{nt} = (\phi_1 + \phi_2) = (192.84 + 192.84) = 385.7 \quad (7)$$

ANS $\varepsilon_{nn} = -70.19\mu$; $\varepsilon_{tt} = -529.81\mu$; $\gamma_{nt} = 385.7\mu$

---

9. 9

Solution $\varepsilon_{yy} = -1000\ \mu$ $\varepsilon_{nn} = ?$ $\varepsilon_{tt} = ?$ $\gamma_{nt} = ?$

From geometry in Figure (b) we obtain the following.

$$\Delta y = \Delta n\sin 70 \qquad PP_1 = |\varepsilon_{yy}|\Delta y = 1000(10^{-6})\Delta n\ \sin 70 = 939.69(\Delta n)(10^{-6}) \quad (1)$$

$$PP_n = PP_1\cos 20 = 883.02(\Delta n)(10^{-6}) \qquad P_nP_1 = PP_1\sin 20 = 321.39(\Delta n)(10^{-6}) \qquad \varepsilon_{nn} = -PP_n/OP = -883.02(10^{-6}) \quad (2)$$

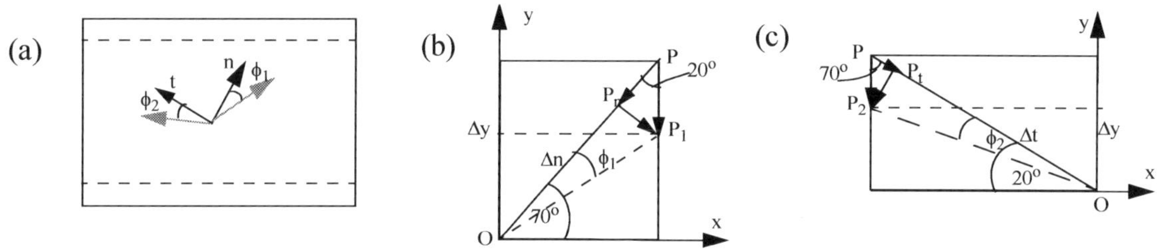

$$OP_n = OP - PP_n = (\Delta n)[1 - 883.02(10^{-6})] \approx (\Delta n) \qquad \tan\phi_1 \approx \phi_1 = P_nP_1/OP_n \approx 321.39(10^{-6}) = 321.39\mu rads \quad (3)$$

From geometry in Figure (c) we obtain the following.

$$\Delta y = \Delta t\sin 20 \qquad PP_2 = |\varepsilon_{yy}|\Delta y = 1000(10^{-6})\Delta t\ \sin 20 = 342.02(\Delta t)(10^{-6}) \quad (4)$$

$$PP_t = PP_2\cos 70 = 116.98(\Delta t)(10^{-6}) \qquad P_tP_2 = PP_2\sin 70 = 321.39(\Delta t)(10^{-6}) \qquad \varepsilon_{tt} = -PP_t/OP = -116.98(10^{-6}) \quad (5)$$

$$OP_t = OP - PP_t = (\Delta t)[1 - 116.98(10^{-6})] \approx (\Delta t) \qquad \tan\phi_2 \approx \phi_2 = P_tP_2/OP_t \approx 321.39(10^{-6}) = 321.39\mu rads \quad (6)$$

$$\gamma_{nt} = -(\phi_1 + \phi_2) = -(321.39 + 321.39) = -642.8\mu \quad (7)$$

ANS $\varepsilon_{nn} = -883\mu$; $\varepsilon_{tt} = -117\mu$; $\gamma_{nt} = -642.8\mu$

---

9. 10

Solution $\gamma_{xy} = 500\ \mu$ $\varepsilon_{nn} = ?$ $\varepsilon_{tt} = ?$ $\gamma_{nt} = ?$

From geometry in Figure (b) we obtain the following.

$$\Delta y = \Delta n\sin 70 \qquad PP_1 = AP\tan|\gamma_{xy}| \approx \Delta y|\gamma_{xy}| = 500(10^{-6})\Delta n\ \sin 70 = 469.85(\Delta n)(10^{-6}) \quad (1)$$

$$PP_n = PP_1\cos 70 = 160.7(\Delta n)(10^{-6}) \qquad P_nP_1 = PP_1\sin 70 = 441.51(\Delta n)(10^{-6}) \qquad \varepsilon_{nn} = PP_n/OP = 160.7(10^{-6}) \quad (2)$$

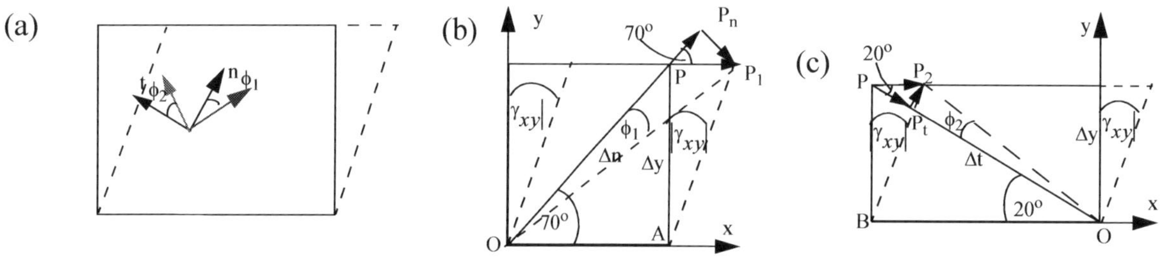

$$OP_n = OP + PP_n = (\Delta n)[1 + 160.7(10^{-6})] \approx (\Delta n) \qquad \tan\phi_1 \approx \phi_1 = P_nP_1/OP_n \approx 441.51(10^{-6}) = 441.51\mu rads \quad (3)$$

From geometry in Figure (c) we obtain the following.

$$\Delta y = \Delta t\sin 20 \qquad PP_2 = BP\tan|\gamma_{xy}| \approx \Delta y|\gamma_{xy}| = 500(10^{-6})\Delta t\ \sin 20 = 171.01(\Delta t)(10^{-6}) \quad (4)$$

$$PP_t = PP_2\cos 20 = 160.7(\Delta t)(10^{-6}) \qquad P_tP_2 = PP_2\sin 20 = 58.49(\Delta t)(10^{-6}) \qquad \varepsilon_{tt} = -PP_t/OP = -160.7(10^{-6}) \quad (5)$$

$$OP_t = OP - PP_t = (\Delta t)[1 - 160.7(10^{-6})] \approx (\Delta t) \qquad tan\phi_2 \approx \phi_2 = P_tP_2/OP_t \approx 58.49(10^{-6}) = 58.49\mu rads \quad (6)$$

$$\gamma_{nt} = -(\phi_1 - \phi_2) = -(441.51 - 58.49) = -383 \quad (7)$$

ANS $\varepsilon_{nn} = 160.7\mu$; $\varepsilon_{tt} = -160.7\mu$; $\gamma_{nt} = -383\mu rad$

---

9. 11

Solution $\varepsilon_{xx} = 600\ \mu$ $\quad \varepsilon_{nn} = ?$ $\quad \varepsilon_{tt} = ?$ $\quad \gamma_{nt} = ?$

From geometry in Figure (b) we obtain the following.

$$\Delta x = \Delta n sin 40 \qquad PP_1 = |\varepsilon_{xx}|\Delta x = 600(10^{-6})\Delta n\ sin 40 = 385.67(\Delta n)(10^{-6}) \quad (1)$$

$$PP_n = PP_1 cos 50 = 247.91(\Delta n)(10^{-6}) \qquad P_nP_1 = PP_1 sin 50 = 295.44(\Delta n)(10^{-6}) \qquad \varepsilon_{nn} = PP_n/OP = 247.91(10^{-6}) \quad (2)$$

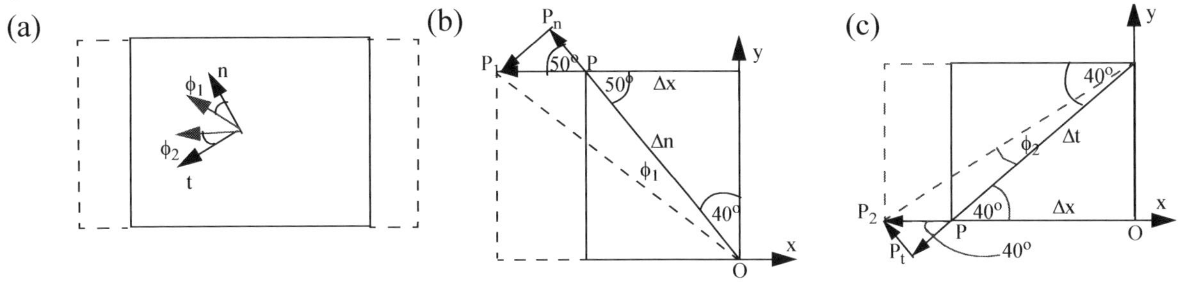

$$OP_n = OP + PP_n = (\Delta n)[1 + 247.91(10^{-6})] \approx (\Delta n) \qquad tan\phi_1 \approx \phi_1 = P_nP_1/OP_n \approx 295.44(10^{-6}) = 295.44\mu rads \quad (3)$$

From geometry in Figure (c) we obtain the following.

$$\Delta x = \Delta t cos 40 \qquad PP_2 = |\varepsilon_{xx}|\Delta x = 600(10^{-6})\Delta t\ cos 40 = 459.63(\Delta t)(10^{-6}) \quad (4)$$

$$PP_t = PP_2 cos 40 = 352.09(\Delta t)(10^{-6}) \qquad P_tP_2 = PP_2 sin 40 = 295.44(\Delta t)(10^{-6}) \qquad \varepsilon_{tt} = PP_t/OP = 352.09(10^{-6}) \quad (5)$$

$$OP_t = OP - PP_t = (\Delta t)[1 - 100(10^{-6})] \approx (\Delta t) \qquad tan\phi_2 \approx \phi_2 = P_tP_2/OP_t \approx 295.44(10^{-6}) = 295.44\mu rads \quad (6)$$

$$\gamma_{nt} = (\phi_1 + \phi_2) = (295.44 + 295.44) = 590.9 \quad (7)$$

ANS $\varepsilon_{nn} = 247.9\mu$; $\varepsilon_{tt} = 352.1\mu$; $\gamma_{nt} = 590.9\mu$

---

9. 12

Solution $\varepsilon_{yy} = 600\ \mu$ $\quad \varepsilon_{nn} = ?$ $\quad \varepsilon_{tt} = ?$ $\quad \gamma_{nt} = ?$

From geometry in Figure (b) we obtain the following.

$$\Delta y = \Delta n cos 40 \qquad PP_1 = |\varepsilon_{yy}|\Delta y = 600(10^{-6})\Delta n\ cos 40 = 459.63(\Delta n)(10^{-6}) \quad (1)$$

$$PP_n = PP_1 cos 40 = 352.09(\Delta n)(10^{-6}) \qquad P_nP_1 = PP_1 sin 40 = 295.44(\Delta n)(10^{-6}) \qquad \varepsilon_{nn} = PP_n/OP = 352.09(10^{-6}) \quad (2)$$

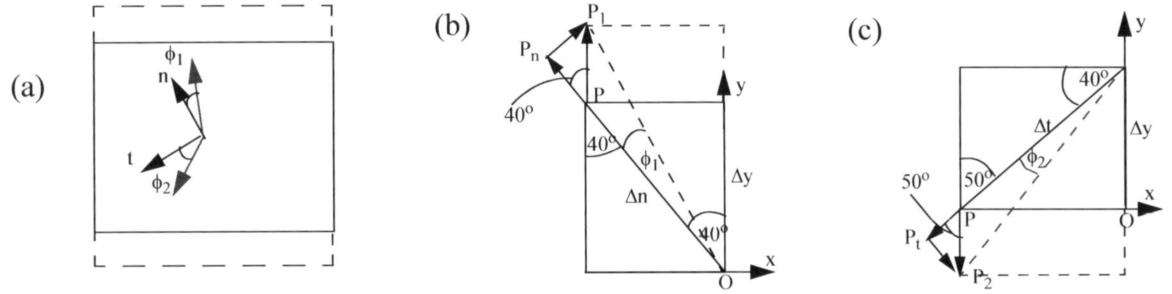

$$OP_n = OP + PP_n = (\Delta n)[1 + 352.09(10^{-6})] \approx (\Delta n) \qquad tan\phi_1 \approx \phi_1 = P_nP_1/OP_n \approx 295.44(10^{-6}) = 295.44\mu rads \quad (3)$$

From geometry in Figure (c) we obtain the following.

$$\Delta y = \Delta t sin 40 \qquad PP_2 = |\varepsilon_{yy}|\Delta y = 600(10^{-6})\Delta t\ sin 40 = 385.67(\Delta t)(10^{-6}) \quad (4)$$

$$PP_t = PP_2 cos 50 = 247.91(\Delta t)(10^{-6}) \qquad P_tP_2 = PP_2 sin 50 = 295.44(\Delta t)(10^{-6}) \qquad \varepsilon_{tt} = PP_t/OP = 247.91(10^{-6}) \quad (5)$$

$$OP_t = OP + PP_t = (\Delta t)[1 + 247.91(10^{-6})] \approx (\Delta t) \qquad tan\phi_2 \approx \phi_2 = P_tP_2/OP_t \approx 295.44(10^{-6}) = 295.44\mu rads \quad (6)$$

$$\gamma_{nt} = -(\phi_1 + \phi_2) = -(295.44 + 295.44) = -590.9 \quad (7)$$

ANS $\varepsilon_{nn} = 352.1\mu$; $\varepsilon_{tt} = 247.9\mu$; $\gamma_{nt} = -590.9\mu$

---

9. 13

Solution $\gamma_{xy} = 600\ \mu$ $\varepsilon_{nn} = ?$ $\varepsilon_{tt} = ?$ $\gamma_{nt} = ?$

From geometry in Figure (b) we obtain the following.

$$\Delta x = \Delta n \sin 40 \qquad PP_1 = AP\tan|\gamma_{xy}| \approx \Delta x|\gamma_{xy}| = 600(10^{-6})\Delta n \ \sin 40 = 385.67(\Delta n)(10^{-6}) \quad \textbf{(1)}$$

$$PP_n = PP_1\cos 40 = 295.44(\Delta n)(10^{-6}) \qquad P_nP_1 = PP_1\sin 40 = 247.91(\Delta n)(10^{-6}) \qquad \varepsilon_{nn} = -PP_n/OP = -295.44(10^{-6}) \quad \textbf{(2)}$$

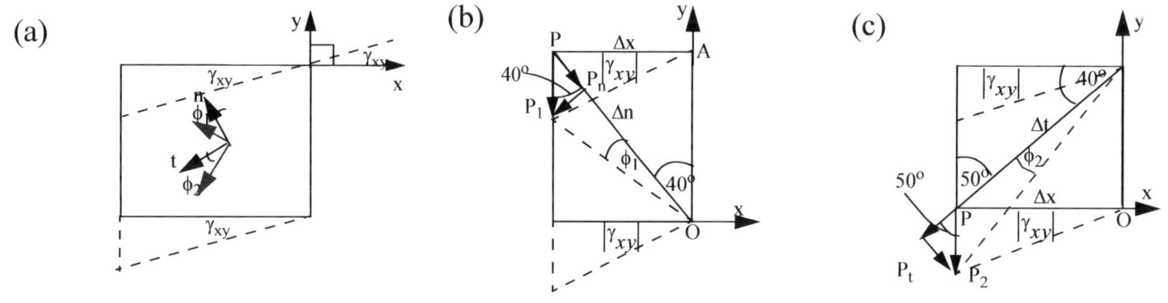

$$OP_n = OP - PP_n = (\Delta n)[1 - 295.44(10^{-6})] \approx (\Delta n) \qquad \tan\phi_1 \approx \phi_1 = P_nP_1/OP_n \approx 247.91(10^{-6}) = 247.91\mu rads \quad \textbf{(3)}$$

From geometry in Figure (c) we obtain the following.

$$\Delta x = \Delta t\cos 40 \qquad PP_2 = OP\tan|\gamma_{xy}| \approx \Delta x|\gamma_{xy}| = 600(10^{-6})\Delta t \ \cos 40 = 459.63(\Delta t)(10^{-6}) \quad \textbf{(4)}$$

$$PP_t = PP_2\cos 50 = 295.44(\Delta t)(10^{-6}) \qquad P_tP_2 = PP_2\sin 50 = 352.09(\Delta t)(10^{-6}) \qquad \varepsilon_{tt} = PP_t/OP = 295.44(10^{-6}) \quad \textbf{(5)}$$

$$OP_t = OP + PP_t = (\Delta t)[1 + 295.44(10^{-6})] \approx (\Delta t) \qquad \tan\phi_2 \approx \phi_2 = P_tP_2/OP_t \approx 352.09(10^{-6}) = 352.09\mu rads \quad \textbf{(6)}$$

$$\gamma_{nt} = -(\phi_2 - \phi_1) = -(352.09 - 247.91) = -104.2 \quad \textbf{(7)}$$

ANS $\varepsilon_{nn} = -295.4\mu$; $\varepsilon_{tt} = 295.4\mu$; $\gamma_{nt} = -104.2\mu rad$

Section 9.2-9.4

9. 14

Solution $\varepsilon_{xx} = -400\ \mu$ $\varepsilon_{yy} = 600\ \mu$ $\gamma_{xy} = -500\ \mu$ principal directions?

ANS Principal direction one will be in sector 3 and principal direction two will be in sector 5
Principal direction one will be in sector 7 and principal direction two will be in sector 1

---

9. 15

Solution $\varepsilon_{xx} = -600\ \mu$ $\varepsilon_{yy} = -800\ \mu$ $\gamma_{xy} = 500\ \mu$ principal directions?

ANS Principal direction one will be in sector 1 and principal direction two will be in sector 3
Principal direction one will be in sector 5 and principal direction two will be in sector 7

---

9. 16

Solution $\varepsilon_{xx} = 800\ \mu$ $\varepsilon_{yy} = 600\ \mu$ $\gamma_{xy} = -1000\ \mu$ principal directions?

ANS Principal direction one will be in sector 4 and principal direction two will be in sector 6
Principal direction one will be in sector 8 and principal direction two will be in sector 2

---

9. 17

Solution $\varepsilon_{xx} = 0$ $\varepsilon_{yy} = 600\ \mu$ $\gamma_{xy} = -500\ \mu$ principal directions?

ANS Principal direction one will be in sector 3 and principal direction two will be in sector 5
Principal direction one will be in sector 7 and principal direction two will be in sector 1

---

9. 18

Solution $\varepsilon_{xx} = -1000\ \mu \quad \varepsilon_{yy} = -500\ \mu \quad \gamma_{xy} = 700\ \mu$ principal directions?

ANS Principal direction one will be in sector 2 and principal direction two will be in sector 4

Principal direction one will be in sector 6 and principal direction two will be in sector 8

---

9. 19

Solution

The solution proceeds as follows.

$$\varepsilon_{nn} = (\varepsilon_{xx}+\varepsilon_{yy})/2 + \{(\varepsilon_{xx}-\varepsilon_{yy})/2\}cos2\theta + (\gamma_{xy}/2)sin2\theta \tag{1}$$

$$\frac{d\varepsilon_{nn}}{d\theta} = \frac{(\varepsilon_{xx}-\varepsilon_{yy})}{2}(-2sin2\theta) + \frac{\gamma_{xy}}{2}(2cos2\theta)\Big|_{\theta=\theta_p} \quad or \quad tan2\theta_p = \frac{\gamma_{xy}}{\varepsilon_{xx}-\varepsilon_{yy}} \tag{2}$$

Above equation is the desired result.

---

9. 20

Solution

We have the angle at which the normal strain is maximum or minimum.In the solution below R_{eq} is defined for derivation and it is different from the radius calculated from the Mohr's circle.

$$sin2\theta_p = \pm\gamma_{xy}/R_{eq} \qquad cos2\theta_p = \pm(\varepsilon_{xx}-\varepsilon_{yy})/R_{eq} \qquad R_{eq} = \sqrt{(\varepsilon_{xx}-\varepsilon_{yy})^2+\gamma_{xy}^2} \tag{1}$$

$$\varepsilon_{nn} = \frac{(\varepsilon_{xx}+\varepsilon_{yy})}{2} + \frac{(\varepsilon_{xx}-\varepsilon_{yy})}{2}cos2\theta + \frac{\gamma_{xy}}{2}sin2\theta = \frac{(\varepsilon_{xx}+\varepsilon_{yy})}{2} + \frac{(\varepsilon_{xx}-\varepsilon_{yy})}{2}\left[\pm\frac{(\varepsilon_{xx}-\varepsilon_{yy})}{R_{eq}}\right] + \frac{\gamma_{xy}}{2}\left[\pm\frac{\gamma_{xy}}{R_{eq}}\right] \text{ or}$$

$$\varepsilon_{nn} = \frac{(\varepsilon_{xx}+\varepsilon_{yy})}{2} \pm \frac{(\varepsilon_{xx}-\varepsilon_{yy})^2+\gamma_{xy}^2}{2R_{eq}} = \frac{(\varepsilon_{xx}+\varepsilon_{yy})}{2} \pm \frac{R_{eq}^2}{2R_{eq}} = \frac{(\varepsilon_{xx}+\varepsilon_{yy})}{2} \pm \frac{\sqrt{(\varepsilon_{xx}-\varepsilon_{yy})^2+\gamma_{xy}^2}}{2} = \frac{(\varepsilon_{xx}+\varepsilon_{yy})}{2} \pm \sqrt{\left(\frac{\varepsilon_{xx}-\varepsilon_{yy}}{2}\right)^2 + \frac{\gamma_{xy}}{2}^2} \tag{2}$$

Above equation is the desired result.

---

9. 21

Solution

In the solution below R_{eq} is defined for derivation and it is different from the radius calculated from the Mohr's circle.

The shear strain can be re-written in double angles as:

$$\gamma_{nt} = -(\varepsilon_{xx}-\varepsilon_{yy})sin2\theta + \gamma_{xy}cos2\theta \tag{1}$$

We have the angle at which the normal strain is maximum or minimum.We can write the following:

$$\text{'}2\theta_p = \pm\gamma_{xy}/R_{eq} \qquad cos2\theta_p = \pm(\varepsilon_{xx}-\varepsilon_{yy})/R_{eq} \qquad R_{eq} = \sqrt{(\varepsilon_{xx}-\varepsilon_{yy})^2+} \tag{2}$$

$$\gamma_{nt} = -(\varepsilon_{xx}-\varepsilon_{yy})[\pm\gamma_{xy}/R_{eq}] + \gamma_{xy}[\pm(\varepsilon_{xx}-\varepsilon_{yy})/R_{eq}] = 0 \tag{3}$$

$$\varepsilon_{nn} = \frac{(\varepsilon_{xx}+\varepsilon_{yy})}{2} \pm \frac{(\varepsilon_{xx}-\varepsilon_{yy})^2+\gamma_{xy}^2}{2R_{eq}} = \frac{(\varepsilon_{xx}+\varepsilon_{yy})}{2} \pm \frac{R_{eq}^2}{2R_{eq}} = \frac{(\varepsilon_{xx}+\varepsilon_{yy})}{2} \pm \frac{\sqrt{(\varepsilon_{xx}-\varepsilon_{yy})^2+\gamma_{xy}^2}}{2} = \frac{(\varepsilon_{xx}+\varepsilon_{yy})}{2} \pm \sqrt{\left(\frac{\varepsilon_{xx}-\varepsilon_{yy}}{2}\right)^2 + \frac{\gamma_{xy}}{2}^2} \tag{4}$$

Above equation is the desired result.

---

9. 22

Solution

The principal angle is given by $tan2\theta_p = \gamma_{xy}/(\varepsilon_{xx}-\varepsilon_{yy})$

Differentiating γ_{nt} with respect θ and assuming the slope is zero at $\theta = \theta_s$ we obtain:

$$\frac{d\gamma_{nt}}{d\theta} = -(\varepsilon_{xx}-\varepsilon_{yy})(2cos2\theta) + \gamma_{xy}(-2sin2\theta)\Big|_{\theta=\theta_s} \quad or \quad tan2\theta_s = -\frac{(\varepsilon_{xx}-\varepsilon_{yy})}{\gamma_{xy}} \tag{1}$$

$$tan2\theta_s tan2\theta_p = -1 \quad or \quad tan2\theta_s = -cot2\theta_p = tan(2\theta_p \pm \pi/2) \quad or \quad \theta_s = \theta_p \pm \pi/4 \tag{2}$$

Above equation shows that the plane of maximum shear stress is 45° to the principal planes.

---

9. 23

Solution

In the solution below R_{eq} is defined for derivation and it is different from the radius calculated from the Mohr's circle.

We have the following:

$$tan2\theta_s = -(\varepsilon_{xx}-\varepsilon_{yy})/\gamma_{xy} \qquad cos2\theta_s = \pm\gamma_{xy}/R_{eq} \qquad sin2\theta_s = \mp(\varepsilon_{xx}-\varepsilon_{yy})/R_{eq} \qquad R_{eq} = \sqrt{(\varepsilon_{xx}-\varepsilon_{yy})^2+\gamma_{xy}^2} \quad (1)$$

$$\gamma_{nt} = -(\varepsilon_{xx}-\varepsilon_{yy})sin2\theta+\gamma_{xy}cos2\theta = -(\varepsilon_{xx}-\varepsilon_{yy})\left[\mp\frac{(\varepsilon_{xx}-\varepsilon_{yy})}{R_{eq}}\right]+\gamma_{xy}\left[\pm\frac{\gamma_{xy}}{R_{eq}}\right] = \pm[(\varepsilon_{xx}-\varepsilon_{yy})^2+\gamma_{xy}^2]/R_{eq} = R_{eq} = |\varepsilon_1-\varepsilon_2| \quad (2)$$

Above equation is same as equation for maximum inplane shear strain.

---

9. 24

Solution

The normal & shear strain transformation equations in terms of double angle are as given below.

$$\varepsilon_{nn} = \frac{(\varepsilon_{xx}+\varepsilon_{yy})}{2}+\frac{(\varepsilon_{xx}-\varepsilon_{yy})}{2}cos2\theta+\frac{\gamma_{xy}}{2}sin2\theta \quad or \quad \varepsilon_{nn}-\frac{\varepsilon_{xx}+\varepsilon_{yy}}{2} = \frac{(\varepsilon_{xx}-\varepsilon_{yy})}{2}cos2\theta+\frac{\gamma_{xy}}{2}sin2\theta \quad (1)$$

$$\gamma_{nt} = -(\varepsilon_{xx}-\varepsilon_{yy})sin2\theta+\gamma_{xy}cos2\theta \quad or \quad \frac{\gamma_{nt}}{2} = -\frac{(\varepsilon_{xx}-\varepsilon_{yy})}{2}sin2\theta+\frac{\gamma_{xy}}{2}cos2\theta \quad (2)$$

Squaring and adding we obtain:

$$\left(\varepsilon_{nn}-\frac{\varepsilon_{xx}+\varepsilon_{yy}}{2}\right)^2+\left(\frac{\gamma_{nt}}{2}\right)^2 = \left(\frac{\varepsilon_{xx}-\varepsilon_{yy}}{2}\right)^2[cos^22\theta+sin^22\theta]+\left(\frac{\gamma_{xy}}{2}\right)^2[sin^22\theta+cos^22\theta] \quad (3)$$

$$\left(\varepsilon_{nn}-\frac{\varepsilon_{xx}+\varepsilon_{yy}}{2}\right)^2+\left(\frac{\gamma_{nt}}{2}\right)^2 = \left(\frac{\varepsilon_{xx}-\varepsilon_{yy}}{2}\right)^2+\left(\frac{\gamma_{xy}}{2}\right)^2 \quad (4)$$

Above equation is the equation for the Mohr's circle for strain.

---

9. 25

Solution $\varepsilon_{xx} = -400\ \mu$. $\varepsilon_{nn} = ?$ $\varepsilon_{tt} = ?$ $\gamma_{nt} = ?$ $\theta = -30^o$

Substituting $\varepsilon_{xx} = -400\ \mu$., $\varepsilon_{yy} = 0$, $\gamma_{xy} = 0$ and $\theta = -30^o$ into strain transformation equations we obtain the following.

$$\varepsilon_{nn} = -400cos^2(-30) = -300\mu \qquad \varepsilon_{tt} = -400sin^2\theta = -100\mu \qquad \gamma_{nt} = -2(-400)sin(-30)cos(-30) = -346.4\mu \quad (1)$$

ANS $\varepsilon_{nn} = -300\mu$; $\varepsilon_{tt} = -100\mu$; $\gamma_{nt} = -346.4\mu$

---

9. 26

Solution $\varepsilon_{xx} = -400\ \mu$. $\varepsilon_{nn} = ?$ $\varepsilon_{tt} = ?$ $\gamma_{nt} = ?$ $\theta = -30^o$

The center of the Mohr's circle is located at -200 μ, the radius is R = 200μ. Points V, H, N, and T are shown below.

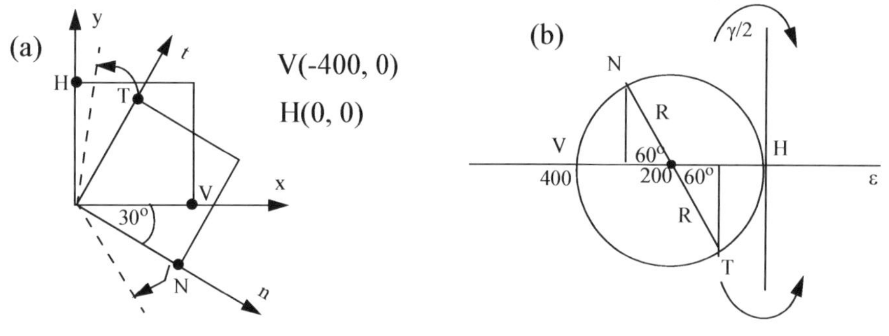

$$\varepsilon_{nn} = -200-Rcos60 = -300 \qquad \varepsilon_{tt} = -200+Rcos60 = -100 \qquad |\gamma_{nt}/2| = Rsin60 \quad or \quad |\gamma_{nt}| = 346.4\mu \quad (1)$$

The shear strain is negative as the angle between *n* and *t* increases.

ANS $\varepsilon_{nn} = -300\mu$; $\varepsilon_{tt} = -100\mu$; $\gamma_{nt} = -346.4\mu$

---

9. 27

Solution $\varepsilon_{yy} = 600\mu$. $\varepsilon_{nn} = ?$ $\varepsilon_{tt} = ?$ $\gamma_{nt} = ?$ $\theta = -30^o$

Substituting $\varepsilon_{xx} = 0$, $\varepsilon_{yy} = 600\mu$, $\gamma_{xy} = 0$ and $\theta = -30^o$ into strain transformation equations we obtain the following.

$$\varepsilon_{nn} = 600sin^2(-30) = 150 \qquad \varepsilon_{tt} = 600cos^2(-30) = 450 \qquad \gamma_{nt} = 2(600)sin(-30)cos(-30) = -519.6 \quad (1)$$

ANS $\varepsilon_{nn} = 150\mu$; $\varepsilon_{tt} = 450\mu$; $\gamma_{nt} = -519.6\mu$

---

9. 28

Solution $\varepsilon_{yy} = 600\mu.$ $\varepsilon_{nn} = ?$ $\varepsilon_{tt} = ?$ $\gamma_{nt} = ?$ $\theta = -30^o$

The center of the Mohr's circle is located at 300 μ, the radius is R = 300μ. Points V, H, N, and T are shown below

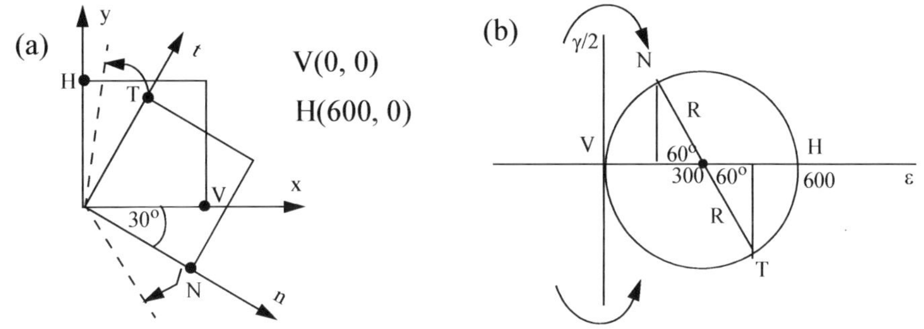

$$\varepsilon_{nn} = 300 - R\cos 60 = 150 \qquad \varepsilon_{tt} = 300 + R\cos 60 = 450 \qquad |\gamma_{nt}/2| = R\sin 60 \quad or \quad |\gamma_{nt}| = 519.62\mu \qquad \textbf{(1)}$$

The shear strain is negative as the angle between *n* and *t* increases.

ANS $\varepsilon_{nn} = 150\mu;\ \varepsilon_{tt} = 450\mu;\ \gamma_{nt} = -519.6\mu$

---

9. 29

Solution $\gamma_{xy} = -500\mu.$ $\varepsilon_{nn} = ?$ $\varepsilon_{tt} = ?$ $\gamma_{nt} = ?$ $\theta = -30^o$

Substituting $\varepsilon_{xx} = 0$, $\varepsilon_{yy} =$, $\gamma_{xy} = -500\mu$ and $\theta = -30^o$ into strain transformation equations we obtain the following.

$$\varepsilon_{nn} = -500\sin(-30)\cos(-30) = 216.5 \qquad \varepsilon_{tt} = -(-500)\sin(-30)\cos(-30) = -216.5 \qquad \textbf{(1)}$$

$$\gamma_{nt} = (-500)(\cos^2(-30) - \sin^2(-30)) = -250 \qquad \textbf{(2)}$$

ANS $\varepsilon_{nn} = 216.5\mu;\ \varepsilon_{tt} = -216.5\mu;\ \gamma_{nt} = -250\mu$

---

9. 30

Solution $\gamma_{xy} = -500\mu.$ $\varepsilon_{nn} = ?$ $\varepsilon_{tt} = ?$ $\gamma_{nt} = ?$ $\theta = -30^o$

The center of the Mohr's circle is located at the origin, the radius is R = 250μ. Points V, H, N, and T are shown below.

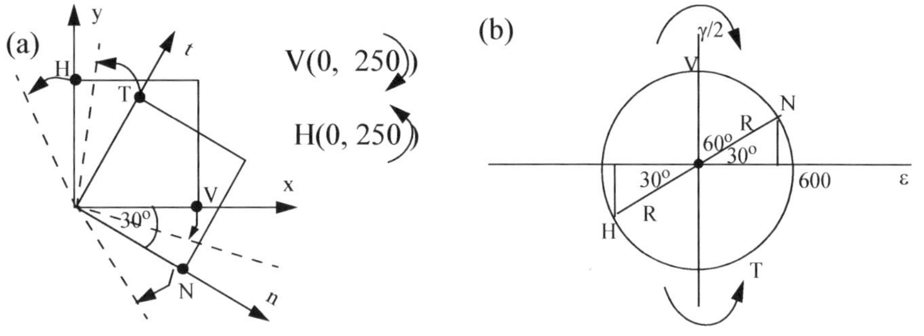

$$\varepsilon_{nn} = R\cos 30 = 216.51 \qquad \varepsilon_{tt} = -R\cos 30 = -216.51 \qquad |\gamma_{nt}/2| = R\sin 30 \quad or \quad |\gamma_{nt}| = 250 \qquad \textbf{(1)}$$

The shear strain is negative as the angle between *n* and *t* increases.

ANS $\varepsilon_{nn} = 216.5\mu;\ \varepsilon_{tt} = -216.5\mu;\ \gamma_{nt} = -250\mu$

---

9. 31

Solution Plane Strain $\varepsilon_{xx} = -400\ \mu$ $\varepsilon_{yy} = 600\ \mu$ $\gamma_{xy} = -500\ \mu$

$\varepsilon_1 = ?$ $\varepsilon_2 = ?$ $\varepsilon_3 = ?$ $\theta_1 = ?$ $\gamma_{max} = ?$ $\varepsilon_{nn} = ?$ $\varepsilon_{tt} = ?$ $\gamma_{nt} = ?$

(a) The principal angle θ_p is calculated and substituted into the normal strain equation to determine one principal strain.

$$\theta_p = 0.5\ atan\left(\frac{-500}{-400-600}\right) = 13.28^o \qquad \textbf{(1)}$$

$$\varepsilon_p = (-400)\cos^2(13.28) + (600)\sin^2(13.28) + (-500)\sin(13.28)\cos(13.28) = -459\mu \qquad \textbf{(2)}$$

The other principal strain is: $-400 + 600 - (-459) = 659\mu$. As this strain is greater than the above strain, therefore this is principal strain one. Thus principal angle 1 is $\theta_1 = \theta_p \pm 90$.

$$\varepsilon_1 = 659\mu \qquad \varepsilon_2 = -459\mu \qquad \varepsilon_3 = 0 \qquad \theta_1 = 103.28^o \ or\ -76.72^o \qquad \textbf{(3)}$$

(b) The difference between ε_1 and ε_2 is the largest difference between principal strains. The maximum shear strain is

$$\gamma_{max}/2 = |(\varepsilon_1 - \varepsilon_2)/2| = |(659 + 459)/2| = 1118\mu \quad (4)$$

(c) Substituting the given strains and θ = -70° into strain transformation equations we obtain the following.

$$\varepsilon_{nn} = (-400)cos^2(-70) + (600)sin^2(-70) + (-500)sin(-70)cos(-70) = 643.7\mu \quad (5)$$

$$\varepsilon_{tt} = (-400)cos^2(-70) + (600)sin^2(-70) - (-500)sin(-70)cos(-70) = -443.7\mu \quad (6)$$

$$\gamma_{nt} = -2(-400 - 600)cos(-70)sin(-70) + (-500)(cos^2(-70) - sin^2(-70)) = -259.8\mu \quad (7)$$

ANS $\varepsilon_1 = 659\mu$; $\varepsilon_2 = -459\mu$; $\varepsilon_3 = 0$; $\theta_1 = 103.28^o$ *or* -76.72^o ; $\gamma_{max} = 1118\mu$; $\varepsilon_{nn} = 643.7\mu$; $\varepsilon_{tt} = -443.7\mu$; $\gamma_{nt} = -259.8\mu$

---

9. 32

Solution Plane Strain $\varepsilon_{xx} = -600\ \mu$ $\varepsilon_{yy} = -800\ \mu$ $\gamma_{xy} = 500\ \mu$

$\varepsilon_1 = ?$ $\varepsilon_2 = ?$ $\varepsilon_3 = ?$ $\theta_1 = ?$ $\gamma_{max} = ?$ $\varepsilon_{nn} = ?$ $\varepsilon_{tt} = ?$ $\gamma_{nt} = ?$

(a) The principal angle θ_p is calculated and substituted into the normal strain equation to determine one principal strain.

$$\theta_p = 0.5\ atan\left(\frac{(500)}{(-600 + 800)}\right) = 34.1^o \quad (1)$$

$$\varepsilon_p = (-600)cos^2(34.1) + (-800)sin^2(34.1) + (500)sin(34.1)cos(34.1) = -430.7\mu \quad (2)$$

The other principal strain is: $-600 - 800 - (-430.7) = -969.3\mu$. As this strain is smaller than the above strain, therefore the above strain is principal strain one. Thus principal angle 1 is $\theta_1 = \theta_p$.

$$\varepsilon_1 = -430.7\mu \qquad \varepsilon_2 = -969.3\mu \qquad \varepsilon_3 = 0 \qquad \theta_1 = 34.1^o \text{ or } -145.9^o \quad (3)$$

(b) The difference between ε_2 and ε_3 is the largest difference between principal strains. The maximum shear strain is

$$\gamma_{max}/2 = |(\varepsilon_3 - \varepsilon_2)/2| = |(0 + 969.3)/2| = 969.3\mu \quad (4)$$

(c) Substituting the given strains and θ = 20° into strain transformation equations we obtain the following.

$$\varepsilon_{nn} = (-600)cos^2(20) + (-800)sin^2(20) + (500)sin(20)cos(20) = -462.7\mu \quad (5)$$

$$\varepsilon_{tt} = (-600)cos^2(20) + (-800)sin^2(20) - (500)sin(20)cos(20) = -937.3\mu \quad (6)$$

$$\gamma_{nt} = -2(-600 - (-800))cos(20)sin(20) + (500)(cos^2(20) - sin^2(20)) = 254.5\mu \quad (7)$$

ANS $\varepsilon_1 = -430.7\mu$; $\varepsilon_2 = -969.3\mu$; $\varepsilon_3 = 0$; $\theta_1 = 34.1^o$ or -145.9^o ; $\gamma_{max} = 969.3\mu$; $\varepsilon_{nn} = -462.7\mu$; $\varepsilon_{tt} = -937.3\mu$; $\gamma_{nt} = 254.5\mu$

---

9. 33

Solution Plane Strain $\varepsilon_{xx} = 250\ \mu$ $\varepsilon_{yy} = 850\ \mu$ $\gamma_{xy} = 1600\ \mu$

$\varepsilon_1 = ?$ $\varepsilon_2 = ?$ $\varepsilon_3 = ?$ $\theta_1 = ?$ $\gamma_{max} = ?$ $\varepsilon_{nn} = ?$ $\varepsilon_{tt} = ?$ $\gamma_{nt} = ?$

(a) The principal angle θ_p is calculated and substituted into the normal strain equation to determine one principal strain.

$$\theta_p = 0.5\ atan\left(\frac{1600}{250 - 850}\right) = -34.7^o \quad (1)$$

$$\varepsilon_p = (250)cos^2(13.28) + (850)sin^2(13.28) + (1600)sin(13.28)cos(13.28) = -304.4\mu \quad (2)$$

The other principal strain as: $250 + 850 - (-304.4) = 1404.4\mu$. As this strain is greater than the above strain, therefore this is principal strain one. Thus principal angle 1 is $\theta_1 = \theta_p \pm 90$.

$$\varepsilon_1 = 1404.4\mu \qquad \varepsilon_2 = -304.4\mu \qquad \varepsilon_3 = 0 \qquad \theta_1 = 55.3^o \text{ or } 124.7^o \quad (3)$$

(b) The difference between ε_1 and ε_2 is the largest difference between principal strains. The maximum shear strain is

$$\gamma_{max}/2 = |(\varepsilon_1 - \varepsilon_2)/2| = |(1404.4 + 304.4)/2| = 1708.8 \quad (4)$$

(c) Substituting the given strains and θ = 115° into strain transformation equations we obtain the following.

$$\varepsilon_{nn} = (250)cos^2(115) + (850)sin^2(115) + (1600)sin(115)cos(115) = 130 \quad (5)$$

$$\varepsilon_{tt} = (250)cos^2(115) + (850)sin^2(115) - (1600)sin(115)cos(115) = 970 \quad (6)$$

$$\gamma_{nt} = -2(250-850)cos(115)sin(115)+(1600)(cos^2(115)-sin^2(115)) = -1488 \quad (7)$$

ANS $\varepsilon_1 = 1404.4\mu$; $\varepsilon_2 = -304.4\mu$; $\varepsilon_3 = 0$; $\theta_1 = 55.3^o$ or 124.7^o ; $\gamma_{max} = 1708.8\mu$; $\varepsilon_{nn} = 130\mu$; $\varepsilon_{tt} = 970\mu$; $\gamma_{nt} = -1488\mu$

---

9. 34

Solution Plane Strain $\varepsilon_{xx} = -1800\ \mu$ $\varepsilon_{yy} = -3600\ \mu$ $\gamma_{xy} = -1500\ \mu$

$\varepsilon_1 = ?$ $\varepsilon_2 = ?$ $\varepsilon_3 = ?$ $\gamma_{max} = ?$ $\varepsilon_{nn} = ?$ $\varepsilon_{tt} = ?$ $\gamma_{nt} = ?$

(a) The principal angle θ_p is calculated and substituted into the normal strain equation to determine one principal strain.

$$\theta_p = 0.5\ atan\left(\frac{(-1500)}{(-1800+3600)}\right) = -19.9^o \quad (1)$$

$$\varepsilon_p = (-1800)cos^2(-19.9)+(-3600)sin^2(-19.9)+(-1500)sin(-19.9)cos(-19.9) = -1528.5\mu \quad (2)$$

The other principal strain is: $-1800-3600-(-1528.5) = -3871.5\mu$. As this strain is smaller than the above strain, therefore the above strain is principal strain one. Thus principal angle 1 is $\theta_1 = \theta_p \pm 90$.

$$\varepsilon_1 = -1528.5\mu \qquad \varepsilon_2 = -3871.5\mu \qquad \varepsilon_3 = 0 \qquad \theta_1 = -19.9^o \text{ or } 160.1^o \quad (3)$$

(b) The difference between ε_2 and ε_3 is the largest difference between principal strains. The maximum shear strain is

$$\gamma_{max}/2 = |(\varepsilon_3-\varepsilon_2)/2| = |(0+3871.5)/2| = 3871.5 \quad (4)$$

(c) Substituting the given strains and $\theta = 205^o$ into strain transformation equations we obtain the following.

$$\varepsilon_{nn} = (-1800)cos^2(205)+(-3600)sin^2(205)+(-1500)sin(205)cos(205) = -2695.5 \quad (5)$$

$$\varepsilon_{tt} = (-1800)sin^2(205)+(-3600)cos^2(205)-(-1500)sin(205)cos(205) = -2704.5 \quad (6)$$

$$\gamma_{nt} = -2(-1800-(-3600))cos(205)sin(205)+(-1500)(cos^2(205)-sin^2(205)) = -2343 \quad (7)$$

ANS $\varepsilon_1 = -1528.5\mu$; $\varepsilon_2 = -3871.5\mu$; $\varepsilon_3 = 0$; $\theta_1 = -19.9^o$ or 160.1^o ; $\gamma_{max} = 3871.5\mu$; $\varepsilon_{nn} = -2695.5\mu$; $\varepsilon_{tt} = -2704.5\mu$; $\gamma_{nt} = -2343\mu$

---

9. 35

Solution Plane Strain $\varepsilon_{nn} = 2000\ \mu$ $\varepsilon_{tt} = -800\ \mu$ $\gamma_{nt} = 750\ \mu$

$\varepsilon_1 = ?$ $\varepsilon_2 = ?$ $\varepsilon_3 = ?$ $\gamma_{max} = ?$ $\varepsilon_{xx} = ?$ $\varepsilon_{yy} = ?$ $\gamma_{xy} = ?$

(a) We can use *n* and *t* as the reference frame as principal strains, and maximum shear strains are unique at a point.

$$\varepsilon_{1,2} = (2000-800)/2 \pm \sqrt{\{(2000-(-800))/2\}^2+(375)^2} = 600 \pm 1449.4 \qquad \varepsilon_1 = 2049\mu \qquad \varepsilon_2 = -849\mu \qquad \varepsilon_3 = 0 \quad (1)$$

(b) The difference between ε_1 and ε_2 is the largest difference between principal strains. The maximum shear strain is

$$\gamma_{max}/2 = |(\varepsilon_1-\varepsilon_2)/2| = |(2049+849)/2| = 2898 \quad (2)$$

(c) Substituting the given strains in place of ε_{xx}, ε_{yy}, γ_{xy} and $\theta=55^o$ into strain transformation equations we obtain the following.

$$\varepsilon_{xx} = (2000)cos^2(55)+(-800)sin^2(55)+(750)sin(55)cos(55) = 2898 \quad (3)$$

$$\varepsilon_{yy} = (2000)cos^2(55)+(-800)sin^2(55)-(750)sin(55)cos(55) = 726.4 \quad (4)$$

$$\gamma_{xy} = -2(2000-(-800))cos(55)sin(55)+(750)(cos^2(55)-sin^2(55)) = -2888 \quad (5)$$

ANS $\varepsilon_1 = 2049\mu$; $\varepsilon_2 = -849\mu$; $\varepsilon_3 = 0$; $\gamma_{max} = 2898\mu$; $\varepsilon_{xx} = 473.6\mu$; $\varepsilon_{yy} = 726.4\mu$; $\gamma_{nt} = -2888\mu$

---

9. 36

Solution Plane Strain $\varepsilon_{nn} = -2000\ \mu$ $\varepsilon_{tt} = -800\ \mu$ $\gamma_{nt} = -600\ \mu$

$\varepsilon_1 = ?$ $\varepsilon_2 = ?$ $\varepsilon_3 = ?$ $\gamma_{max} = ?$ $\varepsilon_{xx} = ?$ $\varepsilon_{yy} = ?$ $\gamma_{xy} = ?$

(a) We can use *n* and *t* as the reference frame as principal strains, and maximum shear strains are unique at a point.

$$\varepsilon_{1,2} = (-2000-800)/2 \pm \sqrt{\{(-2000-(-800))/2\}^2+(-300)^2} = -1400 \pm 670.8 \qquad \varepsilon_1 = -729\mu \qquad \varepsilon_2 = -2071\mu;\ \varepsilon_3 = 0 \quad (1)$$

(b) The difference between ε_3 and ε_2 is the largest difference between principal strains. The maximum shear strain is

$$\gamma_{max}/2 = |(\varepsilon_3-\varepsilon_2)/2| = |(0+2070.8)/2| = 2071 \quad (2)$$

(c) Substituting the given strains in place of ε_{xx}, ε_{yy}, γ_{xy} and $\theta= -35^o$ into strain transformation equations we obtain the following

$$\varepsilon_{xx} = (-2000)cos^2(-35)+(-800)sin^2(-35)+(-600)sin(-35)cos(-35) = -1323 \quad (3)$$

$$\varepsilon_{yy} = (-2000)cos^2(-35) + (-800)sin^2(-35) - (-600)sin(-35)cos(-35) = -1477 \quad (4)$$

$$\gamma_{xy} = -2(-2000-(-800))cos(-35)sin(-35) + (-600)(cos^2(-35) - sin^2(-35)) = -1333 \quad (5)$$

ANS $\varepsilon_1 = -729\mu$; $\varepsilon_2 = -2071\mu$; $\varepsilon_3 = 0$; $\gamma_{max} = 2071\mu$; $\varepsilon_{xx} = -1323\mu$; $\varepsilon_{yy} = -1477\mu$; $\gamma_{nt} = -1333\mu$

---

9. 37

Solution Plane Strain $\varepsilon_{nn} = 350\ \mu$ $\varepsilon_{tt} = 700\ \mu$ $\gamma_{nt} = 1400\ \mu$

$\varepsilon_1 = ?$ $\varepsilon_2 = ?$ $\varepsilon_3 = ?$ $\gamma_{max} = ?$ $\varepsilon_{xx} = ?$ $\varepsilon_{yy} = ?$ $\gamma_{xy} = ?$

(a) We can use *n* and *t* as the reference frame as principal strains, and maximum shear strains are unique at a point.

$$\varepsilon_{1,2} = (350+700)/2 \pm \sqrt{\{(350-700)/2\}^2 + (700)^2} = 525 \pm 721.5 \qquad \varepsilon_1 = 1246.5\mu \qquad \varepsilon_2 = -196.5\mu \qquad \varepsilon_3 = 0 \quad (1)$$

(b) The difference between ε_1 and ε_2 is the largest difference between principal strains. The maximum shear strain is

$$\gamma_{max}/2 = |(\varepsilon_1 - \varepsilon_2)/2| = |(1246.5 + 196.5)/2| = 1443 \quad (2)$$

(c) Substituting the given strains in place of ε_{xx}, ε_{yy}, γ_{xy} and θ=-105° into strain transformation equations we obtain the following

$$\varepsilon_{xx} = 350cos^2(-105) + 700sin^2(-105) + 1400sin(-105)cos(-105) = 1027 \quad (3)$$

$$\varepsilon_{yy} = 350cos^2(-105) + 700sin^2(-105) - 1400sin(-105)cos(-105) = 23 \quad (4)$$

$$\gamma_{xy} = -2(350-700)cos(-105)sin(-105) + 1400(cos^2(-105) - sin^2(-105)) = -1037 \quad (5)$$

ANS $\varepsilon_1 = 1246.5\mu$; $\varepsilon_2 = -196.5\mu$; $\varepsilon_3 = 0$; $\gamma_{max} = 1443\mu$; $\varepsilon_{xx} = 1027\mu$; $\varepsilon_{yy} = 23\mu$; $\gamma_{xy} = -1037\mu$

---

9. 38

Solution Plane Strain $\varepsilon_{nn} = -3600\ \mu$ $\varepsilon_{tt} = 2500\ \mu$ $\gamma_{nt} = -1000\mu$

$\varepsilon_1 = ?$ $\varepsilon_2 = ?$ $\varepsilon_3 = ?$ $\theta_1 = ?$ $\gamma_{max} = ?$ $\varepsilon_{xx} = ?$ $\varepsilon_{yy} = ?$ $\gamma_{xy} = ?$

(a) We can use *n* and *t* as the reference frame as principal strains, and maximum shear strains are unique at a point.

$$\varepsilon_{1,2} = \left(\frac{-3600+2500}{2}\right) \pm \sqrt{\left(\frac{-3600-2500}{2}\right)^2 + (-500)^2} = -550 \pm 3090.7 \qquad \varepsilon_1 = 2541\mu \qquad \varepsilon_2 = -3641\mu \qquad \varepsilon_3 = 0 \quad (1)$$

(b) The difference between ε_1 and ε_2 is the largest difference between principal strains. The maximum shear strain is

$$\gamma_{max}/2 = |(\varepsilon_1 - \varepsilon_2)/2| = |(2540.7 + 3640.7)/2| = 6181 \quad (2)$$

(c) Substituting the given strains in place of ε_{xx}, ε_{yy}, γ_{xy} and θ=162° into strain transformation equations we obtain the following

$$\varepsilon_{xx} = (-3600)cos^2(162) + 2500sin^2(162) + (-1000)sin(162)cos(162) = -2724 \quad (3)$$

$$\varepsilon_{yy} = (-3600)cos^2(162) + 2500sin^2(162) - (-1000)sin(162)cos(162) = 1624 \quad (4)$$

$$\gamma_{xy} = -2(-3600-2500)cos(162)sin(162) + (-1000)(cos^2(162) - sin^2(162)) = -4395 \quad (5)$$

ANS $\varepsilon_1 = 2541\mu$; $\varepsilon_2 = -3641\mu$; $\varepsilon_3 = 0$; $\gamma_{max} = 6181\mu$; $\varepsilon_{xx} = -2724\mu$; $\varepsilon_{yy} = 1624\mu$; $\gamma_{nt} = -4395\mu$

---

9. 39

Solution $\varepsilon_1 = 1200\ \mu$ $\varepsilon_2 = 300\ \mu$ $\theta_1 = 27.5^0$ $\varepsilon_{xx} = ?$ $\varepsilon_{yy} = ?$ $\gamma_{xy} = ?$

Figure (a) shows the Mohr's circle for the given state of strain. The center is at 750 μ and the radius is R = 450 μ. Figure (b) shows the orientation of principal directions P_1 and P_2 at 27.5° counter-clockwise from the x axis as per the value of θ_1. To get to points V and H, we must rotate clockwise from the principal direction points P_1 and P_2. Starting from P_1 and P_2 on the Mohr's circle we rotate 55° clockwise to get points V and H as shown in Figure (a).

(a)

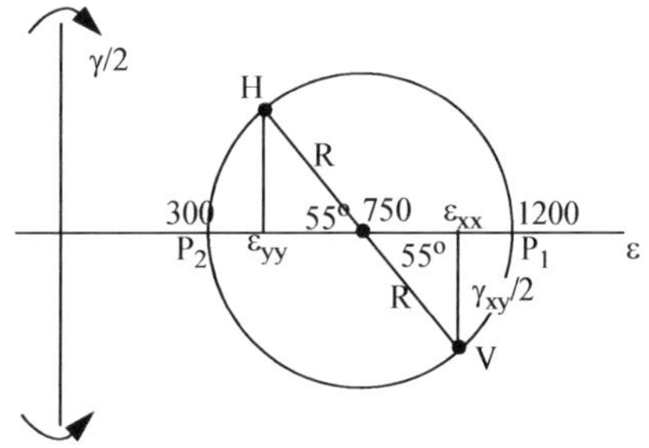

(b)

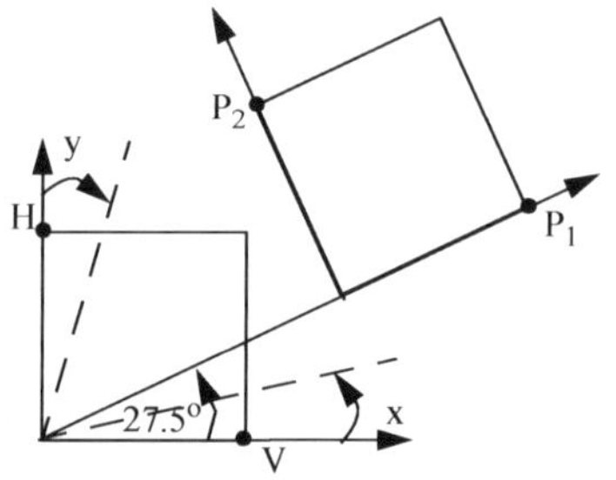

The normal strains in the x and y coordinate system are:

$\varepsilon_{xx} = 750 + R\cos 55 = 1008.11$ $\quad \varepsilon_{yy} = 750 - R\cos 55 = 491.89$ $\quad |\gamma_{xy}/2| = R\sin 55 = 368.62$ $\quad or \quad |\gamma_{xy}| = 737.2$ **(6)**

We note from Figure (b) that angle decreases between *n* and *t*, hence shear strain is positive.

ANS $\varepsilon_{xx} = 1008.1\mu$; $\varepsilon_{yy} = 491.9\mu$; $\gamma_{xy} = 737.2\mu$

---

9. 40

Solution $\quad \varepsilon_1 = 900\ \mu \quad \varepsilon_2 = -600\ \mu \quad \theta_1 = -20^0 \quad \varepsilon_{xx} = ? \quad \varepsilon_{yy} = ? \quad \gamma_{xy} = ?$

Figure (a) shows the Mohr's circle for the given state of strain. The center is at 150 μ and the radius is R = 750 μ. Figure (b) shows the orientation of principal directions P_1 and P_2 at 27.5^o counter-clockwise from the x axis as per the value of θ_1. To get to points V and H, we must rotate clockwise from the principal direction points P_1 and P_2. Starting from P_1 and P_2 on the Mohr's circle we rotate 55° clockwise to get points V and H as shown in Figure (a).

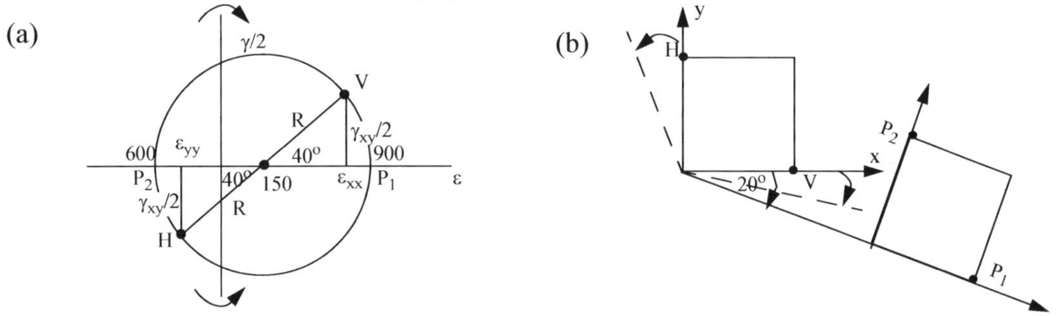

The normal strains in the x and y coordinate system are:

$\varepsilon_{xx} = 150 + R\cos 40 = 724.5$ $\quad \varepsilon_{yy} = 150 - R\cos 40 = -424.5$ $\quad |\gamma_{xy}/2| = R\sin 40 = 482.09$ $\quad or \quad |\gamma_{xy}| = 964.2$ **(1)**

We note from Figure (b) that angle between *n* and *t* increases, hence shear strain is negative.

ANS $\varepsilon_{xx} = 724.5\mu$; $\varepsilon_{yy} = -424.5\mu$; $\gamma_{xy} = -964.2\mu$

---

9. 41

Solution $\quad \varepsilon_1 = -200\ \mu \quad \varepsilon_2 = -2000\ \mu \quad \theta_1 = 105^0 \quad \varepsilon_{xx} = ? \quad \varepsilon_{yy} = ? \quad \gamma_{xy} = ?$

Figure (a) shows the Mohr's circle for the given state of strain. The center is at -1100 μ and the radius is R = 900 μ. Figure (b) shows the orientation of principal directions P_1 and P_2 at 105° counter-clockwise from the x axis as per the value of θ_1. To get to points V and H, we must rotate clockwise from the principal direction points P_1 and P_2. Starting from P_1 and P_2 on the Mohr's circle we rotate 210° clockwise to get points V and H as shown in Figure (a).

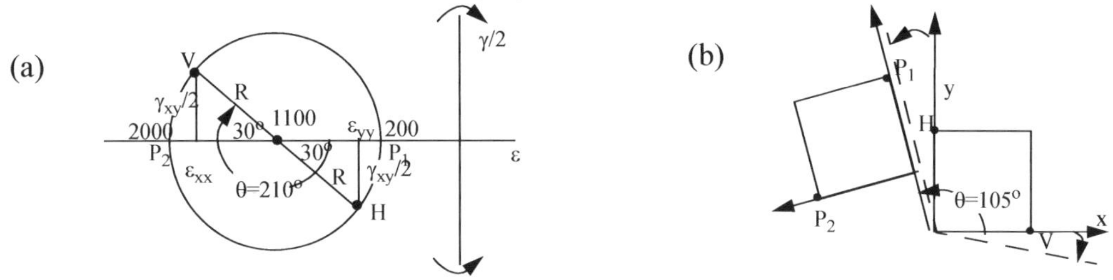

The normal strains in the x and y coordinate system are:

$\varepsilon_{xx} = -1100 - R\cos 30 = -1879.4$ $\quad \varepsilon_{yy} = -1100 + R\cos 30 = -320.6$ $\quad |\gamma_{xy}/2| = R\sin 30 = 450$ $\quad or \quad |\gamma_{xy}| = 900$ **(1)**

We note from Figure (b) that angle between *n* and *t* increases, hence shear strain is negative.

ANS $\varepsilon_{xx} = -1879\mu$; $\varepsilon_{yy} = -320.6\mu$; $\gamma_{xy} = -900\mu$

---

9. 42

Solution $\quad \varepsilon_1 = 1400\ \mu \quad \varepsilon_2 = -600\ \mu \quad \theta_1 = -75^0 \quad \varepsilon_{xx} = ? \quad \varepsilon_{yy} = ? \quad \gamma_{xy} = ?$

Figure (a) shows the Mohr's circle for the given state of strain. The center is at 400 μ and the radius is R = 1000 μ. FigureFigure (b) shows the orientation of principal directions P_1 and P_2 at 75° clockwise from the x axis as per the value of θ_1. To get to points V and H, we must rotate counter-clockwise from the principal direction points P_1 and P_2. Starting from P_1 and P_2 on the Mohr's circle we rotate 150° counter-clockwise to get points V and H as shown in Figure (a). The normal strains in the x and y coordinate system are:

$\varepsilon_{xx} = 400 - R\cos 30 = -466.03 \qquad \varepsilon_{yy} = 400 + R\cos 30 = 1266.03 \qquad |\gamma_{xy}/2| = R\sin 30 = 500 \quad or \quad |\gamma_{xy}| = 1000$ **(1)**

We note from Figure (b) that the angle between *n* and *t* increases, hence shear strain is negative.

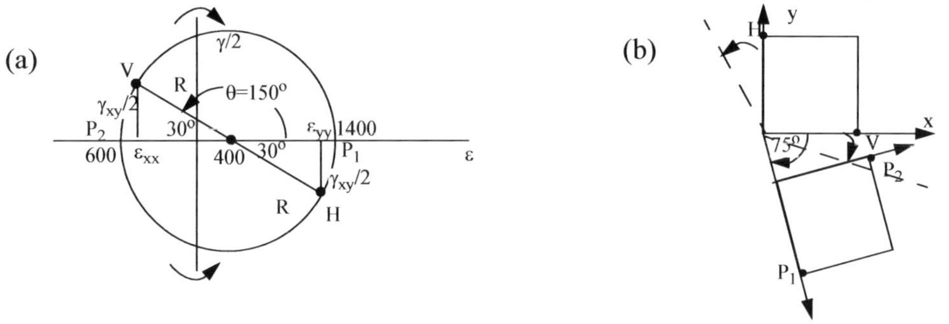

ANS $\varepsilon_{xx} = -466\mu$; $\varepsilon_{yy} = 1266\mu$; $\gamma_{xy} = -1000\mu$

---

9. 43

Solution Plane Stress $E = 70$ GPa $\nu = 0.2$ $\varepsilon_1 = ?$ $\varepsilon_2 = ?$ $\varepsilon_3 = ?$ $\theta_1 = ?$ $\gamma_{max} = ?$

From the given stress cube we have the following stresses.

$$\sigma_{xx} = -30MPa \qquad \sigma_{yy} = 60MPa \qquad \tau_{xy} = 40MPa \tag{1}$$

The principal angle θ_p is calculated and substituted into the normal stress equation to determine one principal stress.

$$\theta_p = 0.5\ atan\left(\frac{(2)(40)}{((-30)-60)}\right) = -20.82^o \tag{2}$$

$$\sigma_p = (-30)cos^2(-20.82) + (60)sin^2(-20.82) + 2(40)sin(-20.82)cos(-20.82) = -45.2MPa \tag{3}$$

The other principal stress is $-30 + 60 - (-45.2) = 75.2MPa$. As this stress is greater than the above stress, therefore this is principal stress one.

$$\sigma_1 = 75.2MPa \qquad \sigma_2 = -45.2MPa \qquad \sigma_3 = 0 \qquad \theta_1 = \theta_p \pm 90 \quad or \quad \theta_1 = -69.2^o \text{ or } 110.8^o \tag{4}$$

From Hooke's law in principal coordinates we obtain the following.

$$\varepsilon_1 = \frac{[\sigma_1 - \nu(\sigma_2 + \sigma_3)]}{E} = \frac{[75.2 - 0.25(-45.2)](10^6)}{70(10^9)} \qquad \varepsilon_2 = \frac{[\sigma_2 - \nu(\sigma_3 + \sigma_1)]}{E} = \frac{[(-45.2) - 0.25(75.2)](10^6)}{70(10^9)} \tag{5}$$

$$\varepsilon_3 = \frac{[\sigma_3 - \nu(\sigma_1 + \sigma_2)]}{E} = \frac{[0 - 0.25(75.2 - 45.2)](10^6)}{70(10^9)} \qquad \frac{\gamma_{max}}{2} = \left|\frac{\varepsilon_1 - \varepsilon_2}{2}\right| = \left|\frac{1236 + 914.4}{2}\right| \quad or \quad \gamma_{max} = 2150 \tag{6}$$

ANS $\varepsilon_1 = 1236\mu$; $\varepsilon_2 = -914.4\mu$; $\varepsilon_3 = -107.1\mu$; $\theta_1 = -69.2^o$ or 110.8^o ; $\gamma_{max} = 2150\mu$

---

9. 44

Solution Plane Stress $E = 70$ GPa$\varepsilon_1 = ?$ $\nu = 0.25\varepsilon_2 = ?$ $\varepsilon_3 = ?$ $\theta_1 = ?$ $\gamma_{max} = ?$

From the given stress cube we have the following stresses.

$$\sigma_{xx} = 45MPa \qquad \sigma_{yy} = 15MPa \qquad \tau_{xy} = -20MPa \tag{1}$$

The principal angle θ_p is calculated and substituted into the normal stress equation to determine one principal stress.

$$\theta_p = 0.5\ atan\left(\frac{(2)(-20)}{(45-15)}\right) = -26.57^o \tag{2}$$

$$\sigma_p = (45)cos^2(-26.57) + (15)sin^2(-26.57) + 2(-20)sin(-26.57)cos(-26.57) = 55MPa \tag{3}$$

The other principal stress is: $45 + 15 - 55 = 5MPa$ is smaller than the stress above. The above stress is principal stress one.

$$\sigma_1 = 55MPa \qquad \sigma_2 = 5MPa \qquad \sigma_3 = 0 \qquad \theta_1 = -26.57^o \ or\ 153.4^o \tag{4}$$

From Hooke's law in principal coordinates we obtain the following.

$$\varepsilon_1 = \frac{[\sigma_1 - \nu(\sigma_2 + \sigma_3)]}{E} = \frac{[55 - 0.25(5)](10^6)}{70(10^9)} \qquad \varepsilon_2 = \frac{[\sigma_2 - \nu(\sigma_3 + \sigma_1)]}{E} = \frac{[5 - 0.25(55)](10^6)}{70(10^9)} \tag{5}$$

$$\varepsilon_3 = \frac{[\sigma_3 - \nu(\sigma_1 + \sigma_2)]}{E} = \frac{[0 - 0.25(5 + 55)](10^6)}{70(10^9)} \qquad \frac{\gamma_{max}}{2} = \left|\frac{\varepsilon_1 - \varepsilon_3}{2}\right| = \left|\frac{767.9 + 214.3}{2}\right| \quad or \quad \gamma_{max} = 982.2 \tag{6}$$

ANS $\varepsilon_1 = 767.9\mu$; $\varepsilon_2 = -125.0\mu$; $\varepsilon_3 = -214.3\mu$; $\theta_1 = -26.57^o$ *or* 153.4^o ; $\gamma_{max} = 982.2\mu$

---

9. 45

Solution Plane Stress E = 30,000 ν = 0.28 $\varepsilon_1 = ?$ $\varepsilon_2 = ?$ $\varepsilon_3 = ?$ $\theta_1 = ?$ $\gamma_{max} = ?$

From the given stress cube we have the following stresses.

$$\sigma_{xx} = -10 ksi \qquad \sigma_{yy} = -20 ksi \qquad \tau_{xy} = 30 ksi \tag{1}$$

The principal angle θ_p is calculated and substituted into the normal strain equation to determine one principal strain.

$$\theta_p = 0.5\ atan\left(\frac{(2)(30)}{(-10-(-20))}\right) = 40.27^o \tag{2}$$

$$\sigma_p = (-10)cos^2(-40.27) + (-20)sin^2(-40.27) + 2(30)sin(-40.27)cos(-40.27) = 15.4 ksi \tag{3}$$

The other principal stress $-10-20-15.4 = -45.4 ksi$ is smaller than the stress above. The above stress is principal stress one.

$$\sigma_1 = 15.4 ksi \qquad \sigma_2 = -45.4 ksi \qquad \sigma_3 = 0 \qquad \theta_1 = 40.27^o \text{ or } -139.73^o \tag{4}$$

$$\varepsilon_1 = \frac{[\sigma_1 - \nu(\sigma_2+\sigma_3)]}{E} = \frac{[15.4-0.28(-45.4)]}{30000} \qquad \varepsilon_2 = \frac{[\sigma_2 - \nu(\sigma_3+\sigma_1)]}{E} = \frac{[(-45.4)-0.28(15.4)]}{30000} \tag{5}$$

$$\varepsilon_3 = \frac{[\sigma_3 - \nu(\sigma_1+\sigma_2)]}{E} = \frac{[0-0.28(15.4-45.4)]}{30000} \qquad \frac{\gamma_{max}}{2} = \left|\frac{\varepsilon_1-\varepsilon_2}{2}\right| = \left|\frac{937.7+1657.7}{2}\right| = 2595 \tag{6}$$

ANS $\varepsilon_1 = 938\mu$; $\varepsilon_2 = -1658\mu$; $\varepsilon_3 = 280\mu$; $\theta_1 = 40.27^o$ or -139.73^o; $\gamma_{max} = 2595\mu$

---

9. 46

Solution Plane Strain E = 105 GPa ν = 0.35 $\varepsilon_1 = ?$ $\varepsilon_2 = ?$ $\varepsilon_3 = ?$ $\theta_1 = ?$ $\gamma_{max} = ?$

From the given stress cube we have the following stresses.

$$\sigma_{xx} = -20 MPa \qquad \sigma_{yy} = 40 MPa \qquad \tau_{xy} = -40 MPa \tag{1}$$

The principal angle θ_p is calculated and substituted into the normal strain equation to determine one principal strain.

$$\theta_p = 0.5\ atan\left(\frac{(2)(-40)}{(-20-40)}\right) = 26.57^o \qquad \sigma_p = -20cos^2(26.57) + 40sin^2(26.57) + 2(-40)sin(26.57)cos(26.57) = -40 MPa \tag{2}$$

The other principal stress $-20+40-(-40) = 60 MPa$ is greater than the above stress, therefore this stress is principal stress one. Thus principal angle 1 is $\theta_1 = \theta_p \pm 90$. For plane strain $\varepsilon_{zz} = 0$ and is used to find σ_3 as shown below.

$$\sigma_1 = 60 MPa \qquad \sigma_2 = -40 MPa \qquad \sigma_3 = \sigma_{zz} = \nu(\sigma_{xx}+\sigma_{yy}) = 0.35(-20+40) = 7 MPa \qquad \theta_1 = 116.5^o \text{ or } -63.43^o \tag{3}$$

$$\varepsilon_1 = \frac{[\sigma_1 - \nu(\sigma_2+\sigma_3)]}{E} = \frac{[60-0.35(-40+7)](10^6)}{105(10^9)} \qquad \varepsilon_2 = \frac{[\sigma_2 - \nu(\sigma_3+\sigma_1)]}{E} = \frac{[-40-0.35(60+7)](10^6)}{105(10^9)} \tag{4}$$

$$\gamma_{max}/2 = |(\varepsilon_1-\varepsilon_2)/2| = |(681.4+604.3)/2| = 1285.7 \tag{5}$$

ANS $\varepsilon_1 = 681.4\mu$; $\varepsilon_2 = -604.3\mu$; $\varepsilon_3 = 0$; $\theta_1 = 116.5^o$ or -63.43^o; $\gamma_{max} = 1285.7\mu$

---

9. 47

Solution Plane Strain E = 70 GPa ν = 0.25 $\varepsilon_1 = ?$ $\varepsilon_2 = ?$ $\varepsilon_3 = ?$ $\theta_1 = ?$ $\gamma_{max} = ?$

From the given stress cube we have the following stresses.

$$\sigma_{xx} = 35 MPa \qquad \sigma_{yy} = 25 MPa \qquad \tau_{xy} = -20 MPa \textbf{ st} \tag{1}$$

The principal angle θ_p is calculated and substituted into the normal stress equation to determine one principal stress.

$$\theta_p = 0.5\ atan\left(\frac{(2)(-20)}{(35-25)}\right) = -37.98^o \tag{2}$$

$$\sigma_p = (35)cos^2(-37.98) + (25)sin^2(-37.98) + 2(-20)sin(-37.98)cos(-37.98) = 50.6 MPa \tag{3}$$

The other principal stress is: $35+25-50.6 = 9.4 MPa$. As this stress is smaller than the above stress, the above stress is principal stress one. Thus principal angle 1 is $\theta_1 = \theta_p$. For plane strain $\varepsilon_{zz} = 0$ and is used to find σ_3 as shown below.

$$\sigma_1 = 50.6 MPa \qquad \sigma_2 = 9.4 MPa \qquad \sigma_3 = \sigma_{zz} = \nu(\sigma_{xx}+\sigma_{yy}) = 0.25(35+25) = 15 MPa \qquad \theta_1 = -37.98^o \text{ or } 142.2^o \tag{4}$$

$$\varepsilon_1 = \frac{[\sigma_1 - \nu(\sigma_2+\sigma_3)]}{E} = \frac{[50.6-0.25(9.4+15)](10^6)}{70(10^9)} \qquad \varepsilon_2 = \frac{[\sigma_2 - \nu(\sigma_3+\sigma_1)]}{E} = \frac{[9.4-0.25(50.6+15)](10^6)}{70(10^9)} \tag{5}$$

$$\gamma_{max}/2 = |(\varepsilon_1-\varepsilon_3)/2| = |(636+100.3)/2| = 736.3 \tag{6}$$

ANS $\varepsilon_1 = 636\mu$; $\varepsilon_2 = -100.3\mu$; $\varepsilon_3 = 0$; $\theta_1 = -37.98^o$ or 142.2^o; $\gamma_{max} = 736.3\mu$

---

9. 48

Solution Plane Strain E = 30,000 ksi ν = 0.28 $\varepsilon_1 = ?$ $\varepsilon_2 = ?$ $\varepsilon_3 = ?$ $\theta_1 = ?$ $\gamma_{max} = ?$

From the given stress cube we have the following stresses.

$$\sigma_{xx} = -25ksi \qquad \sigma_{yy} = -15ksi \qquad \tau_{xy} = -20ksi \tag{1}$$

The principal angle θ_p is calculated and substituted into the normal stress equation to determine one principal stress.

$$\theta_p = 0.5\,atan\left(\frac{2(-20)}{-25-(-15)}\right) = 37.98^o \qquad \sigma_p = -25cos^2(37.98)+(-15)sin^2(37.98)+2(-20)sin(37.98)cos(37.98) = -40.62ksi \tag{2}$$

The other principal stress is: $-25+(-15)-(-40.62) = 0.62ksi$. As this stress is greater than the above stress, this stress is principal stress one. Thus principal angle 1 is $\theta_1 = \theta_p \pm 90$. For plane strain $\varepsilon_{zz} = 0$ and is used to find σ_3 as shown below.

$$\sigma_1 = 0.62ksi \qquad \sigma_2 = -40.62ksi \qquad \sigma_3 = \sigma_{zz} = \nu(\sigma_{xx}+\sigma_{yy}) = 0.28(-25-15) = -11.2ksi \qquad \theta_1 = -52.02^o \text{ or } 127.98^o \tag{3}$$

$$\varepsilon_1 = \frac{[\sigma_1 - \nu(\sigma_2+\sigma_3)]}{E} = \frac{[0.62-0.28(-40.62-11.2)]}{30000} \qquad \varepsilon_2 = \frac{[\sigma_2 - \nu(\sigma_3+\sigma_1)]}{E} = \frac{[-40.62-0.28(0.62-11.2)]}{30000} \tag{4}$$

$$\gamma_{max}/2 = |(\varepsilon_1-\varepsilon_3)/2| = |(504+1255)/2| = 1759 \tag{5}$$

ANS $\varepsilon_1 = 504\mu$; $\varepsilon_2 = -1255\mu$; $\varepsilon_3 = 0$; $\theta_1 = -52.02^o$ or 127.98^o; $\gamma_{max} = 1759\mu$

---

9. 49

Solution E_x=7,500ksi E_y= 2,500 ksi G_{xy} = 1,250 ksi ν_{xy}= 0.25. $\theta_1 = ?$ for strains $\bar{\theta}_1 = ?$for stresses

The principal angle θ_p is calculated and substituted into the normal strain equation to determine one principal strain.

$$\theta_p = 0.5\ atan\left(\frac{(-500)}{(-400-600)}\right) = 13.3^o \qquad \varepsilon_p = (-400)cos^2(13.3)+(600)sin^2(13.3)+(-500)sin(13.3)cos(13.3) = -459\mu \tag{1}$$

The other principal strain is: $-400+600-(-459) = 659\mu$. As this strain is greater than the above strain , therefore this strain is principal strain one. Thus principal angle 1 is $\theta_1 = \theta_p \pm 90$ or $\theta_1 = 103.3^o$ or -76.7^o

From the given material constants we have: $\nu_{yx} = (E_y/E_x)\nu_{xy} = (2,500/(7,500))(0.25) = 0.0833$

Substituting the given strain values and the constants in the given equation, we obtain the following

$$\varepsilon_{xx} = \frac{\sigma_{xx}}{E_x} - \frac{\nu_{yx}}{E_y}\sigma_{yy} = \frac{\sigma_{xx}}{7,500} - \frac{0.0833}{2,500}\sigma_{yy} = (-400)(10^{-6}) \qquad or \qquad \sigma_{xx} - 0.25\sigma_{yy} = -3ksi \tag{2}$$

$$\varepsilon_{yy} = \frac{\sigma_{yy}}{E_y} - \frac{\nu_{xy}}{E_x}\sigma_{xx} = \frac{\sigma_{yy}}{2,500} - \frac{(0.25)}{7500}(\sigma_{xx}) = 600(10^{-6}) \qquad or \qquad 3\sigma_{yy} - 0.25\sigma_{xx} = 4.5ksi \tag{3}$$

$$\sigma_{xx} = -2.681ksi \qquad \sigma_{yy} = 1.277ksi \qquad \tau_{xy} = G_{xy}\gamma_{xy} = (1250)(-500)(10^{-6}) = -0.625ksi \tag{4}$$

The principal angle θ_p is calculated and substituted into the normal stress equation to determine one principal stress.

$$\bar{\theta}_p = 0.5\,atan\left(\frac{2(-0.625)}{-2.681-1.277}\right) = 8.76^o \qquad \sigma_p = -2.681\,cos^2(8.76)+1.277\,sin^2(8.76)+2(-0.625)sin(8.76)cos(8.76) = -2.78ksi$$

The other principal stress is: $-2.68+1.28-(-2.78) = 1.38ksi$. As this stress is greater than the above stress in Eq. 7, this stress is principal stress one. Thus principal angle 1 is $\bar{\theta}_1 = \bar{\theta}_p \pm 90$

ANS Strain: 103.3^o or -76.7^o; Stress: ($\bar{\theta}_1 = 98.76^o$ or -81.24^o)

---

9. 50

Solution E_x=7,500ksi E_y= 2,500 ksi G_{xy} = 1,250 ksi ν_{xy}= 0.25. $\theta_1 = ?$ for strains $\bar{\theta}_1 = ?$for stresses

The principal angle $\bar{\theta}_p$ is calculated and substituted into the normal stress equation to determine one principal stress

$$\bar{\theta}_p = 0.5\ atan\left(\frac{(2)(5)}{10-(-7)}\right) = 15.23^o \qquad \sigma_p = 10cos^2(15.23)+(-7)sin^2(15.23)+2(5)sin(15.23)cos(15.23) = 11.36ksi \tag{1}$$

The other principal stress as: $10-7-11.36 = -8.36ksi$. As this stress is smaller than the above stress, the above stress is principal stress one. Thus principal angle 1 is $\bar{\theta}_1 = \bar{\theta}_p$ or $\bar{\theta}_1 = 15.23^o$ or -164.77^o.

From the given material constants we have: $\nu_{yx} = (E_y/E_x)\nu_{xy} = (2,500/(7,500))(0.25) = 0.0833$

Substituting the given stress values and the constants above in the given equation we obtain the following

$$\varepsilon_{xx} = \frac{\sigma_{xx}}{E_x} - \frac{\nu_{yx}}{E_y}\sigma_{yy} = \frac{10}{7,500} - \frac{0.0833}{2,500}(-7) = 1567(10^{-6}) = 1567\mu \tag{2}$$

$$\varepsilon_{yy} = \frac{\sigma_{yy}}{E_y} - \frac{\nu_{xy}}{E_x}\sigma_{xx} = \frac{(-7)}{2,500} - \frac{(0.25)}{7500}(10) = -3133(10^{-6}) = -3133\mu \quad (3)$$

$$\gamma_{xy} = \frac{\tau_{xy}}{G_{xy}} = \frac{5}{1250} = 4000(10^{-6}) = 4000\mu \quad (4)$$

The principal angle θ_p is calculated and substituted into the normal strain equation to determine one principal strain.

$$\theta_p = 0.5\ atan\left(\frac{4000}{1567-(-3133)}\right) = 20.2^o \qquad \varepsilon_p = 1567 cos^2(20.2) + (-3133)sin^2(20.2) + 4000 sin(20.2)cos(20.2) = 2302\mu \quad (5)$$

The other principal strain is: $1567 - 3133 - 2302 = -3868\mu$. As this strain is smaller than the above strain, the above strain is principal strain one. Thus principal angle 1 is $\theta_1 = \theta_p$

ANS Strain: $\theta_1 = 20.2^o$ or -159.8^o $\quad$ *Stress*: $\bar{\theta}_1 = 15.23^o$ or -164.77^o

---

9. 51

Solution E_x=50 MPa $\quad$ E_y= 18 MPa $\quad$ G_{xy} = 9 MPa $\quad$ ν_{xy}= 0.25. θ_1 = ? for strains $\bar{\theta}_1$ = ?for stresses

The principal angle θ_p is calculated and substituted into the normal strain equation to determine one principal strain.

$$\theta_p = 0.5\ atan\left(\frac{300}{800-200}\right) = 13.3^o \qquad \varepsilon_p = 800 cos^2(13.3) + 200 sin^2(13.3) + 300 sin(13.3)cos(13.3) = 835.4\mu \quad (1)$$

The other principal strain is: $800 + 200 - 835.4 = 164.6\mu$. As this strain is smaller than the above strain, the above strain is the principal strain one. Thus principal angle 1 is $\theta_1 = \theta_p$ or $\theta_1 = 13.3^o$ or -166.7^o.

From the given material constants we have: $\nu_{yx} = (E_y/E_x)\nu_{xy} = (18/50)(0.25) = 0.09$

Substituting the given strain values and the constants in the given equation we obtain the following

$$\varepsilon_{xx} = \frac{\sigma_{xx}}{E_x} - \frac{\nu_{yx}}{E_y}\sigma_{yy} = \frac{\sigma_{xx}}{50(10^9)} - \frac{0.09}{18(10^9)}\sigma_{yy} = 800(10^{-6}) \quad or \quad \sigma_{xx} - 0.25\sigma_{yy} = 40MPa \quad (2)$$

$$\varepsilon_{yy} = \frac{\sigma_{yy}}{E_y} - \frac{\nu_{xy}}{E_x}\sigma_{xx} = \frac{\sigma_{yy}}{18(10^9)} - \frac{(0.25)}{50(10^9)}(\sigma_{xx}) = 200(10^{-6}) \quad or \quad 2.7778\sigma_{yy} - 0.25\sigma_{xx} = 10MPa \quad (3)$$

$$\sigma_{xx} = 41.84MPa \qquad \sigma_{yy} = 7.37MPa \qquad \tau_{xy} = G_{xy}\gamma_{xy} = (9)(10^9)(300)(10^{-6}) = 2.70MPa \quad (4)$$

The principal angle $\bar{\theta}_p$ is calculated and substituted into the normal stress equation to determine one principal stress.

$$\bar{\theta}_p = 0.5\ atan\left(\frac{2(2.70)}{41.84-7.37}\right) = 4.45^o \qquad \sigma_p = 41.84 cos^2(4.45) + 7.37 sin^2(4.45) + 2(2.70) sin(4.45)cos(4.45) = 42.05MPa \quad (5)$$

The other principal stress is: $41.84 + 7.37 - 42.05 = 7.16MPa$. As this stress is smaller than the above stress, the above stress is principal stress one. Thus principal angle 1 is $\bar{\theta}_1 = \bar{\theta}_p$.

ANS Strain: $\theta_1 = 13.3^o$ or -166.7^o $\quad$ Stress: $\bar{\theta}_1 = 4.45^o$ or -175.55^o

---

9. 52

Solution E_x=50 MPa $\quad$ E_y= 18 MPa $\quad$ G_{xy} = 9 MPa $\quad$ ν_{xy}= 0.25. $\quad$ θ_1 = ? for strains $\bar{\theta}_1$ = ? for stresses

The principal angle $\bar{\theta}_p$ is calculated and substituted into the normal stress equation to determine one principal stress.

$$\bar{\theta}_p = 0.5\ atan\left(\frac{(2)(-30)}{70-(-49)}\right) = 35.35^o \quad (1)$$

$$\sigma_p = (-70)cos^2(35.35) + (-49)sin^2(35.35) + 2(-30)sin(35.35)cos(35.35) = -91.3MPa \quad (2)$$

The other principal stress as: $-70 - 49 - (-91.3) = -27.7MPa$. As this stress is greater than the stress above, this stress is principal stress one. Thus principal angle 1 is $\bar{\theta}_1 = \bar{\theta}_p \pm 90^o$ or $\bar{\theta}_1 = 125.35^o$ or -54.65^o.

From the given material constants we have: $\nu_{yx} = (E_y/E_x)\nu_{xy} = (18/50)(0.25) = 0.09$

Substituting the given stress values and the constants above in the given we obtain the following.

$$\varepsilon_{xx} = \frac{\sigma_{xx}}{E_x} - \frac{\nu_{yx}}{E_y}\sigma_{yy} = \frac{(-70)(10^6)}{50(10^9)} - \frac{0.09}{18(10^9)}(-49)(10^6) = -1155(10^{-6}) = -1155\mu \quad (3)$$

$$\varepsilon_{yy} = \frac{\sigma_{yy}}{E_y} - \frac{\nu_{xy}}{E_x}\sigma_{xx} = \frac{(-49)(10^6)}{18(10^9)} - \frac{0.25}{50(10^9)}(-70) = -2372(10^{-6}) = -2372\mu(10^6) \quad (4)$$

$$\gamma_{xy} = \frac{\tau_{xy}}{G_{xy}} = \frac{(-30)}{9(10^9)} = -3333(10^{-6}) = -3333\mu \quad (5)$$

The principal angle θ_p is calculated and substituted into the strain equation to determine one principal strain.

$$\theta_p = 0.5\ atan\left(\frac{(-3333)}{-1155-(-2372)}\right) = -34.97^o \quad (6)$$

$$\varepsilon_p = (-1155)cos^2(-34.97) + (-2372)sin^2(-34.97) + (-3333)sin(-34.97)cos(-34.97) = 10.7\mu \quad (7)$$

The other principal strain is: $-1155-2372-10.7 = -3537.7\mu$. As this strain is smaller than the strain above, the above strain is principal strain one. Thus principal angle 1 is $\theta_1 = \theta_p$.

ANS Stress: $\bar{\theta}_1 = 125.35^o$ or -54.65^o Strain: $\theta_1 = -34.97^o$ or 145.03^o

---

Section 9.5

9. 53

Solution

Substituting $(\theta \pm 180)$ in place of θ in strain transformation equation and noting that $sin(\theta \pm 180) = -sin\theta$ and $cos(\theta \pm 180) = -cos\theta$ we obtain:

$$\varepsilon_{nn} = \varepsilon_{xx}cos^2(\theta \pm 180) + \varepsilon_{yy}sin^2(\theta \pm 180) + \gamma_{xy}sin(\theta \pm 180^0)cos(\theta \pm 180) \text{ or}$$

$$\varepsilon_{nn} = \varepsilon_{xx}[-cos\theta]^2 + \varepsilon_{yy}[-sin\theta]^2 + \gamma_{xy}[-sin\theta][-cos\theta] = \varepsilon_{nn} = \varepsilon_{xx}cos^2\theta + \varepsilon_{yy}sin^2\theta + \gamma_{xy}sin\theta cos\theta \quad (1)$$

The above equation is the desired result.

---

9. 54

Solution $\varepsilon_{xx} = 400\mu\ in/in$ $\varepsilon_{yy} = -200\mu\ in/in$ $\gamma_{xy} = 500\mu\ rad$ $\varepsilon_a = ?$ $\varepsilon_b = ?$ $\varepsilon_c = ?$

By inspection of the given figure $\varepsilon_b = \varepsilon_{xx}$. Substituting the given strains and $\theta_a = -30^o$ and $\theta_c = 60^o$ in the strain transformation equation we obtain

$$\varepsilon_a = (400)cos^2(-30) + (-200)sin^2(-30) + 500sin(-30)cos(-30) = 33.5 \quad (1)$$

$$\varepsilon_c = (400)cos^2(60) + (-200)sin^2(60) + 500sin(60)cos(60) = 166.5 \quad (2)$$

ANS $\varepsilon_a = 33.5\mu\ in/in$; $\varepsilon_b = 400\mu\ in/in$; $\varepsilon_c = 166.5\mu\ in/in$

---

9. 55

Solution $\varepsilon_a = 200\mu\ in/in$ $\varepsilon_b = 100\mu\ in/in$ $\varepsilon_c = -400\mu\ in/in$ $\varepsilon_{xx} = ?$ $\varepsilon_{yy} = ?$ $\gamma_{xy} = ?$

By inspection of the given figure $\varepsilon_{xx} = \varepsilon_b$

Substituting $\theta_a = -30^o$ and $\theta_c = 60^o$ and the given strain values into the strain transformation equation we obtain:

$$\varepsilon_a = 100cos^2(-30) + \varepsilon_{yy}sin^2(-30) + \gamma_{xy}sin(-30)cos(-30) = 200 \quad or \quad 0.25\varepsilon_{yy} - 0.433\gamma_{xy} = 125 \quad (1)$$

$$\varepsilon_c = 100cos^2(60) + \varepsilon_{yy}sin^2(60) + \gamma_{xy}sin(60)cos(60) = -400 \quad or \quad 0.75\varepsilon_{yy} + 0.433\gamma_{xy} = -425 \quad (2)$$

Solving the two equations we obtain the strain values shown below.

ANS $\varepsilon_{xx} = 100\mu$ in./in. ; $\varepsilon_{yy} = -300\mu$ in./in. $\gamma_{xy} = -461.9\mu$ rad

---

9. 56

Solution $\sigma_{xx} = 22\ ksi(T)$ $\sigma_{yy} = 15\ ksi(C)$ $\tau_{xy} = -10\ ksi$

E = 10,000 ksi and G =4,000 ksi $\varepsilon_a = ?$ $\varepsilon_b = ?$ $\varepsilon_c = ?$ Plane stress

We can calculate the following.

$$\gamma_{xy} = \tau_{xy}/G = (-10)/4000 = -2500\mu \qquad \nu = E/(2G) - 1 = 10000/[2(4000)] - 1 = 0.25 \quad (1)$$

From Generalized Hooke's law, for plane stress ($\sigma_{zz} = 0$) we obtain:

$$\varepsilon_{xx} = \frac{\sigma_{xx}}{E} - \frac{\nu}{E}(\sigma_{yy} + \sigma_{zz}) = \frac{22}{10000} - \frac{0.25}{10000}(-15) = 2575\mu \qquad \varepsilon_{yy} = \frac{\sigma_{yy}}{E} - \frac{\nu}{E}(\sigma_{xx} + \sigma_{zz}) = \frac{(-15)}{10000} - \frac{0.25}{10000}(22) = -2050\mu \quad (2)$$

By inspection we obtain $\varepsilon_a = \varepsilon_{xx}$

Substituting $\theta_b = 60^o$, $\theta_c = 135^o$ in normal strain transformation equation we obtain:

$$\varepsilon_b = 2575cos^2(60) + (-2050)sin^2(60) + (-2500)sin(60)cos(60) = -1976.3 \quad (3)$$

$$\varepsilon_c = 2575cos^2(135) + (-2050)sin^2(135) + (-2500)sin(135)cos(135) = 1512.5 \quad (4)$$

ANS $\varepsilon_a = 2575\mu$ in./in. ; $\varepsilon_b = -1976\mu$ in./in. ; $\varepsilon_c = 1513\mu$ in./in.

---

9. 57

Solution $\varepsilon_a = -600\ \mu\ in/in$ $\quad \varepsilon_b = 500\ \mu\ in/in$ $\quad \varepsilon_c = 400\ \mu\ in/in$

E = 10,000 ksi G =4,000 ksi $\sigma_{xx} = ?$ $\sigma_{yy} = ?$ $\tau_{xy} = ?$

By inspection we obtain: $\varepsilon_{xx} = -600\mu\ in/in$

Substituting $\theta_b = 60^o$ and $\theta_c = 135^o$ and the given strain values into strain transformation equation we obtain:

$$\varepsilon_b = -600cos^2(60) + \varepsilon_{yy}sin^2(60) + \gamma_{xy}sin(60)cos(60) = 500 \quad or \quad 0.75\varepsilon_{yy} + 0.433\gamma_{xy} = 650 \quad (1)$$

$$\varepsilon_c = -600cos^2(135) + \varepsilon_{yy}sin^2(135) + \gamma_{xy}sin(135)cos(135) = 400 \quad or \quad 0.5\varepsilon_{yy} - 0.5\gamma_{xy} = 700 \quad (2)$$

$$\varepsilon_{yy} = 1061.9\mu\ in/in \qquad \gamma_{xy} = -338.1\mu\ rad \qquad \tau_{xy} = G\gamma_{xy} = (4000)(-338.1)(10^{-6}) = -1.352 \quad (3)$$

The Poisson's ratio is: $\nu = E/(2G) - 1 = 10000/[2(4000)] - 1 = 0.25$.

From Generalized Hooke's law, for plane stress ($\sigma_{zz} = 0$) we obtain:

$$\varepsilon_{xx} = \frac{\sigma_{xx}}{E} - \frac{\nu}{E}(\sigma_{yy} + \sigma_{zz}) \quad or \quad \sigma_{xx} - 0.25\sigma_{yy} = E\varepsilon_{xx} = (10000)(-600)(10^{-6}) = -6ksi \quad (4)$$

$$\varepsilon_{yy} = \frac{\sigma_{yy}}{E} - \frac{\nu}{E}(\sigma_{xx} + \sigma_{zz}) \quad or \quad \sigma_{yy} - 0.25\sigma_{xx} = E\varepsilon_{yy} = (10000)(1061.9)(10^{-6}) = 10.62ksi \quad (5)$$

Solving we obtain the stresses as shown below.

ANS $\sigma_{xx} = 3.57$ ksi (C) ; $\sigma_{yy} = 9.73$ ksi (T) ; $\tau_{xy} = -1.35$ ksi

---

9. 58

Solution $\sigma_{xx} = 50\ MPa(T)$ $\quad \sigma_{yy} = 20\ MPa(C)$ $\quad \tau_{xy} = 96\ MPa$ E = 80 GPa G = 32 GPa $\varepsilon_a = ?$ $\varepsilon_b = ?$ $\varepsilon_c = ?$

We can calculate the following.

$$\gamma_{xy} = \frac{\tau_{xy}}{G} = \frac{(96)(10^6)}{32(10^9)} = 3000\mu \qquad \nu = \frac{E}{2G} - 1 = \frac{80}{2(32)} - 1 = 0.25 \quad (1)$$

From Generalized Hooke's law, for plane stress ($\sigma_{zz} = 0$) we obtain:

$$\varepsilon_{xx} = \frac{\sigma_{xx}}{E} - \frac{\nu}{E}(\sigma_{yy} + \sigma_{zz}) = \frac{(50)}{(80)(10^9)} - \frac{0.25}{(80)(10^9)}(-20) = 687.5\mu \quad (2)$$

$$\varepsilon_{yy} = \frac{\sigma_{yy}}{E} - \frac{\nu}{E}(\sigma_{xx} + \sigma_{zz}) = \frac{(-20)}{(80)(10^9)} - \frac{0.25}{(80)(10^9)}(50) = -406.3\mu \quad (3)$$

By inspection of the given figure we obtain: $\varepsilon_a = 687.5\mu$ m/m and $\varepsilon_b = -406.3\mu$ m/m

Substituting $\theta_c = -25^o$ and the strains into strain transformation equation we obtain.

$$\varepsilon_c = 687.5cos^2(-25) + (-406.3)sin^2(-25) + (3000)sin(-25)cos(-25) = -656.9 \quad (4)$$

ANS $\varepsilon_a = 687.5\mu$ m/m ; $\varepsilon_b = -406.3\mu$ m/m ; $\varepsilon_c = -656.9\mu$ m/m

---

9. 59

Solution $\varepsilon_a = 1000\ \mu\ m/m$ $\quad \varepsilon_b = 1500\mu\ m/m$ $\quad \varepsilon_c = -450\ \mu\ m/m$ E = 80GPa G =32GPa $\sigma_{xx} = ?$ $\sigma_{yy} = ?$ $\tau_{xy} = ?$

We can write the following by inspection of the given figure. $\varepsilon_{xx} = 1000\mu\ m/m$ $\quad \varepsilon_{yy} = 1500\mu\ m/m$

Substituting $\theta_c = -25^o$ into normal strain transformation equation we obtain:

$$\varepsilon_c = 1000cos^2(-25) + 1500sin^2(-25) + \gamma_{xy}sin(-25)cos(-25) = -450 \quad or \quad \gamma_{xy} = 4018.8\mu\ rad \quad (1)$$

$$\tau_{xy} = G\gamma_{xy} = (32)(10^9)(4018.8)(10^{-6}) = 128.6(10^6)N/m^2 \qquad \nu = E/(2G) - 1 = 80/[2(32)] - 1 = 0.25 \quad (2)$$

From Generalized Hooke's law, for plane stress ($\sigma_{zz} = 0$) we obtain:

$$\varepsilon_{xx} = \frac{\sigma_{xx}}{E} - \frac{\nu}{E}(\sigma_{yy} + \sigma_{zz}) \quad or \quad \sigma_{xx} - 0.25\sigma_{yy} = E\varepsilon_{xx} = (80)(10^9)(1000)(10^{-6}) = 80MPa \quad (3)$$

$$\varepsilon_{yy} = \frac{\sigma_{yy}}{E} - \frac{\nu}{E}(\sigma_{xx} + \sigma_{zz}) \quad or \quad \sigma_{yy} - 0.25\sigma_{xx} = E\varepsilon_{yy} = (80)(10^9)(1500)(10^{-6}) = 120MPa \quad (4)$$

ANS $\sigma_{xx} = 117.3$ MPa ; $\sigma_{yy} = 149.3$ MPa ; $\tau_{xy} = 128.6$ MPa

---

9. 60

Solution: $\varepsilon_a = -800\ \mu\ m/m$, $\varepsilon_b = -300\ \mu\ m/m$, $\varepsilon_c = -700\ \mu\ m/m$ $\varepsilon_1 = ?$ $\varepsilon_2 = ?$ $\varepsilon_3 = ?$ $\theta_1 = ?$ $\gamma_{max} = ?$

By inspection of the given figure we obtain: $\varepsilon_{xx} = -800\mu\ m/m$ and $\varepsilon_{yy} = -300\mu\ m/m$

Substituting $\theta_c = 135^o$ into the normal strain transformation equation we obtain:

$$\varepsilon_c = (-800)cos^2(135) + (-300)sin^2(135) + \gamma_{xy}sin(135)cos(135) = -700 \quad or \quad \gamma_{xy} = 300\mu\ rad \quad (1)$$

The principal angle θ_p is calculated and substituted into the normal strain transformation equation to determine one principal strain as shown below.

$$\theta_p = 0.5\ atan\left(\frac{300}{-800+300}\right) = -15.48^o \quad (2)$$

$$\varepsilon_p = (-800)cos^2(-15.48) + (-300)sin^2(-15.48) + (300)sin(-15.48)cos(-15.48) = -841.5\mu \quad (3)$$

The other principal strain is: $-800 - 300 - (-841.5) = -258.5\mu$. As this strain is greater than the above strain, this strain is principal strain one. Thus principal angle 1 is $\theta_1 = \theta_p \pm 90$.

$$\varepsilon_1 = -258.5\mu \qquad \varepsilon_2 = -841.5\mu \qquad \theta_1 = 74.52^o \text{ or } -105.48^o \quad (4)$$

For plane stress: $\varepsilon_3 = \varepsilon_z = -\frac{\nu}{(1-\nu)}(\varepsilon_{xx} + \varepsilon_{yy}) = -\frac{0.28}{(1-0.28)}(-800-300) = 427.8$

The maximum shear strain is given by:

$$\gamma_{max}/2 = |(\varepsilon_3 - \varepsilon_2)/2| = |(427.8 + 841.5)/2| \quad or \quad \gamma_{max} = 1269 \quad (5)$$

ANS $\varepsilon_1 = -258.5\mu$; $\varepsilon_2 = -841.5\mu$; $\varepsilon_3 = 427.8\mu$; $\theta_1 = 74.52^o$ or -105.48^o ; $\gamma_{max} = 1269\mu$

---

9. 61

Solution $\varepsilon_a = 400\ \mu\ m/m$ $\varepsilon_b = 200\ \mu\ m/m$ $\varepsilon_c = 0$ E = 210 GPa, ν = 0.28 $\sigma_1 = ?$ $\sigma_2 = ?$ $\sigma_3 = ?$ $\theta_1 = ?$ $\tau_{max} = ?$

By inspection of the given figure we obtain: $\varepsilon_{xx} = 400\mu\ m/m$ and $\varepsilon_{yy} = 200\mu\ m/m$

Substituting $\theta_c = 135^o$ into the normal strain transformation equation we obtain:

$$\varepsilon_c = (400)cos^2(135) + (200)sin^2(135) + \gamma_{xy}sin(135)cos(135) = 0 \quad or \quad \gamma_{xy} = 600\mu\ rad \quad (1)$$

$$G = \frac{E}{2(1+\nu)} = \frac{210}{2(1+0.28)} = 82.03GPa \qquad \tau_{xy} = G\gamma_{xy} = (82.03)(10^9)(600)(10^{-6}) = 49.22MPa \quad (2)$$

From Generalized Hooke's law, for plane stress ($\sigma_{zz} = 0$) we obtain:

$$\varepsilon_{xx} = \frac{\sigma_{xx}}{E} - \frac{\nu}{E}(\sigma_{yy} + \sigma_{zz}) \quad or \quad \sigma_{xx} - 0.28\sigma_{yy} = E\varepsilon_{xx} = (210)(10^9)(400)(10^{-6}) = 84MPa \quad (3)$$

$$\varepsilon_{yy} = \frac{\sigma_{yy}}{E} - \frac{\nu}{E}(\sigma_{xx} + \sigma_{zz}) \quad or \quad \sigma_{yy} - 0.28\sigma_{xx} = E\varepsilon_{yy} = (210)(10^9)(200)(10^{-6}) = 42MPa \quad (4)$$

$$\sigma_{xx} = 103.9MPa \qquad \sigma_{yy} = 71.1MPa \quad (5)$$

The principal angle θ_p is calculated and substituted into the stress equation to determine one principal stress as shown below.

$$\theta_p = 0.5\ atan\left(\frac{(2)(49.22)}{103.9-71.1}\right) = 35.79^o \quad (6)$$

$$\sigma_p = (103.9)cos^2(35.79) + (71.1)sin^2(35.79) + 2(49.22)sin(35.79)cos(35.79) = 139.38MPa \quad (7)$$

The other principal stress is: $103.9 + 71.1 - 139.38 = 35.62MPa$. As this stress is smaller than the above stress, the above stress is principal stress one. The principal angle 1 is $\theta_1 = \theta_p$ and the maximum shear stress is $\tau_{max} = |(\sigma_1 - \sigma_3)/2| = 69.7$ MPa

ANS $\sigma_1 = 139.4$ MPa (T) ; $\sigma_2 = 35.6$ MPa (T) ; $\sigma_3 = 0$; $\tau_{max} = 69.7$ MPa ; $\theta_1 = 35.79^o$ or -144.2^o

---

9. 62

Solution $\varepsilon_a = -100\ \mu\ in/in$ $\varepsilon_b = 200\ \mu\ in/in$ $\varepsilon_c = 300\ \mu\ in/in$ E = 10,000 ksi ν = 0.25

$\varepsilon_1 = ?$ $\varepsilon_2 = ?$ $\varepsilon_3 = ?$ $\theta_1 = ?$ $\gamma_{max} = ?$

By inspection of the given figure we obtain: $\varepsilon_{xx} = 200\mu\ m/m$ o and $\varepsilon_{yy} = 300\mu\ m/m$

Substituting $\theta_a = -60^o$ into strain transformation equation we obtain:

$$\varepsilon_a = (200)cos^2(-60) + (300)sin^2(-60) + \gamma_{xy}sin(-60)cos(-60) = -100 \quad or \quad \gamma_{xy} = 866\mu\ rad \qquad \textbf{(1)}$$

The principal angle θ_p is calculated and substituted into the stress equation to determine one principal stress as shown below.

$$\theta_p = 0.5\ atan\left(\frac{866}{200-300}\right) = -41.71^o \qquad \textbf{(2)}$$

$$\varepsilon_p = (200)cos^2(-41.71) + (300)sin^2(-41.71) + (866)sin(-41.71)cos(-41.71) = -185.9\mu \qquad \textbf{(3)}$$

The other principal strain is: $200 + 300 - (-185.9) = 685.9\mu$. As this strain is greater than the above strain, this strain is principal strain one. Thus principal angle 1 is $\theta_1 = \theta_p \pm 90$

$$\varepsilon_1 = 685.9\mu \qquad \varepsilon_2 = -185.9\mu \qquad \theta_1 = 48.29^o \text{ or } -131.7^o \qquad \textbf{(4)}$$

$$\varepsilon_3 = \varepsilon_z = -\frac{\nu}{(1-\nu)}(\varepsilon_{xx} + \varepsilon_{yy}) = -\frac{0.25}{(1-0.25)}(200+300) \qquad \frac{\gamma_{max}}{2} = \left|\frac{\varepsilon_1 - \varepsilon_2}{2}\right| = \left|\frac{685.9 + 185.9}{2}\right| \quad or \quad \gamma_{max} = 871.8\mu$$

ANS $\varepsilon_1 = 685.9\mu$; $\varepsilon_2 = -185.9\mu$; $\varepsilon_3 = -166.7\mu$; $\theta_1 = 48.29^o$ or -131.7^o; $\gamma_{max} = 871.8\mu$

---

9. 63

Solution $\varepsilon_a = 500\ \mu\ in/in$ $\varepsilon_b = 500\ \mu\ in/in$ $\varepsilon_c = 500\ \mu\ in/in$ E = 10,000 ksi ν = 0.25

$\sigma_1 = ?$ $\sigma_2 = ?$ $\sigma_3 = ?$ $\theta_1 = ?$ $\tau_{max} = ?$

By inspection of the given figure we obtain: $\varepsilon_{xx} = 500\mu\ m/m$ and $\varepsilon_{yy} = 500\mu\ m/m$

Substituting $\theta_a = -60^o$ s into the normal stress transformation equation we obtain:

$$\varepsilon_a = (500)cos^2(-60) + (500)sin^2(-60) + \gamma_{xy}sin(-60)cos(-60) = 500 \quad or \quad \gamma_{xy} = 0 \qquad \textbf{(1)}$$

From Generalized Hooke's law, for plane stress ($\sigma_{zz} = 0$) we obtain:

$$\varepsilon_{xx} = \frac{\sigma_{xx}}{E} - \frac{\nu}{E}(\sigma_{yy} + \sigma_{zz}) \quad or \quad \sigma_{xx} - 0.25\sigma_{yy} = E\varepsilon_{xx} = (10000)(500)(10^{-6}) = 5ksi \qquad \textbf{(2)}$$

$$\varepsilon_{yy} = \frac{\sigma_{yy}}{E} - \frac{\nu}{E}(\sigma_{xx} + \sigma_{zz}) \quad or \quad \sigma_{yy} - 0.25\sigma_{xx} = E\varepsilon_{yy} = (10000)(500)(10^{-6}) = 5ksi \qquad \textbf{(3)}$$

$$\sigma_{xx} = 6.67ksi \qquad \sigma_{yy} = 6.67ksi \qquad G = \frac{E}{2(1+\nu)} = \frac{10000}{2(1+0.25)} = 4000ksi \qquad \tau_{xy} = G\gamma_{xy} = (4000)(0) = 0 \qquad \textbf{(4)}$$

As shear stress is zero, σ_{xx} and σ_{yy} are the principal stress. As these two stress components are the same, all planes have the same normal stress. The maximum shear stress is $\tau_{max} = |(\sigma_1 - \sigma_3)/2| = |6.67/2| = 3.34$ ksi.

ANS $\sigma_1 = 6.67$ ksi (T); $\sigma_2 = 6.67$ ksi (T); $\sigma_3 = 0$; $\theta_1 =$ all directions; $\tau_{max} = 3.34$ ksi

---

9. 64

Solution 50 mm x 50 mm F = 100 kN E = 70 GPa G = 28 GPa $\varepsilon_{30} = ?$

The axial stress and strain, the transverse strain, and strain in the gage can be calculated as shown below.

$$\sigma_{xx} = \frac{100(10^3)}{(0.05)(0.05)} = 40(10^6)N/m^2 \qquad \varepsilon_{xx} = \frac{\sigma_{xx}}{E} = \frac{40(10^6)}{70(10^9)} = 571.4\mu \qquad \textbf{(1)}$$

$$\nu = E/(2G) - 1 = 70/[2(28)] - 1 = 0.25 \qquad \varepsilon_{yy} = -\nu\varepsilon_{xx} = -0.25(571.4) = -142.85\mu \qquad \gamma_{xy} = 0 \qquad \textbf{(2)}$$

$$\varepsilon_{30} = (571.4)cos^2(30) + (-142.85)sin^2(30) = 392.9 \qquad \textbf{(3)}$$

ANS $\varepsilon_{30} = 392.9\mu$

---

9. 65

Solution 50 mm x 50 mm E = 70 GPa G = 28 GPa $\varepsilon_{30} = 200\ \mu$. F = ?

The axial stress and strain, the transverse strain, and strain in the gage can be calculated as shown below.

$$\sigma_{xx} = F/[(0.05)(0.05)] = 400F\ N/m^2 \qquad \varepsilon_{xx} = \sigma_{xx}/E = 400F/[70(10^9)] = 5.7154F(10^{-9}) \qquad (1)$$

$$\nu = E/(2G)-1 = 70/[2(28)]-1 = 0.25 \qquad \varepsilon_{yy} = -\nu\varepsilon_{xx} = -0.25(5.7154F)(10^{-9}) = -1.428F(10^{-9}) \qquad \gamma_{xy} = 0 \qquad (2)$$

$$\varepsilon_{30} = (5.7154F)(10^{-9})cos^2(30) + (-1.428F)(10^{-9})sin^2(30) = 3.924F(10^{-9}) = 200(10^{-6}) \quad or \quad F = 50.9(10^3)\text{ N} \qquad (3)$$

ANS $F = 50.9$ kN

---

9. 66

Solution d = 2 in E = 30000 ksi ν = 0.3 F = 100 kips ε_{45} =?

The axial stress and strain, the transverse strain, and strain in the gage can be calculated as shown below.

$$\sigma_{xx} = \frac{100}{(\pi 2^2)/4} = 31.83\ ksi \qquad \varepsilon_{xx} = \frac{\sigma_{xx}}{E} = \frac{31.83}{30000} = 1061\mu \qquad \varepsilon_{yy} = -\nu\varepsilon_{xx} = -0.3(1061) = -318.3\mu \qquad \gamma_{xy} = 0 \qquad (1)$$

$$\varepsilon_{45} = [1061\,cos^2(-45) + (-318.3)sin^2(-45)] = 371.6\mu \qquad (2)$$

ANS $\varepsilon_{45} = 371.6\mu$

---

9. 67

Solution d = 2 in E = 30000 ksi ν = 0.3 ε_{45} = 1000 μ F = ?

The axial stress and strain, the transverse strain, and strain in the gage can be calculated as shown below.

$$\sigma_{xx} = \frac{F}{(\pi 2^2)/4} = 0.3183F \qquad \varepsilon_{xx} = \frac{\sigma_{xx}}{E} = \frac{0.3183F}{30000} = 10.61F\mu \qquad \varepsilon_{yy} = -\nu\varepsilon_{xx} = -0.3(10.61F) = -3.183F\mu \qquad \gamma_{xy} = 0 \qquad (1)$$

$$\varepsilon_{45} = [(10.61F)cos^2(-45) + (-3.183F)sin^2(-45)](10^{-6}) = 3.716F(10^{-6}) = 1000(10^{-6}) \qquad F = 269\text{ kips} \qquad (2)$$

---

9. 68

Solution d = 2 in E = 30000 ksi ν = 0.3 ε_{60} = ?

Using the figure below we obtain the following.

$$J = \pi(2^4)/32 = 1.571 in^4 \qquad \tau_{x\theta} = 30(1)/1.571 = 19.1ksi = -\tau_{xy} \quad or \quad \tau_{xy} = -19.1ksi \qquad (1)$$

(a)

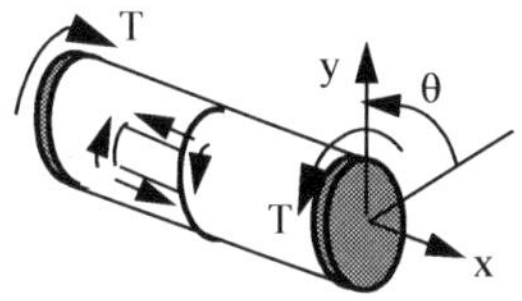

$$G = \frac{E}{2(1+\nu)} = \frac{30000}{2(1.3)} = 11538.5ksi \qquad \gamma_{xy} = \frac{\tau_{xy}}{G} = -\frac{19.1}{11538.5} = -1655.3\mu \qquad \varepsilon_{xx} = 0 \qquad \varepsilon_{yy} = 0 \qquad (2)$$

Substituting the strains and θ = -60°, we obtain

$$\varepsilon_{60} = (-1655.3)sin(-60)cos(-60) = 716.77 \qquad (3)$$

ANS $\varepsilon_{60} = 716.8\mu$

---

9. 69

Solution d = 50 mm E = 70 GPa G = 28 GPa T =500 N-m ε_{40} = ?

Using the figure below we obtain the following.

$$\tau_{x\theta} = \frac{(500)(0.025)}{\pi(0.05^4)/32} = 20.372(10^6)N/m^2 = -\tau_{xy} \quad or \quad \tau_{xy} = -20.372(10^6)N/m^2 \qquad (1)$$

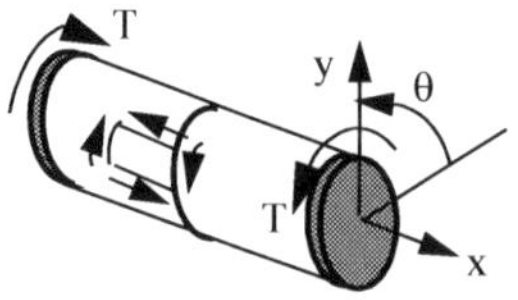

$$\gamma_{xy} = \tau_{xy}/G = -20.372(10^6)/[28(10^9)] = -727.55\mu \qquad \varepsilon_{xx} = 0 \qquad \varepsilon_{yy} = 0 \qquad (2)$$

Substituting the strains and θ = 40°, we obtain

$$\varepsilon_{40} = (-727.55T)sin(40)cos(40) = -358.25\mu \qquad (3)$$

ANS $\varepsilon_{40} = -358.3\mu$

---

9. 70

Solution d = 50 mm E = 70 GPa G = 28 GPa ε_{40} = -600 μ T =?

Using the figure below we obtain the following.

$$\tau_{x\theta} = \frac{T(0.025)}{\pi(0.05^4)/32} = 40.743T(10^3)N/m^2 = -\tau_{xy} \quad or \quad \tau_{xy} = -40.743T(10^3)N/m^2 \tag{1}$$

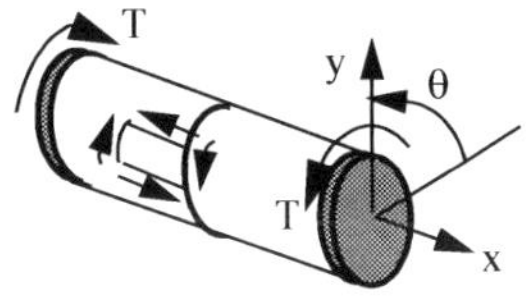

$$\gamma_{xy} = \tau_{xy}/G = -(40.743T(10^3))/28(10^9) = -1.4551T\mu \qquad \varepsilon_{xx} = 0 \qquad \varepsilon_{yy} = 0 \tag{2}$$

Substituting the strains and θ = 40°, we obtain

$$\varepsilon_{40} = (-1.4551T)sin(40)cos(40) = -600 \quad or \quad T = 837.4 \text{ N-m} \tag{3}$$

ANS $T = 837.4$ N-m

---

9. 71

Solution d = 1000 mm t = 10 mm p = 200 kPa E = 210GPa ν= 0.28 ε_a = ? ε_b = ?

The solution proceeds as follows.

$$\sigma_{xx} = \frac{pr}{2t} = \frac{(200)(10^3)(0.5)}{2(0.01)} = 5MPa \qquad \sigma_{yy} = \frac{pr}{t} = \frac{(200)(10^3)(0.5)}{(0.01)} = 10MPa \tag{1}$$

$$\varepsilon_{xx} = \frac{\sigma_{xx}}{E} - \frac{\nu}{E}(\sigma_{yy} + \sigma_{zz}) = \frac{5(10^6)}{210(10^9)} - \frac{0.28(10)(10^6)}{210(10^9)} = 10.476\mu \tag{2}$$

$$\varepsilon_{yy} = \frac{\sigma_{yy}}{E} - \frac{\nu}{E}(\sigma_{xx} + \sigma_{zz}) = \frac{(10)(10^6)}{210(10^9)} - \frac{0.28(5)(10^6)}{210(10^9)} = 40.952\mu \tag{3}$$

The shear stress τ_{xy} = 0, hence γ_{xy} = 0. Substituting the strains and θ_a = 40° and θ_b = - 50° we obtain the following.

$$\varepsilon_a = (10.476)cos^2 40 + (40.952)sin^2 40 = 23.1 \qquad \varepsilon_b = (10.476)cos^2(-50) + (40.952)sin^2(-50) = 28.4 \tag{4}$$

ANS $\varepsilon_a = 23.1\mu$; $\varepsilon_b = 28.4\mu$

---

9. 72

Solution E = 70 GPa, ν=0.25 P= 10kN M = 5kN-m ε_a= ? ε_b= ?

By equilibrium of forces and moment we obtain:

$$\boldsymbol{V}_y = P = 10 \text{ kN} \qquad \boldsymbol{M}_z = M + 0.5P = 5 + 5 = 10 \text{ kN-m} \tag{1}$$

(a)

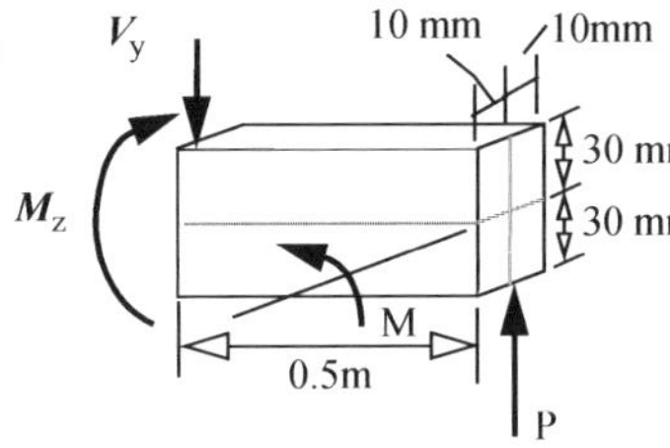

(b)

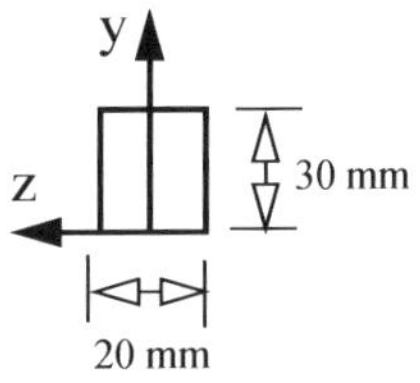

$$I_{zz} = \frac{(0.02)(0.06^3)}{12} = 0.36(10^{-6})m^4 \qquad (\sigma_{xx})_a = 0 \qquad (\sigma_{xx})_b = \frac{-\boldsymbol{M}_z y_b}{I_{zz}} = \frac{-(10)(10^3)(-0.03)}{0.36(10^{-6})} = 833.3(10^6) \tag{2}$$

As the shear force $\boldsymbol{V}_y$ is positive, the bending shear stress τ_{xy} will be positive at the neutral axis.

$$(Q_z)_a = (0.02)(0.03)(0.015) = 9.0(10^{-6})m^3 \qquad (\tau_{xy})_a = \left|\frac{V_y(Q_z)_a}{I_{zz}t}\right| = \frac{(10)(10^3)(9.0)(10^{-6})}{0.36(10^{-6})(0.02')} = 12.5(10^6) \qquad (\tau_{xy})_b = 0 \tag{3}$$

$$G = \frac{70}{2(1+0.25)} = 28GPa \qquad (\gamma_{xy})_a = \frac{(\tau_{xy})_a}{G} = \frac{12.5(10^6)}{28(10^9)} = 446.4(10^{-6}) = 446.4\mu \qquad (\varepsilon_{xx})_a = 0 \qquad (\varepsilon_{yy})_a = 0 \tag{4}$$

$$(\varepsilon_{xx})_b = \frac{833.3(10^6)}{70(10^9)} = 11.904(10^{-3}) = 11904\mu \qquad (\varepsilon_{yy})_b = -0.25(\varepsilon_{xx})_b = -2976\mu \qquad (\gamma_{xy})_a = 0 \tag{5}$$

Substituting θ = - 25° and the above strains we obtain the following.

$$\varepsilon_a = 446.4\sin(-25)\cos(-25) = -170.99\mu \qquad \varepsilon_b = 11904\cos^2(-25) + [-2976]\sin^2(-25) = 9246.3\mu \qquad (6)$$

ANS $\varepsilon_a = -171\mu$; $\varepsilon_b = 9246\mu$

---

9. 73

Solution $\varepsilon_a = -386$ μm/m and $\varepsilon_b = 4092$ μm/m. E – 70 GPa ν=0.25 P= ? M= ?

By equilibrium of forces and moment we obtain:

$$V_y = P \qquad M_z = M + 0.5P \qquad (1)$$

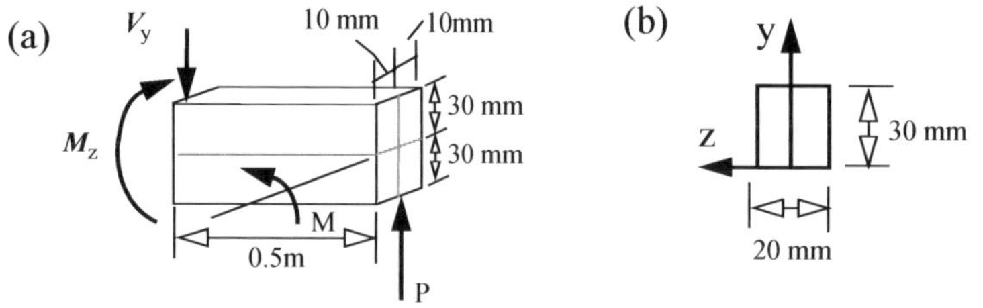

$$I_{zz} = \frac{(0.02)(0.06^3)}{12} = 0.36(10^{-6})m^4 \qquad (\sigma_{xx})_a = 0 \qquad (\sigma_{xx})_b = \frac{-M_z y_b}{I_{zz}} = \frac{-(M+0.5P)(-0.03)}{0.36(10^{-6})} = 83.33(M+0.5P)(10^3) \qquad (2)$$

As the shear force V_y is positive, the bending shear stress τ_{xy} will be positive at the neutral axis.

$$(Q_z)_a = (0.02)(0.03)(0.015) = 9.0(10^{-6})m^3 \qquad (\tau_{xy})_a = \left|\frac{V_y(Q_z)_a}{I_{zz}t}\right| = \frac{(P)(9.0)(10^{-6})}{0.36(10^{-6})(0.02')} = 1250P \qquad (\tau_{xy})_b = 0 \qquad (3)$$

$$G = \frac{70}{2(1+0.25)} = 28GPa \qquad (\varepsilon_{xx})_a = 0 \qquad (\varepsilon_{yy})_a = 0 \qquad (\gamma_{xy})_a = \frac{(\tau_{xy})_a}{G} = \frac{1250P}{28(10^9)} = 44.643P(10^{-9}) \qquad (4)$$

$$(\varepsilon_{xx})_b = \frac{(\sigma_{xx})_b}{E} = \frac{83.33(M+0.5P)(10^3)}{70(10^9)} = 1.1904(M+0.5P)(10^{-6}) \qquad (\varepsilon_{yy})_b = -\nu(\varepsilon_{xx})_b = -0.25(\varepsilon_{xx})_b \qquad (\gamma_{xy})_a = 0 \qquad (5)$$

Substituting θ = - 30° and the strains we obtain:

$$\varepsilon_a = 44.643P(10^{-9})\sin(-30)\cos(-30) = -19.331P(10^{-9}) = -386(10^{-6}) \qquad or \qquad P = 19.97(10^3)\text{ N} \qquad (6)$$

$$\varepsilon_b = (\varepsilon_{xx})_b\cos^2(-30) + [-0.25(\varepsilon_{xx})_b]\sin^2(-30) = 0.6875(\varepsilon_{xx})_b = 0.6875[1.1904(M+0.5P)(10^{-6})] \qquad (7)$$

$$\varepsilon_b = 0.8184(M+0.5P)(10^{-6}) = 4092(10^{-6}) \qquad or \qquad M+0.5P = 5000 \qquad or \qquad M = -4985N-m \qquad (8)$$

ANS $P = 19.97$ kN ; $M = -4.985$ kN-m

---

9. 74

Solution E = 210 GPa ν=0.28 d = 50 mm P = 100 kN ε_{20} =?

By equilibrium of forces and noting that the deformation of the two bars is the same, we obtain:

$$N_{AB} + N_{BC} = 2P = 200(10^3) \qquad \delta_{AB} = \delta_{BC} \qquad (1)$$

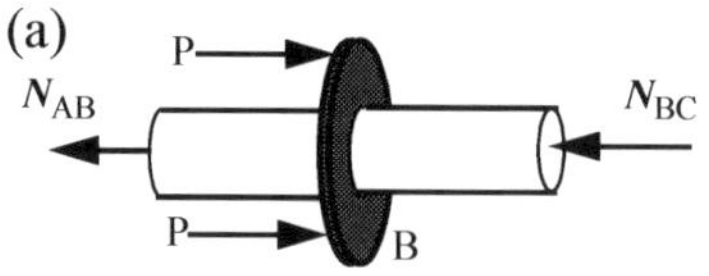

The deformation of the bars can be written as:

$$A = \pi(0.05)^2/4 = 1.963(10^{-3}) \qquad \delta_{AB} = \frac{N_{AB}L_{AB}}{EA} = \frac{N_{AB}(0.75)}{210(10^9)(1.963)(10^{-3})} = 1.8189N_{AB}(10^{-9}) \qquad (2)$$

$$\delta_{BC} = \frac{N_{BC}L_{BC}}{EA} = \frac{N_{BC}(2)}{210(10^9)(1.963)(10^{-3})} = 4.8504N_{BC}(10^{-9}) \qquad (3)$$

$$\delta_{AB} = 1.8189N_{AB}(10^{-9}) = \delta_{BC} = 4.8504N_{BC}(10^{-9}) \qquad or \qquad N_{BC} = 0.375N_{AB} \qquad N_{AB}(1+0.375) = 200(10^3) \qquad (4)$$

$$N_{AB} = 145.45 10^3 \qquad N_{BC} = 0.5455P = 54.55(10^3) \qquad (5)$$

$$\sigma_{xx} = N_{BC}/A = (54.55(10^3))/[1.963(10^{-3})] = 27.79(10^6)\ \frac{N}{m^2}(C) \qquad (6)$$

$$\varepsilon_{xx} = \frac{\sigma_{xx}}{E} = \frac{-27.79(10^6)}{210(10^9)} = -0.1323(10^{-3}) \qquad \varepsilon_{yy} = -\nu\varepsilon_{xx} = -0.28[-0.1323(10^{-3})] = 0.03705(10^{-3}) \qquad \gamma_{xy} = 0 \quad \textbf{(7)}$$

$$\varepsilon_{20} = [-0.1323(10^{-3})]cos^2(20) + [0.03705(10^{-3})]sin^2(20) = -112.51(10^{-3}) \quad \textbf{(8)}$$

ANS $\varepsilon_{20} = -112.5\mu$

---

9. 75

Solution E = 210 GPa ν=0.28 d = 50 mm $\varepsilon_{20} = -214\mu$ P=?

From previous problem we have the following.

$$\boldsymbol{N}_{AB} + \boldsymbol{N}_{BC} = 2P \qquad \delta_{AB} = \delta_{BC} \qquad A = 1.963(10^{-3})m^2 \qquad \delta_{AB} = 1.8189\boldsymbol{N}_{AB}(10^{-9}) \qquad \delta_{BC} = 4.8504\boldsymbol{N}_{BC}(10^{-9}) \quad \textbf{(1)}$$

Solution proceeds as follows.

$$\delta_{AB} = 1.8189\boldsymbol{N}_{AB}(10^{-9}) = \delta_{BC} = 4.8504\boldsymbol{N}_{BC}(10^{-9}) \quad or \quad \boldsymbol{N}_{BC} = 0.375\boldsymbol{N}_{AB} \qquad \boldsymbol{N}_{AB}(1 + 0.375) = 2P \quad \textbf{(2)}$$

$$\boldsymbol{N}_{AB} = 1.4545P \qquad \boldsymbol{N}_{BC} = 0.5455P \qquad \sigma_{xx} = \boldsymbol{N}_{BC}/A = 0.5455P/[1.963(10^{-3})] = 0.2779P(10^3)\ \frac{N}{m^2}(C) \quad \textbf{(3)}$$

$$\varepsilon_{xx} = \frac{\sigma_{xx}}{E} = -\frac{0.2779P(10^3)}{210(10^9)} = -1.3232P(10^{-9}) \qquad \varepsilon_{yy} = -\nu\varepsilon_{xx} = -0.28[-1.3232P(10^{-9})] = 0.3705P(10^{-9}) \qquad \gamma_{xy} = 0 \quad \textbf{(4)}$$

$$\varepsilon_{20} = [-1.3232P(10^{-9})]cos^2(20) + [0.3705P]sin^2(20) = -1.1251P(10^{-9}) = -214(10^{-6}) \quad or \quad P = 190.2(10^3)\text{ N} \quad \textbf{(5)}$$

ANS $P = 190.2$ kN

---

9. 76

Solution E = 210 GPa ν=0.28 d = 50 mm T = 10 kN-m ε_{20}=?

By moment equilibrium and geometry we obtain the following.

$$\boldsymbol{T}_{AB} = T_A \qquad \boldsymbol{T}_{BC} = T_A - T = T_A - 10(10^3) \qquad J = \pi(0.05)^4/32 = 0.6136(10^{-6})m^4 \quad \textbf{(1)}$$

(a)

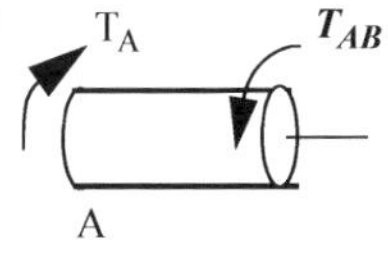

(b)

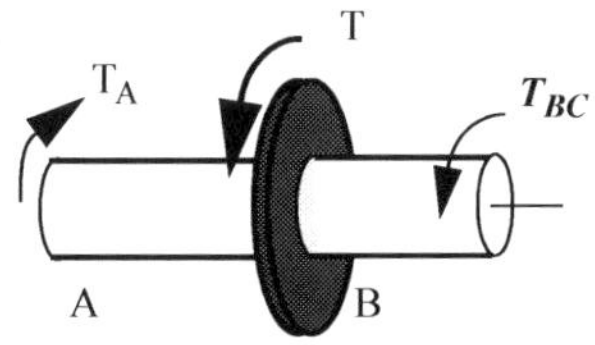

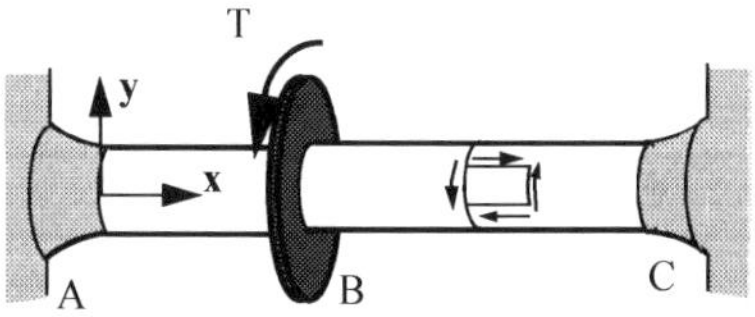

The relative rotation of the ends can be written and added as shown below.

$$\phi_B - \phi_A = \frac{\boldsymbol{T}_{AB}(x_B - x_A)}{GJ} = \frac{(T_A)(0.75)}{GJ} \qquad \phi_C - \phi_B = \frac{\boldsymbol{T}_{BC}(x_C - x_B)}{GJ} = \frac{(T_A - 10(10^3))(2)}{GJ} \quad \textbf{(2)}$$

$$\phi_C - \phi_A = \frac{T_A(0.75)}{GJ} + \frac{(T_A - T)2}{GJ} = 0 \quad or \quad T_A = \frac{2T}{2.75} = 0.727T \qquad \boldsymbol{T}_{BC} = -2727T = -2.727(10^3) \quad \textbf{(3)}$$

$$\tau_{x\theta} = \left|\frac{\boldsymbol{T}_{BC}\rho_{max}}{J}\right| = \frac{2.727(10^3)(0.025)}{0.6136(10^{-6})} = 111.1(10^6)\frac{N}{m^2} = \tau_{xy} \quad or \quad \tau_{xy} = 111.1(10^6)\frac{N}{m^2} \quad \textbf{(4)}$$

$$G = \frac{E}{2(1+\nu)} = \frac{210}{2(1+0.28)} = 82.03GPa \qquad \gamma_{xy} = \frac{\tau_{xy}}{G} = 111.1(10^6)[82.03(10^9)] = 1.354(10^{-3}) = 1354\mu \quad \textbf{(5)}$$

$$\varepsilon_{xx} = 0 \qquad \varepsilon_{yy} = 0 \qquad \varepsilon_{20} = 1354cos(20)sin(20) = 435.16\mu \quad \textbf{(6)}$$

ANS $\varepsilon_{20} = 435.2\mu$

---

9. 77

Solution E = 210 GPa ν=0.28 d = 50 mm $\varepsilon_{20} = 1088\mu$ T= ?

From previous problem we have

$$\boldsymbol{T}_{BC} = -2727T \qquad J = 0.6136(10^{-6})m^4 \qquad G = 82.03GPa \quad \textbf{(1)}$$

The torsional shear stress on the surface of segment BC is:

$$\tau_{x\theta} = \left|\frac{\boldsymbol{T}_{BC}\rho_{max}}{J}\right| = \frac{(0.2727T)(0.025)}{0.6136(10^{-6})} = 11.11T(10^3)\frac{N}{m^2} = \tau_{xy} \quad or \quad \tau_{xy} = 11.11T(10^3)\frac{N}{m^2} \quad \textbf{(2)}$$

$$\gamma_{xy} = \tau_{xy}/G = 11.11T(10^3)/[82.03(10^9)] = 0.1354T(10^{-6}) = 0.1354T\mu \quad (3)$$

$$\varepsilon_{xx} = 0 \qquad \varepsilon_{yy} = 0 \qquad \varepsilon_{20} = 0.1354T\cos(20)\sin(20) = 0.04352T\mu = 1088\mu \qquad or \qquad T = 25(10^3)\text{ N-m} \quad (4)$$

ANS $T = 25$ kN-m

---

9. 78

Solution

From the given transformation equations we can write the following.

$$x = n\cos\theta - t\sin\theta \qquad y = n\sin\theta + t\cos\theta \qquad \frac{\partial x}{\partial n} = \cos\theta \qquad \frac{\partial y}{\partial n} = \sin\theta \quad (1)$$

$$\frac{\partial u_n}{\partial x} = \frac{\partial u}{\partial x}\cos\theta + \frac{\partial v}{\partial x}\sin\theta \qquad \frac{\partial u_n}{\partial y} = \frac{\partial u}{\partial y}\cos\theta + \frac{\partial v}{\partial y}\sin\theta \quad (2)$$

By chain rule we have

$$\varepsilon_{nn} = \frac{\partial u_n}{\partial n} = \left(\frac{\partial u_n}{\partial x}\right)\frac{\partial x}{\partial n} + \left(\frac{\partial u_n}{\partial y}\right)\frac{\partial y}{\partial n} = \left(\frac{\partial u}{\partial x}\cos\theta + \frac{\partial v}{\partial x}\sin\theta\right)\cos\theta + \left(\frac{\partial u}{\partial y}\cos\theta + \frac{\partial v}{\partial y}\sin\theta\right)\sin\theta = \quad \text{or}$$

$$\varepsilon_{nn} = \frac{\partial u}{\partial x}\cos^2\theta + \frac{\partial v}{\partial y}\sin^2\theta + \left(\frac{\partial u}{\partial y} + \frac{\partial v}{\partial x}\right)\cos\theta\sin\theta = \varepsilon_{xx}\cos^2\theta + \varepsilon_{yy}\sin^2\theta + \gamma_{xy}\cos\theta\sin\theta \quad (3)$$

Above equation is the desired result.

---

9. 79

Solution

From the given transformation equations we can write the following.

$$x = n\cos\theta - t\sin\theta \qquad y = n\sin\theta + t\cos\theta \qquad \frac{\partial x}{\partial t} = -\sin\theta \qquad \frac{\partial y}{\partial t} = \cos\theta \quad (1)$$

$$\frac{\partial u_t}{\partial x} = -\frac{\partial u}{\partial x}\sin\theta + \frac{\partial v}{\partial x}\cos\theta \qquad \frac{\partial u_t}{\partial y} = -\frac{\partial u}{\partial y}\sin\theta + \frac{\partial v}{\partial y}\cos\theta \quad (2)$$

By chain rule we have

$$\varepsilon_{tt} = \frac{\partial u_t}{\partial t} = \left(\frac{\partial u_t}{\partial x}\right)\frac{\partial x}{\partial t} + \left(\frac{\partial u_t}{\partial y}\right)\frac{\partial y}{\partial t} = \left(-\frac{\partial u}{\partial x}\sin\theta + \frac{\partial v}{\partial x}\cos\theta\right)(-\sin\theta) + \left(-\frac{\partial u}{\partial y}\sin\theta + \frac{\partial v}{\partial y}\cos\theta\right)(\cos\theta) \quad \text{or}$$

$$\varepsilon_{tt} = \frac{\partial u}{\partial x}\sin^2\theta + \frac{\partial v}{\partial y}\cos^2\theta - \left(\frac{\partial u}{\partial y} + \frac{\partial v}{\partial x}\right)\cos\theta\sin\theta = \varepsilon_{xx}\sin^2\theta + \varepsilon_{yy}\cos^2\theta - \gamma_{xy}\cos\theta\sin\theta \quad (3)$$

Above equation is the desired result.

---

9. 80

Solution

From the given transformation equations we can write the following.

$$x = n\cos\theta - t\sin\theta \qquad y = n\sin\theta + t\cos\theta \qquad \frac{\partial x}{\partial n} = \cos\theta \qquad \frac{\partial y}{\partial n} = \sin\theta \qquad \frac{\partial x}{\partial t} = -\sin\theta \qquad \frac{\partial y}{\partial t} = \cos\theta \quad (1)$$

$$\frac{\partial u_t}{\partial x} = -\frac{\partial u}{\partial x}\sin\theta + \frac{\partial v}{\partial x}\cos\theta \qquad \frac{\partial u_t}{\partial y} = -\frac{\partial u}{\partial y}\sin\theta + \frac{\partial v}{\partial y}\cos\theta \qquad \frac{\partial u_n}{\partial x} = \frac{\partial u}{\partial x}\cos\theta + \frac{\partial v}{\partial x}\sin\theta \qquad \frac{\partial u_n}{\partial y} = \frac{\partial u}{\partial y}\cos\theta + \frac{\partial v}{\partial y}\sin\theta \quad (2)$$

By chain rule we have

$$\frac{\partial u_t}{\partial n} = \left(\frac{\partial u_t}{\partial x}\right)\frac{\partial x}{\partial n} + \left(\frac{\partial u_t}{\partial y}\right)\frac{\partial y}{\partial n} = \left(-\frac{\partial u}{\partial x}\sin\theta + \frac{\partial v}{\partial x}\cos\theta\right)\cos\theta + \left(-\frac{\partial u}{\partial y}\sin\theta + \frac{\partial v}{\partial y}\cos\theta\right)\sin\theta = \left(-\frac{\partial u}{\partial x} + \frac{\partial v}{\partial y}\right)\sin\theta\cos\theta + \frac{\partial v}{\partial x}\cos^2\theta - \frac{\partial u}{\partial y}\sin^2\theta \quad (3)$$

$$\frac{\partial u_n}{\partial t} = \left(\frac{\partial u_n}{\partial x}\right)\frac{\partial x}{\partial t} + \left(\frac{\partial u_n}{\partial y}\right)\frac{\partial y}{\partial t} = \left(\frac{\partial u}{\partial x}\cos\theta + \frac{\partial v}{\partial x}\sin\theta\right)(-\sin\theta) + \left(\frac{\partial u}{\partial y}\cos\theta + \frac{\partial v}{\partial y}\sin\theta\right)\cos\theta = \left(-\frac{\partial u}{\partial x} + \frac{\partial v}{\partial y}\right)\sin\theta\cos\theta + \frac{\partial u}{\partial y}\cos^2\theta - \frac{\partial v}{\partial x}\sin^2\theta \quad (4)$$

$$\gamma_{nt} = \frac{\partial u_t}{\partial n} + \frac{\partial u_n}{\partial t} = 2\left(-\frac{\partial u}{\partial x} + \frac{\partial v}{\partial y}\right)\sin\theta\cos\theta + \left(\frac{\partial u}{\partial y} + \frac{\partial v}{\partial x}\right)(\cos^2\theta - \sin^2\theta) = 2(\varepsilon_{yy} - \varepsilon_{xx})\sin\theta\cos\theta + \gamma_{xy}(\cos^2\theta - \sin^2\theta) \quad (5)$$

Above equation is the desired result.

---

9. 81

Solution

Solution proceeds as follows.

$$\varepsilon_{xx} = \frac{\sigma_{xx}}{E_x} - \frac{\nu_{yx}}{E_y}\sigma_{yy} \qquad \varepsilon_{yy} = \frac{\sigma_{yy}}{E_y} - \frac{\nu_{xy}}{E_x}\sigma_{xx} \qquad \gamma_{xy} = \frac{\tau_{xy}}{G_{xy}} \qquad \frac{\nu_{yx}}{E_y} = \frac{\nu_{xy}}{E_x} \tag{1}$$

Substituting the following stresses $\sigma_{xx} = \sigma$ $\sigma_{yy} = -\sigma$ $\tau_{xy} = 0$ into normal stress transformation equations

$$\sigma_{nn} = \sigma cos^2\theta + (-\sigma)sin^2\theta = \sigma cos2\theta \qquad \sigma_{tt} = \sigma sin^2\theta + (-\sigma)cos^2\theta = -\sigma cos2\theta \tag{2}$$

$$\varepsilon_{xx} = \left(\frac{1}{E_x} + \frac{\nu_{yx}}{E_y}\right)\sigma \qquad \varepsilon_{yy} = -\left(\frac{1}{E_y} + \frac{\nu_{xy}}{E_x}\right)\sigma \qquad \gamma_{xy} = 0 \tag{3}$$

$$\varepsilon_{nn} = \varepsilon_{xx}cos^2\theta + \varepsilon_{yy}sin^2\theta + \gamma_{xy}cos\theta sin\theta = \left[\left(\frac{1}{E_x} + \frac{\nu_{yx}}{E_y}\right)\sigma\right]cos^2\theta + \left[-\left(\frac{1}{E_y} + \frac{\nu_{xy}}{E_x}\right)\sigma\right]sin^2\theta \text{ or}$$

$$\varepsilon_{nn} = \left[\left(\frac{1}{E_x} + \frac{\nu_{yx}}{E_y}\right)\sigma\right]\frac{(1+cos2\theta)}{2} + \left[-\left(\frac{1}{E_y} + \frac{\nu_{xy}}{E_x}\right)\sigma\right]\frac{(1-cos2\theta)}{2} = \left(\frac{1}{E_x} - \frac{1}{E_y} + \frac{\nu_{yx}}{E_y} - \frac{\nu_{xy}}{E_x}\right)\frac{\sigma}{2} + \left[\frac{1}{E_x} + \frac{1}{E_y} + \frac{\nu_{yx}}{E_y} + \frac{\nu_{xy}}{E_x}\right]\frac{\sigma cos2\theta}{2}$$

Substituting into Generalized Hooke's law for isotropic materials in plane stress we obtain:

$$\varepsilon_{nn} = \frac{\sigma_{nn}}{E} - \frac{\nu\sigma_{tt}}{E} = \frac{(1+\nu)}{E}\sigma cos2\theta = \left(\frac{1}{E_x} - \frac{1}{E_y} + \frac{\nu_{yx}}{E_y} - \frac{\nu_{xy}}{E_x}\right)\frac{\sigma}{2} + \left[\frac{1}{E_x} + \frac{1}{E_y} + \frac{\nu_{yx}}{E_y} + \frac{\nu_{xy}}{E_x}\right]\frac{\sigma cos2\theta}{2} \tag{4}$$

If above equation is to be valid for any angle θ, then the following conditions must be satisfied.

$$\left(\frac{1}{E_x} - \frac{1}{E_y} + \frac{\nu_{yx}}{E_y} - \frac{\nu_{xy}}{E_x}\right) = 0 \qquad \frac{1}{2}\left[\frac{1}{E_x} + \frac{1}{E_y} + \frac{\nu_{yx}}{E_y} + \frac{\nu_{xy}}{E_x}\right] = \frac{(1+\nu)}{E} \tag{5}$$

Noting that $\nu_{yx}/E_y = \nu_{xy}/E_x$, the above equation implies $E_x = E_y$, which in turn implies $\nu_{yx} = \nu_{xy}$.

Substituting these identities we obtain $\frac{1}{2}\left[\frac{1}{E_x} + \frac{1}{E_y} + \frac{\nu_{yx}}{E_y} + \frac{\nu_{xy}}{E_x}\right] = \frac{2}{2}\left[\frac{1}{E_x} + \frac{\nu_{xy}}{E_x}\right] = \frac{(1+\nu)}{E}$ or $\frac{E}{(1+\nu)} = \frac{E_x}{(1+\nu_{xy})}$,

Substituting the following stresses $\sigma_{xx} = 0$ $\sigma_{yy} = 0$ $\tau_{xy} = \tau$ and corresponding strains $\varepsilon_{xx} = 0$ $\varepsilon_{yy} = 0$ $\gamma_{xy} = \tau/G_{xy}$ into shear stress and shear strain transformation equation, we obtain:

$$\tau_{nt} = \tau(cos^2\theta - sin^2\theta) = \tau cos2\theta \qquad \gamma_{nt} = \tau(cos^2\theta - sin^2\theta)/G_{xy} = \tau cos2\theta/G_{xy} \tag{6}$$

Substituting into the Generalized Hooke's law for isotropic material, we obtain:

$$\tau_{nt} = G\gamma_{nt} = \frac{E}{2(1+\nu)}\gamma_{nt} \quad \text{or} \quad \tau cos2\theta = \frac{E}{2(1+\nu)}\left[\frac{\tau}{G_{xy}}cos2\theta\right] \quad \text{or} \quad G_{xy} = \frac{E}{2(1+\nu)} = \frac{E_x}{(1+\nu_{xy})} \tag{7}$$

Above equations are the desired results.

---

9. 82

Solution Plane Strain $\varepsilon_1 = ?$ $\varepsilon_2 = ?$ $\varepsilon_3 = ?$ $\theta_1 = ?$ $\gamma_{max} = ?$

For plane strain the third principal strain is: $\varepsilon_3 = 0$

We obtain the following strains from the given displacements.

$$\varepsilon_{xx} = \frac{\partial u}{\partial x} = [0.5(2x) + 0.5y + 0.25](10^{-3}) = [x + 0.5y + 0.25](10^{-3}) \tag{1}$$

$$\varepsilon_{yy} = \frac{\partial v}{\partial y} = [0.25(-2y) - x](10^{-3}) = -[x + 0.5y](10^{-3}) \tag{2}$$

$$\gamma_{xy} = \frac{\partial u}{\partial y} + \frac{\partial v}{\partial x} = [0.5(-2y) + 0.5x + 0.25(2x) - y](10^{-3}) = [x - 2y](10^{-3}) \tag{3}$$

θ (°)	x (mm)	y (mm)	ε_{xx} μ	ε_{yy} μ	γ_{xy} μ	θ_p (rads)	ε_p μ	ε_{p2} μ	ε_1 μ	ε_2 μ	θ_1 (°)
0	1.000	0.000	1250	-1000	1000	0.209	1356	-1106	1356	-1106	11.98
30	0.866	0.500	1366	-1116	-134	-0.027	1368	-1118	1368	-1118	-1.54
60	0.500	0.866	1183	-933	-1232	-0.264	1349	-1099	1349	-1099	-15.11
90	0.000	1.000	750	-500	-2000	-0.506	1304	-1054	1304	-1054	-29.00
120	-0.500	0.866	183	67	-2232	-0.759	1243	-993	1243	-993	-43.51
150	-0.866	0.500	-366	616	-1866	0.543	-929	1179	1179	-929	121.12
180	-1.000	0.000	-750	1000	-1000	0.260	-883	1133	1133	-883	104.87
210	-0.866	-0.500	-866	1116	134	-0.034	-868	1118	1118	-868	88.07
240	-0.500	-0.866	-683	933	1232	-0.326	-891	1141	1141	-891	71.34
270	-0.000	-1.000	-250	500	2000	-0.606	-943	1193	1193	-943	55.28
300	0.500	-0.866	317	-67	2232	0.700	1257	-1007	1257	-1007	40.12
330	0.866	-0.500	866	-616	1866	0.450	1316	-1066	1316	-1066	25.77

Starting with $\theta = 0$ in steps of 30° we find the following on a spreadsheet as shown in the table above.

1. The coordinates: $x = r\cos\theta = \cos\theta \qquad y = r\sin\theta = \sin\theta$
2. We find the strains from the above equations.
3. The principal angle: $\theta_p = atan[\gamma_{xy}/(\varepsilon_{xx}-\varepsilon_{yy})]/2$
4. The normal strain on principal plane: $\varepsilon_p = \varepsilon_{xx}\cos^2\theta_p + \varepsilon_{yy}\sin^2\theta_p + \gamma_{xy}\sin\theta_p\cos\theta_p$.
5. The other principal strain: $\varepsilon_{p2} = \varepsilon_{xx}+\varepsilon_{yy}-\varepsilon_p$
6. If $\varepsilon_p \ge \varepsilon_{p2}$ then $\theta_1 = \theta_p$ and $\varepsilon_1 = \varepsilon_p$; $\varepsilon_2 = \varepsilon_{p2}$ otherwise $\theta_1 = \theta_p + 90$ and $\varepsilon_1 = \varepsilon_{p2}$; $\varepsilon_2 = \varepsilon_p$.

---

9. 83

Solution E = 70 GPa $\nu = 0.25$ P= ? w = ?

By equilibrium of forces and moment we obtain:

$$V_y = -(P+0.4w) \qquad M_z = -[0.4P + (0.4w)(0.2)] = -0.4(P+0.2w) \tag{1}$$

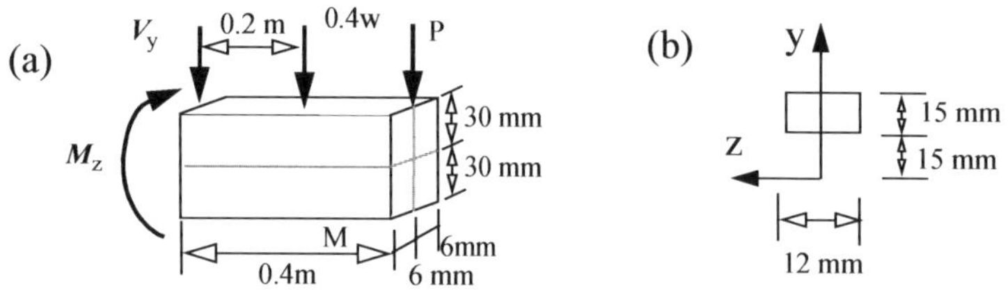

The bending normal stress and strain at the two gage locations are

$$I_{zz} = (0.012)(0.06^3)/12 = 0.216(10^{-6}) \qquad \sigma_{xx} = \frac{-M_z y}{I_{zz}} = -\frac{[-0.4(P+0.2w)](0.015)}{(0.216)(10^{-6})} = 27.777(P+0.2w)(10^3) \tag{2}$$

$$\varepsilon_{xx} = \frac{\sigma_{xx}}{E} = \frac{27.777(P+0.2w)(10^3)(10^3)}{70(10^9)} = 0.3968(P+0.2w)(10^{-6}) \tag{3}$$

As the shear force V_y is negative, the bending shear stress τ_{xy} will be negative.

$$Q_z = (0.012)(0.015)(0.0225) = 4.05(10^{-6})m^3 \qquad \tau_{xy} = -\left|\frac{V_y(Q_z)_a}{I_{zz}t}\right| = -\frac{(P+0.4w)(4.05)(10^{-6})}{(0.216)(10^{-6})(0.012)} = -1562.5(P+0.4w) \tag{4}$$

$$G = E/[2(1+\nu)] = 70/[2(1+0.25)] = 28GPa \qquad \gamma_{xy} = \frac{\tau_{xy}}{G} = \frac{-1562.5(P+0.4w)}{28(10^9)} = -0.0558(P+0.4w)(10^{-6}) \tag{5}$$

Substituting $\varepsilon_{yy} = -\nu\varepsilon_{xx}$ and $\theta_a = 45^o$ and $\theta_b = -45^o$ into normal strain transformation equation, we obtain:

$$\varepsilon_a = \varepsilon_{xx}\cos^2(45) + (-0.28\varepsilon_{xx})\sin^2(45) + \gamma_{xy}\sin(45)\cos(45) = \frac{(1-\nu)\varepsilon_{xx}}{2} + \frac{\gamma_{xy}}{2} \tag{6}$$

$$\varepsilon_b = \varepsilon_{xx}\cos^2(-45) + (-\nu\varepsilon_{xx})\sin^2(-45) + \gamma_{xy}\sin(-45)\cos-(45) = \frac{(1-\nu)\varepsilon_{xx}}{2} - \frac{\gamma_{xy}}{2} \tag{7}$$

Solving we obtain:

$$\varepsilon_{xx} = \frac{1}{(1-\nu)}(\varepsilon_a+\varepsilon_b) = 0.3968(P+0.2w)(10^{-6}) \quad or \quad P+0.2w = 3.36(\varepsilon_a+\varepsilon_b)(10^6) \qquad \gamma_{xy} = (\varepsilon_a-\varepsilon_b) \tag{8}$$

$$\gamma_{xy} = -0.0558(P+0.4w)(10^{-6}) = (\varepsilon_a-\varepsilon_b) \quad or \quad P+0.4w = -17.92(\varepsilon_a-\varepsilon_b)(10^6) \tag{9}$$

$$P = [6.72(\varepsilon_a+\varepsilon_b)+17.92(\varepsilon_a-\varepsilon_b)](10^6) \qquad w = -[16.8(\varepsilon_a+\varepsilon_b)+89.6(\varepsilon_a-\varepsilon_b)] \tag{10}$$

The various values of P and w can be found on a spreadsheet from the reading of ε_a and ε_b as shown in the table below.

ε_a (μ)	ε_b (μ)	P N	w N/m
1501	2368	10463	12297
1433	2276	9818	12851
1385	2193	9565	11929
1483	2336	10378	11888
1470	2331	10114	12909
1380	2191	9464	12316
1448	2282	10120	11689
1496	2366	10362	12684
1398	2223	9549	12725
1411	2228	9813	11704

CHAPTER 10

Section 10.1

10. 1

Solution d = 2in $\theta = 60^o$ σ_{nn} =? τ_{nt} =?

By equilibrium of forces and moment we obtain the following.

$$N = 50\ kips \qquad T = 30\ in-kips \tag{1}$$

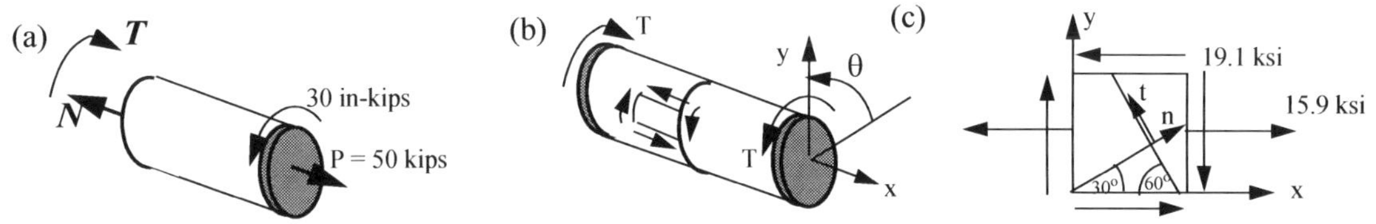

$$A = \frac{\pi(2)^2}{4} = 3.1419in^2 \qquad \sigma_{xx} = \frac{N}{A} = \frac{50}{3.1419} = 15.92ksi \qquad J = \frac{\pi(2)^4}{32} = 1.5708in^4 \qquad \tau_{x\theta} = \frac{T\rho}{J} = \frac{30(1)}{1.5708} = 19.1ksi \tag{2}$$

Fig (c) shows the stress cube at point A. We note

$$\sigma_{xx} = 15.92ksi \qquad \sigma_{yy} = 0ksi \qquad \tau_{xy} = -19.1ksi \qquad \theta = 30°, \tag{3}$$

$$\sigma_{nn} = 15.19cos^2 30 + 2(-19.1)cos30sin30 = -4.6ksi \qquad \tau_{nt} = -15.19(cos30sin30) + (-19.1)[cos^2 30 - sin^2 30] = -16.4ksi$$

ANS $\sigma_{nn} = 4.6ksi(C)$; $\tau_{nt} = -16.4ksi$

---

10. 2

Solution: d = 4 in σ_{nn} =? τ_{nt} =?

By equilibrium of forces and moments we obtain the following.

$$N = 120\ kips \qquad T = -300\ in-kips \tag{1}$$

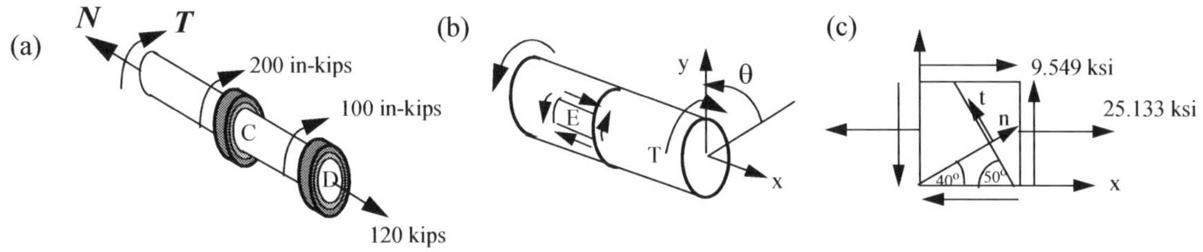

$$A = \frac{\pi(4)^2}{4} = 12.57in^2 \qquad \sigma_{xx} = \frac{N}{A} = \frac{120}{12.57} = 9.55ksi \qquad J = \frac{\pi(4)^4}{32} = 25.13in^4 \qquad \tau_{x\theta} = \frac{T\rho}{J} = \frac{-300(2)}{25.13} = -23.87ksi \tag{2}$$

Fig (c) shows the stress cube at point E. We note

$$\sigma_{xx} = 9.549ksi \qquad \sigma_{yy} = 0 \qquad \sigma_{xy} = 23.872ksi \qquad \theta = 40° \tag{3}$$

$$\sigma_{nn} = 9.549cos^2 40 + (2)(23.872)cos40sin40 = 29.11ksi \qquad \tau_{nt} = -9.549cos40sin40 + 23.872[cos^2 40 - sin^2 40] = -0.557ksi \tag{4}$$

ANS σ_{nn} = 29.11 ksi (T) ; τ_{nt} = –0.557 ksi

---

10. 3

Solution: d = 75mm $\theta = 20^o$ E = 250 GPa $\nu = 0.3$

By equilibrium of forces and moments in Fig (a) we obtain the following.

$$T = T\ kN-m \qquad N = -P\ kN \qquad A = \pi(0.075)^2/4 = 4.418(10^{-3})m^2 \qquad J = \pi(0.075)^4/32 = 3.106(10^{-6})m^4 \tag{1}$$

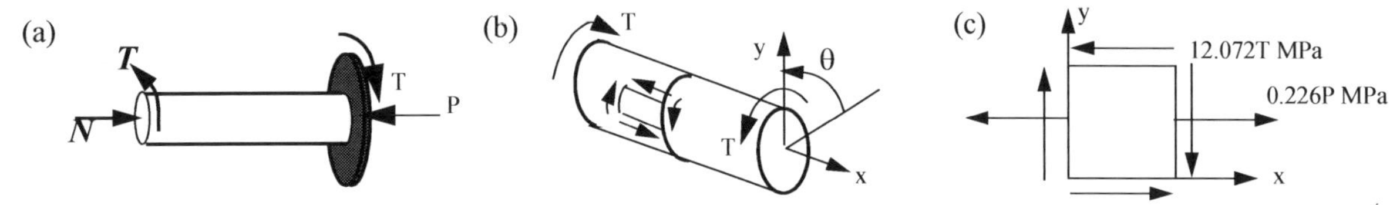

$$\sigma_{xx} = \frac{N}{A} = \frac{-P(10^3)}{4.418(10^{-3})} = -0.226P(10^6)\frac{N}{m^2} = -11.3(10^6)\frac{N}{m^2} \qquad \varepsilon_{xx} = \frac{\sigma_{xx}}{E} = -\frac{11.3(10^6)}{250(10^9)} = -45.2\ \mu \tag{2}$$

$$\tau_{x\theta} = \frac{T\rho}{J} = \frac{T(10^3)(37.5)(10^{-3})}{3.106(10^{-6})} = 12.072T(10^6)\frac{N}{m^2} = 241.44(10^6)\frac{N}{m^2} \qquad \tau_{xy} = -241.44(10^6)\frac{N}{m^2} \tag{3}$$

$$\varepsilon_{yy} = -\nu\varepsilon_{xx} = 13.56\mu \qquad G = \frac{E}{2(1+\nu)} = 96.15GPa \qquad \gamma_{xy} = \frac{\tau_{xy}}{G} = \frac{-241.44(10^6)}{96.15(10^9)} = -2511.1\ \mu \qquad (4)$$

$$\varepsilon_{20} = (-45.2)(\cos 20)^2 + (13.56)(\sin 20)^2 + (-2511.1)\cos 20\sin 20 = -845\mu \qquad (5)$$

ANS $\varepsilon_{20} = -845\mu$

---

10. 4

Solution: d = 75mm $\theta = 20^o$ E = 250 GPa $\nu = 0.3$ $\varepsilon_{20} = -450\mu$ T = 10 kN-m P=?

From previous problem we have the following.

$$\sigma_{xx} = -0.226P(10^6)\frac{N}{m^2} \qquad \tau_{xy} = -12.072T(10^6)\frac{N}{m^2} \qquad G = 96.15GPa \qquad (1)$$

In the above equation P is in kN and T is in kN-m. The strains are

$$\varepsilon_{xx} = \frac{\sigma_{xx}}{E} = -\frac{0.226P(10^6)}{250(10^9)} = -0.9054P\mu \qquad \varepsilon_{yy} = -\nu\varepsilon_{xx} = 0.2716P\mu \qquad \gamma_{xy} = \frac{\tau_{xy}}{G} = \frac{-12.072T(10^6)}{96.15(10^9)} = -125.5T\mu \qquad (2)$$

$$\varepsilon_{20} = (-0.9054P)(\cos 20)^2 + (0.2716P)(\sin 20)^2 + (-125.5T)\cos 20\sin 20 = (-0.7677P - 40.335T)\mu \text{ or}$$

$$\varepsilon_{20} = -(0.7677)(P) - (40.335)(10) = -450 \qquad or \qquad P = 60.76\text{ kN} \qquad (3)$$

ANS $P = 60.76$ kN

---

10. 5

Solution: d = 75mm $\theta = 20^o$ E = 250 GPa $\nu = 0.3$ $\varepsilon_{20} = -300\mu$ P = 55kN T=?

From previous problem we have

$$\varepsilon_{20} = (-0.7677P - 40.335T)\mu \qquad (1)$$

In the above equation P is in kN and T is in kN-m.. We are given $\varepsilon_{20} = -300\mu$ and P = 55kN. Substituting we obtain

$$\varepsilon_{20} = -(0.7677)(55) - (40.335)T = -300 \qquad or \qquad T = 6.39\text{ kN-m} \qquad (2)$$

ANS $T = 6.39$ kN-m

---

10. 6

Solution: d=2 E=30,000 $\nu = 0.3$ P = 70 kips T = 50 in-kips. ε_a=? ε_b=?

By equilibrium of forces and moments and geometry we obtain the following.

$$N = P = 70kips \qquad \boldsymbol{T} = T = 50in-kips \qquad A = \pi(2)^2/4 = 3.1416in^2 \qquad J = \pi(2)^4/32 = 1.5708in^4 \qquad (1)$$

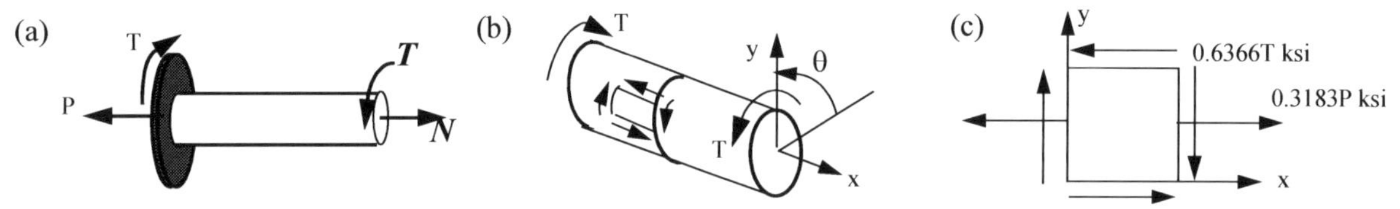

The axial and torsional stresses and strains are as shown below.

$$\sigma_{xx} = \frac{N}{A} = \frac{P}{3.1416} = 0.3183Pksi \qquad \varepsilon_{xx} = \frac{\sigma_{xx}}{E} = -\frac{0.3183P}{30000} = 10.61P\mu \qquad \varepsilon_{yy} = -\nu\varepsilon_{xx} = -3.183P\mu \qquad (2)$$

$$\tau_{x\theta} = \frac{\boldsymbol{T}\rho}{J} = \frac{\boldsymbol{T}(10^3)(1)}{1.5708} = 0.6366Tksi = -\tau_{xy} \qquad G = \frac{E}{2(1+\nu)} = 11538ksi \qquad \gamma_{xy} = \frac{\tau_{xy}}{G} = \frac{-0.6366T}{11538} = -55.174T\mu \qquad (3)$$

Substituting $\theta_a = 150^o$, $\theta_b = 60^o$, and the value of strains in normal strain transformation equation we obtain the following.

$$\varepsilon_a = (10.61P)(\cos 150)^2 + (-3.183P)(\sin 150)^2 + (-55.174T)\cos 150\sin 150 = (7.1617P + 23.891T)\mu \qquad (4)$$

$$\varepsilon_b = (10.61P)(\cos 60)^2 + (-3.183P)(\sin 60)^2 + (-55.174T)\cos 60\sin 60 = (0.26525P - 23.891T)\mu \qquad (5)$$

Substituting P = 70 kips and T = 50 in-kips we obtain the strain gage readings as shown below.

$$\varepsilon_a = (7.1617)(70) + (23.891)(50) = 1696\mu \qquad \varepsilon_b = (0.26525)(70) - (23.891)(50) = -1176\mu \qquad (6)$$

ANS $\varepsilon_a = 1696\mu$; $\varepsilon_b = -1176\mu$

10.7

Solution: d=2 E=30,000 $\nu = 0.3$ $\varepsilon_a = 2078\ \mu$ $\varepsilon_b = -1410\mu$ P= ? T=?

From previous problem we have the following.

$$\varepsilon_a = (7.1617P + 23.891T)\mu \qquad \varepsilon_b = (0.26525P - 23.891T)\mu \tag{1}$$

Equating the above equations to the given strains we obtain two equations in two unknowns that are solved for P & T.

$$(7.1617)P + (23.891)T = 2078 \qquad (0.26525)P - (23.891)T = -1410 \qquad or \qquad P = 89.9\text{ kips} \qquad T = 60.0\text{ in-kips} \tag{2}$$

ANS $P = 89.9$ kips ; $T = 60.0$ in-kips

10.8

Solution: $E_s = 200$ GPa $G_s = 80$ GPa $E_{br} = 100$ GPa $G_{br} = 40$ GPa σ_{max} =? τ_{max} =?

$$N_{AB} = R_A \qquad T_{AB} = -T_A \qquad N_{BC} = R_A - 80 \qquad T_{BC} = -T_A + 10 \qquad N_{ND} = R_A - 80 \qquad T_{CD} = -T_A + 10 \tag{1}$$

(a)

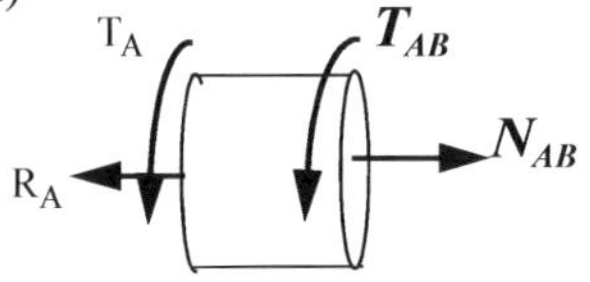

(b)

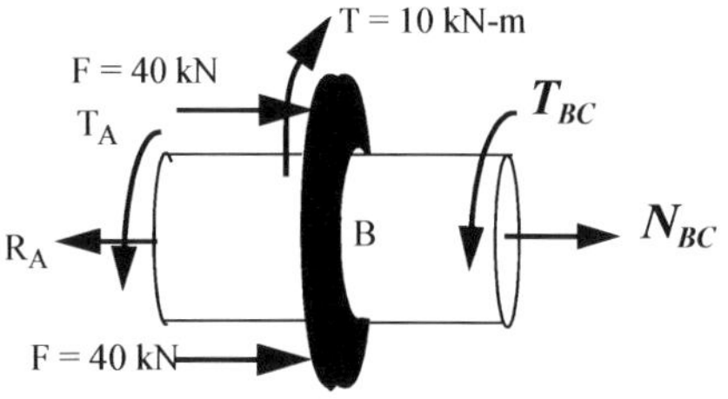

The relative displacement of ends of each segment can be written and added as shown below.

$$A = \frac{\pi}{4}(0.1^2) = 7.854(0.1^{-3})m^2 \qquad u_B - u_A = \frac{N_{AB}(x_B - x_A)}{E_{AB}A_{AB}} = \frac{R_A(5)}{(200)(10^9)(7.854)(10^{-3})} = 3.183R_A(10^{-9}) \tag{2}$$

$$u_C - u_B = \frac{N_{BC}(x_C - x_B)}{E_{BC}A_{BC}} = \frac{(R_A - 80)(3)}{(200)(10^9)(7.854)(10^{-3})} = 1.9099(R_A - 80)(10^{-9}) \tag{3}$$

$$u_D - u_C = \frac{N_{CD}(x_D - x_C)}{E_{CD}A_{CD}} = \frac{(R_A - 80)(4)}{(100)(10^9)(7.854)(10^{-3})} = 5.0929(R_A - 80)(10^{-9}) \tag{4}$$

$$u_D - u_A = [3.183R_A + 1.9099(R_A - 80) + 5.0929(R_A - 80)](10^{-9}) = 0 \qquad or \qquad R_A = \frac{[1.9099 + 5.0929](80)}{(3.183 + 1.9099 + 5.0929)} = 55kN \tag{5}$$

The relative rotation of ends of each segment can be written and added as shown below.

$$J = \frac{\pi}{32}(0.1^4) = 9.817(0.1^{-6})m^4 \qquad \phi_B - \phi_A = \frac{T_{AB}(x_B - x_A)}{G_{AB}J_{AB}} = \frac{(-T_A)(5)}{(80)(10^9)(9.817)(10^{-6})} = -6.37T_A(10^{-6}) \tag{6}$$

$$\phi_C - \phi_B = \frac{T_{BC}(x_C - x_B)}{G_{BC}J_{BC}} = \frac{(-T_A + 10)(3)}{(80)(10^9)(9.817)(10^{-6})} = 3.812(-T_A + 10)(10^{-6}) \tag{7}$$

$$\phi_D - \phi_C = \frac{T_{CD}(x_D - x_C)}{G_{DC}J_{DC}} = \frac{(-T_A + 10)(4)}{(40)(10^9)(9.817)(10^{-6})} = 10.19(-T_A + 10)(10^{-6}) \tag{8}$$

$$\phi_D - \phi_A = [-6.37T_A + 3.82(-T_A + 10) + 10.19(-T_A + 10)](10^{-6}) = 0 \qquad or \qquad T_A = \frac{[3.82 + 10.19](10)}{(6.37 + 3.82 + 10.19)} = 6.875kN - m \tag{9}$$

Substituting we obtain

$$N_{AB} = 55kN \qquad T_{AB} = -6.88kN - m \qquad N_{BC} = -25kN \qquad T_{BC} = 3.12kN - m \qquad N_{CD} = -25kN \qquad T_{CD} = 3.12kN - m$$

As diameter is the same in all segments the maximum axial and torsional shear stress will exist in segments AB,

$$\sigma_{xx} = \frac{N}{A} = \frac{55(10^3)}{7.854(10^{-3})} = 7(10^6)\frac{N}{m^2} = 7MPa(T) \qquad \tau_{x\theta} = \frac{T_{AB}\rho}{J} = \frac{(-6.88)(10^3)(0.05)}{9.817(10^{-6})} = 35(10^6)\frac{N}{m^2} = -35.MPa(T) \tag{10}$$

Noting that $\sigma_{yy} = 0$, the principal stresses and maximum shear stress are as shown below.

$$\sigma_{1,2} = \frac{7+0}{2} \pm \sqrt{\left(\frac{7}{2}\right)^2 + (35)^2} = 3.5 \pm 35.17 \qquad \sigma_1 = 38.67MPa \qquad \sigma_2 = -31.67MPa \qquad \tau_{max} = \frac{\sigma_1 - \sigma_2}{2} = 35.17 \tag{11}$$

ANS $\sigma_{max} = 38.7$ MPa ; $\tau_{max} = 35.2$ MPa

10.9

Solution: $E_s = 200$ GPa $G_s = 80$ GPa $E_{br} = 100$ GPa $G_{br} = 40$ GPa σ_{nn} =? τ_{nt} =?

From previous problem and the figures below we obtain the following.

$$\sigma_{xx} = 7MPa(T) \qquad \tau_{x\theta} = -35.MPa(T) = -\tau_{xy} \qquad \theta = 55°. \tag{1}$$

Substituting in stress transformation equations we obtain the following

$$\sigma_{nn} = 7cos^2 55 + 2(35)cos55sin55 = 35.19MPa \qquad \tau_{nt} = (-7)cos55sin55 + (35)[cos^2 55 - sin^2 55] = -15.26MPa \qquad (2)$$

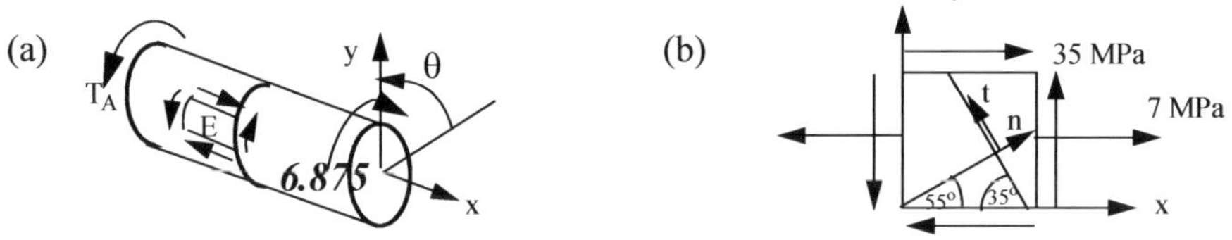

ANS $\sigma_{nn} = 35.2$ MPa (T) ; $\tau_{nt} = -15.3$ MPa

---

10. 10

Solution: $t = (1/2)in$, $P_1 = 72$ kips, $P_2 = 0$, $P_3 = 6$ kips, $\sigma_A = ?$, $\tau_A = ?$, $\sigma_B = ?$, $\tau_B = ?$

By equilibrium of forces and moments we obtain

$$N = -P_1 \qquad V_y = P_2 \qquad V_z = -P_3 \qquad M_z = 60P_2 \qquad M_y = -60P_3 \qquad (1)$$

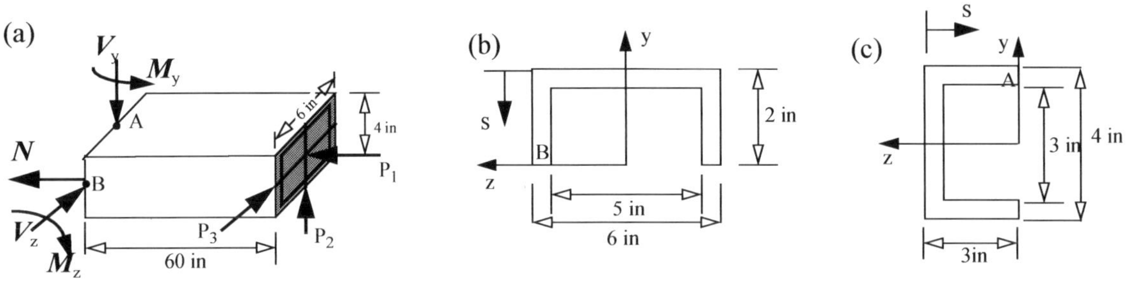

With $P_2 = 0$, we have axial stress, and stresses due to bending about y-axis.

$$\text{Axial:} \quad A = (6)(4) - (5)(3) = 9in^2 \qquad (\sigma_{xx})_{A,B} = \frac{N}{A} = \frac{-72}{9} = -8ksi \qquad (2)$$

$$\text{Bending about y-axis} \quad I_{yy} = \frac{1}{12}(4)(6^3) - \frac{1}{12}(3)(5^3) = 40.75in^4 \qquad \sigma_A = 0 \qquad \sigma_B = \frac{-M_y z_B}{I_{yy}} = \frac{-(-60)(6)(3)}{40.75} = 26.5ksi \quad (3)$$

$$(\tau_{xy})_B = 0 \qquad (Q_y)_A = 2(3)(0.5)(1.5) + 3(0.5)(2.75) = 8.625in^3 \qquad (\tau_{xs})_A = \frac{-V_z(Q_y)_A}{I_{yy}t} = \frac{-(-6)8.625}{40.75(1)} = 1.27ksi = -(\tau_{xz})_A \quad (4)$$

The total stresses at points A and B are found and shown on a stress cube below.

$$(\sigma_{xx})_A = -8ksi \qquad (\sigma_{xx})_B = -8 + 26.5 = 18.5ksi \qquad (5)$$

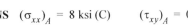

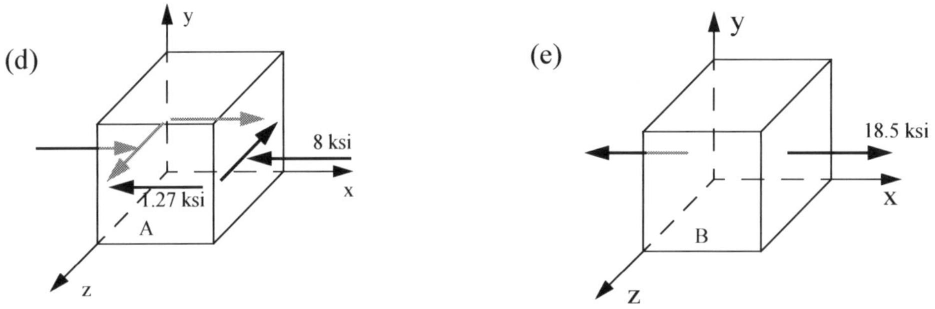

ANS $(\sigma_{xx})_A = 8$ ksi (C) $\quad (\tau_{xy})_A = 0 \quad (\tau_{xz})_A = -1.27$ ksi ; $(\sigma_{xx})_B = 18.5$ ksi (T) $\quad (\tau_{xy})_B = 0 \quad (\tau_{xz})_B = 0$

---

10. 11

Solution: $P_1 = 72$ kips, $P_2 = 3$ kips, $P_3 = 0$, $(\sigma_1)_{A,B} = ?$, $(\sigma_2)_{A,B} = ?$, $(\tau_{max})_{A,B} = ?$

From previous problem we have the following.

$$N = -P_1 \qquad V_y = P_2 \qquad V_z = -P_3 \qquad M_z = 60P_2 \qquad M_y = -60P_3 \qquad A = 9in^2 \qquad (1)$$

With $P_3 = 0$ we have axial and stresses due to bending about z -axis.

$$\text{Axial} \quad (\sigma_{xx})_{A,B} = \frac{N}{A} = \frac{-72}{9} = -8ksi \qquad (2)$$

$$\text{Bending about z-axis} \quad I_{zz} = \frac{6(4^3)}{12} - \frac{5(3^3)}{12} = 20.75in^4 \qquad (\sigma_{xx})_A = \frac{-M_z y_B}{I_{zz}} = \frac{-(180)(2)}{20.75} = -17.35ksi \qquad (\sigma_{xx})_B = 0 \quad (3)$$

$(\tau_{xz})_A = 0 \qquad (Q_z)_B = 2(2)(0.5)(1) + 5(0.5)(1.75) = 6.375 in^4 \qquad (\tau_{xs})_B = \dfrac{-V_y(Q_z)_B}{I_{zz}t} = \dfrac{-(3)6.375}{20.75(1)} = -0.922 ksi = -(\tau_{xy})_B$ **(4)**

The total stresses at points A and B are found and shown on a stress cube below.

$(\sigma_{xx})_A = -8 - 17.35 = -25.35 ksi \qquad (\tau_{xy})_A = (\tau_{xz})_A = 0 \qquad (\sigma_{xx})_B = -8 ksi \qquad (\tau_{xy})_B = 0.922 ksi \qquad (\tau_{xz})_B = 0$ **(5)**

We note that shear stresses are zero at point A.Thus the principal stresses and maximum shear stresses are

$$(\sigma_1)_A = 0 \qquad (\sigma_2)_A = 25.35 \text{ ksi (C)} \qquad (\tau_{max})_A = \left|\frac{\sigma_1 - \sigma_2}{2}\right| = \frac{25.35}{2} = 12.7 \text{ ksi} \tag{6}$$

The principal stresses and maximum shear stress at point B are found as shown below.

$$(\sigma_{1,2})_B = \left(\frac{-8}{2}\right) \pm \sqrt{\left(\frac{-8}{2}\right)^2 + (0.922)^2} = -4 \pm 4.1 \qquad or \qquad (\sigma_1)_B = 0.1 \text{ ksi (T)} \qquad (\sigma_2)_B = 8.1 \text{ ksi (C)} \tag{7}$$

$$(\tau_{max})_B = \left|\frac{\sigma_1 - \sigma_2}{2}\right| = \left|\frac{0.1 + 8.1}{2}\right| = 4.1 \text{ ksi} \tag{8}$$

ANS $(\sigma_1)_A = 0$; $(\sigma_2)_A = 25.35$ ksi (C); $(\tau_{max})_A = 12.7$ ksi; $(\sigma_1)_B = 0.1$ ksi (T); $(\sigma_2)_B = 8.1$ ksi (C); $(\tau_{max})_B = 4.1$ ksi

---

10. 12

Solution: E = 200 GPa $\nu = 0.3$ P_1= 3 kN P_2= 40 kN $\varepsilon_a = ?$ $\varepsilon_b = ?$

By equilibrium of forces and moments and geometry we obtain the following

$$N = P_2 \; kN \qquad M_z = -0.4P_1 \; kN-m \qquad A = (0.01)(0.06) = 0.6(10^{-3})m^2 \qquad I_{zz} = \frac{(0.01)(0.06^3)}{12} = 0.18(10^{-6})m^4 \tag{1}$$

(a)

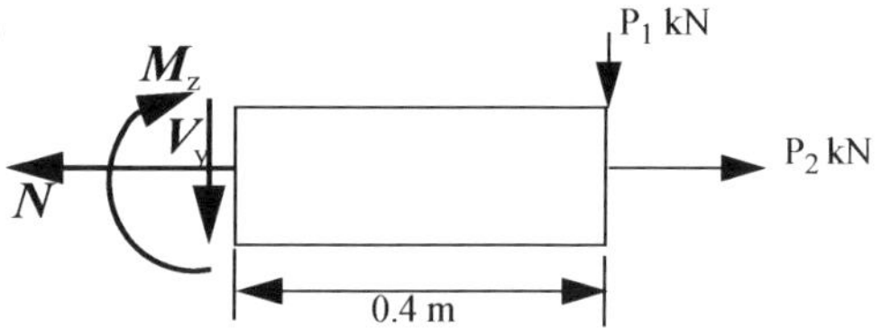

The normal stresses at points A and B are as shown below.

$$\text{Axial:} \quad (\sigma_{xx})_{a,b} = \frac{N}{A} = \frac{P_2(10^3)}{(0.6)(10^{-3})} = 1.667P_2(10^6)\frac{N}{m^2} \tag{2}$$

$$\text{Bending:} \quad (\sigma_{xx})_{a,b} = \frac{-M_z y_{a,b}}{I_{zz}} = \frac{-(-0.4P_1)(10^3)(\pm 30)(10^{-3})}{0.18(10^{-6})} = \pm 66.67P_1(10^6)\frac{N}{m^2} \tag{3}$$

$$\text{Total:} \quad (\sigma_{xx})_a = (1.667P_2 + 66.67P_1)(10^6)\frac{N}{m^2} \qquad (\sigma_{xx})_b = (1.667P_2 - 66.67P_1)(10^6)\frac{N}{m^2} \tag{4}$$

$$\varepsilon_a = (\varepsilon_{xx})_a = \frac{(\sigma_{xx})_a}{E} = (8.333P_2 + 333.3P_1)\mu = 8.333(40) + 333.3(3) = 1333.33 \tag{5}$$

$$\varepsilon_b = (\varepsilon_{xx})_b = \frac{(\sigma_{xx})_b}{E} = (8.333P_2 - 333.3P_1)\mu = 8.333(40) - (333.3)(3) = -666.66 \tag{6}$$

ANS $\varepsilon_a = 1333\mu$; $\varepsilon_b = -666.66\mu$

---

10. 13

Solution: E = 200 GPa $\nu = 0.3$ $\varepsilon_a = 1000\ \mu$; $\varepsilon_b = -750\ \mu$ P_1= ? kN, P_2= ?

From previous problem and the given data we have the following.

$$\varepsilon_a = (8.333P_2 + 333.3P_1)\mu = 1000\mu \qquad \varepsilon_b = (8.333P_2 - 333.3P_1)\mu = -750\mu \qquad or \qquad P_1 = 2.625 \text{ kN} \qquad P_2 = 15 \text{ kN} \tag{1}$$

ANS $P_1 = 2.625$ kN; $P_2 = 15$ kN

---

10. 14

Solution: E = 200 GPa $\nu = 0.3$ P_1= 3 kN; P_2= 40 kN $\varepsilon_a = ?$ $\varepsilon_b = ?$

From earlier problem we have the following.

$$N = P_2 \; kN \qquad V_y = P_1 \; kN \qquad M_z = -0.4P_1 \; kN-m \qquad A = 0.6(10^{-3})m^2 \qquad I_{zz} = 0.18(10^{-6})m^4 \tag{1}$$

$$\text{Axial} \quad (\sigma_{xx})_{a,b} = 1.667P_2(10^6)\frac{N}{m^2} \qquad \text{Bending} \quad (\sigma_{xx})_{a,b} = 0 \tag{2}$$

In finding the bending shear stress we note that s and y are in opposite direction.

$$Q_z = (0.01)(0.03)(0.015) = 4.5(10^{-6})m^3 \qquad \tau_{xs} = \frac{-V_yQ_z}{I_zt} = \frac{-(-P_1)(10^3)(4.5)(10^{-6})}{(0.18)(10^{-6})(0.01)} = 2.5P_1(10^6)\frac{N}{m^2} = -\tau_{xy} \quad (3)$$

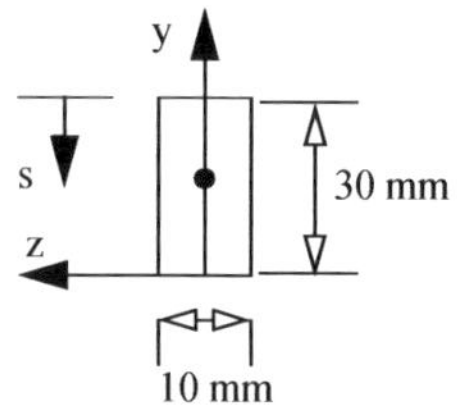

$$\varepsilon_{xx} = \frac{1.667P_2(10^6)}{200(10^9)} = 8.333P_2\mu \qquad \varepsilon_{yy} = -\nu\sigma_{xx} = -2.5P_2\mu \qquad G = \frac{E}{2(1+\nu)} = 76.923GPa \qquad \gamma_{xy} = \frac{\tau_{xy}}{G} = -32.5P_1\mu \quad (4)$$

Substituting $\theta_a = 35^o$ and $\theta_b = -35^o$ and the above strains into strain transformation equations we obtain the following.

$$\varepsilon_a = 8.333P_2cos^235 - 2.5P_2sin^235 - 32.5P_1cos35sin35 = (4.769P_2 - 15.27P_1) = 4.769(40) - 15.27(3) = 145 \quad (5)$$

$$\varepsilon_b = 8.333P_2cos^2(-35) - 2.5P_2sin^2(-35) - 32.5P_1cos(-35)sin(-35) = (4.769P_2 + 15.27P_1) = 4.769(40) + 15.27(3) = 236.6 \quad (6)$$

ANS $\varepsilon_a = 145\mu$; $\varepsilon_a = 236.6\mu$

---

10. 15

Solution: E = 200 GPa $\nu = 0.3$ $\varepsilon_a = 133\ \mu$ $\varepsilon_b = -159\ \mu$ P_1= ? kN, P_2= ?

From previous problem and the given data we have the following.

$$\varepsilon_a = 4.769P_2 - 15.27P_1 = 133 \qquad \varepsilon_b = 4.769P_2 + 15.27P_1 = 159 \qquad or \qquad P_1 = 0.85 \text{ kN} \qquad P_2 = 30.61 \text{ kN} \quad (1)$$

ANS $P_1 = 0.85$ kN ; $P_2 = 30.61$ kN

---

10. 16

Solution: E = 30,000 $\nu = 0.3$ ε_A =? ε_B =?

By equilibrium of forces and moments awe obtain the following.

$$N = 40\ kips \qquad V_y = -2\ kips \qquad V_z = -3\ kips \qquad M_z = -2(24) = -48\ in-kips \qquad M_y = -3(24) = -72\ in-kips \quad (1)$$

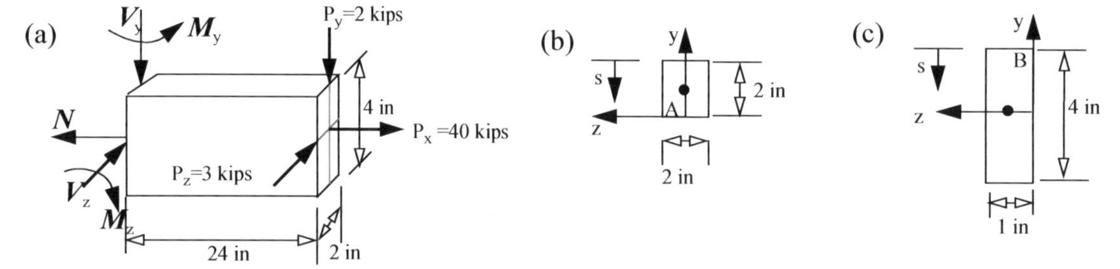

Axial: $A = (2)(4) = 8in^2 \qquad (\sigma_{xx})_{A,B} = N/A = 40/8 = 5ksi$ (2)

Bending about z axis: $I_{zz} = \frac{1}{12}(2)(4^3) = 10.667in^4 \qquad (\sigma_{xx})_A = 0 \qquad (\sigma_{xx})_B = \frac{-M_zy_B}{I_{zz}} = \frac{-(-48)(2)}{10.667} = 9ksi$ (3)

Bending about y axis: $I_{yy} = \frac{1}{12}(4)(2^3) = 2.667in^4 \qquad (\sigma_{xx})_A = \frac{-M_yz_A}{I_{yy}} = \frac{-(-72)(1)}{2.667} = 27ksi \qquad (\sigma_{xx})_B = 0$ (4)

Total: $(\sigma_{xx})_A = 5 + 27 = 32ksi \qquad (\sigma_{xx})_B = 5 + 9 = 14ksi$ (5)

We note that s and y direction are opposite in Fig. (b) and x and z direction are opposite in Fig. (c).

$$(Q_z)_A = 2(2)(1) = 4in^3 \qquad -(\tau_{xy})_A = (\tau_{xs})_A = \frac{-V_y(Q_z)_A}{I_{zz}t} = \frac{-(-2)(4)}{(10.667)(2)} = 0.375ksi \qquad or \qquad (\tau_{xy})_A = (-0.375)ksi \quad (6)$$

$$(Q_y)_B = 4(1)(0.5) = 2in^3 \qquad -(\tau_{xz})_B = (\tau_{xs})_B = \frac{-V_z(Q_y)_B}{I_{yy}t} = \frac{-(-3)(2)}{(2.667)(4)} = 0.5625ksi \qquad or \qquad (\tau_{xz})_B = (-0.5625)ksi \quad (7)$$

$$(\varepsilon_{xx})_A = \frac{(\sigma_{xx})_A}{E} = \frac{32}{30000} = 1067\mu \qquad (\varepsilon_{yy})_A = -\nu(\varepsilon_{xx})_A = (-320)\mu \quad (8)$$

$$(\varepsilon_{xx})_B = \frac{(\sigma_{xx})_B}{E} = \frac{14}{30000} = 467\mu \qquad (\varepsilon_{zz})_B = -\nu(\varepsilon_{xx})_B = (-140)\mu \quad \textbf{(9)}$$

$$G = \frac{E}{2(1+\nu)} = 11538 \qquad (\gamma_{xy})_A = \frac{(\tau_{xy})_A}{G} = -32.5\mu \qquad (\gamma_{xz})_B = \frac{(\tau_{xz})_B}{G} = -48.75\mu \quad \textbf{(10)}$$

Substituting $\theta_A = 30^o$ and $\theta_B = 30^o$ and the corresponding strains at A and B in strain transformation equation we obtain

$$\varepsilon_a = 1067(cos30)^2 - 320(sin30)^2 - 32.5cos30sin30 = 706.2 \qquad \varepsilon_b = 467(cos30)^2 - 140(sin30)^2 - 48.75cos30sin30 = 294.1 \quad \textbf{(11)}$$

ANS $\varepsilon_a = 706.2\mu$; $\varepsilon_b = 294.1\mu$

---

10. 17

Solution: $(\sigma_{xx})_A$ =? $(\tau_{xy})_A$ =? $(\tau_{xz})_A$ =? $(\sigma_{xx})_B$ =? $(\tau_{xy})_B$ =? $(\tau_{xz})_B$ =?

By equilibrium of forces and moments we obtain

$$\boldsymbol{N} = 100kN \qquad \boldsymbol{V}_y = 15kN \qquad \boldsymbol{V}_z = -12kN \quad \textbf{(1)}$$

$$T = 2kN-m \qquad \boldsymbol{M}_z = -12(0.75) = -9kN-m \qquad \boldsymbol{M}_y = 15(0.75) = 11.25kN-m \quad \textbf{(2)}$$

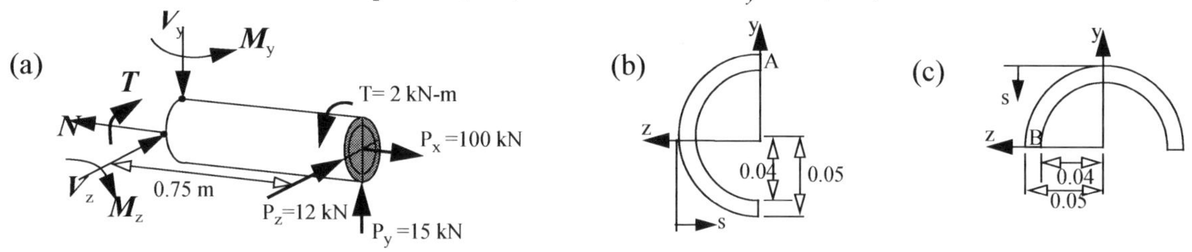

$$A = \frac{\pi}{4}[(0.1)^2 - (0.08)^2] = 2.827(10^{-3})m^2 \qquad J = \frac{\pi}{32}[(0.1)^4 - (0.08)^4] = 5.796(10^{-6})m^4 \qquad I_{yy} = I_{zz} = \frac{J}{2} = 2.898(10^{-6})m^4 \quad \textbf{(3)}$$

$$(Q_y)_A = \left[\left(\frac{4R_o}{3\pi}\right)\left(\frac{\pi R_o^2}{2}\right) - \left(\frac{4R_i}{3\pi}\right)\left(\frac{\pi R_i^2}{2}\right)\right] = \frac{2}{3}(R_o^3 - R_i^3) = \frac{2}{3}(0.05^3 - 0.04^3) = 40.67(10^{-6})m^3 \qquad (Q_z)_B = 40.67(10^{-6})m^3 \quad \textbf{(4)}$$

$$\text{Axial:} \quad (\sigma_{xx})_{A,B} = \frac{\boldsymbol{N}}{A} = \frac{100(10^3)}{2.827(10^{-3})} = 35.37(10^6)N/m^2 \quad \textbf{(5)}$$

$$\text{Bending about z axis:} \quad (\sigma_{xx})_A = \frac{-\boldsymbol{M}_z y_A}{I_{zz}} = \frac{-(11.25)(10^3)(0.05)}{2.898(10^{-6})} = -194.1(10^6)N/m^2 \qquad (\sigma_{xx})_B = 0 \quad \textbf{(6)}$$

$$\text{Bending about y axis:} \quad (\sigma_{xx})_A = 0 \qquad (\sigma_{xx})_B = \frac{-\boldsymbol{M}_y z_B}{I_{yy}} = \frac{-(-9)(10^3)(0.05)}{2.898(10^{-6})} = 155.3(10^6)N/m^2 \quad \textbf{(7)}$$

$$\text{Total:} \quad (\sigma_{xx})_A = (35.37 - 194.1)(10^6)N/m^2 = -158.7 \text{ MPa} \qquad (\sigma_{xx})_B = (35.37 + 155.3)(10^6)N/m^2 = 190.7 \text{ MPa} \quad \textbf{(8)}$$

Note s and z direction in Fig. (b) are opposite and s and y direction in Fig. (c) are opposite.

$$-(\tau_{xz})_A = (\tau_{xs})_A = \frac{-\boldsymbol{V}_z(Q_y)_A}{I_{yy}t} = \frac{-(-12)(10^3)(40.67)(10^{-6})}{(2.898)(10^{-6})(0.02)} = 8.42(10^6)\frac{N}{m^2} \qquad (\tau_{xz})_A = -8.42(10^6)\frac{N}{m^2} \qquad (\tau_{xz})_B = 0 \quad \textbf{(9)}$$

$$-(\tau_{xy})_B = (\tau_{xs})_B = \frac{-\boldsymbol{V}_y Q_z}{I_{zz}t} = \frac{-(15)(10^3)(40.67)(10^{-2})}{(2.898)(0.02)} = -10.52(10^6)N/m^2 \qquad (\tau_{xy})_B = 10.52(10^6)N/m^2 \qquad (\tau_{xy})_A = 0 \quad \textbf{(10)}$$

$$\text{Torsion:} \quad \tau_{x\theta} = \frac{\boldsymbol{T}\rho}{J} = \frac{(2)(10^3)(0.05)}{5.796(10^{\ 6})} = 17.25(10^6)N/m^2 \qquad (\tau_{xz})_A = 17.25(10^6)N/m^2 \qquad (\tau_{xy})_B = -17.25(10^6)N/m^2 \quad \textbf{(11)}$$

$$\text{Total:} \quad (\tau_{xz})_A = (-8.42 + 17.25)(10^6)N/m^2 = 8.83 \text{ MPa} \qquad (\tau_{xy})_B = (10.52-17.25)(10^6)N/m^2 = -6.73 \text{ MPa} \quad \textbf{(12)}$$

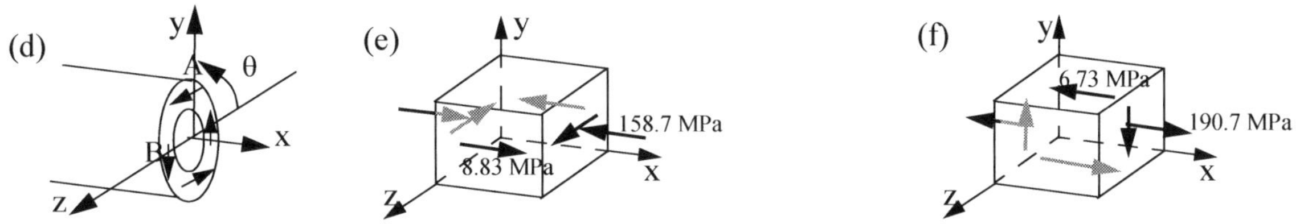

ANS $(\sigma_{xx})_A = 158.7$ MPa (C) ; $(\sigma_{xx})_B = 190.7$ MPa (T) ; $(\tau_{xy})_A = 0$; $(\tau_{xz})_A = 8.83$ MPa ; $(\tau_{xy})_B = -6.73$ MPa ; $(\tau_{xz})_B = 0$

---

10. 18

Solution: $(\sigma_1)_B$ =? $(\sigma_2)_B$ =? $(\tau_{max})_B$ =?

From previous problem we have the following non-zero stress components at B.

$$(\sigma_{xx})_B = 190.7 MPa(T) \qquad (\tau_{xy})_B = -6.73 MPa \tag{1}$$

The third principal stress is zero. The principal stresses and maximum shear can be found as shown below.

$$\sigma_{1,2} = \frac{190.7}{2} \pm \sqrt{\left(\frac{190.7}{2}\right)^2 + (-6.73)^2} = 95.35 \pm 95.87 \qquad \sigma_1 = 190.9 \qquad \sigma_2 = 0.24 \qquad \sigma_3 = 0$$

$$\tau_{max} = [190.9 - (-0.24)]/2 = 95.87$$

ANS $\sigma_1 = 190.9$ MPa (T) ; $\sigma_2 = 0.24$ MPa (C) ; $\sigma_3 = 0$; $\tau_{max} = 95.87$ MPa

---

10. 19

Solution $P_x = 9$ kN $P_y = 0$ $P_z = 0$ Stresses at points A and B by inspection.

All shear stresses are zero. By inspection, the normal stresses and internal force and moment can be written as follows.

$$(\sigma_{xx})_A = \sigma_{axial} \qquad (\sigma_{xx})_B = \sigma_{axial} + \sigma_{bend-y} \qquad |N| = 9kN \qquad |M_y| = (9)(1.5) = 13.5\ kN-m \tag{1}$$

$$A = \frac{\pi}{4}(0.12^2 - 0.1^2) = 3.4558(10^{-3})\ m^2 \qquad \sigma_{axial} = \frac{(9)(10^3)}{3.4558(10^{-3})} = 2.6044(10^6)(N/m^2) = 2.6044 MPa \tag{2}$$

$$I_{yy} = \frac{\pi}{64}(0.12^4 - 0.1^4) = 5.270(10^{-6})\ m^4 \qquad \sigma_{bend-y} = \frac{(13.5)(10^3)(0.06)}{5.270(10^{-3})} = 153.7(10^6)(N/m^2) = 153.7 MPa \tag{3}$$

$$(\sigma_{xx})_A = 2.60 \text{ MPa (T)} \qquad (\sigma_{xx})_B = 2.6044 + 153.7 = 156.3 \text{ MPa (T)} \tag{4}$$

(a) Point A

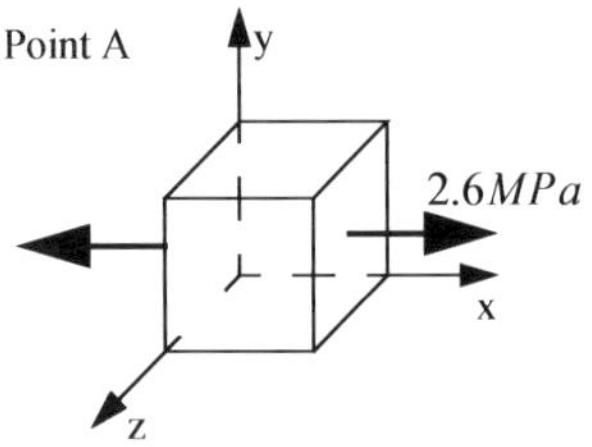

(b) Point B

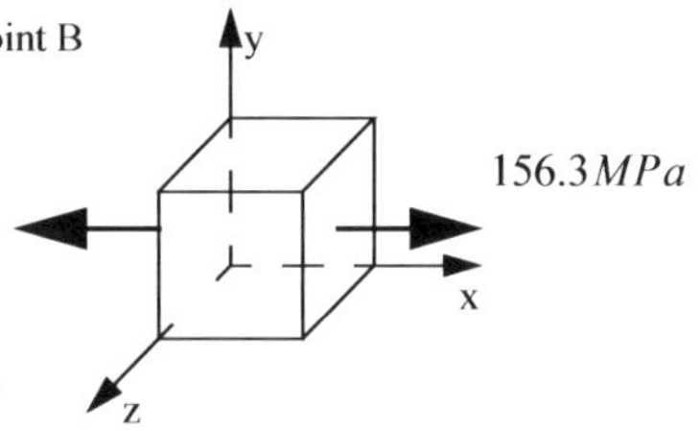

ANS $(\sigma_{xx})_A = 2.60$ MPa (T) ; $(\sigma_{xx})_B = 2.6044 + 153.7 = 156.3$ MPa (T)

---

10. 20

Solution $P_x = 0$ $P_y = 12$ kN $P_z = 0$ Stresses at points A and B by inspection.

For only P_y non-zero, by inspection, the magnitude of internal forces, moments, and stresses can be written as follows.

$$|V_y| = 12kN \qquad T = (12)(1.5) = 18\ kN-m \qquad |M_z| = (12)(1.2) = 14.4\ kN-m \tag{1}$$

$$(\sigma_{xx})_A = -\sigma_{bend-z} \qquad (\sigma_{xx})_B = 0 \qquad (\tau_{xz})_A = -\tau_{tor} \qquad (\tau_{xy})_B = \tau_{tor} + \tau_{bend-z} \tag{2}$$

$$J = \frac{\pi}{32}(0.12^4 - 0.1^4) = 10.54(10^{-6})\ m^4 \qquad \tau_{tor} = \frac{(18)(10^3)(0.06)}{10.54(10^{-6})} = 102.47(10^6)N/m^2 = 102.47 MPa \tag{3}$$

$$I_{zz} = \frac{\pi}{64}(0.12^4 - 0.1^4) = 5.270(10^{-6})\ m^4 \qquad \sigma_{bend-z} = \frac{(14.4)(10^3)(0.06)}{5.270(10^{-6})} = 163.95(10^6)N/m^2 = 163.95 MPa \tag{4}$$

$$)_z)_B = \frac{4}{3}[(0.06)^3 - (0.05)^3] = 60.67(10^{-6})\ m^3 \qquad \tau_{bend-z} = \frac{(12)(10^3)(60.67)(10^{-6})}{5.270(10^{-6})(0.02)} = 6.907(13.814)(10^6)N/m^2 = 6.91 MI \tag{5}$$

$$(\tau_{xy})_B = \tau_{tor} + \tau_{bend-z} = 102.47 + 6.91 = 109.39 \text{ MPa} \tag{6}$$

(a) Due to torsion

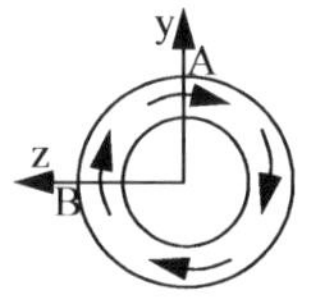

(b) Due to bending about z-axis

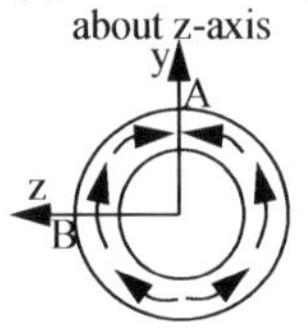

(c) Point A

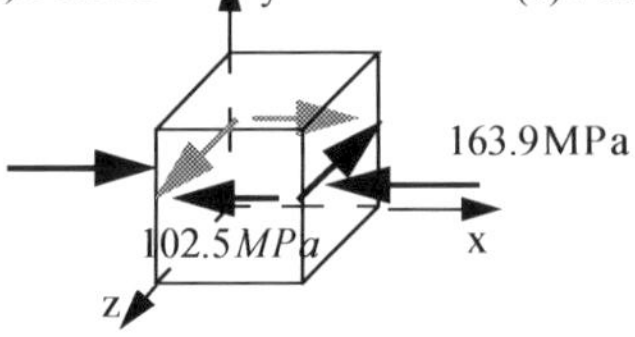

(d) Point B

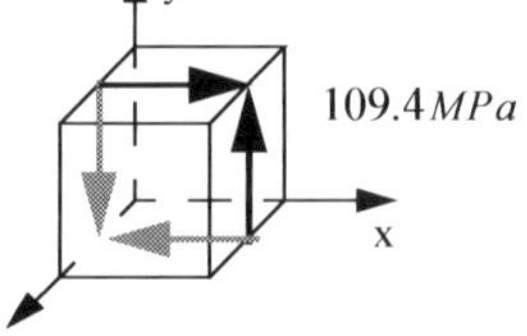

ANS $(\sigma_{xx})_A = 163.95$ MPa (C) ; $(\sigma_{xx})_B = 0$; $(\tau_{xz})_A = 0$; $(\tau_{xy})_B = 109.4$ MPa

---

10. 21

Solution $P_x = 0$ $P_y = 0$ $P_z = 15$ kN Stresses at points A and B by inspection.

For only P_z non-zero, by inspection, the magnitude of internal forces, moments, and stresses can be written as follows.

$$|V_z| = 15kN \qquad |M_y| = (15)(1.2) = 18 \ kN-m \tag{1}$$

$$(\sigma_{xx})_A = 0 \qquad (\sigma_{xx})_B = -\sigma_{bend-y} \qquad (\tau_{xz})_A = \tau_{bend-y} \qquad (\tau_{xy})_B = 0 \tag{2}$$

$$I_{yy} = \frac{\pi}{64}(0.12^4 - 0.1^4) = 5.270(10^{-6}) \ m^4 \qquad \sigma_{bend-y} = \frac{(18)(10^3)(0.06)}{5.270(10^{-6})} = 204.9(10^6)N/m^2 = 204.9MPa \tag{3}$$

$$(Q_z)_B = \frac{4}{3}[(0.06)^3 - (0.05)^3] = 60.67(10^{-6}) \ m^3 \qquad \tau_{bend-y} = \frac{(15)(10^3)(60.67)(10^{-6})}{5.270(10^{-6})(0.02)} = 8.668(10^6)N/m^2 = 8.67MPa \tag{4}$$

$$(\sigma_{xx})_A = 0 \qquad (\tau_{xz})_A = 8.67 \text{ MPa} \qquad (\sigma_{xx})_B = 204.9 \text{ MPa (C)} \qquad (\tau_{xy})_B = 0 \ . \tag{5}$$

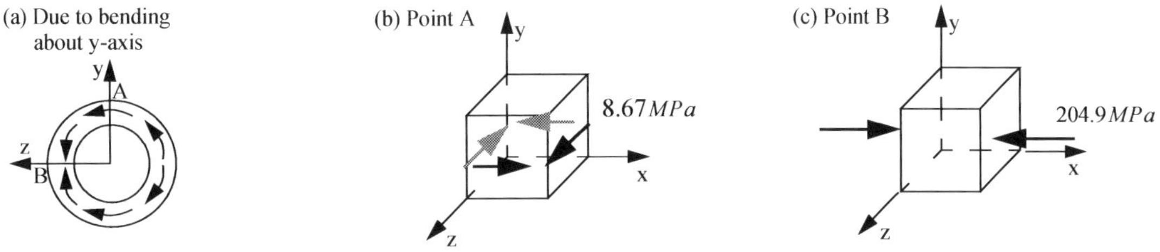

ANS $(\sigma_{xx})_A = 0$; $(\tau_{xz})_A = 8.67$ MPa ; $(\sigma_{xx})_B = 204.9$ MPa (C) ; $(\tau_{xy})_B = 0$

---

10. 22

Solution $P_x = 9$kN $P_y = 0$ $P_z = 0$ Stresses at points A and B by inspection.

For only P_x non-zero, by inspection, the magnitude of internal forces, moments, and stresses can be written as follows. All shear stresses are zero.

$$|N| = 9kN \qquad |M_z| = (9)(0.3) = 2.7 \ kN-m \qquad (\sigma_{xx})_A = \sigma_{axial} + \sigma_{bend-z} \qquad (\sigma_{xx})_B = \sigma_{axial} \tag{1}$$

$$A = \frac{\pi}{4}(0.12^2 - 0.1^2) = 3.4558(10^{-3}) \ m^2 \qquad \sigma_{axial} = \frac{(9)(10^3)}{3.4558(10^{-3})} = 2.604(10^6)(N/m^2) = 2.604MPa \tag{2}$$

$$I_{zz} = \frac{\pi}{64}(0.12^4 - 0.1^4) = 5.270(10^{-6}) \ m^4 \qquad \sigma_{bend-z} = \frac{(2.7)(10^3)(0.06)}{5.270(10^{-3})} = 30.7(10^6)(N/m^2) = 30.74MPa \tag{3}$$

$$(\sigma_{xx})_A = 2.604 + 30.74 = 33.34 \text{ MPa (T)} \qquad (\sigma_{xx})_B = 2.604 \text{ MPa (T)}. \tag{4}$$

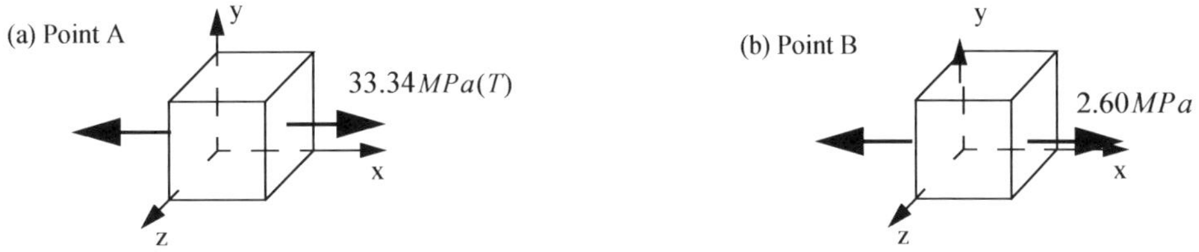

ANS $(\sigma_{xx})_A = 33.34$ MPa (T) ; $(\sigma_{xx})_B = 2.604$ MPa (T)

---

10. 23

Solution $P_x = 0$ $P_y = 12$kN $P_z = 0$ Stresses at points A and B by inspection.

For only P_y non-zero, by inspection, the magnitude of internal forces, moments, and stresses can be written as follows.

$$|V_z| = 12kN \qquad |M_z| = (12)(0.5) = 6 \ kN-m \tag{1}$$

$$(\sigma_{xx})_A = \sigma_{bend-z} \qquad (\sigma_{xx})_B = 0 \qquad (\tau_{xz})_A = 0 \qquad (\tau_{xy})_B = -\tau_{bend-z} \tag{2}$$

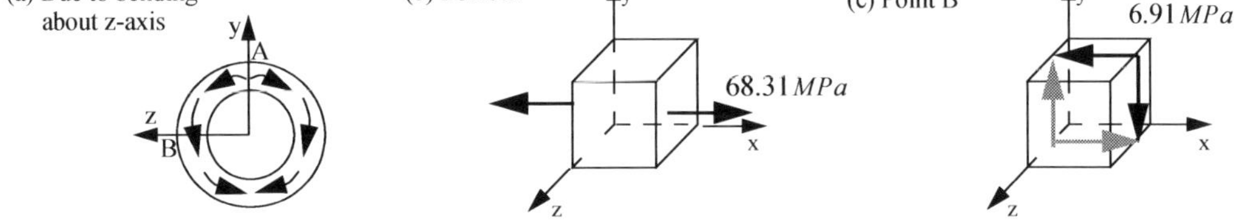

$$I_{yy} = \frac{\pi}{64}(0.12^4 - 0.1^4) = 5.270(10^{-6}) \ m^4 \qquad \sigma_{bend-y} = \frac{(6)(10^3)(0.06)}{5.270(10^{-6})} = 68.31(10^6)N/m^2 = 68.31MPa \tag{3}$$

$$(Q_z)_B = \frac{4}{3}[(0.06)^3 - (0.05)^3] = 60.67(10^{-6}) \ m^3 \qquad \tau_{bend-y} = \frac{(12)(10^3)(60.67)(10^{-6})}{5.270(10^{-6})(0.02)} = 6.91(10^6)N/m^2 = 6.91MPa \tag{4}$$

$(\sigma_{xx})_A = 68.31 \text{ MPa (T)} \qquad (\tau_{xz})_A = 0 \qquad (\sigma_{xx})_B = 0 \qquad (\tau_{xy})_B = -6.91 \text{ MPa}$ **(5)**

ANS $(\sigma_{xx})_A = 68.31 \text{ MPa (T)}$; $(\tau_{xz})_A = 0$; $(\sigma_{xx})_B = 0$; $(\tau_{xy})_B = -6.91 \text{ MPa}$

---

10. 24

Solution $P_x = 0$ $P_y = 0$ $P_z = 15\text{kN}$ Stresses at points A and B by inspection.

For only P_z non-zero, by inspection, the magnitude of internal forces, moments, and stresses can be written as follows.

$$|V_z| = 15kN \qquad T = (15)(0.3) = 4.5 \ kN-m \qquad |M_y| = (15)(0.5) = 7.5 \ kN-m \tag{1}$$

$$(\sigma_{xx})_A = 0 \qquad (\sigma_{xx})_B = -\sigma_{bend-y} \qquad (\tau_{xz})_A = \tau_{tor} + \tau_{bend-y} \qquad (\tau_{xy})_B = -\tau_{tor} \tag{2}$$

$$J = \frac{\pi}{32}(0.12^4 - 0.1^4) = 10.54(10^{-6}) \ m^4 \qquad \tau_{tor} = \frac{(4.5)(10^3)(0.06)}{10.54(10^{-6})} = 25.62(10^6) N/m^2 = 25.62 MPa \tag{3}$$

$$I_{yy} = \frac{\pi}{64}(0.12^4 - 0.1^4) = 5.270(10^{-6}) \ m^4 \qquad \sigma_{bend-z} = \frac{(7.5)(10^3)(0.06)}{5.270(10^{-6})} = 85.39(10^6) N/m^2 = 85.39 MPa \tag{4}$$

$$(Q_y)_B = \frac{4}{3}[(0.06)^3 - (0.05)^3] = 60.67(10^{-6}) \ m^3 \qquad \tau_{bend-z} = \frac{(15)(10^3)(60.67)(10^{-6})}{5.270(10^{-6})(0.02)} = 8.668(10^6) N/m^2 = 8.67 MPa \tag{5}$$

$$(\sigma_{xx})_A = 0 \qquad (\tau_{xz})_A = 25.62 + 8.67 = 34.29 \text{ MPa} \qquad (\sigma_{xx})_B = 85.39 \text{ MPa (C)} \qquad (\tau_{xy})_B = -25.62 \text{ MPa} . \tag{6}$$

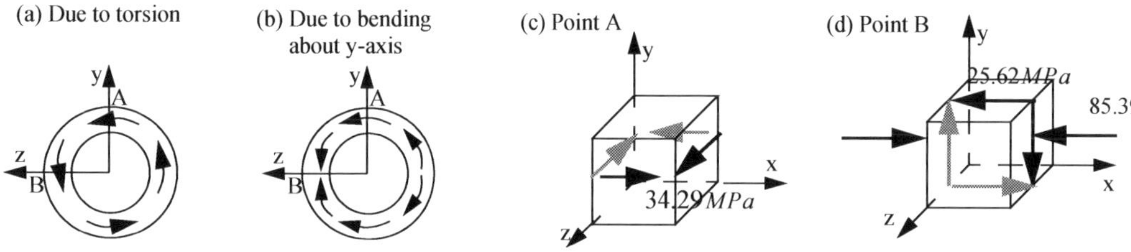

ANS $(\sigma_{xx})_A = 0$; $(\tau_{xz})_A = 34.29 \text{ MPa}$; $(\sigma_{xx})_B = 85.39 \text{ MPa (C)}$; $(\tau_{xy})_B = -25.62 \text{ MPa}$

---

10. 25

Solution $P_x = 9\text{kN}$ $P_y = 0$ $P_z = 0$ Stresses at points A and B by inspection.

For only P_x non-zero, by inspection, the magnitude of internal forces, moments, and stresses can be written as follows. All shear stresses are zero.

$$|N| = 9kN \qquad |M_y| = (9)(0.7) = 6.3 \ kN-m \qquad (\sigma_{xx})_A = \sigma_{axial} \qquad (\sigma_{xx})_B = \sigma_{axial} + \sigma_{bend-y} \tag{1}$$

$$A = \frac{\pi}{4}(0.12^2 - 0.1^2) = 3.4558(10^{-3}) \ m^2 \qquad \sigma_{axial} = \frac{(9)(10^3)}{3.4558(10^{-3})} = 2.604(10^6)(N/m^2) = 2.604 MPa \tag{2}$$

$$I_{yy} = \frac{\pi}{64}(0.12^4 - 0.1^4) = 5.270(10^{-6}) \ m^4 \qquad \sigma_{bend-z} = \frac{(6.3)(10^3)(0.06)}{5.270(10^{-3})} = 71.73(10^6)(N/m^2) = 71.73 MPa \tag{3}$$

$$(\sigma_{xx})_A = 2.604 \text{ MPa (T)} \qquad (\sigma_{xx})_B = 2.604 + 71.73 = 74.33 \text{ MPa (T)} . \tag{4}$$

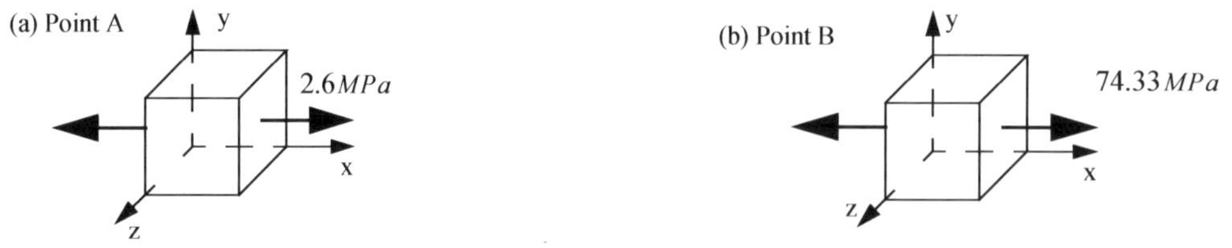

ANS $(\sigma_{xx})_A = 2.604 \text{ MPa (T)}$; $(\sigma_{xx})_B = 74.33 \text{ MPa (T)}$

---

10. 26

Solution $P_x = 0$ $P_y = 12 \text{ kN}$ $P_z = 0$ Stresses at points A and B by inspection.

For only P_y non-zero, by inspection, the magnitude of internal forces, moments, and stresses can be written as follows.

$$|V_y| = 12kN \qquad T = (15)(0.7) = 10.5 \ kN-m \qquad |M_z| = (12)(1.3) = 15.6 \ kN-m \tag{1}$$

$$(\sigma_{xx})_A = \sigma_{bend-z} \qquad (\sigma_{xx})_B = 0 \qquad (\tau_{xz})_A = \tau_{tor} \qquad (\tau_{xy})_B = -\tau_{tor} - \tau_{bend-z} \tag{2}$$

$$J = \frac{\pi}{32}(0.12^4 - 0.1^4) = 10.54(10^{-6}) \ m^4 \qquad \tau_{tor} = \frac{(10.5)(10^3)(0.06)}{10.54(10^{-6})} = 59.77(10^6) N/m^2 = 59.77 MPa \tag{3}$$

$$I_{zz} = \frac{\pi}{64}(0.12^4 - 0.1^4) = 5.270(10^{-6})\ m^4 \qquad \sigma_{bend-z} = \frac{(15.6)(10^3)(0.06)}{5.270(10^{-6})} = 177.6(10^6)N/m^2 = 177.6MPa \quad (4)$$

$$(Q_z)_B = \frac{4}{3}[(0.06)^3 - (0.05)^3] = 60.67(10^{-6})\ m^3 \qquad \tau_{bend-z} = \frac{(12)(10^3)(60.67)(10^{-6})}{5.270(10^{-6})(0.02)} = 6.91(10^6)N/m^2 = 6.91MPa \quad (5)$$

$$(\sigma_{xx})_A = 177.6\text{ MPa (T)} \qquad (\tau_{xz})_A = 59.77\text{ MPa} \qquad (\sigma_{xx})_B = 0 \qquad (\tau_{xy})_B = -59.77 - 6.91 = -66.68\text{ MPa} \quad (6)$$

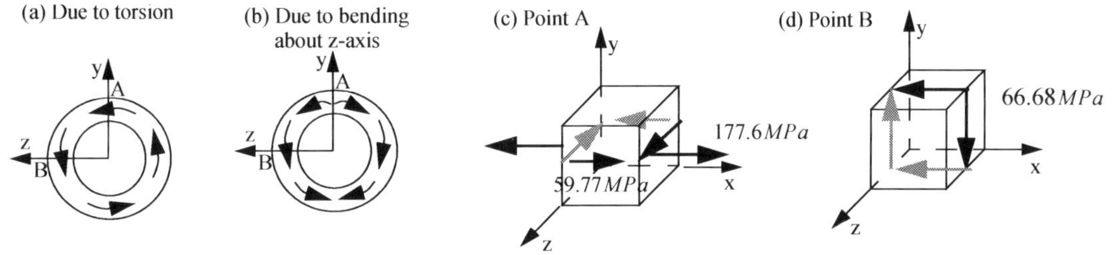

ANS $(\sigma_{xx})_A = 177.6$ MPa (T) ; $(\tau_{xz})_A = 59.77$ MPa ; $(\sigma_{xx})_B = 0$; $(\tau_{xy})_B = -66.68$ MPa

---

10. 27

Solution $P_x = 0$ $P_y = 0$ $P_z = 15$ kN Stresses at points A and B by inspection.

For only P_z non-zero, by inspection, the magnitude of internal forces, moments, and stresses can be written as follows.

$$|V_z| = 15kN \qquad |M_y| = (15)(1.3) = 19.5\ kN-m \quad (1)$$

$$(\sigma_{xx})_A = 0 \qquad (\sigma_{xx})_B = -\sigma_{bend-y} \qquad (\tau_{xz})_A = \tau_{bend-y} \qquad (\tau_{xy})_B = 0 \quad (2)$$

$$I_{yy} = \frac{\pi}{64}(0.12^4 - 0.1^4) = 5.270(10^{-6})\ m^4 \qquad \sigma_{bend-y} = \frac{(19.5)(10^3)(0.06)}{5.270(10^{-6})} = 222(10^6)N/m^2 = 222MPa \quad (3)$$

$$(Q_y)_B = \frac{4}{3}[(0.06)^3 - (0.05)^3] = 60.67(10^{-6})\ m^3 \qquad \tau_{bend-y} = \frac{(15)(10^3)(60.67)(10^{-6})}{5.270(10^{-6})(0.02)} = 8.668(10^6)N/m^2 = 8.67MPa \quad (4)$$

$$(\sigma_{xx})_A = 0 \qquad (\tau_{xz})_A = 8.67\text{ MPa} \qquad (\sigma_{xx})_B = 222\text{ MPa (C)} \qquad (\tau_{xy})_B = 0\ . \quad (5)$$

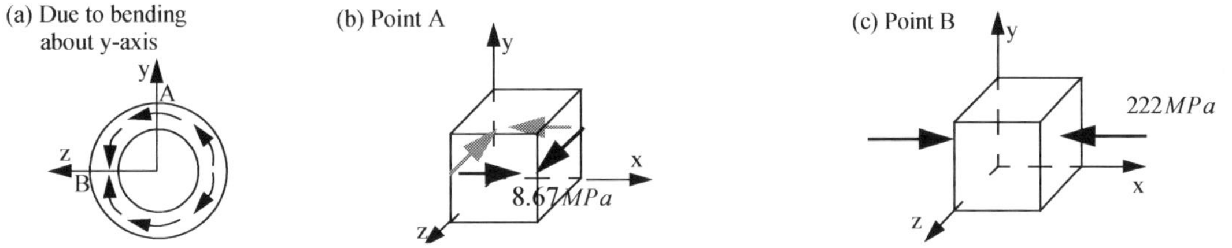

ANS $(\sigma_{xx})_A = 0$; $(\tau_{xz})_A = 8.67$ MPa ; $(\sigma_{xx})_B = 222$ MPa (C) ; $(\tau_{xy})_B = 0$

---

10. 28

Solution $P_x = 10$ kN $P_y = 0$ $P_z = 0$ Stresses at points A and B=?

For only P_x non-zero, by inspection, the magnitude of internal forces, moments, and stresses can be written as follows. All shear stresses are zero.

$$|N| = 10\ kN \qquad |M_z| = (10)(0.7) = 7\ kN-m \qquad (\sigma_{xx})_A = \sigma_{axial} + \sigma_{bend-z} \qquad (\sigma_{xx})_B = \sigma_{axial} \quad (1)$$

$$A = \frac{\pi}{4}(0.12^2 - 0.1^2) = 3.4558(10^{-3})\ m^2 \qquad \sigma_{axial} = \frac{(10)(10^3)}{3.4558(10^{-3})} = 2.894(10^6)(N/m^2) = 2.894MPa \quad (2)$$

$$I_{zz} = \frac{\pi}{64}(0.12^4 - 0.1^4) = 5.270(10^{-6})\ m^4 \qquad \sigma_{bend-z} = \frac{(7)(10^3)(0.06)}{5.270(10^{-3})} = 79.70(10^6)(N/m^2) = 79.70MPa \quad (3)$$

$$(\sigma_{xx})_A = 2.894 + 79.70 = 84.11MPa(T) \qquad (\sigma_{xx})_B = 79.70MPa(T) \quad (4)$$

The maximum normal stresses are the principal stress 1. The maximum shear stress is half the principal stress 1

ANS $(\sigma_{max})_A = 84.11$ MPa (T) $(\tau_{max})_A = 40.61$ MPa $(\sigma_{max})_B = 79.70$ MPa (T) $(\tau_{max})_B = 39.85$ MPa

---

10. 29

Solution $P_x = 0$ $P_y = 15$ kN $P_z = 0$ Stresses at points A and B=?

For only P_y non-zero, by inspection, the magnitude of internal forces, moments, and stresses can be written as follows.

$$|V_y| = 15kN \qquad |M_z| = (15)(0.9-0.5) = 6 \ kN-m \tag{1}$$

$$(\sigma_{xx})_A = -\sigma_{bend-z} \qquad (\sigma_{xx})_B = 0 \qquad (\tau_{xz})_A = 0 \qquad (\tau_{xy})_B = \tau_{bend-z} \tag{2}$$

$$I_{zz} = \frac{\pi}{64}(0.12^4 - 0.1^4) = 5.270(10^{-6}) \ m^4 \qquad \sigma_{bend-z} = \frac{(6)(10^3)(0.06)}{5.270(10^{-6})} = 68.31(10^6)N/m^2 = 68.31MPa \tag{3}$$

$$(Q_z)_B = \frac{4}{3}[(0.06)^3 - (0.05)^3] = 60.67(10^{-6}) \ m^3 \qquad \tau_{bend-z} = \frac{(15)(10^3)(60.67)(10^{-6})}{5.270(10^{-6})(0.02)} = 8.668(10^6)N/m^2 = 8.67MPa \tag{4}$$

(a) Due to bending about z-axis

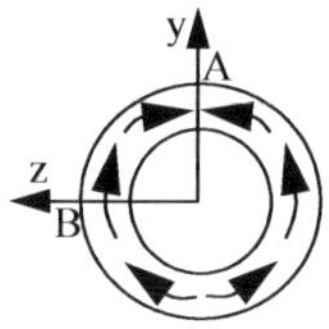

$$(\sigma_{xx})_A = 68.31MPa(C) \qquad (\tau_{xz})_A = 0 \qquad (\sigma_{xx})_B = 0 \qquad (\tau_{xy})_B = 8.67MPa \tag{5}$$

The maximum normal stresses are the principal stresses. These with maximum shear stress can be determined as shown below.

ANS $(\sigma_{max})_A = 68.3$ MPa (C) $\quad (\tau_{max})_A = 34.15$ MPa $\quad (\sigma_{max})_B = 17.27$ MPa (T) or (C) $\quad (\tau_{max})_B = 8.67$ MPa

---

10. 30

Solution $\quad P_x = 0 \quad P_y = 0 \quad P_z = 20$ kN $\quad$ Stresses at points A and B=?

For only P_z non-zero, by inspection, the magnitude of internal forces, moments, and stresses can be written as follows.

$$|V_z| = 20kN \qquad T = (20)(0.7) = 14 \ kN-m \qquad |M_y| = (20)(0.9-0.5) = 8 \ kN-m \tag{1}$$

$$(\sigma_{xx})_A = 0 \qquad (\sigma_{xx})_B = -\sigma_{bend-y} \qquad (\tau_{xz})_A = \tau_{tor} + \tau_{bend-y} \qquad (\tau_{xy})_B = -\tau_{tor} \tag{2}$$

(a) Due to torsion

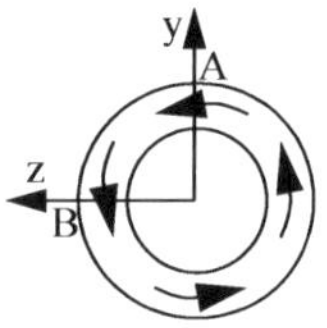

(b) Due to bending about y-axis

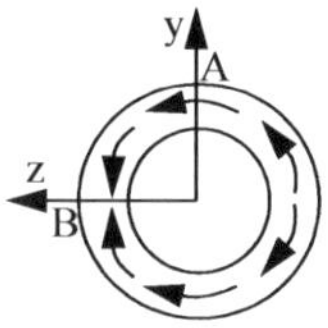

$$I_{yy} = \frac{\pi}{64}(0.12^4 - 0.1^4) = 5.270(10^{-6}) \ m^4 \qquad \sigma_{bend-y} = \frac{(8)(10^3)(0.06)}{5.270(10^{-6})} = 91.08(10^6)N/m^2 = 91.08MPa \tag{3}$$

$$J = \frac{\pi}{32}(0.12^4 - 0.1^4) = 10.54(10^{-6}) \ m^4 \qquad \tau_{tor} = \frac{(14)(10^3)(0.06)}{10.54(10^{-6})} = 79.70(10^6)N/m^2 = 79.70MPa \tag{4}$$

$$(Q_y)_A = \frac{2}{3}[(0.06)^3 - (0.05)^3] = 60.67(10^{-6}) \ m^3 \qquad \tau_{bend-y} = \frac{(20)(10^3)(60.67)(10^{-6})}{5.270(10^{-6})(0.02)} = 11.51(10^6)N/m^2 = 11.51MPa \tag{5}$$

$$(\sigma_{xx})_A = 0 \qquad (\tau_{xz})_A = 79.70 + 11.51 = 91.21MPa \qquad (\sigma_{xx})_B = 91.08MPa(C) \qquad (\tau_{xy})_B = -79.70MPa \tag{6}$$

The maximum normal stresses are the principal stresses. These with maximum shear stress can be determined as shown below.

$$(\sigma_{max})_A = 91.21 \text{ MPa (T) or (C)} \qquad (\tau_{max})_A = 102.7 \text{ MPa} \tag{7}$$

$$(\sigma_2)_B = -\frac{91.08}{2} - \sqrt{\left(\frac{91.08}{2}\right)^2 + 79.70^2} = -45.54 - 91.79 = -137.33MPa \qquad (\sigma_{max})_B = 137.33 \text{ MPa (C)} \tag{8}$$

$$(\tau_{max})_B = 91.79 \text{ MPa} \tag{9}$$

ANS $(\sigma_{max})_A = 102.7$ MPa (T) or (C) ; $(\tau_{max})_A = 91.21$ MPa ; $(\sigma_{max})_B = 137.33$ MPa (C) ; $(\tau_{max})_B = 91.79$ MPa

---

10. 31

Solution $\quad d_0 = 2$ in $\quad t = (1/4)in$, $\quad$ Stresses at points A and B =?

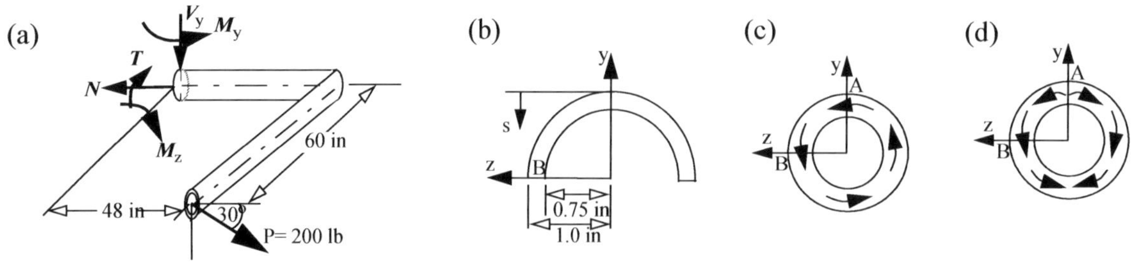

By equilibrium of forces and moments we obtain.

$$N = 200cos(30) = 173.2lb \qquad V_y = (-200)sin(30) = -100\ lb \qquad T = [200sin(30)](60) = 6000in-lb \quad (1)$$

$$M_y = [(-200)cos(30)](60) = -10392.3in-lb \qquad M_z = [(-200)sin(30)](48) = -4800in-lb \quad (2)$$

$$\text{Axial:} \quad A = \frac{\pi}{4}[(2)^2-(1.5)^2] = 1.374in^2 \qquad (\sigma_{xx})_{A,B} = \frac{N}{A} = \frac{173.2}{1.374} = 126psi(T) \quad (3)$$

$$\text{Bending about z-axis} \quad I_{zz} = \frac{J}{2} = 0.5369in^4 \qquad (\sigma_{xx})_A = \left|\frac{M_z y_A}{I_{zz}}\right| = \frac{(4800)(1)}{0.5369} = 8940psi(T) \qquad (\sigma_{xx})_B = 0 \quad (4)$$

$$\text{Bending about y-axis} \quad I_{yy} = \frac{J}{2} = 0.5369in^4 \qquad (\sigma_{xx})_A = 0 \qquad (\sigma_{xx})_B = \left|\frac{M_y z_A}{I_{yy}}\right| = \frac{(10392.3)(1)}{0.5369} = 19356psi(T) \quad (5)$$

$$\text{Total:} \quad (\sigma_{xx})_A = 126+8940 = 9066\text{ psi (T)} \qquad (\sigma_{xx})_B = 126+19356 = 19482\text{ psi (T)} \quad (6)$$

$$(Q_z)_B = \frac{2}{3}[(1)^3-(0.75)^3] = 0.3854in^3 \qquad (\tau_{xz})_A = 0 \qquad (\tau_{xy})_B = -\left|\frac{V_y(Q_z)_B}{I_{zz}t}\right| = \frac{-(100)(0.3854)}{(0.5369)(0.5)} = -143.6psi \quad (7)$$

$$J = \frac{\pi}{32}[2^4-1.5^4] = 1.074in^4 \qquad \tau_{tor} = \left|\frac{T\rho_{max}}{J}\right| = \frac{6000(1)}{1.074} = 5587.6psi \qquad (\tau_{xz})_A = 5587.6psi \qquad (\tau_{xy})_B = -5587.6psi \quad (8)$$

$$\text{Total:} \quad (\tau_{xz})_A = 5588\text{ psi} \qquad (\tau_{xy})_B = (-143.6-5587.6) = -5731.2\text{ psi} \quad (9)$$

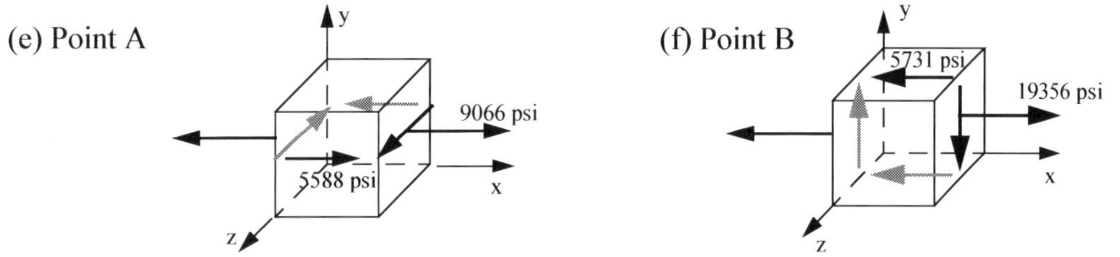

ANS $(\sigma_{xx})_A = 9066$ psi (T); $(\tau_{xz})_A = 5588$ psi; $(\sigma_{xx})_B = 19482$ psi (T); $(\tau_{xy})_B = -5587.6psi$

---

10. 32

Solution: σ_{max} =? τ_{max} =?

From previous problem we have the following.

$$(\sigma_{xx})_B = 19482psi(T) \qquad (\tau_{xy})_B = -5731psi \quad (1)$$

The principal stresses and maximum shear stress are found asshown below.

$$\sigma_{1,2} = \left(\frac{19482}{2}\right) \pm \sqrt{\left(\frac{19482}{2}\right)^2 + (5731)^2} = 9741 \pm 11302 \qquad \sigma_{max} = \sigma_1 = 21043psi \qquad \tau_{max} = \left|\frac{\sigma_1-\sigma_2}{2}\right| = 11302\text{ psi} \quad (2)$$

ANS $\sigma_{max} = 21043$ psi (T); $\tau_{max} = 11302$ psi

---

10. 33

Solution: $d_0 = 40$ mm, t = 10 mm, stresses at A and B =?

We determine the sign of stresses by inspection. By equilibrium of forces and moments we obtain

$$N = 10cos(15) = 9.659kN \qquad V_y = 10sin(15) = 2.588kN \qquad T = (-10sin(15))(0.1) = -0.2588kN-m \quad (1)$$

$$M_y = (-10cos(15))(0.1) = -0.9659kN-m \qquad M_z = (10)sin(15)(0.65) = 1.6823kN-m \quad (2)$$

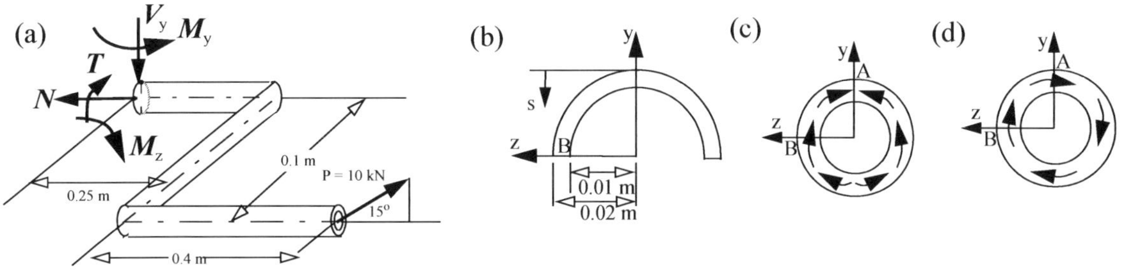

$$J = \frac{\pi[0.04^4-0.02^4]}{32} = 0.2356(10^{-6})m^4 \qquad I_{yy} = I_{zz} = \frac{J}{2} = 0.1178(10^{-6})m^4 \qquad (Q_z)_B = \frac{2}{3}(0.02^3-0.01^3) = 4.667(10^{-6})m^3 \quad (3)$$

$$\text{Axial} \quad A = \frac{\pi}{4}[0.04^2-0.02^2] = 0.9425(10^{-3})m^2 \qquad (\sigma_{xx})_{A,B} = \left|\frac{N}{A}\right| = \frac{9.659(10^3)}{0.9425(10^{-3})} = 10.25(10^6)\frac{N}{m^2} = 10.25MPa(T) \quad (4)$$

Bending about z-axis $(\sigma_{xx})_A = \left|\dfrac{M_z y_A}{I_{zz}}\right| = \dfrac{(1.6823)(10^3)(0.02)}{(0.1178)(10^{-6})} = 285.6(10^6)\dfrac{N}{m^2} = 285.6MPa(C)$ $\quad (\sigma_{xx})_B = 0$ (5)

Bending about the y-axis $(\sigma_{xx})_A = 0 \quad (\sigma_{xx})_B = \left|\dfrac{M_y z_B}{I_{yy}}\right| = \dfrac{(0.9659)(10^3)(0.02)}{(0.1178)(10^{-6})} = 163.98(10^6)\dfrac{N}{m^2} = 164MPa(T)$ (6)

Total: $(\sigma_{xx})_A = 10.25 - 285.6 = -275.35$ MPa) $\quad (\sigma_{xx})_B = 10.25 + 164 = 174.25$ MPa (7)

$$(\tau_{xz})_A = 0 \qquad (\tau_{xy})_B = \left|\frac{V_y(Q_t)_B}{I_{zz}t}\right| = \frac{(2.588)(10^3)(4.667)(10^{-6})}{(0.1178)(10^{-6})(0.02)} = 5.13(10^6)\frac{N}{m^2} = 5.13MPa \tag{8}$$

$$\tau_{tor} = \left|\frac{T\rho_{max}}{J}\right| = \frac{(0.2588)(10^3)(0.02)}{0.2356(10^{-6})} = 21.97(10^6)\frac{N}{m^2} = 21.97MPa \qquad (\tau_{xz})_A = -21.97MPa \qquad (\tau_{xy})_B = 21.97MPa \tag{9}$$

Total $(\tau_{xz})_A = -22$ MPa $\quad (\tau_{xy})_B = (5.13 + 21.97) = 27.1$ MPa . (10)

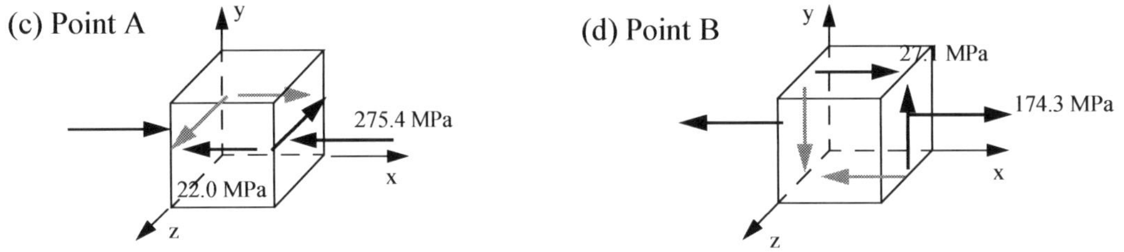

ANS $(\sigma_{xx})_A = 275.35$ MPa (C) ; $(\tau_{xz})_A = -22$ MPa ; $(\sigma_{xx})_B = 174.25$ MPa (T) ; $(\tau_{xy})_B = 27.1$ MPa

---

10. 34

Solution: σ_{max} =? $\qquad \tau_{max}$ =?

From previous problem we have the following.

$$(\sigma_{xx})_B = 174.3MPa \qquad (\tau_{xy})_B = 27.1MPa \tag{1}$$

The principal stresses and maximum shear stress can be found as shown below.

$$\sigma_{1,2} = \frac{174.3}{2} \pm \sqrt{\frac{174.3^2}{4} + 27.1^2} = 87.15 \pm 91.27 \qquad \sigma_{max} = \sigma_1 = 178.42MPa \qquad \tau_{max} = \left|\frac{\sigma_1 - \sigma_2}{2}\right| = 91.3 \text{ MPa} \tag{2}$$

ANS $\sigma_{max} = 178.4$ MPa (T) ; $\tau_{max} = 91.3$ MPa

---

10. 35

Solution $d_0 = 2$ in, $t = (1/4)in$, Stresses at point A =?

We determine the sign of stresses by inspection. By equilibrium of forces and moments we obtain

$$N = 1000lb \qquad V_y = 200lb \qquad V_z = 800lb \tag{1}$$

$$T = (800)(10) = 8000in-lb \qquad M_z = (200)(16) - (1000)(10) = (-6800)in-lb \qquad M_y = (800)(16) = 12800in-lb \tag{2}$$

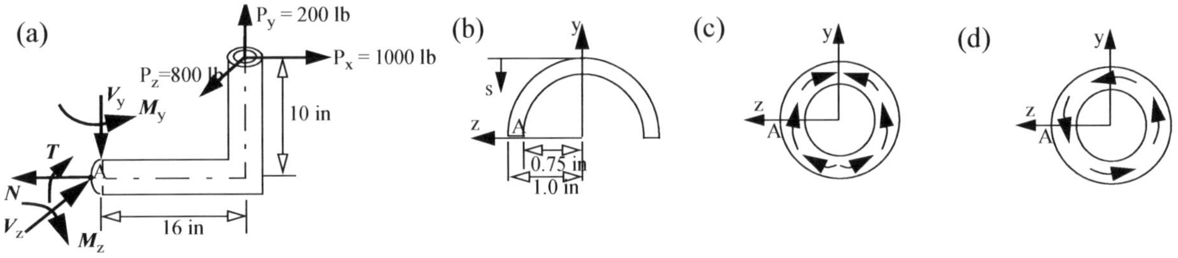

$$J = \frac{\pi[2^4 - 1.5^4]}{32} = 1.0738in^4 \qquad I_{yy} = I_{zz} = 0.5369in^4 \qquad (Q_z)_A = \frac{2}{3}(1^3 - 0.75^3) = 0.3854in^3 \tag{3}$$

$$\text{Axial} \quad A = \frac{\pi[2^2 - 1.5^2]}{4} = 1.3745in^2 \qquad (\sigma_{xx})_A = \left|\frac{N}{A}\right| = \frac{1000}{1.3745} = 728psi \tag{4}$$

$$\text{Bending about the z-axis} \quad (\sigma_{xx})_A = \left|\frac{M_y z_A}{I_{yy}}\right| = \frac{12800(1)}{0.5369} = 23841psi(C) \tag{5}$$

$$\text{Total} \quad (\sigma_{xx})_A = -23841 + 728 = -23113psi \tag{6}$$

$$(\tau_{xy})_A = \left|\frac{V_y(Q_z)_A}{I_{zz}t}\right| = \frac{(200)(0.3854)}{(0.5369)(0.5)} = 287psi \qquad (\tau_{xy})_A = -\left|\frac{T\rho}{J}\right| = \frac{-(8000)(1)}{1.0738} = -7450psi \tag{7}$$

Total $(\tau_{xy})_A = 287 - 7450 = -7163psi$ **(8)**

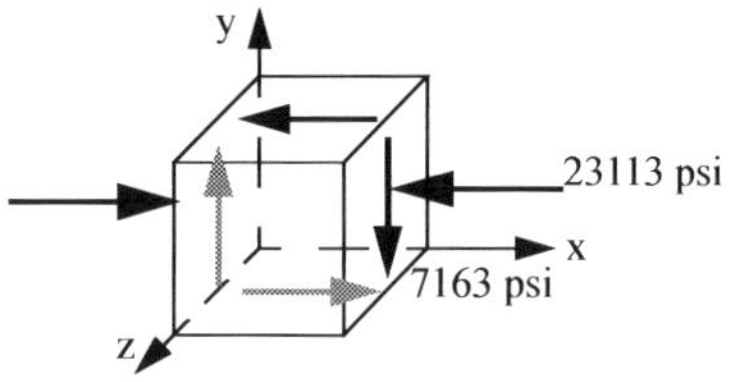

ANS $(\sigma_{xx})_A = 23113psi(C)$; $(\tau_{xy})_A = -7163psi$

---

10. 36

Solution σ_{nn} =? τ_{nt} =?

From previous problem we have the stresses shown below on a stress cube with the inclined plane.

$$(\sigma_{xx})_A = -23113psi \qquad (\tau_{xy})_A = -7163psi \quad \textbf{(1)}$$

(a)

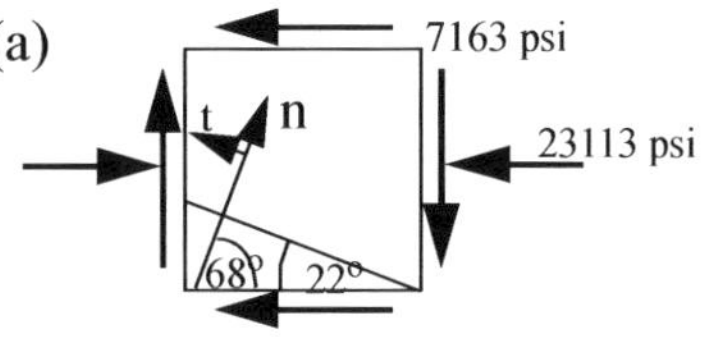

Substituting the stresses and the angle of 68° into stress transformation equations we obtain the following.

$\sigma_{nn} = -23113cos^2 68 + 2(-7163)cos68sin68 = -8219psi$ $\tau_{nt} = 23113cos68sin68 + (-7163)[cos^2 68 - sin^2 68] = 13180psi$

ANS $\sigma_{nn} = 8219psi(C)$; $\tau_{nt} = 13180psi$

---

10. 37

Solution: d_o = 4 in, d_i = 3 in, σ_{max} =?, τ_{max} =?

By symmetry the reaction force at each wall is half the total force i.e. $R_A = R_B$ = 1600lb. The magnitude of the torque on each pulley is: $T_{ext} = (1200)(12) - (400)(12) = 9600in-lb$. Figure below shows the shaft with equivalent loads and the shear forcebending moment diagram.

(a)

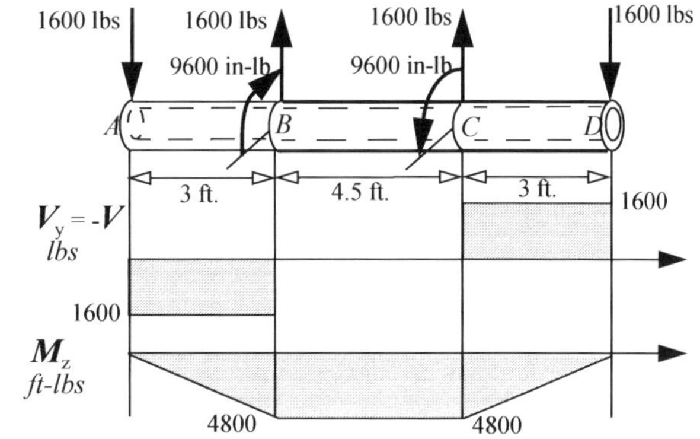

$$J = \frac{\pi}{32}[4^4 - 3^4] = 17.18in^4 \qquad \tau_{tor} = \left|\frac{T\rho_{max}}{J}\right| = \frac{(9600)(2)}{17.18} = 1118psi \quad \textbf{(2)}$$

$$I_{zz} = \frac{J}{2} = 8.59in^4 \qquad \sigma_{bend} = \left|\frac{M_z y_{max}}{I_{zz}}\right| = \frac{(4800)(12)(2)}{8.59} = 13411psi \quad \textbf{(3)}$$

$$\sigma_{1,2} = \frac{13411}{2} \pm \sqrt{\left(\frac{13411}{2}\right)^2 + 1118^2} = 6705.5 \pm 6798.1 \qquad \sigma_{max} = \sigma_1 = 13504 \text{ psi} \qquad \tau_{max} = \left|\frac{\sigma_1 - \sigma_2}{2}\right| = 6798.1 \text{ psi} \quad \textbf{(4)}$$

ANS σ_{max} = 13504 psi (T); τ_{max} = 6798 psi

---

10. 38

Solution d_o = 20 mm t = 15 mm Stresses at points A and B =?

By equilibrium we obtain:

$$\boldsymbol{T} = 7 - 9 = -2 \text{ kN-m} \qquad V_y = -20 + 10 = -10 \text{ kN} \qquad \boldsymbol{M_z} + 20(0.8) - 10(0.4) = 0 \quad or \quad \boldsymbol{M_z} = -12 \text{ kN-m} \quad \textbf{(1)}$$

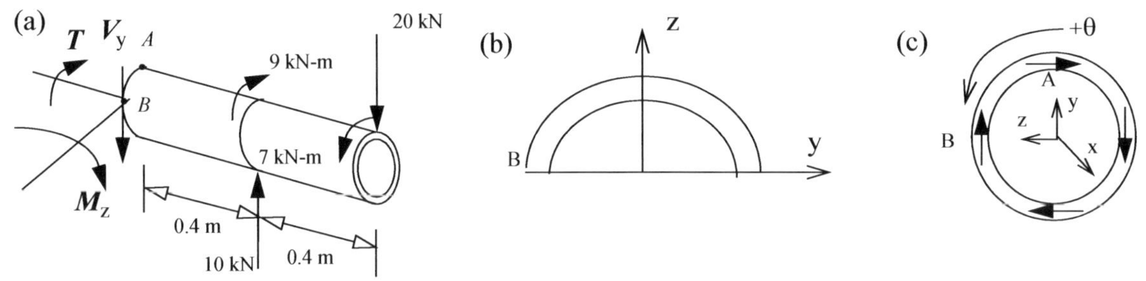

$$J = \frac{\pi}{32}[0.12^4 - 0.09^4] = 13.916(10^{-6})\ \text{m}^4 \qquad I_{zz} = \frac{J}{2} = 6.958(10^{-6})\ \text{m}^4 \qquad (Q_z)_B = \frac{2}{3}(0.06^3 - 0.045^3) = 83.25(10^{-6})\ \text{m}^3 \quad \textbf{(2)}$$

$$\tau_{x\theta} = \frac{T\rho}{J} = \frac{(-2)(10^3)(0.06)}{13.916(10^{-6})} = (-8.623)(10^6)N/m^2 \qquad (\tau_{xz})_A = -8.623\ \text{MPa} \qquad (\tau_{xy})_B = 8.623\ \text{MPa} \quad \textbf{(3)}$$

$$(\sigma_{xx})_A = \frac{-M_z y_A}{I_{zz}} = \frac{-(-12)(10^3)(0.06)}{6.958(10^{-6})} = 103.48(10^6)N/m^2 = 103.48\ \text{MPa(T)} \qquad (\sigma_{xx})_B = 0 \quad \textbf{(4)}$$

$$-(\tau_{xy})_B = (\tau_{xs})_B = \frac{-V_y Q_z}{I_{zz}t} = \frac{-(-10)(10^3)(83.25)(10^{-6})}{[6.958(10^{-6})](0.03)} = 3.988(10^6)N/m^2 \quad or \quad (\tau_{xy})_B = -3.988\ \text{MPa} \quad \textbf{(5)}$$

$$\text{Total} \qquad (\tau_{xz})_A = -8.62\ \text{MPa} \qquad (\tau_{xy})_B = (8.623 - 3.988)\ \text{MPa} = 4.635\ \text{MPa} \quad \textbf{(6)}$$

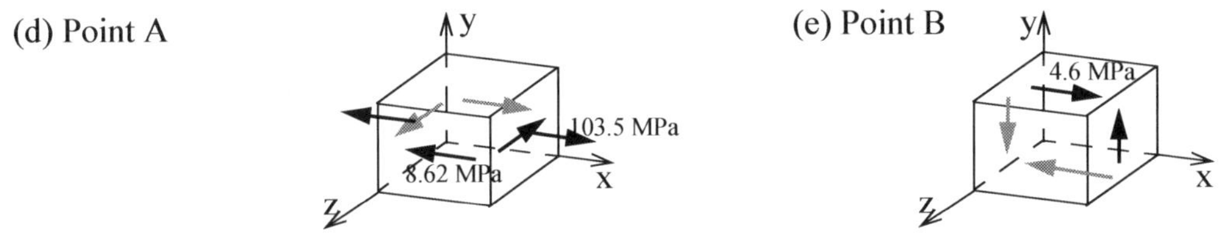

ANS $(\sigma_{xx})_A = 103.5\ \text{MPa (T)}$; $(\sigma_{xx})_B = 0$; $(\tau_{xz})_A = -8.62\ \text{MPa}$; $(\tau_{xy})_B = 4.6\ \text{MPa}$

---

10. 39

Solution 6 in. x 1 in $\sigma_{max} = ?$ $\tau_{max} = ?$

We draw the axial force and the shear-moment diagrams as shown below..

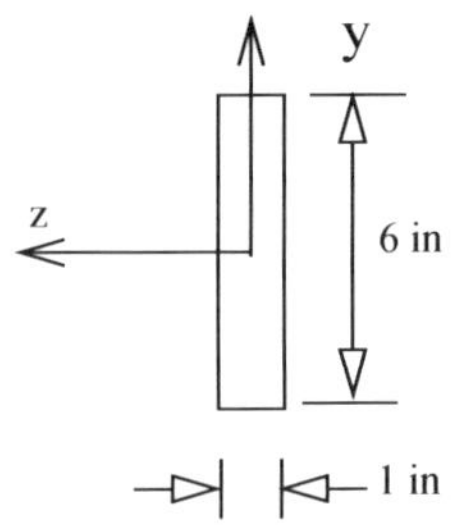

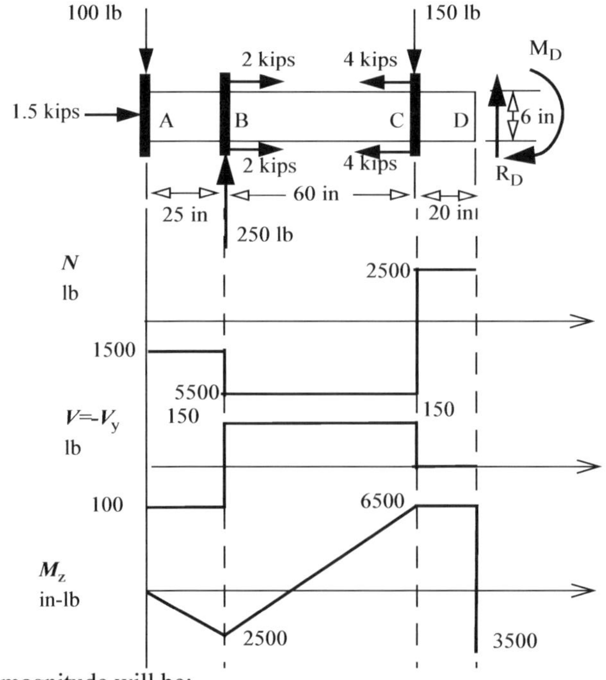

The maximum σ_{xx} will be just before C and its magnitude will be:

$$(\sigma_{xx})_{max} = \left|\frac{N}{A}\right| + \left|\frac{M_z y_{max}}{I_{zz}}\right| = \frac{5500}{(6)(1)} + \frac{6500(3)}{(1)(6^3)/12} = 2000\ \text{psi} \quad \textbf{(1)}$$

As $(\sigma_{xx})_{max} \gg (\tau_{xy})_{max}$ in bending, the maximum normal and shear stress will be $(\sigma_{xx})_{max}$ and $(\sigma_{xx})_{max}/2$, respectively.

ANS $\sigma_{max} = 2$ ksi ; $\tau_{max} = 1$ ksi

---

10. 40

Solution p = 300 psi, $R_0 = 10$ in, $t = 0.25$, Stresses at A =?

By equilibrium of moment $\boldsymbol{T}$ = -200 in-kips.

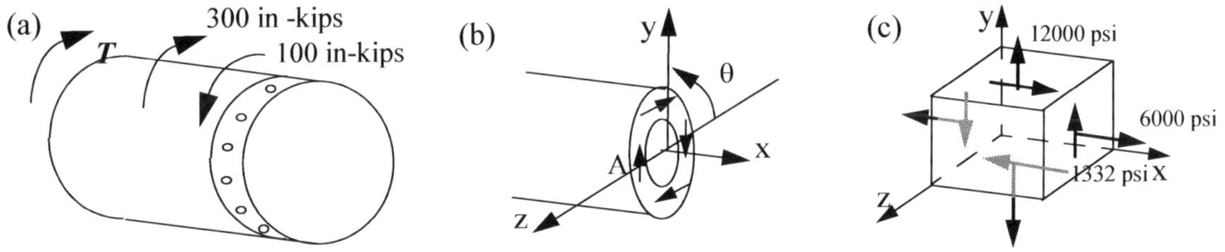

$$J = \frac{\pi}{2}[10^4 - 9.75^4] = 1512.9 in^4 \qquad \tau_{x\theta} = \left|\frac{T\rho}{J}\right| = \frac{(-200)(10)}{1512.9} = -1.322 ksi = -1322 \text{ psi} \qquad (\tau_{xy})_A = 1322 psi \qquad \textbf{(1)}$$

$$\sigma_{\theta\theta} = \frac{pr}{t} = \frac{(300)(10)}{0.25} = 12000 \text{ psi} \qquad \sigma_{xx} = \frac{pr}{2t} = 6000 \text{ psi} \qquad \textbf{(2)}$$

Fig. (c) shows the stress cube at point A.

ANS $\sigma_{xx} = 6000$ psi ; $\sigma_{yy} = 12000$ psi ; $\tau_{xy} = 1322$ psi

---

Section 10.2

10. 41

Solution $d_o = 5$ ft t = 6 in $\gamma = 120$ lb/ft^3 H= 30 ft p_{max} =? for no tensile stress.

The maximum tensile bending stress will be at the bottom of the chimney. By equilibrium

$$\boldsymbol{N} = W = \gamma A H \qquad \boldsymbol{M}_z = \frac{p d_o H^2}{2} = \frac{(p)(5)(30)^2}{2} = 2250p \text{ ft-lb} \qquad \textbf{(1)}$$

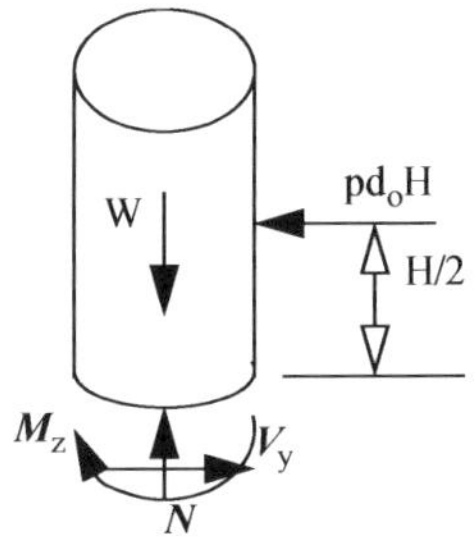

$$A = \frac{\pi}{4}(d_o^2 - d_i^2) = \frac{\pi}{4}(5^2 - 4^2) = 7.0686 \text{ ft}^2 \qquad \sigma_{axial} = \frac{\boldsymbol{N}}{A} = \frac{\gamma A H}{A} = \gamma H = (120)(30) = 3600 \text{ lb/ft}^2 (C) \qquad \textbf{(2)}$$

$$I = \frac{\pi}{64}(d_o^4 - d_i^4) = \frac{\pi}{64}(5^4 - 4^4) = 18.11 \text{ ft}^4 \qquad \sigma_{bend} = \frac{\boldsymbol{M}_z y_{max}}{I} = \frac{[2250p](5/2)}{18.11} = 310.54p \text{ lb/ft}^2 (T) \qquad \textbf{(3)}$$

$$\sigma_{axial} > \sigma_{bend} \quad or \quad 3600 > 310.54p \quad or \quad p < \frac{3600}{310.54} \quad or \quad p < 11.59 \text{ lb/ft}^2 \qquad \textbf{(4)}$$

ANS $p_{max} = 11.5$ lb/ft^2

---

10. 42

Solution $d_o = 1.5$ m t = 150 mm $\gamma = 1800$ kg/m^3 $p = 800$ Pa H_{max} =? for no tensile stress.

From previous problem we have the following.

$$\boldsymbol{N} = W = \gamma A H \qquad \boldsymbol{M}_z = \frac{p d_o H^2}{2} = \frac{(800)(1.5)H^2}{2} = 600H^2 \qquad \textbf{(1)}$$

$$\sigma_{axial} = \frac{\boldsymbol{N}}{A} = \frac{\gamma A H}{A} = \gamma H = (1800)(9.81)H = 17658H \text{ N}/m^2 (C) \qquad \textbf{(2)}$$

$$I = \frac{\pi}{64}(d_o^4 - d_i^4) = \frac{\pi}{64}(1.5^4 - 1.2^4) = 0.1467 \text{ m}^4 \qquad \sigma_{bend} = \frac{\boldsymbol{M}_z y_{max}}{I} = \frac{[600H^2](1.5/2)}{0.1467} = 3067.1H^2 \text{ N}/m^2 (T). \qquad \textbf{(3)}$$

$$\sigma_{axial} > \sigma_{bend} \quad or \quad 17658H > 3067.1H^2 \quad or \quad H < \frac{17658}{3067.1} \quad or \quad H < 5.757 \text{ m} \qquad \textbf{(4)}$$

ANS $H_{max} = 5.75$ m

---

10. 43

Solution: $d_o = 100$ mm, $d_i = 50$ mm, $\sigma_{max} \le 200$MPa, $\tau_{max} \le 115$ MPa, E = 200 GPa G = 80 GPa, $\nu = 0.25$ T_{max} =?, ε_A =?,

The axial and torsional shear stress can be found as shown below.

$$A = \frac{\pi}{4}[(0.1)^2 - (0.05)^2] = 5.89(10^{-3})m^2 \qquad \sigma_{xx} = \frac{800(10^3)}{5.89(10^{-3})} = 135.8(10^6)\frac{N}{m^2} = 135.8MPa(C) \quad (1)$$

$$J = \frac{\pi}{32}[(0.1)^4 - (0.05)^4] = 9.204(10^{-6})m^4 \qquad \tau_{tor} = \left|\frac{T\rho}{J}\right| = \frac{T(10^3)(0.05)}{9.204(10^{-6})} = 5.432(T)MPa = (\tau_{xy})_A \quad (2)$$

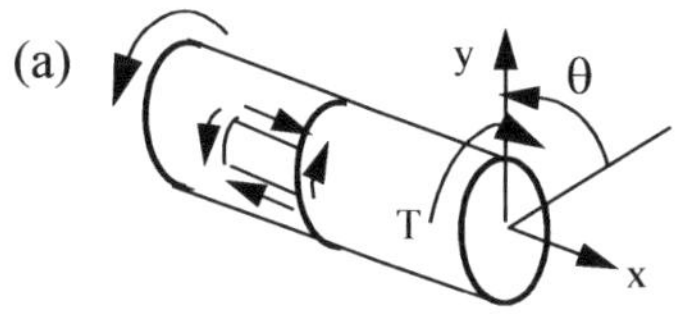

$$\sigma_{max} = \left|\left(\frac{-135.8}{2}\right) - \sqrt{\left(\frac{-135.8}{2}\right)^2 + (5.432T)^2}\right| \le 200 \quad or \quad \sqrt{\left(\frac{135.8}{2}\right)^2 + (5.432T)^2} \le 132.1 \quad (3)$$

$$\tau_{max} = \sqrt{\left(\frac{135.8}{2}\right)^2 + (5.432T)^2} \le 115 \quad or \quad \left(\frac{135.8}{2}\right)^2 + (5.432T)^2 \le 13225 \quad or \quad T \le 17.1 \quad T_{max} = 17.0 \text{ in.-kips} \quad (4)$$

$$\varepsilon_{xx} = \frac{(-135.8)(10^6)}{200(10^9)} = -679\mu \qquad \varepsilon_{yy} = -\nu\varepsilon_{xx} = 169.75\mu \qquad \gamma_{xy} = \frac{\tau_{xy}}{G} = \frac{(5.432)(17)(10^6)}{(80)(10^9)} = 1154\mu \quad (5)$$

Substituting the above strains and $\theta = -35^o$ in strain transformation equation we obtain

$$\varepsilon_A = -(679)(cos(-35))^2 + (169.75)(sin(-35))^2 + 1154(cos(-35)sin(-35)) = -942\mu \quad (6)$$

ANS $T_{max} = 17.0$ in.-kips ; $\varepsilon_A = -942\mu$

---

10. 44

Solution: $\sigma_T \le 160$ MPa (T) $\sigma_C \le 120$ MPa (C)

By equilibrium of forces and moments and geometry we obtain the following.

$$\boldsymbol{N} = P \qquad \boldsymbol{M}_z = -0.054P \qquad A = (12)(6) + (18)(6) + (18)(6) = 288mm^2 \quad (1)$$

$$I_{zz} = \frac{1}{12}(12)(6^3) + (12)(6)(16.5-3)^2 + \frac{1}{12}(6)(18^3) + (6)(18)(15-16.5)^2 + \frac{1}{12}(18)(6^3) + (18)(6)(27-16.5)^2 = 28.728(10^3)mm^4 \quad (2)$$

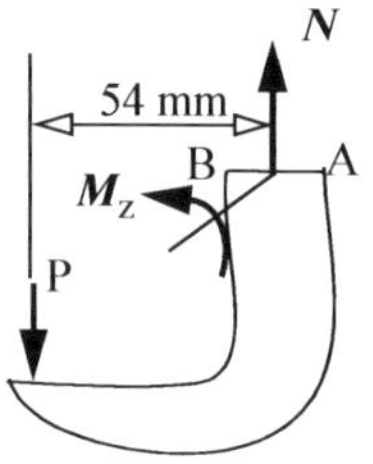

$$(\sigma_{xx})_{A,B} = \frac{\boldsymbol{N}}{A} = \frac{P}{288(10^{-6})} = 3.472P(10^3)\frac{N}{m^2}(T) \qquad (\sigma_{xx})_A = \left|\frac{\boldsymbol{M}_z y_A}{I_{zz}}\right| = \frac{(0.054P)(16.5)(10^{-3})}{28.728(10^{-9})} = 31.01P(10^3)\frac{N}{m^2}(C) \quad (3)$$

$$(\sigma_{xx})_B = \left|\frac{\boldsymbol{M}_z y_B}{I_{zz}}\right| = \frac{(0.054P)(30-16.5)(10^{-3})}{28.728(10^{-9})} = 25.376P(10^3)\frac{N}{m^2}(T) \quad (4)$$

$$(\sigma_{xx})_A = (31.01P - 3.472P)(10^3) = 27.538P(10^3) \le 120(10^6) \quad or \quad P \le 4.357(10^3)N \quad (5)$$

$$(\sigma_{xx})_B = (25.376P + 3.472P)(10^3) = 28.848P(10^3) \le 160(10^6) \quad or \quad P \le 5.546(10^3)N \quad (6)$$

ANS $P_{max} = 4.3kN$

---

10. 45

Solution: $\sigma_G \le 20$MPa, $\tau_G \le 12$ MPa, P_{max} =?

By equilibrium of forces and moment and geometry we obtain the following.

$$N = P\cos 55 \qquad V_y = -P\sin 55 \qquad M_z = (-P\sin 55)(3) = -2.457P \tag{1}$$

$$\eta_c = \frac{(50)(250)(125) + (50)(250)(275)}{(50)(250) + (50)(250)} = 200mm \qquad A = (50)(250) + (50)(250) = 25(10^3)mm^2 \tag{2}$$

$$I_{zz} = \frac{1}{12}(50)(250^3) + (50)(250)(200 - 125)^2 + \frac{1}{12}(250)(50^3) + (50)(250)(275 - 200)^2 = 208.3(10^6)mm^4 \tag{3}$$

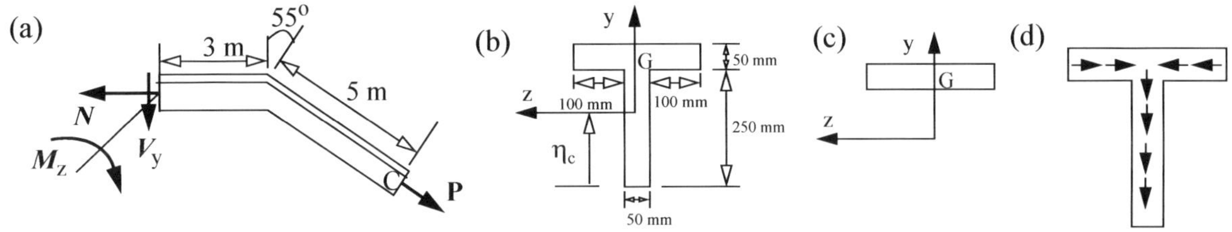

The stresses in glue at section AA can be found as shown below.

$$(\sigma_{xx})_G = \frac{N}{A} + \left|\frac{M_z y_G}{I_{zz}}\right| = \frac{0.5736P}{25(10^{-3})} + \frac{(2.457P)(250 - 200)(10^{-3})}{208.3(10^{-6})} = 22.943P + 589.77P = 612.72P \tag{4}$$

$$(Q_z)_G = (250)(50)(75) = 937.5(10^3)mm^3 \qquad (\tau_{xy})_G = \left|\frac{V_y(Q_t)_G}{I_{zz}t}\right| = \frac{(0.8191P)(937.5)(10^{-6})}{(208.3)(10^{-6})(50)(10^{-3})} = 73.73P \tag{5}$$

$$\sigma_{max} = \left(\frac{612.72P}{2}\right) + \sqrt{\left(\frac{612.72P}{2}\right)^2 + (73.73P)^2} = (306.4 + 315.1)P = 621.5P \le 20(10^6) \qquad or \qquad P \le 32.18(10^3)N \tag{6}$$

$$\tau_{max} = \sqrt{\left(\frac{612.72P}{2}\right)^2 + (73.73P)^2} = 315.1P \le 12(10^6) \qquad or \qquad P \le 38.08(10^3)N \tag{7}$$

ANS $P_{max} = 32.1$ kN

---

10. 46

Solution: $\tau_{max} < 90$ MPa, P = 40 kN, d_{BD} =?, d_{AB} =? to nearest 5 mm

By equilibrium of forces and moment about point A, we obtain the following.

$$(N_{BD}\sin\theta)(1) - P(1.5) = 0 \qquad or \qquad N_{BD} = 78.10kN \qquad A_y = N_{BD}\sin\theta - P = 20kN \qquad A_x = N_{BD}\cos\theta = 50kN \tag{1}$$

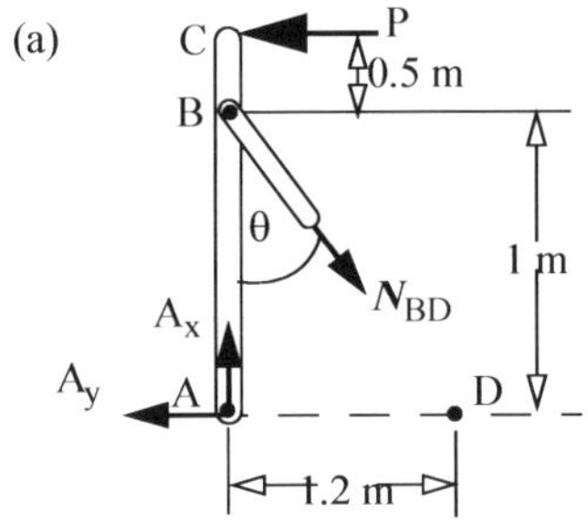

$\tan\theta = 1.2$ or $\theta = 50.194°$

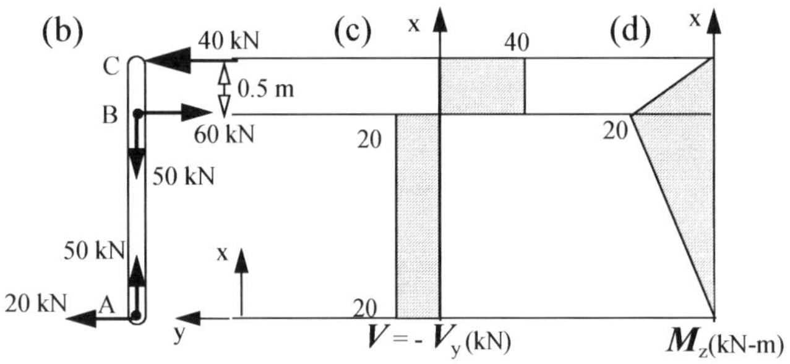

$$(\tau_{BD})_{max} = \frac{\sigma_{BD}}{2} = \frac{1}{2}\frac{N_{BD}}{A_{BD}} = \frac{1}{2}\frac{(78.10)(10^3)}{\pi d_{BD}^2/4} \le 90(10^6) \qquad or \qquad d_{BD} \ge 0.0235m \qquad d_{BD} = 25 \text{ mm} \tag{2}$$

The minimum diameter to the nearest 5 mm is

Fig (b) shows the forces in component form that are acting on member ABC. Figs.(b) and (c) show the shear force and bending moment diagrams. Noting that the axial stress in segment AB is compressive, the maximum normal stress is σ_{xx} is sum of maximum bending normal stress and axial stress.

$$\sigma_{AB} = \left|\frac{N_{AB}}{A_{AB}}\right| + \left|\frac{M_z(y_{AB})_{max}}{I_{AB}}\right| = \frac{(50)(10^3)}{\pi d^2_{AB}/4} + \frac{(20)(10^3)(d_{AB}/2)}{\pi d^4_{AB}/64} = \frac{63.66(10^3)}{d^2_{AB}} + \frac{203.72(10^3)}{d^3_{AB}} \tag{3}$$

The normal stress in Eq. (4) is on the left surface of segment AB, where the bending shear stress is zero. The maximum shear stress is thus

$$(\tau_{AB})_{max} = \frac{\sigma_{AB}}{2} = \frac{31.83(10^3)}{d^2_{AB}} + \frac{101.86(10^3)}{d^3_{AB}} \le 90(10^6) \tag{4}$$

For an initial guess we neglect the first term and obtain: $d^3_{AB} \ge (101.86/90)(10^{-3})$ or $d_{AB} \ge 0.104m$. We consider values of d_{AB} = 0.105 m and obtain τ_{max} = 90.93 (10^6) which exceeds the allowable stress. We next consider d_{AB} = 0.11 and obtain τ_{max} = 79.2.93 (10^6) which is below the allowable stress value.

ANS $d_{BD} = 25\text{ mm}$; $d_{AB} = 110\text{ mm}$

---

10. 47

Solution $d_{AB} = 0.75\text{in}$, $A_{BC} = 2\text{ in x }2\text{ in}$, $\sigma_{AB} = ?$, $\sigma_{BC} = ?$,

By equilibrium of moment about point C and by equilibrium of forces we obtain the following.

$$N_{AB}(66 sin60) - (5280)(33) = 0 \qquad or \qquad N_{AB} = 3048.4 lbs \qquad (1)$$

$$C_y - 5280 + N_{AB} sin60 = 0 \qquad or \qquad C_y = 2640 lb \qquad C_x - N_{AB} cos60 = 0 \qquad or \qquad C_x = 1524.2 lb \qquad (2)$$

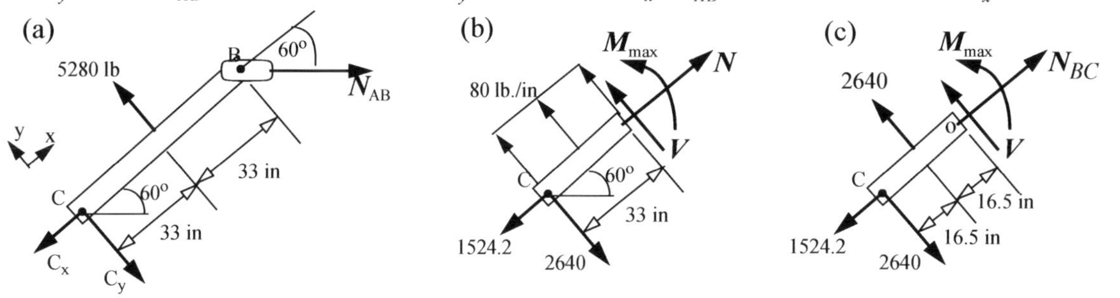

$$\sigma_{AB} = \frac{N_{AB}}{A_{AB}} = \frac{3048.4}{\pi(0.75)^2/4} = \frac{3048.4}{0.4418} psi = 6899.9 psi \qquad (3)$$

The maximum bending moment will occur at the center of CB. By equilibrium of force and moment about point O we obtain.

$$N_{BC} = 1524.2 lb \qquad M_{max} + (2640)(33) - (2640)(16.5) = 0 \qquad or \qquad M_{max} = 43560 in-lb \qquad (4)$$

$$\sigma_{BC} = \frac{N_{BC}}{A_{BC}} + \left|\frac{M_{max}(y_{BC})_{max}}{I_{BC}}\right| = \frac{1524.2}{(2)(2)} + \frac{(43560)(1)}{(2)(2^3)/12} = (381 + 32670) psi = 33051 psi \qquad (5)$$

ANS $\sigma_{AB} = 6.9\text{ ksi (T)}$; $\sigma_{BC} = 33.05\text{ ksi (T)}$

---

10. 48

Solution: $A_{ABC} = 100\text{ mm x }150\text{ mm}$, $A_{CDE} = 100\text{ mm x }200\text{ mm}$, $A_{BD} = 100\text{ mm x }50\text{ mm}$, $|\sigma_{max}| \le 20 MPa$, $w_{max} = ?$

By equilibrium of moment about point A in figure (a) and by equilibrium of forces we obtain the following.

$$E_x(3) - (2.5)(5w) = 0 \qquad or \qquad E_x = 4.1667w \qquad A_x = E_x = 4.1667w \qquad A_y = 5w \qquad (1)$$

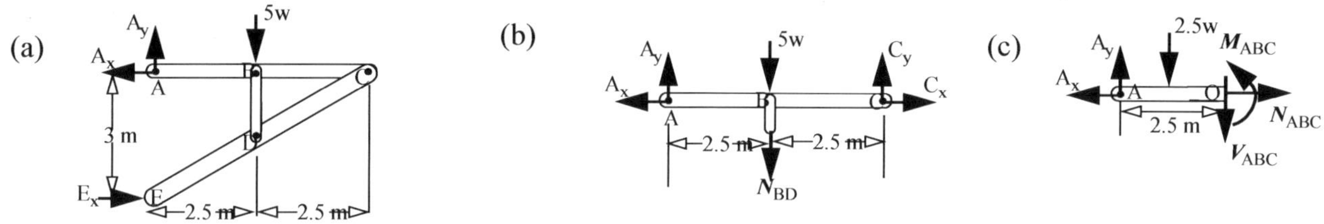

By equilibrium of moment about C in figure (b) and equilibrium of forces we obtain the following.

$$A_y(5) - 5w(2.5) - N_{BD}(2.5) = 0 \qquad or \qquad N_{BD} = 5w \qquad C_x = A_x = 4.1667w \qquad C_y = 5w + N_{BD} - A_y = 5w \qquad (2)$$

$$\sigma_{BD} = \frac{N_{BD}}{A_{BD}} = \frac{5w}{(0.1)(0.05)} = 1000w \le 20(10^6) \qquad w \le 20(10^3)\frac{N}{m} \qquad (3)$$

The maximum moment will be in the center of member ABC. By equilibrium of force and moment figure (c) we obtain.

$$N_{ABC} = A_x = 4.1667w \qquad M_{ABC} = A_y(2.5) - (2.5w)(1.25) = 9.375w \qquad (4)$$

$$\sigma_{ABC} = \frac{N_{ABC}}{A_{ABC}} + \frac{M_{ABC}(y_{ABC})_{max}}{I_{ABC}} = \frac{4.1667w}{(0.1)(0.15)} + \frac{(9.375w)(0.075)}{(0.1)(0.15^3)/12} = 25278w \le 20(10^6) \qquad or \qquad w \le 791.2\frac{N}{m} \qquad (5)$$

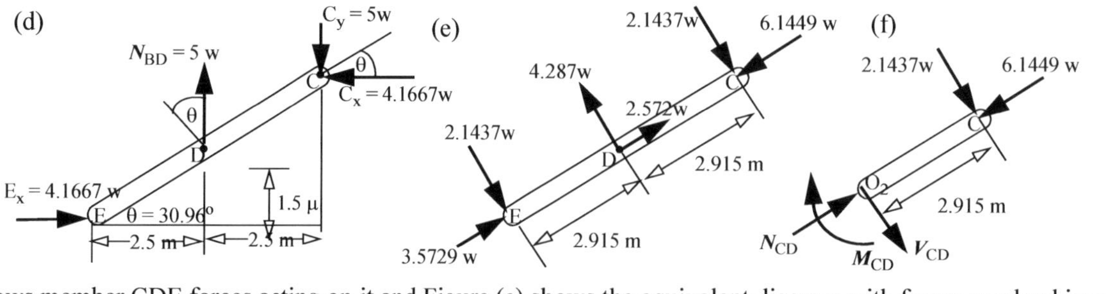

Figure (d) shows member CDE forces acting on it and Figure (e) shows the equivalent diagram with forces resolved in axial and transverse direction.The maximum bending moment will be in the middle of member CDE. By equilibrium of forces and moment about point O_2, we obtain.

$$N_{CD} = 6.1449w \qquad M_{CD} = (-2.1437w)(2.915) = -6.25w \qquad (6)$$

$$\sigma_{CD} = \frac{N_{CD}}{A_{CD}} + \frac{M_{CD}(y_{CD})_{max}}{I_{CD}} = \frac{6.1449w}{(0.1)(0.2)} + \frac{(6.25w)(0.1)}{(0.1)(0.2^3)/12} = 9682.1w \le 20(10^6) \quad or \quad w \le 2.06(10^3)\frac{N}{m} \qquad (7)$$

ANS $w = 791.2$ N/m

---

10. 49

Solution: W = 300lb, $\sigma_w \le 1.2$ psi, $\tau_w \le 6$ ksi, Lumber =?, d_b =? nearest 1/8 inch

By equilibrium of moment about point A in figure {a) and equilibrium of forces we obtain the following.

$$F_y(9) = (600)(3) \quad or \quad F_y = 200lb \quad A_y = 600 - F_y = 400lb \quad A_x = 0 \qquad (1)$$

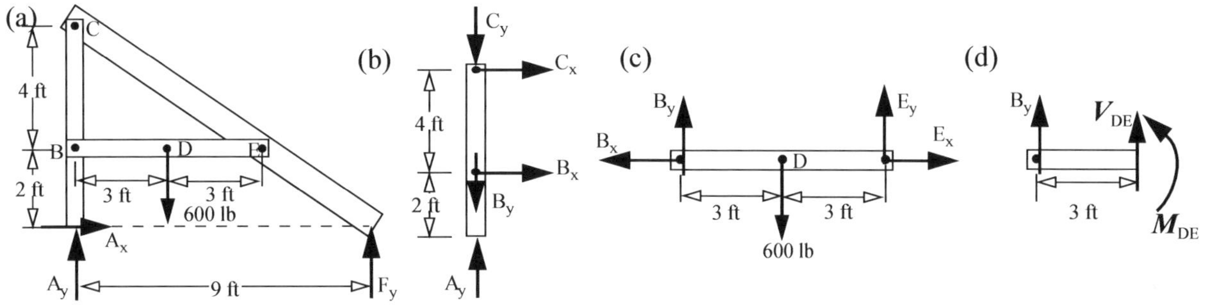

By equilibrium of moment about point B in figure (b) and equilibrium of force in the x - direction we obtain.

$$C_x(4) = 0 \quad or \quad C_x = 0 \quad B_x = -C_x = 0 \quad C_y = A_y - B_y = 400 - B_y \qquad (2)$$

By equilibrium of moment about point E in figure (c) and force equilibrium we obtain.

$$B_y(6) - (600)(3) = 0 \quad or \quad B_y = 300lb \quad E_y = 600 - B_y = 300lb \quad E_x = B_x = 0 \quad C_y = 400 - B_y = 100lb \qquad (3)$$

The largest force acts on pin D.The shear stress in pin D is.

$$\tau_A = \frac{600}{\pi d_b^2/4} \le 6(10^3) \quad or \quad d_b \ge 0.356in \quad or \quad d_b = \frac{3}{8}\text{ in} \qquad (4)$$

Segment AB of member ABC will have the largest compressive axial stress.

$$\sigma_{AB} = \frac{400}{A_{AB}} \le 1200psi \quad or \quad A_{AB} \ge 0.333in^2 \quad \text{Use 2 in x 4 in for ABC.} \qquad (5)$$

The maximum bending moment in member BDE will be at the center. By equilibrium of moment in figure (d) we obtain

$$M_{DE} = (3)(B_y) = 900ft-lb = 10800in-lb \qquad (6)$$

$$\sigma_{BDE} = \frac{M_{DE}}{S_{BDE}} = \frac{10800}{S_{BDE}} \le 1200 \quad or \quad S_{DE} \ge 9in^3 \quad \text{Use 2 in x 6 in for BDE} \qquad (7)$$

Figure (e) shows the forces acting on member CEF. Figure (f) shows the equivalent diagram with forces resolved in axial and transverse direction.The maximum axial stress will be in segment EF. The maximum bending moment will be at E in member CEF. By equilibrium of forces and moment in figure (g) we obtain.

$$N_{EF} = 110.94lb \qquad M_{EF} = (166.4)(3.61) = 600ft-lb = 7200in-lb \qquad (8)$$

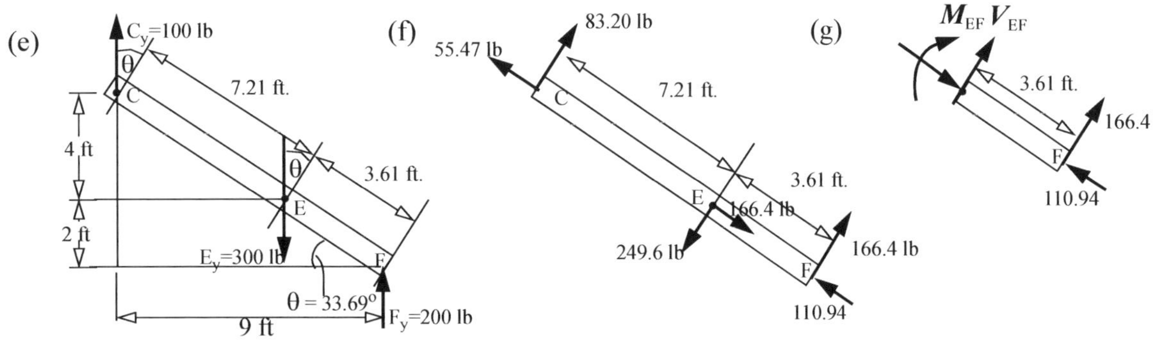

$$\sigma_{CEF} = \frac{N_{EF}}{A_{CEF}} + \frac{M_{EF}}{S_{CEF}} = \frac{110.94}{A_{CEF}} + \frac{7200}{S_{CEF}} \le 1200 \qquad (9)$$

The 2 in. x 4 in. cross-section has as area $A_{CEF} = 8$ and section modulus $S_{CEF} = 5.3$. Substituting these values in the above equation, we obtain, $\sigma_{CEF} = 1372$ psi, which exceeds 1200 psi. We nest consider 2 in. x 6 in. cross-section for which $A_{CEF} = 12$ and $S_{EF} = 12$. Substituting these values in the above equation we obtain $\sigma_{CEF} = 609$ psi which is less than 1200 psi.

ANS $d_b = (3/8)$ in. **;** Use 2 in x 4 in for ABC. .; Use 2 in x 6 in for BDE ; Use 2 in x 6 in for CEF

---

10. 50

Solution $A_{AB} = 4$ in. x 8 in., $d_{al} = (1/2)in$, $\sigma_w \le 1.5$ ksi, $\tau_{al} \le 8$ ksi, $E_w = 1800$ ksi, $E_{al} = 10{,}000$ $P_{max} = ?$

Fig (a) and (b) shows the cantilever beam and the axial rod.

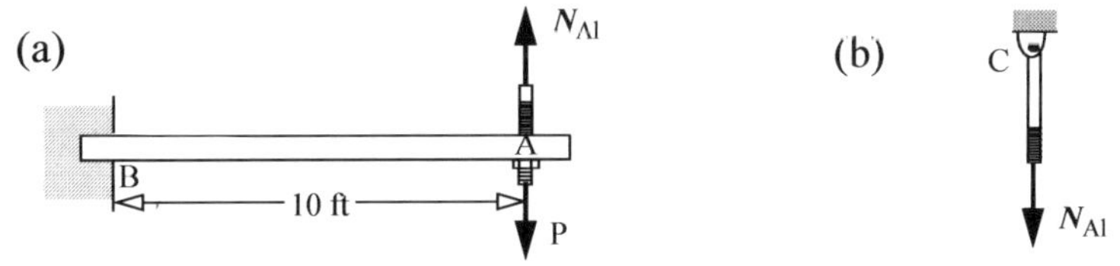

Comparing figure (a) to case 1 in Table C.3 we obtain the following.

$$a = 120in \qquad b = 0 \qquad P = P - N_{al} \qquad I = \frac{1}{12}(4)(8^3) = 170.67in^4 \qquad v_A = \frac{(P - N_{al})(120^3)}{(3)(1800)(170.67)} = 1.875(P - N_{al}) \quad \textbf{(1)}$$

The extension of aluminum rod can be found as shown below.

$$\delta_{al} = \frac{N_{al}L_{al}}{E_{al}A_{al}} = \frac{N_{al}(54)}{(10000)(\pi)(1/2)^2/4} = 0.0275N_{al} = v_A = 1.875(P - N_{al}) \quad or \quad N_{al} = \frac{1.875}{1.9025}P = 0.9855P \quad \textbf{(2)}$$

$$\sigma_{al} = \frac{N_{al}}{A_{al}} = \frac{0.9855P}{\pi(1/2)^2/4} = 5.0193P \qquad \tau_{al} = \sigma_{al}/2 = 2.5097P \le 8 \quad or \quad P \le 3.19kips \quad \textbf{(3)}$$

The maximum bending moment in the beam will be at the left wall.

$$M_{max} = (P - N_{al})(120) = 1.735P \qquad \sigma_w = \left|\frac{M_{max}y_W}{I_W}\right| = \frac{(1.735P)(4)}{(170.67)} \le 1.5 \quad or \quad P \le 36.9kips \quad \textbf{(4)}$$

ANS $P_{max} = 3.1$ kips

---

10. 51

Solution: $d_o = 1.5$ in. $t = (1/4)in$ $E = 30{,}000$ ksi, $\nu = 0.28$ Stresses at A and B.

In figure (a) support at D is replaced by a reaction force. The beam ED is subjected to a couple of 30 in-kips and a force R_D comparing beam in figure (a) to cases (1) and (2) in Table C.3, we have $M = -30$ in-kips, $P = R_D$, $L = 60$, $E = 30{,}000$ ksi.

$I = \pi[1.5^4 - 1^4]/64 = 0.1994in^4$. The deflection of D is the superposition of deflection in cases 1 and 2 and should equal to zero as shown below.

$$v_D = \frac{(-30)(60^2)}{(2)(30000)(0.1994)} + \frac{(R_o)(60^3)}{(2)(30000)(0.1994)} = 0 \quad or \quad R_D = \frac{3}{2}\frac{(30)}{60} = 0.75kip \quad \textbf{(1)}$$

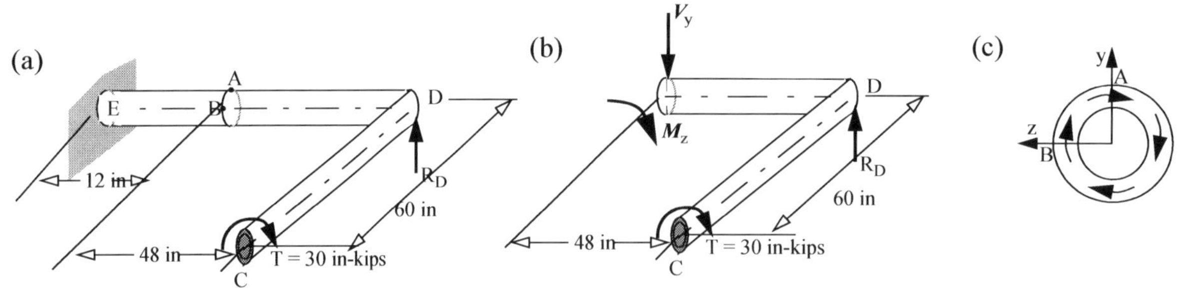

By equilibrium of forces and moment in figure (b)

$$V_y = R_D = 0.75kip \qquad M_z + 30 - R_D(48) = 0 \quad or \quad M_z = 6in - kips \quad \textbf{(2)}$$

$$(\sigma_{xx})_A = -\left(\frac{M_z y_A}{I}\right) = -\left[\frac{6(0.75)}{0.1994}\right] = -22.6ksi = 22.6 \text{ ksi (C)} \qquad (\sigma_{xx})_B = 0 \quad \textbf{(3)}$$

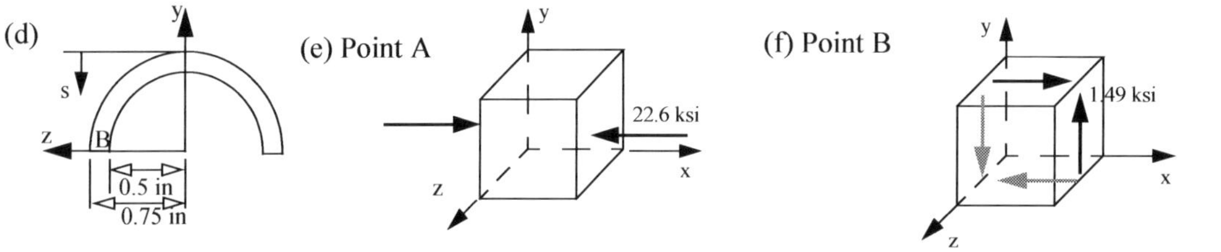

Figure (c) shows the direction of shear stress. At point A the shear stress is zero and at point B it is in the positive y - direction.

$$(Q_z)_B = \frac{2}{3}\left[\left(\frac{1.5}{2}\right)^3 - \left(\frac{1}{2}\right)^3\right] = 0.1979in^3 \qquad (\tau_{xy})_B = \left|\frac{V_y(Q_z)_B}{It_B}\right| = \frac{(0.75)(0.1979)}{(0.1994)(0.5)} = 1.49\text{ ksi} \qquad (\tau_{xy})_A = 0 \tag{4}$$

ANS $(\sigma_{xx})_A = 22.6$ ksi (C); $(\tau_{xy})_A = 0$; $(\sigma_{xx})_B = 0$; $(\tau_{xy})_B = 1.49$ ksi

---

10. 52

Solution $d_{al} = 8$ mm, $\sigma_w \le 14$ MPa, $\sigma_s \le 140$ MPa, $\tau_{al} \le 60$ MPa,
$E_w = 12.6$ GPa, $E_s = 200$ GPa $E_{al} = 70$ GPa, $w_{max} = ?$

The area moment of inertia and the bending rigidity can be found as shown below.

$$(I_{zz})_w = (40)(100)^3/12 = 3.333(10^6)mm^4 \qquad (I_{zz})_s = (40)(6)^3/12 + (40)(6)(53^2) = 0.6749(10^6)mm^4 \tag{1}$$

$$\sum E_j(I_{zz})_j = (12.6)(10^9)(3.333)(10^{-6}) + (2)(200)(10^9)(0.6749)(10^{-6}) = 311.95(10^3)N-m^2 \qquad 1$$

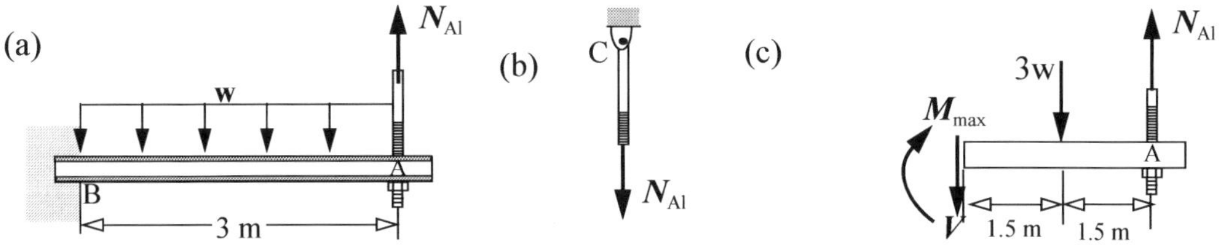

The cross sectional area of aluminum rod is: $A_{al} = \pi(8)^2/4 = 50.26mm^2$. Comparing figure (a) to case 1 and 3 in Table C.3, we obtain a = 3m, b = 0, P = $-N_{al}$, p_o = w. The bending rigidity (EI) is replaced by the value given above. We obtain

$$v_A = \frac{-N_{al}(3^3)}{(3)(311.95)(10^3)} + \frac{w(3^4)}{(8)(311.95)(10^3)} = [-28.95N_{al} + 32.46w](10^{-6}) \tag{2}$$

The extension of aluminum rod can be found and equated to the above displacement.

$$\delta_{al} = \frac{N_{al}(1.3)}{(70)(10^9)(50.26)(10^{-6})} = 0.3695N_{al}(10^{-6}) = v_A = [-28.95N_{al} + 32.46w](10^{-6}) \qquad or \qquad N_{al} = 1.107W \qquad 3$$

$$\sigma_{al} = \frac{N_{al}}{A_{al}} = \frac{1.107w}{(50.26)(10^{-6})} = 22.02w(10^3) \qquad \tau_{max} = \frac{\sigma_{al}}{2} = 11.01w(10^3) \le 60(10^6) \qquad w \le 5.45(10^3)\frac{N}{m} \tag{4}$$

The maximum bending moment will be at the wall as shown in figure (c).

$$M_{max} + 3w(1.5) - N_{al}(3) = 0 \qquad or \qquad M_{max} = -1.179w \tag{5}$$

The maximum normal stress in wood will be at $y_w = \pm 0.05m$ and in steel at $y_s = \pm 0.056m$.

$$\sigma_w = \left|-\frac{E_w M_z y_w}{\sum E_j(I_{zz})_j}\right| = \left|\frac{(12.6)(10^9)(1.179w)(0.05)}{(311.95)(10^3)}\right| = 2.381w(10^3) \le 14(10^6) \qquad or \qquad w \le 5.88(10^3)\frac{N}{m} \tag{6}$$

$$\sigma_s = \left|-\frac{E_s M_z y_s}{\sum E_j(I_{zz})_j}\right| = \left|\frac{(200)(10^9)(1.179w)(0.05)}{(311.95)(10^3)}\right| = 42.33w(10^3) \le 140(10^6) \qquad or \qquad w \le 3.307(10^3)\frac{N}{m} \tag{7}$$

ANS $w_{max} = 3.3$ kN/m

---

10. 53

Solution: $A_{BD} = A_{CE} = 6$ in x 6in, $A_{AB} = A_{AC} = A_{BC} = 2$ in x 8 in, $\sigma_{max} = ?$ in each member

From equilibrium in figures (a) and (b) we obtain the following.

$$D_x = 0 \qquad D_y = E_y = 184.75 \qquad N_{BD} = D_y = 184.75 \tag{1}$$

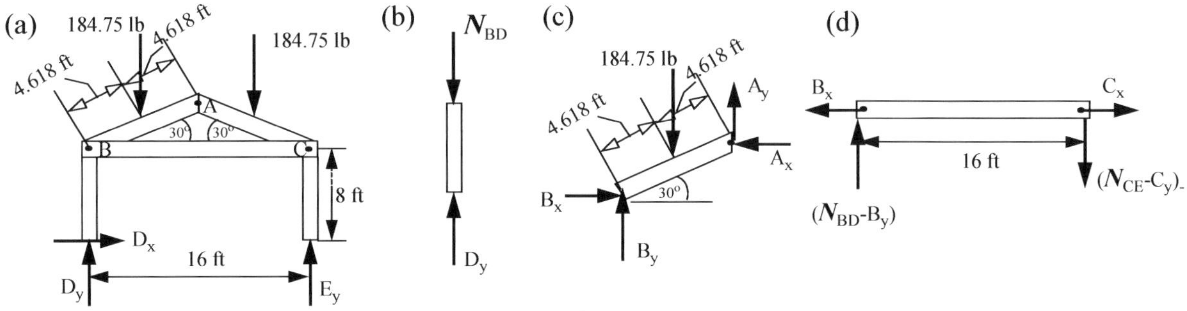

$$\sigma_{BD} = \sigma_{CE} = \frac{184.75}{(6)(6)} = 5.13\text{ psi (C)} \tag{2}$$

By equilibrium of moment about point C in figure (d) and about point A in figure (c) we obtain

$$(N_{BD} - B_y)(16) = 0 \qquad B_y = 184.75 \tag{3}$$

$$B_x(9.236\sin 30) - B_y(9.236\cos 30) + 184.75(4.68\cos 30) = 0 \quad or \quad B_x = 159.78lb \tag{4}$$

$$A_x = B_x = 159.78 \qquad A_y = 184.75 - B_y = 0 \tag{5}$$

$$\sigma_{BC} = \frac{B_x}{A_{BC}} = \frac{159.78}{(2)(8)} = 9.986psi \tag{6}$$

Fig. (e) shows the coordinates x, y and coordinates x_1, y_1 that are in axial and in transverse direction to member *AB*. The resolution of forces can be written as: $F_{x1} = F_x\cos 30 + F_y\sin 30$ and $F_{y1} = -F_x\sin 30 + F_y\cos 30$. Using these a equations the forces in Figure (c) are transformed to those shown in Figure (f). The bending moment will be maximum at the center of member AB. By equilibrium of forces and moments in Figure (h) we obtain the following.

$$N_{AB} = 230.9 - 46.2 = 184.7lb \qquad M_{AB} = (80)(4.618) - (80)(2.309) = 184.72lb - ft \tag{7}$$

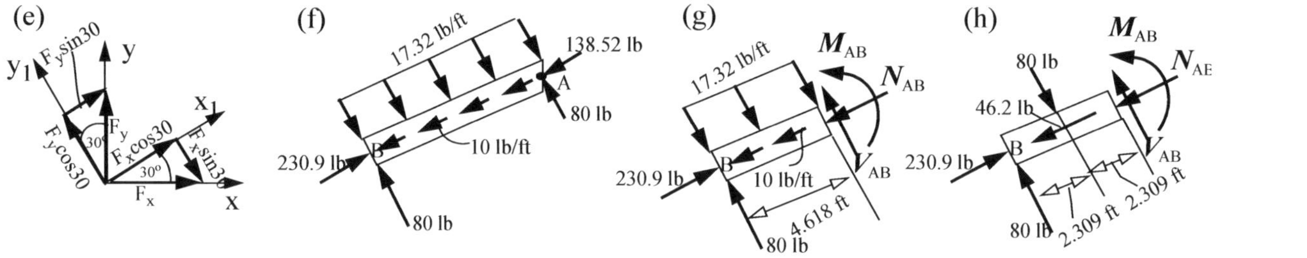

$$\sigma_{AB} = \frac{N_{AB}}{A_{AB}} + \frac{M_{AB}y_{AB}}{I_{AB}} = \frac{184.7}{(2)(8)} + \frac{(184.72)(12)(4)}{(2)(8)^3/12} = 11.5 + 103.91 = 115.5psi \tag{8}$$

ANS $\sigma_{BD} = \sigma_{CE} = 5.13$ psi (C) ; $\sigma_{AB} = 115.5psi(C)$; $\sigma_{BC} = 10$ psi (T)

---

10. 54

Solution: $(d_{CD})_0 = 16$ in, $(d_{EF})_0 = (d_{PG})_0 = 12$ in, $t = 1$, $p = 20$ lb/ft^2 Stresses at A and B=?

The pressure on the pipes and signboards are replaced by statically equivalent forces. By equilibrium of forces and moment we obtain the following.

$$V_z = -[(2)(600) + (2)(20) + 2(160) + 533.3] = -2093.3lb \tag{1}$$

$$T = -\left[(2)(20)\left(14.5 + \frac{8}{12}\right) + (2)(160)\left(4 + \frac{8}{12}\right) + (600)\left(18 + \frac{8}{12}\right) + (600)\left(11 + \frac{8}{12}\right)\right] = -20300\ ft - lb = -243600\ in - lb \tag{2}$$

$$M_y = (20 + 160)(20) + (20 + 160)(17) + 2(600)(18.5) + (533.3)(10) = 34193.3\ ft - lb = 410320\ in - lb \tag{3}$$

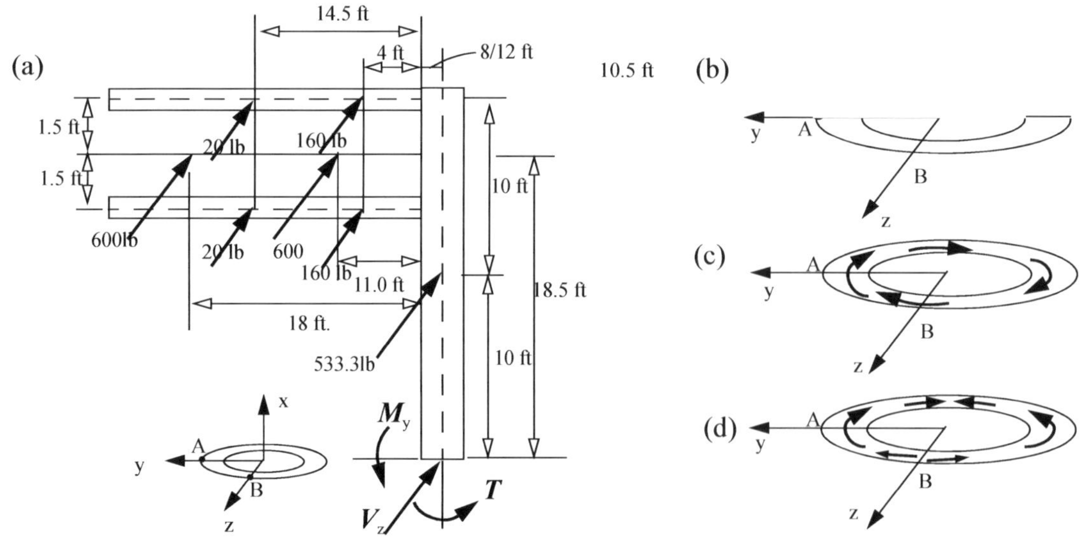

The polar moment of inertia for pipe CDE is $J = \pi[(16)^4 - (14)^4]/32 = 2662.5in^4$ and $I = J/2 = 1331.3in^4$

The normal stress at point A is zero and at point B is tensile.

$$(\sigma_{xx})_A = 0 \qquad (\sigma_{xx})_B = \frac{(410320)(8)}{1331.3} = 2465.7psi \tag{4}$$

The torsional shear stress, and bending shear stresses are calculated and sign determined using figures (c) and (d).

$$\tau_{x\theta} = \frac{T\rho}{J} = \frac{(-243600)(8)}{2662.5} = -731.94psi \qquad (\tau_{xz})_A = -731.94psi \qquad (\tau_{xy})_B = 731.94psi \qquad \textbf{(5)}$$

$$Q_y = \frac{2}{3}[8^3 - 7^3] = 112.67\ in^3 \qquad (\tau_{xz})_A = -\frac{(2093.3)(112.67)}{(1331.3)(2)} = -88.579psi \qquad (\tau_{xy})_B = 0 \qquad \textbf{(6)}$$

$$\text{Total} \quad (\tau_{xz})_A = (-731.94 - 88.579)psi = -820.5psi \qquad (\tau_{xy})_B = 731.94psi \qquad \textbf{(7)}$$

ANS $(\sigma_{xx})_B = 2465.7psi(T)$; $(\tau_{xz})_A = -820.5$ psi ; $(\tau_{xy})_B = 731.9$ psi

---

10. 55

Solution: $d_o = 1$ in., $t = (1/16)in$, $\sigma_{max} \le 12$ksi, $\tau_{max} \le 8$ ksi, W_{max} =?, to nearest lb.

By equilibrium of moment about B and equilibrium of forces, we obtain the following.

$$\frac{W}{2}(23) + \frac{W}{2}(15) - (T_{CD}sin38.6)(12) - (T_{CD}cos38.6)(12) = 0 \qquad or \qquad T_{CD} = 1.126W \qquad \textbf{(1)}$$

$$B_x = T_{CD}cos38.6 = 0.88W \qquad B_y = W - T_{CD}sin38.6 = 0.298W \qquad \textbf{(2)}$$

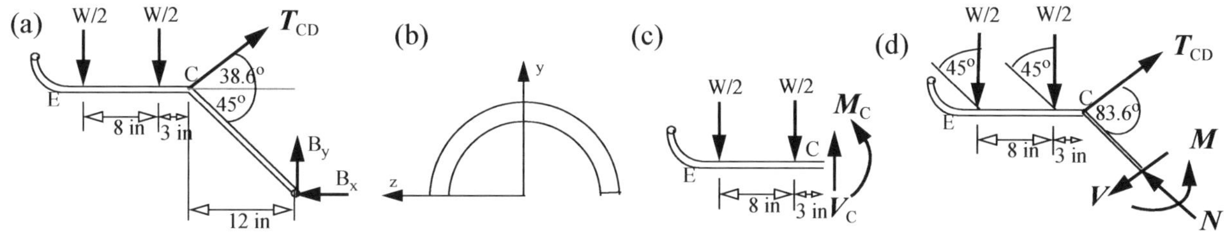

In EC, the maximum moment and shear force will be just before C as shown in Fig. (c).

$$M_c = -(W/2)(11) - (W/2)(3) = -7W \qquad V_c = W \qquad \textbf{(3)}$$

$$A = \pi(1^2 - 0.875^2)/4 = 0.184in^2 \qquad I = \pi(1^4 - 0.875^4)/64 = 20.31(10^{-3})in^4 \qquad Q_z = (2/3)(1^3 - 0.875^3) = 0.22in^3 \qquad \textbf{(4)}$$

The magnitude of maximum bending normal and shear stress can be found as

$$|(\sigma_{xx})_{max}| = \left|\frac{(7W)(0.5)}{20.31(10^{-3})}\right| = 172.31W \le 12(10^5) \qquad or \qquad W \le 69.6lb \qquad \textbf{(5)}$$

$$|(\tau_{xy})_{max}| = \left|\frac{(W)(0.22)}{20.31(10^{-3})(0.125)}\right| = 86.67W \le 8(10^3) \qquad or \qquad W \le 92.3lb \qquad \textbf{(6)}$$

The maximum shear stress on the top and bottom surface on a plane 45° to the axis of EC is

$$\tau_{max} = (\sigma_{xx})_{max}/2 = 86.15W \le 8(10^3) \qquad or \qquad W \le 92.86lb \qquad \textbf{(7)}$$

In BC, the maximum moment will be next to C. Figure (d). shows the free body diagram after an imaginary cut is made just next to C in member BC. The moment value is same as before. By equilibrium of forces we obtain.

$$N = T_{CD}(cos83.6) + 2\left(\frac{W}{2}\right)(cos45) = 0.833W \qquad V = T_{CD}(sin83.6) - 2\left(\frac{W}{2}\right)(sin45) = 0.412W \qquad M = M_c = -7W \qquad \textbf{(8)}$$

The magnitude of maximum normal stress is

$$|\sigma_{BC}| = \left|\frac{0.833W}{0.184}\right| + \left|\frac{(7W)(0.5)}{(20.31)(10^{-3})}\right| = 176.86W \le 12(10^5) \qquad or \qquad W \le 67.86lb \qquad \textbf{(9)}$$

ANS $W_{max} = 67\ \ lb$

---

10. 56

Solution: $\sigma_{max} \le 18$ksi W_{max} =?

By equilibrium of moment about point A and equilibrium of forces, we obtain the following.

$$(N_{BC}sin41.63^o)(9) + \frac{W}{2}(2) - \frac{W}{2}(11.5) = 0 \qquad or \qquad N_{BC} = 0.794W \qquad \textbf{(1)}$$

$$A_y = N_{BC}sin41.63^o - W/2 = 0.027W \qquad A_x = N_{BC}cos41.63^o - W/2 = 0.093W \qquad \textbf{(2)}$$

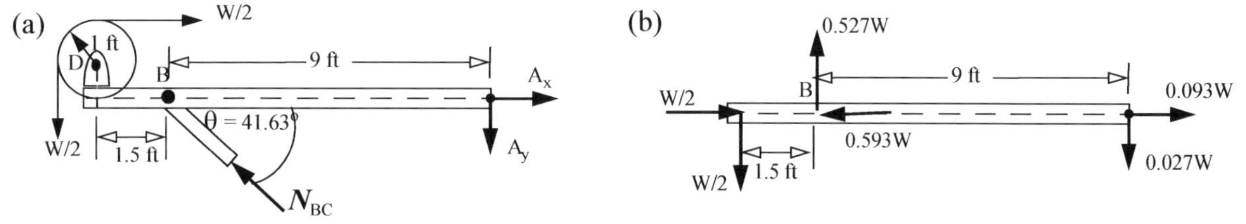

$$A_{AA} = 2\left[(2)\left(\frac{1}{8}\right)+\left(2-\frac{1}{8}\right)\left(\frac{1}{8}\right)\right] = 0.969in^2 \qquad \sigma_{BC} = \frac{0.794W}{0.969} = 0.8196W \le 18ksi \qquad W \le 21.96kip \quad (1)$$

Fig. (b) shows the forces acting in axial and transverse direction on member ABD. The maximum moment and axial force will be just left of B. The magnitudes are

$$\boldsymbol{N} = 0.5W \qquad \boldsymbol{M} = (0.5W)(1.5) = 0.75Wft-kips = 9Win-kips \quad (2)$$

$$A_{BB} = [4(0.25)+(2)(1.75)(0.25)] = 1.875in^2 \qquad I_{zz} = \frac{(0.25)(4^3)}{12}+2\left[\frac{(1.75)(0.25^3)}{12}+(1.75)(0.25)(1.875^2)\right] = 4.414in^4 \quad (3)$$

The maximum compressive stress in segment BD is

$$(\sigma_{BD})_{max} = \frac{(9W)(2)}{4.414}+\frac{(0.5)W}{1.875} = 4.375W \le 18ksi \qquad W \le 4.14kip \quad (4)$$

ANS $W_{max} = 4.1$ kips

---

10. 57

Solution d = 50 mm $\sigma_{max} \le 100$ MPa Failure Envelope =?

Solution proceeds as follows. Below P is in kN and T is in N-m.

$$A = \frac{\pi}{4}(0.05)^2 = 1.963(10^{-3})mm^2 \qquad \sigma_{xx} = \frac{P(10^3)}{1.963(10^{-3})} = 0.5094P(10^6)\frac{N}{m^2} = 0.5094P\ MPa \quad (1)$$

$$J = \frac{\pi}{32}(0.05)^4 = 0.613(10^{-6})mm^4 \qquad \tau_{x\theta} = \frac{T(0.025)}{0.613(10^{-6})} = 0.0408T(10^6)\frac{N}{m^2} = 0.0408TMPa \quad (2)$$

$$\sigma_1 = \left(\frac{0.5094P}{2}\right)+\sqrt{\left(\frac{0.5094P}{2}\right)^2+(0.0408T)^2} = 0.2547P+\sqrt{(0.2547P)^2+(0.0408T)^2} \le 100 \quad (3)$$

Failure envelope is shown in figure below.

P (kN)	T (N-m)
0.00	2450.98
15.00	2355.48
30.00	2255.94
45.00	2151.80
60.00	2042.36
75.00	1926.71
90.00	1803.66
105.00	1671.58

P (kN)	T (N-m)
120.00	1528.12
135.00	1369.72
150.00	1190.43
165.00	978.83
180.00	706.46
195.00	200.17
196.31	0.00

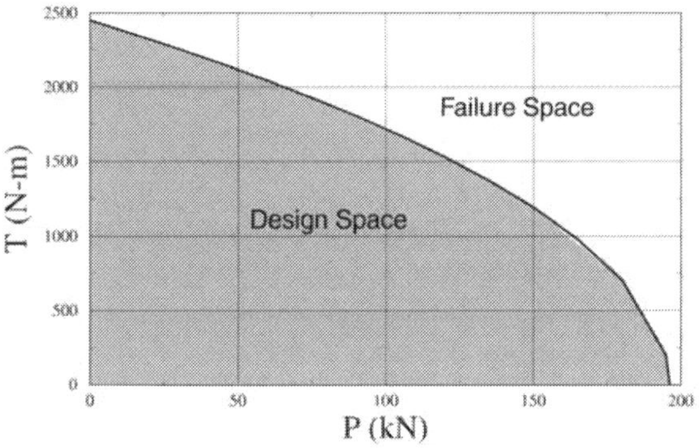

---

10. 58

Solution: $\tau_{max} \le 75$ MPa, Failure Envelope =?

From previous problem we have the following.

$$\sigma_1 = 0.2547P+\sqrt{(0.2547P)^2+(0.0408T)^2} \qquad \sigma_2 = 0.2547P-\sqrt{(0.2547P)^2+(0.0408T)^2} \quad (1)$$

$$\tau_{max} = \left|\frac{\sigma_1-\sigma_2}{2}\right| = \sqrt{(0.2547P)^2+(0.0408T)^2} \le 75 \quad (2)$$

when $T = 0$, the value of $P = 294.46$ kN. The table below shows the values of P and T that satisfy the equality above. Failure envelope is shown in the figure below.

P (kN)	T (N-m)
0.00	1840.94
20.00	1836.69
40.00	1823.89
60.00	1802.34
80.00	1771.73
100.00	1731.58
120.00	1681.21
140.00	1619.66

P (kN)	T (N-m)
160.00	1545.60
180.00	1457.13
200.00	1351.40
220.00	1223.97
240.00	1067.08
260.00	864.85
280.00	570.98
294.52	0.07

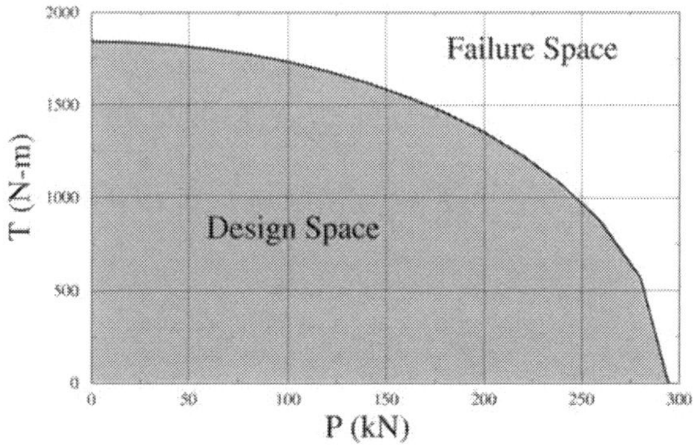

10. 59

Solution L = 1.5 m E = 200 GPa G = 80 GPa $\delta \le 0.5$ mm $\Delta\phi \le 3^o$ $\sigma_{max} \le 100$ MPa, Failure Envelope =?

From earlier problem, we have the following.

$$A = 1.963(10^{-3})mm^2 \qquad J = 0.613(10^{-6})mm^4 \qquad \sigma_1 = 0.2547P + \sqrt{(0.2547P)^2 + (0.0408T)^2} \le 100 \qquad (1)$$

The deformation and rotation can be written as shown below.

$$\delta = \frac{P(10^3)1.5}{(200)(10^9)(1.963)(10^{-3})} = 3.820P(10^{-6}) \le 0.5(10^{-3}) \qquad P \le 130.9kN \qquad (2)$$

$$\Delta\phi = 3° = 52.36(10^{-3})rads \qquad \Delta\phi = \frac{T(1.5)}{(80)(10^9)(0.613)(10^{-6})} = 30.59T(10^{-6}) \le 52.36(10^{-3}) \quad or \quad T \le 1.712(10^3) \qquad (3)$$

Table in earlier problem shows the values of T and P corresponding to the stress limitation in the first equation.. Figure below shows the failure envelope corresponding to all the equality signs.

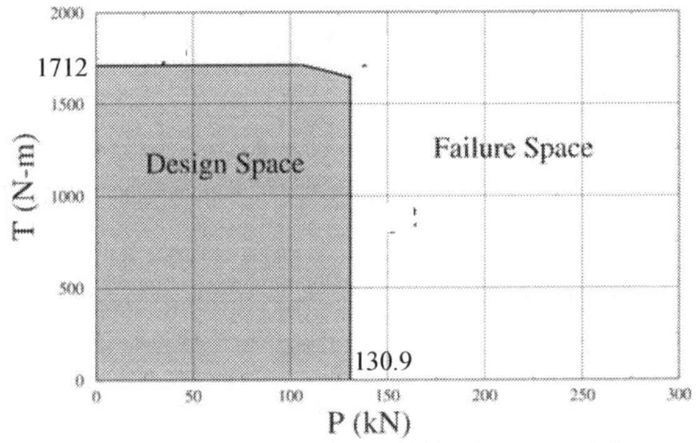

10. 60

Solution: d_o = 40 mm. $t = 10mm$, $\tau_{max} \le 60$MPa, Failure Envelope =?

By equilibrium of forces and moment we obtain:

$$N = P_1 \qquad V_y = P_2 \qquad M_z = 0.65P_2 \qquad T = (-0.1)P_1 \qquad M_y = (-0.1)P_1 \qquad (1)$$

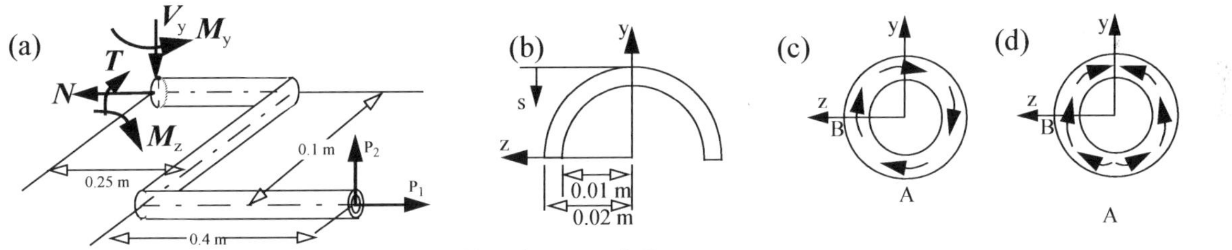

The area of cross section and the area moment of inertia are as follows.

$$A = \pi[0.04^2 - 0.02^2]/4 = 0.9425(10^{-3})m^2 \qquad J = \pi[0.04^4 - 0.02^4]/32 = 0.2356(10^{-6})m^4 \qquad I = J/2 = 0.1178(10^{-6})m^4$$

We assume that P_1 and P_2 are in kN. The normal stresses at points A and B are

$$(\sigma_{xx})_A = \frac{P_1(10^3)}{0.9425(10^{-3})} + \frac{(0.65P_2)(10^3)(0.02)}{0.1178(10^{-6})} = (1.061P_1 + 110.36P_2)(10^6) = (1.061P_1 + 110.36P_2)MPa \qquad (2)$$

$$(\sigma_{xx})_B = \frac{P_1(10^3)}{0.9425(10^{-3})} + \frac{(0.1P_1)(10^3)(0.02)}{0.1178(10^{-6})} = 18.039P_1(10^6) = 18.039P_1MPa \qquad (3)$$

Fig. (c) shows the direction of torsional shear stress. Fig. (d) shows direction of shear stress due bending about z - axis. The shear stress at points A and B are

$$Q_z = \frac{2}{3}[(0.02)^3 - (0.01)^3] = 4.67(10^{-6})m^3 \qquad (\tau_{xz})_A = \frac{(0.1)P_2(10^3)(0.02)}{0.2356(10^{-6})} = 8.489P_2(10^6) = 8.489P_2MPa \qquad (4)$$

$$(\tau_{xz})_B = \frac{(0.1)P_2(10^3)(0.02)}{0.2356(10^{-6})} + \frac{P_2(10^3)(4.67)(10^{-6})}{0.1178(10^{-6})(0.02)} = 10.471P_2(10^6) = 10.471P_2MPa \qquad (5)$$

The maximum shear stress at points A and B are

$$(\tau_{max})_A = \sqrt{\left(\frac{1.061P_1 + 110.36P_2}{2}\right)^2 + (8.489P_2)^2} \le 60 \quad or \quad P_1 = [\sqrt{(60)^2 - (8.489P_2)^2} - 55.18P_2]/0.5305 \qquad (6)$$

$$(\tau_{max})_B = \sqrt{\left(\frac{18.039P_1}{2}\right)^2 + (10.471P_2)^2} \le 60 \qquad P_1 = [\sqrt{(60)^2 - (10.471P_2)^2}]/9.0195 \tag{7}$$

Table below can be constructed on a spread sheet and failure envelope constructed as shown below

P_2 Eq.(6)	P_1 Eq. (6)	P_2 Eq. (7)	P_1 Eq. (7)
0.000	113.10	0.00	6.65
0.100	102.69	0.50	6.63
0.200	92.25	1.00	6.55
0.300	81.79	1.50	6.42
0.400	71.31	2.00	6.23
0.500	60.81	2.50	5.99
0.600	50.28	3.00	5.67
0.700	39.73	3.50	5.27
0.800	29.16	4.00	4.76
0.900	18.57	4.50	4.12
1.000	7.95	5.00	3.25
1.075	0.00	5.73	0.00

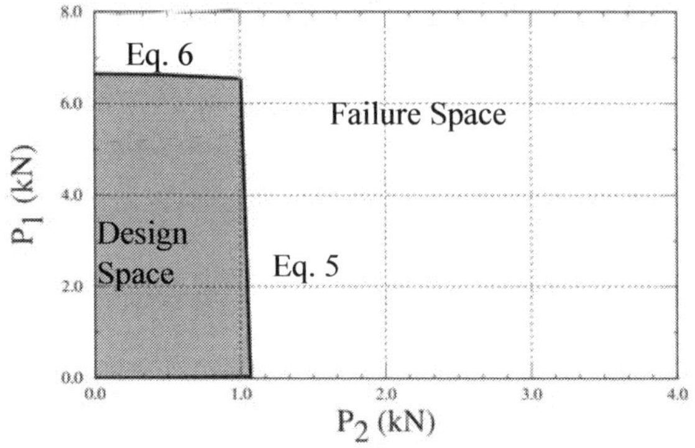

10. 61

Solution: $d_o = 2$ in $\quad t = 1/4in$, $\quad \tau_{max} \le 24$ ksi, $\quad$ Failure Envelope =?

By equilibrium of forces and moment and geometry we obtain the following.

$$N = P_1 \qquad V_z = P_2 \qquad T = 10P_2 \qquad M_y = 16P_2 \qquad M_z = -10P_1 \tag{1}$$

$$A = \pi[(2)^2 - (1.5)^2]/4 = 1.374in^2 \qquad J = \pi[(2)^4 - (1.5)^4]/32 = 1.074in^4 \qquad I = J/2 = 0.5369in^4 \tag{2}$$

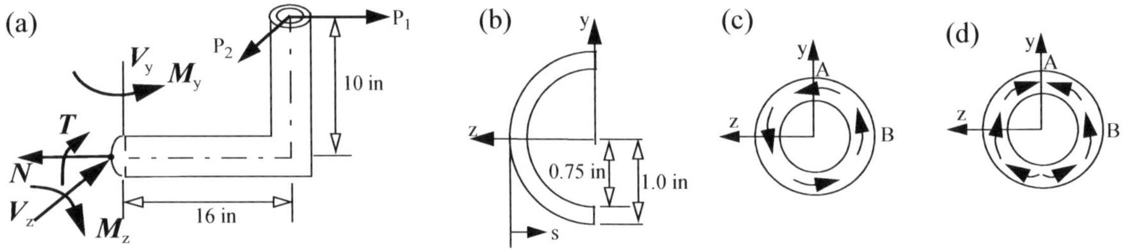

$$(\sigma_{xx})_A = \frac{P_1}{1.374} + \frac{(10P_1)(1)}{0.5369} = 19.353P_1 \qquad (\sigma_{xx})_B = \frac{P_1}{1.374} + \frac{(16P_2)(1)}{0.5369} = 0.7278P_1 + 29.801P_2 \tag{3}$$

Figures (c) and (d) shows the direction of torsional and bending shear stress.

$$Q_y = \frac{2}{3}[1^3 - 0.75^3] = 0.3854in^3 \qquad (\tau_{xz})_A = \frac{(10)P_2(1)}{1.074} + \frac{P_2(0.3854)}{(0.5369)(0.5)} = 10.746P_2 \qquad (\tau_{xz})_B = \frac{(10)P_2(1)}{1.074} = 9.311P_2 \tag{4}$$

$$(\tau_{max})_A = \sqrt{\left(\frac{19.353P_1}{2}\right)^2 + (10.746P_2)^2} \le 24 \qquad or \qquad P_1 = \frac{1}{9.676}[\sqrt{(24)^2 - (10.746P_2)^2}] \tag{5}$$

$$(\tau_{max})_B = \sqrt{\left(\frac{0.7278P_1 + 29.801P_2}{2}\right)^2 + (9.311P_2)^2} \le 24 \qquad or \qquad P_1 = \frac{1}{0.3637}[\sqrt{(24)^2 - (9.311P_2)^2} - 14.90P_2] \tag{6}$$

Table below can be constructed on a spread sheet and failure envelope constructed as shown below

P_2 Eq.(6)	P_1 Eq. (6)	P_2 Eq. (7)	P_1 Eq. (7)
0.000	2.480	0.000	65.952
0.200	2.470	0.150	59.699
0.400	2.440	0.300	53.220
0.600	2.389	0.450	46.514
0.800	2.316	0.600	39.573
1.000	2.218	0.750	32.390
1.200	2.092	0.900	24.951
1.400	1.933	1.050	17.240
1.600	1.731	1.200	9.235
1.800	1.468	1.350	0.907
2.000	1.104	1.360	0.339
2.200	0.427	1.365	0.055
2.233	0.000	1.366	0.000

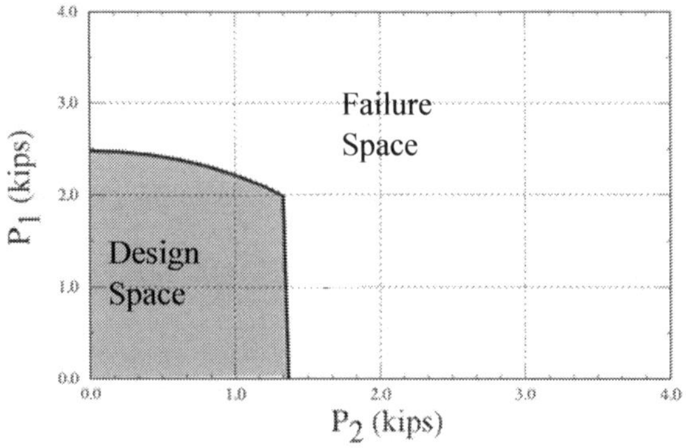

10. 62

Solution: L = 5 ft. T = 200 in-kips, P = 100 kips, $R_i = 1$, $\tau_{max} \le 10$ ksi R_o =? lightest

The axial stress and maximum torsional shear stress can be written as

$$\sigma_{xx} = \frac{100}{\pi(R^2{}_o - 1)} = \frac{31.831}{(R^2{}_o - 1)} \qquad \tau_{x\theta} = \frac{200(R_o)}{\pi(R^4{}_o - 1)} = \frac{127.323(R_o)}{(R^4{}_o - 1)} \tag{1}$$

(a)

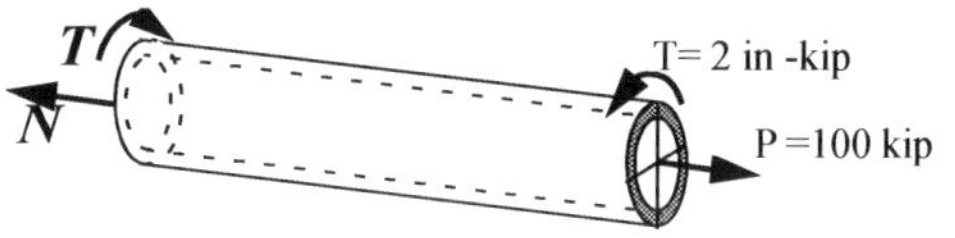

The maximum shear stress is

$$\tau_{max} = \sqrt{\left(\frac{15.915}{(R^2{}_o - 1)}\right)^2 + \left(\frac{127.323(R_o)}{(R^4{}_o - 1)}\right)^2} \le 10 \qquad \text{Define } f(R_o) = \sqrt{\left(\frac{15.915}{(R^2{}_o - 1)}\right)^2 + \left(\frac{127.323(R_o)}{(R^4{}_o - 1)}\right)^2} - 10 \qquad f(R_o) \le 0 \tag{1}$$

The root of $f(R_o)$ corresponding to the equality sign can be found on a spreadsheet as shown in table below. $R_o = 2.405in$

R_o	$f(R_o)$	R_o	$f(R_o)$
2.000	7.786	2.4000	0.068
2.100	5.227	2.4025	0.036
2.200	3.160	2.4050	0.004
2.300	1.469	2.4075	-0.027
2.400	0.068	2.4100	-0.059
2.500	-1.105	2.4125	-0.090
2.600	-2.095	2.4150	-0.121
2.700	-2.938	2.4175	-0.152

---

10. 63

Solution t = 15 mm $\sigma_{max} \le 150\ MPa$ R_o = ? nearest millimeter

By force and moment equilibrium and geometry we obtain the following.

$$N = 100\ kN \qquad V_y = 15\ kN \qquad T = 2\ kN-m \qquad M_z = (15)(1.2) = 18\ kN-m \tag{1}$$

$$A = \pi[R_o^2 - (R_o - 0.015)^2]\ m^2 \qquad I = \frac{\pi}{4}[R_o^4 - (R_o - 0.015)^4]\ m^4 \qquad J = 2I \qquad Q_z = \frac{2}{3}[R_o^3 - (R_o - 0.015)^3]\ m^3$$

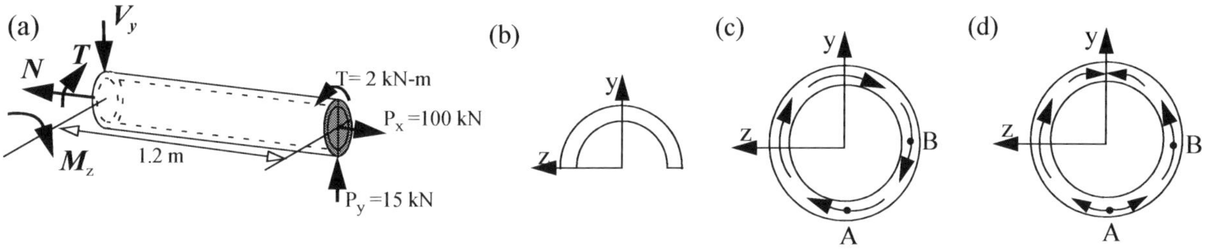

R_o m	A m^2	I m^4	Q_z m^3	$(\sigma_{xx})_A$ MPa	$(\tau_{xz})_A$ MPa	$f_1(R_o)$ MPa	$(\sigma_{xx})_B$ MPa	$(\tau_{xy})_B$ MPa	$f_2(R_o)$ MPa
0.055	4.477E-03	5.176E-06	6.825E-05	213.6	10.6	64.1	22.3	6.6	-125.9
0.056	4.571E-03	5.505E-06	7.113E-05	205.0	10.2	55.5	21.9	6.5	-126.4
0.057	4.665E-03	5.847E-06	7.407E-05	196.9	9.7	47.4	21.4	6.3	-126.8
0.058	4.760E-03	6.203E-06	7.707E-05	189.3	9.4	39.8	21.0	6.2	-127.3
0.059	4.854E-03	6.573E-06	8.013E-05	182.2	9.0	32.6	20.6	6.1	-127.7
0.060	4.948E-03	6.958E-06	8.325E-05	175.4	8.6	25.8	20.2	6.0	-128.2
0.061	5.042E-03	7.358E-06	8.643E-05	169.1	8.3	19.5	19.8	5.9	-128.6
0.062	5.137E-03	7.773E-06	8.967E-05	163.0	8.0	13.4	19.5	5.8	-129.0
0.063	5.231E-03	8.203E-06	9.297E-05	157.4	7.7	7.7	19.1	5.7	-129.3
0.064	5.325E-03	8.649E-06	9.633E-05	152.0	7.4	2.3	18.8	5.6	-129.7
0.065	5.419E-03	9.111E-06	9.975E-05	146.9	7.1	-2.8	18.5	5.5	-130.0
0.066	5.513E-03	9.589E-06	1.032E-04	142.0	6.9	-7.6	18.1	5.4	-130.4
0.067	5.608E-03	1.008E-05	1.068E-04	137.4	6.6	-12.3	17.8	5.3	-130.7

Figures (c) and (d) shows the direction of torsional and bending shear stress.

$$(\sigma_{xx})_A = \frac{100(10^3)}{A} + \frac{18(10^3)R_o}{I} \qquad (\sigma_{xx})_B = \frac{100(10^3)}{A} \qquad (\tau_{xz})_A = \frac{2(10^3)(R_o)}{J} + \frac{15(10^3)(Q_z)}{I(0.015)} \qquad (\tau_{xy})_B = \frac{2(10^3)(R_o)}{J}$$

The principal stress one should be less than or equal to 150 MPa. The value can be written as:

$$(\sigma_{max})_A = \frac{(\sigma_{xx})_A}{2} + \sqrt{\left[\frac{(\sigma_{xx})_A}{2}\right]^2 + (\tau_{xz})_A^2} \le 150(10^6) \qquad or \qquad f_1(R_o) = \frac{(\sigma_{xx})_A}{2} + \sqrt{\left[\frac{(\sigma_{xx})_A}{2}\right]^2 + (\tau_{xz})_A^2} - 150(10^6) \le 0 \tag{2}$$

$$(\sigma_{max})_B = \frac{(\sigma_{xx})_B}{2} + \sqrt{\left[\frac{(\sigma_{xx})_B}{2}\right]^2 + (\tau_{xy})_B^2} \le 150(10^6) \quad or \quad f_2(R_o) = \frac{(\sigma_{xx})_B}{2} + \sqrt{\left[\frac{(\sigma_{xx})_B}{2}\right]^2 + (\tau_{xy})_B^2} - 150(10^6) \le 0 \quad (3)$$

The calculation of the various quantities as a function of R_o can be done on a spread sheet. The results are shown in the table above. $f_2(R_o)$ is always less than zero for values of R_o shown in column one. $f_1(R_o)$ changes sign at the value of $R_o = 0.064$ m.

ANS $R_o = 64$ mm

---

10. 64

Solution T = 30 kN-m P = 100 kN σ_{max} and τ_{max} vs. x

The axial, torsional, maximum normal stress, and maximum shear stress can be found as shown below.

$$\sigma_{xx} = \frac{100(10^3)}{\pi R^2} = \frac{31.83(10^3)}{R^2} \qquad \tau_{x\theta} = \frac{(30)(10^3)(R)}{(\pi R^4/2)} = \frac{19.098(10^3)}{R^3} \quad (1)$$

$$\sigma_{max} = \frac{\sigma_{xx}}{2} + \sqrt{\left(\frac{\sigma_{xx}}{2}\right)^2 + \tau_{x\theta}^2} \qquad \tau_{max} = \sqrt{\left(\frac{\sigma_{xx}}{2}\right)^2 + \tau_{x\theta}^2} \quad (2)$$

The table below shows the calculation of the stresses and the figure shows the plot of σ_{max} and τ_{max} vs. x.

x (m	R (mm)	σ_{xx} (MPa)	$\tau_{x\theta}$ (MPa)	σ_{max} (MPa)	τ_{max} (MPa)
0.00	100.60	3.15	18.76	20.40	18.82
0.10	92.70	3.70	23.98	25.90	24.05
0.20	82.60	4.67	33.89	36.30	33.97
0.30	79.60	5.02	37.87	40.46	37.95
0.40	75.90	5.53	43.68	46.53	43.77
0.50	68.80	6.72	58.65	62.10	58.74
0.60	68.00	6.88	60.74	64.28	60.84
0.70	65.90	7.33	66.73	70.50	66.83
0.80	60.10	8.81	87.98	92.50	88.09
0.90	60.30	8.75	87.11	91.59	87.22
1.00	59.10	9.11	92.52	97.19	92.63
1.10	54.00	10.92	121.29	126.87	121.41
1.20	54.80	10.60	116.05	121.47	116.17
1.30	54.10	10.88	120.62	126.18	120.74
1.40	49.40	13.04	158.42	165.08	158.56
1.50	50.60	12.43	147.42	153.76	147.55

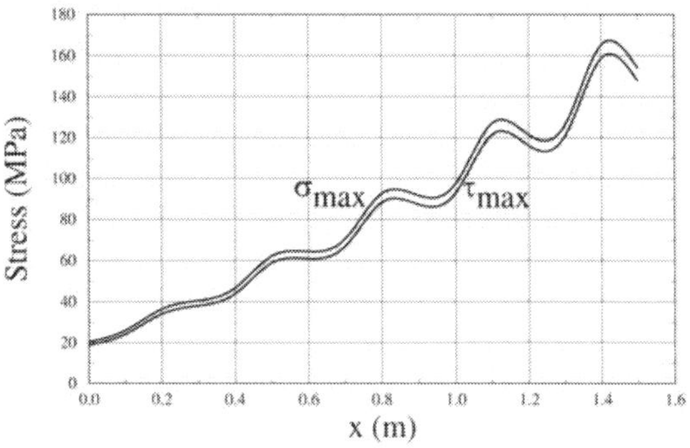

Section 10.3

10. 65

Solution $\sigma_{xx} = 10P$ MPa $\sigma_{yy} = -20P$ MPa $\tau_{xy} = 5P$ MPa $\sigma_{yield} = 160$MPa
$P_{max} = ?$ Max. Shear $P_{max} = ?$ Max. Distortion . Plane stress

The principal stresses, maximum shear stress, von-Misses stress can be found as shown below.

$$\sigma_{1,2} = \frac{(10P - 20P)}{2} \pm \sqrt{\left(\frac{10P + 20P}{2}\right)^2 + (5P)^2} = -5P \pm 15.81P \quad or \quad \sigma_1 = 10.81P \quad \sigma_2 = -20.81P \quad \sigma_3 = 0 \quad (1)$$

$$\tau_{max} = |\sigma_1 - \sigma_2/2| = 15.81P \le \sigma_{yield}/2 \quad or \quad P \le 5.06 \quad (2)$$

$$\sigma_{von} = \frac{1}{\sqrt{2}}\sqrt{(\sigma_1-\sigma_2)^2 + (\sigma_2-\sigma_3)^2 + (\sigma_3-\sigma_1)^2} = \frac{P}{\sqrt{2}}\sqrt{(31.62)^2 + (20.81)^2 + (10.81)^2} = 27.836P \le 160 \quad or \quad P \le 5.75 \quad (3)$$

ANS (a) $P_{max} = 5kN$ (b) $P_{max} = 5.75kN$

---

10. 66

Solution: $\sigma_{xx} = -4P$ $\sigma_{yy} = 3P$ $\tau_{xy} = -5P$ $\sigma_T < 18$ ksi $\sigma_c < 32$ ksi $P_{max} = ?$

The principal stresses can be found and Mohr's modified theory used as shown below.

$$\sigma_{1,2} = \frac{(-4P + 3P)}{2} \pm \sqrt{\left(\frac{-4P - 3P}{2}\right)^2 + (-5P)^2} = -0.5P \pm 6.1P \quad or \quad \sigma_1 = 5.6P\ ksi(T) \quad \sigma_2 = -6.6P = 6.6P\ ksi(C) \quad (1)$$

$$\frac{\sigma_1}{\sigma_T} - \frac{\sigma_2}{\sigma_C} \le 1 \quad or \quad \frac{5.6P}{18} - \frac{6.6P}{32} \le 1 \quad or \quad 0.1049P \le 1 \quad or \quad P \le 9.54 \quad (2)$$

ANS $P_{max} = 9.5kips$

10. 67

Solution: $\sigma_{xx} = 9$ ksi $\sigma_{yy} = -6$ ksi $\tau_{xy} = -4$ ksi $\sigma_T = 18$ ksi $\sigma_c = 32$ ksi k = ?

The principal stresses can be found and Mohr's modified theory used as shown below.

$$\sigma_{1,2} = \frac{(9-6)}{2} \pm \sqrt{\left(\frac{9+6}{2}\right)^2 + (-4)^2} = 1.5 \pm 8.5 \quad or \quad \sigma_1 = 10ksi(T) \quad \sigma_2 = -7ksi = 7ksi(C) \quad (1)$$

$$F = \frac{\sigma_1}{\sigma_T} - \frac{\sigma_1}{\sigma_C} = \frac{10}{18} - \frac{7}{32} = 0.3368 \quad k = \frac{1}{F} = \frac{1}{0.3368} = 2.969 \quad (2)$$

ANS $k = 2.969$

---

10. 68

Solution: E = 10,000 ksi, ν= 0.25, $\sigma_{yeild} = 24$ ksi, k = ?

From the given figure we have $\varepsilon_{xx} = -600\mu$, $\theta_b = 60^o$ and $\theta_c = 135^O$. Substituting into strain transformation equations we obtain the following.

$$\varepsilon_b = -600(cos 60)^2 + \varepsilon_{yy}(sin 60)^2 + \sigma_{xy}(cos 60)(sin 60) = 500 \quad or \quad 0.75\varepsilon_{yy} + 0.433\gamma_{xy} = 650 \quad (1)$$

$$\varepsilon_c = -600(cos 135)^2 + \varepsilon_{yy}(sin 135)^2 + \gamma_{xy}(cos 135)(sin 135) = 400 \quad or \quad 0.5\varepsilon_{yy} - 0.5\gamma_{xy} = 700 \quad (2)$$

Solving we obtain: $\varepsilon_{yy} = 1061.8\mu$ *and* $\gamma_{xy} = -338.1\mu$

From Generalized Hooke's law we obtain the following.

$$G = E/[2(1+\nu)] = 10000/[2(1.25)] = 4000ksi \quad \tau_{xy} = G\gamma_{xy} = -1.352ksi \quad (3)$$

$$\sigma_{xx} - \nu\sigma_{yy} = E\varepsilon_{xx} = (10000)(-600)(10^{-6}) = -6ksi \quad \sigma_{yy} - \nu\sigma_{xx} = E\varepsilon_{xy} = (10000)(1061.8)(10^{-6}) = 10.62ksi \quad (4)$$

Solving with ν = 0.25 we obtain: $\sigma_{xx} = -3.568ksi$ *and* $\sigma_{yy} = 9.728ksi$

$$\sigma_{1,2} = \frac{(-3.568+9.728)}{2} \pm \sqrt{\left(\frac{-3.568-9.728}{2}\right)^2 + (-1.352)^2} = 3.080 \pm 6.784 \quad or \quad \sigma_1 = 9.864ksi \quad \sigma_2 = -3.704ksi \quad (5)$$

$$\tau_{max} = |(\sigma_1 - \sigma_2)/2| = 6.784ksi \quad k = \frac{(\sigma_{yield}/2)}{\tau_{max}} = \frac{12}{6.784} = 1.76 \quad (6)$$

ANS $k = 1.76$

---

10. 69

Solution: E = 200 GPa, ν= 0.28, $\sigma_{yield} = 210$ MPa, k = ?

From the given figure we have $\varepsilon_{xx} = -800\mu$, $\varepsilon_{yy} = -300\mu$, and $\theta_c = 135^O$. Substituting in strain transformation equation we obtain the following.

$$\varepsilon_c = -800(cos 135)^2 + (-300)(sin 135)^2 + \gamma_{xy}(cos 135)(sin 135) = -700 \quad or \quad \gamma_{xy} = 300\mu \quad (1)$$

From Generalized Hooke's law we have the following.

$$G = 200/[2(1+0.28)] = 78.125GPa \quad \tau_{xy} = 300(10^{-6})(78.125)(10^9) = 23.44(10^6)N/m^2 = 23.44MPa \quad (2)$$

$$\sigma_{xx} - \nu\sigma_{yy} = E\varepsilon_{xx} = (200)(10^9)(-800)(10^{-6}) = -160(10^6)N/m^2 = -160MPa \quad (3)$$

$$\sigma_{yy} - \nu\sigma_{xx} = E\varepsilon_{yy} = (200)(10^9)(-300)(10^{-6}) = -60(10^6)N/m^2 = -60MPa \quad (4)$$

Solving with ν = 0.28 we obtain: $\sigma_{xx} = -191.8MPa$ and $\sigma_{yy} = -113.7MPa$

$$\sigma_{1,2} = \frac{-191.8-113.7}{2} \pm \sqrt{\left(\frac{-191.8+113.7}{2}\right)^2 + 23.44^2} = -152.75 \pm 45.55 \quad \sigma_1 = -107.2 \quad \sigma_2 = -198.3 \quad \sigma_3 = 0 \quad (5)$$

$$\sigma_{von} = \frac{1}{\sqrt{2}}\sqrt{(\sigma_1-\sigma_2)^2 + (\sigma_2-\sigma_3)^2 + (\sigma_3-\sigma_1)^2} = \frac{1}{\sqrt{2}}\sqrt{(-107.2+198.3)^2 + (107.2)^2 + (198.3)^2} = 171.9MPa \quad (6)$$

$$k = \sigma_{yield}/\sigma_{von} = 210/171.9 = 1.22 \quad (7)$$

ANS $k = 1.22$

---

10. 70

Solution: r = 3 ft., t = 1/2 in., $\sigma_{yield} = 30$ ksi, $p_{max} = ?$

There are no shear stresses. The principal stresses are the axial and hoop stresses.

$$\sigma_1 = \sigma_{\theta\theta} = p(36)/(0.5) = 72p \quad \sigma_2 = \sigma_{xx} = p(36)/[2(0.5)] = 36p \quad \sigma_3 = 0 \quad (1)$$

$$\sigma_{von} = \frac{1}{\sqrt{2}}\sqrt{(\sigma_1-\sigma_2)^2+(\sigma_2-\sigma_3)^2+(\sigma_3-\sigma_1)^2} = \frac{P}{\sqrt{2}}\sqrt{(36p)^2+(36p)^2+(72p)^2} = 62.35p \le 30 \quad or \quad p \le 0.481 ksi \quad \textbf{(2)}$$

ANS $p = 481 psi$

---

10. 71

Solution: 18in < R < 36in   p = 750 psi σ_{yield} = 60 ksi   k = 1.5   Failure envelope: R and t

The allowable, axial, hoop, principal , and von Mises stresses can be found as shown below.,

$$\sigma_{max} = 60/1.5 = 40psi \qquad \sigma_1 = pR/t = 750R/t \qquad \sigma_2 = pR/(2t) = 375R/t \qquad \sigma_3 = 0 \quad \textbf{(1)}$$

$$\sigma_{von} = \frac{1}{\sqrt{2}}\sqrt{(750-375)^2+(750)^2+(375)^2}\left(\frac{R}{t}\right) = 649.5\left(\frac{R}{t}\right) \le 40(10^3) \quad or \quad (R/t) \le 61.58 \quad \textbf{(2)}$$

The value of R and t that satisfy the above equation and are in the range for r that is given can be found using a spread sheet as shown in table below. The failure envelope is shown in the figure below.

R (in)	t (in)
18	0.292
20	0.325
22	0.357
24	0.390
26	0.422
28	0.455
30	0.487
32	0.520
34	0.552
36	0.585

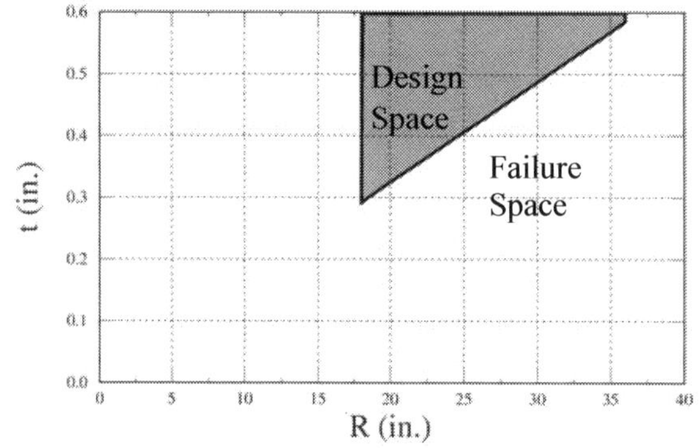

---

10. 72

Solution

The solution proceeds as follows.

$$\sigma_1 = \frac{(\sigma_{xx}-\sigma_{yy})}{2} + \sqrt{\left(\frac{\sigma_{xx}+\sigma_{yy}}{2}\right)^2+(\tau_{xy})^2} \qquad \sigma_2 = \frac{(\sigma_{xx}-\sigma_{yy})}{2} - \sqrt{\left(\frac{\sigma_{xx}+\sigma_{yy}}{2}\right)^2+(\tau_{xy})^2} \qquad \sigma_3 = 0 \quad \textbf{(1)}$$

$$(\sigma_1-\sigma_2)^2 = 4\left[\frac{(\sigma_{xx}-\sigma_{yy})^2}{2}+(\tau_{xy})^2\right] = (\sigma_{xx})^2+(\sigma_{yy})^2-2(\sigma_{xx})(\sigma_{yy})+4(\tau_{xy})^2 \quad \textbf{(2)}$$

$$(\sigma_1)^2+(\sigma_2)^2 = 2\left[\frac{(\sigma_{xx}+\sigma_{yy})^2}{2}+\frac{(\sigma_{xx}-\sigma_{yy})^2}{2}+(\tau_{xy})^2\right] = 2\left[\frac{2(\sigma_{xx})^2}{4}+\frac{2(\sigma_{yy})^2}{4}+(\tau_{xy})^2\right] = (\sigma_{xx})^2+(\sigma_{yy})^2+2(\tau_{xy})^2 \quad \textbf{(3)}$$

$$\sigma_{von} = \frac{\sqrt{(\sigma_1-\sigma_2)^2+(\sigma_1)^2+(\sigma_2)^2}}{\sqrt{2}} = \frac{\sqrt{2(\sigma_{xx})^2+2(\sigma_{yy})^2-2(\sigma_{xx})(\sigma_{yy})+6(\tau_{xy})^2}}{\sqrt{2}} = \sqrt{(\sigma_{xx})^2+(\sigma_{yy})^2-(\sigma_{xx})(\sigma_{yy})+3(\tau_{xy})^2} \quad \textbf{(4)}$$

Above equation is the desired result.

---

10. 73

Solution:

For plain strain we have the following.

$$\varepsilon_{zz} = 0 = [\sigma_{zz}-\nu(\sigma_{xx}+\sigma_{yy})]/E \quad or \quad \sigma_{zz} = \nu(\sigma_{xx}+\sigma_{yy}) \qquad \tau_{zy} = 0 \qquad \tau_{zx} = 0 \quad \textbf{(1)}$$

Substituting the stress values in the given equation we obtain the following.

$$\sigma_{von} = \sqrt{\sigma_{xx}^2+\sigma_{yy}^2+\sigma_{zz}^2-\sigma_{xx}\sigma_{yy}-\sigma_{zz}(\sigma_{xx}+\sigma_{yy})+3\tau_{xy}^2} = \sqrt{\sigma_{xx}^2+\sigma_{yy}^2-\sigma_{xx}\sigma_{yy}+\nu^2(\sigma_{xx}+\sigma_{yy})^2-\nu(\sigma_{xx}+\sigma_{yy})+3\tau_{xy}^2} \text{ or}$$

$$\sigma_{von} = \sqrt{\sigma_{xx}^2+\sigma_{yy}^2-\sigma_{xx}\sigma_{yy}+(\nu^2-\nu)[\sigma_{xx}^2+\sigma_{yy}^2+2\sigma_{xx}\sigma_{yy}]+3\tau_{xy}^2} = \sqrt{(\sigma_{xx}^2+\sigma_{yy}^2)(1+\nu^2-\nu)-\sigma_{xx}\sigma_{yy}(1+2\nu-2\nu^2)+3\tau_{xy}^2} \quad \textbf{(2)}$$

Above equation is the desired result.

---

10. 74

Solution: σ_{von} = ?   at θ = 0 and θ = π / 2

Substituting θ = 0 in the given equations we obtain the following.

$$\sigma_{xx} = \frac{K_I}{\sqrt{2\pi r}} \qquad \sigma_{yy} = \frac{K_I}{\sqrt{2\pi r}} \qquad \tau_{xy} = 0 \qquad \sigma_1 = \sigma_2 = \frac{K_I}{\sqrt{2\pi r}} \qquad \sigma_3 = 0 \tag{1}$$

$$\sigma_{von} = \frac{1}{\sqrt{2}}\sqrt{(\sigma_1-\sigma_2)^2+(\sigma_1)^2+(\sigma_2)^2} = \frac{1}{\sqrt{2}}\frac{K_I}{\sqrt{2\pi r}}\sqrt{2} = \frac{0.7071K_I}{\sqrt{\pi r}} \tag{2}$$

Substituting $\theta = \pi / 2$ in the given equations we obtain the following.

$$\sigma_{xx} = \frac{K_I}{4\sqrt{\pi r}} \qquad \sigma_{yy} = \frac{3K_I}{4\sqrt{\pi r}} \qquad \tau_{xy} = \frac{-K_I}{4\sqrt{\pi r}} \qquad \sigma_{1,2} = \frac{K_I}{2\sqrt{\pi r}} \pm \sqrt{\left(\frac{K_I}{4\sqrt{\pi r}}\right)^2+\left(\frac{-K_I}{4\sqrt{\pi r}}\right)^2} = \frac{K_I}{\sqrt{\pi r}}\left(\frac{1}{2} \pm \frac{\sqrt{2}}{4}\right) \tag{3}$$

$$\sigma_1 = \frac{0.85355K_I}{\sqrt{\pi r}} \qquad \sigma_2 = \frac{0.14645K_I}{\sqrt{\pi r}} \qquad \sigma_3 = 0 \tag{4}$$

$$\sigma_{von} = \frac{1}{\sqrt{2}}\sqrt{(\sigma_1-\sigma_2)^2+(\sigma_1)^2+(\sigma_2)^2} = \frac{K_I}{\sqrt{2}\sqrt{\pi r}}\sqrt{(0.7071)^2+(0.8535)^2+(0.1464)^2} = \frac{0.7071K_I}{\sqrt{\pi r}} \tag{5}$$

ANS (a) $\sigma_{von} = \dfrac{0.7906K_I}{\sqrt{\pi r}}$ (b) $\sigma_{von} = \dfrac{0.7071K_I}{\sqrt{\pi r}}$

CHAPTER 11

Section 11.1

11. 1

Solution: $P_{cr} = f(k,L) = ?$

The deformation of the spring and moment at point A are as shown below.

$$\delta = L\sin\theta \qquad M_A = PL\sin\theta - (kL\sin\theta)L\cos\theta = L\sin\theta(P - kL\cos\theta) \tag{1}$$

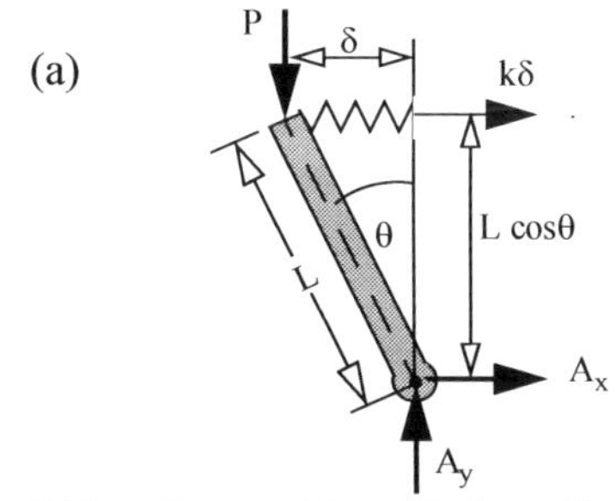

When P = kL then equilibrium condition of $M_A = 0$ cannot be met at any θ except θ = 0.

ANS $P_{cr} = kL$

11. 2

Solution: $P_{cr} = f(k,L) = ?$

The deformation of the springs and the moment at point A are as shown below.

$$\delta_1 = L\sin\theta \qquad \delta_2 = (L/2)\sin\theta \qquad M_A = PL\sin\theta - (kL\sin\theta)L\cos\theta - \left(\frac{kL}{2}\sin\theta\right)\frac{L}{2}\cos\theta = L\sin\theta\left(P - \frac{5kL}{4}\cos\theta\right) \tag{1}$$

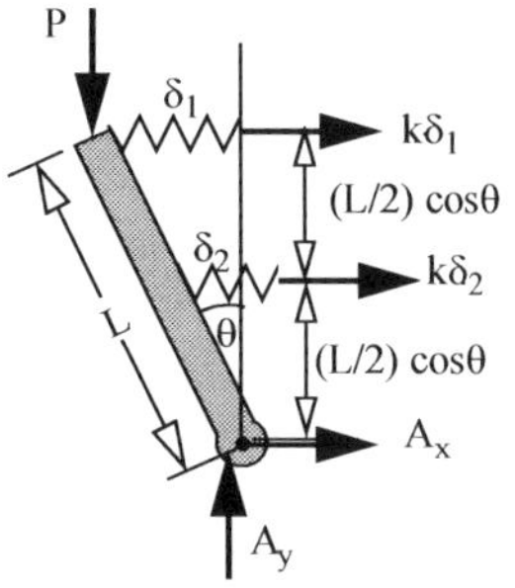

When P=5kL/4 the equilibrium condition of $M_A = 0$ cannot be met at any θ except θ = 0.

ANS $P_{cr} = 5kL/4$

11. 3

Solution: $P_{cr} = f(k,L) = ?$

The deformation of the spring, equilibrium of forces in Figure (b) and the moment at point A in Figure (a) are as follows.

$$\delta = (L/2)\sin\theta \qquad N\cos\theta = P \qquad R = N\sin\theta = P\tan\theta \tag{1}$$

$$M_A = R\left(\frac{L}{2}\cos\theta + \frac{L}{2}\cos\theta\right) - \left(\frac{kL}{2}\sin\theta\right)\frac{L}{2}\cos\theta = P\tan\theta(L\cos\theta) - \frac{kL^2}{4}\sin\theta\cos\theta = L\sin\theta\left(P - \frac{kL}{4}\cos\theta\right) \tag{2}$$

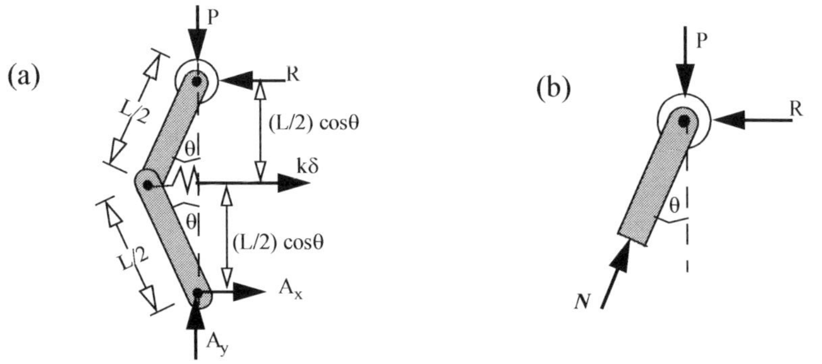

When P= kL/4 the equilibrium condition of $M_A = 0$ cannot be met at any θ except θ = 0.

ANS $P_{cr} = kL/4$

11.4

Solution: P_{cr} =?

The deformation of the linear spring and moment at point A can be written as follows.

$$:\delta = 1.2 sin\theta \qquad M_A = P(1.2)sin\theta - 25(1.2sin\theta)(1.2cos\theta) - 30\theta \tag{1}$$

(a)

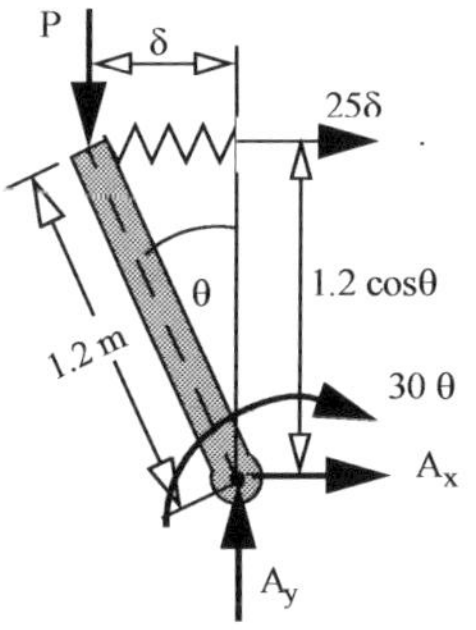

For small perturbations from the equilibrium position of $\theta = 0$, we can approximate $sin\theta \approx \theta$ and $cos\theta \approx 1$ to obtain

$$M_A = 1.2P\theta - 25(1.2)^2\theta - 30 = (1.2P - 36 - 30)\theta = (1.2P - 66)\theta \tag{2}$$

For P= 66/1.2= 55kN, the equilibrium condition of $M_A = 0$ is met at every θ, implying the system is in neutral equilibrium.

ANS $P_{cr} = 55$ kN

11.5

Solution: P_{cr} =?

The deformation of the linear springs and moment at point A can be written as shown below.

$$\delta_1 = 60 sin\theta \qquad \delta_2 = 30 sin\theta \qquad M_A = P(60)sin\theta - 8(60sin\theta)(60cos\theta) - 8(30sin\theta)(30cos\theta) - 2000\theta \tag{1}$$

(a)

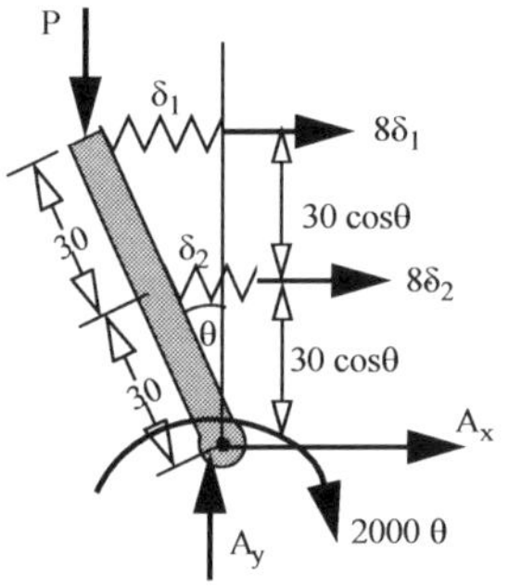

For small perturbations from the equilibrium position of $\theta = 0$, we can approximate $sin\theta \approx \theta$ and $cos\theta \approx 1$ to obtain

$$M_A \cong 60P\theta - 28800\theta - 7200\theta - 2000\theta = (60P - 38000)\theta \tag{2}$$

For P= 38000/60 = 633.3 lb., the equilibrium condition of $M_A = 0$ is met at every θ, implying the system is in neutral equilibrium.

ANS $P_{cr} = 633.3 lb$

11.6

Solution: P_{cr} =?

The deformation of the spring, the force equilibrium in Figure (b) and the moment at point A in Figure (a) can be written as follows.

$$\delta = 30 sin\theta \qquad N = P cos\theta \qquad R = N sin\theta = P tan\theta \tag{1}$$

$$\mathbf{M}_A = R(30cos\theta + 30cos\theta) - 8(30sin\theta)(30cos\theta) - 2000\theta = 60P sin\theta - 7200 sin\theta cos\theta - 2000\theta \tag{2}$$

(a)

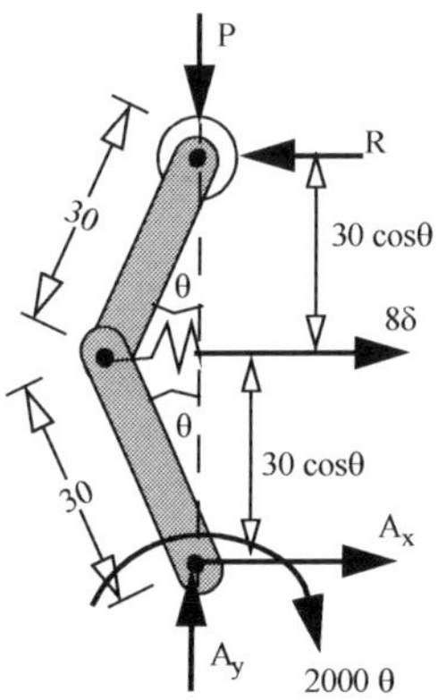

(b)

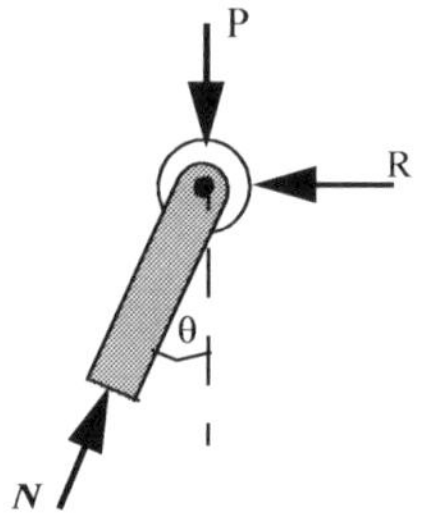

For small perturbations from the equilibrium position of $\theta = 0$, we can approximate $sin\theta \approx \theta$ and $cos\theta \approx 1$ to obtain

$$M_A = 60P\theta - 7200\theta - 2000\theta = (60P - 9200)\theta \quad \textbf{(3)}$$

For P= 9200/60 = 153.3 lb., the equilibrium condition of $M_A = 0$ is met at every θ, implying the system is in neutral equilibrium.

ANS $P_{cr} = 153.3 \text{ lb}$

---

11. 7

Solution k = 25 kN/m K= 30 kN/rad. P_{cr} =?

By geometry and equilibrium of forces in Figure (b) we obtain the following.

$$BD = 1.2sin\theta_1 \approx 1.2\theta_1 \qquad CE = 1.2sin\theta_2 \approx 1.2\theta_2 \qquad BE = 1.2cos\theta_2 \approx 1.2 \qquad AD = 1.2cos\theta_1 \approx 1.2 \quad \textbf{(1)}$$

$$\delta = CF = CE + BD = 1.2(\theta_1 + \theta_2) \quad \textbf{(2)}$$

$$B_x = k\delta = (25)[1.2(\theta_1 + \theta_2)] = 30(\theta_1 + \theta_2) \qquad B_y = P \quad \textbf{(3)}$$

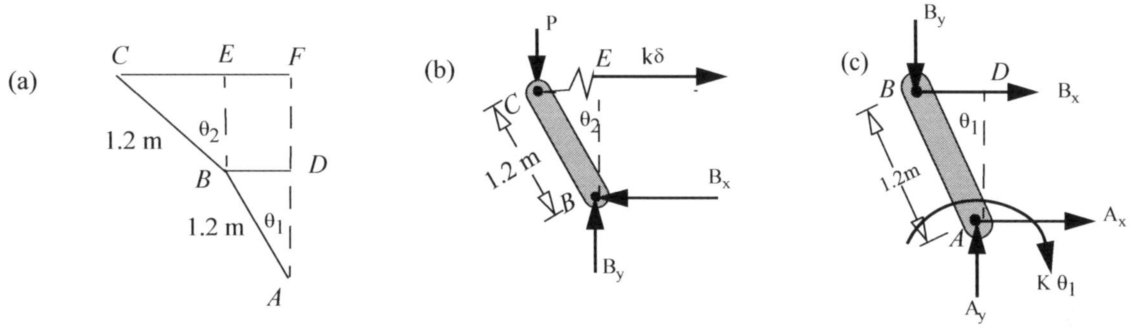

By equilibrium of moments about point *B* and *A* we obtain two equations which can be written in matrix form as shown below.

$$B_y(BD) - B_x(AD) - K\theta_1 = 0 \quad or \quad P(1.2\theta_1) - 30(\theta_1 + \theta_2)(1.2) - 30\theta_1 = 0 \quad or \quad (P - 55)\theta_1 - 30\theta_2 = 0 \quad \textbf{(4)}$$

$$P(CE) - (k\delta)(BE) = P(1.2\theta_2) - (25)[1.2(\theta_1 + \theta_2)](1.2) = 0 \quad or \quad (P - 30)\theta_2 - 30\theta_1 = 0 \quad \textbf{(5)}$$

$$\begin{bmatrix} (P-55) & -30 \\ -30 & (P-30) \end{bmatrix} \begin{Bmatrix} \theta_1 \\ \theta_2 \end{Bmatrix} = \begin{Bmatrix} 0 \\ 0 \end{Bmatrix} \quad \textbf{(6)}$$

$$(P - 55)(P - 30) - 900 = 0 \quad or \quad P^2 - 85P + 750 = 0 \quad or \quad P_{1,2} = \frac{85 \pm \sqrt{85^2 - 4(750)}}{2} = \frac{85 \pm 65}{2} \quad \textbf{(7)}$$

$$P_1 = 75 \text{ kN} \qquad P_2 = 10 \text{ kN} \quad \textbf{(8)}$$

ANS $P_{cr} = 10 \text{ kN}$

---

11. 8

Solution k = 8 lb/in K= 2000 in.-lb/rad P_{cr} =?

By geometry and equilibrium of forces in Figure (b) we obtain the following.

$$BD = 30sin\theta_1 \approx 30\theta_1 \qquad CE = 30sin\theta_2 \approx 30\theta_2 \qquad BE = 30cos\theta_2 \approx 30 \qquad AD = 30cos\theta_1 \approx 30 \quad \textbf{(1)}$$

$$\delta_1 = BD = 30\theta_1 \qquad \delta_2 = CF = CE + BD = 30(\theta_1 + \theta_2) \quad \textbf{(2)}$$

$$B_x = k\delta = (8)[30(\theta_1 + \theta_2)] = 240(\theta_1 + \theta_2) \qquad B_y = P \quad \textbf{(3)}$$

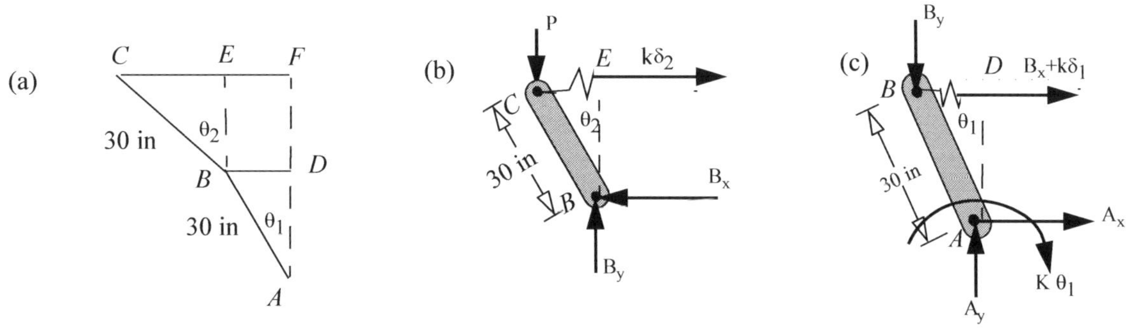

By equilibrium of moments about point *B* and *A* we obtain two equations which can be written in matrix form as shown below.

$$B_y(BD) - (B_x + k\delta_1)(AD) - K\theta_1 = 0 \quad or \quad P(30\theta_1) - [240(\theta_1 + \theta_2) + 240\theta_1](30) - 2000\theta_1 = 0 \text{ or}$$

$$(30P - 7200 - 7200 - 2000)\theta_1 - 7200\theta_2 = 0 \quad or \quad (P - 546.7)\theta_1 - 240\theta_2 = 0 \quad \textbf{(4)}$$

$$P(CE) - (k\delta_2)(BE) = P(30\theta_2) - (8)[30(\theta_1 + \theta_2)](30) = 0 \quad or \quad (P - 240)\theta_2 - 240\theta_1 = 0 \quad \textbf{(5)}$$

$$\begin{bmatrix} (P-546.7) & -240 \\ -240 & (P-240) \end{bmatrix} \begin{Bmatrix} \theta_1 \\ \theta_2 \end{Bmatrix} = \begin{Bmatrix} 0 \\ 0 \end{Bmatrix} \quad (6)$$

$$(P-546.7)(P-240)-240^2 = 0 \quad or \quad P^2 - 786.67P + 73600 = 0 \quad (7)$$

$$P_{1,2} = \frac{786.67 \pm \sqrt{786.67^2 - 4(73600)}}{2} = \frac{786.67 \pm 569.6}{2} \qquad P_1 = 678.13 \text{ lb} \qquad P_2 = 108.53 \text{ lb} \quad (8)$$

ANS $P_{cr} = 108.5$ lb

Section 11.2

11. 9

Solution: E =200 GPa L=5m d_i=75 mm d_o=100 mm L/r =? P_{cr} =? σ_{cr} =? P_{cr2} =?

The solution proceeds as follows.

$$A = \frac{\pi}{4}[0.1^2 - 0.075^2] = 3.436(10^{-3})m^2 \qquad I = \frac{\pi}{64}[0.1^4 - 0.075^4] = 3.356(10^{-6})m^4 \quad (1)$$

$$r = \sqrt{\frac{I}{A}} = 31.25(10^{-3})m \qquad \frac{L}{r} = \frac{5}{31.25(10^{-3})} = 160 \quad (2)$$

$$(P_{cr}) = \frac{\pi^2 EI}{L^2} = \frac{\pi^2(200)(10^9)(3.356)(10^{-6})}{5^2} = 264.9(10^3)N \qquad \sigma_{cr} = \frac{P_{cr}}{A} = \frac{264.9(10^3)}{(3.436)(10^{-3})} = 77.1(10^6)N/m^2 \quad (3)$$

$$P_{cr2} = 2^2(\pi^2 EI/L^2) = 4P_{cr} = 1059.6(10^3)N \quad (4)$$

ANS $L/r = 160$; $P_{cr} = 265$ kN ; $\sigma_{cr} = 77.1$ MPa (T) ; $P_{cr2} = 1060$ kN

---

11. 10

Solution: L = 30 ft E = 30000 ksi t= 0.5 in 4in x 4in built in ends L_{eff}/r =? P_{cr} =? σ_{cr} =?

The solution proceeds as follows.

$$A = (4)(4)-(3)(3) = 7in^2 \qquad I = [(4)(4)^3 - (3)(3)^3]/12 = 14.583in^4 \quad (1)$$

$$r = \sqrt{\frac{I}{A}} = 1.443in \qquad L_{eff} = 0.5L = (0.5)(30)(12) = 180in \qquad \frac{L_{eff}}{r} = \frac{180}{1.443} \quad (2)$$

$$(P_{cr}) = \frac{4\pi^2 EI}{L^2} = \frac{4\pi^2(30000)(14.583)}{(360)^2} \qquad \sigma_{cr} = \frac{P_{cr}}{A} = \frac{133.3}{7} = 19.04ksi \quad (3)$$

ANS $L_{eff}/r = 124.7$; $P_{cr} = 133.3$ kips ; $\sigma_{cr} = 19.0$ ksi (C)

---

11. 11

Solution L= 10 ft E = 1,800 ksi 4 in. x 6 in $(P_{cr})_1$ =? $(P_{cr})_2$ =? case 1

$$I_{min} = (6)(4^3)/12 = 32 \text{ in}^4 \qquad (P_{cr})_1 = \pi^2 EI_{min}/L^2 = \pi^2(1800)(32)/120^2 = 39.47 \text{ kips} \quad (1)$$

$$(P_{cr})_2 = 4(P_{cr})_1 = 157.88 \text{ kips} \quad or \quad I_{max} = \frac{1}{12}(6^3)(4) = 72 \text{ in}^4 \qquad (P_{cr})_2 = \frac{\pi^2 EI_{max}}{L^2} = \frac{\pi^2(1800)(72)}{120^2} = 88.83 \text{ kips} \quad (2)$$

ANS $(P_{cr})_1 = 39.47$ kips ; $(P_{cr})_2 = 88.83$ kips

---

11. 12

Solution L= 4 m E = 210 GPa 120 mm x 80 mm t=10 mm $(P_{cr})_1$ =? $(P_{cr})_2$ =? case 2

$$I_{min} = [0.12(0.08^3) - 0.10(0.06^3)]/12 = 3.32(10^{-6}) \text{ m}^4 \qquad I_{max} = [0.08(0.12^3) - 0.06(0.10^3)]/12 = 6.52(10^{-6}) \text{ m}^4 \quad (1)$$

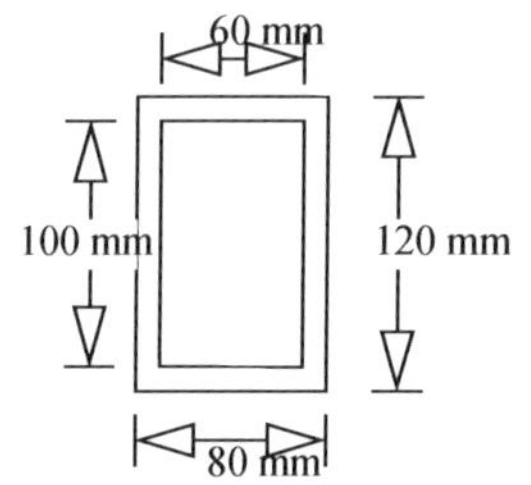

$$(P_{cr})_1 = \frac{\pi^2 EI_{min}}{4L^2} = \frac{\pi^2(210)(10^9)(3.32)(10^{-6})}{4(4^2)} = 107.5(10^3)\text{ N} = 107.5\text{ kN} \quad (1)$$

$$(P_{cr})_2 = 9(P_{cr})_1 = 967.6\text{ kN} \quad or \quad (P_{cr})_2 = \frac{\pi^2 EI_{max}}{4L^2} = \frac{\pi^2(210)(10^9)(6.52)(10^{-6})}{4(4^2)} = 211.15(10^3)\text{ N} = 211.15\text{ kN} \quad (2)$$

ANS $(P_{cr})_1 = 107.5\text{ kN}$; $(P_{cr})_2 = 211.2\text{ kN}$

---

11. 13

Solution L= 12 ft E = 1,800 ksi 6 in. x 8 in $(P_{cr})_1$ =? $(P_{cr})_2$ =? case 3

$$I_{min} = 8(6^3)/12 = 144\text{ in}^4 \qquad I_{max} = 6(8^3)/12 = 256\text{ in}^4 \quad (1)$$

$$(P_{cr})_1 = \frac{20.13EI_{min}}{L^2} = \frac{20.13(1800)(144)}{144^2} = 251.6\text{ kips} \quad (2)$$

$$(P_{cr})_2 = \frac{59.68EI_{min}}{L^2} = \frac{59.68(1800)(144)}{144^2} = 746\text{ kips} \quad or \quad (P_{cr})_2 = \frac{20.13EI_{max}}{L^2} = \frac{\pi^2(1800)(256)}{144^2} = 447.3\text{ kips} \quad (3)$$

ANS $(P_{cr})_1 = 251.6\text{ kips}$; $(P_{cr})_2 = 447.3\text{ kips}$

---

11. 14

Solution L= 4 m E = 210 GPa 120 mm x 90 mm t=15 mm $(P_{cr})_1$ =? case 4

$$I_{min} = [0.12(0.09^3) - 0.90(0.06^3)]/12 = 5.67(10^{-6})\text{ m}^4 \qquad (P_{cr})_1 = \frac{4\pi^2 EI_{min}}{L^2} = \frac{\pi^2(210)(10^9)(5.67)(10^{-6})}{5^2} = 1857(10^3)\text{ N} \quad (1)$$

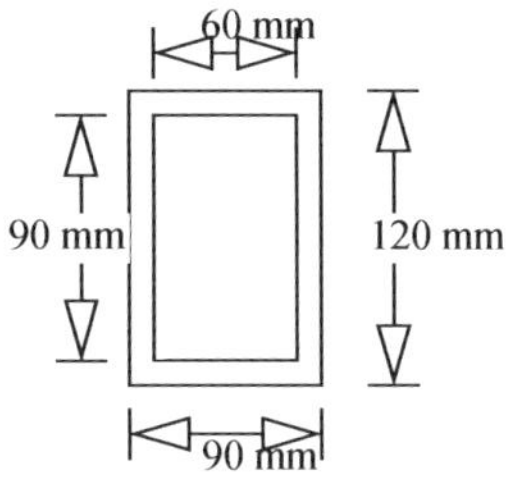

ANS $(P_{cr})_1 = 1857\text{ kN}$

---

11. 15

Solution: L = 20 ft E = 1800 ksi 8in x 8in built and simple support L_{eff}/r =? P_{cr} =? σ_{cr} =?

The solution proceeds as follows.

$$A = (8)(8) = 64in^2 \qquad I = \frac{1}{12}(8)(8)^3 = 341.33in^4 \quad (1)$$

$$r = \sqrt{\frac{I}{A}} = 2.309in \qquad L_{eff} = 0.7L = (0.7)(240) = 168in \qquad \frac{L_{eff}}{r} = \frac{168}{2.309} = 72.7 \quad (2)$$

$$P_{cr} = \frac{(20.19)EI}{L^2} = \frac{(20.19)(1800)(341.33)}{(240)^2} \qquad \sigma_{cr} = \frac{P_{cr}}{A} = \frac{215.4}{64} = 3.365ksi \quad (3)$$

ANS $L_{eff}/r = 72.7$; $P_{cr} = 215.4kip$; $\sigma_{cr} = 3.36ksi(C)$

---

11. 16

Solution W12x35 L = 21 ft E = 30000 ksi simple support L/r =? P_{cr} =? σ_{cr} =? P_{cr3} =?

From Section C.6 and definition of slenderness ratio we have the following.

$$A = 10.3\ in^2 \qquad I = 24.5\ in^4 \qquad r = 1.54\ in \qquad L/r = (21)(12)/1.54 = 163.6 \quad (1)$$

$$P_{cr} = \frac{\pi^2 EI}{L^2} = \frac{\pi^2(30000)(24.5)}{(252)^2} = 114.2kip \qquad \sigma_{cr} = \frac{P_{cr}}{A} = \frac{114.2}{10.3} = 11.09ksi \quad (2)$$

ANS $L/r = 163.6$; $P_{cr} = 114.2\text{ kip}$; $\sigma_{cr} = 11.1\text{ ksi (C)}$

---

11. 17

Solution S200 x 34 L = 6 m E = 200 GPa built in ends L_{eff}/r =? P_{cr} =? σ_{cr} =?

From Section C.6 and definition of slenderness ratio we have the following.

$$A = 4368\ mm^2 \qquad I = 1.794(10^6)\ mm^4 \qquad r = 20.3\ mm \qquad L_{eff} = 0.5L = 3\ m \qquad \frac{L_{eff}}{r} = \frac{3}{(20.3)(10^{-3})} = 147.78 \quad (1)$$

$$P_{cr} = \frac{4\pi^2 EI}{L^2} = \frac{4\pi^2(200)(10^9)(1.794)(10^{-6})}{6^2} = 393.47(10^3)\text{ N} \qquad \sigma_{cr} = \frac{P_{cr}}{A} = \frac{393.47(10^3)}{4368(10^{-6})} = 90.07(10^6)\text{ N/m}^2 \quad (2)$$

ANS $L_{eff}/r = 147.8$; $P_{cr} = 393.5$ kN $\sigma_{cr} = 90.07$ MPa (C)

---

11. 18

Solution t = 0.125 in E = 90 ksi σ_{yield} = 90 ksi Long and Short Columns = ?

The solution proceeds as follows.

$$A = a^2 - (a-0.25)^2 \qquad I = \frac{a^4 - (a-0.25)^4}{12} \qquad \sigma_{cr} = \frac{P_{cr}}{A} = \frac{\pi^2 E}{(L/r)_{cr}^2} = 90 \quad or \quad (L/r)_{cr} = \sqrt{\frac{\pi^2(9000)}{90}} = 31.41 \quad (1)$$

From the given values of L and a, we can calculate I, A, r, and L/r using spreadsheet as shown in the table below. Comparing L/r to $(L/r)_{cr}$ we can determine short and long column as shown below.

L (ft)	a (in)	I (in^4)	A (in^2)	r (in)	L/r	Column Type
1.0	1.125	0.085	0.500	0.411	29.167	Short
1.5	1.500	0.218	0.688	0.564	31.934	Long
2.0	1.750	0.360	0.813	0.665	36.071	Long
2.5	2.750	1.511	1.313	1.073	27.962	Short
3.0	3.000	1.984	1.438	1.175	30.643	Short
3.5	3.000	1.984	1.438	1.175	35.750	Long
4.0	3.000	1.984	1.438	1.175	40.857	Long

---

11. 19

Solution t = 10 mm E = 100 GPa σ_{yield} =600 MPa Long and Short Columns = ?

The solution proceeds as follows.

$$A = \pi[d^2 - (d-20)^2]/4\ mm^2 \qquad I = \pi[d^4 - (d-20)^4]/64\ mm^4 \quad (1)$$

$$\sigma_{cr} = \frac{P_{cr}}{A} = \frac{4\pi^2 E}{(L/r)_{cr}^2} = 600(10^6)\ N/m^2 \quad or \quad (L/r)_{cr} = \sqrt{\frac{4\pi^2(100)(10^9)}{600(10^6)}} = 81.11 \quad (2)$$

From the given values of L and d, the area moment of inertia I, the cross-sectional area A, the radius of gyration r, and the slenderness ratio can be found using as shown in the table below. Columns with slenderness ratio less than the critical slenderness ratio in Eq. 1 are short columns and those greater than the critical slenderness ratio are long columns as shown in the table below.

L (m)	d (mm)	I (mm^4)	A (mm^2)	r (mm)	L/r	Column Type
1	60	5.11E+05	1.57E+03	18.0	55.5	Short
2	80	1.37E+06	2.20E+03	25.0	80.0	Short
3	100	2.90E+06	2.83E+03	32.0	93.7	Long
4	150	1.08E+07	4.40E+03	49.6	80.6	Short
5	200	2.70E+07	5.97E+03	67.3	74.3	Short
6	225	3.91E+07	6.75E+03	76.1	78.8	Short
7	250	5.44E+07	7.54E+03	84.9	82.4	Long

---

11. 20

Solution: $P_{cr1}:P_{c2}:P_{cr3}$ =?

The dimensions for the three shapes in terms of cross sectional areas are calculated as shown below.

$$\text{Square} \quad a_s^2 = A \qquad a_s = \sqrt{A} \qquad I_s = \frac{1}{12}a_s^4 = \frac{A^2}{12} = 0.0833A^2 \quad (1)$$

$$\text{Circle} \quad \pi a_c^2 = A \qquad a_c = \sqrt{\frac{A}{\pi}} \qquad I_c = \frac{\pi}{12}a_c^4 = \frac{A^2}{4\pi} = 0.07958A^2 \quad (2)$$

$$\text{Triangle} \quad \frac{a_t^2 \sin 60}{2} = A \qquad a_t = \sqrt{\frac{4A}{\sqrt{3}}} \qquad I_t = \frac{1}{36}a_t(a_t \sin 60)^3 = \frac{3\sqrt{3}}{288}a_t^4 = \frac{A^2}{6\sqrt{3}} = 0.0962A^2 \quad (3)$$

Noting that $P_{cr} = \pi^2 EI/L^2$ we obtain the following.

$$P_{cr1}:P_{cr2}:P_{cr3} = I_s:I_c:I_t = 0.0833:0.07958:0.0962 = 1:0.9549:1.1547 \quad (4)$$

The ratio of critical loads will not change with end conditions.

ANS $P_{cr1}:P_{cr2}:P_{cr3} = 1:0.9549:1.1547$

---

11. 21

Solution F=750 lb E = 30000 ksi d = 1/4 in σ_{yield}=30 ksi L_{AP}= 8 in L_{BP} = 10 in K=?

By equilibrium of forces we obtain

$$N_{BP} sin 70 = 750 sin 40 \quad or \quad N_{BP} = 513.03 lb \qquad N_{AP} = 750 cos 40 - N_{BP} cos 70 \quad or \quad N_{AP} = 399.07 lb \quad \textbf{(1)}$$

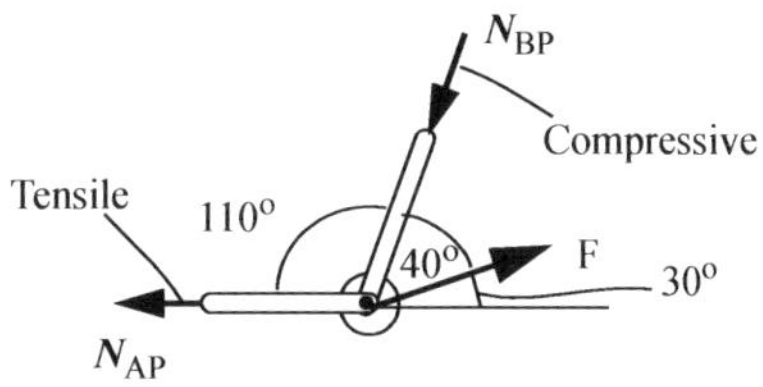

$$A = \frac{\pi}{4}\left(\frac{1}{4}\right)^2 = 0.0491 in^2 \qquad \sigma_{BP} = \frac{N_{BP}}{A} = \frac{513.03}{0.0491} = 10448.7 psi \quad \textbf{(2)}$$

$$I = \frac{\pi}{64}\left(\frac{1}{4}\right)^4 = 191.7(10^{-6}) in^4 \qquad (P_{cr})_{BP} = \frac{\pi^2 EI}{L_{BP}^2} = \frac{\pi^2(30)(10^6)(191.7)(10^{-6})}{10^2} = 567.6 lb \quad \textbf{(3)}$$

$$K_y = \frac{\sigma_{yield}}{\sigma_{BP}} = \frac{30000}{10448.7} = 2.87 \qquad K_{cr} = \frac{(P_{cr})_{BP}}{N_{BP}} = \frac{567.6}{513.03} = 1.106 \quad \textbf{(4)}$$

ANS $K = 1.106$

---

11. 22

Solution F=600 lb E = 30000 ksi d = 1/4 in σ_{yield}=30 ksi L_{AP}= 7in L_{BP} = 10 in K=?

By equilibrium of forces we obtain the following.

$$N_{BP} sin 60 = 600 sin 25 \quad or \quad N_{BP} = 292.8 lb \qquad N_{AP} = 600 cos 25 + N_{BP} cos 60 \quad or \quad N_{AP} = 690.18 lb \quad \textbf{(1)}$$

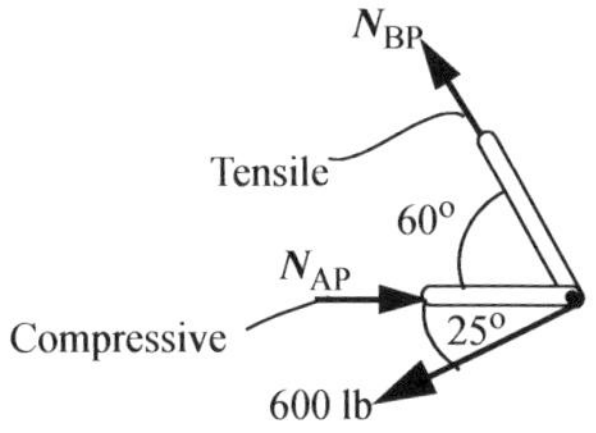

$$A = \frac{\pi}{4}\left(\frac{1}{4}\right)^2 = 0.0491 in^2 \qquad \sigma_{AP} = \frac{N_{AP}}{A} = \frac{690.18}{0.0491} = 14060 psi(C) = 14.06 ksi(C) \quad \textbf{(2)}$$

$$I = \frac{\pi}{64}\left(\frac{1}{4}\right)^4 = 191.7(10^{-6}) in^4 \qquad (P_{cr})_{AP} = \frac{\pi^2 EI}{L_{AP}^2} = \frac{\pi^2(30)(10^6)(191.7)(10^{-6})}{7^2} = 1158.4 lb \quad \textbf{(3)}$$

$$K_y = \frac{\sigma_{yield}}{\sigma_{AP}} = \frac{30}{14.06} = 2.13 \qquad K_{cr} = \frac{(P_{cr})_{AP}}{N_{AP}} = \frac{1158.4}{690.18} = 1.68 \quad \textbf{(4)}$$

ANS $K = 1.68$

---

11. 23

Solution F=750 lb E = 15000 ksi d = 1/4 in σ_{yield}=12 ksi L_{AP}= 7in L_{BP} = 9in K=?

By equilibrium of forces we obtain the following.

$$N_{AP} cos 75 + N_{BP} cos 30 = 500 cos 40 = 383.02 \qquad N_{AP} sin 75 - N_{BP} sin 30 = 500 sin 40 = 321.39 \quad \textbf{(1)}$$

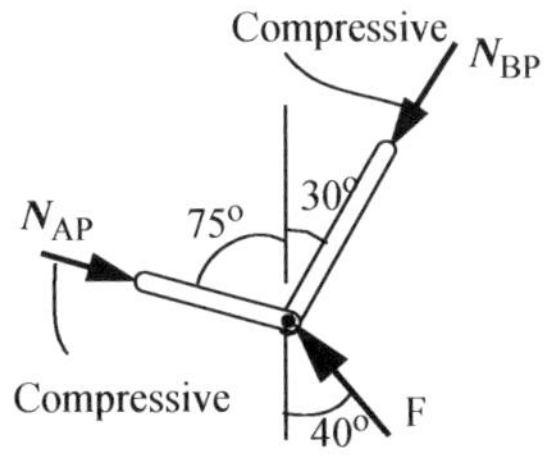

$$N_{AP} = 486.42 lb \qquad N_{BP} = 296.91 lb \quad \textbf{(2)}$$

$$A = \frac{\pi}{4}\left(\frac{1}{4}\right)^2 = 0.0491 in^2 \qquad \sigma_{AP} = \frac{N_{AP}}{A} = \frac{486.42}{0.0491} = 9906.7psi(C) \qquad (3)$$

$$I = \frac{\pi}{64}\left(\frac{1}{4}\right)^4 = 191.7(10^{-6})in^4 \qquad (P_{cr})_{AP} = \frac{\pi^2 EI}{L_{AP}^2} = \frac{\pi^2(15)(10^6)(191.7)(10^{-6})}{7^2} = 579.2lb \qquad (4)$$

$$(P_{cr})_{BP} = \pi^2 EI/L_{BP}^2 = \pi^2(15)(10^6)(191.7)(10^{-6})/9^2 = 350.37lb \qquad (5)$$

$$K_y = \frac{\sigma_{yield}}{\sigma_{AP}} = \frac{12}{9906.7} = 1.211 \qquad K_{cr1} = \frac{(P_{cr})_{AP}}{N_{AP}} = \frac{579.2}{486.42} = 1.191 \qquad K_{cr2} = \frac{(P_{cr})_{BP}}{N_{BP}} = \frac{350.37}{296.91} = 1.180 \qquad (6)$$

ANS $K = 1.18$

---

11. 24

Solution F=10 kN E = 200 GPa d = 10 m σ_{yield} = 200 MPa L_{AP}= 200 mm L_{BP}= 300 mm K = ?

By force equilibrium in the x-direction and from geometry we obtain the following.

$$N_A + N_B cos70 = F \quad or \quad N_A + 0.342N_B = 10(10^3) \qquad \delta_A = \delta_P \qquad \delta_B = \delta_P cos70 = 0.342\delta_P \qquad (1)$$

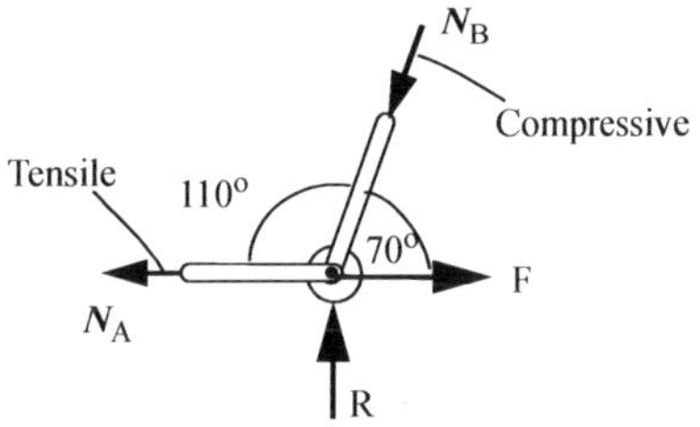

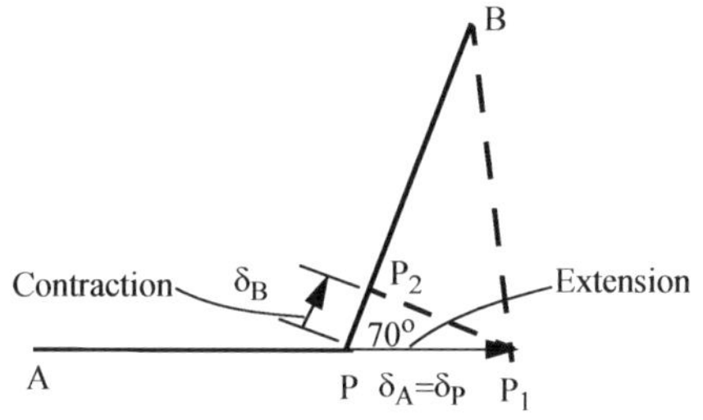

$$A = \frac{\pi}{4}0.01^2 = 78.54(10^{-6})m^2 \qquad \delta_A = \frac{N_A L_A}{EA} = \frac{N_A(0.2)}{(200)(10^9)(78.54)(10^{-6})} = \delta_P \quad or \quad N_A = 78.54\delta_P(10^6) \qquad (2)$$

$$\delta_B = \frac{N_B L_B}{EA} = \frac{N_B(0.35)}{(200)(10^9)(78.54)(10^{-6})} = 0.342\delta_P \quad or \quad N_B = 15.35\delta_P(10^6) \qquad (3)$$

$$N_A + 0.342N_B = [(78.54\delta_P) + 0.342(15.35\delta_P)](10^6) = 10(10^3) \quad or \quad \delta_P = 0.1193(10^{-3})m \qquad (4)$$

$$N_A = (78.54)(0.1193)(10^3) = 9.373(10^3)\ N \qquad N_B = (15.35)(0.1193)(10^3) = 1.832(10^3)\ N \qquad (5)$$

$$\sigma_A = \frac{N_A}{A} = \frac{9.373(10^3)}{(78.54)(10^{-6})} = 119.34(10^6)\ N/m^2 = 119.3\ MPa(T) \qquad (6)$$

$$I = \frac{\pi}{64}0.01^4 = 0.4909(10^{-9})m^4 \qquad (P_{cr})_B = \frac{\pi^2 EI}{L_B^2} = \frac{\pi^2(200)(10^9)(0.4909)(10^{-9})}{0.35^2} = 7.910(10^3)\ N \qquad (7)$$

$$K_y = \frac{\sigma_{yield}}{\sigma_A} = \frac{200}{119.3} = 1.676 \qquad K_{cr} = \frac{(P_{cr})_B}{N_B} = \frac{7.910(10^3)}{1.832(10^3)} = 4.32 \qquad (8)$$

ANS $K = 1.67$

---

11. 25

Solution F=10 kN E = 200 GPa d = 10 m σ_{yield} = 360 MPa L_{AP}= 200 mm L_{BP}= 300 mm K = ?

By force equilibrium in the x-direction and from geometry we obtain the following. the followg

$$N_A sin30 + N_B sin60 = F \quad or \quad 0.5N_A + 0.866N_B = 10(10^3) \qquad \delta_A = \delta_P cos60 = 0.5\delta_P \qquad \delta_B = \delta_P sin60 = 0.866\delta_P \qquad (1)$$

$$A = \frac{\pi}{4}0.01^2 = 78.54(10^{-6})m^2 \qquad \delta_A = \frac{N_A L_A}{EA} = \frac{N_A(0.2)}{(200)(10^9)(78.54)(10^{-6})} = 0.5\delta_P \quad or \quad N_A = 39.27\delta_P(10^6) \qquad (2)$$

$$\delta_B = \frac{N_B L_B}{EA} = \frac{N_B(0.3)}{(200)(10^9)(78.54)(10^{-6})} = 0.866\delta_P \quad or \quad N_B = 45.34\delta_P(10^6) \qquad (3)$$

$$0.5N_A + 0.866N_B = [0.5(39.27\delta_P) + 0.866(45.34\delta_P)](10^6) = 10(10^3) \quad or \quad \delta_P = 0.1698(10^{-3})m \qquad (4)$$

$$N_A = (39.27)(0.1698)(10^3) = 6.667(10^3)\ N \qquad N_B = (45.34)(0.1698)(10^3) = 7.697(10^3)\ N \qquad (5)$$

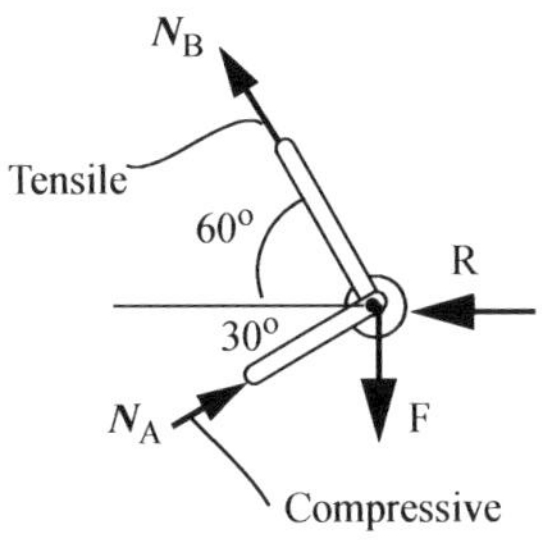

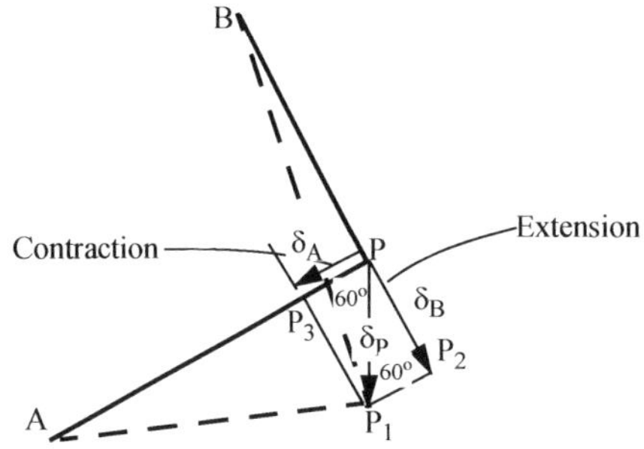

$$\sigma_B = \frac{N_B}{A} = \frac{7.697(10^3)}{(78.54)(10^{-6})} = 98.01(10^6)N/m^2 = 98.01 \quad MPa(T) \tag{6}$$

$$I = \frac{\pi}{64}0.01^4 = 0.4909(10^{-9})m^4 \qquad (P_{cr})_A = \frac{\pi^2 EI}{L_A^2} = \frac{\pi^2(200)(10^9)(0.4909)(10^{-9})}{0.2^2} = 24.22(10^3) \ N \tag{7}$$

$$K_y = \frac{\sigma_{yield}}{\sigma_B} = \frac{360}{98.01} = 3.673 \qquad K_{cr} = \frac{(P_{cr})_A}{N_A} = \frac{24.22(10^3)}{6.667(10^3)} = 3.633 \tag{8}$$

ANS $K = 3.63$

---

11. 26

Solution F=10 kN E = 200 GPa d = 10 m σ_{yield} = 360 MPa L_{AP}= 200 mm L_{BP}= 300 mm K = ?

By force equilibrium in the y-direction and by geometery we obtain the following.

$$N_A cos 75 + N_B cos 30 = F \qquad or \qquad 0.2588 N_A + 0.8660 N_B = 10(10^3) \tag{1}$$

$$\delta_A = \delta_P cos 75 = 0.2588\delta_P \qquad \delta_B = \delta_P cos 30 = 0.8660\delta_P \tag{2}$$

(a)

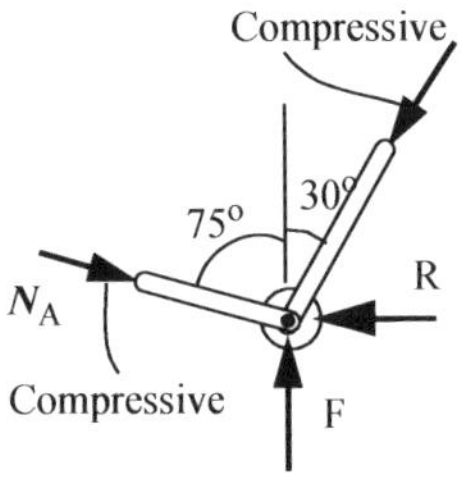

(b)

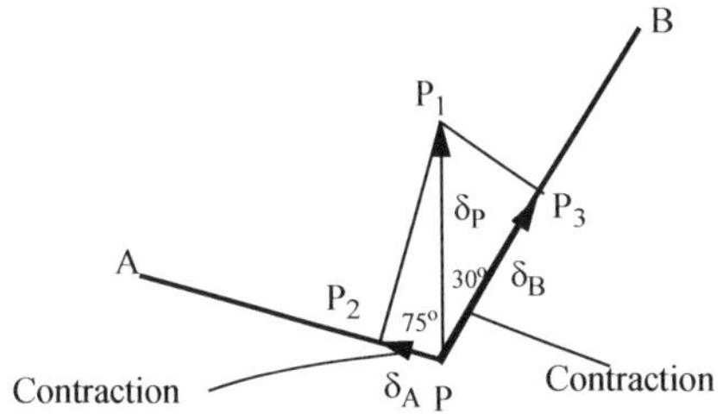

$$A = \frac{\pi}{4}0.01^2 = 78.54(10^{-6})m^2 \qquad \delta_A = \frac{N_A L_A}{EA} = \frac{N_A(0.2)}{(200)(10^9)(78.54)(10^{-6})} = 0.2588\delta_P \qquad or \qquad N_A = 20.326\delta_P(10^6) \tag{3}$$

$$\delta_B = \frac{N_B L_B}{EA} = \frac{N_B(0.3)}{(200)(10^9)(78.54)(10^{-6})} = 0.866\delta_P \qquad or \qquad N_B = 45.343\delta_P(10^6) \tag{4}$$

$$0.2588 N_A + 0.8660 N_B = [0.2588(20.326\delta_P) + 0.866(45.343\delta_P)](10^6) = 10(10^3) \qquad or \qquad \delta_P = 0.2246(10^{-3})m \tag{5}$$

$$N_A = (20.326)(0.2246)(10^3) = 4.564(10^3) \ N \qquad N_B = (45.343)(0.2246)(10^3) = 10.183(10^3) \ N \tag{6}$$

$$\sigma_B = \frac{N_B}{A} = \frac{10.183(10^3)}{(78.54)(10^{-6})} = 129.65(10^6)N/m^2 = 129.65 \quad MPa(C) \tag{7}$$

$$I = \frac{\pi}{64}0.01^4 = 0.4909(10^{-9})m^4 \qquad (P_{cr})_A = \frac{\pi^2 EI}{L_A^2} = \frac{\pi^2(200)(10^9)(0.4909)(10^{-9})}{(0.2)^2} = 24.22(10^3) \ N \tag{8}$$

$$(P_{cr})_B = \frac{\pi^2 EI}{L_B^2} = \frac{\pi^2(200)(10^9)(0.4909)(10^{-9})}{0.3^2} = 10.77(10^3) \ N \tag{9}$$

$$K_y = \frac{\sigma_{yield}}{\sigma_B} = 1.543 \qquad K_{cr_1} = \frac{(P_{cr})_A}{N_A} = 5.31 \qquad K_{cr_2} = \frac{(P_{cr})_B}{N_B} = 1.057 \tag{10}$$

ANS $K = 1.057$

---

11. 27

Solution P_{cr1} = ? P_{cr2} = ? for Case 2

The boundary value problem for Case 2 is:

$$EI\frac{d^2v}{dx^2} + Pv = Pv(L) \quad or \quad \frac{d^2v}{dx^2} + \lambda^2 v = \lambda^2 v(L) \quad v(0) = 0 \quad \frac{dv}{dx}(0) = 0 \tag{1}$$

The solution is as follows.

$$v_h(x) = A\cos\lambda x + B\sin\lambda x \quad v_p(x) = v(L) \quad v(x) = v_h(x) + v_p(x) = A\cos\lambda x + B\sin\lambda x + v(L) \tag{2}$$

$$v(0) = A + v(L) = 0 \quad or \quad A = -v(L) \tag{3}$$

$$\frac{dv}{dx}(x) = -\lambda A\sin\lambda x + \lambda B\cos\lambda x \quad \frac{dv}{dx}(0) = \lambda B = 0 \quad or \quad B = 0 \tag{4}$$

$$v(x) = v(L)[-\cos\lambda x + 1] \quad or \quad v(L) = v(L)[-\cos\lambda L + 1] \quad or \quad \cos\lambda L = 1 \quad or \quad \lambda L = (2n+1)\frac{\pi}{2} \tag{5}$$

$$For\ n = 0 \quad \sqrt{\frac{P_{cr1}}{EI}} = \frac{\pi}{2L} \quad or \quad P_{cr1} = \frac{\pi^2 EI}{4L^2} \quad For\ n = 1 \quad \sqrt{\frac{P_{cr2}}{EI}} = \frac{3\pi}{2L} \quad or \quad P_{cr2} = \frac{9\pi^2 EI}{4L^2} \tag{6}$$

ANS $P_{cr1} = \frac{\pi^2 EI}{4L^2}$; $P_{cr2} = \frac{9\pi^2 EI}{4L^2}$

---

11. 28

Solution $P_{cr1} = ?$ $P_{cr2} = ?$ for Case 3

The boundary value problem for Case 3 is:

$$EI\frac{d^2v}{dx^2} + Pv = R_B(L-x) \quad or \quad \frac{d^2v}{dx^2} + \lambda^2 v = \frac{R_B}{EI}(L-x) \quad v(0) = 0 \quad \frac{dv}{dx}(0) = 0 \quad v(L) = 0 \tag{1}$$

The solution is as follows.

$$v_h(x) = A\cos\lambda x + B\sin\lambda x \quad v_p(x) = \frac{R_B}{\lambda^2 EI}(L-x) \quad v(x) = v_h(x) + v_p(x) = A\cos\lambda x + B\sin\lambda x + \frac{R_B}{\lambda^2 EI}(L-x) \tag{2}$$

$$v(0) = A + \frac{R_B L}{\lambda^2 EI} = 0 \quad or \quad A = -\frac{R_B L}{\lambda^2 EI} \tag{3}$$

$$\frac{dv}{dx}(x) = -\lambda A\sin\lambda x + \lambda B\cos\lambda x - \frac{R_B}{\lambda^2 EI} \quad or \quad \frac{dv}{dx}(0) = \lambda B - \frac{R_B}{\lambda^2 EI} = 0 \quad or \quad B = \frac{R_B}{\lambda^3 EI} \tag{4}$$

$$v(L) = A\cos\lambda L + B\sin\lambda L = \left[-\frac{R_B L}{\lambda^2 EI}\cos\lambda L + \frac{R_B}{\lambda^3 EI}\sin\lambda L\right] = \frac{R_B L}{\lambda^2 EI}(-\lambda L\cos\lambda L + \sin\lambda L) = 0 \quad or \quad \tan\lambda L = \lambda L \tag{5}$$

$$\lambda_1 L = 4.4934 \quad or \quad \sqrt{\frac{P_{cr1}}{EI}} = \frac{4.4934}{L} \quad or \quad P_{cr1} = \frac{20.19EI}{L^2} \tag{6}$$

$$\lambda_2 L = 7.7253 \quad or \quad \sqrt{\frac{P_{cr2}}{EI}} = \frac{7.7253}{L} \quad or \quad P_{cr2} = \frac{59.68EI}{L^2} \tag{7}$$

ANS $P_{cr1} = \frac{20.19EI}{L^2}$; $P_{cr2} = \frac{59.68EI}{L^2}$

---

11. 29

Solution $P_{cr1} = ?$ $P_{cr2} = ?$ for Case 4

The boundary value problem for Case 4 is:

$$EI\frac{d^2v}{dx^2} + Pv = R_B(L-x) + M_B \quad or \quad \frac{d^2v}{dx^2} + \lambda^2 v = \frac{R_B(L-x)}{EI} + \frac{M_B}{EI} \tag{1}$$

$$v(0) = 0 \quad \frac{dv}{dx}(0) = 0 \quad v(L) = 0 \quad \frac{dv}{dx}(L) = 0 \tag{2}$$

The solution is as follows.

$$v_h(x) = A\cos\lambda x + B\sin\lambda x \quad v_p(x) = \frac{R_B(L-x)}{\lambda^2 EI} + \frac{M_B}{\lambda^2 EI} \quad v(x) = v_h(x) + v_p(x) = A\cos\lambda x + B\sin\lambda x + \frac{R_B(L-x)}{\lambda^2 EI} + \frac{M_B}{\lambda^2 EI} \tag{3}$$

$$\frac{dv}{dx}(x) = -\lambda A\sin\lambda x + \lambda B\cos\lambda x - \frac{R_B}{\lambda^2 EI} \quad \frac{dv}{dx}(0) = \lambda B - \frac{R_B}{\lambda^2 EI} = 0 \quad or \quad \frac{R_B}{\lambda^2 EI} = \lambda B \tag{4}$$

$$v(0) = A + \frac{R_B L}{\lambda^2 EI} + \frac{M_B}{\lambda^2 EI} = A + \lambda LB + \frac{M_B}{\lambda^2 EI} = 0 \quad or \quad \frac{M_B}{\lambda^2 EI} = -(A + \lambda LB) \quad (5)$$

$$v(L) = A\cos\lambda L + B\sin\lambda L + \frac{M_B}{\lambda^2 EI} = 0 \quad or \quad A(\cos\lambda L - 1) + B(\sin\lambda L - \lambda L) = 0 \quad (6)$$

$$\frac{dv}{dx}(L) = -\lambda A\sin\lambda L + \lambda B\cos\lambda L - \frac{R_B}{\lambda^2 EI} = 0 \quad or \quad -A\sin\lambda L + B(\cos\lambda L - 1) = 0 \quad (7)$$

For non-trivial solution the determinant of the above equation must be zero.

$$\begin{vmatrix} (\cos\lambda L - 1) & (\sin\lambda L - \lambda L) \\ -\sin\lambda L & (\cos\lambda L - 1) \end{vmatrix} = (\cos\lambda L - 1)^2 + \sin\lambda L(\sin\lambda L - \lambda L) = 2(1-\cos\lambda L) - \lambda L\sin\lambda L = 0 \text{ or}$$

$$2[2\sin^2(\lambda L/2)] - \lambda L[2\sin(\lambda L/2)\cos(\lambda L/2)] = 0 \quad or \quad \sin(\lambda L/2)[\sin(\lambda L/2) - \lambda L/2\cos(\lambda L/2)] = 0 \quad (8)$$

Above equation is satisfied if the following conditions are satisfied.

$$\sin(\lambda L/2) = 0 \quad or \quad \lambda L/2 = n\pi \quad \lambda_1 L/2 = \pi = 3.1415 \quad \lambda_2 L/2 = 2\pi = 6.283$$

$$or \quad (9)$$

$$\tan(\lambda L/2) = \lambda L/2 \quad or \quad \lambda_1 L/2 = 4.4934 \quad \lambda_2 L/2 = 7.7253$$

Using the two lowest roots we obtain the following.

$$\lambda_1 L/2 = \pi \quad or \quad \sqrt{\frac{P_{cr1}}{EI}} = \frac{2\pi}{L} \quad or \quad P_{cr1} = \frac{4\pi^2 EI}{L^2} \quad (10)$$

$$\lambda_2 L/2 = 4.4934 \quad or \quad \sqrt{\frac{P_{cr2}}{EI}} = \frac{2(4.4934)}{L} \quad or \quad P_{cr2} = \frac{80.76EI}{L^2} \quad (11)$$

ANS $P_{cr1} = \frac{4\pi^2 EI}{L^2}$; $P_{cr2} = \frac{80.76EI}{L^2}$

---

11. 30

Solution Characteristic Equation = ? $K = 0$ is Case 1 and $K = \infty$ is Case 3

By equilibrium of moment about point A in Figure (a) and about point O we obtain the following.

$$R_B L = K\frac{dv}{dx}(0) \quad or \quad R_B = \frac{K}{L}\frac{dv}{dx}(0) \quad and \quad \boldsymbol{M}_z + Pv = R_B(L-x) = \frac{K(L-x)}{L}\frac{dv}{dx}(0) \quad (1)$$

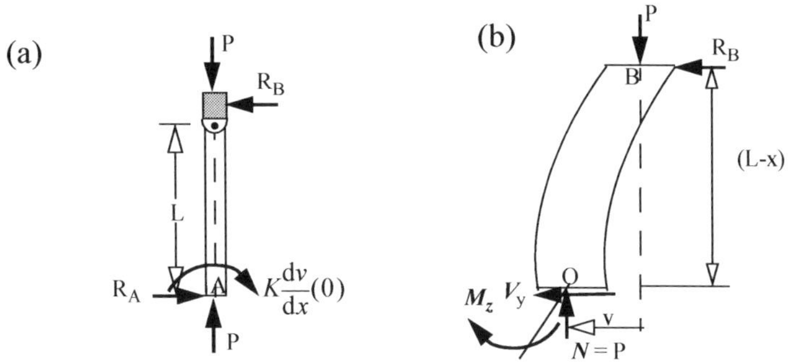

The boundary value problem can be written as:

$$EI\frac{d^2v}{dx^2} + Pv = \frac{K(L-x)}{L}\left[\frac{dv}{dx}(0)\right] \quad or \quad \frac{d^2v}{dx^2} + \lambda^2 v = \frac{K(L-x)}{EIL}\left[\frac{dv}{dx}(0)\right] \quad v(0) = 0 \quad v(L) = 0 \quad (2)$$

The solution proceeds as follows.

$$v_h(x) = A\cos\lambda x + B\sin\lambda x \quad v_p(x) = \frac{K(L-x)}{\lambda^2 EIL}\frac{dv}{dx}(0) \quad v(x) = v_h(x) + v_p(x) = A\cos\lambda x + B\sin\lambda x + \frac{K(L-x)}{\lambda^2 EIL}\left[\frac{dv}{dx}(0)\right] \quad (3)$$

$$\frac{dv}{dx}(x) = -\lambda A\sin\lambda x + \lambda B\cos\lambda x - \frac{K}{\lambda^2 EIL}\frac{dv}{dx}(0) \quad (4)$$

$$\frac{dv}{dx}(0) = \lambda B - \frac{K}{\lambda^2 EIL}\frac{dv}{dx}(0) \quad or \quad B = \frac{1}{\lambda}\left[\frac{dv}{dx}(0)\right]\left(1 + \frac{K}{\lambda^2 EIL}\right) = \frac{(\lambda^2 EIL + K)}{\lambda^3 EIL}\left[\frac{dv}{dx}(0)\right] \quad (5)$$

$$v(0) = A + \frac{K}{\lambda^2 EI}\frac{dv}{dx}(0) = 0 \quad or \quad A = -\frac{K}{\lambda^2 EI}\left[\frac{dv}{dx}(0)\right] \quad (6)$$

$$v(L) = A\cos\lambda L + B\sin\lambda L = 0 \quad or \quad B\tan\lambda L = -A \quad or \quad \frac{(\lambda^2 EIL + K)}{\lambda^3 EIL}\left[\frac{dv}{dx}(0)\right]\tan\lambda L = \frac{K}{\lambda^2 EI}\left[\frac{dv}{dx}(0)\right] \text{ or}$$

$$\tan\lambda L = \frac{K\lambda L}{K + \lambda^2 EIL} \tag{7}$$

As $K \to 0$ we obtain $\tan\lambda L = 0$ or $\sin\lambda L = 0$, which is the Characteristic equation of case 1.

The above equation can be written as: $\tan\lambda L = \dfrac{\lambda L}{1 + (\lambda^2 EIL)/K}$. As $K \to \infty$ we obtain $\tan\lambda L = \lambda L$ which is the Characteristic equation of case 3.

ANS $\tan\lambda L = \dfrac{K\lambda L}{K + \lambda^2 EIL}$

---

11. 31

Solution Characteristic Equation = ? $K = 0$ is Case 2 and $K = \infty$ is?

By equilibrium of moment about point O we obtain:

$$M_z + P(v - v(L)) - K\frac{dv}{dx}(L) = 0 \tag{1}$$

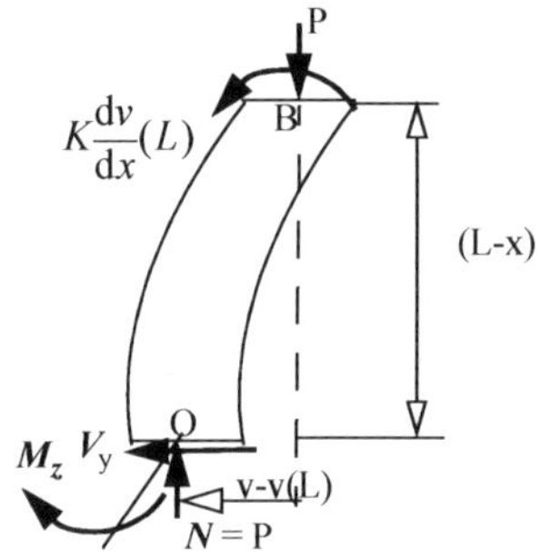

The boundary value problem can be written as:

$$EI\frac{d^2v}{dx^2} + Pv = Pv(L) + K\frac{dv}{dx}(L) \quad or \quad \frac{d^2v}{dx^2} + \lambda^2 v = \lambda^2 v(L) + \frac{K}{EI}\frac{dv}{dx}(L) \quad v(0) = 0 \quad \frac{dv}{dx}(0) = 0 \tag{2}$$

$$v_h(x) = A\cos\lambda x + B\sin\lambda x \quad v_p(x) = v(L) + \frac{K}{\lambda^2 EI}\frac{dv}{dx}(L) \quad v(x) = v_h(x) + v_p(x) = A\cos\lambda x + B\sin\lambda x + v(L) + \frac{K}{\lambda^2 EI}\frac{dv}{dx}(L) \tag{3}$$

The solution proceeds as follows.

$$v(0) = A + v(L) + \frac{K}{\lambda^2 EI}\frac{dv}{dx}(L) = 0 \quad or \quad A = -\left[v(L) + \frac{K}{\lambda^2 EI}\frac{dv}{dx}(L)\right] \tag{4}$$

$$\frac{dv}{dx}(x) = -\lambda A\sin\lambda x + \lambda B\cos\lambda x \quad \frac{dv}{dx}(0) = \lambda B = 0 \quad or \quad B = 0 \tag{5}$$

$$v(L) = A\cos\lambda L + v(L) + \frac{K}{\lambda^2 EI}\frac{dv}{dx}(L) \quad or \quad \frac{dv}{dx}(L) = -A\left(\frac{\lambda^2 EI}{K}\right)\cos\lambda L = -\lambda A\sin\lambda L \quad or \quad \tan\lambda L = \lambda\frac{EI}{K} \tag{6}$$

We can write the above equation as $K\sin\lambda L = \lambda EI\cos\lambda L$. As $K \to 0$ we obtain $\cos\lambda L = 0$ which is the Characteristic equation of case 2. The above equation can also be written as: $\sin\lambda L = (\lambda EI\cos\lambda L)/K$. As $K \to \infty$ we obtain $\sin\lambda L = 0$ which is the characteristic equation of case 1.

ANS $\tan\lambda L = \lambda EI/K$

---

11. 32

Solution $P_{cr} = f(E, I, L, \alpha) = ?$ For $\alpha = 0.5$ $P_{cr} = ?$

Using singularity functions we can write the moment equilibrium equation as shown below.

$$M_z + Pv + R\langle x - \alpha L\rangle^1 = 0 \tag{1}$$

The boundary value problem can be written as:

$$EI\frac{d^2v}{dx^2} + Pv = -R\langle x - \alpha L\rangle^1 \quad or \quad \frac{d^2v}{dx^2} + \lambda^2 v = -\frac{R}{EI}\langle x - \alpha L\rangle^1 \quad v(0) = 0 \quad v(L) = 0 \quad v(\alpha L) = 0 \tag{2}$$

$$v_h(x) = A\cos\lambda x + B\sin\lambda x \quad v_p(x) = -\frac{R}{\lambda^2 EI}\langle x - \alpha L\rangle^1 \quad v(x) = v_h(x) + v_p(x) = A\cos\lambda x + B\sin\lambda x - \frac{R}{\lambda^2 EI}\langle x - \alpha L\rangle^1 \tag{3}$$

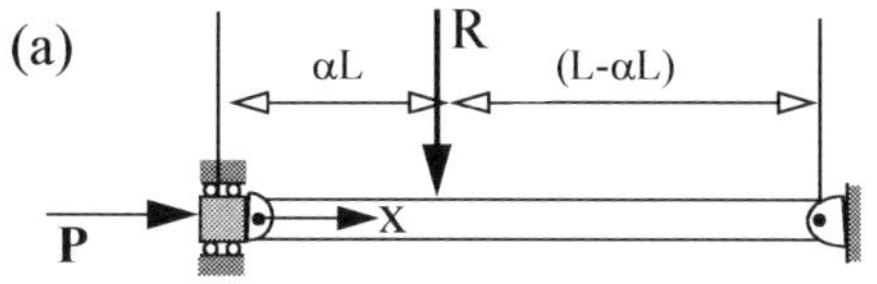

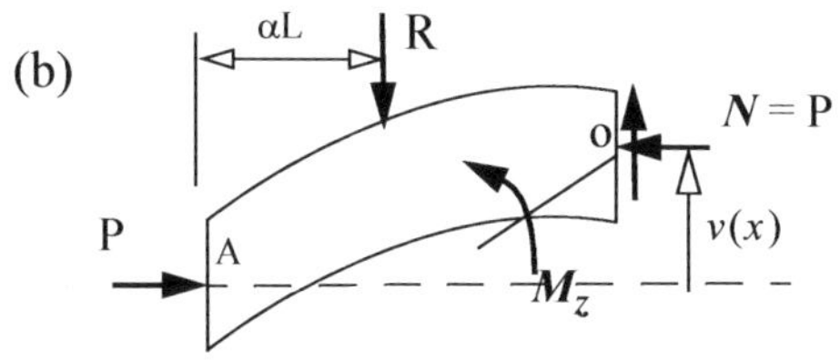

$$v(0) = A - 0 = 0 \quad or \quad A = 0 \qquad v(L) = B\sin\lambda L - \frac{R}{\lambda^2 EI}\langle L - \alpha L\rangle^1 = 0 \quad or \quad B\sin\lambda L = \frac{RL}{\lambda^2 EI}(1-\alpha) \qquad \textbf{(4)}$$

$$v(\alpha L) = B\sin\lambda\alpha L = 0 \quad or \quad \sin\lambda\alpha L = 0 \quad or \quad \lambda\alpha L = n\pi \quad n = 1, 2\ldots\ldots \qquad \textbf{(5)}$$

$$\lambda_{cr} = \sqrt{\frac{P_{cr}}{EI}} = \frac{\pi}{\alpha L} \quad or \quad P_{cr} = \frac{\pi^2 EI}{\alpha^2 L^2} \quad \text{For } \alpha = 0.5 \quad P_{cr} = \frac{4\pi^2 EI}{L^2} \qquad \textbf{(6)}$$

ANS $P_{cr} = \dfrac{\pi^2 EI}{\alpha^2 L^2}$

---

11. 33

Solution $v(L) = f_1(E, L, I, \alpha) = ?$ $P_{cr} = f_2(E, L, I, \alpha) = ?$

By equilibrium of moment about point O we obtain the following.

$$M_z - P(v(L) - v) + \alpha P(L - x) = 0 \qquad \textbf{(1)}$$

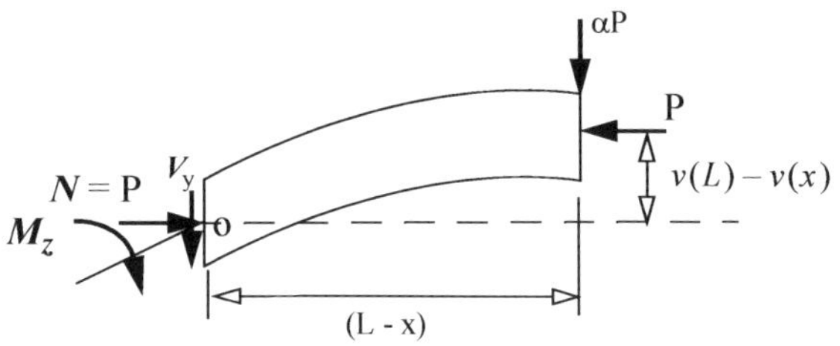

The boundary value problem can be written as follows.

$$EI\frac{d^2v}{dx^2} + Pv = Pv(L) - \alpha P(L-x) \quad or \quad \frac{d^2v}{dx^2} + \lambda^2 v = \lambda^2 v(L) - \lambda^2\alpha(L-x) \qquad v(0) = 0 \qquad \frac{dv}{dx}(0) = 0 \qquad \textbf{(2)}$$

The solution proceeds as follows.

$$v_h(x) = A\cos\lambda x + B\sin\lambda x \qquad v_p(x) = v(L) - \alpha(L-x) \qquad v(x) = v_h(x) + v_p(x) = A\cos\lambda x + B\sin\lambda x + v(L) - \alpha(L-x) \qquad \textbf{(3)}$$

$$v(0) = A + v(L) - \alpha L = 0 \quad or \quad A = \alpha L - v(L) \qquad \textbf{(4)}$$

$$\frac{dv}{dx}(x) = -\lambda A\sin\lambda x + \lambda B\cos\lambda x - \alpha \qquad \frac{dv}{dx}(0) = \lambda B - \alpha = 0 \quad or \quad B = \frac{\alpha}{\lambda} \qquad \textbf{(5)}$$

$$v(L) = A\cos\lambda L + B\sin\lambda L + v(L) = (\alpha L - v(L))\cos\lambda L + \frac{\alpha}{\lambda}\sin\lambda L + v(L) \quad or \quad v(L) = \alpha(\lambda L + \tan\lambda L) \qquad \textbf{(6)}$$

The tangent function, hence displacement v(L), becomes unbounded at $\lambda L = \pi/2$ thus, $P_{cr} = \pi^2 EI/4L^2$

ANS $v(L) = \alpha(\lambda L + \tan\lambda L)$; $P_{cr} = \pi^2 EI/4L^2$

---

11. 34

Solution $v(L) = f_1(E, L, I, \alpha) = ?$ $P_{cr} = f_2(E, L, I, \alpha) = ?$

By equilibrium of moment about point O we obtain the following.

$$M_z - P(v(L) - v) - \alpha PL = 0 \qquad \textbf{(1)}$$

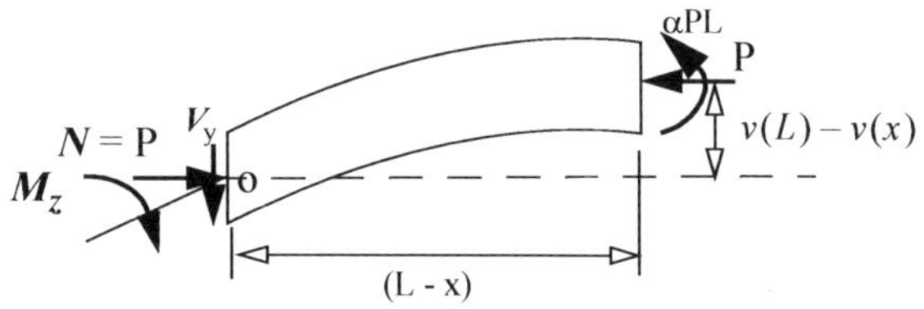

The boundary value problem can be written as:

$$EI\frac{d^2v}{dx^2} + Pv = Pv(L) + \alpha PL \quad or \quad \frac{d^2v}{dx^2} + \lambda^2 v = \lambda^2 v(L) + \lambda^2 \alpha L \quad v(0) = 0 \quad \frac{dv}{dx}(0) = 0 \quad (2)$$

The solution proceeds as follows.

$$v_h(x) = A\cos\lambda x + B\sin\lambda x \quad v_p(x) = v(L) + \alpha L \quad v(x) = v_h(x) + v_p(x) = A\cos\lambda x + B\sin\lambda x + v(L) + \alpha L \quad (3)$$

$$v(0) = A + v(L) + \alpha L = 0 \quad or \quad A = -\alpha L - v(L) \quad (4)$$

$$\frac{dv}{dx}(x) = -\lambda A\sin\lambda x + \lambda B\cos\lambda x \quad \frac{dv}{dx}(0) = \lambda B = 0 \quad or \quad B = 0 \quad (5)$$

$$v(x) = [-\alpha L - v(L)]\cos\lambda x + v(L) + \alpha L = [\alpha L + v(L)](1 - \cos\lambda x) \quad (6)$$

$$v(L) = [\alpha L + v(L)](1 - \cos\lambda L) \quad or \quad v(L) = \frac{\alpha L(1 - \cos\lambda L)}{\cos\lambda L} \quad (7)$$

$$v(x) = \left[\alpha L + \frac{\alpha L(1 - \cos\lambda L)}{\cos\lambda L}\right](1 - \cos\lambda x) = \frac{\alpha L}{\cos\lambda L}(1 - \cos\lambda x) \quad (8)$$

The denominator is zero when $\lambda L = \pi/2$ thus, $P_{cr} = \pi^2 EI/(4L^2)$

ANS $v(L) = \alpha L(1 - \cos\lambda L)/(\cos\lambda L)$; $P_{cr} = \pi^2 EI/(4L^2)$

---

11. 35

Solution $v(L) = f_1(E, L, I, \alpha) = ?$ $P_{cr} = f_2(E, L, I, \alpha) = ?$

Figure shows the column of a previous problem and the associated result of the deflection at x = L below.

$$v(L) = \alpha_1(\lambda_1 L + \tan\lambda_1 L) \quad \lambda_1 = \sqrt{P_1/EI} \quad (1)$$

(a)

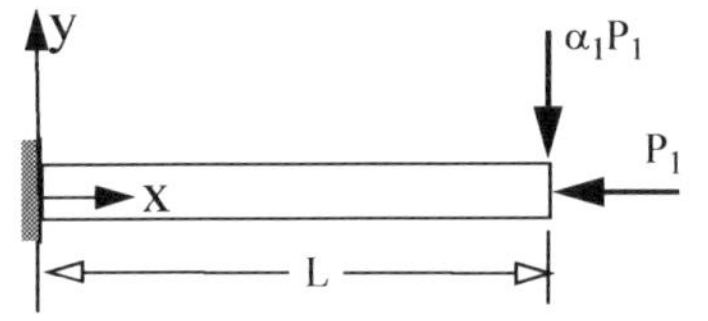

$$P_1 = P\cos\alpha \quad \alpha_1 P_1 = P\sin\alpha \quad \text{therefore} \quad \alpha_1 = \tan\alpha \quad \lambda_1 = \sqrt{P\cos\alpha/EI} = \lambda\sqrt{\cos\alpha} \quad (2)$$

$$v(L) = \tan\alpha[\lambda L\sqrt{\cos\alpha} + \tan(\lambda L\sqrt{\cos\alpha})] \quad (3)$$

The tangent function, hence displacement v(L), becomes unbounded at $(\lambda L\sqrt{\cos\alpha}) = \pi/2$ thus, $P_{cr} = (\pi^2 EI)/(4L^2\cos\alpha)$

ANS $v(L) = \tan\alpha[\lambda L\sqrt{\cos\alpha} + \tan(\lambda L\sqrt{\cos\alpha})]$; $P_{cr} = (\pi^2 EI)/(4L^2\cos\alpha)$

---

11. 36

Solution E = 210 GPa 15 mm x 25 mm $P_{max} = ?$

By geometry, equilibrium of moment about B and equilibrium of forces we obtain the following.

$$\tan\theta = 0.7/1 \quad or \quad \theta = 34.99° \quad L_{AC} = 0.7/\sin\theta = 1.221m \quad I_{min} = (0.025)(0.015^3)/12 = 7.031(10^{-9})m^4 \quad (1)$$

$$N_{AC}\cos\theta(0.7) = P(1) \quad or \quad N_{AC} = 1.744P \quad (2)$$

$$N_{BD}\sin\theta = P \quad or \quad N_{BD} = 1.744P \quad N_{CD} = N_{BD}\cos\theta = 1.429P \quad (3)$$

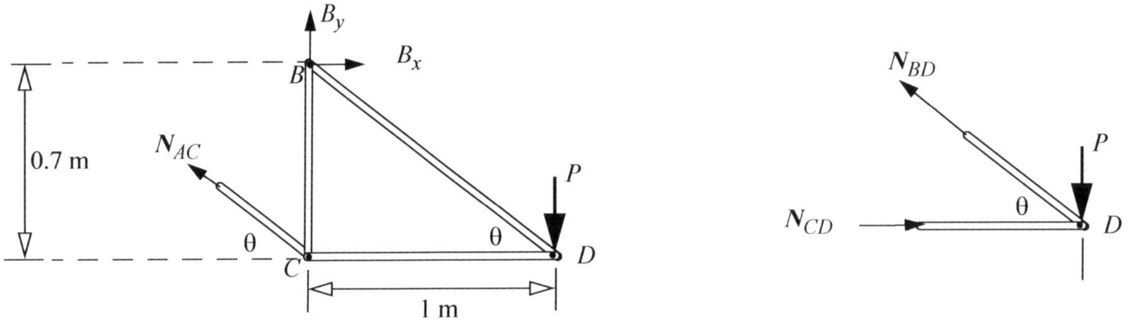

$$(N_{AC})_{cr} = \pi^2 EI_{min}/L_{AC}^2 = [\pi^2(210)(10^9)(7.031)(10^{-9})]/1.221^2 = 9780.2 \geq 1.744P \quad or \quad P \leq 5513.1 \text{ N} \quad (4)$$

$$(N_{CD})_{cr} = \pi^2 EI_{min}/L_{CD}^2 = [\pi^2(210)(10^9)(7.031)(10^{-9})]/1^2 = 14572.5 \geq 1.429P \quad or \quad P \leq 10197 \text{ N} \quad (5)$$

ANS $P_{max} = 5513$ N

---

11. 37

Solution E = 210 GPa 15 mm x 25 mm $P_{max} = ?$

By geometry, equilibrium of moment about B and equilibrium of forces we obtain the following.

$\theta_1 = atan(0.6) = 30.97°$ $\theta_2 = atan(0.6/1.4) = 23.2°$ $L_{AC} = 0.6/sin\theta_1 = 1.166m$ $I_{min} = 7.031(10^{-9})m^4$ **(1)**

$N_{AC}cos\theta_1(0.7) = P(1.4)$ or $N_{AC} = 2.332P$ **(2)**

$N_{BD}sin\theta_2 = P$ or $N_{BD} = 2.538P$ $N_{CD} = N_{BD}cos\theta_2 = 2.333P$ **(3)**

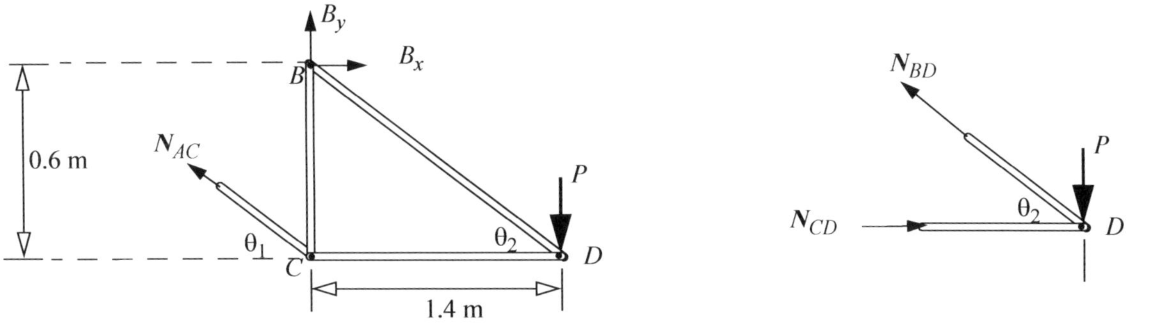

$$(N_{AC})_{cr} = \frac{\pi^2 EI_{min}}{L_{AC}^2} = \frac{\pi^2(210)(10^9)(7.031)(10^{-9})}{1.166^2} = 10715.1 \geq 2.332P \quad or \quad P \leq 4594.8 \text{ N} \quad \textbf{(4)}$$

$$(N_{CD})_{cr} = \frac{\pi^2 EI_{min}}{L_{CD}^2} = \frac{\pi^2(210)(10^9)(7.031)(10^{-9})}{1.4^2} = 7434.98 \geq 2.333P \quad or \quad P \leq 3186.87 \text{ N} \quad \textbf{(5)}$$

ANS $P_{max} = 3186.9$ N

---

11. 38

Solution E= 30,000 ksi 1/2 in.x 1in $P_{max} = ?$

By geometry and equilibrium of forces we obtain the following.

$\theta = atan(36/48) = 36.87°$ $L_{AC} = 36/sin\theta = 60in$ $I_{min} = (1)(1/2)^3/12 = 10.4167(10^{-3})in^4$ **(1)**

$N_{BC} = Psin45 = 0.707P$ or $N_{AB} = Pcos45 = 0.707P$ $N_{AC}sin\theta = N_{BC}$ or $N_{AC} = 1.178P$ **(2)**

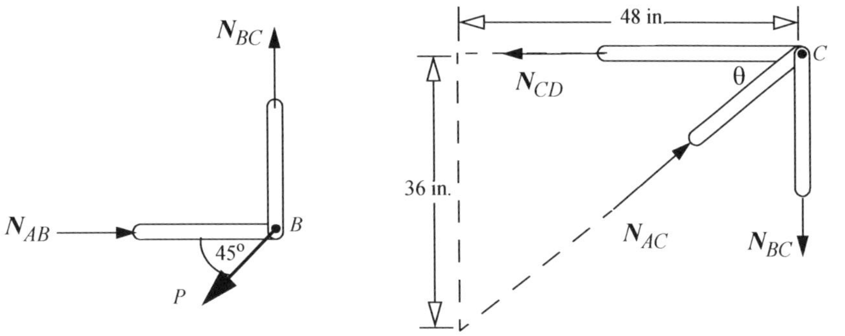

$$(N_{AC})_{cr} = \frac{\pi^2 EI_{min}}{L_{AC}^2} = \frac{\pi^2(30000)(10.4167)(10^{-3})}{60^2} = 0.857kips \geq 1.178P \quad or \quad P \leq 0.728 \text{ kips} \quad \textbf{(3)}$$

$$(N_{AB})_{cr} = \frac{\pi^2 EI_{min}}{L_{AB}^2} = \frac{\pi^2(30000)(10.4167)(10^{-3})}{48^2} = 1.339 \geq 0.707P \quad or \quad P \leq 1.893 \text{ kips} \quad \textbf{(4)}$$

ANS $P_{max} = 728$ kips

---

11. 39

SolutionE= 30,000 ksi 1/2 in.x 1in $P_{max} = ?$

By geometry and equilibrium of forces we obtain the following.

$\theta = atan(36/48) = 36.87°$ $L_{BD} = 36/sin\theta = 60in$ $I_{min} = (1)(1/2)^3/12 = 10.4167(10^{-3})in^4$ **(1)**

$N_{BC} = 0 \qquad N_{BD}sin\theta = Psin45 \qquad or \qquad N_{BD} = 1.178P \qquad N_{AB} = N_{BD}cos\theta + Pcos45 = 1.65P$ **(2)**

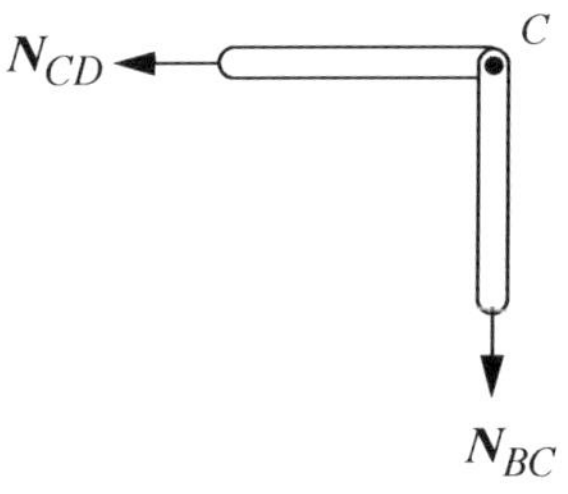

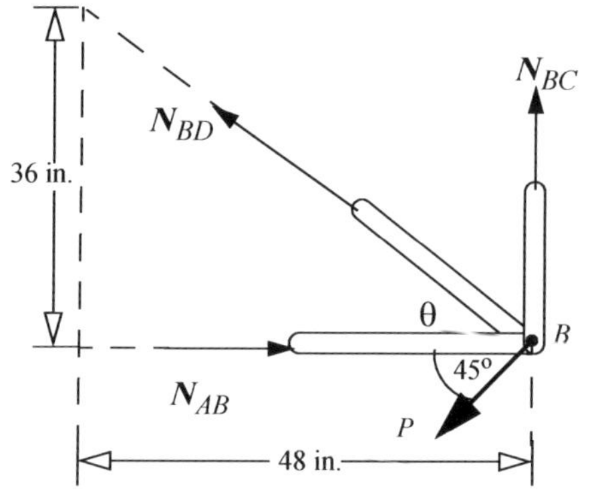

$$(N_{AB})_{cr} = \frac{\pi^2 EI_{min}}{L_{AB}^2} = \frac{\pi^2(30000)(10.4167)(10^{-3})}{48^2} = 1.339 \geq 1.65P \qquad or \qquad P \leq 0.812 \text{ kips} \qquad \textbf{(3)}$$

ANS $P_{max} = 812$ kips

---

11. 40

Solution: W = 5 kip , E = 1880 ksi , σ_{max} = 3 ksi , L_{max} =? to nearest inch

By the equilibrium of forces we obtain.

$N_{AC}sin30 = 10 \qquad or \qquad N_{AC} = 20kip \qquad N_{BC} = N_{AC}cos30 \qquad or \qquad N_{BC} = 17.32kip$ **(1)**

(a)

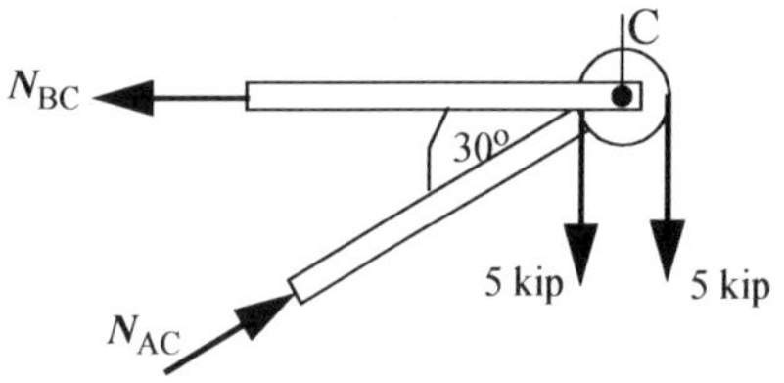

(b)

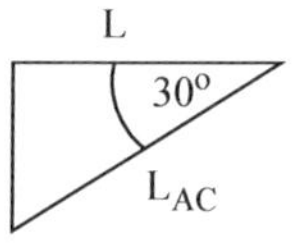

$A = (2)(4) = 8in^2 \qquad I_{min} = (4)(2)^3/12 = 2.667in^4 \qquad L_{AC} = L/(cos30)$ **(2)**

The maximum normal stress in AC is: $\sigma_{max} = 20/8 = 2.5ksi$ irrespective of the length and is less than allowable value.

$$P_{cr} = \frac{\pi^2 EI_{min}}{L_{AC}^2} = \frac{\pi^2(1800)(2.667)}{L_{AC}^2} \geq N_{AC} \qquad or \qquad L_{AC} = \frac{L}{cos30} \leq \sqrt{\frac{\pi^2(1800)(2.667)}{20}} \qquad L \leq 42.15in \qquad \textbf{(3)}$$

ANS $L = 42in$

---

11. 41

Solution: E = 30000 psi σ_{yield} = 30ksi k = 2 P_{max} =?

By equilibrium of forces and geometry we obtain the following.

$N_{CD} = R \qquad N_{AB} = 2P - R \qquad \delta_{AB} = \delta_{CD} \qquad A_{AB} = \pi(2)^2/4 = 3.141in^2 \qquad A_{CD} = \pi(3)^2/4 = 7.068in^2$ **(1)**

(a)

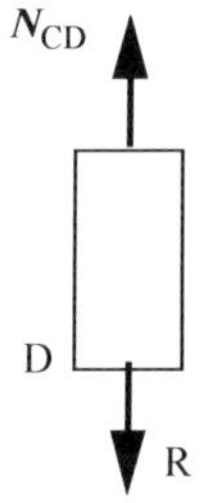

(b)

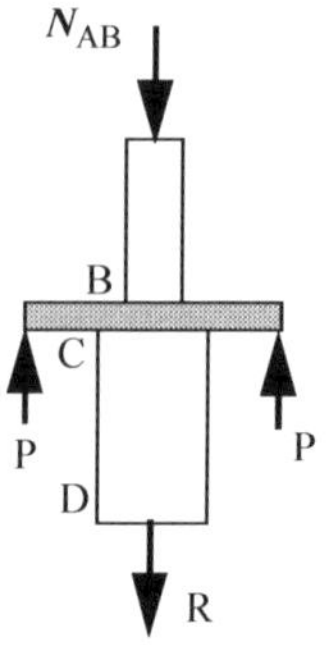

$$\delta_{AB} = \frac{N_{AB}L_{AB}}{EA_{AB}} = \delta_{CD} = \frac{N_{CD}L_{CD}}{EA_{CD}} \qquad or \qquad \frac{R(120)}{3.141} = \frac{(2P-R)(96)}{7.068} \qquad or \qquad R = 0.5246P \qquad \textbf{(2)}$$

$$N_{CD} = 0.5246P \qquad or \qquad N_{AB} = 1.4754P \qquad \textbf{(3)}$$

$$\sigma_{CD} = \frac{N_{CD}}{A_{CD}} = \frac{0.5246P}{7.068} \leq 15 \qquad or \qquad P \leq 202.1kip \qquad \sigma_{AB} = \frac{N_{AB}}{A_{AB}} = \frac{(1.4754P)}{3.141} \leq 15 \qquad or \qquad P \leq 31.94kips \qquad \textbf{(4)}$$

$$I_{AB} = \frac{\pi}{64}(2)^4 = 0.7854 in^4 \qquad P_{cr} = \frac{20.13 EI_{AB}}{L_{AB}^2} = \frac{20.13(30000)(0.7854)}{(120)^2} = 32.94 \qquad (5)$$

$$N_{AB} \le \frac{P_{cr}}{2} \quad or \quad 1.4754P \le \frac{32.94}{2} \quad or \quad P \le 11.162 kip \qquad (6)$$

ANS $P_{max} = 11162$ lb

---

11. 42

Solution: E = 10000 psi t = 1/8in d_0= 2in $\sigma_{max} \le 40 ksi$ L =?

By equilibrium of forces we obtain the following.

$$N = 2T\cos 30 \qquad F = 2T\cos 60 \qquad or \qquad N = \frac{F\cos 30}{\cos 60} = 1.732F \qquad (1)$$

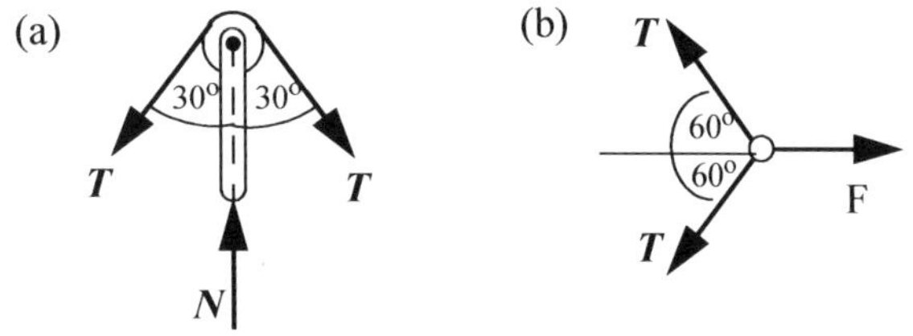

$$A = \pi(2^2 - 1.75^2)/4 = 0.7363 in^2 \qquad \sigma = \frac{N}{A} = \frac{1.732F}{0.7363} = 2.352F \le 40 \qquad or \qquad F \le 17 kips \qquad (2)$$

$$I = \pi(2^4 - 1.75^4)/64 = 0.325 in^4 \qquad P_{cr} = \frac{\pi^2 EI}{L^2} = \frac{\pi^2(10000)(0.325)}{L^2} = \frac{32076.2}{L^2} \qquad (3)$$

$$N \le P_{cr} \qquad or \qquad 1.732F \le \frac{32076.2}{L^2} \qquad or \qquad F \le \frac{18519}{L^2} \qquad (4)$$

The table below shows the values of F calculated from equation above for values of L varying between 4 ft to 8 ft in increments of 6 inches. As the value of F in the table is always less than 17 kips, the limit on stress does not affect the solution.

L (ft)	4.0	4.5	5.0	5.5	6.0	6.5	7.0	7.5	8.0
F (kip)	8.04	6.35	5.14	4.25	3.57	3.04	2.62	2.29	2.01

---

11. 43

Solution: E = 12.6GPa σ_{max} = 18MPa 200mm x 50mm w_{max} =? k_{BD} =?

By equilibrium of moment about point A and equilibrium of forces we obtain the reactions and draw the shear-moment diagrams as shown below.

$$N(2.25) - (3.25w)(1.625w) = 0 \qquad or \qquad N = 2.3472w \qquad R_A = 3.25w - N = 0.9028w \qquad (1)$$

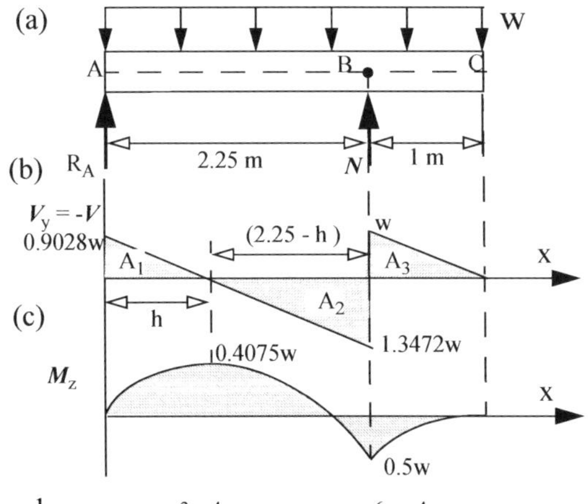

$$\frac{h}{0.9028w} = \frac{(2.25 - h)}{1.3472w}$$

$$h = 0.9028$$

$$A_1 = \frac{1}{2}(0.9028w)(0.9028) = 0.4075w$$

$$A_2 = \frac{1}{2}(1.3472w)(1.3472) = 0.9075w$$

$$A_3 = \frac{1}{2}(1)(w) = 0.5w$$

$$I_{zz} = \frac{1}{12}(0.05)(0.2)^3 m^4 = 33.33(10^{-6})m^4 \qquad \sigma_{max} = \left|\frac{M_{max} y_{max}}{I_{zz}}\right| = \frac{(0.5w)(0.1)}{33.33(10^{-6})} = 1500w \le 18 \qquad or \qquad w \le 12(10^3)\frac{N}{m} \qquad (2)$$

$$I_{min} = \frac{1}{12}(0.2)(0.05)^3 = 2.083(10^{-6})m^4 \qquad P_{cr} = \frac{\pi^2 EI_{min}}{L_{BD}^2} = \frac{\pi^2(12.6)(10^9)(2.083)(10^{-6})}{2^2} = 64.77 \qquad (3)$$

$$N \le P_{cr} \quad or \quad 2.3472w \le 64.76(10^3) \quad or \quad w \le 27.59(10^3)\frac{N}{m} \tag{4}$$

$$\sigma_{BD} = \frac{N}{A} = \frac{2.3472w}{(0.05)(0.2)} = 234.7w \le 18(10^6) \quad or \quad w \le 76.7(10^3)\frac{N}{m} \tag{5}$$

$$w_{max} = (12)(10^3)\text{ N/m} \qquad N = (2.3472)(12)(10^3) = 28.17(10^3) \qquad k_{BD} = \frac{P_{cr}}{N} = \frac{64.76(10^3)}{28.17(10^3)} = 2.298 \tag{6}$$

ANS $w_{max} = 12$ kN/m ; $k_{BD} = 2.3$

---

11. 44

Solution: 200mm x 50mm E = 12.6GPa $\sigma_{max} \le 18MPa$ w_{max} =? k_{BD} =?

Beam AB can be considered as the sum of two beams below. The deflection of point B can be written as superposition of deflection in case 1 and 3 of Table C.3.

$$a = L_{AB} \qquad b = 0 \qquad v_B = \frac{wL_{AB}^4}{8EI_{AB}} - \frac{NL_{AB}^3}{3EI_{AB}} \tag{1}$$

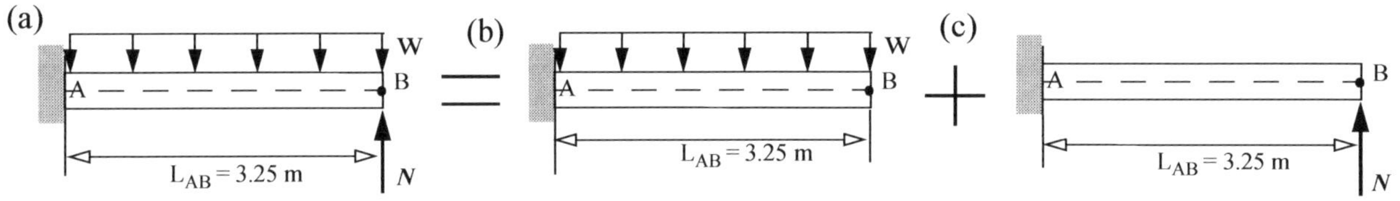

$$A_{BC} = (0.05)(0.2) = 0.01m^2 \qquad I_{AB} = \frac{1}{12}(0.05)(0.2)^3 = 33.33(10^{-6})m^4 \qquad \delta_{BC} = \frac{NL_{BC}}{EA_{BC}} = v_B \text{ or}$$

$$\frac{w(3.25)^4}{8(12.6)(10^9)(33.33)(10^{-6})} - \frac{N(3.25)^3}{3(12.6)(10^9)(33.33)(10^{-6})} = \frac{N(2)}{(12.6)(10^9)(0.01)} \quad or \quad N = 1.218W \tag{2}$$

Figure (d) shows the free body diagram of the beam AB. By force and moment equilibrium we obtain the wall reactions and draw the shear-moment diagram as shown below,

$$R_A = 3.25w - 1.218w = 2.032w \qquad M_A - 3.25w\left(\frac{3.25}{2}\right) + 1.218w(3.25) = 0 \quad or \quad M_A = 1.3228w \tag{3}$$

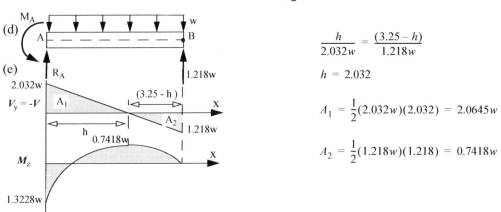

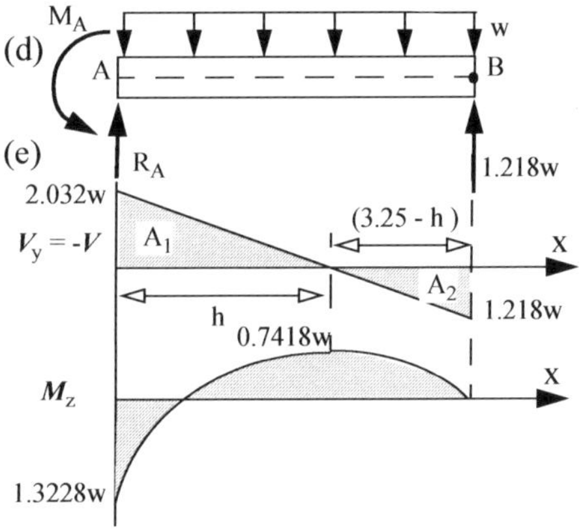

$$\frac{h}{2.032w} = \frac{(3.25 - h)}{1.218w}$$

$$h = 2.032$$

$$A_1 = \frac{1}{2}(2.032w)(2.032) = 2.0645w$$

$$A_2 = \frac{1}{2}(1.218w)(1.218) = 0.7418w$$

$$\sigma_{max} = \left|\frac{M_{max}y_{max}}{I_{AB}}\right| = \frac{(1.3228w)(0.1)}{33.33(10^{-6})} = 3.9684w(10^3) \le 18(10^6) \quad or \quad w \le 4.538(10^3) \tag{4}$$

$$I_{min} = \frac{1}{12}(0.2)(0.05)^3 = 2.083(10^{-6})m^4 \qquad P_{cr} = \frac{\pi^2 EI_{min}}{L_{BD}^2} = \frac{\pi^2(12.6)(10^9)(2.083)(10^{-6})}{2^2} = 64.77 \tag{5}$$

$$N \le P_{cr} \quad or \quad 1.218w \le 64.77(10^3) \quad or \quad w \le 53.18(10^3) \qquad w_{max} = 4.538(10^3)\text{ N/m} \tag{6}$$

$$N = 1.218W = (1.218)(4.5)(10^3) = 5.481(10^3) \qquad k_{BD} = [64.77(10^3)]/[5.481(10^3)] = 11.81 \tag{7}$$

ANS $w_{max} = 4.5$ kN/m ; $k_{BD} = 11.8$

---

11. 45

Solution: E = 30000ksi $\sigma_{max} \le 25ksi$ $A_A = 1\text{in}^2$ $A_B = 2\text{in}^2$ P_{max}=?

From earlier problem in chapter 4 we have compressive axial force in members A and B as

$$N_A = (0.085P + 1.556)kip \qquad N_B = (0.2988P - 0.6483)kip \tag{1}$$

$$\sigma_A = \frac{N_A}{A_A} = \frac{(0.083P + 1.556)}{1} \le 25 \quad or \quad P \le 282.4 kips \tag{2}$$

$$\sigma_B = \frac{N_B}{A_B} = \frac{(0.2988P - 0.6483)}{2} \le 25 \quad or \quad P \le 169.5 kips \tag{3}$$

$$A_A = \pi d_A^4/4 = 1 \quad or \quad d_A = 1.128 \quad I_A = \pi d_A^{\ 4}/64 = \pi(1.128)^4/64 = 0.0796 in^4 \tag{4}$$

$$A_B = \pi d_B^4/4 = 2 \quad or \quad d_B = 1.595 \quad I_B = \pi d_B^{\ 4}/64 = \pi(1.595)^4/64 = 0.3183 in^4 \tag{5}$$

Member A can be modeled as pin connected and member B as built in on one end. The critical buckling loads are as follows.

$$(P_{cr})_A = \frac{\pi^2 EI_A}{L_A^2} = \frac{\pi^2(30000)(0.0796)}{(36)^2} = 18.18 kip \qquad (P_{cr})_B = \frac{\pi^2 EI_B}{4L_B^2} = \frac{\pi^2(30000)(0.3183)}{4(48)^2} = 10.22 kip \tag{6}$$

$$N_A \le (P_{cr})_A \quad or \quad 0.083P + 1.556 \le 18.18 \quad or \quad P \le 200.3 kips \tag{7}$$

$$N_B \le (P_{cr})_B \quad or \quad 0.2988P - 0.6483 \le 10.22 \quad or \quad P \le 36.37 kips \tag{8}$$

ANS $P_{max} = 36.3$ kip

---

11. 46

Solution

By equilibrium of moment about O and equilibrium of forces in y-direction we obtain the following.

$$(V_y + dV_y)dx + M_z + dM_z - M_z - (p_y dx)\frac{dx}{2} + P dv = 0 \quad or \quad V_y dx + dM_z + P dv = 0 \quad or \quad V_y = -\left(\frac{dM_z}{dx} + P\frac{dv}{dx}\right) \tag{1}$$

$$V_y + dV_y - V_y + p_y dx = 0 \quad or \quad \frac{dV_y}{dx} = -p_y \quad or \quad -\frac{d^2 M_z}{dx^2} - P\frac{d^2 v}{dx^2} = -p_y \tag{2}$$

(a)

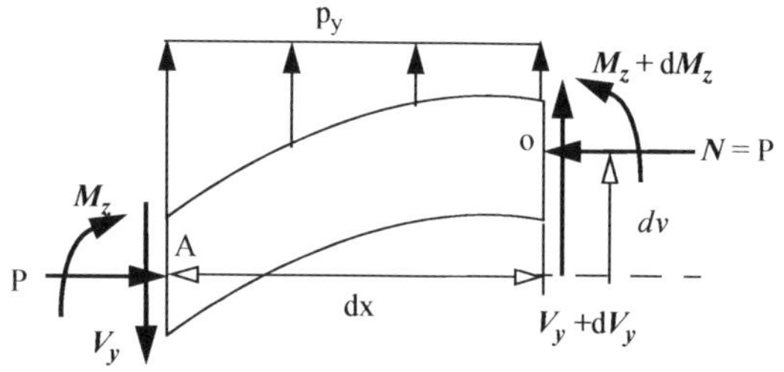

$$M_z = EI\frac{d^2 v}{dx^2} \qquad \frac{d^2}{dx^2}\left(EI\frac{d^2 v}{dx^2}\right) + P\frac{d^2 v}{dx^2} = p_y \quad or \quad EI\frac{d^4 v}{dx^4} + P\frac{d^2 v}{dx^2} = p_y \tag{3}$$

The above equation is the desired result.

---

11. 47

Solution: v_{max} =?

Substituting p_y = -w into the results of previous problem and using the boundary conditions of a the simple supported column, we obtain the following boundary value problem

$$EI\frac{d^4 v}{dx^4} + P\frac{d^2 v}{dx^2} = -w \quad or \quad \frac{d^4 v}{dx^4} + \lambda^2\frac{d^2 v}{dx^2} = -\frac{w}{EI} \tag{1}$$

$$v(0) = 0 \quad M_z(0) = EI\frac{d^2 v}{dx^2}(0) = 0 \quad v(L) = 0 \quad M_z(L) = EI\frac{d^2 v}{dx^2}(L) = 0 \tag{2}$$

The solution proceeds as follows.

$$\frac{d^2 v}{dx^2} + \lambda^2 v = -\left(\frac{w}{EI}\right)\frac{x^2}{2} + C_1 x + C_2 \tag{3}$$

$$\frac{d^2 v}{dx^2}(0) = C_2 = 0 \quad and \quad \frac{d^2 v}{dx^2}(L) = -\frac{w}{EI}\left(\frac{L^2}{2}\right) + C_1 L = 0 \quad or \quad C_1 = \frac{wL}{2EI} \tag{4}$$

Substituting C_1 and C_2 we can write the boundary value problem statement as shown below. obtain

$$\frac{d^2 v}{dx^2} + \lambda^2 v = \frac{wLx}{2EI} - \frac{wx^2}{2EI} \qquad v(0) = 0 \qquad v(L) = 0 \tag{5}$$

The boundary value problem statement above is the same as given in the cited Example, thus maximum deflection is as given in the Example.

ANS $v_{max} = -\frac{wEI}{P^2}\left[sec\left(\frac{L}{2}\sqrt{\frac{P}{EI}}\right) - 1\right] + \frac{wL}{8P}$

11.48

Solution: $\Delta T_{crit} = ?$

The compressive axial reaction force P can be found by noting that the total axial strain is zero

$$\varepsilon = \sigma/E + \alpha\Delta T = 0 \qquad or \qquad \sigma = -E\alpha\Delta T = -P/A \qquad or \qquad P = E\alpha\Delta TA \tag{1}$$

At critical temperature P would equal to the critical buckling load. Thus

$$P_{cr} = E\alpha\Delta T_{crit}A = \pi^2 EI/L^2 \qquad or \qquad \Delta T_{crit} = \pi^2(I/A)/(\alpha L^2) = (\pi^2 r^2)/(\alpha L^2) \tag{2}$$

ANS $\Delta T_{crit} = \pi^2/[\alpha(L/r)^2]$

11.49

Solution

By equilibrium of moment about point O and equilibrium of forces in the y-direction we obtain the following.

$$M_z + dM_z - M_z + V_y dx - (kdx)v\left(\frac{dx}{2}\right) + Pdv = 0 \qquad or \qquad V_y = -\left[\frac{dM_z}{dx} + P\frac{dv}{dx}\right] \tag{1}$$

$$V_y + dV_y - V_y - (kdx)v = 0 \qquad or \qquad \frac{dV_y}{dx} - kv = 0 \tag{2}$$

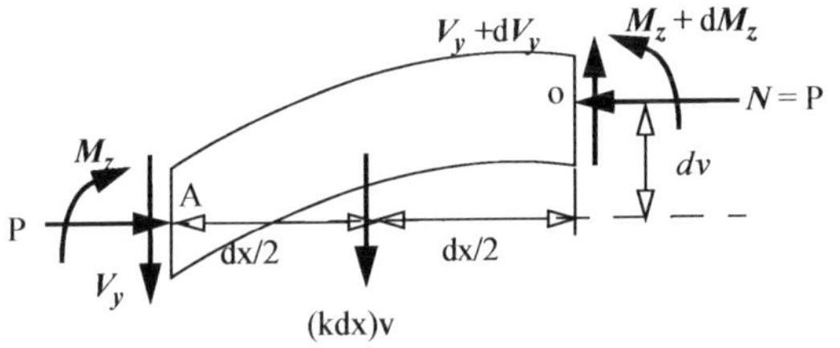

Substituting the first equation into the second equation, we obtain the following.

$$-\left(\frac{d^2M_z}{dx^2} + P\frac{d^2v}{dx^2}\right) - kv = 0 \qquad or \qquad EI\frac{d^4v}{dx^4} + P\frac{d^2v}{dx^2} + kv = 0 \tag{3}$$

The above equation is the desired result.

11.50

Solution

Consider the solution below with its derivatives to obtain the following.

$$v(x) = A\cos\Lambda x + B\sin\Lambda x \qquad \frac{dv}{dx} = -A\Lambda\sin\Lambda x + B\Lambda\cos\Lambda x \tag{1}$$

$$\frac{d^2v}{dx^2} = -A\Lambda^2\cos\Lambda x - B\Lambda^2\sin\Lambda x = -\Lambda^2(A\cos\Lambda x + B\sin\Lambda x) = -\Lambda^2 v \tag{2}$$

$$\frac{d^4v}{dx^4} = \frac{d^2}{dx^2}\left[\frac{d^2v}{dx^2}\right] = \frac{d^2}{dx^2}[-\Lambda^2 v] = -\Lambda^2\frac{d^2v}{dx^2} = \Lambda^4 v \tag{3}$$

Substituting the above equations into the differential equation obtained in the previous problem we obtain the following.

$$EI\frac{d^4v}{dx^4} + P\frac{d^2v}{dx^2} + kv = EI[\Lambda^4 v] + P[-\Lambda^2 v] + kv = 0 \qquad or \qquad (EI\Lambda^4 - P\Lambda^2 + k)v = 0 \qquad or \qquad EI\Lambda^4 - P\Lambda^2 + k = 0 \tag{4}$$

For simply supported column we have the following boundary conditions.

$$v(0) = A = 0 \qquad and \qquad v(L) = B\sin\Lambda L = 0 \qquad or \qquad \sin\Lambda L = 0 \qquad or \qquad \Lambda L = n\pi \qquad or \qquad \Lambda = n\pi/L \tag{5}$$

Substituting Λ

$$EI\left(\frac{n\pi}{L}\right)^4 - P_n\left(\frac{n\pi}{L}\right)^2 + k = 0 \qquad or \qquad P_n\left(\frac{n^2\pi^2}{L^2}\right) = EI\left(\frac{n^4\pi^4}{L^4}\right) + k \qquad or \qquad P_n = \frac{\pi^2 EI}{L^2}\left[n^2 + \frac{1}{n^2}\left(\frac{kL^4}{\pi^4 EI}\right)\right] \tag{6}$$

ANS $P_n = \frac{\pi^2 EI}{L^2}\left[n^2 + \frac{1}{n^2}\left(\frac{kL^4}{\pi^4 EI}\right)\right]$

---

11. 51

Solution

By equilibrium of moment about point A we obtain:

$$M_z + Pv = 0 \quad (1)$$

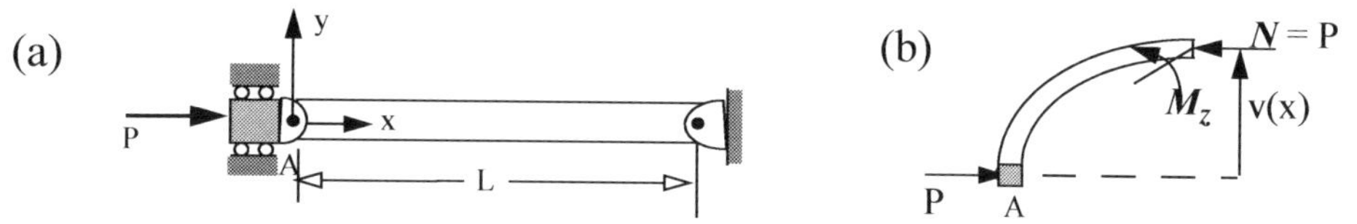

Substituting the moment-curvature relationship $M_z = \sum_{j=1}^{n} E_j(I_{zz})_j$ for composite beam derived in an earlier problem in chapter 6 earlier we obtain the differential equation below.

$$\left(\sum_{j=1}^{n} E_j(I_{zz})_j\right)\frac{d^2v}{dx^2} + Pv = 0 \quad (2)$$

The above differential equation differs from the case 1 of Table 11.1 in the bending rigidity which is a constant. We may thus replace EI the bending rigidity in the Euler buckling load formula with the bending rigidity for the composite to obtain the results below.

ANS $P_{cr} = [\pi^2 \sum_{i=1}^{n} E_i I_i]/L_{eff}^2$

---

11. 52

Solution $P_{cr} = f(E, d, L)$

The bending rigidity of the cross-section can be found as shown below.

$$I_i = \pi d^4/64 \qquad I_o = \pi((2d)^4 - d^4)/64 = (15\pi)d^4/64 \qquad \sum_{j=1}^{n} E_j(I_{zz})_j = EI_i + 2EI_o = E\left(\frac{\pi}{64}d^4\right) + 2E\left(\frac{15\pi}{64}d^4\right) = \frac{31\pi}{64}d^4 \quad (1)$$

Substituting the bending rigidity and $L_{eff} = L$ into the results of previous problem we obtain the following.

$$P_{cr} = \left[\pi^2\left(\frac{31\pi}{64}d^4\right)\right]/L^2 \quad (2)$$

ANS $P_{cr} = \dfrac{31\pi^3 d^4}{64L^2}$

---

11. 53

Solution $P_{cr} = f(E, d, L)$

The bending rigidity of the cross-section can be found as shown below.

$$I_1 = \frac{1}{12}a^4 \qquad I_2 = \frac{1}{12}(0.25a)(a^3) = \frac{1}{48}a^4 \qquad \sum_{j=1} E_j(I_{zz})_j = EI_1 + 2[2EI_o] = E\left(\frac{1}{12}a^4\right) + 4E\left(\frac{1}{48}a^4\right) = \frac{E}{6}a^4 \quad (1)$$

Substituting the bending rigidity and $L_{eff} = 2L$ into results of earlier problem we obtain the following.

$$P_{cr} = \left[\pi^2\left(\frac{E}{6}a^4\right)\right]/(2L)^2 \quad (2)$$

ANS $P_{cr} = \pi^2 a^4 E/(24L^2)$

Section 11.3

11. 54

Solution $v(x) = ?$

The figure in problem 11.34 is shown as Figure (a) and Figure (b) shows an equivalent diagram to the one given in the problem.

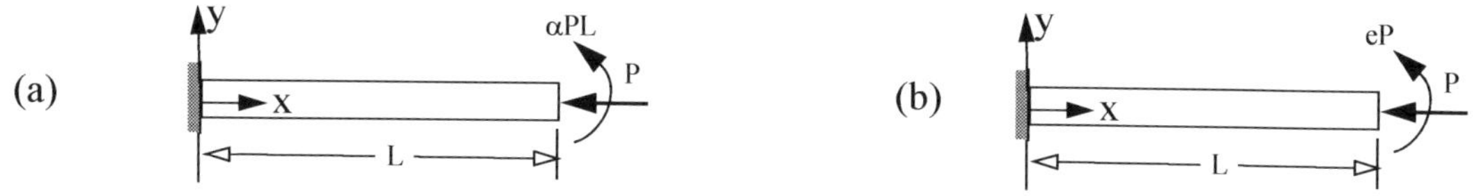

The following was one of the equation in problem 11.34.

$$v(x) = \frac{\alpha L}{\cos\lambda L}(1 - \cos\lambda x) \quad (1)$$

Comparing Figures (a) and (b) we obtain: $\alpha L = e$. which we substitute into the above equation to obtain the following.

$$v(x) = \frac{e}{\cos\lambda L}(1-\cos\lambda x) \quad (2)$$

ANS $v(x) = \frac{e(1-\cos\lambda x)}{\cos\lambda L}$

---

11. 55

Solution P = 3 kips E = 30000 psi L= 5 ft $\varepsilon = 0.25$ σ_{max} =? v_{max} =?

The solution proceeds as follows.

$$A = \pi[2.5^2 - 2^2]/4 = 1.767in^2 \qquad I = \pi[2.5^4 - 2^4]/64 = 1.132in^4 \qquad r = \sqrt{I/A} = 0.8004in \quad (1)$$

$$\frac{ec}{r^2} = \frac{(0.25)(1.25)}{(0.8004)^2} = 0.4878 \qquad \frac{L_{eff}}{r} = \frac{(2)(5)(12)}{(0.8004)} = 149.92 \qquad P_{cr} = \frac{\pi^2 EI}{4L^2} = \frac{\pi^2(30000)(1.132)}{4(60)^2} = 23.37kip \quad (2)$$

$$v_{max} = e\left[\sec\left(\frac{\pi}{2}\sqrt{\frac{P}{P_{cr}}}\right) - 1\right] = 0.25\left[\sec\left(\frac{\pi}{2}\sqrt{\frac{3}{23.37}}\right) - 1\right] \quad (3)$$

$$\sigma_{max} = \frac{P}{A}\left[1 + \frac{ec}{r^2}\sec\left(\frac{L_{eff}}{2r}\sqrt{\frac{P}{EA}}\right)\right] = \frac{3}{1.767}\left[1 + 0.4878\sec\left(\frac{149.92}{2}\sqrt{\frac{3}{(30000)(1.767)}}\right)\right] \quad (4)$$

ANS $v_{max} = 0.0458in$; $\sigma_{max} = 2.68ksi(C)$

---

11. 56

Solution: P = 3 kips E = 30000 ksi σ_{max} = 8 ksi L_{max} =?

Solution proceeds as follows.

$$A = \pi[2.5^2 - 2^2]/4 = 1.767in^2 \qquad I = \pi[2.5^4 - 2^4]/64 = 1.132in^4 \qquad r = \sqrt{I/A} = 0.8004in \quad (1)$$

$$\sigma_{max} = \frac{P}{A}\left[1 + \frac{ec}{r^2}\sec\left(\frac{L_{eff}}{2r}\sqrt{\frac{P}{EA}}\right)\right] = \frac{3}{1.767}\left[1 + 0.4878\sec\left(\frac{L_{eff}}{2(0.8004)}\sqrt{\frac{3}{(30000)(1.767)}}\right)\right] \text{ or}$$

$$\sigma_{max} = 1.6978[1 + 0.4878\sec(4.699L_{eff}(10^{-3}))] = 8 \quad or \quad \sec(4.699L_{eff}(10^{-3})) = 7.6097 \text{ or}$$

$$4.699L_{eff}(10^{-3}) = 1.439 \quad or \quad L_{eff} = \frac{1.439}{4.699(10^{-3})} = 2L_{max} \quad or \quad L_{max} = 153.11 \text{ in } \textbf{or} \quad (2)$$

ANS $L_{max} = 153.1$ in

---

11. 57

Solution: L = 5 ft σ_{yield} = 30 ksi E = 30000 ksi e = 0.25 P_{max} =?

Solution proceeds as follows.

$$A = \pi[2.5^2 - 2^2]/4 = 1.767in^2 \qquad I = \pi[2.5^4 - 2^4]/64 = 1.132in^4 \qquad r = \sqrt{I/A} = 0.8004in \quad (1)$$

From figure of failure envelopes for steel we obtain the following estimate.

$$\frac{L_{eff}}{r} = \frac{(2)(5)(12)}{(0.8004)} = 149.92 \qquad \frac{ec}{r^2} = \frac{(0.25)(1.25)}{(0.8004)^2} = 0.4878 \qquad \frac{P}{A\sigma_{yield}} = 0.32 \qquad P = (0.32)(30)(1.767) = 16.963 \text{ kip} \quad (2)$$

ANS $P_{max} = 16.96$ kip

---

11. 58

Solution: P = 100 kN E = 70 GPa L= 2 m e = 0.009 σ_{max} =? v_{max} =?

The solution proceeds as follows.

$$A = (0.05)^2 - (0.03)^2 = 1.6(10^{-3})m^2 \qquad I = \frac{[0.05^4 - 0.03^4]}{12} = 0.453(10^{-6})m^4 \qquad r = \sqrt{\frac{I}{A}} = 0.01683m \quad (1)$$

$$\frac{ec}{r^2} = \frac{(0.009)(0.025)}{(0.01683)^2} = 0.7942 \qquad P_{cr} = \frac{20.13EI}{L^2} = \frac{(20.13)(70)(10^9)(0.453)(10^{-6})}{2^2} = 159.69kN \quad (2)$$

$$v_{max} = e\left[\sec\left(\frac{\pi}{2}\sqrt{\frac{P}{P_{cr}}}\right) - 1\right] = 0.009\left[\sec\left(\frac{\pi}{2}\sqrt{\frac{100}{159.69}}\right) - 1\right] = 18.956(10^{-3}) \text{ m} \quad (3)$$

$$\frac{L_{eff}}{r} = \frac{(0.7)(2)}{(0.01683)} = 83.172$$

$$\sigma_{max} = \frac{P}{A}\left[1+\frac{ec}{r^2}sec\left(\frac{L_{eff}}{2r}\sqrt{\frac{P}{EA}}\right)\right] = \frac{100(10^3)}{1.6(10^{-3})}\left[1+0.7942sec\left(\frac{83.172}{2}\sqrt{\frac{100(10^3)}{(70)(10^9)(1.6)(10^{-3})}}\right)\right] = 216.5(10^6)\text{ N/m}^2 \quad \textbf{(4)}$$

ANS $v_{max} = 18.96$ mm ; $\sigma_{max} = 216.5$ MPa (C)

---

11. 59

Solution: P = 100 kN E = 70 GPa $\varepsilon = 0.009$ L_{max} =? $\sigma_{max} = 250$ MPa

From previous problem we have the following.

$$A = 1.6(10^{-3})m^2 \qquad I = 0.453(10^{-6})m^4 \qquad r = 0.01683m \qquad \frac{ec}{r^2} = 0.7942 \quad \textbf{(1)}$$

$$\sigma_{max} = \frac{P}{A}\left[1+\frac{ec}{r^2}sec\left(\frac{L_{eff}}{2r}\sqrt{\frac{P}{EA}}\right)\right] = \frac{100(10^3)}{1.6(10^{-3})}\left[1+0.7942sec\left(\frac{L_{eff}}{2(0.01683)}\sqrt{\frac{100(10^3)}{(70)(10^9)(1.6)(10^{-3})}}\right)\right] \text{ or}$$

$$\sigma_{max} = 62.5(10^6)[1+0.7942sec(0.8877L_{eff})] = 250(10^6) \quad or \quad sec(0.8877L_{eff}) = 3.777 \text{ or}$$

$$0.8877L_{eff} = 1.3028 \quad or \quad L_{eff} = 1.4676 = 0.7L_{max} \quad or \quad L_{max} = 2.0965\text{m} \quad \textbf{(2)}$$

ANS $L_{max} = 2.1$m

---

11. 60

Solution: L = 2 m $\sigma_{yield} = 280$ MPa E = 70 GPa e = 0.009m P_{max} =?

From earlier problem we have the following.

$$A = 1.6(10^{-3})m^2 \qquad (ec)/r^2 = 0.7942 \qquad L_{eff}/r = 83.172 \quad \textbf{(1)}$$

We note that $\sigma_{yield}/E = 0.004$. From failure envelope for Aluminum we obtain the following estimate.

$$\frac{P}{A\sigma_{yield}} = 0.23 \quad or \quad P_{max} = (0.23)(280)(10^6)(1.6)(10^{-3}) = 103.04(10^3)\text{ N} \quad \textbf{(2)}$$

ANS $P_{max} = 103$ kN

---

11. 61

Solution: W8 X 18 P = 20kips E = 30000 psi L= 9ft e = 0.3 in σ_{max} =? v_{max} =?

From the table in Section C.6 we have the following.

$$A = 5.26\text{ in}^2 \qquad I = 7.97\text{ in}^4 \qquad r = 1.23\text{ in} \qquad c = \frac{5.25}{2} = 2.65\text{ in} \quad \textbf{(1)}$$

$$\frac{ec}{r^2} = \frac{(0.3)(2.625)}{(1.23)^2} = 0.5205 \qquad \frac{L_{eff}}{r} = \frac{(2)(9)(12)}{(1.23)} = 175.61 \qquad P_{cr} = \frac{\pi^2 EI}{4L^2} = \frac{\pi^2(30000)(7.97)}{4(108)^2} = 50.579kip \quad \textbf{(2)}$$

$$v_{max} = e\left[sec\left(\frac{\pi}{2}\sqrt{\frac{P}{P_{cr}}}\right)-1\right] = 0.3\left[sec\left(\frac{\pi}{2}\sqrt{\frac{20}{50.579}}\right)-1\right] = 0.245\text{ in} \quad \textbf{(3)}$$

$$\sigma_{max} = \frac{P}{A}\left[1+\frac{ec}{r^2}sec\left(\frac{L_{eff}}{2r}\sqrt{\frac{P}{EA}}\right)\right] = \frac{20}{5.26}\left[1+0.5305sec\left(\frac{175.61}{2}\sqrt{\frac{20}{(30000)(5.26)}}\right)\right] = 7.4\text{ ksi} \quad \textbf{(4)}$$

ANS $v_{max} = 0.245$ in ; $\sigma_{max} = 7.4$ ksi (C)

---

11. 62

Solution: W8 x 18 P = 20kips E = 30000 psi e = 0.3 in $\sigma_{max} = 24$ ksi L_{max} =?

From the previous problem we have the following.

$$A = 5.26\text{ in}^2 \qquad I = 7.97\text{ in}^4 \qquad r = 1.23\text{ in} \qquad \frac{ec}{r^2} = 0.5205 \quad \textbf{(1)}$$

$$\sigma_{max} = \frac{P}{A}\left[1+\frac{ec}{r^2}sec\left(\frac{L_{eff}}{2r}\sqrt{\frac{P}{EA}}\right)\right] = \frac{20}{5.26}\left[1+0.5205sec\left(\frac{L_{eff}}{2(1.23)}\sqrt{\frac{20}{(30000)(5.26)}}\right)\right] \text{ or}$$

$$\sigma_{max} = 3.802[1+0.5205sec(4.576L_{eff}(10^{-3}))] = 24 \qquad sec(4.576L_{eff}(10^{-3})) = 10.205 \text{ or}$$

$$4.576L_{eff}(10^{-3}) = 1.473 \quad or \quad L_{eff} = 321.8 = 2L_{max} \quad or \quad L_{max} = 160.9\text{ in.} \quad \textbf{(2)}$$

ANS $L_{max} = 160.9$ in.

---

11. 63

Solution: W 8x 18 L = 9 ft E = 30000 psi e = 0.3 in σ_{yield} = 30ksi P_{max} =?

From previous problems we have:

$$A = 5.26 in^2 \qquad (ec)/r^2 = 0.5205 \qquad L_{eff}/r = 175.61 \tag{1}$$

We note that $\sigma_{yield}/E = 0.001$. From failure envelope for steel we obtain the following estimate.

$$P/(A\sigma_{yield}) = 0.25 \qquad or \qquad P_{max} = (0.3)(30)(5.26) = 39.45 \text{ kips} \tag{2}$$

ANS $P_{max} = 39.45$ kips

---

11. 64

Solution: L = 6 ft d_0 = 3in t = 1/8in P_{cr} =?

The cross-sectional area, the area moment of inertia, and the tangent modulus of elasticity for the three regions of stress- strain curve are as shown below.

$$A = \frac{\pi}{4}(3^2 - 2.75^2) = 1.129 in^2 \qquad I = \frac{\pi}{64}(3^4 - 2.75^4) = 1.168 in^4 \tag{1}$$

$$E_1 = \frac{14}{0.001} = 14000 ksi \qquad E_2 = \frac{37-14}{0.004-0.001} = 7666.7 ksi \qquad E_3 = \frac{43-37}{0.007-0.004} = 2000 ksi \tag{2}$$

$$P_{cr} = \frac{\pi^2 E_t(1.168)}{72^2} = 2.224 E_t(10^{-3}) \qquad \sigma_{cr} = \frac{P_{cr}}{A} = \frac{2.224 E_t(10^{-3})}{1.129} = 1.9696 E_t \tag{3}$$

Substituting $E_t = E_1$ we obtain: $\sigma_{cr} = (1.9696)(14000)(10^{-3}) = 27.57 ksi$. As the stress value is greater than 14 ksi we t try $E_t = E_2$ and obtain $\sigma_{cr} = (1.9696)(7666.7)(10^{-3}) = 15.1 ksi$. The stress value is in the second linear region. Substituting $E_t = E_2$ in we obtain

$$P_{cr} = (2.224)(7666.7)(10^{-3}) = 17.05 \text{ kip} \tag{4}$$

ANS $P_{cr} = 17.0$ kip

---

11. 65

Solution: t = 10mm 75mm x 75mm L = 0.75m P_{cr} =?

The cross-sectional area, the area moment of inertia, and the tangent modulus of elasticity for the three regions of stress- strain curve are as shown below.

$$A = (0.075^2 - 0.055^2) = 2.6(10^{-3}) m^2 \qquad I = \frac{1}{12}(0.075^4 - 0.055^4) = 1.874(10^{-6}) m^4 \tag{1}$$

$$E_1 = \frac{280(10^6)}{0.004} = 70 GPa \qquad E_2 = \frac{(330-280)(10^6)}{0.005-0.004} = 50 GPa \qquad E_3 = \frac{(390-330)(10^6)}{0.007-0.005} = 30 GPa \tag{2}$$

$$P_{cr} = \pi^2 E_t(1.874)(10^{-6})/0.75^2 = 32.88 E_t(10^{-6}) \qquad \sigma_{cr} = P_{cr}/A = (32.88 E_t)/[2.6(10^{-3})] = 12.65 E_t(10^{-3}) \tag{3}$$

Substituting $E_t = E_1$ we obtain: $\sigma_{cr} = (12.65)(70)(10^9)(10^{-3}) = 885 MPa$. As this stress is greater than 280MPa, we try $E_t = E_2$ and obtain: $\sigma_{cr} = (12.65)(50)(10^9)(10^{-3}) = 632.5 MPa$. Once more the stress level is greater than the upper limit of region two i.e greater than 330MPa. We next try $E_t = 30 MPa$ and obtain: $\sigma_{cr} = (12.65)(30)(10^9)(10^{-3}) = 379.5 MPa$ The stress is in the third linear region. Substituting $E_t = E_3$ we obtain

$$P_{cr} = (32.88)(30)(10^9)(10^{-6}) = 986(10^3) \text{ N} \tag{4}$$

ANS $P_{cr} = 986$ kN

---

11. 66

Solution

By equilibrium of moment about point O we obtain:

$$M_z + P\left(v - v_o \sin\frac{\pi x}{L}\right) = 0 \qquad or \qquad EI\frac{d^2 v}{dx^2} + Pv(x) = Pv_o \sin\frac{\pi x}{L} \qquad or \qquad \frac{d^2 v}{dx^2} + \frac{P}{EI}v(x) = \frac{P}{EI}v_o \sin\frac{\pi x}{L} \tag{1}$$

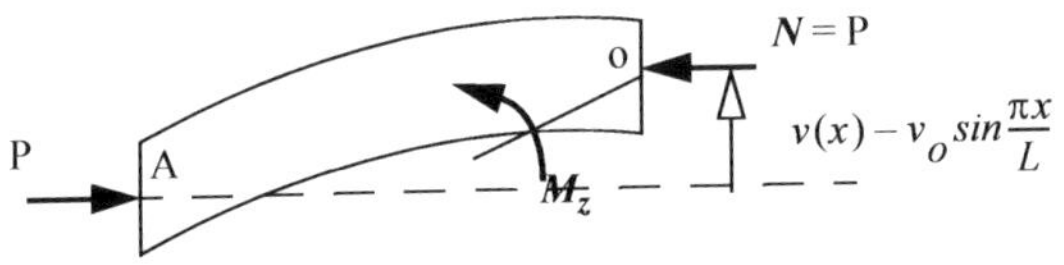

The boundary value problem can be written as shown below. **(1)**

$$\frac{d^2v}{dx^2}+\lambda^2 v = \lambda^2 v_o sin\frac{\pi x}{L} \qquad v(0) = 0 \qquad v(L) = 0 \qquad where \qquad \lambda^2 = \frac{P}{EI} \tag{2}$$

To find the particular solution we substitute $v_p(x) = C sin\pi(x/L)$ into the differential equation to obtain the following.

$$\left[-C\left(\frac{\pi}{L}\right)^2+\lambda^2 C\right]sin\frac{\pi x}{L} = \lambda^2 v_o sin\frac{\pi x}{L} \qquad or \qquad C = \frac{\lambda^2 v_o}{\lambda^2-(\pi/L)^2} = \frac{v_o}{1-P/P_{cr}} \tag{3}$$

$$v_h(x) = A cos\lambda x + B sin\lambda x \qquad v(x) = v_h(x)+v_p(x) = A cos\lambda x + B sin\lambda x + \frac{v_o}{1-P/P_{cr}}sin\frac{\pi x}{L} \tag{4}$$

$$v(0) = A = 0 \qquad or \qquad A = 0 \qquad v(L) = B sin\lambda L = 0 \qquad or \qquad B = 0 \tag{5}$$

ANS $v(x) = \left(\frac{v_o}{1-P/P_{cr}}\right)sin\frac{\pi x}{L}$

---

11. 67

Solution

We write the bending rigidity for composite beam as:

$$\sum_{j=1}^{2} E_j(I_{zz})_j = E_t I_1 + E I_2 = E_r I \qquad or \qquad E_r = E_t(I_1/I)+E(I_2/I) \tag{1}$$

$$P_{cr} = (\pi^2 E_r I)/L_{eff}^2 \tag{2}$$

The above equations are the desired results.

---

11. 68

Solution: d = 2ft E = 8000 ksi L= 20ft e = 2in $\sigma_{max} \le 20ksi$ P_{max} =? (a) pinned (b) built in

From geometry we have the following..

$$A = \pi(12)^2 = 452.4in^2 \qquad I = \frac{\pi}{4}(12)^4 = 16286in^4 \qquad r = \sqrt{\frac{I}{A}} = 6in \qquad \frac{ec}{r^2} = \frac{(2)(12)}{6^2} = 0.6667 \tag{1}$$

(a) For pinned column L_{eff} = L = 240in. Let σ = P/A

$$\sigma_{max} = \sigma\left[1+\frac{ec}{r^2}sec\left(\frac{L_{eff}}{2r}\sqrt{\frac{\sigma}{A}}\right)\right] = \sigma\left[1+0.6667sec\left(20\sqrt{\frac{\sigma}{8000}}\right)\right] = 20ksi \qquad f_1(\sigma) = \sigma[1+0.6667sec(0.2236\sqrt{\sigma})]-20 = 0 \tag{2}$$

Table below shows the calculation of the root σ in the above equation. From the table we obtain the root as σ = 10.56 ksi.

$$P_{max} = (10.56)(452.4) = 4777.3 \text{ kip} \tag{3}$$

(b) For built in ends L_{eff} = 0.5L = 120in.

$$\sigma_{max} = \sigma\left[1+0.6667sec\left(10\sqrt{\frac{\sigma}{8000}}\right)\right] = 20ksi \qquad f_2(\sigma) = \sigma[1+0.6667sec(0.1118\sqrt{\sigma})]-20 = 0 \tag{4}$$

Table below shows the calculation of the root in the above equation. From the table we obtain the root as σ =11.64 ksi.

$$P_{max} = (11.64)(452.4) = 5265.9 \text{ kip} \tag{5}$$

ANS (a) P_{max} = 4777 kip (b) P_{max} = 5266 kip

σ	$f_1(\sigma)$	σ	$f_1(\sigma)$	σ	$f_1(\sigma)$
1.00	-18.32	10.00	-1.23	10.50	-0.15
2.00	-16.60	10.10	-1.02	10.51	-0.13
3.00	-14.84	10.20	-0.80	10.52	-0.11
4.00	-13.04	10.30	-0.58	10.53	-0.09
5.00	-11.20	10.40	-0.37	10.54	-0.06
6.00	-9.31	10.50	-0.15	10.55	-0.04
7.00	-7.38	10.60	0.07	10.56	-0.02
8.00	-5.39	10.70	0.29	10.57	0.00
9.00	-3.34	10.80	0.50	10.58	0.02
10.00	-1.23	10.90	0.72	10.59	0.05
11.00	0.95	11.00	0.95	10.60	0.07

σ	$f_2(\sigma)$	σ	$f_2(\sigma)$	σ	$f_2(\sigma)$
4.00	-13.27	11.00	-1.13	11.60	-0.07
5.00	-11.56	11.10	-0.95	11.61	-0.05
6.00	-9.84	11.20	-0.78	11.62	-0.03
7.00	-8.12	11.30	-0.60	11.63	-0.02
8.00	-6.39	11.40	-0.42	11.64	0.00
9.00	-4.65	11.50	-0.25	11.65	0.02
10.00	-2.89	11.60	-0.07	11.66	0.04
11.00	-1.13	11.70	0.11	11.67	0.05
12.00	0.64	11.80	0.29	11.68	0.07
13.00	2.42	11.90	0.46	11.69	0.09
14.00	4.21	12.00	0.64	11.70	0.11

---

11. 69

Solution E = 70 GPa P_{max} = ?

From earlier problem we have thye following.

$$A = 1.6(10^{-3})\ m^2 \qquad \frac{ec}{r^2} = 0.7942 \qquad \frac{L_{eff}}{r} = 83.172 \tag{1}$$

$$\sigma_{max} = \frac{P}{A}\left[1+\frac{ec}{r^2}sec\left(\frac{L_{eff}}{2r}\sqrt{\frac{P}{EA}}\right)\right] = \sigma(10^6)\left[1+0.7942\,sec\left(\frac{83.172}{2}\sqrt{\frac{\sigma(10^6)}{70(10^9)}}\right)\right] = 280(10^6) \qquad where \qquad \sigma = \frac{P}{A} \text{ or}$$

$$f(\sigma) = \sigma\left[1+0.7942\,sec\left(\frac{83.172}{2}\sqrt{\frac{\sigma}{70000}}\right)\right]-280 = 0 \tag{2}$$

The roots of the above equation can be solved on a spreadsheet as shown in Appendix B and are reported in the table below.

σ=P/A (MPa)	f(σ)	σ=P/A (MPa)	f(σ)	σ=P/A (MPa)	f(σ)	σ=P/A (MPa)	f(σ)
50	-140.10	65	-41.65	69	-1.10	69.09	-0.08
55	-113.75	65.5	-37.06	69.01	-0.99	69.091	-0.06
60	-81.75	66	-32.35	69.02	-0.87	69.092	-0.05
65	-41.65	66.5	-27.50	69.03	-0.76	69.093	-0.04
70	10.61	67	-22.52	69.04	-0.65	69.094	-0.03
75	82.39	67.5	-17.40	69.05	-0.53	69.095	-0.02
80	188.44	68	-12.12	69.06	-0.42	69.096	-0.01
85	363.33	68.5	-6.69	69.07	-0.30	69.097	0.00
90	711.71	69	-1.10	69.08	-0.19	69.098	0.02
95	1767.79	69.5	4.66	69.09	-0.08	69.099	0.03
		70	10.61	69.1	0.04	69.1	0.04

$$\sigma = \frac{P}{A} = 69.1\ MPa \qquad or \qquad P = (69.097)(10^6)(1.6)(10^{-3}) = 110555.2\ N \tag{3}$$

ANS $P = 110555$ N

11. 70

E = 30000 ksi $\qquad P_{max} = ?$

Solution

From earlier problem we have the following.

$$A = 5.26\ in^2 \qquad \frac{ec}{r^2} = 0.5205 \qquad \frac{L_{eff}}{r} = 175.61 \tag{1}$$

$$\sigma_{max} = \frac{P}{A}\left[1+\frac{ec}{r^2}sec\left(\frac{L_{eff}}{2r}\sqrt{\frac{P}{EA}}\right)\right] = \sigma\left[1+0.5205\,sec\left(\frac{175.61}{2}\sqrt{\frac{\sigma}{30000}}\right)\right] = 30 \qquad where \qquad \sigma = \frac{P}{A} \text{ or}$$

$$f(\sigma) = \sigma\left[1+0.5205\,sec\left(\frac{175.61}{2}\sqrt{\frac{\sigma}{30000}}\right)\right]-30 = 0 \tag{2}$$

The roots of the above equation can be solved on a spreadsheet as shown in Appendix B. The results are reported in the table below.

σ=P/A (ksi)	f(σ)	σ=P/A (ksi)	f(σ)	σ=P/A (ksi)	f(σ)	σ=P/A (ksi)	f(σ)
1	-28.40	7.000	-6.99	7.500	-1.00	7.560	-0.09
2	-26.62	7.100	-5.97	7.510	-0.85	7.561	-0.08
3	-24.56	7.200	-4.87	7.520	-0.70	7.562	-0.06
4	-22.06	7.300	-3.69	7.530	-0.55	7.563	-0.05
5	-18.85	7.400	-2.40	7.540	-0.40	7.564	-0.03
6	-14.34	7.500	-1.00	7.550	-0.25	7.565	-0.02
7	-6.99	7.600	0.54	7.560	-0.09	7.566	-0.00
8	8.48	7.700	2.22	7.570	0.06	7.567	0.01
9	72.62	7.800	4.09	7.580	0.22	7.568	0.03
10	-181.65	7.900	6.16	7.590	0.38	7.569	0.05
11	-70.95	8.000	8.48	7.600	0.54	7.570	0.06

$$\sigma = P/A = 7.567\ ksi \qquad or \qquad P = (7567)(5.26) = 39802\ lb \tag{3}$$

ANS $P = 39802$ lb

SOLUTION TO STATIC REVIEW EXAM 2

SR. 1

(a) As the y-axis is the axis of symmetry, the centroid will lie on the y-axis. Thus $z_c = 0$ **1 point**.

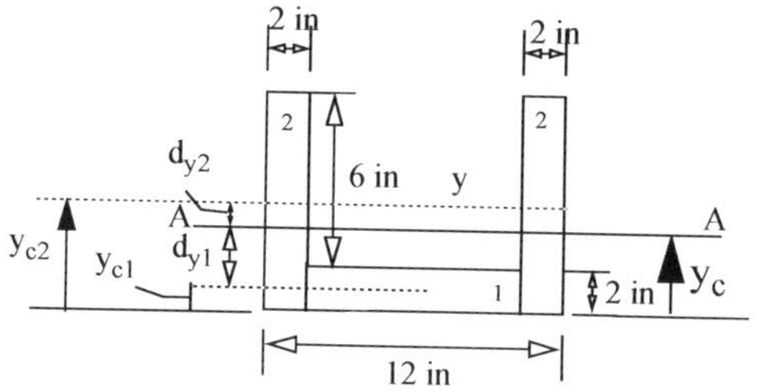

Calculation of the centroid.

section	y_{ci} in	A_i in^2	$y_{ci}A_i$ in^3
1	1	(8)(2)=16	16
2	4	2(8)(2)=32	128
Total		48	144

For each correct entry 1 point for a total of 8 points.

We obtain: $y_c = 144/48$ or $\boxed{y_c = 3in}$ **1 point for the correct answer and units.**

section	$d_{zi} = y_c - y_{ci}$ in	$I_{zizi} = \frac{1}{12}a_i b_i^3$ in^4	$I_{zizi} + A_i d_{zi}^2$ mm^4
1	2 in	$(8)(2^3)/12=5.333$	69.333
2	1 in	$(2+2)(8^3)/12=170.66$	202.667
	1 point for each correct entry	**2 points for each correct entry**	**1 point for each correct entry**

We obtain $\boxed{I_{AA} = 272\ in^4}$ **1 point for the correct answer**
1 point for the correct units.

SR. 2

We can replace uniform and linear loading by an equivalent forces as shown in Figure A.8, then replace it by a single force as shown in Figure (a). The magnitude of the forces are:

$$F_1 = (8000)(10)(3) = 240000 lbs = 240 kip \qquad F_2 = \left[\frac{1}{2}(8000)(9)(3)\right] = 108000 lb = 108 kip \tag{1}$$

For correct location of force F_1 and F_2, 3 points/force. For correct magnitudes of F_1 and F_2, 3 points/force.--Total 12

(a)

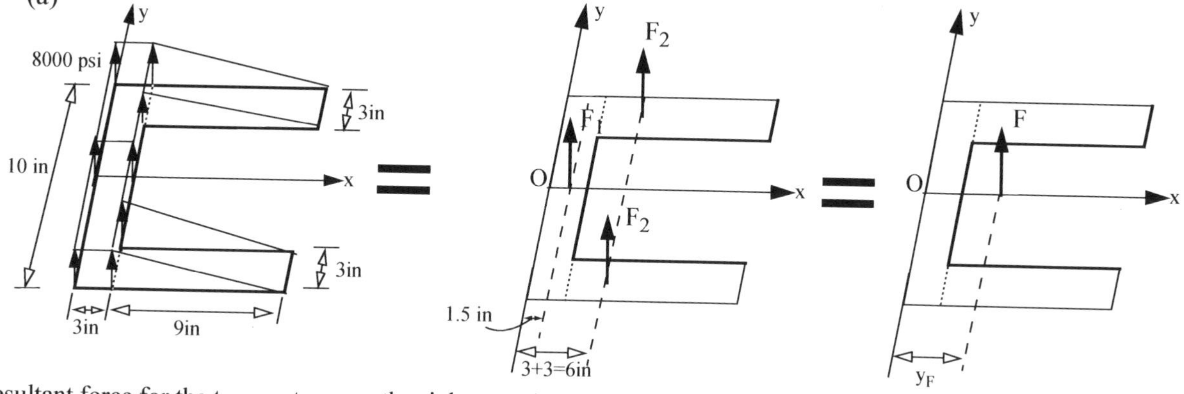

The resultant force for the two systems on the right must be the same. We obtain: $F = F_1 + 2F_2 = 240 + 2(108)$ or

$\boxed{F = 456\ kips}$ **2 points for the correct answer**
1 point for the correct units.

The resultant moment about any point (point O) for the two systems on the right must be the same. We obtain: $1.5(F_1) + (2)(6)(F_2) = x_F(F)$ or $x_F = (1656)/456$ or $\boxed{x_F = 3.63\ in}$ **1 point for the correct answer**
1 point for the correct units.

1 point for each correct entry in this equation.

SR. 3

We make imaginary cut at E and draw free body diagrams as shown below
From Figure (a) we obtain:

$N_E - 64 + 180 = 0$ **1 point for the correct equation.** $\boxed{N_E = 116kN(C)}$ **1 point for correct answer.**
1 point for correct units.
1 point for reporting compression

(a) Internal axial force calculations

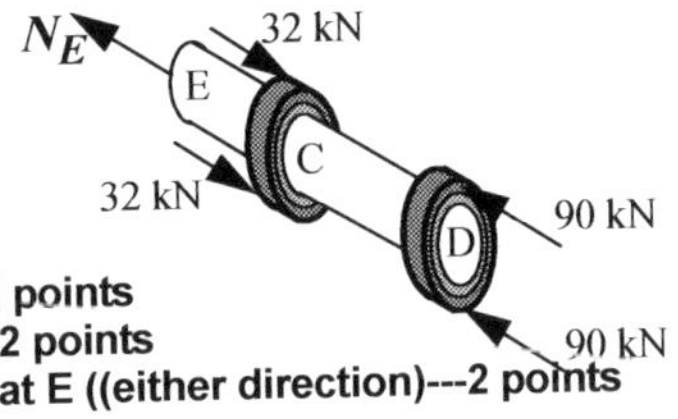

For drawing180 kN at D--2 points
64 kN at C ----2 points
Normal force at E ((either direction)---2 points

(b) Internal torque calculations

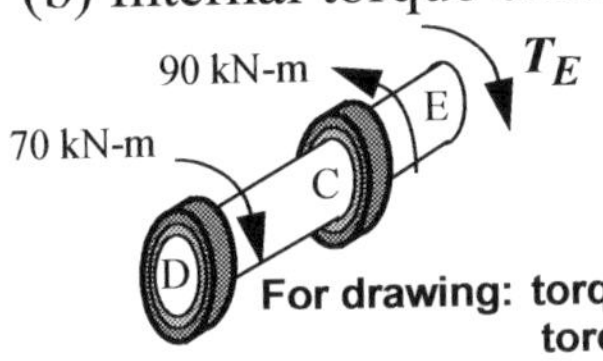

For drawing: torque at D--2 points
torque at C ----2 points
torque at E (either direction) ---2

From Figure (b) we obtain:

$T_E - 90 + 70 = 0$ **2 points for the correct equation.**

$\boxed{T_E = 20\ kN-m}$ **1 point for correct answer.**
1 point for correct units.

SR. 4

Figure (a) shows the free body diagram of the entire beam.

(a)

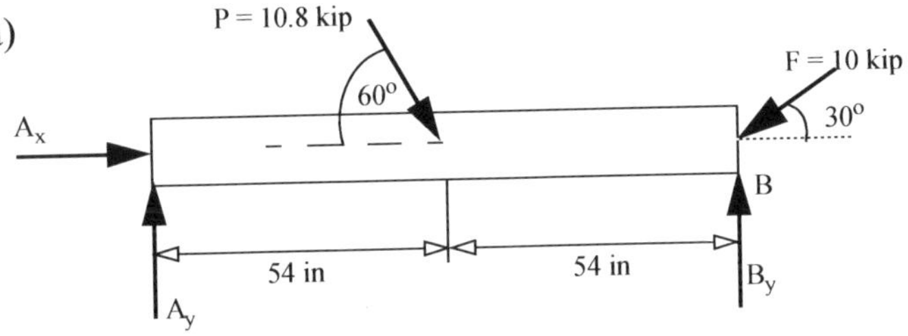

0.5 point for each force besides P
2 points for force magnitude and direction of
1 point for correct location of P
Total 5 points.

By equilibrium of moment about A we obtain:

$B_y(108) - (10 sin 30)(108) - (10.8 sin 60)(54) = 0$ or $B_y = 9.676 kip$ **0.5 point for each term in the equation for a total of 1.5 points.**

By equilibrium of forces in the y direction we obtain:

$A_y + B_y - 10.8 sin 60 - 10 sin 30 = 0$ or $A_y = 4.676 kip$ **0.5 point for each term in the equation for a total of 1.5 points.**

By equilibrium of forces in the x-direction we obtain:

$A_x - 10 cos 30 + 10.8 cos 60 = 0$ or $A_x = 3.26 kip$ **0.5 point for each term in the equation for a total of 1.5 points.**

We make an imaginary cut at C and draw the free body diagram as shown in Figure (b) and replace the distributed load to get Figure (c)

(b)

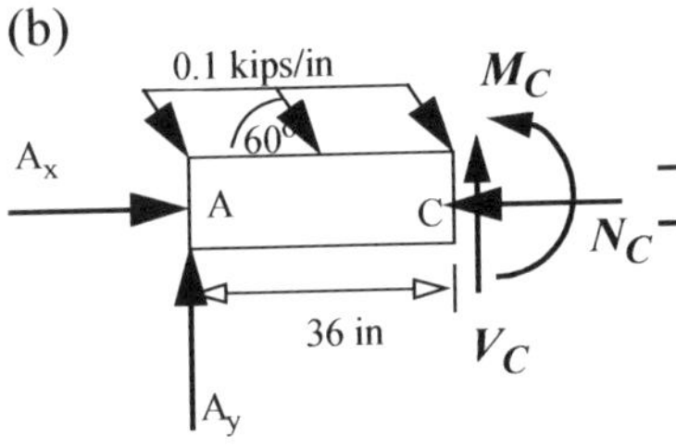

(c)

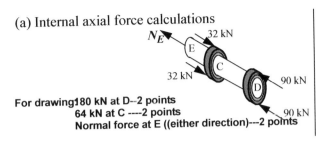

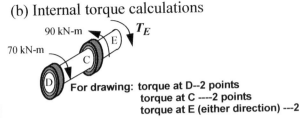

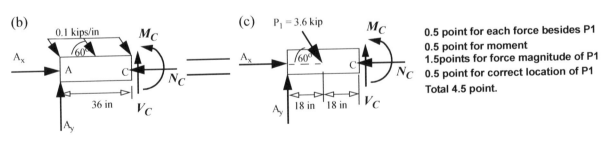

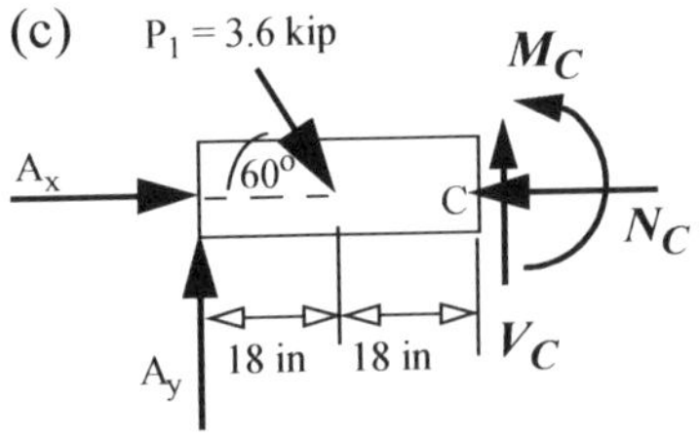

0.5 point for each force besides P1
0.5 point for moment
1.5points for force magnitude of P1
0.5 point for correct location of P1
Total 4.5 point.

By equilibrium of forces in the x-direction we obtain:

$N_C - 3.6 cos 60 - A_x = 0$ or $\boxed{N_C = 5.06 kip(C)}$ **0.5 point for each term in the equation for a total of 1.5 points. 0.5 point for compression**

By equilibrium of forces in the y direction we obtain:

$A_y + V_C - 3.6 sin 60 = 0$ or $\boxed{V_C = 1.558 kip\ downward}$ **0.5 point for each term in the equation for a total of 1.5 points. 0.5 point for downward**

By equilibrium of moment about C we obtain:

$M_C + (3.6 sin 60)(18) - A_y(36) = 0$ or $\boxed{M_C = 112.2 in-kip\ CCW}$ **0.5 point for each term in the equation for a total of 1.5 points.0.5 point for CCW.**

SR. 5

By inspection we can write the following answers.

Internal Force/Moment	Section AA zero/non-zero	Section BB zero/non-zero
Axial Force	zero	non-zero
Shear force	non-zero in x direction.	non-zero in z direction.
Shear force	non-zero in z direction.	zero in y direction.
Torque	non-zero	non-zero
Bending Moment	non-zero in x-direction.	non-zero in y direction.
Bending Moment	non-zero in z direction.	non-zero in z direction.

1 point for each correct zero/non-zero entry.

1 point for each correct direction.

Total 20 points.

33562745R00224

Made in the USA
San Bernardino, CA
24 April 2019